Lecture Notes in Computer Science 5163

Commenced Publication in 1973
Founding and Former Series Editors:
Gerhard Goos, Juris Hartmanis, and Jan van Leeuwen

Editorial Board

Věra Kůrková
Roman Neruda
Jan Koutník (Eds.)

Artificial Neural Networks – ICANN 2008

18th International Conference
Prague, Czech Republic, September 3-6, 2008
Proceedings, Part I

Springer

Volume Editors

Věra Kůrková
Roman Neruda
Institute of Computer Science
Academy of Sciences of the Czech Republic
Pod Vodarenskou vezi 2
182 07 Prague 8, Czech Republic
E-mail: {vera, roman}@cs.cas.cz

Jan Koutník
Department of Computer Science
Czech Technical University in Prague
Karlovo nam. 13
121 35 Prague 2, Czech Republic
E-mail: koutnij@fel.cvut.cz

Library of Congress Control Number: 2008934470

CR Subject Classification (1998): F.1, I.2, I.5, I.4, G.3, J.3, C.2.1, C.1.3

LNCS Sublibrary: SL 1 – Theoretical Computer Science and General Issues

ISSN   0302-9743
ISBN-10   3-540-87535-2 Springer Berlin Heidelberg New York
ISBN-13   978-3-540-87535-2 Springer Berlin Heidelberg New York

Springer is a part of Springer Science+Business Media

springer.com

© Springer-Verlag Berlin Heidelberg 2008
Printed in Germany

Typesetting: Camera-ready by author, data conversion by Scientific Publishing Services, Chennai, India
Printed on acid-free paper   SPIN: 12520565   06/3180   5 4 3 2 1 0

Preface

This volume is the first part of the two-volume proceedings of the 18th International Conference on Artificial Neural Networks (ICANN 2008) held September 3–6, 2008 in Prague, Czech Republic. The ICANN conferences are annual meetings supervised by the European Neural Network Society, in cooperation with the International Neural Network Society and the Japanese Neural Network Society. This series of conferences has been held since 1991 in various European countries and covers the field of neurocomputing and related areas. In 2008, the ICANN conference was organized by the Institute of Computer Science, Academy of Sciences of the Czech Republic together with the Department of Computer Science and Engineering from the Faculty of Electrical Engineering of the Czech Technical University in Prague. Over 300 papers were submitted to the regular sessions, two special sessions and two workshops. The Program Committee selected about 200 papers after a thorough peer-review process; they are published in the two volumes of these proceedings. The large number, variety of topics and high quality of submitted papers reflect the vitality of the field of artificial neural networks.

The first volume contains papers on the mathematical theory of neurocomputing, learning algorithms, kernel methods, statistical learning and ensemble techniques, support vector machines, reinforcement learning, evolutionary computing, hybrid systems, self-organization, control and robotics, signal and time series processing and image processing.

The second volume is devoted to pattern recognition and data analysis, hardware and embedded systems, computational neuroscience, connectionistic cognitive science, neuroinformatics and neural dynamics. It also contains papers from two special sessions, "Coupling, Synchronies, and Firing Patterns: From Cognition to Disease," and "Constructive Neural Networks," and two workshops, New Trends in Self-Organization and Optimization of Artificial Neural Networks, and Adaptive Mechanisms of the Perception-Action Cycle.

It is our pleasure to express our gratitude to everyone who contributed in any way to the success of the event and the completion of these proceedings. In particular, we thank the members of the Board of the ENNS who uphold the tradition of the series and helped with the organization. With deep gratitude we thank all the members of the Program Committee and the reviewers for their great effort in the reviewing process. We are very grateful to the members of the Organizing Committee whose hard work made the vision of the 18th ICANN reality. Zdeněk Buk and Eva Pospíšilová and the entire Computational Intelligence Group at Czech Technical University in Prague deserve special thanks for preparing the conference proceedings. We thank to Miroslav Čepek for the conference website administration. We thank Milena Zeithamlová and Action M Agency for perfect local arrangements. We also thank Alfred Hofmann, Ursula

Barth, Anna Kramer and Peter Strasser from Springer for their help with this demanding publication project. Last but not least, we thank all authors who contributed to this volume for sharing their new ideas and results with the community of researchers in this rapidly developing field of biologically motivated computer science. We hope that you enjoy reading and find inspiration for your future work in the papers contained in these two volumes.

June 2008

Věra Kůrková
Roman Neruda
Jan Koutník

Organization

Conference Chairs

General Chair	Věra Kůrková, Academy of Sciences of the Czech Republic, Czech Republic
Co-Chairs	Roman Neruda, Academy of Sciences of the Czech Republic, Czech Republic
	Jan Koutník, Czech Technical University in Prague, Czech Republic
	Milena Zeithamlová, Action M Agency, Czech Republic
Honorary Chair	John Taylor, King's College London, UK

Program Committee

Włodzisław Duch	Nicolaus Copernicus University in Torun, Poland
Luis Alexandre	University of Beira Interior, Portugal
Bruno Apolloni	Università Degli Studi di Milano, Italy
Timo Honkela	Helsinki University of Technology, Finland
Stefanos Kollias	National Technical University in Athens, Greece
Thomas Martinetz	University of Lubeck, Germany
Guenter Palm	University of Ulm, Germany
Alessandro Sperduti	Università Degli Studi di Padova, Italy
Michel Verleysen	Université catholique de Louvain, Belgium
Alessandro E.P. Villa	Universite jouseph Fourier, Grenoble, France
Stefan Wermter	University of Sunderland, UK
Rudolf Albrecht	University of Innsbruck, Austria
Peter Andras	Newcastle University, UK
Gabriela Andrejková	P.J. Šafárik University in Košice, Slovakia
Bartlomiej Beliczynski	Warsaw University of Technology, Poland
Monica Bianchini	Università degli Studi di Siena, Italy
Andrej Dobnikar	University of Ljubljana, Slovenia
José R. Dorronsoro	Universidad Autónoma de Madrid, Spain
Péter Érdi	Hungarian Academy of Sciences, Hungary
Marco Gori	Università degli Studi di Siena, Italy
Barbora Hammer	University of Osnabrück, Germany

Tom Heskes	Radboud University Nijmegen, The Netherlands
Yoshifusa Ito	Aichi-Gakuin University, Japan
Janusz Kacprzyk	Polish Academy of Sciences, Poland
Paul C. Kainen	Georgetown University, USA
Mikko Kolehmainen	University of Kuopio, Finland
Pavel Kordík	Czech Technical University in Prague, Czech Republic
Vladimír Kvasnička	Slovak University of Technology in Bratislava, Slovakia
Danilo P. Mandic	Imperial College, UK
Erkki Oja	Helsinki University of Technology, Finland
David Pearson	Université Jean Monnet, Saint-Etienne, France
Lionel Prevost	Université Pierre et Marie Curie, Paris, France
Bernadete Ribeiro	University of Coimbra, Portugal
Leszek Rutkowski	Czestochowa University of Technology, Poland
Marcello Sanguineti	University of Genova, Italy
Kateřina Schindler	Austrian Academy of Sciences, Austria
Juergen Schmidhuber	TU Munich (Germany) and IDSIA (Switzerland)
Jiří Šíma	Academy of Sciences of the Czech Republic, Czech Republic
Peter Sinčák	Technical University in Košice, Slovakia
Miroslav Skrbek	Czech Technical University in Prague, Czech Republic
Johan Suykens	Katholieke Universiteit Leuven, Belgium
Miroslav Šnorek	Czech Technical University in Prague, Czech Republic
Ryszard Tadeusiewicz	AGH University of Science and Technology, Poland

Local Organizing Committee

Zdeněk Buk	Czech Technical University in Prague
Miroslav Čepek	Czech Technical University in Prague
Jan Drchal	Czech Technical University in Prague
Paul C. Kainen	Georgetown University
Oleg Kovářík	Czech Technical University in Prague
Rudolf Marek	Czech Technical University in Prague
Aleš Pilný	Czech Technical University in Prague
Eva Pospíšilová	Academy of Sciences of the Czech Republic
Tomáš Siegl	Czech Technical University in Prague

Referees

S. Abe
R. Adamczak
R. Albrecht
E. Alhoniemi
R. Andonie
G. Angelini
D. Anguita
C. Angulo-Bahón
C. Archambeau
M. Atencia
P. Aubrecht
Y. Avrithis
Ľ. Beňušková
T. Beran
Z. Buk
G. Cawley
M. Čepek
E. Corchado
V. Cutsuridis
E. Dominguez
G. Dounias
J. Drchal
D. A. Elizondo
H. Erwin
Z. Fabián
A. Flanagan
L. Franco
D. François
C. Fyfe
N. García-Pedrajas
G. Gnecco
B. Gosselin
J. Grim
R. Haschke
M. Holena
J. Hollmen
T. David Huang

D. Húsek
A. Hussain
M. Chetouani
C. Igel
G. Indiveri
S. Ishii
H. Izumi
J.M. Jerez
M. Jiřina
M. Jiřina, jr.
K.T. Kalveram
K. Karpouzis
S. Kasderidis
M. Koskela
J. Kubalík
M. Kulich
F.J. Kurfess
M. Kurzynski
J. Laaksonen
E. Lang
K. Leiviskä
L. Lhotská
A. Likas
C. Loizou
R. Marek
E. Marchiori
M. A. Martín-Merino
V. di Massa
F. Masulli
J. Mańdziuk
S. Melacci
A. Micheli
F. Moutarde
R. Cristian Muresan
M. Nakayama
M. Navara
D. Novák

M. Olteanu
D. Ortiz Boyer
H. Paugam-Moisy
K. Pelckmans
G. Peters
P. Pošík
D. Polani
M. Porrmann
A. Pucci
A. Raouzaiou
K. Rapantzikos
M. Rocha
A. Romariz
F. Rossi
L. Sarti
B. Schrauwen
F. Schwenker
O. Simula
A. Skodras
S. Slušný
A. Stafylopatis
J. Šťastný
D. Štefka
G. Stoilos
A. Suárez
E. Trentin
N. Tsapatsoulis
P. Vidnerová
T. Villmann
Z. Vomlel
T. Wennekers
P. Wira
B. Wyns
Z. Yang
F. Železný

Table of Contents – Part I

Mathematical Theory of Neurocomputing

Learning Algorithms

Kernel Methods, Statistical Learning, and Ensemble Techniques

Support Vector Machines

Reinforcement Learning

Evolutionary Computing

Control and Robotics

Signal and Time Series Processing

Image Processing

Image Processing – Recognition Systems

Table of Contents – Part II

Pattern Recognition and Data Analysis

Hardware, Embedded Systems

Computational Neuroscience

Connectionistic Cognitive Science

Neuroinformatics

Neural Dynamics

Special Session: Coupling, Synchronies and Firing Patterns: from Cognition to Disease

Special Session: Constructive Neural Networks

Workshop: New Trends in Self-organization and Optimization of Artificial Neural Networks

Workshop: Adaptive Mechanisms of the Perception-Action Cycle

Dimension Reduction for Mixtures of Exponential Families

Shotaro Akaho

Neuroscience Research Institute, AIST, Tsukuba 3058568, Japan

Abstract. Dimension reduction for a set of distribution parameters has been important in various applications of datamining. The exponential family PCA has been proposed for that purpose, but it cannot be directly applied to mixture models that do not belong to an exponential family. This paper proposes a method to apply the exponential family PCA to mixture models. A key idea is to embed mixtures into a space of an exponential family. The problem is that the embedding is not unique, and the dimensionality of parameter space is not constant when the numbers of mixture components are different. The proposed method finds a sub-optimal solution by linear programming formulation.

1 Introduction

In many applications, dimension reduction is important for many purposes such as visualization and data compression. Traditionally, the principal component analysis (PCA) has been widely used as a powerful tool for dimension reduction in the Euclidean space. However, data are often given as binary strings or graph structures that have very different nature from Euclidean vectors.

One approach that we take here is to regard such a data as a parameter of a probability distribution. Information geometry[1] gives a mathematical framework of the space of probability distributions, and a dimension reduction method has been proposed for a class of exponential family[2,3,4,5]. There are mainly two advantages of information geometrical approach to other conventional methods: one is that the information geometrical projection from data point always lies on the support of parameters, and the other is that the projection is defined more naturally for a distribution than the conventional Euclidean projection.

In this paper, we focus on the mixture models[6], which are very flexible and are often used for clustering. However, we cannot apply the exponential family PCA to the mixture models, because they are not members of an exponential family. Our main idea is to embed mixture models into the space of an exponential family. However, that is not straightforward because the embedding is not unique and the dimensionality of parameter space is not constant when the numbers of mixture components are different. Those problems can be resolved by solving some combinatorial optimization problem, which is computationally intractable. Therefore, we propose a method that finds a sub-optimal solution by separating the problems into subproblems, each of which can be optimized easier.

V. Kůrková et al. (Eds.): ICANN 2008, Part I, LNCS 5163, pp. 1–10, 2008.

2   S. Akaho

The proposed framework is not only useful for visualization and data compression, but also for applications which have been developed recently in the field of datamining, privacy preserving datamining[7] and distributed datamining[8]. In the distributed datamining, raw data are collected in many distributed sites. Those datae are not directly sent to the center, but processed into statistical data in each site in order to preserve privacy as well as to reduce communication costs, and then those statistical data are sent to the center. Similar framework has begun to be studied in the field of sensor networks[9].

2   e-PCA and m-PCA: Dual Dimension Reduction

2.1   Information Geometry of Exponential Family

In this section, we review the exponential family PCA called e-PCA and m-PCA[4]. Exponential family is defined as a class of distributions given by

$$p(\boldsymbol{x}; \boldsymbol{\theta}) = \exp\{\sum_{i=1}^{d} \theta_i F_i(\boldsymbol{x}) + C(\boldsymbol{x}) - \psi(\boldsymbol{\theta})\}, \tag{1}$$

with a random variable $\boldsymbol{x}$ and a parameter $\boldsymbol{\theta} = (\theta_1, \ldots, \theta_d)^\top$. The whole set of distribution $p(\boldsymbol{x}; \boldsymbol{\theta})$ by changing $\boldsymbol{\theta}$ forms a space (manifold) $\mathcal{S}$. The structure of the manifold is determined by introducing a Riemannian metric and an affine connection. Statistically natural metric is a Fisher information matrix $g_{jk}(\boldsymbol{\theta}) = \mathbb{E}_{\boldsymbol{\theta}}[\{\partial \log p(\boldsymbol{x}; \boldsymbol{\theta})/\partial \theta_j\}\{\partial \log p(\boldsymbol{x}; \boldsymbol{\theta})/\partial \theta_k\}]$ and the natural connection is α-connection specified by one real valued parameter α. In particular, $\alpha = \pm 1$ is important, because $\mathcal{S}$ becomes a flat manifold. When $\alpha = 1$, the space is called e-flat[1] with respect to an affine coordinate (e-coordinate) $\boldsymbol{\theta}$. When $\alpha = -1$, the exponential family is also flat with respect to another affine coordinate (m-coordinate) $\boldsymbol{\eta} = (\eta_1, \ldots, \eta_d)^\top$ defined by $\eta_i = \mathbb{E}_{\boldsymbol{\theta}}[F_i(\boldsymbol{x})]$. The coordinates $\boldsymbol{\theta}$ and $\boldsymbol{\eta}$ are dually related and transformed each other by Legendre transform, and we write this coordinate transform by $\boldsymbol{\theta}(\boldsymbol{\eta}), \boldsymbol{\eta}(\boldsymbol{\theta})$.

2.2   e-PCA and m-PCA

Since the manifold of an exponential family is flat in the e- and m- affine coordinate, there are two kinds of flat submanifolds for dimension reduction accordingly. The e-PCA (and m-PCA) is defined by finding the e-flat (m-flat) submanifold that fits to samples given as a set of points of the exponential family. Here we describe only e-PCA, because m-PCA is completely dual to e-PCA that is given by exchanging e- and m- in the description of e-PCA.

Let us define the h dimensional e-flat subspace $\mathcal{M}$. The points on $\mathcal{M}$ can be expressed by

$$\boldsymbol{\theta}(\boldsymbol{w}; U) = \sum_{j=1}^{h} w_j \boldsymbol{u}_j + \boldsymbol{u}_0, \tag{2}$$

---

[1] 'e' stands for 'exponential' and 'm' stands for 'mixture'.

where $U = [\boldsymbol{u}_0, \boldsymbol{u}_1, \ldots, \boldsymbol{u}_h] \in \mathbb{R}^{d \times h}$ is a matrix containing basis vectors of the subspace and $\boldsymbol{w} = (w_1, \ldots, w_h)^\top \in \mathbb{R}^h$ is a local coordinate on $\mathcal{M}$.

Suppose we have a set of parameters $\boldsymbol{\theta}^{(1)}, \ldots, \boldsymbol{\theta}^{(n)} \in \mathcal{S}$ as sample points. For dimension reduction, we need to consider the projection of the sample points onto $\mathcal{M}$, which is defined by a geodesic that is orthogonal to $\mathcal{M}$ with respect to the Fisher information. According to the two kinds of geodesic, we can define e-projection and m-projection.

Amari[1] has proved that the m-projection onto an e-flat submanifold is unique, and further it is given by the point that minimizes the m-divergence,

$$K_m(p, q) = \int p(x)\{\log p(x) - \log q(x)\}dx, \tag{3}$$

hence we take m-projection for the e-PCA[2].

As a cost function of fitting of the sample points to a submanifold, it is convenient to take sum of the m-divergence

$$L(U, W) = \sum_{i=1}^{n} K_m(\boldsymbol{\theta}^{(i)}, \boldsymbol{\theta}(\boldsymbol{w}^{(i)}; U)), \tag{4}$$

where $W = (\boldsymbol{w}^{(1)}, \ldots, \boldsymbol{w}^{(n)})$, and e-PCA is defined by finding U and W that minimize $L(U, W)$. Note that even when data are given as values of a random variable instead of parameters, the random variable can be related to a parameter, thus we can apply the same framework[10].

2.3   Alternating Gradient Descent Algorithm

Although it is difficult to optimize $L(U, W)$ with respect to U and W simultaneously, the optimization becomes easier by alternating procedures in which optimization is performed for one variable with fixing the other variable. If we fix the basis vectors U, the projection onto an e-flat space from a sample point is unique as mentioned above. On the other hand, optimizing U with fixing W is also an m-projection to the e-flat subspace determined by W that is a submanifold of the product space of $\mathcal{S}^n$. Therefore, it also has a unique solution.

In each optimization step, we can apply a Newton-like method[4], but we only use an simple gradient descent in this paper. Note that whatever algorithm we use, the algorithm does not always converge to the global solution, even if each alternating step is globally optimized, like in EM and variational Bayes.

The gradient descent algorithm is given by

$$\Delta w_j^{(i)} = -\epsilon_w \boldsymbol{u}_j^\top \Delta \boldsymbol{\eta}^{(i)}, \quad \Delta \boldsymbol{u}_j = -\epsilon_u \sum_{i=1}^{n} w_j^{(i)} \Delta \boldsymbol{\eta}^{(i)}, \quad \Delta \boldsymbol{u}_0 = -\epsilon_u \sum_{i=1}^{n} \Delta \boldsymbol{\eta}^{(i)} \tag{5}$$

---

[2] By duality, we take e-projection for m-PCA, and e-divergence is defined by $K_e(p, q) = K_m(q, p)$.

where $\Delta\boldsymbol{\eta}^{(i)} = \tilde{\boldsymbol{\eta}}^{(i)} - \boldsymbol{\eta}^{(i)}$ is a difference of m-coordinates between the point specified by the current estimate $\boldsymbol{w}^{(i)}$ and the sample point. As a general tendency, the problem is more sensitive against U than against W. Thus we take a learning constant to be $\epsilon_w > \epsilon_u$.

Further, the basis vectors U has a redundancy of linear transformation, that is, when U is transformed by any non-singular matrix A, the same solution is obtained by transforming W to WA^{-1}. It also happens that two different bases $\boldsymbol{u}_i$ and $\boldsymbol{u}_j$ converge to the same direction if they are optimized by ordinary gradient descent without any constraints. Therefore, we restrict U to be an orthogonal frame (i.e. $UU^{\top} = I_d$). Such a space is called Grassmann manifold. The optimization in Grassmann manifold is often used for finding principal components or minor components [11]. The natural gradient for U is given by

$$\Delta U_{\mathrm{nat}} = \Delta U - UU^{\top}\Delta U \tag{6}$$

where ΔU be the matrix whose columns are $\Delta\boldsymbol{u}_j$ in (5). Since this update rule does not preserve the orthogonal constraint strictly, we need to orthogonalize it (we apply this in the experiment), or update U along the geodesic.

2.4   e-Center and m-Center

An important special case of the e-PCA and m-PCA is a zero dimensional subspace that corresponds to a point. The only parameter in that case is $\boldsymbol{u}_0$ that is given in a closed form

$$\boldsymbol{\theta}_c^e = \boldsymbol{\theta}\left(\frac{1}{n}\sum_{i=1}^{n}\boldsymbol{\eta}(\boldsymbol{\theta}^{(i)})\right), \quad \boldsymbol{\eta}_c^m = \boldsymbol{\eta}\left(\frac{1}{n}\sum_{i=1}^{n}\boldsymbol{\theta}(\boldsymbol{\eta}^{(i)})\right). \tag{7}$$

We call them e-center and m-center respectively.

2.5   Properties of e-PCA and m-PCA

In this subsection, we summarize several points that the e-PCA and m-PCA are different from the ordinary PCA.

The first thing is about the hierarchical relation between different dimensions. Since e-PCA and m-PCA includes nonlinear part in its formulation, an optimal low dimensional subspace is not always included in a higher dimensional one. In some applications, hierarchical structures are necessary or convenient. In such cases, we can construct an algorithm that finds the optimal subspace by constraining the search space.

The second thing is about the domain (or support) of $\mathcal{S}$. The parameter set of exponential family is a local coordinate. That means $\boldsymbol{\theta}$ does not define the probability distribution for all values of $\mathbb{R}^d$. In general, it forms a convex region for e- and m- coordinate systems. It is known that the m-projection for e-PCA is guaranteed to be included in that region. However, when we apply the gradient-type algorithm, too large step size causes the excess of the candidate solution from the domain. In our implementation, the candidate solution is checked to be included in each learning step, and the learning constant is adaptively changed in the case of excess.

The third thing is about the initialization problem. Since the alternating algorithm only gives a local optimum, it is important to find a good initial solution. The naive idea is to use the conventional PCA using Euclidean metric, and $\boldsymbol{u}_0$ is initialized by its e-center. However, the initialization problem is related to the domain problem above, i.e., the initialization points have to lie in the domain region. For simplicity, we take $W = 0$ in our numerical simulation, which corresponds to the initial projection point being always $\boldsymbol{u}_0$.

3 Embedding of Mixture Models

Now let us move on to our main topic, the dimension reduction of mixture models. A major difficulty is that mixture models are not members of an exponential family. If we add a latent variable z representing which component $\boldsymbol{x}$ is generated from, $p(\boldsymbol{x}, z; \boldsymbol{\theta})$ belongs to the exponential family.

3.1 Latent Variable Model

Mixture of exponential family is written as

$$p(\boldsymbol{x}) = \sum_{i=0}^{k} \pi_i f_i(\boldsymbol{x}; \boldsymbol{\xi}_i), \quad f_i(\boldsymbol{x}; \boldsymbol{\xi}_i) = \exp(\boldsymbol{\xi}_i \cdot \boldsymbol{F}_i(\boldsymbol{x}) - \psi_i(\boldsymbol{\xi}_i)), \quad i = 0, \ldots, k. \quad (8)$$

Since the number of freedom $\{\pi_i\}$ is k, we regard $\pi_1, \ldots, \pi_k$ as parameters and define π_0 by $\pi_0 = 1 - \sum_{i=1}^{k} \pi_i$.

When $z \in \{0, 1, 2, \ldots, k\}$ is a latent variable representing which component of mixture $\boldsymbol{x}$ is generated from, the distribution of $(\boldsymbol{x}, z)$ is an exponential family [12] as written down below.

$$p(\boldsymbol{x}, z) = \pi_z f_z(\boldsymbol{x}; \boldsymbol{\xi}_z) \exp \left[\sum_{i=1}^{k} \boldsymbol{\xi}_i \cdot \boldsymbol{F}_i(\boldsymbol{x}) \delta_i(z) \right.$$

$$\left. + \boldsymbol{\xi}_0 \cdot \boldsymbol{F}_0(\boldsymbol{x}) \left(1 - \sum_{i=1}^{k} \delta_i(z) \right) + \sum_{i=1}^{k} \nu_i \delta_i(z) - \psi_* \right], \quad (9)$$

where $\delta_i(z) = 1$ when $z = i$, and 0 otherwise, and

$$\nu_i = \log \pi_i - \psi_i(\boldsymbol{\xi}_i) - (\log \pi_0 - \psi_0(\boldsymbol{\xi}_0)), \quad \psi_* = -\log \pi_0 + \psi_0(\boldsymbol{\xi}_0). \quad (10)$$

The e-coordinate of this model is $\boldsymbol{\theta} = \nu_1, \ldots, \nu_k, \boldsymbol{\xi}_0, \boldsymbol{\xi}_1, \ldots, \boldsymbol{\xi}_k$, and the m-coordinate $\boldsymbol{\eta}$ is $\mathbb{E}_{\boldsymbol{\theta}}[\delta_i(z)] = \pi_i$ corresponding to ν_i, and $\mathbb{E}_{\boldsymbol{\theta}}[\boldsymbol{F}_i(\boldsymbol{x})\delta_i(z)] = \pi_i \boldsymbol{\gamma}_i$ corresponding to $\boldsymbol{\xi}_i$, where $\boldsymbol{\gamma}_i = \mathbb{E}_{\boldsymbol{\theta}}[\boldsymbol{F}_i(\boldsymbol{x})]$ is the m-coordinate of each component distribution $f_i(\boldsymbol{x}; \boldsymbol{\xi}_i)$.

3.2   Problems of the Embedding

There are two problems in the embedding described above. The first one is that the embedding is not unique, because the mixture distribution is invariant when components are exchanged. The other happens when there are different numbers of mixture components. In such a case, we cannot embed them directly into one common space, because the dimensions of mixture components are different.

For the first problem, we will find the embedding so that embedded distributions are located as closely as possible. Once the embedding is completed, the e-PCA (or m-PCA) can be applied directly. For the second problem, we will split the components to adjust the dimensions between different numbers of components.

3.3   Embedding for the Homogeneous Mixtures

Firstly, we consider the homogeneous case in which the numbers of components are the same for all mixtures $\boldsymbol{\theta}^{(i)}$.

A naive way to resolve the problem is that we perform e-PCA (or m-PCA) for any possible embeddings and take the best one. However, it is not practical because the number of possible embeddings increase exponentially with respect to the number of components and the number of mixtures. Instead, we try to find a configuration by which mixtures get as close together as possible. The following proposition shows the divergence between two mixtures in the embedded space is given in a very simple form.

Proposition 1. *Suppose there are two mixture distributions with the same numbers of components, and their distributions with latent variables be*

$$p_1(\boldsymbol{x}, z) = \alpha_z f_z(\boldsymbol{x}; \xi_z), \qquad p_2(\boldsymbol{x}, z) = \beta_z f_z(\boldsymbol{x}; \zeta_z). \tag{11}$$

The m-divergence between p_1 and p_2 is given by

$$K_m(p_1, p_2) = \sum_{i=0}^{k} \alpha_i [K_m(f_i(\boldsymbol{x}; \xi_i), f_i(\boldsymbol{x}; \zeta_i)) + \log \frac{\alpha_i}{\beta_i}]. \tag{12}$$

This means that the divergence is separated into sum of functions each of which depends only on pairwise components of the two mixtures. Note that the divergence between the original mixtures is not so simple.

Based on this fact, we can derive the optimal embedding for two mixtures that minimizes the divergence. It should be noted that the optimality is not invariant with respect to the order of p_1 and p_2 because the divergence is not a symmetric function. For the general 'n mixtures' case, we apply the following greedy algorithm based on the pairwise optimality.

[Embedding algorithm (for e-PCA, homogeneous)]

   1. Embed $\boldsymbol{\theta}^{(1)}$ in any configuration
   2. Repeat the following procedures for $i = 2, 3, \ldots, n$
      (a) Let $\boldsymbol{\theta}_c^e$ be the e-center of already embedded mixtures for $j = 1, \ldots, i-1$.
      (b) Embed $\boldsymbol{\theta}^{(i)}$ so as to minimize the m-divergence between $\boldsymbol{\theta}^{(i)}$ and $\boldsymbol{\theta}_c^e$ in the embedded space (see next subsection).

Fig. 1. Matching of distributions. Left: Homogeneous case. The sum of weights is minimized. Right: Heterogeneous case. In this example, the k-th component of the left group is split and matched with two components $(0, k'$-th) of the right group.

3.4   The Optimal Matching Solution

In this subsection, we give an optimization method to find a matching between two mixtures so as to minimize the cost function (12) that is the sum of component-wise functions (Left of fig.1).

Letting the weight values be

$$\omega_{ij} = K_m(p_1, p_2) = \alpha_i \left[K_m(f_i(\boldsymbol{x}; \xi_i), f_j(\boldsymbol{x}; \zeta_j)) + \log \frac{\alpha_i}{\beta_j} \right], \tag{13}$$

we obtain the optimization problem in terms of the linear programming,

$$\min_{a_{ij}} \sum_{i=0}^{k} \sum_{j=0}^{k} \omega_{ij} a_{ij} \quad \text{s.t.} \quad a_{ij} \geq 0, \quad \sum_{i=0}^{k} a_{ij} = \sum_{j=0}^{k} a_{ij} = 1 \tag{14}$$

The solution a_{ij} takes binary values (0 or 1) by the following integrality theorem.

Proposition 2 (Integrality theorem[13]).   *In the transshipment problem, $\min_{a_{ij}} \sum_{i=0}^{k} \sum_{j=0}^{k'} \omega_{ij} a_{ij}$ s.t. $a_{ij} \geq 0, \sum_{i=0}^{k} a_{ij} = s_j, \sum_{j=0}^{k'} a_{ij} = t_i$, a_{ij} has an integer optimal solution when the problem has at least one feasible solution and s_j, t_i are all integers. In particular, the solution given by the simplex method always gives the integer solution.*

3.5   General Case: Splitting Components

When numbers of components of mixtures are different (heterogeneous case), we can adjust the numbers by splitting components. Splitting the components of mixtures have played an important role in different situations, for example, to find an optimal number of components for fitting the mixture[14].

Let $\alpha f(\boldsymbol{x})$ be one of the components of a mixture, it can be split into $k+1$ components like

$$\alpha_i f(\boldsymbol{x}; \xi), \quad i = 0, \ldots, k, \quad \sum_{i=0}^{k} \alpha_i = \alpha, \quad \alpha_i > 0 \tag{15}$$

We need to determine two things: which component should be split and how large weights of splitting α_i should be. However, since it is hard to optimize them simultaneously, we solve the problem sequentially, that is, first we determine the component to be split based on the optimal assignment problem in the previous subsection, and then we optimize the weights of splitting.

3.6   Component Selection

Suppose we have two mixtures p_1 and p_2 given in (11). When their numbers of components are different (heterogeneous case), we need to find matching one-to-many. Here let $z = 0, 1, \ldots, k$ for p_1 and $z = 0, 1, \ldots, k'$ for p_2. In order to find the one-to-many matching, we extend the optimization problem of the homogeneous case to the heterogeneous case in a natural way

$$\min_{a_{ij}} \sum_{i=0}^{k} \sum_{j=0}^{k'} \omega_{ij} a_{ij} \quad \text{s.t.} \quad a_{ij} \geq 0, \quad \sum_{i=0}^{k} a_{ij} \geq 1, \quad \sum_{j=0}^{k'} a_{ij} = 1, \tag{16}$$

where ω_{ij} is defined by (13), and we assumed p_1 has a smaller number of components than p_2 ($k \leq k'$) and some equality constraints are replaced by inequality constraints to deal with one-to-many matching (the right of fig.1).

Note that this problem only gives a sub-optimal matching for the entire problem, because the splitting weights are not taken into account. However, from the computational point of view, the integrality property of the solution is preserved and all weights are guaranteed to be binary values, and further virtue of this formulation is that the homogeneous case is included as a special case of the heterogeneous case.

3.7   Optimal Weights

After the matching is performed, we split the component $\alpha f(\boldsymbol{x}; \xi)$ into $k + 1$ components given by (15) and find the optimal correspondence to the components $\beta_i f_i(\boldsymbol{x}; \zeta_i), (i = 0, \ldots, k)$. This can be given by the following proposition.

Proposition 3. *The optimal splitting that minimizes the sum of m-divergence between $\alpha_i f(\boldsymbol{x}; \xi)$ and $\beta_i f_i(\boldsymbol{x}; \zeta_i), (i = 0, \ldots, k)$ is given by*

$$\alpha_i^e = \frac{\beta_i}{Z} \exp(-K_m(f(\boldsymbol{x}; \xi), f(\boldsymbol{x}; \zeta_i))), \tag{17}$$

where Z is a normalization constant. The splitting for e-divergence is given by

$$\alpha_i^m = \frac{\beta_i}{Z}. \tag{18}$$

Now we summarize the embedding method in the general case including both homogeneous and heterogeneous.

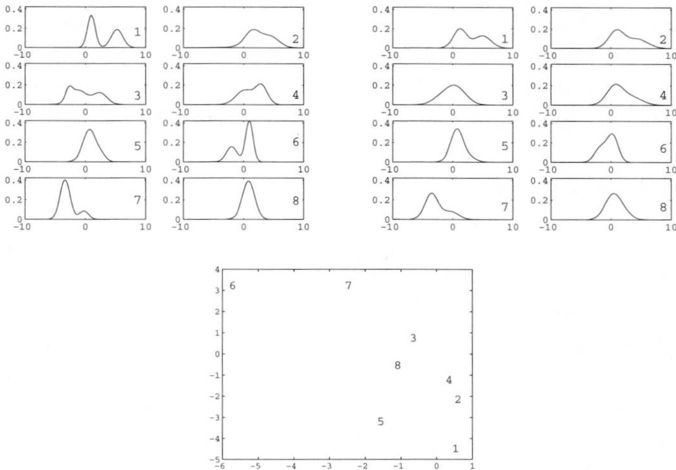

Fig. 2. Up-left: Original mixtures, Up-right: Mixtures with reduced dimension, Down: Two dimensional scatter plots of mixtures

[Embedding algorithm (for e-PCA, general)]

1. Sort $\boldsymbol{\theta}^{(1)}, \ldots, \boldsymbol{\theta}^{(n)}$ in the descending order of the numbers of components.
2. Embed $\boldsymbol{\theta}^{(1)}$ in any configuration
3. Repeat the following (a),(b),(c) for $i = 2, 3, \ldots, n$
   (a) Let $\boldsymbol{\theta}_c^e$ be e-center of already embedded mixtures $j = 1, \ldots, i - 1$.
   (b) Solve (16) to find the correspondence between $\boldsymbol{\theta}^{(i)}$ and $\boldsymbol{\theta}_c^e$.
   (c) If the number of components of $\boldsymbol{\theta}^{(i)}$ is smaller than $\boldsymbol{\theta}_c^e$, then split the components by (17).

4   Numerical Experiments

We applied the proposed method to a synthetic data set of one dimensional Gaussian mixtures. First, Gaussian mixtures are generated (total 8 = 4 mixtures with 3 components + 3 mixtures with 2 components + 1 mixture with 1 component), where the parameters of those mixtures (mixing weight, mean and variance of each component) are determined at random. (the upper left of fig.2).

The learning coefficient of e-PCA is taken to be $\epsilon_w = 0.1, \epsilon_u = 0.01$ except the cases that the parameter exceeds the domain boundary or the objective function increases exceptionally (in such cases the learning rate is decreased adaptively). The update of U is performed 20 steps, each of which follows after 50 step updates of W for the sake of stable convergence.

The down figure of fig. 2 shows the result of dimension reduction (e-PCA) to 2 dimensional subspace from 8 dimensional original space (= the number of parameters of Gaussian mixtures with 3 components). The objective function of $L(U, W)$ is about 6.4 in the initial solution (the base function is initialized Euclidean PCA) that decreased to about 1.9. The upper right of fig. 2 shows the projected distributions obtained by e-PCA. We see their original shapes are well-preserved even in 2-D subspace, but the shapes are little smoothed. We also applied m-PCA as well, a similar but different results are obtained.

5   Concluding Remarks

We have proposed a dimension reduction method of parameters of mixture distributions. There are two important problems to be solved: One is to find a good initial solution because the final solution is not a global optimum, though the optimum solution is obtained in each step. The other is to develop a stable and fast algorithm. As for the embedding, there is a lot of possibilities to be improved from the proposed greedy algorithm. The application of the real world data and extensions to other structured models like HMM and other types of methods like clustering are all left as future works.

References

1. Amari, S.: Differential Geometrical Methods in Statistics. Springer, Heidelberg (1985)
2. Amari, S.: Information Geometry on Hierarchy of Probability Distributions. IEEE Trans. on Information Theory 41 (2001)
3. Collins, M., Dasgupta, S., Schapire, R.: A Generalization of Principal Component Analysis to the Exponential Family. In: Advances in NIPS, vol. 14 (2002)
4. Akaho, S.: The e-PCA and m-PCA: dimension reduction by information geometry. In: IJCNN 2004, pp. 129–134 (2004)
5. Watanabe, K., Akaho, S., Okada, M.: Clustering on a Subspace of Exponential Family Using Variational Bayes Method. In: Proc. of Worldcomp2008/Information Theory and Statistical Learning (2008)
6. McLachlan, G., Peel, D.: Finite Mixture Models. Wiley, Chichester (2000)
7. Agrawal, R., Srikant, R.: Privacy-Preserving Data Mining. In: Proc. of the ACM SIGMOD, pp. 439–450 (2000)
8. Kumar, A., Kantardzic, M., Madden, S.: Distributed Data Mining: Framework and Implementations. IEEE Internet Computing 10, 15–17 (2006)
9. Chong, C.Y., Kumar, S.: Sensor networks: evolution, opportunities, and challenges. Proc. of the IEEE 91, 1247–1256 (2003)
10. Buntine, W.: Variational extensions to EM and multinomial PCA. In: Elomaa, T., Mannila, H., Toivonen, H. (eds.) ECML 2002. LNCS (LNAI), vol. 2430. Springer, Heidelberg (2002)
11. Edelman, A., Arias, T., Smith, S.: The geometry of algorithms with orthogonality constraints. SIAM J. Matrix Anal. Appl. 20(2), 303–353 (1998)
12. Amari, S.: Information geometry of the EM and em algorithms for neural networks. Neural Networks 8(9), 1379–1408 (1995)
13. Chvátal, V.: Linear Programming. W.H. Freeman and Company, New York (1983)
14. Fukumizu, K., Akaho, S., Amari, S.: Critical lines in symmetry of mixture models and its application to component splitting. In: Proc. of NIPS15 (2003)

Several Enhancements to Hermite-Based Approximation of One-Variable Functions

Bartlomiej Beliczynski[1] and Bernardete Ribeiro[2]

Warsaw University of Technology,
Institute of Control and Industrial Electronics,
ul. Koszykowa 75, 00-662 Warszawa, Poland
B.Beliczynski@ee.pw.edu.pl
Department of Informatics Engineering, Center for Informatics and Systems,
University of Coimbra,
Polo II, P-3030-290 Coimbra, Portugal
bribeiro@dei.uc.pt

Abstract. Several enhancements and comments to Hermite-based one-variable function approximation are presented. First of all we prove that a constant bias extracted from the function contributes to the error decrease. We demonstrate how to choose that bias. Secondly we show how to select a basis among orthonormal functions to achieve minimum error for a fixed dimension of an approximation space. Thirdly we prove that loss of orthonormality due to truncation of the argument range of the basis functions does not effect the overall error of approximation and the expansion coefficients. We show how this feature can be used. An application of the obtained results to ECG data compression is presented.

1 Introduction

A set of Hermite functions forming an orthonormal basis is naturally attractive for various approximation, classification and data compression tasks. These basis functions are defined on the real numbers set $\mathbb{R}$ and they can be recursively calculated. The approximating function coefficients can be determined relatively easily to achieve the best approximation property. Since Hermite functions are eigenfunctions of the Fourier transform, time and frequency spectra are simultaneously approximated. Each subsequent basis function extends frequency bandwidth within a limited range of well concentrated energy; see for instance [1]. By introducing scaling parameter we may control the bandwidth influencing at the same time the dynamic range of the input argument, till we strike a desirable balance.

If Hermite one-variable functions are generalized to two variables, they retain the same useful property and turn out to be very suitable for image compression tasks.

Recently in several publications (see for instance [2], [3]) it was suggested to use Hermite functions as activation functions in neural schemes. In [3], a so called "constructive" approximation scheme is used. It is a type of incremental approximation developed in [4], [5]. The novelty of this approach is that contrary to the

V. Kůrková et al. (Eds.): ICANN 2008, Part I, LNCS 5163, pp. 11–20, 2008.

traditional neural architecture, every node in the hidden layer has a different activation function. It gains several advantages of the Hermite functions. However, in such approach orthogonality of Hermite functions is not really exploited.

In this paper we return to the basic tasks of one-variable function approximation. For this classical problem we are offering two enhancements and one proof of correctness.

For fixed basis functions in a Hilbert space, there always exists the best approximation. If the basis is orthonormal, the approximation can relatively easily be calculated in the form of expansion coefficients. Those coefficients represent the original function approximated in the Hermite basis. The coefficients usually require less space than the original data. At first glance there seems to be little room for improvement. However one may slightly reformulate the problem. Instead of approximating the function f, one may approximate $f - f_0$, where f_0 is a fixed chosen function. After the approximation is done, f_0 is added to the approximant of $f - f_0$. From approximation and data compression point of view, this procedure makes sense if additional efforts put into the representation of f_0 are compensated by reduction of the approximation error.

In a typically stated approximation problem a basis of $n+1$ functions $\{e_0, e_1, .., e_n\}$ is given and we are looking for their expansion coefficients. We may however reformulate that problem in the following way. Let us search for any $n+1$ Hermite basis functions, not necessarily with consecutive indices, ensuring the smallest error of approximation. This is the second issue.

The third problem which is stated and discussed here is the problem of loosing orthonormality property by basis functions if the set $\mathbb{R}$ is replaced by its subset. When the approximating basis is orthonormal, the expansion coefficients are calculated easily. Otherwise these calculations are more complicated. However we prove that despite of loss of orthonormality, we may determine the Hermite expansion coefficients as before.

In this paper we are focusing on Hermite basis, however many of the studied properties are applicable to any orthonormal basis. Our enhancements were tested and demonstrated with ECG data compression, a well known application area.

This paper is organized as follows. In Section 2 basic facts about approximation needed for later use are recalled. In Section 3 Hermite functions are shortly described. Then we present our results in Section 4: bias extraction, basis functions selection and proof of correctness for expansion coefficients calculation despite the lack of basis orthonormality. In Section 5 certain practicalities are presented and an application of our improvements to ECG data compression is demonstrated and discussed. In Section 6 conclusions are drawn.

2   Approximation Framework

Some selected facts on function approximation useful for this paper will be recalled. Let us consider the following function

$$f_{n+1} = \sum_{i=0}^{n} w_i g_i,\tag{1}$$

where $g_i \in \mathcal{G} \subset \mathcal{H}$, and $\mathcal{H}$ is a Hilbert space $\mathcal{H} = (\mathcal{H}, ||.||)$, $i = 0, ..., n$, and $w_i \in \mathbb{R}$, $i = 0, \ldots, n$.

For any function f from a Hilbert space $\mathcal{H}$ and a closed (finite dimensional) subspace $\mathcal{G} \subset \mathcal{H}$ with basis $\{g_0, ..., g_n\}$ there exists a unique best approximation of f by elements of $\mathcal{G}$ ([6]). Let us denote it by g_b. Because the error of the best approximation is orthogonal to all elements of the approximation space $f - g_b \bot \mathcal{G}$, the coefficients w_i may be calculated from the set of linear equations

$$\langle g_i, f - g_b \rangle = 0 \text{ for } i = 0, ..., n \tag{2}$$

where $\langle .,. \rangle$ denotes inner product.

The formula (2) can also be written as $\langle g_i, f - \sum_{k=0}^{n} w_k g_k \rangle = \langle g_i, f \rangle - \sum_{k=0}^{n} w_k \langle g_i, g_k \rangle = 0$ for $i = 0, ..., n$ or in the matrix form

$$\Gamma w = G_f \tag{3}$$

where $\Gamma = [\langle g_i, g_j \rangle]$, $i, j = 0, ..., n$, $w = [w_0, ..., w_n]^T$, $G_f = [\langle g_0, f \rangle, ..., \langle g_n, f \rangle]^T$ and "T" denotes transposition.

Because there exists a unique best approximation of f in a $n+1$ dimensional space $\mathcal{G}$ with basis $\{g_0, ..., g_n\}$, the matrix Γ is nonsingular and $w_b = \Gamma^{-1} G_f$.

For any basis $\{g_0, ..., g_n\}$ one can find such orthonormal basis $\{e_0, ..., e_n\}$, $\langle e_i, e_j \rangle = 1$ when $i = j$ and $\langle e_i, e_j \rangle = 0$ when $i \neq j$ that $span\{g_0, ..., g_n\} = span\{e_0, ..., e_n\}$. In such a case, Γ is a unit matrix and

$$w_b = \left[\langle e_0, f \rangle, \langle e_2, f \rangle, . .., \langle e_n, f \rangle \right]^T. \tag{4}$$

Finally (1) will take the form

$$f_{n+1} = \sum_{i=0}^{n} \langle e_i, f \rangle e_i, \quad i = 0, 1, ..., n. \tag{5}$$

The squared error $error_{n+1} = <f - f_n, f - f_n>$ of the best approximation of a function f in the basis $\{e_0, ..., e_n\}$ is thus expressible by

$$||error_{n+1}||^2 = ||f||^2 - \sum_{i=0}^{n} w_i^2. \tag{6}$$

3   Hermite Functions

We will be looking at an orthonormal set of functions in the form of Hermite functions. Their expansion coefficients are easily and independently calculated from (4). Let us consider a space of a great practical interest $L^2(-\infty, +\infty)$

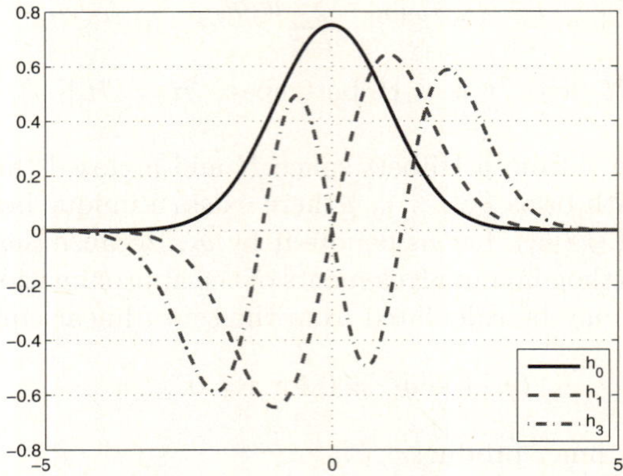

Fig. 1. Hermite functions h_0, h_1, h_3

with the inner product defined $< x, y >= \int\limits_{-\infty}^{+\infty} x(t)y(t)dt$. In such space a sequence of linearly independent and bounded functions could be defined as follows $h_0(t) = w(t) = e^{-t^2/2}$, $\quad h_1(t) = tw(t)$,..., $h_n(t) = t^n w(t)$,..This basis could be orthonormalized by using the well known and efficient Gram-Schmidt process (see for instance [6]). Finally a new now orthonormal basis spanning the same space is obtained

$$h_0(t), h_1(t), ..., h_n(t), ... \tag{7}$$

where

$$h_n(t) = c_n e^{-\frac{t^2}{2}} H_n(t); \;\; H_n(t) = (-1)^n e^{t^2} \frac{d^n}{dt^n}(e^{-t^2}); \;\; c_n = \frac{1}{(2^n n! \sqrt{\pi})^{1/2}}. \tag{8}$$

The polynomials $H_n(t)$ are called Hermite polynomials and the functions $e_n(t)$ Hermite functions. According to (8) the first several Hermite functions could be calculated

$$h_0(t) = \frac{1}{\pi^{1/4}} e^{-\frac{t^2}{2}}; \qquad h_1(t) = \frac{1}{\sqrt{2}\pi^{1/4}} e^{-\frac{t^2}{2}} 2t;$$

$$h_2(t) = \frac{1}{2\sqrt{2}\pi^{1/4}} e^{-\frac{t^2}{2}}(4t^2 - 2); \;\; h_3(t) = \frac{1}{4\sqrt{3}\pi^{1/4}} e^{-\frac{t^2}{2}}(8t^3 - 12t)$$

A plot of several functions of the Hermite basis are shown in Fig.1.

4   Main Results

4.1   Extracting of Bias

In this section our first enhancement is introduced. Let f be any function from a Hilbert space $\mathcal{H}$. Instead of approximating function f, we suggest to approximate the function $f - f_0$, where $f_0 \in \mathcal{H}$ is a known function. Later f_0 is added to the approximant of $f - f_0$. Now a modification of (5) will be the following

$$f_{n+1}^{f_0} = f_0 + \sum_{i=0}^{n} < f - f_0, e_i > e_i, \qquad (9)$$

Then the approximation error will be expressed as

$$e_n^{f_0} = f - f_{n+1}^{f_0} = f - f_0 - \sum_{i=0}^{n} < f - f_0, e_i > e_i,$$

and similarly to (6) its squared norm

$$||e_{n+1}^{f_0}||^2 = ||f - f_0||^2 - \sum_{i=0}^{n} < f - f_0, e_i >^2 \qquad (10)$$

Theorem 1. *Let $\mathcal{H}$ be a Hilbert space of functions on a subset of $\mathbb{R}$ containing the interval $[a,b]$, let f be a function from $\mathcal{H}$, $f \in \mathcal{H}$, $\{e_0, e_1, .. ., e_n\}$ be an orthonormal set in $\mathcal{H}$, c be a constant $c \in \mathbb{R}$. Let $f_0 = c 1_{[a,b]}$ where $1_{[a,b]}$ denotes a function of value 1 in the range $[a,b]$ and 0 elsewhere, and the approximation formula be the following*

$$f_{n+1}^{f_0} = f_0 + \sum_{i=0}^{n} < f - f_0, e_i > e_i$$

then the norm of the approximation error is minimized for $c = c_0$ and

$$c_0 = \frac{< f, 1_{[a,b]} > - \sum_{i=0}^{n} < f, e_i > < e_i, 1_{[a,b]} >}{(b-a) - \sum_{i=0}^{n} < e_i, 1_{[a,b]} >^2} \qquad (11)$$

Proof. The squared error formula (10) could be expressed as follows $||e_{n+1}^{f_0}||^2 = ||f||^2 + ||f_0||^2 - 2 < f, f_0 > - \sum_{i=0}^{n} (\langle f, e_i \rangle - \langle e_i, f_0 \rangle)^2 = ||f||^2 + c^2(b-a) - 2c < f, 1_{[a,b]} > - \sum_{i=0}^{n} (\langle f, e_i \rangle^2 + c^2 \langle e_i, 1_{[a,b]} \rangle^2 - 2c \langle f, e_i \rangle \langle e_i, 1_{[a,b]} \rangle)$. Now differentiating the squared error formula in respect of c and equating it to zero one obtains (11).

Along the Theorem we are suggesting the two step approximation. First f_0 should be calculated and then the function $f - f_0$ will be approximated in a usual way.

Remark 1. One may notice that in many applications c_0 of (11) could well be approximated by

$$c_0 \simeq \frac{< f, 1_{[a,b]} >}{(b-a)} \qquad (12)$$

The right hand side of (12) expresses the mean value of the approximated function f in the range $[a,b]$. A usual choice of $[a,b]$ is such as an actual function f argument range.

4.2 Basis Selection

In a typically stated approximation problem there is a function to be approximated f and a basis $\{e_0, e_1, ..., e_n\}$ of approximation. We are looking for the function expansion coefficients related to the basis functions.

The problem may however be reformulated in the following way. Let search for any $n + 1$ Hermite-basis functions, not necessarily with consecutive indices, ensuring the smallest error of approximation. In practice this easily can be done. Since for any orthonormal basis an indicator of the error reduction associated with the basis function e_i is $|w_i| = |<f, e_i>|$, one may calculate sufficiently many coefficients and order them accordingly to their absolute values. Then using as many coefficients as needed, one ensures the fastest decrease of error with respect to the number of basis functions. A theoretical question might however be stated. How many coefficients should initially be calculated as to achieve the $n + 1$ most significant ones in the whole set?

4.3 Truncated Basis

According to equation (3), the best approximation coefficients could easily be calculated if Γ is a unit matrix. This happens if the approximation basis is orthonormal. The coefficients are given by (4). While implementating, the basis functions are represented via their values in a finite range of discrete arguments. In practice the Γ matrix is approximated as follows

$$\Gamma \simeq B(r)^T B(r)$$

where $B(r)^{M,n+1} = [e_0(r), e_1(r), ..., e_n(r)]$, and r is a vector of M real entries from the range $[a, b]$. Adopting Matlab notation, $e_0(r)$ means a vector of e_0 function values calculated in the arguments specified by the vector r. Thus r contains the samples of a real axis $(-\infty, \infty)$. Obviously it must be truncated, so it also means that $B(r)^T B(r) \neq I_{n+1}$. The basis represented by $B(r)$ is not orthonormal any more. However formula (4) is still valid and can be used. More precisely this is stated in the Proposition 1.

Proposition 1. *Let $B = \{e_0, e_1, ..., e_n\}$ be a set of orthonormal functions, $e_i \in \mathcal{L}^2(-\infty, \infty)$, $i = 0, 1, ..., n$. Let the function f be such that $f \subset \mathcal{L}^2(-\infty, \infty)$ and $f(t) = 0$ when $t \notin [a, b]$ then*

$$w_b = \left[\langle e_0^t, f \rangle, \langle e_1^t, f \rangle, ..., \langle e_n^t, f \rangle \right]^T$$

where $e_i^t(t)$, $i = 0, 1, ..., n$ are $e_i(t)$ truncated to $[a, b]$.

Proof. According to (3) $w_b = \Gamma^{-1} G_f$ and because orhonormality of basis $w_b = \left[\langle e_0, f \rangle, \langle e_1, f \rangle, ..., \langle e_n, f \rangle \right]^T$, but $f(t) = 0$ when $t \notin [a, b]$, thus $\langle e_i, f \rangle = \langle e_i^t, f \rangle$ for every $i = 0, 1, ..., n$.

Remark 2. If a function to be approximated is defined over a limited range of its argument, the orthonormal basis of approximation could be limited to that range and despite the loss of orthonormality, the best approximation is calculated in the same way.

5   Practicalities and an Example

5.1   Calculating Hermite Functions

Hermite functions could directly be calculated from (8). However for large n, their components, the Hermite polynomials reach very large values and those calculations are error prone. Better way to calculate Hermite functions is to use the following well known recursive formula

$$h_{n+1}(t) = \sqrt{\frac{2}{n+1}} t h_n(t) - \sqrt{\frac{n}{n+1}} h_{n-1}(t); \ \ n = 1, 2, 3, ... \tag{13}$$

see for instance [1].

If a function to be approximated is represented by a set of pairs $\{(t_i, f_i)\}_{i=-N}^{N}$, in the range $[-t_{\max}, t_{\max}]$ then starting from $h_0(t)$ and $h_1(t)$ one obtains Hermite-basis from (13) and the expansion coefficients from (4).

If sampling is regular with sampling time denoted as T, then the inner product in (4) could be approximated as follows

$$\int_{-\infty}^{+\infty} h_n(t) f(t) dt \approx \sum_{i=-N}^{N} h_n(t_i) f(t_i) T \tag{14}$$

5.2   Scaling Basis

From engineering point of view, to meet various approximation requirements, it is important to adjust the time and frequecy scales of the basis. One can modify the basis (7) by scaling t variable via $\sigma \in (0, \infty)$ as a parameter. So if one substitutes $t := \frac{t}{\sigma}$ into (8) and modifies c_n to ensure orthonormality, then

$$h_n(t, \sigma) = c_{n,\sigma} e^{-\frac{t^2}{2\sigma^2}} H_n(\frac{t}{\sigma}) \ \ \text{where} \ c_{n,\sigma} = \frac{1}{(\sigma 2^n n! \sqrt{\pi})^{1/2}} \tag{15}$$

and

$$h_n(t, \sigma) = \frac{1}{\sqrt{\sigma}} h_n(\frac{t}{\sigma}) \ \text{and} \ \widetilde{h}_n(\omega, \sigma) = \sqrt{\sigma} \widetilde{h}_n(\sigma \omega) \tag{16}$$

Note that h_n as defined by (15) is the two arguments function whereas h_n as defined by (8) has only one argument. These functions are related by (16).

Thus by introducing the scaling parameter σ into (15) one may adjust both the dynamic range of the input argument $h_n(t, \sigma)$ and its frequency bandwidth

$$t \in [-\sigma\sqrt{2n+1}, \sigma\sqrt{2n+1}] \ ; \ \ \omega \in [-\frac{1}{\sigma}\sqrt{2n+1}, \frac{1}{\sigma}\sqrt{2n+1}] \tag{17}$$

Suppose that a one-variable function f defined over the range of its argument $t \in [-t_{\max}, t_{\max}]$ has to be approximated by using Hermite expansions. Assume that the retained function angular frequency should at least be ω_r, then according to (17), the following two conditions should be fulfilled

$$\sigma\sqrt{2n+1} \geq t_{\max} \ \ \text{and} \ \ \frac{1}{\sigma}\sqrt{2n+1} \geq \omega_r \tag{18}$$

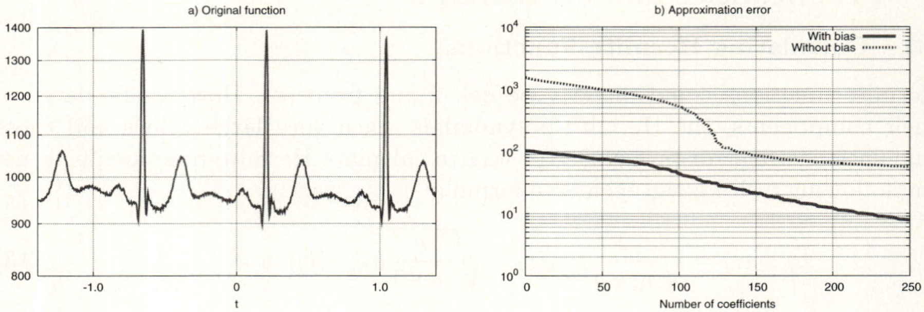

Fig. 2. a) The original function to be approximated, b) Approximation error with and without estimation of bias

or

$$\sigma \in [\sigma_1, \sigma_2] \quad \text{where} \quad \sigma_1 = \frac{t_{\max}}{\sqrt{2n+1}} \quad \text{and} \quad \sigma_2 = \frac{\sqrt{2n+1}}{\omega_r} \tag{19}$$

5.3   An Illustrative Example

This is a realistic practical example of an approximation and compression of ECG signal. Such compressed signal could be transmitted via communication channel providing mobile ECG based monitoring. The original signal collected in 1000 data points selected from the MIT-BIH Database[1] is shown in Fig.2a. It should be emphasized that the process of calculating Hermite expansion coefficients does not require any division operation, instead only multiplications and additions are used. In such a case, high accuracy of the approximation is expected. The only question is how to obtain it with as few coefficients as possible. The first step is to estimate initially only the bias and then to calculate the expansion coefficients according to (9). This method allows to decrease the approximation error more than 10 times as can be gleaned from Fig.2b.

It is interesting to find out which Hermite basis functions contribute the most to the decrease of the error. For this example the first 10 indices chosen out of 500 are the following

$$\begin{aligned}
\text{With bias:} \quad &(32, 1, 23, 30, 2, 49, 7, 4, 22, 28, ...) \\
\text{Without bias:} \quad &(0, 2, 4, 6, 8, 12, 10, 16, 14, 22, ...)
\end{aligned}$$

One can see that if the bias is not estimated initially, the most significant improvement to the error decrease comes from even functions. Indeed, the even Hermite functions have mean values different from zero.

Yet another improvement comes by selecting Hermite basis functions from a larger set, according to how they diminish the error. This is measured by the absolute value of the expansion coefficients. The error versus the number of

---

[1] http://www.physionet.org/physiobank/database/html/mitdbdir/mitdbdir.hm

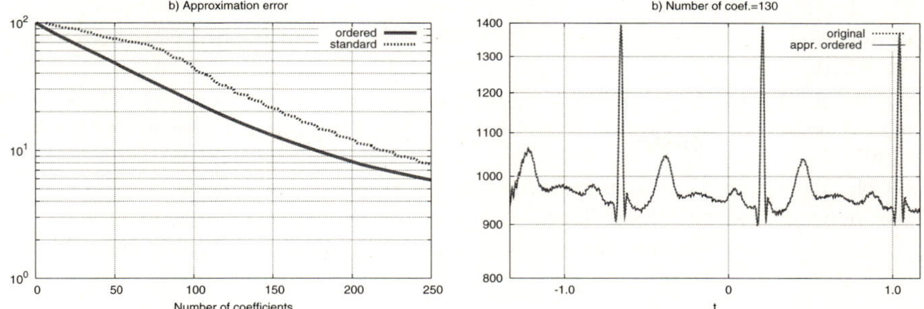

Fig. 3. a) Approximation error for the standard and ordered basis with bias estimation, b) The original and approximating function for the ordered basis of 130 coefficients out of 500, and the bias estimation. The differences are hardly seen.

coefficients, for the standard and ordered basis with bias estimation, is shown in Fig. 3a. In Fig.3b, the original and approximating function, with bias estimation for the ordered basis of 130 coefficients out of 500, are presented. The approximation error can be directly read from Fig. 3a.

The modification of the approximation equation as in (9) is the most important factor of the error decrease. Further improvement can be achieved by selecting such Hermite functions which contribute the most to the error decrease. If we consider this process from a compression point of view, one has also to take into account the indices of Hermite functions used in the basis, which should as well be stored or transmitted.

6   Conclusions

We described two enhancements and one proof of correctness for the Hermite function based approximation. The results have been proved in an abstract way and applied to the well known problem of ECG data compression. The algorithm is very efficient and numerically robust. We have run approximation with several thousand coefficients. These good numerical properties are due to the orthogonal basis, the recursive formula for the basis calculation and avoidance of searching methods. Significant improvement is achieved if one initially estimates the bias and then approximates the function with the extracted bias.

References

1. Beliczynski, B.: Properties of the Hermite activation functions in a neural approximation scheme. In: Beliczynski, B., Dzielinski, A., Iwanowski, M., Ribeiro, B. (eds.) ICANNGA 2007. LNCS, vol. 4432, pp. 46–54. Springer, Heidelberg (2007)
2. Linh, T.H.: Modern Generations of Artificial Neural Networks and their Applications in Selected Classification Problems. Publishing House of the Warsaw University of Technology (2004)

3. Ma, L., Khorasani, K.: Constructive feedforward neural networks using Hermite polynomial activation functions. IEEE Transactions on Neural Networks 16, 821–833 (2005)
4. Kwok, T., Yeung, D.: Constructive algorithms for structure learning in feedforward neural networks for regression problems. IEEE Trans. Neural Netw. 8(3), 630–645 (1997)
5. Kwok, T., Yeung, D.: Objective functions for training new hidden units in constructive neural networks. IEEE Trans. Neural Networks 8(5), 1131–1148 (1997)
6. Kreyszig, E.: Introductory functional analysis with applications. J. Wiley, Chichester (1978)

Multi-category Bayesian Decision by Neural Networks

Yoshifusa Ito[1], Cidambi Srinivasan[2], and Hiroyuki Izumi[3]

[1] School of Medicine, Aichi Medical University
Nagakute-cho, Aichi-ken, 480-1195 Japan
ito@aichi-med-u.ac.jp
[2] Department of Statistics, University of Kentucky
Patterson Office Tower, Lexington, Kentucky 40506, USA
srini@ms.uky.edu
[3] Department of Policy Science, Aichi-Gakuin University
Nisshin-shi, Aichi-ken, 470-0195 Japan
hizumi@psis.aichi-gakuin.ac.jp

Abstract. For neural networks, learning from dichotomous random samples is difficult. An example is learning of a Bayesian discriminant function. However, one-hidden-layer neural networks with fewer inner parameters can learn from such signals better than ordinary ones. We show that such neural networks can be used for approximating multi-category Bayesian discriminant functions when the state-conditional probability distributions are two dimensional normal distributions. Results of a simple simulation are shown as examples.

1   Introduction

The goal of this paper is to show that a one-hidden-layer neural network having a small number of inner parameters can learn multi-category Bayesian discriminant functions better than ordinary neural networks.

In 1998, Funahashi [2] proposed a neural network which may learn Bayesian discriminant functions in the case of two categories with normal distributions. If the patterns to be classified are in $\mathbf{R}^d$, his network has $2d$ hidden units. However, we have realized that this number can be decreased to $d + 1$ [3,5]. Since it is a general belief that a network having a smaller number of units works better, we used a network having $d + 1$ hidden units in simulations [4,6,7]. However, the network was successful only when the state-conditional probability distributions are just one-dimensional.

The constants in the approximation formula are called parameters, if they are optimized when the network is trained. Among the parameters, we call those the inner parameters if they are included in the activation functions. If the coefficients of the activation functions and the constant to be summed up with them in the approximation formula are to be optimized, then they are called the outer parameters.

V. Kůrková et al. (Eds.): ICANN 2008, Part I, LNCS 5163, pp. 21–30, 2008.

We observed that the difficulty in training of our networks arose from the optimization of the inner parameters. In the case of learning Bayesian discriminant functions, the teacher signals are dichotomous random variables. Learning with such teacher signals is difficult because the approximation cannot be realized by simply bringing the output of the network close to the target function. It may be a general method to simplify the parameter space when we meet difficulty in training neural networks. But it can be proved that the number of hidden units cannot be decreased from $d + 1$ [3]. Hence, we had to shift to a new policy to reduce the number of inner parameters of the network though it causes increment of the outer parameters. A special feature of the new network is that all the constants included in the activation functions are not to be optimized. This policy is theoretically detailed in [3] and, in this paper, we use neural networks based on an approximation theorem in [3].

In Ito et al. [4,5], we have discussed on the extension of the network to the multi-category case, where the method of obtaining the posterior probabilities is a little complicated. In this paper, we show that a set of Bayesian discriminant functions in the multi-category cases can be obtained by a simple method. They are not posterior probabilities in the multi-category Bayesian decision, but are those in the two-category cases. The multi-category Bayesian decision is reduced to plural two-category decisions in this paper.

Thus, use of a one-hidden-layer neural network having a small number of inner parameters and extension to the multi-category case are the main points of this paper. Though we still meet problems of local minima and over-learning in simulations with the network, the performance of the networks is considerably better. The network succeeds in the case of two-dimensional normal distributions, in which ordinary networks have never succeeded in our laboratory. The simulation illustrated in this paper is on the case of three categories with two dimensional normal distributions.

In below we start with historical results, because we use some of their ingredients and they include our experience which led us to the new type of the networks.

2   Preliminaries

The categories are denoted by $\theta_1, \cdots, \theta_n$ respectively and we set $\Theta = \{\theta_1, \cdots, \theta_n\}$. Let $\mathbf{R}^d$ be the d-dimensional Euclidean space ($\mathbf{R} = \mathbf{R}^1$) and let $x \in \mathbf{R}^d$ be the patterns to be classified. Denote by $N(\mu_i, \Sigma_i)$, $i = 1, \cdots, n$, the normal distributions, where μ_i and Σ_i are the mean vectors and the covariance matrices. They are the distributions of the patterns x from the respective categories. The prior probabilities are denoted by $P(\theta_i)$. Hence, $\sum_{i=1}^{n} P(\theta_i) = 1$. We suppose that $P(\theta_i)$ are all non-zero. The state-conditional probability density functions are denoted by $p(x|\theta_i)$. Here,

$$p(x|\theta_i) = \frac{1}{\sqrt{(2\pi)^d \Sigma_i}} e^{-\frac{1}{2}(x-\mu_i)^t \Sigma_i^{-1}(x-\mu_i)}, i = 1, \cdots, n. \tag{1}$$

Throughout the covariance matrices are assumed to be non-degenerate and, hence, Σ_i as well as Σ_i^{-1} are positive definite. Denote by $P(\theta_i|x)$ the posterior probabilities. We have

$$P(\theta_i|x) = \frac{P(\theta_i)p(x|\theta_i)}{p(x)}, \tag{2}$$

where $p(x) = \sum_{i=1}^{n} P(\theta_i)p(x|\theta_i)$.

Consider the case $n = 2$. Then, the network has a single output unit (see [2], [5], [11]). Let $F(x, w)$ be the output, where w is the weight vector. Let $\xi(x, \theta)$ be a function on $\mathbf{R}^d \times \Theta$, $\Theta = \{\theta_1, \theta_2\}$ such that $\xi(x, \theta_1) = 1$ and $\xi(x, \theta_2) = 0$. Let $E[\xi(x, \cdot)|x]$ and $V[\xi(x, \cdot)|x]$ be the conditional mean and variance of ξ respectively. In [10], [11] the proposition below is proved.

Proposition 1. If we set

$$\mathcal{E}(w) = \int_{\mathbf{R}^d} \sum_{i=1}^{2} (F(x, w) - \xi(x, \theta_i))^2 P(\theta_i)p(x|\theta_i)dx, \tag{3}$$

then

$$\mathcal{E}(w) = \int_{\mathbf{R}^d} (F(x, w) - E[\xi(x, \cdot)|x])^2 p(x) + \int_{\mathbf{R}^d} V[\xi(x, \cdot)|x]p(x)dx. \tag{4}$$

The second term of the right-hand side of (4) is independent of w. The conditional mean coincides with the posterior probability: $E[\xi(x, \cdot)|x] = P(\theta_i|x)$. Hence, if the output can approximate the posterior probability, we can expect that $F(x, w)$ approximates the posterior probability $E[\xi(x, \cdot)|x]$ when $\mathcal{E}(w)$ is minimized. Accordingly, learning of the network is carried out by minimizing

$$\mathcal{E}_n(w) = \frac{1}{n} \sum_{t=1}^{n} (F(x^{(k)}, w) - \xi(x^{(k)}, \theta^{(k)}))^2, \tag{5}$$

where $\{(x^{(k)}, \xi(x^{(k)}, \theta^{(k)}))\}_{k=1}^{n}$, $(x^{(k)}, \theta^{(k)}) \subset \mathbf{R}^d \times \Theta$, is the training set. Minimization of (5) can be realized by sequential learning. This method of training has actually been stated by many authors [2,4-11]. In this paper, this proposition plays an important role.

3 Two Traditional Networks

3.1 Funahashi's Network

Funahashi [2] has proposed a neural network which may learn a Bayesian discriminant function for the case of two categories involving the normal distribution. In this case, the posterior probability $P_1(\theta_1|x)$, the ratio $P_1(\theta_1|x)/P_2(\theta_2|x)$ and the log ratio $\log P_1(\theta_1|x)/P_2(\theta_2|x)$ can all be used as Bayesian discriminant functions [1].

Since the state-conditional probability distributions are normal, the log ratio is a quadratic function:

$$\log \frac{P(\theta_1|x)}{P(\theta_2|x)} = \log \frac{P(\theta_1)P(x|\theta_1)}{P(\theta_2)P(x|\theta_2)} = \log \frac{P(\theta_1)}{P(\theta_2)} - \frac{1}{2}\log \frac{|\Sigma_1|}{|\Sigma_2|}$$

$$- \frac{1}{2}\left\{ (x-\mu_1)^t \Sigma_1^{-1}(x-\mu_1) - (x-\mu_2)^t \Sigma_2^{-1}(x-\mu_2) \right\}. \tag{6}$$

He remarked that if the inner potential of the output unit of a one-hidden-layer neural network approximates this log ratio and the activation function of the output unit is the logistic function $\sigma(t) = (1 + e^{-t})^{-1}$, then the output of the network is the posterior probability:

$$\sigma\left(\log \frac{P(\theta_1|x)}{P(\theta_2|x)} \right) = \frac{P(\theta_1|x)}{P(\theta_1|x) + P(\theta_2|x)} = P(\theta_1|x). \tag{7}$$

Funahashi [2] also proved that if the activation function g satisfies a certain condition, a linear sum

$$\varphi_1(x) = \sum_{i=1}^{2d} a_i g(w_i \cdot x + t_i) + a_0, \tag{8}$$

can approximate any quadratic form, say $Q(x)$, on any compact set in $\mathbf{R}^d$. Since σ maps $\mathbf{R}$ onto $(0,1)$, $\sigma(\varphi_1(x))$ can approximate $\sigma(Q(x))$ in the sense of $L^2(\mathbf{R}^d, p)$, where p is the probability density defined in Section 2.

3.2   Previous Networks

Funahashi's formula (8) can approximate any quadratic form in the sense of $L^2(\mathbf{R}^d, p)$ [2]. However, it is proved in [3] that the number of the activation functions in the approximation formula can be reduced to $d + 1$:

$$\varphi_2(x) = \sum_{i=0}^{d} a_i g(w_i \cdot x + t_i) + a_0. \tag{9}$$

Moreover, it is established in [3] that this number is necessary and sufficient for the network to be capable of approximating any quadratic formula on $\mathbf{R}^d$. Hence, the activation functions in (9) cannot be decreased anymore. The approximation can also be in the sense of $L^2(\mathbf{R}^d, p)$.

If the approximation formula can include linear terms, one of the activation functions can be replaced by an inner product:

$$\varphi_3(x) = \sum_{i=1}^{d} a_i g(w_i \cdot x + t_i) + w_0 \cdot x + a_0. \tag{10}$$

This approximation formula can be realized by a neural network of the form illustrated in Fig.1. In Introduction, we say that (9) as well as (10) have $d + 1$ activation functions, for simplicity.

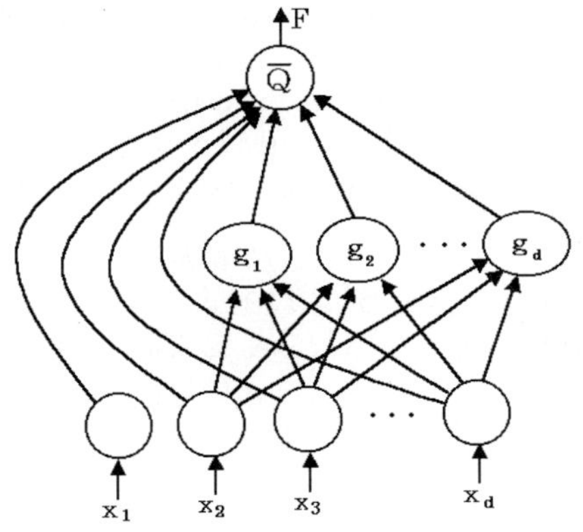

Fig. 1. *A* single-hidden-layer neural network having direct connections between the input layer and output unit

We have used neural networks based on the approximation formula (10) in the case of two categories with normal distributions [6,7]. But learning was successful only in the case where the probability distributions are simple: it was successful only in one-dimensional case. In order to overcome the difficulty, a general strategy is to decrease the number of hidden units (activation functions in the approximation formula). But the hidden units cannot be further decreased [3]. Observing the learning process, we have concluded that the difficulty comes from training of the inner parameters w_i and t_i. So we looked for the neural networks having a small number of inner parameters.

4   Networks with Fewer Inner Parameters

Learning with the dichotomous teacher signals may be a difficult task for the networks. In Fig.2, learning with dichotomous random teacher signals is compared to that with deterministic teacher signals.

The dots in Fig.1 are teacher signals. They are deterministic in the left and dichotomous in the right. In the latter they are from $\{(x, \xi(x, \theta))\}$, where $\xi(x, \theta)$ is defined above Proposition 1, and the target function is the posterior probability $P(\theta_1|x) = E[\xi(x, \theta_1)|x]$. For simplicity, the one-dimensional examples are illustrated. In the case of learning with random dichotomous teacher signals, the network cannot approximate the function just by simply decreasing the distance of the output from the target.

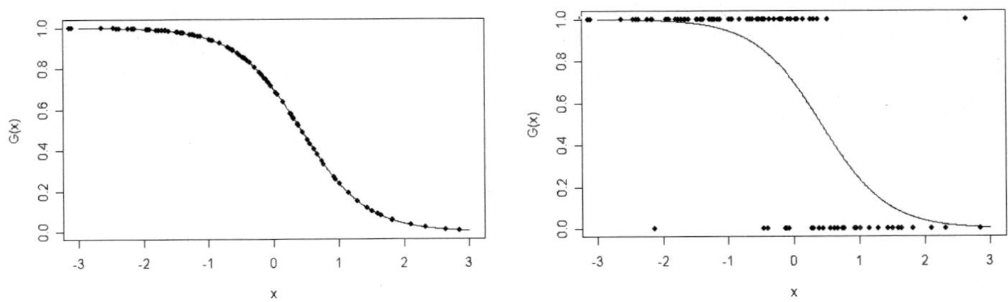

Fig. 2. Curves are target functions. Teacher signals are deterministic (Left) and dichotomous (Right).

The lemma and the theorem in this section are proved in [3]. Though we omit details to avoid repetition, the notation are self explanatory. Among the notation, μ_u is the projection of the probability measure μ onto the line $\{ut : -\infty < t < \infty\}$, where $u \in \mathbf{S}^{d-1}$ is a unit vector. If $g^{(i)}(0) \neq 0$ $(1 \leq i \leq n)$ and $g^{(i)}$ are bounded, the condition of the theorem is satisfied for most of familiar probability measure.

Lemma 1. *For any $n \in \mathbf{N}_0$, there are unit vectors $u_k \in \mathbf{S}^{d-1}$ $(1 \leq k \leq {}_{n+d-1}C_n)$ such that, for any i $(0 \leq i \leq n)$, the functions $(u_k \cdot x)^i$ $(1 \leq k \leq {}_{i+d-1}C_i)$ are linearly independent.*

For $n = 2$, $u_1 = (1, 0, \cdots, 0), \cdots, u_d = (0, \cdots, 0, 1), \cdots, u_{d+1} = (\frac{1}{\sqrt{2}}, \frac{1}{\sqrt{2}}, 0, \cdots, 0), \cdots, u_{\frac{1}{2}d(d+1)} = (0, \cdots, 0, \frac{1}{\sqrt{2}}, \frac{1}{\sqrt{2}})$ are such unit vectors.

Theorem 2. *Let μ be a probability measure on $\mathbf{R}^d$ and let $g \in C^N(\mathbf{R})$ $(N \in \mathbf{N}_0)$ such that $g^{(i)}(0) \neq 0$ $(1 \leq i \leq n \leq N)$. Suppose that there exists a function g_0 on $\mathbf{R}$ such that , for any $\delta(|\delta| < 1)$,*

$$|g^{(i)}(\delta t)t^i| < g_0(t), \qquad 0 \leq i \leq n,$$

and, for any unit vectors $u \in \mathbf{S}^{d-1}$,

$$g_0 \in L^p(\mathbf{R}, \mu_u)(1 \leq p \leq \infty).$$

Then, for any polynomial P of degree n and any $\xi > 0$, there are coefficients a_{ik} and constants $\delta_i(0 \leq i \leq n, 1 \leq k \leq {}_{n+d-1}C_n)$ for which

$$\left\| \frac{\partial^\alpha}{\partial x^\alpha}P - \frac{\partial^\alpha}{\partial x^\alpha}\bar{P} \right\|_{L^p(\mathbf{R}^d,\mu)} < \varepsilon, \qquad |\alpha| \leq N,$$

where

$$\bar{P}(x) = \sum_{i=0}^{n} \sum_{k=1}^{{}_{n+d-1}C_n} a_{ik}g(\delta_i u_k \cdot x)$$

and u_k are unit vectors satisfying the condition of the lemma above.

We use this theorem for $n = 2$. Since $a_{01}g(\delta_0 u_1 \cdot x)$ in the approximation formula in the theorem is to approximate a constant and $\sum_{k=1}^{d} a_{1k}g(\delta_1 u_k \cdot x)$ to approximate an inner product of a vector and x, we can use an approximation formula of the form

$$\varphi_4(x) = \sum_{k=1}^{d(d+1)/2} a_{2k}g(\delta u_k \cdot x) + \sum_{k=1}^{d} a_{1k}x_k + a_0 \tag{11}$$

to approximate quadratic forms on $\mathbf{R}^d$ [8,9]. This approximation formula has only a single inner constant δ, but has more outer parameters a_0, a_{1k} and a_{2k}. The approximation formula (11) can be realized by a neural network of the form illustrated in Fig.2 having $\frac{1}{2}d(d+1)$ hidden units. Since $\sigma''(0) = 0$, the logistic function does not satisfy the condition of the theorem. However, its shift does.

5   Extension to Multi-category Cases

Now we treat the multi-category Bayesian decision problem. The posterior probabilities $P(\theta_i|x)$, $i = 1, \cdots, n$, can be used as Bayesian discriminant functions [1]. However, these cannot be obtained in the same way as in the case of two categories, where (7) plays an important role. In order to approximate $P(\theta_i|x)$ in the same way, the inner potential of the output unit must approximate a log ratio

$$\log \frac{P(\theta_i)p(x|\theta_i)}{\sum_{j\neq i} P(\theta_j)p(x|\theta_j)}.$$

This is not generally a quadratic form for $n \geq 3$. Theoretically, a neural network having many units can approximate this log ratio, but learning of this log ratio may be very difficult.

We reduce the multi-category case to the two-category case. If we choose (x, θ_i) and (x, θ_n), $i \neq n$, from an n category teacher sequence, we obtain a two-category teacher sequence. We can treat this sequence as a teacher sequence from a two category source. Then, the prior probabilities of patterns from the respective categories are

$$P_{in}(\theta_i) = \frac{P(\theta_i)}{P(\theta_i) + P(\theta_n)}, \qquad P_{in}(\theta_n) = \frac{P(\theta_n)}{P(\theta_i) + P(\theta_n)}. \qquad (12)$$

Hence, the posterior probabilities are

$$P_{in}(\theta_i|x) = \frac{P_{in}(\theta_i)p(x|\theta_i)}{p_{in}(x)}, \, P_{in}(\theta_n|x) = \frac{P_{in}(\theta_n)p(x|\theta_n)}{p_{in}(x)}, \qquad (13)$$

where $p_{in}(x)$ is the probability density function of x:

$$p_{in}(x) = P_{in}(\theta_i)p(x|\theta_i) + P_{in}(\theta_n)p(x|\theta_n). \qquad (14)$$

We have

$$P_{in}(\theta_i|x) = \sigma\left(\log \frac{P_{in}(\theta_i|x)}{P_{in}(\theta_n|x)}\right) = \sigma\left(\log \frac{P_{in}(\theta_i)p(x|\theta_i)}{P_{in}(\theta_n)p(x|\theta_n)}\right)$$

$$= \sigma\left(\log \frac{P(\theta_i)p(x|\theta_i)}{P(\theta_n)p(x|\theta_n)}\right) = \sigma\left(\log \frac{P(\theta_i|x)}{P(\theta_n|x)}\right).$$

Accordingly,

$$P_{in}(\theta_i|x) = \frac{P(\theta_i|x)}{P(\theta_i|x) + P(\theta_n|x)}. \qquad (15)$$

Since the log ratio

$$\log \frac{P_{in}(\theta_i|x)}{P_{in}(\theta_n|x)} \qquad (16)$$

is a quadratic form, this posterior probability can be approximated by a neural network based on (11). For $i = n$, we have $P_{nn}(\theta_n|x) = \dfrac{1}{2}$. Hence, using $n - 1$ neural networks we can obtain (15) for $i = 1, \cdots, n$.

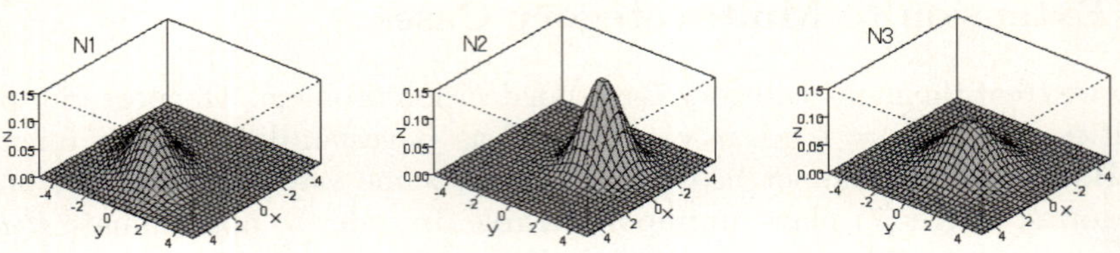

Fig. 3. State-conditional probability density functions of Categories 1 (N1), 2 (N2) and 3 (N3)

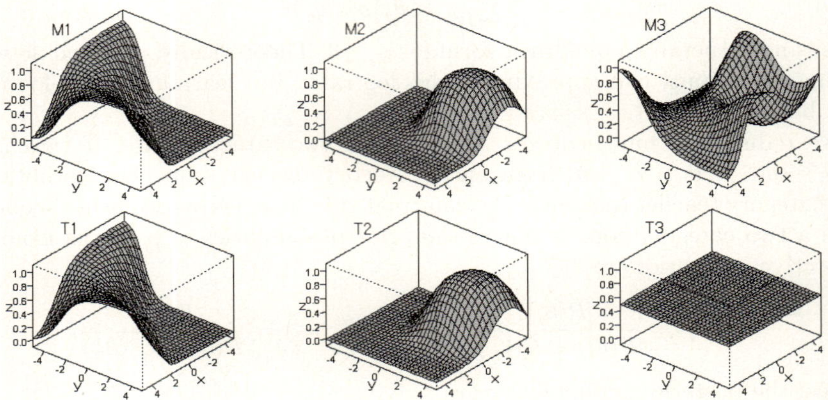

Fig. 4. Posterior probabilities $P(\theta_i|x)$ (upper) and $P_{i3}(\theta_i|x)$ (lower), $i=1,2,3$ (see Text)

Duda et al. [1] remarked that if the discriminant functions h_i are replaced by $f(h_i)$, where f is a monotonically increasing function, the resulting classification is unchanged [1]. We further remark that even if the monotone function f is parameterized by $x \in \mathbf{R}^d$, say f_x, the classification is unchanged. Consequently, Bayesian classification by $P(\theta_i|x)$, $i = 1, \cdots, n$ is unchanged even if they are respectively replaced by (15).

6   Simulations

Our prior experience had shown that an ordinary one-hidden layer neural network can rarely learn the Bayesian discriminant function if the dimension of the space of patterns is higher than or equal to 2 [6,7]. The present network performs better in higher dimensions. In the simulation below, the state-conditional probability distributions are two-dimensional normal distributions: $N(\mu_1, \Sigma_1)$, $N(\mu_2, \Sigma_2)$ and $N(\mu_3, \Sigma_3)$, where $\mu_1 = (1,\text{-}1)$, $\mu_2 = (\text{-}1,1)$, $\mu_3 = (0,0)$ and

$$\Sigma_1 = \begin{pmatrix} 2 & 1 \\ 1 & 2 \end{pmatrix}, \Sigma_2 = \begin{pmatrix} 1 & -0.1 \\ -0.1 & 1 \end{pmatrix}, \Sigma_3 = \begin{pmatrix} 2 & 0 \\ 0 & 2 \end{pmatrix}, \text{ and } P(\theta_1) = P(\theta_2) = P(\theta_3) =$$

1/3. In Fig.3, N1, N2 and N3 illustrate these normal distributions respectively.

Fig.4 illustrates the discriminant functions: M1, M2 and M3 are the posterior probabilities $P(\theta_1|x)$, $P(\theta_2|x)$ and $P(\theta_3|x)$, and T1, T2 and T3 are posterior

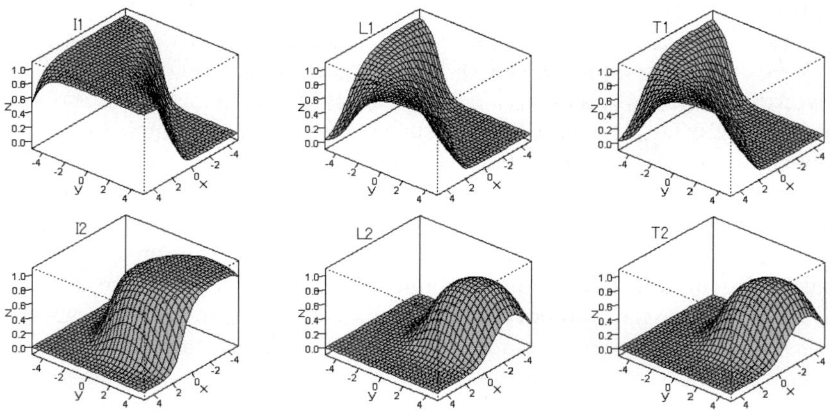

Fig. 5. Learning results of the two networks compared with theoretically obtained posterior probabilities

Table 1. Allocation results in a simulation

	$P(\theta_i|x)$	$P_{i3}(\theta_i|x)$	$SP_{i3}(\theta_i|x)$
Alloc. to θ_1	395	395	391
Alloc. to θ_2	394	394	378
Alloc. to θ_3	211	211	231
Correct Alloc.	646	646	650

probabilities $P_{13}(\theta_1|x)$, $P_{23}(\theta_2|x)$ and $P_{33}(\theta_3|x)$ (see Sections 3 and 5 for the notation). In the figure T1 and T2 look similar to M1 and M2 respectively. However, the surfaces in T1 and T2 are above those in M1 and M2 respectively. Their values at origin are $M1(0,0) = 0.19$, $T1(0,0) = 0.30$, $M2(0,0) = 0.36$ and $T2(0,0) = 0.44$ as can be expected from their definitions: for $i \neq n$, $P_{in}(\theta_i|x) > P(\theta_i|x)$ by (15).

Two networks are constructed based on (11), where a shifted logistic function is used. They are trained with 2000 pairs (x, θ) of teacher signals. In Fig.5 performances of the two neural networks are illustrated. The outputs at the starting point of learning are I1 and I2, those at the final stage are L1 and L2, and their targets are T1 and T2 respectively. The targets T1 and T2 are posterior probabilities $P_{13}(\theta_1|x)$ and $P_{23}(\theta_2|x)$ respectively, which are also shown in Fig.4.

In an example, the trained networks are tested with 1000 test signals, among which 337, 341 and 322 are from the categories θ_1, θ_2 and θ_3 respectively. The columns of Table 1 show the allocation results by the three sets of discriminant functions, $\{P(\theta_i|x),\ i = 1, 2, 3\}$, $\{P(\theta_{i3}|x),\ i = 1, 2, 3\}$ and $\{SP_{i3}(\theta_i|x),\ i = 1, 2, 3\}$, where $SP_{i3}(\theta_i|x),\ i = 1, 2$ are $L1$ and $L2$ in Fig. 5 and $SP_{33}(\theta_3|x) = \frac{1}{2}$. The numbers of correctly allocated patterns by the respective sets of discriminant functions are 646, 646 and 650.

7   Discussions

If the log ratio (16) is a polynomial of low degree, our method can be applied and Bayesian discriminant functions $P_{in}(\theta_i|x)$ in multi-category cases can be obtained. In the case of our familiar distributions such as binomial, polynomial, exponential distributions and others, the log ratio is linear. Hence, the method can be applied.

In our network, the number of inner parameters is decreased but this inevitably causes increment of hidden units and, hence, the number of outer parameters is increased. However, the total number of the parameters is not so increased. In the approximation formulae (10) and (11), there are $2d + 1$ and $\frac{1}{2}d^2 + \frac{3}{2}d + 1$ outer parameters and $d^2 + d$ and 1 inner parameters respectively. Hence, the total numbers of parameters in (10) and (11) are $d^2 + 3d + 1$ and $\frac{1}{2}d^2 + \frac{3}{2}d + 2$ respectively. The latter is smaller than the former. In these counting, note that the unit vectors u_k in (11) are fixed and not regarded as parameters.

Training of the outer parameters is easier than the inner parameters. Accordingly, training of the network based on (11) is easier than that based on (10). However, the problems of local minima and overlearning are still to be solved. Furthermore, we still have to choose the initial values of the network parameters carefully in simulations.

References

1. Duda, R.O., Hart, P.E.: Pattern classification and scene analysis. John Wiley & Sons, New York (1973)
2. Funahashi, K.: Multilayer neural networks and Bayes decision theory. Neural Networks 11, 209–213 (1998)
3. Ito, Y.: Simultaneous approximations of polynomials and derivatives and their applications to neural networks. Neural Computation (in press)
4. Ito, Y., Srinivasan, C.: Multicategory Bayesian decision using a three-layer neural network. In: Proceedings of ICANN/ICONIP 2003, pp. 253–261 (2003)
5. Ito, Y., Srinivasan, C.: Bayesian decision theory on three-layer neural networks. Neurocomputing 63, 209–228 (2005)
6. Ito, Y., Srinivasan, C., Izumi, H.: Bayesian learning of neural networks adapted to changes of prior probabilities. In: Duch, W., Kacprzyk, J., Oja, E., Zadrożny, S. (eds.) ICANN 2005. LNCS, vol. 3697, pp. 253–259. Springer, Heidelberg (2005)
7. Ito, Y., Srinivasan, C., Izumi, H.: Discriminant analysis by a neural network with Mahalanobis distance. In: Kollias, S.D., Stafylopatis, A., Duch, W., Oja, E. (eds.) ICANN 2006. LNCS, vol. 4132, pp. 350–360. Springer, Heidelberg (2006)
8. Ito, Y., Srinivasan, C., Izumi, H.: A neural network having fewer inner constants to be trained and Bayesian decision. In: Proceedings of IJCNN 2007, pp. 1783–1788 (2007)
9. Ito, Y., Srinivasan, C., Izumi, H.: Learning of Bayesian discriminant functions by a layered neural network. In: ICONIP 2007 Proceedings, Springer, Heidelberg (2007)
10. Richard, M.D., Lipmann, R.P.: Neural network classifiers estimate Bayesian a posteriori probabilities. Neural Computation 3, 461–483 (1991)
11. Ruck, M.D., Rogers, S., Kabrisky, M., Oxley, H., Sutter, B.: The multilayer perceptron as approximator to a Bayes optimal discriminant function. IEEE Transactions on Neural Networks 1, 296–298 (1990)

Estimates of Network Complexity and Integral Representations

Paul C. Kainen[1] and Věra Kůrková[2]

[1] Department of Mathematics, Georgetown University, Washington,
D.C. 20057-1233, USA
kainen@georgetown.edu
[2] Institute of Computer Science, Academy of Sciences of the Czech Republic
Pod Vodárenskou věží 2, Prague 8, Czech Republic
vera@cs.cas.cz

Abstract. Upper bounds on rates of approximation by neural networks are derived for functions representable as integrals in the form of networks with infinitely many units. The bounds are applied to perceptron networks.

1 Introduction

The most straightforward measure of network complexity is the number of network units. Estimates of this number can be obtained by using tools from approximation theory. Assuming that training data are chosen from some multivariable function or considering all functions interpolating such data, one can investigate rates of approximations of these functions by neural networks. Comparison of such rates for various types of computational units may suggest what types of functions interpolating the data and what properties of the units themselves permit approximation with fewer units.

One can estimate rates of neural network approximation using a result by Maurey [1], Jones [2] and Barron [3] on approximation by "variable-basis functions". Maurey-Jones-Barron's theorem implies an upper bound on rates of approximation by one-hidden-layer networks with an increasing number n of units. The bound is of the form $n^{-1/2}$ times a certain variational norm tailored to the class of computational units of the function to be approximated [4]. Similar bounds in terms of variational norms were derived for speed of convergence of infima of empirical and expected error functionals over networks with an increasing number of units [5,6]. Thus, properties of variational norms corresponding to various types of network units can provide some insight into the impact of a choice of units on network complexity.

Integral transformations (such as Fourier, Gabor or Radon) play an important role in applied science. Also in neurocomputing, various integral transformations in the form of "networks with infinitely many units" have been considered. Originally, they were used to prove the universal approximation property [7,8] and later, in combination with the Maurey-Jones-Barron theorem, integral transforms were applied to estimate network complexity. Barron [3] used functions

V. Kůrková et al. (Eds.): ICANN 2008, Part I, LNCS 5163, pp. 31–40, 2008.
© Springer-Verlag Berlin Heidelberg 2008

representable as weighted Fourier transforms, Girosi and Anzelotti [9] studied convolutions with suitable kernels, Kůrková et al. [10] derived an integral representation in the form of a network with infinitely many Heaviside perceptrons for smooth compactly supported functions, and Kainen et al. [11,12] extended this representation to functions vanishing sufficiently rapidly at infinity.

In this paper, we present a quite general framework for estimation of variational norms for functions representable as integrals of the form of networks with infinitely many computational units. Our approach is based on the Bochner integral (e.g., [22, pp. 130–132]) which extends the concept of Lebesgue integrals to mappings into Banach spaces; the value of a Bochner integral is a function, not a number. The Bochner integral is suitable for dealing with arbitrary computational units. One can represent them as mappings from the spaces of parameters (weights, biases, centroids, etc.) to Banach spaces of functions computable by units with such parameters; see Kainen [19]. A special case of application of the Bochner to estimation of rates of approximation by neural networks was sketched by Girosi and Anzellotti [9].

The paper is organized as follows. Section 2 recalls Maurey-Jones-Barron's theorem and variational norms. Section 3 gives upper bounds on variational norms for functions representable as integrals of the form of networks with infinitely many hidden units. In Section 4, these estimates are applied to perceptron networks. Section 5 is a brief discussion.

2 Rates of Variable-Basis Approximation and Variational Norm

One-hidden-layer feedforward neural networks belong to a class of computational models which can mathematically be described as *variable-basis* schemas. Such models compute functions from sets of the form

$$\operatorname{span}_n G = \left\{ \sum_{i=1}^n w_i g_i \,|\, w_i \in \mathbb{R},\, g_i \in G \right\},$$

where G is a set (or *dictionary*) of functions. For example, G could be the set of trigonometric polynomials, the set of functions computable by perceptron or radial-basis function units, etc. The number n expresses the *model complexity* (in the case of one-hidden-layer neural networks, it is the number of units in the hidden layer).

The distance of an element f in a normed linear space $(\mathcal{X}, \|.\|_{\mathcal{X}})$ from a subset A is denoted

$$\|f - A\|_{\mathcal{X}} = \inf_{g \in A} \|f - g\|_{\mathcal{X}}.$$

An upper bound on approximation by

$$\operatorname{conv}_n G = \left\{ \sum_{i=1}^n a_i g_i \,|\, a_i \in [0,1], \sum_{i=1}^n a_i = 1,\, g_i \in G \right\}$$

was derived by Maurey [1], rediscovered by Jones [2] (with an iterative construction), and refined by Barron [3].

Theorem 1 (Maurey-Jones-Barron). *Let G be a bounded nonempty subset of a Hilbert space $(X, \|.\|)$ and $s_G = \sup_{g \in G} \|g\|$. Then for every $f \in \mathrm{cl}\,\mathrm{conv}\,G$ and for every positive integer n,*

$$\|f - \mathrm{conv}_n G\|^2 \le \frac{s_G^2 - \|f\|^2}{n}.$$

Theorem 1 can be reformulated in terms of a norm called *G-variation*. This norm is defined for any bounded nonempty subset G of any normed linear space $(X, \|.\|)$ as the Minkowski functional of the closed convex symmetric hull of G, i.e.,

$$\|f\|_G = \inf \left\{ c > 0 \mid c^{-1} f \in \mathrm{cl}\,\mathrm{conv}\,(G \cup -G) \right\}, \tag{1}$$

where the closure cl is taken with respect to the topology generated by the norm $\|.\|$ and conv denotes the convex hull. It is a norm on the subspace of $\mathcal{X}$ formed by those $f \in \mathcal{X}$ for which $\|f\|_G < \infty$. G-variation depends on the norm on the ambient space, but as this is implicit, we omit it in the notation.

Variational norms were introduced by Barron [13] for characteristic functions of certain families of subsets of $\mathbb{R}^d$, in particular, for the set of characteristic functions of closed half-spaces corresponding to the set of functions computable by Heaviside perceptrons. The general concept was defined by Kůrková [14]. The following reformulation of Theorem 1 in terms of variational norm from [14] (see also [4]) allows one to give an upper bound on variable-basis approximation for all functions in a Hilbert space.

Theorem 2. *Let $(\mathcal{X}, \|.\|)$ be a Hilbert space, G its bounded nonempty subset, $s_G = \sup_{g \in G} \|g\|$. Then for every $f \in \mathcal{X}$ and every positive integer n,*

$$\|f - \mathrm{span}_n G\|^2 \le \frac{s_G^2 \|f\|_G^2 - \|f\|^2}{n}.$$

A similar result to Theorem 2 can be obtained in $\mathcal{L}^q$-spaces with $q \in (1, \infty)$ using a result by Darken et al. [15]; for a slightly simplified argument see also [6].

Theorem 3. *Let G be a bounded subset of $\mathcal{L}^q(\Omega, \rho)$, $q \in (1, \infty)$, and ρ a Radon measure. Then for every $f \in \mathrm{cl}\,\mathrm{conv}\,G$ and every positive integer n,*

$$\|f - \mathrm{conv}_n G\|_{\mathcal{L}^q(\Omega, \rho)} \le \frac{2^{1+1/r} s_G \|f\|_G}{n^{1/s}},$$

where $1/q + 1/p = 1$, $r = \min(p, q)$, $s = \max(p, q)$.

Thus, investigation of variational norms can provide some insight into when multivariable functions can be efficiently approximated by various computational models. The following proposition states basic properties of the variational norm: (i) follows directly from the definition and for (ii) see [10] and [18].

Proposition 1. *Let $(\mathcal{X}, \|.\|)$ be a normed linear space, G its nonempty bounded subset. Then:*
(i) for all $f \in \mathcal{X}$ representable as $f = \sum_{i=1}^{k} w_i g_i$ with all $g_i \in G$ and $w_i \in \mathbb{R}$, $\|f\|_G \leq \sum_{i=1}^{k} |w_i|$;
(ii) for all $f \in \mathcal{X}$, $\{f_i\}_{i=1}^{\infty} \subset \mathcal{X}$ satisfying $b_i = \|f_i\|_G < \infty$ for all i, $\lim_{i \to \infty} \|f_i - f\|_{\mathcal{X}} = 0$ and $b = \lim_{i \to \infty} b_i$, one has $\|f\|_G \leq b$.

3 Variational Norm of Functions Representable as Networks with Infinitely Many Units

Typically, sets of functions computable by neural-network units are parameterized; that is they are of the form

$$G_\phi = \{\phi(.,y) \,|\, y \in Y\},$$

where $\phi : \Omega \times Y \to \mathbb{R}$, Y is the set of parameters and Ω is the set of variables. Such a parameterized set of functions can be represented by a mapping

$$\Phi : Y \to \mathcal{X},$$

where $\mathcal{X}$ is a suitable function space; Φ is defined for all $y \in Y$ as

$$\Phi(y)(x) = \phi(x,y).$$

We denote

$$\Phi(Y) = G_\phi = \{\phi(.,y) \,|\, y \in Y\} \qquad \text{and} \qquad s_\Phi = \sup_{y \in Y} \|\phi(.,y)\|_{\mathcal{X}}. \qquad (2)$$

For example, the set of functions computable by *perceptrons with an activation function* $\sigma : \mathbb{R} \to \mathbb{R}$ can be described by a mapping Φ_σ on $\mathbb{R}^{d+1}$ defined for $(v, b) \in \mathbb{R}^d \times \mathbb{R} = \mathbb{R}^{d+1}$ as

$$\Phi_\sigma(v,b)(x) = \sigma(v \cdot x + b).$$

Similarly, the sets of *radial-basis functions* with the radial function $\psi : \mathbb{R} \to \mathbb{R}$ can be described by a mapping Φ_ψ on $\mathbb{R}^{d+1}$ defined for $(v, b) \in \mathbb{R}^d \times \mathbb{R} = \mathbb{R}^{d+1}$ as

$$\Phi_\psi(v,b)(x) = \psi(b\|x - v\|).$$

In this paper, we consider parameterized sets of functions belonging to the space $(\mathcal{C}(\Omega), \|.\|_{\sup})$ of all continuous functions on $\Omega \subseteq \mathbb{R}^d$ with the supremum norm or to a reproducing kernel Hilbert space or to an $\mathcal{L}^q$-space with $q \in [1, \infty)$. A Hilbert space $\mathcal{X}$ of point-wise defined real-valued functions on an arbitrary set Ω is called a *reproducing kernel Hilbert space (RKHS)* when all evaluation functionals on $\mathcal{X}$ are bounded [16]. For ρ a measure on Ω and $q \in [1, \infty)$, we denote by $\mathcal{L}^q(\Omega, \rho)$ the space of all real-valued functions h satisfying $\int_\Omega |h(y)|^q d\rho < \infty$. When ρ is the Lebesgue measure, we sometimes write merely $\mathcal{L}^q(\Omega)$.

To estimate rates of approximation by neural networks we estimate $\Phi(Y)$-variation of functions representable as one-hidden-layer networks with infinitely many hidden units from $\Phi(Y)$. More precisely, we consider functions which can be expressed for a suitable measure μ on Y and μ a. e. $x \in \Omega$ as Lebesgue integrals of the form

$$f(x) = \int_Y w(y)\phi(x,y)d\mu(y). \tag{3}$$

Such functions are images of the corresponding weight functions w under the integral operator L_ϕ defined as

$$L_\phi(w)(x) = \int_Y w(y)\phi(x,y)d\mu(y).$$

Analogy with Proposition 1(i) suggests that for f representable as on (3) one should expect

$$\|f\|_{\Phi(Y)} \leq \int_Y |w(y)|d\mu(y). \tag{4}$$

Various special cases of integral representations of the form (3) have been used in combination with Maurey-Jones-Barron's theorem. For example, Barron [3] proved that a function f representable as a weighted Fourier transform belongs to a convex hull of trigonometric units and thus Theorem 1 can be used to estimate rates of approximation of f by perceptrons networks. Girosi and Anzelotti [9] proved a similar estimate for convolutions with suitable kernels. Explicitly as an upper bound on variation, the estimate in terms of the $\mathcal{L}^1(Y)$-norm of the weight function w was derived in [10] for integral representations $\int_Y w(y)\phi(x,y)dy$ with both Ω and Y compact and ϕ continuous in both variables. It was extended in [17] to integral representations with Heaviside kernel.

However, many functions of interest are defined on non-compact domains Ω, their integral representations may have parameters in non-compact sets Y, and computational units need not be continuous. The following theorems include these cases.

Theorem 4. *Let $(\mathcal{X}, \|.\|_{\mathcal{X}})$ be a Banach space of real-valued functions on a set $\Omega \subseteq \mathbb{R}^p$ such that all evaluation functionals on $\mathcal{X}$ are bounded and suppose $f \in \mathcal{X}$ can be represented for all $x \in \Omega$ as*

$$f(x) = \int_Y w(y)\phi(x,y)d\mu(y),$$

where Y, w, ϕ and μ satisfy both following conditions:

(i) $Y \subseteq \mathbb{R}^p$, p is a positive integer, $Y \setminus Y_0 = \cup_{m=1}^\infty Y_m$, where $\mu(Y_0) = 0$ and for all $m \geq 1$, Y_m is compact and $Y_m \subseteq Y_{m+1}$, and $(Y, \mathcal{S}, \mu)$ is a Radon measure space,
(ii) $\Phi(Y)$ is a bounded subset of $\mathcal{X}$, $w \in \mathcal{L}^1(Y, \mu)$, and $w\Phi : Y \setminus Y_0 \to \mathcal{X}$ is continuous.

Then $\|f\|_{\Phi(Y)} \leq \|w\|_{\mathcal{L}^1(Y,\mu)}$ and for all positive integers n,

$$\|f - \text{span}_n \Phi(Y)\|_{\mathcal{X}}^2 \leq \frac{s_\Phi^2 \|w\|_{\mathcal{L}^1(Y,\mu)}^2 - \|f\|_{\mathcal{X}}^2}{n}.$$

Theorem 5. *Let $\mathcal{X} = \mathcal{L}^q(\Omega, \rho)$, $q \in [1, \infty)$, where $\Omega \subseteq \mathbb{R}^d$ and ρ is a σ-finite measure. Let $f \in \mathcal{X}$ satisfy for ρ-a.e. $x \in \Omega$*

$$f(x) = \int_Y w(y)\phi(x, y)d\mu(y),$$

where Y, w, ϕ, and μ satisfy the following three conditions:
(i) $Y \subseteq \mathbb{R}^p$, p is a positive integer, $Y \setminus Y_0 = \cup_{m=1}^{\infty} Y_m$, where $\mu(Y_0) = 0$ and for all $m \geq 1$, Y_m is compact and $Y_m \subseteq Y_{m+1}$,
(ii) $\Phi(Y)$ is a bounded subset of $\mathcal{X}$, $w \in \mathcal{L}^1(Y, \mu)$, and $w\Phi : Y \setminus Y_0 \to \mathcal{X}$ is continuous,
(iii) $(Y, \mathcal{S}, \mu)$ is a Radon measure space and $\phi : \Omega \times Y \to \mathbb{R}$ is $\rho \times \mu$-measurable. Then for all positive integers n, for all $q \in [1, \infty)$

$$\|f\|_{\Phi(Y)} \leq \|w\|_{\mathcal{L}^1(Y,\mu)},$$

for all $q \in (1, \infty)$ and q' satisfying $1/q + 1/q' = 1$, $r = \min(q, q')$, $s = \max(q, q')$,

$$\|f - \text{span}_n \Phi(Y)\|_{\mathcal{X}} \leq \frac{2^{1/r} 2 s_\Phi \|w\|_{\mathcal{L}^1(Y,\mu)}}{n^{1/s}},$$

and for $q = 2$,

$$\|f - \text{span}_n \Phi(Y)\|_{\mathcal{X}}^2 \leq \frac{s_\Phi^2 \|w\|_{\mathcal{L}^1(Y,\mu)}^2 - \|f\|_{\mathcal{X}}^2}{n}.$$

Rigorous derivations of these two theorems are rather lengthy and require some technical lemmas; here we only sketch the main ideas (for full proofs see [18]). The proofs are based on properties of the *Bochner integral $I(w\Phi)$* of the mapping $w\Phi : Y \to \mathcal{X}$ defined as $w\Phi(y) = w(y)\Phi(y)$. Using Lebesgue dominated convergence, one can show that $\Phi(Y)$-variation of $I(w\Phi)$ is bounded by $\|w\|_{\mathcal{L}^1(Y,\mu)}$. A crucial step in proving that the Bochner integral $I(w\Phi)$ is equal to f is the equality of evaluations of the Bochner integral at almost all x to the Lebesgue integrals $f(x) = \int_Y w(y)\phi(x, y)d\mu(y)$. This equality holds for all Banach spaces $\mathcal{X}$ of pointwise defined functions such that all evaluation functionals on $\mathcal{X}$ are bounded; this class includes all RKHSs and the space $(\mathcal{C}(\Omega), \|.\|_{\sup})$. It was shown in [19] that this equality also holds for $\mathcal{L}^q$-spaces with $q \in [1, \infty)$. Theorems 4 and 5 are proven for compact sets of parameters first and then extended to non-compact parameter sets using convergence properties of variational norm from Proposition 1 and Lebesgue dominated convergence.

Theorems 4 and 5 show that for functions representable as networks with infinitely many units, the growth of model complexity with increasing accuracy depends on the $\mathcal{L}^1$-norm of the output weight function.

4   Approximation by Perceptron Networks

To apply results from Section 3 to neural networks, we use the following straightforward corollary of Theorem 5 about parameterized families in $\mathcal{L}^2(\Omega) = \mathcal{L}^2(\Omega, \lambda)$, where λ denotes the Lebesgue measure.

Corollary 1. *Let $\Omega \subseteq \mathbb{R}^d$ be Lebesgue measurable and $f \in \mathcal{L}^2(\Omega)$ be such that for a.e. $x \in \Omega$,*

$$f(x) = \int_Y w(y)\phi(x, y)dy,$$

where Y, w, and ϕ satisfy the following three conditions:
(i) $Y \subseteq \mathbb{R}^p$ is Lebesgue measurable, p is a positive integer, $Y \setminus Y_0 = \cup_{m=1}^\infty Y_m$, where $\lambda(Y_0) = 0$ and for all positive integers m, Y_m is compact and $Y_m \subseteq Y_{m+1}$,
(ii) $\Phi(Y)$ is a bounded subset of $\mathcal{L}^2(\Omega)$, $w \in \mathcal{L}^1(Y)$, and $w\Phi : Y \setminus Y_0 \to \mathcal{X}$ is continuous,
(iii) $\phi : \Omega \times Y \to \mathbb{R}$ is Lebesgue measurable.
Then $\|f\|_{\Phi(Y)} \leq \|w\|_{\mathcal{L}^1(Y)}$ and for all positive integers n,

$$\|f - \operatorname{span}_n \Phi(Y)\|_{\mathcal{L}^2(\Omega)}^2 \leq \frac{s_\Phi^2 \|w\|_{\mathcal{L}^1(Y)}^2 - \|f\|_{\mathcal{L}^2(\Omega)}^2}{n}.$$

A function $\sigma : \mathbb{R} \to \mathbb{R}$ is called *sigmoidal* when it is nondecreasing and $\lim_{t \to -\infty} \sigma(t) = 0$ and $\lim_{t \to \infty} \sigma(t) = 1$. For every compact $\Omega \subset \mathbb{R}^d$, the mapping

$$\Phi_\sigma : \mathbb{R}^d \times \mathbb{R} \to \mathcal{L}^2(\Omega),$$

which is defined for all $x \in \Omega$ as $\Phi_\sigma(v, b)(x) = \sigma(v \cdot x + b)$, maps parameters (input weights v and biases b) to functions computable by perceptrons with the activation function σ. Let $\phi_\sigma : \Omega \times S^{d-1} \times \mathbb{R} \to \mathbb{R}$ be defined as $\phi_\sigma(x, e, b) = \sigma(x \cdot e + b)$.

We denote by $\vartheta : \mathbb{R} \to \mathbb{R}$ the *Heaviside function*, i.e., $\vartheta(t) = 0$ for $t < 0$ and $\vartheta(t) = 1$ for $t \geq 0$, and S^{d-1} denote the unit sphere in $\mathbb{R}^d$. It is easy to see that for any bounded subset Ω of $\mathbb{R}^d$, $\Phi_\vartheta(S^{d-1} \times \mathbb{R}) = \Phi_\vartheta(\mathbb{R}^d \times \mathbb{R})$. It was shown in [14] that for every $\Omega \subset \mathbb{R}^d$ compact and every continuous sigmoidal function σ, $\Phi_\sigma(\mathbb{R}^d \times \mathbb{R})$-variation in $\mathcal{L}^2(\Omega)$ is equal to $\Phi_\vartheta(S^{d-1} \times \mathbb{R})$-variation. Thus to apply Theorem 2 to estimate rates of approximation by sigmoidal perceptron networks, it suffices to estimate Φ_ϑ-variation.

For Ω compact, $\Phi_\vartheta : S^{d-1} \times \mathbb{R} \to \mathcal{L}^2(\Omega)$ is continuous, $\Phi_\vartheta(S^{d-1} \times \mathbb{R})$ is a bounded subset of $\mathcal{L}^2(\Omega)$ and $\phi_\vartheta : \Omega \times S^{d-1} \times \mathbb{R} \to \mathbb{R}$ is Lebesgue measurable. Thus, by Corollary 1, for a function $f \in \mathcal{L}^2(\Omega)$ representable for almost all $x \in \Omega$ as $f(x) = \int_{S^{d-1} \times \mathbb{R}} w(e, b)\vartheta(e \cdot x + b)dedb$ with $w \in \mathcal{L}^1(S^{d-1} \times \mathbb{R})$, $\Phi_\vartheta(S^{d-1} \times \mathbb{R})$-variation of f is bounded from above by $\|w\|_{\mathcal{L}^1(S^{d-1} \times \mathbb{R})}$.

Corollary 1 can be applied to all sufficiently differentiable functions which are either compactly supported or decay fast at infinity. Such functions can be expressed as networks with infinitely many Heaviside perceptrons. More precisely, for d odd, all d-variable sufficiently differentiable real-valued functions which are either compactly supported [14] or are of a weakly controlled decay [12] have a representation of the form

$$f(x) = \int_{S^{d-1}\times\mathbb{R}} w_f(e,b)\vartheta(e\cdot x + b)\,de\,db, \tag{5}$$

where the weight function $w_f(e,b)$ is a product of a function $a(d)$ of the number of variables d converging with d increasing exponentially fast to zero and a "flow of the order d through the hyperplane" $H_{e,b} = \{x \in \mathbb{R}^d \mid x \cdot e + b = 0\}$. More precisely,

$$w_f(e,b) = a(d)\int_{H_{e,b}} (D_e^{(d)}(f))(y)\,dy,$$

where

$$a(d) = (-1)^{(d-1)/2}(1/2)(2\pi)^{1-d}$$

and $D_e^{(d)}$ denotes the directional derivative of the order d in the direction e. The class of functions with weakly controlled decay contains all functions from $\mathcal{C}^d(\mathbb{R}^d)$ with compact support as well as all functions from the Schwartz class $\mathcal{S}(\mathbb{R}^d)$ (for its definition see [20, p.251]; in particular, it contains the Gaussian function $\gamma_d(x) = \exp(-\|x\|^2)$).

Applying Corollary 1 to the integral representation (5) we get for a large class of functions an upper bound on rates of approximation by perceptron networks. The $\mathcal{L}^1$-norm of the weight function w_f can be further estimated by the maximal value of the $\mathcal{L}^1$-norm of partial derivatives of the function to be approximated. In [17], *Sobolev seminorm* $\|.\|_{d,1,\infty}$, was defined as

$$\|f\|_{d,1,\infty} = \max_{|\alpha|=d} \|D^\alpha f\|_{\mathcal{L}^1(\mathbb{R}^d)}$$

and it was shown that for all d odd and all f of a weakly controlled decay

$$\|w_f\|_{\mathcal{L}^1(S^{d-1}\times\mathbb{R})} \leq k(d)\|f\|_{d,1,\infty},$$

where $k(d) \sim \left(\frac{4\pi}{d}\right)^{1/2} \left(\frac{e}{2\pi}\right)^{d/2}$. To avoid complicated notation, in the upper bound in next theorem we assume that all functions are restricted to the set Ω.

Theorem 6. *Let $\sigma : \mathbb{R} \to \mathbb{R}$ be a continuous sigmoidal function or the Heaviside function, d be an odd positive integer, $f \in \mathcal{C}^d(\mathbb{R}^d)$ be either compactly supported with $\Omega = \mathrm{supp}(f)$ or f be of a weakly controlled decay and Ω be any compact subset of $\mathbb{R}^d$. Then for all positive integers n,*

$$\|f - \mathrm{span}_n\Phi_\sigma(\mathbb{R}^{d+1})\|^2_{\mathcal{L}^2(\Omega)} \leq \frac{\lambda(\Omega)^2\|w_f\|^2_{\mathcal{L}^1(S^{d-1}\times\mathbb{R})} - \|f\|^2_{\mathcal{L}^2(\Omega)}}{n}$$

$$\leq \frac{k(d)^2\lambda(\Omega)^2\|f\|^2_{d,1,\infty} - \|f\|^2_{\mathcal{L}^2(\Omega)}}{n},$$

where $w_f(e,b) = a(d)\int_{H_{e,b}}(D_e^{(d)}(f))(y)\,dy$, $a(d) = (-1)^{(d-1)/2}(1/2)(2\pi)^{1-d}$, and $k(d) \sim \left(\frac{4\pi}{d}\right)^{1/2}\left(\frac{e}{2\pi}\right)^{d/2}$.

Hence, rates of approximation by perceptron networks of sufficiently smooth functions depend on the maximum of the $\mathcal{L}^1$-norms of their partial derivatives and on a function of the number of variables d which converges exponentially fast to zero as d increases.

5 Discussion

We have provided a unifying framework to estimate neural-network complexity. Our proof technique, based on Bochner integration, enabled us to derive upper bounds on rates of approximation for a wide class of functions, including those representable by integrals over non-compact sets of parameters.

Applying our results to perceptron networks, we estimated the growth of network complexity for smooth multivariable functions. Our theorems hold under mild assumptions on type of network units and so they can be combined with many other integral representations, for example with convolutions with Gaussian and Bessel kernels, which were studied in [9], [21].

Acknowledgements

V. K. was partially supported by the Ministry of Education of the Czech Republic, project Center of Applied Cybernetics 1M684077004 (1M0567). P. C. K. was partially supported by the Georgetown University.

References

1. Pisier, G.: Remarques sur un resultat non publié de B. Maurey. Seminaire d'Analyse Fonctionelle 1980-1981 I(12) (1981)
2. Jones, L.K.: A simple lemma on greedy approximation in Hilbert space and convergence rates for projection pursuit regression and neural network training. Annals of Statistics 24, 608–613 (1992)
3. Barron, A.R.: Universal approximation bounds for superpositions of a sigmoidal function. IEEE Transactions on Information Theory 39, 930–945 (1993)
4. Kůrková, V.: High-dimensional approximation and optimization by neural networks. In: Suykens, J., Horváth, G., Basu, S., Micchelli, C., Vandewalle, J., (eds.) Advances in Learning Theory: Methods, Models and Applications, ch.4, pp. 69–88. IOS Press, Amsterdam (2003)
5. Kůrková: Minimization of error functionals over perceptron networks. Neural Computation 20(1), 252–270 (2008)
6. Kůrková, V., Sanguineti, M.: Error estimates for approximate optimization by the extended Ritz method. SIAM J. Optimization 2(15), 461–487 (2005)
7. Carroll, S.M., Dickinson, B.W.: Construction of neural nets using the Radon transform. In: Proceedings of IJCNN, vol. I, pp. 607–611. IEEE Press, New York (1989)
8. Ito, Y.: Representation of functions by superpositions of a step or sigmoid function and their applications to neural network theory. Neural Networks 4, 385–394 (1991)
9. Girosi, F., Anzellotti, G.: Rates of convergence for radial basis functions and neural networks. In: Mammone, R.J. (ed.) Artificial Neural Networks for Speech and Vision, London, pp. 97–113. Chapman and Hall, Boca Raton (1993)

10. Kůrková, V., Kainen, P.C., Kreinovich, V.: Estimates of the number of hidden units and variation with respect to half-spaces. Neural Networks 10, 1061–1068 (1997)
11. Kainen, P.C., Kůrková, V., Vogt, A.: An integral formula for Heaviside neural networks. Neural Network World 10, 313–320 (2002)
12. Kainen, P.C., Kůrková, V., Vogt, A.: Integral combinations of Heavisides. Mathematische Nachrichten (to appear, 2008)
13. Barron, A.R.: Neural net approximation. In: Proc. 7th Yale Workshop on Adaptive and Learning Systems, pp. 69–72. Yale University Press (1992)
14. Kůrková, V.: Dimension–independent rates of approximation by neural networks. In: Warwick, K., Kárný, M. (eds.) Computer–Intensive Methods in Control and Signal Processing: Curse of Dimensionality, pp. 261–270. Birkhauser, Boston (1997)
15. Darken, C., Donahue, M., Gurvits, L., Sontag, E.: Rate of approximation results motivated by robust neural network learning. In: Proceedings of the 6th Annual ACM Conference on Computational Learning Theory, pp. 303–309. ACM, New York (1993)
16. Aronszajn, N.: Theory of reproducing kernels. Transactions of AMS 68, 337–404 (1950)
17. Kainen, P.C., Kůrková, V., Vogt, A.: A Sobolev-type upper bound for rates of approximation by linear combinations of Heaviside plane waves. Journal of Approximation Theory 147, 1–10 (2007)
18. Kainen, P.C., Kůrková, V.: An integral upper bound for neural-network approximation. Neural Computation, Research Report V-1023, Institute of Computer Science, Prague (submitted, 2008), http://www.cs.cas.cz/research
19. Kainen, P.C.: On evaluation of Bochner integrals. Research Report GU-08-30-07
20. Adams, R.A., Fournier, J.J.F.: Sobolev Spaces. Academic Press, Amsterdam (2003)
21. Kainen, P.C., Kůrková, V., Sanguineti, M.: Estimates of approximation rates by Gaussian radial-basis functions. In: Beliczynski, B., Dzielinski, A., Iwanowski, M., Ribeiro, B. (eds.) ICANNGA 2007. LNCS, vol. 4432, pp. 11–18. Springer, Heidelberg (2007)
22. Zaanen, A.C.: An Introduction to the Theory of Integration. North Holland, Amsterdam (1961)

Reliability of Cross-Validation for SVMs in High-Dimensional, Low Sample Size Scenarios

Sascha Klement, Amir Madany Mamlouk, and Thomas Martinetz

Institute for Neuro- and Bioinformatics, University of Lübeck

Abstract. A Support-Vector-Machine (SVM) learns for given 2-class-data a classifier that tries to achieve good generalisation by maximising the minimal margin between the two classes. The performance can be evaluated using cross-validation testing strategies. But in case of low sample size data, high dimensionality might lead to strong side-effects that can significantly bias the estimated performance of the classifier. On simulated data, we illustrate the effects of high dimensionality for cross-validation of both hard- and soft-margin SVMs. Based on the theoretical proofs towards infinity we derive heuristics that can be easily used to validate whether or not given data sets are subject to these constraints.

1   Introduction

Learning from examples in cases where many degrees of freedom but only few examples are available is commonly considered a challenging problem. Due to massively parallel data acquisition systems, such as microarray gene analysis [1,2], it easily happens that there are very few data points described by thousands of features. In such cases, theoretical and practical issues such as generalisation bounds, run-time, or memory-footprint considerations require sophisticated validation methods to measure the performance of machine learning algorithms in high-dimensional feature spaces.

Two commonly used methods in this context are support vector machines (SVM, [3,4]) as a classification system and cross-validation (CV, [5]) as a validation tool. Both methods are closely connected. Typically, when using the SVM there is a tendency to increase the data dimensionality as the classification problem is simplified in higher dimensions; on the other hand, CV is the method of choice in scenarios with relatively few data points compared to the dimensionality.

There is a well-known effect that if dimensionality is increased towards infinity, a finite set of points will lose more and more of its spatial topology. In the limit, the points will be located on the vertices of a regular simplex [6], i.e. all samples have nearly the same distances to the origin as well as among each other, and they are pairwise orthogonal. These properties were shown for multivariate standard normal distributions with zero mean and identity covariance matrix but hold under much weaker assumptions [7].

V. Kůrková et al. (Eds.): ICANN 2008, Part I, LNCS 5163, pp. 41–50, 2008.

Usually, the dimensionality will be finite, but we will show that even comparatively low dimensional data will behave as if being infinitely dimensional. So, especially for low sample size data, *infinity* is rather *small*.

First, we show that the leave-one-out CV error for hard-margin SVMs will approach 1 in high-dimensional feature spaces for equal-sized classes drawn from the same distribution – despite the expected error rate of 0.5, which would be the outcome for the same setting in low dimensions.

This first observation is generalised to two classes drawn from different distributions. Hall [6] showed that whenever these classes are too close together, a hard-margin SVM will vote along the majority rule alone for a dimension towards infinity. We will show empirically that this will occur even in quite finite dimensions when the given sample size is small.

Finally, we show the soft-margin approach to make things even worse. One might think that only simple hard-margin SVMs are prone to severe overfitting. Soft-margin SVMs increase the margin to reduce the fat-shattering dimension and should therefore reduce overfitting by allowing training errors. Unfortunately, this does not increase the generalisation performance, again due to the counterintuitive geometric properties of only few samples in high-dimensional space and the asymmetries of a resampling scheme such as leave-one-out cross-validation. In the soft-margin case infinity becomes even *smaller*.

It should be emphasised that especially dealing with high-dimensional but small sample size data leads to various counterintuitive and unfamiliar side-effects, which can significantly impact training and validation. In tasks such as the analysis of genetic microarrays, practical and financial issues strictly limit the maximum sample size but all the same rely on the analysis of high-dimensional content. Therefore, we want to elucidate the constraints on dimensionality, sample size, and class distribution, which might help to make training with SVMs still feasible in such scenarios.

2   Leave-One-Out and Hard-Margin SVM

Leave-one-out cross-validation is commonly used to estimate the generalisation performance and to fine-tune training parameters for various machine learning algorithms. The computational complexity limits the usage to small sample size data, but there it is commonly regarded to give good approximations of the true generalisation performance.

Still, it will fail in certain scenarios such as in the following example. Assume a balanced two class random dataset, i.e. samples drawn from an arbitrary distribution with randomly assigned class labels. The best classifier for completely random datasets is simply the majority voting rule [5]. Unfortunately, leave-one-out cross-validation will indicate very poor performance, since the original balanced dataset becomes unbalanced in each and every validation step. As the left-out pattern reduces the size of one class, the majority classifier will always vote for the other larger class, but the left-out pattern belongs to the smaller class. Thus, the classifier will always vote wrong. This behaviour is independent

of dimensionality or training set size but requires the prior knowledge of dealing
with completely random datasets.

In general, we do not have this knowledge and want to apply a generic classification framework such as linear SVMs for separation. To examine this framework
in detail, we define a very simple unlearnable scenario with the training set

$$\mathcal{D} = \{\boldsymbol{x}_i, y_i\}_{i=1}^{n} \tag{1}$$

consisting of feature vectors $\boldsymbol{x}_i \in \mathbb{R}^d$ where each entry is drawn from the standard normal distribution and corresponding class labels $y_i \in \{-1, +1\}$. Without
loss of generality we set

$$y_1 = \ldots = y_{\frac{n}{2}} = -1 \quad \text{and} \tag{2}$$

$$y_{\frac{n}{2}+1} = \ldots = y_n = +1 \ , \tag{3}$$

i.e. the training set represents two equal-sized classes drawn from the same distribution. For high-dimensional low sample size data $d \gg n$ holds, and therefore
a separating hyperplane exists in general, except for cases with three or more
collinear data points having alternating class labels.

We used leave-one-out cross validation to approximate the generalisation performance on this degenerated dataset by training a linear SVM n times using
the MinOver algorithm [4], each time on a different subset of size $n - 1$. The
resulting classification functions $f_i(\boldsymbol{x})$ are then used to classify the remaining
pattern and the leave-one-out error E is determined by

$$E = \frac{1}{2n} \sum_{i=1}^{n} |f_i(\boldsymbol{x}_i) - y_i| \ .$$

This procedure was repeated 100 times for each $n \in \{4, 6, \ldots, 60\}$ and fixed
$d = 1000$ (see Fig. 1, left) as well as for $n = 20$ and logarithmic-spaced

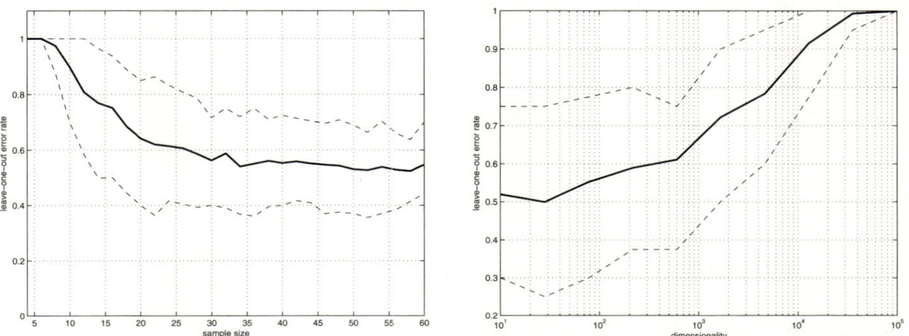

Fig. 1. The leave-one-out error rate of a hard-margin SVM tends to 1 for fixed dimensionality d and decreasing sample size n (left) as well as for fixed sample size and
increasing dimensionality (right). Solid lines depict the mean of 100 trials, while dashed
lines mark the 5th and 95th percentiles.

$d \in [10, 10^5]$ (see Fig. 1, right). Obviously, E depends on the training set size and tends to 1 for $n \to 4$ and $d \to \infty$. For larger training sets E tends to 50%, which is what we would intuitively assume. In order to explain error rates of close to 1 for small sample sizes, we switch from finite dimensionality to the remarkable case of finite datasets and infinite dimensionality. We will give a proof for this observation only using the geometric properties of high-dimensional small sample size data and simple vector algebra.

Theorem 1. *For any dataset* $\mathcal{D} = \{\boldsymbol{x}_i, y_i\}_{i=1}^n$ *with* $\boldsymbol{x}_i \in \mathrm{I\!R}^d$ *drawn from the multivariate standard normal distribution,* $y_1 = \ldots = y_{\frac{n}{2}} = -1$ *and* $y_{\frac{n}{2}+1} = \ldots = y_n = +1$ *with* n *fixed the leave-one-out error-rate of a hard-margin SVM is 1 for* $d \to \infty$.

Proof. For $d \to \infty$ all $\boldsymbol{x}_i \in \mathcal{D}$ will lie on the vertices of a regular n-simplex as well as all pairwise angles will be orthogonal [6]. The total variability of $\mathcal{D}$ is provided in the rotation of this simplex. Without loss of generality we set $\boldsymbol{x}_i = \sqrt{d}\,\boldsymbol{e}_i$, so that for $d \to \infty$

$$\|\boldsymbol{x}_i\|_2 = \sqrt{d} \qquad \forall\, i$$
$$\|\boldsymbol{x}_i - \boldsymbol{x}_j\|_2 = \|\sqrt{d}\,\boldsymbol{e}_i - \sqrt{d}\,\boldsymbol{e}_j\|_2 = \sqrt{2\,d} \qquad \forall\, i \neq j$$
$$\boldsymbol{x}_i^T \boldsymbol{x}_j = d\,\boldsymbol{e}_i^T \boldsymbol{e}_j = 0 \qquad \forall\, i \neq j \ .$$

Again without loss of generality we select $\boldsymbol{x}_1, \ldots, \boldsymbol{x}_{n-1}$ for training. Now, we can analytically determine the maximum margin classifier

$$f(\mathbf{x}) = \mathrm{sgn}\left(\boldsymbol{w}^T \boldsymbol{x} + b\right)$$
$$\text{that minimises} \quad \boldsymbol{w}^T \boldsymbol{w}$$
$$\text{subject to} \quad y_i\left(\boldsymbol{w}^T \boldsymbol{x}_i + b\right) \geq 1 \qquad \forall\, i$$

with simple vector algebra. Since all samples are pairwise orthogonal also the centroids of both classes are orthogonal. Thus, the separating hyperplane with maximum margin is orthogonal to the straight line through the centroids (see Fig. 2), i.e.

$$\boldsymbol{w} = \bar{\boldsymbol{x}}^+ - \bar{\boldsymbol{x}}^- \quad \text{with} \quad \bar{\boldsymbol{x}}^+ = \frac{1}{\frac{n}{2}} \sum_{i=1}^{\frac{n}{2}} \sqrt{d}\,\boldsymbol{e}_i \quad \text{and} \quad \bar{\boldsymbol{x}}^- = \frac{1}{\frac{n}{2} - 1} \sum_{i=\frac{n}{2}+1}^{n-1} \sqrt{d}\,\boldsymbol{e}_i \ .$$

With $\boldsymbol{p}$ as the centre of this line we get the bias through

$$b = -\boldsymbol{w}^T \boldsymbol{p} = -\left(\bar{\boldsymbol{x}}^+ - \bar{\boldsymbol{x}}^-\right)^T \frac{1}{2}\left(\bar{\boldsymbol{x}}^+ + \bar{\boldsymbol{x}}^-\right) = \frac{2\,d}{n\,(n-2)} \ .$$

The left-out data point $\boldsymbol{x}_n$ gets misclassified, because

$$f(\boldsymbol{x}_n) = \boldsymbol{w}^T \sqrt{d}\,\boldsymbol{e}_n + b = b > 0 \ .$$

This will be the case in each and every validation step, so the total leave-one-out classification error is 1. $\qquad\qquad\square$

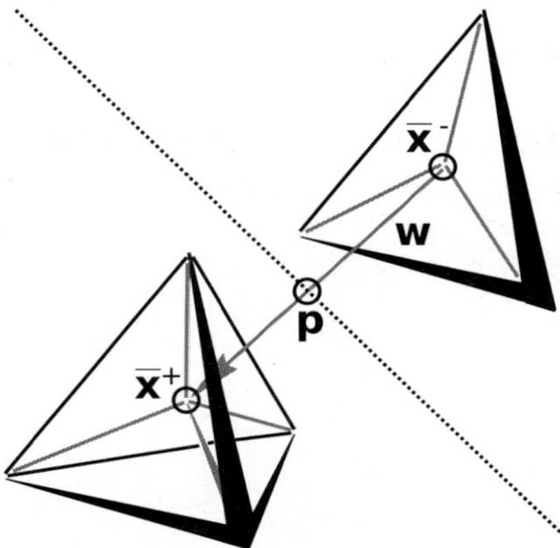

Fig. 2. Geometric sample configuration used in the proof of Theorem 1. Here, the case of 7 samples is visualised. The samples from the larger class form a 4-simplex while those from the smaller class form a 3-simplex. The centroids of both simplices are denoted by $\bar{x}^+$ and $\bar{x}^-$ respectively. The separating hyperplane with maximum margin has the normal vector w and contains the point p.

As dimensionality decreases, the data points differ more and more from the vertices of a regular simplex and are no longer orthogonal. So the conditions of the above proof are only approximately fulfilled. Therefore, probability increases for data points to be correctly classified, and the leave-one-out error-rate will be less than 1 and will converge to the intuitive error-rate of 0.5 for $n \to \infty$.

3   Practical Bounds for Critical Scenarios

So far, we examined the leave-one-out error in the limit for $d \to \infty$ which seems to be unrealistic in practical cases. But for high dimensional low sample size data infinity is rather *small* as our next experiment will stress. We show empirically which size dimensionality needs to have to behave as if it were infinite. Again, we sampled two classes independently and identically distributed from the standard normal distribution and trained support vector machines for $n \in \{4, 6, \ldots, 40\}$ and logarithmic-spaced $d \in [10, 10^5]$. For each configuration the procedure was repeated 10 times and the mean leave-one-out error was determined.

The colour-coded results are shown in Fig. 3. For any fixed number of data points n (e.g. $n = 20$) the leave-one-out error reaches 1 for a specific dimensionality (in this case at about $d = 10000$). Due to the probabilistic behaviour of the dataset, a precise borderline does not exist, but the tendency is obvious. We illustrate this by means of linear, quadratic and exponential extrapolation derived from the approximate border points of $n \in \{4, 6, 8\}$. A linear border is obviously too low while exponential behaviour is too steep – a slightly super-quadratic tendency is most promising. However, this heuristic suits only for illustrating

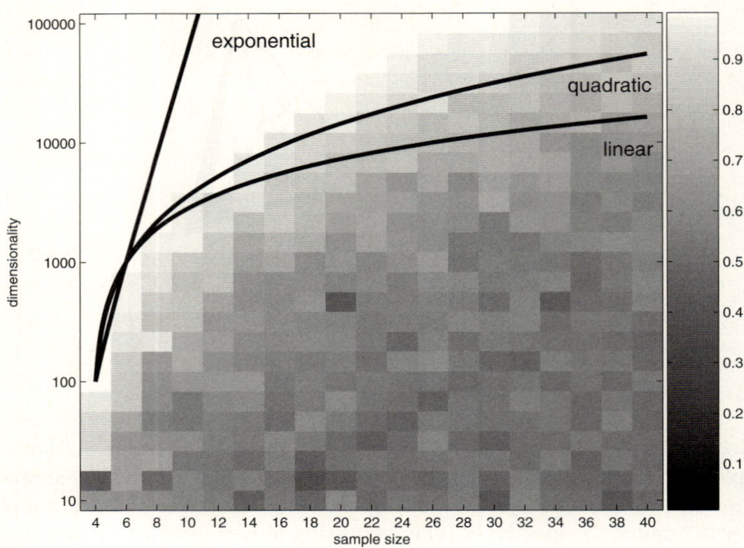

Fig. 3. Colour-coded error-rates depending on sample size and dimensionality. White colour indicates an error-rate of 1, i.e. the corresponding parameter sets behave as if they had infinite dimensionality. Additionally, three candidates for the border function are shown.

the asymptotic behaviour and should not be taken as an exact mathematical coherence.

3.1   Real Two-Class Scenarios

Now, we will generalise the random scenario from Sect. 2 to the case where both classes are drawn from different distributions. Assume the samples of the two classes to be distributed as $\boldsymbol{\mathcal{X}} = \left(\mathcal{X}^{(1)}, \ldots, \mathcal{X}^{(d)}\right)^{T}$ and $\boldsymbol{\mathcal{Y}} = \left(\mathcal{Y}^{(1)}, \ldots, \mathcal{Y}^{(d)}\right)^{T}$. With $d \to \infty$ the following conditions shall hold:

$$\sigma^2 = \lim_{d \to \infty} \frac{1}{d} \sum_{i=1}^{d} \mathrm{var}(\mathcal{X}^{(i)})$$

$$\tau^2 = \lim_{d \to \infty} \frac{1}{d} \sum_{i=1}^{d} \mathrm{var}(\mathcal{Y}^{(i)})$$

$$\mu^2 = \lim_{d \to \infty} \frac{1}{d} \sum_{i=1}^{d} \left(E(\mathcal{X}^{(i)}) - E(\mathcal{Y}^{(i)})\right)^2 \ .$$

Any Gaussian or rectangular distributions in d dimensions fulfil these conditions. Let k and l be the number of data points of the data sets X and Y drawn from $\boldsymbol{\mathcal{X}}$ and $\boldsymbol{\mathcal{Y}}$, respectively. If we then train a hard-margin SVM on these data point this leads to [6]:

Theorem 2. *Assume that $\frac{\sigma^2}{k} \geq \frac{\tau^2}{l}$ (otherwise interchange X and Y).*

If $\mu^2 > \frac{\sigma^2}{k} - \frac{\tau^2}{l}$, then the probability of a new datum from either population to be correctly classified by the SVM converges to 1 as $d \to \infty$. Otherwise any new sample will be classified as belonging to class Y as $d \to \infty$.

Again, we want to visualise how *small* infinite dimensionality in this case is. We sampled two equal-sized classes each from the multivariate standard normal distribution so that $\sigma^2 = \tau^2 = 1$ with the total number of data samples $n \in \{4, 6, \ldots, 40\}$ and logarithmic-spaced $d \in [10, 10^5]$. We illustrate exemplarily the dependency of sample size and leave-one-out error for an interclass distance of $\mu^2 = \frac{1}{30}$. The mean colour-coded results of 10 independent runs are shown in Fig. 4. As stated in Theorem 2, a certain threshold exists – separating convergence to zero error from convergence to an error rate of one. Obviously, the threshold corresponds in this case to a total sample size of 12, i.e. 5 samples from one class are trained vs. 6 samples from the other class within each validation step. Then, the proposed ratio exceeds the critical interclass distance at

$$\frac{\sigma^2}{k} - \frac{\tau^2}{l} = \frac{1}{5} - \frac{1}{6} = \frac{1}{30} = \mu^2 \ ,$$

which is perfectly the initially chosen value.

As a consequence of Theorem 2, it follows that two classes having different means in one dimension and having the same mean in all other dimension will not be separable, since μ is close to zero. So, the leave-one-out error rate will *always* converge to exactly 1 for any sample size as dimensionality goes to infinity.

3.2 Soft Margin and the Fat-Shattering Dimension

Hard-margin SVMs are prone to overfitting since one single outlier will strongly affect the separating hyperplane and reduce generalisation performance.

In the simplest case of two equal-sized classes drawn from the same distribution – defined as in (1)–(3) – we can explicitly derive an upper bound for the fat-shattering dimension *fat* according to [8] by

$$fat(\gamma) \leq \min\left\{\left\lceil \frac{R^2}{\gamma^2} \right\rceil, d\right\} + 1 \tag{4}$$

where R is the radius of the smallest enclosing sphere containing all samples and γ denotes the margin of separation. For randomly distributed datasets with $d \to \infty$ as used before the margin derives to

$$\gamma = \frac{1}{2}\|\boldsymbol{w}\| = \sqrt{\frac{d\,(n-1)}{n\,(n-2)}} \ .$$

Using (4) and $R = \sqrt{d}$ the fat-shattering dimension is upper bounded by

$$fat(\gamma) \leq \min\left\{\left\lceil \frac{d\,n\,(n-2)}{d\,(n-1)} \right\rceil, d\right\} + 1 = \min\{n-1, d\} + 1 = n \ .$$

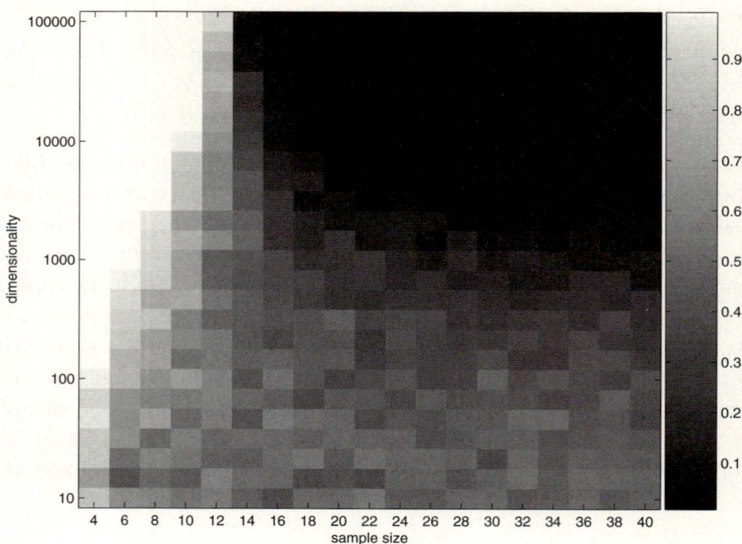

Fig. 4. Colour-coded error-rates depending on sample size and dimensionality for $\mu^2 = \frac{1}{30}$. White indicates an error-rate of 1, black an error rate of 0. The parameter sets corresponding to black or white behave as if they had infinite dimensionality.

Here the bounds $(n-2) < \frac{n\,(n-2)}{n-1} < (n-1)$ are used. Since $d \gg n$ and all samples are pairwise orthogonal, $fat(\gamma)$ is also lower bounded by n. Thus, in the hard-margin case the fat-shattering dimension indicates that no generalisation is possible, as expected. Considering a 2-norm soft-margin SVM, the following kernel has to be used instead of the dot product in the dual representation [3]:

$$K(\boldsymbol{x}_i, \boldsymbol{x}_j) = \Phi(\boldsymbol{x}_i)^T \Phi(\boldsymbol{x}_j) = \boldsymbol{x}_i^T \boldsymbol{x}_j + \frac{\delta_{ij}}{C}$$

where δ_{ij} is Kronecker's delta. In the kernel space the data points are again located on the edges of a regular simplex since

$$||\Phi(\boldsymbol{x}_i)|| = \sqrt{K(\boldsymbol{x}_i, \boldsymbol{x}_i)} = \sqrt{\boldsymbol{x}_i^T \boldsymbol{x}_i + \frac{1}{C}} = \sqrt{d + \frac{1}{C}},$$

$$||\Phi(\boldsymbol{x}_i) - \Phi(\boldsymbol{x}_j)|| = \sqrt{(\Phi(\boldsymbol{x}_i) - \Phi(\boldsymbol{x}_j))^T (\Phi(\boldsymbol{x}_i) - \Phi(\boldsymbol{x}_j))}$$

$$= \sqrt{K(\boldsymbol{x}_i, \boldsymbol{x}_i) - 2K(\boldsymbol{x}_i, \boldsymbol{x}_j) + K(\boldsymbol{x}_j, \boldsymbol{x}_j)}$$

$$= \sqrt{\boldsymbol{x}_i^T \boldsymbol{x}_i + \boldsymbol{x}_j^T \boldsymbol{x}_j + \frac{2}{C}} = \sqrt{2\left(d + \frac{1}{C}\right)}$$

and all samples are pairwise orthogonal:

$$\Phi(\boldsymbol{x}_i)^T \Phi(\boldsymbol{x}_j) = K(\boldsymbol{x}_i, \boldsymbol{x}_j) = \boldsymbol{x}_i^T \boldsymbol{x}_j = 0 \ .$$

So, when training a 2-norm soft margin SVM we intrinsically deal with a hard-margin SVM of dimensionality $(d + \frac{1}{C})$. For $C \gg 1$ the softness term is negligible but for $C < 1$ dimensionality will increase dramatically. Thus, with the soft-margin approach we implicitly increase dimensionality so that the leave-one-out error rate will be close to 1 already for a smaller number of dimensions.

At first glance, this is counterintuitive since soft margin approaches increase the margin and therefore reduce the fat-shattering dimension so that overfitting is reduced and generalisation performance improves. But in case of high-dimensional low sample size data the asymmetries of leave-one-out cross-validation stronger affect generalisation performance than overfitting problems do.

For 1-norm soft margin approaches we expect the same behaviour but a closed mathematical formulation cannot be derived as simply as above.

4   Conclusions

Machine learning methods provide good results easily as long as large sample sizes in connection with comparatively low dimensionality are given. However, in practical applications such as the analysis of microarray data or other high-dimensional data mining problems, we often have a reverse situation: Extremely few points, for which it is not possible to increase the sample size significantly and often a high-dimensional feature space resulting from massively parallel data acquisition.

Increasing the dimensionality of artificial randomly distributed small sample size data can be shown to result in a surprising reorganisation of the data on the vertices of a regular simplex (see Theorem 1). Exactly the same is happening for non-random data, i.e. for true 2-class problems whenever the two classes fulfil the proximity criterion of Theorem 2. Thus, for $d \rightarrow \infty$, random and non-random data scenarios are not to distinguish anymore – by any metric-based measure.

We examined the practical impact of $d \rightarrow \infty$, i.e. how *large* infinity really is. Our empirical simulations suggest that there might be a sub-exponential relation between sample size and dimensionality. For an identically distributed equal-sized two-class dataset consisting of 20 samples in 10000 dimensions we already have an infinity-like behaviour. Here, we cannot learn anything from the data by metric-based methods at all as distances are becoming approximately equal. Apparently, infinite dimensionality is not that *large*.

In general, we do not know the true variances and true mean values of the class distributions, so these values have to be estimated from the data. Again, due to the counterintuitive properties of high-dimensional low sample size data, the data might degenerate in just the same way for $d \rightarrow \infty$. Thus, whenever acquiring degenerated data we can only speculate whether we have random or non-random data. Although we focused on scenarios that are not learnable by definition, the results are truly important in real-world scenarios. Avoiding an asymmetric resampling scheme such as leave-one-out cross-validation may not be possible due to the small sample size.

We showed that a soft-margin approach does not improve the generalisation performance of the SVM on high-dimensional low sample size data. We gave a proof for the first data scenario that introducing softness has the same effect as an increase of dimensionality. Thus, the soft-margin technique is making infinity even *smaller*.

Especially biometric problems may suffer from artifacts of high-dimensional low sample size data as we showed exemplarily for leave-one-out cross-validation with support vector classification. Typically, these problems are supposed to have a much lower intrinsic dimension than the observed dimension; thus, it should be possible to reduce dimensionality to avoid the mentioned effects. Future work should cover critical validation techniques of a widespread range of machine learning methods to further investigate counterintuitive artifacts and to find solutions to overcome these effects.

References

1. Golub, T., Slonim, D., Tamayo, P., Huard, C., Gaasenbeek, M., Mesirov, J., Coller, H., Loh, M., Downing, J., Caligiuri, M., Bloomfield, C., Lander, E.: Molecular Classification of Cancer: Class Discovery and Class Prediction by Gene Expression Monitoring. Science 286(5439), 531–537 (1999)
2. Lockhart, D.J., Winzeler, E.: Genomics, Gene Expression and DNA Arrays. Nature 405, 827–836 (2000)
3. Cristianini, N., Shawe-Taylor, J.: Support Vector Machines and other kernel-based learning methods. Cambridge University Press, Cambridge (2000)
4. Martinetz, T., Labusch, K., Schneegaß, D.: SoftDoubleMinOver: A Simple Procedure for Maximum Margin Classification. In: Duch, W., Kacprzyk, J., Oja, E., Zadrożny, S. (eds.) ICANN 2005. LNCS, vol. 3697, pp. 301–306. Springer, Heidelberg (2005)
5. Kohavi, R.: A study of cross-validation and bootstrap for accuracy estimation and model selection. In: IJCAI, pp. 1137–1145 (1995)
6. Hall, P., Marron, J.S., Neeman, A.: Geometric representation of high dimension, low sample size data. J. R. Statist. Soc. 67(3), 427–444 (2005)
7. Ahn, J., Marron, J.S., Muller, K.M., Chi, Y.Y.: The high-dimension, low-sample-size geometric representation holds under mild conditions. Biometrika 94(3), 760–766 (2007)
8. Bartlett, P., Shawe-Taylor, J.: Generalization Performance of Support Vector Machines and Other Pattern Classifiers. In: Advances in Kernel Methods: Support Vector Learning, pp. 43–54. MIT Press, Cambridge (1999)

Generalization of Concave and Convex Decomposition in Kikuchi Free Energy

Yu Nishiyama[1] and Sumio Watanabe[2]

[1] Department of Computational Intelligence and Systems Science,
Tokyo Institute of Technology,
4259, Nagatuta, Midori-ku, Yokohama, 226-8503 Japan
`nishiyudesu@cs.pi.titech.ac.jp`
[2] Precision and Intelligence Laboratory, Tokyo Institute of Technology,
4259, Nagatsuta, Midori-ku, Yokohama, 226-8503 Japan
`swatanab@pi.titech.ac.jp`

Abstract. Computing marginal probabilities from a high-dimensional distribution is generally an intractable task which arises in various probabilistic information processing. The minimum of Kikuchi free energy is known to yield more accurate marginals than that of Bethe free energy. Concave convex procedure (CCCP) proposed by Yuille is known to be an algorithm which guarantees to monotonically decrease both free energies.

In this paper, we generalize CCCP algorithm for Kikuchi free energy and present a new CCCP (NCCCP) algorithm which includes a high-dimensional free vector. The free vector enables NCCCP to be more stable and to reduce enormous computational cost underlying CCCP. Moreover it seems that there exists an optimal series of free vectors that can minimize the huge computational cost. This means that there exists an optimal concave and convex decomposition schedule in Kikuchi free energy.

1 Introduction

Computing marginal probabilities from a high-dimensional distribution is generally an intractable task which arises in various probabilistic information processing. One famous task is to compute maximizer of the posterior marginal (MPM) in Bayesian inference. Belief propagation (BP) [1] is one of the efficient algorithms which provide marginals and has been studied in various fields such as inference for Bayesian networks [2], error correcting codes [3],[4], code division multiple access (CDMA) in mobile communication [5], and probabilistic image processing [6]. Loopy belief propagation (LBP), which is the BP applied to the distribution defined on a cyclic graph, is known not to compute precise marginals and does not guarantee the convergence. However, since the marginals computed by BP are equivalent to those giving extrema of Bethe free energy [7],[8], the central problem is changed to minimization of Bethe free energy and various optimization techniques have been proposed [9],[10]. Kikuchi free energy [11] is an extension of Bethe free energy and the minimum of Kikuchi free energy is known

V. Kůrková et al. (Eds.): ICANN 2008, Part I, LNCS 5163, pp. 51–60, 2008.

to yield more accurate marginals than those of Bethe. The belief propagation which corresponds to Kikuchi free energy was proposed by Yedidia [8] and is known to be generalized BP (GBP) algorithm.

For both Bethe and Kikuchi free energies, the convergent algorithm named concave convex procedure (CCCP), which guarantees to monotonically decrease the free energies, was proposed by Yuille [12]. Although CCCP algorithm offers better performances for convergence and bit error rate, CCCP is known to need huge computational cost.

In this paper, we generalize the CCCP algorithm for Kikuchi free energy and present a new CCCP (NCCCP) algorithm. Since NCCCP includes a lot of free parameters which one can decide in advance, NCCCP can be more flexible and can freely change the dynamics of CCCP till a GBP fixed-point. After we present NCCCP algorithm for Kikuchi free energy, we show the numerical experiments of NCCCP for Bethe free energy when Bethe free energy is applied to multi-dimensional Gaussian distributions. We note that Bethe free energy is a special case of Kikuchi free energy.

This paper is organized as follows. In the next section, we review the mathematical frameworks of Bethe and Kikuchi free energies and the CCCP algorithm. In section **3**, we present NCCCP algorithm for Kikuchi free energy. In section **4**, **5**, to confirm the effectiveness of NCCCP, we apply NCCCP to simple Gaussian distributions and demonstrate the NCCCP algorithm. In section **6**, conclusion and future works follow.

2   Bethe and Kikuchi Free Energies

2.1   Target Distribution

Let the (high-dimensional) target distribution whose marginals we wish to compute be the form of

$$p(x_1, x_2, \ldots, x_d) = \frac{1}{Z} \prod_{\{ij\} \in \mathcal{B}} W_{ij}(x_i, x_j), \tag{1}$$

where $\{x_i\}$ are finite discrete variables, function $W_{ij}(x_i, x_j)$ is strictly positive and exists if a pair of two variables x_i, x_j is included in set $\mathcal{B}$, and Z is the normalization constant. We regard the distribution as that defined on an undirected graph $\mathcal{G} = (\mathcal{V}, \mathcal{E})$ where vertex or node set is given by $\mathcal{V} = \{1, \ldots, d\}$ that labels variables $\{x_i\}$ and edge set is given by $\mathcal{E} = \mathcal{B}$. Let α be a subset of vertex set $\mathcal{V}$. Then the marginal distributions $p_\alpha(\mathbf{x}_\alpha)$ is defined by

$$p_\alpha(\mathbf{x}_\alpha) = \sum_{\mathbf{x} \backslash \mathbf{x}_\alpha} \frac{1}{Z} \prod_{\{ij\} \in \mathcal{B}} W_{ij}(x_i, x_j),$$

where $\mathbf{x}_\alpha$ is the subvector of $\mathbf{x}$ corresponding to subset $\alpha(\subset \mathcal{V})$ and $\sum_{\mathbf{x} \backslash \mathbf{x}_\alpha}$ is the summation over $x_i, (i \in \mathcal{V} \backslash \alpha)$. The direct calculation of marginals is intractable since the terms of the summation increase exponentially.

2.2   Bethe and Kikuchi Free Energies

If the graph $\mathcal{G}$ is a singly-connected graph (i.e., acyclic graph), the target distribution $p(\mathbf{x})$ is exactly rewritten, using the marginals of $p(\mathbf{x})$, as

$$p(\mathbf{x}) = \prod_{k=1}^{d} p_k(x_k) \prod_{\{ij\}\in\mathcal{B}} \frac{p_{ij}(x_i, x_j)}{p_i(x_i)p_j(x_j)}. \tag{2}$$

When one considers to obtain the true marginals $\{p_i\}$, $\{p_{ij}\}$ by minimizing a certain functional, one prepares a trial distribution $r(\mathbf{x})$ replacing $\{p_i\}$, $\{p_{ij}\}$ in eq.(2) by trial marginals $\{r_i\}$, $\{r_{ij}\}$ and minimizes Kullback Leibler distance from the trial distribution $r(\mathbf{x})$ to the target distribution (1). The minimization is equivalent to that of the functional with respect to trial marginals $\{r_i\}$, $\{r_{ij}\}$:

$$F_{Bethe}(\{r_i\}, \{r_{ij}\}) \equiv -\sum_{\{ij\}\in\mathcal{B}} S_{ij}(r_{ij}) + \sum_{i=1}^{d}(|\mathcal{N}_i| - 1)S_i(r_i) + \sum_{\{ij\}\in\mathcal{B}} \langle E_{ij}(x_i, x_j)\rangle_{r_{ij}}. \tag{3}$$

Here $E_{ij}(x_i, x_j) = -\log W_{ij}(x_i, x_j)$ and $\langle\cdot\rangle_{r_{ij}}$ denotes the expectation value over r_{ij}. $\mathcal{N}_i$ is the set of nodes directly connecting to node i and $|\mathcal{N}_i|$ is the number of elements of the set $\mathcal{N}_i$. A single entropy $S_i(r_i)$ and joint entropy $S_{ij}(r_{ij})$ are defined by $-\sum_{x_i} r_i \ln r_i$ and $-\sum_{x_i,x_j} r_{ij} \ln r_{ij}$, respectively. If graph $\mathcal{G}$ is a singly-connected graph, the minimum of Bethe free energy is equal to Helmholtz free energy $-\log Z$ and the minimum is given at exact marginals $\{p_i\}$, $\{p_{ij}\}$. The minimization of Bethe free energy is considered even when the graph $\mathcal{G}$ represents a cyclic graph. The trial marginals that give the minimum of Bethe free energy are expected to be close to the true marginals $\{p_i\}$, $\{p_{ij}\}$.

Kikuchi free energy [11] is an extension of Bethe free energy. Examples of Kikuchi approximation are shown in [8],[12]. Let *basic clusters* be the clusters of nodes, which yield set of regions $\mathcal{R}$ on which Kikuchi free energy is constructed. Let $\mathcal{R}$ be the set of regions made by basic clusters, the intersections of basic clusters, the intersections of the intersections, and so on, till the intersections do not yield new regions. Let $sup(\alpha)$ be the set of regions ($\subset \mathcal{R}$) which contain subset α (*super-regions of* α). Similarly, let $sub(\alpha)$ be the set of regions ($\subset \mathcal{R}$) which are contained in subset α (*sub-regions of* α). Under the definitions, Kikuchi free energy is defined by

$$F_K(\{r_\alpha\}) \equiv -\sum_{\alpha\in\mathcal{R}} c_\alpha S_\alpha(r_\alpha) + \sum_{\alpha\in\mathcal{R}} c_\alpha \langle E_\alpha(\mathbf{x}_\alpha)\rangle_{r_\alpha}. \tag{4}$$

Here $E_\alpha(\mathbf{x}_\alpha)$ denotes $E_\alpha(\mathbf{x}_\alpha) = -\sum_{\{ij\}\subset\alpha} \log W_{ij}(x_i, x_j)$ and $\langle\cdot\rangle_{r_\alpha}$ is the expectation value over r_α. The summation $\sum_{\{ij\}\subset\alpha}$ is taken over all pairs (in $\mathcal{B}$) of nodes contained in subset α. Entropy $S_\alpha(r_\alpha)$ is defined by $-\sum_{\mathbf{x}_\alpha} r_\alpha(\mathbf{x}_\alpha) \ln r_\alpha(\mathbf{x}_\alpha)$. c_α denotes *over-counting number* of region α and is calculated according to $c_\alpha = 1 - \sum_{\beta\in sup(\alpha)} c_\beta$, where $c_\alpha = 1$ for basic clusters α. The energy term in eq.(4) is exactly the same as that of Bethe free energy (3) but the entropy term in eq.(4) is different and generalized. When one considers basic clusters to be all pairs of nodes indicated by edge set $\mathcal{B}$, the entropy term of eq.(4) is reduced to that of eq.(3) and Kikuchi free energy is the same as Bethe free energy. Kikuchi free energy is the functional with respect to trial marginals $\{r_\alpha\}$ defined on set $\mathcal{R}$. In what follows, we often denote $\{r_\alpha\}$ by $\mathbf{r}$.

2.3 CCCP for Kikuchi Free Energy

Concave convex procedure (CCCP) is a general iterative algorithm, proposed by Yuille [12], which computes extrema of a function or a functional if the objective function(al) consists of the summation of concave and convex function(al)s. Let $F(\mathbf{r}) = F_{vex}(\mathbf{r}) + F_{cave}(\mathbf{r})$ be the function(al) of vector $\mathbf{r}$ that has minima. Here $F_{vex}(\mathbf{r})$ and $F_{cave}(\mathbf{r})$ are convex and concave, respectively. Iterations of $\mathbf{r}^{(t)}$ according to $\nabla F_{vex}(\mathbf{r}^{(t+1)}) \leftarrow -\nabla F_{cave}(\mathbf{r}^{(t)})$ guarantees to monotonically decrease function(al) $F(\mathbf{r})$ (i.e., $F(\mathbf{r}^{(t+1)}) \leq F(\mathbf{r}^{(t)})$). The proof is shown in [12]. The above iterative equation is obtained by the extremum condition of new objective function(al) $F(\mathbf{r}^{(t+1)}|\mathbf{r}^{(t)}) = F_{vex}(\mathbf{r}^{(t+1)}) + \mathbf{r}^{(t+1)} \cdot \nabla F_{cave}(\mathbf{r}^{(t)})$ with respect to $\mathbf{r}^{(t+1)}$.

Kikuchi free energy straightforwardly consists of summation of concave and convex functionals. CCCP guarantees to monotonically decrease Kikuchi free energy. For Kikuchi free energy, convex and concave parts $F_{K,vex}$, $F_{K,cave}$ are chosen to be (Yuille pointed out that many others are possible.)

$$F_{K,vex}(\{r_\alpha\}) = -\sum_{\alpha \in \mathcal{R}} c_{max} S_\alpha(r_\alpha) + \sum_{\alpha \in \mathcal{R}} c_{max} \langle E_\alpha(\mathbf{x}_\alpha) \rangle_{r_\alpha},$$

$$F_{K,cave}(\{r_\alpha\}) = -\sum_{\alpha \in \mathcal{R}} (c_\alpha - c_{max}) S_\alpha(r_\alpha) + \sum_{\alpha \in \mathcal{R}} (c_\alpha - c_{max}) \langle E_\alpha(\mathbf{x}_\alpha) \rangle_{r_\alpha}, \quad (5)$$

where c_{max} denotes $c_{max} = \max_{\alpha \in \mathcal{R}} c_\alpha$. Since entropy S_α is concave, $c_{max} > 0$, and $c_\alpha - c_{max} \leq 0$, above decomposition is one of reasonable decompositions.

Under the decomposition, extremum condition of a new functional (CCCP free energy)

$$F_{K,CCCP}(\mathbf{r}^{(t+1)}|\mathbf{r}^{(t)}) = F_{K,vex}(\mathbf{r}^{(t+1)}) + \mathbf{r}^{(t+1)} \cdot \nabla F_{K,cave}(\mathbf{r}^{(t)}) \quad (6)$$

with respect to trial marginals $\mathbf{r}^{(t+1)}$ yields *outer loop* of CCCP, which is the update from $\mathbf{r}^{(t)}$ to $\mathbf{r}^{(t+1)}$. Since marginals satisfy the normalization and consistency properties, one has to consider the Lagrange functional $L_{K,CCCP}$ made by $F_{K,CCCP}$. Normalization and consistency properties of marginals are given by

$$\sum_{\mathbf{x}_\alpha} r_\alpha(\mathbf{x}_\alpha) = 1, \ \ \alpha \in \mathcal{R}, \qquad \sum_{\mathbf{x}_{\alpha \backslash \beta}} r_\alpha(\mathbf{x}_\alpha) = r_\beta(\mathbf{x}_\beta), \ \ \alpha \in \mathcal{R}, \ \ \beta \in sub_d(\alpha), \quad (7)$$

where the set $sub_d(\alpha)$ is defined by all sub-regions of α which have no super-regions in $sub(\alpha)$ ($sub_d(\alpha) \subset sub(\alpha)$). For later use, we also define the set $sup_d(\alpha)$ to be all super-regions of α which have no sub-regions in $sup(\alpha)$ ($sup_d(\alpha) \subset sup(\alpha)$). We define the Lagrange functional $L_{K,CCCP}$ to be

$$L_{K,CCCP}(\mathbf{r}^{(t+1)}|\mathbf{r}^{(t)}) = F_{K,CCCP}(\mathbf{r}^{(t+1)}|\mathbf{r}^{(t)}) + \sum_{\alpha \in \mathcal{R}} v_\alpha \Big\{ \sum_{\mathbf{x}_\alpha} r_\alpha^{(t+1)}(\mathbf{x}_\alpha) - 1 \Big\}$$

$$+ \sum_{\alpha \in \mathcal{R}} \sum_{\beta \in sub_d(\alpha)} \sum_{\mathbf{x}_\beta} \eta_{\alpha,\beta}(\mathbf{x}_\beta) \Big\{ \sum_{\mathbf{x}_{\alpha \backslash \beta}} r_\alpha^{(t+1)}(\mathbf{x}_\alpha) - r_\beta^{(t+1)}(\mathbf{x}_\beta) \Big\},$$

where v_α and $\eta_{\alpha,\beta}(\mathbf{x}_\beta)$ are Lagrange multipliers for normalization and consistency, respectively. The extremum condition yields outer loop of CCCP. The outer loop for Kikuchi free energy is given by

$$r_k^{(t+1)} = \{r_k^{(t)}\}^{\frac{c_{max}-c_k}{c_{max}}} \exp\left\{ -\tilde{v}_k - \sum_{\beta \in sub_d(k)} \frac{\eta_{k,\beta}(\mathbf{x}_\beta)}{c_{max}} + \sum_{\gamma \in sup_d(k)} \frac{\eta_{\gamma,k}(\mathbf{x}_k)}{c_{max}} - \frac{c_k E_k(\mathbf{x}_k)}{c_{max}} \right\},$$

$$k \in \mathcal{R}, \quad (8)$$

where we have changed the Lagrange multipliers v_k to $\tilde{v}_k$. Lagrange multipliers $\tilde{v}_k$ and $\eta_{\alpha,\beta}(\mathbf{x}_\beta)$, which are necessary for computing outer loop, are given by following *inner loop*:

$$\exp(\tilde{v}_k^{(\tau+1)}) = \sum_{\mathbf{x}_k} \{r_k^{(t)}\}^{\frac{c_{max}-c_k}{c_{max}}} \exp\left\{ - \sum_{\beta \in sub_d(k)} \frac{\eta_{k,\beta}^{(\tau)}(\mathbf{x}_\beta)}{c_{max}} + \sum_{\gamma \in sup_d(k)} \frac{\eta_{\gamma,k}^{(\tau)}(\mathbf{x}_k)}{c_{max}} - \frac{c_k E_k(\mathbf{x}_k)}{c_{max}} \right\},$$

$$k \in \mathcal{R},$$

$$\exp\left(\frac{2\eta_{k,l}^{(\tau+1)}(\mathbf{x}_l)}{c_{max}} \right) =$$

$$\sum_{\mathbf{x}_{k \setminus l}} \{r_k^{(t)}\}^{\frac{c_{max}-c_k}{c_{max}}} \exp\left\{ -\tilde{v}_k^{(\tau)} - \sum_{\beta \in sub_d(k) \setminus l} \frac{\eta_{k,\beta}^{(\tau)}(\mathbf{x}_\beta)}{c_{max}} + \sum_{\gamma \in sup_d(k)} \frac{\eta_{\gamma,k}^{(\tau)}(\mathbf{x}_k)}{c_{max}} - \frac{c_k E_k(\mathbf{x}_k)}{c_{max}} \right\}$$

$$\Bigg/ \left[\{r_l^{(t)}\}^{\frac{c_{max}-c_l}{c_{max}}} \exp\left\{ -\tilde{v}_l^{(\tau)} - \sum_{\beta \in sub_d(l)} \frac{\eta_{l,\beta}^{(\tau)}(\mathbf{x}_\beta)}{c_{max}} + \sum_{\gamma \in sup_d(l) \setminus k} \frac{\eta_{\gamma,l}^{(\tau)}(\mathbf{x}_l)}{c_{max}} - \frac{c_l E_l(\mathbf{x}_l)}{c_{max}} \right\} \right],$$

$$k \in \mathcal{R}, \quad l \in sub_d(k), \quad (9)$$

which are the updates from time τ to $\tau + 1$ and obtained by the conditions (7). CCCP converges when one considers asynchronous update of inner loop (i.e., only one Lagrange multiplier is updated at each time t) but does not when synchronous update in general. NCCCP can converge even for the synchronous update by choosing a free vector $\mathbf{u}$ properly.

The fixed-point of outer loop r_k^* corresponds to the (approximate) marginals of $p(\mathbf{x})$ that give a extremum of Kikuchi free energy.

3   New CCCP Algorithm for Kikuchi Free Energy

In this section, we generalize CCCP algorithm for Kikuchi free energy and present a new CCCP (NCCCP) algorithm. The extension is based on the general decomposition of concave and convex functionals in Kikuchi free energy. NCCCP algorithm contains a new $|\mathcal{R}|$-dimensional vector $\mathbf{u}$ which one can decide freely in advance. However, NCCCP guarantees to monotonically decrease Kikuchi free energy. Conventional CCCP algorithm corresponds to the point $\mathbf{u} = \mathbf{c}_{max} - \mathbf{c}$, where $\mathbf{c}_{max}$ is a vector whose components are all c_{max} and $\mathbf{c}$ is a vector $\mathbf{c} = \{c_\alpha\}$. The $|\mathcal{R}|$-dimensional free vector $\mathbf{u}$ enables NCCCP to converge even if update manner of inner loop is either asynchronous, synchronous, or combination of those and change the dynamics till GBP fixed-points. As a result, NCCCP algorithm is more stable and can eliminate huge computational cost till GBP fixed-points.

In stead of the one fixed decomposition given by eqs.(5), we generally decompose Kikuchi free energy as follows.

$$F_{K,vex}(\mathbf{r}) = F_K(\mathbf{r}) + f(\mathbf{r}), \qquad F_{K,cave}(\mathbf{r}) = -f(\mathbf{r}),$$

where $F_K(\mathbf{r})$ is the Kikuchi free energy and $f(\mathbf{r})$ is an arbitrary convex functional with respect to $\mathbf{r}$. We have used the trivial pair creation $0 = f(\mathbf{r}) - f(\mathbf{r})$, where if $f(\mathbf{r})$ is convex, negative $f(\mathbf{r})$ is concave. Particularly, we consider $f(\mathbf{r})$ to be $f(\mathbf{r}) = \sum_{\alpha \in \mathcal{R}} u_{E,\alpha} \langle E_\alpha(\mathbf{x}_\alpha)\rangle_{r_\alpha} - \sum_{\alpha \in \mathcal{R}} u_{S,\alpha} S_\alpha(r_\alpha)$, where $u_{E,\alpha}$ and $u_{S,\alpha}$ are the free parameters. Then, $F_{K,vex}(\mathbf{r})$ and $F_{K,cave}(\mathbf{r})$ are given by as follows.

$$F_{K,vex}(\mathbf{r}, \mathbf{u}) = - \sum_{\alpha \in \mathcal{R}} (c_\alpha + u_{S,\alpha}) S_\alpha(r_\alpha) + \sum_{\alpha \in \mathcal{R}} (c_\alpha + u_{E,\alpha}) \langle E_\alpha(\mathbf{x}_\alpha)\rangle_{r_\alpha},$$

$$F_{K,cave}(\mathbf{r}, \mathbf{u}) = \sum_{\alpha \in \mathcal{R}} u_{S,\alpha} S_\alpha(r_\alpha) - \sum_{\alpha \in \mathcal{R}} u_{E,\alpha} \langle E_\alpha(\mathbf{x}_\alpha)\rangle_{r_\alpha}. \tag{10}$$

In order for $F_{K,vex}(\mathbf{r})$ and $F_{K,cave}(\mathbf{r})$ to be convex and concave, respectively, free parameters have to satisfy

$$u_{S,\alpha} > \max\{0, -c_\alpha\}, \quad {}^{\forall}u_{E,\alpha} \in R, \qquad \alpha \in \mathcal{R}.$$

Since $\langle E_\alpha(\mathbf{x}_\alpha)\rangle_{r_\alpha}$ is linear with respect to r_α, free parameter $u_{E,\alpha}$ does not affect the convexity and concavity and can take an arbitrary real number.

Under the general decomposition with the parameters, CCCP free energy (6) with the parameters is newly constructed and the extremum condition yields new CCCP (NCCCP) algorithm including many free parameters. We note that though NCCCP free energy depends on the parameters, Kikuchi free energy and the extremum do not. In fact, since free parameters $\{u_{E,\alpha}\}$ do not affect convexity and concavity, parameters $\{u_{E,\alpha}\}$ do not appear in following NCCCP algorithm. Hence, we simplify the notation $u_{S,\alpha}$ as u_α. we write $\mathbf{u} = \{u_\alpha\}$. We define the feasible set $\mathcal{U}$ as

$$\mathcal{U} = \{\mathbf{u} \in R^{|\mathcal{R}|} | u_\alpha > \max\{0, -c_\alpha\}, {}^{\forall}\alpha \in \mathcal{R}\}.$$

The NCCCP algorithm is given by as follows.

Theorem 1. *The outer loop of NCCCP algorithm for minimizing Kikuchi free energy is given by the update equations from time t to $t+1$ as follows:*

$$r_k^{(t+1)} = \{r_k^{(t)}\}^{\frac{u_k}{c_k + u_k}} \exp\left\{ -\tilde{v}_k - \sum_{\beta \in sub_d(k)} \frac{\eta_{k,\beta}(\mathbf{x}_\beta)}{c_k + u_k} + \sum_{\gamma \in sup_d(k)} \frac{\eta_{\gamma,k}(\mathbf{x}_k)}{c_k + u_k} - \frac{c_k E_k(\mathbf{x}_k)}{c_k + u_k} \right\},$$

$$k \in \mathcal{R}.$$

Corresponding inner loop of NCCCP is given by the update equations from time τ to $\tau + 1$ as follows:

$$\exp(\tilde{v}_k^{(\tau+1)}) = \sum_{\mathbf{x}_k} \{r_k^{(t)}\}^{\frac{u_k}{c_k + u_k}} \exp\left\{ -\sum_{\beta \in sub_d(k)} \frac{\eta_{k,\beta}^{(\tau)}(\mathbf{x}_\beta)}{c_k + u_k} + \sum_{\gamma \in sup_d(k)} \frac{\eta_{\gamma,k}^{(\tau)}(\mathbf{x}_k)}{c_k + u_k} - \frac{c_k E_k(\mathbf{x}_k)}{c_k + u_k} \right\},$$

$$k \in \mathcal{R},$$

$$\exp\left\{\left(\frac{1}{c_k+u_k}+\frac{1}{c_l+u_l}\right)\eta_{k,l}^{(\tau+1)}(\mathbf{x}_l)\right\}=$$

$$\sum_{\mathbf{x}_{k\setminus l}}\{r_k^{(t)}\}^{\frac{u_k}{c_k+u_k}}\exp\left\{-\tilde{v}_k^{(\tau)}-\sum_{\beta\in sub_d(k)\setminus l}\frac{\eta_{k,\beta}^{(\tau)}(\mathbf{x}_\beta)}{c_k+u_k}+\sum_{\gamma\in sup_d(k)}\frac{\eta_{\gamma,k}^{(\tau)}(\mathbf{x}_k)}{c_k+u_k}-\frac{c_k E_k(\mathbf{x}_k)}{c_k+u_k}\right\}$$

$$\Bigg/\left[\{r_l^{(t)}\}^{\frac{u_l}{c_l+u_l}}\exp\left\{-\tilde{v}_l^{(\tau)}-\sum_{\beta\in sub_d(l)}\frac{\eta_{l,\beta}^{(\tau)}(\mathbf{x}_\beta)}{c_l+u_l}+\sum_{\gamma\in sup_d(l)\setminus k}\frac{\eta_{\gamma,l}^{(\tau)}(\mathbf{x}_l)}{c_l+u_l}-\frac{c_l E_l(\mathbf{x}_l)}{c_l+u_l}\right\}\right],$$

$$k\in\mathcal{R},\quad l\in sub_d(k),$$

where u_k satisfies $u_k>\max\{0,-c_k\}$ for $^\forall k\in\mathcal{R}$.

Thus, we have obtained NCCCP algorithm which includes a $|\mathcal{R}|$-dimensional free vector $\mathbf{u}\in\mathcal{U}$. We note that when the free vector $\mathbf{u}$ is the point $\mathbf{u}=\mathbf{c}_{max}-\mathbf{c}$, above outer and inner loops of NCCCP are equivalent to eqs.(8),(9), respectively. For the NCCCP algorithm, the following remark is important.

Remark 1. *NCCCP algorithm guarantees to monotonically decrease Kikuchi free energy even if free vector $\mathbf{u}$ is changed within $\mathcal{U}$ at every outer loop.*

From remark 1, free vector $\mathbf{u}$ depends on time t and can be written as $\mathbf{u}(t)$. This intuitively means that one can vary the decomposition manner at every time t and can minimize Kikuchi free energy according to a decomposition schedule specified by $\mathbf{u}(t)$.

To choose appropriate free vectors $\mathbf{u}(t)$ enables NCCCP to be more stable even if update manner of inner loop is asynchronous, synchronous, or combination of those and to reduce huge computational cost underlying CCCP.

In what follows, we show the effectiveness of NCCCP algorithm through a simple case of Gaussian distributions. We consider the NCCCP for Bethe free energy, which is a special case of Kikuchi free energy, and apply the algorithm to simple multi-dimensional Gaussian distributions. In the numerical experiments, we fixed the free vector $\mathbf{u}$ to be a constant for all outer loops. However, by choosing appropriate value of free parameter, we know that NCCCP can be more stable and eliminate expensive computational cost.

4   NCCCP Algorithm in Gaussian Distributions

In this section, for the purpose of showing the effectiveness of NCCCP, we implement the NCCCP algorithm to multi-dimensional Gaussian distributions and give the outer loop and inner loop in a simple case. The numerical results are shown in the next section. We consider the NCCCP for Bethe free energy, which is the particular case where basic clusters are chosen to be the set $\mathcal{B}$. Hence, the set of regions $\mathcal{R}$ is $\mathcal{R}=\{\mathcal{B},\mathcal{V}\}$. We consider the target distribution (1) to be Gaussian distributions with mean vector $\mathbf{0}$ and inverse covariance matrix S^1. We write $p(\mathbf{x})\sim N(\mathbf{0},S)$ and the components of the matrix S as $(S)_{ij}=s_{i,j}$.

---

[1] Though, in this paper, we defined the target distribution (1) to be a probability mass function, we can also consider that the distribution and the subsequent algorithm are continuous.

We set the marginals $\{r_i\}$, $\{r_{ij}\}$ and Lagrange multipliers $\{\eta_{ij,i}(x_i)\}$, $\{\eta_{ij,j}(x_j)\}$ computed by NCCCP to be Gaussian distributions given by

$$r_i(x_i) \sim N(0, \Lambda_i), \qquad\qquad i \in \{1, \cdots, d\},$$
$$r_{ij}(x_i, x_j) \sim N(\mathbf{0}, \tilde{S}_{ij}),$$
$$\exp\{-\eta_{ij,i}(x_i)\} \sim N(0, \tilde{\eta}_{ij,i}),$$
$$\exp\{-\eta_{ij,j}(x_j)\} \sim N(0, \tilde{\eta}_{ij,j}), \quad \{ij\} \in \mathcal{B}. \tag{11}$$

We define $\bar{\Lambda}_i \equiv \Lambda_i/s_{i,i}$, $\bar{\eta}_{ij,i} \equiv \tilde{\eta}_{ij,i}/s_{i,i}$, $\bar{\eta}_{ij,j} \equiv \tilde{\eta}_{ij,j}/s_{j,j}$, and $\bar{s}_{i,j} \equiv s_{i,j}/\sqrt{s_{i,i}s_{j,j}}$.

By substituting eqs.(11) into theorem 1 and considering basic clusters are chosen to be $\mathcal{B}$ and the target distribution is Gaussian distributions, we obtain the outer and inner loops in multi-dimensional Gaussian distributions.

In this paper, for simplicity, we consider more special case where inverse covariance matrix S satisfies

$$|\mathcal{N}_i| = |\mathcal{N}|, \quad i \in \{1, \cdots, d\}, \qquad \bar{s}_{i,j} = \bar{s}, \quad \{ij\} \in \mathcal{B}, \tag{12}$$

where $|\mathcal{N}_i|$ is the number of edges directly connecting to the node i. Then, if we take initial values $\{\bar{\eta}_{ij,i}^{(0)}\}$ and $\{\bar{\Lambda}_i^{(0)}\}$ properly, all nodes in $\mathcal{V}$ can not be distinguished from one another and indices of $\{\bar{\eta}_{ij,i}^{(\tau)}\}$, $\{\bar{\Lambda}_i^{(t)}\}$, $\{u_i\}$ can all be removed as $\bar{\eta}^{(\tau)}$, $\bar{\Lambda}^{(t)}$, u, respectively. The abbreviated outer and inner loops of the NCCCP are given by as follows.

Theorem 2. *Outer loop of NCCCP in Gaussian distributions under the conditions (12) is given by*

$$\bar{\Lambda}^{(t+1)} = \frac{u}{u - |N| + 1}\bar{\Lambda}^{(t)} - \frac{|\mathcal{N}|}{u - |N| + 1}\bar{\eta}^{(\tilde{\tau})},$$

where $\tilde{\tau}$ is the number of iterations of inner loop. Corresponding inner loop of the NCCCP is given by

$$\bar{\eta}^{(\tau+1)} = \frac{u - |\mathcal{N}| + 1}{u - |\mathcal{N}| + 2}\left\{ -\frac{1}{|\mathcal{N}|} + \frac{\bar{s}^2}{\frac{1}{|\mathcal{N}|} + \bar{\eta}^{(\tau)}} + \frac{u}{u - |\mathcal{N}| + 1}\bar{\Lambda}^{(t)} - \frac{|\mathcal{N}| - 1}{u - |\mathcal{N}| + 1}\bar{\eta}^{(\tau)} \right\},$$

where free parameter u satisfies $u > |\mathcal{N}| - 1$.

5   Numerical Experiments of NCCCP

In this section, we show the numerical results of NCCCP subject to the outer and inner loops in theorem 2. For simplicity, we fixed the free parameter u to be a constant. We consider four different Gaussian distributions with inverse covariance matrix S as follows:

$$S_1 = S(10, 0.1, 4), \quad S_2 = S(10, 0.1, 5), \quad S_3 = S(10, 2, 5), \quad S_4 = S(10, 0.1, 10),$$

where $S(s_0, s, d)$ denotes a $d \times d$ inverse covariance matrix in which all diagonal and off-diagonal components are given by s_0 and s, respectively. We examined

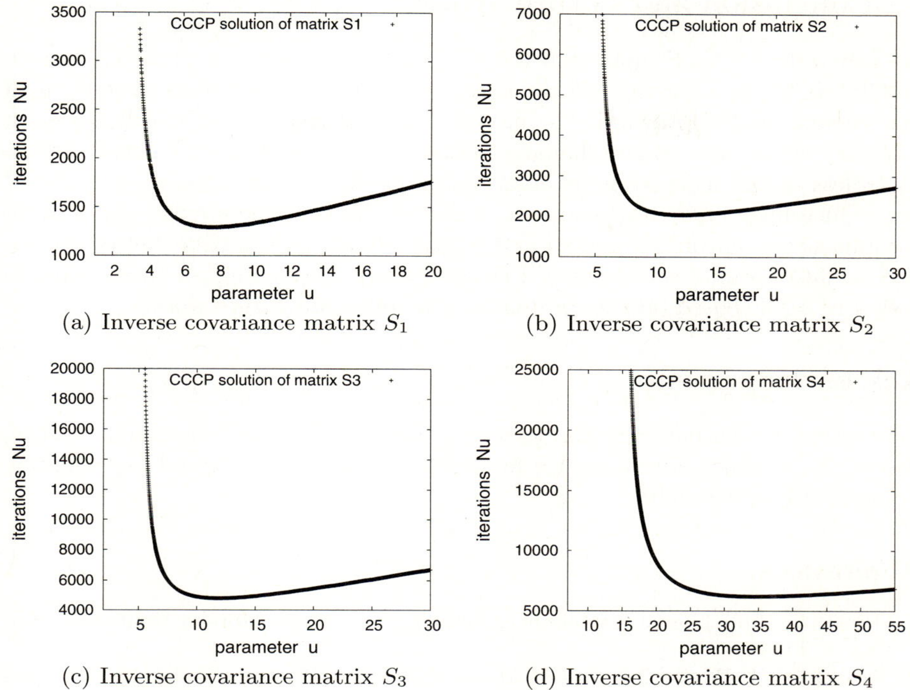

(a) Inverse covariance matrix S_1 (b) Inverse covariance matrix S_2

(c) Inverse covariance matrix S_3 (d) Inverse covariance matrix S_4

Fig. 1. Numerical experiments of NCCCP for four different Gaussian distributions $N(\mathbf{0}, S)$. Vertical axes are the number of iterations till convergence. Horizontal axes are the values of given parameter u.

number of iterations which requires till convergence with changing the value of free parameter u. We define the number of iterations Nu to be total counts of inner loop till convergence. Fig. 1.(a)(b)(c)(d) show the results, where vertical axes are the number of iterations Nu and horizontal axes are the values of u. In each figure, the fixed-points computed by NCCCP at any values of parameter u were the same and equivalent to that of loopy belief propagation. The conventional CCCP corresponds to the point $u = |\mathcal{N}|$, that is, $u = 3$, $u = 4$, $u = 4$, and $u = 9$ for each figure, respectively. We know that the number of iterations Nu diverges at $u = |\mathcal{N}|$ for four figures. This means that CCCP does not converge for the Gaussian distributions. However, the number Nu can be finite and NCCCP converges by making free parameter u sufficiently large in advance. This means that NCCCP is more stable. Also, we know that we can eliminate enormous computational cost underlying CCCP by choosing value of u properly. The saving appears prominently when one considers higher-dimensional target distributions. We observe that there exists an optimal value of parameter u that can minimize the huge computational cost.

6   Conclusion and Future Works

We generalized CCCP algorithm for Kikuchi free energy and presented a new CCCP (NCCCP) algorithm which guarantees to monotonically decrease Kikuchi free energy. NCCCP includes a high-dimensional free vector which takes an arbitrary vector in a certain domain and is more flexible. NCCCP can be stable regardless of the asynchronous or synchronous update of inner loop. NCCCP can reduce huge computational cost and it seems that there exists an optimal decomposition schedule of convex and concave functionals in Kikuchi free energy. To evaluate optimal schedules of free vector $\mathbf{u}(t)$ depending on time t and to design NCCCP based on the optimality are important future works.

Acknowledgement

This research was partially supported by the Ministry of Education, Science, Sports and Culture, Grant-in-Aid for JSPS Fellows 19005165 and for Scientific Research 18079007, 2008.

References

1. Pearl, J.: Probabilistic Reasoning in Intelligent Systems. Morgan Kaufmann, San Mateo (1988)
2. Murphy, K.P., Weiss, Y., Jordan, M.I.: Loopy Belief Propagation for Approximate Inference: An Empirical Study. In: 15th conference on Uncertainty in Artificial Intelligence, pp. 467–475 (1999)
3. McEliece, R.J., MacKay, D.J.C., Cheng, J.: Turbo Decoding as an Instance of Pearl's "Belief Propagation" Algorithm. IEEE Journal on Selected Areas in Communications 16(2), 140–152 (1998)
4. MacKay, D.J.C.: Good Error-Correcting Codes Based on Very Sparse Matrices. IEEE Transactions on Information Theory 45(2), 399–431 (1999)
5. Kabashima, Y.: A CDMA multiuser detection algorithm on the basis of belief propagation. J. Phys. A, Math. Gen. 36(43), 11111–11121 (2003)
6. Tanaka, K.: Statistical-mechanical approach to image processing. J. Phys. A, Math. Gen. 35(37), R81–R150 (2002)
7. Kabashima, Y., Saad, D.: The Belief in TAP. Neural Information Processing Systems 11 (1998)
8. Yedidia, J., Freeman, W., Weiss, Y.: Generalized Belief Propagation. Neural Information Processing Systems 13 (2000)
9. Welling, M., Teh, Y.W.: Belief Optimization for Binary Networks: A Stable Alternative to Loopy Belief Propagation. In: 17th conference on Uncertainty in Artificial Intelligence, pp. 554–561 (2001)
10. Tonosaki, Y., Kabashima, Y.: Sequential Minimization of the Bethe Free Energy of Ising Spin Systems. Interdisciplinary Information Sciences 13(1), 57–64 (2007)
11. Kikuchi, R.: A Theory of Cooperative Phenomena. Phys. Rev. 81(6), 988–1003 (1951)
12. Yuille, A.L.: CCCP Algorithms to Minimize the Bethe and Kikuchi Free Energies: Convergent Alternatives to Belief Propagation. Neural Computation 14(7), 1691–1722 (2002)

Analysis of Chaotic Dynamics Using Measures of the Complex Network Theory

Yutaka Shimada[1,2], Takayuki Kimura[1], and Tohru Ikeguchi[1]

[1] Graduate school of Science and Engineering, Saitama University,
255 Shimo-Ohkubo Saitama 338–8570, Japan
[2] `sima@nls.ics.saitama-u.ac.jp`

Abstract. Complex phenomena are observed in various situations. These complex phenomena are produced from deterministic dynamical systems or stochastic systems. Then, it is an important issue to clarify what is a source of the complex phenomena and to analyze what kind of response will emerge. Then, in this paper, we analyze deterministic chaos from a new aspect. The analysis method is based on the idea that attractors of nonlinear dynamical systems and networks are characterized by a two-dimensional matrix: a recurrence plot and an adjacent matrix. Then, we transformed the attractors to the networks, and evaluated the clustering coefficients and the characteristic path length to the networks. As a result, the networks constructed from the chaotic systems show a small world property.

1   Introduction

Complex phenomena often appear in real worlds, for example, fluctuation of financial in-dices in a stock market, change of weather conditions, electrical activities of neural systems, such as brain waves and nerve response, seismic activities of earthquakes and so on. These complex phenomena might be produced from a deterministic nonlinear dynamical system. Even if the complex phenomena are produced by a deterministic but nonlinear rule, it is not so easy to predict or control them, because its dynamics often exhibits chaos. This fact indicates that it is an important issue to clarify what is a source of the complex phenomena to develop an effective strategy towards prediction and control.

To measure the chaotic dynamics, several methods have already been proposed. The Lyapunov exponent is a measure for orbital instability due to sensitive dependence on initial condition. If the Lyapnov exponents are positive, corresponding dynamical system exhibits chaos. The Kolmogorov-Sinai entropy evaluates predictability of chaotic motions[1]. If the Kolmogorov-Sinai entropy takes a positive value, the system exhibits long-term unpredictability. The fractal dimension quantifies a geometrical structure of chaotic attractors. Chaotic attractors often have fractal structure, then the dimension would be a fractional value. Although these indices quantify chaotic attractors or chaotic motions from several view points,there exists a common feature. These measures use interpoint distances between two points on the attractors produced from the dynamical systems.

V. Kůrková et al. (Eds.): ICANN 2008, Part I, LNCS 5163, pp. 61–70, 2008.

In this sense, the recurrence plot[2,3] is also an important methodology to analyze chaotic attractors, because it represents the interpoint distances through a binary image. Then, it can visualize how recurrences occur on the attractors. It can often be used for analyzing nonstationarity of the dynamical systems. Using the recurrence plots, various hidden information of the systems can be revealed, such as local temporal correlation.

Different from an impact given by the nonlinear dynamical system theory which shows that a low dimensional nonlinear dynamical system can produce complex phenomena, recent drastic progress has been made on the network theory. In particular, from 1998, a new emerging field, called a complex network theory is widely distributed through several areas, such as biology, sociology, physics and so on. Using the complex network theory, complex phenomena in structural properties in various real networks have been clarified. These properties are a small-world property discovered by D. Watts and S. Strogatz[4] and a scale-free property discovered by A.-L. Barabási and R.Albert[5]. In addition, various measures have been proposed to analyze the structure and the dynamics of such complex networks[6].

The recurrence plot is a two-dimensional image which reflects dynamical and geometrical information of an attractor produced from nonlinear, possibly chaotic dynamical systems. Namely, the recurrence plot can be represented by a two-dimensional square matrix with ones or zeros. On the other hand, connections between nodes in networks are described by an adjacent matrix, which is also a two-dimensional square matrix. Thus, it means that the recurrence plots of the dynamical systems and the adjacent matrices of the network can be related each other. Then, we can apply the measures of the complex network theory to analyze the nonlinear dynamical systems that produce deterministic chaos. In this sense, J. Zhang and M. Small have already reported several results on characterization of the complicated dynamics in observed phenomena using the measures of the complex networks, such as the clustering coefficients, the characteristic path length, and the degree distributions[7]. Although their results are interesting, they mainly evaluated and analyzed pseudo-periodic time series. Then, in this paper, we applied the method to fully developed chaotic time series produced from a mathematical model and a real time series. In case of analyzing the mathematical model, we also evaluated numerical results with bifurcation phenomenon. As a result, we confirmed that the small world property emerges from the fully developed chaotic time series, and that the orbital instability, one of essential characteristic of deterministic chaos, produces a small-world property. In case of analyzing the real time series, we found that the time series also exhibits a small-world property.

2   Characterization of Deterministic Chaos Using the Complex Network Theory

2.1   Construction of a Network from an Attractor

To characterize the chaotic behavior using the complex network theory, we first generate an attractor from a dynamical system or reconstruct an attractor from a time series. If we have an explicit information of the dynamical system, an attractor is obtained as an asymptotic set of all the trajectories of the dynamical system. On the other hand, if we only have an observed time series, an attractor is reconstructed by transforming the observed time series to a reconstructed state space by a time delay coordinate[8].

Next, we construct a network from the attractor. To construct the networks, we take the same way as that of the recurrence plots from the attractor.

Let $\boldsymbol{x}(i)$ $(i = 1, \ldots, N)$ be the ith point on the attractor from the dynamical system. Then, a distance between $\boldsymbol{x}(i)$ and $\boldsymbol{x}(j)$ is defined by $d_{ij} = \mid \boldsymbol{x}(i) - \boldsymbol{x}(j) \mid$. Let $R(i, j)$ be the (i, j)th entry of a two-dimensional $N \times N$ square matrix R. Then, the algorithm for producing the recurrence plots is described as follows.

1. Select the ith point $\boldsymbol{x}(i)$ $(i = 1, \ldots, N)$, and calculate d_{ij} $(j = 1, \ldots, N)$.
2. Select M near neighbor points $(\boldsymbol{x}(k_1), \ldots, \boldsymbol{x}(k_M))$ of $\boldsymbol{x}(i)$. Thus, $\boldsymbol{x}(k_1)$ is the nearest neighbor of $\boldsymbol{x}(i)$.
3. Set $R(i, j) = 1$ if $j \in I_i$ and $R(i, j) = 0$ if $j \notin I_i$, where I_i is a set of indices of the selected M points $(k_1, \ldots, k_M)$.
4. Repeat the steps 1 to 3 for all $\boldsymbol{x}(i)$.
5. Symmetrization of the matrix R. Namely, for all i and j, if $R(i, j) = 1$, then $R(j, i) = 1$.

Now, we have a two-dimensional $N \times N$ square matrix R produced by the above algorithm. This matrix reflects the information of interpoint distances of attractor as an adjacent matrix of the network. The (i, j)th entry of the matrix R represents whether the ith node and the jth node are connected($R(i, j) = 1$) or not($R(i, j) = 0$). If the ith node and the jth node are connected, we consider that the ith node and the jth node are adjacent each other. The network generated by the above algorithm reflects the interpoint distances of the attractor. Namely, it contains a relationship between two points on the attractor.

2.2   Measures of the Complex Network Theory

The clustering coefficient and the characteristic path length[4] are basic measures to evaluate the property of the network in the complex network theory[4]. The small world property of the networks can be evaluated by the clustering coefficient and the characteristic path length.

The clustering coefficient is defined as follows:

$$C = \frac{1}{N} \sum_{i=1}^{N} C_i, \qquad C_i = \frac{l_i}{k_i \, {}_{\mathrm{C}_2}}, \tag{1}$$

where k_i is the number of adjacent nodes of the ith node, l_i is the number of connections between the adjacent nodes of the ith node, and N is the number of nodes. The clustering coefficient shows a local connectivity among any three nodes in the network.

The characteristic path length is defined as follows:

$$L = \frac{1}{N(N-1)} \sum_{i=1}^{N} \sum_{j=1, i \neq j}^{N} d_{ij}, \tag{2}$$

where d_{ij} is the shortest path length between the ith node and the jth node. The characteristic path length shows a global accessibility of the network.

If a network has a high clustering coefficient and a low characteristic path length, the network has the small world property. Such a network is called a small world network.

3   Experiments

We evaluated our proposed measure for the mathematical models and a real time series. In these experiments, we calculated the clustering coefficient C and the characteristic path length L for networks transformed from several attractors. The clustering coefficient C and the characteristic path length L of those networks were normalized by C_R and L_R, which were calculated from a randomized network. The randomized networks were generated by randomization of the connections of the constructed networks.

3.1   Evaluation with a Mathematical Model

First, to check the validity of the method, we used an attractor produced from the Rössler system[9]. The Rössler system is given by the following three-dimensional ordinary differential equations:

$$\dot{x} = -(y+z), \quad \dot{y} = x + 0.2y, \quad \dot{z} = 0.2 + z(x - \alpha). \tag{3}$$

We can generate not only periodic but also chaotic attractors by changing the parameter α in Eq.(3). Then, we conducted the following two experiments **(1)** and **(2)** to evaluate the networks produced from the periodic and chaotic attractors.

(1) We calculated the clustering coefficient C and the characteristic path length L by increasing of the number of adjacent nodes M (Fig.2). In this experiment, we fixed the number of nodes, $N = 15,000$,

(2) We calculated the clustering coefficient C and the characteristic path length L by increasing of the number of nodes N (Fig.3). In this experiment, we fixed the number of adjacent nodes $M = 20$,

In the above experiments **(1)** and **(2)**, we first used the following four values: $\alpha = 2.5$ (a period-1 attractor), $\alpha = 3.5$ (a period-2 attractor), $\alpha = 4.0$ (a period-4 attractor), and $\alpha = 5.0$ (a chaotic attractor).

Further, to evaluate bifurcation structure of the networks transformed from attractors of the Rössler system, we conducted the following experiment **(3)**.

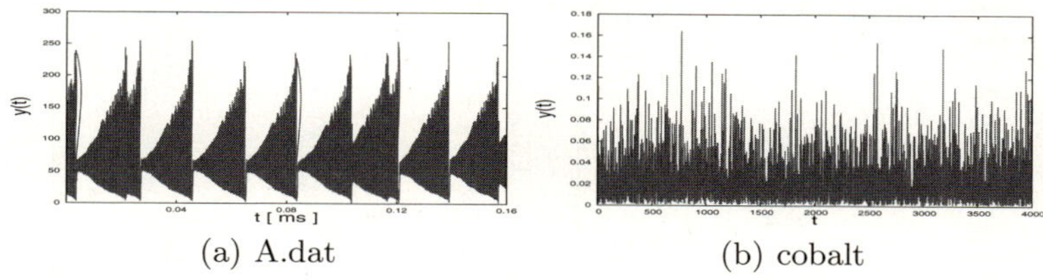

(a) A.dat   (b) cobalt

Fig. 1. A time series of (a) A.dat and (b) gamma ray emission of cobalt

(3) We changed the value of α in Eq.(3) from 2.5 to 8.0, and calculated the clustering coefficient C and the characteristic path length L of the networks produced from attractors $(2.5 \leq \alpha \leq 8.0)$ (Fig.4). In this experiment, we fixed the number of nodes, $N = 10,000$, and the number of adjacent nodes, $M = 20$,

In this experiment, we divided C and L by the maximum value of C and L for the bifurcation parameter α. Then, we defined C^* and L^* by the following equations:

$$C^*(\alpha) = \frac{C(\alpha)/C_R(\alpha)}{\max_{\alpha}(C(\alpha)/C_R(\alpha))}, \text{ and } L^*(\alpha) = \frac{L(\alpha)/L_R(\alpha)}{\max_{\alpha}(L(\alpha)/L_R(\alpha))}.$$

3.2 Evaluation for a Time Series

We evaluated three reconstructed attractors from a time series. The first one is reconstructed from the first component (x) of the chaotic Lorenz system[10] given by the following three dimensional ordinary differential equations:

$$\dot{x} = -10x + 10y, \ \dot{y} = -xz + 28x - y, \ \dot{z} = xy - \frac{8}{3}z. \tag{4}$$

The second one is a real time series, A.dat[11](Fig.1(a)). The time series A.dat is one of the data set which was used in the competition of time series prediction organized by the Santa Fe Institute in the 1992. This time series is reported to exhibit chaotic behavior[11].

The third one is a random time series: an interval time series of gamma ray emission of cobalt(Fig.1(b)). This time series shows very erratic behavior and is considered to be a truly random one.

To analyze these time series, we reconstruct an attractor by a time delay coordinate[8].

4 Results and Discussions

4.1 Mathematical Model

Results for four attractors of the Rössler system are shown in Figs.2 and 3. From Fig.2(a), the clustering coefficients (C) in all attractors are large, if the number

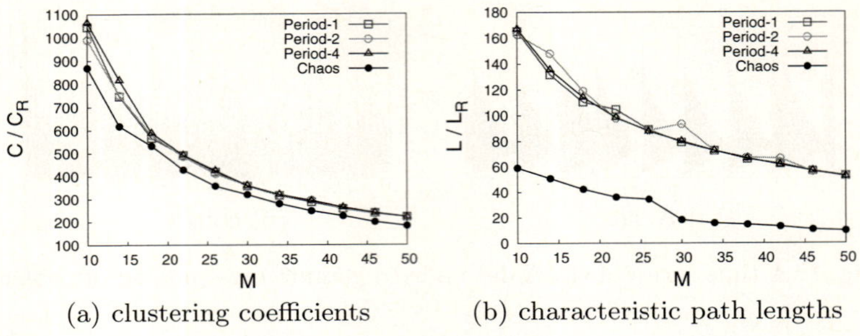

(a) clustering coefficients   (b) characteristic path lengths

Fig. 2. Results of (a) clustering coefficients C and (b) characteristic path lengths L in case of increasing M. In these results, the number of points on the attractors is fixed: $N = 15,000$.

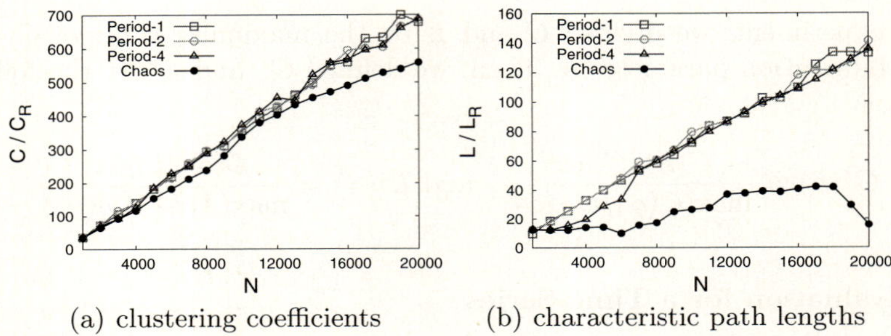

(a) clustering coefficients   (b) characteristic path lengths

Fig. 3. Results of (a) clustering coefficients C and (b) characteristic path lengths L in case of increasing N. In these results, the number of adjacent nodes is fixed: $M = 20$.

of the adjacent nodes (M) is small. In addition, from Fig.2(b), although the characteristic path lengths (L) of the periodic attractors are large, these of the chaotic attractor are small when the number of adjacent nodes M is small. From Fig.2, the chaotic attractor has large C and small L. Thus, the chaotic attractor shows the small world property.

Next, we analyzed the Rössler system by increasing the size of the network, N. In Fig.3(a), although the clustering coefficient (C) of the periodic attractors shows a similar tendency if the number of nodes (N) increases, the chaotic attractor shows smaller values of C than the periodic attractors. In addition, in Fig.3(b), although the periodic attractors show large characteristic path length (L) when the number of nodes (N) increases, the chaotic attractor keeps small values of L. From Figs.3(a) and (b), the chaotic attractor converged to have a small world property when the size of networks becomes large.

In Fig.4, we evaluated the clustering coefficient C and the characteristic path length L with an analysis on bifurcations of the Rössler system (Eq.(3)). In this figure, we changed the parameter α. The upper figure of Fig.4 is a bifurcation diagram. To construct the bifurcation diagram, we extracted local minima of the variable x of the Rössler system. We plotted the nth local minimum X_n on

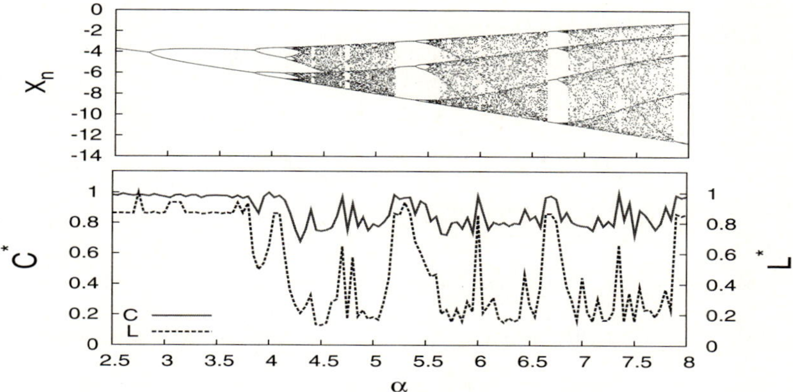

Fig. 4. A bifurcation diagram (upper) and values of clustering coefficients C and characteristic path lengths L (lower), when the parameter α in Eq.(3) is changed. In the bifurcation diagram the nth local minimum X_n of the x component of the Rössler system are extracted and plotted on the vertical axis.

the vertical axis. Then, we increased α from 2.5 to 8.0. In the lower figure of Fig.4, we can find that the clustering coefficient C and the characteristic path length L become large, when α is to a value with which periodic attractors are produced. On the other hand, the clustering coefficients C and the characteristic path length L of chaotic attractors become small. These results clearly show that when the attractors become chaotic, the small world property emerges.

4.2   Time Series

Results of the time series of the Lorenz system are shown in Fig.5, those of A.dat are shown in Fig.6, and those of cobalt are shown in Fig.7.

In the case of the chaotic Lorenz system and the A.dat, both C (Fig.6(a)), and L (Fig.6(b)) exhibit almost the same tendency. From Figs.5(a) and 6(a), if the reconstructed dimension increases, the clustering coefficients (C) converge to almost the same value and the values of C are large. In addition, from Figs.5(b) and 6(b), if the reconstructed dimension increases, the characteristic path lengths (L) also converged. The values of L (without normalization) are about 20 which is sufficiently smaller than the network size N. In Figs.5 and 6, both of these results show that even if we only observed a single variable chaotic time series, we can identify the small world properties from reconstructed attractors. On the other hand, in the case of analyzing cobalt data the results show different tendency from the chaotic Lorenz system and A.dat. From Fig.7, if the reconstructed dimension increases, the clustering coefficients (C) and the characteristic path lengths (L) do not converge but decrease.

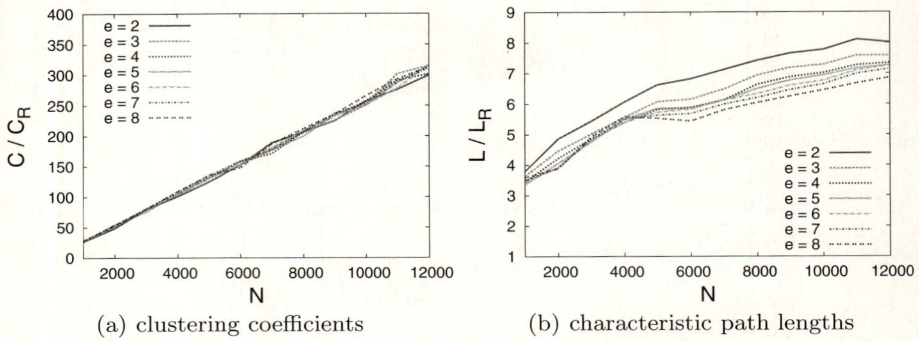

(a) clustering coefficients   (b) characteristic path lengths

Fig. 5. Results of (a) clustering coefficients, and (b) characteristic path lengths for the x component of the Lorenz system. In these figures, $M = 20$. The reconstructed dimension e is 2–8.

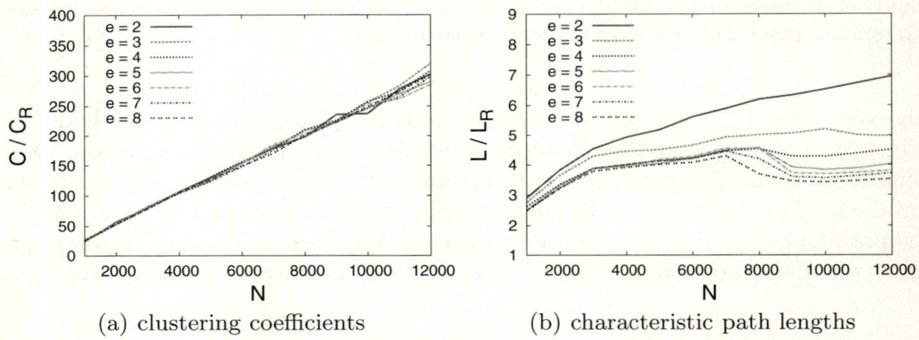

(a) clustering coefficients   (b) characteristic path lengths

Fig. 6. Results of (a) clustering coefficients, and (b) characteristic path lengths for the A.dat. In these figures, $M = 20$. The reconstructed dimension e is 2–8.

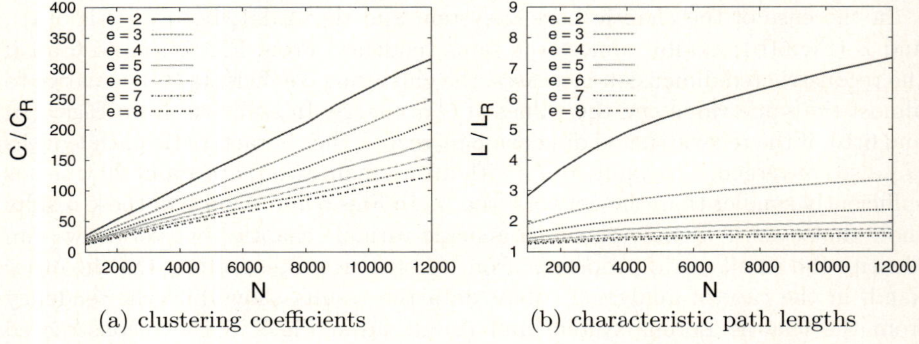

(a) clustering coefficients   (b) characteristic path lengths

Fig. 7. Results of (a) clustering coefficients, and (b) characteristic path lengths for the cobalt. In these figures, $M = 20$. The reconstructed dimension e is 2–8.

In the above analysis method, the nodes in the constructed network correspond to the points on the attractor, and the connections of the nodes are decided by spatial distribution between the points on the attractor. In other words, if the spatial distribution between the nodes becomes specific, the attractor has intrinsic local structure. Thus, the clustering coefficient and the characteristic path length for the constructed network may reflect information of a local structure of the points on the attractor. Thus, the results obtained in this paper suggest that the orbital instability of the deterministic chaos could have be represented as a small world property.

5   Conclusion

In this paper, we investigated how attractors of nonlinear dynamical systems are translated by the complex network theory. The key idea of our method is based on a relationship between the recurrence plot of the dynamical systems and the adjacent matrix of the network theory. Using this idea, we characterized the Rössler system and chaotic time series using the clustering coefficient and the characteristic path length. From the results, the periodic attractor shows the regular network property, which means that the networks has higher clustering coefficients and higher characteristic path lengths. On the other hand, the chaotic attractor shows the small world property, because the networks has high clustering coefficient, but smaller characteristic path length.

From these results, we claim that the clustering coefficients measure short-term predictability, and the characteristic path lengths do long-term predictability. Thus, we will investigate the above conjecture quantitatively from the view point of the complex network theory. In addition, in this paper, we only evaluated undirected and unweighted networks for the sake of simplicity. Thus, it is an important future problem to analyze directional and weighted networks. It is also important to apply the method to growing networks.

The research of TI is partially supported by Grant-in-Aid for Exploratory Research (No.20650032) from JSPS.

References

1. e.q. Ruelle, D.: Chaotic evolution and strange attractors. Cambridge University Press, Cambridge (1989)
2. Thiel, M., Romano, M.C., Kurths, J., Rolfs, M., Kliegl, R.: Twin surrogates to test for complex synchronisation. Europhysics Letters 74, 535–541 (2006)
3. Thiel, M., Romano, M.C.: Estimation of dynamical invariants without embedding by recurrence plots. Chaos: An Interdisciplinary Journal of Nonlinear Science 14(2), 234–243 (2004)
4. Watts, D.J., Strogatz, S.H.: Collective dynamics of 'small-world' networks. Nature 393, 440–442 (1998)
5. Barabási, A.L., Albert, R.: Emergence of scaling in random networks. Science 286, 509–512 (1999)

6. Costa, L.D.F., Rodrigues, F.A., Travieso, G., Villas Boas, P.R.: Characterization of complex networks: A survey of measurements. Advances in Physics 56, 167–242 (2007)
7. Zhang, J., Small, M.: Complex network from pseudoperiodic time series: Topology versus dynamics. Physical Review Letters 96(238701) (2006)
8. Packard, N.H., Crutchfield, J.P., Farmar, J.D., Shaw, R.S.: Geometry from a time series. Physical Review Letters 45(9), 712–716 (1980)
9. Rossler, O.E.: An equation for continuous chaos. Physics Letters A 57(5), 397–398 (1976)
10. Lorenz, E.N.: Deterministic nonperiodic flow. Journal of The Atmospheric Sciences 20, 131–141 (1963)
11. Weigend, A.S., Gershenfeld, N.A.: Time series prediction: Forecasting the future and understanding the past. Santa Fe Institute (1992)

Global Dynamics of Finite Cellular Automata

Martin Schüle[1,*], Thomas Ott[2], and Ruedi Stoop[1]

[1] Institute of Neuroinformatics, ETH and University of Zurich, 8057 Zurich,
Switzerland
[2] Institute of Applied Simulation, ZHAW Zurich, 8820 Wädenswil, Switzerland

Abstract. A novel algebraic and dynamic systems approach to finite
elementary cellular automata is presented. In particular, simple algebraic expressions for the local rules of elementary cellular automata are
deduced and the cellular automata configurations are represented via
Fourier analysis. This allows for a further analysis of the global dynamics
of cellular automata by the use of tools derived from functional analysis
and dynamical system theory.

1 Introduction

Cellular automata (CA) models have been widely studied and applied in physics,
biology and computer science. They are amongst the simplest systems which exhibit self-organisation and are closely related to neural networks. CA have also
been suggested as the generic discrete model of pattern formation and decentralised computation. Despite their simple appearance there are a number of
important open theoretical questions [1,2,3,4].

Contrary to the usual mathematical treatment we consider here *finite* cellular
automata. This starting point ensures a close analogy to actual biological systems
and may eventually be generalised in the infinite size limit.
A cellular automaton is specified by a d-dimensional regular discrete lattice L
with boundary conditions and a finite set Σ of states x_i assigned to each node
or cell i of the lattice. A *local rule* f acting on the cells in the range k of
the *neighbourhood* N_k^i of each cell i determines the dynamics of the cellular
automaton in discrete time steps starting from an initial condition. In this study
we consider only *finite* CA, that is CA with a finite number N of cells.

Unless stated otherwise, only *elementary* cellular automata will be considered, that is cellular automata with $d = 1$, $\Sigma = \{0,1\}$ and nearest neighbourhood $k = 3$. In this case, there are 256 different possible local rules $x_i^{t+1} =
f(x_{i-1}^t, x_i^t, x_{i+1}^t)$. The N cells are subject to *periodic boundary conditions* and
their states x_i are updated synchronously by the local rule. Local rules are given
by a *rule table*.

Example 1 (rule 110). The rule table of CA rule 110 is $(f(111) = 0, f(110) =
1, f(101) = 1, f(100) = 0, f(011) = 1, f(010) = 1, f(001) = 1, f(000) = 0)$.

* This work was supported by ETH Research Grant TH-04 07-2.

V. Kůrková et al. (Eds.): ICANN 2008, Part I, LNCS 5163, pp. 71–78, 2008.

It is customary to assign a decimal number to such rule tables. One speaks of *rule 110* as the binary expansion of the decimal number 110, that is $110 = 01101110$, encodes the rule table when read from left to right.

A configuration $x^t = (x_0^t, x_1^t, ..., x_{N-1}^t)$ of a CA is the string of the states of the N cells at the time t. Starting from an initial configuration x^0, the global function or map F then maps configurations x^t to subsequent configurations x^{t+1} thereby generating a space-time pattern. The global map F is only indirectly given through the local rule f. Figure 1 shows examples of space-time patterns generated by the CA rules 90 and 110.

Understanding the global dynamics induced by the global map F is the long-standing challenge of CA theory. CA with a finite number N of cells will eventually always become periodic (after at most 2^N steps). However, already the rather small CA of figure 1 with 65 cells could yield an intractable $2^{65} \sim 3.7 \cdot 10^{19}$ time steps. In this paper we present a novel approach in order to further analyse the global dynamics of finite CA. The outline of the paper is as follows: First, we introduce algebraic expressions for the local CA rules. Second, a continuous representation of the CA configuration space by Fourier analysis is presented. We then discuss the implications and possibilities of this approach for finite CA in view of functional analysis and dynamic system theory.

2   Algebraic Expressions of Local Rules

If a CA rule table is viewed as a truth table from propositional logic it is immediately clear that every rule table represents a *Boolean function*, which must be expressible as a *disjunctive normal form* (DNF) [5]. The DNF is a disjunction of clauses, where a clause is a conjunction of Boolean variables.

Example 2 (DNF of rule 110). CA rule 110 written as a DNF is $(X_{i-1} \wedge X_i \wedge \neg X_{i+1}) \vee (X_{i-1} \wedge \neg X_i \wedge X_{i+1}) \vee (\neg X_{i-1} \wedge X_i \wedge X_{i+1}) \vee (\neg X_{i-1} \wedge X_i \wedge \neg X_{i+1}) \vee (\neg X_{i-1} \wedge \neg X_i \wedge X_{i+1})$ with the Boolean variables X_i, the disjunction denoted by $\vee$, the conjunction by $\wedge$ and the negation by $\neg$.

The representation of cellular automata rules as Boolean functions has in fact been used by Wolfram to program cellular automata rules [6]. However, this line of research seems to have not yielded any further insights into CA rule dynamics [2]. We propose to translate the DNF of CA rules to algebraic expressions which will allow for a further analysis of CA dynamics. A similar idea but within a different framework was also proposed by Chua [8]. In the representation chosen here, the conjunction $X_i \wedge X_j$ is expressed by the algebraic multiplication, $x_i \cdot x_j$, the disjunction $X_i \vee X_j$ is expressed by the algebraic relation $x_i + x_j - x_i \cdot x_j$ and the negation $\neg X_i$ is expressed by $1 - x_i$. All CA rules $x_i^{t+1} = f(x_{i-1}{}^t, x_i{}^t, x_{i+1}{}^t)$ expressible as Boolean functions can be written in such a form, where the state of the cell i at time $t+1$ is given by an algebraic expression of the neighbourhood states at time t.

Example 3 (Algebraic expressions of rule 110 and rule 90). The algebraic approach proposed here is discussed by example of the prototypic rules 110 and

Fig. 1. Space-time patterns generated by CA rules 90 and 110. The vertical axis is the time axis. **a** Space-time pattern of rule 90 applied on a initial configuration of 65 cells of which all are in state 0 except one being in state 1 for 250 iterations. **b** Space-time pattern of rule 90 applied on a random initial configuration of 65 cells for 250 iterations. **c** Space-time pattern of rule 110 applied on a initial configuration of 65 cells of which all are in state 0 except one being in state 1 for 250 iterations. **d** Space-time pattern of rule 90 applied on a random initial configuration of 65 cells for 250 iterations.

90. Rule 110 is amongst the most complex elementary CA rules as it has been proven that it is capable of universal computation in the Turing sense [14]. Rule 90 is among the simplest elementary CA rules as one of the so-called additive CA rules [7].

The Boolean formula (DNF) for CA rule 110 becomes

$$x_i{}^{t+1} = x_i{}^t + x_{i+1}{}^t - x_i{}^t \cdot x_{i+1}{}^t - x_{i-1}{}^t \cdot x_i{}^t \cdot x_{i+1}{}^t \tag{1}$$

and for CA rule 90 (=01011010)

$$x_i{}^{t+1} = x_{i-1}{}^t + x_{i+1}{}^t - 2x_{i-1}{}^t \cdot x_{i+1}{}^t. \tag{2}$$

Elementary CA rules are grouped according to the 0-1-transformation and the *left-right*-transformation of the rules yielding 88 independent groups of rules which essentially display the same global dynamics [9]. Within the algebraic approach proposed here these transformations are simple algebraic operations on the CA rules. The $0-1$-transformation T_{0-1} is given by $T_{0-1}f(x_{i-1}, x_i, x_{i+1}) = 1 - f(1 - x_{i-1}, 1 - x_i, 1 - x_{i+1})$ and the left-right transformation $T_{left-right}$ is given by $T_{left-right}f(x_{i-1}, x_i, x_{i+1}) = f(x_{i+1}, x_i, x_{i-1})$.

Example 4 (The group of rules equivalent to rule 110). The group of rules equivalent to rule 110 under the 0-1-transformation and the left-right-transformation:
Rule 110 (=01101110)

$$x_i^{t+1} = x_i^t + x_{i+1}^t - x_i^t x_{i+1}^t - x_{i-1}^t x_i^t x_{i+1}^t \tag{3}$$

Rule 137 (=10001001) (0-1-transformation)

$$x_i^{t+1} = 1 - x_{i-1}^t - x_i^t - x_{i+1}^t + x_{i-1}^t x_i^t + x_{i-1}^t x_{i+1}^t + 2x_i^t x_{i+1}^t - x_{i-1}^t x_i^t x_{i+1}^t \tag{4}$$

Rule 124 (=01111100) (left-right-transformation)

$$x_i^{t+1} = x_{i-1}^t + x_i^t - x_{i-1}^t x_i^t - x_{i-1}^t x_i^t x_{i+1}^t \tag{5}$$

Rule 193 (=11000001) (0-1-transformation and left-right-transformation)

$$x_i^{t+1} = 1 - x_{i-1}^t - x_i^t - x_{i+1}^t + 2x_{i-1}^t x_i^t + x_{i-1}^t x_{i+1}^t + x_i^t x_{i+1}^t - x_{i-1}^t x_i^t x_{i+1}^t \tag{6}$$

Note that all CA rules with $\Sigma = \{0, 1\}$ can be assigned by this method to an algebraic expression. In fact, this approach allows to derive an algebraic expression for any kind of network which operates with Boolean functions.

If the variables x_i are drawn from a probability distribution we get algebraic expressions for probabilistic elementary CA which yield in the mean field approach the familiar mean field equations of probabilistic elementary CA [4,6].
To the best of our knowledge, we have here for the first time simple algebraic expressions for *all* elementary CA rules. Previous studies either examine only restricted classes of elementary CA rules, mostly *additive* CA rules [7] or cannot yield the same simple forms as our approach [8]. Of course, the algebraic expressions derived here apply equally to infinite elementary cellular automata.

In general, the algebraic approach proposed here will, through the equations of the form (1), give a *system of coupled nonlinear Boolean difference equations* for the time evolution of the global configurations x^t. We will not pursue this line further in this paper but instead investigate a continuous representation of CA configurations in order to analyse global CA behaviour.

3   Global Functions and Dynamics of Finite Cellular Automata

The global dynamics of a cellular automaton is determined by repeatedly applying the global map F to some initial configuration x^0. The global map F is only indirectly given by the local map f. One might be tempted to assign rational or real (for infinite configurations) numbers to the configurations x^t [6,10].

This approach has the drawback that it induces a wrong topology in the CA configuration space. The set of all possible CA configurations of length N forms a *N-dimensional hypercube* with the *Hamming distance* as the natural metric. The Hamming distance counts the number of bits for which two binary strings differ and is here defined by $d_H(x,y) = \frac{1}{N}\sum_{n=0}^{N-1}(x[n]-y[n])^2$ for two CA configurations, i.e. two binary strings $x[n]$ and $y[n]$ of length N. We are looking for a transformation which preserves the topology induced by the Hamming metric. In the usual dynamic system approach to CA, infinite size CA in d dimensions are treated as continuous functions on the compact metric space Σ^{Z^d} equipped with the product topology [1]. This topology for CA as dynamical system has been criticized in [11]. Here, we suggest a different approach for the topological dynamics of finite CA.

We propose to treat the CA configurations $x = x[n], n = 0, ..., N-1$ as a sample of a continuous periodic signal $x(s)$. A continuous signal with period 1 has a Fourier series

$$x(s) = \sum_{k=-\infty}^{\infty} \hat{x}_c(k)e^{2\pi iks} \tag{7}$$

with $\hat{x}_c(k)$ being the Fourier coefficients. If the signal is band limited and sampling is at a rate higher than the Nyquist rate the Fourier coefficients will up to a constant factor be equal to the *discrete* Fourier transform values $\hat{x}(k)$[12]. That is $\hat{x}(k) = N\hat{x}_c(k)$ with the discrete Fourier transform

$$\hat{x}(k) = \sum_{n=0}^{N-1} x[n]e^{-\frac{2\pi ikn}{N}}, 0 \le k \le N-1. \tag{8}$$

We thus have in (7) only a finite summation yielding a continuous function $x(s)$ which returns the initial configuration $x[n]$ when sampled at the rate $\frac{1}{N}$.

The global dynamics of CA crucially depend on the initial configurations. Our approach takes this fact into account as simple or periodic initial configuration will automatically yield simple forms through the discrete Fourier transform (8).

The functions $x(s)$ of equation (7) with finite N form a finite Hilbert space $L_2[0,1]$ with the inner product $(x,y) = \int_0^1 x(s)^*y(s)ds$, where $x(s)^*$ denotes the complex conjugate of $x(s)$. The distance metric is given by the inner product $\int_0^1 (x(s)-y(s))^*(x(s)-y(s))ds$ which results in

$$d(x,y) = \int_0^1 (x(s)-y(s))^*(x(s)-y(s))ds \tag{9}$$

$$= \frac{1}{N^2} \sum_{k=0}^{N-1} (\hat{x}(k)-\hat{y}(k))^*(\hat{x}(k)-\hat{y}(k)) \tag{10}$$

$$= \frac{1}{N} \sum_{n=0}^{N-1} (x[n]-y[n])^2 \tag{11}$$

$$= d_H(x,y)$$

We thus have through the discrete Fourier transform a distance-preserving isomorphism between two metric spaces, that is between a finite Hamming space (hypercube) and a finite Hilbert space $L_2[0,1]$. The distance measure in Fourier space is given by (10). All dynamical properties of CA (e.g. fixed-points) can accordingly be studied in Fourier space. In general, the dynamics will be given by a nonlinear map in the Fourier space which will be the focus of future work. Additive CA rules [6] however result in simple closed forms irrespective of the initial conditions.

Example 5 (Global dynamics of rule 90). For the additive rule 90, the algebraic expression of the local rule in equation (2) can be shortened to $x_i^{t+1} = (x_{i-1}^t + x_{i+1}^t) \ mod \ 2$. Inserting x_i^{t+1} into the Fourier series expression (7) yields

$$x(s)^T = \frac{1}{N} \sum_{k=0}^{N-1} (2\cos(\frac{2\pi k}{N}))^T \hat{x}(k) e^{2\pi iks} mod 2 \tag{12}$$

for the temporal evolution of the CA with T being an arbitrary time step. Note that this equation applies to any initial configuration and is therefore more general and simpler than previous results [7]. It might be surprising that the rather complicated looking dynamics of figure 1 (b) is generated by such a simple form.

Our approach yields through (7) the well behaved functions of a finite Hilbert space $L_2[0,1]$. Accordingly, results from functional analysis can readily be applied. As a first example consider the *Banach fixed-point theorem.*

Theorem (Banach fixed-point theorem for finite CA). *If the global function* $F : X \to X$ *is k-contractive:*

$$d(Fx, Fy) \le kd(x, y), \forall x, y \in X \tag{13}$$

and fixed $k, 0 \le k \le 1$, the global function F has exactly one fixed point $x*$ on the closed set X. That is there is a unique fixed point, a single homogeneous state the cellular automaton evolves to.

Proof. The classic Banach fixed-point theorem from functional analysis holds because all necessary conditions are satisfied within our approach [13].   □

Only the trivial rules 0 and 255 are contractive and therefore yield single homogeneous states. In ongoing work we study subsets of configurations which yield unique fixed points for certain rules through a (in this regard) restricted Banach fixed-point theorem. The crucial point is proving that these subsets are closed. As the functions $x(s)$ live in a finite Hilbert space, which is in a certain regard the most convenient space for analysis, we expect that more results from functional analysis will turn out to be useful in the analysis of global CA dynamics.

4   Concluding Remarks

Despite the simplicity of the local update rules for elementary cellular automata, the prediction of the global dynamics is a difficult problem, in its generality unsolved to date. In fact, in the infinite case, it has been proven that rule 110 is capable of universal computation and its global dynamics are therefore, in general, intractable [14].

Here, the idea is to tackle the problem by associating the configurations of a finite cellular automaton with states of a continuous state space, allowing for an analysis from a viewpoint of dynamic system theory. Previous approaches following this idea suffered from the fact that, first, there were no simple algebraic expressions for the local CA rules and, second, the association map did not preserve the topology of the original configuration space, the N-dimensional hypercube induced by the Hamming distance. This led to irreconcilable problems in the analysis of the global dynamics. In this contribution we showed:

1. that the rules of a elementary cellular automaton can be translated into algebraic expressions, yielding the possibility of an interpretation as continuous maps,
2. that on the basis of Fourier series, a continuous representation of the configurations can be introduced, which preserves the topology of the configuration space. This leads to a description of the global dynamics of a elementary cellular automaton in terms of the dynamics in a finite Hilbert space $L_2[0,1]$.

This preliminary contribution focused on illustrating the method characterised by these two main points. We believe that this approach opens up various new possibilities in the analysis of global dynamics of CA. We believe that the problems emerging within our approach can be amenable to a rigorous mathematical treatment, this is however outside of the scope of this contribution. Ongoing and future work will focus on studying non-additive CA from a dynamic system perspective. In addition, the Fourier series representation in the Hilbert space $L_2[0,1]$ stipulates further analysis of the dynamics from a functional analysis perspective. Finally, our approach shall in some weaker sense be generalised in the thermodynamic limit, that is the infinite size limit $N \to \infty$.

References

1. Kari, J.: Theory of Cellular Automata: A Survey. Theoretical Computer Science 334, 3–33 (2005)
2. Wolfram, S.: A New Kind of Science. B&T (2002)
3. Garzon, M.: Models of Massive Parallelism: Analysis of Cellular Automata and Neural Networks. Springer, Heidelberg (1995)
4. Deutsch, A., Dormann, S.: Cellular Automaton Modeling of Biological Pattern Formation. Birkhäuser, Basel (2004)
5. Mendelson, E.: Introduction to Mathematical Logic. Chapman & Hall, Boca Raton (1997)
6. Wolfram, S.: Statistical Mechanics of Cellular Automata. Reviews of Modern Physics 51, 601–644 (1984)
7. Martin, O., Odlyzko, A.M., Wolfram, S.: Algebraic Properties of Cellular Automata. Communications in Mathematical Physics 93, 219–258 (1984)
8. Chua, L.O.: A Nonlinear Dynamics Perspective of Wolframs New Kind of Science, Part I. International Journal of Bifurcation and Chaos 12, 2655–2766 (2002)

9. Li, W., Packard, N.: The Structure of the Elementary Cellular Automata Rule Space. Complex Systems 4, 281–297 (1990)
10. Chua, L.O.: A Nonlinear Dynamics Perspective of Wolframs New Kind of Science, Part IV. International Journal of Bifurcation and Chaos 15, 1045–1183 (2005)
11. Blanchard, F., Formenti, E., Kurka, P.: Cellular automata in the Cantor, Besicovitch and Weyl topological spaces. Complex Systems 11, 107–123 (1999)
12. Porat, B.: A Course in Digital Signal Processing. John Wiley &Sons, Inc., Chichester (1997)
13. Zeidler, E.: Applied Functional Analysis. Springer, Heidelberg (1995)
14. Cook, M.: Universality in Elementary Cellular Automata. Complex Systems 15, 1–40 (2004)

Semi-supervised Learning of Tree-Structured RBF Networks Using Co-training

Mohamed F. Abdel Hady*, Friedhelm Schwenker, and Günther Palm

Institute of Neural Information Processing
University of Ulm
D-89069 Ulm, Germany
{mohamed.abdel-hady,friedhelm.schwenker,guenther.palm}@uni-ulm.de

Abstract. Supervised learning requires a large amount of labeled data but the data labeling process can be expensive and time consuming. Co-training is a semi-supervised learning method that reduces the amount of required labeled data through exploiting the available unlabeled data in supervised learning to boost the accuracy. It assumes that the patterns are described by multiple independent feature sets and each feature set is sufficient for classification. On the other hand, most of the real-world pattern recognition tasks involve a large number of categories which may make the task difficult. The tree-structured approach is a multi-class decomposition strategy where a complex multi-class problem is decomposed into tree structured binary sub-problems. In this paper, we propose a framework that combines the tree-structured approach with Co-training. We show that our framework is especially useful for classification tasks involving a large number of classes and a small amount of labeled data where the tree-structured approach does not perform well by itself and when combined with Co-training, the unlabeled data boosts its accuracy.

1 Introduction

Many real world pattern recognition problems involve a large number of classes that can be assigned to any pattern. As the number of classes increases, the overall classification task becomes more difficult. Multi-class decomposition schemes [1], such as the tree-structured approach, have been considered where instead of using only a single classifier, an ensemble of classifiers is developed, and their results are combined. A multi-class decomposition scheme is a three-stage procedure where in the first stage the problem is decomposed into simpler subproblems. The second stage solves these sub-problems separately and then their solutions are combined to yield the final result. Advantages of problem decomposition are efficiency in learning, interpretability and scalability [2].

Although past decomposition schemes were focused on supervised classification, many multi-class pattern recognition problems are characterized by difficulty of collecting a large amount of labeled data. For example, in remote sensing

* This work has been supported by the German Science Foundation DFG under grant SCHW623/4-3 and a scholarship of the German Academic Exchange Service DAAD.

V. Kůrková et al. (Eds.): ICANN 2008, Part I, LNCS 5163, pp. 79–88, 2008.
© Springer-Verlag Berlin Heidelberg 2008

image classification, it was pointed out that the large number of spectral bands of modern sensors and the large number of land-cover classes require the labeling of a number of training examples that is an expensive or tedious process. At the same time, the collection of unlabeled data in these multi-class applications is easy and inexpensive. Hence, such multi-class applications require a multi-class decomposition scheme that achieve high accuracy using only a few labeled examples and many unlabeled data.

Semi-supervised learning approaches such as Co-training (for an exhaustive review, see [3]) provide a way to use a large amount of unlabeled data to boost the accuracy of a given supervised learning algorithm and to reduce the amount of labeled data required for learning. Despite of the practical relevance of these approaches within the multi-class decomposition context, there is not much work on integrating unlabeled data into multi-class decomposition [4], in the machine learning literature.

The main contribution of this paper is the combination of the tree-structured approach with the Co-training algorithm. We show that our method is especially useful for classification tasks involving a large number of classes and a small amount of labeled data where the tree-structured framework does not perform well by itself but when used in combination with Co-training setting, the unlabeled data will improve its performance.

The paper is organized as follows: The Co-training method is discussed in Section 2. The tree-structured multi-class decomposition scheme will be explained in Section 3. In Section 4, we introduce the semi-supervised learning of tree-structured RBF networks using Co-training and in Sections 5 and 6 experimental results of applying the proposed method to 2D and 3D object recognition from images are presented. Section 7 offers conclusion and future directions.

2   Co-training Algorithm

Co-training is a semi-supervised learning method that relies on the assumption that a given task is described by two redundant feature sets (views) where each feature set is sufficient for classification and the two feature sets of each instance are conditionally independent given the class. It incrementally uses the unlabeled data to a build classifier over each view [5]. An initial classifier in each view is created using the initial available labeled training examples. For each iteration, each classifier is applied to all unlabeled examples, and for each class it ranks the examples by confidence and selects a predefined number of the most confident examples. These high-confidence examples are labeled with the estimated class labels and added to the labeled training set. Based on the new training set, a new classifier is learned in each view, and the whole process is repeated for several iterations. At the end, a combined classifier is created by combining the prediction of the classifiers learned in each view. That is, each classifier extracts information from the unlabeled data and exchanges it with the other one. In web page classification [6], Co-training builds two classifiers given two independent feature sets, one uses the words appearing on the page itself and the other uses

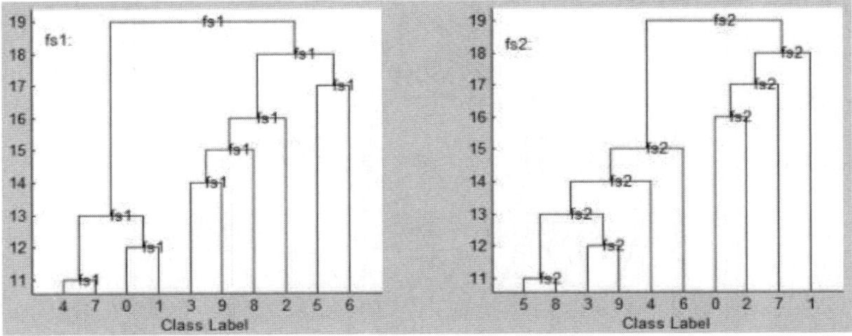

Fig. 1. Two different trees generated for the handwritten digits dataset

the words appearing in the hyperlinks pointing to that page. The results in [6] show that Co-training outperforms Expectation-Maximization (EM) and that the combination of the two embedded classifiers outperforms either of the two individually, because of the independence of the feature sets.

3   Tree-Structured Approach Using RBF Networks

A tree-structured approach decomposes a given K-class problem into a tree of K-1 binary problems, each associated with an RBF network (see Figure 2) using a single feature set (Single-View Tree) or a group of feature sets (Multi-View Tree). The strategy is based on creating a framework of tree structure through decomposing the set of classes into two disjoint subsets recursively until each subset contains one class. In [7], we have constructed an ensemble of tree-structured RBF networks using only labeled data.

3.1   Tree Training Phase

In the first step, the tree structure of the classifier is generated. The performance of a tree-structured classifier is sensitive to the tree structure. There are many approaches to split a set of classes into two disjoint meta-classes at each node. For example, in the handwritten digits recognition problem, we can construct two meta-classes by separating the even digits from the odd digits or we can put the digits from 0 to 4 in one meta-class and the rest in the other meta-class. It is recommended to generate the tree structure based on the similarity among the data points of the classes either automatically or manually if the number of classes is not very large. As in [7], the tree structure will be generated using 2-means clustering of the centroids of the classes (see Figure 1). In the second step of the training phase, an RBF network is assigned to each node (see Figure 2) where the Gaussian function in (1) is used as the radial basis function

$$h_j(\|x - c_j\|) = exp(-\frac{\|x - c_j\|^2}{2\sigma_j^2})$$

(1)

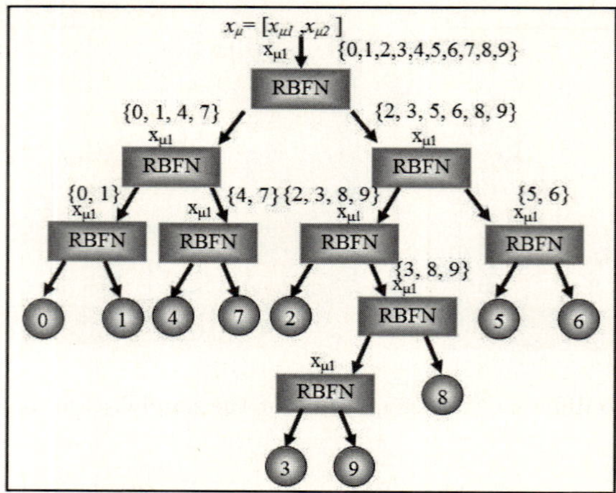

Fig. 2. Tree-structured RBF Networks corresponding to the left tree of Figure 1

For training such an RBF network a two-phase learning procedure [8] is used. In the first phase, the RBF centers $c_1, \ldots, c_k$ are determined by performing class-specific k-means clustering and the distance between c_j and the nearest prototype is used as the width of the jth RBF neuron where α has to be set heuristically (in our experiments $\alpha=1.5$).

$$\sigma_j = \alpha \min \{\|c_j - c_i\| : i = 1, \ldots, k, i \neq j\} \tag{2}$$

Then, in the second phase the output layer weights W are computed from

$$HW = Y \tag{3}$$

where Y is the matrix of target outputs of the m training examples and H is the activation matrix defined by

$$H = h_j(\|x - c_j\|)_{j=1,\ldots,k}^{i=1,\ldots,m} \tag{4}$$

Therefore, calculating the pseudoinverse of H provides a least squares solution to the system of linear equations in (3). This direct computation is faster than error backpropagation learning algorithm and yields good classification results.

3.2 Tree Classification Phase

Two different strategies can be used to compute the final decision of a tree-structured classifier: decision-tree-like evaluation and a global tree evaluation scheme known as the rule of combination in the Dempster-Shafer theory of evidence[9,10]. A simple and fast method to get the class label of a given sample is to traverse the tree, starting from the root node to a leaf node, as in the decision tree approach but this approach does not estimate class probabilities for the given samples.

Dempster-Shafer theory of evidence is a framework for representing and combining evidences. The main reason for using this theory in the combination of tree-structured binary classifiers is its ability to easily represent and combine evidences at different levels of abstraction. The Dempster-Shafer theory starts by assuming a universe of discourse consisting of a finite set of mutually exclusive atomic hypotheses θ. Let 2^θ denote the set of all subsets of θ. Then a function $m : 2^\theta \rightarrow [0, 1]$ is called basic probability assignment (bpa) if it satisfies

$$m(\phi) = 0 \quad \text{and} \quad \sum_{A \subseteq \theta} m(A) = 1 \tag{5}$$

It is possible to combine the basic probability assignments produced by n independent sources $m_1, \ldots, m_n$ using the orthogonal sum defined in (6) and (7) where the normalization factor K indicates how much $m_1, \ldots, m_n$ are contradictory.

$$m(A) = K \sum_{\cap A_i = A} \prod_{1 \leq i \leq n} m_i(A_i) \tag{6}$$

where

$$K^{-1} = 1 - \sum_{\cap A_i = \phi} \prod_{1 \leq i \leq n} m_i(A_i) = \sum_{\cap A_i \neq \phi} \prod_{1 \leq i \leq n} m_i(A_i) \tag{7}$$

In our study, we assume that θ is the set of class labels and the output of each RBF network is transformed into *bpa* and then the resulting *bpas* of all the classifiers are combined using the orthogonal sum without normalization. In contrast to the decision-tree-like approach, all the classifiers within the tree structure, are participating in classification and the final output estimates class probabilities for the given samples.

4 Combining Tree-Structured Approach and Co-training

We propose a new framework that aims to combine the advantages that tree-structured multi-class decomposition approach offers for classification tasks using labeled data and that of Co-training for combining unlabeled data and labeled data in learning. The pseudo-code of the framework is described in Algorithm 1. Let L be the set of labeled training examples and U be the set of unlabeled examples. The algorithm begins by constructing a multi-view forest $F^{(0)}$ consisting of two different tree classifiers $TC_1^{(0)}$ and $TC_2^{(0)}$ using any learning algorithm (in this study, two-phase learning procedure of an RBF network). The following process will then be repeated for T times or until U becomes empty.

For each example $x = (x_1, x_2)$ in U, the multi-view forest $F^{(t-1)}$ predicts a label for x where the average of the class probabilities produced by the tree classifiers $TC_1^{(t-1)}$ and $TC_2^{(t-1)}$ is used to measure the confidence in the forest prediction. For each class j, the n_j most confident examples assigned to class j are added to L and removed from U. Then, $F^{(t)}$ is trained using the updated training set. For classification phase, the forest $F^{(T)}$ resulting from the last iteration is used.

Algorithm 1. Co-Training by tree-structured classifiers algorithm

---

Require: L: set of m labeled training examples
    U: set of unlabeled examples
    V_1 and V_2: two feature sets(views) representing the examples
    T: maximum number of co-training iterations
    $\{n_j\}_{i=1}^{J}$: number of examples to select per class per iteration that must be directly proportional with the class prior probability where J is the number of classes
 1: Train tree classifiers $TC_1^{(0)}$ from L based on V_1 and $TC_2^{(0)}$ from L based on V_2
 2: Create an initial multi-view forest, $F^{(0)} = \{TC_1^{(0)}, TC_2^{(0)}\}$
 3: **for** $t = 1$ to T **do**
 4:   **if** U is empty **then**
 5:     Set $T = t$-1 and abort loop
 6:   **end if**
 7:   Apply the forest $F^{(t-1)}$ on U. (use the average of the class probabilities produced by classifiers $TC_1^{(t-1)}$ and $TC_2^{(t-1)}$)
 8:   For each class j, Select the n_j most confident examples assigned to class j
 9:   Move the most confident examples from U to L
10:   Train tree classifiers $TC_1^{(t)}$ and $TC_2^{(t)}$ using the new L
11:   Create a multi-view forest, $F^{(t)} = \{TC_1^{(t)}, TC_2^{(t)}\}$
12: **end for**
13: Return multi-view forest $F^{(T)}$

---

5   Datasets

5.1   Coil-20 Dataset

This benchmark dataset is a set of object images obtained from Columbia Object Image Library [11]. The dataset contains the images of 20 different objects, for each object 72 training samples are available. Each image was divided into 3×3 overlapped sub-images. Colour histogram and orientation histogram utilising Sobel edge detection are extracted from each sub-image [12]. Then the colour histograms are concatenated into one input feature set for one tree classifier and the orientation histograms forming another input feature set for the other.

5.2   Fruits Dataset

The fruits images data set, collected at the University of Ulm, consists of seven different objects with 120 images per object (see Figure 3). Each image was divided into 3×3 overlapped sub-images. Colour histogram and orientation histogram utilising Canny edge detection are extracted from each sub-image [12]. Then the 9 colour histograms are concatenated into one input feature set for one tree classifier and the 9 orientation histograms are used as a feature set for the other.

5.3   Handwritten Digits Dataset

The handwritten STATLOG digits data set [8] consists of 10,000 images (1,000 images per class) and each image is represented by a 16×16 matrix containing

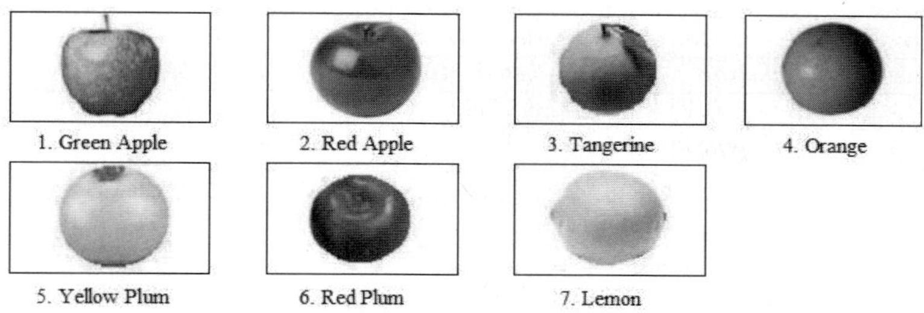

1. Green Apple   2. Red Apple   3. Tangerine   4. Orange

5. Yellow Plum   6. Red Plum   7. Lemon

Fig. 3. A sample of the images in the fruits dataset

the 8-bit grey values of each pixel (see Figure 4). In our study we used only 200 images per class. Each sample is represented by two feature vectors: a 40-dim vector results from reshaping the 16×16 image matrix into 256-dim vector then performing feature extraction using PCA and projecting the 256-dim vector onto the first 40 principal components. The second is a 144-dim vector representing 9 orientation histograms where an image matrix has been divided into 3×3 overlapped sub-images and an orientation histogram is created for each.

Fig. 4. A sample of the handwritten digits

6   Experimental Results

To estimate the accuracy 10 runs of 4-fold cross validation for all the datasets have been performed; consequently, each test set has 360, 210 and 500 while each training set consists of 1080, 630 and 1500 for COIL-20, fruits and digits, respectively. Table 1 summarizes the results. First, we assume that the whole training set is 100% labeled, to provide an upper bound to evaluate our framework (second column). Then, the training sets are split randomly into two sets L and U: 10% of the training examples per class (N_j) are used as L (10, 5 and 15 for fruits, COIL-20 and digits, respectively), while the remaining are U.

Co-Training has been performed until all the unlabeled examples in U are labeled and added to L. At each iteration, a probabilistic label is assigned to each unlabeled example. This label results from averaging the probabilistic labels computed by the two tree classifiers using Dempster-Schafer based combiner, then it ranks the examples by confidence and select the 4 most confident examples per class (n_j) and move them to L.

For COIL-20 dataset as shown in Figure 5, after 12 iterations the co-trained multi-view forest reaches a performance of 94.4% starting from 89.58%. Although the accuracy for Fruits dataset was 90.11% using only 10% labeled data, the

Table 1. The classification accuracy and the corresponding absolute gain

Dataset	*Baseline*	*10% Labeled*	*10% Labeled + Unlabeled Data*	Gain
COIL-20	99.6% ± 0.19	89.58% ± 1.56	94.4% ± 0.99	4.82%
Fruits	98.08% ± 0.26	90.11% ± 1.47	92.95% ± 1.43	2.84%
Digits	96.88% ± 0.14	90.78% ± 0.53	95.48% ± 0.31	4.7%

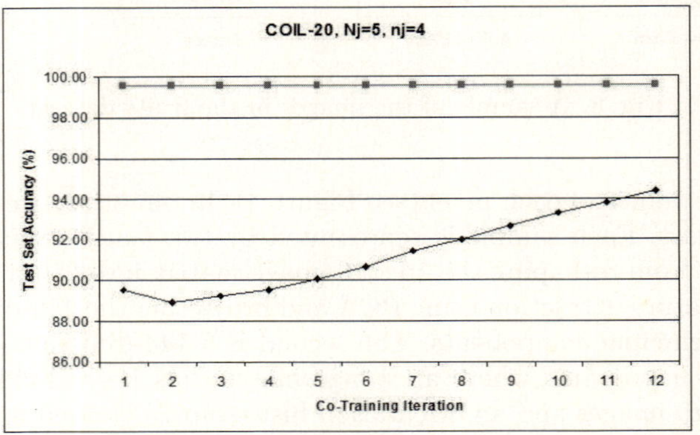

Fig. 5. The performance of Co-training for the COIL-20 data set as it labels more and more unlabeled examples. The horizontal line shows the performance of the baseline.

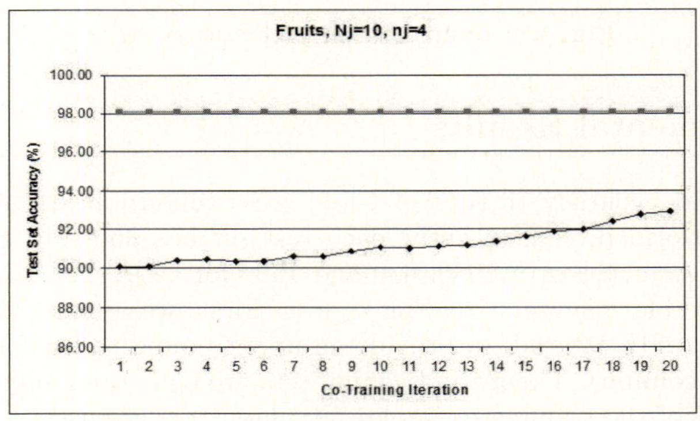

Fig. 6. The recognition rate of Co-training for fruits data set as the number of labeled examples increases. The horizontal line shows the performance of the baseline.

co-trained multi-view forest achieves an accuracy of 92.95% (see Figure 6) after 20 iterations. Therefore, the improvement was not satisfactory because the randomly selected ten samples were not enough to discriminate between red apple and red plum. Figure 7 shows that handwritten digits recognition rate was 90.78% using the 10% selected data and reaches 95.48% after 33 rounds.

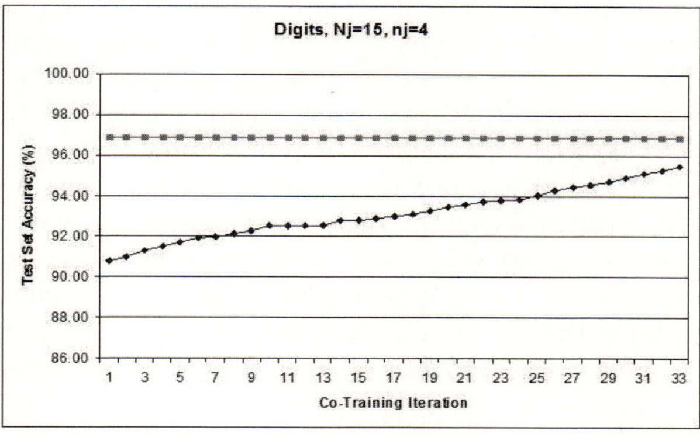

Fig. 7. The accuracy of Co-training for digits data set as it labels more and more unlabeled examples. The horizontal line shows the performance of the baseline.

In [4], the combination of Error-Correcting Output Codes (ECOC) and Co-training are shown to outperform both ECOC and Co-training using Naive Bayes as binary classifier and the approach was applied to text classification. Hence, we expect that using other semi-supervised learning algorithms such as self-training, EM and Co-EM [6] to integrate unlabeled data into multi-class decomposition schemes such as one-versus-one and one-versus-others would also improve their performance. Finding the best combination of semi-supervised learning and multi-class decomposition is beyond the scope of this paper.

Our main contribution is to show that integrating semi-supervised learning and multi-class decomposition leads to an improvement in performance. Another contribution is to prove that the probabilistic labels produced by Dempster-Schafer based tree evaluation method can be used by Co-training to measure the confidence in the tree classifier prediction.

7 Conclusions and Future Work

The results described in this paper are encouraging and show that the combination of tree-structured approach and Co-training is useful for learning with labeled and unlabeled data. We have shown that the proposed framework outperforms the supervised tree-structured approach when the number of labeled data is small and the amount of unlabeled data is large. In this study, the proposed framework was evaluated using RBF network as binary classifier type but in the future, we will investigate another types of binary classifiers such as SVM and k-NN. In addition, the reported results show that the accuracy after Co-training iterations does not reach the level achieved using 100% of the training set labeled. Hence, further investigation is required to minimize this gap.

There are two different architectures to combine the tree-structured approach and Co-training. The first architecture, proposed in this paper, is to train two tree-structured classifiers based on the two feature sets separately and to combine them using Co-training. The second architecture is to decompose the given k-class problem into $k-1$ binary problems using the tree-structured approach then learning each of these binary problems through training two binary classifiers based on the two feature sets separately and combining them using Co-training. It would be beneficial to study the second architecture in future work.

The Dempster-Shafer based combiner provides not only the resulting class but also estimates class probabilities for the presented samples which is required in the Co-training setting to measure confidence. But it is time consuming because all binary classifiers within the tree are evaluated and additional calculations are needed for combining the individual classification results. As a future work, we will investigate alternatives for faster estimation of class probabilities.

References

1. Kahsay, L., Schwenker, F., Palm, G.: Comparison of multiclass SVM decomposition schemes for visual object recognition. In: Kropatsch, W.G., Sablatnig, R., Hanbury, A. (eds.) DAGM 2005. LNCS, vol. 3663, pp. 334–341. Springer, Heidelberg (2005)
2. Hrycej, T.: Modular Learning in Neural Networks: A Modularized Approach to Neural Network Classification. John Wiley & Sons, Inc., NY (1992)
3. Seeger, M.: Learning with labeled and unlabeled data, Technical report (2002)
4. Ghani, R.: Using error-correcting codes with co-training for text classification with a large number of categories. In: Proceedings of Workshop on Text Mining at the First IEEE Conference on Data Mining (2001)
5. Blum, A., Mitchell, T.: Combining labeled and unlabeled data with co-training. In: Proceedings of the the 11th annual conference on Computational Learning Theory, pp. 92–100. Morgan Kaufmann Publishers, San Francisco (1998)
6. Nigam, K., Ghani, R.: Analyzing the effectiveness and applicability of co-training. In: Proceedings of the ninth international conference on Information and knowledge management, pp. 86–93. ACM, New York (2000)
7. Abdel Hady, M.F., Palm, G., Schwenker, F.: Multi-view forests of radial basis function networks based on dempster-shafer evidence theory. In: Verleysen, M. (ed.) Proceedings of the 16th European Symposium on Artificial Neural Networks (ESANN 2008), Bruges, Belgium, pp. 307–312. D-side publications (April 2008)
8. Schwenker, F., Kestler, H., Palm, G.: Three learning phases for radial basis function networks. Neural Networks 14, 439–458 (2001)
9. Dempster, A.P.: A generalization of bayesian inference. Journal of the Royal Statistical Society, 205–247 (1968)
10. Shafer, G.: A Mathematical Theory of Evidence. Princeton University Press, Princeton (1976)
11. Nene, S., Nayar, S., Murase, H.: Columbia object image library: Coil (1996)
12. Fay, R.: Feature selection and information fusion in hierarchical neural networks for iterative 3D-object recognition. PhD thesis, Ulm University (2007)

A New Type of ART2 Architecture and Application to Color Image Segmentation

Jiaoyan Ai[1], Brian Funt[2], and Lilong Shi[2]

[1] Guangxi University, China
shinin@vip.163.com
[2] Simon Fraser University, Canada

Abstract. A new neural network architecture based on adaptive resonance theory (ART) is proposed and applied to color image segmentation. A new mechanism of similarity measurement between patterns has been introduced to make sure that spatial information in feature space, including both magnitude and phase of input vector, has been taken into consideration. By these improvements, the new ART2 architecture is characterized by the advantages: (i) keeping the traits of classical ART2 network such as self-organizing learning, categorizing without need of the number of clusters, etc.; (ii) developing better performance in grouping clustering patterns; (iii) improving pattern-shifting problem of classical ART2. The new architecture is believed to achieve effective unsupervised segmentation of color image and it has been experimentally found to perform well in a modified $L^*u^*v^*$ color space in which the perceptual color difference can be measured properly by spatial information.

Keywords: ART2, similarity measurement, unsupervised segmentation, color image.

1 Introduction

With highly developed applications of color information in recent decades, the segmentation of color image has been given more and more attention. Various segmentation techniques have been proposed in the literature, amongst which the application of artificial neural networks (ANNs) comes to be attractive. ANNs have several advantages over many conventional computational algorithms, e.g., high degree of parallelism, nonlinear mapping, adaptivity, and error tolerance etc. Guo Dong [1] summarized different types of neural networks proposed for the segmentation of color image.

Both based on competitive learning there are two prevalent self-organizing network models: self-organizing map (SOM) of Kohonen [2] and adaptive resonance theory (ART) of Carpenter and Grossberg [3]. Although it has been commonly used in the unsupervised segmentation of color image [1], [4], SOM has following defects: (i) the map size should be defined in advance and can't be changed during learning; (ii) the learning is time-consuming. In comparison with it, ART is characterized by advantages: (i) no any prior knowledge of cluster

V. Kůrková et al. (Eds.): ICANN 2008, Part I, LNCS 5163, pp. 89–98, 2008.

number is needed; (ii) fast learning algorithm can be used. Adaptive resonance theory was introduced by Carpenter G. A. in 1976 who subsequently proposed three main types of ART network together with Grossberg S. [3], [5], [6].

In this paper, we cited ART2 network which can perform real-time unsupervised learning to analog input pattern, and overcome the disadvantage of most forward network that is easily running into local minimum [7], [8]. But the similarity measurement mechanism of ART2 is based on phase information of patterns that results in the loss of pattern magnitude and inhibits its application in segmentation of color image. Besides, pattern-shifting problem exists because the learning algorithm of classical ART2 is insensitive to gradually changed input patterns. In order to overcome these defects, various methods have been proposed in the literature [9], [10], [11], [12], [13], [14], but these improvements usually add more complexity to network in structure or in algorithms. By introducing a new mechanism of similarity measurement and a quick learning algorithm, we propose a new ART2 architecture without introducing complexity neither in structure nor in algorithms, and it is proved to be able to achieve effective color segmentation in perceptually uniform color space.

2   Structure and Algorithms of Classical ART2 Architecture

A classical ART2 architecture is shown in Fig. 1.

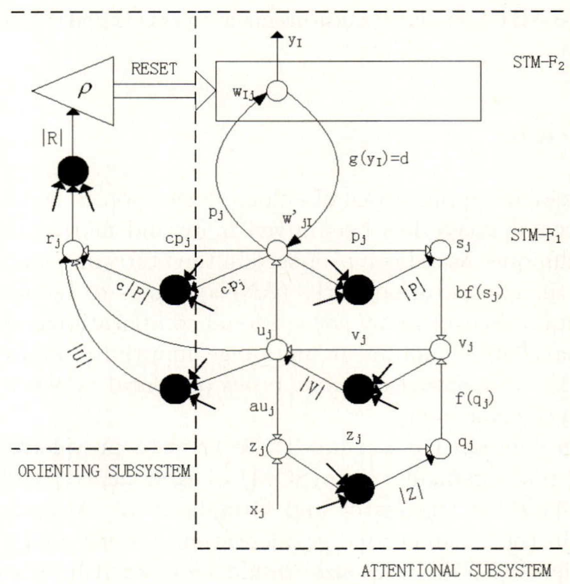

Fig. 1. A classical ART2 architecture

The network is partitioned by dashed lines into two subsystems. In attentional subsystem, STM-F_1 layer preprocesses input pattern to denoise and enhance contrast; the STM-F_2 layer keeps nodes to represent pattern prototypes, and $y_I = 1$ when node I is activated. Orienting subsystem executive similarity vigilance-testing between input pattern and activated pattern prototype. The bigger black circles mean calculation of modulus. Main algorithms of the classical ART2 network are as following [7]:

$$z_j = x_j + au \, , \tag{1}$$

$$q_j = \frac{z_j}{e + ||Z||} \, , \tag{2}$$

$$v_j = f(q_j) + bf(s_j) \, , \tag{3}$$

$$u_j = \frac{v_j}{e + ||V||} \, , \tag{4}$$

$$p_j = u_j + \sum_{i=0}^{m-1} g(y_i)w'_{ji} \, , \tag{5}$$

$$s_j = \frac{p_j}{e + ||P||} \, , \tag{6}$$

where,

$$f(x) = \begin{cases} x & if \ x \geq \theta \\ 0 & if \ x < \theta \end{cases}, \ \text{and} \ g(y_j) = \begin{cases} d & if \ j = I \\ 0 & if \ j \neq I \end{cases} . \tag{7}$$

Activation mechanism runs under competitive selection described as:

$$T_I = \max\{T_i = \sum_{j=1}^{n} p_j w_{ij} \mid i = 1, 2, \ldots, m\} \, . \tag{8}$$

m is the maximal allowable number of categories; n is the dimension of input pattern; I is the category number of current active node; e is error parameter. System will carry out unsupervised learning when the pattern prototype stored in the current active node is successfully matched to the input pattern. With quick learning, instar and outstar connection weights w_{Ij} and w'_{jI} update by

$$w_{Ij} = w'_{jI} = \frac{u_j}{1 - d} \, . \tag{9}$$

The similarity vigilance-testing of orienting subsystem processes with vector r described by

$$r_i = \frac{U_i + cPi}{||U|| + c||P||}, \ \text{and} \ R = ||r|| \, . \tag{10}$$

If $R \geq \rho + e$, update the instar and outstar weights of current active node, otherwise, the active node should be reset and inhibited, and system should continue to search the most matching pattern prototype among the residual nodes. If no node passes vigilance-testing, create a new node, that is, create a new category.

According to the algorithms, there is $||U|| = 1$, and $||W_I|| = ||W'_I|| = ||U/(1-d)|| = 1(1-d)$. Consequently no magnitude information of original input patterns has been included during the similarity vigilance-testing. To some applications (for example, speech recognition) in which magnitude is not the invariant feature, classical ART2 network has been widely applied. But in some other cases, the magnitude must been considered. For example, in image recognition, sometimes we must adopt gray intensity or color chroma etc. Some transform algorithms maybe allow alternatively to get invariant feature independent of magnitude, but usually cost is huge. Therefore it is significant to modify classical ART2 so as to make magnitude an important feature. Following sections will describe in detail how to modify structure and algorithms to carry out recognition and classification with combination of phase and magnitude.

3   Modification of Structure and Algorithms

Modified ART2 architecture is shown in Fig. 2.

Compared with classical ART2, there are following modifications:

(1) Extracting magnitude of pattern before the original input pattern into F_1.
(2) Adapting minimum-win competitive rule in activation mechanism of F_2 nodes,

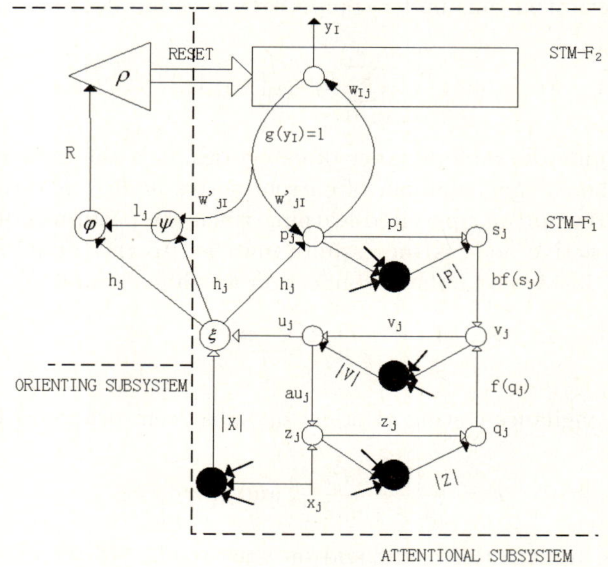

Fig. 2. The new ART2 architecture

$$D_i = ||P - W_i|| = \left(\sum_j (p_j - w_{ij})^2 \right)^{\frac{1}{2}} , \text{ and } D_I = \min(D_i) . \tag{11}$$

(3) Introducing three special functions to assistant similarity measurement between input pattern and stored pattern:

$$H = \xi(||X||, U) = ||X|| * U , \tag{12}$$

$$L = \psi(W_I', H) = \begin{cases} W_I' & if \ ||W_I'|| \neq 0 \\ H & if \ ||W_I'|| = 0 \end{cases} , \tag{13}$$

$$R = \phi(H, L, \delta) = \exp(-||H - L||/\delta) , \tag{14}$$

where, R is the final similarity measurement between two patterns for vigilance-testing.

(4) Quick weight adjustment (learning) algorithms:

$$w_{jI}'^{new} = w_{Ij}^{new} = \frac{1}{2}(||X|| \bullet u_j + w_{jI}'^{old}) . \tag{15}$$

By above modifications, the magnitude of original input pattern has been considered both during similarity vigilance-testing and weight adjustment. In addition, the stored pattern which is the closest to input pattern by Euclidian-distance will be activated.

4 Analysis of the New ART2 Network

The important modifications of the new ART2 architecture are similarity measurement and matching based on Euclidian-distance between pattern vectors. Then all modifications work together to appropriately apply spatial information to measure similarity, but keep the same foundation of classical ART2 network.

4.1 System Initialization

The processing nodes and downward connection weight are initialized to zero (same as classical ART2), but upward connection weight should be discussed forward. Initialization of upward connection weight aims to enable selecting or creating a new category when there is no existing category matching input pattern, but avoids to do so when there is matching category. So the initialization must make initial weight vectors far away from input pattern. In classical ART2 network, because competitive mechanism chooses the node of maximum weighted summation input as winner, upward channel vectors are usually initialized to evenly distributed random values inside a small zone, so new category's weighted summation is small and difficult to be chosen. Differently in the new ART2 network, competitive mechanism chooses the node of minimum Euclidian-distance as winner, so upward channel vectors should be initialized to be close to zeros in order to make the Euclidian-distance far enough between initial upward vector and input vector,

$$0 \le w_{ij} \le \beta \,, \tag{16}$$

where β is a mini constant values or mini random values inside a small zone. Notice that zeros are allowed here, because the Euclidian-distance is not zero between nonzero vector and zero vector.

4.2 STM-F_1 Self-stabilization Stage

The function of STM-F_1 in the new ART2 network is same as classical one that is to denoise and enhance contrast, but their operating process is somehow different mainly lying in the top pattern P:

$$p_j = h_j + \sum_{i=0}^{m-1} g(y_i)w'_{ji}, \text{ and } h_j = ||X|| \bullet u_j \,. \tag{17}$$

This top pattern P restores the magnitude information of input pattern. When there is no node in F_2 active, that is $g(y_i) = 0$, P represents middle pattern H.

After STM-F_1 self-stabilization, top pattern P is sent to feature representation field through upward channel for similarity matching in order to search the node with the shortest Euclidian-distance to pattern P. Calculate all Euclidian-distances of instar connection weight vectors to P. If $D_I = \min(D_i)$, then activate node I. There are some instances which should be considered:

(1) If define the maximum number of categories in advance, and no node has been executed that is never learned, when all topward channel vectors have the same initial values, system may chose to active one node randomly or according to index number, otherwise minimum-win rule still works.

(2) If define the maximum number of categories in advance, and all learned nodes have been inhibited because of not passing vigilance-testing, execute as instance (1) among residual unlearned nodes.

(3) If not define the maximum number of categories in advance, and all new nodes are dynamically added, then there is just one node initially. It is chosen when input the first pattern, and create new node dynamically if the following pattern does not match it, the rest may be deduced by analogy. Because new node will be created only when all learned nodes have been inhibited, input pattern may be directly stored into the new category.

(4) Quick searching: directly chose or create a new node will greatly shorten searching time. It means that when the node with the minimum Euclidian-distance among all learned nodes don't pass vigilance-testing the others will also not pass.

4.3 Vigilance Testing

When a node in F_2 is activated, its downward channel vector will be sent to orienting subsystem for similarity vigilance-testing together with middle-level pattern of input. If node I is activated, there are several possibilities:

(1) It is an unexecuted node. $W_I' = 0$, such that by Eq. (13) and (14) there are

$$L = H \text{ and } R = 1 . \tag{18}$$

That means two patterns are the same, so undoubtedly the node will pass vigilance-testing
(2) It is an executed node. $W_I' \neq 0$, by Eq. (12), (13), (14) there is

$$R = \exp\left(-||W_I' - ||X|| * U||/\delta\right) . \tag{19}$$

By defining

$$D_I' = ||W_I' - ||X|| * U|| , \tag{20}$$

and vigilance parameter ρ , the requirement for passing vigilance-testing is

$$\exp(-D_I'/\delta) \geq \rho \Rightarrow D_I' \leq (-\ln\rho)\delta . \tag{21}$$

With a determinate ρ, all those input patterns located inside a super-sphere with W_I' as center and $(-\ln\rho)\delta$ as radius will pass vigilance-testing. It is obvious that super-sphere's radius linearly varies with δ , therefore the selection of δ directly decides clustering precision. In addition, with defined δ the super-sphere's radius logarithmically varies with vigilance parameter, so selection of ρ can also regulate classifying precision. Based on that, δ and ρ should be appropriately selected according to actual application background.

4.4 Weight Adjustment (Learning)

Weight learning in this paper is a kind of quick algorithms of approximate mean vector. By analysis of articles [9], [10] the learning algorithms of classical ART2 usually result in pattern shifting, and it can be improved by adapting varied K-means as weight learning algorithm. In article [10] the activated times K of each node is recorded, and new weight update by

$$W^{new} = \frac{1}{k}((k-1)W^{old} + input) \tag{22}$$

By this method, the new weight vector more correctly represent clustering center of a category with combination of all learned patterns and the new input pattern. The disadvantage is having to add a mechanism in network to record K, and it becomes not important after enough learning. We adapt approximate mean defined by Eq. (15) that avoids recording mechanism while sufficiently simulate the idea of varied K-mean algorithm.

5 Experimental Results and Analysis

To describe the design and application of the new ART2 architecture, we carried out some experiments in this paper.

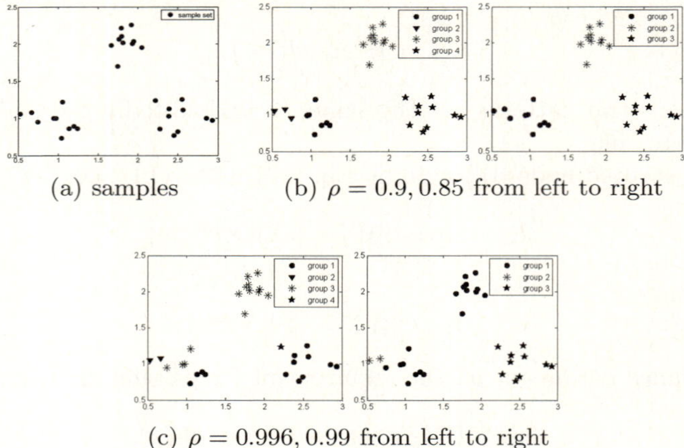

(a) samples (b) $\rho = 0.9, 0.85$ from left to right

(c) $\rho = 0.996, 0.99$ from left to right

Fig. 3. Sample set and classifying results. (a) is samples distribution; (b)showed classifying results of the new ART2 network given parameters $a = b = 10, \theta = 0.01, \delta = 5, e = 0$; (c) showed classifying results of the classical ART2 network given parameters $a = b = 10, \theta = 0.01, c = 0.1, d = 0.9, e = 0$.

5.1   Performance in Classifying Clustering Groups

Sample set includes thirty randomly distributed data points which are expected at points $(1, 1)$, $(2, 2)$, and $(2.5, 1)$ respectively disturbed by noise of Gaussian distribution. The samples and classifying results are shown in Fig. 3. In Fig. 3(a) we can visually perceive that all samples are distributed as three clustering groups. The experimental results have shown that the new ART2 is more suitable for group clustering data than the classical ART2 architecture.

5.2   Segmentation of Color Image

With the advantages described above, it is believed that the new ART2 architecture should be able to achieve effective segmentation of color image in perceptually uniform color space. In this paper, we conducted all experiments in a modified CIE$L^*u^*v^*$ color space in which the perceptual color difference can be measured properly by spatial information [15]. In [1], Gu Dong has given detailed steps for the conversion from RGB to modified CIE$L^*u^*v^*$.

A. Segmentation of Artificial Color Image

Given an artificial color image in which several perceptually close colors are nested, experiments were designed to test the classifying ability of the new ART2 architecture. Results have been shown in Fig. 4.

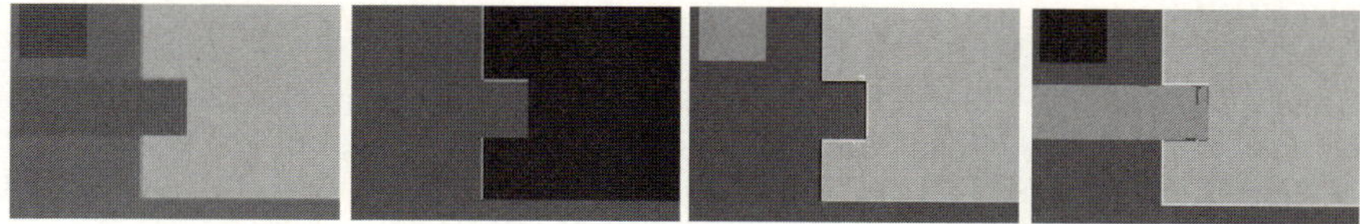

Fig. 4. Artificial color image and segmentation results with the new ART2 network. From left to right they are the original image and segmentation results given parameters $a = b = 5, \theta = 0.1, \delta = 1, e = 0$ with different vigilance testing $\rho = 0.85, 0.9$ and 0.95 respectively.

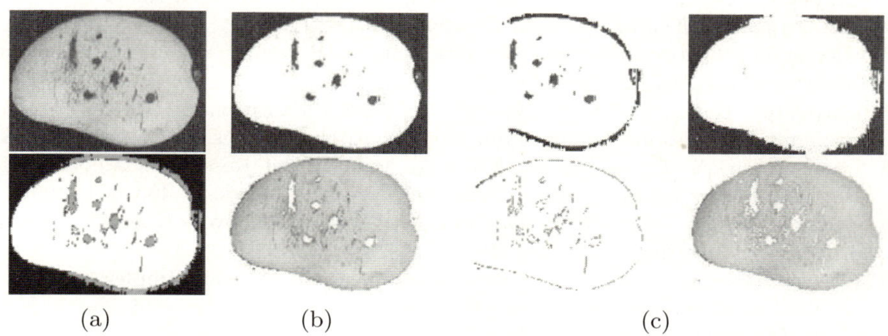

(a)   (b)   (c)

Fig. 5. Natural color image and segmentation results with the new ART2 network. By column, (a) showed original image and final segmentation; given parameters $a = b = 5, \theta = 0.1, \delta = 1, e = 0.01$, (b) showed segmentation results with $\rho = 0.8$; (c) showed the subsegmentation of (b) with $\rho = 0.9$ and $\rho = 0.85$ from top to bottom.

B. Segmentation of Natural Color Image

Fig. 5 has shown the segmentation results of natural color image with the new ART2 network.

6   Conclusion

The new ART2 architecture proposed in this paper introduces a new similarity mechanism without changing the classical ART2 foundation, and the modifications of structure and corresponding algorithms work together to improve network's performance in grouping clustering data and have been experimentally found effective in color image segmentation in uniform color space. Of course there still is need for more work to make the new network generally applicable. For example: how to preprocess data to make sure the spatial compactness; and how to select parameters appropriately? That will be our next goal.

Acknowledgments. The work is supported by Guangxi Science Foundation (0832058) and Computational Vision Lab of School of Computing Science in Simon Fraser University, Canada.

References

1. Dong, G., Xie, M.: Color Clustering and Learning for Image Segmentation Based on Neural Networks. IEEE Transactions on Neural Networks 16(4), 925–936 (2005)
2. Kohonen, T.: The Self-Organizing Map. IEEE Proceedings 78, 1464–1477 (1990)
3. Carpenter, G.A., Grossberg, S.: A Massively Parallel Architecture for a Self-Organizing Neural Pattern Recognition Machine. Computer Vision, Graphics, and Image Processing 37, 54–115 (1987)
4. Yeo, N.C., Lee, K.H., Venkatesh, Y.V., Ong, S.H.: Colour Image Segmentation Using the Self-Organizing Map and Adaptive Resonance Theory. Image and Vision Computing 23, 1060–1079 (2005)
5. Carpenter, G.A., Grossberg, S.: The ART of Adaptive Pattern Recognition by a Self-Organizing Neural Network. Computer, 77–88 (1988)
6. Huang, D.: Theory of Neural Network Pattern Recognition System. Electronic Industry Press, Beijing (1996)
7. Carpenter, G.A., Grossberg, S.: ART-2: Self-Organization of Stable Category Recognition Codes for Analog Input Pattern. Applied Optics 26, 4919–4930 (1987)
8. Carpenter, G.A., Grossberg, S.: ART2-A: An Adaptive Resonance Algorithm for Rapid Category Learning and Recognition. Neural Networks 4, 493–504 (1991)
9. Shen, A., Yu, B., Guan, H.: Research on ART-2 Neural Network Classifier. Journal of Northern Jiaotong University 20(2), 146–151 (1996)
10. Cong, S., Zheng, Y., Wang, Y.: The Improvement and Modeling Implementation of ART-2 Neural Network. Computer Engineering and Applications 14, 25–27 (2002)
11. Tang, H., Sang, N., Cao, Z., Zhang, T.: Research and Improvements of ART-2 Neural Networks. Infrared and Laser Engineering 33(1), 101–106 (2004)
12. Xu, Y., Deng, H., Li, Y.: An Improved ART2 Neural Network Clustering Algorithm. Journal of Computer Applications 26(3), 659–662 (2006)
13. Gu, M., Ge, L.: Improved Algorithm for ART2 Neural Network. Journal of Computer Applications 27(4), 945–947 (2007)
14. Gu, M., Ge, L.: Improvements of ART2 Neural Network Based on Amplitude Component. Computer Engineering and Applications 43(13), 52–54 (2007)
15. Color Metric, http://www.compuphase.com/cmetric.htm

BICA: A Boolean Indepenedent Component Analysis Approach

B. Apolloni[1], S. Bassis[1], and A. Brega[2]

[1] Dipartimentimento di Scienze dell'Informazione
via Comelico 39/41, 20135 Milano, Italy
{apolloni,bassis}@dsi.unimi.it
[2] Dipartimento di Matematica Federigo Enriques
via Saldini 50, 20133 Milano, Italy
andrea.brega@unimi.it

Abstract. We analyze the potentialities of an approach to represent general data records through Boolean vectors in the philosophy of ICA. We envisage these vectors at an intermediate step of a clustering procedure aimed at taking decisions from data. With a "divide et conquer" strategy we first look for a suitable representation of the data and then assign them to clusters. We assume a Boolean coding to be a proper representation of the input of the discrete function computing assignments. We demand the following of this coding: to preserve most information so as to prove appropriate independently of the particular clustering task; to be concise, in order to get understandable assignment rules; and to be sufficiently random, to prime statistical classification methods. In the paper we toss these properties in terms of entropic features and connectionist procedures, whose validation is checked on a series of benchmarks.

1 The Representation Problem

A general wisdom in clustering a set of objects is: the better you group them, the better you understand what they are. On the other hands, you may claim to have understood the true nature of the objects you are handling only when the decisions you have taken about them prove suitable. This is the essence of the connectionist philosophy [1] where representation and related symbols and meaning are a corollary of the proved suitability. In the statistical assumption that the framework where objects and decisions live is repetitive enough that probability distributions around boiling up events make operational sense, we may expect that we can define general features of a suitable representation of data, independently of their subsequent clustering at the basis of decisions. In very abstract terms, we identify the goal of a clustering by the minimization of the conditional entropy of objects given the clusters c to which they are assigned. To concretize this quantity we need a pattern y_i through which the objects are identified. Denoting with capital letters (such as Y) random variables and small letters (y) their realizations – as usual – and calling the above entropy $H(Y|C)$, we realize that in the case of the univocal assignment of objects to clusters, this entropy is given by the difference between the entropies $H(Y)$ of Y and $H(C)$ of C, so that a suitable representation Z of the objects may locate its entropy at an intermediate level between the two. With this aim, on the

V. Kůrková et al. (Eds.): ICANN 2008, Part I, LNCS 5163, pp. 99–108, 2008.

one hand we may prove the optimality of representations with independent components. In turn, this fits the usual entropic criterion of maximizing the mutual information between the components [2]. On the other, we decide to require these components to be Boolean as well, in the direction of taking the discrete decisions underlying clustering. We synthesize these two goals into a Boolean Independent Component Analysis, BICA for short [3].

Since we are not giving any semantic value to these variables – for the moment – we commit their computation to a connectionist device. Since we want to preserve the information content of the data, we set up a neural network with double duty: i) extracting a compressed version of the data in a tight hidden layer, and ii) mapping it onto Boolean vectors that minimize a specific cost function. On the one hand we prove the correctness of the cost function with respect to our goal. On the other, we are left with a binary coding that has no specific meaning and depends on the network initialization. Hence, with a given probability density of the initialization parameters we obtain a generator of random codes which may be used as the primer of random base learners within learning ensemble schemes [4]. We have experienced BICA coding in various contexts and different clustering tools [5,3]. At the close of the paper we synthesize some of these experiments in order to highlight the application value of our approach.

2   The Ntropic Question

A common translation of a clustering problem in entropic terms lies in the minimization of the mutual information between the pattern vector $Y = (Y_1, \ldots, Y_\nu)$ through which we characterize the objects and the class to which we attribute them, in the idea that a same object could be generated with a different likelihood to more than one class [6]. In particular, the k-means clustering rule: attribute the object o_i to the j-th among a set of clusters $\{c_1, \ldots, c_k\}$ if $\nu_r(\boldsymbol{y}_i - \boldsymbol{v}_j)$ is the minimum of the set $\{\nu_r(\boldsymbol{y}_i - \boldsymbol{v}_t), t = 1, \ldots, k\}$ with ν_r a suitable norm[1]. and $\boldsymbol{v}_j$ the template of the j-th cluster, matches an object distribution law

$$f_{\boldsymbol{Y}|C=c_j}(\boldsymbol{y}_i) \propto \mathrm{e}^{-\nu_r(\boldsymbol{y}_i-\boldsymbol{v}_t)^r} \tag{1}$$

and an *a priori* probability distribution of the clusters that is uniform on the set of C specifications. By definition

$$H(\boldsymbol{Y}|C) = -\sum_{c_i} p_{c_i} H(\boldsymbol{Y}|c_i) \tag{2}$$

where p_{c_i} is the probability measure of cluster c_i. With this notation

$$f_{\boldsymbol{Y}}(\boldsymbol{y}) = \sum_{c_i} p_{c_i} f_{\boldsymbol{Y}|C=c_i}(\boldsymbol{y}) = \sum_{c_i} p_{c_i} \frac{1}{(2\pi)^{n/2}} \mathrm{e}^{-(\boldsymbol{y}-\boldsymbol{v}_{c_i})^T (\boldsymbol{y}-\boldsymbol{v}_{c_i})} \tag{3}$$

Hence

$$H(\boldsymbol{Y}|C) = \sum_{c_i} p_{c_i} \mathrm{E}[(\boldsymbol{Y}-\boldsymbol{v}_{c_i})^T (\boldsymbol{Y}-\boldsymbol{v}_{c_i})] + a \tag{4}$$

---

[1] Say $\nu_r(\boldsymbol{x}) = \left(\sum_{i=1}^n |x_i|^r\right)^{\frac{1}{r}}$.

where a is a constant and E denotes the expectation operator. By denoting with $I(X,Y)$ the mutual information between X and Y and re-reading in true entropic terms our clustering strategy, the general goal of any useful mapping from Y to C is to ensure a high mutual information [7]. Now, in the case where the mapping realizes a partition of Y range, hence in the case of univocal mapping from the above range to $\{c_1, \ldots, c_k\}$ – an operation that we denote as a *labeling* of the y_is with flags possibly having the same name of the clusters, by abuse – we have the following expression of the mutual information.

Lemma 1. *For any univocal mapping from Y to C,*

$$H(Y|C) = H(Y) - H(C) \tag{5}$$

$$I(Y,C) = H(C) \tag{6}$$

Proof. Let us prove the claim in the case of discrete Y, hence specifying into patterns indexed by an integer i, and C having only two clusters, hence inducing a label L ranging in $\{0,1\}$, the extension to the general case being trivial. With these conditions we have

$$H(Y) = -\sum_i p_i \ln p_i = -\sum_{i \in A} p_i \ln p_i - \sum_{i \in \overline{A}} p_i \ln p_i$$

$$= -\left(\sum_{i \in A} p_i\right) \sum_{i \in A} \frac{p_i}{\left(\sum_{i \in A} p_i\right)} \ln \left(\frac{p_i}{\sum_{i \in A} p_i} \sum_{i \in A} p_i\right)$$

$$- \left(\sum_{i \in \overline{A}} p_i\right) \sum_{i \in \overline{A}} \frac{p_i}{\sum_{i \in \overline{A}} p_i} \ln \left(\frac{p_i}{\sum_{i \in \overline{A}} p_i} \sum_{i \in \overline{A}} p_i\right)$$

$$= -p_A \sum_{i \in A} p_{i|A} \ln p_{i|A} - p_{\overline{A}} \sum_{i \in \overline{A}} p_{i|\overline{A}} \ln p_{i|\overline{A}}$$

$$- p_A \ln p_A - p_{\overline{A}} \ln p_{\overline{A}} = H(Y|L) + H(L) \tag{7}$$

where A is the set of the indices of Y patterns labeled by 1, $\overline{A}$ its complement, and p_i is the probability of the pattern or set denoted by the index i. By solving the equality in $H(Y|L)$ we obtain the first claim. Coupling this result with the definition $I(Y,L) = H(Y) - H(Y|L)$ we obtain $I(Y,L) = H(L)$ which brings us to the second claim.

Claim (6) says that a clustering so more preserves information of patterns the more the entropy of clusters is higher, i.e. the more they are detailed. Claim (5) denotes the gap between entropy of patterns and entropy of clusters that is managed by the clustering algorithm. As $H(Y)$ is not up to us, the algorithm may decide to group the patterns so as to increase $H(C)$, in line with the claim (6) suggestion.

   In a case such as ours where the labels of the patterns are given and no other information about them is granted, we are harmless with respect to the $H(C)$ management. Rather, we focus on an algorithm that reproduces the given labeling. If the task is hard, we may try to split it into two steps: i) improve the patterns representation so that their labels may be more understandable, and ii) find a clustering algorithm with this new input. This corresponds to dividing the gap between $H(Y)$ and $H(C)$ in two steps and, ultimately, looking for an encoding Z of Y minimizing $H(Z|C)$, i.e. the residual gap $H(Z) - H(C)$. We have two constraints to this minimization. One is to maintain the partitioning property of the final mapping. Hence

Definition 1. *Consider a set A of Y patterns, each affected by a label $c_i \in C$. We will say that an encoding Z of Y is correct if it never happens that two patterns of A with different labels receive the same codeword.*

The second constraint is strategic: as we do not know the algorithm we will invent to cluster the patterns, we try to preserve almost all information that could be useful to a profitable running of the algorithm.

This goal is somehow fuzzy, since *vice versa* our final goal is to reduce the mean information of the patterns exactly to its lower bound $H(C)$. A property that is operationally proof against the mentioned fuzzyness is the independence of the components of Z. The following claim is connected to a special way of minimizing $H(Z)$. This is why we speak of operationally proofness. It is however a very natural way in absence of any further constraint on Z.

Lemma 2. *Consider a set A of Y patterns and a probability distribution p over the patterns. Assume that for any mapping from Y to Z entailing an entropy $H(Z)$ there exists a mapping from Y to Z' with $H(Z') = H(Z)$ such that the Z' components are independent. Then the function*

$$\widetilde{H}(Z) = -\sum_{i=1}^{n}\sum_{j=1}^{r_i} \pi_{j,i} \ln \pi_{j,i} \tag{8}$$

where n is the number of components, each assuming r_i values with marginal probability $\pi_{j,i}$, has minima over the above mappings in Zs having independent components.

Proof. $H(Z) = \widetilde{H}(Z) - I(Z)$ where $I(Z)$ denotes the overall mutual information between Z components, a term that is always positive and 0 only when the components are independent. Hence for each Z with non independent components there exists a Z' with independent components and $\widetilde{H}(Z') < \widetilde{H}(Z)$ by hypothesis.

Finally, with the aim of lowering the vagueness of our entropic goal, in many applications we decide to have Boolean Z, as a way of forcing some reduction of data redundancy and information as well, in the direction of taking a discrete decision, ultimately, the clustering. This produces the side benefit of a concise description of both patterns and clustering formulas as a nice premise for a semantic readability of them. To this end, and limiting ourselves to a binary partition of the patterns, we assume as cost function of the single pattern s in our coding problem the following Schur-concave function [8] which we call *edge pulling* function:

$$E_s = \ln\left(\prod_{i=1}^{n} z_{s,i}^{-z_{s,i}}(1 - z_{s,i})^{-(1-z_{s,i})}\right) \tag{9}$$

where $z_{s,i}$ is the i-th component of the encoding of the pattern s. In line with the general capability of Schur-concave functions of leading to independent component located on the boundaries of their domain [9], we may prove that minimizing over the possible encodings the sum $\breve{H}$ of the logarithm of E_s over all patterns leads us to a representation of the patterns that is binary and with independent bits, which we call BICA representation.

Lemma 3. *Any mapping $\mathbf{Z}$, that is correct according to Definition 1 on a set of m training patterns and meets assumptions in Lemma 2 while minimizing the sum of the* edge pulling *function (9) on the patterns, is expected to produce Boolean independent components minimizing (2) as well.*

Proof. This function has a minimum when each $z_{s,i}$ is Boolean. This is why we call it edge pulling *function. Moreover, from Jensen inequality [10] on the function $g(x) = x \ln x$ we have that*

$$\breve{H} \equiv \frac{1}{m} \sum_{s=1}^{m} E_s \leq \sum_{i=1}^{n} \left[-\sum_{s=1}^{m} \frac{z_{s,i}}{m} \ln \left(\sum_{s=1}^{m} \frac{z_{s,i}}{m} \right) \right.$$
$$\left. - \left(1 - \sum_{s=1}^{m} \frac{z_{s,i}}{m} \right) \ln \left(1 - \sum_{s=1}^{m} \frac{z_{s,i}}{m} \right) \right] \equiv \widehat{H} \quad (10)$$

where $\widehat{H}$ is an estimate of $H(\mathbf{Z})$, i.e. the entropy of the $\mathbf{Z}$ patterns we want to cluster computed with the empirical distribution law [11] in the assumption that $\mathbf{Z}$ components are independent. Finally, for $z_{s,i}$ close to either 0 or 1 and binary vectors almost orthogonal (so that also $\sum_s \frac{z_{s,i}}{m}$ is close to 0), $g(x)$ behaves in (10) almost linearly, making the inequality almost an equality. Thus, within this range of values we are minimizing $\widehat{H}$ as well, getting an encoding with components that are independent by Lemma 2 and Boolean as well.

The above lemma is only part of the history. In view of minimizing $\widehat{H}$, another lever is represented by the *number of bits* into which we want to encode the patterns. In the next section we will see that this number is fixed at a breakeven point between different utilities. Here we highlight that the evident appeal of a small number of bits contrasts their independence, a property that is variously evoked in the above lemmas and is generally promoted by the sparseness of the data representation [12].

3   The Application Versant

In the previous chapter we summed up in a systematic way the theory underlying BICA coding. We started its implementation some years ago, even before the full assessment of the theory [5] drawn by the need to state Boolean rules as a conclusion of complete processing leading from continuous sensory data to their interpretation [13]. The benefits we met in the above and analogous applications are not linked with the accuracy of the produced rules *per se*, but, maintaining it in a somehow equivalence range, with a profitable balance between the manageability and the understandability of their representation, as a counterpart of the intermediate positioning of $H(\mathbf{Z})$ between $H(\mathbf{C})$ and

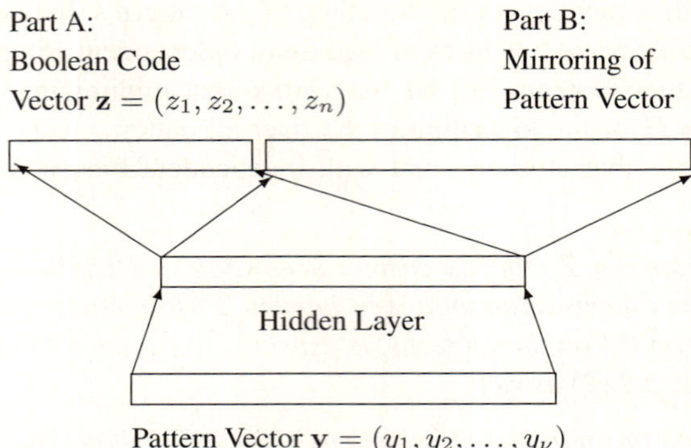

Fig. 1. Layout of the neural network mapping features to symbols. The same hidden layer neurons feed both the Boolean variables generation in part A and the input mirroring in part B.

$H(\boldsymbol{Y})$. In the following we discuss two different ways of tossing this balance. In both cases we just quote as marginal tools the clustering algorithms fed by the encoded data and their competitors, since our interest is focused on the information management.

3.1 The Encoder

We standardized the coding of the patterns through the use of an hourglass network. This is the merge of two neural networks as in Fig. 1. The former is a mirroring feed-forward network made of equal sized input and output layers and a narrower hidden layer. The features fed in the input must be exactly reproduced in the output. Since hidden neurons are in smaller number, they must compress the information carried by the input data without losing all that is necessary for their mirroring. This is a common task [14] that can be achieved by the network trained with back-propagation (for short BP) algorithm rooted on a quadratic error.

The latter network shares input and hidden neurons with the former. But its output layer has so many neurons as the number of bits with which we decide to code the patterns, each devoted to code a single bit. The training algorithm is the same, so that the whole training is through BP, where the error computed in the second output layer is given by the edge pulling function (9). We used some special expedients to direct the training that can be found in [3].

3.2 The Benefits of BICA Coding

We are left with a Boolean vector as a counterpart of a pattern of features collected to classify a set of objects. We have neither the explicit meaning of the bits, nor special criteria to prefix their number. We are not puzzled in principle by the former lack, since we may assume the meaning as a statistical utility of the symbol deriving precisely from their use [15]. In this perspective problems like feature selection [16] find a solution as a corollary of the clustering rather than a premise of it. Rather, we will

Table 1. Trade-off between compression and accuracy with BICA contrasted with a Neural network classification on dataset Sonar. μ: mean value σ: standard deviation. Correct %: percentage of correctly classified patterns of the test set. n.a. denotes not available data.

BICA					Neural Network				
Length			Correct %		Length			Correct %	
Signal	Formula				Signal	Formula			
	μ	σ	μ	σ		μ	σ	μ	σ
50	101.2	9.8	75.3	19.9	240	n.a.	n.a.	84.7	5.7
60	83.3	2.3	73.3	15.4					

see that some randomness of the association of features to bits will be beneficial to new learning techniques. As for the sizing of the Boolean vector, we base it mainly on trial and error in search of a breakeven point between the shortness of the vector (which calls for the shortness of formulas exploiting it as well) and the correctness of the clustering, pragmatically leaving the above considerations on the bits' independence as a side effect. This is another way of conjugating the structural complexity minimization problem [17], here read in terms of balance between understandability and accuracy of the clustering rules [18].

We tossed the above balance on the Sonar dataset [19]. It contains a total of 208 labelled patterns, 111 items obtained by bouncing sonar signals off metal cylinders, and the rest off rocks. Each pattern is a set of 60 numbers in the range 0.0 to 1.0 representing the rebounded time integrated energy within a particular frequency band. We contrast our two-phase clustering procedure directly with a neural network classifier to claim the understandability of the produced formulas. These formulas are extracted through a quick heuristic called StatEx in [3] that, like with bin packing greedy algorithms, groups most positive points (labelled as metal) in the support of a first consistent monomial, then adds a further monomial to group the most remaining points, and so on. The neural network implemented by Gorman and Sejnowski is a three-layer feedforward network with 12 hidden nodes [20]. With both methods we followed the cross-validation scheme proposed by these authors, where the patterns are grouped in 13 random sets of 16 patterns each. 12 of these sets are used for training and the remaining set for testing. This is repeated 13 times, changing the test set each time. Table 1 reports the numerical results in terms of length of input, length of formulas and classification accuracy. As for the latter the connectionist classifier wins, as expectable. However we reduce to $1/5$ the length of the data representation paying less than 10 percentage points, although with a higher variance. We reckon this rate by conventionally attributing, in a conservative way, a length of 4 bits to the original real signals. Actually the classification problem is difficult *per se* as denoted by the length of the disjunction of monomials discovered by our method, on the one hand, and the high number of hidden nodes on the other. Also, the goal to hit is not univocal. As shown in the table we may compress the data either into 50 Boolean variables, getting an optimal accuracy, or into 60 Boolean variables, getting an optimal formula concision.

With the second experiment we propose, we may appreciate the random nature of the binary encoding coming from the random initialization of the hourglass network in Section 3.1. The benchmark is a microarray database referring to the Colon adenocarcinoma pathology [21]. It is composed of 62 samples: 40 tumoral and 22 normal colon tissue samples. Each sample reports the fluorescence of 2000 genes that we compress into 20 bit long patterns. BICA vectors are not a univocal representation of the original data. Given the randomness of the network parameters we use for initializations, we expect that different Boolean vectors may enhance different features of the data to be clustered, so that some features are more appropriate for discriminating a certain subset of them, while other features serve the same purpose for other subsets. Hence on each BICA representation we train a Support Vector Machine (SVM) [22], considered as a *base learner* of an ensemble learning procedure [23]. Then we suitably combine the classification proposed by each SVM. Namely, as a variant of majority voting rule, we compute for each example the frequency with which base learners answer 1, and we gather frequencies corresponding to either positive or negative samples. In the loose assumption that frequencies in each group follow a Gaussian distribution we locate a threshold at the cross of the cumulative distribution function (c.d.f.) of the right distribution with the complement to 1 of the c.d.f. of the left distribution [6], i.e.

$$t = \frac{\hat{\mu}_- \hat{\sigma}_+ + \hat{\mu}_+ \hat{\sigma}_-}{\hat{\sigma}_- + \hat{\sigma}_+}$$

where $\hat{\mu}_-$ and $\hat{\sigma}_-$ are, respectively, the sample estimates of parameters μ_- and σ_- of the negative distribution; idem for the positive distribution. With this threshold we classify new records.

The general scheme of experiments is the following: according to the multiple hold-out scheme, we randomly split the dataset in two equally-sized training and test sets. We repeat this process 50 times.

We train 50 hourglass networks for every partition, obtaining 50 BICA encodings. On each encoding of a training set we train an SVM getting an ensemble of 50 SVMs decreeing the cluster of each pattern in the training set. Hence we pick up the most efficient subset of SVM capable of jointly classifying the whole training set correctly. With these SVMs, in number of 3, we obtain the classification accuracy in Table 2. As a technical detail, we use a polynomial kernel to have the single learners working as non-linear separators. In particular, Fig. 2 compares the error percentages on the instances of the multiple hold-out scheme. They have been obtained on each record of the data set with three different methods, namely: Random Subspace Ensemble (RSE) [24] exploiting random projections to lower dimensional subspaces, Genet [25] wisely adapting the training of a single SVM in a semi-automatic mode, and our ensemble learner. We also adopted a leave-one-out cross-validation technique to contrast our results with the literature, getting similar classification error rates around 10% [26]. The pictures in Fig. 2 show that there are records for which the error rate is high for every method we consider. This matches the biological experience [21] that most normal samples are enriched in muscle cells, while tumor samples are enriched in epithelial cells, while the above samples consistently misclassified by all methods present an "inverted" tissue composition: normal samples are rich in epithelial cells, tumor samples are rich in muscle cells.

As a conclusion, the two phase classifier is comparable to other methods. The additional benefit arises from: i) the short description again of the classification rule: three hyperplanes on 20 Boolean variables; ii) the possibility of reducing the latter number by isolating the variables that really identify the hyperplanes; and iii) coming back along

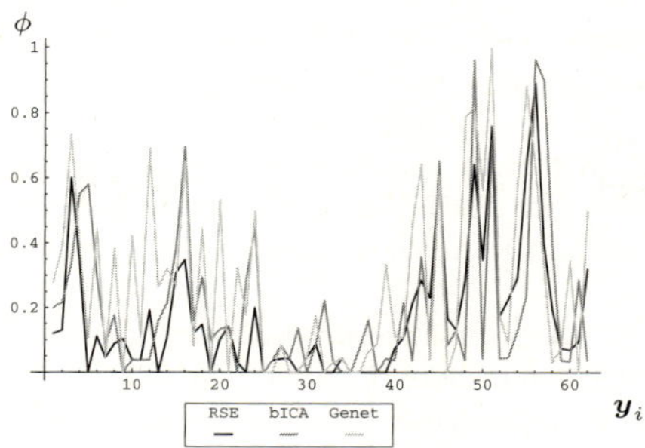

Fig. 2. Frequencies ϕ of error of the patterns y_i within the multiple hold-out scheme. Curve parameter: the classification procedure.

Table 2. Classification accuracy on Colon dataset split over training sets (Train) and test sets (Test)

Train accuracy		Test accuracy		Input bits
μ	σ	μ	σ	
0.98	0.02	0.792	0.09	60

the hourglass, identifying the features, among the original 2000 that really compute these variables.

4   Conclusions

Boolean variables seem to be the most elementary way to categorize data. Hence they are output of primary encoding systems and output of rules managing the former as well. We may assume the reason for their wide use to lie in the simplicity of the reasoning they support, hence of the formulas through which the latter are expressed. In this paper we discuss these features from both entropic and a semantic perspectives. The former is relatively easy, since it involves the syntax of the underlying logic. As for the semantic one, we exhibit some benefits in terms of length of the formulas that we broadly identify with their understandability. We also introduce some prospective strategies for attributing meaning to these variables through their links to relevant features of the objects they refer to. A concrete exploitation of this research line will form a part of our future work.

References

1. Rumelhart, D.E., Hinton, G.E., Williams, R.J.: Learning internal representations by error propagation. In: Parallel Distrib. Proc., vol. 1, pp. 318–362. MIT Press, Cambridge (1987)
2. Kreutz-Delgado, K., Murray, J.F., Rao, B.D., Engan, K., Lee, T.W., Sejnowski, T.J.: Dictionary learning algorithms for sparse representation. Neural Computation 15(2), 349–396 (2003)

3. Apolloni, B., Malchiodi, D., Brega, A.: BICA: a Boolean Independent Component Analysis Algorithm. Hybrid Intelligent System, 131–136 (2005)
4. Dietterich, T.: Ensemble methods in machine learning. In: Kittler, J., Roli, F. (eds.) MCS 2000. LNCS, vol. 1857, pp. 1–15. Springer, Heidelberg (2000)
5. Apolloni, B., Esposito, A., Malchiodi, D., Orovas, C., Palmas, G., Taylor, J.: A general framework for learning rules from data. IEEE Transactions on Neural Networks 15(6) (2004)
6. Duda, R., Hart, P.: Pattern Classification and Scene Analysis. John Wiley & Sons, New York (1973)
7. Bell, A.J., Sejnowski, T.J.: An information-maximization approach to blind separation and blind deconvolution. Neural Computation 7, 1129–1159 (1995)
8. Kreutz-Delgado, K., Rao, B.: Application of concave/Schur-concave functions to the learning of overcomplete dictionaries and sparse representations. In: Signals, Systems & Computers, 1998. Conference Record of the Thirty-Second Asilomar Conference, vol. 1, pp. 546–550 (1998)
9. Rockafellar, R.: Convex Analysis. Princeton University Press, Princetion (1970)
10. Cover, T., Thomas, J.: Elements of Information Theory. Wiley & Sons, New York (1938)
11. Glivenko, V.: Sulla determinazione empirica delle leggi di probabilità. Giornale dell'Istituto Italiano degli Attuari 3, 92–99 (1933)
12. Delgado, K., Murray, J., Rao, B., Engan, K., Lee, T., Sejnowski, T.: Dictionary learning algorithms for sparse representation. Neural Computation 15, 349–396 (2003)
13. Apolloni, B., Malchiodi, D., Orovas, C., Palmas, G.: From synapses to rules. Cognitive Systems Research 3, 167–201 (2002)
14. Pollack, J.: Recursive distributed representation. Artificial Intelligence 46, 77–105 (1990)
15. Koenig, S., Holte, R.C.: PAC Meditation on Boolean Formulas. In: Koenig, S., Holte, R.C. (eds.) Abstraction, Reformulation and Approximation. Springer, Berlin (2002)
16. Amaldi, E., Kann, V.: On the approximation of minimizing non zero variables or unsatisfied relations in linear systems. Theoretical Computer Science 209, 237–260 (1998)
17. Vapnik, V.: The Nature of Statistical Learning Theory. Springer, New York (1995)
18. Apolloni, B., Brega, A., Malchiodi, D., Palmas, G., Zanaboni, A.: Learning rule representations from data. IEEE Trans. on Systems, Man and Cybernetics, Part A 36(5), 1010–1028 (2006)
19. Asuncion, A., Newman, D.: UCI machine learning repository (2007)
20. Gorman, R.P., Sejnowski, T.J.: Analysis of hidden units in a layered network trained to classify sonar targets. Neural Networks 1, 75–89 (1988)
21. Alon, U., et al.: Broad patterns of gene expressions revealed by clustering analysis of tumor and normal colon tissues probed by oligonucleotide arrays. PNAS 96, 6745–6750 (1999)
22. Cristianini, N., Shawe-Taylor, J.: An introduction to Support Vector Machines and other kernel-based learning methods. Cambridge University Press, Cambridge (2000)
23. Dietterich, T.G.: Ensemble methods in machine learning. In: Kittler, J., Roli, F. (eds.) MCS 2000. LNCS, vol. 1857, pp. 1–15. Springer, Heidelberg (2000)
24. Ho, T.: The random subspace method for constructing decision forests. IEEE Transactions on Pattern Analysis and Machine Intelligence 20(8), 832–844 (1998)
25. Bobrowski, L., Lukaszuk, T.: Selection of the linearly separable feature subsets. In: Rutkowski, L., Siekmann, J.H., Tadeusiewicz, R., Zadeh, L.A. (eds.) ICAISC 2004. LNCS (LNAI), vol. 3070, pp. 544–549. Springer, Heidelberg (2004)
26. Furey, T., Cristianini, N., Duffy, N., Bednarski, D., Schummer, M., Haussler, D.: Support vector machine classification and validation of cancer tissue samples using microarray expression data. Bioinformatics 16(10), 906–914 (2000)

Improving the Learning Speed in 2-Layered LSTM Network by Estimating the Configuration of Hidden Units and Optimizing Weights Initialization

Débora C. Corrêa[1], Alexandre L.M. Levada[2], and José H. Saito[1]

[1] Federal University of São Carlos, Computer Department, São Paulo, Brazil
[2] University of São Paulo, Physics Institute of São Carlos, São Paulo, Brazil
`debora_correa@dc.ufscar.br`, `alexandreluis@ursa.ifsc.usp.br`,
`saito@dc.ufscar.br`

Abstract. This paper describes a method to initialize the LSTM network weights and estimate the configuration of hidden units in order to improve training time for function approximation tasks. The motivation of this method is based on the behavior of the hidden units and the complexity of the function to be approximated. The results obtained for 1-D and 2-D functions show that the proposed methodology improves the network performance, stabilizing the training phase.

1 Introduction

Recurrent Neural Networks (RNNs) represent a non-algorithmic computation and can be viewed as mathematical abstraction of human brain structure and processing. They can perform complex mappings from input to output sequences and they learn their behavior extracting the necessary features to represent the given information from a training set of correct examples sequences [5] [7] [8] [11]. The training samples are fed to the network and the training process try to minimize the error between the actual and the desired network output using gradient descent by which the connection weights are progressively adjusted in the direction that reduces this error most rapidly. However, previous gradient based methods [5][11][15][16] share a problem: over time, as gradient information is passed backward to update weights whose values affect later outputs, the error/gradient information is continually decreased or increased by weight-update scalar values [2][4][6][12][16]. This means that the temporal evolution of the path integral over all error signals flowing back in time exponentially depends on the magnitude of the weights [4]. For this reason, standard recurrent neural networks fail to learn when in presence of long time lags between relevant input and target events [4] [6][12].

The Long-Short-Term Memory (LSTM) [3] [4] [7] [12] [13] network is a novel recurrent network architecture that implements an appropriate gradient-based learning algorithm. It was designed to overcome the vanishing-gradient problem

V. Kůrková et al. (Eds.): ICANN 2008, Part I, LNCS 5163, pp. 109–118, 2008.

[6] by enforcing constant error flow through constant error carrousels (CECs) within special units, permitting then non-decaying error flow back into time [4][7]. This enables the network to learn important data and store them without degradation of long period of time. The LSTM networks have been used in many applications, such as speech recognition, function approximation, music composition, among other applications.

Our motivation is first to extend Nguyen and Widrows [10] adaptive initialization method by describing the behavior of the hidden units in a LSTM network in function approximation tasks and propose a method for initialization of its weights in order to reduce the training time. It can be observed that besides improving training speed, the proposed method offers stabilization that is very important and desired feature during the learning process. We also propose to combine this weight initialization method with an estimative of the ideal number of hidden units necessary to perform function approximations based on the number of local minima/maxima points.

This paper is organized as follows: section 2 presents the traditional LSTM network architecture, including its forward training algorithm; section 3 discusses the behavior of hidden units for one input and multiple input networks, and presents the proposed method of initialization; section 4 describes how to estimate the number of hidden units; section 5 shows the obtained results and section 6 presents the conclusions and future works.

2 Traditional LSTM Neural Networks

The memory block is the basic unit in the hidden layer of a LSTM network and it substitutes the hidden unit in a RNN neural network (Figure 1(a)) [4].

A memory block is formed by one or more memory cells and a pair of adaptive, multiplicative gating units which gate input and output to all cells in the block. Figure 1(b) shows a detailed memory block with one memory cell. Each memory cell has a recurrently self-connected linear unit called Constant Error Carrousel

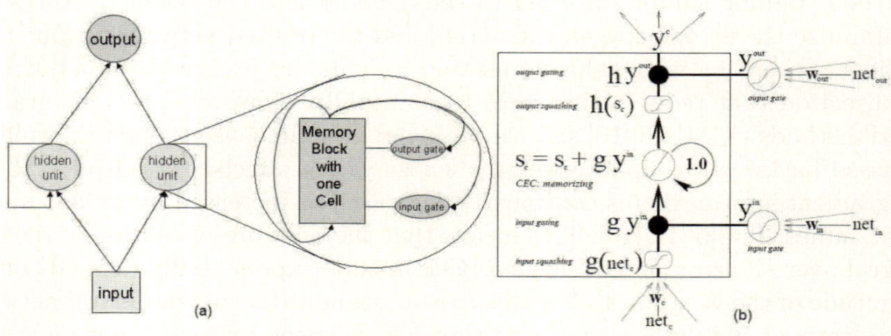

Fig. 1. (a) Left: RNN with one recurrent hidden layer. Right: LSTM with memory blocks in the hidden layer (only one is shown) [4, pp.11]. (b) A memory block with one cell [4, pp.12].

(CEC) [4][7], representing the cell state. The CEC helps to solve the vanishing error problem. For a detailed explanation of the LSTM network forward and backward pass see reference [4] and the work of Hochreiter & Schmidhuber [7].

3   The Behavior of LSTM Hidden Nodes in Approximating Non-linear 1-D Function

Neural networks with more then one layer can be used to approximate any arbitrary functions [8][10], given that they have enough number of hidden units. Basically, during training in order to approximate a desired function $d(x)$, the neural network builds piece-wise linear approximations $y(x)$ to the function $d(x)$. The pieces are then summed to generate the resulting function. The idea is that each hidden unit is responsible for a single piece-wise linear approximation.

In this approach, the adaptive weights and bias control the caracteristics of these approximations [10]. During learning process, the weights must move in such a manner that the region of interest is divided into sub-intervals, so that each sub-interval is responsible for a small piece of the desired function $d(x)$, that is, each hidden unit should provide a different contribution. Therefore, it is reasonable to admit that the training process can be reduced by setting the initial weights of the hidden units so that each hidden unit is associated to a different sub-interval at the start of training. The network is then trained normally, with each hidden unit still having the freedom to adjust its weights during training. However, most of these adjustments will probably be small, once part of these movements were eliminated by the proposed weight initialization method.

In the example, $d(x)$ is a real function to be approximated by the neural network over the region $[-1, 1]$. So, the length of the interval is 2. Let H be the number of memory blocks and M the number or memory cells in each block used to approximate the function $d(x)$. Therefore, each memory block will be responsible for a sub-interval of length $2/HM$ on the average.

According to [10], the $f(x)$, $g(x)$ and $h(x)$ sigmoid functions used in LSTM networks can be considered approximately linear for $x \in [-1, 1]$, but saturate as x becomes larger in magnitude. From the equations that provide the relationship between the LSTM output for one memory block with one memory cell and a one-dimensional input x, and considering the first iteration step, where there is no feedback from the recurrent connections yet, and adopting a linear approximation to the sigmoid function over the range $[-1, 1]$, we can say that:

$$-1 < w_{c_1^1} b_{in_1} b_{out_1} x < 1 \tag{1}$$

where $w_{c_1^1} b_{in_1} b_{out_1}$ is the output approximation to the j-th memory block output in the network initialization. This implies that:

$$\frac{-1}{w_{c_1^1} b_{in_1} b_{out_1}} < x < \frac{1}{w_{c_1^1} b_{in_1} b_{out_1}} \tag{2}$$

resulting in an interval with length $\dfrac{2}{w_{c_1^1} b_{in_1} b_{out_1}}$.

As previously described, it is expected that each memory block be responsible, in average, for a sub-interval of length $2/HM$. Therefore:

$$HM = w_{c_1^1} b_{in_1} b_{out_1} \tag{3}$$

Adopting the bias initialization first proposed by GERS [6], we have:

$$\begin{cases} b_{in_1} = b_{out_1} = -0.5 \\ b_{in_2} = b_{out_2} = -1.0 \\ \vdots \\ b_{in_j} = b_{out_j} = -(0.5 * j) \end{cases} \tag{4}$$

As a result, we propose a general expression for the $w_{c_j^v 1}$ weights initialization for a LSTM network with one input, given by the following equation:

$$w_{c_j^v 1} = \frac{HM}{(0.5 * j)^2} * \alpha \tag{5}$$

As we found in [10], it is preferable to have the sub-intervals overlap slightly, thus $\alpha \in [0, 1]$ represents the coefficient that controls this overlapping between sub-intervals. Along the experiments in this work, we will consider $\alpha = 0.7$ as suggested in [10]. Note that $w_{c_1^1 1} = ... = w_{c_1^M 1}$ when there are M memory cells in the block.

The interpretation of a N-dimensional function ($N > 1$) approximation is a little more complicated. Basically, it is used the same concepts adopted in computerized tomography image reconstruction [9]. The fundamental mathematical tool is the Fourier Slice Theorem. The idea consists of trying to approximate the Fourier Transform $D(U)$ of the desired function $d(x)$ from 1-D slices $D_i(U)$. Using the inverse transform in the result, it is possible to define an approximation $\tilde{d}(x)$ to the desired function $d(x)$. Considering a neural network with 2 external inputs, a typical output of a hidden unit possess as its Fourier Transform (FT), a slice, denoted by $D_i(U)$, of the 2-D FT $D(U)$ [10]. The time domain version of $D_i(U)$, $d_i(x)$, is a one variable function and can be approximated by the neural network as described in the previous section. This motives us to approximate S slices (1-D functions) with I intervals each using H memory blocks with M memory cells each.

As before training it is not possible to know how many slices the network will produce, it is suggested in [10], that the number of slices should be $S = I^{N-1}$, where N is the number of network inputs. Besides, as each element of input vector $\mathbf{X}$ belongs to the interval $[-1, 1]$ this means that each interval is approximately $2/I$ long. So, we have,

$$\begin{aligned} HM &= SI \\ HM &= I^N \\ I &= (HM)^{1/N} \end{aligned} \tag{6}$$

Similarly to equation (8), we can say that:

$$\frac{2}{(HM)^{I/N}} = \frac{2}{w_{c_j^v m} b_{in_j} b_{out_j}} \tag{7}$$

which results in:

$$w_{c_j^v m} = \frac{(HM)^{1/N}}{(0.5 * j)^2} * \alpha \qquad (8)$$

with $\alpha \in [0, 1]$. In the same way, $\alpha = 0.7$ is used to provide some overlap between sub-intervals.

4   Estimating the Number of Hidden Units

In order to estimate the number of memory cells of the LSTM neural network in function approximations, we propose to use the number of local minima/maxima points. Since each memory cell is responsible for a piecewise linear approximation, it is reasonable to assume that the minimum number of memory cells should be the same number of subintervals between the local minima/maxima points, as illustrates the function $y^2(x) = \sin^2(\pi x) + \exp(x^5)$ in Figure 2(a) and Figure 2(b). Connecting the extreme points provides a good approximation for the function (Figure 2(c)).

The motivation for this approach relies on a major result from shape analysis theory: points of high curvature concentrate geometrical information [1], providing good shape descriptors. These points can be well approximated by local extreme values. We propose to use the information presented in the target function $d(x)$, more precisely, the number of local minima/maxima and the number of network input units to calculate an initial estimate to the number of hidden units. Note that in this method we assign a specific network configuration depending on the observed target function. We detect the local minima/maxima

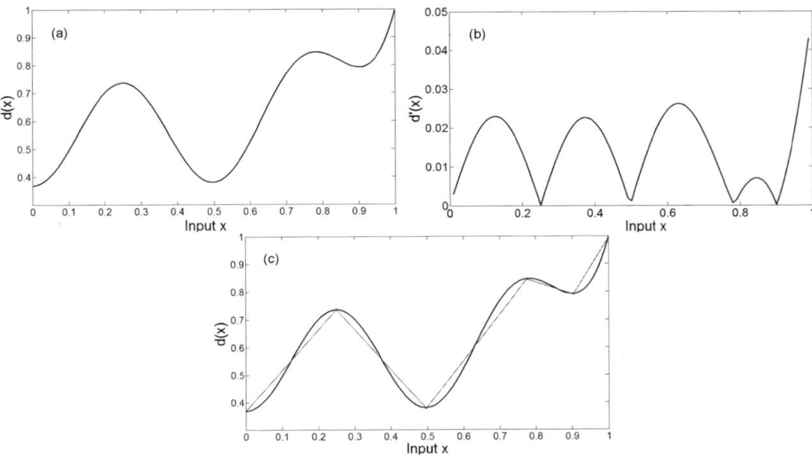

Fig. 2. (a) 1-D Function (b) Detected extreme points (c) Linear approximation by extreme points

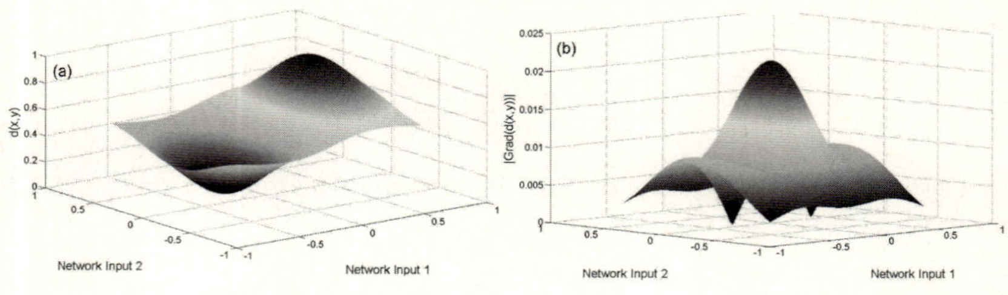

Fig. 3. (a) 2-D function (b) Detected extreme points

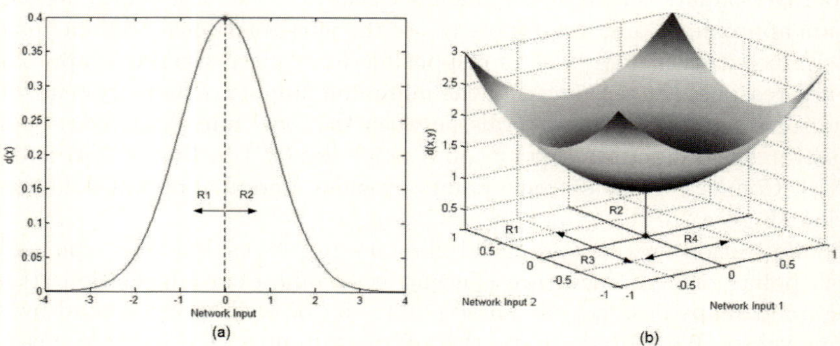

Fig. 4. Division of the function domain by a local extreme point (a) Two regions (b) Four regions

points by using finite difference approximations for first derivatives. There are three commonly finite difference approximations described in literature: forward, backward and central differences. The central difference yields the most accurate approximation in terms of Taylor expansion. The expression for the first derivative, in a point x, using the central difference is given by [14]:

$$\frac{d}{dx}f(x) \approx \frac{f(x+h) - f(x-h)}{2} \tag{9}$$

In case of 2-D real functions, the gradient of $f(x, y)$ can be approximated by the central differences in directions x and y. Figure 3(a) shows the 2-D function $f(x, y) = 0.25 + x \exp\{-(x/0.5)^2 - (y/0.5)^2\}$ and Figure 3(b) shows the detected local minima/maxima points (condition $\nabla f(x, y) = \mathbf{0}$).

For 1-D functions, the definition of one local minimum/maximum point divides the function domain in 2 non-overlapping regions (Figure 4(a)). In case of functions defined on $\Re^2$, the plan is subdivided in 4 non-overlapping quadrants (Figure 4(b)). This fact motivates the use of a heuristic to set the number of hidden units (HM) needed to approximate a function. Actually, we define HM to be proportional to $(n^2 K) + 1$, where n denotes the number of network input units (function dimensionality) and K is the number of function extreme points in a given interval of interest.

5   Experiments

To test and evaluate the proposed methodologies, we trained a LSTM network to approximate the 1-D and 2D functions illustrated in Figures 2 and 3. Table 1 shows the obtained mean square errors for the random initialization, where the initial values of the weights $w_{c_j^v}$ are chosen randomly from a uniform distribution between -0.2 e 0.2, as suggested in [4] and the mean square error using our optimized initialization of weights. We have trained various configurations of memory blocks and memory cells per block using different learning rates. We trained each network configuration during 300 epochs, N times (N=10). For each configuration we present the best and worst errors in Table 1. Errors with magnitude less than 10^{-4} were considered as zero. The improvement in LSTM training is clear, since in all cases the optimized initialization provides significantly smaller errors. Also, the proposed method reduces the difference between the best and worst cases, reflecting a more stable behavior. Note that network configurations near to our estimative (5,1) present the best results, suggesting that the criterion adopted in the proposed method for estimation of the number of hidden nodes is valid.

As illustrative examples, we plotted the network output and mean square error for the best and worst performances, using random and optimized initialization of weights and with the "ideal" configuration of five memory blocks and one memory cell per block. For the worse performance, Figure 5(a) shows the desired 1-D function $y^2\left(x\right) = \sin^2\left(\pi x\right) + \exp\left(x^5\right)$ and the network output after training

Table 1. Mean Square Error for 1-D function approximation training using randomly and optimized initialization

H	M	Mean Square Error (Randomly Initialization)				Mean Square Error (Optimized Initialization)			
		$\alpha{=}0.5$	$\alpha{=}1$	$\alpha{=}2$	$\alpha{=}3$	$\alpha{=}0.5$	$\alpha{=}1$	$\alpha{=}2$	$\alpha{=}3$
2	1	0.0076	0.0022	0.0017	0.0013	0.0030	0.0020	0.0017	0.0014
		0.0087	0.0048	0.0093	0.0093	0.0031	0.0020	0.0023	0.0014
3	1	0.0070	0.0020	0.0010	0.0010	0.0021	0.0014	**0**	**0**
		0.0074	0.0097	0.0086	0.0091	0.0024	0.0020	0.0012	0.0010
4	1	0.0060	0.0028	0.0023	0.0017	0.0018	0.0011	**0**	**0**
		0.0084	0.0086	0.0082	0.0089	0.0034	0.0026	0.0011	0.0010
5	1	**0.0031**	**0.0012**	**0.0010**	**0**	**0.0010**	**0**	**0**	**0**
		0.0076	0.0088	0.0070	0.0086	0.0030	0.0010	**0**	**0**
1	2	0.0081	0.0087	0.0053	0.0037	0.0093	0.0081	0.0048	0.0032
		0.0097	0.0097	0.0098	0.0098	0.0094	0.0097	0.0049	0.0034
1	3	0.0065	0.0081	0.0089	0.0020	0.0058	0.0044	0.0023	0.0018
		0.0097	0.0093	0.0094	0.0061	0.0069	0.0045	0.0024	0.0018
1	4	0.0068	0.0031	0.0020	0.0024	0.0052	0.0029	0.0018	0.0015
		0.0096	0.0090	0.0092	0.0098	0.0052	0.0029	0.0018	0.0015
1	5	0.0050	0.0028	0.0020	0.0016	0.0022	0.0021	**0**	**0**
		0.0086	0.0084	0.0097	0.0088	0.0034	0.0023	0.0016	0.0013

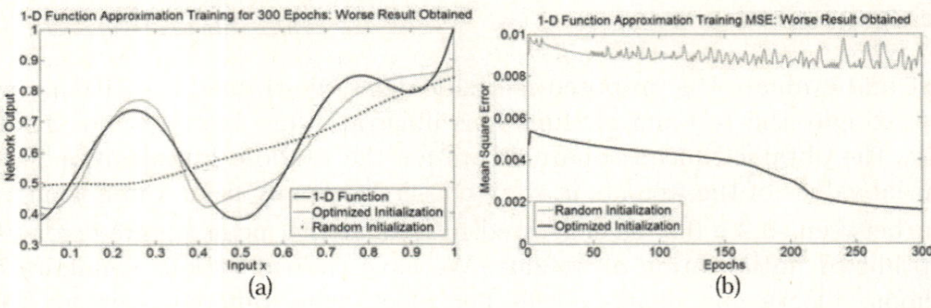

(a)   (b)

Fig. 5. (a)1-D function and network output after training with random initialization and optimized initialization - worse case (b)Mean square error for the training in (a)

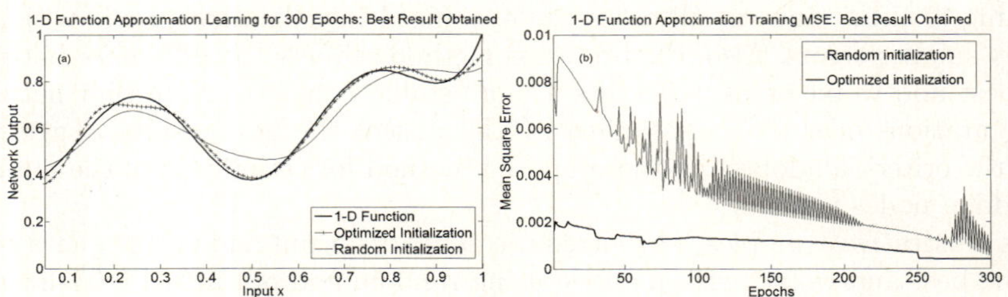

Fig. 6. (a)1-D function and network output after training with random initialization and optimized initialization - best case (b)Mean square error for the training in (a)

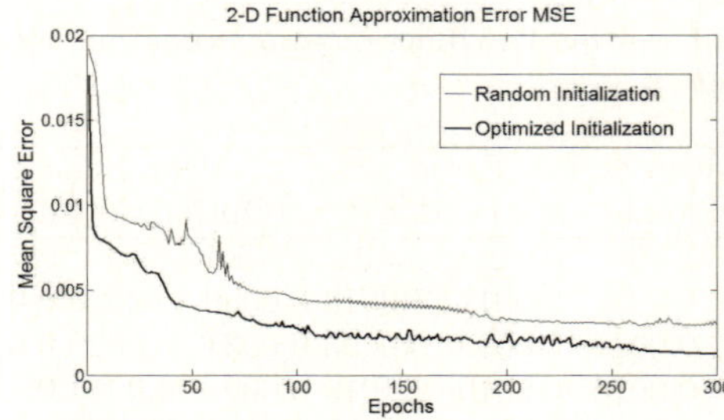

Fig. 7. Mean square error for the 2-D function described above

for randomly and optimized initialization and Figure 5(b) shows the mean square error as a function of training time for both training cases.

Similarly, Figure 6(a) shows the same desired 1-D function (solid curve) and the network outputs for the best case using random and optimized initialization Again, Figure 6(b) shows the mean square error for both training.

We trained the network to approximate the 2-D function illustrated in Figure 3 with different configurations similar to the 1-D case. The results indicated that, similar to the 1-D function training, the close the configuration is to the ideal one (in this case, 9 hidden units) the smaller is the mean square error. Figure 7 shows the mean square errors in the best cases for random and optimized initialization.

6   Conclusions

This paper described the behavior of the hidden units in a LSTM neural network in approximating a non-linear desired function. The approximation is done by forming a union of piece-wise linear segments. We proposed to improve LSTM network training by estimating the ideal number of hidden units in conjunction with an automated method for network weight initialization.

With the proposed initialization scheme, some initial weights are pre-defined close to their final desired location after training, increasing network's learning speed. The proposed methodology was also useful in other different training problems. As observed in [10], we also noted that the larger is the number of hidden units used to approximate a complicated desired response function, the better is the performance of the initialization method.

The obtained results show that, in the randomly initialization, an increase of the number of memory blocks do not guarantee a better performance, once the additional memory blocks may not be well initialized (initial values are randomly defined) and their initial configurations are very distant from the final desired one. Besides, two or more memory blocks can be initialized very similar one to another and remain close after the training process, not providing different significant contributions, once their outputs are highly correlated (two or more memory blocks may be almost equivalent). In addition, the random initialization often provides larger differences between the best and the worst performances, reflecting a very unstable behavior during the training stage.

Differently, the proposed weight initialization approach combined with the detection of local minima/maxima points offer a good estimative of the ideal configuration of the network. In other words, the closer a network configuration is from the ideal configuration (reference), the better were the obtained results, suggesting the accuracy of the proposed methodology. Another contribution is that our method also minimizes the problem of instability that generally occurs when training neural networks. It seems plausible to have a training process less dependent of initial conditions. We also observed that, although large learning rates not necessary means better performance, the LSTM network training with large learning rates ($\alpha > 1$), often present good results.

Future works may include an attempt to optimize the initialization of other LSTM network parameters, as the bias recurrent parameters. We also aim to train and test the network with more complex N-dimensional ($N > 2$) functions and with different application domains, for instance, classification tasks and musical composition, in order to compare the network learning performance with and without the proposed initialization method in real problems.

References

1. Costa, L.F., Cesar, R.M.: Shape Analysis and Classification: Theory and Practice. CRC Press, Boca Raton (2001)
2. Franklin, J.A.: Recurrent Neural Networks for Music Computation. Informs Journal on Computing 18(3), 321–328 (2006)
3. Gaves, A., Schmidhuber, J.: Framewise phoneme classification with bidirectional LSTM and other neural network architectures. Neural Networks 18(5-6), 602–610 (2005)
4. Gers, F.: Long Short-Term Memory in Recurrent Neural Networks. PhD Thesis (2001)
5. Haykin, S.: Neural Networks: A comprehensive Foundation, 2nd edn. Prentice Hall, Englewood Cliffs (1998)
6. Hochreiter, S., Bengio, Y., Frasconi, P., Schmidhuber, J.: Gradient flow in recurrent nets: The difficulty of learning long-term dependencies. In: Kolen, J., Kremer, S.C. (eds.) A Field Guide to Dynamical Recurrent Networks. IEEE Press, New York (2001)
7. Hochreiter, S., Schmidhuber, J.: Long Short-Term Memory. Neural Computation 9(8), 1735–1780 (1997)
8. Irie, B., Miyake, S.: Capabilities of three-layerd perceptrons. In: Proceedings of the IEEE International Conference on Neural Networks p. I–641 (1998)
9. Kak, A.C., Malcolm, S.: Principles of Computerized Tomographic Imaging. IEEE Press, New York (2001)
10. Nguyen, D., Widrow, B.: Improving the learning speed of 2-layer neural networks by choosing initial values of adaptive weights. Proc. IJCNN 3, 21–26 (1990)
11. Pearlmutter, B.A.: Gradient calculations for dynamic recurrent neural networks: A survey. IEEE Transactions on Neural Network 6(5), 1212–1228 (1995)
12. Perez-Ortiz, J.A., Gers, F.A., Eck, D., Schmidhuber, J.: Kalman Filters improve LSTM network performance in problems unsolvable by traditional recurrent nets. Neural Networks 16(2), 241–250 (2003)
13. Schmidhuber, J., Wierstra, D., Gagliolo, M., Gomez, F.: Training Recurrent Networks by Evolino. Neural Computation 19, 757–779 (2007)
14. Smith, G.D.: Numerical Solution of Partial Differential Equations: Finite Difference Methods, 3rd edn. Clarendon Press (1985)
15. Werbos, P.J.: Generalization of backpropagation with application to a recurrent gas market model. Neural Networks 1, 339–356 (1988)
16. Williams, R.J., Zipser, D.: Gradient-Based Learning Algorithms for Recurrent Networks and Their Computational Complexity. In: Chauvin, Y., Rumelhart, D.E. (eds.) Back-Propagation: Theory, Architectures and Applications, Hillsdale, NJ (1992)

Manifold Construction Using the Multilayer Perceptron

Wei-Chen Cheng and Cheng-Yuan Liou

Department of Computer Science and Information Engineering
National Taiwan University
Republic of China
Supported by National Science Council
cyliou@csie.ntu.edu.tw

Abstract. We present a training method which adjusts the weights of
the MLP (Multilayer Perceptron) to preserve the distance invariance in
a low dimensional space. We apply visualization techniques to display
the detailed representations of the trained neurons.

1   Introduction

The MDS (Multidimensional Scaling) [1] searches for a low dimensional representation that preserves the distance relations among patterns. The Isomap [2] extends the MDS by using the geodesic distance to construct the nonlinear manifold.
The LLE (Locally Linear Embedding) [3] fits certain linear coefficients to maintain the local geometric surface in the manifold. The SOM [4] and PCA have broad
applications, such as text mining, restoration of space-dependent blurred images
[5] and economics analysis [6]. The conformal SOM [7][8] is further devised for the
geometrical surface mapping, biological morphology and monitoring continuous
states. In this work, we develop a training method for the MLP [9] using the SIR
algorithm [10][11][12] to construct the distance preservation mapping.

2   Method

Suppose the MLP network has $L + 1$ layers. It contains $L - 1$ hidden layers,
an input layer and an output layer. Let the ith layer have n_i neurons or n_i
dimensions. The dimension of the first hidden layer is n_1. The dimension of the
output layer is n_L. As an example, the MLP shown in Figure 1 has an input
layer, two non-linear hidden layers and a linear output layer. The output layer
contains two neurons and $n_3 = 2$.

Let $\mathbf{x}^p$ denote the pth data pattern. $\mathbf{x}^p$ is a column vector and $\mathbf{x}^p \in X =
\{\mathbf{x}^p | p = 1, .., P\}$. Let $\mathbf{y}^{(p,m)} = [y_1^{(p,m)}, y_2^{(p,m)}, ..., y_{n_m}^{(p,m)}]^T$ denote the output vec-
tor of the mth layer of the MLP when the pth pattern, $\mathbf{x}^p$, is fed in to the input
layer. Let $w_{ij}^{(m)}$ be the weight connecting the ith neuron in the mth layer and

V. Kůrková et al. (Eds.): ICANN 2008, Part I, LNCS 5163, pp. 119–127, 2008.

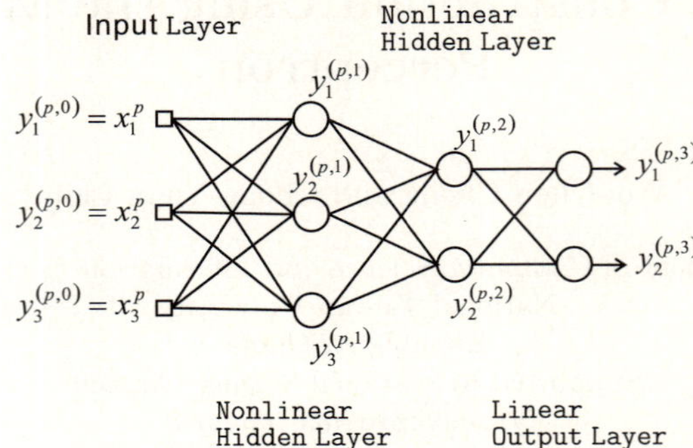

Fig. 1. The MLP network has one input layer, two non-linear hidden layers and a linear output layer

the jth neuron in the $(m-1)$th layer. The output, $y_i^{(p,m)}$, of the ith neuron in the mth layer is calculated by

$$v_i^{(p,m)} = \sum_j w_{ij}^{(m)} y_j^{(p,m-1)}, \; m = 1, \ldots, L-1 \tag{1}$$

and

$$y_i^{(p,m)} = \tanh\left(v_i^{(p,m)}\right), \; m = 1, \ldots, L-1. \tag{2}$$

The output of the ith neuron in the output layer is calculated by

$$y_i^{(p,L)} = \sum_j w_{ij}^{(L)} y_j^{(p,L-1)}. \tag{3}$$

The $\mathbf{y}^{(p,0)}$ is the input pattern $\mathbf{x}^p$,

$$\mathbf{y}^{(p,0)} = \mathbf{x}^p. \tag{4}$$

For any two patterns, $\mathbf{x}^p$ and $\mathbf{x}^q$, the SIR algorithm tries to minimize the energy, $E_{pq}(r)$, to preserve the invariance [10] in the mapping,

$$E_{pq}(r) = \frac{1}{4} f\left(\|\mathbf{x}^p - \mathbf{x}^q\|, r\right) \left(\left\|\mathbf{y}^{(p,L)} - \mathbf{y}^{(q,L)}\right\|^2 - \|\mathbf{x}^p - \mathbf{x}^q\|^2\right)^2, \tag{5}$$

where f is a threshold function,

$$f(t,r) = \begin{cases} 0 & \text{if } t > r \\ 1 & \text{if } t \leq r \end{cases}. \tag{6}$$

We call f a control function. By reducing the energy in (5), we expect that the output of the MLP can preserve the distance relations among local pattern pairs. This reduction can be accomplished by the delta rule toward the gradient descent direction, $\frac{\partial E_{pq}(r)}{\partial w_{ij}^{(m)}}$. The local gradients, $\delta_i^{(p,L)}$ and $\delta_i^{(q,L)}$, of the ith linear output neuron are

$$\delta_i^{(p,L)} = f\left(\|\mathbf{x}^p - \mathbf{x}^q\|, r\right) \left(\left\|\mathbf{y}^{(p,L)} - \mathbf{y}^{(q,L)}\right\|^2 - \|\mathbf{x}^p - \mathbf{x}^q\|^2\right) \left(y_i^{(p,L)} - y_i^{(q,L)}\right) \tag{7}$$

and

$$\delta_i^{(q,L)} = f\left(\|\mathbf{x}^p - \mathbf{x}^q\|, r\right) \left(\left\|\mathbf{y}^{(p,L)} - \mathbf{y}^{(q,L)}\right\|^2 - \|\mathbf{x}^p - \mathbf{x}^q\|^2\right) \left(y_i^{(p,L)} - y_i^{(q,L)}\right). \tag{8}$$

The local gradients, $\delta_i^{(p,m)}$ and $\delta_i^{(q,m)}$, of the ith hidden neuron in mth layer are

$$\delta_i^{(p,m)} = ((1 - y_i^{(p,m)})(1 + y_i^{(p,m)})) \sum_k \delta_k^{(p,m+1)} w_{ki}^{(m+1)}, \ m = 1, \ldots, L-1 \tag{9}$$

and

$$\delta_i^{(q,m)} = ((1 - y_i^{(q,m)})(1 + y_i^{(q,m)})) \sum_k \delta_k^{(q,m+1)} w_{ki}^{(m+1)}, \ m = 1, \ldots, L-1. \tag{10}$$

Differentiating (5) with respect to $w_{ij}^{(L)}$ yields

$$\frac{\partial E_{pq}(r)}{\partial w_{ij}^{(L)}} = \delta_i^{(p,L)} y_j^{(p,L-1)} - \delta_i^{(q,L)} y_j^{(q,L-1)}. \tag{11}$$

Differentiating (5) with respect to $w_{ij}^{(m)}$ of the mth hidden layer yields

$$\frac{\partial E_{pq}}{\partial w_{ij}^{(m)}} = \delta_i^{(p,m)} y_j^{(p,m-1)} - \delta_i^{(q,m)} y_j^{(q,m-1)}, \ m = 1, \ldots, L. \tag{12}$$

The updated equation for the weights [9] is

$$w_{ij}^{(m)} = w_{ij}^{(m)} - \eta \left(\delta_i^{(p,m)} y_j^{(p,m-1)} - \delta_i^{(q,m)} y_j^{(q,m-1)}\right), \ m = 1, \ldots, L, \tag{13}$$

where η is the training rate.

2.1 Training Method

The parameter r in (5) controls the size of the neighborhood region of each pattern in the input space. At the beginning of the training process, the parameter r is set to be a large number, a large size region. Those patterns in this large region will be collected and used to train the MLP. When the training epoch increases, the value of r is decreased linearly. At the end of the training process, only those pattern pairs which have very small distance between them will be used. The SIR training algorithm is briefly described below.

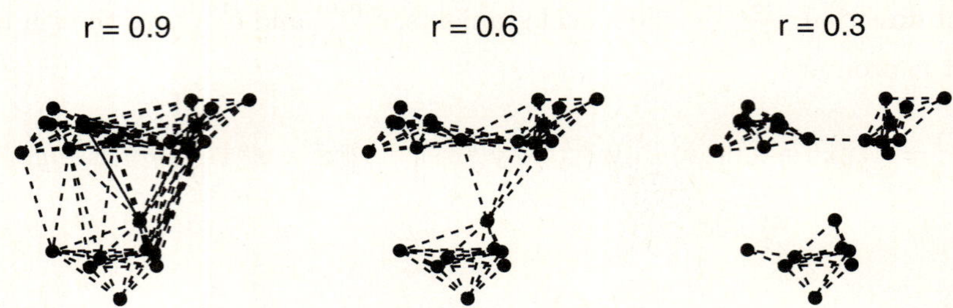

Fig. 2. This diagram shows the shrunk size r during the training process. (a) When r is large, many far neighborhood patterns are included in the training set. (b)(c) When r is decreased to a small value, few near neighborhood patterns are included in the training set.

1. Select two patterns, $\mathbf{x}^p$ and $\mathbf{x}^q$. The distance between them is less than r, $\|\mathbf{x}^p - \mathbf{x}^q\| < r$.
2. Adjust weights of the MLP by (13).
3. Gradually reduce the value of r and repeat step 1 and step 2.

The SIR repeatedly selects two patterns and checks whether their distance is smaller than r. If so, the SIR algorithm adjusts its weights by using the equation (13). If not, the SIR selects another pair of patterns. The region size r is illustrated in Figure 2. One may prepare a list that contains all qualified neighborhood patterns for each pattern for a given size r. With this list, the SIR can be operated in a batch mode.

3   Simulations and Summary

In the first simulation, we show the mapping capability. Figure 3 shows the patterns on a curved surface. The training patterns are sampled from the surface in Figure 3(a). In each training epoch, we randomly select 30 pattern points from the surface. The MLP used in this simulation has four layers, $L = 3$, including a three dimensional input layer, $n_0 = 3$, two hidden layers, $n_1 = 3$ and $n_2 = 2$, and a two dimensional linear output layer, $n_3 = 2$. Figure 3 displays detailed trained results. The mapped surface, from $\mathbf{y}^{(p,0)}$ to $\mathbf{y}^{(p,1)}$, of the first hidden layer, $n_1 = 3$, is plotted in Figure 3(b). From this Figure, one can see that the round surface in the input space is transformed into a bended plane. We uniformly sample a set of patterns from the curved surface in Figure 3(a) and map them to the second hidden layer, $\mathbf{y}^{(p,2)}$, to display the mapping relations in the Figure 3(c). Figure 3(d) shows the output of the linear layer, $n_3 = 2$. From this Figure 3, the SIR twists the input space to minimize the energy function in each layer.

In the second simulation, we show the latent manifold of an image. We randomly sample two 31×31 sub-images from the Lena image in Figure 4(a). These two 31×31 sub-images are re-arranged into two 961×1 column vectors, $n_0 = 961$.

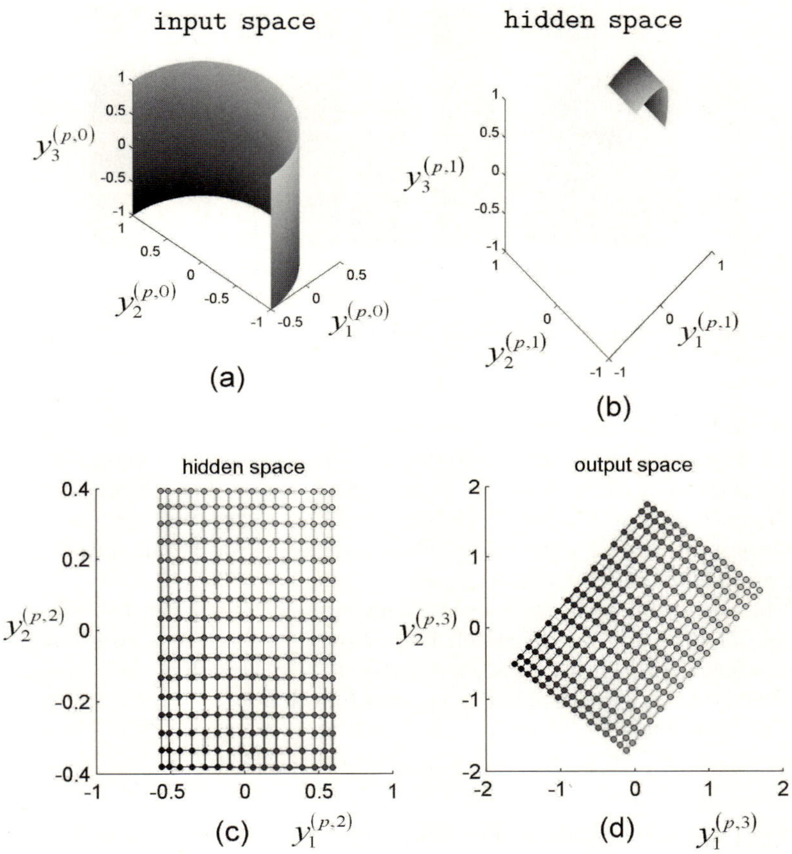

Fig. 3. (a) The curved surface in the 3D input space. (b) The hidden representation of the first hidden layer. (c) The output of the second hidden layer. (d) The output of the linear output layer.

Then we use these two vectors in the SIR. Figure 4(b) shows 20 sampled sub-images. All initial weights are set to random values in the range $[-2, 2]$. The training rate is $\eta = 0.005$. The number of neurons is $n_1 = n_2 = 30$. The output layer has three neurons, $n_3 = 3$. Note that we will apply the PCA to the output of the linear output layer, $\mathbf{y}^{(p,3)}$, to rotate the output in the 3D space for comparison purpose. Initially, the parameter r is set to 1500 and is gradually reduced during the training process. We also provide the results obtained by using the SOM in Figure 5(a).

The result of SIR is shown in Figure 5(b). We display the three output values, $y_1^{(p,3)}$, $y_2^{(p,3)}$, $y_3^{(p,3)}$, of the linear output layer by the three RGB colors. The left image displays the output in RGB color and the other three images is the

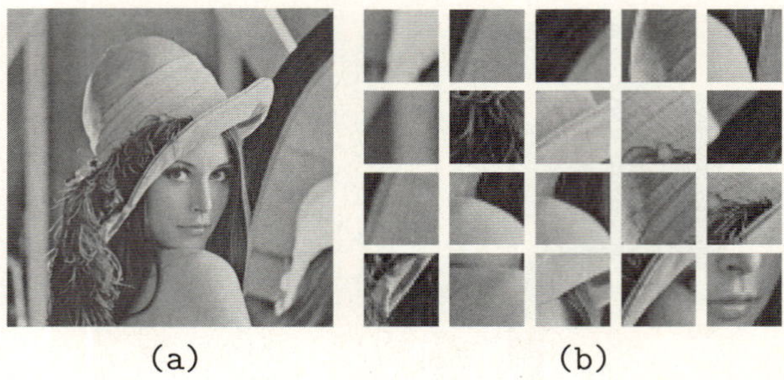

(a)   (b)

Fig. 4. (a) The original Lena image. (b) Twenty examples of 31×31 sub-images.

output of each neuron. Figures 5(a)(c)(d) show results of the SOM. We use $9 = 3 \times 3$ neurons in the SOM. Figure 5(a) is the result of using the nine color representations in Figure 5(d) for each neuron. Figure 5(c) shows the sensitivity of each neuron to sub-images. Both SIR and SOM accomplish vector quantization of those sub-images. The proposed SIR provides a continuous output space and SOM provides a discrete output space.

In the last simulation, we use the dataset in [1]. The dataset consists of many records of different animals. Each record pattern has 17 features including 15 Boolean features and two numerical features. We normalize each feature dimension within the range $[-1, 1]$. The control function is

$$f(t, r) = e^{\frac{-t}{r}}. \tag{14}$$

The f is plotted in Figure 6. The MLP has eight hidden neurons in first hidden layer, $n_1 = 8$, and two hidden neurons in second hidden layer, $n_2 = 2$. The last layer is a two dimensional output layer, $n_3 = 2$. The trained output is shown in Figure 7. Results of other dimension reduction algorithms are shown together for comparison. We use the evaluation function in [1] to calculate the performance error ε,

$$\varepsilon = \left(\frac{\sum\limits_{p,q} \left(\|\mathbf{x}^p - \mathbf{x}^q\| - \|\mathbf{y}^{(p,L)} - \mathbf{y}^{(q,L)}\| \right)^2}{\sum\limits_{p,q} \|\mathbf{x}^p - \mathbf{x}^q\|^2} \right)^{\frac{1}{2}}. \tag{15}$$

Table 1 shows the error of different dimension reduction algorithms. The parameters k of LLE and Isomap which denote the number of neighborhood are set to be 25 and 20, respectively. The MDS gives larger error, $\varepsilon = 0.3480$, than that obtained by the proposed SIR, $\varepsilon = 0.2590$. The error of the SIR is 25.57% less than the error of the MDS. The advantage of proposed SIR is that it finds a set of two dimensional representations to represent the high dimensional patterns and returns a set of weights to describe the input-output mapping.

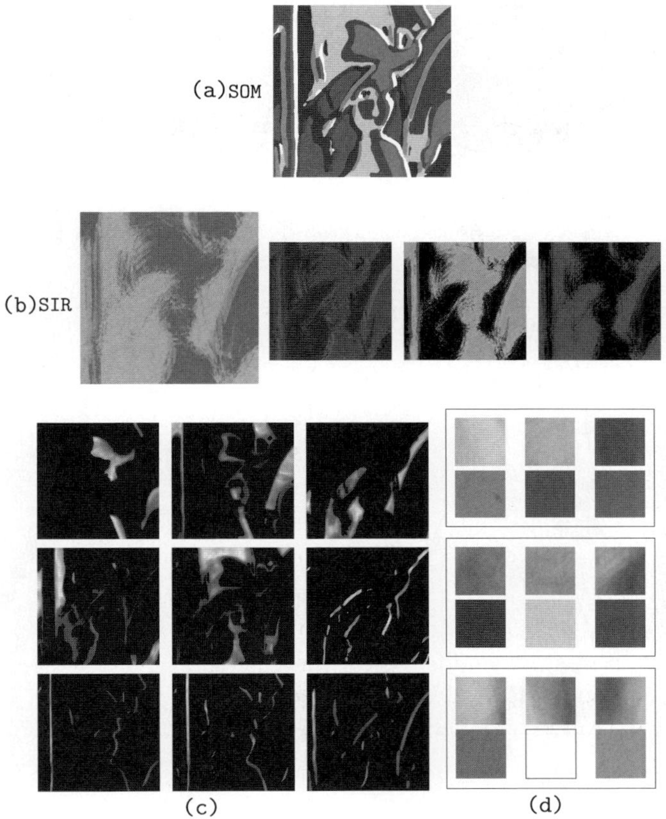

Fig. 5. (a) The quantization result of the 3×3 SOM. (b) Results of the SIR. (c) The segmentation by the SOM. Each plot shows the activated region of each neuron. (d) The nine colors for displaying the neurons in the SOM.

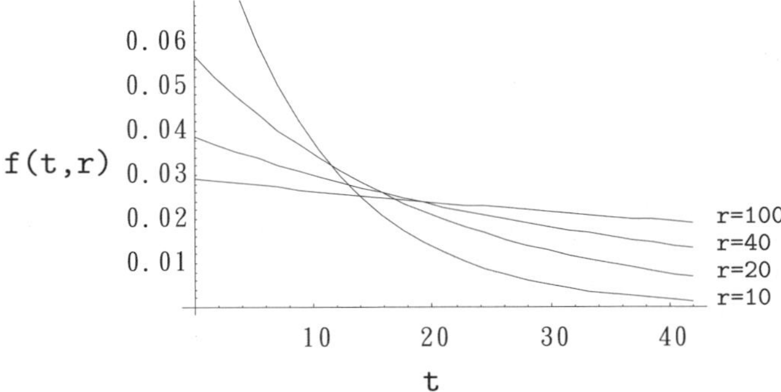

Fig. 6. Plot of the function in (14)

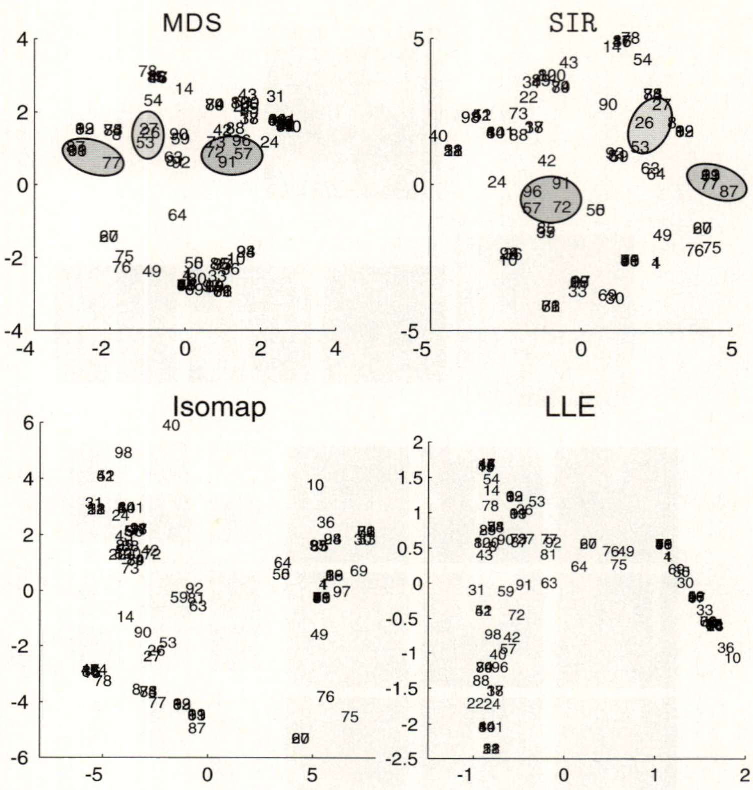

Fig. 7. The results of MDS, SIR, LLE, and Isomap. Label numbers denote patterns. Color ellipses mark the corresponding regions.

Table 1. The ε generated by different mthods

MDS	SIR	LLE	Isomap
0.3480	0.2590	0.6451	0.6344

References

1. Garrido, L., Góez, S., Roca, J.: Improved multidimensional scaling analysis using neural networks with distance-error backpropagation. Neural Networks 11(3), 595–600 (1999)
2. Tenenbaum, J., Silva, V., Langford, J.: A global geometric framework for nonlinear dimensionality reduction. Science 290, 2319–2323 (2000)
3. Roweis, S., Saul, L.: Nonlinear dimensionality reduction by locally linear embedding. Science 290, 2323–2326 (2000)
4. Kohonen, T.: Self-organized formation of topologically correct feature maps. Biological Cybernetics 43(1), 59–69 (1982)

5. Liou, C., Tai, W.: Maximum a posteriori restoration of blurred image using self-organization. Journal of Electronic Imaging, Special Issue on Computational Intelligence in Image Processing 7, 86–93 (1998)
6. Liou, C.Y., Kuo, Y.T.: Economic state indicator on neuronic map. In: The 9th International Conference on Neural Information Processing (ICONIP), Singapore, pp. 787–791 (2002)
7. Liou, C., Tai, W.: Conformal self-organization for continuity on a feature map. neural networks. Journal of Electronic Imaging, Special Issue on Computational Intelligence in Image Processing 12, 893–905 (1999)
8. Liou, C.Y., Kuo, Y.T., Huang, J.C.: Smooth seamless surface construction based on conformal self-organizing map. In: King, I., Wang, J., Chan, L.-W., Wang, D. (eds.) ICONIP 2006. LNCS, vol. 4232, pp. 1012–1021. Springer, Heidelberg (2006)
9. Rumelhart, D., Hinton, G., Williams, R.: Learning internal representations by error propagation. In: Rumelhart, D., McClelland, J. (eds.) Parallel Distributed Processing: Explorations in the Microstructure of Cognition, pp. 318–362. MIT Press, Cambridge (1986)
10. Liou, C.Y., Chen, H.T., Huang, J.C.: Separation of internal representations of the hidden layer. In: Proceedings of the International Computer Symposium, ICS, Workshop on Artificial Intelligence, Chiayi, Taiwan, pp. 26–34 (2000)
11. Liou, C.Y., Cheng, W.C.: Resolving hidden representations. In: Ishikawa, M., Doya, K., Miyamoto, H., Yamakawa, T. (eds.) ICONIP 2007, Part II. LNCS, vol. 4985, pp. 254–263. Springer, Heidelberg (2008)
12. Liou, C.Y., Cheng, W.C.: Manifold construction by local neighborhood preservation. In: Ishikawa, M., Doya, K., Miyamoto, H., Yamakawa, T. (eds.) ICONIP 2007, Part II. LNCS, vol. 4985, pp. 683–692. Springer, Heidelberg (2008)

Improving Performance of a Binary Classifier by Training Set Selection

Cezary Dendek[1] and Jacek Mańdziuk[2]

[1] Warsaw University of Technology, Faculty of Mathematics and Information Science,
Plac Politechniki 1, 00-661 Warsaw, Poland
`dendekc@student.mini.pw.edu.pl`
[2] Warsaw University of Technology, Faculty of Mathematics and Information Science,
Plac Politechniki 1, 00-661 Warsaw, Poland
(Phone: (48 22) 621 93 12; Fax: (48 22) 625 74 60)
`mandziuk@mini.pw.edu.pl`
`http://www.mini.pw.edu.pl/~mandziuk/`

Abstract. In the paper a method of training set selection, in case of low data availability, is proposed and experimentally evaluated with the use of k-*NN* and neural classifiers. Application of proposed approach visibly improves the results compared to the case of training without postulated enhancements.

Moreover, a new measure of distance between events in the pattern space is proposed and tested with k-*NN* model. Numerical results are very promising and outperform the reference literature results of k-*NN* classifiers built with other distance measures.

1   Introduction

The problem of binary classification based on small corpus of data, which implies the use of simple classifier models, is quite often encountered in several application fields. In order to improve classifier's accuracy in such cases the training set should be selected with special care due to significant influence of each element on the classification quality (caused by a small size of the data set).

In this paper a new approach to the problem of training set selection is proposed in Sect. 2 and experimentally evaluated in the context of supervised training with feedforward neural networks and k-*NN* models. The idea relies on the iterative training set creation based on the test of the hypothesis that elements of the training set fulfill the model assumptions and can be properly classified by chosen (target) classifier. Presented algorithm provides a generalization of approach described in [1] in the context of iterative reevaluation of typicality of each observation.

Additionally, a new measure of distance between elements of the pattern space is proposed in Sect. 3 and evaluated with k-*NN* model. Proposed distance function expresses the empirical probability that two vectors are representations of the same event.

The benchmark data sets and results of numerical evaluation of proposed training set construction method and distance measure efficacy are presented in Sections 4 and 5, resp. Conclusions and directions for future research are placed in the last section.

Presented measure of distance is a continuation of authors' previous works [2,3] related to properties of metrical structure of pattern space.

V. Kůrková et al. (Eds.): ICANN 2008, Part I, LNCS 5163, pp. 128–135, 2008.

2 Outlier Extraction in the Case of Binary Classification

2.1 Introduction

The usual approach during the task of fitting a model belonging to the class of models
to given set of observations is to select a model that performs best on this set.

This approach however can be misleading in case of modeling phenomena of complex structure when only a small data set is available, which naturally limits the possible degree of complexity of the model. In such case some non-typical observations
from data set can hinder the quality of a fit by their influence during the training phase
(these observations should be regarded as outliers in the context of selected model and
the whole data set).

Hence, in order to improve the performance of the chosen model one may consider
to eliminate these non-typical, misleading observations from the training set.

2.2 Algorithm of Non-typical Observation Extraction in Case of *k-NN* Classifiers

Having a distance measure defined on pattern space it is possible to extract patterns
that are non-typical for the considered models. In the case of binary classification the
following algorithm can be proposed utilizing the *k-NN* approach:

1. Let TS be a set of all observations.
2. $\forall x \in TS$ find k closest observations from the remaining set $TS \setminus \{x\}$. The distance
   is calculated only based on the parts of patterns that do not belong to dimensions
   being predicted, since these (predicted) dimensions will be unknown during the test
   phase. Hence, a part of a pattern belonging to predicted dimensions is set to 0.
3. Among those k observations, a fraction of them belonging to the same class as x
   defines a *confidence coefficient* of x.
4. If k is reasonably large and a confidence coefficient of x is close to 0, then x should
   be considered as non-typical observation for the *k-NN* classifier.

In described algorithm given element x is conditionally classified under the condition
of its neighborhood (defined by parameter k and distance function). The whole process
can be considered as *k-NN* leave-one-out crossvalidation (denoted by *1-CV*). Generally
speaking, typicality of observation is a function of data set and the considered classification algorithm. These observations lead to generalization of the algorithm described
above to other binary classifier classes.

2.3 Algorithm of Non-typical Observation Extraction in General Case of Binary Classifiers

1. Let TS be a set of all observations.
2. Perform *1-CV* using given class of models m times (independence of each particular model with respect to the training set is important as well as m being reasonably
   large; m cannot however be higher than number of models in class being evaluated).
   Let $C_{i,TS \setminus \{x\}}(x)$ be the binary correctness of classification of element x in step i,
   $i \in \{1, \ldots, m\}$.

3. A fraction of proper classifications of element x defines a *confidence coefficient* $(CC_{CV(TS)}(x))$ of given pattern (under the condition of model used and the data set):

$$CC_{CV(TS)}(x) = \frac{\sum_{i=1}^{m} C_{i,TS\setminus\{x\}}(x)}{m}, \tag{1}$$

where $CC_{CV(TS)}(x)$ is calculated using *1-CV* on the TS set.

4. $\forall x$ if m is reasonably large and $CC_{CV(TS)}(x)$ is smaller than predefined threshold α the observation is considered as a non-typical one, being a member of non-typical observation set (NTS) defined as follows:

$$NTS = \{x \in TS : CC_{CV(TS)}(x) < \alpha\}, \tag{2}$$

$$TS := TS \setminus NTS. \tag{3}$$

In case of classifiers with time-consuming training phase, application of presented approach can be hindered by overall calculation time. In such a case, instead of application of *1-CV*, a whole data set can be used as both training and testing set. In most cases the sets generated by both methods should be highly positively correlated.

Non-typicality of the elements of the set created with the use of the algorithm described by equations (1)-(3) depends on the other elements of the data set (including other non-typical observations), which may cause improper categorization of some observations as non-typical.

In order to reduce this effect re-categorization of non-typical observations (NTS set) with respect to typical ones (TS set) is proposed. The procedure can be expressed by the following algorithm.

2.4   Algorithm of Outliers Extraction in General Case of Binary Classifiers

1. $\forall x \in NTS$ calculate its confidence coefficient $(CC_{TS}(x))$ by building m classifiers with use of TS as training set. Observe that in this case the calculation is performed *without applying the* 1-CV *procedure*.
2. Move elements of $CC_{TS}(x) > \beta, \beta \geq 0$ from NTS to TS, for a given, pre-defined threshold β.

$$TTS = \{x \in NTS : CC_{TS}(x) > \beta\}, \tag{4}$$

$$NTS := NTS \setminus TTS, \tag{5}$$

$$TS := TS \cup TT. \tag{6}$$

3. If any elements have been moved (i.e. $|TTS| > 0$), goto step 1. Otherwise, finish the procedure. NTS is the set containing outliers.

2.5   Remarks

Algorithm presented in sections 2.3 and 2.4 can be regarded as *forward* ($TS \to NTS$) and *backward* ($NTS \to TS$) steps of outlier selection procedure. Those steps can be repeated iteratively until stabilization of NTS set.

The confidence coefficient of elements belonging to NTS set, calculated during the phase of outlier extraction can be used to estimate the efficiency of cooperative voting classifier as well as a number of classifiers used in cooperative model.

3   Distance in the Training Patterns Space

Pattern space has naturally defined structure of metrical space which is obtained by its immersion into $\mathbb{R}^n$. This approach however does not preserve structure of probability space, which can be used to improve accuracy of estimators.

Improved immersion can be obtained with the use of *Cumulative Density Functions* (CDF) by transformation of pattern space. Let CDF_i denotes univariate CDF calculated on i-th dimension of pattern space. Transformation of pattern is defined as follows:

$$(CDF(x))_i := CDF_i(x_i)$$

CDF transformation results in uniform distribution of patterns in each dimension. In order to provide an effective measure of distance between two events correlation has to be taken into consideration, by applying *Mahalanobis-derived distance*. Let Σ denotes covariance matrix of $CDF_i(x_i)$. Distance between events A and B is defined as:

$$d_{cdf}(A, B) := \sqrt{\left(CDF(A) - CDF(B)\right)^T \Sigma^{-1}\left(CDF(A) - CDF(B)\right)}$$

where $CDF(A)$ and $CDF(B)$ are representations of these events according to CDF. A classical Mahalanobis distance defined as a measure of distance between a vector and a set of all observations can be obtained as follows:

$$D_{cdf}(A) := \sqrt{\left(CDF(A) - \frac{1}{2}\right)^T \Sigma^{-1}\left(CDF(A) - \frac{1}{2}\right)}$$

due to uniform distribution of CDF_i values in the set $[0, 1]$, for each i.

4   Data Sets

In order to provide experimental support of presented method validity and generate results that can be compared to other sources BUPA Liver Disorders and Pima Indians Diabetes data sets available at *UCI Machine Learning Repository* [4] has been chosen. Both of them contain observations of rather complex medical phenomena of catastrophic nature and provide only correlated predictors. Size of the sample is limited to 345 and 768 elements, respectively. Both sets are rather challenging tasks for classification methods.

In the following experiments either the "raw" data sets were used or the transformed ones obtained by application of *Empirical Cumulative Distribution Function* (ECDF) which resulted in normalization of data and uniformity of marginal distributions.

5   Results

Both data sets has been evaluated with use of a neural network and *k-NN* classifier. Application of outlier removal algorithm described in Sect. 2 resulted in significant increase of classifier quality in both cases.

5.1   k-NN Classifier

During evaluation *k-NN* models with standard distance function and distance defined in Sect. 3 has been used. Discrimination levels has been set as follows: $\alpha = 1$, $\beta = 0.5$.

Observe that in case of classifiers using the ECDF transformation, the metric function was modified by repeating the calculation of the matrix Σ based *exclusively on elements of the set* $TS \setminus NTS$. The re-calculation of Σ after outliers removal from TS enhances classification capabilities. This property is owed to the fact that after removing a non-empty set NTS from TS the correlation matrix has changed and the "old" correlation matrix calculated for TS does not fit the current model for the set $TS \setminus NTS$.

Numerical results of classifiers quality (misclassification rates in per cent) for *1-CV* estimation are presented in Tables 1, 2 and 3. Table 1 presents misclassification results in the case of standard Euclidean metric (i.e. without ECDF transformation) in the "raw" *1-CV* case and with the use of outlier removal algorithm described in Sect. 2.4. Table 2 provides a similar comparison in the case of ECDF transformation and standard distance. The results of the combination of the outlier removal algorithm with ECDF transformation with (column 3) and without (column 2) recalculation of Mahalanobis-based distance metric (Σ) vs standard *1-CV* (column 1) in ECDF space are presented in Table 3.

The results presented in the Tables fully confirm the efficacy of the proposed outliers removal algorithm. Moreover, the advantage of using ECDF transformation is also clear. Finally, the results support application of the Mahalanobis-derived metric - in all cases except for *3-NN* and Pima Indians Diabetes set, presented in Table 3 the use of this

Table 1. *k-NN* misclassification rate, standard distance in Euclidean space

k	BUPA Liver Disorders		Pima Indians Diabetes	
	initial *1-CV*	*1-CV* after outlier removal	initial *1-CV*	*1-CV* after outlier removal
3	36.52	33.62	30.60	26.30
5	33.62	32.75	28.52	27.99

Table 2. *k-NN* misclassification rate, standard distance in ECDF image space

k	BUPA Liver Disorders		Pima Indians Diabetes	
	initial *1-CV*	*1-CV* after outlier removal	initial *1-CV*	*1-CV* after outlier removal
3	35.07	30.14	27.73	23.96
5	31.88	29.28	26.56	25.26

Table 3. *k-NN* misclassification rate, Mahalanobis-derived distance in ECDF image space

k		BUPA Liver Disorders			Pima Indians Diabetes	
	initial *1-CV*	*1-CV* after outlier removal	*1-CV* after outlier Σ removal	initial *1-CV*	*1-CV* after outlier removal	*1-CV* after outlier Σ removal
3	33.91	28.12	26.96	27.60	24.09	24.09
5	31.59	30.43	28.70	26.56	25.00	24.87

Table 4. k-NN misclassification rate comparison

distance maeasure	estimation method	BUPA Liver Disorders	Pima Indians Diabetes
Mahalanobis-based with outlier removal (Table 3)	leave-one-out CV	26.96	24.09
Mahalanobis-based (Table 3)	leave-one-out CV	31.59	26.56
weighted distances [5]	100 x 5–CV	36.22	27.39
adaptive distance measure [6]	10–CV	32.94	28.16
boosting distance estimation [7]	100 x 20/80	33.58	28.91
cam weighted distance [8]	leave-one-out CV	35.3	24.7

metric improved the results compared to the cases of standard metric in the ECDF space
(Table 2) and Cartesian metric on the untransformed data (Table 1). The advantage is
especially visible in the case of Liver Disorders data.

In summary, the use of the Mahalonobis-based distance metric combined with the
algorithm of outliers removal allowed, within the *k-NN* models, to achieve comparable
or better results than the best ones presented in the literature. In particular the following
comparisons with other definitions of the metric functions within the considered data
sets has been taken into account: weighted distances [5], adaptive distance measure [6],
boosting distance estimation [7] and cam weighted distance [8]. Table 4 presents the
summary of results accomplished by the *k-NN* classifiers.

5.2   Neural Classifier

Both data sets has been evaluated with use of a neural network classifier (MLP with
one hidden layer), using various sizes of hidden layer and backpropagation as a training
method.

Classifier quality has been estimated with use of leave-one-out crossvalidation performed 10 times (the results presented in teh paper are the average values over these 10
trials). The initial estimation of the network's classification quality concerned the full
data sets. In the final estimation, the observations classified as outliers were removed
from the training set and appeared only in the test set.

In case of Pima Indians Diabetes data set initial non-typical observation set has been
obtained with the use of training and testing performed on full data set, due to high

Table 5. Neural network misclassification rate

number of hidden units	BUPA Liver Disorders		Pima Indians Diabetes	
	initial 1-CV	1-CV after outlier removal	initial 1-CV	1-CV after outlier removal
2	28.79	26.32	24.84	22.27
3	29.37	24.54	24.77	21.59
4	28.60	25.39	24.85	21.43
5	29.08	24.46	24.22	21.35

computational cost of full crossvalidation and in order to provide an empirical support for a claim presented in Sect. 2.5.

For both data sets values of discriminant coefficients has been set to $\alpha = 0.5$, $\beta = 0.5$. Discrimination has been made with a support of $m = 20$ independent (with respect to the training set) classifiers of a given architecture.

The results of *1-CV* estimation of the neural classifiers with and without application of the outliers removal algorithm are presented in Table 5. Similarly to the case of *k-NN* classifiers the use of the removal of non-typical observations from the training set clearly enhances the efficacy of neural classifiers. In effect, the MLP-based classifiers outperform the *k-NN* ones further decreasing the misclassification figures to the levels of 24.46% and 21.35%, respectively for Liver Disorders and Pima Indians Diabetes data sets.

Based on the comprehensive comparison of various classification methods presented in [9] for the case of Pima Indians Diabetes set the results accomplished by the algorithm using ECDF transformation of the data combined with outliers removal method presented in this paper can be regarded as very promising.

6   Conclusions

The problem of classifier construction based on small corpus of data is quite common in real-life situations. In case of fitting a general model (constructed without additional knowledge about modelled phenomena) training set selection can improve overall classifier's efficacy.

In the paper a method of training set selection is proposed and experimentally evaluated with *k-NN* and neural network classifiers. It is shown that proposed approach produces in average better results than training without its use in case of two sample problems of small corpus of data.

Additionally, a new measure of distance between events in the pattern space is proposed and evaluated with *k-NN* model. Crossvalidational estimate of resulting model quality has been compared with numerical results provided by other researchers, concerning the *k-NN* classifiers built with other distance measures.

Tests in other problem domains are under research. Other possible uses of presented distance measure (especially in the context of training sequence construction presented in [2]) are considered as future research plans.

Acknowledgement

This work was supported by the Warsaw University of Technology grant number 503G 1120 0007 007.

References

1. Sane, S.S., Ghatol, A.A.: Use of instance typicality for efficient detection of outliers with neural network classifiers. In: Mohanty, S.P., Sahoo, A. (eds.) ICIT, pp. 225–228. IEEE Computer Society, Los Alamitos (2006)
2. Dendek, C., Mandziuk, J.: Including metric space topology in neural networks training by ordering patterns. In: Kollias, S.D., Stafylopatis, A., Duch, W., Oja, E. (eds.) ICANN 2006. LNCS, vol. 4132, pp. 644–653. Springer, Heidelberg (2006)
3. Mańdziuk, J., Shastri, L.: Incremental class learning approach and its application to handwritten digit recognition. Inf. Sci. Inf. Comput. Sci. 141(3-4), 193–217 (2002)
4. Asuncion, A., Newman, D.J.: UCI machine learning repository (2007)
5. Pardes, R., Vidal, E.: Learning weighted metrics to minimize nearest-neighbor classification error. IEEE Transactions on Pattern Analysis and Machine Intelligence 28(7), 1100–1110 (2006)
6. Wang, J., Neskovic, P., Cooper, L.N.: Improving nearest neighbor rule with a simple adaptive distance measure. Pattern Recogn. Lett. 28(2), 207–213 (2007)
7. Amores, J., Sebe, N., Radeva, P.: Boosting the distance estimation. Pattern Recogn. Lett. 27(3), 201–209 (2006)
8. Zhou, C.Y., Chen, Y.Q.: Improving nearest neighbor classification with cam weighted distance. Pattern Recogn. 39(4), 635–645 (2006)
9. Tsakonas, A.: A comparison of classification accuracy of four genetic programming-evolved intelligent structures. Inf. Sci. 176(6), 691–724 (2006)

An Overcomplete ICA Algorithm by InfoMax and InfoMin

Yoshitatsu Matsuda[1] and Kazunori Yamaguchi[2]

[1] Department of Integrated Information Technology,
Aoyama Gakuin University,
5-10-1 Fuchinobe, Sagamihara-shi, Kanagawa, 229-8558, Japan
`matsuda@it.aoyama.ac.jp`
`http://www-haradalb.it.aoyama.ac.jp/~matsuda`
[2] Department of General Systems Studies,
Graduate School of Arts and Sciences, The University of Tokyo,
3-8-1, Komaba, Meguro-ku, Tokyo, 153-8902, Japan
`yamaguch@graco.c.u-tokyo.ac.jp`

Abstract. It is known that the number of the edge detectors significantly exceeds that of input signals in the visual system of the brains. This phenomenon has been often regarded as overcomplete independent component analysis (ICA) and some generative models have been proposed. Though the models are effective, they need to assume some ad-hoc prior probabilistic models. Recently, the InfoMin principle was proposed as a comprehensive framework with minimal prior assumptions for explaining the information processing in the brains and its usefulness has been verified in the classic non-overcomplete cases. In this paper, we propose a new ICA contrast function for overcomplete cases, which is deductively derived from the the InfoMin and InfoMax principles without any prior models. Besides, we construct an efficient fixed-point algorithm for optimizing it by an approximate Newton's method. Numerical experiments verify the effectiveness of the proposed method.

1 Introduction

The information theoretical approach has been regarded as a promising framework for elucidating the fundamental mechanism of the brains [1,2]. Especially, it has been shown that the InfoMax approach with nonlinear neurons is useful for explaining the generation of edge detectors in the visual system [3]. The InfoMax approach is strongly related to independent component analysis (ICA) [4], and many works on ICA models explaining the visual system have been published [5,6,7,8]. Though the effectiveness of the previously-proposed models has been verified, the derivation of these models is not plausible. The derivation essentially depends on the nonlinearity of neurons and the property of the contrast function (objective function) of these ICA models is quite sensitive to the curvature of a given ad-hoc nonlinear function. In order to avoid this problem, the InfoMin principle was recently proposed as a simple framework for explaining the information processing in the brains. The combination of the InfoMin principle and

V. Kůrková et al. (Eds.): ICANN 2008, Part I, LNCS 5163, pp. 136–144, 2008.

the ordinary InfoMax one with some minimal assumptions can comprehensively explain various objective functions for topographic mappings [9], prewhitening process and the classical ICA model [10], topographic ICA model [10,11], and the generation of hierarchical long edge detectors [12].

The purpose of this paper consists of two part. The first one is to derive an overcomplete ICA contrast function deductively from the InfoMin principle [10]. It gives a comprehensive framework to overcomplete ICA. The second one is to construct an efficient algorithm optimizing the contrast function. It gives a new efficient technique to solve overcomplete ICA. "Overcompleteness" means that the number of edge detectors significantly exceeds that of input signals, and it is one of the most important properties in the visual system. Some overcomplete ICA models have been proposed by assuming a prior probabilistic model. For example, in [13], an overcomplete ICA algorithm was constructed as a simple extension of a classical non-overcomplete ICA by giving a probabilistic distribution of the source signals. In [14], an overcomplete ICA contrast function was proposed, which measures the pseudo-orthogonality of separated signals and is derived from a prior probabilistic model of the mixing matrix. On the other hand, in this paper, we derive an overcomplete contrast function by only the InfoMin and InfoMax principles with no ad-hoc prior probabilistic models except for large gaussian noise. Besides, we construct an efficient algorithm for optimizing the function by an approximate Newton's method.

This paper is organized as follows. In Section 2, some previously-proposed knowledge on the InfoMin principles is briefly described and the new contrast function for overcomplete ICA is proposed. In Section 3, an efficient fixed-point algorithm for optimizing the function is constructed. In Section 4, some numerical experiments show that the proposed algorithm is effective. Lastly, this paper is concluded in Section 5.

2   New Contrast Function for Overcomplete ICA

2.1   Brief Introduction to InfoMin and InfoMax

We briefly explain the InfoMin and InfoMax principles here for the reader's convenience. In the information processing of the brain, it seems to be natural to require that *"the information through the whole neurons must be maximized."* This is the *InfoMax* principle. As far as InfoMax is sustained, the second principle, *"the information transfer through each neuron must be minimized,"* seems desirable. This principle, named *InfoMin*, was first proposed as a unifying framework for forming topographic mappings [9]. Roughly speaking, InfoMin asserts that a neuron which rarely fires is more desirable than the one which does with a probability of 50 percent.

In order to derive the closed form of objective functions for the above InfoMax and InfoMin principles, we assume only that (1) the third cumulants are negligible; (2) there exists additive gaussian noise whose covariance is $\Delta^2 I$ (*Delta^2* is a constant variance of the noise, and I is the identity matrix); (3) the variance of every signal is the same constant (which is set to 1 without losing generality). In consequence, the InfoMax principle is equivalent to the minimization of

$$\phi_{InfoMax} = \sum_{i,j\neq i} \left(\kappa^{ij}\right)^2 + O\left(\frac{1}{\Delta^2}\right) \tag{1}$$

where $\kappa^{ij} = E\{z_i z_j\}$ is the second-order cumulant of the i-th and j-th signals. Here, $\boldsymbol{z} = (z_i)$ is the vector of signals and $E\{\}$ is the expectation operator. In the linear model, $\boldsymbol{z}$ is given as a linear mixture of observed signals (see Section 2.2). Besides, note that the term of $O\left(\frac{1}{\Delta^2}\right)$ is negligible for sufficiently large Δ^2. On the other hand, the InfoMin principle is equivalent to the minimization of

$$\phi_{InfoMin} = -\sum_{i} \left(\kappa^{iiii}\right)^2 + O\left(\frac{1}{\Delta^2}\right) \tag{2}$$

where $\kappa^{iiii} = E\{z_i^4\} - 3$ is the kurtosis of the i-th signal. Because it is natural to assume that all the signals are super-gaussian in the visual system, it is further simplified to

$$\phi_{InfoMin} = -\sum_{i} \kappa^{iiii} \tag{3}$$

See [10] for the detailed derivation. In non-overcomplete cases, the minimization of $\phi_{InfoMax}$ corresponds to any prewhitening process. Then, the minimization of $\phi_{InfoMin} = -\sum_i \kappa^{iiii}$ is equivalent to the classical kurtosis-based ICA.

2.2   Derivation of New Contrast Function

In overcomplete cases, it is obvious that $\phi_{InfoMax}$ in Eq. (1) measures the degree of the pseudo-orthogonality of the signals. Therefore, as a simple contrast function for overcomplete cases, we adopt the linear combination of $\phi_{InfoMax}$ and $\phi_{InfoMin}$ as follows:

$$\phi_{over} = \phi_{InfoMax} + (\alpha+1)\,\phi_{InfoMin} = \sum_{i,j\neq i} \left(E\{z_i z_j\}\right)^2 - (\alpha+1)\sum_{i} E\{z_i^4\}, \tag{4}$$

where $(\alpha+1)$ is the weight of $\phi_{InfoMin}$, and the constant subtracter -3 ($= \kappa^{iiii} - E\{z_i^4\}$) is omitted. $(\alpha+1)$ must be positive and should be adjusted empirically. The optimization of Eq. (4) is difficult because it includes a term of the square of the expectation operator. In order to overcome this problem, we propose to use the following contrast function as a substitute for ϕ_{over}:

$$\phi = \sum_{i,j\neq i} E\{z_i^2 z_j^2\} - \alpha \sum_{i} E\{z_i^4\}. \tag{5}$$

The replacement of ϕ_{over} with ϕ is justified by the following two points. First, it is easily proved that $E\{z_i^2 z_j^2\} \geq \left(E\{z_i z_j\}\right)^2$ and $E\{z_i^4\} > 0$ if $\boldsymbol{z}$ is given as a linear mixture of super-gaussian signals. So, the following inequality always holds in super-gaussian cases:

$$\phi > \phi_{over}. \tag{6}$$

It guarantees that ϕ gives an upper bound of ϕ_{over}. It supports the use of ϕ. Roughly speaking, it means that the minimization of $(E\{z_i z_j\})^2$ (relatively difficult) can be replaced by that of $E\{z_i^2 z_j^2\}$ if the kurtoses also has to be minimized. Second, we pay attention to that $\sum_i E\{z_i^4\} + \sum_{i,j\neq i} E\{z_i^2 z_j^2\}$ is a constant if the orthogonality of z holds. Therefore, if z is near to the optimum and the pseudo-orthogonality holds ($\sum_{i,j\neq i} (E\{z_i^2 z_j^2\})^2$ is quite small), the following approximation holds:

$$\sum_{i,j\neq i} E\{z_i^2 z_j^2\} \simeq K - \sum_i E\{z_i^4\}, \tag{7}$$

where K is a constant. Then, the following approximation holds if the pseudo-orthogonality holds:

$$\phi \simeq K - \sum_i E\{z_i^4\} - \alpha \sum_i E\{z_i^4\} = K - (\alpha + 1) E\{z_i^4\}. \tag{8}$$

On the other hand, if the pseudo-orthogonality holds, $(E\{z_i z_j\})^2 \simeq 0$ for each i and $j\,(\neq i)$. So, the following approximation holds:

$$\phi_{over} \simeq -(\alpha + 1) \sum_i E\{z_i^4\}. \tag{9}$$

Therefore, around the optimum of z, the following approximation holds:

$$\mathrm{argmin}_{z}\,\phi \simeq \mathrm{argmin}_{z}\,\phi_{over}. \tag{10}$$

It shows that the global optimum of ϕ is quite near to that of ϕ_{over} if the pseudo-orthogonality is achievable. It also supports the use of ϕ.

In expectation of future extendibility, we use the following contrast function generalized from Eq. (5):

$$\phi_{general} = E\left\{ \sum_{i,j\neq i} F(z_i) F(z_i) - \alpha \sum_i G(z_i) \right\}. \tag{11}$$

$\phi_{general}$ for $F(u) = u^2\ G(u) = u^4$ is equivalent to Eq. (5). The form is the sum of the simple contrast function of each signal and the nonlinear covariance between each pair of signals.

3   Algorithm

Here, we assume that z is given as a linear mixture of observed signals x by

$$z = Wx \tag{12}$$

where W is an $M \times N$ separating matrix, M is the number of overcomplete signals, and N is that of observed ones. Here, only the observed signals x are known. Both

z and W must be estimated. Besides, we assume that the observed signals x are prewhitened ($E\left\{xx^T\right\}$ is the $N \times N$ identity matrix I) and the variance of each z_i is 1 ($E\left\{z_i^2\right\} = 1$). Then, a fixed-point algorithm minimizing $\phi_{general}$ in Eq. (11) with respect to W is derived in the same way as in fast ICA [15].

The update equation for each z_i is derived as follows. Let w_i be the i-th row of W. According to the Kuhn-Tucker conditions in the optima of $\phi_{general}$ under the constraint $E\left\{z_i^2\right\} = w_i w_i^T = 1$, the optimal w_i needs to satisfy

$$E\left\{x\left(\sum_{j\neq i} F'(z_i) F(z_j) - \alpha G'(z_i)\right)\right\} - \beta_i w_i^T = 0, \qquad (13)$$

where β_i is a Lagrange multiplier, which is given as

$$\beta_i = E\left\{\left(\sum_{j\neq i} z_i F'(z_i) F(z_j)\right) - \alpha z_i G'(z_i)\right\} \qquad (14)$$

by multiplying both terms by w_i. Besides, its Jacobian matrix J w.r.t. w_i is given as

$$J = E\left\{xx^T\left(\sum_{j\neq i} F''(z_i) F(z_j) - \alpha G''(z_i)\right)\right\} - \beta_i I. \qquad (15)$$

Now, the crucial approximation is given as

$$E\left\{xx^T\left(\sum_{j\neq i} F''(z_i) F(z_j) - \alpha G''(z_i)\right)\right\}$$
$$\simeq E\left\{xx^T\right\} E\left\{\sum_{j\neq i} F''(z_i) F(z_j) - \alpha G''(z_i)\right\}$$
$$= E\left\{\sum_{j\neq i} F''(z_i) F(z_j) - \alpha G''(z_i)\right\} I \qquad (16)$$

where xx^T is assumed to be independent of the other terms. Note that the same assumption was used in the derivation of fast ICA. Therefore, the inversion of J is easily calculated and $\phi_{general}$ is minimized by Newton's method. The update equation is given as

$$w_i := w_i - \frac{E\left\{x^T\left(\sum_{j\neq i} F'(z_i) F(z_j) - \alpha G'(z_i)\right)\right\} - \beta_i w_i}{|E\left\{\sum_{j\neq i} F''(z_i) F(z_j) - \alpha G''(z_i)\right\} - \beta_i|} \qquad (17)$$

Though Eqs. (14) and (17) involve heavy computational costs of the summation $\sum_{j\neq i}$ for each i, they can be rewritten as

$$\beta_i = E\left\{z_i F'(z_i) \gamma - z_i F'(z_i) F(z_i) - \alpha z_i G'(z_i)\right\} \qquad (18)$$

and

$$\boldsymbol{w}_i := \boldsymbol{w}_i - \frac{E\left\{\boldsymbol{x}^T \left(F'\left(z_i\right)\gamma - F'\left(z_i\right)F\left(z_i\right) - \alpha G'\left(z_i\right)\right)\right\} - \beta_i \boldsymbol{w}_i}{\left|E\left\{F''\left(z_i\right)\gamma - F''\left(z_i\right)F\left(z_i\right) - \alpha G''\left(z_i\right)\right\} - \beta_i\right|}, \tag{19}$$

where γ is given as

$$\gamma = \sum_i F\left(z_i\right). \tag{20}$$

Because γ does not depend on i, the update equation Eq. (19) can be carried out efficiently. Besides, in practice, the following normalization for each $\boldsymbol{w}$ is done at each update:

$$\boldsymbol{w}_i := \frac{\boldsymbol{w}_i}{\sqrt{\boldsymbol{w}_i \boldsymbol{w}_i^T}}. \tag{21}$$

For ϕ in Eq. (5) where $F\left(u\right) = u^2$ and $G\left(u\right) = u^4$, the update equation Eq. (19) is simplified further to

$$\boldsymbol{w}_i := \boldsymbol{w}_i - \frac{E\left\{\boldsymbol{x}^T \left(2z_i\gamma - \left(2 + 4\alpha\right)z_i^3\right)\right\} - \beta_i \boldsymbol{w}_i}{\left|2M - 2 - 12\alpha - \beta_i\right|}, \tag{22}$$

where β and γ are given as

$$\beta_i = E\left\{2z_i^2\gamma - \left(2 + 4\alpha\right)z_i^4\right\} \tag{23}$$

and

$$\gamma = \sum_i z_i^2, \tag{24}$$

respectively.

Table 1. Results for artificial data: 100,000 samples of M-dimensional Laplace-distributed random variables were used as the source signals. Then, the observed signals were constructed by multiplying the source signals by a randomly-generated $N \times M$ mixing matrix $\boldsymbol{A}$. The separating matrix $\boldsymbol{W}$ was estimated by Eqs. (22), (23), and (24). N was set to 50. M was set to 50, 100, and 200. α was set to 10,000. The table shows $\sum_{i,j \neq i} \left(\kappa^{ij}\right)^2 / N\left(N-1\right)$, $\sum_i \kappa^{iiii}/N$, the reduction rate of Amari's errors [16] compared to initial random conditions, and the update times for convergence. They were averaged over 10 runs.

	$\dfrac{\sum_{i,j\neq i}\left(\kappa^{ij}\right)^2}{N(N-1)}$	$\dfrac{\sum_i \kappa^{iiii}}{N}$	Reduction rate of Amari's error	Update times for convergence
$N = 50,\ M = 50$	0.019	3.02	0.22	9.6
$N = 50,\ M = 100$	0.024	0.93	0.42	19.4
$N = 50,\ M = 200$	0.033	0.33	0.57	121

4   Results

Here, the proposed fixed-point algorithm is applied to some data.

First, some experiments on artificial data of Laplace-distributed random variables were carried out. As a result, Table 1 shows the mean of the squares of covariances $(\sum_{i,j\neq i}\left(\kappa^{ij}\right)^2/N\left(N-1\right))$, the mean of kurtoses $(\sum_i \kappa^{iiii}/N)$, and the reduction rate of a separating error. In the non-overcomplete case ($M = 50$), the proposed algorithm could generate rather good results. The separating error was reduced to less than one quarter, and the pseudo-orthogonality was attained (in other words, the mean of the squares of covariances was quite small). In addition, the mean of kurtoses was near to 3.0 which is the theoretical value of the kurtosis of the Laplace-distribution. Though the reduction rate of

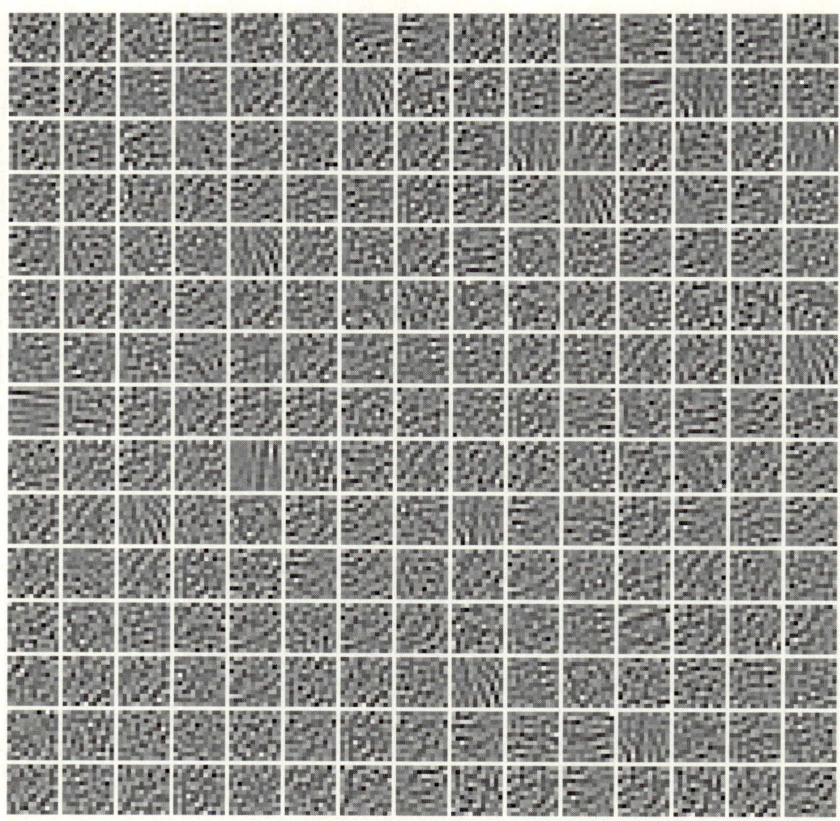

Fig. 1. Formation of overcomplete edge detectors from natural scenes: Here, 100 PCA-whitened components of 30,000 samples of natural scenes of 12×12 pixels were used as $\boldsymbol{x}$. The number of overcomplete signals was set to 225. α was set to 10,000. $\boldsymbol{W}$ was estimated by Eqs. (22), (23), and (24). Each small image corresponds to a row of $\boldsymbol{W}$.

the error deteriorated in the overcomplete cases ($M = 100$ and $M = 200$), it was nevertheless near to 0.5. It is more important that the pseudo-orthogonality was always satisfied even in the overcomplete cases. In addition, it shows that the convergence is quite fast especially for $M = 50$ and $M = 100$. So, the results show the effectiveness of the proposed method.

Second, this algorithm is applied to natural scenes. At the convergence state, $\sum_{i,j\neq i} \left(\kappa^{ij}\right)^2 /N\left(N-1\right)$ was 0.029, $\sum_i \kappa^{iiii}/N$ was 20.1, and the update times was 41. It shows that the pseudo-orthogonality was attained also for natural scenes. Fig. 1 displays the generated components $\boldsymbol{W}$. Though they are noisy, several clear edge detectors can be found. It is interesting that the cumulant-based method could generate some edge detectors. But, in order to generate more clear edge detectors, we may have to replace $F\left(u\right)$ and $G\left(u\right)$ with some robust functions.

5   Conclusion

In this paper, we proposed a new contrast function for overcomplete ICA, which is deductively derived from the InfoMax and InfoMin principles. In addition, we constructed an efficient fixed-point algorithm for optimizing the function. The numerical experiment showed that this algorithm can generate overcomplete edge detectors from natural scenes.

Though the proposed algorithm is based on Newton's method, it is just an approximation. We are now trying to analyze the convergence property of the algorithm and clarify the validity of this approximation. In addition, the parameter α must be empirically set in the current algorithm. It is an important parameter combining the InfoMax and InfoMin principles, so we are planning to examine its property. We are also planning to try non-cumulant-based functions as $F\left(u\right)$ and $G\left(u\right)$ in order to obtain more robust results and generate clearer edge detectors. Besides, we are planning to compare the proposed algorithm with other overcomplete ICA methods statistically. Lastly, we are planning to extend the InfoMin framework for explaining more general properties of the visual system. This work is partially supported by Grant-in-Aid for Young Scientists (KAK-ENHI) 19700267.

References

1. Linsker, R.: Self-organization in a perceptual network. Computer 21(3), 105–117 (1988)
2. Nadal, J.P., Parga, N.: Nonlinear neurons in the low-noise limit: a factorial code maximizes information-transfer. Network: Computation in Neural Systems 5(4), 565–581 (1994)
3. Bell, A.J., Sejnowski, T.J.: The independent components of natural scenes are edge filters. Vision Research 37(23), 3327–3338 (1997)
4. Bell, A.J., Sejnowski, T.J.: An information-maximization approach to blind separation and blind deconvolution. Neural Computation 7, 1129–1159 (1995)

5. van Hateren, J.H., van der Schaaf, A.: Independent component filters of natural images compared with simple cells in primary visual cortex. Proceedings of the Royal Society of London: B 265, 359–366 (1998)
6. Hyvärinen, A., Hoyer, P.O., Inki, M.: Topographic independent component analysis. Neural Computation 13(7), 1527–1558 (2001)
7. Hoyer, P.O., Hyvärinen, A.: A multi-layer sparse coding network learns contour coding from natural images. Vision Research 42(12), 1593–1605 (2002)
8. Karklin, Y., Lewicki, M.S.: Learning higher-order structures in natural images. Network: Computation in Neural Systems 14, 483–499 (2003)
9. Matsuda, Y., Yamaguchi, K.: The InfoMin criterion: an information theoretic unifying objective function for topographic mappings. In: Kaynak, O., Alpaydın, E., Oja, E., Xu, L. (eds.) ICANN 2003 and ICONIP 2003. LNCS, vol. 2714, pp. 401–408. Springer, Heidelberg (2003)
10. Matsuda, Y., Yamaguchi, K.: The InfoMin principle for ICA and topographic mappings. In: Rosca, J.P., Erdogmus, D., Príncipe, J.C., Haykin, S. (eds.) ICA 2006. LNCS, vol. 3889, pp. 958–965. Springer, Heidelberg (2006)
11. Matsuda, Y., Yamaguchi, K.: A fixed-point algorithm of topographic ICA. In: Kollias, S.D., Stafylopatis, A., Duch, W., Oja, E. (eds.) ICANN 2006. LNCS, vol. 4132, pp. 587–594. Springer, Heidelberg (2006)
12. Matsuda, Y., Yamaguchi, K.: A connection-limited neural network by infomax and infomin. In: Proceedings of IJCNN 2008, Hong Kong. IEEE, Los Alamitos (in press, 2008)
13. Lewicki, M.S., Sejnowski, T.J.: Learning overcomplete representations. Neural Computation 12(2), 337–365 (2000)
14. Hyvärinen, A., Hoyer, P.O., Inki, M.: Estimating overcomplete independent component bases for image windows. Journal of Mathematical Imaging and Vision 17, 139–152 (2002)
15. Hyvärinen, A.: Fast and robust fixed-point algorithms for independent component analysis. IEEE Transactions on Neural Networks 10(3), 626–634 (1999)
16. Amari, S., Cichocki, A.: A new learning algorithm for blind signal separation. In: Touretzky, D., Mozer, M., Hasselmo, M. (eds.) Advances in Neural Information Processing Systems, vol. 8, pp. 757–763. MIT Press, Cambridge (1996)

OP-ELM: Theory, Experiments and a Toolbox

Yoan Miche[1,2], Antti Sorjamaa[1], and Amaury Lendasse[1]

[1] Department of Information and Computer Science, HUT, Finland
[2] Gipsa-Lab, INPG, France

Abstract. This paper presents the Optimally-Pruned Extreme Learning Machine (OP-ELM) toolbox. This novel, fast and accurate methodology is applied to several regression and classification problems. The results are compared with widely known Multilayer Perceptron (MLP) and Least-Squares Support Vector Machine (LS-SVM) methods. As the experiments (regression and classification) demonstrate, the OP-ELM methodology is considerably faster than the MLP and the LS-SVM, while maintaining the accuracy in the same level. Finally, a toolbox performing the OP-ELM is introduced and instructions are presented.

1 Introduction

The amount of information is increasing rapidly in many fields of science. It creates new challenges for storing the massive amounts of data as well as to the methods, which are used in the data mining process. In many cases, when the amount of data grows, the computational complexity of the used methodology also increases.

Feed-forward neural networks are often found to be rather slow to build, especially on important datasets related to the data mining problems of the industry. For this reason, the nonlinear models tend not to be used as widely as they could, even considering their overall good performances. The slow building of the networks comes from a few simple reasons; many parameters have to be tuned, by slow algorithms, and the training phase has to be repeated many times to make sure the model is proper and to be able to perform model structure selection (number of hidden neurons in the network, regularization parameters tuning...).

Guang-Bin Huang *et al.* in [1] propose an original algorithm for the determination of the weights of the hidden neurons called Extreme Learning Machine (ELM). This algorithm decreases the computational time required for training and model structure selection of the network by hundreds. Furthermore, the algorithm is rather simplistic, which makes the implementation easy.

In this paper, a methodology called Optimally-Pruned ELM (OP-ELM), based on the original ELM, is proposed. The OP-ELM methodology, presented in Section 2, is compared in Section 3 using several experiments and two well-known methods, the Least-Squares Support Vector Machine (LS-SVM) and the Multilayer Perceptron (MLP). Finally, a toolbox for performing the OP-ELM is introduced in Appendix.

V. Kůrková et al. (Eds.): ICANN 2008, Part I, LNCS 5163, pp. 145–154, 2008.

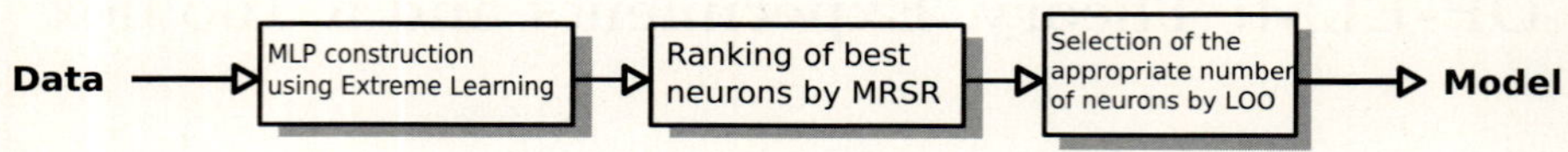

Fig. 1. The three steps of the OP-ELM algorithm

2   OP-ELM

The OP-ELM is made of three main steps summarized in Figure 1.

2.1   Extreme Learning Machine (ELM)

The first step of the OP-ELM algorithm is the core of the original ELM: the building of a single-layer feed-forward neural network. The idea of the ELM has been proposed by Guang-Bin Huang *et al.* in [1], even the idea of such network was already proposed in [2].

In the context of a single hidden layer perceptron network, let us denote the weights between the hidden layer and the output by $\mathbf{b}$. Activation functions proposed in the OP-ELM Toolbox differ from the original ELM choice since linear, sigmoid and gaussian functions are proposed here. For the output layer, a linear function is used.

The main novelty introduced by the ELM is in the determination of the kernels, initialized randomly. While the original ELM used only sigmoid kernels, gaussian, sigmoid and linear are proposed in OP-ELM: gaussian ones have their centers taken randomly from the data points and a width randomly drawn between percertile 20 percent and percentile 80 percent of the distance distribution of the input space; sigmoids weights are drawn at random from a uniform distribution in the interval $[-5, 5]$. A theorem proposed in [1] states that with the additional hypothesis of infinite differentiability of the activation functions, output weights $\mathbf{b}$ can be computed from the hidden layer output matrix $\mathbf{H}$: the columns $\mathbf{h_i}$ of $\mathbf{H}$ are computed by $\mathbf{h_i} = \mathtt{Ker}(\mathbf{x_i}^T)$, where $\mathtt{Ker}$ stands for either linear, sigmoid or gaussian activation functions (including multiplication by first layer weights). Finally, the output weights $\mathbf{b}$ are computed by $\mathbf{b} = \mathbf{H}^{\dagger}\mathbf{y}$, where $\mathbf{H}^{\dagger}$ stands for the Moore-Penrose inverse [3] and $\mathbf{y} = (y_1, \ldots, y_M)^T$ is the output.

The only remaining parameter in this process is the number of neurons N of the hidden layer. From a practical point of view, it is advised to set the number of neurons clearly above the number of the variables in the dataset, since the next step aims at pruning the useless neurons from the hidden layer.

2.2   Multiresponse Sparse Regression (MRSR)

For the removal of the useless neurons of the hidden layer, the Multiresponse Sparse Regression proposed by Timo Similä and Jarkko Tikka in [4] is used. It is mainly an extension of the Least Angle Regression (LARS) algorithm [5] and hence is actually a variable ranking technique, rather than a selection one. The main idea of this algorithm is the following: denote by $\mathbf{T} = [\mathbf{t}_1 \ldots \mathbf{t}_p]$ the $n \times p$ matrix of targets, and by $\mathbf{X} = [\mathbf{x}_1 \ldots \mathbf{x}_m]$ the $n \times m$ regressors matrix. MRSR adds each regressor one by one to the model $\mathbf{Y}^k = \mathbf{X}\mathbf{W}^k$, where $\mathbf{Y}^k = [\mathbf{y}_1^k \ldots \mathbf{y}_p^k]$ is the target approximation by the model. The $\mathbf{W}^k$ weight matrix has k nonzero rows at kth step of the MRSR. With each new step a new nonzero row, and a new regressor to the total model, is introduced.

An important detail shared by the MRSR and the LARS is that the ranking obtained is exact in the case, where the problem is linear. In fact, this is the case, since the neural network built in the previous step is linear between the hidden layer and the output. Therefore, the MRSR provides the exact ranking of the neurons for our problem.

Details on the definition of a cumulative correlation between the considered regressor and the current model's residuals and on the determination of the next regressor to be added to the model can be found in the original paper about the MRSR [4].

MRSR is hence used to rank the kernels of the model: the target is the actual output y_i while the "variables" considered by MRSR are the outputs of the kernels h_i.

2.3   Leave-One-Out (LOO)

Since the MRSR only provides a ranking of the kernels, the decision over the actual best number of neurons for the model is taken using a Leave-One-Out method. One problem with the LOO error is that it can get very time consuming if the dataset tends to have a high number of samples. Fortunately, the PRESS (or PREdiction Sum of Squares) statistics provide a direct and exact formula for the calculation of the LOO error for linear models. See [6,7] for details on this formula and implementations:

$$\epsilon^{\mathrm{PRESS}} = \frac{y_i - \mathbf{h}_i\mathbf{b}}{1 - \mathbf{h}_i\mathbf{P}\mathbf{h}_i^T}, \tag{1}$$

where $\mathbf{P}$ is defined as $\mathbf{P} = (\mathbf{H}^T\mathbf{H})^{-1}$ and $\mathbf{H}$ the hidden layer output matrix defined in subsection 2.1.

The final decision over the appropriate number of neurons for the model can then be taken by evaluating the LOO error versus the number of neurons used (properly ranked by MRSR already).

In the end, a single-layer neural network possibly using a mix of linear, sigmoid and gaussian kernels is obtained, with a highly reduced number of neurons, all within a small computational time (see section 3 for comparisons of performances and computational times between MLP, LSSVM and OP-ELM).

2.4   Discussion on the Advantages of the OP-ELM

In order to have a very fast and still accurate algorithm, each of the three presented steps have a special importance in the whole OP-ELM methodology.

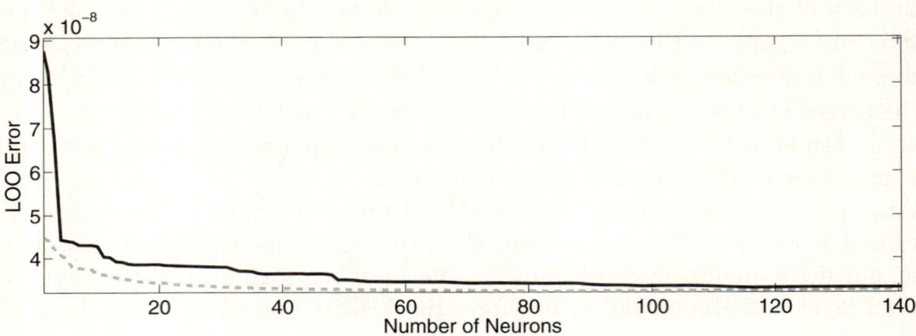

Fig. 2. Comparison of LOO error with and without the MRSR ranking. The solid blue line represents the LOO error without and the dashed orange one with the MRSR.

Indeed, the ELM is very fast, as can be seen in the original ELM paper [1] and in the experiments in Section 3. The ELM also has the advantage of being rather simple, for the process of training and initializing the neural network weights.

The variable ranking by the MRSR is also one of the fastest ranking methods providing the exact best ranking of the variables, since the problem is linear, when creating the neural network using ELM.

The linearity also enables the model structure selection step using the Leave-One-Out, which is usually very time-consuming. Thanks to the PRESS statistics formula for the LOO error calculation, the structure selection can be done in a reasonable time.

The final model structure selection for the OP-ELM model using the Ailerons dataset (see Section 3) is shown in Figure 2.

It can be seen from Figure 2 that the OP-ELM benefits greatly from the MRSR ranking step of its methodology. The convergence is faster and the LOO error gets smaller with respect to the number of neurons when the MRSR is used than when it is not.

3   Experiments

This section demonstrates the speed and accuracy of the OP-ELM method using several different regression and classification datasets. For the comparison, Section 3.2 provides also the performances using a well-known MultiLayer Perceptron (MLP) [8] and Least-Squares Support Vector Machine (LSSVM) [9] implementations. Following subsection shows a toy example to illustrate the performance of OP-ELM on a simple case that can be plotted.

3.1   Toy Example: Sum of Two Sines

A set of 1000 training points are generated following a sum of two sines. This gives a one-dimensional example, where no feature selection has to be performed. Figure 3 plots the obtained model on top of the training data.

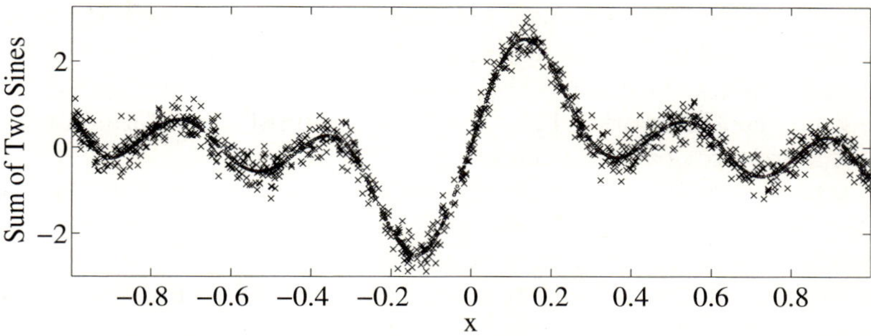

Fig. 3. Plot of a one dimensional sum of sines as black crosses and the model obtained by OP-ELM as blue circles

The model approximates the data very nicely.

3.2   Real Measurement Datasets

For the comparison of the three methods, we selected several different datasets: eight regression and four classification problems. For each dataset, all three methods are applied and the performances compared.

Each dataset is divided into two sets, train and test sets. The trainset includes two thirds of the data, selected randomly without replacement, and the testset one third. Table 1 shows some key information about the datasets and the selected hyperparameters for the LS-SVM and the MLP methods.

Table 1. Key information about the selected datasets and the selected hyperparameters for the LS-SVM and the MLP. For the classification, the variables column also includes the number of classes in the dataset.

| | | Samples | | LS-SVM | | MLP |
Regression	Variables	Train	Test	Gamma	Sigma	Neurons
Abalone	8	2784	1393	3.23	7.73	2
Ailerons	5	4752	2377	3.78	5.06	9
Elevators	6	6344	3173	2.2	7.27	10
Auto Price	15	106	53	621.84	5.43	7
Servo	4	111	56	211.94	4.19	7
Breast Cancer	32	129	65	18.99	4.94	2
Bank	8	2999	1500	1099	4.92	6
Stocks	9	633	317	67.39	2.3	12
Classification						
Iris	4 / 3	100	50	151.53	1.24	4
Wisconsin Breast Cancer	30 / 1	379	190	1169	6.07	1
Pima Indians Diabetes	8 / 1	512	256	57.88	2.38	1
Wine	13 / 3	118	60	14.54	7.21	7

The hyperparameters for the LS-SVM and the MLP are selected using a 10-fold Cross-Validation. The LS-SVM is performed using the LS-SVM toolbox [9] with the default settings for the hyperparameters and the grid search. The MLP is performed using a Neural Network toolbox, which is a part of the Matlab software from the Mathworks. The training of the MLP is performed using the Levenberg-Marquardt backpropagation.

In order to decrease the possibility of local minima with the MLP, the training is repeated 10 times for each fold and the best network according to the training error is selected for validation. For example, in order to validate the MLP network using 12 hidden neurons, we have to train a total of 100 MLP networks with 12 hidden neurons to evaluate the validation error. This procedure is done for each number of hidden nodes from 1 to 20 and the selected number according to the validation MSE is selected.

Table 2 shows the results of the validation for each method. Also included is the respective calculation time consumed when using similar computer systems in calculations.

Table 3 shows the results of the test for each method.

Table 2. Validation errors for each method and the calculation times in seconds

	Validation Error			Calculation Time		
Regression	LS-SVM	MLP	OP-ELM	LS-SVM	MLP	OP-ELM
Abalone	4,42	**4,16**	4,35	9,36E+04	2640	25
Ailerons	2,80E-08	**2,76E-08**	2,77E-08	1,40E+05	6360	500
Elevators	2,92E-06	**1,99E-06**	2,00E-06	1,04E+06	4080	1250
Auto Price	2,13E+07	8,46E+06	**5,21E+06**	660	720	0,015
Servo	0,598	**0,361**	0,506	480	480	0,34
Breast Cancer	1301	1514	**1164**	900	1500	0,22
Bank	2,45E-03	**8,93E-04**	1,00E-03	1,21E+06	3360	54
Stocks	**0,485**	0,878	0,819	720	1320	3,8
Classification						
Iris	0,923	**0,980**	0,950	300	540	0,13
Wisconsin Breast Cancer	0,953	**0,966**	0,958	960	2340	2,82
Pima Indians Diabetes	0,744	**0,777**	0,775	600	600	172
Wine	0,972	**0,983**	**0,983**	420	900	0,41

From Tables 3 and 3 we can see that in general, the OP-ELM is on the same performance level than the other methods. On some datasets, the method performs worse and on some, better than the LS-SVM or the MLP.

On all the datasets, however, the OP-ELM method is clearly the fastest, with several orders of magnitude. For example, in the Abalone dataset using the OP-ELM is more than 3700 times faster than the LS-SVM and roughly 100 times faster than the MLP.

Finally, it should be noted that considering the very long computational time, which relates to the complexity of the problem in the LS-SVM case, some of the bad results obtained when using LS-SVM can be due to the fact that the

Table 3. Test errors for each method

Regression	LS-SVM	MLP	OP-ELM
Abalone	4,45	**4,34**	4,58
Ailerons	2,82E-08	**2,64E-08**	2,69E-08
Elevators	2,87E-06	**2,11E-06**	**2,11E-06**
Auto Price	2,25E+07	**1,02E+07**	5,91E+06
Servo	0,644	**0,565**	0,589
Breast Cancer	907	1033	**670**
Bank	2,47E-03	**9,05E-04**	1,00E-03
Stocks	**0,419**	0,764	0,861
Classification			
Iris	0,870	0,940	**0,980**
Wisconsin Breast Cancer	0,927	0,947	**0,968**
Pima Indians Diabetes	0,687	**0,769**	0,754
Wine	0,950	**0,967**	0,950

algorithm might not have converged properly. This might explain the results on
Bank and Elevators datasets.

4   Conclusions

In this paper we have demonstrated the speed and accuracy of the OP-ELM
methodology. Comparing to two well-known methodologies, the LS-SVM and
the MLP, the OP-ELM achieves roughly the same level of accuracy with several
orders of magnitude less calculation time.

Our goal is not to prove that the OP-ELM provides the best results in terms
of the MSE, but instead to show that it provides very accurate results very fast.
This makes it a valuable tool for applications, which need a small response time
and a good accuracy. Indeed, the ratio between the accuracy and the calculation
time is very good.

For further work, the comparisons with other methodologies are performed
in order to verify the applicability and accuracy of the OP-ELM with different
datasets.

References

1. Huang, G.B., Zhu, Q.Y., Siew, C.K.: Extreme learning machine: Theory and applications. Neurocomputing 70(1–3), 489–501 (2006)
2. Miller, W.T., Glanz, F.H., Kraft, L.G.: Cmac: An associative neural network alternative to backpropagation. Proceedings of the IEEE 70, 1561–1567 (1990)
3. Rao, C.R., Mitra, S.K.: Generalized Inverse of Matrices and Its Applications. John Wiley & Sons, Chichester (1972)
4. Similä, T., Tikka, J.: Multiresponse sparse regression with application to multidimensional scaling. In: Duch, W., Kacprzyk, J., Oja, E., Zadrożny, S. (eds.) ICANN 2005. LNCS, vol. 3697, pp. 97–102. Springer, Heidelberg (2005)

5. Efron, B., Hastie, T., Johnstone, I., Tibshirani, R.: Least angle regression. Annals of Statistics 32, 407–499 (2004)
6. Myers, R.: Classical and Modern Regression with Applications, 2nd edn. Duxbury, Pacific Grove (1990)
7. Bontempi, G., Birattari, M., Bersini, H.: Recursive lazy learning for modeling and control. In: European Conference on Machine Learning, pp. 292–303 (1998)
8. Haykin, S.: Neural Networks: A Comprehensive Foundation, 2nd edn. Prentice Hall, Englewood Cliffs (1998)
9. Suykens, J., Gestel, T.V., Brabanter, J.D., Moor, B.D., Vanderwalle, J.: Least-Squares Support-Vector Machines. World Scientific, Singapore (2002)
10. Lendasse, A., Sorjamaa, A., Miche, Y.: OP-ELM Toolbox, `http://www.cis.hut.fi/projects/tsp/index.php?page=research&subpage=downloads`
11. Whitney, A.W.: A direct method of nonparametric measurement selection. IEEE Transactions on Computers C-20, 1100–1103 (1971)

Appendix: Toolbox

A small overview of the OP-ELM Toolbox [10] is given in this appendix. Users can also refer to the toolbox documentation provided within the OP-ELM Toolbox package at the address: `http://www.cis.hut.fi/projects/tsp/index.php?page=research&subpage=downloads`. Main functions with their use and arguments are first listed, followed by a small example.

`gui_OPELM`

This command invokes the Graphical User Interface (GUI) for the OP-ELM Toolbox commands (quickly reviewed in the following). It can be used for typical problems. Figure 4 shows a snapshot of the GUI.

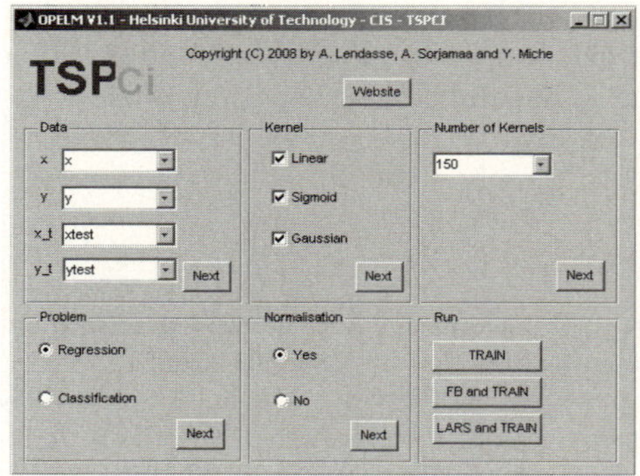

Fig. 4. Snapshot of the OP-ELM toolbox GUI

train_OPELM

This function trains a model using OP-ELM and the proper inputs.

```
function [model]=train_OPELM(data,[kernel],[maxneur],[problem],
                             [normal],[KM])
Inputs:    data        the data (can be multi-output)
           [kernel]    (optional) is the type of kernels to use.
                       Either 'l' (linear), 's' (sigmoid), 'g'
                       (gaussian), 'ls' (linear+sigmoid), 'lg'
                       (linear+gaussian) or 'lsg' (lin+sig+gauss).
           [maxneur]   (optional) maximum number of neurons model.
           [problem]   (optional) Either 'r' (regression) or 'c'
                       (classification).
           [normal]    (optional) Normalize data or not.
           [KM]        (optional) specifies a previously computed
                       Kernel Matrix to be used as initialization
                       of the model.
Output:    [model]     the obtained model.
```

sim_OPELM

This function uses the model computed by **train_OPELM** on a test dataset.

```
function [yh,error]=sim_OPELM(model,datatest) Inputs:    model
is the model previously obtained by the
                     train_OPELM function.
                     datatest  is the test data.
Outputs:   yh          the estimated output by the model.
           error       the mean square error (for regression problem)
                       or classification error with confusion matrix
                       (for classification problem).
                   (if real output is known).
```

LARS_Selection_OPELM

This function uses the MRSR algorithm [4] with OP-ELM algorithm.

```
function myinputs=LARS_Selection_OPELM(data,[kernel],[maxneur],
                              [problem],[normal])
Inputs:   same as train_OPELM Output:   myinputs    a 1xd matrix of
'1' (for selected variables)
                 and '0' (for the unselected).
```

FB_OPELM

This function uses the Forward-Backward [11] algorithm with OP-ELM algorithm.

```
function myinputs=FB_OPELM(data,[input_init],[kernel],[maxneur],
                                      [problem],[normal])
Inputs: same as train_OPELM
        [input_init] (optional) is an initialization of the input
                     selection to be used for the Forward-Backward
                     algorithm. Specified as a 1xd matrix of '0'
                     (if the considered variable is not to
                     be taken) and '1' (if it is to be taken).
Ouput:  myinputs     a 1xd matrix of '1' (for selected variables)
                     and '0' (for the unselected).
```

Example of use

```
>> data.x = (-1:0.001:1)';
>> data.y = 2*sin(8*data.x)+5*sin(3*data.x)+0.5*randn(2001,1);
>> datatest.x = (-1:0.0001:1)';
>> datatest.y = 2*sin(8*datatest.x)+5*sin(3*datatest.x)+0.5*randn(20001,1);
```

Create some data...

Train a model on the data with default parameters...

```
>> [model] = train_OPELM(data);
Warning: normalization unspecified...
-------> Switching to forced normalization.
Warning: problem unspecified...
-------> Switching to regression problem.
Warning: kernel type unspecified...
-------> Switching to lsg kernel.
Warning: maximum number of neurons unspecified...
-------> Switching to 100 maximum neurons.
```

Display some info about the built model...

```
>> show_model(model)
Model for regression, build on 2001x1 data, one dimensional
output. Uses 16 neurons; LOO error is 2.522819e-01.
```

Use the model on test data...

```
>> [yth,error] = sim_OPELM(model,datatest);
>> error
error = 0.2522
```

Robust Nonparametric Probability Density Estimation by Soft Clustering

Ezequiel López-Rubio, Juan Miguel Ortiz-de-Lazcano-Lobato,
Domingo López-Rodríguez, and María del Carmen Vargas-Gonzalez

School of Computing
University of Málaga
Campus de Teatinos, s/n. 29071 Mlaga
Spain
{ezeqlr,jmortiz}@lcc.uma.es, dlopez@ctima.uma.es

Abstract. A method to estimate the probability density function of multivariate distributions is presented. The classical Parzen window approach builds a spherical Gaussian density around every input sample. This choice of the kernel density yields poor robustness for real input datasets. We use multivariate Student-t distributions in order to improve the adaptation capability of the model. Our method has a first stage where hard neighbourhoods are determined for every sample. Then soft clusters are considered to merge the information coming from several hard neighbourhoods. Hence, a specific mixture component is learned for each soft cluster. This leads to outperform other proposals where the local kernel is not as robust and/or there are no smoothing strategies, like the manifold Parzen windows.

1   Introduction

The estimation of the unknown probability density function(PDF) of a continuous distribution from a set of input data forming a representative sample drawn from the underlying density is a problem of fundamental importance to all aspects of machine learning and pattern recognition (see [2],[11] and [14]).

Parametric methods make a priori assumptions about the unknown distribution. They consider a particular functional form for the PDF and reduce the problem to the estimation of the required functional parameters. On the other hand, nonparametric approaches make less rigid assumptions. Popular nonparametric methods include the histogram, kernel estimation, nearest neighbour methods and restricted maximum likelihood methods, as can be found in [4], [6] and [3].

The kernel density estimator, also commonly referred to as the Parzen window estimator, [9], places a local Gaussian kernel on each data point of the training set. Then, the PDF is approximated by summing all the kernels, which are multiplied by a normalizing factor. Thus, this model can be viewed as a finite mixture model (see [7]) where the number of mixture components equals the number of points in the data sample. Parzen windows estimates are usually built using a

V. Kůrková et al. (Eds.): ICANN 2008, Part I, LNCS 5163, pp. 155–164, 2008.

'spherical Gaussian' with a single scalar variance parameter, which spreads the density mass equally along all input space directions and gives too much probability to irrelevant regions of space and too little along the principal directions of variance of the distribution. This drawback is partially solved in Manifold Parzen Windows algorithm [15], where a different covariance matrix is calculated for each component. The covariance matrix is estimated by considering a hard neighbourhood of each input sample. We propose in Section 2 to build soft clusters to share the information among neighbourhoods. This leads to filter the input noise by smoothing the estimated parameters. Furthermore, we use multivariate Student-t distributions, which have heavier tails than the Gaussians, in order to achieve robustness in the presence of outliers ([10], [12], [16]).

We present in section 3 the mixture of multivariate Student-t distributions which is learnt from the soft clusters. The asymptotical convergence of the proposed method is formally proven in Section 4. We show some experimental results in section 5, where our method produces more precise density estimations than the Manifold Parzen Windows and other approaches. Finally, Section 6 is devoted to conclusions.

2   The Smooth Parzen Windows Method

Let x be a D-dimensional real-valued random variable and $p()$ an arbitrary probability density function over x which is unknown and we want to estimate. The training set of the algorithm is formed by N observations of the random variable. For each training sample x_i we build a hard Q-neighbourhood H_i with the Q nearest neighbours of x_i, including itself. Hence H_i is interpreted as a random event which happens iff the input belongs to that neighbourhood. The knowledge about the local structure of the distribution around x_i is obtained when we calculate the mean vector $\boldsymbol{\mu}$ and the correlation matrix $\boldsymbol{R}$:

$$\boldsymbol{\mu}(H_i) = E[\boldsymbol{x}|H_i] = \frac{1}{Q} \sum_{x_j \in H_i} x_j \tag{1}$$

$$\boldsymbol{R}(H_i) = E[\boldsymbol{x}\boldsymbol{x}^T|H_i] = \frac{1}{Q} \sum_{x_j \in H_i} x_j x_j^T \tag{2}$$

Now we present a smoothing procedure to merge the information from different hard neighbourhoods. A soft cluster i is defined by a random event named S_i, which verifies when the input belongs to cluster i. Each hard neighbourhood H_j contributes to S_i with a normalized weight w_{ij}:

$$w_{ij} = P[H_j|S_i] \tag{3}$$

So, we have

$$\forall i \in \{1, 2, \ldots, M\}, \sum_{j=1}^{N} w_{ij} = 1 \tag{4}$$

where the number of soft clusters M may be different from the number of hard neighbourhoods N. We can infer the structure of the soft cluster by merging the information from the hard neighbourhoods:

$$\boldsymbol{\mu}(S_i) = E[\boldsymbol{x}|S_i] = \sum_j P[H_j|S_i]E[\boldsymbol{x}|H_j] = \sum_{j=1}^{N} w_{ij}\boldsymbol{\mu}(H_j) \tag{5}$$

$$\boldsymbol{R}(S_i) = E[\boldsymbol{x}\boldsymbol{x}^T|S_i] = \sum_j P[H_j|S_i]E[\boldsymbol{x}\boldsymbol{x}^T|H_j] = \sum_{j=1}^{N} w_{ij}\boldsymbol{R}(H_j) \tag{6}$$

In order to define a multivariate Student-t distribution we need the estimation of the covariance matrix C for each soft cluster:

$$\boldsymbol{C}(S_i) = E[(\boldsymbol{x} - \boldsymbol{\mu}(S_i))\,(\boldsymbol{x} - \boldsymbol{\mu}(S_i))^T\,|S_i] = \boldsymbol{R}(S_i) - \boldsymbol{\mu}(S_i)\boldsymbol{\mu}(S_i)^T \tag{7}$$

Finally, we need a method to determine the merging weights w_{ij}. We propose two approaches:

a) If $M = N$, we can perform the smoothing by replacing the 'hard' model at the data sample $\boldsymbol{x}_i$ by a weighted average of its neighbours ranked by their distance to $\boldsymbol{x}_i$. Here the model at $\boldsymbol{x}_i$ has the maximum weight, and their neighbours $\boldsymbol{x}_j$ have a weight which is a decreasing function of the distance from $\boldsymbol{x}_i$ to $\boldsymbol{x}_j$:

$$\boldsymbol{\omega}_{ij} = \exp\left(-\frac{\|\boldsymbol{x}_i - \boldsymbol{x}_j\|^2}{\psi^2}\right) \tag{8}$$

$$w_{ij} = \frac{\boldsymbol{\omega}_{ij}}{\sum_{k=1}^{N} \boldsymbol{\omega}_{ik}} \tag{9}$$

where ψ is a parameter to control the width of the smoothing. Please note that $\boldsymbol{\omega}_{ii} = 1$.

b) We may use the fuzzy c-means algorithm [1] to perform a soft clustering. This algorithm partitions the set of training data into M clusters so it minimizes the distance within the cluster. The objective function is:

$$J = \sum_{i=1}^{M}\sum_{j=1}^{N} m_{ij}^{\phi}d_{ij}^{2} \tag{10}$$

where ϕ is the fuzzy exponent which determines the degree of fuzzyness, and d_{ij} is the distance between training sample $\boldsymbol{x}_j$ and the centroid of cluster i. The degrees of membership of training sample j to soft cluster i are obtained as m_{ij}, which can be regarded as the probability of training sample j belonging to cluster i. In this approach the weights w_{ij} of the local models that we merge to yield the model of cluster i are computed as follows:

$$w_{ij} = \frac{m_{ij}}{\sum_{i=1}^{N} m_{ik}} \tag{11}$$

3 Robust Density Model

Once we have the estimations of the mean vectors $\boldsymbol{\mu}(S_i)$ and covariance matrices $\boldsymbol{C}(S_i)$ for each soft cluster S_i, it is needed to obtain a multivariate Student-t distribution from them. First we define our probability model, which is a mixture of multivariate Student-t distributions. Then we discuss how to make it learn from the data.

3.1 Mixture Model

The proposed algorithm is designed to estimate an unknown density distribution $p()$ from which the N samples of the training dataset are generated. The generated estimator will be formed by a mixture of M multivariate Student-t distributions, one for each soft cluster:

$$\hat{p}(\boldsymbol{x}) = \frac{1}{M} \sum_{i=1}^{M} K_i(\boldsymbol{x}) \tag{12}$$

$$K_i(\boldsymbol{x}) = \frac{\Gamma(\frac{\gamma_i+D}{2})|\boldsymbol{\Sigma}_i|^{-1/2}}{\left(\Gamma(\frac{1}{2})\right)^D \Gamma(\frac{\gamma_i}{2})\gamma_i^{D/2}} \left(1 + \frac{(\boldsymbol{x} - \boldsymbol{\mu}_i)^T \boldsymbol{\Sigma}_i^{-1} (\boldsymbol{x} - \boldsymbol{\mu}_i)}{\gamma_i} \right)^{\frac{\gamma_i+D}{2}} \tag{13}$$

where Γ is the gamma function, D is the dimension of the training samples and γ_i is the degrees of freedom parameter of the i-th Student-t kernel. We require $\gamma_i > 2$ in order that both the mean and the covariance matrix exist.

3.2 Model Learning

As known [16], when $\gamma_i > 1$ the mean of the i-th Student-t distribution exists and is $\boldsymbol{\mu}_i$. We estimate $\boldsymbol{\mu}_i$ by the mean of the i-th soft cluster. On the other hand, if $\gamma_i > 2$ then the covariance matrix exists and is given by $\gamma_i(\gamma_i - 2)^{-2}\boldsymbol{\Sigma}_i$. We estimate $\boldsymbol{\Sigma}_i$ with the help of the covariance matrix of the i-th soft cluster. Hence we have:

$$\boldsymbol{\mu}_i = \boldsymbol{\mu}(S_i) \tag{14}$$

$$\boldsymbol{\Sigma}_i = \frac{(\gamma_i - 2)^2}{\gamma_i} \boldsymbol{C}(S_i) \tag{15}$$

Finally we need to estimate the degrees of freedom parameter γ_i. Since there is no closed formula to obtain its value (see [10]), we follow a maximum likelihood approach here. We choose the value of γ_i which maximizes the log-likelihood of the training set:

$$L = \sum_{j=1}^{N} log\ \hat{p}(\boldsymbol{x}_j) = \sum_{j=1}^{N} log \left(\frac{1}{M} \sum_{i=1}^{M} K_i(\boldsymbol{x}) \right) \tag{16}$$

We have only M unknowns to optimize, namely $\gamma_i, i = 1, \ldots, M$. So, the maximization of L with respect to the γ_i's is done by a standard method such as Levenberg-Marquardt. The constraint $\gamma_i > 2$ is enforced for all i (see for example [5]).

3.3   Summary

The training algorithm can be summarized as follows:

1. For each training sample, compute the mean vector $\boldsymbol{\mu}(H_i)$ and correlation matrix $\boldsymbol{R}(H_i)$ of its hard neighbourhood H_i with equations (1) and (2).
2. Estimate the merging weights w_{ij} either by the distance method (9) or the fuzzy c-means algorithm (11).
3. Compute the mean vectors $\boldsymbol{\mu}(S_i)$ and covariance matrices $\boldsymbol{C}(S_i)$ of each soft cluster S_i following (5) and (7).
4. Obtain the optimal values of the degrees-of-freedom parameters γ_i by maximizing the log-likelihood (16). Note that the parameters $\boldsymbol{\mu}_i$ and $\boldsymbol{\Sigma}_i$ are not subject to optimization, because they are computed by equations (14) and (15), respectively.

4   Convergence Proof

In this section we prove that our estimator $\hat{p}()$ converges to the true density function $p()$ in the limit $N \to \infty$ and $M \to \infty$.

Lemma 1. *Every local Student-t kernel $K_i(\boldsymbol{x})$ tends to the D-dimensional Dirac delta function $\delta(\boldsymbol{x} - \boldsymbol{\mu}(S_i))$ as $N \to \infty$ and $M \to \infty$.*

Proof. In the limit $N \to \infty$ and $M \to \infty$ the clusters S_i reduce their volume to zero. This means that $\gamma_i^p \to 0$ for all i and p, where γ_i^p is the p-th eigenvalue of the Mahalanobis distance matrix Σ_i. Hence the kernels $K_i(\boldsymbol{x})$ are confined to a shrinking volume centered at $\boldsymbol{\mu}(S_i)$, because the variances in each direction are γ_i^p , but they continue to integrate to 1. So, we have that $K_i(\boldsymbol{x}) \to \delta(\boldsymbol{x} - \boldsymbol{\mu}(S_i))$. It should be noted that if we had allowed $\gamma_i \to 0$, the tails of the kernels could be so heavy that this property would have not hold, but this is not the case because we require $\gamma_i > 2$.

Theorem 1. *The expected value of the proposed estimation tends to the true probability density function as $N \to \infty$ and $M \to \infty$.*

Proof. The expectation is w.r.t. the underlying distribution of the training samples, which is the true probability density function $p()$:

$$E[\hat{p}(\boldsymbol{x})] = \frac{1}{M} \sum_{i=1}^{M} E[K_i(\boldsymbol{x})] \tag{17}$$

Since $K_i(\boldsymbol{x})$ are independent and identically distributed random variables we get

$$E[\hat{p}(\boldsymbol{x})] = E[K_i(\boldsymbol{x})] = \int p(\boldsymbol{y}) K_{\boldsymbol{y}}(\boldsymbol{x}) d\boldsymbol{y} \tag{18}$$

where $K_{\boldsymbol{y}}()$ is a multivariate Student-t centered in $\boldsymbol{y}$. Then, by Lemma 1, if $N \to \infty$ and $M \to \infty$ then $K_{\boldsymbol{y}}()$ shrinks to a Dirac delta:

$$E[\hat{p}(\boldsymbol{x})] = \int p(\boldsymbol{y}) \delta(\boldsymbol{x} - \boldsymbol{y}) d\boldsymbol{y} \tag{19}$$

So, the expectation of the estimation converges to a convolution of the true density with the Dirac delta function. Then,

$$E[\hat{p}(\boldsymbol{x})] \to p(\boldsymbol{x}) \tag{20}$$

Theorem 2. *The variance of the proposed estimation tends to zero as* $N \to \infty$ *and* $M \to \infty$.

Proof. The variance is w.r.t. the underlying distribution of the training samples, which is the true probability density function $p()$:

$$var[\hat{p}(\boldsymbol{x})] = var\left[\frac{1}{M} \sum_{i=1}^{M} K_i(\boldsymbol{x})\right] \tag{21}$$

Since $K_i(\boldsymbol{x})$ are independent and identically distributed random variables we get

$$var[\hat{p}(\boldsymbol{x})] = var\left[\frac{1}{M} K_i(\boldsymbol{x})\right] \tag{22}$$

By the properties of variance and (18) we obtain

$$var[\hat{p}(\boldsymbol{x})] = \frac{1}{M}\left(E[(K_i(\boldsymbol{x}))^2] - E[K_i(\boldsymbol{x})]^2\right) = \frac{1}{M}\left(E[(K_i(\boldsymbol{x}))^2] - E[\hat{p}(\boldsymbol{x})]^2\right) \tag{23}$$

By definition of expectation

$$var[\hat{p}(\boldsymbol{x})] = \frac{1}{M}\left(\int p(\boldsymbol{y})\,(K_{\boldsymbol{y}}(\boldsymbol{x}))^2\,d\boldsymbol{y} - E[\hat{p}(\boldsymbol{x})]^2\right) \tag{24}$$

where again $K_{\boldsymbol{y}}()$ is a multivariate Student-t centered in $\boldsymbol{y}$. We can bound the integral of the above equation with the help of (18), and so we get

$$var[\hat{p}(\boldsymbol{x})] \leq \frac{\sup(N(\cdot))E[\hat{p}(\boldsymbol{x})]}{M} \to 0 \quad \text{as } N \to \infty \text{ and } M \to \infty \tag{25}$$

5   Experimental Results

This section shows some experiments we have designed in order to study the quality of the density estimation achieved by our method. We call it SmoothTDist when the distance weighting is used, and SmoothTFuzzy when we use fuzzy c-means. Vincent and Bengio's method is referred as MParzen, the original Parzen windows method (with isotropic Gaussian kernels) is called OParzen, and finally the Mixtures of Probabilistic PCA model of Tipping and Bishop [13] is called MPPCA. For this purpose the performance measure we have chosen is the average negative log likelihood

$$ANLL = -\frac{1}{T} \sum_{i=1}^{T} log\ \hat{p}(\boldsymbol{x}_i) \tag{26}$$

where $\hat{p}()$ is the estimator, and the test dataset is formed by T samples $\boldsymbol{x}_i$.

5.1   Experiment on 2D Artificial Data

We have considered two artificial 2D datasets. The first dataset consists of a training set of 100 points, a validation set of 100 points and a test set of 10000 points, which are generated from the following distribution of two dimensional (x, y) points:

$$x = 0.04t\sin(t) + \epsilon_x, \qquad y = 0.04t\cos(t) + \epsilon_y \tag{27}$$

where $t \sim U(3, 15)$, $\epsilon_x \sim N(0, 0.01)$, $\epsilon_y \sim N(0, 0.01)$, $U(a, b)$ is uniform in the interval (a, b) and $N(\mu, \sigma)$ is a normal density. The second dataset is a capital letter 'S'.

We have optimized separately all the parameters of the five competing models with disjoint training and validation sets. The performance of the optimized models has been computed by 10-fold cross-validation, and the results are shown in Table 1, with the best result marked in bold. It can be seen that our models outperform the other three in density distribution estimation (note that lower is better).

Figures 1 and 2 show density distribution plots corresponding to the five models. Darker areas represent zones with high density mass and lighter ones indicate the estimator has detected a low density area.

Table 1. Quantitative results on the artificial datasets (standard deviations in parentheses)

Method	ANLL on test set (espiral)	ANLL on test set (capital 'S')
SmoothTDist	**-1.0901 (1.1887)**	-1.9384 (1.2157)
SmoothTFuzzy	-1.0880 (1.3858)	**-2.1176 (1.0321)**
OParzen	1.0817 (1.3357)	-0.6929 (0.4003)
MParzen	-0.9505 (0.3301)	-1.0956 (0.0657)
MPPCA	0.2473 (0.0818)	-0.1751 (0.6204)

We can see in the plots that our models have less density holes (light areas) and less 'bumpiness'. This means that our model represents more accurately the true distribution, which has no holes and is completely smooth. We can see that the quantitative ANLL results agree with the plots, because the lowest values of ANLL match the best-looking plots. So, our model outperforms clearly the other three considered approaches.

5.2   Density Estimation Experiment

A density estimation experiment has been designed, where we have chosen nine datasets from the UCI Repository of Machine Learning Databases [8]. As before, we have optimized all the parameters of the five competing models with disjoint training and validation sets. The parameters for the density estimator of each dataset have been optimized separately. Table 2 shows the results of the 10-fold

Fig. 1. Density estimation for the 2D espiral dataset. From left to right and from top to bottom: SmoothDist, SmoothFuzzy, OParzen, MParzen and MPPCA.

Table 2. ANLL on test set (lower is better). The standard deviations are shown in parentheses).

Database	SmoothTDist	SmoothTFuzzy	OrigParzen	ManifParzen	MPPCA
BrCancerWis	**-48.17 (2.48)**	9.54 (1.68)	9.35 (1.51)	10.22 (1.50)	13.19 (0.99)
Glass	**-197.18 (72.35)**	-8.22 (4.90)	-1.75 (5.87)	-1.04 (5.74)	-3.98 (6.24)
Ionosphere	**-21.92 (9.73)**	-8.01 (8.58)	-15.68 (3.24)	-14.83 (3.21)	-1.21 (5.15)
Liver	**17.53 (3.56)**	21.41 (0.63)	20.27 (3.45)	21.09 (3.31)	22.25 (0.83)
Pima	**-304.37 (36.48)**	27.95 (0.31)	32.21 (3.91)	32.85 (3.56)	29.87 (0.89)
Segmentation	65.96 (3.08)	**-71.22 (10.95)**	51.18 (3.93)	51.97 (3.86)	18.77 (4.29)
TAE	**-2.57 (11.70)**	7.59 (1.99)	8.17 (0.55)	9.09 (0.54)	11.98 (0.12)
Wine	**-59.84 (1.58)**	62.78 (18.53)	29.16 (4.55)	31.14 (7.01)	18.98 (0.63)
Yeast	**-319.02 (22.86)**	-9.50 (0.34)	-17.85 (0.33)	-16.96 (0.31)	-11.99 (0.31)

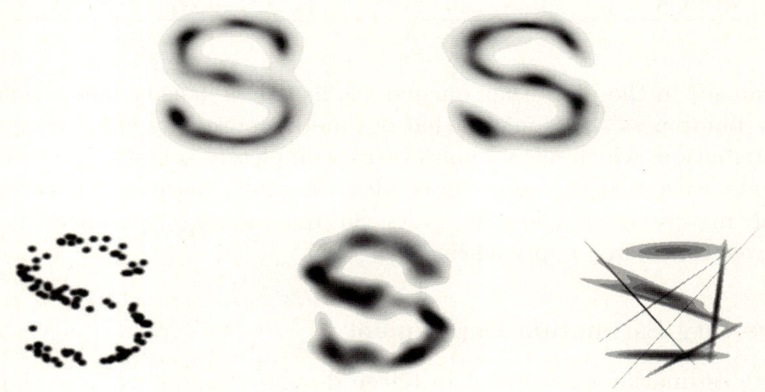

Fig. 2. Density estimation for the 2D capital 'S' dataset. From left to right and from top to bottom: SmoothDist, SmoothFuzzy, OParzen, MParzen and MPPCA.

cross-validation, with the winning models in bold. Our two proposals show a superior performance.

We have used the T-test to check the statistical significance of the difference between the two best performing models for each database. We have considered that the difference is statistically significant if we have less than 0.05 probability that the difference between the means is caused by chance. It has been found that the difference is statistically significant for all the considered databases.

6   Conclusions

We have presented a probability density estimation model. It is based in the Parzen window approach. Our proposal builds local models for a hard neighbourhood of each training sample. Then soft clusters are obtained by merging these local models, and local multivariate Student-t kernels are introduced. This allows our method to represent input distributions more faithfully than three well-known density estimation models. Computational results show the superior performance of our method.

Acknowledgements

The authors acknowledge support from CICYT (Spain) through grant TIN2005-02984 (including FEDER funds). This work is also partially supported by Junta de Andalucía (Spain) through grant TIC-01615.

References

1. Bezdek, J.C.: Numerical taxonomy with fuzzysets. J. Math. Biol. 1, 57–71 (1974)
2. Bishop, C.: Neural Networks for Pattern Recognition. Oxford University Press, Oxford (1995)
3. Hjort, N.L., Jones, M.C.: Locally Parametric Nonparametric Density Estimation. Annals of Statistics 24(4), 1619–1647 (1996)
4. Izenman, A.J.: Recent developments in nonparametric density estimation. Journal of the American Statistical Association 86(413), 205–224 (1991)
5. Kanzow, C., Yamashita, N., Fukushima, M.: Levenberg-Marquardt methods for constrained nonlinear equations with strong local convergence properties. Journal of Computational and Applied Mathematics 172, 375–397 (2004)
6. Lejeune, M., Sarda, P.: Smooth estimators of distribution and density functions. Computational Statistics and Data Analysis 14, 457–471 (1992)
7. McLachlan, G., Peel, D.: Finite Mixture Models. Wiley, Chichester (2000)
8. Newman, D.J., Hettich, S., Blake, C.L., Merz, C.J.: UCI Repository of machine learning databases. Department of Information and Computer Science, University of California, Irvine (1998),
   `http://www.ics.uci.edu/~mlearn/MLRepository.html`
9. Parzen, E.: On the Estimation of a Probability Density Function and Mode. Annals of Mathematical Statistics 33, 1065–1076 (1962)

10. Shoham, S.: Robust clustering by deterministic agglomeration EM of mixtures of multivariate t-distributions. Pattern Recognition 35, 1127–1142 (2002)
11. Silverman, B.: Density Estimation for Statistics and Data Analysis. Chapman and Hall, New York (1986)
12. Svensén, M., Bishop, C.M.: Robust Bayesian mixture modeling. Neurocomputing 64, 235–252 (2005)
13. Tipping, M.E., Bishop, C.M.: Mixtures of Probabilistic Principal Components Analyzers. Neural Computation 11, 443–482 (1999)
14. Vapnik, V.N.: Statistical Learning Theory. John Wiley and Sons, New York (1998)
15. Vincent, P., Bengio, Y.: Manifold Parzen Windows. Advances in Neural Information Processing Systems 15, 825–832 (2003)
16. Wang, H., Zhang, Q., Luo, B., Wei, S.: Robust mixture modelling using multivariate t-distribution with missing information. Pattern Recognition Letters 25, 701–710 (2004)

Natural Conjugate Gradient on Complex Flag Manifolds for Complex Independent Subspace Analysis

Yasunori Nishimori[1], Shotaro Akaho[1], and Mark D. Plumbley[2]

[1] Neuroscience Research Institute, National Institute of Advanced Industrial Science
and Technology (AIST),
AIST Central2, 1-1-1, Umezono, Tsukuba, Ibaraki 305-8568, Japan
`y.nishimori@aist.go.jp, s.akaho@aist.go.jp`
[2] Department of Electronic Engineering, Queen Mary University of London,
Mile End Road, London E1 4NS, UK
`mark.plumbley@elec.qmul.ac.uk`

Abstract. We study the problem of complex-valued independent subspace analysis (ISA). We introduce complex flag manifolds to tackle this problem, and, based on Riemannian geometry, propose the natural conjugate gradient method on this class of manifolds. Numerical experiments demonstrate that the natural conjugate gradient method yields better convergence compared to the natural gradient geodesic search method.

1 Introduction

The main aim of this paper is to explain how a new class of complex manifolds, *complex flag manifolds* can be used to solve the complex independent subspace analysis (complex ISA) task. We investigate two geodesic search-type optimization methods on complex flag manifolds. One is the natural gradient geodesic search method (NGS), and the other is the natural conjugate gradient method (NCG). The former method was proposed on the orthogonal group to solve the non-negative ICA task by Plumbley [15], and the latter was proposed by Edelman et al. [5] for the case of real Stiefel and Grassmann manifolds. Although several authors have recently proposed optimization methods on complex manifolds such as the unitary group [1],[6], complex Stiefel and Grassmann manifolds [10], complex flag manifolds have not been explored so far in the signal processing literature. Also, most authors have concentrated on either the natural gradient or the Newton's method on complex manifolds, however, the behavior of the conjugate gradient method over complex manifolds has not been reported.

The organization of this paper is as follows. In section 2 the problem of complex-valued independent subspace analysis is formulated, and how complex flag manifolds naturally arise for solving this problem is explained; subsection 2.1 reviews the Riemannian geodesic line search method as well as the Riemannian conjugate gradient method, subsection 2.2 derives the update rules for both methods on complex flag manifolds. Section 3 illustrates the comparison

V. Kůrková et al. (Eds.): ICANN 2008, Part I, LNCS 5163, pp. 165–174, 2008.

of the performance of both methods as applied to a complex ISA problem on synthetic data sets.

2   Complex Independent Subspace Analysis

Complex ISA is a straightforward generalization of real ISA [8]. We observe a linear mixture of source signals $s \in \mathbf{C}^N$:

$$x = As, \text{ where } A \in \mathbf{C}^{N \times N} \text{ is nondegenerate.} \tag{1}$$

We assume that s is decomposed into d_j-tuples $(j = 1, \ldots, r + 1)$ as

$$s = (s_1, \ldots, s_{d_1}, s_{d_1+1}, \ldots, s_{d_1+d_2}, \ldots, s_n, s_{n+1}, \ldots, s_N)^\top, \sum_{j=1}^{r+1} d_j = N \tag{2}$$

such that signals belonging to different tuples are statistically independent, while signals belonging to the same tuple are allowed to be statistically dependent, and $\top$ denotes the (real) transpose operator. Further, the $(r + 1)$st tuple is assumed to be composed of Gaussian noises and $d_{r+1} = N - n$. Then we define the task of complex ISA as estimating n original sources $s_1, \ldots, s_n$, not including $(N-n)$ Gaussian noises $s_{n+1}, \ldots .s_N$ from this statistical assumption (up to ambiguity). Often we deal with the noiseless case $N = n$.

To solve the complex ISA problem, as with ordinary real ICA, we first center and pre-whiten the data:

$$z = Bx, \text{ such that } E_z\left[zz^H\right] = I_N, \tag{3}$$

where H denotes the Hermitian transpose operator. Then solving the complex ISA task reduces to finding a rectangular unitary matrix

$$W = [W_1, W_2, \ldots, W_r] \in \mathbf{C}^{N \times n}, \text{ where } W^H W = I_n, W_j \in \mathbf{C}^{N \times d_j} \tag{4}$$

such that the output vector

$$y = W^H z = W^H BAs \tag{5}$$

satisfies the statistical assumption of the original sources. Therefore the complex ISA task can be solved by optimization on the complex Stiefel manifold:

$$\mathrm{St}(N, n\,;\mathbf{C}) = \left\{ W \in \mathbf{C}^{N \times n} | W^H W = I_n \right\}. \tag{6}$$

Note that the complex Stiefel manifold $\mathrm{St}(N, n\,;\mathbf{C})$ includes the unitary group

$$U(N) = \left\{ W \in \mathbf{C}^{N \times N} | W^H W = I_N \right\} \tag{7}$$

as a special case when $N = n$.

For a cost function, we use

$$f(W) = \sum_{j=1}^{r} E_z \left[\log p(\sum_{l=1}^{d_j} |w_{jl}^H z|^2) \right], \tag{8}$$

where E_z denotes the empirical mean with respect to z, and p is a probability density function of the power of the projection of the observed signals to the subspace spanned by w_{jl}'s ($W_j = w_{j1}, \ldots, w_{jd_j}$), that generalizes the maximum likelihood-based parametric cost function for real ISA proposed by Hyvärinen and Hoyer [8]. However, this cost function is invariant to the unitary transformations of each W_j $(j = 1, \ldots, r)$:

$$W \mapsto W \operatorname{diag}[U_1, U_2, \ldots, U_r], \ U_j \in U(d_j),$$

so an additional structure is present in the target manifold over which a solution matrix is searched. The target manifold is obtained by identifying those points on $\mathrm{St}(N, n\,;\,\mathbf{C})$ related by the following equivalence relation $\sim$

$$W_1 \sim W_2 \iff \exists U_i \in U(d_i) \text{ s.t. } W_2 = W_1 \operatorname{diag}[U_1, U_2, \ldots, U_r]. \tag{9}$$

This quotient space is called the *complex flag manifold,* and we denote it as $\mathrm{Fl}(n, d_1, \ldots, d_r\,;\,\mathbf{C})$. $\mathrm{Fl}(n, d_1, \ldots, d_r\,;\,\mathbf{C})$ is isomorphic to $U(N)/U(d_1) \times \cdots \times U(d_r) \times U(N - n)$ as a homogeneous space, and includes the complex Grassmann manifold as a special case when $r = 1$. We will exploit this geometric structure for the optimization of the complex ISA cost function. For instance, the Riemannian Hessian of the ISA cost function on the tangent space of simpler complex Stiefel manifold is degenerate, therefore we cannot directly apply the Riemannian Newton's method on the Stiefel manifold to ISA.

2.1 Geodesic Search Method

The update rules of both geodesic search type methods for minimizing a real-valued smooth cost function f over a smooth manifold M go as follows. We denote the equation of the geodesic emanating from a point $W \in M$ in the direction of V as $\varphi_M(W, V, t)$, where t is a time parameter s.t.

$$\varphi_M(W, V, 0) = W, \quad \frac{d}{dt}\varphi_M(W, V, 0)|_{t=0} = V \tag{10}$$

hold. Although a geodesic is a generalization of a straight line to a manifold and is computed by a Riemannian metric g on the manifold, we use one metric for each manifold in this paper, and do not refer which metric to use with this notation. The formula [12]

$$\varphi_{\mathrm{St}(N,n)}(W, V, t) = \exp(t(DW^\top - WD^\top))W, \text{ where } D = (I - \frac{1}{2}WW^\top)V \tag{11}$$

is used in the subsequent part. We extend both methods such that they are applicable to complex flag manifolds. At each iteration, both methods search for a point where the cost function takes its minimum along the geodesic in the direction of each search direction as follows. The search direction for NGS is the natural gradient of the cost function, while the NCG search direction is determined by using the parallel transportation of the one-step former search direction to the current position as well as the natural gradient of the cost function.

– the natural gradient geodesic search method (NGS)

$$t^k_{\min} = \arg\min_t f(\varphi_M(W_k, G_k, t)) \tag{12}$$

$$G_k = -\operatorname{grad}^M_{W_k} f(W_k) \tag{13}$$

$$W_{k+1} = \varphi_M(W_k, G_k, t^k_{\min}) \tag{14}$$

– natural conjugate gradient method (NCG)

$$H_0 = -\operatorname{grad}^M_{W_0} f(W_0), \tag{15}$$

$$H_k = -\operatorname{grad}^M_{W_k} f(W_k) + \gamma_k \tau H_{k-1} \tag{16}$$

$$t^k_{\min} = \arg\min_t f(\varphi_M(W_k, H_k, t)) \tag{17}$$

$$W_{k+1} = \varphi_M(W_k, H_k, t^k_{\min}) \tag{18}$$

Here τH_{k-1} denotes the parallel transportation of H_k along $\varphi_M(W_{k-1}, H_{k-1}, t)$ to W_k, and[1]

$$\gamma_k = \frac{g_{W_k}(\operatorname{grad}^M_{W_k} f(W_k) - \operatorname{grad}^M_{W_{k-1}} f(W_{k-1}), \operatorname{grad}^M_{W_k} f(W_k))}{g_{W_{k-1}}(\operatorname{grad}^M_{W_{k-1}} f(W_{k-1}), \operatorname{grad}^M_{W_{k-1}} f(W_{k-1}))}. \tag{19}$$

Since $\varphi_{\mathrm{St}(N,n)}(W_k, H_k, t)$ is a geodesic, the parallel transportation vector τH_k is equal to the velocity vector

$$\frac{d}{dt}\varphi_{\mathrm{St}(N,n)}(W_k, H_k, t) = (D_k W_k^\top - W_k^\top D_k)\exp(t(D_k W_k^\top - W_k^\top D_k)) \tag{20}$$

at $t = t^k_{\min}$. Therefore

$$\tau H_k = (D_k W_k^\top - W_k^\top D_k)W_{k+1}, \tag{21}$$

where $D_k = (I - \frac{1}{2}W_k W_k^\top)H_k$.

2.2   Geometry of Complex Flag Manifolds

We first consider an optimization problem over a complex Stiefel manifold:

$$\mathrm{St}(N, n\,;\mathbb{C}) \ni W = W^\Re + iW^\Im \mapsto F(W) \in \mathbb{R}. \tag{22}$$

---

[1] There are several versions of selecting this γ_k for the conjugate gradient method. Here we use the Fletcher-Reeves rule. More precisely, $\tau\operatorname{grad}_{W_{k-1}} f(W_{k-1})$ should be used instead of $\operatorname{grad}_{W_{k-1}} f(W_{k-1})$ in the numerator, however, it is not known how to compute it cheaply, so we use this approximated version as is used in [5].

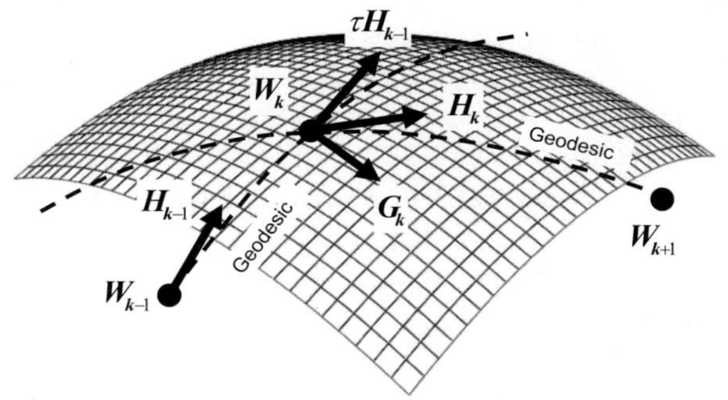

Fig. 1. Natural conjugate gradient update

Since the cost function F is real-valued, we regard $\mathrm{St}(N, n;\mathbb{C})$ as a real manifold; taking the real and imaginary parts of the defining equation

$$W^H W = (W^{\Re\top} - iW^{\Im\top})(W^{\Re} + iW^{\Im}) = I_n \tag{23}$$

separately, we get a real manifold M underlying $\mathrm{St}(N, n;\mathbb{C})$, which is a submanifold in $\mathbb{R}^{2N\times n}$ defined by the constraints:

$$M := \{W' = \begin{pmatrix} W^{\Re} \\ W^{\Im} \end{pmatrix} \in \mathbb{R}^{2N\times n}|$$
$$W^{\Re\top}W^{\Re} + W^{\Im\top}W^{\Im} = I_n, W^{\Re\top}W^{\Im} - W^{\Im\top}W^{\Re} = O_n\}. \tag{24}$$

We further embed M by the following map:

$$v : W' \mapsto \tilde{W} = \begin{pmatrix} W^{\Re} & -W^{\Im} \\ W^{\Im} & W^{\Re} \end{pmatrix}. \tag{25}$$

From the relation (24), $\tilde{W}^\top \tilde{W} = I_{2n}$ holds. In fact, this can be elaborated; $\tilde{M} = v(M)$ is a *totally geodesic*, or *autoparallel* submanifold of $\mathrm{St}(2N, 2n ; \mathbf{R})$ relative to the normal metric. That is, every geodesic on $\tilde{M}$ becomes a geodesic on $\mathrm{St}(2N, 2n ; \mathbf{R})$ as well, as follows:

$$\mathrm{St}(N, n ; \mathbf{C}) \to M \xrightarrow{v} \tilde{M} \to \mathrm{St}(2N, 2n ; \mathbf{R}) \tag{26}$$
$$W \qquad \mapsto W' \mapsto \tilde{W}$$

Thus, minimizing F over $\mathrm{St}(N, n ; \mathbb{C})$ can be solved through the minimization of $f(\tilde{W}) := F(W)$ on $\tilde{M}$. Though we start from this real manifold $\tilde{M}$ for the optimization, the update rule can be expressed by the original complex variables using the following nice properties the map $u : W \mapsto \tilde{W}$ possesses:

1. $u(I_m) = I_{2m}$
2. $u(W_1 W_2) = u(W_1)u(W_2)$, where $W_1 \in \mathbf{C}^{l\times m}, W_2 \in \mathbf{C}^{m\times n}$
3. $u(W_1 + W_2) = u(W_1) + u(W_2)$, where $W_1, W_2 \in \mathbf{C}^{m\times n}$

4. $u(W^H) = u(W)^\top$, where $W \in \mathbf{C}^{m \times m}$
5. $u(\exp(W)) = \exp(u(W))$, where $W \in \mathbf{C}^{m \times m}$

For instance, the geodesic

$$\varphi_{\mathrm{St}(2N,2n;\mathbf{R})}(\tilde{W}, \tilde{V}, t) = \exp(t(\tilde{D}\tilde{W}^\top - \tilde{W}\tilde{D}^\top))\tilde{W}, \quad \tilde{D} = (I - \frac{1}{2}\tilde{W}\tilde{W}^\top)\tilde{V}, \quad (27)$$

on $\tilde{M}$ is readily expressed in the complex form by switching the real transpose operator $\top$ to the Hermitian transpose operator H :

$$\varphi_{\mathrm{St}(N,2;\mathbf{C})}(W, V, t) = \exp(t(DW^H - WD^H))W, \tag{28}$$

where $D = (I - \frac{1}{2}WW^H)V, V = u^{-1}\tilde{V}$.

Next we derive another ingredient necessary for the optimization: the natural gradient. Since $\tilde{M}$ is a submanifold of $\mathrm{St}(2N, 2n\,;\mathbf{R})$, the natural gradient of f on $\tilde{M}$ is obtained by projecting the natural gradient of f on $\mathrm{St}(2N, 2n;\mathbf{R})$ at $\tilde{W}$ to the tangent space of $\tilde{M}$ at $\tilde{W}$. It is easily verified that the natural gradient of f on $\mathrm{St}(2N, 2n\,;\mathbf{R})$ at $\tilde{W}$ always lies in the tangent space of $\tilde{M}$ at $\tilde{W}$, hence we get

$$\mathrm{grad}^{\tilde{M}}_{\tilde{W}} f(\tilde{W}) = F(\tilde{W}) = \tilde{\nabla}F - \tilde{W}^\top\tilde{\nabla}F\tilde{W}^\top, \text{where}\, \tilde{\nabla}F = \begin{pmatrix} \frac{\partial F}{\partial W^{\Re}} & -\frac{\partial F}{\partial W^{\Im}} \\ \frac{\partial F}{\partial W^{\Im}} & \frac{\partial F}{\partial W^{\Re}} \end{pmatrix}. \quad (29)$$

$\tilde{\nabla}F$ comes from the Euclidean gradient of F on M;

$$\tilde{\nabla}F = v\left(\left(\frac{\partial F}{\partial W^{\Re}}, \frac{\partial F}{\partial W^{\Im}}\right)^\top\right), \tag{30}$$

and the corresponding complex gradient becomes[2]

$$2\frac{\partial F}{\partial \overline{W}} := \frac{\partial F}{\partial W^{\Re}} + i\frac{\partial F}{\partial W^{\Im}}. \tag{31}$$

This complex gradient has been used for the optimization on complex manifolds by many authors including [1], [2], [4], [6], [9], [10], [17]. Finally, the following formula for the complex form of the natural gradient is obtained by replacing

$$\tilde{\nabla}F \longmapsto 2\frac{\partial F}{\partial \overline{W}}, \qquad \text{and} \qquad \top \longmapsto H; \tag{32}$$

$$\mathrm{grad}^{\mathrm{St}(N,n;\mathbf{C})}_W F = 2\left(\frac{\partial F}{\partial \overline{W}} - W\frac{\partial F}{\partial \overline{W}}^H W\right). \tag{33}$$

---

[2] This notation using $\overline{W}$ (the complex conjugate of W) is due to the Wirtinger calculas [9].

In the same manner as for the complex Stiefel manifold, we can take the complex flag manifold $\mathrm{Fl}(n, d_1, \ldots, d_r\,;\mathbb{C})$ as a totally geodesic submanifold by using the following embedding

$$v : W = [W_1, W_2, \ldots, W_r] \mapsto \tilde{W}$$

$$= \begin{pmatrix} W_1^{\Re} & -W_1^{\Im} & W_2^{\Re} & -W_2^{\Im} & \cdots & W_r^{\Re} & -W_r^{\Im} \\ W_1^{\Im} & W_1^{\Re} & W_2^{\Im} & W_2^{\Re} & \cdots & W_r^{\Im} & W_r^{\Re} \end{pmatrix}, \quad (34)$$

where $W_k = W_k^{\Re} + i W_k^{\Im} \in \mathbf{C}^{N \times d_k}, \sum_{k=1}^{r} d_r = n$, of $\mathrm{Fl}(n, 2d_1, \ldots, 2d_r\,;\mathbb{R})$. Thus, the formula for the natural gradient and geodesics of real flag manifolds we obtained in [14] can also be complexified by replacing the real Euclidean gradient by the complex gradient and the transpose operator by the Hermitian transpose operator;

$$\mathrm{grad}_W^{\mathrm{Fl}(n,d_1,\ldots,d_r;\mathbb{C})} F = \left[V_1^{\mathbf{C}}, V_2^{\mathbf{C}}, \ldots, V_r^{\mathbf{C}} \right], \quad (35)$$

$$V_j^{\mathbf{C}} = X_j^{\mathbf{C}} - (W_j W_j^{H} X_j^{\mathbf{C}} + \sum_{l \neq j} W_l X_l^{\mathbf{C}^{H}} W_j), \quad (36)$$

where $X_j^{\mathbf{C}} = \frac{\partial F}{\partial W_j^{\Re}} + i \frac{\partial F}{\partial W_j^{\Im}}$.

$$\varphi_{\mathrm{Fl}(n,d_1,\ldots,d_r;\mathbb{C})}(W, V, t) = \exp(t(DW^{H} - WD^{H}))W, \quad (37)$$

where $D = (I - \frac{1}{2}WW^{H})V$.

3   Numerical Experiments

We applied the Riemannian conjugate gradient method to the complex-valued ISA problem using the following synthetic data sets and compared its performance with that of the natural gradient geodesic search method. Let $s = (s_1, s_2, \ldots, s_8)^{\top} = (\mathbf{s}_1, \mathbf{s}_2, \mathbf{s}_3, \mathbf{s}_4)^{\top} \in \mathbb{C}^8$ be a source signals, where $\mathbf{s}_i \in \mathbb{C}^2$ and $\mathbf{s}_1, \mathbf{s}_2$ are sampled from a generalization of PSK8 signal, $\mathbf{s}_3$, sampled from a generalization of QAM(4) signal, and $\mathbf{s}_4$, sampled from a two dimensional complex Gaussian noise.[3]$\mathbf{s}_1, \mathbf{s}_2$ take either of the following 24 values; $(\pm 1, 0), (0, \pm 1), (\pm i, 0), (0, \pm i), (\pm 1 \pm i, \pm 1 \pm i)/2$ independently at a time. The codomain of a PSK8 signal lies on a circle equiangularly in the complex plane $\simeq \mathbb{R}^2$, while that of this generalized signal lies on a hypersphere in $\mathbb{C}^2 \simeq \mathbb{R}^4$ equiangularly. QAM(4) signal takes either of 4 values $(\pm 1 \pm i)$ independently, while $\mathbf{s}_3 \in \mathbb{C}^2$ takes either of the following 16 values $(\pm 1 \pm i, \pm 1 \pm i)$ independently. We observe an instantaneous linear mixture of the signals $x = As$, where A is a regular 8×8 complex matrix. After whitening $z = Bx$, the complex ISA task in the sense of this paper can be solved by optimization of a cost function over the complex flag manifold $\mathrm{Fl}(8, 2, 2, 2;\mathbb{C})$ such that $y = W^H z$ give recovered signals, where $W \in \mathrm{Fl}(8, 2, 2, 2\,;\mathbb{C})$. We use the following Kurtosis-like higher order statistics for a cost function.

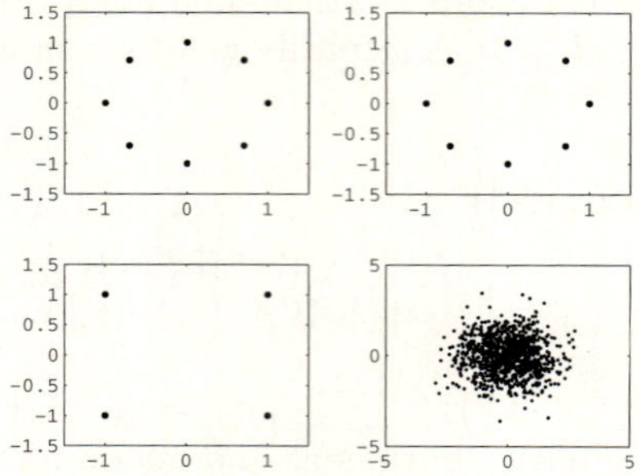

Fig. 2. Source signals of complex ISA (schematic)

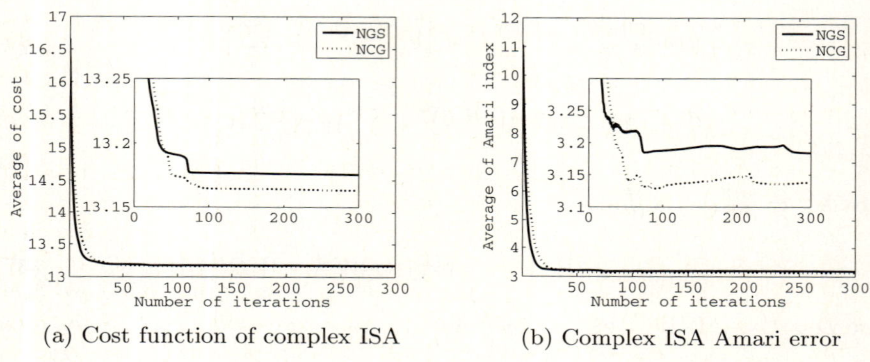

(a) Cost function of complex ISA   (b) Complex ISA Amari error

Fig. 3. Numelical simulation results

$$f(W) = \sum_{i=1}^{3} E_z(\|\mathbf{y}_i\|^4),\tag{38}$$

where $y = (y_1, y_2, \ldots, y_6)^\top = (\mathbf{y}_1^\top, \mathbf{y}_2^\top, \mathbf{y}_3^\top)^\top, \mathbf{y}_i \in \mathbb{C}^2$. Thus we get the natural gradient of f by substituting

$$\frac{\partial f}{\partial W_i^\Re} = E_z(\|\mathbf{y}_i\|^2(z\mathbf{y}_i^H + \bar{z}\mathbf{y}_i^\top))\tag{39}$$

$$\frac{\partial f}{\partial W_i^\Im} = E_z(\|\mathbf{y}_i\|^2(z\mathbf{y}_i^H - \bar{z}\mathbf{y}_i^\top))\tag{40}$$

into (35). Also, to evaluate the performance, the following generalized Amari error was used. Let us put $C = W^H BA \in \mathbb{C}^{6\times8}$, decompose it into 3×4 block matrices $D_{ij} \in \mathbb{C}^2 (i = 1, \ldots, 3, j = 1, \ldots, 4)$, and make a new 4×4 matrix $\tilde{D}$ such that $\tilde{D}_{ij} = \max_{\|x\|=1} \|D_{ij}x\|, (i = 1, \ldots, 3, j = 1, \ldots, 4), \tilde{D}_{4,4} = 1, \tilde{D}_{i,j} = 0,$

---

[3] Note that the data sets s_i actually take value in $\mathbb{C}^2$ instead of $\mathbb{C}$. Simplified versions are drawn in Figure 2 for visualization.

otherwise. The Amari error applied to this matrix yields an appropriate quantity to measure how close C is block diagonal up to scaling and permutation.

$$E(\tilde{D}) = \sum_{i=1}^{4} \sum_{j=1}^{4} \frac{\tilde{D}_{ij}}{\max_k \tilde{D}_{ik}} + \frac{\tilde{D}_{ij}}{\max_k \tilde{D}_{kj}} - 2 \times 4 \qquad (41)$$

Figure 3 show average behaviors of two algorithms over 100 trials. Note that a zoomed figure around the convergence point is superimposed for each of the figures. We observed that the natural gradient geodesic search (NGS) decreased the cost function more quickly at early stages of learning, yet the natural conjugate gradient method (NCG) exhibited faster and better convergence around the convergence points. One possible reason for this difference is because, around the convergence point, the cost function can be well approximated by a quadratic function. This assumption is exploited to derive the conjugate gradient method. Similar observation in favour of the natural conjugate gradient on a manifold of probability distributions was recently reported by Honkela et al.[7]. Also the effectiveness of the conjugate gradient method in real ICA was shown by Martin-Clemente et al.[11].

Lastly we would like to mention that the problem of complex ISA was also discussed by Szabo et al.[16]. Their proposed learning algorithm consists of 2 steps: first seek an ordinary complex ICA solution, then grouping dependent components by swapping two basis vectors belonging to different subspaces until the swapping does not lower the cost function. Here we didn't compare their algorithm with ours because we consider the case when the effect of the local minima of the cost function is not so serious. When the effect is serious, we also needed to swap basis vectors to get out of the local minima in [13]. From a computational complexity point of view, the optimization of a parametric ICA cost function is similar to the optimization of a parametric ISA cost function; the second greedy search step in their algorithm costs more expensive.

4   Conclusions

We introduced complex flag manifolds to tackle complex version of independent subspace analysis and proposed the Riemannian conjugate gradient method for the optimization on this class of complex manifolds. The Riemannian conjugate gradient method outperformed simpler natural gradient geodesic search. This Riemannian method is applicable to other signal processing problems that are formalized as optimization problems on complex flag manifolds.

Acknowledgements

This work is supported in part by JSPS Grant-in-Aid for Young Researchers 19700229, and MEXT Grant-in-Aid for Scientific Research on Priority Areas 17022033.

References

1. Abrudan, T., Eriksson, J., Koivunen, V.: Steepest Descent Algorithms for Optimization under Unitary Matrix Constraint (preprint)
2. Adali, T., Li, H.: A practical formulation for computation of complex gradients and its application to maximum likelihood. In: Proc. IEEE Int. Conf. Acoust., Speech, Signal Processing (ICASSP), Honolulu, Hawaii (April 2007)
3. Amari, S.: Natural gradient works efficiently in learning. Neural Computation 10, 251–276 (1998)
4. Brandwood, D.H.: A Complex Gradient Operator and its Application in Adaptive Array Theory. IEE Proceedings H (Microwaves, Optics, and Antennas) 130(1), 11–16 (1983)
5. Edelman, A., Arias, T.A., Smith, S.T.: The geometry of algorithms with orthogonality constraints. SIAM Journal on Matrix Analysis and Applications 20(2), 303–353 (1998)
6. Fiori, S.: A Study on Neural Learning on Manifold Foliations: The Case of the Lie Group SU(3). Neural Computation 4, 1091–1197 (2008)
7. Honkela, A., Tornio, M., Raiko, T., Karhunen, J.: Natural Conjugate Gradient in Variational Inference. In: Ishikawa, M., Doya, K., Miyamoto, H., Yamakawa, T. (eds.) ICONIP 2007, Part II. LNCS, vol. 4985. Springer, Heidelberg (to appear, 2008)
8. Hyvärinen, A., Hoyer, P.O.: Emergence of phase and shift invariant features by decomposition of natural images into independent feature subspaces. Neural Computation 12(7), 1705–1720 (2000)
9. Kreutz-Delgado, K.: The Complex Gradient Operator and the CR -Calculus, Technical report, University of California, San Diego (2003)
10. Manton, J.: Optimization algorithms exploiting unitary constraints. IEEE Transactions on Signal Processing 50(3), 635–650 (2002)
11. Clemente, R.M., Puntonet, C.G., Catalina, J.I.A.: Blind Signal Separation Based on the Derivatives of the Output Cumulants and a Conjugate Gradient Algorithm. In: Proceedings of Int. Conf. Independent Component Analysis Signal Separation, San Diego, pp. 390–393 (2001)
12. Nishimori, Y., Akaho, S.: Learning algorithms utilizing quasi-geodesic flows on the Stiefel manifold. Neurocomputing 67, 106–135 (2005)
13. Nishimori, Y., Akaho, S., Plumbley, M.D.: Riemannian optimization method on the flag manifold for independent subspace analysis. In: Rosca, J.P., Erdogmus, D., Príncipe, J.C., Haykin, S. (eds.) ICA 2006. LNCS, vol. 3889, pp. 295–302. Springer, Heidelberg (2006)
14. Nishimori, Y., Akaho, S., Abdallah, S., Plumbley, M.D.: Flag manifolds for subspace ICA problems. In: Proceedings of the International Conference on Acoustics, Speech, and Signal Processing (ICASSP 2007), Honolulu, Hawaii, USA, April 15-20, vol. IV, pp. 1417–1420 (2007)
15. Plumbley, M.D.: Algorithms for non-negative independent component analysis. IEEE Transactions on Neural Networks 14(3), 534–543 (2003)
16. Szabo, Z., Lorincz, A.: Real and Complex Independent Subspace Analysis by Generalized Variance, arXiv:math/0610438v1
17. van den Bos, A.: Complex Gradient and Hessian. IEE Proc.-Vis. Image Signal Processing 141(6), 380–382 (1994)

Quadratically Constrained Quadratic Programming for Subspace Selection in Kernel Regression Estimation

Marco Signoretto, Kristiaan Pelckmans, and Johan A.K. Suykens

K.U. Leuven, ESAT-SCD, Kasteelpark Arenberg 10, B-3001 Leuven (Belgium)
{Marco.Signoretto,Kristiaan.Pelckmans,Johan.Suykens}@esat.kuleuven.be

Abstract. In this contribution we consider the problem of regression estimation. We elaborate on a framework based on functional analysis giving rise to structured models in the context of reproducing kernel Hilbert spaces. In this setting the task of input selection is converted into the task of selecting functional components depending on one (or more) inputs. In turn the process of learning with embedded selection of such components can be formalized as a convex-concave problem. This results in a practical algorithm that can be implemented as a quadratically constrained quadratic programming (QCQP) optimization problem. We further investigate the mechanism of selection for the class of linear functions, establishing a relationship with LASSO.

1 Introduction

Problems of model selection constitute one of the major research challenges in the field of machine learning and pattern recognition. In particular, when modelling in the presence of multivariate data, the problem of input selection is largely unsolved for many classes of algorithms [1]. An interesting convex approach was found in the use of l_1-norm, resulting in sparseness amongst the optimal coefficients. This sparseness is then interpreted as an indication of non-relevance. The *Least Absolute Shrinkage and Selection Operator* (LASSO) [2] was amongst the first to advocate this approach, but also the literature on basis pursuit [3] employs a similar strategy. However, these procedures mainly deal with the class of linear models. On the other hand, when the functional dependency is to be found in a broader class, there is a lack of principled methodologies for modeling with embedded approaches for pruning irrelevant inputs.

On a different track, recent advances in convex optimization have been exploited to tackle general problems of model selection and many approaches have been proposed [4], [5]. The common denominator of the latter is conic programming, a class of convex problems broader than linear and quadratic programming [6].

In the present paper we focus on regression estimation. By taking a functional analysis perspective we present an optimization approach, based on a QCQP problem, that permits to cope with the problem of selection in a principled way. In our framework the search of a functional dependency is performed in an hypothesis space formed by the direct sum of orthogonal subspaces. In particular,

V. Kůrková et al. (Eds.): ICANN 2008, Part I, LNCS 5163, pp. 175–184, 2008.

functional ANOVA models, a popular way of modelling in presence of multivariate data, are a special case of this setting. Roughly speaking, since each subspace is composed of functions which depend on a subset of inputs, these models provide a natural framework to deal with the issue of input selection. Subsequently we present our optimization approach, inspired by [4]. In a nutshell, it provides at the same time a way to optimize the hypothesis space and to find in it the minimizer of a regularized least squares functional. This is achieved by optimizing a family of parametrized reproducing kernels each corresponding to an hypothesis space equipped with an associated inner-product. This approach leads to sparsity in the sense that it corresponds to select a subset of subspaces forming the class of models. This paper is organized as follows. In Section 2 we introduce the problem of regression estimation. Section 3 presents the abstract class of hypothesis spaces we deal with. Section 4 presents our approach for learning the functional dependency with embedded selection of relevant components. Before illustrating the method on a real-life dataset (Section 6), we uncover in Section 5 the mechanism of selection for a simple case, highlighting the relation with LASSO.

2   Regression Estimation in Reproducing Kernel Hilbert Spaces

In the following we use boldface for vectors and capital letters for matrices, operators and functionals. We further denote with $[\boldsymbol{w}]_i$ and $[A]_{ij}$ respectively the $i-$th component of $\boldsymbol{w}$ and the element (i, j) of A.

The general search for functional dependency in the context of regression can be modelled as follows [7]. It is generally assumed that for each $\boldsymbol{x} \in \mathcal{X}$, drawn from a probability distribution $p(\boldsymbol{x})$, a corresponding $y \in \mathcal{Y}$ is attributed by a *supervisor* according to an unknown conditional probability $p(y|\boldsymbol{x})$. The typical assumption is that the underlying process is deterministic but the output measurements are affected by (Gaussian) noise. The learning algorithm is then required to estimate the *regression* function i.e. the expected value $h(\boldsymbol{x}) = \int y p(y|\boldsymbol{x})dy$, based upon a training set $\mathcal{Z}^N = \{(\boldsymbol{x}_1, y_1) \times \ldots \times (\boldsymbol{x}_N, y_N)\}$ of N i.i.d. observations drawn according to $p(\boldsymbol{x}, y) = p(y|\boldsymbol{x})p(\boldsymbol{x})$. This is accomplished by finding from an appropriate set of functions $\mathcal{H}$, the minimizer $\hat{f}$ of the *empirical risk* associated to $\mathcal{Z}^N$: $R_{\text{emp}}f \triangleq \frac{1}{N}\sum_{i=1}^{N}(f(\boldsymbol{x}_i) - y_i)^2$. When the learning process takes place in a *reproducing kernel Hilbert space* $\mathcal{H}$ and, more specifically, in $\mathcal{I}_r = \{f \in \mathcal{H} \; : \; \|f\| \leq r\}$ for some $r \geq 0$, the minimizer of the empirical risk is guaranteed to exist and the crucial choice of r can be guided by a number of probabilistic bounds [8]. For fixed r the minimizer of $R_{\text{emp}}f$ in $\mathcal{I}_r$ can be equivalently found by minimizing the regularized risk functional:

$$M_\lambda f \triangleq R_{\text{emp}}f + \lambda\|f\|_{\mathcal{H}}^2 \tag{1}$$

where λ is related to r via a decreasing global homeomorphism for which there exists a precise closed form [8]. In turn the unique minimizer $\hat{f}_\lambda$ of (1) can be easily computed. Before continuing let's recall some aspects of reproducing kernel Hilbert spaces. For the general theory on this subject we make reference to the literature [9],[10],[8],[11]. In a nutshell reproducing kernel Hilbert spaces are spaces of functions[1], in the following denoted by $\mathcal{H}$, in which for each $\boldsymbol{x} \in \mathcal{X}$ the evaluation functional $L_{\boldsymbol{x}} : f \mapsto f(\boldsymbol{x})$ is a bounded linear functional. In this case the Riesz theorem guarantees that there exists $k_{\boldsymbol{x}} \in \mathcal{H}$ which is the representer of the evaluation functional $L_{\boldsymbol{x}}$ i.e. it satisfies $L_{\boldsymbol{x}} f = \langle f, k_{\boldsymbol{x}} \rangle$ where $\langle \cdot, \cdot \rangle$ denotes the inner-product in $\mathcal{H}$. We then call *reproducing kernel* (r.k.) the symmetric bivariate function $k(\boldsymbol{x}, \boldsymbol{y})$ which for fixed $\boldsymbol{y}$ is the representer of $L_{\boldsymbol{y}}$. We state explicitly a well-known result which can be found in one of its multiple forms e.g. in [8].

Theorem 1. *Let $\mathcal{H}$ be a reproducing kernel Hilbert space with reproducing kernel $k : \mathcal{X} \times \mathcal{X} \mapsto \mathbb{R}$. For a given sample $\mathcal{Z}^N$ the regularized risk functional (1) admits over $\mathcal{H}$ a unique minimizer $\hat{f}_\lambda$ that can be expressed as*

$$\hat{f}_\lambda = \sum_{i=1}^{N} [\hat{\boldsymbol{\alpha}}]_i k_{\boldsymbol{x}_i} \tag{2}$$

where $\hat{\boldsymbol{\alpha}}$ is the unique solution of the well-posed linear system:

$$(\lambda N I + K)\boldsymbol{\alpha} = \boldsymbol{y}, \tag{3}$$

$[K]_{ij} = \langle k_{\boldsymbol{x}_i}, k_{\boldsymbol{x}_j} \rangle$ *and $\langle \cdot, \cdot \rangle$ refers to the inner-product defined in $\mathcal{H}$.*

Corollary 1. *The function (2) minimizes (1) if and only if $\hat{\boldsymbol{\alpha}}$ is the solution of*

$$\max_{\boldsymbol{\alpha}} \; 2\boldsymbol{\alpha}^\top \boldsymbol{y} - \boldsymbol{\alpha}^\top K \boldsymbol{\alpha} - \lambda N \boldsymbol{\alpha}^\top \boldsymbol{\alpha}. \tag{4}$$

Notice that minimizing (1) is an optimization problem in a possibly infinite dimensional Hilbert Space. The crucial aspect of theorem 1 is that it actually permits to compute the solution $\hat{f}$ by resorting to standard methods in finite dimensional Euclidean spaces. We now turn into the core of structured hypothesis space which represents the theoretical basis for our modeling with embedded selection approach.

3   Structured Hypothesis Space

3.1   General Framework

Assume $\{\mathcal{F}^{(i)}\}_{i=1}^d$ are orthogonal subspaces of a RKHS $\mathcal{F}$ with reproducing kernel $k : \mathcal{X} \times \mathcal{X} \mapsto \mathbb{R}$. Denoted with $P^{(i)}$ the projection operator mapping $\mathcal{F}$ onto $\mathcal{F}^{(i)}$, their r.k. is given by $k^{(i)} = P^{(i)} k^2$. It can be shown [11],[12],[13] that the

---

[1] In the following we always deal with real functions.

[2] Projection operators here generalize the notion of the ordinary projection operators in Euclidean spaces. By Pk we mean that P is applied to the bivariate function k as a function of one argument keeping the other one fixed. More precisely if $k_{\boldsymbol{x}}$ is the representer of $L_{\boldsymbol{x}}$ then $Pk_{\boldsymbol{x}} = Pk(\cdot, \boldsymbol{x})$.

space $\mathcal{H} = \mathcal{F}^{(1)} \oplus \mathcal{F}^{(2)} \oplus \ldots \oplus \mathcal{F}^{(d)}$ of functions $\{f = f^{(1)} + \ldots + f^{(d)}, \; f^{(i)} \in \mathcal{F}^{(i)}\}$ equipped with the inner product:

$$\langle f, g \rangle_{\mathcal{H}, \gamma} = [\gamma]_i^{-1} \langle f^{(1)}, g^{(1)} \rangle_{\mathcal{F}} + \ldots + [\gamma]_d^{-1} \langle f^{(d)}, g^{(d)} \rangle_{\mathcal{F}}, \; \gamma \in \mathbb{R}^d : \gamma \succ \mathbf{0} \quad (5)$$

is a RKHS with reproducing kernel $k(\boldsymbol{x}, \boldsymbol{y}) = \sum_{i=1}^{d} [\gamma]_i k^{(i)}(\boldsymbol{x}, \boldsymbol{y})$. Indeed for each $\boldsymbol{x}$ the evaluation functional is bounded and by applying the definition of r.k., which is based on the inner product and the Riesz theorem, it is easy to check that $k_{\boldsymbol{x}}$ as above is its representer.

Notice that, even if the inner-product (5) requires $\gamma \succ \mathbf{0}$, for any choice of $\gamma \succeq \mathbf{0}$ in the kernel expansion k there always exists a corresponding space such that if $[\gamma]_j = 0, \forall j \in \mathcal{I}$ then $k = \sum_{j \notin \mathcal{I}} [\gamma]_j k^{(j)}$ is the reproducing kernel of $\mathcal{H} = \oplus_{j \notin \mathcal{I}} \mathcal{F}^{(j)}$ equipped with the inner-product $\langle f, g \rangle_{\mathcal{H}, \gamma} = \sum_{j \notin \mathcal{I}} [\gamma]_j^{-1} \langle f^{(j)}, g^{(j)} \rangle_{\mathcal{F}}$.

Let $\hat{f}_\lambda = \sum_{i=1}^{N} [\hat{\boldsymbol{\alpha}}]_i k_{\boldsymbol{x}_i}$ be the solution of (1) in $\mathcal{H}$. Notice that its projection in $\mathcal{F}^{(j)}$ is $P^{(j)} \hat{f}_\lambda = P^{(j)}(\sum_{i=1}^{N} [\hat{\boldsymbol{\alpha}}]_i k_{\boldsymbol{x}_i}) = \sum_{i=1}^{N} [\hat{\boldsymbol{\alpha}}]_i P^{(j)} \left(\sum_{l=1}^{d} [\gamma]_l k_{\boldsymbol{x}_i}^{(l)} \right) = [\gamma]_j \sum_{i=1}^{N} [\hat{\boldsymbol{\alpha}}]_i k_{\boldsymbol{x}_i}^{(j)}$ and for the norm induced by (5) we easily get $\|\hat{f}_\lambda\|_{\mathcal{H}, \gamma}^2 = \sum_{j=1}^{d} [\gamma]_j^{-1} \|P^{(j)} \hat{f}_\lambda\|_{\mathcal{F}}^2$ where $\|P^{(j)} \hat{f}_\lambda\|_{\mathcal{F}}^2 = [\gamma]_j^2 \hat{\boldsymbol{\alpha}}^\top K^{(j)} \hat{\boldsymbol{\alpha}}, \; [K^{(j)}]_{lk} = k^{(j)}(\boldsymbol{x}_l, \boldsymbol{x}_k)$.

Besides the tensor product formalism of ANOVA models, which we present in Subsection 3.3, the present ideas naturally apply to the class of linear functions.

3.2 Space of Linear Functions

Consider the Euclidean space $\mathcal{X} = \mathbb{R}^d$. The space of linear functions $f(\boldsymbol{x}) = \boldsymbol{w}^\top \boldsymbol{x}$ forms its dual space $\mathcal{X}^\star$. Since $\mathcal{X}$ is finite dimensional with dimension d, also its dual has dimension d and the dual base is simply given by[3]:

$$e_i(\boldsymbol{e}_j) = \begin{cases} 1 \, , \, j = i \\ 0 \, , \, j \neq i \end{cases} . \quad (6)$$

where $\{\boldsymbol{e}_j\}_{j=1}^{d}$ denotes the canonical basis of $\mathbb{R}^d$. We can further turn $\mathcal{X}^\star$ into an Hilbert space by posing:

$$\langle e_j, e_j \rangle_{\mathcal{X}^\star} = \begin{cases} 1 \, , \, j = i \\ 0 \, , \, j \neq i \end{cases} \quad (7)$$

which defines the inner-product for any couple $(f, g) \in \mathcal{X}^\star \times \mathcal{X}^\star$. Since $\mathcal{X}^\star$ is finite dimensional, it is a RKHS with r.k.[4]: $k(\boldsymbol{x}, \boldsymbol{y}) = \sum_{i=1}^{d} e_i(\boldsymbol{x}) e_i(\boldsymbol{y}) = \boldsymbol{x}^\top \boldsymbol{y}$ where for the last equality we exploited the linearity of e_i and the representation of vectors of $\mathbb{R}^d$ in terms of the canonical basis.

Consider now the spaces $\mathcal{X}_j^\star$ spanned by e_j for $j = 1, \ldots, d$ and denote with $P^{(j)}$ the projection operator mapping $\mathcal{X}^\star$ into $\mathcal{X}_j^\star$. For their r.k. we simply have:

---

[3] Notice that, as an element of $\mathcal{X}^\star$, e_i is a linear function and (6) defines the value of $e_i(\boldsymbol{x})$ for all $\boldsymbol{x} \in \mathcal{X}$. Since any element of $\mathcal{X}^\star$ can be represented as a linear combination of the basis vectors e_j, (6) also defines the value of $f(\boldsymbol{x})$ for all $f \in \mathcal{X}^\star$, $\boldsymbol{x} \in \mathcal{X}$.

[4] Indeed, any finite dimensional Hilbert space of functions $\mathcal{H}$ is a RKHS (see e.g. [10]).

$k^{(j)}(\boldsymbol{x}, \boldsymbol{y}) = P^{(j)}k(\boldsymbol{x}, \boldsymbol{y}) = \boldsymbol{x}^\top A^{(j)}\boldsymbol{y}$ where $A^{(j)} = \boldsymbol{e}_j \boldsymbol{e}_j^\top$. Finally, redefining the standard dot-product according to (5), it is immediate to get $k(\boldsymbol{x}, \boldsymbol{y}) = \boldsymbol{x}^\top A\boldsymbol{y}$ where $A = \mathrm{diag}(\boldsymbol{\gamma})$.

3.3   Functional ANOVA Models

ANOVA models represent an attempt to overcome the curse of dimensionality in non-parametric multivariate function estimation. Before introducing them we recall the concept of tensor product of RKHSs.

We refer the reader to the literature for the basic definitions and properties about tensor product Hilbert spaces [14],[15],[16],[17]. Their importance is mainly due to the fact that any reasonable multivariate function can be represented as a tensor product [15]. We just recall the following. Let $\mathcal{H}_1, \mathcal{H}_2$ be RKHSs defined respectively on $\mathcal{X}_1$ and $\mathcal{X}_2$ and having r.k. k_1 and k_2. Then the tensor product $\mathcal{H} = \mathcal{H}_1 \otimes \mathcal{H}_2$ equipped with the inner-product $\langle f_1 \otimes f_1', f_2 \otimes f_2', \rangle \triangleq \langle f_1, f_1' \rangle_{\mathcal{H}_1} \langle f_2, f_2' \rangle_{\mathcal{H}_2}$ is a reproducing kernel Hilbert space with reproducing kernel: $k_1 \otimes k_2 : ((\boldsymbol{x}_1, \boldsymbol{y}_1), (\boldsymbol{x}_2, \boldsymbol{y}_2)) \to k_1(\boldsymbol{x}_1, \boldsymbol{y}_1)k_2(\boldsymbol{x}_2, \boldsymbol{y}_2)$ where $(\boldsymbol{x}_1, \boldsymbol{y}_1) \in \mathcal{X}_1 \times \mathcal{X}_1, \;\; (\boldsymbol{x}_2, \boldsymbol{y}_2) \in \mathcal{X}_2 \times \mathcal{X}_2$ [9]. Notice that the previous can be immediately extended to the case of tensor product of a finite number of spaces.

Let now $\mathcal{I}$ be an index set and $\mathcal{W}_j$, $j \in \mathcal{I}$ be reproducing kernel Hilbert spaces. Consider the tensor product $\mathcal{F} = \otimes_{j \in \mathcal{I}} \mathcal{W}_j$. An orthogonal decomposition in the factor spaces $\mathcal{W}_j$ induces an orthogonal decomposition of $\mathcal{F}$ [13],[12]. Consider e.g. the twofold tensor product $\mathcal{F} = \mathcal{W}_1 \otimes \mathcal{W}_2$. Assume now $\mathcal{W}_1 = \mathcal{W}_1^{(1)} \oplus \mathcal{W}_1^{(2)}$, $\mathcal{W}_2 = \mathcal{W}_2^{(1)} \oplus \mathcal{W}_2^{(2)}$. Then $\mathcal{F} = \left(\mathcal{W}_1^{(1)} \otimes \mathcal{W}_2^{(1)}\right) \oplus \left(\mathcal{W}_1^{(2)} \otimes \mathcal{W}_2^{(1)}\right) \oplus \left(\mathcal{W}_1^{(1)} \otimes \mathcal{W}_2^{(2)}\right) \oplus \left(\mathcal{W}_1^{(2)} \otimes \mathcal{W}_2^{(2)}\right)$ and each element $f \in \mathcal{F}$ admits a unique expansion:

$$f = f^{(11)} + f^{(21)} + f^{(12)} + f^{(22)}. \tag{8}$$

Each factor: $\mathcal{F}^{(jl)} \triangleq \mathcal{W}_1^{(j)} \otimes \mathcal{W}_2^{(l)}$ is a RKHS with r.k. $k^{(jl)} = P_1^{(j)}k_1 \otimes P_2^{(l)}k_2$ where $P_i^{(j)}$ denotes the projection of $\mathcal{W}_i$ onto $\mathcal{W}_i^{(j)}$. In the context of functional ANOVA models [13],[12] one considers the domain $\mathcal{X}$ as the Cartesian product of sets: $\mathcal{X} = \mathcal{X}_1 \times \mathcal{X}_2 \times \cdots \times \mathcal{X}_d$ where typically $\mathcal{X}_i \subset \mathcal{R}^{d_i}$, $d_i \geq 1$. In this context for $i = 1, \ldots, d$, $\mathcal{W}_i$ is a space of functions on $\mathcal{X}_i$, $\mathcal{W}_i^{(1)}$ is typically the subspace of constant functions and $\mathcal{W}_i^{(2)}$ is its orthogonal complement (i.e. the space of functions having zero projection on the constant function). In such a framework equation (8) is typically known as ANOVA decomposition.

Concrete constructions of these spaces are provided e.g. in [10] and [11]. Due to their orthogonal decomposition, they are a particular case of the general framework presented in the beginning of this section.

4   Subspace Selection Via Quadratically Constrained Quadratic Programming

Consider an hypothesis space $\mathcal{H}$ defined as in the previous section and fix $\lambda > 0$ and $\boldsymbol{\gamma} \succeq \boldsymbol{0}$. It is immediate to see, in view of theorem 1, that the minimizer of the regularized risk functional $M_{\lambda,\gamma}f = R_{\mathrm{emp}}f + \lambda \sum_{j=1}^{d}[\gamma]_j^{-1}\|P^{(j)}f\|_{\mathcal{F}}^2$ admits a representation as $\sum_{i=1}^{N}[\hat{\alpha}]_i \left(\sum_{j=1}^{d}[\gamma]_j k_{\boldsymbol{x}_i}^{(j)}\right)$ where $\hat{\alpha}$ solves the unconstrained optimization problem given by corollary 1. Adapted to the present setting (4) becomes: $\max\limits_{\alpha} 2\boldsymbol{\alpha}^\top \boldsymbol{y} - \sum_{j=1}^{d}[\gamma]_j \boldsymbol{\alpha}^\top K^{(j)}\boldsymbol{\alpha} - \lambda N \boldsymbol{\alpha}^\top \boldsymbol{\alpha}$.

Consider the following optimization problem, consisting in an outer and an inner part:

$$\left[\begin{array}{c} \min\limits_{\gamma \in \mathbb{R}^d} \max\limits_{\alpha} \ 2\boldsymbol{\alpha}^\top \boldsymbol{y} - \sum_{j=1}^{d}[\gamma]_j \boldsymbol{\alpha}^\top K^{(j)}\boldsymbol{\alpha} - \lambda N \boldsymbol{\alpha}^\top \boldsymbol{\alpha} \\ \text{s.t.} \quad \begin{array}{c} \mathbf{1}^\top \gamma = p \\ \gamma \succeq \mathbf{0} \end{array} \end{array}\right]. \tag{9}$$

The idea behind it is that of optimizing simultaneously both γ and α. Correspondingly, as we will show, we get at the same time the selection of subspaces and the optimal coefficients $\hat{\alpha}$ of the minimizer (2) in the actual structure of the hypothesis space. In the meantime notice that, if we pose $\hat{f} = \sum_{i=1}^{N}[\hat{\alpha}]_i \left(\sum_{j=1}^{d}[\hat{\gamma}]_j k_{\boldsymbol{x}_i}^{(j)}\right)$ where $\hat{\gamma}$, $\hat{\alpha}$ solve (9), $\hat{f}$ is clearly the minimizer of $M_{\lambda,\hat{\gamma}}f$ in $\mathcal{H} = \oplus_{j \in \mathcal{I}^+}\mathcal{F}^{(j)}$ where $\mathcal{I}^+$ is the set corresponding to non-zero $[\hat{\gamma}]_j$. The parameter p in the linear constraint controls the total sum of parameters $[\gamma]_j$. While we adapt the class of hypothesis spaces according to the given sample, the capacity, which critically depends on the training set, is controlled by the regularization parameter λ selected outside the optimization. Problem (9) was inspired by [4] where similar formulations were devised in the context of transduction i.e. for the task of completing the labeling of a partially labelled dataset. By dealing with the more general search for functional dependency, our approach starts from the minimization of a risk functional in a Hilbert space. In this sense problem (9) can also be seen as a convex relaxation to the model selection problem arising from the design of the hypothesis space. See [13],[11] for an account on different non-linear non-convex heuristics for the search of the best γ.

The solution of (9) can be carried out solving a QCQP problem. Indeed we have the following result that, due to space limitations, we state without proof.

Proposition 1. *Let p and λ be positive parameters and denote with $\hat{\alpha}$ and $\hat{\nu}$ the solution of problem:*

$$\left[\begin{array}{c} \min\limits_{\alpha \in \mathbb{R}^N,\ \nu \in \mathbb{R},\ \beta \in \mathbb{R}} \ \lambda N \beta - 2\boldsymbol{\alpha}^\top \boldsymbol{y} + \nu p \\ \text{s.t.} \quad \begin{array}{c} \boldsymbol{\alpha}^\top I \boldsymbol{\alpha} \leq \beta \\ \boldsymbol{\alpha}^\top K^{(j)}\boldsymbol{\alpha} \leq \nu \ j = 1, \ldots, d \end{array} \end{array}\right]. \tag{10}$$

Define the sets: $\hat{\mathcal{I}}^+ = \{i : \hat{\boldsymbol{\alpha}}^\top K^{(i)}\hat{\boldsymbol{\alpha}} = \hat{\nu}\}$, $\hat{\mathcal{I}}^- = \{i : \hat{\boldsymbol{\alpha}}^\top K^{(i)}\hat{\boldsymbol{\alpha}} < \hat{\nu}\}$ and let h be a bijective mapping between index sets: $h : \hat{\mathcal{I}}^+ \to \{1, \ldots, |\hat{\mathcal{I}}^+|\}$. Then:

$$[\hat{\gamma}]_i = \begin{cases} [\boldsymbol{b}]_{h(i)}, & i \in \hat{\mathcal{I}}^+ \\ 0, & i \in \hat{\mathcal{I}}^- \end{cases}, \quad [\boldsymbol{b}]_{h(i)} > 0 \ \forall \ i \in \hat{\mathcal{I}}^+, \quad \sum_{i \in \hat{\mathcal{I}}^+}[\boldsymbol{b}]_{h(i)} = p \tag{11}$$

and $\hat{\alpha}$ are the solutions of (9).

Basically, once problem (9) is recognized to be convex [4], it can be converted into (10) which is a QCQP problem [6]. In turn the latter can be efficiently solved e.g. with CVX [18] and the original variables $\hat{\boldsymbol{\alpha}}$ and $\hat{\boldsymbol{\gamma}}$ can be computed. Notice that, while one gets the unique solution $\hat{\boldsymbol{\alpha}}$, the non-zero $[\hat{\boldsymbol{\gamma}}]_{i\in\hat{\mathcal{I}}^+}$ are just constrained to be strictly positive and to sum up to p. This seems to be an unavoidable side effect of the selection mechanism. Any choice (11) is valid since for each of them the optimal value of the objective function in (9) is attained. Despite of that one solution among the set (11) corresponds to the optimal value of the dual variables associated to the constraints $\boldsymbol{\alpha}^\top K^{(j)}\boldsymbol{\alpha} \leq \nu\ j = 1,\ldots,d$ in (10). Since interior point methods solve both the primal and dual problem, by solving (10) one gets at no extra price an optimal $\hat{\boldsymbol{\gamma}}$.

5   Relation with LASSO and Hard Thresholding

Concerning the result presented in proposition 1 we highlighted that the solution $\hat{f}$ has non-zero projection only on the subspaces whose associated quadratic constraints are active i.e. for which the boundary $\hat{\nu}$ is attained. This is why in the following we refer to (10) as *Quadratically Constraint Quadratic Programming Problem for Subspace Selection* (QCQPSS). Interestingly this mechanism of selection is related to other thresholding approaches that were studied in different contexts. Among them the most popular is LASSO, an estimate procedure for linear regression that shrinks some coefficients to zero [2]. If we pose $X = \begin{bmatrix} \boldsymbol{x}_1 \ldots \boldsymbol{x}_N \end{bmatrix}^5$ the LASSO estimate $[\hat{\boldsymbol{w}}]_j$ of the coefficients corresponds to the solution of problem:

$$\begin{bmatrix} \min_{\boldsymbol{w}\in\mathbb{R}^d} \|\boldsymbol{y} - X^\top \boldsymbol{w}\|_2^2 \\ \text{s.t.} \quad \|\boldsymbol{w}\|_1 \leq t \end{bmatrix}.$$

It is known [2] that, when $XX^\top = I$, the solution of LASSO corresponds to $[\hat{\boldsymbol{w}}]_j = \text{sign}([\tilde{\boldsymbol{w}}]_j)\,(|[\tilde{\boldsymbol{w}}]_j| - \xi_t)^+$ where $\tilde{\boldsymbol{w}}$ is the least squares estimate $\tilde{\boldsymbol{w}} \triangleq \arg\min_{\boldsymbol{w}} \|\boldsymbol{y} - X^\top \boldsymbol{w}\|_2^2$ and ξ_t depends upon t. Thus in the orthogonal case LASSO selects (and shrinks) the coefficients corresponding to the LS estimate that are bigger than the threshold. This realizes the so called *soft thresholding* which represents an alternative to the hard threshold estimate (subset selection) [19]: $[\hat{\boldsymbol{w}}]_j = [\tilde{\boldsymbol{w}}]_j\,I(|[\tilde{\boldsymbol{w}}]_j| > \gamma)$ where γ is a number and $I(\cdot)$ denotes here the indicator function.

When dealing with linear functions as in Subsection 3.2 it is not too hard to carry out explicitly the solution of (9) when $XX^\top = I$. We were able to demonstrate the following. Again we do not present here the proof for space limitations.

Proposition 2. *Assume* $XX^\top = I$, $k^{(j)}(\boldsymbol{x},\boldsymbol{y}) = \boldsymbol{x}^\top A^{(j)}\boldsymbol{y}$, $A^{(j)} = \boldsymbol{e}_j\boldsymbol{e}_j^\top$ *and let* $[K^{(j)}]_{lm} = k^{(j)}(\boldsymbol{x}_l,\boldsymbol{x}_m)$. *Let* $\hat{\boldsymbol{\gamma}}$, $\hat{\boldsymbol{\alpha}}$ *be the solution of* (9) *for fixed values of*

---

⁵ It is typically assumed that the observations are normalized i.e. $\sum_i [\boldsymbol{x}_i]_j = 0$ and $\sum_i [\boldsymbol{x}_i]_j^2 = n\ \forall\ j$.

p and λ. Then the function $\hat{f}_{\lambda,p} = \sum_{i=1}^{N} [\hat{\boldsymbol{\alpha}}]_i \left(\sum_{j=1}^{d} [\hat{\boldsymbol{\gamma}}]_j k_{\boldsymbol{x}_i}^{(j)} \right)$ can be equivalently restated as:

$$\hat{f}_{\lambda,p}(\boldsymbol{x}) = \hat{\boldsymbol{w}}^\top \boldsymbol{x}, \ [\hat{\boldsymbol{w}}]_j = I([\tilde{\boldsymbol{w}}]_j > \xi) \frac{[\tilde{\boldsymbol{w}}]_j}{1 + \frac{N\lambda}{[\hat{\boldsymbol{\gamma}}]_j}}$$

where $\xi = \lambda N \sqrt{\hat{\nu}}$, $\hat{\nu}$ being the solution of (10).

In comparison with l_1−norm selection approaches, when dealing with linear functions, problem (9) provides a principled way to combine sparsity and l_2 penalty. Notice that the coefficients associated to the set of indices $\hat{\mathcal{I}}^+$ are a shrinked version of the corresponding LS estimate and that the amount of shrinkage depends upon the parameters λ and $\hat{\boldsymbol{\gamma}}$.

6 Experimental Results

We illustrate here the method for the class of linear functions.

In near-infrared spectroscopy absorbances are measured at a large number of evenly-spaced wavelengths. The publicly available Tecator data set (`http://lib.stat.cmu.edu/datasets/tecator`) donated by the Danish Meat Research Institute, contains the logarithms of the absorbances at 100 wavelengths of finely chopped meat recorded using a Tecator Infrared Food and Feed Analyzer. It consists of sets for training, validation and testing. The logarithms of absorbances are used as predictor variables for the task of predicting the percentage of fat. The result on the same test set of QCQPSS as well as the LASSO, were compared with the results obtained by other subset selection algorithms[6] as reported

Table 1. Best RSS on test dataset for each cardinality of subset of variables. QCQPSS achieves very good results for models depending on small subsets of variables. The results for the first 5 methods are taken from [20].

Num. of vars.	Forward sel.	Backward elim.	Sequential replac.	Sequential 2-at-a-time	Random 2-at-a-time	QCQPSS	LASSO
1	14067.6	14311.3	14067.6	14067.6	14067.6	1203100	6017.1
2	2982.9	3835.5	2228.2	2228.2	2228.2	**1533.8**	1999.5
3	1402.6	1195.1	1191.0	1156.3	1156.3	**431.0**	615.8
4	1145.8	1156.9	833.6	799.7	799.7	**366.1**	465.6
5	1022.3	1047.5	711.1	610.5	610.5	402.3	391.4
6	913.1	910.3	614.3	475.3	475.3	390.5	388.7
7	−	−	−	−	−	358.9	356.1
8	852.9	742.7	436.3	417.3	406.2	376.0	351.7
9	−	−	−	−	−	369.4	345.3
10	746.4	553.8	348.1	348.1	340.1	−	344.3
12	595.1	462.6	314.2	314.2	295.3	−	322.9
15	531.6	389.3	272.6	253.3	252.8	−	331.5

---

[6] In these cases after the selection, the parameters were fitted according to least squares.

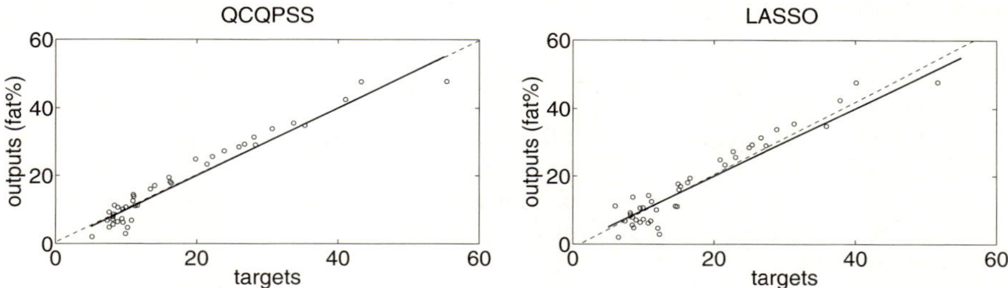

Fig. 1. Targets vs. outputs (test dataset) for the best-fitting 3 wavelengths models. The best linear fit is indicated by a dashed line. The perfect fit (output equal to targets) is indicated by the solid line. The correlation coefficients are respectively .97 (QCQPSS - left panel) and .959 (LASSO - right panel).

(with discussion) in [20]. For both QCQPSS and the LASSO the data were normalized. The values of p and λ in (10) were taken from a predefined grid. For each value of $|\hat{\mathcal{I}}^{+}_{p,\lambda}|$ as a function of (p, λ), a couple $(\hat{p}, \hat{\lambda})$ was selected if $\hat{f}_{\hat{p},\hat{\lambda}}$ achieved the smallest residual sum of squares (RSS) computed on the base of the validation set. A corresponding criterion was used in the LASSO for the choice of $\hat{t}$ associated to any cardinality of the selected subsets. QCQPSS selected at most 9 regressors while by tuning the parameter in the LASSO one can select subsets of variables of an arbitrary cardinality. However very good results were achieved for models depending on small subsets of variables, as reported on Table 1.

7 Conclusions

We have presented an abstract class of structured reproducing kernel Hilbert spaces which represents a broad set of models for multivariate function estimation. Within this framework we have elaborated on a convex approach for selecting relevant subspaces forming the structure of the approximating space. Subsequently we have focused on the space of linear functions, in order to gain a better insight on the selection mechanism and to highlight the relation with LASSO.

Acknowledgment. This work was sponsored by the Research Council KUL: GOA AMBioRICS, CoE EF/05/006 Optimization in Engineering(OPTEC), IOF-SCORES-/4CHEM, several PhD/postdoc and fellow grants; Flemish Government: FWO: PhD postdoc grants, projects G.0452.04, G.0499.04, G.0211.05, G.0226.06, G.0321.06, G.03-/02.07, G.0320.08, G.0558.08, G.0557.08, research communities (ICCoS, ANMMM, MLDM); IWT: PhD Grants, McKnow-E, Eureka-Flite+; Belgian Federal Science Policy Office: IUAP P6/04 (DYSCO, Dynamical systems, control and optimization, 2007-2011) ; EU: ERNSI; Contract Research: AMINAL.

References

1. Pelckmans, K., Goethals, I., De Brabanter, J., Suykens, J., De Moor, B.: Componentwise Least Squares Support Vector Machines. In: Wang, L. (ed.) Support Vector Machines: Theory and Applications, pp. 77–98. Springer, Heidelberg (2005)
2. Tibshirani, R.: Regression Shrinkage and Selection via the Lasso. Journal of the Royal Statistical Society. Series B (Methodological) 58(1), 267–288 (1996)
3. Chen, S.: Basis Pursuit. PhD thesis, Department of Statistics, Stanford University (November 1995)
4. Lanckriet, G., Cristianini, N., Bartlett, P., El Ghaoui, L., Jordan, M.: Learning the Kernel Matrix with Semidefinite Programming. The Journal of Machine Learning Research 5, 27–72 (2004)
5. Tsang, I.: Efficient hyperkernel learning using second-order cone programming. IEEE transactions on neural networks 17(1), 48–58 (2006)
6. Ben-Tal, A., Nemirovski, A.: Lectures on Modern Convex Optimization: Analysis, Algorithms, and Engineering Applications. Society for Industrial Mathematics (2001)
7. Vapnik, V.: Statistical Learning Theory. Wiley, New York (1998)
8. Cucker, F., Zhou, D.: Learning Theory: An Approximation Theory Viewpoint (Cambridge Monographs on Applied & Computational Mathematics). Cambridge University Press, New York (2007)
9. Aronszajn, N.: Theory of reproducing kernels. Transactions of the American Mathematical Society 68, 337–404 (1950)
10. Berlinet, A., Thomas-Agnan, C.: Reproducing Kernel Hilbert Spaces in Probability and Statistics. Kluwer Academic Publishers, Dordrecht (2004)
11. Wahba, G.: Spline Models for Observational Data. CBMS-NSF Regional Conference Series in Applied Mathematics, vol. 59. SIAM, Philadelphia (1990)
12. Gu, C.: Smoothing spline ANOVA models. Series in Statistics (2002)
13. Chen, Z.: Fitting Multivariate Regression Functions by Interaction Spline Models. J. of the Royal Statistical Society. Series B (Methodological) 55(2), 473–491 (1993)
14. Light, W., Cheney, E.: Approximation theory in tensor product spaces. Lecture Notes in Math., vol. 1169 (1985)
15. Takemura, A.: Tensor Analysis of ANOVA Decomposition. Journal of the American Statistical Association 78(384), 894–900 (1983)
16. Huang, J.: Functional ANOVA models for generalized regression. J. Multivariate Analysis 67, 49–71 (1998)
17. Lin, Y.: Tensor Product Space ANOVA Models. The Annals of Statistics 28(3), 734–755 (2000)
18. Grant, M., Boyd, S., Ye, Y.: CVX: Matlab Software for Disciplined Convex Programming (2006)
19. Donoho, D., Johnstone, J.: Ideal spatial adaptation by wavelet shrinkage. Biometrika 81(3), 425–455 (2003)
20. Miller, A.: Subset Selection in Regression. CRC Press, Boca Raton (2002)

The Influence of the Risk Functional in Data Classification with MLPs

Luís M. Silva[1], Mark Embrechts[2], Jorge M. Santos[1,4], and J. Marques de Sá[1,3]

[1] INEB -Instituto de Engenharia Biomédica, Divisão de Sinal e Imagem
Campus FEUP, Rua Dr. Roberto Frias, 4200 - 465 Porto - Portugal
[2] Rensselaer Polytechnic Institute, Troy, New York, USA
[3] Faculdade de Engenharia da Universidade do Porto, Porto, Portugal
[4] Instituto Superior de Engenharia do Porto, Porto, Portugal
{lmsilva,jmsa}@fe.up.pt, embrem@rpi.edu, jms@isep.ipp.pt

Abstract. We investigate the capability of multilayer perceptrons using specific risk functionals attaining the minimum probability of error (optimal performance) achievable by the class of mappings implemented by the multilayer perceptron (MLP). For that purpose we have carried out a large set of experiments using different risk functionals and datasets. The experiments were rigorously controlled so that any performance difference could only be attributed to the different risk functional being used. Statistical analysis was also conducted in a careful way. From the several conclusions that can be drawn from our experimental results it is worth to emphasize that a risk functional based on a specially tuned exponentially weighted distance attained the best performance in a large variety of datasets. As to the issue of attaining the minimum probability of error we also carried out classification experiments using non-MLP classifiers that implement complex mappings and are known to provide the best results until this date. These experiments have provided evidence that at least in many cases, by using an adequate risk functional, it will be possible to reach the optimal performance.

1   Introduction

We consider a classification problem with a set of classes $\Omega = \{\omega\}$ and a parametric machine (parameter set $W = \{w\}$) such as the multilayer perceptron (MLP) performing a mapping $Y = \varphi(X) = \varphi_w(X)$ from the input variable X into the output variable Y. The machine is trained by some algorithm in order to minimize a risk functional on a parameter set $W = \{w\}$ of the function class $\Phi = \{\varphi_w\}$ implemented by the classifier, attempting to approximate some target variable T. The risk (written here only for the continuous case) is often an expected distance between T and Y,

$$R_\Phi = \sum_\Omega P(\omega) \int_{X,T} L(t,y)dF(t,x|\omega) \quad \text{with } y = \varphi_w(x) \tag{1}$$

V. Kůrková et al. (Eds.): ICANN 2008, Part I, LNCS 5163, pp. 185–194, 2008.

where the so-called cost or loss function $L(\cdot)$ can be chosen in various ways. For the popular mean square error (MSE), $L(t, y) = (t - y)^2$ is a quadratic distance; for cross-entropy and two-class problems with $Y \in [0, 1]$ and $T \in \{0, 1\}$, $L(t, y) = t \ln y + (1 - t) \ln(1 - y)$, is a special case of the Kullback-Leibler divergence; etc. Recently, a generalized exponential loss function enjoying the property of being able to emulate the behavior of a whole family of loss functions, by single parameter tuning, has also been proposed [1]. It is also possible to choose the risk as a functional of the error probability density function (pdf), whenever it exists, $f(e) \equiv f_E(e)$ with $E = T - Y$, particularly the Shannon's entropy functional,

$$R_\Phi = - \int_E \ln f(e) dF(e), \tag{2}$$

or Rényi's entropy functional

$$R_\Phi = \frac{1}{1 - \alpha} \ln \int_E [f(e)]^{\alpha - 1} dF(e). \tag{3}$$

Note that $E \equiv E(\varphi)$. This approach corresponds to the Minimum of Error Entropy (MEE) principle [2,3,4]. These are more sophisticated risk functionals in the sense that they reflect the whole pdf of $T - Y$, instead of a single distance between T and Y. The main problem in data classification is the possibility of attaining the minimum probability of error, $\min_W Pe_\Phi$, afforded by the machine architecture for some w^*, the so-called *optimal solution*. For instance, if hypothetically a certain $\min_W R_\Phi$ does not lead to $\min_W Pe_\Phi$, one has to conclude that a risk functional is being used which fails to adequately take into account the whole complexity available in the family set Φ. One should then turn to another risk functional. The practical problem is that usually $\min_W Pe_\Phi$ is unknown and some risk functionals - equivalently, some learning algorithms may exhibit a better behavior than competing ones in some datasets and worse behavior in other datasets. (In fact, this must necessarily happen, taking into account the well-known results of the no-free-lunch theorems (see e.g. [5]).) In order to acquire experimental evidence on how different risk functionals behave in different datasets, we performed an extensive comparison of MLP classification results involving several types of risk functionals and datasets. Since we use the same type of classifier (MLP with the same architecture) and the same training and test sets the only influence in the results is from the loss functions. The experiments were conducted in a rigorously controlled way. The results were thoroughly evaluated.

2   The Datasets

We restricted ourselves to artificial and real-world two-class datasets. As artificial datasets we used checkerboard datasets such as the one shown in Figure 1. Checkerboard datasets are complex, controllable and unbalanced datasets. We used two different configurations: 2×2 and 4×4 checkerboards. For each one of the configurations we built three datasets with different numbers of elements

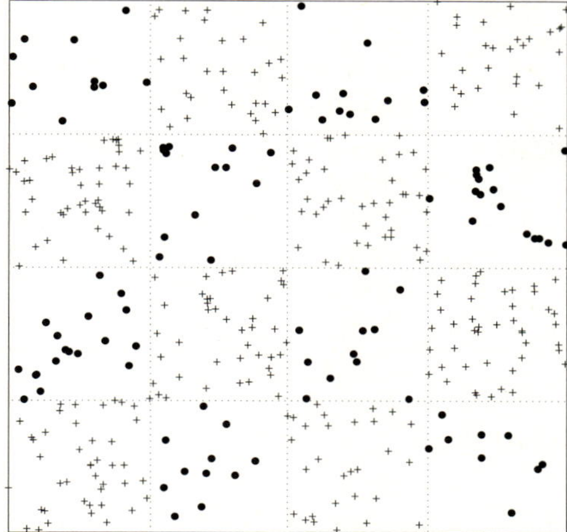

Fig. 1. An example of the 4×4 checkerboard dataset with 400 points (100 elements in the minority class: dots). Dotted lines are for visualization purpose only.

(points) but with a common characteristic: a fixed number of elements belonging to the minority class. The percentage of elements of this minority class is 50, 25 and 10% of the total number of elements. In Table 1 we show the different characteristics of the checkerboard datasets. CB2×2(200, 50) means "checkerboard 2×2 dataset with a total of 200 cases, 50% of them of the minority class"; likewise for the other checkerboard datasets.

We used several real-world datasets summarized in Table 2. All datasets can be found in the UCI repository [6]. All these datasets differ a lot among them specially in what concerns the number of features and their topology. As a matter of fact and regarding this last aspect, the Wilks' lambda, a $[0, 1]$-separability measure related to the within-class over between-class sum of squares, varies - as shown in Table 2 - from such a low value as 0.23 (Wdbc) indicating a quite good separability to such a high value as 0.86 (Liver) corresponding to a large class overlap.

Table 1. The artificial checkerboard datasets

Data set	number of elements	number of elements per class
CB2×2(200,50)	200	100-100
CB2×2(400,25)	400	300-100
CB2×2(1000,10)	1000	900-100
CB4×4(200,50)	200	100-100
CB4×4(400,25)	400	300-100
CB4×4(1000,10)	1000	900-100

Table 2. The 2-class real-world datasets

Data set	number of elements	number of features	number of el. per class	Wilks' lambda
Cleveland Heart Disease 2	297	13	160-137	0.47
Diabetes	768	8	500-268	0.70
Ionosphere	351	34	225-126	0.38
Liver	345	6	200-145	0.86
Sonar	208	60	111-97	0.38
Wdbc	569	30	357-212	0.23

3   MLP Architectures

In all experiments with a given dataset we used the same MLP architectures, with as many inputs as the number of features, one hidden layer and a single output. The number of hidden neurons, n_h, was chosen in order to assure a not too complex network with acceptable generalization. For that purpose we used some criteria as guidelines and performed some preliminary experiments. As criteria we took into account the minimum number of lines needed to separate the checkerboard classes and the well-known rule of thumb $n_h = w/\epsilon$ (based on a formula given in [7]), where w is the number of weights and the expected error rate. The number of hidden neurons is given in Table 3. All neurons used the hyperbolic tangent as activation function. As risk functionals we used:

1. Mean square error (MSE):

$$R_\Phi = \sum_\Omega P(\omega) \int_{X,T} (t - y)^2 dF(t, x|\omega) \qquad (4)$$

2. Cross-entropy (CE):

$$R_\Phi = \sum_\Omega P(\omega) \int_{X,T} (t \ln y + (1 - t) \ln(1 - y)) dF(t, x|\omega) \qquad (5)$$

Table 3. Number of hidden neurons, n_h

Data set	n_h	Data set	n_h
Checkerboard 2×2	4	Ionosphere	4
Checkerboard 4×4	16	Sonar (10)*	10
Cleveland Heart Disease 2	2	Sonar (20)*	20
Diabetes	9	Liver	8
Wdbc	3		

(*)Same dataset but used in two n_h-different series of experiments.

3. Generalized exponential (EXP):

$$R_{\Phi} = \sum_{\Omega} P(\omega) \int_{X,T} \tau e^{(t-y)^2/\tau} dF(t, x|\omega) \tag{6}$$

4. Shannon's entropy of the errors (HS):

$$R_{\Phi} = - \int_{E} \ln f(e) dF(e) \tag{7}$$

5. Quadratic Rényi's entropy of the errors (HR2):

$$R_{\Phi} = - \ln \int_{E} [f(e)] dF(e) \tag{8}$$

All these risks depend on the family of functions $\Phi = \{\varphi_w\}$ implemented by the MLP, where $W = \{w\}$ is the set of MLP weights.

4   The Experiments

As learning algorithm we used backpropagation (BP) of the errors. The practical implementations of this algorithm for all the risk functionals are described in the cited references. Note that for MEE the only available learning algorithm is BP; therefore, we had to use it for a fair comparison among all risk functionals. Regularization was performed by early stopping according to the same criterion, as follows: for each data set 10 runs were performed in order to determine the optimal number of epochs, Ep, (as well as the optimal smoothing parameter h for HS and HR2) for each method. The optimal Ep (h) were chosen as those values achieving the mean test error over the 10 runs. Initial values for h were selected using the formula given in [8]. The inputs were all pre-processed in order to standardize them to zero mean and unit variance. The support of the output target variable was $\{-1, 1\}$. In all experiments we used the 2-fold cross validation method. In this method in each run half of the data set is randomly chosen for training and the other half for testing. The datasets are next used with reverse roles (the original training set becomes the test set and vice-versa). Each experiment consisted of 20 runs of the algorithm. For this purpose twenty different random splits of the data sets were generated and stored. The *same* twenty different random splits were used as inputs for all MLPs with different risk functionals. This guaranteed that no differences in the results were due to different splits of the data sets. After the 20 runs the mean and standard deviation of the following performance measures were computed:

AUC: The area in percentage under the Receiver Operating Characteristic (ROC) curve, which measures the trade-off between sensitivity and specificity in two-class decision tables (for details, see e.g. [9]). The higher the area the better is the decision rule.

BCR: The balanced correct defined as $50 \times TN/(TN+FP) + 50 \times TP/(FN+TP)$ in percentage (T=true; F=false; P=positive; N=negative).

COR: The classification correct rate in percentage.

The first two measures are based on the resulting 2×2 decision table, considering as "abnormal" class the one with lesser cases. They are specially suitable for unbalanced datasets, as the artificial checkerboard datasets, where an optimistically high COR could arise from a too high sensitivity or specificity. AUC and BCR give an adequate picture in those situations.

5   The Results

All results were ranked and subject to "multiple comparison among groups" (post-hoc one-way anova tests) statistical tests, using Tukey's least significant difference criterion when the test probability was less that the specified significance level (0.05), i.e., when the test was significant (rejecting the null hypothesis of equal means), and the more strict Tukey's honestly significant difference criterion, otherwise. Based on the statistical tests we were able to decide whether or not a functional that performed better was indeed significantly better (at that significance level) than others with worst performance. The results obtained for the 2×2 and 4×4 checkerboard datasets are shown in Table 4. In all tables average values are followed by standard deviations between parentheses. In all the tables the best average results are in bold; the statistically significant best are underlined. For the 2×2 checkerboard datasets the EXP and HS functionals performed better in general than all other functionals (both in average and variance). The sum of the ranks for the BCR performance index disclosed the following order from best to worst: EXP, HS, HR2, CE and MSE (*exaequo*). For the 4×4 checkerboard datasets the CE and EXP functionals performed better than all other functionals. The sum of the ranks for the BCR performance index disclosed the following order from best to worst: CE, EXP, MSE, HS and HR2 (*exaequo*). Table 5 shows the results for real-world datasets. These datasets are more challenging in terms of number of features, but less challenging in terms of class unbalance and class topology. For these datasets we chose to only look to the COR performance index. The sum of the ranks for the COR performance index disclosed the following order from best to worst: CE, EXP, HS, MSE, HR2. In general, taking into account the sum of ranks for the COR performance index for all datasets, the best functional was EXP.

6   How Far from Optimality

We have of course no way to know the $\min_W Pe_\Phi$ value for the real-world datasets. However, we may find it instructive to compare the MLP results with the best results achievable by a battery of more sophisticated classifiers that do not depend on a stochastic iterative process as BP. For that purpose the same datasets were submitted to the following classifying algorithms:

1. PLS - Partial Least Squares algorithm described in [10, 11] with five latent variables.

Table 4. Results for 2×2 and 4×4 checkerboard datasets. Significantly best results underlined.

Dataset	CE	EXP	MSE	HS	HR2
CB2×2(200,50)					
AUC	97.58 (1.26)	97.89 (2.85)	**<u>98.41</u>** (1.08)	84.85 (17.07)	97.13 (4.51)
BCR	91.25 (2.50)	**<u>92.94</u>** (4.30)	92.87 (2.48)	86.96 (3.17)	92.48 (3.17)
CB2×2(400,25)					
AUC	98.21 (2.34)	98.89 (1.64)	96.41 (8.30)	**<u>99.03</u>** (1.32)	91.88 (10.82)
BCR	92.87 (2.35)	93.29 (3.61)	92.47 (4.73)	**<u>94.36</u>** (1.80)	90.70 (5.45)
CB2×2(1000,10)					
AUC	97.97 (2.77)	**<u>98.94</u>** (2.50)	62.72 (15.92)	95.07 (7.07)	96.15 (6.22)
BCR	83.40 (3.69)	**<u>94.14</u>** (5.16)	76.97 (6.51)	90.22 (5.66)	91.80 (4.87)
CB4×4(200,50)					
AUC	**<u>85.01</u>** (3.13)	83.89 (3.39)	80.89 (4.96)	77.32 (8.42)	74.39 (6.61)
BCR	**<u>79.40</u>** (3.20)	78.54 (3.57)	77.94 (4.39)	73.58 (4.96)	75.59 (5.45)
CB4×4(400,25)					
AUC	**<u>89.95</u>** (1.65)	84.11 (6.21)	76.96 (6.88)	71.41 (5.56)	70.63 (6.78)
BCR	**<u>82.98</u>** (1.78)	80.21 (3.05)	76.56 (3.96)	70.93 (4.03)	71.20 (3.69)
CB4×4(1000,10)					
AUC	**<u>91.28</u>** (3.06)	89.72 (4.13)	70.94 (5.37)	68.15 (3.23)	75.44 (5.45)
BCR	80.49 (1.98)	**<u>81.47</u>** (3.64)	70.77 (3.63)	67.98 (3.95)	73.04 (2.71)

Table 5. Results for real-world datasets. Significantly best results underlined.

Dataset	CE	EXP	MSE	HS	HR2
Clev. Heart Dis. 2					
COR	**<u>83.33</u>** (1.07)	81.72 (1.29)	82.42 (1.08)	82.72 (1.11)	81.77 (1.81)
Diabetes					
COR	**<u>76.82</u>** (0.77)	76.76 (0.79)	76.58 (0.88)	76.66 (0.85)	75.84 (0.78)
Ionosphere					
COR	88.04 (1.46)	88.37 (1.88)	87.81 (1.21)	87.71 (1.37)	**88.50** (1.33)
Liver					
COR	69.04 (2.01)	69.80 (1.27)	68.52 (1.97)	69.08 (1.86)	**<u>70.32</u>** (1.54)
Sonar (10)					
COR	77.93 (2.86)	**<u>78.75</u>** (2.84)	77.91 (2.96)	77.76 (2.57)	75.50 (4.35)
Sonar (20)					
COR	78.70 (2.60)	78.82 (2.92)	78.82 (2.51)	**79.18** (2.50)	77.43 (3.01)
Wdbc					
COR	**<u>97.44</u>** (0.55)	97.21 (0.68)	97.39 (0.67)	97.36 (0.66)	96.89 (0.42)

2. K-PLS - Kernel Partial Least Squares algorithm described in [10, 11] with five latent variables.
3. LS-SVM - Least Squares Support Vector Machine algorithm, described in [12, 13, 14].
4. SVMR - The classical Support Vector Machine algorithm applied in regression mode, described in [15, 16, 17].
5. SVMc - The classical Support Vector Machine algorithm, described in [15, 16, 17].

For the K-PLS and SVM algorithms the Gaussian kernel was used. Table 6 allows comparing the best MLP BCR result with the best BCR result obtained with the above algorithms (table header "Other") for the checkerboard datasets (with the algorithm number between brackets). Table 7 does the same for the real-world datasets and for COR.

Table 6. MLP versus "Other" (checkerboard datasets). Significantly best results underlined.

Dataset	MLP	Other
CB2×2(200,50)	**92.94** (4.30)	92.43 (2.22) (2)
CB2×2(400,25)	**94.36** (1.80)	92.44 (1.88) (2)
CB2×2(1000,10)	**94.14** (5.16)	92.55 (2.42) (5)
CB4×4(200,50)	79.40 (3.20)	**83.31** (3.02) (4)
CB4×4(400,25)	82.98 (1.78)	**85.24** (1.71) (2)
CB4×4(1000,10)	81.47 (3.64)	**82.89** (2.64) (3)

Table 7. MLP versus "Other" (real-world datasets). Significantly best results underlined.

Dataset	MLP	Other
Clev. Heart Dis. 2	**83.33** (1.06)	83.31 (0.97) (5)
Diabetes	76.82 (0.77)	**76.97** (0.87) (1)
Ionosphere	88.50 (1.33)	**95.00** (0.62) (3)
Liver	70.32 (1.54)	**70.84** (1.63) (3)
Sonar (10)	78.75 (2.84)	**85.07** (1.93) (3)
Sonar (20)	79.18 (2.50)	**85.07** (1.93) (3)
Wdbc	**97.44** (0.55)	97.28 (0.45) (3)

7   Conclusions

In the checkerboard experiments the EXP and CE functionals were the ones with better achievement, with high values of AUC and BCR even in presence of severe data unbalance. The HS functional performed very well in the 2×2 cases but poorly in the 4×4. The best functionals for the real-world datasets were CE and EXP. HS also behaved quite well achieving in general better performance than MSE. The bottom line conclusions drawn from our experiments are as follows: EXP is a very performing functional in a large variety of datasets, which can be explained for its capability of emulating a large class of functionals [1]; EXP, CE and HS seem definitely well performing algorithms only superseded in rare cases by MSE or HR2; the entropic functionals have sometimes a remarkable performance (e.g., for CB2×2(400,25), Ionosphere, Liver, Sonar20), although the characterization of such situations is for now unclear. When comparing the best classification results attained with MLPs with those attained with (generally considered) more sophisticated algorithms we conclude from our experiments that often MLPs will compete and outperform those ones. This may come as a surprise for those who perhaps thought that MLPs was an outdated approach. On the other hand, if we regard the best performance obtained as a sort of upper bound of the unknown $\min_W Pe_\Phi$, then our results provide some evidence that an MLP with a suitable risk functional will often be able to reach $\min_W Pe_\Phi$. (We say "upper bound" because the non-MLP algorithms correspond to more complex Φ-function families.) The suitability of algorithms to datasets is a tricky subject, given the many factors that come into play and the many points of view one may have on the matter. As to the former one may for instance ask whether or not it is fair to compare all algorithms with the same architecture and features; it may happen that one algorithm will have better performance in more complex architectures than other competing ones, or that it will cope better with redundant features than competing ones. As to the latter, one could consider being interested in looking to other performance indexes than the ones we have used. Nevertheless, the suitability of algorithms to datasets is an interesting and useful subject and a central topic in metalearning issues. The existing evaluation and comparison studies carried out with care and with a statistically sound background are scarce. One of the reasons of such scarcity is also that the experiments involved are much time consuming and demand a variety of resources that are not often available. The present paper is just one contribution to this topic that we intend to widen in the future.

Acknowledgments. This work was supported by the Portuguese FCT-Fundação para a Ciência e a Tecnologia (project POSC/EIA/56918/2004). First author is also supported by FCT's grant SFRH/BD/16916/2004.

References

1. Silva, L.M., Alexandre, L.A., Marques de Sá, J.: Data classification with multilayer perceptrons using a generalized error function. Neural Networks (2008) doi:10.1016/j.neunet.2008.04,004
2. Erdogmus, D., Príncipe, J.C.: Generalized information potential criterion for adaptive system training. IEEE Transactions on Neural Networks 13(5), 1035–1044 (2002)
3. Erdogmus, D., Príncipe, J.C.: An Error-Entropy Minimization Algorithm for Supervised Training of Nonlinear Adaptive Systems. IEEE Transactions on Signal Processing 50(7), 1780–1786 (2002)

4. Santos, J.M., Alexandre, L.A., Marques de Sá, J.: The Error Entropy Minimization Algorithm for Neural Network Classification. In: Int. Conf. on Recent Advances in Soft Computing, Nottingham, United Kingdom (2004)
5. Duda, R., Hart, P., Stork, D.: Pattern Classification. Wiley-Interscience, Chichester (2001)
6. Newman, D., Hettich, S., Blake, C., Merz, C.: UCI repository of machine learning databases (1998)
7. Baum, E.B., Haussler, D.: What size net gives valid generalization? Neural Computation 1(1), 151–160 (1990)
8. Santos, J.M., Marques de Sá, J., Alexandre, L.A.: Neural Networks Trained with the EEM Algorithm: Tuning the Smoothing Parameter. WSEAS Transactions on Systems 4(4), 295–300 (2005)
9. Marques de Sá, J.: Using SPSS, STATISTICA, MATLAB and R, 2nd edn. Springer, Heidelberg (2007)
10. Embrechts, M.J., Szymanski, B., Sternickel, M.: Introduction to Scientific Data Mining: Direct Kernel Methods and Applications. In: Computationally Intelligent Hybrid Systems, ch. 10, pp. 317–363. Wiley Interscience, Chichester (2004)
11. Bennett, K.P., Embrechts, M.J.: An Optimization Perspective on Kernel Partial Least Squares Regression. In: Advances in Learning Theory: Methods, Models and Applications. NATO Science Series, Series III: Computer and System Sciences, vol. 190, pp. 227–249. IOS Press, Amsterdam (2003)
12. Suykens, J.A.K., Vandewalle, J.: Least squares support vector machine classifiers. Neural Processing Letters 9(3), 293–300 (1999)
13. Suykens, J.A.K., Gestel, T.V., Brabanter, J.D., Moor, B.D., Vandewalle, J.: Least Squares Support Vector Machines. World Scientific Pub. Co., Singapore (2003)
14. Keerthi, S.S., Shevade, S.K.: SMO algorithm for least squares SVM formulations. Neural Computation 15, 487–507 (2003)
15. Cristianini, N., Shawe-Taylor, J.: Support Vector Machines and Other Kernel-Based Learning Methods. Cambridge University Press, Cambridge (2000)
16. Vapnik, V.: Statistical Learning Theory. Wiley-Interscience, Chichester (1998)
17. Schölkopf, B., Smola, A.J.: Learning with Kernels. MIT Press, Cambridge (2002)

Nonnegative Least Squares Learning for the Random Neural Network

Stelios Timotheou[*]

Intelligent Systems and Networks Group
Department of Electrical and Electronic Engineering
Imperial College
London SW7 2BT, UK
`stelios.timotheou@imperial.ac.uk`

Abstract. In this paper, a novel supervised batch learning algorithm for the Random Neural Network (RNN) is proposed. The RNN equations associated with training are purposively approximated to obtain a linear Nonnegative Least Squares (NNLS) problem that is strictly convex and can be solved to optimality. Following a review of selected algorithms, a simple and efficient approach is employed after being identified to be able to deal with large scale NNLS problems. The proposed algorithm is applied to a combinatorial optimization problem emerging in disaster management, and is shown to have better performance than the standard gradient descent algorithm for the RNN.

1   Introduction

One of the most prominent features of neural networks is their ability to learn from examples. Supervised learning has been extensively employed in a plethora of neural networks models, including the random neural network (RNN), which is of interest in this paper. RNN is a recurrent neural network inspired by the pulsed behaviour of natural neuronal networks [1,2]. In contrast to other recurrent neural models, its steady state can be described by a product form probability distribution of the neuron states, whereas its signal flow equations are analytically solvable. Moreover, as is the case for conventional artificial neural networks, RNN is able to approximate continuous and bounded functions [3,4]. A gradient descent supervised learning algorithm for RNN that sequentially updates the weights of the network for each pattern was introduced in [5]. The algorithm has been successfully applied in many problems including image processing [6], magnetic resonance imaging [7] and wafer surface reconstruction [8]; a survey of applications can be found in [6].

Apart from its usefulness as a neural network, RNN can be used as a modelling tool to capture the behaviour of interactive agents in several contexts. The RNN

---

[*] This research was undertaken as part of the ALADDIN (Autonomous Learning Agents for Decentralised Data and Information Networks) project and is jointly funded by a BAE Systems and EPSRC strategic partnership (EP/C548051/1).

V. Kůrková et al. (Eds.): ICANN 2008, Part I, LNCS 5163, pp. 195–204, 2008.

has a direct analogy to queueing networks and extends them by introducing the notion of "negative" customers for work removal. This notion has inspired a great amount of scientific work resulting in numerous extensions of queueing networks called *G-networks* [9]. Recently, a variation of RNN was employed to model both the interactions taking place in gene regulatory networks [10] and the synchronized firing effects in neuronal networks [11,12].

Therefore, introducing efficient learning algorithms for the RNN is important not only for applications where the RNN is used as a neural network tool but also in cases where a combination of modelling and learning is required. In this paper, a new batch learning algorithm for the RNN based on an approximation of the network's steady-state equations, is proposed. The obtained system is linear with nonnegativity constraints and can be solved using linear Nonnegative Least Squares (NNLS) techniques.

Linear least squares techniques have been utilized in feedforward connectionist neural networks [13,14]. The learning process becomes linear by either neglecting the scaling effect or approximating the nonlinear activation function. Least squares solution is only applied to single layers of neurons and not the whole network. On the contrary, the proposed approach is developed for a recurrent network with least squares applied to the whole network.

The remaining of the paper is structured as follows: In section 2, a concise description of the RNN and its gradient descent learning algorithm is provided. In section 3, supervised learning in RNN is formulated as a NNLS problem which in turn is solved using the RNN-NNLS algorithm developed in section 4. Following, in section 5, is the comparison of its performance with that of the standard gradient algorithm when applied to a combinatorial optimization problem arising in disaster management. The conclusions are summarized in section 6.

2   The Random Neural Network

The RNN model is comprised of N neurons that interact with each other sending excitatory and inhibitory signals. Neuron i is excited if its potential $k_i(t) \in \mathbb{N}$ is positive and it is quiescent or idle if $k_i(t) = 0$. The probability of neuron i being excited is equal to q_i. Neuron i receives excitatory or inhibitory spikes from the outside world according to Poisson processes of rates Λ_i and λ_i. Excitatory spikes increase the internal state of the receiving neuron by 1, whereas inhibitory spikes reduce this by 1 if neuron i is excited. If a neuron i is active, it can fire a spike at time t with an independent exponentially distributed firing rate r_i. If this happens then its potential $k_i(t)$ is reduced by 1. The resulting spike reaches neuron j as an excitatory spike with probability $p^+(i,j)$ or as an inhibitory spike with probability $p^-(i,j)$, or it departs from the network with probability $d(i)$. Because the probabilities associated with neuron i must sum up to 1 it is true that: $\sum_{j=1}^{N} [p^+(i,j) + p^-(i,j)] + d(i) = 1$.

The signal-flow equations of the network are described by the following system of nonlinear equations:

$$q_i = \min\{1, \frac{\lambda^+(i)}{r_i + \lambda^-(i)}\}, \quad 0 \le q_i \le 1$$

$$\lambda^+(i) = \Lambda_i + \sum_{j=1}^{N} r_j q_j p^+(j,i) \tag{1}$$

$$\lambda^-(i) = \lambda_i + \sum_{j=1}^{N} r_j q_j p^-(j,i)$$

The above system of nonlinear equations (1) always exists and it is unique [1,2,5], whereas the steady-state probability distribution of the neuron potentials is in product form: $\pi(\underline{k}) = \prod_{i=1}^{N}(1 - q_i)q_i^{k_i}$, if $q_i < 1$.

Similar to connectionist models, RNN has weights, $w^+(i,j)$ and $w^-(i,j)$, that are used for learning. These weights represent the firing rates of excitatory and inhibitory signals from neuron i to neuron j respectively, and are given by $w^+(i,j)=r_i p^+(i,j) \ge 0$ and $w^-(i,j) = r_i p^-(i,j) \ge 0$. From the definition of the RNN weights, an expression for the firing rate r_i is derived:

$$r_i = \sum_{j=1}^{N} [w^+(i,j) + w^-(i,j)]/(1 - d(i)) \tag{2}$$

Supervised learning in RNN can be performed by mapping the positive and negative values of the input patterns to the parameters $\underline{\Lambda}_k = [\Lambda_{1k}, \ ... \ , \Lambda_{Nk}]$, $\underline{\lambda}_k = [\lambda_{1k}, \ ... \ , \lambda_{Nk}]$ respectively, and the output patterns to the parameters $\underline{q}_k = [q_{1k}, \ ... \ , q_{Nk}]$. In order to update the RNN weights $w^{\pm}(u,v)$ according to the gradient descent rule, is is required to find expressions for the derivatives $\partial q_i/\partial w^{\pm}(u,v)$. Although this is challenging task because the system of equations (1) is nonlinear, as proven in [5], these terms can be expressed as a linear system of equations with unknowns $\partial \underline{q}/\partial w^{\pm}(u,v)$ that can be efficiently solved.

The general steps of the RNN gradient descent supervised learning algorithm are as follows [5]:

1. Initialize the weights of the network
2. For each input-output pattern do the following:
    - Initialize appropriately the parameters $\underline{\Lambda}_k$, $\underline{\lambda}_k$ and $\underline{q}_k$
    - Solve the system of nonlinear equations (1)
    - Use the results obtained to calculate the terms $\partial \underline{q}/\partial w^{\pm}(u,v)$
    - Update the weights according to the gradient descent rule
3. Repeat the procedure in step 2 until convergence

3   NNLS Formulation of RNN Supervised Learning

The purpose of supervised learning is to find an optimal set of weights so that for a set K of input patterns, the network's output closely represents the desired one. As already mentioned, the output of the k-th pattern is associated with the parameters $\underline{q}_k$, which are derived from the system of equations (1). Assuming that $0 \leq \lambda^+(i)/(r_i + \lambda^-(i)) \leq 1, \forall i, k$ and that $d(i) = 0, \ \forall i$ then combining (1) with (2) for all patterns yields:

$$q_{ik} \sum_{j=1}^{N} w^-(i,j) + q_{ik} \sum_{j=1}^{N} q_{jk} w^-(j,i) +$$

$$+ q_{ik} \sum_{j=1}^{N} w^+(i,j) - \sum_{j=1}^{N} q_{jk} w^+(j,i) = \Lambda_{ik} - q_{ik}\lambda_{ik}, \ \forall i, \ k \tag{3}$$

$$\text{and } w^+(i,j) \geq 0, \ w^-(i,j) \geq 0 \ \forall \ i, \ j$$

The above system is comprised of NK equations. If all q_{ik} are known then the only unknowns are the $2N^2$ weights $w^+(i,j)$ and $w^-(i,j)$ and the resulting system is linear with nonnegativity constraints. The assumption that all q_{ik} are known is valid when there is a large number of output neurons which are set to their desired output values; for the other neurons, q_{ik} can be set to random values.

Because of the nonnegativity constraints and the fact that $K \neq 2N$ a unique solution to the system (4) may not exist. In fact, the system can be either over-determined or under-determined depending on the value of K. The best that can be done is to minimize the least squares error, which can be accomplished by formulating the system of equations (4) as a linear NNLS problem:

$$\min_{\underline{w} \geq 0} f(\underline{w}) \ = \ \min_{\underline{w} \geq 0} 0.5 \|\mathbb{A}\underline{w} - \underline{b}\|_2^2, \quad \mathbb{A} \in \mathbb{R}^{NK \times 2N^2}, \ \underline{b} \in \mathbb{R}^{NK \times 1}, \ \underline{w} \in \mathbb{R}^{2N^2 \times 1} \tag{4}$$

where $\| \cdot \|_2$ is the L2-norm. The gradient of $f(\underline{w})$ is $\nabla f(\underline{w}) = \mathbb{A}^T(\mathbb{A}\underline{w}) - \mathbb{A}^T\underline{b}$. If $\mathbb{A}$ is a full rank matrix, NNLS is a strictly convex quadratic optimization problem.

Let row ik of matrix $\mathbb{A}$ be the one associated with equation ik of system (4) and ij^+ and ij^- columns be the ones corresponding to the variables $w^+(i,j)$ and $w^-(i,j)$ of $\underline{w}$. Then, $\mathbb{A}(ik, ij^+) = \mathbb{A}(ik, ij^-) = q_{ik}$, $\mathbb{A}(ik, ji^+) = -q_{jk}$ and $\mathbb{A}(ik, ji^-) = q_{ik}q_{jk} \ \forall j \neq i$, while for $j = i$ $\mathbb{A}(ik, ii^-) = q_{ik} + q_{ik}^2$. The other elements of row ik of $\mathbb{A}$ are equal to zero. In addition, the elements of $\underline{b}$ are given by:

$$b(ik) = \Lambda_{ik} - q_{ik}\lambda_{ik}, \quad \forall \ i, k \tag{5}$$

Notice that $\mathbb{A}$ is of high dimensionality ($NK \times 2N^2$ elements). However, it is highly sparse since each of its $2N^2$ element rows, contain only $4N - 3$ nonzero elements. Moreover, the nonzero values of $\mathbb{A}$ can be easily calculated using the q_{ik} values. Consequently, vector $\underline{q} \in \mathbb{R}^{NK \times 1}$ can be stored instead of $\mathbb{A}$ and used to calculate simple operations involving $\mathbb{A}$ such as matrix-vector products. Consequently, the developed approach should not require the storage of $\mathbb{A}$ and involve simple operations with $\mathbb{A}$.

4   Nonnegative Least Squares Learning

The algorithms for the solution of NNLS problems can be divided into two broad classes: *active set algorithms* and *iterative approaches.*

Active set algorithms [15,16] rely on the fact that any variables that take negative or zero values when the unconstrained least squares problem is solved, do not contribute to the solution of the constrained problem. These constraints form a set called *active,* while the set of constraints corresponding to positive variables is called *passive.* All constraints are initially inserted into the active set. Afterwards, an iterative procedure is followed to identify and remove variables from the particular set. This involves the solution of the unconstrained problem formed by the variables in the passive set. Consequently, active set methods require matrix inversion operations and are not appropriate.

Iterative methods are based on standard techniques used in nonlinear optimization to update $\underline{w}$ such as gradient descent methods and interior point methods. A popular approach for NNLS problems is to update the current solution with the use of *projected gradient methods* (6). These methods can identify several active set constraints at a single iteration and do not require sophisticated matrix operations. In the projected gradient methods, the update takes place towards the steepest descent direction. However, by using the projection operation (7), it is ensured that the new point is within the feasible region.

$$\underline{w}^{\tau+1} = P[\underline{w}^{\tau} - s^{\tau}\mathbb{D}^{\tau}\nabla f(\underline{w})], \quad s^{\tau} \geq 0, \ \mathbb{D}^{\tau} \in \mathbb{R}^{2N^2 \times 2N^2} \tag{6}$$

$$P[w_i] = \begin{cases} w_i, & w_i > 0 \\ 0, & w_i \leq 0 \end{cases} \tag{7}$$

Many different projected gradient methods have been developed which employ different techniques for selecting the step size s^{τ} and the gradient scaling matrix $\mathbb{D}^{\tau}$.

Lin proposed an algorithm based on first order projected gradient information that makes no use of $\mathbb{D}^{\tau}$ [17]. Lin introduced an efficient modification of the "Armijo rule along the projection arc" (APA) [18] to update the value for the step size s^{τ}. In APA, $s^{\tau+1}$ is found by starting from a large value for s^{τ} and exponentially decreasing it until condition (8) is satisfied.

$$f(\underline{w}^{\tau+1}) - f(\underline{w}^{\tau}) \leq \sigma\nabla f(\underline{w})^T(\underline{w}^{\tau+1} - \underline{w}^{\tau}) \tag{8}$$

In Lin's approach, $s^{\tau+1}$ is obtained by starting the search for the new step-size from the previous optimal value and appropriately increasing or decreasing it until condition (8) is satisfied (Algorithm 1).

This algorithm is not only simple and efficient but also it is appropriate for the purposes of this paper since it requires only matrix-vector multiplications involving $\mathbb{A}$ and there is no need to store the latter. Specifically, the calculations involving $\mathbb{A}$ are $\mathbb{A}\underline{w}$ to find $f(\underline{w})$ as well as $\underline{z} = \mathbb{A}\underline{w}$, $\mathbb{A}^T\underline{z}$ and $\mathbb{A}^T\underline{b}$ to get $\nabla f(\underline{w})$.

A weakness of the above approach is that it relies on first order gradient information and has slow convergence.

Algorithm 1. Projected Gradient Algorithm for the NNLS problem

---

$\quad$ Set $\sigma = 0.01$, $\beta = 0.1$
$\quad$ Set $s^0 = 1$ and $\underline{w}^1 = 0$
$\quad$ **repeat**
$\quad\quad$ $s^\tau \leftarrow s^{\tau-1}$
$\quad\quad$ **if** $\{(8)$ is satisfied$\}$ **then**
$\quad\quad\quad$ **repeat**
$\quad\quad\quad\quad$ $s^\tau \leftarrow s^\tau / \beta$
$\quad\quad\quad\quad$ Set $\underline{w}^{\tau+1} \leftarrow P[\underline{w}^\tau - s^\tau \nabla f(\underline{w}^\tau)]$
$\quad\quad\quad$ **until** $\{((8)$ is not satisfied$)$ or $(\underline{w}(s^\tau/\beta) = \underline{w}(s^\tau))\}$
$\quad\quad$ **else**
$\quad\quad\quad$ **repeat**
$\quad\quad\quad\quad$ $s^\tau \leftarrow s^\tau \beta$
$\quad\quad\quad\quad$ Set $\underline{w}^{\tau+1} \leftarrow P[\underline{w}^\tau - s^\tau \nabla f(\underline{w}^\tau)]$
$\quad\quad\quad$ **until** $\{(8)$ is satisfied$\}$
$\quad\quad$ **end if**
$\quad\quad$ Set $\underline{w}^{\tau+1} \leftarrow P[\underline{w}^\tau - s^\tau \nabla f(\underline{w}^\tau)]$
$\quad$ **until** $\{$Convergence$\}$

A quasi-Newton method that uses the gradient scaling matrix $\mathbb{D}^\tau$ achieving fast convergence was proposed in [19]. This algorithm not only exploits the active set constraints but it also utilizes an approximation of the Hessian matrix to update $\mathbb{D}^\tau$. Nevertheless, as the algorithm progresses, $\mathbb{D}^\tau$ becomes dense and requires storage making the algorithm prohibitive for the problem examined in this paper.

Taking everything into consideration, the projected gradient method proposed in [17] has been employed. The slow convergence of the algorithm is not a major issue because obtaining a solution close to optimum is sufficient if the fact that the problem itself is an approximation is regarded.

The proposed RNN-NNLS learning algorithm for obtaining the weights of RNN according to the input-output pairs is outlined in Algorithm 2. Two significant issues related to the algorithm are highlighted. Firstly, the NNLS algorithm is executed without matrix $\mathbb{A}$ as input; $\underline{q}$ is sufficient to perform all the matrix-vector product operations associated with $\mathbb{A}$. Hence, the order of memory required is the same as for the standard RNN learning algorithm. Secondly, when solving the nonlinear system of equations (1) in the main for-loop of the algorithm, only the non-output neurons are updated; the values of the output neurons are kept constant to their desired values. The procedure is repeated for a number of iterations and the best acquired solution is exploited.

5　Experimental Results

For the evaluation of the proposed NNLS learning algorithm, a problem emerging in disaster management is considered. The problem deals with the allocation of emergency units to locations of injured civilians which is a hard combinatorial

---

Algorithm 2. Supervised RNN-NNLS Batch Learning Algorithm

Based on the input patterns initialize Λ_{ik} and λ_{ik} $\forall$ i, k

Set $0 < q_{i_{out}k} < 1$ $\forall$, k, i_{out}, according to the desired output of neuron i of the k-th pattern. Index i_{out} denotes an output neuron

Initialize the remaining neurons $0 < q_{\overline{i_{out}}k} < 1$ $\forall$ i, k randomly and based on $q_{i_{out}k}$, $q_{\overline{i_{out}}k}$ form $\underline{q} \in \mathbb{R}^{NK \times 1}$

for a number of iterations **do**

    Update $\underline{b}$ according to (5)

    $\underline{w} \leftarrow ProjectedGradientNNLS(\underline{q}, \underline{b})$

    for all k **do**

        Update $q_{\overline{i_{out}}k}$ by solving the nonlinear system of equations (1)

    end for

end for

---

optimization problem. Problems of this nature require fast and close to optimal decisions. A heuristic method that could be employed to solve such problems is to train a neural network off-line with instances of the optimization problem in the same physical context as the disaster and then exploit the trained neural network to solve any problem instances that arise during the emergency. The developed algorithm is applied to such a problem for training and decision and its performance is compared to that obtained when training is performed with the standard gradient descent RNN learning algorithm [5].

In the emergency problem considered, a number of civilians I_j is injured at incident j with a total of N_L such simultaneous and spatially distributed incidents. N_U emergency units are positioned at different locations and need to collect the injured. Unit i can collect $c_i > 0$ injured and has response time to incident j, $T_{ij} > 0$. Under the assumption that one unit can be allocated to only one incident and that the total capacity of the emergency units, $c_t = \sum_i (c_i)$, is sufficient to collect all the injured, the goal is to find an allocation matrix $\underline{x}$ with elements x_{ij} which minimizes total response time $f(\underline{x})$:

$$\min \ f(\underline{x}) = \sum_{i=1}^{N_U}\sum_{j=1}^{N_L} T_{ij}x_{ij} \tag{9}$$

subject to the constraints:

$$\sum_{j=1}^{N_L} x_{ij} = 1 \ \forall \ i, \qquad \sum_{i=1}^{N_U} c_i x_{ij} \geq I_j \ \forall \ j, \qquad x_{ij} \in \{0,1\} \forall \ i, \ j \tag{10}$$

The binary variables x_{ij} represent whether emergency unit i is allocated to incident j. The first constraint designates that an emergency unit must be allocated to exactly one incident, whilst the second indicates that the total capacity of the units allocated to an incident must be at least equal to the number of civilians injured there. The above problem is NP-hard.

Assuming that the locations and capacities of the emergency units as well as the locations of the incidents are fixed the problem can be mapped to a

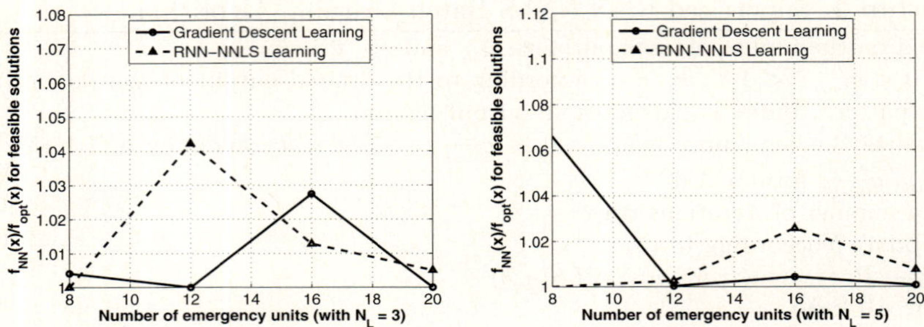

Fig. 1. Average of $f_{NN}(\underline{x})/f_{opt}(\underline{x})$ for the solutions where the units are able to remove all the injured civilians (i.e. the "feasible" ones)

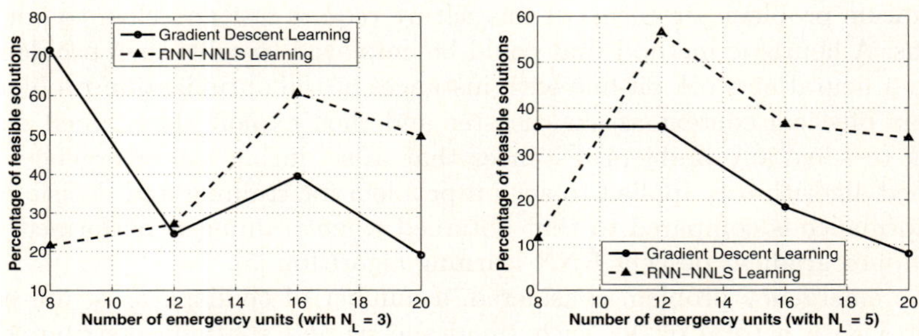

Fig. 2. Percentage of solutions in which all injured civilians are evacuated; these solutions are called "feasible" in the graphs, for want of a better term

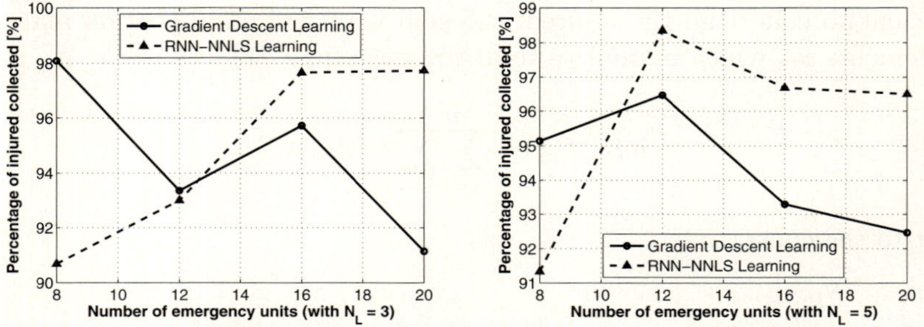

Fig. 3. Percentage of injured civilians that are collected

supervised learning context, where the inputs to the network are the parameters I_j and the outputs correspond to the binary variables x_{ij}. To represent the output variables two neurons are used, a "positive" and a "negative". If $x_{ij} = 1$, then the excitation level of the corresponding "positive" output neuron is high (close to 1) and the excitation level of the "negative" output neuron is low (close to 0). If $x_{ij} = 0$, then the reverse is true. Thus, the corresponding neural network architecture is comprised of a fully recurrent neural network with N_L input neurons and $2N_U N_L$ output neurons.

Training was performed for values of N_U ranging from 8 to 20 and $N_L = \{3, 5\}$ using 250 optimally solved training optimization instances. For every set of training instances, the parameters $T_{ij} \in (0, 1)$ and $c_i \in \{1, 2, 3, 4\}$ were fixed to uniformly distributed random values, while for each of the training pattern $I_j \in \mathbb{N}$ were also randomly generated from the uniform distribution in the range $[0.5 \cdot c_t/N_L, 1.1 \cdot c_t/N_L]$.

Testing was performed using a distinct generated set of 200 test cases, with the same probability distributions for all parameters, so that the training and testing sets were disjoint. To emphasize the fact that $\mathbb{A}$ cannot be stored notice that for $N_U = 20$ and $N_L = 5$ a network of $N = 205$ neurons is produced and $\mathbb{A}$ is a 51250×84050 matrix.

The results produced by the two methods were evaluated based on:

1. The ratio $f_{NN}(\underline{x})/f_{opt}(\underline{x})$ that shows how close to optimality the solutions are (Fig.1)
2. The percentage of instances solved so that all of the injured were evacuated (feasible solutions)(Fig. 2)
3. The percentage of injured civilians collected (Fig. 3)

All metrics were averaged over the testing instances.

As illustrated in Fig. 1 both algorithms approach the optimal solutions equally well. Concerning the percentage of injured civilians collected and the percentage of instances where all civilians are collected, Figures 2 and 3 imply that for $N_U = 8$ the gradient descent algorithm performs better than the RNN-NNLS one. This may be the case because for small networks there are less local minima and gradient descent can reach a good solution. However, for the rest of the cases, the RNN-NNLS learning algorithm performs substantially better that the gradient descent one. This shows that the developed algorithm can work quite well for large-scale networks and that the approximation is more accurate as the network grows.

6   Conclusions

A novel algorithm for learning in the RNN, the RNN-NNLS, has been presented. The algorithm is based on approximating the system of equations describing RNN by a linear system with nonnegative variables. For the solution of the formulated NNLS problem an efficient and simple projected gradient algorithm, shown to be able to solve the problem at hand without the storage of any large matrices, is employed. RNN-NNLS algorithm appears to have better performance than the gradient descent RNN algorithm when applied to a hard combinatorial optimization problem associated with the allocation of emergency units to locations of injured civilians.

References

1. Gelenbe, E.: Random neural networks with positive and negative signals and product form solution. Neural Computation 1(4), 502–510 (1989)
2. Gelenbe, E.: Stability of the random neural network. Neural Computation 2(2), 239–247 (1990)
3. Gelenbe, E., Mao, Z.H., Li, Y.D.: Function approximation with spiked random networks. IEEE Trans. on Neural Networks 10(1), 3–9 (1999)
4. Gelenbe, E., Mao, Z.H., Li, Y.D.: Function approximation by random neural networks with a bounded number of layers. J. Differential Equations and Dynamical Systems 12(1&2), 143–170 (2004)
5. Gelenbe, E.: Learning in the recurrent random network. Neural Computation 5, 154–164 (1993)
6. Bakircioglu, H., Kocak, T.: Survey of random neural network applications. European Journal of Operational Research 126(2), 319–330 (2000)
7. Gelenbe, E., Feng, T., Krishnan, K.R.R.: Neural network methods for volumetric magnetic resonance imaging of the human brain. Proceedings of the IEEE 84(1), 1488–1496 (1996)
8. Gelenbe, E., Kocak, T.: Wafer surface reconstruction from top-down scanning electron microscope images. Microlectronic Engineering 75, 216–233 (2004)
9. Artalejo, J.R.: G-networks: A versatile approach for work removal in queueing networks. European Journal of Operational Research 126(2), 233–249 (2000)
10. Gelenbe, E.: Steady-state solution of probabilistic gene regulatory networks. Physical Review E 76(1), 031903 (2007)
11. Gelenbe, E., Timotheou, S.: Random Neural Networks with Synchronised Interactions. In: Neural Computation. MIT Press, Cambridge (2008) (accepted for publication)
12. Gelenbe, E., Timotheou, S.: Synchronised Interactions in Spiked Neuronal Networks. The Computer Journal (2008) (Oxford Journals)
13. Biegler-Konig, F., Barmann, F.: A learning algorithm for multilayered neural networks based on linear least squares problems. Neural Networks 6(1), 127–131 (1993)
14. Fontenla-Romero, O., Erdogmus, D., Principe, J.C., Alonso-Betanzos, A., Castillo, E.: Linear least-squares based methods for neural networks learning. In: Artificial Neural Networks and Neural Information Processing ICANN/ICONIP 2003, pp. 84–91 (2003)
15. Lawson, C.L., Hanson, R.J.: Solving Least Squares Problems. Prentice Hall, Englewood Cliffs (1987)
16. Bro, R., Long, S.D.: A fast non-negativity-constrained least squares algorithm. Journal of Chemometrics 11(5), 393–401 (1997)
17. Lin, C.J.: Projected gradient methods for nonnegative matrix factorization. Neural Computation 19(10), 2756–2779 (2007)
18. Bertsekas, D.: Nonlinear Programming. Athena Scientific (1995)
19. Kim, D., Sra, S., Dhillon, I.: A new projected quasi-newton approach for the nonnegative least squares problem. Technical report, Dept. of Computer Sciences, The University of Texas at Austin (December 2006)

Sparse Bayes Machines for Binary Classification

Daniel Hernández-Lobato

Escuela Politécnica Superior,
Universidad Autónoma de Madrid,
C/ Francisco Tomás y Valiente, 11, Madrid 28049 Spain
`daniel.hernandez@uam.es`

Abstract. In this paper we propose a sparse representation for the Bayes Machine based on the approach followed by the Informative Vector Machine (IVM). However, some extra modifications are included to guarantee a better approximation to the posterior distribution. That is, we introduce additional refining stages over the set of active patterns included in the model. These refining stages can be thought as a backfitting algorithm that tries to fix some of the mistakes that result from the greedy approach followed by the IVM. Experimental comparison of the proposed method with a full Bayes Machine and a Support Vector Machine seems to confirm that the method is competitive with these two techniques. Statistical tests are also carried out to support these results.

1   Introduction

The Bayes Machine (BM) is a kernel machine that is based on a full Bayesian approach to binary classification [1]. Initially inspired on the Bayes Point Machine [2], the BM is able to take into account the whole set of hypotheses consistent with the data while at the same time allowing for mislabeled instances. Because the model is described within a full Bayesian approach it is able to provide a posterior distribution over the model parameters including a noise level parameter. This means that the BM is able to learn the intrinsic noise in the class labeling of the training data. Exact inference in such a model is unfeasible and approximation techniques have to be used. In [1], approximate inference is carried out by means of the algorithm Expectation Propagation (EP) [3]. The computational cost of this algorithm, and hence the cost of training a BM, is $\mathcal{O}(n^3)$, where n is the number of training instances. Finally, the BM has shown to be competitive with other famous kernel based classifiers like Support Vector Machines (SVMs) [4] and Gaussian Process Classifiers (GPCs) [5]. In this work we aim to obtain a sparse representation for the BM based on previous work on the Informative Vector Machine (IVM) [6,7]. The IVM approach is based on finding a set of active patterns of size $d < n$ in such a way that the posterior approximation obtained by making use of those patterns only is as close as possible to the one obtained when all the available patterns are used for this purpose. The active set of patterns is computed by means of a greedy algorithm that uses the Kullback-Liebler (KL) divergence as a measure of dissimilarity. Nevertheless, it could be

V. Kůrková et al. (Eds.): ICANN 2008, Part I, LNCS 5163, pp. 205–214, 2008.

the case that this greedy approach includes some patterns at the early stages of the algorithm that should not actually be included and vice versa. We aim to solve this problem by means of the inclusion of further refining stages that allow for the active set of patterns to change over later iterations.

2   Bayes Machines

Assume we are given an i.i.d. sample $\mathcal{D} = \{(\mathbf{x}_1, y_1), \ldots, (\mathbf{x}_n, y_n)\}$, $\mathbf{x}_i \in \mathcal{X}$ and $y_i \in \mathcal{Y} = \{-1, 1\}$ (hence, we will focus on binary classification problems). The BM is a full Bayesian approach to linear binary classification [1]. Given a hyperplane $\mathbf{w}$ and a noise level ϵ, the likelihood of the BM for this data set $\mathcal{D}$ is given by

$$\mathcal{P}(\mathcal{D}|\mathbf{w}, \epsilon) = \prod_i \mathcal{P}(y|\mathbf{x}, \mathbf{w}, \epsilon) = \prod_i \left[\epsilon + (1 - 2\epsilon)\Theta(y_i \mathbf{w}^t \mathbf{x}_i) \right] , \tag{1}$$

where Θ is a step function and $\mathbf{w}$ is required to go through the origin. Note that this can easily be achieved by extending each instance with one extra component that always takes value 1. This likelihood turns out to be uniform over all hyperplanes $\mathbf{w}$ that linearly separate the data with a fixed error rate ϵ. For a fixed error rate ϵ, the likelihood for a hyperplane $\mathbf{w}$ is just $\epsilon^r (1 - \epsilon)^{n-r}$, where r is the number of misclassified instances in the training set. Note that this last quantity is maximum at $\epsilon = r/n$, which is precisely the error rate of the hyperplane $\mathbf{w}$.

Given the likelihood function (1) a posterior distribution can be computed by making use of Bayes theorem, namely

$$\mathcal{P}(\mathbf{w}, \epsilon|\mathcal{D}) = \frac{\mathcal{P}(\mathcal{D}|\mathbf{w}, \epsilon)\mathcal{P}(\mathbf{w}, \epsilon)}{\mathcal{P}(\mathcal{D})} , \tag{2}$$

where $\mathcal{P}(\mathbf{w}, \epsilon)$ is some prior distribution over the model parameters and $\mathcal{P}(\mathcal{D})$ is just a normalization constant. Because of the special form of the likelihood (1), computing (2) is not tractable and hence, approximation techniques are required. Particularly, an approximating distribution $\mathcal{Q}(\mathbf{w}, \epsilon)$ can be computed by means of EP [3].

EP proceeds to approximate the numerator in the right hand side of (2) by a product of simple terms $\tilde{t}_i$ that belong to the exponential family of probability distributions, although they do not have to integrate one. That is,

$$\mathcal{P}(\mathcal{D}|\mathbf{w}, \epsilon)\mathcal{P}(\mathbf{w}, \epsilon) = \prod_i t_i(\mathbf{w}, \epsilon) \approx \prod_i \tilde{t}_i(\mathbf{w}, \epsilon) , \tag{3}$$

where each $\tilde{t}_i$ is chosen so that $\tilde{t}_i \prod_{j \neq i} \tilde{t}_j$ is as close as possible to $t_i \prod_{j \neq i} \tilde{t}_j$ in terms of the Kullback-Liebler (KL) divergence [8]. The normalization constant $\mathcal{P}(\mathcal{D})$ can then be approximated by the following integral

$$\mathcal{P}(\mathcal{D}) \approx \int \prod_i \tilde{t}_i(\mathbf{w}, \epsilon) d\mathbf{w} d\epsilon . \tag{4}$$

Note that because the terms $\tilde{t}_i$ belong to the exponential family of probability distributions so it does their product and hence, this last integral is very easy to compute. In the EP implementation of the BM, the terms $\tilde{t}_i$ are set to be the product of a Gaussian distribution and a beta distribution as follows

$$\tilde{t}_i(\mathbf{w}, \epsilon) = s_i exp\left(-\frac{1}{2v_i}(\mathbf{w}^t y_i \mathbf{x}_i - m_i)^2\right) \epsilon^{a_i}(1 - \epsilon)^{b_i}, \tag{5}$$

where s_i is just a constant value. In this way, because of the closure under multiplication property of the exponential family, the approximating distribution Q is also the product of a Gaussian distribution and a beta distribution, namely

$$Q(\mathbf{w}, \epsilon) = \mathcal{N}(\mathbf{w}|\mathbf{m_w}, \mathbf{V_w})\mathcal{B}(\epsilon|a_\epsilon, b_\epsilon), \tag{6}$$

where

$$\mathbf{V_w} = \left(\mathbf{I} + \sum_i v_i^{-1} \mathbf{x}_i \mathbf{x}_i^t\right)^{-1}, \tag{7}$$

$$\mathbf{m_w} = \mathbf{V_w} \sum_i m_i v_i^{-1} y_i \mathbf{x}_i, \tag{8}$$

$$a_\epsilon = \sum_i a_i + 1, \tag{9}$$

$$b_\epsilon = \sum_i b_i + 1. \tag{10}$$

In a similar way, we can compute (4) quite easily.

The class label of a new arbitrary instance $\mathbf{x}$ is computed by means of the predictive distribution of the model, where the true posterior distribution is approximated by Q. That is,

$$P(y|\mathcal{D}, \mathbf{x}) \approx \int P(y|\mathbf{x}, \mathbf{w}, \epsilon) Q(\mathbf{w}, \epsilon) d\mathbf{w} d\epsilon. \tag{11}$$

Because of the simple form of the posterior approximation Q, this last integral is relatively easy to solve.

Note that the whole EP algorithm can be written in terms of inner products between the training instances. This means that feature expansion by means of the *kernel trick* can be straight forward implemented. This procedure allows for the BM to address non linear classification problems quite easily. However, for this purpose we need to keep up to date the matrix

$$\mathbf{A} = (\mathbf{C}^{-1} + \mathbf{\Lambda}^{-1})^{-1}, \tag{12}$$

and the vector

$$\mathbf{h} = \mathbf{A}\mathbf{\Lambda}^{-1}\mathbf{m}, \tag{13}$$

where $\mathbf{C}$ is the inner product matrix or Gram matrix, and $\mathbf{\Lambda}$ is a diagonal matrix whose elements are $v_1, \ldots, v_n$.

Finally, the most compelling feature of the BM is its ability to learn the intrinsic noise in the class labeling and its major drawback is that the training time scales like $\mathcal{O}(n^3)$, where n is the number of training instances.

3   Sparse Bayes Machines

In this section we will try to obtain a *sparse* representation for the BM (SBM) by following the same procedure as in the IVM [6,7]. However, we will introduce some extra modifications in order to obtain a better approximation.

The easiest way to obtain a sparse representation for the BM is just to leave some of the v_i, a_i and b_i parameters of the $\tilde{t}_i$ terms fixed to $+\infty$ and 0 respectively. By doing this, we will set those terms to be uniform and hence, they will not contribute to the posterior approximation. This effect is exactly the same as the one produced by removing from the training set those patterns associated to these terms. For the sake of clarity we will maintain an active set of elements $\mathcal{I} \subset \{1,\ldots,n\}$, $|\mathcal{I}| = d < n$ so that $\forall i \notin \mathcal{I}\ v_i = +\infty$ and $a_i = b_i = 0$. The key point is how to select $\mathcal{I}$ so that the posterior approximation obtained is as similar as possible to the one that results when $\mathcal{I} = \{1,\ldots,n\}$. For this purpose we follow the greedy approach proposed in [6] with further refining stages. We will start with a selection set $\mathcal{I}$ that includes a random set of d elements (recall that at the beginning of the EP algorithm all terms $\tilde{t}_i$ are set to be uniform). Until convergence of the active set $\mathcal{I}$ and the terms $\tilde{t}_i$, we will iterate over all the elements in $\mathcal{I}$. For each element i in $\mathcal{I}$, we will first update $\mathcal{I}$ by removing that element form it (i.e. the new active set is fixed to $\mathcal{I} \setminus \{i\}$). Then, we will select a new pattern j to be included in the active set $\mathcal{I}$ from the set of available patterns $\mathcal{J} = \{1,\ldots,n\} \setminus \mathcal{I}$. Please, note that after the first removal $i \in \mathcal{J}$. The selection of j is made by computing a score function on each pattern in $\mathcal{J}$ and then picking the winner. Particularly, we will score the patterns in $\mathcal{J}$ by the KL divergence between the posterior approximation $\mathcal{Q}$ before and after each pattern inclusion, as suggested in [7]. Because our posterior approximation (6) factorizes, the KL divergence is the sum of the KL divergences between marginals.

Performing only one iteration of this algorithm over the active set $\mathcal{I}$ is equivalent to the approach followed by the IVM. Nevertheless, we expect that the later refining stages will actually significantly improve the approximation. We would like to point out that convergence of the proposed algorithm is not guaranteed at all. However, we expect it to work in a similar way as the EP algorithm. As a matter of fact, experimental results seem to confirm this. Finally, we would like to remark that exchange moves are described in [7] although they have not been analyzed. Please, note that the procedure just described is different from computing the active set $\mathcal{I}$ at the first iteration of the algorithm and then running EP over the terms $\tilde{t}_i$ included in $\mathcal{I}$. The difference lies in the fact that the proposed approach allows for $\mathcal{I}$ to change over the later refining stages.

In order to rank points efficiently we have to describe the matrix $\mathbf{A}$, as suggested in [6], by means of the Woodbury formula. That is,

$$\mathbf{A} = \mathbf{C} - \mathbf{M}^t\mathbf{M}\,, \tag{14}$$

$$\mathbf{M} = \mathbf{L}^{-1}\boldsymbol{\Lambda}_{\mathcal{I}}^{-\frac{1}{2}}\mathbf{C}_{\mathcal{I},\cdot} \in \mathbb{R}^{d,n}\,, \tag{15}$$

where $\mathbf{L}$ is the lower-triangular Cholesky factor of

$$\mathbf{B} = \mathbf{I} + \boldsymbol{\Lambda}_{\mathcal{I}}^{-\frac{1}{2}}\mathbf{C}_{\mathcal{I}}\boldsymbol{\Lambda}_{\mathcal{I}}^{-\frac{1}{2}} \in \mathbb{R}^{d,d}\,. \tag{16}$$

Note that in the EP algorithm some of the v_i parameters can be negative [3] and hence $\Lambda^{-\frac{1}{2}}$ is not well defined. This is just solved by making use of complex numbers in the implementation.

Let $\mathcal{Q}^{\backslash i}$ be the posterior approximation after removing training pattern i from the active set $\mathcal{I}$ and then, let $\mathcal{Q}^{\backslash i \cup j}$ be the posterior approximation after including training pattern j. We have that the KL divergence between $\mathcal{Q}^{\backslash i \cup j}$ and $\mathcal{Q}^{\backslash i}$ is given by

$$KL\left(\mathcal{Q}^{\backslash i \cup j}|\mathcal{Q}^{\backslash i}\right) = \frac{1}{2}log(1 + v_j^{-1}\lambda_j^{\backslash i}) - \frac{1}{2}\frac{v_j^{-1}\lambda_j^{\backslash i}}{1 + v_j^{-1}\lambda_j^{\backslash i}} + \frac{1}{2}(\alpha_j^{\backslash i})^2\lambda_j^{\backslash i}$$

$$+ log\left(\frac{\beta(a_\epsilon^{\backslash i}, b_\epsilon^{\backslash i})}{\beta(a_\epsilon^{\backslash i \cup j}, b_\epsilon^{\backslash i \cup j})}\right) + (1 - \phi(z_j))\psi(a_\epsilon^{\backslash i \cup j})$$

$$+ \phi(z_j)\psi(b_\epsilon^{\backslash i \cup j}) - \psi(a_\epsilon^{\backslash i} + b_\epsilon^{\backslash i} + 1)\,, \qquad (17)$$

where

$$\lambda_j^{\backslash i} = A_{jj} + A_{ji}^2/(v_i - A_{ii})\,, \qquad (18)$$

$$h_j^{\backslash i} = h_j + \frac{A_{ji}}{v_i - A_{ii}}(h_i - m_i)\,, \qquad (19)$$

$$z_j = h_j^{\backslash i}/\sqrt{\lambda_j^{\backslash i}}\,, \qquad (20)$$

$$\epsilon = a_\epsilon^{\backslash i}/(a_\epsilon^{\backslash i} + b_\epsilon^{\backslash i})\,, \qquad (21)$$

$$\alpha_j^{\backslash i} = 1/\sqrt{\lambda_j^{\backslash i}}(1 - 2\epsilon)\mathcal{N}(z_j|0,1)/(\epsilon + (1 - 2\epsilon)\phi(z_j))\,, \qquad (22)$$

$$h_j^{\backslash i \cup j} = h_j^{\backslash i} + \lambda_j^{\backslash i}\alpha_j^{\backslash i}\,, \qquad (23)$$

$$v_j = \lambda_j^{\backslash i}\left(\frac{1}{\alpha_j^{\backslash i}h_j^{\backslash i \cup j}} - 1\right)\,, \qquad (24)$$

$$a_\epsilon^{\backslash i \cup j} = a_\epsilon^{\backslash i} + 1 - \phi(z_j)\,, \qquad (25)$$

$$b_\epsilon^{\backslash i \cup j} = b_\epsilon^{\backslash i} + \phi(z_j)\,, \qquad (26)$$

ϕ is the cumulative probability function of a standard Gaussian distribution and ψ is the digamma function. Recall that (17) has to be computed for all the remaining patterns, i.e. $\forall j \in \mathcal{J}$. Note that for this purpose we need a column of the matrix $\mathbf{A}$. This column can be computed in $\mathcal{O}(nd)$ steps from $\mathbf{M}$ and $\mathbf{C}$.

Once that a new training pattern j has been selected for inclusion, we have to update the matrixes $\mathbf{M}$ and $\mathbf{L}$, and the vector $\mathbf{h}$ ($\tilde{t}_j$ is updated as in the standard BM). Let l be the position of the element i in $\mathcal{I}$. Then, as introduced in [7] the new $\mathbf{B}$ matrix turns out to be

$$\mathbf{B}_{new} = \mathbf{B} + \left(\tilde{\boldsymbol{\delta}}_l + \tilde{\mathbf{u}}\right)\left(\tilde{\boldsymbol{\delta}}_l + \tilde{\mathbf{u}}\right)^t - \tilde{\mathbf{u}}\tilde{\mathbf{u}}^t\,, \qquad (27)$$

where $\tilde{\boldsymbol{\delta}}_l = \eta^{\frac{1}{2}}\boldsymbol{\delta}_l$, $\tilde{\mathbf{u}} = \eta^{-\frac{1}{2}}\mathbf{u}$, $\eta = v_j^{-1}C_{jj} + v_i^{-1}C_{ii} - 2v_j^{-\frac{1}{2}}v_i^{-\frac{1}{2}}C_{ij} > 0$, $\mathbf{u} = \Lambda^{-\frac{1}{2}}(v_j^{-\frac{1}{2}}\mathbf{C}_{\mathcal{I},j} - v_i^{-\frac{1}{2}}\mathbf{C}_{\mathcal{I},i})$ and $\boldsymbol{\delta}_l$ is a vector whose components are all zeros except the *lth* component that takes value 1. These are two rank 1 updates, and hence, the Cholesky factor $\mathbf{L}$ can be updated in $\mathcal{O}(d^2)$ steps because for each rank 1 update, $\mathbf{L}_{new} = \mathbf{L}\tilde{\mathbf{L}}$, where $\tilde{\mathbf{L}}$ has a special form which allows for a fast multiplication algorithm (actually $\tilde{\mathbf{L}}$ only depends on $\mathcal{O}(d)$ parameters) [9].

Regarding the update of the matrix $\mathbf{M}$ we would like to point out that the update rule description that appears in the appendix of [7] is incorrect because it does not take into account the changes in $\mathbf{C}_{\mathcal{I}}$ after the update of $\mathcal{I}$. The right update rule for the matrix $\mathbf{M}$, for each rank 1 update of $\mathbf{B}$, is

$$\mathbf{M}_{new} = \tilde{\mathbf{L}}^{-1}\left(\mathbf{M} + (v_j^{-\frac{1}{2}} - v_i^{-\frac{1}{2}})\mathbf{L}^{-1}\boldsymbol{\delta}_l\mathbf{C}_{i,.} + v_j^{-\frac{1}{2}}\mathbf{L}^{-1}\boldsymbol{\delta}_l\left(\mathbf{C}_{j,.} - \mathbf{C}_{i,.}\right)^t\right). \tag{28}$$

By means of the Woodbury formula, a multiplication by the factor $\tilde{\mathbf{L}}^{-1}$ can be easily computed in terms of $\tilde{\mathbf{L}}$, so in general these products, and hence the update rule defined by (28), can be computed in $\mathcal{O}(dn)$. Note that (28) points out that we actually need to know $\mathbf{L}^{-1}$. This is no problem because it can be updated as follows $\mathbf{L}_{new}^{-1} = \tilde{\mathbf{L}}^{-1}\mathbf{L}^{-1}$.

Now, in order to update the vector $\mathbf{h}$, we have to make use of its definition, given by (13), and the following relationship that arises from (12) and the Woodbury formula

$$\mathbf{A}_{new} = \mathbf{A} - \mathbf{A}\mathbf{I}_{.,\{ji\}}\boldsymbol{\Delta}^{\frac{1}{2}}\mathbf{P}^{-1}\boldsymbol{\Delta}^{\frac{1}{2}}\mathbf{I}_{\{ji\},.}\mathbf{A}, \tag{29}$$

where $\mathbf{P} = \mathbf{I} + \boldsymbol{\Delta}^{\frac{1}{2}}\mathbf{A}_{\{ji\}}\boldsymbol{\Delta}^{\frac{1}{2}}$ and $\boldsymbol{\Delta} = diag(v_j^{-1}, v_i^{-1})$ is a matrix denoting the changes in $\boldsymbol{\Lambda}$. Let $\boldsymbol{\gamma} = (v_j^{-1}m_j, v_i^{-1}m_i)^t$ be a vector denoting the changes in $\boldsymbol{\Delta}\mathbf{m}$. Then, we have that

$$\mathbf{h}_{new} = \mathbf{h} + \mathbf{A}_{.,\{ji\}}\boldsymbol{\Delta}^{\frac{1}{2}}\mathbf{P}^{-1}\left(\boldsymbol{\Delta}^{-\frac{1}{2}}\boldsymbol{\gamma} - \boldsymbol{\Delta}^{\frac{1}{2}}\mathbf{h}_{\{ji\}}\right), \tag{30}$$

which can be computed in $\mathcal{O}(nd)$ steps. See [7] for further details.

Hence, the preprocessing of a pattern in the active set $\mathcal{I}$ requires a total cost of $\mathcal{O}(nd)$ steps. Because $|\mathcal{I}| = d$, an iteration of the whole algorithm takes $\mathcal{O}(nd^2)$ steps. The storage requirements are dominated by the matrix $\mathbf{M}$ which demands $\mathcal{O}(nd)$ memory units. Please, note that the elements of the kernel matrix $\mathbf{C}$ can be computed when required. Once training is completed, we can evaluate the marginal log likelihood of the training data (which can be used for kernel selection) as in a standard BM by computing $\mathbf{A}_{\mathcal{I}}$, $\boldsymbol{\Lambda}_{\mathcal{I}}$ and $\mathbf{C}_{\mathcal{I}}$. Prediction over new arbitrary instances can be achieved in a similar way. The only problem of this sparse approach is how to select the size d of the active set $\mathcal{I}$. In general this is a difficult task. However, the average log likelihood of the current predictor on the remaining patterns $\mathcal{J}$ can be in principle evaluated for this purpose, as suggested in [6].

4   Experiments

In this section we try to empirically assess the performance of the sparse approach described in the previous section over some classification problems. First,

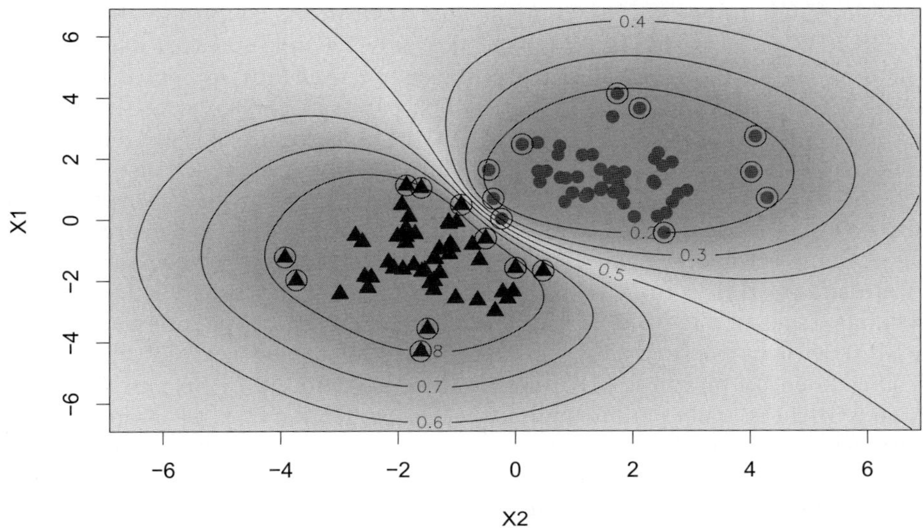

Fig. 1. Decision boundary obtained by training a SBM over the toy problem. Patterns included in the model are those that are highlighted with black circumferences. In this case the parameter d was set to 20% of the number of available patterns. Contour curves of the predictive distribution are also plotted.

we start with a toy problem to visually check which patterns the SBM actually selects. We generated a simple 2-class 2-dimensional data set whose input features x_1 and x_2 are drawn from two Gaussian distributions with unit covariance matrix and mean vectors $\boldsymbol{\mu}_1 = (1,1)^t$ and $\boldsymbol{\mu}_2 = (-1,-1)^t$ respectively. The class label of each instance, either 1 or -1, is assigned according to which Gaussian distribution the instance belongs to. This means that the optimal decision boundary of the problem is just the hyperplane defined by the vector $\mathbf{w} = (1,1,0)^t$. In this way, we built a training set of 100 instances and trained a SBM on it. We set a Gaussian Kernel in the SBM algorithm like the following one

$$C(\mathbf{x}_i, \mathbf{x}_j) = exp\left(-\gamma|\mathbf{x}_i - \mathbf{x}_j|^2\right) , \tag{31}$$

where in this case we fixed $\gamma = 0.1$ and $d = 20$ (i.e. we only retain 20% of the initial patters). After fifteen iterations the algorithm converged to a stable configuration, though just after the first six iterations the active set $\mathcal{I}$ had already converged. Fig. 1 displays the decision boundary of the resulting SBM. The patterns that were selected after the training phase are highlighted with a black circumference. Note that only those patterns that surround each cluster of points are included in the model. This is because the kernel employed allows for non linear decision boundaries. If a linear kernel had been used, only those patterns near the decision boundary would have been included.

We also carried out experiments over the MNIST handwritten digits data base [10] comparing our method with the SVM [4]. For each digit i so that $i \neq 9$, we considered binary tasks of the form i against 9. We down-sampled the bitmaps of size 28×28 pixels to size 14×14 pixels (i.e. only 196 attributes are considered). For each one of those classification problems we randomly extracted from the MNIST training set of 60000 instances, 500 patterns associated to the digit i, and 500 patterns associated to the digit 9. Then, each one of these datasets of 1000 instances is randomly split 50 times into a training set of 100 instances and a test set of 900 instances. The attributes were normalized to have zero mean and unit standard deviation on the training set. Each method is then trained on the training set and evaluated on the test set. Particularly, we compare results of a standard BM, a SBM and a SVM. In order to check whether the further refining stages of the SBM are beneficial or not, we also compare results with a SBM without these stages. This SBM implementation actually follows the assumed density filtering (ADF) approach suggested in [6,7]. Because the SBM cannot fix the sparsity parameter d, for each classification task, we initialized this parameter with the average number of support vectors obtained by the SVM. In all methods the same kernel is used, namely the Gaussian kernel described in (31). The γ parameter is fixed in all methods by a grid search procedure, where the logarithm of the grid is fixed to ten equally spaced points in the interval $[-10, -4]$. In the case of the SVM a cross validation procedure is used to estimate which parameter value performs best. In the other methods, the marginal log likelihood of the training data is used for this purpose. The cost parameter C of the SVM is fixed in a similar way, by a cross validation grid search procedure where the log_{10} of the grid is fixed to ten equally spaced points in the interval $[-3, 3]$.

Table 1 displays the average error of each method, for each problem, alongside with the average number of support vectors returned by the SVM. The table points out that the BM obtains the best error rates while the SBM seems to provide similar results. On the other hand, error rates of the SBM without the refining stages (fourth column in the table) are slightly higher than those of the previous methods. Finally, the error rates of the SVM are the highest ones. These experimental results remark the beneficial properties of the further refining stages that have been implemented in this work.

Table 1. Error values of each method and average number of support vectors

Problem	BM	SBM	SBM$_{\text{ADF}}$	SVM	# SV
0 vs 9	1.5±0.6	1.5±0.6	2.0±0.7	3.2±1.1	54.2±16.4
1 vs 9	1.7±0.9	1.7±0.9	3.1±1.4	3.5±1.3	56.1±17.9
2 vs 9	2.4±0.6	2.4±0.5	2.8±0.7	4.5±1.6	59.3±16.2
3 vs 9	3.5±1.0	4.0±1.2	4.9±1.4	6.1±1.6	54.4±15.8
4 vs 9	8.9±1.6	10.7±1.6	11.3±1.9	11.6±2.1	61.9±13.9
5 vs 9	4.2±0.9	5.1±1.3	5.2±1.3	6.8±2.1	59.2±12.8
6 vs 9	1.2±0.7	1.2±0.8	2.2±0.9	4.2±2.0	72.0±17.3
7 vs 9	8.9±1.4	10.6±1.9	11.0±1.7	10.6±2.0	56.1±13.1
8 vs 9	4.0±0.7	4.7±1.0	4.9±0.9	6.4±2.0	59.7±13.2

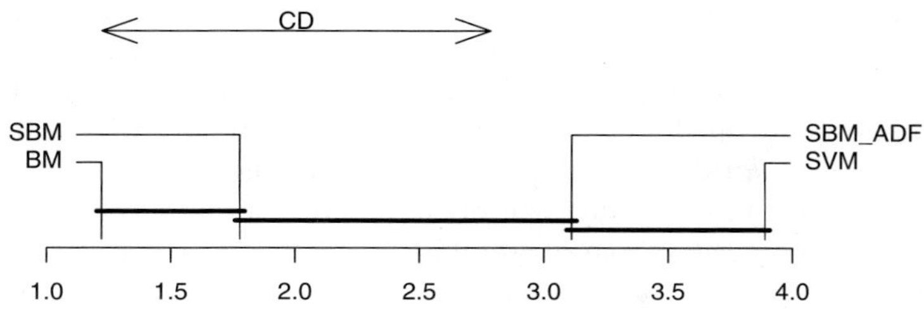

Fig. 2. Graphical result of a Nemenyi test between the average ranks of each algorithm in Table 1. The critical difference (CD) between average ranks is displayed at the top.

In order to assess whether or not these results are statistically significant, we follow the approach suggested in [11]. A Friedman test over the average rank values of each algorithm states that it is very likely that there is a difference between their performance (a p-value below 0.05 was obtained). As a result, we applied a Nemenyi test over those average ranks to check for differences between pairs of algorithms (an α-value of 0.05 is employed for this purpose). Fig. 2 displays graphically the results of this test, as suggested in [11]. Algorithms that belong to the same group are linked together with a solid black line. The critical difference (CD) between ranks is also displayed at the top. The picture points out that there is statistical evidence supporting that the BM and the SBM are better than the SVM over this set of classification problems. Nevertheless, there is not enough evidence for a difference between the performances of the BM and SBM. Finally, the picture also shows that the SBM without the further refining stages performs worse than the BM.

5   Conclusion

In this paper we have proposed a *sparse* representation for the BM. We call this model sparse Bayes machine (SBM). The approach we have followed is based on the one developed in the IVM [6,7]. Nevertheless, we have introduced some extra modifications in order to obtain a better posterior approximation. Particularly, we have included further refining stages into the algorithm which allow for the active set $\mathcal{I}$ of selected patterns to change over later iterations. These refining stages can be thought as a backfitting algorithm that tries to fix some of the mistakes that result from the greedy approach followed by the IVM. They actually work in a similar way as the EP algorithm, which improves the approximation obtained by ADF [3]. The training cost of the proposed method is $\mathcal{O}(d^2 n)$, where n is the number of training patterns available and $d < n$ is some sparsity parameter that has to be fixed in some way. This is an important improvement over the training cost of the BM, which is $\mathcal{O}(n^3)$. Prediction time is also improved in a similar way. In order to assess the accuracy of the proposed technique, experiments were carried out over the MNIST data base of handwritten digits [10], comparing its performance with those of a SVM [4], a full BM, and a SBM without the further refining stages. The proposed method outperfoms the SVM and obtains similar error rates to those of a full BM. However, only some of the available patters are used after training for prediction. In addition, these experiments pointed out the beneficial properties of implementing further refining stages in the algorithm. Statistcal tests were also carried out to support these results. Finally, one of the major drawbacks of the proposed approach is that the sparsity parameter d has to be fixed in some way. In general this is a difficult task. However, we also believe that the average log likelihood of the model on the remaining patterns $\mathcal{J}$ can be evaluated in principle for this purpose, as suggested in [6].

Acknowledgment

This work has been supported by *Consejería de Educación de la Comunidad Autónoma de Madrid, European Social Fund, Universidad Autónoma de Madrid* and *Ministerio de Educación y Ciencia, Dirección General de Investigación* under project TIN2007-66862-C02-02.

References

1. Hernández-Lobato, D., Hernández-Lobato, J.M.: Bayes machines for binary classification. Pattern Recognition Letters 29, 1466–1473 (2008)
2. Herbrich, R., Graepel, T., Campbell, C.: Bayes point machines. Journal of Machine Learning Research 1, 245–279 (2001)
3. Minka, T.: A Family of Algorithms for approximate Bayesian Inference. PhD thesis, Massachusetts Institute of Technology (2001)
4. Vapnik, V.N.: The nature of statistical learning theory. Springer, New York (1995)
5. Kim, H.C., Ghahramani, Z.: Bayesian gaussian process classification with the em-ep algorithm. IEEE Transactions on Pattern Analisys Machine Intelligence 28(12), 1948–1959 (2006)
6. Lawrence, N., Seeger, M., Herbrich, R.: Fast sparse gaussian process methods: The informative vector machine. In: Becker, S., Thrun, S., Obermayer, K. (eds.) Advances in Neural Information Processing Systems, vol. 15, pp. 609–616. MIT Press, Cambridge (2003)
7. Seeger, M.: Bayesian Gaussian Process Models: PAC-Bayesian Generalisation Error Bounds and Sparse Approximations. PhD thesis (2003)
8. Bishop, C.M.: Pattern Recognition and Machine Learning (Information Science and Statistics). Springer, Heidelberg (August 2006)
9. Gill, P.E., Golub, G.H., Murray, W., Saunders, M.A.: Methods for modifying matrix factorizations. Mathematics of Computation 28(126), 505–535 (1974)
10. LeCun, Y., Bottou, L., Bengio, Y., Haffner, P.: Gradient-based learning applied to document recognition. Proceedings of the IEEE 86(11), 2278–2324 (1998)
11. Demšar, J.: Statistical comparisons of classifiers over multiple data sets. Journal of Machine Learning Research 7, 1–30 (2006)

Tikhonov Regularization Parameter in Reproducing Kernel Hilbert Spaces with Respect to the Sensitivity of the Solution

Kateřina Hlaváčková-Schindler

Commission for Scientific Visualization, Austrian Academy of Sciences Donau-City
Str. 1, A-1220 Vienna, Austria
`katerina.schindler@assoc.oeaw.ac.at`

Abstract. In our paper, we consider Tikhonov regularization in the reproducing Kernel Hilbert Spaces. In this space we derive upper and lower bound of the interval which contains the optimal value of Tikhonov regularization parameter with respect to the sensitivity of the solution without computing the singular values of the corresponding matrix. For the case of normalized kernel, we give an explicit formula for the optimal regularization parameter with respect to the sensitivity of the solution which needs only the knowledge of the minimal singular value of the corresponding matrix.

1 Introduction

We will deal with the following problem: Given sparse, noisy samples of a function f, how do we recover f as accurately as possible? Infinitely many curves pass through the samples which makes the problem to be ill-posed. A prior knowledge about the function must be introduced to make the solution unique. A theoretical framework to do this is regularization. It constraints the solution by "stabilizers" and establishes the tradeoff between fidelity and smoothness of the solution. Poggio et al. in [13] introduced this problem to neural networks. They showed that regularization leads to approximation schemes that are equivalent to networks with one layer of hidden units, called regularization networks. The Tikhonov regularization, which they used, identifies the approximating function as the minimizer of a cost functional which includes an error term and a smoothness functional, usually called a stabilizer. The most important step of Tikhonov regularization is to determinate parameter γ, multiplying the stabilizer. In practice is this parameter used mostly heuristically (e.g. the discrepancy principle [10], generalized cross-validation [4], the L-curve [5] etc.). More analytical approach to the determination of γ was suggested by Regińska, Cucker and Smale and by Kůrková. Regińska in [14] investigated the relationship between the regularization parameter and the sensitivity of the regularized solution of a discrete ill-posed problem (for the sensitivity definition, see Section 5). Similarly as Regińska, Cucker and Smale in [2] and Kůrková in [9] investigated the influence of the regularization parameter on the ill-posedness of the problem.

V. Kůrková et al. (Eds.): ICANN 2008, Part I, LNCS 5163, pp. 215–224, 2008.

All the mentioned approaches to find a regularization parameter require the knowledge of singular values of the matrix corresponding to the problem. The singular values are most often computed by means of singular value decomposition (SVD), having high computational demands. In addition, SVD finds only approximative values of the singular values with a certain stability of the solution of the method. For symmetric (kernel) matrices, one usually applies various numerical optimization methods, among which probably the most known one is the Rayleigh quotient iteration [11]. However, even for positive definite kernels, only asymptotic or probabilistic results for their singular values are in general known (i.e. [3]). In some practical problems where is the computational time important (neural networks including), an approximative value of the Tikhonov parameter can be sufficient. In our paper, we consider Tikhonov regularization in the reproducing Kernel Hilbert Spaces. We derive upper and lower bound of the interval which contains the optimal value of Tikhonov regularization parameter with respect to the sensitivity of the solution without computing the singular values. For the case of normalized kernel, we give an explicit formula for the optimal regularization parameter with respect to the sensitivity of the solution which needs only the knowledge of the minimal singular value of the corresponding matrix.

The paper is organized as follows. Section 2 explains the minimization of the empirical error by pseudo-inversion. Inverse problems in Reproducing Kernel Hilbert Spaces Operators (RKHS) are discussed in Section 3. Tikhonov regularization in RKHS is discussed in Section 4. To optimal values of the regularization parameter in Hilbert spaces is devoted Section 5. Our main results are concentrated in Section 6, followed by conclusions in Section 7.

2 Minimization of the Empirical Error by Pseudo-inversion

We will shortly review the theoretical analysis of the minimization of the empirical cost function based on pseudo-inversion from Cucker and Smale [2] and Kůrková in [9], [7] and [8] and apply it to the regularization parameter determination. Consider a data set of pairs of data $Z = \{(x_i, y_i) \in \Omega \times \mathcal{R}, i = 1, \ldots, m\}$, $\Omega \subset R^d$. The empirical error function with the square loss function is defined as

$$E_Z(f) = \frac{1}{t} \sum_{i=1}^{t} (f(x_i) - y_i)^2. \tag{1}$$

The problem of minimization of the empirical error functional (1) can be studied in the framework of theory of inverse problems and in multidimensional spaces. Having an operator $A : (X, \|.\|) \rightarrow (Y, \|.\|))$ between two Banach spaces, an *inverse problem* defined by A is to find for $g \in Y$ some $f \in X$ such that

$$A(f) = g \tag{2}$$

(Bertero [1]). If for every $g \in Y$ there exists a unique solution $f \in X$, i.e. a unique inverse operator $A^{-1} : Y \rightarrow X$, then the inverse problem is called well-posed.

As a measure of the stability of the solutions of an inverse problem, one can use the condition number defined for a well-posed problem given by an operator A as $cond(A) = \|A\|\|A^{-1}\|$. Inverse problems are often ill-posed or ill-conditioned. When a solution does not exist, one can search for best approximate solution f^o, called a *pseudo-solution*, defined by $\|A(f^o) - g\|_Y = \min_{f \in X} \|A(f) - g\|_Y$ and a normal pseudo-solution f^+, which is a pseudo-solution of the minimal norm, i.e., $\|f^+\|_X = \min\{\|f^o\|_X : f^o \in S(g)\}$, where $S(g)$ is the set of all pseudo-solutions of the inverse problem given by an operator A and data g. When for every $g \in Y$ there exists a normal pseudo-solution f^+, then a pseudo-inverse operator $A^+ : Y \to X$ can be defined as $A^+(g) = f^+$. Similarly as in the case of well-posed problems, the *condition number of an operator* A with a pseudo-inverse A^+ is defined as $cond(A) = \|A\|\|A^+\|$.

3   Inverse Problem in Reproducing Kernel Hilbert Spaces

For X and Y finite dimensional, the pseudo-solution can be described in terms of Moore-Penrose pseudo-inverse of the matrix corresponding to the operator A. The concept of Moore-Penrose pseudo-inversion has been extended to the case of linear continuous operators between Hilbert spaces. Kůrková in [7] expressed as the *inverse problem* the problem of minimization of the empirical error E_Z in a Hilbert space X of functions on some Ω. Let $Z = \{z = (x, y)\}$, where $x = (x_1, \ldots, x_m) \in \Omega^m$ and $y = (y_1, \ldots, y_m) \in R^m$, be a sample set defining the empirical error functional E_Z. Consider an operator $L_x : X \to R^m$ defined as $L_x(f) = (\frac{f(x_1)}{\sqrt{m}}, \ldots, \frac{f(x_m)}{\sqrt{m}})$. Then E_Z can be represented as

$$E_Z = \|L_x - \frac{y}{\sqrt{m}}\|_2^2, \tag{3}$$

where $\|.\|_2$ denotes the l_2 norm on R^m. We will now limit to a class of Hilbert spaces, on which are operators L_x continuous, a reproducing kernel Hilbert space RKHS. A reproducing kernel Hilbert space RKHS is a Hilbert space formed by functions defined on a nonempty set such that for every $x \in \Omega$ the evaluation functional $\mathcal{F}_x$, defined for any f in the Hilbert space as $\mathcal{F}_x(f) = f(x)$ is bounded. RKHS can be characterized in terms of kernels, which are symmetric positive semi-definite functions $K : \Omega \times \Omega \to R$ i.e., functions satisfying for all m, all $(w_1, \ldots, w_m) \in R^m$, and all $(x_1, \ldots, x_m) \in \Omega^m$, $\sum_{i,j=1}^m w_i w_j K(x_i, x_j) \geq 0$. A kernel is positive definite if $\sum_{i,j=1}^m w_i w_j K(x_i, x_j) = 0$ for any distinct $x_1, \ldots, x_m$ implies that for all $i = 1, \ldots, m, w_i = 0$. To every RKHS one can associate a unique kernel $K : \Omega \times \Omega \to R$ such that for every f in the RKHS and $x \in \Omega$ $f(x) = \langle f, K_x \rangle_K$ (*reproducing property*), where $K_x : \Omega \to R$ is defined as $K_x(y) = K(x, y)$ for all $y \in \Omega$.

For a kernel K, a positive integer m and a vector $x = (x_1, \ldots, x_m)$, by $K[x]$ we denote the $m \times m$ matrix defined as $K[x]_{i,j} = K(x_i, x_j)$, called the *Gram matrix of the kernel K with respect to the vector x.* The most common kernel is the *Gaussian kernel* $G_x^\rho(y) = e^{-\rho\|x-y\|^2}$ on $R^d \times R^d$. For this kernel, the space $\mathcal{H}_{G^\rho}(R^d)$ (the hypothesis space) contains all functions computable by RBF networks with a fixed width equal to ρ.

It was proven in [2] and in [9] that for every kernel K and every sample of empirical data set Z, there exists a solution of the problem of minimization of the empirical error functional E_Z (3) over the space $\mathcal{H}_K(\Omega)$. This function is formed by a linear combination of functions $K_{x_1}, \ldots, K_{x_m}$:

$$f^+ = L_x^+\left(\frac{y}{\sqrt{m}}\right) = \sum_{i=1}^{m} c_i K_{x_i}. \tag{4}$$

Function f^+ can be interpreted as an input/output function of a neural network with one hidden layer of kernel units and a single linear output unit. The coefficients of the linear combination $c = (c_1, \ldots, c_m)$ (output weights) can be computed applying the pseudo-inverse of the Gram matrix of the kernel K with respect to the input data vector x on the output data vector: $c = K[x]^+ y$. The set of such solutions $argmin(\mathcal{H}_K(\Omega); E_z)$ is a closed convex set of the form $\sum_{i=1}^{m} c_i K_{x_i} + N(L_x)$, where $c = K[x]^+ y$ and $N(L_x)$ is the null space of the operator L_x. For K positive definite, the solution interpolates the data and minimum of E_Z over $\mathcal{H}_K(\Omega)$ is equal to zero. We will study the sensitivity of the solution $f^+ = \sum_{i=1}^{m} c_i K_{x_i}$ with respect to a small perturbation of the vector of output data y by means of the condition number of the matrix $K[x]$, which we will denote $\kappa(K)$. The *condition number of a symmetric positive definite matrix* is the ratio between the largest and the smallest singular value.

4   Tikhonov Regularization in RKHS

The function $f^+ = \sum_{i=1}^{m} c_i K_{x_i}$ with $c = K[x]^+ y$ guarantees the best fit to the sample data z that can be achieved using functions from the space $(\mathcal{H}_K(\Omega); E_Z)$. By choosing as a hypothesis space to be a RKHS, we impose a condition on oscillations of potential solutions. The restriction on potential solutions can be by penalizing the size of the norm $\|.\|_K$ of the solution. This approach to constrain the solutions of ill-posed inverse problems is called Tikhonov regularization [16].

Tikhonov's regularization replaces the (functional minimization) problem

$$\min_{f \in X} \|A(f) - g\|_Y^2 \tag{5}$$

by the problem

$$\min_{f \in X} \|A(f) - g\|_Y^2 + \gamma \|f\|_X^2 \tag{6}$$

where the regularization parameter γ plays the role of a trade-off between fitting the empirical and conceptual data. It was proven by [1] that even if the original inverse problem is ill-posed (i.e. it does not have a unique solution), for every $\gamma > 0$, the regularized problem has a unique solution (due to the uniform convexity of the functional $\|.\|_Y^2$). With γ going to zero, the solutions of the regularized problem converge to the pseudo-solution with the minimal norm $A^+(g)$. Let

$I : R^m \to R^m$ be the identity operator and $I[x]_{i,j}$ corresponding $m \times m$ matrix corresponding to it. The following theorem is a shortened version of the theorem from Kůrková [9].

Theorem 1. *[9] Let $K : \Omega \times \Omega \to R$ be a kernel, m be a positive integer, $z = (x, y)$, where $x = (x_1, \ldots, x_m) \in \Omega^m$, $x_1, \ldots x_m$ are distinct, $y = (y_1, \ldots, y_m) \in R^m$ and $\gamma > 0$, then:*

(i) There exists a unique solution f^γ of the problem (6) in $\mathcal{H}_K(\Omega)$;
(ii) $f^\gamma = \sum_{i=1}^m c_i K_{x_i}$, where $c = (K[x] + \gamma m I)^{-1} y$;
(iii) When K is positive definite, then

$$\kappa(K_\gamma) := \kappa(K[x] + \gamma m I)) = 1 + \frac{(\kappa(K[x]) - 1)\lambda_{min}}{\lambda_{min} + \gamma m} \tag{7}$$

where λ_{min} is the minimal singular value of $K[x]$.

Theorem 1 (iii) shows how much ill-conditioning of the problem (5) in a RKHS can be improved by regularization. As $\lim_{\gamma m \to \infty}(1 + \frac{((\kappa(K[x]) - 1)\lambda_{min})}{\lambda_{min} + \gamma m}) = 1$, for sufficiently large γm, the condition number of the matrix $(K[x] + \gamma m I)$ is close to 1. In case the size of the data set m is fixed, the condition number of the regularized matrix from (6) is influenced only by the Tikhonov regularization parameter of γ, to which we will devote this paper. The following corollary shows explicitly, how the regularization influences the size of the condition number of the corresponding matrix.

Corollary 1. *Let $K : \Omega \times \Omega \to R$ be a kernel, m be a positive integer and its corresponding Gram matrix $K[x]$ be positive definite and normalized. Then for every $\gamma > 0$ regularization parameter holds*

(i) $\kappa(K_\gamma) \leq 1 + \frac{1}{\gamma m}$.
(ii) If $m \geq 2$ and $\gamma \geq \frac{1}{m}$ then $\kappa(K_\gamma) \leq 2$.
(iii) $\kappa(K[x] + \gamma m I) < \kappa(K[x])$.

Proof: (i) Since the Gram matrix is symmetric, from Theorem 1 (iii) we have $\kappa(K_\gamma) = 1 + \frac{(\frac{1}{\lambda_{min}} - 1)\lambda_{min}}{\lambda_{min} + \gamma m} = 1 + \frac{1 - \lambda_{min}}{\lambda_{min} + \gamma m} = \frac{1 + \gamma m}{\lambda_{min} + \gamma m}$. We want to investigate for which $\gamma > 0$ is $\frac{1 + \gamma m}{\lambda_{min} + \gamma m} < 1 + \frac{1}{\gamma m}$. This unequation leads to $\gamma m < \lambda_{min} + \gamma m$ which holds for any $\lambda_{min} > 0$. (ii) is obvious. (iii) Since $\frac{\lambda_{min}}{\lambda_{min} + \gamma m} < 1$; if both sides of this unequation are multiplied by $\kappa(K[x]) - 1$, we get then $1 + \frac{(\kappa(K[x]) - 1)\lambda_{min}}{\lambda_{min} + \gamma m} < \kappa(K[x])$ which leads to $\kappa(K[x] + \gamma m I) < \kappa(K[x])$.

5   Sensitivity of Solutions of Ill-Posed Problems in General Hilbert Spaces

Reginska in [14] investigated discrete ill-posed problem for a general operator in a Hilbert Space. She proved that the condition number can be decreased only to the square root of that for the non-regularized problem. Comparing this result to Corollary 1 from Section 4, one can see how advantageous for condition numbers was to limit the minimization problem on RKHS.

Consider problem (2) where A is a compact operator in a Hilbert space X. As it was already said in Section 2, if the range of operator A is not finite dimensional, the inverse of A as well as the pseudoinverse A^+ are not bounded and consequently the problem (2) is ill-posed in the space X also in the least squares sense (problem (1)). So, for slightly perturbed right-hand side g_δ, $\|g - g_\delta\| \leq \delta$, the solution $A^+ g_\delta$ does not necessarily exist, or, if it exists, the distance between these two solutions may be arbitrary large. For achieving a stable solution of ill-posed problems a regularization method can be applied. In practice a finite dimensional approximation of (2) is considered, i.e.

$$A_m f_m = g_m \tag{8}$$

where A_m is a linear operator in a finite dimensional space. The problem (8) is called *a discrete ill-posed problem* ([14]). The spaces X_m are not assumed to be subspaces of X, so one can apply the Temam approach ([15]) where an approximation of X is a sequence of triples $\{X_m, r_m, p_m\}_{m=1}^{\infty}$ where X_m is a Hilbert space and $r_m : X \to X_m$, and $p_m : X_m \to X$ are linear bounded restriction and prolongation operators, respectively. The *approximation is convergent* if $p_m r_m$ converges point-wisely to the identity in X as $m \to 0$. If r_m and p_m are uniformly bounded, then the *approximation* is said to be *stable*. We will assume that the considered approximation of X is convergent and stable and $r_m p_m$ is the identity operator on X_m for any m. We will assume that $g_m := r_m g$. Let $g_{m,\delta} = r_m g_\delta$ and f_m^δ be the solution of (8) with the right hand side $g_{m,\delta}$. Let us consider the Tikhonov regularization method which replaces the least squares problem (1) by the following functional minimization problem:

$$\min_{\{f \in X_m\}} \{\|A_m f_m - g_{m,\beta}\|^2 + \beta \|f\|^2\}. \tag{9}$$

The solution of (9) is called a regularized solution and satisfies the equation

$$(A_m^* A_m + \beta I) f_{m,\gamma}^\beta = A_m^* g_{n,\beta}. \tag{10}$$

Notice, that we are using a different notation for the Tikhonov parameters here and in (7), since the regularization matrices differ. It holds $\beta = m\gamma$.

Discrete ill-posed problems are ill-conditioned. Similarly as it was already mentioned in Section 3 for symmetric positive definite matrices, the condition number $\kappa(A_m)$ is defined as $\kappa(A_m) = \frac{\lambda_{max}(A_m)}{\lambda_{min}(A_m)}$ where $\lambda_{max}(A_m)$ is the largest singular value of A_m and $\lambda_{min}(A_m)$ the smallest one. Roughly said, the condition number characterizes "the sensitivity of problem (8) on data perturbation". Let $f_{m,\beta}$ and $f_{m,\beta}^\beta$ be solutions of (10) for the right-hand side g_m and $g_{m,\beta} := r_m g_\beta$. If $\|f_{m,\beta}\| > 0$ then the influence of the right hand side error on the relative solution error can be described by

$$\frac{\|f_{m,\beta} - f_{m,\beta}^\beta\|}{\|f_{m,\beta}\|} \leq \kappa(A_{m,\beta}) \frac{\|g_m - g_{m,\beta}\|}{\|g_m\|}. \tag{11}$$

A parameter $\beta > 0$ will be called *optimal with respect to the sensitivity of problem* (8) if for all $\beta > 0$ achieves the right hand side of (11) in β its minimum. Regińska proved in [14] the following theorem.

Theorem 2. *[14] The optimal choice of regularization parameter γ with respect to sensitivity of problem is*

$$\beta_m = \lambda_{min}(A_m)\lambda_{max}(A_m) \tag{12}$$

and gives

$$\kappa_m^{opt} = \frac{1}{2}(\kappa(A_m)^{1/2} + \kappa(A_m)^{-1/2}). \tag{13}$$

6   Optimal Regularization Parameter with Respect to the Sensitivity of the Solution in RKHS

In this section we will apply the results from previous sections to the regularization in finite dimensional RKHS. In RKHS holds that $A^+ = A^*$ (in the formula (10)). The following theorem gives the value of the regularization parameter γ for which the solution of (9) in RKHS has optimal sensitivity.

Theorem 3. *Let $K : \Omega \times \Omega \to R$ be a kernel, m be a positive integer, $Z = \{z = (x, y)\}$, where $x = (x_1, \ldots, x_m) \in \Omega^m$, $x_1, \ldots x_m$ are distinct, $y = (y_1, \ldots, y_m) \in R^m$. Let the corresponding Gram matrix $K[x]$ be positive definite. Then the optimal value of parameter $\gamma_{opt} > 0$ with respect to the sensitivity of the regularized solution of the problem $(\mathcal{H}_K(\Omega); E_Z + \gamma\|.\|_K^2)$ is*

$$\gamma_{opt} = \frac{\lambda_{min}(\kappa(K) - \frac{\kappa(K)+1}{2\sqrt{\kappa(K)}})}{(\frac{\kappa(K)+1}{2\sqrt{\kappa(K)}} - 1)}. \tag{14}$$

Proof: Asserting formula (13) into (7) one gets $\lambda_{min}\kappa(K) + \beta = \lambda_{min}\frac{\kappa(K)+1}{2\sqrt{\kappa(K)}} + \beta\frac{\kappa(K)+1}{\sqrt{\kappa(K)}}$ which leads to $\lambda_{min}(\kappa(K) - \frac{\kappa(K)+1}{2\sqrt{\kappa(K)}}) = \beta(\frac{\kappa(K)+1}{2\sqrt{\kappa(K)}} - 1)$ and consequently to formula (14).

As it has been already said, if K is a symmetric positive definite matrix, its condition number is $\kappa(K) = \frac{\lambda_{max}}{\lambda_{min}}$, where λ_{max} and λ_{min} are the largest and the smallest singular values, respectively. Suppose still that the matrix norm $\|K\|_2 = \max \frac{\|Kx\|_2}{\|x\|_2}$ equals to 1, i.e. the matrix K is normalized, where $\|x\|_2$ is the Euclidean (vector) norm. This can be always achieved by multiplying the set of equations $Kx = b$ by a suitable constant (or see Remark 1 in Section 6.2). Then we get $\kappa(K) = \frac{1}{|\lambda_{min}|}$. A normalized symmetric positive definite matrix is thus ill-conditioned if and only if $\|K^{-1}\|_2$ is large, that is if λ_{min} is small [18].

Corollary 2. *Let $K : \Omega \times \Omega \to R$ be a kernel, m be a positive integer, $z = (x, y)$, where $x = (x_1, \ldots, x_m) \in \Omega^m$, $x_1, \ldots x_m$ are distinct, $y = (y_1, \ldots, y_m) \in R^m$.*

Let the corresponding Gram matrix $K[x]$ be positive definite and normalized. Then the optimal value of parameter $\gamma_{opt} > 0$ with respect to the sensitivity of the regularized solution f_{opt} of the problem $(\mathcal{H}_K(\Omega); E_z + \gamma\|.\|_K^2)$ is

$$\gamma_{opt} = \frac{1}{(1 + \lambda_{min})} - \frac{\lambda_{min}(2 - \sqrt{\lambda_{min}})}{2}. \tag{15}$$

6.1 An Upper Bound on γ_{opt} for the Case of Unknown λ_{min}

The computation of singular values of a matrix by a full singular value decomposition for real-time applications is usually very time-consuming. In some practical problems (neural networks including), where time is an important aspect, even an upper estimate for γ_{opt} can suffice. Consider a kernel whose corresponding Gram matrix K is positive definite and normalized. Let $\|K\|_F = \sqrt{\sum_{i=1}^m \sum_{j=1}^m |k_{ij}|^2}$ denote the Frobenius norm of matrix K. Then avoiding the direct singular values computation of K and using only the (easily computable) information of the values of K, the determinant of K and the Frobenius norm of K, one can get the upper bound for γ_{opt} by the following theorem.

Theorem 4. *Let $K : \Omega \times \Omega \to R$ be a kernel whose corresponding $m \times m$ Gram matrix K is positive definite and $\|K\|_2 = 1$. Let $\|K\|_F$ denotes the Frobenius norm of matrix K and $\det K$ the determinant of K. Denote $k_{ij} = k(x_i, x_j)$. Let the matrix $K' = (k'_{ij})$ be obtained by deleting from K any arbitrary row or column. Denote $K^{(0)} = K, K^{(s)} = (k_{ij}^{(s)}) = (K^{(s-1)})'$, where $s = 1, \ldots, m - 1$. Denote $r_i(K) = \sum_{j=1}^m |k_{ij}|$, $i = 1, \ldots, m$. Then*

$$\gamma_{opt} \le \frac{1}{1 + \theta} - \theta + \frac{\psi\sqrt{\psi}}{2} \tag{16}$$

where $\psi = \max_{i,j,k_{i,j}^{(m-1)} \ne 0} r_i(K^{(m-1)})$ and $\theta = \max\{\frac{|\det K|}{2^{m/2-1}\|K\|_F}, \omega\}$ where $\omega = \min\{\frac{1}{2}\min_{i\ne j,|k_{ij}|\ne 0}\{|k_{ii}| + |k_{jj}| - [(|k_{ii}| - |k_{jj}|)^2 + 4\sum_{s=1,s\ne i}^m |k_{is}| \sum_{r=1,r\ne j}^m |k_{jr}|)]^{\frac{1}{2}}\}, \min_{i=1,\ldots,m} |k_{ii}|\}$.

Proof: We will apply Corollary 2 and some known upper and lower bounds for λ_{min}. For the lower estimate of λ_{min} from the first part of the sum from formula (14) $\frac{1}{(1+\lambda_{min})} - \lambda_{min}$ we used maximum of lower bounds from [12] (page 144) $\frac{|\det K|}{2^{(m/2)-1}\|K\|_F} \le \lambda_{min}$ and the lower bound from Theorem 3.5 from [6] (in Appendix) and applied it for the symmetric positive definite matrices. For the upper bound of λ_{min} in the resting summand of formula (15) $\frac{\lambda_{min}\sqrt{\lambda_{min}}}{2}$ we used Theorem 2.2 from [6] and applied it for the symmetric positive definite matrices.

6.2 A Lower Bound on γ_{opt} for the Case of Unknown λ_{min}

Theorem 5. *Let $K : \Omega \times \Omega \to R$ be a kernel whose corresponding $m \times m$ Gram matrix K is positive definite and $\|K\|_2 = 1$. Let $\|K\|_F$ denotes the Frobenius* *norm of matrix K and $\det K$ the determinant of K. Denote $k_{ij} = k(x_i, x_j)$. Let the matrix $K' = (k'_{ij})$ be obtained by deleting from K any arbitrary row or column. Denote $K^{(0)} = K, K^{(s)} = (k_{ij}^{(s)}) = (K^{(s-1)})'$, where $s = 1, \ldots, m - 1$. Denote $r_i(K) = \sum_{j=1}^{m} |k_{ij}|$, $i = 1, \ldots, m$. Then*

$$\gamma_{opt} \geq \frac{1}{1 + \psi} - \psi + \frac{\theta\sqrt{\theta}}{2} \tag{17}$$

where $\psi = \max_{i,j,k_{i,j}^{(m-1)} \neq 0} r_i(K^{(m-1)})$ and θ and ω were defined in Theorem 4.

Proof: To have a smaller value than the first part of the sum from formula (15) $\frac{1}{(1+\lambda_{min})} - \lambda_{min}$, we use the lower bound from Theorem 2.2. from [6] and to have smaller value than the resting summand from formula (15), we applied the maximum from lower bounds for λ_{min} from Theorem 3.5. from [6] and from [12].

Remark 1: The assumption that the Gram matrix of K is normalized is not at all restrictive. To achieve that we can use for example so called *divisive normalization* defined by Weiss [17], which uses the Laplacian L of the kernel matrix K. For a data set $S = \{x_1, \ldots, x_m\}$, with $x_i \in R^p$, where the $m \times m$ and the symmetric kernel Gram matrix defined as $K = \{k_{ij}\}_{i,j=1}^{m} = k(x_i, x_j)$, the transformation is defined by $D^{-1/2}KD^{-1/2}$, where $D = \{d_{ij}\}_{i,j=1}^{m}$ and $d_{ij} = 0$ for $i \neq j$ and $d_{ij} = \sum_{k=1}^{m} k_{ik}$ for $i = j$, where $K = \{k_{ij}\}_{i,j=1}^{m}$.

7   Conclusion

In this paper, Tikhonov regularization was considered in the reproducing Kernel Hilbert Spaces and upper and lower bound for the optimal value of Tikhonov regularization parameter with respect to the sensitivity of the solution was derived without computing the singular values of the corresponding matrix. For the case of a normalized kernel, we gave an explicit formula for the optimal regularization parameter with respect to the sensitivity of the solution which needs only the knowledge of the minimal singular value of the corresponding matrix. Comparing to the results of Regińska in [14], these values do not require application of the SVD procedure (only for the normalized case, when only the minimal singular value has to be computed), so they can be computed very efficiently.

Acknowledgement

The author thanks to the Czech Grant Agency for the support by project MSMT CR 2C06001 (Bayes).

References

1. Bertero, M.: Linear inverse and ill-posed problems. Advances in Electronic and Electron Physics 75, 1–120 (1989)
2. Cucker, F., Smale, S.: Best Choices for Regularization Parameters in Learning Theory: On the Bias-Variance Problem. Found. Comput. Math. 2, 413–428 (2002)

3. Chang, C.-H., Ha, C.-W.: On eigenvalues of differentiable positive definite kernels. Integr. Equ. Oper. Theory 33, 1–7 (1999)
4. Golub, G.H., Heath, M., Wahba, G.: Generalized cross-validation as a method for choosing a good ridge parameter. Technometrics 21, 215–223 (1979)
5. Hansen, P.C.: Rank-Defficient and Discrete Ill-Posed Problems: Numerical Aspects of Linear Inversion. SIAM, Philadelphia (1997)
6. Kolotilina, L.Y.: Bounds for the singular values of a matrix involving its sparsity patterns. Journal of Mathematical Sciences 137(3) (2006)
7. Kůrková, V.: Learning from data as an inverse problem. In: Antoch, J. (ed.) Proceedings of COMPSTAT 2004, pp. 1377–1384. Physica-Verlag, Heidelberg (2004)
8. Kůrková, V.: Supervised learning with generalization as an inverse problem. Logic Journal of IGPL 13(5), 551–559 (2005)
9. Kůrková, V.: Supervised learning as an inverse problem, Technical report ICS AS CR (2004), http://www.cs.cas.cz/research/library/
10. Morozov, V.A.: On the solution of functional equations by the method of regularization. Soviet Math. Dokl. 7, 414–417 (1966)
11. Parlett, B.N.: The Symmetric Eigenvalue Problem, p. 357. SIAM, Philadelpia (1998)
12. Piazza, G., Politi, T.: An upper bound for the condition number of a matrix in spectral norm. Journal of Computational and Applied Mathematics 143, 141–144 (2002)
13. Poggio, T., Girosi, E.: Regularization algorithms for learning that are equivalent to multilayer networks. Science 247, 978–982 (1990)
14. Reginska, T.: Regularization of discrete ill-posed problems. BIT 44(1), 119–133 (2004)
15. Temam, R.: Numerical Analysis. D. Reidel Publishing Company (1973)
16. Tikhonov, A.N., Arsenin, V.Y.: Solutions of Ill-posed Problems. W.H. Winston, Washington (1977)
17. Weiss, Y.: Segmentation using eigenvectors: A unifying view. In: Proceedings IEEE International Conference on Computer Vision, vol. 9, pp. 975–982. IEEE Press, Piscataway (1999)
18. Wilkinson, J.H.: The algebraic eigenvalue problem. Monographs on Numerical Analysis, p. 191. Oxford University Press, Oxford (1965)

Appendix

Let $\mathcal{C}$ denote the set of complex numbers. Let $A^{m \times n}$ denotes a matrix and $\lambda_1(A) \geq \ldots, \geq \lambda_q(A)$, where $q = \min\{m, n\}$ be its singular values.

Theorem 6. *[6] Let $A = (a_{ij}) \in \mathcal{C}^{m \times n}$, $m, n \geq 1$, and let $q = \min\{m, n\}$. Then for an arbitrary sequence of nested principal $k \times k$ submatrices $A^{[k]}$, $k = n, \ldots, 1$ of the $q \times q$ matrix $\hat{A} = \{a_{ij}\}_{i,j}^q$*

$$\lambda_k(A) \geq \min\{\frac{1}{2} \min_{i \neq j, i,j \in K, |a_{ij}| + |a_{ji}| \neq 0} \{|a_{ii}| + |a_{jj}| - [(|a_{ii}| - |a_{jj}|)^2 \tag{18}$$

$$+ \sum_{s \in K, s \neq i} (|a_{is}| + |a_{si}|) \sum_{r \in K, r \neq j} (|a_{jr}| + |a_{rj}|)]^{1/2}\}, \min_{i \in K} |a_{ii}|\},$$

$k = n, \ldots, 1$, where $A[K] = A^{[k]}$ denotes the a sequence of nested principal $k \times k$ submatrices of A, K is the set of indices specifying the submatrix $A^{[k]}$.

Mixture of Expert Used to Learn Game Play

Peter Lacko and Vladimír Kvasnička

Institute of Applied Informatics
Faculty of Informatics and Information technologies
Slovak University of Technology, Ilkovičova 3, 842 16 Bratislava
{lacko,kvasnicka}@fiit.stuba.sk

Abstract. In this paper, we study an emergence of game strategy in multiagent systems. Symbolic and subsymbolic approaches are compared. Symbolic approach is represented by a backtrack algorithm with specified search depth, whereas the subsymbolic approach is represented by feedforward neural networks that are adapted by reinforcement temporal difference $TD(\lambda)$ technique. We study standard feed-forward networks and mixture of adaptive experts networks. As a test game, we used the game of simplified checkers. It is demonstrated that both networks are capable of game strategy emergence.

1 Introduction

Applications of $TD(\lambda)$ reinforcement learning [1, 2] to computational studies of emergence of game strategies were initiated by Gerald Tesauro [3–5] in 1992. He let a machine-learning program endowed with feed-forward neural network to play against itself a backgammon game. Tesauro has observed that a neural network emerged, which is able to play backgammon on a supreme champion level. These computer experiments were successfully repeated by Pollack et al. [6] who replaced the reinforcement learning adaptation of neural network by a hillclimbing algorithm. He used a couple of neural networks, the currently best network with its mutated version, which played backgammon against each other. Such evolved neural network was playing better and better backgammon game in the course of adaptation. The resulting neural network was not so excellent as the network produced by Tesauro, but it was surprisingly good considering its humble origin from a hill-climbing optimization method. Fogel and Chellapilla [7–9] tried to use an evolutionary algorithm in a population of neural networks, and achieved an emergence of checkers strategies that are competitive with moderate human players (classified as class B). Zaman and Wunsch [10] have used mixture of expert architecture for learning evaluation function of game GO with very good results.

The purpose of this paper is to use $TD(\lambda)$ reinforcement learning method for adaptation of mixture of adaptive experts network that is used as evaluators of next possible positions created from a given position by permitted moves. Neural network evaluates each position by a score number from an open interval $(0,1)$. a position with the largest score is selected as the forthcoming position,

V. Kůrková et al. (Eds.): ICANN 2008, Part I, LNCS 5163, pp. 225–234, 2008.
© Springer-Verlag Berlin Heidelberg 2008

while other moves are ignored. The method is tested on a simplified game of checkers, where that player wins, whose piece first reaches any of the squares on the opposite end of the game board. We compare classical feed-forward neural network with mixture of expert architecture. The networks are trained while playing against MinMax backtrack algorithm with the search depth 2. It is show that mixture of expert architecture can gain better strategy than classical feed-forward neural networks.

2   Simplified Game of Checkers

The game of checkers is played on a square board with sixty-four smaller squares arranged in an 8×8 grid of alternating colors, see Fig. 1. The starting position is with each player having 8 pieces (black, respectively white), on the 8 squares of the same color closest to his edge of the board. Each must make one move per turn. The pieces move one square, diagonally, forward. a piece can only move to a vacant square. One captures an opponent's piece by jumping over it, diagonally, to the adjacent vacant square. If a player can jump, he must. a player wins, if one of his/her pieces reaches asquare on the opponent's edge of the board, or captures the last opponent's piece, or blocks all opponent's moves. These rules do not comply with standard rules of checkers. By introduction of these rules, we avoided a finish, when the board is occupied by asmall number of pieces. Such finish would be very difficult for aneural network to play successfully. The difficulty for a neural network consists probably in the necessity to plan many moves ahead in such a finish.

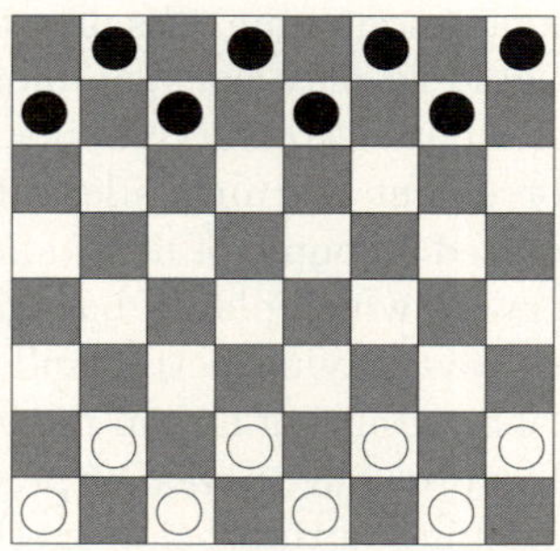

Fig. 1. Initial position of simplified game of checkers

3   Formalization of Game

In this section, we shall deal with aformalization of the game of checkers, which can be applied to all symmetric two player games (chess, go, backgammon, etc.). Let the current position of the game be described by a variable P, this position can be changed by permitted actions moves constituting a set $A(P)$. Using a move $\acute{a} \in A(P)$, the position P shall be transformed into anew position $\acute{P}$,

$P \overset{\acute{a}}{\to} \acute{P}$. An inverse position $\bar{P}$ is obtainable from aposition P by switching the color of all black pieces to white and of all white pieces to black. The term inverse position will be important for aformulation of a common algorithm, identical for both first and second player in a symmetric game. We shall use amultiagent approach, and we shall presume, that the game is played by two agents G1 and G2, which are endowed with cognitive devices, by which they are able to evaluate next positions.The formulation of the algorithm is as follows (or see Fig. 2):

1. The game is started by the first player, $G \leftarrow G1$, from astarting position, $P \leftarrow Pini$.
2. The player G generates from the position P aset of the next permitted positions $A(P) = \{P_1, P_2, ..., P_n\}$. Each such position P_i from the set of these positions is evaluated by a coefficient $0 < z_i < 1$. The player selects as his next position such $\acute{P} \in A(P)$, which is evaluated by themaximum coefficient, $P \leftarrow \acute{P}$. If the position P satisfies condition for victory, then the player G wins and the game continues with the step 4.
3. The other player takes turn in the game, $G \leftarrow G_2$, the position P is generated from the inverse of the current position, $P \leftarrow \bar{P}$, the game continues with the step 2.
4. End of the game.

The described algorithm is not fully deterministic, the first move of the first player is chosen from the set of permitted moves at random, after that the game is fully deterministic. The key role in the algorithm plays the calculation of coefficients $z = z(\acute{P})$ for positions $\acute{P} \in A(P)$. These calculations can be done either by methods of classical artificial intelligence, based on a combination of depth first search and various heuristics, or by soft computing methods. We shall use the second approach, where we shall turn our attention to modern approach of multiagent systems. It is based on a presumption, that the behavior of an agent in his/her environment and/or certain actions he/she performs are fully determined by his/her cognitive device, which demonstrates acertain plasticity (i.e. it is capable of learning). The cognitive device of the agent G is usually numerically realized by a parametric mapping $G(w) : P \to R$, where w is a parameter (parameters) of mapping, which assigns to each position (environment) a real number z, formally $z = G(P, w)$. Achange of w changes also the properties of the cognitive device (i.e. we get different parametric mapping). In this way, we have provided for the plasticity of the cognitive device, which an agent needs to be able to learn in the given environment, where such an agent handles the required task better and better. In general, the parametric mapping $G(w) : P \to R$ can be realized by many various ways, since the only requirement is, that the cognitive device has to be the universal approximator. This general condition is automatically satisfied with arequired precision for power series or for feed-forward neural networks with at least one layer of hidden neurons.

We shall presume that the agent can evaluate the next permitted positions by his/her cognitive device both in the case that he is the first player, as well as if he is the second player. This assumption is plausible for symmetric games, where

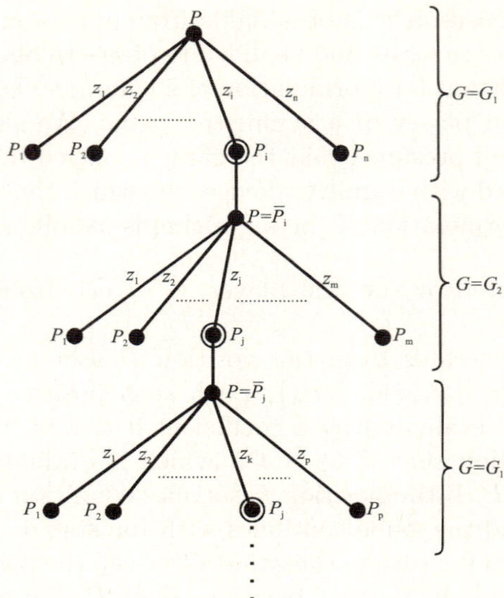

Fig. 2. Diagrammatic representation of symmetric 2-player game. The chosen player G creates from the current position P all the next permitted positions $P1$, $P2$, etc. By application of his cognitive device (represented for example by aneural network), this player evaluates each permitted position by a number z, and he shall chose his next position, one with themaximum evaluation. This elementary act of decision is repeated with the players taking turn.

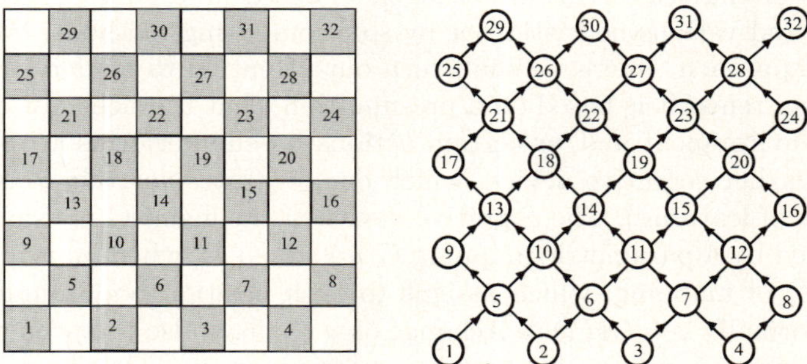

Fig. 3. The left hand part shows indexing of black squares of the game board (the colors are switched in Fig. 1 to see the pieces). The right hand side shows an oriented graph, which simply specifies the permitted moves of single pieces from the current players point of view.

strategic rules both for the first, as well as for the second player are the same, but acertain small difference exists in the fact, that the first player begins the game and therefore can to acertain extent enforce astyle of game to the second player. This requirement for universality of the cognitive device (i.e. the agent

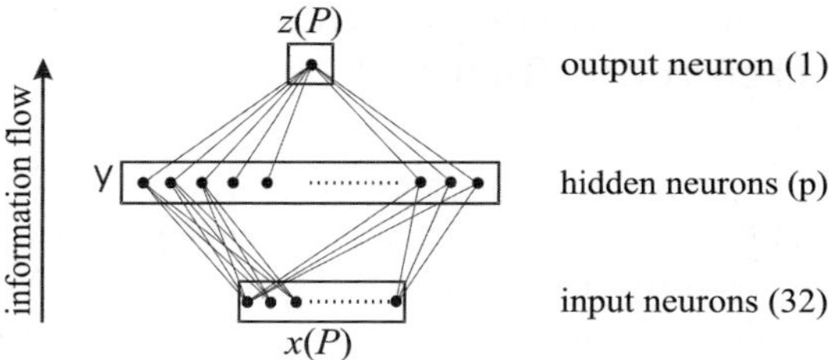

Fig. 4. Feed-forward neural network with one layer of hidden neurons. The input activities are equivalent to 32-dimensional vector x(P), which codes a position of the game. The output activity equals to the real number z(P) from the open interval (0,1), this number is an evaluation of the input position.

does not have aspecial cognitive device for the case, when it is the first player and another cognitive device for the case it is the second player) is realized in the above specified algorithm by the following approach. The second player with his/her cognitive device does not evaluate directly the positions $\acute{P} \in A(P)$ but their inversions $\bar{P}$, i.e. the agent, even while he/she is in the position of the second player, evaluates the next move from the same point of view as if he/she were the first player.

4   The Structure of the Cognitive Device Neural Network

Firstly, before we shall specify the cognitive device of agents, we have to introduce the so-called numerical representation of positions. Position is represented by a 32-dimensional vector (see Fig. 3)

$$x\left(P\right) = (x_1, x_2, ..., x_{32}) \in \{0, 1, -1\}^{32} \tag{1}$$

where single entries specify single squares at the position P of the game

$$x_i = \begin{cases} 0 & (i-\text{th square is free}) \\ 1 & (i-\text{th square is occupied by our piece}) \\ -1 & (i-\text{th square is occupied by the opponent's piece}) \end{cases} \tag{2}$$

The used neural network has the architecture of a feed-forward network with one layer of hidden neurons. The activities of input neurons are determined by anumerical representation x(P) of the given position P, the output activity evaluates the position x(P) (see Fig. 4). The number of parameters of the neural network is $34p + 1$, where p is the number of hidden neurons. The plasticity of cognitive device can be in the case of neural network divided into two parts. It is (a) structural plasticity, which means, that the number of hidden neurons changes (i.e. the structure topology of neural network changes), and (b) parametric plasticity, which corresponds to achange of weights of connections and to achange of threshold coefficients of hidden and output neurons.

5 Adaptation of Cognitive Device of Agent with Temporal Difference TD(λ)-Method with Reward and Punishment

In this section we shall give the basic principles of the used reinforcement learning method, which currently in its temporal difference TD(λ) version belongs to effective algorithmic tools for adaptation of cognitive devices of multiagent systems. The basic principles of reinforcement learning are following: Agent observes the mapping of his/her input pattern to his/her output signal of his/her cognitive device (the output signal is often called an action or control signal). Agent evaluates the quality of the output signal on the basis of the external scalar signal reward. The aim of the learning is such an adaptation of the cognitive organ of the agent, which will modify the output signals toward maximization of external reward signals. In many cases the reward signal is delayed, it arrives only at the end of along sequence of actions and it can be understood as the evaluation of the whole sequence of actions, whether the sequence achieved the desired goal or not. In this case, the agent must solve the problem of temporal credit assignment, where learning is based on differences between temporal credit assignments for single elements of the whole sequence of actions.

In this section we shall outline the construction of TD(λ) method as acertain generalization of astandard method of gradient descent learning of neural networks. Let us presume, that we know the sequence of positions of the given agent - player and their evaluation by areal number z

$$P_1, P_2, ..., P_m, z_{reward} \tag{3}$$

where z_{reward} is an external evaluation of the sequence and corresponds to the fact, that the last position P_m means that the given agent won or lost

$$z_{reward} = \begin{cases} 1 & \text{(sequence of positions won)} \\ 0 & \text{(sequence of positions lost)} \end{cases} \tag{4}$$

From the sequence (3) we shall create m couples of positions and their evaluations by z_{reward}, which shall be used as training set for the following objective function

$$E\left(w\right) = \frac{1}{2}\sum_{t=1}^{m}\left(z_{reward} - G\left(\mathbf{x}_t; w\right)\right)^2 \tag{5}$$

We shall look for such weight coefficients of the neural network cognitive device, which will minimize the objective function. When we would find out such weight coefficients of the network that the function would be zero, then each position from the sequence (3) is evaluated by a number z_{reward}. The recurrent formula for adaptation of the weight coefficients is as follows

$$w = w - \alpha\frac{\partial E}{\partial w} = w + \Delta w \tag{6}$$

$$\Delta w = \alpha \sum_{t=1}^{m} \left(z_{reward} - z_t\right) \frac{\partial z_t}{\partial w} \tag{7}$$

where $z_t = G(P_i, w)$ is an evaluation of t-th position P_t by neural network working as a cognitive device. Our goal will be, that all the positions from the sequence (3) would be evaluated by the same number z_{reward}, which specifies, if outcome of the game consisting from the sequence (3) was a win, draw, or loss for the given player. This approach can be generalized to a formula, which creates the basis of the TD(λ) class of learning methods [1]

$$\Delta w = \sum_{t=1}^{m} \Delta w_t \tag{8}$$

$$\Delta w_t = \alpha \left(z_{t+1} - z_t\right) \sum_{k=1}^{t} \lambda^{t-k} \frac{\partial z_k}{\partial w} \tag{9}$$

where the parameter $0 < \lambda < 1$. Formulas (8) and (9) enable a recurrent calculation of the increment Δw. We shall introduce anew symbol et(λ), which can be easily calculated recurrently

$$e_t\left(\lambda\right) = \sum_{k=1}^{t} \lambda^{t-k} \frac{\partial z_k}{\partial w} \Rightarrow e_{t+1}\left(\lambda\right) = \lambda e_t\left(\lambda\right) + \frac{\partial z_{t+1}}{\partial w} \tag{10}$$

where $e_1\left(\lambda\right) = \partial z_1 / \partial w$. Then the single partial increments Δw_t are determined as follows

$$\Delta w_t = \alpha \left(z_{t+1} - z_t\right) e_t\left(\lambda\right) \tag{11}$$

6   Mixture of Local Experts

When solving complex task, there is almost always that opportunity to split it into smaller subtasks. Such subtask may be better solved by a neural network. We can create a set of neural networks [11], where every network is an expert for some subtask from the original problem. We will use t expert neural networks : $N_i = \left(G^{(i)}, w^{(i)}, \vartheta^{(i)}\right), (i = 1, 2, ..., t)$ and one network which will be used as gating network $N_g = (G^{(g)}, w^{(g)}, \vartheta^{(g)})$. The purpose of gating network is to choose which expert is best suited for solving given problem. We expect that all networks in set N_t have same number of input and output neurons. Gating network have the same number of input neurons like expert networks, but number of output neurons is t (number of expert networks).

For given combination of $\mathbf{x}_I / \hat{\mathbf{x}}_O$ are the outputs of expert neural networks described with a vector $\mathbf{x}_O^{(i)}$, the output of gating network is vector $\mathbf{x}_O^{(g)}$. All output activities are in interval (0,1), we will define proportional coefficients $p_i \in (0, 1)$, they describe the respond of gating network on input vector $\mathbf{x}_I$, proportional coefficients are defined as softmax function:

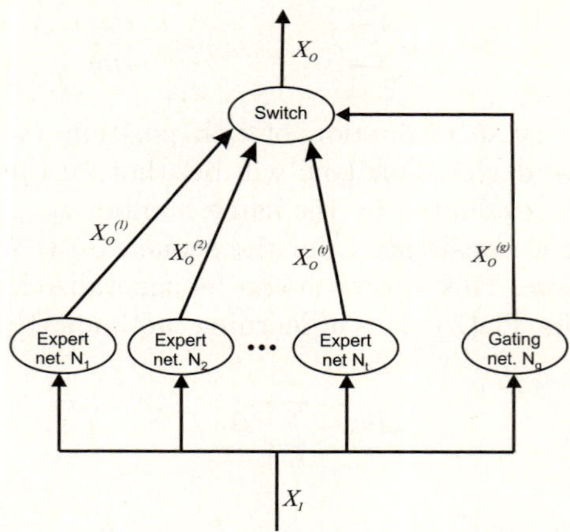

Fig. 5. Mixture of experts network architecture

$$p_i = \frac{\mathbf{x}_i^{(g)}}{\sum_{(j=1)}^{(p)} x_j^{(g)}}, (i = 1, 2, ..., t)$$ (12)

Proportional coefficients will be interpreted as weight of response of corresponding neural network. The output of the whole network will be:

$$\tilde{\mathbf{x}}_O = p_1 \mathbf{x}_O^{(1)} + p_2 \mathbf{x}_O^{(2)} + ... + p_t \mathbf{x}_O^{(t)} = \sum_{i=1}^{t} p_i \mathbf{x}_O^{(i)}$$ (13)

Objective function is then defined as:

$$E = \frac{1}{2} (\tilde{\mathbf{x}}_O - \hat{\mathbf{x}}_O)^2 = \frac{1}{2} \left(\sum_{i=1}^{t} \left(p_i \mathbf{x}_O^{(i)} \right) - \hat{\mathbf{x}}_O \right)^2$$ (14)

We ant to train the mixture of expert architecture with TD(λ) method, so objective function will be defined like this. We are summing up errors from all m moves:

$$E = \frac{1}{2} \sum_{j=1}^{m} \left(\tilde{\mathbf{x}}_O^{(j)} - \hat{\mathbf{x}}_O \right)^2$$ (15)

We then have two formulas for weights change one for gate network:

$$\Delta w_j = \alpha \frac{1}{2} \frac{1}{\left\| \mathbf{x}_O^{(j)} \right\|} \left(\tilde{\mathbf{x}}_O^{(j)} - \hat{\mathbf{x}}_O \right) \sum_{k=1}^{j} \lambda^{j-k} \sum_{i=1}^{t} \left(\frac{\partial \mathbf{x}_O^{(i,k)}}{\partial w} \left(\mathbf{x}_O^{(i,k)} - \tilde{\mathbf{x}}_O^{(k)} \right) \right)$$ (16)

and the second one for expert network weights change:

$$\Delta w_{i,j} = \alpha \frac{1}{2} \left(\tilde{\mathbf{x}}_O^{(j+1)} - \tilde{\mathbf{x}}_O^{(j)} \right) \sum_{k=1}^{j} \lambda^{j-k} p_{i,k} \frac{\partial \mathbf{x}_O^{(i,k)}}{\partial w}$$ (17)

where i is the index of i-th expert.

7   Results

We have tested the mixture of expert architecture and classical feed-forward network. Both used the TD(λ) learning rule while playing against MiniMax algorithm. For training and testing of the feed-forward network we used a two-layered feed-forward network with 8 hidden neurons, the learning rate α=0.1 and the coefficient λ=0.99. The network learned after each game, when it was evaluated by 1 if it won, and 0 evaluated it if it lost. For training and testing of the mixture of experts architecture we used 4 experts. Every expert was a two-layered feed-forward network with 8 hidden neurons. The gating network was two-layered feed-forward network with 8 hidden neurons.

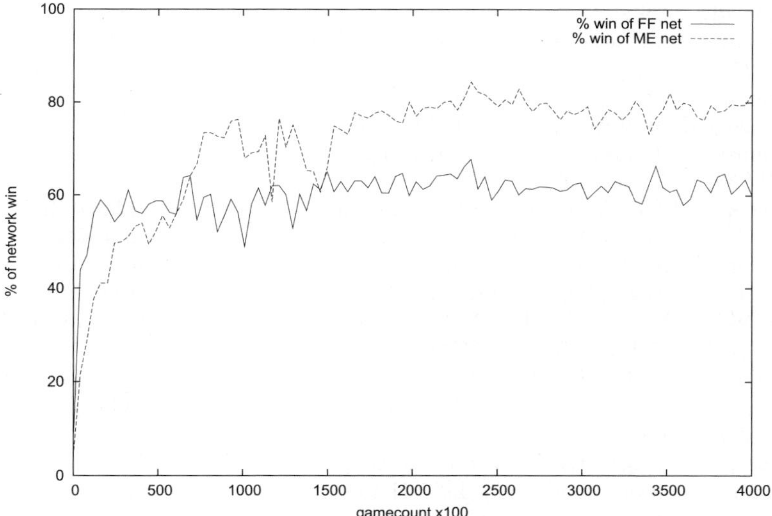

Fig. 6. Comparison of learning between feed-forward network and mixture of experts architecture playing against algorithm MinMax (working to the search depth 2)

In the Fig. 6 there is an evolution of success of neural networks learning against algorithm MinMax (working to the search depth 2). The graph clearly shows that mixture of experts architecture can decompose the problem of game play into smaller parts which are then better solved by local experts.

8   Conclusion

The purpose of this paper is a computational study of game-strategy emergence for a simplified game of checkers. We have compared standard feed-forward network with mixture of experts architecture. Both were trained with approach of reinforcement learning with TD(λ) algorithm. The learning process was execute while playing against MiniMax algorithm. The results clearly shows that mixture of experts architecture can decompose the problem of game play into smaller parts which are then better solved by local experts. Thus this architecture is capable of better game play as standard feed-forward network.

Acknowledgement. This work was supported by Scientic Grant Agency of Slovak Republic under grants No. 1/0804/08 and No. 1/4053/07.

References

1. Sutton, R.S.: Learning to predict by the method of temporal differences. Machine Learning 3, 9–44 (1988)
2. Sutton, R.S., Barto, A.G.: Reinforcement Learning: An Introduction. MIT Press, Cambridge (1998)
3. Tesauro, G.J.: Practical issues in temporal difference learning. Machine Learning 8, 257–277 (1992)
4. Tesauro, G.J.: TD-gammon, a self-teaching backgammon program achieves master-level play. Neural Computation 6(32), 215–219 (1994)
5. Tesauro, G.: Temporal difference learning and td-gammon. Communications of the ACM 38(3), 58–68 (1995)
6. Pollack, J., Blair, A., Land, M.: Coevolution of a backgammon player. In: Langton, C.G. (ed.) Proc. of Artificial Life V, pp. 92–98. MIT Press, Cambridge (1996)
7. Chellapilla, K., Fogel, D.B.: Evolving neural networks to play checkers without expert knowledge. IEEE Trans. Neural Networks 10(6), 1382–1391 (1999)
8. Chellapilla, K., Fogel, D.B.: Evolution, neural networks, games, and intelligence. Proc. IEEE 87(9), 1471–1496 (1999)
9. Chellapilla, K., Fogel, D.B.: Evolving an expert checkers playing program without using human expertise. IEEE Transactions on Evolutionary Computation 5(4), 422–428 (2001)
10. Zaman, R., Wunsch III, D.: TD methods applied to mixture of experts for learning 9x9 go evaluation function. In: IJCNN 1999 vol. 6, pp. 3734–3739 (1999)
11. Jordan, M., Jacobs, R.: Modular and hierarchical learning systems. The Handbook of Brain Theory and Neural Networks. MIT Press, Cambridge (1995)

Unsupervised Bayesian Network Learning for Object Recognition in Image Sequences[*]

Daniel Oberhoff and Marina Kolesnik

Fraunhofer Institut FIT, Schloss Birlinghoven, 53754 Sankt Augustin

Abstract. We present work on the learning of hierarchical Bayesian networks for image recognition tasks. We employ Bayesian priors to the parameters to avoid over-fitting and variational learning. We further explore the effect of embedding Hidden Markov Models with adjusted priors to perform sequence based grouping, and two different learning strategies, one of which can be seen as a first step towards online-learning. Results on a simple data-set show, that the simplest network and learning strategy work best, but that the penalty for the more complex models is reasonable, encouraging work on more complex problems.

1   Introduction

In the face of the increasing amount of data being generated and communicated, the need to automatically process such data becomes ever stronger. A very important form of information, and often the most important form, for us humans is visual information. In the recent decades computers have become capable enough to handle large amounts of visual information, in fact nearly every computer sold today is capable of playing and recording video and transmitting it over the internet. But the handling of this information by the computer is still mostly restricted to storing it or providing interactive editing aids for the human actor. The computer itself is still quite ignorant of the content of this visual data. This stands in contrast to, for example, the automatic text indexing which is integrated into most modern operating systems, and even the emerging automatic music annotation tools.

Many approaches have been developed to tackle this problem with machine learning algorithms and very respectable performances have been achieved, but mostly with fully labeled data, where the task is reduced to "learning" the implicated function from data to labels. But when we think of learning we usually also think of autonomously acquiring new information and learning about the nature of the data even in the absence of a teacher that gives labels to everything. Thus the desire to have a system which will learn everything there is to learn about some visual data, or at least a generally useful subset of "everything" in the form of a structured model, and to apply this model to new data from the same or similar sources and thereby generalize the acquired knowledge. And such a system could still make use of labels by co-generalizing labels applied to

[*] Supported by the EC, Contract No.: 12963 NEST, project MCCOOP.

V. Kůrková et al. (Eds.): ICANN 2008, Part I, LNCS 5163, pp. 235–244, 2008.
© Springer-Verlag Berlin Heidelberg 2008

a subset of data to any amount of new data by making use of the generated model. The mechanism by which this works is analogous to "hashing", where a short descriptor captures the essential information of a data-set or document and this descriptor can be used to "index" it and quickly find related data-sets or documents. This capability to learn without supervision while being able to make use of labels effectively prevents over-fitting. Regularization mechanisms that have been used in the past can be seen as a first step in the direction of unsupervised Bayesian models, where the regularizing mechanisms appear in the form of probabilistic priors.

Following we will first explain the motivation for our work and draw comparisons with existing Bayesian network models reported in the literature. We then describe the overall structure of our probabilistic model with an emphasis on the mathematical details of learning and inference. Finally, we will present results of several experiments in which the model has been applied for recognition of handwritten digits and cars in real world video data. We will conclude with the analysis of the system performance and recommendations for future research.

2   Motivation

Humans seamlessly comprehend visual data, outperforming computers equipped with state-of-the-art algorithms in all but the simplest visual tasks. This spurred a bunch of efforts directed towards understanding, modeling, and replicating the neural apparatus of human visual pathways in the hope to improve the flexibility of recognition approaches to changing visual representations of world objects. The biologically motivated trend has been accelerated by numerous findings concerning the principles underlying the ascendance of neural responses through the hierarchy of visual areas in the human brain.

An important decision when starting composition of such a neural model is the need to decide on the kind and scope of knowledge - as much of it is open to interpretation, that has to be incorporated into the model in details. Early modeling approaches were loaded with over-complicated mechanisms of neural interactions, which led to computationally inefficient schemes (see for example [1]). These, however, helped understanding the emerging principles and paved the way for the construction of more abstract mechanisms, and further on of highly simplified computational schemes, where a number of processing steps have been reduced to the bare minimum (c.f. [2]).

In this work we follow this latest route, in which the hierarchical networked structure of the visual cortex and the mostly local nature of it's computational processes form the basis for the system *topology*, yet the system components are only modeled in their *functionality* so that modeling of unnecessary details is avoided. Specifically, this work is based on a series of models, which formulate the problem as one of discovering the *hidden causes* of the image data, which are modeled by the nodes of the network. Consider for example Fig. 1, *right*: The image of a car is caused by the presence of the car, which reflects light into the eye or camera to form this image. More indirectly, the car is the cause for the presence of individual components, which again can be decomposed into composition of certain image features. This hierarchy of causes is then thought to be modeled by the hierarchical network shown on the left. The appropriate mathematical framework for such a model is that of Bayesian Inference, and has been applied to this problem in at least two other work groups: that of G.E. Hinton et. al. at Toronto University in Canada (see for example [3]), and that of Thomas Dean et. al. at Brown University (see f.e. [4,5]). The presented system is strongly based on the work presented by Thomas Dean et. al. and in the conclusion we will specifically draw a comparison between experimental results and the results presented in [6,5].

3   Probabilistic Model

Fig. 1 shows a cross-section of the Bayesian graphical network for modeling of object images. The underlying probabilistic concept is that a hierarchy of (hidden) nodes is "generating" the perceived image. For example, an image might be a house, which consists of parts like windows, which, in turn, are characterized by a certain combination of edges. The graph is acyclic, but not a tree, as nodes share some of their descendants. This overlap allows to compensate partially for the lack of lateral connectivity and to enforce a certain level of lateral coherency.

Interspersed layers, in which nodes have only single descendants allow the extraction of local invariances from the input. Nodes with multiple descendants should thus faithfully represent the state of their descendants, while nodes with a single descendant should extract information about the presence of particular feature, while neglecting the exact state of that feature (i.e. the presence of a line, but with relatively low precision of the spatial position or rotation of that line). In the Bayesian framework this task is performed by mapping the states of the descendant onto fewer states. Since the objective is invariance we appropriate mechanism to generate this mapping is *sequence based grouping*. With sequence

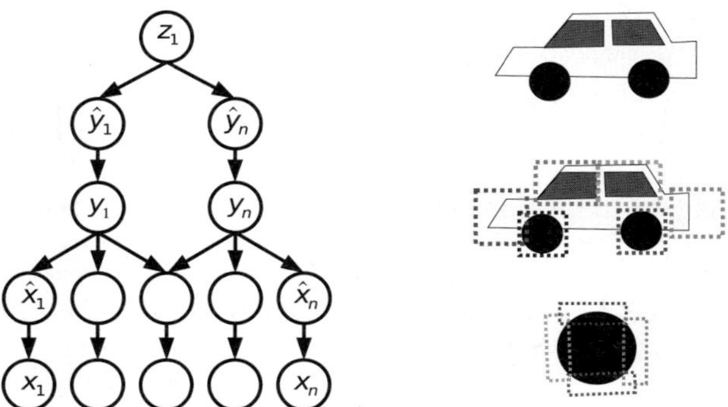

Fig. 1. *left*: 2-D cross-section of the Bayesian network. *right*: Illustration of the "hierarchical composition of parts" model implemented by the graphical model shown on the left.

based grouping we mean a mechanism, that groups together states that tend to follow each other in sequence. The concept of sequence based grouping (also referred to as "trace learning", "slow feature learning", or simply learning of invariances) has received increasing attention in the recent decade [7,5,8,9]. We will discuss our choice of mechanism for sequence based grouping in chapter 3.2.

3.1   Spatial Features

The objective of nodes with multiple descendants is the discovery of spatial patterns which we cast as a mixture modeling problem. In our case the mixture model is a sub-tree of a larger tree of multinomial random variables, the appropriate component model is thus a multivariate multinomial, and the joint probability of the mixture is:

$$p(x, z) = p(x|z)p(z) = \prod_{i,k,m,n} \pi_k^{z_{nk}} \mu_{kim}^{x_{nim} z_{nk}} \tag{1}$$

where the indices i, k, m, n enumerate the descendant, parent states or mixture components, descendant states, and data points, respectively. x is the emitted variable and z is the latent variable of the mixture model. The standard way of learning mixture models is expectation maximization (EM), corresponding to a maximization of the posterior. However EM has various deficiencies. One is that it tends to over-fit the data, the other is a strong dependency of the result on the number of components chosen. This can be avoided by learning a distribution over the parameters, and then optimizing the parameters of this distribution by maximizing the marginal probability. As priors for both the components and the mixture coefficients we choose the Dirichlet distribution, which is the conjugate prior of the multinomial distribution. For the parameters π_k the prior has the form $p(\pi) = \frac{1}{B(\alpha)} \prod_k \pi_k^{\alpha_k - 1}$ with hyper-parameters α_k, where $B(\alpha)$ is the Beta function. The prior over μ is a product of Dirichlet distributions for the componets: $p(\beta) = \prod_k \frac{1}{B(\beta_k)} \prod_{im} (\mu_{ikm}^{\beta_{ikm} - 1})$ with hyper-parameters β_{kim}. The full model with priors becomes:

$$p(x, z, \pi, \mu) = p(x|z, \mu)p(z|\pi)p(\pi)p(\mu) \tag{2}$$

$$= \prod_k \left[\pi_k^{z_k} \prod_n \mu_{kn}^{x_{kn} z_k} \right] \frac{1}{B(\alpha)} \prod_k \left[\pi_k^{\alpha_k - 1} \frac{1}{B(\beta_k)} \prod_{im} (\mu_k^{\beta_{kim} - 1}) \right] \tag{3}$$

The objective is now to maximize the marginal likelihood of the model given the data:

$$L(\alpha, \beta | X) = p(X|\alpha, \beta) \tag{4}$$

$$= \int_{-\infty}^{\infty} p(X|Z, \pi, \mu)p(Z|\pi)p(\mu)p(\pi)d\mu d\pi \tag{5}$$

The direct maximization of this likelihood is generally intractable, but we can approximate the distribution which is marginalized by one that factorizes over the the latent variables and the parameters:

$$p(X, \Theta) \equiv p(X, Z, \pi, \mu) \approx q(Z)q(\pi)q(\mu) \equiv q(\Theta) \tag{6}$$

The complete marginal likelihood can be decomposed into terms dependent on the approximating distributions only and an error term:

$$\ln p(X|\alpha, \beta) = \int q(\Theta) \ln \frac{p(X, \Theta)}{q(\Theta)} d\Theta - \int q(\Theta) \ln \frac{p(\Theta|X)}{q(\Theta)} d\Theta \tag{7}$$

$$\equiv \mathcal{L}(q) + \mathrm{KL}(q \parallel p) \tag{8}$$

where we have identified the second term as the Kullbach-Leibler-divergence, which is always positive. Because of this, $\mathcal{L}(q)$ is a rigorous lower bound on the likelihood, and therefore optimiziation will always improve it (or leave it unchanged), regardless of the choice of approximation $q(\Theta)$. Furthermore when $q(\Theta)$ is a factorization with respect to M subgroups of parameters Θ_i, $i = 1, ..., M$, we can make use of a general solution of this optimization problem ([10], chapter 10.1.1) yielding:

$$\ln q_i^*(\Theta_i) = \mathbb{E}_{j \neq i}\left[\ln p(X, \Theta)\right] \tag{9}$$

Where $\mathbb{E}_{j \neq i}\left[\ldots\right]$ denotes the expectation of the term in brackets with respect to the parameters in all the other groups (in the following we will omit the subscript since it is obvious from the context which group of parameters is being optimized). In the factorized form the maximization can then be efficiently performed over each group of variables separately in an iterative manner using variational calculus. This gives the following solution for optimizing with respect to the latent variables z_{nk}:

$$\ln q^*(z_{kn}) = \rho_{kn}^{z_{kn}} + c \tag{10}$$

$$\Rightarrow \mathbb{E}\left[z_{nk}\right] = \frac{\rho_{nk}}{\sum_k \rho_{nk}} \equiv r_{nk} \tag{11}$$

where:

$$\rho_{kn} = \mathbb{E}\left[\ln p(x_n|z_{kn}, \mu)\right] + \mathbb{E}\left[\ln p(z_{kn}|\pi)\right] \tag{12}$$

$$= \mathbb{E}\left[\ln \pi_k\right] + \sum_i \mathbb{E}\left[x_{ni}\right] \mathbb{E}\left[\ln \mu_{ki}\right] \tag{13}$$

The equivalence relation in (10-11) follows from the fact that $q^*(z_{kn})$ has again the form of a multinomial distribution, and the constant must be the normalization constant. The r_{nk} are called the *responsibilities* because they quantify how much each component is *responsible* for explaining each data point. The occurring expectations are calculated from the form of the Dirichlet distribution:

$$\mathbb{E}\left[\ln \pi_k\right] = \Psi(\alpha_k) - \Psi(\hat{\alpha}) \quad \text{where:} \quad \hat{\alpha} = \sum_k \alpha_k \tag{14}$$

$$\mathbb{E}\left[\ln \mu_{ik}\right] = \Psi(\beta_{ik}) - \Psi(\hat{\beta}_k) \quad \text{where:} \quad \hat{\beta}_k = \sum_i \beta_{ik} \tag{15}$$

We can in turn use this solution to find the optimum with respect to the parameters:

$$\ln q^*(\pi_k) = \ln p(\pi_k) + \sum_n \mathbb{E}\left[\ln p(z_{kn}|\pi_k)\right] = (\alpha_0 - 1)\ln \pi_k + \sum_n r_{kn} \ln \pi_k \quad (16)$$

$$\ln q^*(\mu_{ki}) = \ln p(\mu_k) + \sum_n \mathbb{E}\left[\ln p(x_n|\mu_k)\right] = (\beta_0 - 1)\ln \mu_k + \sum_n r_{kn} x_n \ln \mu_k$$
$$(17)$$

where α_0 and β_0 are the prior hyper-parameter values and the constant c is a normalization constant, which in (13-17) is omitted for brevity. Recognizing the resulting distributions again as Dirichlet distribution we obtain the following parameter updates:

$$\alpha_k = \alpha_0 + \sum_n r_{kn} \quad (18)$$

$$\beta_k = \beta_0 + \sum_n x_n r_{kn} \quad (19)$$

These updated hyper-parameters are then used to calculate the expectations in Eq. (16-17) for the next iteration. This mechanism of iterative approximate Bayesian inference of the optimal parameters bears strong resemblance to the EM-algorithm, and thus is called *variational Bayesian expectation maximization* (VBEM). Note that the form of update equations above is the same for any mixture model from the exponential family with conjugate priors, after x_n is replaced with an appropriate sufficient statistics (see [11], chapter 2.4).

We still have to choose the number of mixture components, but the fact that we have introduced uncertainty about the mixture coefficients protects the model from using the extraneous components to over-fit small subsets of the data. The parameters that remain to be specified prior to learning are the initial values of the priors, α_0 and β_0. These variables can be though of as initial pseudo-event counts, and express the amount by which the distributions are expected to change given the data. For β_0 we would typically choose a relatively small value, suggesting that the components should be determined relatively "sharply", for example 0.01. For α_0 we wish to express our lack of knowledge about the number of needed components. A sensible choice is 0.5, a value suggested by minimum message length (MML) theory.

3.2   Sequence Based Grouping

Sequence based grouping, has the objective to group together components of the mixture model that tend to follow each other in time, and can be termed as the discovery of invariance in the input data [5]. Modeling of temporal dependencies in a Bayesian network amounts to embedding Markov models, effectively turning mixture models into Hidden Markov Models (HMMs). These HMMs are implemented in the grouping nodes with singles descendants in Fig. 1, *left*, which group together states of their descendants, thereby forming compound states. Grouping is performed by the HMM, if the prior expresses the expectation, that transitions from the compound states onto themselves are more frequent than those between compound states. The joint probability of the sequence of compound states of a first order HMM is:

$$p(x^t|_{t\in 0...T}, A, \pi) = p(A)p(\pi)p(x_0|\pi) \prod_{t=1}^{T} p(x^t|x^{t-1}, A). \tag{20}$$

The full joint probability for the compound state sequence together with the state sequence of their descendants is simply the product of the joint probability of the HMM and that of the mixture model, omitting the prior state probabilities. $p(\pi)$ is modeled as for the mixture model above, and $p(A)$ is a dirichlet prior. To encourage grouping we significantly higher initial pseudo-counts to the diagonal entries than to the off-diagonals, namely 100 and 0.5 respectively. The full learning algorithm would amount to performing two passes over the state sequences, one forward and one backward pass, corresponding to a variational form of the Baum-Welch algorithm. But to speed up learning we chose to only perform the forward pass, which prevents us from having to store all internal state for the whole sequence, which is prohibitive in large data-sets. Learning the Markov model thus amounts to learning a multinomial mixture model to model the transitions, in which the responsibilities computed for the previous step are taken as the prior state probabilities. The resulting learning algorithm for the grouping nodes is similar to the above for the mixture models, but the variational responsibilities for the compound states are computed as a normalized product of the responsibilities from the mixture model, and those from the embedded Markov model. The model is thus the same as that employed in [5], but with a modified learning algorithm.

3.3   Network Learning

Above we derived the learning rule for a single multinomial mixture model and outlined how it can be extended to an HMM. To learn the whole network of connected nodes (each of which is an instance of such a mixture model) there are basically two major strategies which can be pursued:

- Subsequent (layer by layer) learning
- Simultaneous learning of all parameters

In the subsequent learning the parameters of the input layer are learned on the input using the above update equations for several iterations over the training data. Next, the parameters of the first layer are fixed and used to generate probabilities of the hidden states for each instance of the training data. These probabilities are then used to learn the parameters of the next higher layer and so on until all parameters have been learned.

The simultaneous learning makes use of a generalization of belief propagation to variational inference called "variational message passing" (VMP) [12]. In VMP expectations are calculated for the whole network by propagating the expected responsibilities, r_{nk}, upwards by using them as input distributions for the next higher layer, and then again downwards in an equivalent manner. The r_{nk} which are passed downwards then act as priors for the lower layers.

For both strategies we choose to tie the parameters of all the nodes in a single layer to speed up learning. Thus the accumulation of pseudo-counts in the update equations (18-19) are summed into a single parameter distribution for each layer.

4   Experiments

To evaluate the system outlined above in practice we conducted an experiment on the MNIST handwritten digit data set[1]. To achieve local shift invariance the input was presented at a series of shifts with the following schedule: first the input was shifted horizontally in small increments, and after the image had been traversed a single increment shift was performed in the vertical direction. Afterwards the same schedule was performed with horizontal and vertical direction exchanged. This procedure was performed for five iterations over the first 1000 images of the prescribed training set. All layers were generally learned simultaneously using variational message passing. In the experiments we used to distinct network layouts: one with, and one without the grouping nodes. Furthermore, we varied the number of states each node can acquire and performed separate experiments in which we omitted the downwards passing of messages during learning, to evaluate if omitting these messages is a viable option to speed up processing. The initial state prior α_0 was generally set to 0.5. The conditional probability prior β_0 was set to 0.01 for the spatial feature nodes. This small value was necessary to prevent especially higher level nodes from under-fitting and collecting all pseudo-counts into a single component with high entropy. Grouping nodes had

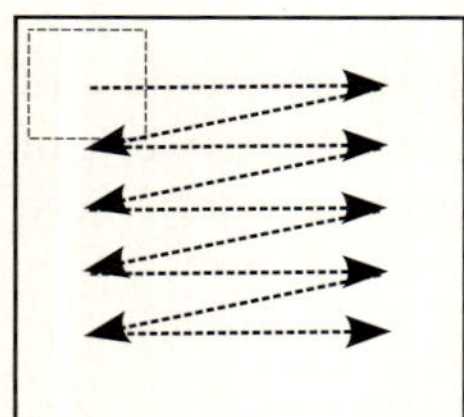

	#(train&test)	#states	% train	% test	comments
1.	1000	100	78.5%	69.1%	
13.	1000	100	68.4%	61.5%	VMP
17.	1000	50	54.8%	49.3%	HMM, VMP
10.	100	100	69%	59%	
12.	100	100	60%	55%	VMP
16.	100	50	39%	40%	HMM, VMP

Fig. 2. *left*: To achieve shift invariance in learning and recognition the input images are presented in the form of a sequence in which a local window is cut out of the whole image which is swept over the image. The illustrated sweeping motion is also repeated with vertical motion. *right*: Recognition results of the various experiments. VMP denotes variational message passing, used to learn all parameters of the network at once. HMM denotes embedded HMMs in the grouping nodes. Where HMMs are not used, the grouping nodes are omitted altogether.

---

[1] http://yann.lecun.com/exdb/mnist/

β_0 set to 100, in order to allow the "fuzziness" resulting from grouping multiple lower level states.

All of the experiments are evaluated in terms of the recognition rate of a supervised multinomial mixture model learned along with the network. Recognition was performed by averaging probabilities of each possible output label over the whole artificial sequence, and then deciding for the label with the highest average probability. We ran three separate experiments, with two training set sizes (100 and 1000), and the same test set of the first 1000 images from the MNIST set. The first experiment was run with a model without the interspersed grouping nodes, learned using subsequent layer-by-layer learning. In the second experiment we used VMP in which variational responsibilities are passed up and down the hierarchy, such that each layer communicates with each other layer. Finally in the third experiment we also introduced the grouping nodes.

The results compare well to those published in [6], even when exact inference was used. Especially, our models do seem to have a smaller tendency to over-fit the data, due to the probabilistic priors attached to the parameters, acting as regularizers. This is even more evident for very small training set sizes, where the performance is still reasonable, and training and test set performance drop equally. A penalty of about 10% is attached to learning the whole network at once. The reason for this should be investigated, but it is possible, that the tendency of the variational approach to under-fit the data is worsened when learning too many parameters simultaneously. Also introducing the embedded HMMs reduced the performance. This could be due to the fact, that the number of parameters is largely increased. In fact, a similar penalty can be noted in [6], where for small training sets (1000 images) introducing lateral connections also decreases the performance, with the opposite effect with a larger training set (10000 images).

5   Conclusion

We have presented a specific implementation of a Bayesian belief network for image recognition with a focus on temporal sequences. Particularly we have presented how variational inference in the form of variational message passing can be used to robustly estimate the parameters of this model and avoid over-fitting. Furthermore VMP reduces the number of times the training data must be visited by allowing to learn all parameters of the network simultaneously instead of subsequent layer by layer learning, and leaves open the possibility of online learning. We also investigated the use of Hidden Markov Models with inertial priors as an addition to our model, with a modification that reduces the number of visitations of the training set and the amount of data that needs to be stored and also leaves open the possibility of online learning.

In comparison with the results presented in [6] our system achieved comparable results, while avoiding over-fitting. We also showed, that all parameters can be learned simultaneously with relatively little loss in performance, opening the way to online learning applications, where all parameters have to be updated in each step. Introducing hidden Markov Models for grouping did not improve the performance. This could be due to the increased number of parameters, but it is also possible, that the sequence based grouping did not help so much in this problem, since the only temporal evolution existed in shifting of the input images. Thus we plan to perform experiments on more complex sequences and real video images.

In terms of computational performance the system as implemented above is relatively slow, due to the large amounts of internal data generated in the form of belief vectors and conditional probability tables (for example processing a single character image in learning mode with 100 possible states per node and using hidden markov models for grouping takes about one third of a second on a single core of a 2.2 GHz Intel Core2Duo processor; that is for a single frame of the generated sequence). But during the above experiments we noticed, that both of these are very sparse, and storing and processing, say, one tenth of the entries would probably not noticeably reduce the performance, so real time performance should be feasible on reasonably sized images by using sparse data representations, even on standard Desktop or Laptop computers.

References

1. Fukushima, K.: Neocognitron: A self-organizing neural network model for a mechanism of pattern recognition unaffected by shift in position. Biol. Cyb. V36(4), 193–202 (1980)
2. Riesenhuber, M., Poggio, T.: 11: How Visual Cortex Recognizes Objects: The Tale of the Standard Model. In: Computational Object Vision, pp. 1640–1653 (2003)
3. Osindero, S., Hinton, G.E.: Modeling image patches with a directed hierarchy of markov random fields. Advances in Neural Processing Systems 20 (2008)
4. Dean, T.: A computational model of the cerebral cortex. In: Proc. 20th Nat. Conf. on Art. Intell. MIT Press, Cambridge (2005)
5. Dean, T.: Learning invariant features using inertial priors. Annals of Mathematics and Artificial Intelligence 47, 223–250 (2006)
6. Dean, T.: Scalable inference in hierarchical generative models. Ann. Art. Intell. & Math.
7. Deco, G., Rolls, E.T.: A neurodynamical cortical model of visual attention and invariant object recognition. Vis. Res. 44 (2004)
8. Masquelier, T., Serre, T., Thorpe, S., Poggio, T.: Learning complex cell invariance from natural videos: a plausibility proof. Technical report, Massachusetts Institute of Technology, Cambridge, MA (2007)
9. Berkes, P., Wiskott, L.: Slow feature analysis yields a rich repertoire of complex cell properties. J. o. Vis. 5/6/9 (2003)
10. Bishop, C.M.: Pattern Recognition and Machine Learning. Springer, Heidelberg (2006)
11. Beal, M.J.: Variational Algorithms or Approximate Bayesian Inference. PhD thesis, University of Cambridge, UK (2003)
12. Winn, J.M., Bishop, C.M.: Variational message passing. J. Mach. L. Res. 6, 558–590 (2005)

Using Feature Distribution Methods in Ensemble Systems Combined by Fusion and Selection-Based Methods

Laura E.A. Santana, Anne M.P. Canuto, and Joã C. Xavier Jr.

Informatics and Applied Mathematics Department
Federal University of RN Natal, RN - Brazil, 59072-970, Telephone: +55-84-3215-3815
lauraemmanuella@yahoo.com.br, anne@dimap.ufrn.br

Abstract. The main prerequisite for the efficient use of ensemble systems is that the base classifiers should be diverse among themselves. One way of increasing diversity is through the use of feature distribution methods in ensemble systems. In this paper, an investigation of the use of feature distribution methods among the classifiers of ensemble systems will be performed. In this investigation, five different methods of data distribution will be used. These ensemble systems will use six existing combination methods, in which four of them are fusion-based methods and the remaining two are selection-based methods. As a result, it is aimed to detect which ensemble systems are more suitable to use feature distribution among the classifier.

Keywords: Ensemble systems, Diversity, Feature distribution in Ensembles.

1 Introduction

The increasing complexity and widening applicability of pattern recognition systems has led to the detailed investigation of many approaches and methodologies. However, it is often found that no single classifier is entirely satisfactory for a particular task, and hence the idea of combining different methods in order to improve overall performance has emerged as potentially very promising [2,6]. The main example of this attempt to combine classifiers is the ensemble (or committees) systems, which exploit the idea that a pool of different classifiers can offer complementary information about patterns to be classified, improving the effectiveness of the overall recognition process.

In the context of ensemble, diversity is one aspect that has been acknowledged as very important [3,9]. For example, there is clearly no accuracy gain in a system that is composed of a set of identical base classifiers. One way of increasing diversity is through the use of feature distribution (feature selection) in ensemble systems. There are several feature distribution methods in the literature, such as in [4,7,11,12,13,16,17].

This paper presents an analysis of the use of data distribution methods in ensemble systems, which are combined by selection and fusion-based methods. The main purpose of this analysis is to detect which combination methods are strongly affected by the use of feature distribution among the classifiers. As a result, it is aimed to help designers in the choice of the ensemble systems for a pattern recognition system.

V. Kůrková et al. (Eds.): ICANN 2008, Part I, LNCS 5163, pp. 245–254, 2008.

2   Related Works

Recently, several authors have investigated the use of feature distribution methods in ensemble systems, such as in [4,11,12,13,16,17]. There are several feature distribution methods, which can be broadly classified as random or nonrandom distribution. In the random distribution, each classifier in the ensemble is built upon a randomly chosen subset of features. In [17], for instance, the random subspace method relies on a pseudorandom procedure to select a small number of dimensions from a given feature space. In each step, such a distribution is made and a subspace is fixed.

There are also the nonrandom feature distribution methods, in which one criterion is used to define the importance of the attributes [4][11][13] [18]. In [18], for instance, the correlation among the errors of the base classifiers is reduced by decoupling the classifiers and training them with different subsets of the input features. It differs from the random subspace as for each class the correlation between each feature and the output of the class is explicitly computed, and the classifier is trained only on the most correlated subset of features.

Recently, several authors have investigated optimization methods to choose subsets of features for classifiers of an ensemble, such as Tabu search, simulated annealing, genetic algorithms, among others [7][10]. In [10], for instance, authors suggest two simple ways to use genetic algorithm to design an ensemble of classifiers. They present two versions of their algorithm. Also, in [7], authors used a simple GA to explore the space of all possible feature subsets, and then create an ensemble based on them.

In all of the aforementioned works, the main focus is the feature distribution methods. However, the accuracy of an ensemble is also strongly affected by the combination methods (rules to combine the outputs of the base classifiers) used by the ensemble. In contrast to the aforementioned works, as previous mentioned, the focus of this work is not on the feature distribution methods, but on the impact of these methods on different combination methods used by the ensembles. In [14,15], an initial investigation of data distribution was performed. These works performed an investigation of data distribution in combination systems was performed, but it was simpler, using less ensemble systems and configurations than the one in this paper. A more extensive analysis will be done in this paper, using five different data distribution methods and four different system sizes applied to two databases.

3   Ensemble Systems

The goal of using ensembles is to improve the performance of a pattern recognition systems [2,9]. There are two main issues in the design of an ensemble: the ensemble components and the combination methods that will be used. With respect to the first issue, the correct choice of the set of base classifiers is fundamental to the overall performance of an ensemble. The ideal situation would be a set of base classifiers with uncorrelated errors - they would be combined in such a way as to minimize the effect of these failures. In other words, the base classifiers should be diverse among themselves. Diversity can be reached in different ways: variations of the parameters of the base classifiers (e.g., initial weights and topology of a neural network model), use of different samples of the dataset as training sets, use of the different types of base classifiers, among others.

Once a set of classifiers has been created, the next step is to choose an effective way of combining their outputs. There are a great number of combination methods reported in the literature [2,6,9]. According to their functioning, there are two main strategies of combination methods: fusion-based, selection-based.

– Fusion-based Methods: it assumes that all base classifiers are equally experienced in the whole feature space. Thus, the decisions of all classifiers are taken into account for any input pattern. There are a vast number of fusion-based methods reported in the literature, such as Voting, Sum, Neural Networks, Na?ve Bayesian, among others;
– Selection-based Methods: unlike the fusion-based methods, only one classifier is used to classify the input pattern. In order to do so, it is important to define a process to choose a member of the ensemble employed to make the decision, which is usually based on the input pattern to be classified. The choice of a classifier used to produce that output is made during the operation phase. This choice is typically based on the certainty of the current decision. One of the main methods is called Dynamic Classifier Selection (DCS) [9].

4   Feature Distribution in Ensemble Systems

In order to reduce redundancy among the attributes of a pattern and to increase the diversity in ensemble systems, data distribution becomes a desired feature [4,11,12,16]. Data distribution can be classified as horizontal and vertical. In horizontal distribution, a classifier of an ensemble system knows all the features of a given pattern, but not all the patterns (pattern distribution). On the other hand, in vertical distribution, a component of an ensemble system knows some features of each pattern, but no component knows all the features of a given example (feature distribution) [11]. In this paper, the vertical distribution approach is investigated, which is also known as feature distribution in ensemble systems. Thus, hereafter, the terms vertical data distribution and feature distribution will be used interchangeably.

Feature distribution methods allocate subsets of features, overlapping or disjoint, for the classifiers of an ensemble system. In this paper, two different approaches of feature distributions will be used, which are:

1. Complete distribution: the classifiers of an ensemble system make their decision based on completely different sets of attributes (disjoint subsets of features). In other words, there is no overlap of attribute among the base components.
2. Partial distribution: the classifiers of an ensemble system make their decision based on partially different sets of attributes (overlapping subsets of features). In this paper, 50% of overlap among the set of attributes will be used. For instance, if an input pattern has 18 attributes, nine of them will be seen by only one component and the remaining nine will be seen by all components.

Also, for completeness, the aforementioned distribution approaches will be compared with systems using no distribution of features among the classifiers. As previously mentioned, there are several feature distribution methods that can be used for ensembles [9], such as random and nonrandom distribution. In the first case, each classifier in the ensemble is built upon a randomly chosen subset of features [8][17]. In contrast, in non-random distribution, a given criterion is used to define the importance of the attributes. Based on this information, the attributes are distributed over the classifiers. There are several criteria that can be used to rank the attributes, such as variance and entropy.

In this paper, five different feature distribution methods will be used, in which one of them is based on the random distribution, four of them are based on the nonrandom distribution (Variance, Entropy, Error and Pearson Correlation). In the next section, the feature distribution methods used in this paper are described.

4.1   Non-random Feature Distribution Methods

All four methods based on a nonrandom feature distribution use same general procedure, changing only the criterion for choosing the features (parameter). The procedure can be described as follows.

1. Rank all T attributes in a decreasing order, based on a given criterion (parameter) applied over a validation set.
2. Put M (M≤T) attributes in a rank list, in the same decreasing order of the previous step (M = T for complete distribution and M = T/2 for partial distribution).
3. For the first N (usually N = number of components) attributes:
   (a) Distribute the attributes randomly among the components in a sense that all components will be assigned only and only one attribute.
4. Remove these attributes from the ranking.
   (a) If the rank list is empty, stop the distribution.
   (b) Otherwise, go to step 3.

The use of the previous procedure is an attempt to make sure that the important attributes (based on a chosen criterion) will be distributed for all components. In applications where the number of attributes is not equally divided into the number of components, the number of attributes of the components will be different.

As already mentioned, the main difference among them is the criterion (parameter) used to rank the attributes. In this paper, four different criteria are used to rank the attributes:

- Statistical variance (method 1). In feature distribution methods, the ranking of the attributes based on variance is a way of measuring the importance of the attributes (attributes with high variance are more important to a component than attributes with low variance);
- Entropy (method 2). Entropy measures the average amount of information necessary to identify the class of a chosen instance. High values of entropy mean that an attribute is capable of correctly classify the input patterns, decreasing the entropy of the set of patterns [4][11].
- Classification Error (method 3). The error parameter defines the relevance of an attribute to the classifiers, by removing this attribute of the set of attribute and

calculating the accuracy of the classifiers. Thus, for a base with N attributes, N classifiers are trained and tested, each one with N-1 attributes. After that, the importance of each attribute is verified by the error presented for the classifier that does not have this attribute in its training set. Error can be used as a parameter in some data distribution methods.

- Pearson Correlation (method 4). Pearson's Product Moment Correlation Coefficient (PMCC) is a value which indicates the strength of linear relationships between variables [5]. As a feature distribution method, it can be used in a sense that only the least correlated features are put together in a subset of features allocated to a base classifier. In this sense, all features were ranked based on the correlation over the other features and distributed over the classifiers.

5   Experimental Setting Up

In this paper, the influence of five feature distribution methods in some ensemble systems is investigated. In other to do this, ensemble systems using six existing combination methods are analyzed, using several system sizes and configurations. In addition, the accuracies of the ensemble systems using complete and partial distributions are compared with systems with no feature distribution. All ensemble systems are applied to two different datasets, which can be described as follows.

- Dataset A - Outdoor Images. This dataset was taken from the UCI repository (segmentation dataset) [1]. The 2.310 instances were drawn randomly from a dataset of 7 outdoor images. The images were hand-segmented to create a classification for every instance, where each instance is a 3x3 region. Eighteen attributes were extracted from each region.
- Dataset B - Proteins. This dataset represents a hierarchical classification, manually detailed, of known structures of proteins. They are organized according to their evolutionary and structural relationship. It has five classes and it is an unbalanced dataset, which has a total of 582 patterns, in which 111 patterns belong to class all-α, 177 patterns to all-β, 203 to α/β, 46 to $\alpha + \beta$ and 45 to small.

Five types of classification methods will be used as base classifiers for the ensemble systems, which are: k-NN (nearest neighbor), C4.5 (decision tree), RBF (radial basis function) network, MLP (multi-layer Perceptron) neural network and fuzzy-MLP network. The choice of the aforementioned classifiers was due to the diversity in the classification criteria used by each method chosen.

Four different system sizes will be used in this investigation, using 3, 5, 7 and 9 classifiers. The choice of a small number of classifiers is due to the fact that diversity has a strong effect in the performance of ensembles when using less than ten classifiers [9]. For each system size, several different configurations will be used. In all these configurations, different base classification methods and/or different topologies of the same base method will be used. For simplicity reasons, values used in this paper are the average accuracies of all used configurations.

In order to compare the impact of using feature distribution methods, the accuracies of the ensembles when using no feature distribution were compared with the ensemble systems using feature distribution (partial and complete). To do this comparison, a statistical test was applied, which is called hypothesis test (t-test) [9]. It is a test which involves testing two learned hypotheses on identical test sets. In order to perform the test, a set of samples (classifier results) from both algorithms should be used to calculate error mean and standard deviation. Based on the information provided, along with the number of samples, the significance of the difference between the two sets of samples, based on a degree of freedom ( ), is defined. In this paper, the confidence level will be 95% (= 0.05).

6   Results and Discussion

Tables 1 to 3 show the results of using feature distribution in ensemble systems. The results obtained in this analysis will be evaluated under three approaches, which are:

- Number of classifiers: in this part of the analysis, the number of classifiers will vary from 3 to 9. It is aimed to investigate how the ensemble systems behave when using feature distribution for different numbers of classifiers (variation in the number of features for each component);
- Data distribution methods: In this part, five different distribution methods will be used. It is aimed to analyze how the ensemble systems behave when using different ways of distributing features among the components;
- The use of feature distribution: In this part, it is investigated the performance of the ensemble systems when using feature distribution is performed It is aimed to comparatively analyze the behavior of the combination methods when using both approaches for feature distribution.

As already mentioned, ensembles with six ensemble methods are used in this investigation. Four of the ensemble rules are fusion-based methods (Sum, Voting, MLP and Naive Bayesian) and two of them are selection-based methods (DCS-DT and DCS-MCB).

6.1   Number of Classifiers

Table 1 shows the results of the ensemble systems when using 3, 5, 7 and 9 components. As already mentioned, all the ensemble systems use all five distribution methods described in Section 4.2. Values in Table 1 are shown in the Win/Loss format. In order to do this, the hypothesis test was applied, comparing ensembles using no feature distribution and ensembles with feature distribution. In Table 1, Win means the number of statistically significant improvements. When the improvement reached by the use of feature distribution is statistically significant, it is considered as a Win. On the other hand, loss means the number of statistically significant decreases. For each cell in Table 1, there are 10 possible comparisons (five feature distribution methods and two datasets). It is important to emphasize that differences that are not statistically significant are not considered (they are not counted in Win or Loss).

From Table 2, it is possible to observe that the use of partial distribution has brought more benefits to the accuracy of the ensemble systems. The relation Win/Loss for partial distribution was 0.72 (55/76), while it was 0.43 (53/123) for complete distribution. It

Table 1. The Win/loss relation of the ensemble systems, when compared with the systems that do not use feature distributions, separated by the number of classifiers

Systems	Partial Distribution				Complete Distribution			
	three	five	Seven	Nine	three	five	Seven	Nine
DT	3/3	2/3	6/3	2/6	3/3	3/5	2/7	1/7
MCB	5/1	1/6	2/5	3/5	5/5	1/5	0/7	0/8
Naive	4/0	3/0	2/1	2/5	4/0	4/4	4/3	1/7
MLP	2/0	2/0	1/0	2/1	3/2	4/5	3/5	3/5
Sum	4/2	3/3	0/6	0/7	4/5	1/5	0/9	0/8
Voting	4/2	1/4	1/6	0/7	5/4	2/4	0/8	0/8

is an expected result since the use of complete distribution leads to the use of a small number of features by each classifier. Mainly because, in this investigation, the datasets have had 126 and 18 attributes. For bigger datasets, the use of complete distribution is expected to have a better performance.

When analyzing the accuracy of the ensemble systems in relation to the increase in the number of classifiers, it can be noticed that the increase in the number of classifiers has caused the increase in the number of losses. The relation Win/Loss for ensembles with three base classifiers is 1.7 (46/27), and it was 0.61 (27/44) for ensembles with five base classifiers, 0.39 (21/54) for seven base classifiers and 0.19 (14/74) for nine base classifiers. Once more it is an expected result since the increase in the number of classifier has caused a decrease in the number of feature seen by the classifiers. For instance, when using complete distribution, the base classifiers in systems with nine components have only two features, which might be too little to do a good classification. These results show that feature distribution is a wished procedure, but a careful choice of the number of features by classifier must be done.

In relation to the increase in the number of classifiers and the combination methods, it can be seen that the increase in the number of classifiers has caused a decrease in the Win/Loss relation, for all combination methods. Nevertheless, the use of MLP combiner has caused the lowest decrease. For instance, when using nine classifiers, the MLP combiner has caused a relation 0.83 (5/6), which was the highest relation of all combination methods and it was close to 1 (one means that the number of wins is equal to the number of losses).

6.2   Data Distribution Methods

Table 2 shows the results of the ensemble systems when using five different distribution methods, which are: random, Pearson correlation, error, variance and entropy. As already mentioned, values in Table 2 consider ensembles with 3, 5, 7 and 9 classifiers. In a general perspective (partial and complete), the highest win/loss relation was obtained by V - variance, which was 1.12 (28/25), followed by Er - error (0.94), Et -entropy (0.42), R - random (0.29) and P - Pearson correlation (0.28).

For the partial distribution, the feature distribution criterion that has provided the highest win/loss relation is variance, which was 4.5 (18/4), followed by error (1.5), random (0.38), entropy (0.25) and Pearson correlation (0.23). On the other hand, for

Table 2. The Win/loss relation of the ensemble systems, when compared with the systems that do not use feature distributions, separated by the feature distribution method

Systems	Partial Distribution					Complete Distribution				
	P	Et	Er	R	V	P	Et	Er	R	V
DT	0/8	1/5	5/1	1/0	6/1	0/8	3/2	2/3	0/7	4/2
MCB	3/3	1/7	4/3	1/3	2/1	1/6	1/4	2/4	1/5	1/6
Naive	2/1	1/2	4/1	0/1	4/1	3/2	3/3	4/1	3/4	0/4
MLP	0/1	1/0	3/0	0/0	5/0	4/4	3/3	0/3	2/4	4/3
Sum	0/4	1/6	2/4	2/4	2/0	1/6	1/5	2/4	1/6	0/6
Voting	0/5	1/4	3/4	1/5	1/1	1/6	2/6	2/6	1/6	1/0

the complete distribution, the highest win/loss relation was obtained by entropy (0.62), followed by error (0.57), variance (0.48), Pearson (0.31) and random (0.25).

It is important to emphasize that the Random technique, which distributes features among the components randomly, provided an average performance for partial distribution. On the other hand, for the complete distribution approach, the random method has provided the lowest win/loss relation. For the complete distribution approach, all attributes are defined by a criterion, while the partial distribution defined only half of the attributes based on this criterion. Because of this, the use of a good criterion to guide the distribution of the attributes is of fundamental importance for the complete distribution. It is shown by the performance of the Random technique.

6.3   The Use of Data Distribution

In this section, the use of data distribution in the ensemble systems is investigated. In order to do this, Table 3 illustrates the win/loss relation of all ensemble systems (for all system sizes and feature distribution method) when using partial and complete distribution.

In a general perspective (partial and complete), the highest Win/Loss relation was reached by Naive 1.2, followed closely by MLP (1.17), by the selection-based methods, DT (0.59) and MCB (0.4), and finally by the non-trainable fusion-based methods, Voting (0.3) and Sum (0.27). In this sense, it is possible to conclude that the use of feature distribution is positive to ensembles combined by Naive and MLP, since they produced more wins than losses (relation higher than 1).

Table 3. It represents statistically significant Win/losses of the ensemble systems, for partial and complete distribution

Systems	Partial Distribution	Complete Distribution
DT	13/15	9/22
MCB	11/17	6/25
Naive	11/6	13/14
MLP	9/1	13/17
Sum	7/18	5/27
Voting	6/19	7/24

For partial distribution, the ensembles systems combined by MLP have provided the highest win/loss relation (9), followed by Naive (1.83), by the selection-based methods, DT (0.87) and MCB (0.65), and by the non-trainable fusion-based methods Sum (0.39) and Voting (0.32). On the other hand, there is a change in the ranking of win/loss relation for complete distribution, in which there is a swap of between the first two places. The first place now belongs to Naive (0.93) and the second one belongs to MLP (0.77). The third place remains unchanged, DT (0.41). However, the voting method jumped from the last to fourth place (0.29). The fifth and six places belong to MCB and Sum, with 0.24 and 0.18, respectively.

Based on these results, it can be concluded that, for partial distribution, the best choice of combination method is MLP. However, for complete distribution, Naive is the best choice for combination method in ensemble systems.

7   Final Remarks

This paper analyzed the influence of important feature distribution methods in ensemble systems combined by six combination methods, in which two of them are selection-based methods and the remaining four methods are fusion-based methods (2 are non-trainable and 2 are trainable). In order to do this, two different approaches of feature distribution (partial and complete) were used for five data distribution methods (random, error, entropy, variance and Pearson correlation). In addition, four different system sizes were used (3, 5, 7 and 9).

Through this analysis, it could be concluded that, in a general perspective, the use of partial distribution was more positive to the performance of the ensemble systems than the complete distribution. It is an expected result since the datasets used in this investigation have a small number of attributes, leading to an insufficient amount of features for the base classifiers. It is believed that the use of bigger datasets leads to better performance of the complete distribution approach. A similar behavior of the ensemble systems was noticed in the variation of number of classifiers, in which the best results were obtained by ensembles with three base classifiers.

In relation to the data distribution method, when considering partial and complete distributions, the highest win/loss relation was obtained by variance, followed by error, entropy, random and Pearson correlation. Finally, in relation to the use of data distribution, for partial distribution, the best choice of combination method is MLP. However, for complete distribution, Naive is the best choice for combination method in ensemble systems.

References

1. Blake, C.L., Merz, C.J.: UCI Repository of machine learning databases. Univ. of California, Dept. of Information and Computer Science,
   `http://www.ics.uci.edu/~mlearn/MLRepository.html`
2. Canuto, A.: Combining neural networks and fuzzy logic for applications in character recognition. PhD thesis, Univ. of Kent. (2001)

3. Canuto, A., Abreu, M., Oliveira, L., Xavier Junior, J., Santos, A.: Investigating the Influence of the Choice of the Ensemble Members in Accuracy and Diversity of Selection-based and Fusion-based Methods for Ensembles. Pattern Recognition Letters 28(4), 472–486 (2007)
4. Caragea, D., Silvescu, A., Honavar, V.: Decision tree induction from distributed, heterogeneous, autonomous data sources. In: Conf. on Int. Systems Design and App. (ISDA) (2003)
5. Chen, P., Popovich, P.: Correlation: Parametric and Nonparametric Measures, 1st edn. Sage Publications, Thousand Oaks (2002)
6. Czyz, J., Sadeghi, M., Kittler, J., Vandendorpe, L.: Decision fusion for face authentication. In: Proc. First Int. Conf. on Biometric Authentication, pp. 686–693 (2004)
7. Gerra-Salcedo, C., Whitley, D.: Genetic approach to feature selection for ensemble creatin. In: Proceedings of the Genetic and Evolutionary Computation Conference, Orlando, USA, pp. 236–243 (1999)
8. Ho, T.K.: The random subspace method for constructing decision forests. IEEE Trans. Pattern Anal. Mach. Intell. 20(8), 832–844 (1998)
9. Kuncheva, L.I.: Combining Pattern Classifiers. Methods and Algorithms. Wiley, Chichester (2004)
10. Kuncheva, L.I., Jain, L.C.: Designing classifier fusion systems by genetic algorithms. IEEE Trans. Evol. Comput. 4(4), 327–336 (2000)
11. Modi, P.J., Tae Kim, P.W.: Classification of Examples by multiple Agents with Private Features. In: Proc. of IEEE/ACM Int. Conf. on Intelligent Agent Technology, pp. 223–229 (2005)
12. Opitz, D.: Feature selection for ensembles. In: Proc. 16th Nat. Conf. on Art. Intelligence, pp. 379–384. AAAI Press, Menlo Park (1999)
13. Rodriguez, J.J., Kuncheva, L.I., Alonso, C.J.: Rotation Forest: A new classifier ensemble method. IEEE Transactions on Pattern Analysis and Machine Intelligence 28(10), 1619–1630 (2006)
14. Santana, L., Oliveira, D., Canuto, A., Souto, M.: A Comparative Analysis of Feature Selection Methods for Ensembles with Different Combination Methods. In: International Joint Conference on Neural Networks (IJCNN), pp. 643–648 (2007)
15. Santana, L., Canuto, A.: An Analysis of Data Distribution Methods in Classifier Combination Systems. In: IJCNN 2008 (accepted, 2008)
16. Tsymbal, A., Pechenizkiy, M., Cunningham, P.: Diversity in search strategies for ensemble feature selection oInformation Fusion 6(1), 83–98 (2005)
17. Tsymbal, A., Puuronen, S., Patterson, D.W.: Ensemble feature selection with the simple Bayesian classification. Inf. Fusion 4, 87–100 (2003)
18. Tumer, K., Oza, N.C.: Input decimated ensembles. Pattern Anal. Appl. 6, 65–77 (2003)

Bayesian Ying-Yang Learning on Orthogonal Binary Factor Analysis

Ke Sun and Lei Xu

Department of Computer Science and Engineering
The Chinese University of Hong Kong, Hong Kong, P.R. China
{ksun,lxu}@cse.cuhk.edu.hk

Abstract. Binary Factor Analysis (BFA) aims to discover latent binary structures in high dimensional data. Parameter learning in BFA suffers from exponential complexity and a large number of local optima. Model selection in BFA is therefore difficult. The traditional approach for model selection is implemented in a two phase procedure. On a prefixed range of model scales, maximum likelihood (ML) learning is performed for each candidate scale. After this enumeration, the optimum scale is selected according to some criterion. In contrast, the Bayesian Ying-Yang (BYY) learning starts from a high dimensional model and automatically deducts the dimension during parameter learning. The enumeration overhead in the two phase approach is saved. This paper investigates a subclass of BFA called Orthogonal Binary Factor Analysis (OBFA). A BYY machine for OBFA is constructed. The harmony measure, which serves as the objective function in the BYY harmony learning, is more accurately estimated by recovering a term that was missing in the previous studies on BYY learning based BFA. Comparison with traditional two phase implementations shows good performance of the proposed approach.

1 Introduction

Latent structures often exist in high dimensional observations. Depending on the characteristics of data, such a structure can be a manifold of much a lower dimension or just a discrete set. Discovering the underlying structure with an appropriate scale of parametric model is critical in parameter learning. In the classical Factor Analysis (FA) [1], the underlying structure is assumed to be a low dimensional Gaussian distribution. A variant model called Binary Factor Analysis (BFA) adopts a vector of independent Bernoulli distribution as the latent model. As in FA, BFA also faces the difficulty of model selection, that is, to determine an appropriate number of binary factors so that the resulting model represents the regularity well but does not overfit the training data. The problem is even more difficult in BFA because of the combinatorial complexity in computation and a huge number of local optima. Research on BFA under the names of latent trait model, item response theory, latent class model or multiple cause model [2,3,4] is widely used in data reduction, psychological measurement, political science, etc.

V. Kůrková et al. (Eds.): ICANN 2008, Part I, LNCS 5163, pp. 255–264, 2008.
© Springer-Verlag Berlin Heidelberg 2008

This paper focuses on a subclass of BFA called Orthogonal Binary Factor Analysis (OBFA) which further restricts the loading matrix in BFA to be orthogonal. OBFA provides a general method to construct a (orthogonal) coordinate system so that the clusters in the training data are separated by certain coordinate planes. This coordinate system is valuable in many practical applications. In a 2-level Orthogonal Experiment Design (OED) [5] with a set of high dimensional structured experiment inputs available, the task is to extract 2^m well separated representative inputs corresponding to m independent factors. Here OBFA can be applied to learn a coordinate system with the representation inputs lying at $\{-1, 1\}^m$. A parallel binary channel with additive noise and a rotation transformation at the output end results naturally in the OBFA model. In an artificial neural network with one hidden layer, the first layer can be regarded as encoding of space regions as binary vectors and thus can be initialized with the equations of such separation planes. Psychological measurement with questionnaires also falls in the category of OBFA if the distribution of answers is linear separable by several orthogonal planes. Moreover, taking advantage of the formal analysis results, OBFA can be utilized to initialize BFA.

The conventional two phase approach for model selection has to enumerate a candidate set of model scales, denoted as $\mathcal{K}$. ML learning is performed for each $k \in \mathcal{K}$ to estimate the parameters Θ. The optimal scale $k^\star$ is selected with

$$\hat{k^\star} = \arg\min_{k \in \mathcal{K}} J(\hat{\Theta}, k), \tag{1}$$

where $J(\hat{\Theta}, k)$ is a model selection criterion, such as Akaike's Information Criterion (AIC) [6], Bozdogan's Consistent Akaike's Information Criterion (CAIC) [7], Schwarz's Bayesian Information Criterion (BIC) [8], Rissanen's Minimum Description Length (MDL) criterion [9]. Huge computation is evolved in this enumeration, especially for problems with a high computational complexity as BFA.

Proposed firstly in 1995 [10] and systematically developed in the past decade [11,12,13], the BYY harmony learning provides a general framework for parameter learning and model selection. Under this framework, model selection can be performed automatically during parameter learning. In this way the computational overhead in the two phase enumeration is saved. The BYY harmony learning is mathematically implemented by maximizing the following harmony measure

$$H(p \,\|\, q) = \int p(\mathbf{R} \,|\, \mathbf{X}) p(\mathbf{X}) \log \left[q(\mathbf{X} \,|\, \mathbf{R}) q(\mathbf{R}) \right] d\mathbf{X} d\mathbf{R}, \tag{2}$$

where $\mathbf{X}$ is the external observations, $\mathbf{R} = \{\mathbf{Y}, \Theta\}$ consists of the inner states $\mathbf{Y}$ and all unknown parameters Θ. The joint distribution $q(\mathbf{X}, \mathbf{R}) = q(\mathbf{X} \,|\, \mathbf{R}) q(\mathbf{R})$ represents the underlying model; the complementary joint distribution $p(\mathbf{X}, \mathbf{R}) = p(\mathbf{R} \,|\, \mathbf{X}) p(\mathbf{X})$ is the posterior model constructed from the observations.

Much work has been dedicated to FA and BFA using the BYY harmony learning [11,12,14]. When the other structures are fixed, several typical choices of $p(\mathbf{Y} \,|\, \mathbf{X})$, an important component in the joint distribution $p(\mathbf{X}, \mathbf{R})$, result in several scenarios with different model selection performances [13]. The paper [14] investigates BFA with $p(\mathbf{Y} \,|\, \mathbf{X})$ in a free structure. A systematical comparison with typical model selection criteria shows the BYY approach has superior performance. As a follow-up work of [14], the present paper studies BFA by considering $p(\mathbf{Y} \,|\, \mathbf{X})$ in a Bayesian structure, restricted to OBFA so that analysis and computation are feasible. Moreover, the harmony measure is more accurately estimated by recovering a term that is missing in [14]. Experiments show good performance of the proposed approach. Also on OBFA, we found that the BYY harmony learning with automatic model selection (BYY-AUTO) is less influenced by local optima, in comparison with the traditional two phase model selection implementations.

The rest of this paper is organized as follows. Section 2 introduces the OBFA model. Section 3 constructs a BYY machine for OBFA and presents the learning algorithms. Section 4 evaluates the proposed approach by comparative experiments and informal analysis. Concluding remarks are made in the last section.

2 Orthogonal Binary Factor Analysis

Factor Analysis in the literature of statistics [1] can be formulated as

$$\mathbf{x} = \mathbf{A}\mathbf{y} + \mathbf{c} + \mathbf{e}, \tag{3}$$

where $\mathbf{x}_{n \times 1}$ represents the observations, $\mathbf{y}_{m \times 1}$ is the corresponding inner representations, $\mathbf{A}_{n \times m}$, $\mathbf{c}_{n \times 1}$ is the loading matrix and the mean offset, $\mathbf{e}_{n \times 1}$ is a zero-mean Gaussian noise independent of $\mathbf{y}$. In the classical FA, $\mathbf{y}$ is Gaussian distributed. Alternatively, BFA (OBFA) assumes $\mathbf{y}_{m \times 1}$ is a binary vector. Let $y_i (i = 1, \ldots, m)$ take values from $\{-1, 1\}$ rather than $\{0, 1\}$ for simplicity. On the independence assumption of $\mathbf{y}$, we get

$$q(\mathbf{y}) = \prod_{i=1}^{m} \theta_i^{(1+y_i)/2}(1 - \theta_i)^{(1-y_i)/2} \quad (0 < \theta_i < 1, \ i = 1, \ldots, m). \tag{4}$$

Geometrically, BFA can be thought as fitting the data with 2^m Gaussians located at the vertices of a m-dimensional hyper-parallelepiped in $\mathbb{R}^n$.

BFA in general is a complex problem. The EM algorithm for ML learning has to estimate the 2^m-point posterior distributions $q(\mathbf{y} \,|\, \mathbf{x}_t)$ for each sample $\mathbf{x}_t$ and thus is in an exponential complexity. Also, the BYY-AUTO algorithm for BFA [14] needs to compute

$$\mathbf{y}_t = \underset{\mathbf{y} \in \{0,1\}^m}{\arg \max} \ \log q(\mathbf{y} \,|\, \mathbf{x}_t). \tag{5}$$

This $\log q(\mathbf{y} \,|\, \mathbf{x})$ is quadratic with respect to $\mathbf{y}$, and thus Eq. (5) is actually a quadratic 0-1 programming problem which falls in the category of NPC.

To make BFA computationally feasible, OBFA further assumes the loading matrix $\mathbf{A}_{n \times m} = \mathbf{Q}_{n \times n}\mathbf{\Lambda}_{n \times m}$, where $\mathbf{Q}$ is orthogonal and $\mathbf{\Lambda}$ is diagonal. Consequently, the posterior distribution $q(\mathbf{y} \,|\, \mathbf{x})$ turns into an independent Bernoulli distribution shown as follows:

$$q(\mathbf{y}\,|\,\mathbf{x}) = \prod_{i=1}^{m} \hat{\theta}_i(\mathbf{x})^{(1+y_i)/2}(1 - \hat{\theta}_i(\mathbf{x}))^{(1-y_i)/2}, \tag{6}$$

$$\hat{\theta}_i(\mathbf{x}) = \frac{\theta_i}{\theta_i + (1 - \theta_i)\exp\left\{-2\mathbf{a}_i^T(\mathbf{x} - \mathbf{c})/\sigma^2\right\}}, \tag{7}$$

where $\mathbf{a}_i$ is the i'th column of $\mathbf{A}$. For the EM-algorithm, the 2^m-point distribution $p(\mathbf{y}\,|\,\mathbf{x}_t)$ can be expressed by an m dimensional vector which is analytically solvable. $E_p(\log q(\mathbf{x}, \mathbf{y})\,|\,\mathbf{y})$, the objective function in the M-step, has avoided the 2^m-term summation due to the posterior independence of $\mathbf{y}$. In this way the EM algorithm for OBFA can avoid the exponential complexity. Also, the BYY-AUTO algorithm for OBFA [14] becomes computationally feasible because the quadratic form in Eq. (5) is diagonal.

3   A BYY Machine for Orthogonal Binary Factor Analysis

The EM algorithm for ML learning can train a model given a prefixed model scale. It is not good to make model selection, i.e., to determine an appropriate number of binary factors for a training dataset. The traditional approach for model selection has to make ML learning by enumerating a range of model scales and then select the best by one of existing typical model selection criteria [6,7,8,9]. In contrast, the BYY harmony learning is able to make model selection during parameter learning and thus saves the computational overhead in the above two phase enumeration. This section will build a BYY machine for OBFA. Readers are referred to [11,12,13] for details about BYY learning.

Two complementary Bayesian decompositions of the joint distribution of the external observations and inner representations feature the basic structure of a BYY Machine. One of its two major components, namely the *Ying Machine*, models the inner representations $\mathbf{Y}$ via the prior distribution $q(\mathbf{Y})$ (Eq. (4)) and describes a transformation from $\mathbf{Y}$ to the external observations $\mathbf{X}$ via a conditional distribution $q(\mathbf{X}\,|\,\mathbf{Y})$. The complementary part, the *Yang Machine*, characterizes the observations $\mathcal{X}_N = \{\mathbf{x}_t\}_{t=1}^{N}$ as the distribution $p(\mathbf{X})$, which can be simply chosen as $p(\mathbf{X}) = \sum_{t=1}^{N} \delta(\mathbf{X} - \mathbf{x}_t)/N$, where $\delta(.)$ is the Dirac delta function; and a posterior distribution $p(\mathbf{Y}\,|\,\mathbf{X})$ represents the projection from the external observations to their inner representations.

For OBFA, instead of a free structure as used in [14], we let $p(\mathbf{Y}\,|\,\mathbf{X}) = q(\mathbf{Y}\,|\,\mathbf{X})$ in the Bayesian structure by Eq. (6). It further follows from Eq. (2) that

$$H(p\|q) = \frac{1}{N}\sum_{t=1}^{N}\left\{\sum_{i=1}^{m}\left[\frac{1 + \hat{y}_i(\mathbf{x}_t)}{2}\log\theta_i + \frac{1 - \hat{y}_i(\mathbf{x}_t)}{2}\log(1 - \theta_i)\right] - \frac{n}{2}\log\sigma^2\right.$$

$$\left. - \frac{1}{2\sigma^2}\|\mathbf{A}\hat{\mathbf{y}}(\mathbf{x}_t) + \mathbf{c} - \mathbf{x}_t\|_2^2 - \frac{1}{2\sigma^2}\sum_{i=1}^{m}\|\mathbf{a}_i\|_2^2(1 - \hat{y}_i^2(\mathbf{x}_t))\right\}, \tag{8}$$

where σ^2 is the spherical error covariance, $\hat{\mathbf{y}}(\mathbf{x}_t) = E(q(\mathbf{y}\,|\,\mathbf{x}_t)) = 2\hat{\theta}(\mathbf{x}_t) - 1$, $\Theta = \{\mathbf{A}, \mathbf{c}, \theta, \sigma\}$ are unknown parameters. The BYY harmony learning on OBFA is implemented by maximizing $H(p\,\|\,q)$ in Eq. (8) for a given dataset $\mathcal{X}_N = \{\mathbf{x}_t\}_{t=1}^{N}$. With a large initial dimension of $\mathbf{y}$, the simplification ability of the BYY harmony learning [11,12,13] will force $\theta_i \to 0$ or 1, $\|\mathbf{a}_i\|_2 \to 0$ for a redundant dimension i. At a certain threshold, we can discard the corresponding dimension. Eventually, the learning algorithm will determine all the unknown parameters as well as an appropriate model scale.

Algorithm 1. BYY-AUTO for OBFA

---

input	: A dataset $\mathcal{X}_N = \{\mathbf{x}_t\}_{t=1}^{N}$ ($\dim(x) = n$)
output	: m (number of binary factors), $\{\mathbf{Q}, \Lambda, \mathbf{c}, \theta, \sigma, \{\hat{\mathbf{y}}(\mathbf{x}_t)\}_{t=1}^{N}\}$

Initialization

$\quad m = m_0$, $\mathbf{Q}_{n \times m} \leftarrow$ a random orthogonal matrix, $\theta_i = 0.5$, $\mathbf{c}_{n \times 1} = \sum_t \mathbf{x}_t / N$,
$\quad \lambda_i^2 = \sum_t \|\mathbf{x}_t - \mathbf{c}\|_2^2 / Nn$, $\sigma = \lambda_i / 2$, $\gamma = \gamma_0$ (γ is the learning rate; $i = 1, \ldots, m$)

Iteration

  Yang Step

    for $t = 1$ **to** N, $i = 1$ **to** m **do**

$$\hat{\mathbf{y}}_i(\mathbf{x}_t) = \begin{cases} 1 & 2(\mathbf{c} - \mathbf{x}_t)^T \mathbf{Q}_i \lambda_i < \sigma^2 \log(\theta_i/(1 - \theta_i)) \\ -1 & \text{otherwise} \end{cases}$$

  Ying Step

    Update the parameters:

$$\mathbf{Q}^{new} = \mathbf{Q}^{old} + \gamma \frac{1}{N\sigma^2} \sum_{t=1}^{N} (\mathbf{x}_t - \mathbf{c})\hat{\mathbf{y}}^T(\mathbf{x}_t)\Lambda, \text{normalize } \mathbf{Q} \text{ s.t. } \mathbf{Q}^T\mathbf{Q} = \mathbf{I}$$

$$\lambda_i = \frac{1}{N} \sum_{t=1}^{N} (\mathbf{x}_t - \mathbf{c})^T \mathbf{Q}_i \hat{y}_i(\mathbf{x}_t), \quad \theta_i = \frac{1}{N} \sum_{t=1}^{N} \frac{1 + \hat{y}_i(\mathbf{x}_t)}{2}, \quad \mathbf{c} = \frac{1}{N} \sum_{t=1}^{N} (\mathbf{x}_t - \mathbf{Q}\Lambda\hat{\mathbf{y}}(\mathbf{x}_t))$$

$$\sigma^2 = \frac{1}{nN} \sum_{t=1}^{N} \left(\|\mathbf{Q}\Lambda\hat{\mathbf{y}}(\mathbf{x}_t) + \mathbf{c} - \mathbf{x}_t\|_2^2 + \sum_{i=1}^{m} \lambda_i^2(1 - \hat{y}_i^2(\mathbf{x}_t)) \right)$$

    Discard the i'th dimension of y if $\lambda_i < \epsilon_\lambda$, $\theta_i < \epsilon_\theta$ or $\theta_i > 1 - \epsilon_\theta$, where ϵ_λ and ϵ_θ are small positive thresholds.

Until Convergence

The BYY-AUTO (gradient descent) is implemented by replacing the loop body in "Iteration" with $\Theta^{new} = \Theta^{old} + \gamma \bigtriangledown H$, where H is the harmony measure in Eq. (8) with $\hat{\mathbf{y}}(\mathbf{x}) = 2\hat{\theta}(\mathbf{x}) - 1$, $\hat{\theta}(\mathbf{x}) = s(\mathbf{Bx} + \mathbf{d})$, $\mathbf{B}$ and $\mathbf{d}$ are newly introduced parameters.

The EM algorithm is implemented by replacing the loop body in "Yang step" with $\hat{\mathbf{y}}(\mathbf{x}_t) = 2\hat{\theta}(\mathbf{x}_t) - 1$, where $\hat{\theta}(\mathbf{x}_t)$ is given by Eq. (7). Although implemented similarly, the EM algorithm for ML learning is not capable of making model selection automatically [11,12,13].

Compared with [14], Eq. (8) is a more accurate expression of the harmony measure. The underlined term resulting from the Bayesian structure of $p(\mathbf{Y}\,|\,\mathbf{X})$ leads to an extra strength for model selection. During the BYY harmony learning, the model scale is being reduced, in the following two scenarios:

1. The prior distribution $q(\mathbf{y})$ deteriorates to an equivalent lower dimensional distribution when some y_i tends to be deterministic. For example, the Bernoulli distribution in Eq. (4) deteriorates when some $\theta_i \to 0$ or 1.
2. The probabilistic transformation $q(\mathbf{x}\,|\,\mathbf{y})$ tends to be singular. For the case $q(\mathbf{x}\,|\,\mathbf{y}) = G(\mathbf{x}\,|\,\mathbf{A}\mathbf{y} + \mathbf{c}, \boldsymbol{\Sigma})^1$ as in BFA, when some $||\mathbf{a}_i||_2 \to 0$, $\mathbf{y}_i$ can be discarded because the information it carries is vanished by a zero multiplier. The underlined term in Eq. (8) just takes this role that pushes $||\mathbf{a}_i||_2 \to 0$ for a redundant dimension i.

We propose two algorithms for OBFA. The first one is gradient descent on the harmony measure (Eq. (8)). Two free parameters $\mathbf{B}_{m \times n}$ and $\mathbf{d}_{m \times 1}$ are introduced and Eq. (7) is replaced by $\hat{\theta}(\mathbf{x}) = s(\mathbf{B}\mathbf{x}+\mathbf{d})^2$ to reduce the dependence of learning on initial conditions. The algorithm can be derived from the gradient expression of Eq. (8) and is omitted in detail due to space limitation. The second one is presented in Algorithm 1 for a fast implementation in help of the approximation $\hat{\mathbf{y}}(\mathbf{x}) = \mathrm{sgn}\left(2\hat{\theta}(\mathbf{x}) - 1\right)^3$. It works in a so-called Ying-Yang iterative procedure similar to the BYY-AUTO algorithm used in [14].

4   Evaluation and Analysis

This section evaluates the proposed approach by comparing it with traditional two phase model selection implementations. Two aspects are investigated, namely, model selection accuracy and effect of local optima. The comparison of efficiency is omitted because of space limitation. As listed in Algorithm 1, the computational cost of EM algorithm is similar to BYY-AUTO for OBFA, while the two phase approach consumes much more time than BYY-AUTO. The candidate range of model scale for the two phase implementation is selected as $[1, \min(\dim(\mathbf{x}), 2\dim(\mathbf{y}) - 1)]$.

To investigate the performance of the algorithms under different situations, we fix the data dimension to 8 and vary the dimensions of $\mathbf{y}$ ($m = 3, 4, 5$), sample sizes ($N = 20, 50, 100$) and noise levels ($\sigma = 0.1, 0.2, 0.4$). We let $\mathbf{y}$ uniformly take values from the set $\mathcal{Y} = \{\mathbf{y} \in \{0,1\}^m : \mod (\sum_{i=1}^{m} y_i,\, 2) = 0\}$ so that the generated data forms 2^{m-1} clusters in $\Re^n$ which can only be well separated by $\geq m$ hyper-planes. The data is randomly generated according to Eq. (3) with the restriction that the columns of $\mathbf{A}$ are orthogonal. The learning methods are evaluated with the distribution of the estimated dimension $\hat{m}$. As shown in Fig. 1(b), 7 series of experiments are resulted. After 100 independent runs per each configuration (m, N, σ), the results of each method are reported on a cube as in Fig. 1(a) by "$\bar{m} \pm \sigma(m)$", where $\bar{m}$ and $\sigma(m)$ stand for the average and standard deviation of $\hat{m}$, respectively. Furthermore, the 8 vertices of each cube are covered with circles to show the relative values of the standard deviation.

---

[1] $G(\cdot\,|\,\mu, \boldsymbol{\Sigma})$ represents a Gaussian distribution with mean μ and covariance $\boldsymbol{\Sigma}$.
[2] $s(.)$ is the sigmoid function.
[3] sgn(.) is the sign function.

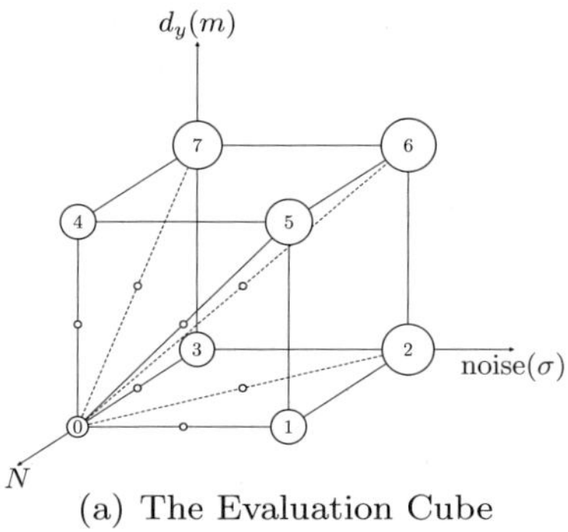

(a) The Evaluation Cube

NO	dim(y)	N	noise(σ)	Vertices
1	3	100	0.1,0.2,0.4	$0 \to 1$
2	3	100,50,20	0.1,0.2,0.4	$0 \to 2$
3	3	100,50,20	0.1	$0 \to 3$
4	3,4,5	100	0.1	$0 \to 4$
5	3,4,5	100	0.1,0.2,0.4	$0 \to 5$
6	3,4,5	100,50,20	0.1,0.2,0.4	$0 \to 6$
7	3,4,5	100,50,20	0.1	$0 \to 7$

(b) Experiment Configuration

Fig. 1. Configuration of Simulated Experiments

All experimental results are presented in Fig. 3. For example, when $\dim(\mathbf{y}) = 3$, $N = 100$ and $\sigma = 0.1$ in Fig. 3(a), the lower left corner shows that the average number of binary factors given by AIC is 4.1 (while the true value is 3) and the standard deviation is 0.80.

In Fig. 3(a) – 3(f), the model selection accuracy of all approaches generally goes down as sample size decreases, hidden dimension increases or noise increase. The two phase implementations tend to overestimate the model size. As noise increases or sample size decreases, this tendency decreases, and then flips to underestimation. In contrast, the results given by the two algorithms of BYY-AUTO are much more accurate. Moreover, the gradient descent implementation of BYY-AUTO is comparatively more accurate.

Next, we use Fig. 2 as an example to intuitively demonstrate the experimental results. The dataset (the clouds) is slightly different from the dataset used in the experiments for illustration. It forms a mixture of four Gaussian distributions with their centers lying on a plane square. Here the task of parameter learning and model selection is to depict the training data with a unknown dimensional cuboid. We first show why the two phase model selection implementations trend to overestimation. On the correct scale ($\dim(\mathbf{y}) = m = 2$), the underlying model

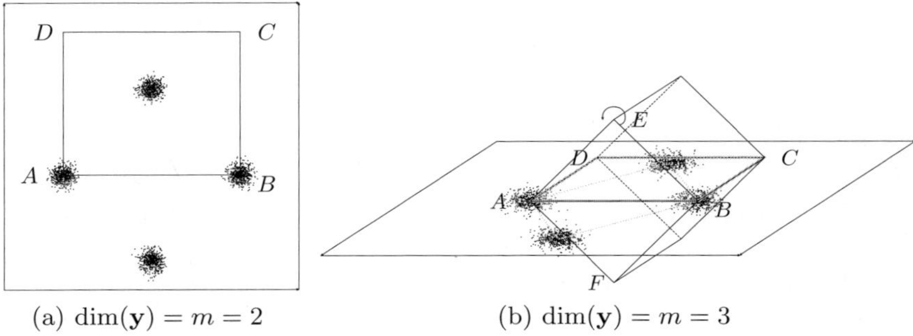

(a) $\dim(\mathbf{y}) = m = 2$ (b) $\dim(\mathbf{y}) = m = 3$

Fig. 2. A Typical Case of Local Optimum

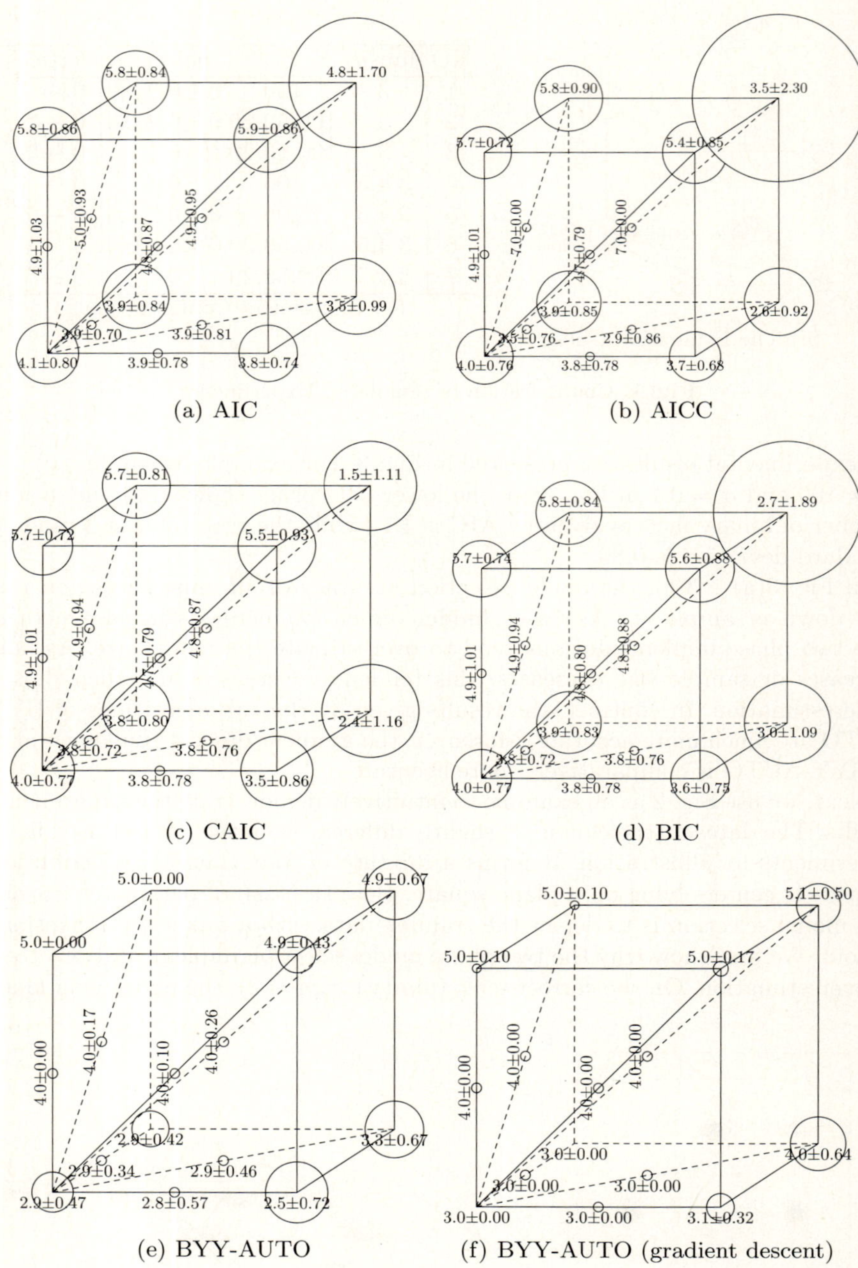

Fig. 3. Model Selection Performance of Different Methods on OBFA

is a rectangle, denoted as $ABCD$. As learning reaches the status shown in Fig. 2(a), a local optimum is likely to occur. It is probably that such a local optimum does not arise on a larger candidate scale, e.g., $\dim(\mathbf{y}) = m = 3$. In this situation a two phase procedure is likely to select the optimal scale as 3 when a local optimum is reached on the correct scale $\dim(\mathbf{y}) = m = 2$. In contrast, the BYY harmony learning starts from a high dimensional cuboid to fit the data then "shrinks" the cuboid to the next lower dimensions. A high dimensional cuboid with more degrees of freedom is less affected by local optima, and we expect this property is kept during dimension reduction. For example, Fig. 2(b) presents a three dimensional cuboid with $ABCD$ in Fig. 2(a) as its four vertices. Another degree of freedom allows the cuboid to be rotated to fit the data. This shows the advantage of initialization with a high dimensional model.

In OBFA, a large number of local optima exist. Two phase model selection implementations are seriously affected by inaccurate estimation of parameters. Better performance can be achieved by repeating the EM training several times for each candidate scale [14], but much more computation overhead will be caused. With almost the same configuration in implementation presented in Algorithm 1, the BYY harmony learning with automatic model selection benefits from initialization with a high dimension model, and is less influenced by local optima according to the experiments. As to the two different implementations of BYY-AUTO, the gradient descent version is more accurate. During gradient descending, an additional strength brought by the underlined term in Eq. (8) will force $\|\mathbf{a}_i\|_2 \to 0$, which is helpful in dimension reduction. This strength is missing in Algorithm 1 because the Yang step makes $\hat{y}_i(\mathbf{x}_t) = \pm 1$ thus vanishes the underlined term in Eq. (8).

5   Concluding Remarks

The presented work investigates model selection on Orthogonal Binary Factor Analysis with the BYY harmony learning. Instead of being free of structure in a previous study [14], the Yang path $p(\mathbf{Y} \mid \mathbf{X})$ is equipped with a Bayesian structure. The harmony measure for BFA is more accurately estimated by recovering a term that can further improve the model selection performance. The proposed approach shows superior performance over the four traditional two phase model selection criteria and is less affected by local optima.

Acknowledgements

The work described in this paper was fully supported by a grant from the Research Grant Council of the Hong Kong SAR (Project No: CUHK4177/07E). We thank Mr. Shikui Tu and Mr. Lei Shi for the helpful discussions.

References

1. Anderson, T.W., Rubin, H.: Statistical inference in factor analysis. In: Proceedings of the Third Berkeley Symposium on Mathematical Statistics and Probability, Berkeley, vol. 5, pp. 111–150 (1956)

2. Bartholomew, D.J., Knott, M.: Latent variable models and factor analysis. Kendallas Library of Statistics, vol. 7. Oxford University Press, New York (1999)
3. Heinen, T.: Latent class and discrete latent trait models: Similarities and differences. Sage, Thousand Oaks (1996)
4. Treier, S., Jackman, S.: Beyond factor analysis: modern tools for social measurement. In: The 2002 Annual Meetings of the Western Political Science Association and the Midwest Political Science Association (2002)
5. Tapchi, G.: System of Experimental Design: Engineering Methods to Optimize Quality and Minimize Costs. UNIPUB/Kraus International Publications (1987)
6. Akaike, H.: A new look at the statistical model identification. IEEE Trans. Automatic Control 19(6), 716–723 (1974)
7. Bozdogan, H.: Model selection and Akaikeas information criterion (AIC): the general theory and its analytical extensions. Psychometrika 52(3), 345–370 (1987)
8. Schwarz, G.: Estimating the dimension of a model. The Annals of Statistics 6(2), 461–464 (1978)
9. Rissanen, J.: Modeling by shortest data description. Automatica 14, 465–471 (1978)
10. Xu, L.: A Unified Learning Scheme: Bayesian-Kullback Ying-Yang Machine. In: Touretzky, D.S., et al. (eds.) Advances in NIPS, vol. 8, pp. 444–450. MIT Press, Cambridge (1996); A preliminary version in Proc. ICONIP 1995, Beijing, pp. 977–988 (1995)
11. Xu, L.: Advances on BYY harmony learning: information theoretic perspective, generalized projection geometry, and independent factor auto-determination. IEEE Trans. Neural Networks 15(5), 885–902 (2004)
12. Xu, L.: Bayesian Ying Yang learning. Scholarpedia 2(3), 1809 (2007),
    http://scholarpedia.org/article/Bayesian_Ying_Yang_Learning
13. Xu, L.: Bayesian Ying Yang System, Best Harmony Learning, and Gaussian Manifold Based Family. In: Zurada, J.M., et al. (eds.) Computational Intelligence: Research Frontiers. LNCS, vol. 5050, pp. 48–78. Springer, Heidelberg (2008)
14. An, Y., et al.: A Comparative Investigation on Model Selection in Binary Factor Analysis. Studies in Computational Intelligence 6, 145–160 (2005)

A Comparative Study on Data Smoothing Regularization for Local Factor Analysis

Shikui Tu, Lei Shi, and Lei Xu

Department of Computer Science and Engineering,
The Chinese University of Hong Kong, Hong Kong, P.R. China
{sktu,shil,lxu}@cse.cuhk.edu.hk

Abstract. Selecting the cluster number and the hidden factor numbers of Local Factor Analysis (LFA) model is a typical model selection problem, which is difficult when the sample size is finite or small. Data smoothing is one of the three regularization techniques integrated in the statistical learning framework, Bayesian Ying-Yang (BYY) harmony learning theory, to improve parameter learning and model selection. In this paper, we will comparatively investigate the performance of five existing formulas to determine the hyper-parameter namely the smoothing parameter that controls the strength of data smoothing regularization. BYY learning algorithms on LFA using these formulas are evaluated by model selection accuracy on simulated data and classification accuracy on real world data. Two observations are obtained. First, learning with data smoothing works better than that without it especially when sample size is small. Second, the gradient method derived from imposing a sample set based improper prior on the smoothing parameter generally outperforms other methods such as the one from Gamma or Chi-square prior, and the one under the equal covariance principle.

1 Introduction

Clustering and dimension reduction have long been considered as two of the fundamental problems in the literature of unsupervised learning. Local Factor Analysis (LFA) model [1,2], also called Mixture of Factor Analyzers (MFA) [3], combines the widely-used Gaussian Mixture Model (GMM) for clustering and the well-known Factor Analysis (FA) model for dimension reduction, and thus it can jointly performs the two tasks, which is different from the scenario that the two tasks are performed sequentially, such as performing dimension reduction after finishing clustering. LFA is practical for high dimensional data analysis by greatly reducing the degree of freedom of covariance matrices. Moreover, LFA also includes mixtures of probabilistic principal component analyzers [4] as its special case. Integrated with these advantages, LFA has wide applications in pattern recognition, bioinformatics, and financial engineering [5,6].

When the LFA model is given, a classical approach for parameter learning is Maximum Likelihood estimation, which is usually implemented by the well-known Expectation Maximization (EM) algorithm [3]. However, due to the limit

V. Kůrková et al. (Eds.): ICANN 2008, Part I, LNCS 5163, pp. 265–274, 2008.

on both the sample size and a prior knowledge of the model, we will often encounter the over-fitting problem. One way to tackle this challenging problem is model selection, which selects an optimal model scale to trade off best fitting and model complexity via minimizing a criterion in a conventional two-phase implementation. There are also incremental (decremental) implementations and automatic approaches to tackle model selection. The former attempt to incorporate as much as possible what already learned as model scale increases (decreases) step by step, while the latter perform model selection automatically during parameter learning, which greatly reduces the computation cost [7]. A comparative investigation of serval approaches for model selection on LFA was given in [1].

Firstly proposed in 1995 [8] and systematically developed in the past decade [2,7,9,10], Bayesian Ying-Yang (BYY) harmony learning theory provides a general statistical learning framework for parameter learning and model selection. It leads to not only a criterion that outperforms typical model selection criteria in a two-phase implementation, but also an automatic approach for model selection. The BYY harmony learning is implemented via maximizing the following harmony measure

$$\max H(p\|q, \mathbf{k}), \ H(p\|q, \mathbf{k}) = \int p(\mathbf{R}|\mathbf{X})p(\mathbf{X}) \ln[q(\mathbf{X}|\mathbf{R})q(\mathbf{R})]d\mathbf{X}d\mathbf{R} \tag{1}$$

where $\mathbf{X}$ is the observation generated from its inner representation $\mathbf{R} = \{\mathbf{Y}, \mathbf{\Theta}\}$ with a parameter set $\mathbf{\Theta}$ collectively representing the underlying structure of $\mathbf{X}$ and $\mathbf{Y}$ correspondingly being the inner representation of $\mathbf{X}$. $p(\mathbf{R}, \mathbf{X}) = p(\mathbf{R}|\mathbf{X})p(\mathbf{X})$ and $q(\mathbf{R}, \mathbf{X}) = q(\mathbf{X}|\mathbf{R})q(\mathbf{R})$ form two types of Bayesian decomposition, which are called Yang machine and Ying machine respectively, and such a Ying-Yang pair is called Bayesian Ying-Yang (BYY) system. Presented in eq.(1) is a best harmony principle which makes a best Ying-Yang matching with least complexity.

In the framework of BYY harmony learning, three new regularization techniques are integrated. First, as $q(\mathbf{R}) = q(\mathbf{Y}|\mathbf{\Theta})q(\mathbf{\Theta})$, prior information of $\mathbf{\Theta}$ can be incorporated via $q(\mathbf{\Theta})$. Second, considering the Yang path $p(\mathbf{R}|\mathbf{X}) = p(\mathbf{\Theta}|\mathbf{X})p(\mathbf{Y}|\mathbf{X}, \mathbf{\Theta}_{y|x})$, if $p(\mathbf{Y}|\mathbf{X}, \mathbf{\Theta}_{y|x})$ is free of structure, by eq.(1) it will lead to an impulse function. To make the Yang machine more regular, we can impose a structure on $p(\mathbf{Y}|\mathbf{X}, \mathbf{\Theta}_{y|x})$ or introduce certain constraints between Ying machine and Yang machine. Third, data smoothing is considered by using a Parzen window density estimator instead of the empirical density in BYY harmony learning. These techniques are all embedded in BYY harmony learning, and take important roles on the performance of model selection. More details are referred to [2,7,9,10]. Due to the space limit, we will focus on data smoothing regularization with help of the first two techniques to determine the smoothing parameter.

It is well known in the literature that adding noise to samples and using Parzen window density estimator can help parameter learning or system modeling in a way closely related to Tikhonov regularization [11], featured by adding to the fitting error with a regularization term controlled by a hyper-parameter. However, it is very difficult to determine such a hyper-parameter. Data smoothing regularization was firstly proposed in [12] by using the Parzen window density estimator as follows for $p(\mathbf{X})$ in eq.(1),

$$p_h(\mathbf{x}) = \frac{1}{N} \sum_{t=1}^{N} G(\mathbf{x}|\mathbf{x}_t, h^2\mathbf{I}) \tag{2}$$

where and through out this paper, $G(\mathbf{x}|\boldsymbol{\mu}, \boldsymbol{\Sigma})$ denotes a Gaussian density with a mean $\boldsymbol{\mu}$ and covariance $\boldsymbol{\Sigma}$. In help of a Taylor expansion like technique in an approximation up to the second order [10], eq.(1) equivalently includes two terms. One is a Tikhonov term, while the other acts to control the value of smoothing parameter h at an appropriate level. BYY harmony learning provides an easy solution to determining h, which plays the same role as the hyper-parameter in Tikhonov regularization. Based on incorporating a prior or adding equal constraints, five formulas have been derived for determining h [2,7,13], which will be further discussed in Section 3.2. Although some formulas have been already used in a comparative study for the model selection problem on LFA [1], there still lack theoretical analysis and systematic comparison among them. Our major contribution is a systematic comparative study on automatic BYY harmony learning algorithms with one of the five formulas for h, via various experiments on a series of simulated data and serval real world data sets. Also, we include those experimental results by the classical two-phase implementation via two typical criteria namely AIC and BIC.

The rest of this paper is organized as follows. In Section 2, we briefly review LFA. The BYY harmony learning on LFA, and the details of data smoothing regularization will be given in Section 3. In Section 4, we design experiments by using various of data sets including simulated data and real world data, followed by concluding remarks in Section 5.

2   Local Factor Analysis

Local Factor Analysis (LFA), also called Mixture of Factor Analyzers (MFA), combines the Gaussian Mixture Model (GMM) with Factor Analysis (FA) for each component's dimension reduction. LFA assumes an observed d-dimensional random variable $\mathbf{x}$ as follows:

$$\begin{aligned}
p(\mathbf{x}) &= \sum_{j=1}^{k} \alpha_j p(\mathbf{x}|j), \sum_{j=1}^{k} \alpha_j = 1, \alpha_j > 0, j = 1, \ldots, k, \\
p(\mathbf{x}|j) &= \int p(\mathbf{x}|\mathbf{y}, j)p(\mathbf{y}|j)d\mathbf{y} = G(\mathbf{x}|\boldsymbol{\mu}_j, \mathbf{U}_j\boldsymbol{\Lambda}_j\mathbf{U}_j^T + \boldsymbol{\Sigma}_j), \\
p(\mathbf{x}|\mathbf{y}, j) &= G(\mathbf{x}|\mathbf{U}_j\mathbf{y} + \boldsymbol{\mu}_j, \boldsymbol{\Sigma}_j), p(\mathbf{y}|j) = G(\mathbf{y}|\mathbf{0}, \boldsymbol{\Lambda}_j), \mathbf{U}_j^T\mathbf{U}_j = \mathbf{I}_{m_j},
\end{aligned} \tag{3}$$

where k is the cluster number and m_j is the hidden dimension of the j-th cluster, and $\boldsymbol{\Lambda}_j, \boldsymbol{\Sigma}_j$ are diagonal. Compared with MFA in [3], the LFA by eq.(3) is expressed differently but equivalently for the convenience of model selection [2].

When the model scale $\mathbf{k} = \{k, \{m_j\}_{j=1}^{k}\}$ is known, the set of parameters $\boldsymbol{\Theta} = \{\alpha_j, \mathbf{A}_j, \boldsymbol{\mu}_j, \boldsymbol{\Sigma}_j\}_{j=1}^{k}$ is often estimated based on a given set of samples $\mathcal{X}_N = \{\mathbf{x}_t\}_{t=1}^{N}$ under the Maximum Likelihood (ML) principle implemented by the famous Expectation Maximization (EM) algorithm [3].

When the model scale is unknown, the task consists of both parameter learning and model selection. Classically, it is tackled in the following two-phase implementation that minimizes one of typical criteria such as Akaike's information criterion (AIC)[14] and Schwarz's Bayesian inference criterion (BIC)[15]:

$$\text{Phase I: } \hat{\boldsymbol{\Theta}}_k = \arg\max_{\Theta} L(\mathcal{X}_N|\boldsymbol{\Theta}_k), \text{ by enumerating each } \mathbf{k} \text{ in a candidate set.}$$

$$\text{Phase II: } \mathbf{k}^* = \arg\max_k \mathbf{J}(\mathbf{k}), \mathbf{J}(\mathbf{k}) = \begin{cases} -2L(\mathcal{X}_N|\boldsymbol{\Theta}_k) + 2D(\mathbf{k}), & \text{for AIC} \\ -2L(\mathcal{X}_N|\boldsymbol{\Theta}_k) + D(\mathbf{k})\ln N, & \text{for BIC} \end{cases} \tag{4}$$

where $L(\mathcal{X}_N|\boldsymbol{\Theta}_k)$ is the log likelihood based on ML estimator $\hat{\boldsymbol{\Theta}}$ under a given model scale $\mathbf{k} = \{k, \{m_j\}_{j=1}^k\}$, and $D(\mathbf{k})$ is the number of free parameters in LFA model, i.e., $D(\mathbf{k}) = k - 1 + 2kd + \sum_{j=1}^{k}[dm_j - m_j(m_j - 1)/2]$.

3   BYY Harmony Learning on LFA with Data Smoothing

3.1   BYY Harmony Learning on LFA

Given $\mathcal{X}_N = \{x_t\}_{t=1}^N$, the parameter set $\boldsymbol{\Theta}$ and model scale $\mathbf{k}$ of LFA can be estimated by maximizing the following harmony function, which is derived according to eq.(1) (see [2,9,10] for details):

$$H(\boldsymbol{\Theta}, h, \mathbf{k}) = H_0(\boldsymbol{\Theta}, \mathbf{k}) - \frac{1}{2}h^2\widetilde{\pi}_q + \mathbf{Z}_q, \mathbf{Z}_q = \ln q(h), \tag{5}$$

$$H_0(\boldsymbol{\Theta}, \mathbf{k}) = \frac{1}{N}\sum_{t=1}^{N}\sum_{j=1}^{k} p_{jt} \ln[\alpha_j G(\mathbf{x}_t|\mathbf{U}_j\mathbf{y}_{jt} + \boldsymbol{\mu}_j, \boldsymbol{\Sigma}_j)G(\mathbf{y}_{jt}|\mathbf{0}, \boldsymbol{\Lambda}_j)]$$

where $\widetilde{\pi}_q = (1/N)\sum_t\sum_j p_{jt}Tr[\boldsymbol{\Sigma}_{x|j}^{-1}] > 0$ and p_{jt}, $\mathbf{y}_{jt}$, $\boldsymbol{\Sigma}_{x|j}$ are given in Tab.(1). For Ying machine, the specific settings of eq.(3) is used, while for Yang machine $p(\mathbf{X})$ is given by the Parzen window estimator by eq.(2) and $p(\mathbf{Y}|\mathbf{X}, \boldsymbol{\Theta}_{y|x})$ is free of structure.

In eq.(5), a Tikhonov-type regularization term $-\frac{1}{2}h^2\widetilde{\pi}_q$ penalizes $H_0(\boldsymbol{\Theta}, \mathbf{k})$ with the regularization strength controlled by h^2. Another term $Z_q = \ln q(h)$ acts to control h at an appropriate level. The learning by maximizing eq.(5) is implemented via a *Ying-Yang-Step* for updating $\boldsymbol{\Theta}$ and a *Smoothing-Step* for updating h. With the *Ying-Yang-Step* similar to those in [1,2,9], an algorithm is sketched in Tab.(1) with the details of *Smoothing-Step* to be further discussed in Section 3.2. Simply letting $h = 0$, the algorithm, denoted as **BYY-h0**, implements the one without data smoothing regularization.

3.2   Data Smoothing Regularization

We compare five previously proposed formulas as the Smoothing-Step in Tab.(1). Actually, the formulas can be summarized as coming from the following two regularization techniques.

Prior Information on Smoothing Parameter. The term $Z_q = \ln q(h)$ in eq.(5) can come from an improper prior for h. One simple choice is $q(h) \propto \frac{1}{h^2}$,

Table 1. Automatic BYY harmony learning algorithm on LFA (see [2,9] for details)

Initialization
Chose a large enough $\mathbf{k}_{init} = \{k_{init}, m_{init}\}$, and randomly initial parameters.

Yang-Step
Randomly pick a sample $\mathbf{x}_t$, and for each $j = 1, \ldots, k$, calculate
$$\mathbf{y}_{jt} = \mathbf{W}_j(\mathbf{x}_t - \boldsymbol{\mu}_j), \mathbf{W}_j = \boldsymbol{\Lambda}_j \mathbf{U}_j^T \boldsymbol{\Sigma}_{x|j}^{-1}, \boldsymbol{\Sigma}_{x|j} = \mathbf{U}_j \boldsymbol{\Lambda}_j U_j^T + \boldsymbol{\Sigma}_j$$
$$p_{jt} = p(j|\mathbf{x}_t) = \begin{cases} 1; & \text{if } j = j_t, j_t = \arg\min_\ell[\varepsilon_t(\theta_\ell) + \varrho_t(\theta_\ell, h)]; \\ 0; & \text{otherwise} \end{cases}$$
$$\varepsilon_t(\theta_j) = -\ln[\alpha_j G(\mathbf{x}_t|\mathbf{U}_j \mathbf{y}_{jt} + \boldsymbol{\mu}_j, \boldsymbol{\Sigma}_j) G(\mathbf{y}_{jt}|\mathbf{0}, \boldsymbol{\Lambda_j})], \varrho_t(\theta_j, h) = h^2 Tr[\boldsymbol{\Sigma}_{x|j}^{-1}]$$

Ying-Step
Update the parameters $\{\boldsymbol{\alpha}_j, \boldsymbol{\mu}_j, \boldsymbol{\Lambda}_j, \mathbf{U}_j, \boldsymbol{\Sigma}_j\}_{j=1}^k$:
$$\alpha_j^{new} = (\alpha_j^{old} + p_{jt}\eta)/(1 + \eta)$$
$$\boldsymbol{\mu}_j^{new} = \boldsymbol{\mu}_j + \eta p_{jt}\mathbf{e}_{jt}, \mathbf{e}_{jt,t} = \mathbf{x}_t - \boldsymbol{\mu}_j, \boldsymbol{\varepsilon}_{j,t} = \mathbf{e}_{j,t} - \mathbf{U}_j \mathbf{y}_{j,\mathbf{x}_t}$$
$$\boldsymbol{\Lambda}_j^{new} = (1 - \eta)\boldsymbol{\Lambda}_j + \eta p_{jt}\{h^2 diag(\mathbf{W}_j \mathbf{W}_j^T) + diag(\mathbf{y}_{jt}\mathbf{y}_{jt}^T)\}$$
$$\boldsymbol{\Sigma}_j^{new} = (1 - \eta)\boldsymbol{\Sigma}_j + \eta p_{jt}\{h^2 diag[(\mathbf{I}_d - \mathbf{U}_j \mathbf{W}_j)(\mathbf{I}_d - \mathbf{U}_j \mathbf{W}_j)^T] + diag[\boldsymbol{\varepsilon}_{j,t}\boldsymbol{\varepsilon}_{j,t}]\}$$
Update $\mathbf{U}_j$ by gradient on the Stiefel manifold (see [13] for details):
$$\mathbf{U}_j^{new} = \mathbf{U}_j + \eta p_{jt}(\mathbf{G}_{U_j} - \mathbf{U}_j \mathbf{G}_{U_j}^T \mathbf{U}_j), \mathbf{G}_{U_j} = \boldsymbol{\Sigma}_{\mathbf{x}|j}^{-1}\mathbf{e}_{jt}\mathbf{y}_{jt}^T + h^2 \boldsymbol{\Sigma}_{\mathbf{x}|j}^{-1}\mathbf{W}_j^T.$$
Model selection:
(1). Discard the j-th component if $\alpha_j \to 0$.
(2). Discard ℓ-th factor of j-th component if $\boldsymbol{\Lambda}_j$'s ℓ-th diagonal entry $\lambda_{j\ell} \to 0$.

Smoothing-Step: (see details in Section 3.2)
A general form: $h^{new} = f(\Theta^{old}, \mathbf{k}, h^{old})$.

Converge: Repeat Yang-Step,Ying-Step,Smoothing-Step, until the parameters
or the harmony function $H(\boldsymbol{\Theta}, h)$ varies very small, where
$$H(\boldsymbol{\Theta}, h) = 0.5 \sum_{j=1}^k \alpha_j\{2\ln\alpha_j - m_j \ln(2\pi) - \ln|\boldsymbol{\Lambda}_j| - \ln|\boldsymbol{\Sigma}_j| - h^2 Tr[\boldsymbol{\Sigma}_{x|j}]\}.$$

---

but it won't work because there is no real root for $\frac{\partial H(\Theta,h,k)}{\partial h} = -h\widetilde{\pi}_q - 2/h = 0$.
Instead, we consider the sample set based improper prior that was firstly proposed in 1997 [12] under the name of data smoothing regularization. Readers are referred to [2] for a recent justification and interpretation. As listed in Tab.(2), we investigate the three formulas for updating h given in [13], denoted by **BYY-h1**.

Also, $q(h)$ is considered in [2] to be a Gamma or Chi-square distribution in form of $q(\zeta) = C^{-1}\zeta^b e^{-a\zeta}$, $a, b > 0, \zeta = h^2$, where C is a normalization constant, and a, b determine its mean $\mu(a, b)$ and variance $\sigma^2(a, b)$. It is subtle to choose (a, b) appropriately. Based on our experience, we simply choose (a, b) such that $\mu(a, b) \in (0, 1)$ and $\sigma^2(a, b)$ is small. The corresponding updating formula is sketched in Tab.(3), denoted as **BYY-h2**.

Equal Covariance. Proposed in [2] under the name of *equal covariance*, h^2 is solved from an equal constraint between Ying and Yang machines:

$$COV_{p(\mathbf{x}|\theta)}(\mathbf{x}) = COV_{p_h(\mathbf{x})}(\mathbf{x}), \tag{6}$$

Table 2. BYY-h1 with an improper prior $q(h) \propto [\sum_t \sum_\tau G(\mathbf{x}_t|\mathbf{x}_\tau, h^2\mathbf{I})/N]^{-1}$ (see [13] for details)

Replace the Smoothing-Step in Tab.(1) with Formula h1(a), h1(b) or h1(c).
Smoothing-Step: Three choices for updating h:

The gradient for h: $\frac{\partial H(\Theta,h,\mathbf{k})}{\partial h} = -h\pi_q + d/h - dh_{x,0}^2/h^3$

where d is the dimension of the observed variable $\mathbf{x}$, and

$\pi_q = \sum_{j=1}^k \alpha_j Tr[\boldsymbol{\Sigma}_{\mathbf{x}|j}^{-1}],\ \boldsymbol{\Sigma}_{\mathbf{x}|j} = \mathbf{U}_j\boldsymbol{\Lambda}_j\mathbf{U}_j^T + \boldsymbol{\Sigma}_j,\ p_{t,\tau} = \frac{\exp\{-d_h(\mathbf{x}_t,\mathbf{x}_\tau)\}}{\sum_{\tau=1}^N \sum_{t=1}^N \exp\{-d_h(\mathbf{x}_t,\mathbf{x}_\tau)\}};$

$h_{x,0}^2 = \frac{1}{d}\sum_{\tau=1}^N \sum_{t=1}^N p_{t,\tau}\|\mathbf{x}_t - \mathbf{x}_\tau\|^2,\ d_h(\mathbf{x}_t,\mathbf{x}_\tau) = \frac{1}{2h^2}(\mathbf{x}_t - \mathbf{x}_\tau)^T(\mathbf{x}_t - \mathbf{x}_\tau).$

Formula h1(a):

$h^{new} = h^{old} + \eta\Delta h,\quad \Delta h = \frac{\partial H(\Theta,h,\mathbf{k})}{\partial h}.$

Formula h1(b):

$h^{new} = h^{old} + \eta\Delta h,\ \Delta h = \begin{cases} 1; & \text{if } H(\Theta, h+\eta, \mathbf{k}) > H(\Theta, h, \mathbf{k}); \\ -1; & \text{if } H(\Theta, h-\eta, \mathbf{k}) > H(\Theta, h, \mathbf{k}); \\ 0; & \text{otherwise} \end{cases}$

Formula h1(c):

Roughly regarding $h_{x,0}$ as a constant to h, and solving $\frac{\partial H(\Theta,h,\mathbf{k})}{\partial h} = 0$ to get

$h^2 = 2h_{x,0}^2/(1 + \sqrt{1 - 4h_{x,0}^2 d^{-1}\pi_q})$

Table 3. BYY-h2 with a proper prior $q(\zeta) = C^{-1}\zeta^b e^{-a\zeta}$, $a, b > 0$, $\zeta = h^2$

Replace the Smoothing-Step in Tab.(1) with Formula h2.
Smoothing-Step:

Formula h2: $h^2 = b/(a + 0.5\pi_q)$, $(a >, b > 0$, prefixed during learning).

Table 4. BYY-h3 with equal covariance $h^2\mathbf{I} + \mathbf{S}_N = \sum_{j=1}^k \alpha_j(\boldsymbol{\mu}_j\boldsymbol{\mu}_j^T + \boldsymbol{\Sigma}_{x|j}) - \boldsymbol{\mu}\boldsymbol{\mu}^T$ (see [2] for details)

Replace the Smoothing-Step in Tab.(1) with Formula h3.
Smoothing-Step:

Formula h3: $h^2 = \frac{1}{d}\left\{\sum_{j=1}^k \alpha_j\|\boldsymbol{\mu}_j\|^2 - \|\sum_{j=1}^k \alpha_j\boldsymbol{\mu}_j\|^2 + Tr[\sum_{j=1}^k \alpha_j\boldsymbol{\Sigma}_{x|j} - \mathbf{S}_N]\right\}$

where $\boldsymbol{\Sigma}_{x|j} = \mathbf{U}_j\boldsymbol{\Lambda}_j\mathbf{U}_j^T + \boldsymbol{\Sigma}_j$, $\boldsymbol{\mu} = \sum_{j=1}^k \alpha_j\boldsymbol{\mu}_j$

$\mathbf{S}_N = \frac{1}{N}\sum_{t=1}^N (\mathbf{x}_t - \bar{\mathbf{x}})(\mathbf{x}_t - \bar{\mathbf{x}})^T$, $\bar{\mathbf{x}} = \frac{1}{N}\sum_{t=1}^N \mathbf{x}_t$

---

where $COV_{u(\mathbf{x})}(\mathbf{x})$ denotes the covariance matrix with respect to a density $u(\mathbf{x})$, and $p(\mathbf{x}|\theta) = \sum_{j=1}^k \int \alpha_j p(\mathbf{x}|\mathbf{y}, j)p(\mathbf{y}|j)d\mathbf{x}$ is the marginal density of LFA, and $p_h(\mathbf{x})$ is the Parzen window density estimator. In this case, we set $Z_q = 0$. The corresponding updating formula, **BYY-h3**, is shown in Tab.(4).

4　Experiments

Two parts of experiments are designed to investigate the performances resulted from using five formulas to estimate the smoothing parameter h in the automatic

BYY harmony learning on LFA in Tab.(1). One examines model selection accuracy on simulated data sets, while the other examines the classification accuracy on serval real world data sets, which characterizes the learning performance at a higher level but still indirectly associated with appropriate model selection for a good generalization ability. We choose learning rate $\eta \in [0.01, 0.1]$, and prior constants $(a, b) = (10, 1)$. Moreover, we also include the classical two-phase implementation with two typical criteria AIC and BIC by eq.(4).

4.1   On Simulated Data

We randomly generate 7 series of datasets from the LFA models with the same component number $k = 3$ and the same local dimensions $m_j = 3$, $j = 1, \ldots, k$. Given in Fig.(1), different experimental environments are set by varying the scale of three basic factors, i.e., noise, dimension and sample size. The environment deteriorates as the noise variance or dimension increases, and as the sample size decreases. The noise variance $\mathbf{\Sigma}_j = \psi^2 I$ is specified based on ζ, $\zeta = \min_j \zeta_j$, where ζ_j is the smallest eigenvalue of $\mathbf{\Lambda_j}$. We initialize $k_{init} = 5$, $m_{j\,init} = 5$, $j = 1, \ldots, k_{init}$ for BYY harmony learning, $1 \le k_{init}, m_{j\,init} \le 5$ for two-phase implementation. For simplicity, we also let $m_j = m$ for all j.

After 100 independent runs for each algorithm on each experimental case, the number of correct selections on model scale $\mathbf{k} = \{k, \{m_j\}_{j=1}^{k}\}$, is listed at the corresponding vertices in the cubes in Fig.(2) arranged all according to Fig.(1), from which we have the following observations:

- All algorithms work well in an environment of good enough. As the situation becomes worse, BYY harmony learning with data smoothing is generally better than that without data smoothing.
- When the sample size is very small, BYY-h1(a) is the best, and BYY-h1(b), BYY-h1(c), BYY-h3 are comparable, followed by BYY-h2.
- BYY algorithms with data smoothing work better than AIC, BIC, especially for a small sample size. Still, BYY algorithm without data smoothing, i.e., BYY-h0, is comparable with AIC and BIC.

4.2   On Real World Data

All methods are tested on four real world data sets collected from UCI repository machine learning database[1], including Pendigits, Optdigits, Segment and Waveform, denoted by PEN, OPT, SEG and WAVE respectively, listed in Tab.(5). The testing method shares an idea similar to the well known k-NN approach. First, training a LFA model M_ℓ for each class $\ell = 1, \ldots, C$; then calculating all the likelihoods $L(\mathbf{x}_t | j, \ell)$ of the test sample $\mathbf{x}_t$ for the j-th component of M_ℓ, where $j = 1, \ldots, k_\ell$, $\ell = 1, \ldots, C$; and assigning $\mathbf{x}_t$ to class $\ell^* = \arg\max_{\ell, j} L(\mathbf{x}_t | j, \ell)$.

The testing results are listed in Tab.(6). BYY harmony learning algorithms with data smoothing are generally better than other methods, while BYY-h1(a) is the best and the most stable one among all.

---

[1] http://www.ics.uci.edu/mlearn/MLRepository.html

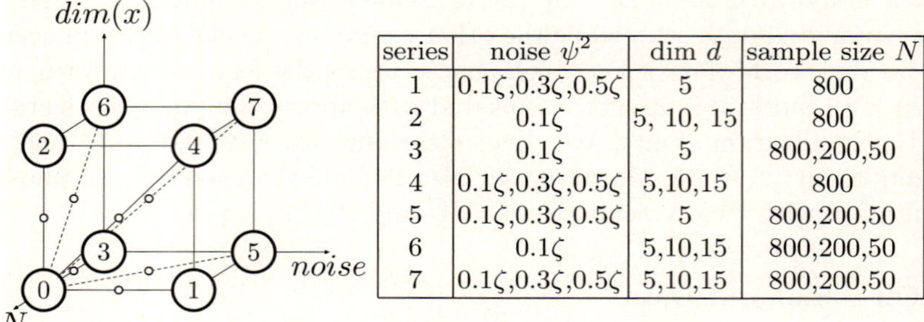

series	noise ψ^2	dim d	sample size N
1	$0.1\zeta,0.3\zeta,0.5\zeta$	5	800
2	0.1ζ	5, 10, 15	800
3	0.1ζ	5	800,200,50
4	$0.1\zeta,0.3\zeta,0.5\zeta$	5,10,15	800
5	$0.1\zeta,0.3\zeta,0.5\zeta$	5	800,200,50
6	0.1ζ	5,10,15	800,200,50
7	$0.1\zeta,0.3\zeta,0.5\zeta$	5,10,15	800,200,50

Fig. 1. Description of 7 series of experimental configurations

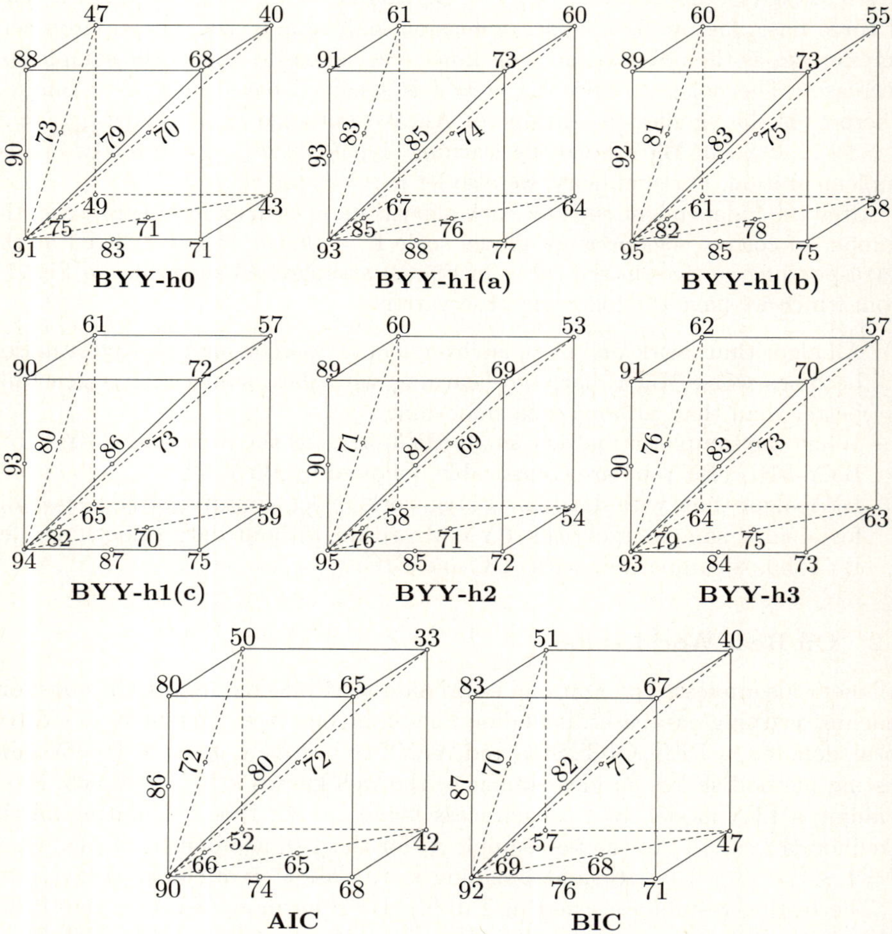

Fig. 2. The correct number of model selection results of 7 series of experimental configurations, collected from 100 independent runs for each method on each case

Table 5. Basic description of real world data sets

Datasets	Training	Testing	Dimension	Classes
PEN	7,494	3,498	16	10
OPT	2,880	1,797	64	10
SEG	700	1,610	14	7
WAVE	300	4,700	21	3

Table 6. The classification accuracies (average $\pm$ standard deviation) on real world data sets, after 20 independent runs for each method

Methods	PEN	OPT	SEG	WAVE
BYY-h0	96.86 ± 0.45	80.82 ± 7.38	85.50 ± 1.44	81.13 ± 1.50
BYY-h1(a)	$\mathbf{98.07 \pm 0.34}$	$\mathbf{98.01 \pm 0.63}$	$\mathbf{87.28 \pm 1.25}$	$\mathbf{83.68 \pm 0.73}$
BYY-h1(b)	97.50 ± 0.19	97.68 ± 0.53	86.98 ± 1.30	82.64 ± 0.78
BYY-h1(c)	97.57 ± 0.31	97.82 ± 0.68	86.56 ± 1.35	83.33 ± 0.69
BYY-h2	97.21 ± 0.47	94.60 ± 1.12	85.01 ± 1.58	83.03 ± 0.78
BYY-h3	97.84 ± 0.17	95.35 ± 1.03	86.74 ± 1.05	83.29 ± 0.97
AIC	93.04 ± 0.27	90.38 ± 0.79	70.33 ± 2.68	69.75 ± 2.06
BIC	94.12 ± 0.24	93.01 ± 0.88	73.80 ± 2.96	81.09 ± 2.51

5   Concluding Remarks

We have comparatively investigated the performance of different formulas for estimating the smoothing parameter in data smoothing regularization during BYY harmony learning, by considering on the model selection problem of LFA modeling. From the experiments, we have at least two observations:

- Data smoothing regularization indeed improves model selection ability considerably, especially when the sample size is small. Among five different formulas for determining smoothing parameter, the sample set based improper prior [12,13] is really promising with the best performances.
- BYY harmony learning with data smoothing regularization outperforms the classical two-phase implementation with the criteria AIC and BIC.

Acknowledgements. The authors would like to thank Mr. Ke Sun for his help in preparing the figures. The work described in this paper was fully supported by a grant from the Research Grant Council of the Hong Kong SAR (Project No: CUHK4177/07E).

References

1. Shi, L.: Bayesian Ying-Yang Harmony Learning for Local Factor Analysis: A Comarative Investigation. In: Tizhoosh, H., Ventresca, M. (eds.) Oppositional Comcepts in Computational Intelligence. Springer Physika-Verlag, Heidelberg (to appear, 2008)

2. Xu, L.: A unified perspective and new results on RHT computing, mixture based learning, and multi-learner based problem solving. Pattern Recognition 40(8), 2129–2153 (2007)
3. Ghahramani, Z., Hinton, G.E.: The EM algorithm for Mixture of Factor Analyzers. Technical Report CRG-TR-96-1, Dept. of Computer Science, University of Toronto (1997)
4. Tipping, M.E., Bishop, C.M.: Mixtures of probabilistic principal component analyzers. Neural Computation 11(2), 443–482 (1999)
5. Ghahramani, Z., Beal, M.J.: Variational Inference for Bayesian Mixtures of Factor Analysers. In: Advances in NIPS, pp. 449–455 (2000)
6. Hinton, G.E., Revow, M., Dayan, P.: Recognizing Handwritten Digits using Mixtures of Linear Models. In: Advances in NIPS, pp. 1015–1022 (1995)
7. Xu, L.: A Trend on Regularization and Model Selection in Statistical Learning: A Bayesian Ying Yang Learning Perspective. In: Duch, W., Mandziuk, J. (eds.) Challenges for Computational Intelligence, pp. 365–406. Springer, Heidelberg (2007)
8. Xu, L.: A Unified Learning Scheme: Bayesian-Kullback Ying-Yang Machine. In: Touretzky, D.S., et al. (eds.) Advances in NIPS, vol. 8, pp. 444–450. MIT Press, Cambridge (1996); A preliminary version in Proc. ICONIP 1995, Beijing, pp. 977–988 (1995)
9. Xu, L.: Bayesian Ying Yang Learning. Scholarpedia 2(3), 1809 (2007), http://scholarpedia.org/article/Bayesian_Ying_Yang_Learning
10. Xu, L.: Bayesian Ying Yang System, Best Harmony Learning, and Gaussian Manifold Based Family. In: Zurada, J.M., Yen, G.G., Wang, J. (eds.) Computational Intelligence: Research Frontiers. LNCS, vol. 5050, pp. 48–78. Springer, Heidelberg (2008)
11. Bishop, C.M.: Training with noise is equivalent to Tikhonov regularization. Neural Computation 7(1), 108–116 (1995)
12. Xu, L.: Bayesian Ying-Yang System and Theory as a Unified Statistical Learning Approach: (I)Unsupervised and Semi-unsupervised Learning. In: Amari, S., Kassabov, N. (eds.) Brain-like Computing and Intelligent Information Systems, pp. 241–274. Springer, Berlin (1997)
13. Xu, L.: Data smoothing regularization, multi-sets-learning, and problem solving strategies. Neural Networks 16(5-6), 817–825 (2003)
14. Akaike, H.: A new look at the statistical model identification. IEEE Transactions on Automatic Control 19(6), 716–723 (1974)
15. Schwarz, G.: Estimating the Dimension of a Model. The Annals of Statistics 6(2), 461–464 (1978)

Adding Diversity in Ensembles of Neural Networks by Reordering the Training Set

Joaquín Torres-Sospedra, Carlos Hernández-Espinosa,
and Mercedes Fernández-Redondo

Departamento de Ingenieria y Ciencia de los Computadores, Universitat Jaume I,
Avda. Sos Baynat s/n, C.P. 12071, Castellon, Spain
{jtorres,espinosa,redondo}@icc.uji.es

Abstract. When an ensemble of neural networks is designed, it is necessary to provide enough diversity to the different networks in the ensemble. We propose in this paper one new method of providing diversity which consists on reordering the training set patterns during the training. In this paper three different algorithms are applied in the process of building ensembles with *Simple Ensemble* and *Cross-Validation*. The first method consists in using the original training set, in the second one the training set is reordered before the training algorithm is applied and the third one consists in reordering the training set at the beginning of each iteration of BackPropagation. With the experiments proposed we want to empirically demonstrate that reordering patterns during the training is a valid source to provide diversity to the networks of an ensemble. The results show that the performance of the original ensemble methods can be improved by reordering the patterns during the training. Moreover, this new source of diversity can be extended to more complex ensemble methods.

1   Introduction

Training an ensemble of neural networks is commonly applied in order to increase the generalization ability with respect to a single neural network. This process consists of training a set of neural networks with different weight initialization or properties in the training process and combining them in a suitable way. The most important methodologies or models to built ensembles and other Multi-Net systems are: *Bagging* [1], *Boosting* [2,3,4], *Stacked Generalization* [5], *Mixture of Experts* [6] among others.

There is no advantage by combining a set of exactly equal networks. The networks in the ensemble should provide slightly different solutions to the problem in order to obtain an additional efficiency by the combination. If we analyse all the methods proposed, we can see that diversity can be provided by modifying some parameters like the initial weight values, the training set, the target equation or network structure. Traditionally, the ensemble methods modify the training set before applying the training algorithm (*Bagging*, *Boosting* or *Cross-Validation*).

In this paper, three different algorithms based on reordering the training set are proposed, i.e., in these new methods we change the order of presentation of the patterns in the training set from one network to another. The first method consists in using the 'original' training set for all the networks in the ensemble. The second one consists in

V. Kůrková et al. (Eds.): ICANN 2008, Part I, LNCS 5163, pp. 275–284, 2008.

reordering the original training set before starting the learning process, i.e., we use a different order of presentation of patterns in the training set for the different networks in the ensemble. Finally, the last one consists in randomly sorting the training set at the beginning of each iteration of *Back-propagation* algorithm. In the two last methods, random sampling without replacement is applied to reorder the training set.

This paper is organized as follows. Firstly, some theoretical concepts are briefly reviewed in section 2. Then, the databases used and the experimental setup are described in section 3. Finally, the experimental results and their discussion are in section 4.

2   Theory

In this section, we describe the training algorithm of *Multilayer Feedforward*, two methods to build ensembles and the proposed methods to reorder the training set.

2.1   The Multilayer Feedforward Network

The *Multilayer Feedforward* architecture is the most known network architecture. This network consists of three layers of computational units. The neurons of the first layer apply the identity function whereas the neurons of the second and third layer apply the sigmoid function. It has been proved that *MF* networks with one hidden layer and threshold nodes can approximate any function with a specified precision [7] and [8]. Figure 1 shows the diagram of the *Multilayer Feedforward* network we have used in our experiments.

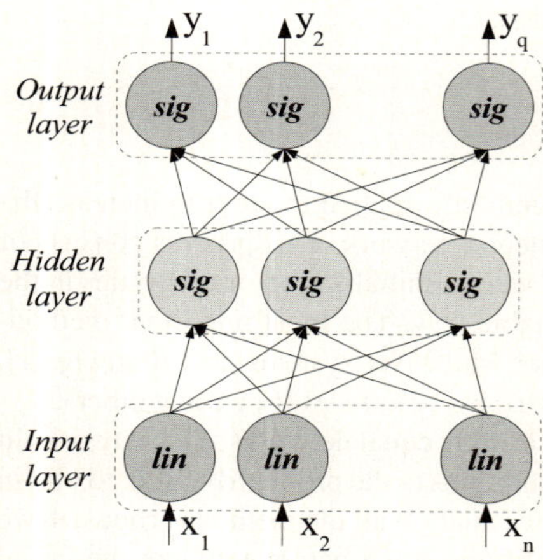

Algorithm 1. ANN Training $\{T, V\}$

Set initial weights values
for $i = 1$ to *iterations* **do**
    Train the network on training set T
    Calculate MSE over validation set V
    Save epoch weights and MSE
end for
Select epoch with minimum MSE
Assign best epoch configuration to the net
Save network configuration

Fig. 1. Multilayer Feedforward Structure

In our experiments we have trained the networks for few iterations. In each iteration, the weights have been adapted with *Back-Propagation* over the training set. At the end of each iteration the Mean Square Error (MSE) has been calculated over the Validation set. When the learning process has finished, we assign the weights of the iteration with lowest MSE of validation to the final network.

For this reason the original learning set L used in the learning process is divided into two subsets: the first set is the training set T which is used to train the networks and the second set is the validation set V which is used to finish the training process.

2.2   Simple Ensemble

A simple ensemble can be constructed by training different networks with the same training set, but with different random initialization [9]. The use of different random initialization is supposed to be applied in the rest of ensemble methods.

Algorithm 2. Simple Ensemble $\{T,V\}$

Generate k different seeds
for $i = 1$ to k **do**
   Set the seed $seed_i$ to the random number generator for generating the weights of the nets
   ANN Training $\{T,V\}$
end for
Save Ensemble Configuration

2.3   Cross Validation

Cross-Validation can be applied to build an ensemble, in this case case we have applied a version called *CVCv2Mod2*.

In this version of *Cross-Validation* the original training set is randomly splitted into 10 subsets, 8 for training and 2 for validation. By changing the subsets used for validation we can create till 45 different training and validation sets.

Algorithm 3. Cross-Validation $\{T,V\}$

Generate 10 different subsets with CV
for $i = 1$ to k **do**
   Select 2 different subsets for Validation V_i
   Select the other subsets for Training T_i
   ANN Training $\{T_i,V_i\}$
end for
Save Ensemble Configuration

2.4   Modifying the Order of Training Set

Original Training Set Order. This method consist on using the original training set T to build the ensembles, all the networks in the ensemble are trained with the exactly the same training set. Each network is trained with the traditional ANN training algorithm, algorithm 1, using the version of the training set generated by the ensemble method.

Static Training Set Order. This method consists on modifying the order of the training set patterns before starting the network learning, the training set T' is drawn by reordering the patterns of the networks training set T. The ANN training algorithm described in algorithm 4 is applied in this case.

Algorithm 4. ANN Training (Static set order) $\{T, V\}$

Set initial weights values
Generate T' by reordering T at random
for $i = 1$ to $iterations$ **do**
    Train the network on the training set T'
    Calculate MSE over validation set V
    Save epoch weights and calculated MSE
end for
Select epoch with minimum MSE
Assign best epoch configuration to the network
Save network configuration

Dynamic Training Set Order. This method consists on modifying the order of the training set patterns at the beginning of each iteration of *Back-Propagation*. The training set T' is reordered at random from the network training set T. The ANN training algorithm described in algorithm 5 is applied in this case.

Algorithm 5. ANN Training (Dynamic set order) $\{T, V\}$

Set initial weights values
for $i = 1$ to $iterations$ **do**
    Generate T' by reordering T at random
    Train the network on the training set T'
    Calculate MSE over validation set V
    Save epoch weights and calculated MSE
end for
Select epoch with minimum MSE
Assign best epoch configuration to the network
Save network configuration

---

3   Experimental Setup

To test the performance of the methods proposed in this paper, we have trained ensembles of 3, 9, 20 and 40 networks with *Simple Ensemble* and *Cross-Validation* using the three different training algorithms described in 2.4. In addition, we repeated ten times the whole process to obtain a mean performance of the ensemble for each database and an error in the performance calculated by standard error theory.

3.1   Databases

We have used the following classification problems from the *UCI repository of machine learning databases* [10] to test the performance of the methods: *Arrhythmia* (aritm), *Balance Scale* (bala), *Cylinder Bands* (band), *BUPA liver disorders* (bupa), *Australian Credit Approval* (cred), *Dermatology* (derma), *Ecoli* (ecoli), *Solar Flares* (flare), *Glass Identification* (glas), *Heart Disease* (hear), *Image segmentation* (img), *Ionosphere*

Database (ionos), *The Monk's Problem 1 and 2* (mok1 and mok2), *Pima Indians Diabetes* (pima), *Haberman's Survival Data* (survi), *Congressional Voting Records* (vote), *Vowel Database* (vowel) and *Wisconsin Breast Cancer* (wdbc).

4   Results and Discussion

4.1   Results

The main results are presented in this subsection. Table 1 shows the results for the ensembles trained with *Simple Ensemble* using the three reordering schemes studied (*traditional*, *static* and *dynamic*) whereas table 2 show the results for the ensembles trained with *Cross-Validation*.

4.2   General Measurements

We have also calculated the Increase of Performance (IoP eq.1) and the Percentage of Error Reduction (PER eq.2) of the results with respect to a single network in order to get a general value for the comparison among all the methods we have studied.

The IoP value is an absolute measurement that denotes the increase of performance of the ensemble with respect to a single network.

$$IoP = Error_{SingleNet} - Error_{Ensemble} \tag{1}$$

The PER value is a relative measurement which ranges from 0%, where there is no improvement by the use of an ensemble method with respect to a single network, to 100%.

$$PER = 100 \cdot \frac{Error_{SingleNet} - Error_{Ensemble}}{Error_{SingleNet}} \tag{2}$$

There can also be negative values in both measurements. A negative value on the IoP and on the PER means that the performance of the ensemble is worse than the performance of the single network.

Finally, we have calculated the mean *Increase of Performance* and the mean *Percentage of Error Reduction* across all databases to get a global measurement to compare the methods presented in the paper. Table 3 shows the results of the mean *Increase of Performance* whereas table 4 shows the results of the mean *Percentage of Error Reduction*. In these tables we have also included the general results of *Adaptive Boosting* (*adaboost*), *ArcX4*, *Adaptive Training Algorithm* (*ata*) and *Evolutionary Ensemble with Negative Correlation Learning* (*eencl*) to compare the training methods studied with traditional ensemble methods.

4.3   Discussion

After analysing the results related to the general measurements (tables 3-4), it can be seen that *Simple Ensemble* and the version included of *Cross Validation* work better

Table 1. Simple Ensemble results

	Traditional Training			
	3 Nets	*9 Nets*	*20 Nets*	*40 Nets*
aritm	73.4±1	73.8±1.1	73.8±1.1	73.8±1.1
bala	96±0.5	95.8±0.5	95.8±0.6	95.9±0.5
band	73.5±1.2	72.9±1.5	73.8±1.3	73.8±1.3
bupa	72.3±1.2	72.4±1.1	72.3±1.1	72.7±1.1
cred	86.5±0.7	86.4±0.7	86.6±0.7	86.5±0.7
derma	97.2±0.7	97.5±0.7	97.3±0.7	97.6±0.7
ecoli	86.6±0.8	86.9±0.8	86.9±0.8	86.9±0.7
flare	81.8±0.5	81.6±0.4	81.5±0.5	81.6±0.5
glas	94±0.8	94±0.7	94±0.7	94.2±0.6
hear	82.9±1.5	83.1±1.5	83.1±1.5	82.9±1.5
img	96.5±0.2	96.7±0.3	96.7±0.2	96.8±0.2
ionos	91.1±1.1	90.3±1.1	90.4±1	90.3±1
mok1	98.3±0.9	98.8±0.8	98.3±0.9	98.3±0.9
mok2	88±2	90.8±1.8	91.1±1.1	91.1±1.2
pima	75.9±1.2	75.9±1.2	75.9±1.2	75.9±1.2
survi	74.3±1.3	74.2±1.3	74.3±1.3	74.3±1.3
vote	95.6±0.5	95.6±0.5	95.6±0.5	95.6±0.5
vowel	88±0.9	91±0.5	91.4±0.8	92.2±0.7
wdbc	96.9±0.5	96.9±0.5	96.9±0.5	96.9±0.5

	Static Training				Dynamic Training			
	3 Nets	*9 Nets*	*20 Nets*	*40 Nets*	*3 Nets*	*9 Nets*	*20 Nets*	*40 Nets*
aritm	74.7±1.6	75.8±1.7	75.8±1.4	75.6±1.6	75.4±1.8	75.3±1.6	75.6±1.7	75.6±1.6
bala	95.4±0.7	95.4±0.8	95.7±0.8	95.8±0.7	95.9±0.8	96.1±0.8	95.9±0.6	95.8±0.7
band	75.1±1.3	74.6±1.3	74±1.2	74±1.4	75.3±1.3	75.6±1.2	75.6±1.1	75.1±1.2
bupa	73.1±1.2	72.3±1.2	72.4±1.2	72.7±1.2	72.6±1.2	72.7±1.1	72.7±1	73.1±1.1
cred	86.6±0.7	87.1±0.6	87±0.7	86.9±0.6	86.5±0.6	86.5±0.6	86.6±0.7	86.7±0.7
derma	97.3±0.7	97.5±0.7	97.6±0.7	97.6±0.7	97.5±0.8	97.6±0.7	97.8±0.6	97.6±0.7
ecoli	86.9±1	86.9±0.9	87.1±1	87.2±1	86.6±0.8	86±0.7	86.6±0.9	86.3±0.8
flare	82.1±0.7	82.1±0.6	82.1±0.6	82.1±0.6	82.2±0.5	82.1±0.3	82.1±0.4	82.1±0.4
glas	92.8±0.9	95.2±0.7	95.4±1	95.4±1	94.6±1	94.6±1	94.4±0.8	94.4±0.8
hear	83.7±1.2	84.2±1.3	83.9±1.3	84.1±1.4	83.1±1.5	83.4±1.6	83.2±1.6	83.2±1.6
img	96.7±0.2	96.9±0.3	96.8±0.3	97±0.3	96.9±0.3	96.8±0.3	97.1±0.2	97.1±0.2
ionos	89.4±1.3	89.4±1.1	89.4±1.2	89.6±1	90.6±1.1	90.1±1	90.1±1	90.1±1
mok1	98.6±0.9	98.6±0.9	98.6±0.9	98.6±0.9	96.6±1.4	97.3±1.4	98.6±0.9	98.6±0.9
mok2	89.9±1.2	91.6±0.8	92.4±0.8	91.6±0.7	91±1.5	91.6±1.4	91±1.2	92±1.1
pima	76.3±1.1	76.1±1	76.7±1	77±1	76.7±1	76.9±1	76.9±1	76.8±1
survi	75.1±1.3	74.3±1.3	74.6±1.5	74.9±1.5	74.9±1.3	74.6±1.4	74.8±1.4	74.6±1.4
vote	96±0.7	95.9±0.6	96±0.6	96.1±0.6	95.9±0.7	95.6±0.7	95.8±0.6	95.8±0.6
vowel	88.6±0.6	90.8±0.9	91.4±0.8	91.6±0.8	89.8±0.7	91.5±0.8	92.3±0.7	92.3±0.7
wdbc	97.1±0.4	97.4±0.4	97.3±0.4	97.3±0.4	97.3±0.5	97.3±0.5	97.3±0.5	97.3±0.5

Table 2. Cross-Validation results

Traditional Training

	3 Nets	9 Nets	20 Nets	40 Nets
aritm	75.4±1.8	75.3±1.6	75.6±1.7	75.6±1.6
bala	95.9±0.8	96.1±0.8	95.9±0.6	95.8±0.7
band	75.3±1.3	75.6±1.2	75.6±1.1	75.1±1.2
bupa	72.6±1.2	72.7±1.1	72.7±1	73.1±1.1
cred	86.5±0.6	86.5±0.6	86.6±0.7	86.7±0.7
derma	97.5±0.8	97.6±0.7	97.8±0.6	97.6±0.7
ecoli	86.6±0.8	86±0.7	86.6±0.9	86.3±0.8
flare	82.2±0.5	82.1±0.3	82.1±0.4	82.1±0.4
glas	94.6±1	94.6±1	94.4±0.8	94.4±0.8
hear	83.1±1.5	83.4±1.6	83.2±1.6	83.2±1.6
img	96.9±0.3	96.8±0.3	97.1±0.2	97.1±0.2
ionos	90.6±1.1	90.1±1	90.1±1	90.1±1
mok1	96.6±1.4	97.3±1.4	98.6±0.9	98.6±0.9
mok2	91±1.5	91.6±1.4	91±1.2	92±1.1
pima	76.7±1	76.9±1	76.9±1	76.8±1
survi	74.9±1.3	74.6±1.4	74.8±1.4	74.6±1.4
vote	95.9±0.7	95.6±0.7	95.8±0.6	95.8±0.6
vowel	89.8±0.7	91.5±0.8	92.3±0.7	92.3±0.7
wdbc	97.3±0.5	97.3±0.5	97.3±0.5	97.3±0.5

Static Training

	3 Nets	9 Nets	20 Nets	40 Nets
aritm	75.2±1.3	76.4±1.5	76.6±1.2	76.6±1.4
bala	96.5±0.8	96.6±0.5	96.1±0.7	96.2±0.6
band	74.7±1.6	75.1±1.1	74.2±1.4	74.7±1.1
bupa	73.4±1.6	74±1.3	73.6±1.4	73.7±1.3
cred	86.7±0.8	86.8±0.8	86.9±0.7	86.9±0.6
derma	97.6±0.7	97±0.6	97.3±0.5	97.5±0.5
ecoli	86.9±0.9	86.8±0.9	87.2±0.9	86.8±0.9
flare	81.9±0.4	82.1±0.5	82.1±0.6	82.4±0.6
glas	94±1.1	95.2±0.8	94.4±0.9	94.4±0.8
hear	84.4±1.5	83.9±1.4	84.4±1.5	84.6±1.5
img	96.71±0.15	96.8±0.3	96.8±0.3	97±0.3
ionos	91.7±0.9	91.1±0.7	91.9±0.8	91.4±0.8
mok1	100±0	99.4±0.6	99.4±0.6	99.4±0.6
mok2	92.6±1.4	94.5±1.6	95.1±1.3	95±1.1
pima	76.5±1.1	76.8±1.1	77±1	76.7±1.1
survi	74.1±1.2	73.9±1.4	73.3±1.5	73.1±1.3
vote	96.3±0.5	96.5±0.5	96.4±0.5	96.4±0.5
vowel	90±1.1	92.3±0.9	93.2±0.8	93.2±0.8
wdbc	96.6±0.4	96.7±0.5	97±0.4	96.9±0.3

Dynamic Training

	3 Nets	9 Nets	20 Nets	40 Nets
aritm	74.5±1.3	76.9±1.3	77.1±1.3	76.6±1.1
bala	95.3±0.8	95.8±0.7	96.1±0.6	95.9±0.6
band	74.7±1	74.6±1.1	75.5±1	75.1±0.9
bupa	73.4±1.7	72.9±1.3	74±1.1	72.9±1.3
cred	86.9±0.7	86.8±0.7	86.6±0.6	87±0.6
derma	97.8±0.4	97.3±0.3	97.2±0.4	97.2±0.4
ecoli	86±1	86.8±1	86.5±0.9	86.5±0.9
flare	82±0.6	82±0.6	81.9±0.6	82.2±0.5
glas	94.4±1.2	94.8±0.9	94.8±1	95±1
hear	84.8±1.4	83.9±1.3	84.1±1.3	84.2±1.5
img	96.5±0.3	97±0.2	97±0.2	97.05±0.18
ionos	90±1.2	90.3±1.2	91.3±0.9	90.7±0.8
mok1	98.8±0.8	99.1±0.7	99.4±0.6	99.4±0.6
mok2	91.8±1.6	93.8±1.4	94±1.4	93.5±1.5
pima	76.8±1.2	77±1.1	77.1±1.2	76.8±1.1
survi	74.1±1.3	74.6±1.4	75.3±1.2	74.9±1.3
vote	96.3±0.4	96.5±0.5	96.5±0.5	96.4±0.5
vowel	91±0.6	92.5±0.9	92.5±0.7	92.7±0.6
wdbc	96.9±0.3	97±0.4	97±0.4	96.9±0.4

Table 3. General Measurements - Mean IoP

method	3 nets	9 nets	20 nets	40 nets
se - traditional	4.97	5.27	5.33	5.42
se - static	5.32	5.67	5.78	5.83
se - dynamic	5.52	5.65	5.79	5.8
cvc - traditional	5.56	5.96	6.17	6.15
cvc - static	5.86	6.18	6.23	6.23
cvc - dynamic	5.66	6.06	6.28	6.13
adaboost	3.68	4.55	4.93	4.98
arcx4	4.09	3.77	4.79	5.24
ata	4.57	4.91	5.01	4.76
eencl	4.41	4.06	4.52	4.2

Table 4. General Measurements - Mean PER

method	3 nets	9 nets	20 nets	40 nets
se - traditional	21.84	23.65	23.72	24.64
se - static	23.78	26.41	27	27.61
se - dynamic	25.71	26.15	27.42	27.16
cvc - traditional	24.05	26.58	28.35	28.67
cvc - static	26.57	27.51	28.82	28.9
cvc - dynamic	25.73	28.03	28.76	28.05
adaboost	15.4	19.5	22.95	24.54
arcx4	18.53	17.69	22.56	24.37
ata	19.44	22.04	22.17	20.21
eencl	16.86	14.92	15.98	17.7

when the propossed static and dynamic reordering algorithms are applied. In most of the cases, they also perform better than traditional ensemble algorithms.

Furthermore, the main results have been resumed in table 5 which shows the performance of a single network and the best performance got among the methods described in section 2. The table also shows the IoP and PER of these results with respect to a single net.

With the extra information derived from table 5, we can see that in 15 databases (78.9% of the cases) the ensemble best result is got with a reordering algorithm (static or dynamic). In addition, the ensemble performance is never lower than the performance of a single network. This fact is very important because there are some traditional methods which performed worse than the single network in some datasets.

Taking into account these facts, the methods analysed in section 2, *Simple Ensemble* and *Cross-Validation*, trained with the *Static Order* and *Dynamic Order* training algorithms should be seriously considered when we want to build an ensemble of neural networks because reordering the patterns during the training adds an extra source of diversity to the ensemble.

Table 5. Best Results

| | Single Net | Ensemble Best Performance | | | | | |
	Performance	Performance	IoP	PER	Method	Algorithm	Nets
aritm	75.6 ± 0.7	77.1 ± 1.3	1.5	6.1	CVC	Dynamic	20
bala	87.6 ± 0.6	96.6 ± 0.5	9	72.6	CVC	Static	9
band	72.4 ± 1.0	75.6 ± 1.2	3.2	11.6	SE	Dynamic	9
bupa	58.3 ± 0.6	74 ± 1.3	15.7	37.6	CVC	Static	9
cred	85.6 ± 0.5	87.2 ± 0.9	1.6	11.1	CVC	Traditional	20
derma	96.7 ± 0.4	97.8 ± 0.4	1.1	33.3	CVC	Dynamic	3
ecoli	84.4 ± 0.7	87.2 ± 0.9	2.8	17.9	CVC	Static	20
flare	82.1 ± 0.3	82.4 ± 0.6	0.3	1.7	CVC	Static	40
glas	78.5 ± 0.9	95.4 ± 1	16.9	78.6	SE	Static	20
hear	82.0 ± 0.9	84.8 ± 1.4	2.8	15.6	CVC	Dynamic	3
img	96.3 ± 0.2	97.1 ± 0.2	0.8	21.6	SE	Dynamic	40
ionos	87.9 ± 0.7	91.9 ± 0.8	4	33.1	CVC	Static	20
mok1	74.3 ± 1.1	100 ± 0	25.7	100	CVC	Static	3
mok2	65.9 ± 0.5	95.8 ± 0.9	29.9	87.7	CVC	Traditional	9
pima	76.7 ± 0.6	77.3 ± 1.1	0.6	2.6	CVC	Traditional	3
survi	74.2 ± 0.8	75.3 ± 1.2	1.1	4.3	CVC	Dynamic	20
vote	95.0 ± 0.4	96.5 ± 0.5	1.5	30	CVC	Static	9
vowel	83.4 ± 0.6	93.5 ± 0.7	10.1	60.8	CVC	Traditional	20
wdbc	97.4 ± 0.3	97.4 ± 0.4	0	0	SE	Static	9

5   Conclusions

In this paper we have proposed and compared two methods to train ensembles of neural networks (*Static Order* and *Dynamic Order*) based on alterations on the order of the training set patterns. For the comparison we have chosen two basic ensemble methods, *Simple Ensemble* and *Cross-Validation*. In the experiments, ensembles of 3, 9, 20 and 40 networks have been trained with *Simple Ensemble* and *Cross-Validation* using the three different training algorithms we have previously mentioned. Two major conclusions can be derived from the results shown in this paper.

The first one is that the use of the *Static order* and *Dynamic order* algorithms highly improves the performance of *Simple Ensemble* and *Cross-Validation*. Sometimes, they provide better results than traditional methods like *Boosting*.

The second conclusion is that applying those training procedures to basic ensemble methods should be seriously considered as an important alternative to more complex traditional methods. If we take into account the accuracy, the computational cost and the simplicity of the models, training a basic ensemble with the *Static order* and *Dynamic order* algorithm is one of the most important ways to build ensembles.

We can conclude by remarking that the best general measurements among the training methods on basic ensemble models have been got with ensembles trained with *Cross-Validation* using the *Dynamic order* training algorithm (20 networks) and using the *Static order* training algorithm (40 networks).

Finally, our research on reordering patterns during training will be extended by applying the two *alternative* algorithms proposed in this paper to complex models in order to get an extra improvement.

References

1. Breiman, L.: Bagging predictors. Machine Learning 24(2), 123–140 (1996)
2. Drucker, H., Cortes, C., Jackel, L.D., LeCun, Y., Vapnik, V.: Boosting and other ensemble methods. Neural Computation 6(6), 1289–1301 (1994)
3. Freund, Y., Schapire, R.E.: Experiments with a new boosting algorithm. In: International Conference on Machine Learning, pp. 148–156 (1996)
4. Raviv, Y., Intratorr, N.: Bootstrapping with noise: An effective regularization technique. Connection Science, Special issue on Combining Estimators 8, 356–372 (1996)
5. Wolpert, D.H.: Stacked generalization. Neural Networks 5(6), 1289–1301 (1994)
6. Jordan, M.I., Jacobs, R.A.: Hierarchical mixtures of experts and the EM algorithm. Neural Computation 6, 181–214 (1994)
7. Bishop, C.M.: Neural Networks for Pattern Recognition. Oxford University Press, New York (1995)
8. Kuncheva, L.I.: Combining Pattern Classifiers: Methods and Algorithms. Wiley-Interscience, Chichester (2004)
9. Fernández-Redondo, M., Hernández-Espinosa, C., Torres-Sospedra, J.: Multilayer feedforward ensembles for classification problems. In: Pal, N.R., Kasabov, N., Mudi, R.K., Pal, S., Parui, S.K. (eds.) ICONIP 2004. LNCS, vol. 3316, pp. 744–749. Springer, Heidelberg (2004)
10. Newman, D.J., Hettich, S., Blake, C.L., Merz, C.J.: UCI repository of machine learning databases (1998), `http://www.ics.uci.edu/~mlearn/MLRepository.html`

New Results on Combination Methods for Boosting Ensembles

Joaquín Torres-Sospedra, Carlos Hernández-Espinosa,
and Mercedes Fernández-Redondo

Departamento de Ingenieria y Ciencia de los Computadores, Universitat Jaume I,
Avda. Sos Baynat s/n, C.P. 12071, Castellon, Spain
`{jtorres,espinosa,redondo}@icc.uji.es`

Abstract. The design of an ensemble of neural networks is a procedure that can
be decomposed into two steps. The first one consists in generating the ensemble,
i.e., training the networks with significant differences. The second one consists in
combining properly the information provided by the networks. *Adaptive Boosting*, one of the best performing ensemble methods, has been studied and improved
by some authors including us. Moreover, *Adaboost* and its *variants* use a specific
combiner based on the error of the networks. Unfortunately, any deep study on
combining this kind of ensembles has not been done yet. In this paper, we study
the performance of some important ensemble combiners on ensembles previously
trained with *Adaboost* and *Aveboost*. The results show that an extra increase of
performance can be provided by applying the appropriate combiner.

1   Introduction

One technique often used to increase the generalization ability of a single neural network consists in training an ensemble of neural networks. This procedure consists of
training a set of neural network with different weight initialization or properties in the
training process and combining them in a suitable way. The two key factors to design
an ensemble are how to train the individual networks and how to combine the outputs
provided by the networks to give a single output.

Adaptive Boosting is a well-known method that construct a sequence of networks in
which the training set of each network is overfitted with hard to learn patterns of the
previous networks. Figure 1 shows the ensemble basic diagram along with the training
diagram. Some authors like Breiman [1], Kuncheva [2] or Oza [3] have deeply studied
and successfully improved *Adaboost*. Unfortunately, any study on combining boosting
methods has not been done. In previouses papers [4,5,6], the combiner *Output average*
was better than the specific *Boosting combiner* in a wide range of cases for the three
new boosting methods proposed.

In this paper, we present a comparison of sixteen different combiners on ensembles
previously trained with *Adaboost* and *Aveboost*, two of the most important boosting
methods, in order to test if the *Boosting combiner* is the most appropriate way to combine boosting ensembles.

This paper is organized as follows. Firstly, some theoretical concepts are reviewed
in section 2. Then, the experimental setup is described in section 3. Finally, the experimental results and their discussion are in section 4.

V. Kůrková et al. (Eds.): ICANN 2008, Part I, LNCS 5163, pp. 285–294, 2008.

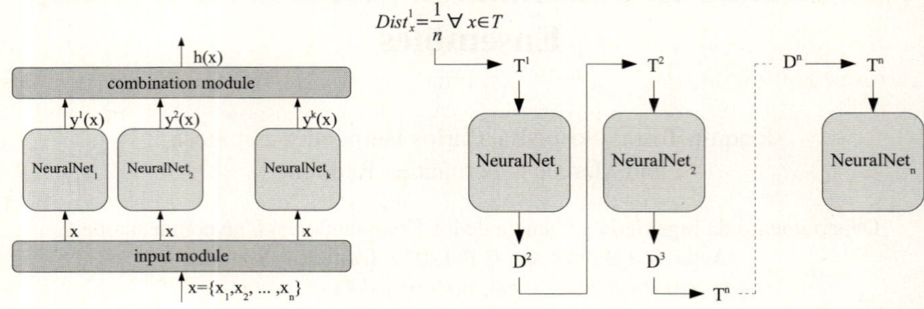

Fig. 1. Boosting diagram

2   Theory

2.1   Adaptive Boosting - Adaboost

In *Adaboost*, the successive networks are trained with a training data set T' selected at random from the original training data set T. The probability of selecting a pattern from T is given by a *sampling distribution* associated to the network $Dist_{net}$. The sampling distribution associated to a network is calculated when the learning process of the previous network has finished. *Adaboost* is described in algorithm 1.

Algorithm 1. AdaBoost $\{T, V, k\}$

Initialize Sampling Distribution: $Dist_x^1 = 1/m \; \forall x \in T$
for $net = 1$ to k **do**
    Create T' sampling from T using $Dist^{net}$
    MF Network Training T', V
    Calculate missclassified vector:
$$miss_x^{net} = \begin{cases} 0 \text{ if } x \text{ is correctly classified} \\ 1 \text{ otherwise} \end{cases}$$
    Calculate error:
$$\epsilon_{net} = \sum_{x=1}^{m} Dist_x^{net} \cdot miss_x^{net}$$
    Update sampling distribution:
$$Dist_x^{net+1} = Dist_x^{net} \cdot \begin{cases} \frac{1}{(2\epsilon_{net})} & \text{if } miss_x^{net} \\ \frac{1}{2(1-\epsilon_{net})} & \text{otherwise} \end{cases}$$
end for

Adaboost and *Aveboost* use a specific method, the *Boosting combiner*, to combine the networks and get the final output or hypothesis eq.1.

$$h(x) = \underset{c=1,...,classes}{\arg\max} \sum_{net:h^{net}(x)=c}^{k} \log \frac{1 - \epsilon_{net}}{\epsilon_{net}} \tag{1}$$

2.2   Averaged Boosting - Aveboost

Oza proposed in [3] *Averaged Boosting, Aveboost. Aveboost* is a method based on *Adaboost* in which the sampling distribution related to a neural network is also based on the number of networks previously trained. The equation to update the sampling distribution is given by:

$$Dist_x^{net+1} = \frac{net \cdot Dist_x^{net} + C_x^{net}}{net + 1} \tag{2}$$

Where:

$$C_x^{net} = Dist_x^{net} \cdot \begin{cases} \frac{1}{(2\epsilon_{net})} & \text{if } miss_x^{net} \\ \frac{1}{2(1-\epsilon_{net})} & \text{otherwise} \end{cases} \tag{3}$$

2.3   Alternative Combiners

In this subsection, the *well-known* alternative combiners studied are briefly reviewed.

Average. This approach simply averages the individual classifier outputs across the different classifiers. The output yielding the maximum of the averaged values is chosen as the correct class.

Majority Vote. Each classifier provides a vote to a class, given by the highest output. The correct class is the one most often voted by the classifiers.

Winner Takes All (WTA). In this method, the class with overall maximum output across all classifier and outputs is selected as the correct class.

Borda Count. For any class c, the *Borda count* is the sum of the number of classes ranked below c by each classifier [7,8]. The class with highest count is selected as correct class.

Bayesian Combination. This combination method was proposed in references [9]. According to this reference a belief value that the pattern x belongs to class c can be approximated by the following equation based on the values of the confusion matrix [8]

$$Bel(c) = \frac{\prod\limits_{net=1}^{k} P(x \in q_c | \lambda_{net}(x) = j_{net})}{\sum\limits_{i=1}^{classes} \prod\limits_{net=1}^{k} P(x \in q_i | \lambda_{net}(x) = j_{net})} \tag{4}$$

Weighted Average. This method introduces weights to the outputs of the different networks prior to averaging. The weights try to minimize the difference between the output of the ensemble and the *desired or true* output. The weights can be estimated from the error correlation matrix. The full description of the method can be found in [10,8].

Choquet Integral. This method is based in the fuzzy integral [11,12] and is the Choquet integral. This combiner is complex and its full description can be found in reference [8].

Fuzzy Integral with Data Dependent Densities. It is another method based on the fuzzy integral and the Choquet integral. But in this case, prior to the application of the method it is performed a partition of the input space into n regions by frequency sensitive learning algorithm (*FSL*). The full description can be found in reference [8].

Weighted Average with Data Dependent weights. This method is the weighted average described above. But in this case, a partition of the space is performed by *FSL* algorithm and the weights are calculated for each partition. We have a different combination scheme for the different partitions of the space. The method is fully described in [8].

BADD Defuzzification Strategy. It is another combination method based on fuzzy logic concepts. The method is complex and the description can also be found in [8].

Zimmermann's Compensatory Operator. This combination method is based in the Zimmermann's compensatory operator described in [13]. The method is complex and can be found in [8].

Dynamically Averaged Networks. Two versions of *Dinamically Averaged Networks* were proposed by Jimenez [14,15]. In these methods instead of choosing static weights derived from the network output on a sample of the input space, we allow the weights to adjust to be proportional to the certainties of the respective network output.

Nash Vote. In this method each voter assigns a number between zero and one for each candidate output. The product of the voter's values is compared for all candidates. The higher is the winner. The method is reviewed in reference [16].

Stacked Combiners (Stacked and Stacked+). Wolpert proposed in 1994 a method to build multiple classifier systems called *Stacked Generalization*. The training in *Stacked Generalization* is divided into two steps. In the first one, the expert networks are trained. In the second one, the combination networks are trained with the outputs provided by the experts.

Stacked Generalization can be adapted to combine ensembles of neural networks if the networks of the ensembles are used as expert networks. In [17], *Stacked* and *Stacked+*, two combiners based on *Stacked Generalization*, were successfully proposed.

3   Experimental Setup

In the experiments, the *Boosting combiner* and the alternative combiners have been applied on ensembles of 3, 9, 20 and 40 *MF* networks previously trained with *Adaptive Boosting* and *Averaged Boosting* on the databases described in subsection 3.1. In the case of *Stacked combiners*, a single *MF* combination network has been applied. Finally, the learning process has been repeated 10 times using different sets in order to get the mean performance and the error calculated by standard error theory.

3.1 Datasets

We have used the following classification problems from the *UCI repository* [18]: *Arrhythmia* (aritm), *Balance Scale* (bala), *Cylinder Bands* (band), *BUPA liver disorders* (bupa), *Australian Credit Approval* (cred), *Dermatology* (derma), *Ecoli* (ecoli), *Solar Flares* (flare), *Glass Identification* (glas), *Heart Disease* (hear), *Image segmentation* (img), *Ionosphere* (ionos), *The Monk's Problem 1&2* (mok1 and mok2), *Pima Indians Diabetes* (pima), *Haberman's Survival Data* (survi), *Congressional Voting Records* (Vote), *Vowel* (vowel) and *Wisconsin Breast Cancer* (wdbc).

The optimal training parameters of the *MF* networks set to train the networks of the ensembles along with the performance of a single network are shown in table 1 whereas the parameters of the combination networks (*Stacked combiners*) are omitted due to the lack of space.

Table 1. MF training parameters

database	hidden	step	mom	ite	accuracy
aritm	9	0.1	0.05	2500	75.6 ± 0.7
bala	20	0.1	0.05	5000	87.6 ± 0.6
band	23	0.1	0.05	5000	72.4 ± 1.0
bupa	11	0.1	0.05	8500	58.3 ± 0.6
cred	15	0.1	0.05	8500	85.6 ± 0.5
derma	4	0.1	0.05	1000	96.7 ± 0.4
ecoli	5	0.1	0.05	10000	84.4 ± 0.7
flare	11	0.6	0.05	10000	82.1 ± 0.3
glas	3	0.1	0.05	4000	78.5 ± 0.9
hear	2	0.1	0.05	5000	82.0 ± 0.9
img	14	0.4	0.05	1500	96.3 ± 0.2
ionos	8	0.1	0.05	5000	87.9 ± 0.7
mok1	6	0.1	0.05	3000	74.3 ± 1.1
mok2	20	0.1	0.05	7000	65.9 ± 0.5
pima	14	0.4	0.05	10000	76.7 ± 0.6
survi	9	0.1	0.2	20000	74.2 ± 0.8
vote	1	0.1	0.05	2500	95.0 ± 0.4
vowel	15	0.2	0.2	4000	83.4 ± 0.6
wdbc	6	0.1	0.05	4000	97.4 ± 0.3

Moreover, the numbers of regions used in *data depend densities* methods was set to $n = 5$. All the parameters have been set after an exhaustive trial and error procedure using the training and validation sets.

4 Results and Discussion

Due to the lack of space, only the general results on combining ensembles trained with *Adaboost* and *Aveboost* are shown in this section instead of showing the complete results. The general measurements used in this paper are described in subsection 4.1.

4.1   General Measurements

In our experiments, we have calculated the *Increase of Performance* (IoP eq.5) and the *Percentage of Error Reduction* (PER eq.6) of the results with respect to a single network in order to perform an exhaustive comparison. The IoP value is an absolute measurement whereas the PER value is a relative measurement. A negative value on these measurements mean that the ensemble performs worse than a single network.

$$IoP = Error_{SingleNet} - Error_{Ensemble} \tag{5}$$

$$PER = 100 \cdot \frac{Error_{SingleNet} - Error_{Ensemble}}{Error_{SingleNet}} \tag{6}$$

Finally, we have calculated the mean IoP and the mean PER across all databases to get a general measurement to compare the methods presented in the paper. The results on combining *Adaboost* are presented in subsection 4.2 whereas the results on combining *Aveboost* are in subsection 4.3.

4.2   Adaboost Results

In this subsection the results of the different combiners on ensembles trained with *Adaptive Boosting* are shown. Table 2 shows the mean IoP and the mean PER for the ensembles trained and combined with the *Boosting combiner* as in the original method *Adaboost* and for the same ensembles combined with the alternative combiners described in subsection 2.3.

Moreover, table 4 shows the best performance for each database on ensembles trained with *Adaboost* among the results provided by combining the ensembles with the *Boosting combiner* and the alternative combiners.

4.3   Aveboost Results

In this subsection the results of the different combiners on ensembles trained with *Averaged Boosting* are shown. Table 3 shows the mean IoP and the mean PER for the ensembles trained and combined with the *Boosting combiner* as in the original method *Aveboost* and for the same ensembles combined with the alternative combiners.

Moreover, table 4 also shows the best performance for each database on ensembles trained with *Aveboost* among the results provided by combining the ensembles with the *Boosting combiner* and the alternative combiners.

4.4   Discussion

According to the general results, the *Boosting combiner* is not the best alternative when it is applied on ensembles trained in *Adaboost* in all the cases. *Stacked combiners* and *Weighted Average with Data-Depend Densities* are better (table 2) when the number of networks is reduced. Analysing table 4, it can be seen that the *boosting combiner* only provides the best result on a few databases (*bupa, survi, vote* and *wdbc*.)

Table 2. Adaptive Boosting - Mean IoP and PER among all databases

Method	Mean IoP				Mean PER			
	3 Nets	9 Nets	20 Nets	40 Nets	3 Nets	9 Nets	20 Nets	40 Nets
adaboost	0.3	0.86	1.15	1.26	1.33	4.26	9.38	12.21
average	−0.18	−0.15	−0.36	−0.35	−20.49	−19.71	−18.81	−22.05
voting	−0.97	−1.48	−1.74	−1.26	−27.18	−27.2	−25.53	−26.21
wta	−1.07	−4.78	−8.22	−11.07	−16.66	−78.22	−132.97	−184.84
borda	−2.74	−2.76	−2.84	−2.07	−50.55	−35.57	−32.71	−32.13
bayesian	−0.36	−1.28	−3.22	−5.46	−6.28	−16.17	−38.15	−65.05
wave	0.59	1.19	0.41	0.44	1.15	6.5	3.7	7.3
choquet	−1.26	−6.23	−	−	−23.68	−107.65	−	−
fidd	−1.37	−6.91	−	−	−27.85	−123.9	−	−
wave dd	0.68	0.72	−	−	4.18	0.67	−	−
badd	−0.18	−0.15	−0.36	−0.35	−20.49	−19.71	−18.81	−22.05
zimm	0.35	−1.16	−13.37	−13.02	−0.28	−28.25	−150.69	−212.35
dan	−12.54	−17.08	−20.23	−19.13	−123.17	−199.31	−244.94	−278.47
dan2	−12.59	−17.45	−20.46	−19.47	−123.5	−202.67	−248.34	−282.06
nash	−1.27	−1.76	−1.71	−1.57	−16.55	−30.14	−28.14	−29.13
stacked	0.69	0.83	0.7	0.51	3.39	2.87	4.7	4.27
stacked+	0.71	0.95	0.64	−0.14	5.43	6.25	4.39	1.81

Table 3. Averaged Boosting - Mean IoP and PER among all databases

Method	Mean IoP				Mean PER			
	3 Nets	9 Nets	20 Nets	40 Nets	3 Nets	9 Nets	20 Nets	40 Nets
aveboost	0.5	1.49	1.83	1.82	1.13	10.46	11.7	10.79
average	0.87	1.61	2	1.8	4.26	11.64	12.93	12.99
voting	0.37	1.54	1.76	1.91	0.28	11.15	12.67	13.01
wta	0.49	0.36	−0.38	−0.88	−0.1	−2.31	−9.2	−10.88
borda	−0.34	1.15	1.57	1.73	−6.73	8.13	11.05	12.12
bayesian	0.02	−0.13	−1.3	−2.87	−4.14	−7.94	−23.39	−40.81
wave	0.85	1.17	1.19	0.32	4.29	8.36	7.65	3.78
choquet	0.21	−0.19	−	−	−3.69	−10.02	−	−
fidd	0.14	−0.35	−	−	−3.76	−11.14	−	−
wave dd	0.92	1.62	−	−	5.62	11.88	−	−
badd	0.87	1.61	2	1.8	4.26	11.64	12.93	12.99
zimm	0.74	0.59	−2.75	−7.53	4.17	5.17	−18.5	−63.01
dan	−2.65	−3.04	−5.06	−5.13	−30.32	−25.63	−34.57	−34.21
dan2	−2.64	−3.13	−5.1	−5.27	−30.37	−26.56	−34.89	−35.38
nash	−0.09	1.03	1.63	1.4	−2.56	7.33	11.34	8.86
stacked	0.9	0.99	0.95	0.96	6.67	7.79	7.15	8.43
stacked+	0.95	1.02	0.83	0.8	6.42	6.22	7.64	6.84

Furthermore, in the case of ensembles trained with *Aveboost*, the *Output average* provides better general results than the *Boosting combiner*. Moreover, analysing table 4, it can be seen that the *boosting combiner* only provides the best result on a few databases (*derma* and *hear*).

Table 4. Adaboost - Best performance

	Adaboost			Aveboost		
Database	Performance	Method	Nets	Performance	Method	Nets
aritm	75.3 ± 0.9	bayes	9	77.0 ± 1.1	average	20
bala	96.8 ± 0.3	stacked	20	97.1 ± 0.3	stacked	3
band	74.9 ± 1.2	average	9	76.4 ± 1.7	nash	20
bupa	73.3 ± 1.4	boosting	9	73.3 ± 1.3	average	40
cred	86.4 ± 0.9	w.ave	3	86.9 ± 0.7	average	9
derma	98.1 ± 0.7	stacked+	3	97.9 ± 0.5	boosting	20
ecoli	87.2 ± 1.0	w.ave	3	87.6 ± 0.9	w.ave	3
flare	82.2 ± 0.6	w.ave	3	82.5 ± 0.6	stacked+	3
glas	96.8 ± 0.7	w.ave	20	97.6 ± 0.5	w.ave	40
hear	84.6 ± 1.4	w.ave.ddd	9	84.9 ± 1.3	boosting	9
img	97.4 ± 0.3	average	20	97.6 ± 0.2	stacked	40
ionos	92 ± 0.9	average	20	92.4 ± 1	zimm	3
mok1	100 ± 0	most combiners	3to40	100 ± 0	alternative combiners	3
mok2	85.9 ± 1.8	w.ave	40	91.6 ± 0.9	w.ave	40
pima	76.6 ± 1.1	average	3	77.1 ± 1	w.ave	3
survi	75.4 ± 1.6	boosting	3	75.1 ± 1.2	voting	3
vote	96.3 ± 0.7	boosting	9	96.4 ± 0.8	stacked+	20
vowel	97.1 ± 0.6	average	40	96.7 ± 0.4	w.ave	40
wdbc	96.7 ± 0.9	boosting	40	96.7 ± 0.3	zimm	9

Although it is not shown in the results tables, the accuracy is highly improved by applying an alternative combiner while the number of networks required to get the best performance is reduced. Sometimes, the boosting combiner provides worse results using more networks compared to the best overall result of table 4.

5   Conclusions

In this paper, we have performed a comparison among sixteen combiners on ensembles trained with *Adaptive Boosting* and *Averaged Boosting*. To carry out our comparison we have used ensembles of 3, 9, 20 and 40 networks previously trained with *Adaptive boosting* and *Averaged boosting* and the accuracy of the ensemble using the *Boosting Combiner*. In our experiments we have selected nineteen databases whose results are not easy to improve with an ensemble of neural networks. Alternatively, we have applied sixteen different combiners on these ensembles to test if the *boosting combiner* is the best method to combine the networks of a *boosting ensemble*.

Finally, we have calculated the mean *Increase of Performance* and the mean *Percentage of Error Reduction* with respect to a single network to compare the combiners.

According the general results, the *Boosting combiner* is not the most effective way to combine an ensemble trained with *Boosting* in a wide number of cases. The original results have been improved with the use of the appropriate alternative combiner. In general, the *Weighted average with or without data depend densities* and the *Stacked combiners* are the best combiners on low sized ensembles trained with *Adaboost*. In a similar way, the *Output average*, the *Voting*, the *Weighted average with/without data depend densities* and the *Borda Count* are the best combiners on ensembles trained with *Aveboost*.

According the best performance for each database (table 4), we can see that the *Output average* and both versions of the *Weighted average* should be seriously considered for combining *boosting ensembles* because the *Boosting combiner* provides the best results only in 15.78% of the cases. In addition, in a some cases not only have the accuracy of the ensembles been improved with the use of an alternative combiner, the numbers of networks to get the best result is also reduced. For instance, the best accuracy for database *ecoli* using the original *Adaboost* (applying the *Boosting combiner*) was got with the 20-network ensembles (86 ± 1.3). The best overall accuracy for this database using *Adaboost* was got by applying the *Weighted average* to the 3-network ensembles (87.2 ± 1.0). The accuracy was improved in 1.2%, the error rate was reduced in 0.3% and the number of networks required were reduced from 20 to 3.

Nowadays, we are extending the comparison we have performed adding more methods and databases. The results we are getting also show that the *Boosting combiner* does not provide either the best general results (best mean IoP or PER) or the best performance for each database.

We can conclude by remarking that the accuracy of a *boosting ensemble* can be improved and its size can be reduced by applying the *Output average* or advanced combiners like the *Weighted average* or the *Stacked combiners* on ensembles previously trained with *Adaboost* or *Aveboost*.

References

1. Breiman, L.: Arcing classifiers. The Annals of Statistics 26(3), 801–849 (1998)
2. Kuncheva, L., Whitaker, C.J.: Using diversity with three variants of boosting: Aggressive. In: Roli, F., Kittler, J. (eds.) MCS 2002. LNCS, vol. 2364. Springer, Heidelberg (2002)
3. Oza, N.C.: Boosting with averaged weight vectors. In: Windeatt, T., Roli, F. (eds.) MCS 2003. LNCS, vol. 2709, pp. 15–24. Springer, Heidelberg (2003)
4. Torres-Sospedra, J., Hernández-Espinosa, C., Fernández-Redondo, M.: Designing a multilayer feedforward ensembles with cross validated boosting algorithm. In: IJCNN 2006 proceedings, pp. 2257–2262 (2006)
5. Torres-Sospedra, J., Hernández-Espinosa, C., Fernández-Redondo, M.: Designing a multilayer feedforward ensemble with the weighted conservative boosting algorithm. In: IJCNN 2007 Proceedings, pp. 684–689. IEEE, Los Alamitos (2007)
6. Torres-Sospedra, J., Hernández-Espinosa, C., Fernández-Redondo, M.: Mixing aveboost and conserboost to improve boosting methods. In: IJCNN 2007 Proceedings, pp. 672–677. IEEE, Los Alamitos (2007)
7. Ho, T.K., Hull, J.J., Srihari, S.N.: Decision combination in multiple classifier systems. IEEE Transactions on Pattern Analysis and Machine Intelligence 16(1), 66–75 (1994)
8. Verikas, A., Lipnickas, A., Malmqvist, K., Bacauskiene, M., Gelzinis, A.: Soft combination of neural classifiers: A comparative study. Pattern Recognition Letters 20(4), 429–444 (1999)
9. Xu, L., Krzyzak, A., Suen, C.Y.: Methods of combining multiple classifiers and their applications to handwriting recognition. IEEE Transactions on Systems, Man, and Cybernetics 22(3), 418–435 (1992)
10. Krogh, A., Vedelsby, J.: Neural network ensembles, cross validation, and active learning. In: Tesauro, G., Touretzky, D., Leen, T. (eds.) Advances in Neural Information Processing Systems, vol. 7, pp. 231–238. The MIT Press, Cambridge (1995)

11. Cho, S.B., Kim, J.H.: Combining multiple neural networks by fuzzy integral for robust classification. IEEE Transactions on System, Man, and Cybernetics 25(2), 380–384 (1995)
12. Gader, P.D., Mohamed, M.A., Keller, J.M.: Fusion of handwritten word classifiers. Pattern Recogn. Lett. 17(6), 577–584 (1996)
13. Zimmermann, H.J., Zysno, P.: Decision and evaluations by hierarchical aggregation of information. Fuzzy Sets and Systems 10(3), 243–260 (1984)
14. Jimenez, D.: Dynamically weighted ensemble neural networks for classification. In: Proceedings of the 1998 International Joint Conference on Neural Networks, IJCNN 1998, pp. 753–756 (1998)
15. Jimenez, D., Darm, T., Rogers, B., Walsh, N.: Locating anatomical landmarks for prosthetics design using ensemble neural networks. In: Proceedings of the 1997 International Conference on Neural Networks, IJCNN 1997 (1997)
16. Wanas, N.M., Kamel, M.S.: Decision fusion in neural network ensembles. In: Proceedings of the 2001 International Joint Conference on Neural Networks, IJCNN 2001, vol. 4, pp. 2952–2957 (2001)
17. Torres-Sospedra, J., Hernández-Espinosa, C., Fernández-Redondo, M.: Combining MF networks: A comparison among statistical methods and stacked generalization. In: Schwenker, F., Marinai, S. (eds.) ANNPR 2006. LNCS (LNAI), vol. 4087, pp. 210–220. Springer, Heidelberg (2006)
18. Newman, D.J., Hettich, S., Blake, C.L., Merz, C.J.: UCI repository of machine learning databases (1998), http://www.ics.uci.edu/~mlearn/MLRepository.html

Batch Support Vector Training
Based on Exact Incremental Training

Shigeo Abe

Graduate School of Engineering
Kobe University
Rokkodai, Nada, Kobe, Japan
abe@kobe-u.ac.jp
http://www2.eedept.kobe-u.ac.jp/~abe

Abstract. In this paper we discuss training support vector machines (SVMs) repetitively solving a set of linear equations, which is an extension of accurate incremental training proposed by Cauwenberghs and Poggio. First, we select two data in different classes and determine the optimal separating hyperplane. Then, we divide the training data set into several chunk data sets and set the two data to the active set, which includes current and previous support vectors. For the combined set of the active set and a chunk data set, we detect the datum that maximally violates the Karush-Kuhn-Tacker (KKT) conditions, and modify the optimal hyperplane by solving a set of equations that constrain the margins of unbounded support vectors and the optimality of the bias term. We iterate this procedure until there are no violating data in any combined set. By the computer experiment using several benchmark data sets, we show that the training speed of the proposed method is comparable with the primal-dual interior-point method combined with the decomposition technique, usually with a smaller number of support vectors.

1 Introduction

Support vector machines (SVMs) are formulated as quadratic programming programs with the number of variables being equal to the number of training data. Therefore, for a large classification problem training becomes slow. There are many methods to accelerate training. The most often used methods are the primal-dual interior point method [1] combined with the decomposition technique [2] or Sequential Minimum Optimization (SMO) [3], which is a special case of the decomposition technique with the working set size of two. Vogt [4] extended SMO when the bias term is not used, i.e., there is no equality constraint in the dual problem.

Some training algorithms are based on the geometrical properties of support vectors [7,8,9,10].

Cauwenberghs and Poggio [11] proposed incremental 1rand decremental training that gives the exact solution. This method is further analyzed in [12]. Along this line Hastie et al. [13] proposed the entire path method to obtain solutions of different values of the margin parameter from the initial solution.

V. Kůrková et al. (Eds.): ICANN 2008, Part I, LNCS 5163, pp. 295–304, 2008.

In this paper we extend the incremental training method discussed in [11] to batch training of L1 and L2 SVMs. Namely, first, we select two data in different classes and determine the optimal separating hyperplane. Then, we divide the training data set into several chunk data sets and generate the active set that includes the current and previous support vectors. For the combined set of the active set and a chunk data set, we detect the datum that maximally violates the Karush-Kuhn-Tacker (KKT) conditions. Then, using a set of linear equations that constrain the margins of the unbounded support vectors and the optimality of the bias term, we modify the optimal hyperplane so that the violating datum satisfies the KKT conditions. After iterating this procedure until no violating data exist in the combined set, we proceed to another combined set until no violating data exist in the entire training data set. Using some benchmark data sets we compare the speeds of batch training and the primal-dual interior-point method with the decomposition technique and demonstrate the usefulness of the proposed method.

In Section 2, we briefly summarize L1 and L2 SVMs, and in Section 3 we discuss the proposed training method. In Section 4, we show the validity of the proposed method by computer experiments.

2   Support Vector Machines

In this section, we summarize support vector machines for two-class problems. In training an SVM, we solve the following optimization problem:

$$\text{minimize} \quad Q(\mathbf{w}, \boldsymbol{\xi}) = \frac{1}{2}||\mathbf{w}||^2 + \frac{C}{p}\sum_{i=1}^{M}\xi_i^p \tag{1}$$

$$\text{subject to} \quad y_i\left(\mathbf{w}^T\boldsymbol{\phi}(\mathbf{x}_i) + b\right) \geq 1 - \xi_i \quad \text{for} \quad i = 1, ..., M, \tag{2}$$

where $\mathbf{w}$ is the weight vector, $\boldsymbol{\phi}(\mathbf{x})$ is the mapping function that maps m-dimensional input vector $\mathbf{x}$ into the feature space, b is the bias term, $(\mathbf{x}_i, y_i)$ $(i = 1, ..., M)$ are M training input-output pairs, with $y_i = 1$ if $\mathbf{x}_i$ belongs to Class 1, and $y_i = -1$ if Class 2, C is the margin parameter that determines the tradeoff between the maximization of the margin and minimization of the classification error, ξ_i is the nonnegative slack variable for $\mathbf{x}_i$, and $p = 1$ for an L1 SVM and $p = 2$ for an L2 SVM.

Introducing the Lagrange multipliers α_i, we obtain the following dual problem for the L1 SVM:

$$\text{maximize} \quad Q(\boldsymbol{\alpha}) = \sum_{i=1}^{M}\alpha_i - \frac{1}{2}\sum_{i,j=1}^{M}\alpha_i\alpha_j\, y_i\, y_j H(\mathbf{x}_i, \mathbf{x}_j) \tag{3}$$

$$\text{subject to} \quad \sum_{i=1}^{M} y_i\,\alpha_i = 0, \quad 0 \leq \alpha_i \leq C \quad \text{for} \quad i = 1, ..., M, \tag{4}$$

and for the L2 SVM:

$$\text{maximize} \quad Q(\boldsymbol{\alpha}) = \sum_{i=1}^{M} \alpha_i - \frac{1}{2} \sum_{i,j=1}^{M} \alpha_i \alpha_j \, y_i \, y_j \left(H(\mathbf{x}_i, \mathbf{x}_j) + \frac{\delta_{ij}}{C} \right) \qquad (5)$$

$$\text{subject to} \quad \sum_{i=1}^{M} y_i \, \alpha_i = 0, \quad \alpha_i \geq 0 \quad \text{for} \quad i = 1, ..., M, \qquad (6)$$

where $H(\mathbf{x}, \mathbf{x}')$ is a kernel function that is given by $H(\mathbf{x}, \mathbf{x}') = \boldsymbol{\phi}^T(\mathbf{x}) \, \boldsymbol{\phi}(\mathbf{x}')$.

In this study we use polynomial kernels with degree d: $H(\mathbf{x}, \mathbf{x}') = (\mathbf{x}^T \mathbf{x}' + 1)^d$ and radial basis function (RBF) kernels: $H(\mathbf{x}, \mathbf{x}') = \exp(-\gamma \|\mathbf{x} - \mathbf{x}'\|^2)$, where γ is a positive parameter for slope control.

Now we define $Q_{ij} = y_i \, y_j \, H(\mathbf{x}_i, \mathbf{x}_j)$ for the L1 SVM and $Q_{ij} = y_i \, y_j \, H(\mathbf{x}_i, \mathbf{x}_j) + \delta_{ij}/C$ for the L2 SVM. Then the KKT complementarity conditions for the L1 SVM are given by

$$\alpha_i \left(\sum_{j=1}^{M} \alpha_j \, Q_{ij} + y_i \, b - 1 + \xi_i \right) = 0 \quad \text{for} \quad i = 1, \dots, M, \qquad (7)$$

$$(C - \alpha_i) \, \xi_i = 0 \qquad \alpha_i \geq 0, \quad \xi_i \geq 0 \quad \text{for} \quad i = 1, \dots, M. \qquad (8)$$

For the solution of (3) and (4), if $\alpha_i > 0$, $\mathbf{x}_i$ are called support vectors; especially if $\alpha_i = C$, bounded support vectors and if $0 < \alpha_i < C$, unbounded support vectors.

The KKT complementarity conditions for the L2 SVM are given by

$$\alpha_i \left(\sum_{j=1}^{M} \alpha_j \, Q_{ij} + y_i \, b - 1 + \xi_i \right) = 0, \quad \alpha_i \geq 0 \qquad \text{for} \quad i = 1, \dots, M. \qquad (9)$$

3 Training by Solving a Set of Linear Equations

3.1 Incremental Training

In the following we discuss incremental training for the L1 SVM based on [11]. Since we use the notation of Q_{ij}, which is common to L1 and L2 SVMs, the only difference of L1 and L2 SVMs is that there is no upper bound in L2 SVMs. Thus, in the following, if we exclude the discussions on the upper bound and consider that the bounded support vectors do not exist for L2 SVMs, we obtain the training method for L2 SVMs.

From the KKT conditions and the optimality of the bias term, the optimal solution of (3) and (4) must satisfy the following set of linear equations:[1]

$$g_i = \sum_{j \in S} Q_{ij} \alpha_j + y_i \, b - 1 = 0 \quad \text{for} \quad i \in S_U, \qquad (10)$$

---

[1] The set of equations is similar to that in [14], which has no mechanism to select support vectors and thus all the training data become support vectors.

$$\sum_{i \in S} y_i \, \alpha_i = 0, \tag{11}$$

where g_i is a margin for $\mathbf{x}_i$ and S_U is the index set of unbounded support vectors.

In incremental training if an added datum satisfies the KKT conditions, we do nothing. But if not, we modify the decision hyperplane under the constraints of (10) and (11). Let $\mathbf{x}_c$ be the added datum. Then if the associated α_c is increased from 0, from (10) and (11), the following equations for the small increment of α_c, $\Delta \alpha_c$, must be satisfied:

$$\sum_{j \in S_U} Q_{ij} \Delta \alpha_j + Q_{ic} \Delta \alpha_c + y_i \, \Delta b = 0 \quad \text{for} \quad i \in S_U, \tag{12}$$

$$\sum_{i \in S_U} y_i \, \Delta \alpha_i + y_c \, \Delta \alpha_c = 0, \tag{13}$$

The above set of equations is valid so long as bounded support vectors remain bounded, unbounded support vectors remain unbounded, and non-support vectors remain non-support vectors. If any of the above conditions is violated, we call it the status change.

From (12) and (13),

$$\Delta b = \beta \Delta \alpha_c, \tag{14}$$
$$\Delta \alpha_j = \beta_j \Delta \alpha_c \quad \text{for} \quad j \in S_U, \tag{15}$$

where

$$\begin{bmatrix} \beta \\ \beta_{s1} \\ \vdots \\ \beta_{sl} \end{bmatrix} = - \begin{bmatrix} 0 & y_{s1} & \cdots & y_{sl} \\ y_{s1} & Q_{s1\,s1} & \cdots & Q_{s1\,sl} \\ \vdots & \vdots & \ddots & \vdots \\ y_{sl} & Q_{sl\,s1} & \cdots & Q_{sl\,sl} \end{bmatrix}^{-1} \begin{bmatrix} y_c \\ Q_{s1\,c} \\ \vdots \\ Q_{sl\,c} \end{bmatrix} \Delta \alpha_c \tag{16}$$

and $\{s1, \ldots, sl\} = S_U$.

For a bounded support vector or a non-support vector, $\mathbf{x}_i$, the margin change due to $\Delta \alpha_c$, Δg_i, is calculated by

$$\Delta g_i = \gamma_i \Delta \alpha_c, \tag{17}$$

where

$$\gamma_i = Q_{ic} + \sum_{j \in S_u} Q_{ij} \, \beta_j + y_i \, \beta. \tag{18}$$

The value of $\Delta \alpha_c$ is determined as the minimum value of $\Delta \alpha_c$ that causes status change.

3.2 Batch Training

Initial Solution. In batch training, we need to start from a solution that satisfy (10) and (11). The simplest one is the solution with one datum for each class, namely, $\mathbf{x}_{s1}$ for Class 1 and $\mathbf{x}_{s2}$ for Class 2:

$$\alpha_{s1} = \alpha_{s2} = \frac{2}{Q_{s1\,s1} + 2Q_{s1\,s2} + Q_{s2\,s2}}, \tag{19}$$

$$b = -\frac{Q_{s1\,s1} - Q_{s2\,s2}}{Q_{s1\,s1} + 2Q_{s1\,s2} + Q_{s2\,s2}}. \tag{20}$$

We choose two data with the minimum distance. If $\alpha_{s1} = \alpha_{s2} \leq C$, $\alpha_{s1} = \alpha_{s2}$ is the solution of the L1 SVM for the training data set consisting of $\mathbf{x}_{s1}$ and $\mathbf{x}_{s2}$. If the solution does not satisfy the upper bound, we select two data with the maximum distance. By this selection of data, however, we cannot obtain the solution for $C = 1$ if we use RBF kernels. This is because $\alpha_{s1} = \alpha_{s2} > 1$ for RBF kernels. In such a situation, we set the value of C that is slightly larger than $\alpha_{s1} = \alpha_{s2}$. But for L2 SVMs there is no such restriction.

Training by the Decomposition Technique. To speedup training, we divide the training data into o chunk data sets. Let the V_i be the index set of ith chunk data set and an active set be A, which includes current and previous support vectors.

Initially $A = \{\mathbf{x}_{s1}, \mathbf{x}_{s2}\}$. One iteration consists of o sub-iterations, in which at the ith sub-iteration the SVM is trained using the training data associated with the combined set of T_i and A: $T_i \cup A$. At the beginning of a sub-iteration we check the KKT conditions using the hyperplane. We divide the combined set $T_i \cup A$ into T_s and T_u, where training data associated with T_s satisfy the KKT conditions and those with T_u do not. We choose $\mathbf{x}_c$ ($c \in T_u$) that maximally violates KKT conditions.

We modify the optimal hyperplane so that $\mathbf{x}_i$ ($i \in T_s$) and $\mathbf{x}_c$ satisfies the KKT conditions. Then we move indices from T_u to T_s whose associated data satisfy the KKT conditions and add to A the indices whose associated data become support vectors. If there is no modification of the hyperplane for the jth iteration, namely, if there is no modification of the hyperplane in any of sub-iteration of the jth iteration, we stop training.

Since previous support vectors may be near the hyperplane, they may resume being support vectors afterwards. To prevent an additional iteration we include in the active set previous support vectors in addition to current support vectors.

3.3   Determining the Minimum Correction

In the following we consider the four cases where the status changes. For each case we calculate the correction of α_c to reach the status change [11]. For the L2 SVM, we need not consider (22) and (23).

1. Correction for $\mathbf{x}_c$.
   Since $\mathbf{x}_c$ violates the KKT conditions, $g_c < 0$. Then if $\gamma_c \geq \varepsilon$, where ε takes a small positive value, and $g_c + \Delta g_c = 0$, $\mathbf{x}_c$ becomes an unbounded support vector. From (17), correction $\Delta\alpha_c$ is calculated by

$$\Delta\alpha_c = -\frac{g_c}{\gamma_c}. \tag{21}$$

But because α_c is upper bounded by C, we need to check this bound; If $\alpha_c + \Delta\alpha_c < C$, $\mathbf{x}_c$ becomes an unbounded support vector. If $\alpha_c + \Delta\alpha_c \geq C$, we bound the correction by

$$\Delta\alpha_c = C - \alpha_c. \tag{22}$$

Then $\mathbf{x}_c$ becomes a bounded support vector.

2. Correction for $\mathbf{x}_{si}$.

If $|\beta_{si}| \geq \varepsilon$, $\mathbf{x}_{si}$ becomes a non-support vector or a bounded support vector. From (15), if $\beta_{si} \geq \varepsilon$, $\mathbf{x}_{si}$ becomes a bounded support vectors for

$$\Delta\alpha_c = \frac{C - \alpha_{si}}{\beta_{si}}. \tag{23}$$

If $\beta_{si} \leq -\varepsilon$, $\mathbf{x}_{si}$ becomes a non-support vectors for

$$\Delta\alpha_c = -\frac{\alpha_{si}}{\beta_{si}}. \tag{24}$$

3. Correction for a bounded support vector $\mathbf{x}_i$ ($g_i < 0$).

If $\gamma_i \geq \varepsilon$, for

$$\Delta\alpha_c = -\frac{g_i}{\gamma_i}, \tag{25}$$

the margin becomes zero and $\mathbf{x}_i$ becomes an unbounded support vector.

4. Correction for a non-support vector $\mathbf{x}_i$ ($g_i > 0$).

The margin becomes zero when $\gamma_i \leq -\varepsilon$ and $g_i + \Delta g_i = 0$. Thus from (21),

$$\Delta\alpha_c = -\frac{g_i}{\gamma_i}. \tag{26}$$

We calculated the smallest collection among Cases 1 to 4. If the correction calculated by Case 1 is not the smallest, this means that the status changes before $\mathbf{x}_c$ becomes a support vector.

In any case, we set up a new set of support vectors and recalculate β, β_j, g_i, and γ_i.

3.4 Inverse Matrix Update and Numerical Instability

If $\mathbf{x}_c$ is added as an unbounded support vector, the $(l+1) \times (l+1)$ matrix Q^{-1} is updated to the $(l+2) \times (l+2)$ matrix Q^{-1} as follows [11]:

$$Q^{-1} \leftarrow \begin{bmatrix} & & & 0 \\ & Q^{-1} & & \vdots \\ & & & 0 \\ 0 & \cdots & 0 & 0 \end{bmatrix} + \frac{1}{\gamma_c} \begin{bmatrix} \beta \\ \beta_{s1} \\ \vdots \\ \beta_{sl} \\ 1 \end{bmatrix} \begin{bmatrix} \beta & \beta_{s1} & \cdots & \beta_{sl} & 1 \end{bmatrix}. \tag{27}$$

If the kth support vector is deleted from the set of support vectors, $\{Q^{-1}\}_{ij}$ is updated as follows [11]:

$$\{Q^{-1}\}_{ij} \leftarrow \{Q^{-1}\}_{ij} - \frac{1}{\{Q^{-1}\}_{kk}}\{Q^{-1}\}_{ik}\{Q^{-1}\}_{kj}. \tag{28}$$

Since Q is positive semi-definite, numerical stability in calculating (27) and (28) is not guaranteed. Namely, in some conditions, because of the round-off errors in calculating (27) and (28) the equalities in (12) and (13) are violated. The occurrence of round-off errors can be detected during or after finishing training by checking the margins of support vectors, the optimality of the bias term and the duality gap.

3.5   Maintenance of Unbounded Support Vectors

There is a case where there is only one unbounded support vector that reaches 0 from non-zero for L1 and L2 SVMs or that changes status from the bounded support vector for the L1 SVM. In such a case $\mathbf{x}_c$ and α_{s1} need to belong to different classes for $\alpha_{s1} = 0$ and the same class for $\alpha_{s1} = C$. If it is not satisfied, we search such a datum that matches with $\mathbf{x}_c$. If so, we change the support vector with the searched one. Otherwise, we search a datum with the minimum margin violation and adjust the bias term so that violation is resolved [12].

4   Performance Evaluation

We evaluated performance of the proposed method using the benchmark data sets [15] shown in Table 1, which lists the number of inputs, classes, training data, test data. The splice data set is the first file of the 100 data sets.[2] The table also shows the parameter values for L1 and L2 SVMs determined by fivefold cross-validation. For instance, $d4$ and $\gamma10$ mean that the kernels are polynomial kernel with degree 4 and RBF kernels with $\gamma = 10$, and $C = 10^5$ means that the value of the margin parameter is 10^5. For multi-class problems we used fuzzy one-against-all SVMs [15] and measured the training time using a workstation (3.6GHz, 2GB memory, Linux operating system). For the obtained solution, we checked the duality gap, the margins of support vectors, and optimality of the bias term to conform that the optimal solution was obtained.

Table 2 shows the effect of the decomposition techniques for the splice data set by L2 SVM and the thyroid data set by the L1 SVM. In the table "Chunk," "Active Set," "Iterations," "Status Changes," and "Time" denote, respectively, the chunk size, the average number of elements in the active set, the number of iterations, the average number of status changes, and the time for training. The average values are per iteration and per decision function. The numerals in parentheses show the time when (27) and (28) were not used. Namely, we calculated the inverse by matrix inversion when needed.

---

[2] http://ida.first.fraunhofer.de/projects/bench/benchmarks.htm

Table 1. Benchmark data sets and parameters for L1 and L2 SVMs

Data	Inputs	Classes	Train.	Test	L1 Parm.	L2 Parm.
Splice	60	2	1000	2175	$\gamma 10C10$	$\gamma 10C51$
Thyroid	21	3	3772	3428	$d4C10^5$	$d4C10^5$
Blood cell	13	12	3097	3100	$d2C3000$	$\gamma 10C100$
Hiragana-50	50	39	4610	4610	$\gamma 10C5000$	$\gamma 10C1000$
Hiragana-13	13	38	8375	8356	$\gamma 10C3000$	$\gamma 10C500$
Hiragana-105	105	38	8375	8356	$\gamma 10C10^4$	$\gamma 10C10^4$

From the table, for using (27) and (28) 20 to 50 times speedup was obtained for the splice data set but for the thyroid data set speed up was about 10%. This is because the number of support vectors for the splice data set was large but that for the thyroid data set was not. The asterisk shows that the solution did not satisfy the equality constraint and the duality gap was large when (27) and (28) were used.

As for the decomposition technique, speedup using (27) and (28) for the splice problem is small. This is because almost all the data become support vectors. For the thyroid data set 18 times speedup was obtained with the chunk size of 50. But there was not much difference among the active set size and the average number of iterations for the change of the chunk size.

Table 3 lists the training results using the proposed method and the primal-dual interior-point method with the decomposition technique. The chunk size for the proposed method was 100 and that for the interior point method was 50. "SVs" and "Rec." denote, respectively, the number of support vectors and recognition rate for the training data and that for the test data in parentheses.

Table 2. Effect of the decomposition technique for the splice data set by the L2 SVM and the thyroid data set by the L1 SVM

Data	Chunk	Active Set	Iterations	Status Changes	Time
Splice	1000	895	2	1171	34 (826)
	400	922	3.0	1648	35 (1398)
	200	924	3.0	1762	33 (1535)
	100	920	3.0	1909	31 (1630)
	50	911	3.0	1848	29 (1529)
	20	903	3.0	1842	29 (1456)
Thyroid	3772	353	2.0	1251	288 (306)
	1000*	340	4.0	1399	67 (70)
	400	295	4.3	1366	36 (39)
	200	275	4.7	1319	20 (23)
	100	268	4.3	1409	16 (18)
	50	259	4.3	1591	14 (16)
	20	264	4.3	2063	21 (23)

Table 3. Comparison of the proposed method and the interior point method with the decomposition technique using L1 and L2 SVMs

SVM	Data	Proposed Method				Interior Point Method		
		SVs	A. Set	Time	Rec.	SVs	Time	Rec.
L1	Splice	771	893	34	90.48 (100)	774	**19**	90.48 (100)
	Thyroid	87	268	16	97.93 (99.97)	141	**8.5**	97.32 (99.76)
	Blood cell	78	197	35	93.13 (98.16)	100	**12**	93.03 (96.77)
	H-50	67*	182	160	99.26 (100)	70	**71**	99.26 (100)
	H-13	39	122	**64**	99.64 (100)	40	134	99.64 (100)
	H-105	91	285	806	100 (100)	91	**256**	100 (100)
L2	Splice	796	920	**31**	90.53 (100)	791	262	90.48 (100)
	Thyroid	99	279	**17**	97.81 (100)	217	40	97.20 (99.79)
	Blood cell	186	310	101	93.71 (97.13)	136	**51**	93.26 (99.55)
	H-50	78	201	184	99.28 (100)	77	**101**	99.26 (100)
	H-13	71	157	**118**	99.70 (99.95)	73	174	99.70 (99.96)
	H-105	91	285	811	100 (100)	91	**295**	100 (100)

The asterisk denotes that we used matrix inversion instead of (27) and (28) because of numerical instability and the bold letters in the Time columns denote the shorter training time. As for the training time the interior point method is faster for 8 data sets and the proposed method for four data sets but except for the hiragana-105 data set both methods are comparable. The number of support vectors by the proposed method is usually smaller. This is because in the proposed method training proceeds by status changes of training data and if the status change occurs simultaneously, we only consider one datum.

5 Conclusions

In this paper we proposed a new training method based on exact incremental training. First, we select two data in different classes and determine the optimal separating hyperplane. Then, for each chunk data and the active set that includes current and previous support vectors, we modify the optimal hyperplane so that the violating datum satisfies the KKT conditions. We iterate the above procedure for each chunk data and the active set until there is no violating data. By the computer experiment using several benchmark data sets, we show that the training speed of the proposed method is comparable with the primal-dual interior-point method combined with the decomposition technique with a smaller number of support vectors.

References

1. Vanderbei, R.J.: LOQO: An Interior Point Code for Quadratic Programming. Technical Report SOR-94-15, Princeton University (1998)
2. Osuna, E., Freund, R., Girosi, F.: An Improved Training Algorithm for Support Vector Machines. In: Proc. NNSP 1997, pp. 276–285 (1997)

3. Platt, J.C.: Fast Training of Support Vector Machines Using Sequential Minimal Optimization. In: Schölkopf, B., et al. (eds.) Advances in Kernel Methods: Support Vector Learning, pp. 185–208. MIT Press, Cambridge (1999)
4. Vogt, M.: SMO Algorithms for Support Vector Machines without Bias Term. Technical report, Institute of Automatic Control, TU Darmstadt, Germany (2002)
5. Kecman, V., Vogt, M., Huang, T.M.: On the Equality of Kernel AdaTron and Sequential Minimal Optimization in Classification and Regression Tasks and alike Algorithms for Kernel Machines. In: Proc. ESANN 2003, pp. 215–222 (2003)
6. Sentelle, C., Georgiopoulos, M., Anagnostopoulos, G., Young, C.: On Extending the SMO Algorithm Sub-Problem. In: Proc. IJCNN 2007, pp. 1249–1254 (2007)
7. Keerthi, S.S., Shevade, S.K., Bhattacharyya, C., Murthy, K.R.K.: A Fast Iterative Nearest Point Algorithm for Support Vector Machine Classifier Design. IEEE Trans. Neural Networks 11(1), 124–136 (2000)
8. Raicharoen, T., Lursinsap, C.: Critical Support Vector Machine without Kernel Function. In: Proc. ICONIP 2002, vol. 5, pp. 2532–2536 (2002)
9. Roobaert, D.: DirectSVM: A Fast and Simple Support Vector Machine Perceptron. In: NNSP 2000, vol. 1, pp. 356–365 (2000)
10. Vishwanathan, S.V.N., Murty, M.N.: SSVM: A Simple SVM Algorithm. In: Proc. IJCNN 2002, vol. 2, pp. 2393–2398 (2002)
11. Cauwenberghs, G., Poggio, T.: Incremental and Decremental Support Vector Machine Learning. In: Leen, T.K., et al. (eds.) Advances in Neural Information Processing Systems, vol. 13, pp. 409–415. MIT Press, Cambridge (2001)
12. Laskov, P., Gehl, C., Kruger, S., Müller, K.-R.: Incremental Support Vector Learning: Analysis, Implementation and Applications. Journal of Machine Learning Research 7, 1909–1936 (2006)
13. Hastie, T., Rosset, S., Tibshirani, R., Zhu, J.: The Entire Regularization Path for the Support Vector Machine. Journal of Machine Learning Research 5, 1391–1415 (2004)
14. Suykens, J.A.K., Vandewalle, J.: Least Squares Support Vector Machine Classifiers. Neural Processing Letters 9(3), 293–300 (1999)
15. Abe, S.: Support Vector Machines for Pattern Classification. Springer, London (2005)

A Kernel Method for the Optimization of the Margin Distribution

F. Aiolli, G. Da San Martino, and A. Sperduti

Dept. of Pure and Applied Mathematics, Via Trieste 63, 35131 Padova - Italy

Abstract. Recent results in theoretical machine learning seem to suggest that nice properties of the margin distribution over a training set turns out in a good performance of a classifier. The same principle has been already used in SVM and other kernel based methods as the associated optimization problems try to maximize the minimum of these margins.

In this paper, we propose a kernel based method for the direct optimization of the margin distribution (KM-OMD). The method is motivated and analyzed from a game theoretical perspective. A quite efficient optimization algorithm is then proposed. Experimental results over a standard benchmark of 13 datasets have clearly shown state-of-the-art performances.

Keywords: Kernel Methods, Margin Distribution, Game Theory.

1 Introduction

Much of last-decade theoretical work on learning machines has been devoted to study the aspects of learning methods that control the generalization performance. In essence, two main features seem to be responsible for the generalization performance of a classifier, namely, keeping low the complexity of the hypothesis space (e.g. by limiting the VC dimension) and producing models which achieve large margin (i.e. confidence in the prediction) over the training set.

The good empirical effectiveness of two of the most popular algorithms, Support Vector Machines (SVM) and AdaBoost, have been in fact explained by the high margin classifiers they are able to produce. Specifically, hard margin SVMs return the hyperplane which keeps all the examples farest away from it, thus maximizing the minimum of the margin over the training set (worst-case optimization of the margin distribution). Similarly, AdaBoost, has been demonstrated to greedily minimize a loss function which is tightly related to the distribution of the margins on the training set. Despite the AdaBoost ability to optimize the margin distribution on the training set, it has been shown in [1] that in certain cases, it can also increase the complexity of the weak hypotheses, thus possibly leading to overfitting phenomena.

The effect of the margin distribution on the generalization ability of learning machines have been studied in [2] and [3], while algorithms trying to optimize explicitly the margin distribution include [4], [5] and [6]. More recently, it has been shown [7] that quite good effectiveness can even be obtained by the optimization of the first moment of the margin distribution (the simple average value over the training set). In this

V. Kůrková et al. (Eds.): ICANN 2008, Part I, LNCS 5163, pp. 305–314, 2008.

case, the problem can be solved very efficiently, since computing the model has time complexity $O(n)$.

In this paper, we propose a kernel machine which explicitly tries to optimize the margin distribution. Specifically, this boils down to an optimization of a weighted combination of margins, via a distribution over the examples, with appropriate constraints related to the entropy (as a measure of complexity) of the distribution.

In Section 1.1 some notation used through the paper is introduced. In Section 2 a game-theoretical interpretation of hard margin SVM is given in the bipartite instance ranking framework (i.e. the problem to induce a separation between positive and negative instances in a binary task) and the problem of optimizing the margin distribution is studied from the same perspective. This game-theoretic analysis leads us to a simple method for optimizing the distribution of the margins. Then, in Section 3, an efficient optimization algorithm is derived. Experimental results are presented in Section 4. Finally, conclusions are drawn.

1.1   Notation and Background

In the context of binary calssification tasks, the aim of a learning algorithm is to return a classifier which minimizes the error on a (unknown) distribution $\mathcal{D}_{\mathcal{X} \times \mathcal{Y}}$ of input/output pairs $(\mathbf{x}_i, y_i)$, $\mathbf{x}_i \in \mathbb{R}^d$, $y_i \in \{-1, +1\}$. The input to the algorithm is a set of pre-classified examples pairs $\mathcal{S} = \{(\mathbf{x}_1, y_1), \ldots, (\mathbf{x}_N, y_N)\}$. With $S^+ = \{\mathbf{x}_1^+, \ldots, \mathbf{x}_p^+\}$ we denote the set of p positive instances, where $\mathbf{x}_i^+$ is the i-th positive instance in S. Similarly $S^- = \{\mathbf{x}_1^-, \ldots, \mathbf{x}_n^-\}$ denotes the set of n negative instances. Clearly, $N = n + p$.

In this paper, we denote by $\Gamma_m \subseteq \mathbb{R}^m$ the set of m-dimensional probability vectors, i.e. $\Gamma_m = \{\gamma \in \mathbb{R}^m | \sum_{i=1}^m \gamma_i = 1, \gamma_i \geq 0\}$. The *convex hull* $\mathbf{ch}(\mathcal{C})$ of a set $\mathcal{C} = \{\mathbf{c}_1, \ldots, \mathbf{c}_m | \mathbf{c}_i \in \mathbb{R}^d\}$, is the set of all affine combinations of points in $\mathcal{C}$ such that the weights γ_i of the combination are non-negative, i.e. $\mathbf{ch}(\mathcal{C}) = \{\gamma_1 \mathbf{c}_1 + \cdots + \gamma_m \mathbf{c}_m | \gamma_i \in \Gamma_m\}$. We also generalize this definition, by defining the η-norm-*convex hull* of a set $\mathcal{C} \in \mathbb{R}^d$ as the subset of $\mathbf{ch}(\mathcal{C})$ which has weights with (squared) norm smaller than a given value η, i.e. $\mathbf{ch}_\eta(\mathcal{C}) = \{\gamma_1 \mathbf{c}_1 + \cdots + \gamma_m \mathbf{c}_m \in \mathbf{ch}(\mathcal{C}) | \, ||\gamma||^2 \leq \eta, \frac{1}{m} \leq \eta \leq 1\}$. Note that, whenever $\eta = \frac{1}{m}$, a trivial set consisting of a single point (the average of points in $\mathcal{C}$), is obtained, while whenever $\eta = 1$ this set will coincide with the convex hull.

2   Game Theory, Learning and Margin Distribution

A binary classification problem can be viewed from two different points of view. Specifically, let $h \in \mathcal{H}$ be an hypothesis space, mapping instances on real values. In a first scenario, let call it *instance classification*, a given hypothesis is said to be consistent with a training set if $y_i h(\mathbf{x}_i) > 0$ (classification constraints) for each example of the training set. In this case, the prediction on new instances can be performed by using the sign of the decision function.

In a second scenario, which we may call *bipartite instance ranking*, a given hypothesis is said consistent with a training set if $h(\mathbf{x}_i^+) - h(\mathbf{x}_j^-) > 0$ (order constraints) for any positive instance $\mathbf{x}_i^+$ and any negative instance $\mathbf{x}_j^-$. Note that when an hypothesis is consistent, then it is always possible to define a threshold which correctly separates positive from negative instances in the training set. In this paper, we mainly focus on this second view, even if a similar treatment can be pursued for the other setting.

In the following, we give an interpretation of the hard-margin SVM as a two players zero-sum game in the bipartite instance ranking scenario presented above. First of all, we recall that, in the classification context, the formulation of the learning problem is based on the maximization of the minimum of the margin in the training set. Then, we propose to slightly modify the pay-off function of the game in order to have a flexible way to control the optimization w.r.t. the distribution of the margin in the training set.

2.1 Hard Margin SVM as a Zero-Sum Game

Consider the following zero-sum game defined for a bipartite instance ranking scenario. Let $\mathcal{P}_{MIN}$ (the nature) and $\mathcal{P}_{MAX}$ (the learner) be the two players. On each round of the game, $\mathcal{P}_{MAX}$ picks an hypothesis h from a given hypotheses space $\mathcal{H}$, while (simultaneously) $\mathcal{P}_{MIN}$ picks a pair of instances of different classes $\mathbf{z} = (\mathbf{x}^+, \mathbf{x}^-) \in \mathcal{S}^+ \times \mathcal{S}^-$. $\mathcal{P}_{MAX}$ wants to maximize its pay-off defined as the achieved margin $\rho_h(\mathbf{z})$ on the pair of examples which, in this particular setting, can be defined by $h(\mathbf{x}^+) - h(\mathbf{x}^-)$. Note that the value of the margin defined in this way is consistent with the bipartite instance ranking setting since it is greater than zero for a pair whenever the order constraint is satisfied for the pair, and less than zero otherwise.

Considering the hypothesis space of hyperplanes defined by unit-length weight vectors

$$\mathcal{H} = \{h(\mathbf{x}) = \mathbf{w}^\top \mathbf{x} - \theta | \ \mathbf{w} \in \mathbb{R}^d s.t. \ ||\mathbf{w}|| = 1, \ \text{and} \ \theta \in \mathbb{R}\},$$

the margin is defined by the difference of the *scores* of the instances, that is

$$\rho_{\mathbf{w}}(\mathbf{x}^+, \mathbf{x}^-) = \mathbf{w}^\top \mathbf{x}^+ - \theta - \mathbf{w}^\top \mathbf{x}^- + \theta = \mathbf{w}^\top (\mathbf{x}^+ - \mathbf{x}^-)$$

Let now be given a mixed strategy for the $\mathcal{P}_{MIN}$ player defined by $\gamma^+ \in \Gamma_p$, the probability of each positive instance to be selected, and $\gamma^- \in \Gamma_n$ the correspondent probabilities for negative instances. We can assume that these probabilities can be marginalized as the associated events are independent. In other words, the probability to pick a pair $(\mathbf{x}_i^+, \mathbf{x}_j^-)$ is simply given by $\gamma_i^+ \gamma_j^-$. Hence, the value of the game, i.e. the expected margin obtained in a game, will be:

$$\begin{aligned}
V((\gamma^+, \gamma^-), \mathbf{w}) &= \textstyle\sum_{i,j} \gamma_i^+ \gamma_j^- \mathbf{w}^\top (\mathbf{x}_i^+ - \mathbf{x}_j^-) \\
&= \mathbf{w}^\top (\textstyle\sum_i \gamma_i^+ \mathbf{x}_i^+ (\sum_j \gamma_j^-) - \sum_j \gamma_j^- \mathbf{x}_j^- (\sum_i \gamma_i^+)) \\
&= \mathbf{w}^\top (\textstyle\sum_i \gamma_i^+ \mathbf{x}_i^+ - \sum_j \gamma_j^- \mathbf{x}_j^-)
\end{aligned}$$

Then, when the player $\mathcal{P}_{MIN}$ is left free to choose its strategy, we obtain the following problem which determines the equilibrium of the game, that is

$$\min_{\gamma^+ \in \Gamma_p, \gamma^- \in \Gamma_n} \ \max_{\mathbf{w} \in \mathcal{H}} \ \mathbf{w}^\top \Big(\sum_i \gamma_i^+ \mathbf{x}_i^+ - \sum_j \gamma_j^- \mathbf{x}_j^- \Big)$$

Now, it easy to realize that a pure strategy is right available to the $\mathcal{P}_{MAX}$ player. In fact, it can maximize its pay-off by setting

$$\hat{\mathbf{w}} = \begin{cases} \text{any } \mathbf{w} \in \mathcal{H} & \text{if } \mathbf{v}(\gamma^+, \gamma^-) = \mathbf{0} \\ \frac{\mathbf{v}(\gamma^+, \gamma^-)}{||\mathbf{v}(\gamma^+, \gamma^-)||} & \text{otherwise} \end{cases} \quad , \text{ where } \mathbf{v}(\gamma^+, \gamma^-) = \sum_i \gamma_i^+ \mathbf{x}_i^+ - \sum_i \gamma_i^- \mathbf{x}_i^-$$

Note that the condition $\mathbf{v}(\gamma^+, \gamma^-) = \mathbf{0}$ implies that the (signed) examples $y_i \mathbf{x}_i$ are not linearly independent, i.e. there exists a linear combination of these instances with not all null coefficients which is equal to the null vector. This condition is demonstrated to be necessary and sufficient for the non linear separability of a set (see [8]).

When the optimal strategy for $\mathcal{P}_{MAX}$ has been chosen, the expected value of the margin according to the probability distributions (γ^+, γ^-), i.e. the value of the game, will be:

$$E[\rho_{\hat{\mathbf{w}}}(\mathbf{x}^+, \mathbf{x}^-)] = \hat{\mathbf{w}}^\top \mathbf{v}(\gamma^+, \gamma^-) = ||\mathbf{v}(\gamma^+, \gamma^-)||$$

Note that the vector $\mathbf{v}(\gamma^+, \gamma^-)$ is defined by the difference of two vectors in the convex hulls of the positive and negative instances respectively, $\mathbf{v}_+ = \sum_i \gamma_i^+ \mathbf{x}_i^+ \in \mathbf{ch}(\mathcal{S}^+)$ and $\mathbf{v}_- = \sum_i \gamma_i^- \mathbf{x}_i^- \in \mathbf{ch}(\mathcal{S}^-)$. Moreover, when γ^+ and γ^- are uniform on their respective sets, the vector $\mathbf{v}(\gamma^+, \gamma^-)$ will be the difference between average points of the sets (a.k.a. their centroids).

Now, we are able to show that the best strategy for $\mathcal{P}_{MIN}$ is the solution obtained by an SVM. For this, let us rewrite the vector $\mathbf{v}(\gamma^+, \gamma^-)$ using a single vector of parameters,

$$\mathbf{v}(\gamma^+, \gamma^-) \equiv \mathbf{v}(\gamma) = \sum_i^N y_i \gamma_i \mathbf{x}_i$$

which can be obtained by a simple change of variables

$$\gamma_i = \begin{cases} \gamma_r^+ & \text{if } \mathbf{x}_i \text{ is the } r\text{-th } \textit{positive} \text{ example} \\ \gamma_r^- & \text{if } \mathbf{x}_i \text{ is the } r\text{-th } \textit{negative} \text{ example} \end{cases}$$

Using the fact that minimizing the squared norm is equivalent to minimize the norm itself, we may formulate the optimization problem to compute the best strategy for $\mathcal{P}_{MIN}$ (which aims to minimize the value of the game):

$$\min_{\gamma^+ \in \Gamma_p, \gamma^- \in \Gamma_n} ||\mathbf{v}(\gamma^+, \gamma^-)|| = \begin{cases} \min_\gamma ||\mathbf{v}(\gamma)||^2 \\ \text{s.t. } \sum_{i:y_i=y} \gamma_i = 1, \forall y \in \{-1, +1\}, \text{ and } \gamma_i \geq 0 \end{cases} \tag{1}$$

As already demonstrated in [9] the problem on the right of Eq. 1 is the same as hard margin SVM when a bias term is present. Specifically, the bias term is chosen as the score of the point standing in the middle between the points $\mathbf{v}_+$ and $\mathbf{v}_-$, i.e.

$$\theta = \frac{1}{2}\hat{\mathbf{w}}^\top(\mathbf{v}_+ + \mathbf{v}_-).$$

Then, the solutions of the two problems are the same. Specifically, the solution maximizes the minimum of the margins in the training set. Clearly, when the training set is not linearly separable, the solution of the problem in Eq.1 will be $\mathbf{v}(\gamma) = \mathbf{0}$.

2.2 Playing with Margin Distributions

The maximization of the minimum margin is not necessarily the optimal choice when dealing with a classification task. In fact, many recent works, including [4,3], have demonstrated that the generalization error depends more properly on the distribution of (lowest) margins in the training set.

Our main idea is then to construct a problem which makes easy to play with the margin distribution. Specifically, we aim at a formulation that allow us to specify a given trade-off between the minimal value and the average value of the margin on the training set.

For this, we extend the previous game considering a further cost for the player $\mathcal{P}_{MIN}$. Specifically, we want to penalize, to a given extent, too pure strategies in such a way to have solutions which are robust with respect to different training example distributions. In this way, we expect to reduce the variance of the best strategy estimation when different training sets are drawn from the true distribution of examples $\mathcal{D}_{\mathcal{X} \times \mathcal{Y}}$. A good measure of the complexity of $\mathcal{P}_{MIN}$ behavior would certainly be the normalized entropy of its strategy which can be defined by

$$E(\gamma) = \frac{1}{2} \left(\frac{1}{\log(p)} \sum_i \gamma_i^+ \log \frac{1}{\gamma_i^+} + \frac{1}{\log(n)} \sum_i \gamma_i^- \log \frac{1}{\gamma_i^-} \right)$$

which has maximum value 1 whenever γ is the uniform distribution on both sets (completely unpredictable startegy) and is 0 when the distribution is picked on a single example per set (completely predictable pure strategy).

However, a (simpler) approximate version of the entropy as defined above, can be obtained by considering the 2-norm of the distribution. In fact, it is well known that, for any distribution $\gamma \in \Gamma_m$ it always holds that $||\gamma||_2 \leq ||\gamma||_1$. Moreover, $||\gamma||_2$ is minimal whenever γ is a uniform distribution and is equal to 1 whenever γ is a pure strategy. Specifically, we can consider the following approximation:

$$E(\gamma) \approx \frac{m}{m-1}(1 - ||\gamma||_2^2)$$

Considering the squared norm of the distribution, we can reformulate the strategy of $\mathcal{P}_{MIN}$ as a trade-off between two objective functions, with a trade-off parameter λ:

$$\min_{\gamma^+ \in \Gamma_p, \gamma^- \in \Gamma_n} (1 - \lambda)||\mathbf{v}(\gamma)||^2 + \lambda||\gamma||^2 \tag{2}$$

It can be shown that the optimal vector $\mathbf{v}(\hat{\gamma})$ which is solution of the problem above, represents the vector joining two points, $\mathbf{v}_+$ into the positive norm-restricted convex hull, i.e. $\mathbf{v}_+ \in \mathbf{ch}_\eta(\mathcal{S}^+)$, and $\mathbf{v}_-$ into the negative norm-restricted convex hull, i.e. $\mathbf{v}_- \in \mathbf{ch}_\eta(\mathcal{S}^-)$, for opportune η.

Similarly to the hard margin SVM, the threshold is defined as the score of the point which is in the middle between these two points, i.e. $\theta = \frac{1}{2}\hat{\mathbf{w}}^\top(\mathbf{v}_+ + \mathbf{v}_-)$.

Finally, it is straightforward to see that this method generalizes (when $\lambda = 1$), the baseline method presented in [10] where the simple difference between the centroid of positives and the centroid of negatives is used as the weight vector, and obviously it generalizes the hard-margin SVM for $\lambda = 0$.

3 Optimization Algorithm

We now propose a very simple method to optimize the problem in Eq. (1). The proposed method optimizes the objective function by a SMO-like procedure which maintains a feasible solution γ at each step, starting from a feasible γ_{init}. Specifically, in order to maintain the solution in the feasible set, at each step, it chooses a pair of variables (γ_r, γ_q) associated to examples of the same class, let say y, and imposes an update of the form:

$$\gamma'_r \leftarrow \gamma_r + \epsilon, \ \gamma'_q \leftarrow \gamma_q - \epsilon$$

With this update, we can exactly measure the extent of the change which occurs to the vector $\mathbf{v}(\gamma)$, i.e.

$$\Delta_{\mathbf{v}(\gamma)} = \mathbf{v}(\gamma') - \mathbf{v}(\gamma) = y\epsilon(\mathbf{x}_r - \mathbf{x}_q)$$

To improve the readability of the following derivations, let us denote by $s_j(\gamma)$ the quantity $s_j(\gamma) = \sum_i y_i \gamma_i \mathbf{x}_i^\top \mathbf{x}_j$. Now, we are able to evaluate how much, and how, the update on the γ's modifies the objective function.

Let us begin from the first term $G(\gamma) = ||\mathbf{v}(\gamma)||^2$. In this case, we have:

$$G(\gamma') = ||\mathbf{v}(\gamma')||^2 = ||\mathbf{v}(\gamma) + \Delta_{\mathbf{v}(\gamma)}||^2 = ||\mathbf{v}(\gamma)||^2 + ||\Delta_{\mathbf{v}(\gamma)}||^2 + 2\mathbf{v}(\gamma)\Delta_{\mathbf{v}(\gamma)}$$

and hence the variation can be obtained by

$$\Delta_{G(\gamma)} = G(\gamma') - G(\gamma) = ||\Delta_{\mathbf{v}(\gamma)}||^2 + 2\mathbf{v}(\gamma)^\top \Delta_{\mathbf{v}(\gamma)}$$

where $||\Delta_{\mathbf{v}(\gamma)}||^2 = \epsilon^2 ||\mathbf{x}_r - \mathbf{x}_q||^2$ and $\mathbf{v}(\gamma)^\top \Delta_{\mathbf{v}(\gamma)} = \epsilon y(s_r(\gamma) - s_q(\gamma))$. Summarizing, we have

$$\Delta_{G(\gamma)} = \epsilon^2 ||\mathbf{x}_r - \mathbf{x}_s||^2 + 2\epsilon y(s_r(\gamma) - s_q(\gamma))$$

For what concerns the second term $H(\gamma) = ||\gamma||^2$, we have

$$H(\gamma') = ||\gamma + \Delta_\gamma||^2 = ||\gamma||^2 + ||\Delta_\gamma||^2 + 2\gamma^\top \Delta_\gamma = ||\gamma||^2 + 2\epsilon^2 + 2\epsilon(\gamma_r - \gamma_q)$$

and thus

$$\Delta_{H(\gamma)} = H(\gamma') - H(\gamma) = 2\epsilon^2 + 2\epsilon(\gamma_r - \gamma_q)$$

Thus the overall objective function $L(\gamma) = (1 - \lambda)||\mathbf{v}(\gamma)||^2 + \lambda||\gamma||^2$ will vary of an amount

$$\begin{aligned}
\Delta_{L(\gamma)} &= L(\gamma') - L(\gamma) = (1 - \lambda)\Delta_{G(\gamma)} + \lambda\Delta_{H(\gamma)} \\
&= (1 - \lambda)(\epsilon^2 ||\mathbf{x}_r - \mathbf{x}_q||^2 + 2y\epsilon(s_r(\gamma) - s_q(\gamma))) + 2\lambda(\epsilon^2 + \epsilon(\gamma_r - \gamma_q)) \\
&= ((1 - \lambda)||\mathbf{x}_r - \mathbf{x}_q||^2 + 2\lambda)\epsilon^2 + 2((1 - \lambda)y(s_r(\gamma) - s_q(\gamma)) + \lambda(\gamma_r - \gamma_q))\epsilon
\end{aligned}$$
$$(3)$$

Finally, setting the derivative of $\Delta_{L(\gamma)}$ w.r.t. to ϵ, to zero, we are able to find the point of minimum (or maximum improvement):

$$\frac{\partial \Delta_{L(\gamma)}(\epsilon)}{\partial \epsilon} = 2\epsilon((1 - \lambda)||\mathbf{x}_r - \mathbf{x}_q||^2 + 2\lambda) + 2((1 - \lambda)(\rho_r - \rho_q) + \lambda(\gamma_r - \gamma_q)) = 0$$

thus obtaining

$$\hat{\epsilon} = \frac{\lambda(\gamma_q - \gamma_r) + (1 - \lambda)y(s_q(\gamma) - s_r(\gamma))}{2\lambda + (1 - \lambda)||\mathbf{x}_r - \mathbf{x}_q||^2} \tag{4}$$

In order to maintain the solution in the feasibility set, the constraints $\hat{\epsilon} < -\gamma_r$, and $\hat{\epsilon} < \gamma_q$ must be enforced. This is made by simply doing a cut with the formula:

$$\hat{\epsilon} = \min\{\hat{\epsilon}, -\gamma_r, \gamma_q\}.$$

3.1   The Algorithm

In the following a pseudo-code of the proposed algorithm is presented.

input:
   training set $S = \{(\mathbf{x}_i, y_i)\}_{i=1,\ldots,N}$
   the convergence tolerance $\delta > 0$
initialize:
   $\gamma_i = \frac{1}{p}$ if $y_i = 1$, $\gamma_i = \frac{1}{n}$ if $y_i = -1$, $\{s_j = \sum_i y_i \gamma_i \mathbf{x}_i^\top \mathbf{x}_j\}_{j=1,\ldots,N}$
repeat
   for each pair (γ_r, γ_q) associated to examples of a same class y
     compute $\hat{\epsilon} = \hat{\epsilon}(\gamma_r, \gamma_q)$ as in Eq. 4
     set $\hat{\epsilon} = \min(\hat{\epsilon}, -\gamma_r)$
     set $\hat{\epsilon} = \min(\hat{\epsilon}, \gamma_q)$
     compute the delta loss $\Delta_{L(\gamma)}$ as in Eq. 3
     if $\Delta_{L(\gamma)}(\hat{\epsilon}) > \delta$ **then** *(update step)*
       $\gamma_r = \gamma_r + \epsilon, \gamma_q = \gamma_q - \epsilon$
       $s_j = s_j + \epsilon(\mathbf{x}_r^\top \mathbf{x}_j - \mathbf{x}_q^\top \mathbf{x}_j), j = 1, \ldots, N$
     end if
   end for
until no update steps have been performed in this iteration
compute the threshold $\theta = \frac{1}{2} \sum_j s_j$
return the vector γ, and the threshold θ.

4   Experiments and Results

The proposed approach has been tested against a popular benchmark consisting of 13 binary datasets[1]. For each dataset, there are 100 different training/test splits of the same data. The classification error is computed as the average error among all these realizations. For a detailed description of the datasets see [11]. In order to have a fair comparison with the results in [11], the same model selection methodology has been used: the best parameter setting for each of the first 5 realizations has been obtained using 5-fold cross validation. Then, the final value of the parameters, namely λ and γ (for the RBF kernel), is selected as the median of the 5 best values obtained in validation. With these parameters, our method has been run again on each training data and then tested for all 100 realizations.

---

[1] All datasets can be downloaded from:
  http://ida.first.fraunhofer.de/projects/bench/benchmarks.htm

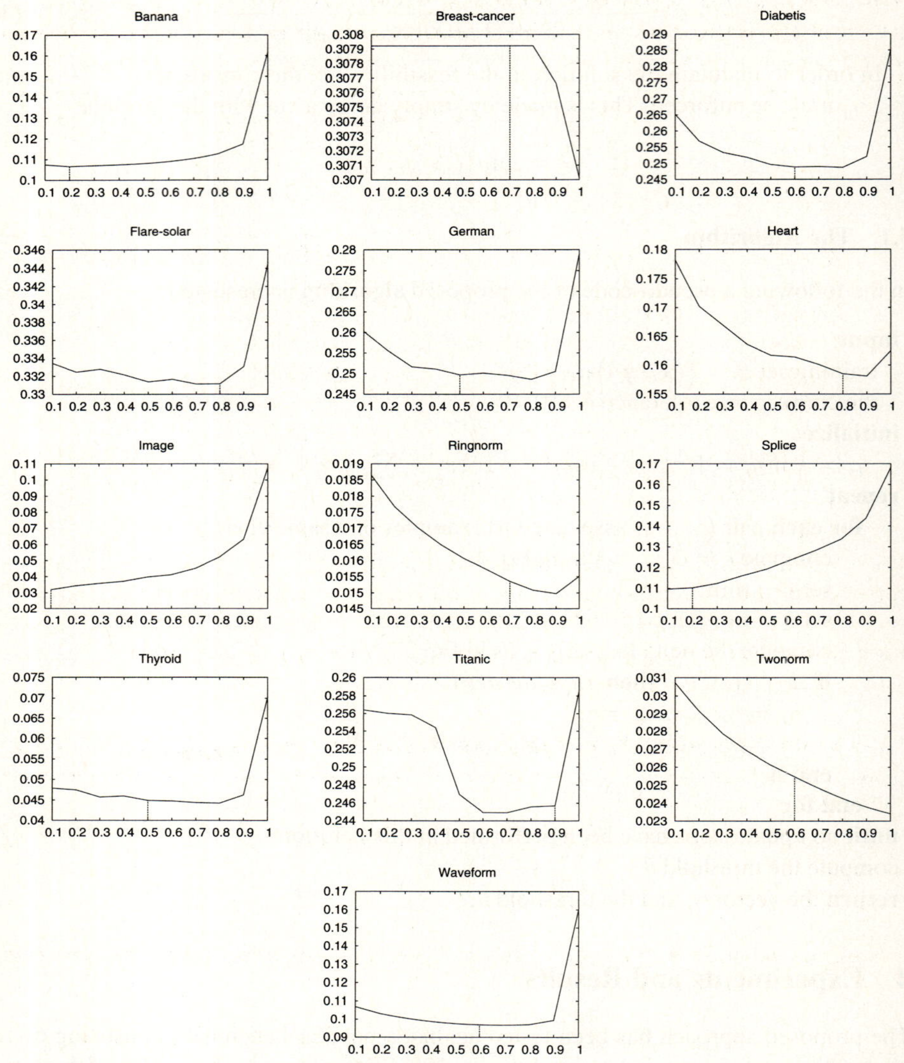

Fig. 1. Classification error (y-axis) with respect to lambda values (x-axis) for each dataset. The vertical dotted line represent the lambda value selected on validation.

Table 1 summarizes the obtained results and compare them against the ones reported in [11] and [12]. Note that the proposed algorithm is better than SVM on 6 datasets. Moreover, on 4 datasets, our method improves over the best performing method.

Figure 1 shows how the error rate varies with respect to λ. Generally speaking, the error rate tends to decrease towards the optimum, and then increase again when λ approaches 1. On the other hand, Figure 2 shows how the training time is affected by λ values. This is only plotted for 2 datasets, namely Image and Waveform, since the of

Table 1. Classification error of SVM and KM-OMD (our method) on 13 datasets. Between brackets the standard deviation. Best method is in bold. When underlined, the method improves the best result, according to [11] and [12], obtained for the dataset (which is reported in the third column).

	SVM	KM-OMD	Best in [11] [12]
banana	0.11530 ($\pm$ 0.0660)	**0.10660** ($\pm$ 0.0150)	0.10730 ($\pm$ 0.0430)
breast-cancer	**0.26040** ($\pm$ 0.0474)	0.30792 ($\pm$ 0.1206)	0.24770 ($\pm$ 0.0463)
diabetis	**0.23530** ($\pm$ 0.0173)	0.24883 ($\pm$ 0.0455)	0.23210 ($\pm$ 0.0163)
flare-solar	**0.32430** ($\pm$ 0.0182)	0.33115 ($\pm$ 0.0489)	0.32430 ($\pm$ 0.0182)
german	**0.23610** ($\pm$ 0.0207)	0.24970 ($\pm$ 0.0570)	0.23610 ($\pm$ 0.0207)
heart	0.15950 ($\pm$ 0.0326)	**0.15850** ($\pm$ 0.0915)	0.15950 ($\pm$ 0.0326)
image	**0.02960** ($\pm$ 0.0060)	0.03198 ($\pm$ 0.0106)	0.02670 ($\pm$ 0.0061)
ringnorm	0.01660 ($\pm$ 0.0012)	**0.01536** ($\pm$ 0.0046)	0.01490 ($\pm$ 0.0012)
splice	**0.10880** ($\pm$ 0.0066)	0.11025 ($\pm$ 0.0142)	0.09500 ($\pm$ 0.0065)
thyroid	0.04800 ($\pm$ 0.0219)	**0.04585** ($\pm$ 0.0475)	0.04200 ($\pm$ 0.0207)
titanic	**0.22420** ($\pm$ 0.0102)	0.24570 ($\pm$ 0.1288)	0.22420 ($\pm$ 0.0102)
twonorm	0.02960 ($\pm$ 0.0023)	**0.02548** ($\pm$ 0.0042)	0.02610 ($\pm$ 0.0015)
waveform	0.09880 ($\pm$ 0.0043)	**0.09685** ($\pm$ 0.0110)	0.09790 ($\pm$ 0.0081)

curves we obtained for the other datasets were similar. The two plots clearly show that learning with low λ values requires more training time, whereas models for higher λ values are faster to compute. It is worth noting that, in most cases, even high λ values (for which the models are much faster to train) give anyway good performances, or at least acceptable when the computational time is an issue.

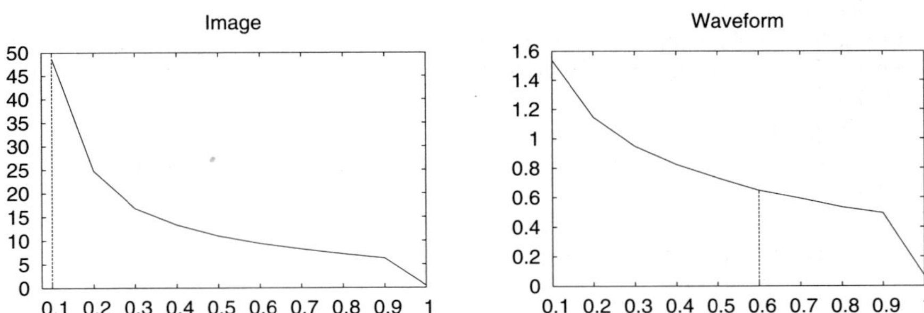

Fig. 2. Training time (y-axis) in seconds with respect to lambda values (x-axis) for Image and Waveform datasets. The vertical dotted line represent the lambda value selected on validation.

5   Conclusions

We have addressed the problem of optimizing the distribution of the margins from a game-theoretical perspective. A simple method has been proposed which consist in optimizing a trade-off between two extreme optimization tasks: the maximization of the minimum margin and the maximization of the average of the margin. The experimental results have shown state-of-the-art performances for some datasets.

In future work, we would like to study under which conditions (e.g. conditions related to the data distribution) our method is to prefer to other state-of-the-art methods. Moreover, the very simple algorithm we have proposed in this paper is not optimized and its optimization could be another direction of our future research. When an optimized version of the algorithm will be available, fair time comparisons with state-of-the-art software for SVM, including SVMLight, will be possible.

References

1. Reyzin, L., Schapire, R.: How boosting the margin can also boost classifier complexity. In: Proceedings of the 23rd International Conference on Machine Learning (ICML) (2006)
2. Garg, A., Har-Peled, S., Roth, D.: On generalization bounds, projection profile, and margin distribution. In: Proceedings of the 11th International Conference on Machine Learning (ICML) (2002)
3. Shawe-Taylor, J., Cristianini, N.: Further results on the margin distribution. In: Proceedings of the 15th International Conference on Machine Learning (ICML) (2003)
4. Garg, A., Roth, D.: Margin distribution and learning algorithms. In: Proceedings of the 12th Conference on Computational Learning Theory (COLT) (1999)
5. Mason, L., Bartlett, P., Baxter, J.: Improved generalization trough explicit optimization of margins. Machine Learning 38(3), 243–255 (2000)
6. Aiolli, F., Sperduti, A.: A re-weighting strategy for improving margins. Artifical Intelligence Journal 137, 197–216 (2002)
7. Pelckmans, K., Suykens, J., Moor, B.D.: A risk minimization principle for a class of parzen estimators. In: Advances in Neural Information Processing Systems (2007)
8. Siu, K.Y., Roychowdhury, V., Kailath, T.: Discrete Neural Computation. Prentice Hall, Englewood Cliffs (1995)
9. Bhattacharyya, C., Keerthi, S., Murthy, K., Shevade, S.: A fast iterative nearest point algorithm for support vector machine classifier design. IEEE Transactions on Neural Networks (2000)
10. Scholkopf, B., Smola, A.: Learning with Kernels. MIT Press, Cambridge (2002)
11. Rätsch, G., Onoda, T., Müller, K.R.: Soft margins for adaboost. Mach. Learn. 42(3), 287–320 (2001)
12. Mika, S., Rätsch, G., Müller, K.R.: A mathematical programming approach to the kernel fisher algorithm. In: NIPS, pp. 591–597 (2000)

A 4–Vector MDM Algorithm for Support Vector Training

Álvaro Barbero, Jorge López, and José R. Dorronsoro[*]

Dpto. de Ingeniería Informática and Instituto de Ingeniería del Conocimiento
Universidad Autónoma de Madrid, 28049 Madrid, Spain

Abstract. While usually SVM training tries to solve the dual of the standard SVM minimization problem, alternative algorithms that solve the Nearest Point Problem (NPP) for the convex hulls of the positive and negative samples have been shown to also provide effective SVM training. They are variants of the Mitchell–Demyanov–Malozemov (MDM) algorithm and although they perform a two vector weight update, they must compute 4 possible updating vectors before deciding which ones to use. In this work we shall propose a 4–vector version of the MDM algorithm that requires substantially less iterations than the previous 2–vector MDM algorithms and leads to faster training if kernel operations are cached.

1   Introduction

Given a linearly separable training sample $\mathcal{S} = \{(X_i, y_i) : i = 1, \ldots, N\}$ with $y_i = \pm 1$, the standard formulation of SVM training seeks [1] to maximize the margin of a separating hyperplane by solving the problem

$$\min \frac{1}{2}\|W\|^2, \ \text{ with } \ y_i(W \cdot X_i + b) \geq 1, i = 1, \ldots, N. \tag{1}$$

There are many algorithms proposed to solve (1) being Platt's SMO or Joachims's SVM–Light the best known (in fact, many other algorithms can be traced to them). Both try to solve the dual problem of (1), that is, to minimize

$$W(\alpha) = \frac{1}{2}\sum_{i,j} \alpha_i\alpha_j y_i y_j X_i \cdot X_j - \sum_i \alpha_i, \ \text{ with } \ \alpha_i \geq 0, \ \sum_i \alpha_i y_i = 0. \tag{2}$$

The optimal weight W^o can be then written in the so–called dual form as $W^o = \sum \alpha_i^o y_i X_i$ and patterns for which $\alpha_i^o > 0$ are called support vectors (SV).

Alternatively, SVM training can be set up [2] to solve the so called Nearest Point Problem (NPP) in which we want to find the nearest points W_+^* and W_-^*

---

[*] All authors have been partially supported by Spain's TIN 2007–66862. The first author is kindly supported by the FPU–MEC grant reference AP2006–02285 and the second by an IIC doctoral fellowship.

V. Kůrková et al. (Eds.): ICANN 2008, Part I, LNCS 5163, pp. 315–324, 2008.
© Springer-Verlag Berlin Heidelberg 2008

of the convex hulls $C(\mathcal{S}_\pm)$ of the positive $\mathcal{S}_+ = \{X_i : y_i = 1\}$ and negative $\mathcal{S}_- = \{X_i : y_i = -1\}$ sample subsets. The maximum margin hyperplane is then $W^* = W_+^* - W_-^*$ and the optimal margin is given by $\|W^*\|/2$. If we write a $W_+ \in C(\mathcal{S}_+)$ as $W_+ = \sum \alpha_p X_p$ with $\sum \alpha_p = 1$ and a $W_- \in C(\mathcal{S}_-)$ as $W_- = \sum \alpha_q X_q$, with $\sum \alpha_q = 1$, we have $W = W_+ - W_- = \sum \alpha_i y_i X_i$ with $X_i \in \mathcal{S} = \mathcal{S}_+ \bigcup \mathcal{S}_-$. We can thus state the NPP problem as follows:

$$\min \frac{1}{2}\|W\|^2, \quad \text{with} \ \ \alpha_i \geq 0, \sum_i \alpha_i y_i = 0, \sum_i \alpha_i = 2, \tag{3}$$

where we assume again a linearly separable training sample. If we restrict ourselves to a zero bias term b, NPP can be further reduced [3] to a Minimum Norm Problem (MNP), where we want to find the vector W in the convex hull $C(\tilde{\mathcal{S}})$ of the set $\tilde{\mathcal{S}} = \{y_i X_i : 1 \geq i \geq N\}$ with the smallest norm. MNP is a much older problem than SVM training and it has received a very large attention, with many different algorithmic proposals. Two of them have been extended to SVM construction, the Gilbert–Schlesinger–Kozinec (GSK) [4,5,3] and the Mitchell–Demyanov–Malozemov (MDM) [6,5]. While the GSK algorithm can be quite slow, the MDM algorithm and some improvements (see [5]) are quite efficient.

The MDM algorithm for MNP iteratively updates $W = \sum \alpha_j y_j X_j$, selecting two vectors $X_L = \arg \min \{y_j W \cdot X_j\}$, $X_U = \arg \max \{y_j W \cdot X_j, \alpha_j > 0\}$ at each step and updating W to $W' = W + \delta(y_L X_L - y_U X_U) = W + \delta Z$, where $Z = y_L X_L - y_U X_U$ and $\delta = \min \{\alpha_U, -W \cdot Z/\|Z\|^2\}$. The rationale for this is that the new vector W' is the one with minimum norm in the line segment $\{W + \lambda Z : \lambda > 0\}$. We can also arrive at the same vector updates following an alternate point of view, in which we select the updating indices L, U so that the new vector W' has essentially the minimum norm among all vectors in $C(\tilde{\mathcal{S}})$ of the form $W' = W + \delta_{i_1} y_{i_1} X_{i_1} + \delta_{i_2} y_{i_2} X_{i_2}$ for any i_1, i_2 choices (see [7]).

This point of view can be also applied when trying to extend the classical MDM algorithm to NPP. More precisely, if we seek to update a current vector W to a new vector of the form $W' = W + \delta_{i_1} y_{i_1} X_{i_1} + \delta_{i_2} y_{i_2} X_{i_2}$ with now X_{i_1}, X_{i_2} in the same class (i.e., $y_{i_1} = y_{i_2}$) and such that the decrease from $\|W\|$ to $\|W'\|$ is maximal among all such choices, we shall show in section 2 that we can select two index pairs (i_{11}, i_{12}), (i_{21}, i_{22}), such that

$$i_{11} = \arg \max_i \{W \cdot X_i : y_i = 1, \alpha_i > 0\}, \quad i_{12} = \arg \min_j \{W \cdot X_j : y_i = 1\},$$
$$i_{21} = \arg \max_i \{W \cdot X_i : y_j = -1\}, \quad i_{22} = \arg \min_j \{W \cdot X_j : y_j = -1, \alpha_j > 0\},$$

compute then $\Delta_1 = W \cdot (X_{i_{11}} - X_{i_{12}})$, $\Delta_2 = W \cdot (X_{i_{21}} - X_{i_{22}})$, and choose $i_1 = i_{11}, i_2 = i_{12}$ if $\Delta_1 > \Delta_2$ and $i_1 = i_{21}, i_2 = i_{22}$ otherwise. These choices, closely related to those in [5,8], result in an update of the form $W' = W + y_{i_1} \delta(X_{i_1} - X_{i_2})$. As the final weight update relies on just two vectors, we shall call the resulting procedure the 2–vector MDM algorithm.

In any case, notice that even if only two vectors are finally used in the MDM updates for NPP, four vectors have to be selected. This suggests the potential interest of another MDM–like procedure where four general vectors are used for the weight updates of the form

$$W' = W + y_1\delta_1 X_1 + y_2\delta_2 X_2 + y_3\delta_3 X_3 + y_4\delta_4 X_4.$$

Moreover, the interpretation of the MDM updating vectors as those achieving a largest minimization of the weight norm suggests that we define weight updates in such a way that the minimization gain in $\|W'\|$ is largest. This is quite difficult in general but the previous 2–vector MDM discussion suggests to try updates of the form

$$W' = W + y_1\delta_1 \left(X_{i_{11}} - X_{i_{12}}\right) + y_2\delta_2 \left(X_{i_{21}} - X_{i_{22}}\right),$$

where we will assume $y_1 = 1 = -y_2$, $X_{i_{11}}, X_{i_{12}} \in \mathcal{S}_+$, $X_{i_{21}}, X_{i_{22}} \in \mathcal{S}_-$. In this setting we shall derive in section 3 a 4–vector MDM–like procedure that gives larger weight norm decreases and margin increases than the 2–vector MDM algorithm and, as a consequence, a procedure for solving NPP requiring substantially less interations while, nevertheless, essentially arriving at the same final SVM models. In any case, the number of iterations is just a measure of the complexity of the final procedure. As usual for SVM algorithms we will also discuss how to kernelize 2– and 4–vector MDM algorithms. When done, it will turn out that 4–vector MDM algorithm needs twice the number of kernel operations per iteration than 2–vector MDM. Thus, unless we cache all the required kernel dot values, 4–vector MDM must require less than half the number of iterations of 2–vector MDM to be competitive, something that may be hard to come by. However, if caching is allowed, we shall argue that the complexity of 4–vector MDM can be estimated at about 1.5 times that of 2–vector MDM. Thus the former would be competitive as long as it requires about 2/3 the number of iterations of the latter, an estimate that even improves if actual running times are considered.

The paper is organized as follows. Section 2 develops the 2–vector MDM algorithm from the viewpoint explained above, section 3 develops the corresponding 4–MDM vector algorithm and they are compared numerically in section 4. The paper ends with a brief discussion.

2   The 2–Vector MDM Algorithm

In this section we will briefly review how to derive a two vector algorithm to solve NPP. Recall that we have to solve (3) assuming an update of the form

$$W' = W + \delta_{i_1} y_{i_1} X_{i_1} + \delta_{i_2} y_{i_2} X_{i_2} \tag{4}$$

The restrictions in (3) imply that $2 = \sum \alpha_i' = \sum \alpha_i + \delta_{i_1} + \delta_{i_2} = 2 + \delta_{i_1} + \delta_{i_2}$ and, also, that $0 = \sum y_i\alpha_i' = \sum y_i\alpha_i + y_{i_1}\delta_{i_1} + y_{i_2}\delta_{i_2} = y_{i_1}\delta_{i_1} + y_{i_2}\delta_{i_2}$. The second one implies that $y_{i_1}\delta_{i_1} = -y_{i_2}\delta_{i_2}$ and, since the first one gives $\delta_{i_2} = -\delta_{i_1}$, we must also have $y_{i_1} = y_{i_2}$. As a consequence, the update (4) becomes $W' = W + \delta_{i_1} y_{i_1} \left(X_{i_1} - X_{i_2}\right) = W + \delta_{i_1} y_{i_1} Z_{i_1,i_2}$, where $Z_{i,j} = X_i - X_j$; thus, once Z_{i_1,i_2} is fixed, $\|W'\|^2$ is a function of δ_{i_1} and we have

$$\Phi(\delta_{i_1}) = \|W'\|^2 = \|W\|^2 + 2\delta_{i_1} y_{i_1} W \cdot Z_{i_1,i_2} + \delta_{i_1}^2 \|Z_{i_1,i_2}\|^2.$$

Hence, $\Phi'(\delta_{i_1}) = 2y_{i_1} W \cdot Z_{i_1,i_2} + 2\|Z_{i_1,i_2}\|^2 \delta_{i_1}$ and solving $\Phi'(\delta_{i_1}^*) = 0$ gives

$$\delta_{i_1}^* = -y_{i_1} \frac{\Delta}{\|Z_{i_1,i_2}\|^2}, \quad \delta_{i_2}^* = -\delta_{i_1}^* = y_{i_1} \frac{\Delta}{\|Z_{i_1,i_2}\|^2}, \tag{5}$$

where we write $\Delta = W \cdot Z_{i_1,i_2} = W \cdot (X_{i_1} - X_{i_2})$ and, in turn,

$$\Phi(\delta_{i_1}^*) = \|W\|^2 - \frac{\Delta^2}{\|Z_{i_1,i_2}\|^2}. \tag{6}$$

If we ignore the $\|Z_{i_1,i_2}\|^2$ denominator, we can maximize the gain by selecting i_1 and i_2 so that Δ is maximized, which we clearly obtain by setting

$$i_{11} = \arg\max_i\{W \cdot X_i : y_i = 1\}, \quad i_{12} = \arg\min_j\{W \cdot X_j : y_j = 1\},$$
$$i_{21} = \arg\max_i\{W \cdot X_i : y_i = -1\}, \quad i_{22} = \arg\min_j\{W \cdot X_j : y_j = -1\},$$

and deciding next which one of the pairs (i_{q1}, i_{q2}), $q = 1, 2$, to choose. To do so we compute the positive values $\Delta_1 = W \cdot (X_{11} - X_{12})$, $\Delta_2 = W \cdot (X_{21} - X_{22})$, and take $i_1 = i_{11}$, $i_2 = i_{12}$ if $\Delta_1 > \Delta_2$ and $i_1 = i_{21}$, $i_2 = i_{22}$ otherwise.

In any case, notice that the resulting updates can be expressed just in terms of the α coefficients as $\alpha'_{i_1} = \alpha_{i_1} + \delta_{i_1}^*$, $\alpha'_{i_2} = \alpha_{i_2} - \delta_{i_1}^*$ while the other α remain unchanged. If we denote the optimal δ for each class as $\delta_{i_{q1}}^*, \delta_{i_{q2}}^*$, $q = 1, 2$, it follows from (5) that $\delta_{i_{11}}^* = -\Delta_1/\|Z_{i_{11},i_{12}}\|^2 < 0$, $\delta_{i_{12}}^* = -\delta_{i_{11}}^* > 0$, and $\delta_{i_{21}}^* = \Delta_2/\|Z_{i_{21},i_{22}}\|^2 > 0$, $\delta_{i_{22}}^* = -\delta_{i_{21}}^* < 0$. Since the new α' multipliers must be positive, it follows that we must ensure that $\alpha_{i_{11}}$ and $\alpha_{i_{22}}$ be positive; in other words, we must refine the previous index choices to

$$i_{11} = \arg\max_i\{W \cdot X_i : y_i = 1, \alpha_i > 0\},$$
$$i_{22} = \arg\min_j\{W \cdot X_j : y_j = -1, \alpha_j > 0\}.$$

Moreover, we may have to clip if necessary the negative increments $\delta_{i_{11}}^*$ and $\delta_{i_{22}}^*$ from below so that we ensure that $\delta_{i_{11}}^* \geq -\alpha_{i_{11}}$ and $\delta_{i_{22}}^* \geq -\alpha_{i_{22}}$. Similarly, we may also have to clip the positive increments $\delta_{i_{12}}^*$ and $\delta_{i_{21}}^*$ from above so that $\delta_{i_{12}}^* < 1 - \alpha_{i_{12}}$ and $\delta_{i_{21}}^* < 1 - \alpha_{i_{21}}$. It is clear from the above discussion that the final procedure only involves dot products and is therefore easy to kernelize. To speed things up we will cache the dot products $W \cdot X_j$ in an N dimensional vector D_j. It is then clear that the D_j can be updated to $D'_j = W' \cdot X_j$ as

$$D'_j = W \cdot X_j + y_{i_1}\delta_{i_1}^* (X_{i_1} \cdot X_j - X_{i_2} \cdot X_j) = D_j + y_{i_1}\delta_{i_1}^* (K(i_1, j) - K(i_2, j)).$$

To initialize the algorithm, we can simply choose two indices i_1, i_2 in the positive and negative classes respectively, set $\alpha_{i_1} = \alpha_{i_2} = 1$ so that we start at $W = y_{i_1}X_{i_1} + y_{i_2}X_{i_2} = X_{i_1} - X_{i_2}$ and initialize the D_j values as $D_j = K(i_1, j) - K(i_2, j)$.

3   The 4–Vector MDM Algorithm

In this section we will explore how to define a four vector extension of 2–vector MDM working with updates of the form

$$W' = W + y_1\delta_1 (X_{i_{11}} - X_{i_{12}}) + y_2\delta_2 (X_{i_{21}} - X_{i_{22}}) = W + y_1\delta_1 Z_1 + y_2\delta_2 Z_2,$$

where $X_{i_{11}}, X_{i_{12}} \in \mathcal{S}_+$, $X_{i_{21}}, X_{i_{22}} \in \mathcal{S}_-$ and we have written $Z_1 = X_{i_{11}} - X_{i_{12}}$, $Z_2 = X_{i_{21}} - X_{i_{22}}$; we will assume from now on $y_1 = 1$, $y_2 = -1$. If we define now $\Phi(\delta_1, \delta_2) = \|W'\|^2$ we have

$$\begin{aligned}
\Phi(\delta_1, \delta_2) &= \|W\|^2 + 2y_1\delta_1 W \cdot Z_1 + 2y_2\delta_2 W \cdot Z_2 + \\
&\quad \delta_1^2\|Z_1\|^2 + 2y_1 y_2 \delta_1 \delta_2 Z_1 \cdot Z_2 + \delta_2^2\|Z_2\|^2 \\
&= \|W\|^2 + 2y_1\delta_1 \Delta_1 + 2y_2\delta_2 \Delta_2 + \delta_1^2 N_1^2 + 2y_1 y_2 \delta_1 \delta_2 N_{12} + \delta_2^2 N_2^2,
\end{aligned}$$

with $\Delta_p = W \cdot Z_p$, $N_p^2 = \|Z_p\|^2$, $p = 1, 2$, and $N_{12} = Z_1 \cdot Z_2$. We then have

$$\frac{\partial \Phi}{\partial \delta_1} = 2\left(y_1\Delta_1 + \delta_1 N_1^2 + y_1 y_2 \delta_2 N_{12}\right), \quad \frac{\partial \Phi}{\partial \delta_2} = 2\left(y_2\Delta_2 + \delta_2 N_2^2 + y_1 y_2 \delta_1 N_{12}\right),$$

and equalling to 0 and solving for the optimal δ_1^*, δ_2^*, we get

$$\begin{aligned}
\delta_1^* &= -y_1\frac{N_2^2\Delta_1 - N_{12}\Delta_2}{N_1^2 N_2^2 - N_{12}^2} = -y_1\frac{C_1}{C} = -\frac{C_1}{C}, \\
\delta_2^* &= -y_2\frac{N_1^2\Delta_2 - N_{12}\Delta_1}{N_1^2 N_2^2 - N_{12}^2} = -y_2\frac{C_2}{C} = \frac{C_2}{C},
\end{aligned} \tag{7}$$

where we have written $C_1 = N_2^2\Delta_1 - N_{12}\Delta_2$, $C_2 = N_1^2\Delta_2 - N_{12}\Delta_1$ and $C = N_1^2 N_2^2 - N_{12}^2$. As a consequence, recalling that $y_{i_1} = 1 = -y_{i_2}$, we have

$$\begin{aligned}
\Phi(\delta_1^*, \delta_2^*) &= \|W\|^2 - 2\frac{\Delta_1 C_1}{C} - 2\frac{\Delta_2 C_2}{C} + \frac{N_1^2 C_1^2}{C^2} + 2\frac{N_{12}C_1 C_2}{C^2} + \frac{N_2^2 C_2^2}{C^2}. \\
&= \|W\|^2 - 2\frac{\Delta_1 C_1 + \Delta_2 C_2}{C} + \frac{N_1^2 C_1^2 + 2N_{12}C_1 C_2 + N_2^2 C_2^2}{C^2}.
\end{aligned}$$

Now it can be easily checked that $\Delta_1 C_1 + \Delta_2 C_2 = N_1^2\Delta_2^2 + N_2^2\Delta_1^2 - 2N_{12}\Delta_1\Delta_2$ and, moreover,

$$\begin{aligned}
C_1^2 &= N_2^4\Delta_1^2 - 2N_2^2 N_{12}\Delta_1\Delta_2 + N_{12}^2\Delta_2^2; \\
C_2^2 &= N_1^4\Delta_2^2 - 2N_1^2 N_{12}\Delta_1\Delta_2 + N_{12}^2\Delta_1^2; \\
C_1 C_2 &= N_1^2 N_2^2\Delta_1\Delta_2 - N_1^2 N_{12}\Delta_2^2 - N_2^2 N_{12}\Delta_1^2 + N_{12}^2\Delta_1\Delta_2 \\
&= \left(N_1^2 N_2^2 + N_{12}^2\right)\Delta_1\Delta_2 - N_1^2 N_{12}\Delta_2^2 - N_2^2 N_{12}\Delta_1^2,
\end{aligned}$$

which, after some easy computations, yield

$$N_1^2 C_1^2 + N_2^2 C_2^2 + 2N_{12}C_1 C_2 = C\left(\Delta_1 C_1 + \Delta_2 C_2\right),$$

and, putting everything together, we arrive at

$$\Phi(\delta_1^*, \delta_2^*) = \|W\|^2 - \frac{N_2^2\Delta_1^2 + N_1^2\Delta_2^2 - 2N_{12}\Delta_1\Delta_2}{C}.$$

Forgetting about the C denominator, we would like thus to maximize

$$J(\delta_1, \delta_2) = N_2^2\Delta_1^2 + N_1^2\Delta_2^2 - 2N_{12}\Delta_1\Delta_2.$$

However, this is a fairly complicated expression for which computing optimal choices of the underlying i_{pq} indices would be quite costly. The option we will take here is to apply the choices made for the 2–vector MDM algorithm applying them here as

$$i_{11} = \min\{W \cdot X_j : y_j = +1\}; \quad i_{12} = \max\{W \cdot X_j : y_j = +1\};$$
$$i_{21} = \max\{W \cdot X_j : y_j = -1\}; \quad i_{22} = \min\{W \cdot X_j : y_j = -1\}.$$

Notice that we have switched the previous i_{11} and i_{12} choices, so that we have now $\Delta_1 < 0$, $\Delta_2 > 0$. The reason for this is to reasonably ensure that both δ_1^*, δ_2^* are positive. In fact, we may expect to have $|\Delta_1| \simeq |\Delta_2|$ and, therefore, we should have $C\delta_1^* \simeq |\Delta_1|(N_2^2 + N_{12}) > 0$ and, similarly, $C\delta_2^* > 0$ (clearly $C > 0$). Although the preceding argument is not a proof, we may expect to arrive at positive increases δ_i^* (in fact, in our experiments a negative increase is extremely rare and if this the case, we revert to a 2–vector MDM update). This, in turn, will simplify the clipping that must also be applied here: the α coefficient updates are now

$$\alpha'_{11} = \alpha_{11} + \delta_1^*, \quad \alpha'_{12} = \alpha_{12} - \delta_1^*, \quad \alpha'_{21} = \alpha_{21} + \delta_2^*, \quad \alpha'_{22} = \alpha_{22} - \delta_2^*, \qquad (8)$$

and to ensure $\alpha_{12} > 0$ and $\alpha_{22} > 0$ so that we can expect $\alpha'_{12} \geq 0$ and $\alpha'_{22} \geq 0$, we refine here the previous selection of i_{12}, i_{22} to

$$i_{12} = \max\{W \cdot X_j : y_j = +1, \alpha_j > 0\}, \quad i_{22} = \min\{W \cdot X_j : y_j = -1, \alpha_j > 0\},$$

and, as before, clip if necessary δ_1^* by the positive α_{12} and δ_2^* by the positive α_{22}. The same initialization of the 2–vector algorithm can be applied here and we can easily kernelize the overall procedure caching again the dot products $D_j = W \cdot X_j$ and updating them as

$$D'_j = D_j + \delta_1^* \left(k(i_{11}, j) - k(i_{12}, j) \right) - \delta_2^* \left(k(i_{21}, j) - k(i_{22}, j) \right).$$

4   Numerical Experiments

We shall compare the performance of the basic 2– and 4–vector MDM algorithms over several datasets taken from the UCI repository but before that we briefly comment some implementation details. To work in non–linearly separable problems we shall allow margin slacks and a quadratic penalty function. That is, we shall solve

$$\min \frac{1}{2}\|W\|^2 + C\sum \xi_i^2 \quad \text{with} \quad y_i(W \cdot X_i + b) \geq 1 - \xi_i, i = 1, \ldots, N.$$

As it is well known [9], the linearly separable case extends easily to the quadratic penalty setting simply working with extended vectors and kernels. In all cases we have used Gaussian kernels $\exp\left(-\|x\|^2/2\sigma^2\right)$ with $2\sigma^2 = 50$ and $C = 10$. While

Table 1. 10–times 10–fold cross–validation average test accuracies for the 2– and 4–vector MDM algorithms with precision $\epsilon = 0.05$

Dataset	2–v. MDM acc.	4–v. MDM acc.
Australian	85.83 ± 4.35	85.83 ± 4.40
Breast-W	96.66 ± 2.18	96.63 ± 2.20
Heartdis	80.47 ± 7.00	80.36 ± 7.13
Ionosphere	95.07 ± 3.41	95.12 ± 3.44
Pima	76.95 ± 4.50	76.87 ± 4.28
Ripley	88.88 ± 2.84	88.89 ± 2.78
Satimage	93.81 ± 0.82	93.82 ± 0.80
Sonar	87.62 ± 7.82	87.57 ± 8.01
Thyroid	97.42 ± 0.49	97.43 ± 0.49
Vehicle	97.52 ± 1.74	97.50 ± 1.72

Table 2. 10–times 10–fold cross–validation average number of iterations for the 2– and 4–vector MDM algorithms with precision $\epsilon = 0.05$ and reduction achieved by 4–vector MDM

Dataset	2–v. MDM # iters.	4–v. MDM # iters.	% 4–v iters.
Australian	1027 ± 36	624 ± 27	60.76
Breast-W	239 ± 18	155 ± 14	64.85
Heartdis	486 ± 20	296 ± 15	60.91
Ionosphere	243 ± 14	150 ± 10	61.73
Pima	1626 ± 35	1100 ± 32	67.65
Ripley	830 ± 20	492 ± 13	59.28
Satimage	5473 ± 103	4176 ± 132	76.30
Sonar	236 ± 8	134 ± 5	56.78
Thyroid	2593 ± 85	1824 ± 67	70.34
Vehicle	492 ± 15	339 ± 15	68.90

possibly not optimal, these parameter choices give reasonable test accuracies and, therefore, are adequate enough for the comparisons to be made here.

To define a stopping condition for both algorithms, notice that the optimal margin m^* coincides with $\|W^*\|/2$. Thus, the values $\|W\|/2$ obtained during training are upper bounds for m^*. Moreover, if b is the bias term associated with W, the margin $m(W, b)$ for W, b can be computed as

$$m(W, b) = \min \left\{ \frac{y_i(W \cdot X_i + b)}{\|W\|} \right\} = \min \left\{ \frac{y_i(D_i + b)}{\|W\|} \right\}$$

and is a lower bound for m^*. We shall thus use the following stopping condition $\frac{1}{2}\|W\| - m(W, b) \leq \frac{\epsilon}{2}\|W\|$ that guarantees that the margin is computed with precision ϵ.

We shall consider two ϵ values, $\epsilon = 0.05$ and the more precise $\epsilon = 0.0005$. Tables 1 and 2 give respectively the average accuracies achieved and the number

Table 3. 10–times 10–fold cross–validation average number of iterations for the 2– and 4–vector MDM algorithms with precision $\epsilon = 0.0005$ and reduction achieved by 4–vector MDM

DATASET	2–V. MDM # ITERS.	4–V. MDM # ITERS.	% RED.
AUSTRALIAN	2637 ± 88	1673 ± 59	63.44
BREAST-W	571 ± 46	396 ± 38	69.35
HEARTDIS	1188 ± 52	746 ± 35	62.79
IONOSPHERE	623 ± 35	416 ± 29	66.77
PIMA	3790 ± 79	2692 ± 61	71.03
RIPLEY	1608 ± 37	1025 ± 28	63.74
SATIMAGE	13432 ± 386	10565 ± 511	78.66
SONAR	539 ± 18	312 ± 11	57.88
THYROID	6409 ± 207	4899 ± 163	76.44
VEHICLE	1331 ± 38	963 ± 38	72.35

of iterations required for the 2– and 4–vector MDM algorithms over a 10 × 10–cross validation at the 0.05 precision. As it can be seen the final accuracies are essentially the same; in particular they are not significantly different under a paired sample t test. On the other hand, the 4–th column in table 2 shows that 4–vector MDM requires substantially less iterations, from about 56.78 % of those of 2–vector MDM in the Sonar problem to 76.30 % for the Satimage dataset. The models obtained at the stricter $\epsilon = 0.0005$ precision are basically the same: their final average accuracies (not shown) are essentially equal, they differ in a very few support vectors and their α coefficients have a correlation essentially of 1. Here, as shown in table 3, 4–vector still requires less iterations, although the reduction is now smaller.

In any case, the number of iterations is not a precise way of comparing algorithm complexity. In fact, the complexities of the 2– and 4– vector algorithms are comparable, except for the D_j update loops over the whole training sample, for which the cost of 4–vector MDM is twice as big, either with respect to kernel operations or, if these are cached, with respect to floating point operations. If no caching is applied, the cost of 4–vector MDM will be twice that of 2–vector MDM, something that the smaller number of iterations that the former needs is not enough to compensate. If kernel operations are cached, the cost of the D_j 4–vector MDM update loop is still twice that of 2–vector MDM, but both also need an extra training sample loop to compute the indices to be used. Taking this into account, the cost of 4–vector MDM would be now about 1.5 times larger than that of 2–vector MDM or, in other words, 4–vector MDM would be faster than the 2–vector algorithm if the former needs less than about 66.6% of the iterations the latter performs. Looking at tables 2 and 3 under this light, we see that, for the 0.05 precision, 4–vector MDM can be considered faster on 6 out of 10 datasets, while for the the 0.05 precision, it would be faster in only 4 datasets. However, when running times are measured, it turns out that 4–vector MDM's advantage increases. Table 4 shows the running time in seconds of 10000 training iterations over the full datasets on a Intel Core 2 processor at 2.00 GHz. As it

Table 4. Running times in seconds of 10000 training iterations over the full datasets

Dataset	2–v. MDM	4–v. MDM	Ratio
Australian	3516	4469	1.271
Breast	3234	4328	1.338
Heart	1437	1891	1.316
Ionosph	1484	2046	1.379
Pima	3781	4719	1.248
Ripley	5609	7375	1.315
Satimage	30473	40475	1.328
Sonar	891	1203	1.350
Thyroid	32814	43286	1.319
Vehicle	4031	5250	1.302

can be seen, a single 4–vector MDM update takes in average 1.317 more time than a 2–vector MDM one. Thus, 4–vector MDM will be faster than 2–vector MDM whenever the former requires 75.9% or less iterations than the latter. When the table values are used for comparison, it turns out that 2–vector MDM is only faster arriving at the 0.05 precision on the Satimage problem and on the Satimage and Thyroid datasets (both are the largest) at the 0.0005 precision.

5   Discussion

In this work we have shown first how to introduce the 2–vector MDM algorithm typically used to solve the Nearest Point Problem (which is equivalent to standard SVM training) when we seek updating patterns that maximize the gain in the NPP norm minimization. While 2–vector MDM only uses two patterns to update the SVM weight vector, it actually selects four, two from each class. We have also considered an extension of standard 2–vector MDM where a 4–pattern weight update takes place. Once the updating patterns are chosen, we have derived the exact update weights that maximize the gain in the NPP norm minimization. We have also derived that gain's value as a function of the updating patterns but the resulting expression seems to be too involved for direct minimization. A reasonably approximation is to choose the desired patterns as those involved in 2–vector SVM. As we have demonstrated numerically, the resulting procedure, that we have called 4–vector MDM, needs substantially less training iterations than 2–vector MDM. However, this may not necessarily translate into a faster procedure, as 4–vector MDM needs twice as many kernel operations than its 2–vector counterpart. On the other hand, if kernel operations are cached, the cost of a single 4–vector update should be about 1.5 times that of a 2–vector one and even less if actual running times are considered. Further work to improve the performance of 4–vector MDM would involve better heuristics to choose the updating vectors or to combine its costlier updates with the faster ones of the 2–vector version.

References

1. Schölkopf, B., Smola, A.: Learning with kernels support vector machines, regularization, optimization, and beyond. MIT Press, Cambridge (2002)
2. Bennett, K., Bredensteiner, E.: Duality and geometry in SVM classifiers. In: Proc. 17th Int. Conf. Machine Learning, pp. 57–64 (2000)
3. Franc, V., Hlaváč, V.: An iterative algorithm learning the maximal margin classiffier. Pattern Recognition 36, 1985–1996 (2003)
4. Gilbert, E.: Minimizing the quadratic form on a convex set. SIAM J. Contr. 4, 61–79 (1966)
5. Keerthi, S., Shevade, S., Bhattacharyya, C., Murthy, K.: A fast iterative nearest point algorithm for support vector machine classifier design. IEEE Transactions on Neural Networks 11(1), 124–136 (2000)
6. Mitchell, B., Dem'yanov, V., Malozemov, V.: Finding the point of a polyhedron closest to the origin. SIAM J. Contr. 12, 19–26 (1974)
7. Barbero, A., López, J., Dorronsoro, J.: An accelerated mdm algorithm for svm training. In: Advances in Computational Intelligence and Learning, Proceedings of ESANN 2008 Conference, pp. 421–426 (2008)
8. Franc, V., Hlaváč, V.: Simple solvers for large quadratic programming tasks. In: Kropatsch, W.G., Sablatnig, R., Hanbury, A. (eds.) DAGM 2005. LNCS, vol. 3663, pp. 75–84. Springer, Heidelberg (2005)
9. Cristianini, N., Shawe-Taylor, J.: An Introduction to Support Vector Machines and Other Kernel-Based Learning Methods. Cambridge University Press, Cambridge (2000)

Implementation Issues of an Incremental and Decremental SVM

Honorius Gâlmeanu[1] and Răzvan Andonie[2]

[1] Electronics and Computers Department
Transilvania University, Braşov, România
galmeanu@vega.unitbv.ro
[2] Computer Science Department
Central Washington University, Ellensburg, USA
andonie@cwu.edu

Abstract. Incremental and decremental processes of training a support vector machine (SVM) resumes to the migration of vectors in and out of the support set along with modifying the associated thresholds. This paper gives an overview of all the boundary conditions implied by vector migration through the incremental / decremental process. The analysis will show that the same procedures, with very slight variations, can be used for both the incremental and decremental learning. The case of vectors with duplicate contribution is also considered. Migration of vectors among sets on decreasing the regularization parameter is given particularly attention. Experimental data show the possibility of modifying this parameter on a large scale, varying it from complete training (overfitting) to a calibrated value.

1 Introduction

In the last decade, datasets have grown to practically infinite sizes. Learning algorithms must process increasing amounts of data using comparatively smaller computing resources. Our learning algorithms do not scale well enough to take advantage of such large datasets. Incremental learning is an attractive solution to this problem. Incremental algorithms may asymptotically outperform learning algorithms that operate by repetitively sweeping over a training set [1]. Systems for incremental learning have been proposed for different standard neural architectures, including support vector machines.

The SVM model has gained a large recognition and two of the most popular implementations are SVMLight ([19]) and LIBSVM([2]). Several authors dealt with the problem of incrementally training a SVM, using various strategies ([17,9,14]). Other authors optimized the SVM parameters by online procedures ([20]). Incremental learning presented by [8] and further extended in [5], opened the opportunity of learning new data while benefiting from the previous knowledge, combined with decremental learning which also allows selective discarding of patterns without losing any additional information.

V. Kůrková et al. (Eds.): ICANN 2008, Part I, LNCS 5163, pp. 325–335, 2008.

The problem of classifying a set of patterns $(x_i, y_i) \in \mathbb{R}^m \times [-1, 1]$ where $i = 1 \ldots N$ for a SVM of the form $g(x_i) = w \cdot \phi(x_i) + w_0$ requires the minimization of

$$\min_{w, w_0, \xi} J(w) = \frac{1}{2} \|w\|^2 + C \sum_{i=1}^{N} \xi_i \tag{1}$$

obeying the constraints

$$y_i(w \cdot \phi(x_i) + w_0) \geq 1 - \xi_i , \quad \text{for } \xi_i \geq 0, \; i = 1 \ldots N \tag{2}$$

The set of slack variables ξ_i are designed to allow the system to cope with incorrectly classified patterns, or patterns that are classified correctly but situated within the separation boundary. The constraint problem is usually expressed using Wolfe representation in its dual form ([18]). Adopting $Q_{ij} = y_i y_j K(x_i, x_j)$ as a more compact way of writing the invariant product, the Karush-Kuhn-Tucker conditions state the following necessary conditions:

$$w = \sum_{i=1}^{N} \lambda_i y_i \phi(x_i) \; , \; \sum_{i=1}^{N} \lambda_i y_i = 1 \; \text{ and } \; 0 \leq \lambda_i \leq C \tag{3}$$

where the regularization parameter C limits the thresholds λ_i. The separating hyperplane given by SVM and the characteristic gradients are:

$$g(x_i) = \sum_j \lambda_j \frac{Q_{ij}}{y_i} + w_0 \; \text{ and } \; h_i = \sum_j \lambda_j Q_{ij} + w_0 y_i - 1 \tag{4}$$

In this context, an exact procedure of adiabatic training has been given in [8]. The procedure is extended in [5], allowing for batch or individual incremental learning or unlearning of patterns. An analysis of an efficient implementation for individual learning of Cauwenberghs&Poggio (CP) algorithm is presented in [15], along with a similar algorithm for one-class learning. The CP algorithm was also extended for regressions ([13,12]).

In all these incremental SVM approaches, the SVM initialization and vector migration through learning / unlearning is not thoroughly discussed, leading to implementation difficulties. In our paper, we try to deal with this issue as follows. In Section 2, we present the relations that should be iteratively evaluated during the learning phase. In Section 3, we present all possible vector migrations between sets, showing that the procedures of vector learning and unlearning can be coupled together. The problem of detecting duplicate contributions of patterns is analyzed in Section 4. Section 5 describes the initial solution and the system start-up. In Section 6 we discuss the details of decrementing the regularization parameter. Sections 7 and 8 contain experimental results and conclusions.

2   Incremental Updates

From the KKT conditions and considering (4), the gradient can be expressed as:

$$h_i = y_i g(x_i) - 1 = \begin{cases} > 0 & \text{for} \quad \lambda_i = 0 \\ = 0 & \text{for} \quad 0 < \lambda_i < C \\ < 0 & \text{for} \quad \lambda_i = C \end{cases} \tag{5}$$

A pattern x from training set D can be accommodated into one of the following categories:

- $x_i \in S \subset D$, where S is the set of *support vectors* situated on the separating hyperplane, for which $h_i = 0$ and associated non-null and unbounded threshold value λ_i;
- $x_i \in O \subset D$, where O is the set of *other vectors*, for which $y_i g(x_i) > 1$, x_i is correctly classified, and has associated threshold value $\lambda_i = 0$;
- $x_i \in E \subset D$, where E is the set of *error vectors*, for which $y_i g(x_i) < 1$, x_i, incorrectly classified or correctly classified but within the boundary region, and having an associated non-null threshold value bounded by the regularization parameter C.

The set $R = \{O \cup E\}$ is the set of *reserve vectors*. Notations of s, r, e are used to refer a specific support, reserve, or error vector.

The CP algorithm relies on the idea of introducing a new vector x_c, and then migrate specific vectors between the sets to have the KKT conditions satisfied again ([5,15]). At first, a new vector to be introduced in the system has initial threshold $\lambda_c = 0$. This threshold is progressively increased, watching, with every increase, the migration of vectors between the sets. Migration is identified by considering the variation of gradients h_i. This variation between initial and final states, that fulfill the KKT conditions, along with thresholds condition, can be written in a compact form as ([15]):

$$
\begin{bmatrix} \Delta h_S \\ \Delta h_R \\ \Delta h_c \\ 0 \end{bmatrix} = \begin{bmatrix} y_S & Q_{SS} \\ y_R & Q_{RS} \\ y_c & Q_{cS} \\ 0 & y_S^T \end{bmatrix} \begin{bmatrix} \Delta w_0 \\ \Delta \lambda_S \end{bmatrix} + \Delta \lambda_c \begin{bmatrix} Q_{cS}^T \\ Q_{cR}^T \\ Q_{cc} \\ y_c \end{bmatrix} \tag{6}
$$

It can be seen that any modification of $\Delta \lambda_c$ should be absorbed by the modification of $\Delta \lambda_s$, Δw_0 and/or the variation of gradients. For support vectors, $\Delta h_s = 0$, since the gradients for support vectors should remain zero, so from the first and the last lines one can solve:

$$
\begin{bmatrix} \Delta w_0 \\ \Delta \lambda_S \end{bmatrix} = - \underbrace{\underbrace{\begin{bmatrix} 0 & y_S^T \\ y_S & Q_{SS} \end{bmatrix}^{-1}}_{P} \begin{bmatrix} y_c \\ Q_{cS}^T \end{bmatrix} \Delta \lambda_c}_{\beta} \quad \text{and} \quad \begin{bmatrix} \Delta h_R \\ \Delta h_c \end{bmatrix} = \underbrace{\begin{bmatrix} y_R & Q_{RS} \\ y_c & Q_{cS} \end{bmatrix} \beta + \begin{bmatrix} Q_{cR}^T \\ Q_{cc} \end{bmatrix}}_{\gamma} \Delta \lambda_c
$$

$$\tag{7}$$

Details can be found in [5,15].

3   Migration between Sets and Learning

The authors previously referred do not perform an exhaustive and specific discussion of the conditions for detecting migration of vectors between sets. We present here a detailed discussion, involving both incremental and decremental algorithms.

1. **Migration of support vectors**
   Consider the relation $\Delta\lambda_s = \beta_s\Delta\lambda_c$, where β_s is the s-th component of vector β. $\Delta\lambda_s$ can reach the following limits:
   - **upper limits** are reached for $\Delta\lambda_s \leq C - \lambda_s$, which means β_s and $\Delta\lambda_c$ having the same sign (and the support vector should migrate to E):

$$
\begin{array}{lll}
\textbf{incremental case} & \beta_s\Delta\lambda_c \leq C - \lambda_s & \\
\Delta\lambda_c > 0 \,,\, \beta_s > 0 & \Delta\lambda_c \leq \frac{C-\lambda_s}{\beta_s} & \delta_{sp} = \min_{s\in S}\left\{\frac{C-\lambda_s}{\beta_s}\right\}
\end{array}
$$

$$
\begin{array}{lll}
\textbf{decremental case} & \beta_s\Delta\lambda_c \leq C - \lambda_s & \\
\Delta\lambda_c < 0 \,,\, \beta_s < 0 & \Delta\lambda_c \geq \frac{C-\lambda_s}{\beta_s} & \delta_{sp} = \max_{s\in S}\left\{\frac{C-\lambda_s}{\beta_s}\right\}
\end{array}
$$

   - **lower limit** is reached for $\Delta\lambda_s \geq -\lambda_s$, which means β_s and $\Delta\lambda_c$ should have opposite signs (and the support vector should migrate to O):

$$
\begin{array}{lll}
\textbf{incremental case} & \beta_s\Delta\lambda_c \geq -\lambda_s & \\
\Delta\lambda_c > 0 \,,\, \beta_s < 0 & \Delta\lambda_c \leq \frac{-\lambda_s}{\beta_s} & \delta_{sm} = \min_{s\in S}\left\{\frac{-\lambda_s}{\beta_s}\right\}
\end{array}
$$

$$
\begin{array}{lll}
\textbf{decremental case} & \beta_s\Delta\lambda_c \geq -\lambda_s & \\
\Delta\lambda_c < 0 \,,\, \beta_s > 0 & \Delta\lambda_c \geq \frac{-\lambda_s}{\beta_s} & \delta_{sm} = \max_{s\in S}\left\{\frac{-\lambda_s}{\beta_s}\right\}
\end{array}
$$

2. **Migration of reserve vectors**
   Consider the relation $\Delta h_r = \gamma_r\Delta\lambda_c$, where γ_r is the r-th component of vector γ. Δh_r can reach zero on the following cases:
   - **other vectors**, $r \in O$, $h_r > 0$, so migration to S takes place when $\Delta h_r < 0$ (γ_r and $\Delta\lambda_c$ of different signs):

$$
\begin{array}{lll}
\textbf{incremental case} & \gamma_r\Delta\lambda_c \geq -h_r & \\
\Delta\lambda_c > 0 \,,\, \gamma_r < 0 & \Delta\lambda_c \leq \frac{-h_r}{\gamma_r} & \delta_{ro} = \min_{r\in O}\left\{\frac{-h_r}{\gamma_r}\right\}
\end{array}
$$

$$
\begin{array}{lll}
\textbf{decremental case} & \gamma_r\Delta\lambda_c \geq -h_r & \\
\Delta\lambda_c < 0 \,,\, \gamma_r > 0 & \Delta\lambda_c \geq \frac{-h_r}{\gamma_r} & \delta_{ro} = \max_{r\in O}\left\{\frac{-h_r}{\gamma_r}\right\}
\end{array}
$$

   - **error vectors**, $r \in E$, $h_r < 0$, so migration to S takes place when $\Delta h_r > 0$ (γ_r and $\Delta\lambda_c$ have the same sign):

$$
\begin{array}{lll}
\textbf{incremental case} & \gamma_r\Delta\lambda_c \leq -h_r & \\
\Delta\lambda_c > 0 \,,\, \gamma_r > 0 & \Delta\lambda_c \leq \frac{-h_r}{\gamma_r} & \delta_{re} = \min_{r\in E}\left\{\frac{-h_r}{\gamma_r}\right\}
\end{array}
$$

$$
\begin{array}{lll}
\textbf{decremental case} & \gamma_r\Delta\lambda_c \leq -h_r & \\
\Delta\lambda_c < 0 \,,\, \gamma_r < 0 & \Delta\lambda_c \geq \frac{-h_r}{\gamma_r} & \delta_{re} = \max_{r\in E}\left\{\frac{-h_r}{\gamma_r}\right\}
\end{array}
$$

3. **Gradient** h_c for the current vector reaches zero. Usually a new vector with positive gradient is classified straightforward as belonging to S, so it will not be trained. When is considered, $h_c < 0$ so $\gamma_c\Delta\lambda_c = \Delta h_c \leq -h_c$, which is positive, so zero can be reached when γ_c and $\Delta\lambda_c$ have the same sign:

$$
\begin{array}{lll}
\textbf{incremental case} & \gamma_c\Delta\lambda_c \leq -h_c & \\
\Delta\lambda_c > 0 \,,\, \gamma_c > 0 & \Delta\lambda_c \leq \frac{-h_c}{\gamma_c} & \delta_c = \left\{\frac{-h_c}{\gamma_c}\right\}
\end{array}
$$

$$
\begin{array}{ll}
\textbf{decremental case} & \text{the objective is not to modify} \\
\Delta\lambda_c < 0 \,,\, \gamma_c < 0 & h_c \text{ but the decrease of } \lambda_c
\end{array}
$$

4. **Threshold λ_c:**

incremental case
λ_c should not overflow C $\delta_{ma} = C - \lambda_c$

decremental case
λ_c should not underrun 0 $\delta_{mi} = \quad -\lambda_c$

The maximum increment / decrement will be calculated as:

incremental case $\Delta\lambda_{max} = \min\{\delta_{sp}, \delta_{sm}, \delta_{ro}, \delta_{re}, \delta_c, \delta_{ma}\}$

decremental case $\Delta\lambda_{min} = \max\{\delta_{sp}, \delta_{sm}, \delta_{ro}, \delta_{re}, \delta_{mi}\}$

Once determined the minimum / maximum, the specific vectors whose s or r indices were this way determined will migrate between the sets as described by various authors ([8,5,15]).

4 Duplicate Contribution of Patterns

The introduction into the support set S of a vector with identical contribution with an existing one would make the matrix P non-invertible. The relations detailed so far would no longer work. We will show that there is no need of an exhaustive search over the set of patterns $\{S \cup R\}$ to find a pattern with identical contribution.

A pattern x_i would have identical contribution with a pattern x_j if we can verify that:

$$Q_{is} = Q_{js}\,, \quad i \neq j\,, \qquad \text{where} \quad s \in S \tag{8}$$

and also that $y_i = y_j$. This would render duplicate lines / columns in the P matrix, which will ruin its invertibility. The best opportunity to detect patterns with identical contribution is to compare the patterns selected to migrate between the sets at the current iteration. We have no concern for patterns with duplicate contribution contained in E or O. We consider the case of two vectors, x_i and x_j trying to enter the set S:

$$\Delta h_i = \gamma_i \Delta\lambda_c = 0 - h_i = \lambda_S Q_{iS} + w_0 y_i - 1 \tag{9}$$

$$\Delta h_j = \gamma_j \Delta\lambda_c = 0 - h_j = \lambda_S Q_{jS} + w_0 y_j - 1 \tag{10}$$

The gradients, $h_i = h_j$, of the vectors that could try to enter simultaneously into the solution could indicate that duplicate contribution vectors are detected. For these, we suggest performing the following verifications during the training phase:

– for a new, yet unlearned vector, if its gradient is determined to be initially zero, check whether it does not have a duplicate contribution with a support vector (which has, by definition, a zero gradient);

- determine if there are more than one minima with the same value, one of which could be δ_c. This means the current vector could have duplicate contribution with that $x_r \in R$ which has the same minima. The current vector should be unlearned, until its λ_c reaches zero again, then disregarded;
- determine if there are more than one minima, containing δ_{ro} or δ_{re} values. If two or more are found, it means that two $x_o \in O$ (or two $x_e \in E$) vectors try to enter the set S. The duplicate x_o vectors will be discarded; for duplicate x_e, current x_c is unlearned until λ_c reaches zero. Then duplicate x_e are unlearned. We would resume learning of current x_c vector afterward.

5   Initial Solution

To start the incremental learning, one would need an initial set of support vectors. Choosing two patterns from opposite classes ($y_1 = -y_2$), we can determine λ_1, λ_2 and w_0 such that:

$$0 = h_1 = Q_{11}\lambda_1 + Q_{12}\lambda_2 + w_0 y_1 - 1 \tag{11}$$
$$0 = h_2 = Q_{21}\lambda_1 + Q_{22}\lambda_2 + w_0 y_2 - 1 \tag{12}$$
$$0 = \lambda_1 y_1 + \lambda_2 y_2 \tag{13}$$

The initial solution can be expressed as:

$$\lambda_1 = \lambda_2 = \frac{2}{Q_{11} + Q_{12} + Q_{21} + Q_{22}} \quad \text{and} \quad w_0 = \frac{Q_{22} - Q_{11}}{Q_{11} + Q_{12} + Q_{21} + Q_{22}} \cdot \frac{1}{y_1}$$

The initial solution could not obey the initial regularization parameter C given by the problem. We will use the decrement procedure given below to re-establish the initial premises.

6   Decrement of Regularization Parameter C

Initial solution puts a constraint of threshold limit parameter C, which should be no less than the calculated λ's. It is possible to start the algorithm with a greater C than the result given by initial solution. The threshold C can also be decreased, as suggested by the original Cauwenberghs and Poggio algorithm. Let us analyze this approach in the following.

The decrement of C can be regarded similar to the previous adiabatic transformations. When trying to maintain the KKT conditions, the variation of gradients can be written as:

$$\Delta h_s = \sum_{j \in S} Q_{sj} \Delta \lambda_j + \sum_{j \in E} Q_{sj} \Delta C + \Delta w_0 y_s \quad s \in S \tag{14}$$

$$\Delta h_r = \sum_{j \in S} Q_{rj} \Delta \lambda_j + \sum_{j \in E} Q_{rj} \Delta C + \Delta w_0 y_r \quad r \in R \tag{15}$$

$$0 = \sum_{s\ inS} \Delta \lambda_s y_s + \sum_{j \in E} \Delta C y_j \tag{16}$$

These relations can be written more compact as:

$$\begin{bmatrix} \Delta h_S \\ \Delta h_R \\ 0 \end{bmatrix} = \begin{bmatrix} y_S & Q_{SS} \\ y_R & Q_{RS} \\ 0 & y_S^T \end{bmatrix} \begin{bmatrix} \Delta w_0 \\ \Delta \lambda_S \end{bmatrix} + \Delta C \begin{bmatrix} \sum_{j \in E} Q_{Sj} \\ \sum_{j \in E} Q_{Rj}^T \\ \sum_{j \in E} y_j \end{bmatrix} \tag{17}$$

For support vectors $\Delta h_s = 0$, from the first and the last lines we can solve:

$$\begin{bmatrix} \Delta w_0 \\ \Delta \lambda_S \end{bmatrix} = - \underbrace{\begin{bmatrix} 0 & y_S^T \\ y_S & Q_{SS} \end{bmatrix}^{-1} \begin{bmatrix} \sum_{j \in E} y_j \\ \sum_{j \in E} Q_{Sj} \end{bmatrix}}_{\beta^e} \Delta C \tag{18}$$

Additionally, the gradients for reserve vectors can be expressed as:

$$\Delta h_R = \underbrace{\left(\begin{bmatrix} y_r & Q_{rS} \end{bmatrix} \beta^e + \sum_{j \in E} Q_{rj} \right)}_{\gamma^e} \Delta C \tag{19}$$

When decrementing C, the same discussion applies, with some slight differences.

1. **Migration of support vectors**

   For the decremental approach ($\Delta C < 0$), $\Delta \lambda_s = \beta_s^e \Delta C$, modifications of support vector thresholds should satisfy:

   $$0 \leq \lambda_s + \Delta \lambda_s \leq C + \Delta C \quad \text{or} \quad 0 \leq \lambda_s + \beta_s^e \Delta C \leq C + \Delta C \tag{20}$$

   – when $\beta_s^e < 0$, the expression is always positive so only superior limit works:

   $$(\beta_s^e - 1)\Delta C \leq C - \lambda_s \ \text{ or } \ \Delta C \geq \frac{C - \lambda_s}{\beta_s^e - 1} \quad \text{so find } \delta_{sp} = \min_{s \in S} \left\{ \frac{C - \lambda_s}{\beta_s^e - 1} \right\} \tag{21}$$

   – when $0 \leq \beta_e < 1$, we have two relations that must be satisfied simultaneously:

   $$\begin{matrix} 0 \leq \Delta \lambda_s + \lambda_s \\ \beta_s^e \Delta C + \lambda_s \geq 0 \end{matrix} \quad \Delta C \geq -\frac{\lambda_s}{\beta_s^e} \quad \text{so find } \delta_{slm} = \min_{s \in S} \left\{ -\frac{\lambda_s}{\beta_s^e} \right\}$$

   $$\begin{matrix} \Delta \lambda_s + \lambda_s \leq C + \Delta C \\ \beta_s^e \Delta C \leq C + \Delta C \end{matrix} \quad \Delta C \geq \frac{C - \lambda_s}{\beta_s^e - 1} \quad \text{so find } \delta_{slp} = \min_{s \in S} \left\{ \frac{C - \lambda_s}{\beta_s^e - 1} \right\} \tag{22}$$

   – when $\beta_s^e > 1$, only the first constraint works, the expression can be written as:

   $$\lambda_s + \beta_s^e \Delta C \geq 0 \ \text{ or } \ \Delta C \geq -\frac{\lambda_s}{\beta_s^e} \quad \text{so find } \delta_{sm} = \min_{s \in S} \left\{ -\frac{\lambda_s}{\beta_s^e} \right\} \tag{23}$$

2. **Migration of reserve vectors**

   The decremental approach means $\Delta C < 0$. Since $\gamma_r^e \Delta C = \Delta h_r$, modifications of gradients h_r should satisfy:

$$\textbf{other vectors } r \in O \quad \gamma_r^e \Delta C \geq -h_r$$
$$\Delta h_r \leq 0 \text{ thus } \gamma_r^e > 0 \quad \Delta C \geq -\frac{h_r}{\gamma_r^e} \quad \delta_{ro} = \min_{r \in O} \left\{ -\frac{h_r}{\gamma_r^e} \right\} \quad (24)$$

$$\textbf{error vectors } r \in E \quad \gamma_r^e \Delta C \leq -h_r$$
$$\Delta h_r \geq 0 \text{ thus } \gamma_r^e < 0 \quad \Delta C \geq -\frac{h_r}{\gamma_r^e} \quad \delta_{re} = \min_{r \in E} \left\{ -\frac{h_r}{\gamma_r^e} \right\} \quad (25)$$

3. **Limit decrease of C**

   Decreasing of C should stop when reaching zero so $C + \Delta C \geq 0$ which brings:

$$\delta_{ma} = -C \tag{26}$$

Determining the first migration means finding the maximum decrease of C:

$$\Delta \lambda_{max} = \min \left\{ \delta_{sp}, \delta_{sm}, \delta_{slp}, \delta_{slm}, \delta_{ro}, \delta_{re}, \delta_{ma} \right\} \tag{27}$$

7 Experimental Results

To illustrate the relations presented so far we constructed an incremental and decremental SVM machine[1] based on CP algorithm[2], where we considered all the relations described. We have trained the system on the USPS ([10]) dataset, which contains 7291 training and 2007 testing patterns, with 256 features. Since this is a multiple-class dataset (10 classes), we have used a randomly selected subset of 1925 training vectors and 557 test vectors for the same two-class problem. For small values of the regularization parameter C, the number of training errors increases and the system is underfitted. For large values, the number of support vectors increases to the extent system performs similar to a hard-margin SVM ([6,19,11]). Practice shows that varying C over a wide range and noting performance progress on a separate validation set can lead to tuning the system for optimal performance. Decrementing C is performed obeying KKT conditions, so each intermediary state is a solution. Train and test accuracies are referred in Table 1, along with the number of migrations.

For each of the configuration obtained, a corresponding system, with the same C and γ (the RBF kernel parameter) values, was trained using the LIBSVM software ([2]). The two very different implementations performed identically, resulting in the same number of misclassified or correctly classified patterns , for each of the C values. It is shown in [3] that obtaining identical solutions is the usual, rather than the exception.

---

[1] The source code of the incremental / decremental machine can be downloaded from http://vega.unitbv.ro/~galmeanu/idsvm.

[2] We have not included a detailed pseudo code of the algorithm, the incremental version can be found in [15] and the decremental version follows straightforward.

Table 1. Classifier performance as constraint C decreases, for USPS subset

USPS training subset, $\sigma = 0.0070922$										
C value	10	7	3	1	0.8	0.5	0.3	0.1	0.08	0.05	0.03
#SV	489	478	377	217	202	162	132	58	44	41	16
#migrations	-	122	149	421	96	221	273	775	179	390	298
Train acc.	1.0000	0.9990	0.9922	0.9745	0.9704	0.9569	0.9496	0.9345	0.9273	0.9075	0.8499
Test acc.	0.9605	0.9623	0.9515	0.9425	0.9408	0.9390	0.9318	0.9300	0.9210	0.8959	0.8492

Table 2. Accuracy performance for Pima Indians subset

Pima Indians Diabetes training subset, $\sigma = 0.125$										
C value	10000	5000	1000	500	100	50	10	5	1	0.5	0.1
#SV	134	122	91	75	51	43	26	19	11	7	6
#migrations	-	77	138	47	101	51	114	44	119	67	88
Train acc.	0.8882	0.8725	0.8294	0.8176	0.8118	0.8078	0.7882	0.7824	0.7745	0.7608	0.6392
Test acc.	0.7882	0.8000	0.8039	0.8078	0.8157	0.8039	0.8039	0.8039	0.7961	0.7201	0.6745

Table 3. Accuracy performance for Reuters subset

Reuters training subset, $\sigma = 0.000100563$										
C value	0.05	0.03	0.01	0.007	0.004	0.002	0.0017	0.0015	0.0012	0.001
#SV	393	367	237	173	96	20	16	12	6	1
#migrations	-	40	213	127	194	186	25	23	21	5
Train acc.	1.000	0.9983	0.9933	0.9866	0.9716	0.8629	0.8211	0.7893	0.7157	0.6522
Test acc.	0.9617	0.9633	0.965	0.96	0.93	0.8533	0.815	0.7933	0.7467	0.6917

For the second experiment, we used the Pima Indians Diabetes dataset ([4]). The set originally has 768 training patterns with 8 features. It is a two-class classification problem, with no test data. We randomly selected 510 training patterns and 255 test patterns. Table (2) shows how the system progressed from over-fitting to an acceptable classification performance when decreasing parameter C. Again, the system presented the same accuracy and classified/miss-classified patterns, for every iteration, when compared with the LIBSVM software.

Finally, we used a polynomial kernel to train a subset of the Reuters collection ([16]). There are 600 train samples and 600 completely different test samples, belonging to two categories. Each vector has a maximum of 9947 features. Table 3 presents the decrease of the classifier's train and test accuracies as C decreases from the initial overfit state.

8   Conclusions

In this paper, we presented the intricacies one should consider when implementing the incremental/decremental SVM learning algorithm. We showed that the regularization parameter can be perturbed by various quantities and this process does not influence the system's expected behavior. As desired and expected, our implementation of the incremental/decremental algorithm lead to the same solution as the LIBSVM implementation of the non-incremental algorithm.

References

1. Bottou, L., Le Cun, Y.: Large Scale Online Learning. In: Thrun, S., Saul, L., Schölkopf, B. (eds.) Advances in Neural Information Processing Systems, vol. 16. MIT Press, Cambridge (2004)
2. Chang, C.C., Lin, C.J.: LIBSVM: a Library for Support Vector Machines (2001), http://www.csie.ntu.edu.tw/~cjlin/libsvm
3. Burges, C.J., Crisp, D.J.: Uniqueness of the SVM Solution. In: Solla, S.A., Leen, T.K., Muller, K.R. (eds.) Advances in Neural Information Processing Systems, vol. 12. Morgan Kaufmann, San Francisco (2000)
4. Blake, C.L., Merz, C.J.: UCI repository of machine learning databases, University of California, Irvine, Dept. of Information and Computer Sciences (1998), http://www.ics.uci.edu/~mlearn/MLRepository.html
5. Diehl, C.P., Cauwenberghs, G.: SVM Incremental Learning, Adaptation and Optimization. In: Proceedings of the IJCNN, vol. 4 (2003)
6. Alpaydin, E.: Introduction to Machine Learning. MIT Press, Cambridge (2004)
7. d'Alche-Buc, F., Ralaivola, L.: Incremental Learning Algorithms for Classification and Regression: local strategies. In: American Institute of Physics Conference Proc., vol. 627, pp. 320–329 (2002)
8. Cauwenberghs, G., Poggio, T.: Incremental and Decremental Support Vector Machine Learning. In: Neural Information Processing Systems, Denver (2000)
9. Fung, G., Mangasarian, O.: Incremental Support Vector Machine Classification. In: Advances in Neural Information Processing Systems. MIT Press, Cambridge (2003)
10. Hull, J.J.: A Database for Handwritten Text Recognition Research. IEEE Transactions on Pattern Analysis and Machine Intelligence 16(5), 550–554 (1994)
11. Taylor, J.S., Cristianini, N.: Kernel Methods for Pattern Analysis. Cambridge University Press, Cambridge (2004)
12. Ma, J., Thelier, J., Perkins, S.: Accurate On-line Support Vector Regression. Neural Computation (2003)
13. Martin, M.: On-line Support Vector Machine Regression. In: Elomaa, T., Mannila, H., Toivonen, H. (eds.) ECML 2002. LNCS (LNAI), vol. 2430. Springer, Heidelberg (2002)
14. Syed, N.A., Liu, H., Sung, K.K.: Incremental Learning with Support Vector Machines. In: Proc. of the Workshop on Support Vector Machines at the International Joint Conference on Artificial Intelligence, IJCAI 1999, Stockholm, Sweden (1999)
15. Laskov, P., Gehl, C., Krüger, S., Müller, K.R.: Incremental Support Vector Learning: Analysis, Implementation and Applications. Journal of Machine Learning Research 7, 1909–1936 (2006)
16. Reuters 21578 Text Categorization Collection, Dataset, http://kdd.ics.uci.edu/databases/reuters21578/reuters21578.html

17. Rüping, S.: Incremental Learning with Support Vector Machines. In: Proceedings of the 2001 IEEE International Conference on Data Mining, ICDM 2001 (2001)
18. Theodoridis, S., Koutroumbas, K.: Pattern Recognition, 2nd edn. Elsevier Academic Press, Amsterdam (2003)
19. Joachims, T.: Learning to Classify Text using Support Vector Machines: Methods, Theory and Algorithms. Kluwer Academic Publishers, Dordrecht (2002)
20. Wang, W., Men, C., Lu, W.: Online Prediction Model Based on Support Vector Machine. Neurocomputing 71, 550–558 (2008)

Online Clustering of Non-stationary Data Using Incremental and Decremental SVM

Khaled Boukharouba[1,2] and Stéphane Lecoeuche[1,2]

[1] LAGIS UMR 8146 - Université des Sciences et Technologies de Lille
Bâtiment P2, 59655 Villeneuve d'Ascq - France
[2] Ecole des Mines de Douai - Département Informatique et Automatique
941, Rue Charles Bourseul, BP838, 59508 Douai - France
`{boukharouba,lecoeuche}@ensm-douai.fr`

Abstract. In this paper we present an online recursive clustering algorithm based on incremental and decremental Support Vector Machine (SVM). Developed to learn evolving clusters from non-stationary data, it is able to achieve an efficient multi-class clustering in a non-stationary environment. With a new similarity measure and different procedures (Creation, Adaptation: incremental and decremental learning, Fusion and Elimination) this classifier can provide optimal updated models of data.

1   Introduction

Clustering analysis is the process of grouping data into clusters such that the patterns in a cluster are similar to each other. Generally, this process is done according to some measures of distance or similarity [1]. In the literature, we find three types of clustering algorithms: offline algorithms, batch algorithms and online algorithms. The first ones are developed for clustering static data where the temporal structure is ignored and require a complete knowledge at the beginning of the learning process. The second ones authorize the introduction of a growing knowledge and the third ones could, in more, take into account model evolution over time.

Recently, various incremental SVM algorithms were introduced, kernel-based classifiers having the advantage of relaxing the assumptions about the data distributions. The work in [2] is the first incremental learning procedure using SVMs. The learning machine is incrementally trained on new data by discarding all previous data except their support vectors. As a generalization of the last strategy, [3] introduced an extension to dynamic samples. They use the iterative sampling in an active manner. [4] proposed a new algorithm based on [2], called SVL incremental algorithm, where the objective function in the optimisation problem is slightly modified by a new regularization term. Kivinen and al. [5] introduced NORMA algorithm which uses stochastic gradient descent in the feature space for classification, regression and novelty detection tasks. Cauwenberghs and al. [6] have developed an on-line recursive algorithm for training support vector machine for a supervised binary setting. It is an exact on-line method to construct the solution recursively one point at a time.

V. Kůrková et al. (Eds.): ICANN 2008, Part I, LNCS 5163, pp. 336–345, 2008.
© Springer-Verlag Berlin Heidelberg 2008

However, most of these incremental algorithms assume that the distribution underlying data does not change over time (i.e. the target concept is assumed to be constant) and then could not deal with the challenge of online learning drifting targets in multi-class environment.

Some authors designing incremental learning procedures have also devoted some efforts to derive algorithms for moving targets frameworks. In general, concept drift is not deeply treated for them. Construction of SVM classifiers for streaming data, a time dependent scenario, is considered in [7] by using a window model based on the fact that old examples may not be a good predictor for future points. In [8], it is performed an experiment where data was generated in such a way that the starting and final separating hyperplanes differ considerably. This means that incoming data is constantly changing and the separating criteria to be learned change dynamically with respect to time. Recently, [9] proposed an online kernel-based algorithm for clustering non-stationary data in multi-class environment. Nevertheless its update rule is based on the technique of stochastic gradient descent which gives only approximate results.

In this paper we present an online clustering algorithm based on incremental and decremental support vector machine which provides exact results. Dedicated to online clustering of non-stationary data, it can achieve an efficient multi-class clustering in an unsupervised context. Beside the previous works the contribution of this paper concerns an exact update of cluster's model whatever the evolution of the data distribution are. Moreover in this work the fusion procedure is improved in order to reduce the complexity.

The paper is organized as follows. The second section presents the formulation of the problem. The concepts of non-stationary data, drifting classes are introduced. In the third section we present the online algorithm with its specific similarity measure and its different adaptation procedures. In section 4, some simulation results and discussions are presented.

2   Problem Statement

In most real-life applications the target concept may change between the learning steps. Before developing the proposed classifier, we will first give the definitions of non-stationary data, evolving cluster and dynamical clustering in the context of our study.

Definition 1. *(Non stationary data) Data are said to be non-stationary; if they have a time-varying distribution [10]. Non-stationary data are also characterized by their drifting concepts, in other words, the underlying data generating mechanism is function of time.*

Two kinds of concept drift that may occur in the real world are normally distinguished in the literature: (1) sudden drift (abrupt, instantaneous changes), and (2) gradual concept drift. This later causes the data-mining model generated from past data to become less accurate in the classification of new data.

Definition 2. *(Evolving cluster) An evolving cluster is defined as a group of similar most recent data. The similarity is assessed from geometric proximity of* *data. When the function that generates data changes over time, the cluster is changing and its model's parameters should be adapted.*

Definition 3. *(Dynamical clustering) From non stationary data, the dynamical clustering task consists in updating the data partition expressed as:* $P^t = \{C_1^t, \ldots, C_m^t, \ldots C_M^t\}$. *where C_m^t is the set of data vectors belonging to the cluster m at time t.*

The problem to be solved in this paper could be formulated as : Knowing a dataset $D = \{X_1, \ldots, X_N\}, X_i \in \Re^d$ of non-stationary data, how to recursively estimate the partition P^t that represents the actual data distribution.

According to some dynamic phenomena such as drifting, creation, fusion or elimination, the partition P^t must be adapted over time using a dynamical clustering approach. Update the partition means update the number of clusters and the cluster's models.

Some of the issues involved are: *Should it be necessary to create a new cluster ? or Which cluster should be updated ? or How much will it be changed? ...* To answer these questions, specific decision rules exploiting position of the recent data in the input space are then required.

In the next section, we present an online classifier dedicated to dynamical clustering of non stationary data. We begin by presenting its model structure followed by the decision rules based on the similarity measure. Then we present the different update procedures.

3   Online Classifier

The key idea used in this paper is to update, according to the data stream, the partition P^t. For this aim, we consider a set of temporal functions expressed as: $\Im^t = \{f_1^t, \ldots, f_m^t, \ldots, f_M^t\}$. Each temporal function f_m^t corresponds to, the mth cluster's decision function defined such as :

$$\forall \left(C_m^t, f_m^t\right) \in P^t \times \Im^t, C_m^t = \left\{X \in \chi, f_m^t\left(X\right) \geq 0, \chi \subset \Re^d\right\} \tag{1}$$

In this paper, the model used is a kernel expansion and it is defined in Hilbert space by:[1]

$$f_m^t\left(x\right) = \sum_i \alpha_{i,m}^t \kappa(x_{i,m}, x) + \rho_m^t \tag{2}$$

where $x_{i,m}$ are the training vectors, the weights $\alpha_{i,m}$ are the Lagrange multiplier weights, obtained by minimizing a convex quadratic objective function [11]:

$$\min_{0 \leq \alpha_{i,m} \leq C} : W_m = \frac{1}{2}\sum_j \sum_i \alpha_{j,m}\alpha_{i,m}\kappa(x_{i,m}, x_{j,m}) + \sum_i \alpha_{i,m}\rho_m - \rho_m \tag{3}$$

ρ_m is the offset of the function and $\kappa(\bullet, \bullet)$ is the kernel, it is defined so that: $\forall X_1, X_2 \in \chi, \kappa(X_1, X_2) = \langle \phi\left(X_1\right), \phi\left(X_2\right)\rangle_\Gamma$
where $\langle \bullet, \bullet \rangle_\Gamma$ is the dot product in feature space.

---

[1] For the sake of simplicity, the time index t will be used only when it is necessary.

In this paper, the chosen kernel is the RBF kernel because of its well-known performances to give an optimal model of data whatever its distribution:

$$\kappa\left(x_1, x_2\right) = \exp\left(\frac{-1}{2\sigma^2}\|x_1 - x_2\|^2\right) \tag{4}$$

where σ is the kernel parameter.

3.1   Classification Principle

At each time t, the KKT conditions of each cluster C_m should be satisfied. These conditions are obtained by computing the gradient of the objective function W_m (eq. 3). These conditions are expressed as:

$$g_{i,m} = \frac{\partial W_m}{\partial \alpha_{i,m}} = \sum_j \alpha_{j,m}\kappa(x_{j,m}, x_{i,m}) + \rho_m \begin{cases} > 0; & \alpha_{i,m} = 0 \\ = 0; & 0 < \alpha_{i,m} < C \\ < 0; & \alpha_{i,m} = C \end{cases} \tag{5}$$

$$\frac{\partial W_m}{\partial \rho_m} = \sum_j \alpha_{j,m} - 1 = 0$$

The KKT conditions divide (classify) data into three categories [12], support vectors $SV_{i,m}$ situated on the boundary function ($g_{i,m} = 0$), error vectors $EV_{i,m}$ outside the boundary ($g_{i,m} < 0$) and the vectors $DV_{i,m}$ inside the boundary ($g_{i,m} > 0$).

This section continues with introducing a new similarity measure computed in the Hilbert space. Next, we present the different procedures for the dynamical classifier : creation, adaptation, fusion and elimination procedure.

3.2   Similarity Measure

Unsupervised clustering algorithms need a criterion used to estimate the membership level of new observations with existing clusters. In this section, we introduce a new similarity measure which only uses cluster's parameters.

Firstly, we check whether the new data X_{new} is situated inside or outside the domains of existing clusters by computing $f_m(X_{new})$. If $f_m(X_{new}) > 0$, X_{new} is inside the cluster C_m therefore the boundary function f_m will not be adapted after the incorporation of the new data.

In the opposite case i.e, $\forall m \in \{1, ..., M\}\,;\, f_m(X_{new}) \leq 0$, we calculate, in the Hilbert space Γ, the angle between the new data X_{new} and the different known vectors $SV_{i,m}$ and $EV_{i,m}$ of different clusters C_m.

The dot product between X_{new} and any X_i is expressed as:

$$\langle\phi\left(X_{new}\right), \phi\left(X_i\right)\rangle_\Gamma = \|\phi\left(X_{new}\right)\|_\Gamma \|\phi\left(X_i\right)\|_\Gamma \cos\left(\phi\left(X_{new}\right), \phi\left(X_i\right)\right)$$

One can note that, using the RBF kernel: $\|\phi\left(X_{new}\right)\|_\Gamma = \|\phi\left(X_i\right)\|_\Gamma = 1$.

Therefore, our similarity measure is defined as the smallest angle $\theta_{nst,m}$ (the angle between X_{new} and the nearest vector for each cluster C_m) and is expressed as:

$$\theta_{nst,m} = \min\left(\cos^{-1}\left(\kappa\left(X_{new}, X_i\right)\right)\right), X_i \in \{\{SV_{i,m}\}, \{EV_{i,m}\}\} \tag{6}$$

Table 1. The decision rules

Case 1	If $\theta_{nst,m} > \theta_{Sim}, \forall m \in \{1, \ldots, M\}$	X_{new} will lead to the **creation** of a new cluster C_{new}.
Case 2	If $card\left(\{m\} \mid \theta_{nst,m} \leq \theta_{Sim}\right) = 1$	X_{new} will be added to C_m as a support or error vector (according to its position regarding C_m) then the boundary function has to be **updated**.
Case 3	If $card\left(\{m\} \mid \theta_{nst,m} \leq \theta_{Sim}\right) \geq 2$	X_{new} is classified as ambiguous; this can lead to the **fusion** of these clusters.

This measure is bounded in $[0, {}^{\pi}/_2]$ and takes ${}^{\pi}/_2$ if X_{new} is infinitely remote from X_i.

Then we compare each smallest angle $\theta_{nst,m}$ to an angle threshold θ_{Sim}. Subject to $f_m(X_{new}) \leq 0$, decision rules are set in table 1.

Remark 1. The threshold θ_{Sim} is a preset parameter that can be defined using statistic properties on an initial dataset.

3.3   Creation and Initialization

At the first acquisition or when the decision rule of our classifier leads to the case 1, the new cluster C_{new} is created with the boundary function f_{new} so that:
$f_{new}(x) = \alpha_{1,new}\kappa\left(x, SV_{1,new}\right) + \rho_{new}$.

where $(SV_{1,new} = X_{new}, \alpha_{1,new} = 1, \rho_{new} = -1)$.

Note that this initialization respects the KKT conditions, since $\sum \alpha_{i,new} = 1$ and $g_{1,new}(SV_{1,new}) = f_{new}(SV_{1,new}) = 0$.

After the creation of C_{new} the sets P^t and $\Im^t$ are incremented: $\mathrm{P}^{t+1} = \mathrm{P}^t \cup \{C_{new}\}$ and $\Im^{t+1} = \Im^t \cup \{f_{new}\}$.

3.4   Updating Procedure

When the decision rule leads to the case 2, this procedure aims at updating the cluster's model after the incorporation of the new data. The classifier update rules are based on incremental SVM [6] by keeping the first-order KKT conditions satisfied. Moreover when a new data X_{new} is added to the cluster C_m these conditions can be expressed differentially as:

$$\Delta g_{i,m} = \kappa(X_{new}, x_{i,m})\Delta\alpha_{new,m} + \sum_j \kappa(SV_{j,m}, x_{i,m})\Delta\alpha_{j,m} + \Delta\rho_m, \forall i \in \{1, \ldots, \tau\}$$
$$0 = \Delta\alpha_{new,m} + \sum_j \Delta\alpha_{j,m} \tag{7}$$

where τ is the cardinality of C_m and $\alpha_{new,m}$ is the coefficient of a new data X_{new} that will be incremented to retain KKT conditions. At the beginning, $\Delta\alpha_{new,m} = \alpha_{new,m} = 0$.

According to (5), $g_{i,m} = 0$ for all the support vectors $SV_{k,m}$, therefore (7) can be expressed $\forall k \in \{1, ..., s_m\}$ as:

$$\kappa(X_{new}, SV_{k,m})\Delta\alpha_{new,m} + \sum_{j=1}^{s_m} \kappa(SV_{j,m}, SV_{k,m})\Delta\alpha_{j,m} + \Delta\rho_m = 0$$

$$\Delta\alpha_{new,m} + \sum_{j=1}^{s_m} \Delta\alpha_{j,m} = 0$$

(8)

With some transformations, we find:

$$\begin{pmatrix} \Delta\rho_m \\ \Delta\alpha_{1,m} \\ \vdots \\ \Delta\alpha_{s_m,m} \end{pmatrix} = - \underbrace{\begin{pmatrix} 0 & 1 & \cdots & 1 \\ 1 & \kappa(SV_{1,m}, SV_{1,m}) & \cdots & \kappa(SV_{1,m}, SV_{s_m,m}) \\ \vdots & \vdots & \ddots & \vdots \\ 1 & \kappa(SV_{s_m,m}, SV_{1,m}) & \cdots & \kappa(SV_{s_m,m}, SV_{s_m,m}) \end{pmatrix}^{-1}}_{R}$$

$$\times \underbrace{\begin{pmatrix} 1 \\ \kappa(SV_{1,m}, X_{new}) \\ \vdots \\ \kappa(SV_{s_m,m}, X_{new}) \end{pmatrix}}_{h} \Delta\alpha_{new,m}$$

$$= \begin{pmatrix} \beta_m & \beta_{1,m} & \cdots & \beta_{s_m,m} \end{pmatrix}^T \Delta\alpha_{new,m}$$

(9)

By substituting $\Delta\rho_m$ and $\Delta\alpha_{i,m}$ in (7), the margin change according to:

$$\Delta g_{i,m} = \left(\kappa(x_{i,m}, X_{new}) + \sum_{j=1}^{s_m} \kappa(SV_{j,m}, x_i)\beta_{j,m} + \beta_m \right) \Delta\alpha_{new,m}, \forall i \in \{1, ..., \tau\}$$

(10)

In order to take into account the drifting target, the cluster size is truncated to τ terms by using a sliding window where the oldest data $X_{old,m}$ has to be left out the cluster. The size of the window is done according to assumptions regarding the dynamic of the non-stationary. When the removed data is situated inside the cluster, it can be left out without any boundary updating; in opposite case, the cluster boundary would change. Both incremental and decremental algorithms (also called learning and unlearning procedures) are proposed in table 2.

3.5   Fusion Procedure

Because of some dynamic phenomena, overlapping can occur between two or more clusters. In the assumption that different clusters are disjoint, this procedure preserves the continuity of cluster region by merging clusters that overlap. Overlapping is detected when two or more different clusters C_{ovp} share ambiguous data. Thus the cluster C_{mrg} is defined as:

$$C_{mrg} = \{X \mid [X \in \{C_i \cap C_j\} \text{ and } card(C_i \cap C_j) \geq N_{ovp}]\}$$

(11)

Fixed to limit the sensibility of the fusion procedure to the noise, the threshold N_{ovp} is defined according to the SNR acting on the data. The clusters set $\mathbf{P}^t$

Table 2. Incremental and Decremental procedures

Incremental Procedure	Decremental procedure

 - $\alpha_{new,m} \leftarrow 0$;
 - Set *step* > 0;
 - $z^t = \left(\rho_m^t\ \alpha_{1,m}^t\ \cdots\ \alpha_{s_m,m}^t \right)^T$;
 - **While** $(g_{new} < 0$ and $\alpha_{new,m} \leq C)$;
 Do:
   • $\alpha_{new,m} = \alpha_{new,m} + step$;
   • $z^{t+1} = z^t - R \times h \times step$; equ.(9)
   • Compute $\Delta g_{i,m}$; equ.(10)
   • $g_{i,m}^{t+1} = g_{i,m}^t + \Delta g_{i,m}, \forall i \in \{1, ..., \tau\}$;
   • Find among $\alpha_{i,m}^{t+1}$ those which are close to zero, corresponding support vectors move inside the margin $(SV_{i,m} \rightarrow DV_{i,m})$. Update h, R and z;
 - **End While**
 - If $\alpha_{new,m} = C$ then X_{new} is an error vector; else it is a support vector.
 - Update R and h.
	
	 - $\alpha_{old,m} \leftarrow$ Weight of $X_{old,m}$;
	 - Set *step* > 0;
	 - $z^t = \left(\rho_m^t\ \alpha_{1,m}^t\ \cdots\ \alpha_{s_m,m}^t \right)^T$;
	 - **While** $(\alpha_{old,m} > 0)$;
	 Do:
	   • $\alpha_{old,m} = \alpha_{old,m} - step$;
	   • $z^{t+1} = z^t + R \times h \times step$; equ.(9)
	   • Compute $\Delta g_{i,m}$; equ.(10)
	   • $g_{i,m}^{t+1} = g_{i,m}^t + \Delta g_{i,m}, \forall i \in \{1, ..., \tau\}$;
	   • Find among $DV_{i,m}$ those which have $g_{i,m}^{t+1} < 0$, these data are candidates to be supports vectors $(DV_{i,m} \rightarrow SV_{i,m})$:
	    - Interrupt unlearning procedure and apply learning procedure on them;
	    - Return to decremental procedure.
	 - **End While**
	 - Leave out $X_{old,m}$ from the cluster.
	 - Update R and h.

and its boundaries functions set $\Im^t$ are then modified: $P^{t+1} = (P^t - \{\cup C_{ovp}\}) \cup \{C_{mrg}\}$ and $\Im^{t+1} = (\Im^t - \{\cup f_{ovp}\}) \cup \{f_{mrg}\}$.

The challenge here is to compute the boundary function f_{mrg}. One solution is to consider the cluster C_{grt} which has the largest number of data and to incorporate data of other clusters one after another and whenever update the boundary function f_{grt}. At the end, $f_{mrg} = f_{grt}$. This solution may be unacceptable in online conditions if the clusters to merge are important. We propose to incorporate to C_{grt} only support vectors SV_i and error vectors EV_i of other clusters and to update f_{grt} for each vector incorporation. This solution provides the same results since support vectors and error vectors are situated on or outside the boundary function. As an example, we have applied the two solutions to merge C_1 and C_2 (figure 1.a). The boundary function f_{mrg} in figure 1.b) was provided with the first solution, and in figure 1.c) with the second solution. These two functions are quite similar.

3.6   Elimination Procedure

The purpose of this procedure is to eliminate clusters that are created because of the noise effect. By comparing the cardinality of all clusters with a cardinality

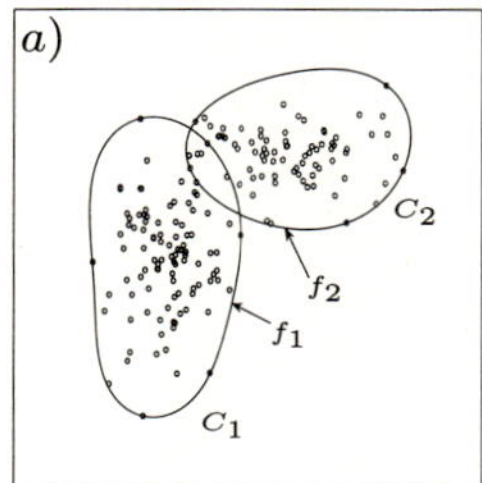

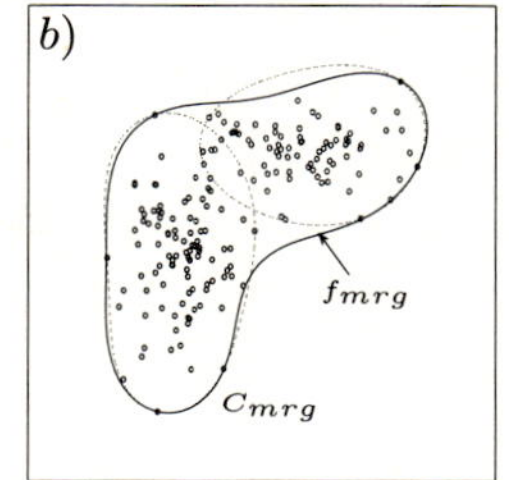

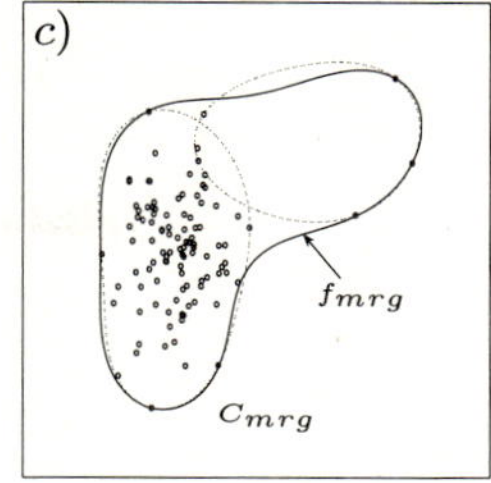

Fig. 1. a) The clusters to be merged. b) Merging using the first method. c) Merging using the second method.

threshold N_c (fixed in a similar way than θ_{Sim}), this procedure eliminates clusters C_{weak} which contain few data after T time (period between two elimination procedures).

So, $\mathbf{P}^{t+1} = \mathbf{P}^t - \{\cup C_{weak}\}$ and $\Im^{t+1} = \Im^t - \{\cup f_{weak}\}$.

4   Experiments and Discussion

In the following experiment, we test the performances of our algorithm, to this end four synthetic datasets have been created in non-stationary environment (2000 time-varying data). Each dataset is an evolving Gaussian density i.e. mean vectors $\overline{X}_i$ and covariance matrices Σ_i change during time:

$$\overline{X}_i = \begin{bmatrix} a \cdot \cos\left(\theta_i(t)\right) + x_{c_i} \\ b \cdot \sin\left(\theta_i(t)\right) + y_{c_i} \end{bmatrix} \quad , \quad \Sigma_i = \delta_i(t) \cdot I_2$$

The different figures, in the table 3, illustrate the evolution of clusters at different stages of online learning. The classifier initializes clusters models from the first data acquisition. Step (a) shows the creation of three clusters, this mechanism is provided by the creation process. Steps (b) and (c) show the capacity of the classifier to update iteratively the densities support of C_1 and C_2 who have migrated and C_3 and C_4 modified through local transformations, thanks to the learning and unlearning procedures. This scenario continues in steps (d) and (e) until cluster C_1 and C_2 become close enough to be merged in step (f) to form $C_{1,2}$ thanks to the fusion procedure. In step (f), the continuity of the cluster C_4 is not preserved, and it seems to be necessary to split this cluster to two different clusters. This could be taken into account in our future research.

The classifier was initialized with: $\sigma = 0.5$, $step = 10^{-4}$, $\theta_{Sim} = \pi/10$, $\tau = 100$, $N_c = 10$ and $N_{ovp} = 10$. Note that the choice of the kernel parameter σ (our knob) influences the decision rule and the updating procedures, a bad choice of this one can yield aberrant results.

The step of incrementation, $step$, influences the speed and the accuracy of the algorithm. If the step is very large, the algorithm becomes faster, but it may not yield too accurate results, even if it is too small it may take a lot of time. $step = 10^{-4}$ seems the optimal choice to get accurate results at an optimal time.

Table 3. Online clustering of non-stationary data

5   Conclusion

We presented an online recursive classifier based on incremental and decremental support vector machine. Dedicated to online clustering in multi-class environment, this classifier is able to give exact results comparing to QP approach. Due to its similarity measure and its recursive nature, it can deal with unsupervised clustering and auto-adapt cluster model over time.

References

1. Jain, A., Dubes, R.: Algorithms for clustering data. Prentice Hall, Englewood Cliffs (1988)
2. Syed, N., Liu, H., Sung, K.: Incremental learning with support vector machines. In: Proceedings of the Workshop on Support Vector Machines at the International Joint Conference on Articial Intelligence (IJCAI 1999) (1999)
3. Mitra, P., Murthy, C., Pal, S.: Data condensation in large databases by incremental learning with support vector machines. In: International Conference on Pattern Recognition (ICPR 2000) (2000)
4. Rüping, S.: Incremental learning with support vector machines. Technical report, Universität Dortmund, Germany (2002)
5. Kivinen, J., Smola, A., Williamson, R.: Online learning with kernels. IEEE Trans. Signal Processing 52(8), 2165–2176 (2004)
6. Cauwenberghs, G., Poggio, T.: Incremental and decremental support vector machine learning. Advances in Neural Information Processing Systems 13, 409–415 (2001)
7. Domeniconi, C., Gunopulos, G.: Incremental support vector machine construction. In: IEEE International Conference on Data Mining (ICDM 2001) (2001)
8. Fung, G., Mangasarian, O.L.: Incremental support vector machine classification. In: International Conference on Data Mining (ICDM 2002) (2002)
9. Amadou, B.H., Lecoeuche, S.: A new kernel-based algorithm for online clustering. In: Duch, W., Kacprzyk, J., Oja, E., Zadrożny, S. (eds.) ICANN 2005. LNCS, vol. 3697, pp. 583–588. Springer, Heidelberg (2005)
10. Angulo, C., Catalá, A.: Online learning with kernels for smart adaptive systems: a review. In: European Network for Intelligent Technologies (ENIT 2003) (2003)
11. Vapnik, V.: The nature of statical learning theory. Springer, New York (1995)
12. Pontil, M., Verri, A.: Properties of support vector machines. Neural Computation 10, 955–974 (1997)

Support Vector Machines for Visualization and Dimensionality Reduction

Tomasz Maszczyk and Włodzisław Duch

Department of Informatics, Nicolaus Copernicus University, Toruń, Poland
`tmaszczyk@is.umk.pl,Google:W.Duch`
`http://www.is.umk.pl`

Abstract. Discriminant functions $g_{\mathbf{W}}(\mathbf{X})$ calculated by Support Vector Machines (SVMs) define in a computationally efficient way projections of high-dimensional data on a direction perpendicular to the discriminating hyperplane. These projections may be used to estimate and display posterior probability densities $p(C|g_{\mathbf{W}}(\mathbf{X}))$. Additional directions for visualization and dimensionality reduction are created by repeating the linear discrimination process in a space orthogonal to already defined projections. This process allows for an efficient reduction of dimensionality and visualization of data, at the same time improving classification accuracy of a single discriminant function. Visualization of real and artificial data shows that transformed data may not be separable and thus linear discrimination will completely fail, but the nearest neighbor or rule-based methods in the reduced space may still provide simple and accurate solutions.

1   Introduction

Many classifiers, including neural networks and support vector machines (SVMs), work as black-box predictors. Their quality is estimated in a global way, on the basis of some accuracy or cost measures. In practical applications it is important to be able to evaluate a specific case, showing the confidence in predictions in the region of space close to this case. From the Bayesian perspective [1] one notes that globally defined priors strongly differ from local priors. Principal Components Analysis (PCA), Independent Component Analysis (ICA), Multidimensional Scaling (MDS) or other methods commonly used for direct visualization of data [2] are very useful for exploratory data analysis, but do not provide any information about reliability of the method used for classification of a specific case. Visualization methods already proved to be very helpful in understanding mappings provided by neural networks [3,4].

This paper shows how to use any linear discriminant analysis (LDA), or SVM classifier in its linear or kernelized version, for dimensionality reduction and data visualization, providing interesting information and improving accuracy at the same time. There is no reason why such systems should be treated as black boxes. In the next section a few linear and non-linear visualization methods are described, and visualization based on linear discrimination is introduced. For

V. Kůrková et al. (Eds.): ICANN 2008, Part I, LNCS 5163, pp. 346–356, 2008.
© Springer-Verlag Berlin Heidelberg 2008

illustration visualization using linear SVM in one and two dimensions for several real and artificial datasets is presented in section 3. This type of visualization is especially interesting because it is fast, projections are easy to understand, and other methods do not seem to find significantly better projections. Conclusions are given in section four.

2 Visualization Algorithms

Visualization methods are discussed in details in many books, for example [2,5]. Below a short description of three popular methods, multidimensional scaling (MDS), principal component analysis (PCA), and Fisher discriminant anlaysis, is given, followed by description of our approach. In the next section empirical comparisons of these four methods are given. Although we have compared our method with many other non-linear and linear methods space limitation do not allow here to present more detailed comparisons.

Multidimensional scaling (MDS) is the only non-linear technique used here. The main idea is to decrease dimensionality while preserving original distances in high-dimensional space. MDS methods need only similarities between objects, so explicit vector representation of objects is not necessary. In metric scaling quantitative evaluation of similarity using numerical functions (Euclidean, cosine or any other measures) is used, while for non-metric scaling qualitative information about the pairwise similarities is sufficient. MDS methods also differ by their cost functions, optimization algorithms, the number of similarity matrices used, and the use of feature weighting. There are many measures of topographical distortions due to the reduction of dimensionality, most of them variants of the stress function:

$$S_T(\mathbf{d}) = \sum_{i>j}^{n} (D_{ij} - d_{ij})^2 \tag{1}$$

where d_{ij} are distances (dissimilarities) in the target (low-dimensional) space, and D_{ij} are distances in the input space, pre-processed or calculated directly using some metric functions. These measures are minimized over positions of all target points, with large distances dominating in the $S_T(\mathbf{d})$. The sum runs over all pairs of objects and thus contributes $O(n^2)$ terms. In the k-dimensional target space there are kn parameters for minimization. For visualization purposes the dimension of the target space is $k =$1-3. The number of objects n may be quite large, making the approximation to the minimization process necessary [6]. MDS cost functions are not easy to minimize, with multiple local minima representing different mappings. Initial configuration is either selected randomly or based on projection of data to the space spanned by principal components. Orientation of axes in the MDS mapping is arbitrary, and the values of coordinates do not have any simple interpretation, as only relative distances are important.

PCA is a linear projection method that finds orthogonal combinations of input features $\mathbf{X} = \{x_1, x_2, ..., x_N\}$ accounting for most variation in the data. Principal components $\mathbf{P}_i$ that result from diagonalization of data covariance matrix guarantee minimal loss of information when position of points are recreated from their low-dimensional projections. Taking 1, 2 or 3 principal components and projecting the data into the space defined by these components $y_{ij} = \mathbf{P}_i \cdot \mathbf{X}_j$ provides for each input vector its representative $(y_{1j}, y_{2j}, ...y_{kj})$ in the target space. For some data distributions such projections show informative structures.

Kernel PCA [7] finds directions of maximum variance for training vectors mapped to an extended space. This space is not constructed in an explicit way, the only condition is that the kernel mapping $K(\mathbf{X}, \mathbf{X}')$ of the original vectors should be a scalar product $\Phi(\mathbf{X}) \cdot \Phi(\mathbf{X}')$ in the extended space. This enables interesting visualization of data, although interpretation of resulting graphs may be rather difficult.

Supervised methods that use information about classes find projections that separate data from different classes in a better way. Fisher Discriminant Analysis (FDA) is a popular algorithm that maximizes the ratio of between-class to within-class scatter, seeking a direction $\mathbf{W}$ such that

$$\max_{\mathbf{W}} J_{\mathbf{W}} = \frac{\mathbf{W}^T \mathbf{S}_B \mathbf{W}}{\mathbf{W}^T \mathbf{S}_I \mathbf{W}} \tag{2}$$

where the scatter matrices $\mathbf{S}_B$ and $\mathbf{S}_I$ are defined by

$$\mathbf{S}_B = \sum_{i=1}^{C} \frac{n_i}{n} (\mathbf{m}_i - \mathbf{m})(\mathbf{m}_i - \mathbf{m})^T; \qquad S_I = \sum_{i=1}^{C} \frac{n_i}{n} \hat{\boldsymbol{\Sigma}}_i \tag{3}$$

where $\mathbf{m}_i$ and $\hat{\boldsymbol{\Sigma}}_i$ are the sample means and covariance matrices of each class and $\mathbf{m}$ is the sample mean [5]. In our implementation pseudoinverse matrix has been used to generate higher FDA directions.

Linear SVM algorithm searches for a hyperplane that provides a large margin of classification, using regularization term and quadratic programming. Non-linear versions are based on a kernel trick [7] that implicitly maps data vectors to a high-dimensional feature space where the best separating hyperplane (the maximum margin hyperplane) is constructed. Linear discriminant function is defined by:

$$g_{\mathbf{W}}(\mathbf{X}) = \mathbf{W}^T \cdot \mathbf{X} + w_0 \tag{4}$$

The best discriminating hyperplane should maximize the distance between decision hyperplane defined by $g_{\mathbf{W}}(\mathbf{X}) = 0$ and the vectors that are nearest to it, $\max_{\mathbf{W}} D(\mathbf{W}, \mathbf{X}^{(i)})$. The largest classification margin is obtained from minimization of the norm $\|\mathbf{W}\|^2$ with constraints:

$$Y^{(i)} g_{\mathbf{W}}(\mathbf{X}^{(i)}) \geq 1 \tag{5}$$

for all training vectors $\mathbf{X}^{(i)}$ that belong to class $Y^{(i)}$. Vector $\mathbf{W}$, orthogonal to the discriminant hyperplane, defines direction on which data vectors are projected, and thus may be used for one-dimensional projections. The same may be done using non-linear SVM based on kernel discriminant:

$$g_{\mathbf{W}}(\mathbf{X}) = \sum_{i=1}^{N_{sv}} \alpha_i K(\mathbf{X}^{(i)}, \mathbf{X}) + w_0 \tag{6}$$

where the summation is over support vectors $\mathbf{X}^{(i)}$ that are selected from the training set. The $x = g_{\mathbf{W}}(\mathbf{X})$ values for different classes may be smoothed and displayed as a histogram, estimating either the $p(x|C)$ class-conditionals or posterior probabilities $p(C|x) = p(x|C)p(C)/p(x)$. Displaying $p(C|x)$ gives some idea about the $\mathbf{X}$ relation to decision border and overlaps between the classes, allowing for immediate estimation of reliability of classification.

The first projection should give $g_{\mathbf{W}_1}(\mathbf{X}) < 0$ for vectors from the first class, and > 0 for the second class. This is obviously possible only for linearly separable data. If this is not the case, a subset of all vectors $\mathcal{D}(\mathbf{W}_1)$ a subset of incorrectly classified vectors projected to $[a(\mathbf{W}_1), b(\mathbf{W}_1)]$ interval that contains the zero point is selected. Visualization may help to separate these $\mathcal{D}(\mathbf{W}_1)$ vectors. The second best direction may be obtained by repeating SVM calculations in the space orthogonalized to the already obtained $\mathbf{W}$ directions, on the subset of $\mathcal{D}(\mathbf{W}_1)$ vectors, as the remaining vectors are already separated in the first dimension. SVM training in its final phase is using anyway mainly vectors from this subset. However, vectors in the $[a(\mathbf{W}_1), b(\mathbf{W}_1)]$ interval do not include some outliers and therefore may lead to significantly different direction.

In two dimensions hierarchical classification rule are obtained:

- If $g_{\mathbf{W}_1}(\mathbf{X}) < a(\mathbf{W}_1)$ Then Class 1
- If $g_{\mathbf{W}_1}(\mathbf{X}) > b(\mathbf{W}_1)$ Then Class 2
- If $g_{\mathbf{W}_2}(\mathbf{X}) < 0$ Then Class 1
- If $g_{\mathbf{W}_2}(\mathbf{X}) > 0$ Then Class 2

where the $[a(\mathbf{W}_1), b(\mathbf{W}_1)]$ interval is determined using estimates of posterior probabilities $p(C|x)$ from smoothed histograms, with the user-determined confidence parameter (for example $p(C|x) > 0.9$ for each class). One could also introduce such confidence intervals for the $\mathbf{W}_2$ direction and reject vectors that are inside this interval. An alternative is to use the nearest neighbor rule after dimensionality reduction.

This process may be repeated to obtain more dimensions. Each additional dimension should help to decrease errors, and the optimal dimensionality is obtained when new dimensions stop decreasing the number of errors in crossvalidation tests. In this way hierarchical system of rules with decreasing reliability is created. Of course it is possible to use other models on the $\mathcal{D}(\mathbf{W}_1)$ data, for example Naive Bayes approach, but we shall not explore this possibility here, concentrating mainly on data visualization.

In case of non-linear kernel $g_{\mathbf{W}}(\mathbf{X})$ provides the first direction, while the second direction may be generated in several ways. The simplest approach is to repeat training on $\mathcal{D}(\mathbf{W})$ subset of vectors that are close to the hyperplane in the extended space using some other kernel, for example a linear kernel.

3 Illustrative Examples

The usefulness of the SVM-based sequential visualization method has been evaluated on a large number of datasets. Here only two artificial binary datasets, and three medical datasets downloaded from the UCI Machine Learning Repository [8] and from [9], are presented as an illustration. A summary of these datasets is presented in Tab. 1; their short description follows:

1. **Parity_8:** 8-bit parity dataset (8 binary features and 256 vectors).
2. **Heart** disease dataset consists of 270 samples, each described by 13 attributes, 150 cases belongs to group "absence" and 120 to "presence of heart disease".
3. **Wisconsin** breast cancer data [10] contains 699 samples collected from patients. Among them, 458 biopsies are from patients labeled as "benign", and 241 are labeled as "malignant". Feature 6 has 16 missing values, removing corresponding vectors leaves 683 examples.
4. **Leukemia:** microarray gene expressions for two types of leukemia (ALL and AML), with a total of 47 ALL and 25 AML samples measured with 7129 probes [9]. Visualization is based on 100 best features from simple feature ranking using FDA index.

For each dataset two-dimensional mappings have been created using MDS, PCA, FDA and SVM-based algorithms (Figs. 1-5).

Table 1. Summary of datasets used for illustrations

Title	#Features	#Samples	#Samples per class		Source
Parity_8	8	256	128 C_0	128 C_1	artificial
Heart	13	270	150 "absence"	120 "presence"	[8]
Wisconsin	10	683	444 "benign"	239 "malignant"	[10]
Leukemia	100	72	47 "ALL"	25 "AML"	[9]

High-dimensional parity problem is very difficult for most classification methods. Many papers have been published on special neural models for parity functions, and the reason is quite obvious, as Fig. 1 illustrates: linear separation cannot be easily achieved because this is a k-separable problem that should be separated into $n+1$ intervals for n bits [11,12]. PCA and SVM find a very useful projection direction $[1, 1..1]$, but the second direction does not help at all. MDS is completely lost, as it is based on preservations of Euclidean distances that in this case do not carry useful information for clustering. FDA shows significant overlaps for projection on the first direction. This is a very interesting example showing that visualization may help to solve a difficult problem in a perfect way even when almost all classifiers fail.

Variations on parity data include random assignment of classes to bit strings with fixed number of 1 bits, creating k-separable ($k \leq n$) data that most methods invented for the parity problem cannot handle [11]. All three linear projections show for such data correct cluster structure along the first direction. However, linear projection has to be followed by a decision tree or the nearest neighbor method, as the data is nonseparable.

For Cleveland Heart data linear SVM gives about $83 \pm 5\%$ accuracy, with the base rate of 54%. Fig. 2 shows nice separation of a significant portion of the

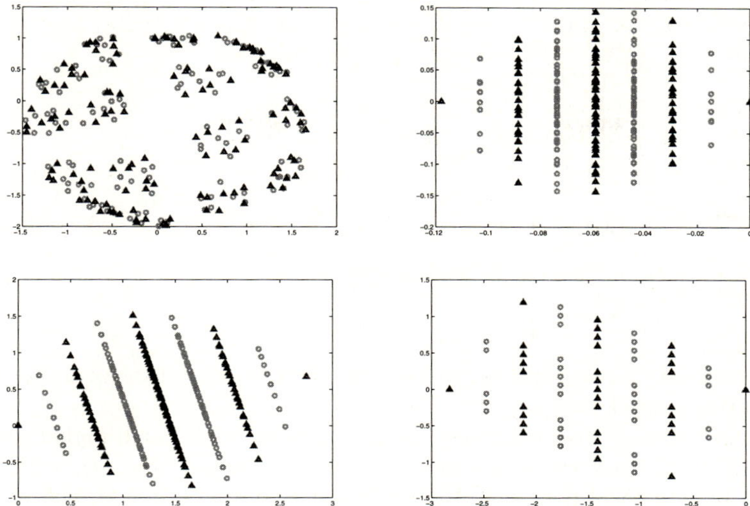

Fig. 1. 8-bit parity dataset, top row: MDS and PCA, bottom row: FDA and SVM

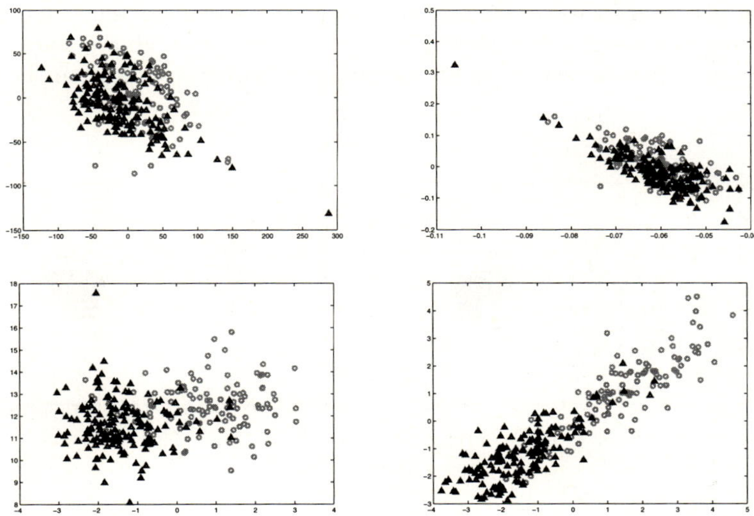

Fig. 2. Heart data set, top row: MDS and PCA, bottom row: FDA and SVM

data, with little improvement due to the second dimension. MDS and PCA are somewhat less useful than FDA and SVM projections.

Displaying class-conditional probability for Parity and Cleveland Heart in the first SVM direction (Fig. 3) may also help to estimate the character of overlaps

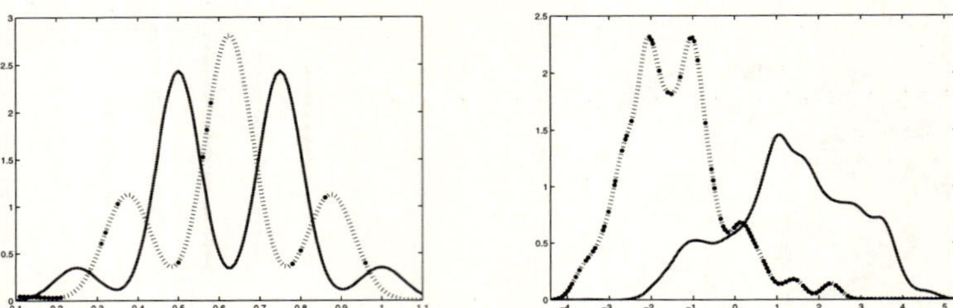

Fig. 3. Estimation of class-conditional probability for Parity and Cleveland Heart in the first SVM direction

and the resulting errors, and help to decide what type of transformation should follow initial projection.

Wisconsin breast cancer dataset can be classified with much higher accuracy, around $97 \pm 2\%$, and therefore shows strong separation (Fig. 4), with benign cancer cases clusterized in one area, and a few outliers that appear far from the main benign cluster, mixing with malignant cases. Most likely these are real misdiagnosed outliers that should in fact be malignant. Only in case of SVM the second direction shows some additional improvement.

Leukemia shows remarkable separation using two-dimensional SVM projection (Fig. 5), thanks to maximization of margin, providing much more interesting projection than other methods. The first direction shows some overlap but in crossvalidation tests it yields significantly better results than the second direction.

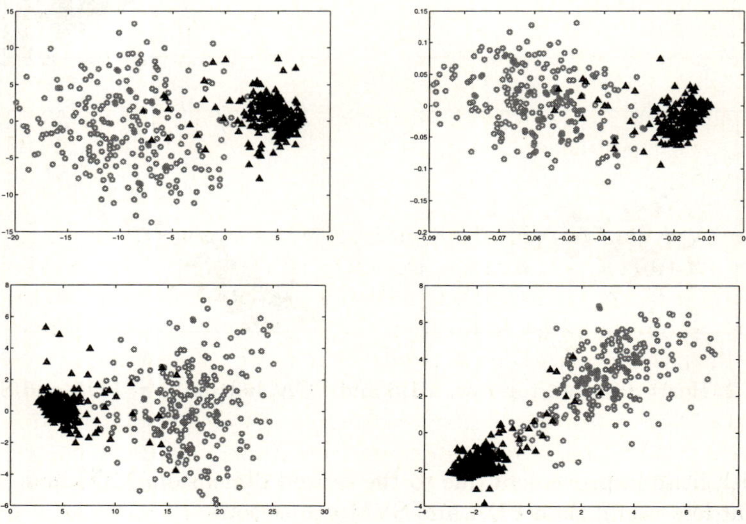

Fig. 4. Wisconsin data set, top row: MDS and PCA, bottom row: FDA and SVM

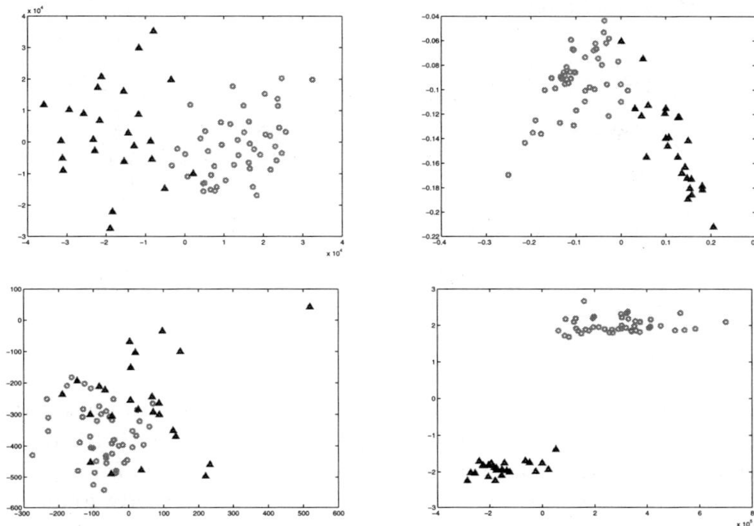

Fig. 5. Leukemia data set, top row: MDS and PCA, bottom row: FDA and SVM

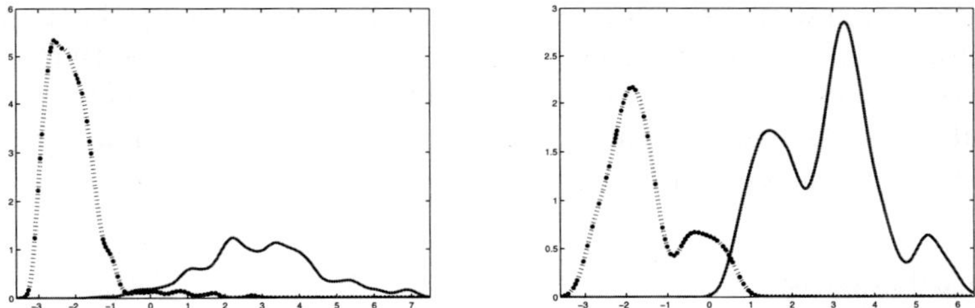

Fig. 6. Estimation of class-conditional probability using linear SVM for Wisconsin (left) and Leukemia (right)

To compare the influence of dimensionality reduction on accuracy of classification for each dataset classification using SVM with linear kernel has been performed in the original and in the reduced two-dimensional space. 10-fold crossvalidation tests have been repeated 10 times and average results collected in Table 2, with accuracies and standard deviations for each dataset. The intention is to illustrate the efficiency of dimensionality reduction; in case of Leukemia starting from pre-selected 100 features from the microarray data does not guarantee correct evaluation of generalization error (feature selection should be done within crossvalidation in order to do it). With such a large number of features and a very few samples, SVM with Gaussian kernel will show nicely separated class-conditional probabilities, but in crossvalidation will perform poorly, showing that strong overfitting occurs.

Table 2. SVM 10-fold crossvalidation accuracy for datasets with reduced features

	# Features	Parity_8	Heart	Wisconsin	Leukemia
PCA	1	41.8±6.2	55.6±8.3	65.0±6.0	65.2±15.6
PCA	2	41.7±5.3	55.6±8.3	65.0±6.0	77.6±19.1
MDS	1	39.7±5.8	60.3±9.3	97.0±2.0	60.2±18.1
MDS	2	38.2±5.4	68.6±9.0	96.7±2.1	94.5± 8.4
FDA	1	40.3±6.5	85.0±6.6	97.2±1.9	75.6±15.4
FDA	2	38.7±7.1	85.2±6.3	97.1±2.0	81.8±14.1
SVM	1	41.9±6.5	84.8±6.5	97.3±1.8	97.2± 5.7
SVM	2	41.8±6.2	84.8±6.5	97.3±1.8	97.2± 5.7
	All	31.4±4.8	83.9±6.3	96.6±2.0	95.4± 7.8

For Heart and Wisconsin data FDA and SVM gives similar, significantly better results than other methods used in this comparison, with SVM achieving much better results for Leukemia, as should also be obvious from data visualization. Adding the second dimensions to the first one already found by SVM obviously does not change the results. However, visualization shows that for highly non-separable data linear projection may still be useful for dimensionality reduction and data preprocessing, combined with various classification or regression algorithms. This is evident in application of SVM with Gaussian kernels to the original and the two-dimensional data. For parity Gaussian kernel SVM fails as badly as linear SVM on the original data, but gives 100% accuracy for the low-dimensional data. The same is true for Leukemia, where crossvalidation results with Gaussian kernel on the original data is quite poor ($79 \pm 10\%$), while on the two-dimensional data it is very accurate($95 \pm 11\%$).

4   Conclusions

There are many methods for data visualization, some of them quite sophisticated [13], with PCA and MDS among the most common. Visualization enables exploratory data analysis, giving much more information than just global information about expected accuracy, or probability of individual cases. In real applications visualization is sometimes used for initial data exploration, but rarely for the evaluation of the mapping implemented by predictors. Visualization can certainly help to understand what black box classifiers really do [3,4]. In industrial, medical or other applications where safety is important evaluation of confidence in predictions, that may be done using visualization methods, is critical.

Sequential dimensionality reduction based on SVM has several advantages: it enables visualization, guarantees dimensionality reduction without loss of accuracy, increases accuracy of the linear discrimination model, is very fast and preserves simple interpretation. Information obtained from unsupervised methods, such as PCA or kernel PCA, provide directions of highest variance, but no information about reliability of classification. There is no reason why SVM decision borders should not be visualized using estimations of class-dependent probabilities, or posterior probabilities $p(C|x)$ in one or two dimensions. This process gives insight into the character of data, helping to construct appropriate predictors by combining linear or non-linear projections with other data models, such as decision trees or nearest neighbor models. For highly non-separable data (symbolic data with inherent complex logic) k-separability approach may be the most appropriate, for very sparse high-dimensional data linear projection on one or two directions may be followed by kernel methods [7,14], prototype-based rules [15], or by the nearest neighbor methods. Such observations allow for implementation of meta-learning as composition of transformations [16], for automatic discovery of the simplest and most reliable models. Visualization will also help to evaluate the reliability of predictions for individual cases, showing them in context of the known cases, and providing information about decision borders of classifiers. We plan to add visualization of probabilities and scatterograms to a few popular SVM packages soon.

References

1. Bishop, C.M.: Pattern Recognition and Machine Learning. Springer, Heidelberg (2006)
2. Pękalska, E., Duin, R.: The dissimilarity representation for pattern recognition: foundations and applications. World Scientific, Singapore (2005)
3. Duch, W.: Visualization of hidden node activity in neural networks: I. In: Rutkowski, L., Siekmann, J.H., Tadeusiewicz, R., Zadeh, L.A. (eds.) ICAISC 2004. LNCS (LNAI), vol. 3070, pp. 38–43. Springer, Heidelberg (2004)
4. Duch, W.: Coloring black boxes: visualization of neural network decisions. In: Int. Joint Conf. on Neural Networks, Portland, vol. I, pp. 1735–1740. IEEE Press, Los Alamitos (2003)
5. Webb, A.: Statistical Pattern Recognition. J. Wiley & Sons, Chichester (2002)
6. Naud, A.: An Accurate MDS-Based Algorithm for the Visualization of Large Multidimensional Datasets. In: Rutkowski, L., Tadeusiewicz, R., Zadeh, L.A., Żurada, J.M. (eds.) ICAISC 2006. LNCS (LNAI), vol. 4029, pp. 643–652. Springer, Heidelberg (2006)
7. Schölkopf, B., Smola, A.: Learning with Kernels. Support Vector Machines, Regularization, Optimization, and Beyond. MIT Press, Cambridge (2001)
8. Merz, C., Murphy, P.: UCI repository of machine learning databases (2007)
9. Golub, T.: Molecular classification of cancer: Class discovery and class prediction by gene expression monitoring. Science 286, 531–537 (1999)
10. Wolberg, W.H., Mangasarian, O.: Multisurface method of pattern separation for medical diagnosis applied to breast cytology. In: PNAS, vol. 87, pp. 9193–9196 (1990)
11. Grochowski, M., Duch, W.: Learning highly non-separable Boolean functions using Constructive Feedforward Neural Network. In: de Sá, J.M., Alexandre, L.A., Duch, W., Mandic, D.P. (eds.) ICANN 2007. LNCS, vol. 4668, pp. 180–189. Springer, Heidelberg (2007)
12. Duch, W.: k-separability. In: Kollias, S.D., Stafylopatis, A., Duch, W., Oja, E. (eds.) ICANN 2006. LNCS, vol. 4131, pp. 188–197. Springer, Heidelberg (2006)

13. van der Maaten, L., Postma, E., van den Herik, H.: Dimensionality reduction: A comparative review (in print, 2008)
14. Cristianini, N., Shawe-Taylor, J.: An Introduction to Support Vector Machines and other Kernel-Based Learning Methods. Cambridge University Press, Cambridge (2000)
15. Duch, W., Blachnik, M.: Fuzzy rule-based systems derived from similarity to prototypes. In: Pal, N.R., Kasabov, N., Mudi, R.K., Pal, S., Parui, S.K. (eds.) ICONIP 2004. LNCS, vol. 3316, pp. 912–917. Springer, Heidelberg (2004)
16. Duch, W., Grudziński, K.: Meta-learning via search combined with parameter optimization. In: Rutkowski, L., Kacprzyk, J. (eds.) Advances in Soft Computing, pp. 13–22. Physica Verlag, Springer, New York (2002)

Multigrid Reinforcement Learning with Reward Shaping

Marek Grześ and Daniel Kudenko

Department of Computer Science
University of York
York, YO10 5DD, UK
{grzes,kudenko}@cs.york.ac.uk

Abstract. Potential-based reward shaping has been shown to be a powerful method to improve the convergence rate of reinforcement learning agents. It is a flexible technique to incorporate background knowledge into temporal-difference learning in a principled way. However, the question remains how to compute the potential which is used to shape the reward that is given to the learning agent. In this paper we propose a way to solve this problem in reinforcement learning with state space discretisation. In particular, we show that the potential function can be learned online in parallel with the actual reinforcement learning process. If the Q-function is learned for states determined by a given grid, a V-function for states with lower resolution can be learned in parallel and used to approximate the potential for ground learning. The novel algorithm is presented and experimentally evaluated.

1 Introduction

Reinforcement learning (RL) is a popular method to design autonomous agents that learn from interactions with the environment. In contrast to supervised learning, RL methods do not rely on instructive feedback, i.e., the agent is not informed what the best action in a given situation is. Instead, the agent is guided by the immediate numerical reward which defines the optimal behaviour for solving the task. This leads to two kinds of problems: 1) the *temporal credit assignment* problem, i.e., the problem of determining which part of the behaviour deserves the reward; 2) slower convergence: conventional RL algorithms employ a delayed approach propagating the final goal reward in a discounted way or assigning a cost to non-goal states. However the back-propagation of the reward over the state space is time consuming.

To speed up the learning process, and to tackle the temporal credit assignment problem, the concept of *shaping reward* has been considered in the field [1,2]. The idea of reward shaping is to give additional (numerical) feedback to the agent in order to improve its convergence rate.

Even though reward shaping has been powerful in many experiments it quickly turned out that, used improperly, it can be also misleading [2]. To deal with such problems potential-based reward shaping $F(s, s')$ was proposed [1] as the

V. Kůrková et al. (Eds.): ICANN 2008, Part I, LNCS 5163, pp. 357–366, 2008.

difference of some potential function Φ defined over a source s and a destination state s':

$$F(s, s') = \gamma\Phi(s') - \Phi(s), \tag{1}$$

where γ is a discount factor. Ng et al. [1] proved that reward shaping defined in this way is necessary and sufficient to learn a policy which is equivalent to the one learned without reward shaping.

One problem with reward shaping is that often detailed knowledge of the potential of states is not available, or very difficult to represent directly in the form of a shaped reward.

In this paper we propose an approach to learn the shaping reward online and use it to enhance basic reinforcement learning. The algorithm starts without any prior knowledge and learns a policy and a shaping reward at the same time. At each step the current approximation of the shaped reward is used to guide the learning process of the target RL. The algorithm operates on two state spaces. The first one is to learn the value function for the ground RL. The second one is obtained by aggregating states from the ground level. This state space has higher generalisation and is used to approximate the potential for the shaping reward. We investigate the case when grid discretisation is used to form a discrete state space. Two types of discretisation with different resolutions naturally allow forming two state spaces: ground states and their aggregation.

When relating our approach to automating shaping [3], the contribution of this paper is two-fold: 1) our algorithm learns reward shaping online through high level reinforcement learning; 2) we present the application of this approach to RL when state aggregation can be applied to form an abstract state space. In particular, we investigate the case when continuous state space is discretised to form a discrete grid where state aggregation can be easily obtained. A review of related research is collected in the final part of the paper.

The principal advantage of our approach is a better convergence rate. Furthermore, overhead computational cost is low, the solution is of general applicability, and knowledge is easily acquired and incorporated. Specifically, knowledge which is necessary to obtain discrete states for ground RL is sufficient to apply our extension.

2   Markov Decision Processes and Reinforcement Learning

A Markov Decision Process (MDP) is a tuple (S, A, T, R), where S is the state space, A is the action space, $T(s, a, s')$ is the probability that action a when executed in state s will lead to state s', $R(s, a, s')$ is the immediate reward received when action a taken in state s results in a transition to state s'. The problem of solving an MDP is to find a policy (i.e., mapping from states to actions) which maximises the accumulated reward. When the environment dynamics (transition probabilities and a reward function) are available, this task becomes a planning problem which can be solved using iterative approaches like policy and value iteration [4].

MDPs represent a modelling framework for RL agents whose goal is to learn an optimal policy when the environment dynamics are not available. For this reason value iteration can not be used. However the concept of an iterative approach in itself is the backbone of the majority of RL algorithms. These algorithms apply so called temporal difference updates to propagate information about values of states, $V(s)$, or state-action pairs, $Q(s, a)$. These updates are based on the difference of the two temporally different estimates of a particular state or state-action value. Model-free SARSA is such a method [4]. It updates state-action values by the formula:

$$Q(s, a) \leftarrow Q(s, a) + \alpha[r + \gamma Q(s', a') - Q(s, a)]. \tag{2}$$

It modifies the value of taking action a in state s, when after executing this action the environment returned reward r, moved to a new state s', and action a' was chosen in state s'.

Immediate reward r which is in the update rule given by Equation 2 represents the feedback from the environment. The idea of reward shaping is to provide an additional reward which will improve the performance of the agent. This concept can be represented by the following formula for the SARSA algorithm:

$$Q(s, a) \leftarrow Q(s, a) + \alpha[r + F(s, a, s') + \gamma Q(s', a') - Q(s, a)],$$

where $F(s, a, s')$ is the general form of the shaping reward which in our analysis is a function $F : S \times S \to \mathbb{R}$. The main focus of this paper is how to approximate this value online in the particular case when it is defined as the difference of potentials of consecutive states s and s' (see Equation 1). This reduces to the problem of how to learn the potential $\Phi(s)$. It is worth noting that the ideal potential would be the one which is equal to the value function, that is, $\Phi(s) = V(s)$, which helps justify why roughly approximating the value function is a promising approach for reward shaping.

3   State Aggregation

State aggregation represents one of the options for tackling the state space explosion which arises in sequential decision making particularly when the mathematical model of the environment needs to have the Markov property [5]. It is based on aggregating states of the original MDP into clusters, and treating these clusters as states of a new MDP. Such an MDP can be more easily solved since its state space can be significantly reduced. States of the ground MDP are grouped according to a given notion of similarity, and in the ideal case states within one cluster should be indistinguishable with regard to the policy (e.g., all states which are in the same abstract state have the same optimal action). When states within aggregates are distinguishable in this sense, such aggregation is approximate. The fact, to what extent the states which belong to one cluster differ, determines the quality of the approximate representation. The trade-off between high quality approximation and tractability of solving an MDP needs to

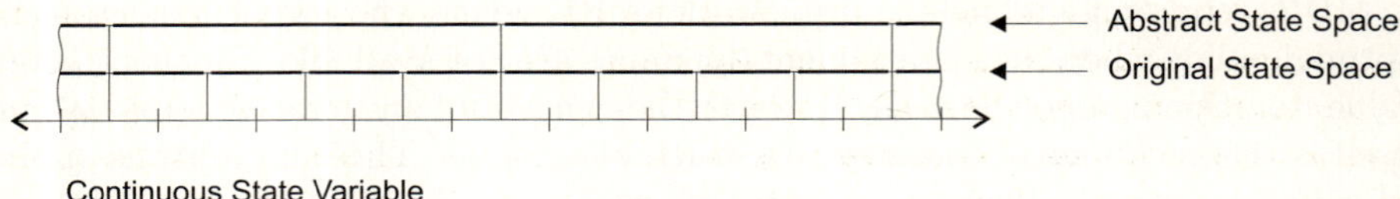

Fig. 1. Multigrid discretisation of the continuous domain variable

be considered. In this paper we propose a RL architecture in which the original RL problem can be represented with a desired precision (i.e., the algorithm can reach the optimal solution) whereas the high level approximation (even though of a low quality) can improve the convergence of ground learning.

In this paper we investigate the particular case when the continuous state space is discretised to form the discrete state space for which the value function can be represented in a tabular form. The state aggregation for high level learning can be easily obtained by applying discretisation with bigger intervals (see Figure 1). Low level discrete states are used to learn the actual solution to the original problem and learning on abstract states with a rough approximation provides useful guidance.

4   Learning Potential for Reward Shaping

We propose a RL architecture with two different discretisations of the state space. The first one is to learn an approximation of the Q-function at the ground RL level. The second one which has lower resolution is to represent an abstract V-function which is used as the potential to calculate the shaping reward (see Equation 1) for the ground level. The algorithm which is proposed here builds on two techniques existing in the field: 1) multigrid discretisation used with MDPs [6], and 2) automatic shaping which was recently proposed [3].

The multigrid discretisation in the MDP setting [6] was used to solve MDPs in a coarse-to-fine manner. While this technique is well suited to dynamic programming methods, (a coarse problem can be solved and used as a starting point for the solution on a finer grid) there was no easy way to merge layers with a different resolution when applied to RL algorithms. First such attempts were made in [7] and this problem was evident there. The need for knowledge of the topology of the state space is necessary in their solution to define how multiple levels are related, but this fact made this approach infeasible for RL tasks. It used a multigrid as a way of obtaining knowledge, but the mechanism to use this knowledge at a ground RL level was missing. We propose potential-based reward shaping as a solution to these problems. The ground RL algorithm does not have to be modified and knowledge can be given in a transparent way via an additional shaping reward.

In the automatic shaping approach [3] an abstract MDP is formulated and solved. In the initial phase of learning, the abstract MDP model is built and after a defined number of episodes the abstract MDP is solved exactly and its value function used as the potential for ground states. In this paper, we propose an algorithm which applies a multigrid strategy when continuous state space is discretised to be represented in a tabular form. Instead of defining an abstract task as dynamic programming for solving an abstract MDP, we use RL to solve the high level task online. Because such an abstract RL does not need to learn the model, a shaping reward can be provided right from the start of learning. Additionally, the multigrid discretisation yields a natural translation between ground and abstract levels. Our method does not require any more knowledge about the environment than necessary to define discretisation at the ground level. We do not need methods to translate abstract to ground states or approximating environment dynamics (transition probabilities) at an abstract level.

Algorithm 1 summarises our approach. It follows the structure of SARSA from [4]. In our case learning at the ground level is the same as in the baseline. The modification is the point where basic SARSA is given shaping reward $F(s, s')$ in line 16 of Algorithm 1 where the temporal difference for ground states is computed. The way in which $F(s, s')$ is evaluated defines our extension.

The shaping reward $F(s, s')$ is computed in line 13 as the difference of the value function of current and previous states visited by the agent. Thus $\Phi(s) = V(z)$ where $V(z)$ is the value function of the abstract level and state s is aggregated by given state z. The value $V(z)$ is learned using temporal difference updates (line 9). All parameters that relate to this task have subscript v in the algorithm. The mapping from state s to a corresponding state z is done in a straightforward way without any special knowledge. Basically, abstract states z aggregate ground states s and subsumption can be easily determined. However with optional, additional knowledge about the problem such a mapping can remove some state variables in the abstract representation and appropriately focus the high level learning.

High level RL is treated as a Semi-MDP [8] since due to higher generalisation an agent can be several time steps within one high level position. For this reason time t is used when the temporal difference in line 9 is computed.

The generic function $reward_v(r)$ shows that high level learning can receive an internally modified reward. The fact which may be considered when defining this function is that the shaping reward should not obscure the reward given by the environment because it may lead to suboptimal solutions when, e.g., the potential is based on non-admissible heuristic.

5   Experimental Methodology

Algorithm 1 was evaluated empirically on the Mountain Car task with a continuous state space [4]. The following values of common RL parameters were used: $\alpha = 0.1$, $\alpha_v = 0.1$, $\gamma = 0.99$ and $\gamma_v = 0.99$. In all experiments ϵ-greedy exploration strategy was used with ϵ decreasing linearly from 0.3 in the first episode to 0.01 in the last episode. All runs were repeated 10 times and average results are presented in graphs. The x-axis represents the number of completed episodes, and the y-axis represents numerical rewards. Following the evaluation

---

Algorithm 1. SARSA-RS: Multigrid SARSA with potential-based reward shaping from the value function of aggregated states

---

1: $\forall s, a\, Q(s,a) = 0$, $\forall z\, V(z) = 0$, $t = 0$
2:    $z \leftarrow$ aggregating state for s
3:    $a \leftarrow$ random action in state s
4: **repeat**
5:        $z' \leftarrow$ aggregating state for s'
6:        $t = t + 1$
7:        $r_v = reward_v(r)$
8:        **if** $r_v \neq 0$ or $z \neq z'$ **then**
9:            $V(z) = (1 - \alpha_v)V(z) + \alpha_v(r_v + \gamma_v^t V(z'))$;
10:           $t = 0$
11:       **end if**
12:       $F(s, s') = \gamma_v V(z') - V(z)$
13:       $z = z'$
14:       $a' \leftarrow$ best action in state s'
15:       with probability ϵ: $a' \leftarrow$ random action in state s'
16:       $Q(s,a) = (1 - \alpha)Q(s,a) + \alpha(r + F(s,s') + \gamma Q(s',a'))$
17:       $s = s'$; $a = a'$
18: **until** state s is terminal

---

process from recent RL competitions, the accumulated reward (part b) of each figure) over all episodes was also used as a measure to compare results in a more readable way. Error bars of the standard error of the mean (SEM) are also presented in graphs for the accumulated reward.

To investigate how well the proposed algorithm scales to big problems, different discretisations for the low-level learning were applied. In this way a range of problems with 231 to 10^4 states was obtained for the analysis. Such an empirical methodology was used, e.g., in [9] and allows for better understanding of properties of evaluated algorithms, especially scalability to bigger problems.

The experiments were performed on the Mountain Car task according to the description in [4]. The agent received a reward of 1 upon reaching the goal state on the right hill and -1 on the left hill. Function $reward_v(r)$ returned zero for negative r and an unchanged value for positive r. An experiment was terminated and the agent placed in a random position after reaching any of the two goal positions or after 10^3 episodes. Three different discretisations were used with 11×21 (these smallest values were taken from [10]), 40×40 and 100×100 intervals on correspondingly the position and velocity of the car. For these discretisations, high level states aggregated 4, 8 and 20 ground states accordingly.

6   Results

The results with three discretisations introduced in the previous section are in Figures 2, 3, and 4 respectively. As these problems gradually become more difficult, the number of episodes was increased in more difficult versions leading to numbers 10^4, $2 \cdot 10^4$, and $5 \cdot 10^4$.

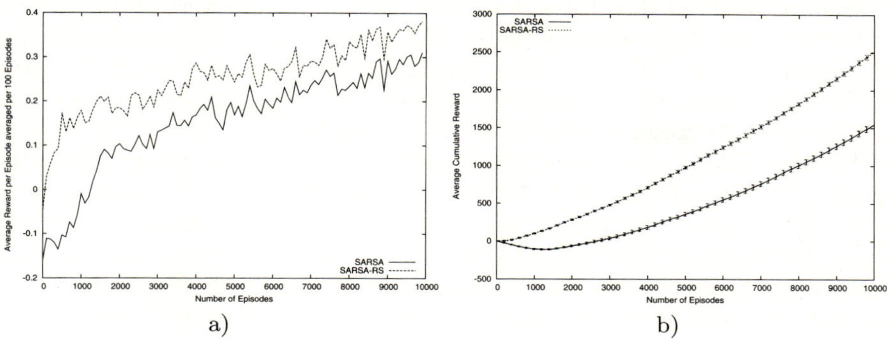

Fig. 2. The average reward per episode a) and the average cumulative reward b) on the Mountain Car task with discretisation 11×21 and aggregation of 4 states

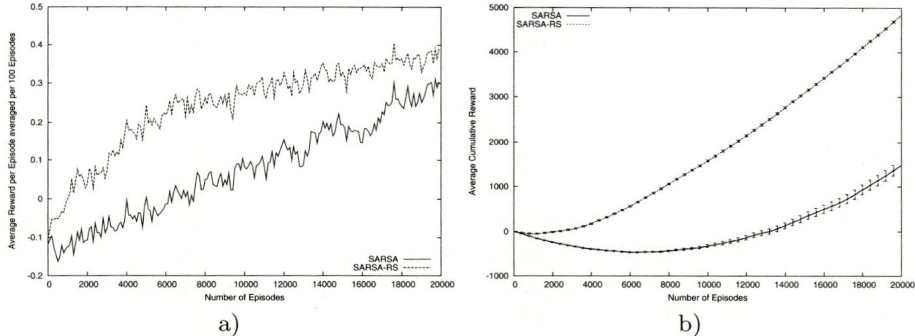

Fig. 3. The average reward per episode a) and the average cumulative reward b) on the Mountain Car task with discretisation 40×40 and aggregation of 8 states

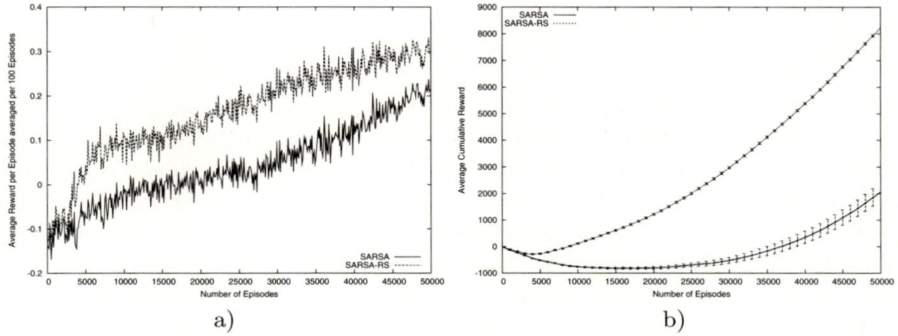

Fig. 4. The average reward per episode a) and the average cumulative reward b) on the Mountain Car task with discretisation 100×100 and aggregation of 20 states

The most essential observation which supports the significance of Algorithm 1 is that the bigger the instance of the problem is, the bigger the positive influence this algorithm has. The particular comparison can be read from the graphs. On the easiest task the accumulated reward of SARSA is 1500 and SARSA-RS 2500 after the same number (10^4) of episodes (Figure 2b). On the 40×40 version when SARSA obtains 1500, SARSA-RS has already scored 4500 after the same number (in this case $2 \cdot 10^4$) of episodes (Figure 3b). The most difficult 100×100 version shows further improvement of reward shaping (Figure 4b). When SARSA obtains 1500, SARSA-RS gains about 7200 after the same number (about $4.7 \cdot 10^4$) of episodes. This relationship can be also found, though in a less clear way, in graphs presenting average reward per episode in the a) part of each figure. This observation clearly shows that the positive influence of our algorithm is more evident when the problem is more difficult (in terms of the number of states).

Learning becomes more difficult when the problem becomes bigger, but the advantage of SARSA-RS when compared with pure SARSA is significant. Error bars placed in graphs for the accumulated reward show statistical significance of this difference.

7   Related Work

The first technique which should be considered in this section is the variable resolution discretisation/abstraction used in the field [11]. The idea is to split some cells (states) and bring a higher resolution to same areas of the state space in order to represent a better policy. The applicability of this idea is however different since it assumes access to environment dynamics. The low-resolution problem is solved as an MDP and then heuristically chosen areas are split to yield more detailed representation when it is predicted to be necessary. In our approach we address a typical RL setting when the environment dynamics are not available in advance.

In our algorithm learning takes place at two levels of abstraction, and so it is worth relating this approach to the general concept of hierarchical machine learning. Stone [12] proposed the universal idea of layered learning where the search space of hypotheses can be reduced by a bottom-up, hierarchical task decomposition into independent subtasks. Each local task is solved separately, and tasks are solved in a bottom-up order. The distinguishing feature of this paradigm is that the learning processes at different layers do not interact with each other and different machine learning algorithms can be used at different layers. In particular, RL was applied to learn in this architecture [12], i.e., to learn at a particular layer. Because tasks are solved independently using results from learning at lower layers, the algorithm proposed in our paper can be seen as a potential choice for selected subtasks.

When relating Algorithm 1 to hierarchical reinforcement learning it is worth taking note of how the hierarchy interacts with reinforcement learning in such algorithms. Regardless of the type of abstraction used to create hierarchy (e.g. state abstraction, hierarchical distance to the goal [13,14], feudal reinforcement learning [15], temporal abstraction [16,8], or both state and temporal abstractions [17]) the hierarchy exists in the final representation of the solution, i.e., the policy is defined on this hierarchy, and learning may take place at all levels of the hierarchy simultaneously. The value function is a function of not only the ground states and actions but also some elements determined by the hierarchy (e.g., in [16] HAMQ-learning maintains an extended Q-table Q([s,m],a) indexed by the pair of environment state s, and machine state m, and an action a at a choice point). In Algorithm 1 the actual RL is not modified and the abstract level learning provides feedback which is given in a transparent way via reward shaping. There is also no need for knowledge about the hierarchal task decomposition, as in the basic case the knowledge which is used to design state representation is sufficient to deploy this algorithm. In particular this can be applied to problems without a clear hierarchy. The state abstraction is in our case used for high level learning to obtain more rapidly an approximate solution which improves the convergence speed of ground learning with a more expressive representation.

8 Conclusion

We propose an algorithm to learn the potential function online at an abstract level, and experimentally evaluate it. The particular case where the continuous state space is discretised and represented in a tabular form shows that simultaneous learning at two levels can converge to a stable solution.

The algorithm proposed in this paper is of potential interest, i.e, may yield significant improvement of the convergence rate, when the bigger instance of the problem is considered with a fine grained policy (i.e., based on a more expressive representation) to be learned.

The strong points of the algorithm are: 1) improved convergence speed at low cost, because there is at most one backup of the V-function for each SARSA backup; 2) separate and external representation of knowledge is obtained; this high level knowledge may be useful for knowledge transfer [18]; 3) no need for explicit domain knowledge; in the basic form the high level learning can be defined using the same knowledge which is used to design the state representation at the ground level.

State aggregation is a general method of obtaining abstract states and can be done by straightforward clustering of neighbouring original states. In the general problem, the application of our algorithm can be considered whenever state aggregation, or in other words grouping of original states, can be applied.

Acknowledgment

This research was sponsored by the United Kingdom Ministry of Defence Research Programme.

References

1. Ng, A.Y., Harada, D., Russell, S.J.: Policy invariance under reward transformations: Theory and application to reward shaping. In: Proceedings of the 16th International Conference on Machine Learning, pp. 278–287 (1999)
2. Randlov, J., Alstrom, P.: Learning to drive a bicycle using reinforcement learning and shaping. In: Proceedings of the 15th International Conference on Machine Learning, pp. 463–471 (1998)
3. Marthi, B.: Automatic shaping and decomposition of reward functions. In: Proceedings of the 24th International Conference on Machine Learning, pp. 601–608 (2007)
4. Sutton, R.S., Barto, A.G.: Reinforcement Learning: An Introduction. MIT Press, Cambridge (1998)
5. Boutilier, C., Dean, T., Hanks, S.: Decision-theoretic planning: Structural assumptions and computational leverage. Journal of Artificial Intelligence Research 11, 1–94 (1999)
6. Chow, C.S., Tsitsiklis, J.N.: An optimal one-way multigrid algorithm for discrete-time stochastic control. IEEE Transactions on Automatic Control 36(8), 898–914 (1991)
7. Anderson, C., Crawford-Hines, S.: Multigrid Q-learning. Technical Report CS-94-121, Colorado State University (1994)
8. Sutton, R.S., Precup, D., Singh, S.P.: Between MDPs and Semi-MDPs: A framework for temporal abstraction in reinforcement learning. Artificial Intelligence 112(1-2), 181–211 (1999)
9. Wingate, D., Seppi, K.D.: Prioritization methods for accelerating MDP solvers. Journal of Machine Learning Research 6, 851–881 (2005)
10. Epshteyn, A., DeJong, G.: Qualitative reinforcement learning. In: Proceedings of the 23rd International Conference on Machine Learning, pp. 305–312 (2006)
11. Munos, R., Moore, A.: Variable resolution discretization in optimal control. Machine Learning 49(2-3), 291–323 (2002)
12. Stone, P., Veloso, M.: Layered learning. In: Proceedings of the 11th European Conference on Machine Learning (2000)
13. Kaelbling, L.P.: Hierarchical learning in stochastic domains: Preliminary results. In: Proceedings of International Conference on Machine Learning, pp. 167–173 (1993)
14. Moore, A., Baird, L., Kaelbling, L.P.: Multi-value-functions: Efficient automatic action hierarchies for multiple goal MDPs. In: Proceedings of the International Joint Conference on Artificial Intelligence, pp. 1316–1323 (1999)
15. Dayan, P., Hinton, G.E.: Feudal reinforcement learning. In: Proceedings of Advances in Neural Information Processing Systems (1993)
16. Parr, R., Russell, S.: Reinforcement learning with hierarchies of machines. In: Proceedings of Advances in Neural Information Processing Systems, vol. 10 (1997)
17. Dietterich, T.G.: Hierarchical reinforcement learning with the MAXQ value function decomposition. Journal of Artificial Intelligence Research 13, 227–303 (2000)
18. Taylor, M.E., Stone, P.: Behavior transfer for value-function-based reinforcement learning. In: Proceedings of the 4th International Joint Conference on Autonomous Agents and Multiagent Systems, pp. 53–59 (2005)

Self-organized Reinforcement Learning
Based on Policy Gradient
in Nonstationary Environments

Yu Hiei[1], Takeshi Mori[2], and Shin Ishii[2,1]

[1] Graduate School of Information Science,
Nara Institute of Science and Technology (NAIST)
Takayama 8916-5, Ikoma, Nara 630-0192, Japan
`yu-h@is.naist.jp`
[2] Graduate School of Informatics,
Kyoto University
Gokasho, Uji, Kyoto 611-0011, Japan
`tak-mori@sys.i.kyoto-u.ac.jp, ishii@i.kyoto-u.ac.jp`

Abstract. In real-world problems, the environment surrounding a controlled system is nonstationary, and the optimal control may change with time. It is difficult to learn such controls when using reinforcement learning (RL) which usually assumes stationary Markov decision processes. A modular-based RL method was formerly proposed by Doya et al., in which multiple-paired predictors and controllers were gated to produce nonstationary controls, and its effectiveness in nonstationary problems was shown. However, learning of time-dependent decomposition of the constituent pairs could be unstable, and the resulting control was somehow obscure due to the heuristical combination of predictors and controllers. To overcome these difficulties, we propose a new modular RL algorithm, in which predictors are learned in a self-organized manner to realize stable decomposition and controllers are appropriately optimized by a policy gradient-based RL method. Computer simulations show that our method achieves faster and more stable learning than the previous one.

1 Introduction

In a nonstationary environment, humans can recognize time-dependent alterations of the environment and make decisions in an adaptive manner. Applying such an adaptive mechanism to robots working in our daily life is important, and machine learning techniques such as reinforcement learning (RL) have been investigated for this purpose in recent years [7] [8]. An important topic in these studies is how to decompose a nonstationary problem into some stationary subproblems.

An effective approach to such a decomposition, called modular selection and identification for control (MOSAIC), has been proposed [1] [2], originally as a computational model of human motor control. The architecture is composed of

V. Kůrková et al. (Eds.): ICANN 2008, Part I, LNCS 5163, pp. 367–376, 2008.

several modules, each of which is the pair of a predictor and a controller, called an inverse model and a forward model, respectively. According to this architecture, the output of each controller is weighted by its responsibility in the current environment, which is calculated by the corresponding predictor. In the learning phase, the learning rate of the parameters in each module is also weighted by the responsibility, so that each module should become adapted to the local environment under the assumptions of temporal continuity and spatial locality of module's activation; even if the environment is nonstationary, it could be regarded to be stationary in a spatially local region. Good controllers cannot be learned without good predictors, but learning of such predictors is not so stable; actually, some global module, which tries to adapt to the whole environment, is inclined to be produced, making the adaptability of MOSAIC to nonstationary environment fail. Doya et al. proposed the multiple model-based reinforcement learning (MMRL) [3], which is an application of the MOSAIC architecture to the RL, and showed its effectiveness in several nonstationary RL problems. Even in this method, however, the instability of the decomposition was inherited from MOSAIC. Furthermore, the probability of selecting control modules was heuristically designed as a combination of the responsibility and the estimated value function, leading the optimality of the resulting control to be obscure.

To overcome these difficulties of the existing techniques, we propose in this study a new modular RL algorithm, whose components, i.e., pairs of a predictor and a controller, are similar to those in the MOSAIC architecture and in MMRL, but its learning scheme and the controllers' structure are more adequately constructed from the points of nonstationary decomposition and RL optimization. For learning of predictors, we employ the idea of learning vector quantization (LVQ) [6], so that the nonstationarity should be decomposed by self-organization of LVQ. For the controllers' structure, the parameters are optimized by the policy gradient RL method [5] based on the reward maximization criterion. Furthermore, we introduce the formulation of the responsibility with regard to the predictor's stabilities, where the responsibility is enlarged when the estimation variance of the corresponding predictor is small, or reduced when the variance is large. Simulation experiments using a problem of random walk in multiple environments (RWME) demonstrate the effectiveness of our proposed scheme. Faster and more stable learning can be achieved by the proposed scheme than by MMRL in both settings of rapid and slow changes of environments.

2   Reinforcement Learning in Nonstationary Environments

RL is a machine learning method, in which the policy is learned through interactions with the environments to maximize the cumulative or the average reward. In many cases, Markov decision processes (MDPs) are assumed. Let the state and the action be s_t and a_t at time t, respectively. The agent selects an action a_t from the probabilistic policy $\pi(a_t|s_t;\theta)$, prescribed by the policy parameter θ, at state s_t, moves to the next state s_{t+1} according to the state transition probability $p(s_{t+1}|s_t, a_t)$ and receives a reward r_{t+1}. In MDPs, the environments surrounding the learning agent are stationary; namely, the probability of s_{t+1} and r_{t+1} conditioned on s_t and a_t is invariant.

In nonstationary environments on which we focus in this study, on the other hand, the transition probability may change with time, and therefore many RL methods formulated in the framework of MDPs cannot be directly applied. In such cases, there are two ways of selecting reasonable actions — switching or combining the stationary policies to construct a nonstationary policy. The former way is valid if the current environment is selected from the finite set of environments and can be recognized, and we only consider which policy is suitable and optimized in each environment. Unfortunately, such a favorable situation rarely happens in many real-world problems, that is, the environment may change with time in a continuous manner or cannot be recognized correctly. The latter way is then more realistic, where the resulting policy is given as a combination of stationary policies, each of which is weighted by the suitability of each policy for the current uncertain environment [3]. Our proposed method, which will be described in the next section, and the previously proposed MMRL [3] are the instances of this latter idea.

3   The Proposed Method

Here, we propose a new modular RL method, in which each module is composed of a pair of a predictor and a controller, which are optimized to perform control in nonstationary environments by self-organized mapping and the policy gradient-based RL method.

In nonstationary environments that can be approximated by combinations of stationary environments, the agent must have some policies, each of which adapts to the corresponding stationary environment, and takes an action by appropriately combining them at different time points. The policy, the probability of selecting an action a_t conditioned on state s_t at time t, is defined as

$$\pi(a_t|s_t; \theta) = \frac{\prod_{i=1}^{M} \pi_i(a_t|s_t; \theta_i)^{\lambda_i}}{\sum_a \prod_{i=1}^{M} \pi_i(a|s_t; \theta_i)^{\lambda_i}}, \tag{1}$$

where $\pi_i(a|s; \theta)$ is the policy, λ_i is the responsibility, which is determined by the corresponding predictor, and θ_i is the policy parameter vector of the ith module. M is the maximum number of modules. If $\lambda_i \equiv 1$ for all i, this model is reduced to a product of experts [9] or conditional random fields [10], which have been extensively studied to deal with complex problems recently. The nonstationary policy (equation (1)) is learned by a variant of the policy gradient RL methods [5], in which the policy parameters are adjusted directly to maximize the average reward. The policy parameters, $\theta_1, \ldots, \theta_M$, are updated as

$$e_i \leftarrow c\gamma e_i + \nabla_{\theta_i} \log \pi_i, \tag{2}$$

$$\theta_i \leftarrow \theta_i + \epsilon e_i r, \tag{3}$$

where e_i is the eligibility trace, which assigns the credit of the current reward r to the current and past scoring functions $\nabla_{\theta_i} \log \pi_i(a_0|s_0; \theta_i), \ldots, \nabla_{\theta_i} \log \pi_i(a_t|s_t; \theta_i)$. $c \in [0, 1]$ denotes the degree of environmental changes where $c = 1$ denotes invariant environment and $c = 0$ denotes large environmental change. γ is the discount rate of the eligibility and ϵ is the learning coefficient.

The responsibility of each module is calculated by the corresponding predictor. We define the observation o_i at time $t + 1$ as

$$o_i \equiv \frac{\exp(-p_i(\hat{s}^i_{t+1}|s_t, a_t)\|s_{t+1} - \hat{s}^i_{t+1}\|^2)}{\sum_j \exp(-p_j(\hat{s}^j_{t+1}|s_t, a_t)\|s_{t+1} - \hat{s}^j_{t+1}\|^2)}, \tag{4}$$

where $\|.\|$ denotes the L2-norm. Note here that the state s is represented as a multinomial vector. The observation o_i is calculated by the difference between the real state s_{t+1} and the state $\hat{s}^i_{t+1}$ drawn from the probability $p_i(\hat{s}^i_{t+1}|s_t, a_t)$ of the ith module, so as to represent a kind of suitability of the ith module for the current environment. Since the behaviors of o_i suffer from large sampling variances, the responsibility λ_i is obtained based on o_i by smoothing out the large variance. For this purpose, we assume a generative model of the observation o_i as a Gaussian distribution, so that the mean μ_i and the variance σ_i^2 are estimated by on-line maximum likelihood estimation:

$$\mu_i \leftarrow (1 - \alpha)\mu_i + \alpha o_i, \tag{5}$$

$$\sigma_i^2 \leftarrow (1 - \alpha)\sigma_i^2 + \alpha(o_i - \mu_i)^2, \tag{6}$$

where α is the learning rate. Online estimation using a constant learning rate suffices for our purpose here. We adopt the moment estimates, μ_i and σ_i^2, to construct the responsibility λ_i as

$$\lambda_i \equiv \beta \frac{\mu_i}{\sqrt{\sigma_i^2}}, \tag{7}$$

where σ_i is used to reduce the responsibility of the modules whose predictors have large variances. β is a constant that represents the total reliability of the modules, and is here set in advance. Weighting the norm $\|.\|^2$ by the probability of $\hat{s}^i_{t+1}$ in equation (4) makes the sampling variance of the suitable module observation small and that of the unsuitable one large, so as to effectively reduce the responsibility of unsuitable modules in equation (7) and hence to result in quick adaptation to changing environments.

For effective decomposition of the nonstationary environment into some approximately stationary ones, we introduce a self-organization mechanism into the estimation of the transition probabilities embedded in the predictors. This mechanism is similar to learning vector quantization (LVQ), which has sometimes been used in supervised learning, and we modify it to be applicable to nonstationary RL problems.

According to our scheme, the ith predictor has an N-dimensional parameter vector w_i, which represents statistics such as stable transition probabilities of the ith environment. These parameters, $w_1, \ldots, w_M$, play crucial roles in decomposing the whole environment, and the features need to be sufficiently different. As their differences are measured by distance, one parameter must be spatially apart from the others for good decomposition. Now we consider that the new observation or the estimation $\hat{w}$ is obtained from the current environment. To decompose well, we must select a parameter w_i to be moved to $\hat{w}$, and the rest of the parameters $w_{j \neq i}$ should be kept apart from $\hat{w}$. The update rule of the parameters is then as follows:

$$w_i \leftarrow w_i + \eta A(\hat{w} - w_i), \qquad\qquad \text{if}\ \ i = \arg\min_j \|A(\hat{w} - w_j)\| \qquad (8)$$

$$w_j \leftarrow w_j - \eta \frac{1}{\|w_i - w_j\|} \frac{A(w_i - w_j)}{\|A(w_i - w_j)\|}, \qquad\qquad \text{otherwise}, \qquad (9)$$

where the weight matrix A is defined as $diag(\delta_1, \ldots, \delta_N)$, which is prescribed by the priority δ_n. η is the learning rate. A concrete value of δ_n is given later (in equations (12) and (13)).

4   Computer Experiments

To examine the effectiveness of the proposed method, we apply it to the RWME task, which is a nonstatinoary RL task, and compare the performance of the proposed and the previous methods.

4.1   Random Walk in Multiple Environments Problem

Here, we introduce the RWME problem, a stochastic shortest path problem composed of 2 absorbing states and 7 non-absorbing states, depicted in Figure 1. In each learning episode, the initial state is uniformly drawn from the state 2, 3, 4, or 5. The agent gets the reward $r_t = -1$ and 1 when it moves to the terminate state $s = 0$ and 8, respectively, otherwise $r_t = 0$.

In this problem, there are three environments whose transition probabilities are totally different. The environment surrounding the learning agent is randomly selected from them or from some combinations of them in each episode. That is, the environment is nonstationary. To solve this problem, time-dependent decomposition of the nonstationary environment into several (probably three) stationary environments is necessary, and each policy must adapt to each decomposed environment.

In this task, the agent has no knowledge of the environment, and then it must estimate the transition probabilities of the current environment. We use the following simple rule to approximate the transition probabilities when the agent takes an action a_t at state s_t:

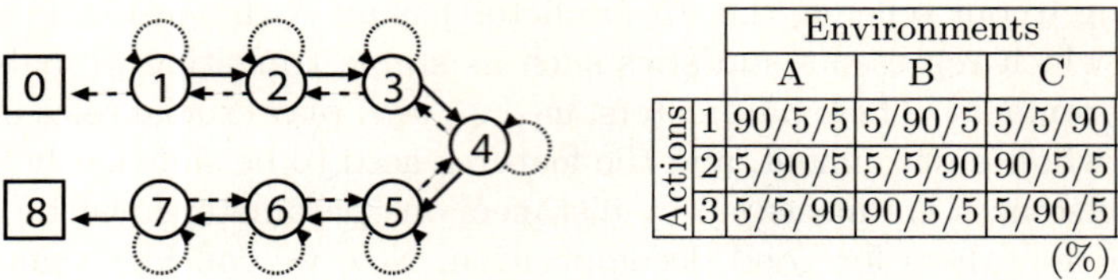

Fig. 1. The problem of random walk in multiple environments. The transition probabilities $p(s_{t+1}|a_t, s_t)$ in three environments (A, B and C) are totally different, as shown in the table. The number in each cell represents the percentage of the transition probability expressed by solid/dashed/dotted arrow in each of the three environments.

$$\hat{p}(s_{t+1}|s_t, a_t) \leftarrow \frac{1}{z}((1 - \eta)\hat{p}(s_{t+1}|s_t, a_t) + \eta), \quad \text{if the next state is } s_{t+1} \tag{10}$$

$$\hat{p}(s_{t+1}|s_t, a_t) \leftarrow \frac{1}{z}(1 - \eta)\hat{p}(s_{t+1}|s_t, a_t), \qquad \text{otherwise,} \tag{11}$$

where η is a constant learning rate and z is a normalization term. The parameters of each predictor are updated by regarding $\hat{p}$ as $\hat{w}$ (equation (8)) in every transition. δ is the parameter of the approximation accuracy in each dimensionality, and is updated as

$$\delta_{s_t,a_t} \leftarrow 1, \tag{12}$$

$$\delta_{s \neq s_t, a \neq a_t} \leftarrow \max(0, \delta_{s,a} - \text{const}). \tag{13}$$

The large value of δ indicates high accuracy. The parameters of each policy are also updated by eligibility traces in each episode.

4.2   Experimental Settings and Results

The experimental settings are as follows.

(a) The current environment is randomly selected from the three (A, B and C) candidates every 20 episodes ($a(k), b(k)$ and $c(k)$ in equation (14) are binary).

(b) Every 40 episodes, the environment surrounding the agent is changed smoothly. The environment is selected from the three candidates and fixed during the earlier 20 episodes, and then the next environment is selected from the rest and the current environment gradually approaches to the new one throughout the later 20 episodes ($a(k), b(k)$ and $c(k)$ in equation (14) are real numbers in [0,1] and smoothly changed).

The transition probability of the current environment in the kth episode is denoted by a convex combination of the state transition matrices, P_A, P_B, and P_C in equation (14) and depicted in Figure 2:

$$P(k) = a(k)P_A + b(k)P_B + c(k)P_C, \qquad a(k) + b(k) + c(k) = 1. \tag{14}$$

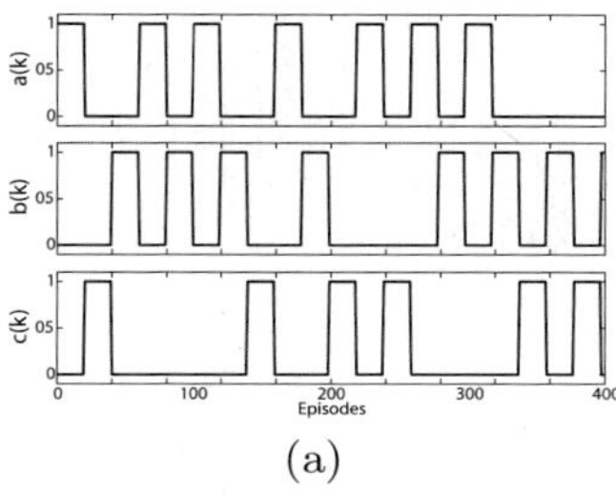

(a)

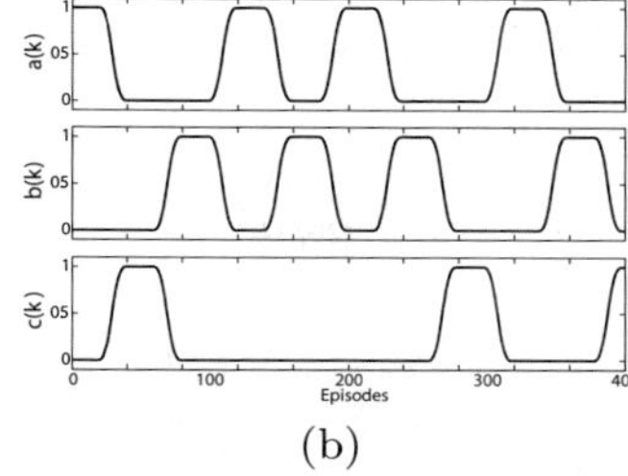

(b)

Fig. 2. An example of the mixture ratio of each environment. In the setting (a), convex combination coefficients are binary and each environmental changes every 20 episodes. In the setting (b), the interval consists of 40 episodes; in the earlier 20 episodes, the environment is selected from the three and fixed, but in the later 20 episodes, it gradually changes to the next one.

In the setting (b), the environment is more ambiguous for the agent than in the setting (a), leading to a more difficult recognition and decomposition of the environment. To prepare one policy for each episode in the gradually changing environment under the setting (b), the number of policies will be infinite.

The maximum number of training episodes was 1000, and each episode terminated when the agent arrived at any terminal state. If the environment did not change, the agent would require a single stationary policy, and the expectation of total rewards per step under the optimal policy was 0.2125. The agent had three modules, each of which had a soft-max policy parameterized by the policy parameters.

Figure 3 shows the learning curves averaged over 10 simulation runs. The vertical axis denotes the average reward in test episodes, each of which was performed after the $(5n + 3)$th learning episode $(n \in \mathbb{N})$ with a fixed initial state being state 4, and the actions were drawn from the corresponding policy. Such selection of test episodes assured the asymmetry of the enviroments, namely $a(k), b(k)$ and $c(k)$ were not equal to 0.5.

Figure 3(a) shows the learning curves under the setting (a). In this setting, the proposed method (solid) and MMRL (dashed) were both successful in adapting to the environmental changes, and appropriately achieved the optimal policy. In contrast, the ordinary policy gradient algorithm (dotted) failed to adapt in this setting. Figure 3(b) shows the learning curves under the setting (b). The upper panel shows the curves in the phase of the fixed environment, and the lower those in the phase of the varying environment. Since the previous method (MMRL) could not well train the predictors, it performed worse in both environments. On the other hand, the performances of the proposed model were good both in the fixed and the changing environments. Furthermore, in the fixed environment, the performance of the proposed model was found to be comparable to that in the setting (a). A major reason for the good performance of the proposed method is the self-organized decomposition of the environments. The MMRL dynamics contains no drive to make the constituent modules as distinct as possible. In the

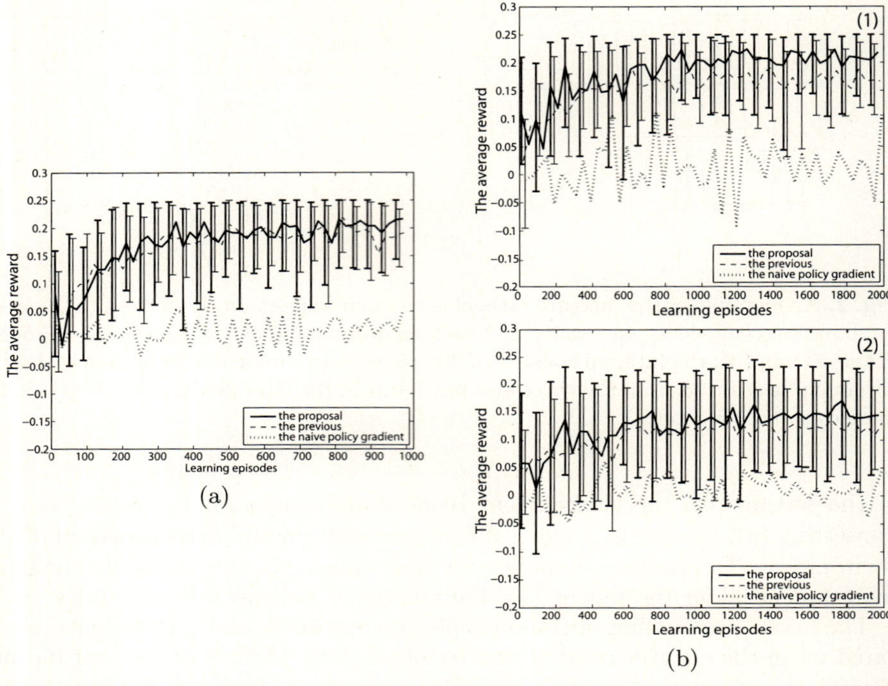

Fig. 3. The learning curves under the settings (a) and (b). A solid line shows the average reward of the proposed method, the dashed line that of MMRL, and the dotted line that of the ordinary policy gradient algorithm (i.e. single module).

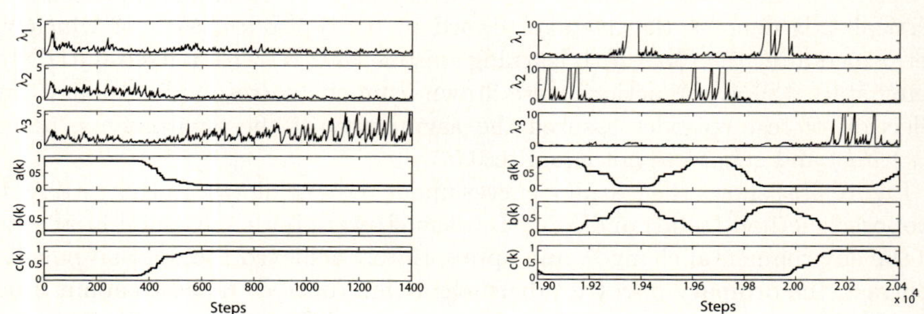

Fig. 4. The responsibility signal

previous model, when the environment was changing, the responsibility values of several modules became almost similar and then the corresponding controllers attempted to adapt to some mixed environments, which produced relatively inferior controllers. On the other hand, the proposed method utilized the self-organized mechanism so that each of the controllers adapted to the ith local subspace well both in the fixed and the changing environments.

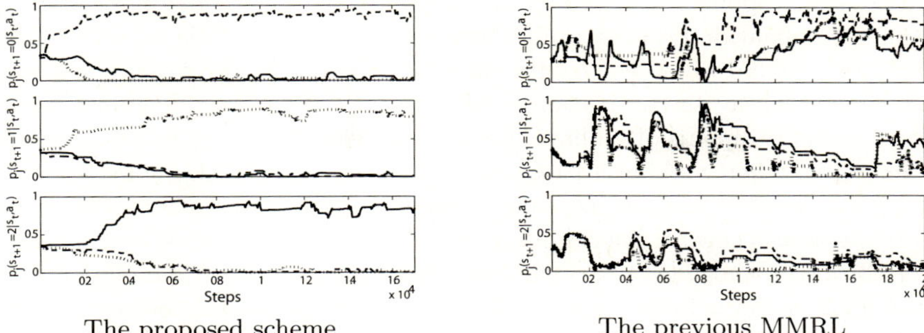

Fig. 5. The decomposition process, which is mainly due to the estimation of the stable transition probability in each module. The top, middle and bottom panels show the probabilities $p_i(s_{t+1} = 0|s_t = 1, a_t = 1), p_i(s_{t+1} = 0|s_t = 1, a_t = 1)$ and $p_i(s_{t+1} = 2|s_t = 1, a_t = 1)$, respectively. The solid, dashed and dotted lines represent the probabilities used in the 1st, 2nd and 3rd modules, respectively. Left panels : our proposed scheme. Right panels : MMRL.

Figure 4 shows the time-series of the responsibility signals before and after learning under the setting (b). Before learning, the next state estimations of all modules were similar, and then the values of the responsibility signals were all comparable. After learning, in contrast, the next state estimation of each module was fitted to its own environment, and then the responsibility signals were different from each other but stable even in the nonstationary environment.

Figure 5 shows the decomposition process under the setting (b), which is represented by the time series of the transition probability estimation $p_i(s_{t+1}|s_t, a_t)$. In the proposed method, as the number of time steps increased, the modules became self-organized. In contrast, the estimations of MMRL closely resembled each other, suggesting that the parameters of the predictors could not be learned appropriately.

5   Discussion

We applied a type of the LVQ to the learning of the predictors and showed that the proposed modular RL scheme was more adaptive and robust than the previous MMRL in temporally changing environments. The predictors' structure (equations (4) and (7)) of our scheme facilitated the self-organization of the modules, which promoted the recognition and decomposition of the surrounding environment. In the self-organization, the repulsive force between the predictors was important to encourage them to keep apart from each other. In the previous MMRL, all the predictors were updated so as to fit the surrounding environment depending only on the responsibility signals. This 'cooperative' feature made the modules develop to be similar to each other especially in rather ambiguous situations.

Next, the responsibility signals in the proposed scheme are not constrained by the normalization process, unlike those in the MMRL. This un-normalization formulation would allow us to introduce some exploration-exploitation mechanism by controlling the activation level (β) of the whole responsibility signals.

Real-world problems often have hierachical local structures in time and space. For example, when making a robot to perform multiple tasks, some coarse motions can be used in various occasions shared by many tasks, whereas they can be decomposed into several finer motions which can also be shared by the tasks. In order to detect and adapt to such hierachically segmented domains effectively, introducing a more hierachical structure to our model will be an interesting extension. In such an extension, some intermediate layers would deal with intermediate localities to extract specific information from the limited sources.

6   Concluding Remarks

In this article, we proposed a novel modular RL method and showed its effectiveness in nonstationary RL tasks. By utilizing its self-organization mechanism, the proposed method performed much better than the existing modular RL method [3] especially when the decomposition of the nonstationary environment was difficult. Moreover, our appropriate formulation with respect to the reward maximization achieved more stable optimization of the constituent controllers.

When considering application of the proposed scheme to large-scale problems, introduction of hierarchical architectures would be necessary. Such an expansion will be shown in our future study.

References

1. Wolpert, D.M., Kawato, M.: Multiple paired forward and inverse models for motor control. Neural Networks 11(7-8), 1317–1329 (1998)
2. Haruno, M., Wolpert, D.M., Kawato, M.: MOSAIC Model for Sensorimotor Learning and Control. Neural Computation 13(10), 2201–2220 (2001)
3. Doya, K., Samejima, K., Katagiri, K., Kawato, M.: Multiple Model-Based Reinforcement Learning. Neural Computation 14(6), 1347–1369 (2002)
4. Sutton, R., Barto, A.: An introduction to reinforcement learning. MIT Press, Cambridge (1998)
5. Williams, R.J.: Simple Statistical Gradient-Following Algorithms for Connectionist Reinforcement Learning. Machine Learning 8, 229–256 (1992)
6. Kohonen, T.: The self-organized map. Proc. IEEE 78(9), 1464–1480 (1990)
7. Singh, S.P.: Transfer of Learning by Composing Solutions of Elemental Sequential Tasks. Machine Learning 8, 323–339 (1992)
8. Sutton, R., Precup, D., Singh, S.: Between MDPs and Semi-MDPs: A Framework for Temporal Abstraction in Reinforcement Learning. Artif. Intell. 112(1-2), 181–211 (1999)
9. Hinton, G.E.: Training Products of Experts by Minimizing Contrastive Divergence. Neural Computation 14(8), 1771–1800 (2002)
10. Lafferty, J., McCallum, A., Pereira, F.: Conditional random fields: Probabilistic model for segmenting and labeling sequence data. In: Proc. 18th International Conf. on Machine Learning (2001)

Robust Population Coding in Free-Energy-Based Reinforcement Learning

Makoto Otsuka[1,2], Junichiro Yoshimoto[1,2], and Kenji Doya[1,2,3]

[1] Initial Research Project, Okinawa Institute of Science and Technology
12-22 Suzaki, Uruma, Okinawa 904-2234, Japan
{otsuka,jun-y,doya}@oist.jp
[2] Graduate School of Information Science, Nara Institute of Science and Technology
8916-5 Takayama, Ikoma, Nara 630-0192, Japan
[3] ATR Computational Neuroscience Laboratories
2-2-2 Hikaridai, "Keihanna Science City," Kyoto 619-0288, Japan

Abstract. We investigate the properties of free-energy-based reinforcement learning using a new experimental platform called the digit floor task. The simulation results showed the robustness of the reinforcement learning method against noise applied in both the training and testing phases. In addition, reward-dependent and reward-invariant representations were found in the distributed activation patterns of hidden units. The representations coded in a distributed fashion persisted even when the number of hidden nodes were varied.

1 Introduction

For the purpose of learning how to behave under unknown environments, reinforcement learning framework can be employed to obtain the optimal behavioral policy from delayed rewards. However, common reinforcement learning algorithms are not good at handling high-dimensional sensory data; therefore, the ad-hoc feature engineering is essential for the successful application of reinforcement learning algorithms [1]. Several methods for finding structures in high-dimensional data have been proposed in the field of reinforcement learning [2,3,4,5]. For fully observable Markov Decision Processes (MDP), most of the proposed methods construct a suitable state space by defining task-independent or task-dependent basis functions. For partially observable Markov Decision Processes (POMDP), the perceptual distinction approach and utile-distinction memory have paved different paths for discovering task-independent and task-dependent hidden structures, respectively. However, none of the approaches that follow the aforementioned paths consider the construction of task-independent and task-dependent structures simultaneously.

Recently, a reinforcement learning algorithm that is capable of handling high-dimensional inputs and actions was proposed by [6,7]. This method exploits the statistical independence realized by the undirected graphical model called the Product of Experts (PoE), and it utilizes the negative free energy of the network

V. Kůrková et al. (Eds.): ICANN 2008, Part I, LNCS 5163, pp. 377–386, 2008.

state to approximate an action value. The proposed method, which we call "free-energy-based reinforcement learning (FE-based RL)," has been demonstrated to be able to solve a temporal credit assignment problem in MDPs with large state and action spaces [6,7]. However, the characteristics of the method with POMDPs have not been fully investigated. In this article, we first demonstrate the noise tolerance obtained by the FE-based RL under the newly proposed task called the "digit floor task." We then analyze the information coding in the activation patterns of hidden nodes in the network.

2   Free-Energy-Based Reinforcement Learning

2.1   Free Energy and Its Derivatives in RBM

The FE-based RL is implemented by the restricted Boltzmann machine (RBM), which is a subtype of the PoE [6,7]. The RBM is an undirected graphical model composed of two layers of binary stochastic nodes. The nodes in a "visible" layer are fully connected to the nodes in a "hidden" layer. The special feature of the RBM is the lack of intra-layer connections between nodes within the same layer. This connectivity ensures that all hidden units are statistically decoupled when visible units are clamped to the observed values.

The stochastic behavior of the RBM network is characterized by the energy of the network. When node V_i in a visible layer $\boldsymbol{V} = \{V_1, \ldots, V_N\}$ is connected to node H_k in a hidden layer $\boldsymbol{H} = \{H_1, \ldots, H_K\}$ by a symmetric weight $w_{ik}{}^1$, the energy of a particular configuration $(\boldsymbol{V} = \boldsymbol{v}, \boldsymbol{H} = \boldsymbol{h})$ is defined by

$$E(\boldsymbol{v}, \boldsymbol{h}) = - \sum_{i,k} v_i w_{ik} h_k \ . \tag{1}$$

The state of each node is stochastically updated so that the probability of the network being in a particular configuration $(\boldsymbol{v}, \boldsymbol{h})$ is given by

$$P(\boldsymbol{v}, \boldsymbol{h}) = \frac{\exp(-E(\boldsymbol{v}, \boldsymbol{h}))}{\sum_{\hat{\boldsymbol{v}}, \hat{\boldsymbol{h}}} \exp(-E(\hat{\boldsymbol{v}}, \hat{\boldsymbol{h}}))} \ . \tag{2}$$

The marginal probability distribution over the visible units is expressed by

$$P(\boldsymbol{v}) = \frac{\sum_{\hat{\boldsymbol{h}}} \exp(-E(\boldsymbol{v}, \hat{\boldsymbol{h}}))}{\sum_{\hat{\boldsymbol{v}}} \sum_{\hat{\boldsymbol{h}}} \exp(-E(\hat{\boldsymbol{v}}, \hat{\boldsymbol{h}}))} = \frac{\exp(-F(\boldsymbol{v}))}{\sum_{\hat{\boldsymbol{v}}} \exp(-F(\hat{\boldsymbol{v}}))} \tag{3}$$

where $F(\boldsymbol{v})$ is called the "equilibrium free energy" of $\boldsymbol{v}$. As shown in [6,7], this quantity is also expressed as the sum of the expected energy and the negative entropy of the hidden nodes

$$F(\boldsymbol{v}) = \sum_{\boldsymbol{h}} P(\boldsymbol{h}|\boldsymbol{v}) E(\boldsymbol{v}, \boldsymbol{h}) + \sum_{\boldsymbol{h}} P(\boldsymbol{h}|\boldsymbol{v}) \log P(\boldsymbol{h}|\boldsymbol{v}) \tag{4}$$

where the hidden node distribution is given by

---

^{1} Each node also possesses a bias weight. However, we do not express these weights explicitly to maintain the notation uncluttered.

$$P(\boldsymbol{h}|\boldsymbol{v}) = \frac{\exp(-E(\boldsymbol{v}, \boldsymbol{h}))}{\sum_{\hat{\boldsymbol{h}}} \exp(-E(\boldsymbol{v}, \hat{\boldsymbol{h}}))} \ . \tag{5}$$

A marked feature of RBM is that the hidden node distribution is factored as $P(\boldsymbol{h}|\boldsymbol{v}) = \prod_k P(h_k|\boldsymbol{v})$, which allows easy computation of the free energy of the network and its derivatives:

$$\frac{\partial F(\boldsymbol{v})}{\partial w_{ik}} = -v_i \langle h_k \rangle, \quad \text{where} \quad \langle h_k \rangle = \frac{1}{1 + \exp(-\sum_i w_{ik} v_i)} \ . \tag{6}$$

2.2   Function Approximation with RBM

If the environmental dynamics of the MDP, given by $(\mathcal{S}, \mathcal{A}, T, R, \gamma)^2$, are known we can obtain the optimal behavioral policy $\pi^* : \mathcal{S} \to \mathcal{A}$ that ensures the highest amount of the discounted return

$$R_t = \sum_{k=1}^{\infty} \gamma^k r_{t+k+1} \ , \tag{7}$$

through the generalized policy iteration. Here, r_t denotes an immediate reward at time t, and γ denotes a discount rate that balances the importance between the immediate and delayed rewards. The generalized policy iteration determines the optimal policy by iterating the following two steps: (1) the policy evaluation step in which the action value function is computed by

$$Q^\pi(s, a) = E_\pi \left\{ \sum_{k=0}^{\infty} \gamma^k r_{t+k+1} | s_t = s, a_t = a \right\}, \tag{8}$$

and (2) the policy improvement step in which the current policy π is updated to π' such that

$$\pi'(s) = \arg\max_a Q^\pi(s, a) \ . \tag{9}$$

In many cases, an agent does not know the dynamics of the environment; therefore, an agent must sample the rewards and state transitions and approximate the action value function by reinforcement learning.

The FE-based RL facilitates the approximation of action values in large state and action spaces. The agent is the RBM network with visible and hidden layers. The visible layer is composed of visible state (observation) nodes and action nodes $\boldsymbol{V} = (\boldsymbol{Z}, \boldsymbol{A})$. Action nodes are imposed with a restriction that only one of the nodes can be active at a time. The hidden layer is purely composed of hidden nodes $\boldsymbol{H}$. An observation node Z_i is connected to the hidden node H_k by the connection weight w_{ik}, and the action node A_j is connected to the hidden node H_k by u_{jk}. Here, the free energy (4) can be expressed as

---

^{2} The $\mathcal{S} \ni s$ and $\mathcal{A} \ni a$ are sets of all possible states and actions, respectively. The T is a state transition probability, and R is a reward function.

$$F(\boldsymbol{z}, \boldsymbol{a}) = -\sum_{k=1}^{K}\left(\sum_{i=1}^{N_s} w_{ik} z_i \langle h_k \rangle + \sum_{j=1}^{N_a} u_{jk} a_j \langle h_k \rangle\right)$$

$$+ \sum_{k=1}^{K} \langle h_k \rangle \log\langle h_k \rangle + \sum_{k=1}^{K} (1 - \langle h_k \rangle) \log(1 - \langle h_k \rangle) \tag{10}$$

for $(\boldsymbol{Z} = \boldsymbol{z}, \boldsymbol{A} = \boldsymbol{a})$. The network is trained so that the negative free energy approximates the action value function, i.e., $Q(\boldsymbol{z}, \boldsymbol{a}) \approx -F(\boldsymbol{z}, \boldsymbol{a})$. By applying SARSA(0) algorithm with a function approximator [1], the particular form of Eq. (10) provides a simple update rule for the network parameters:

$$\Delta w_{ik} = \alpha \left(r_{t+1} - \gamma F(\boldsymbol{z}_{t+1}, \boldsymbol{a}_{t+1}) + F(\boldsymbol{z}_t, \boldsymbol{a}_t)\right) z_{i,t} \langle h_{k,t} \rangle \tag{11}$$

$$\Delta u_{jk} = \alpha \left(r_{t+1} - \gamma F(\boldsymbol{z}_{t+1}, \boldsymbol{a}_{t+1}) + F(\boldsymbol{z}_t, \boldsymbol{a}_t)\right) a_{j,t} \langle h_{k,t} \rangle . \tag{12}$$

Using the estimated action values, an action can be selected from the Boltzmann distribution with the inverse temperature β.

3   Digit Floor Task

The digit floor task is designed to test the noise tolerance of the FE-based RL and to investigate the type of information coded in the activation patterns of a hidden layer.

The goal of an agent in the digit floor task is to discover the "optimal" way to act on a "digit floor" autonomously. The "optimal" behavioral policy leads to the highest discounted return. At each time step, the agent decides the direction of movement depending on previous experiences and the current noisy observation of a hand-written digit painted on the floor. The observations are obtained through a narrow fixed-size "retina;" hence, the agent cannot see the entire floor at once. As a result, the problem is partially observable. After the action is executed, depending on the hidden state (a true class of the hand-written digit painted on a tile) and the selected action, the agent receives a numerical reward and moves toward its intended direction. When the agent reaches the goal position, the agent is randomly transported to a position other than the goal position.

The MNIST database of handwritten digits, which contains 60,000 training images and 10,000 test images, is used for the patterns on the tiles. The original pixel intensity of the images in the MNIST data set is represented by values in the range 0 to 255. The intensity is normalized to the range of $[0, 1]$ in order to provide a probabilistic interpretation to the value. A "tile" is composed of 28×28 pixels of a handwritten digit. A "patch" is composed of 3×3 tiles positioned in a predefined configuration. A "floor" comprises $L \times L$ patches, where for each patch, different tiles are selected randomly from the training data set. The floor has a torus structure; hence, it has no "edge." Fig. 1(c) shows a typical example of floor patterns with $L = 4$.

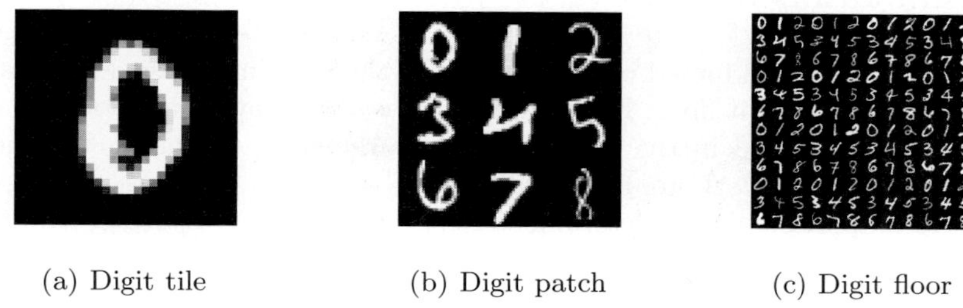

(a) Digit tile   (b) Digit patch   (c) Digit floor

Fig. 1. Building components of a digit floor: (a) is composed of 28×28 pixels; (b) is composed of 3×3 tiles; and (c) is composed of 4×4 patches

To investigate the properties of the FE-based RL, two constraints are imposed in this study.[3] The first constraint is that the unit distance of movement is one tile. The other constraint is that the agent has an "imperfect" vision system with a retina size of 28×28; here, "imperfect" implies that some of pixels observed in the current input are randomly replaced by white noise. Therefore, an agent never observes the exact same visual input even if it has visited the same tile many times. Examples of the noisy inputs are shown in the Fig. 2.

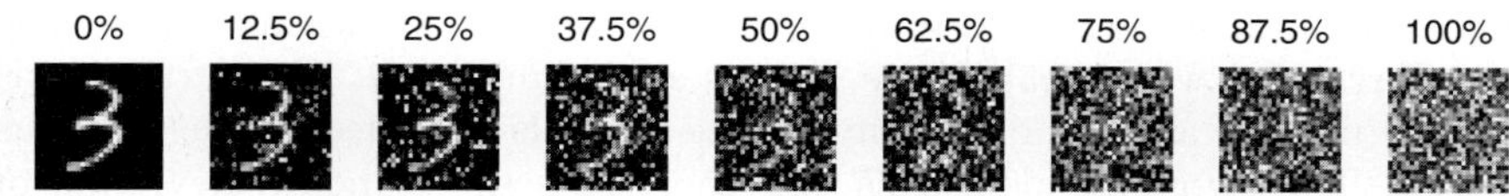

Fig. 2. Examples of observations under noise. The number on the top of each image shows the noise level (the percentage of pixels replaced by white noise).

The digit floor task is formally defined as a POMDP, which is commonly specified with a 7-tuple $(\mathcal{S}, \mathcal{A}, \mathcal{Z}, T, R, P, \gamma)$. A set of hidden states on the floor is given by $\mathcal{S}$, which is a subset of all the possible types of tiles $\mathcal{S}_0 \equiv \{$ "0", ..., "9", "N"$\}$[4]. A set of all the possible actions is a discrete set $\mathcal{A} \equiv \{$'N', 'E', 'S', 'W'$\}$, which indicates "step one tile to the North/East/ South/West." A set of all the possible observations is $\mathcal{Z} \equiv [0, 1]^{784}$. A state transition $T : \mathcal{S} \times \mathcal{A} \to \mathcal{S}$ is deterministic, and the agent always moves toward the intended position. A default reward function $R_0 : \mathcal{S} \times \mathcal{A} \to \mathbb{R}$ is defined as follows: the movements toward the tile "4" from neighboring tiles give a $+10$ reward. Clockwise movements at the corners of each patch result in a $+4$ reward. All other movements incur a -2 reward. The probability distribution over the observation space given the true state $P(z \in \mathcal{Z}|s \in \mathcal{S})$ is expressed on the basis of the exact floor map and noise levels.

---

[3] These constraints can be removed in the general setting.

[4] A double quotation mark is used to indicate a certain digit class. Also "N" indicates a tile is covered with white noise.

4   Simulation Results

The simulation is conducted in two phases. In the training phase, an agent acts on the predefined digit floor. In the testing phase, 400 new images of each digit, which never appeared in the training phase, are shown to the agent, and the behaviors of the network are investigated.

The meta-parameters for learning were set as follows. A learning rate α was exponentially decreased from 0.1 to 0.01 in the course of each epoch. In order to avoid the unwanted influence of the large number of connections on the speed of learning, the learning rate was divided by the square root of the number of total edges in the network. The inverse temperature β, which is a parameter in the Boltzmann action selection rule, was linearly increased from 0 to 2. The discount rate γ was set to 0.9.

4.1   Noise Tolerance

In order to test the influence of noise in both the training and testing phases, images were tainted with different levels of noise. We conducted tests for nine different noise levels incremented by 12.5%, or equivalently 98 noise pixels, from the noiseless image to the purely noisy image.

An agent took 100,000 steps per epoch. The agent had 20 hidden nodes. The optimal action at each reward-invariant hidden state is shown in Fig. 3. The course of training with 12.5% noise level is shown in Fig. 4. The vertical axis shows the moving average of the immediate rewards obtained in the previous 100 steps. The error bar along this axis represents the standard deviation of the moving average over 10 epochs with different initialization. As shown in the figure despite the 12.5% noise level in the training phase, the agent performs at almost the optimal level after 60,000 steps.

Before and after each epoch of training, 400 test images with various noise levels were shown to the agent. Fig. 5(b) shows the count of the greedy actions induced by showing 400 noisy test images to the agent that is trained on a digit floor with 12.5% observation noise. The results for each digit are organized in such a manner that they match the pre-configured position of the digit floor. For example, the upper-left graph of Fig. 5(b) illustrates that most of the test images of "0" triggered the correct optimal action 'E' in all the 10 epochs after training. The same behavioral pattern was observed for all digits used in the task, and therefore the result supports the robustness of the FE-based RL to noise. For comparison, the same 400 noisy test images were shown to the agent before training. The result in Fig. 5(a) indicates that an agent does not have a behavioral tendency for a certain digit.

The agent was able to perform at the satisfactory level even with 37.5% noise level in a test set.[5] However, the performance deteriorated after 50% of noise level in the test set, as shown in Fig. 5(c). When trained with 50% of noise, the performance did not deteriorate even with 50% of test noise, as shown in

---

[5] Data not shown due to the limited space.

Fig. 3. Optimal actions

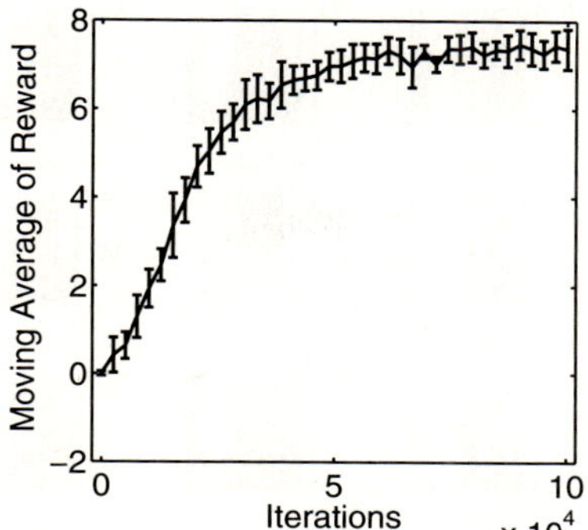

Fig. 4. Moving average of the immediate rewards over 100 steps

Fig. 5(d). The agent could even perform at the satisfactory level with 62.5% test noise level before the performance deteriorated at 75% of noise level. With 75 % noise level, even people find it difficult to detect the correct class of a given digit, as shown in Fig. 2. The results suggest that the tolerance to the noise get better as the noise level in training increases.

4.2   Population Coding of Reward-Dependent and -Invariant States

In order to analyze information coded in the activity patterns of hidden nodes, we modified the previous floor pattern. All the tiles in state "5" in Fig. 1(c) are replaced by purely noisy images. As with other digit tiles, each patch received a different noise tile.

An agent was trained for 30,000 steps per epoch under the 12.5% noise level. Training was repeated for 10 epochs with different initializations for position and weights. For each epoch, the agent was exposed to the 400 unseen hand-written digits in a test set with 12.5% noise level.

Two different reward functions were designed to verify reward-dependent coding. The first reward function was the same as the previously used one, which we call a clockwise (CW) condition. Similarly, the second reward function is called a counterclockwise (CCW) condition. The magnitude of reward in both conditions were the same. The only difference was the optimal action at each corner of the patches.

When 400 test images of each digit were presented to a trained agent, 20 hidden nodes showed differential activations. In order to quantify the similarity between the activation patterns induced by digits, we measured the angles between the median firing patterns of digits,[6] which were collectively represented by a Gram matrix $\boldsymbol{G}$. In order to visualize a structure captured by the Gram matrix,

---

[6] Activation of a hidden node H_k for a given input $\boldsymbol{s}$ is defined as a marginal posterior: $\sum_{\boldsymbol{a}} P(h_k|\boldsymbol{s}, \boldsymbol{a})$.

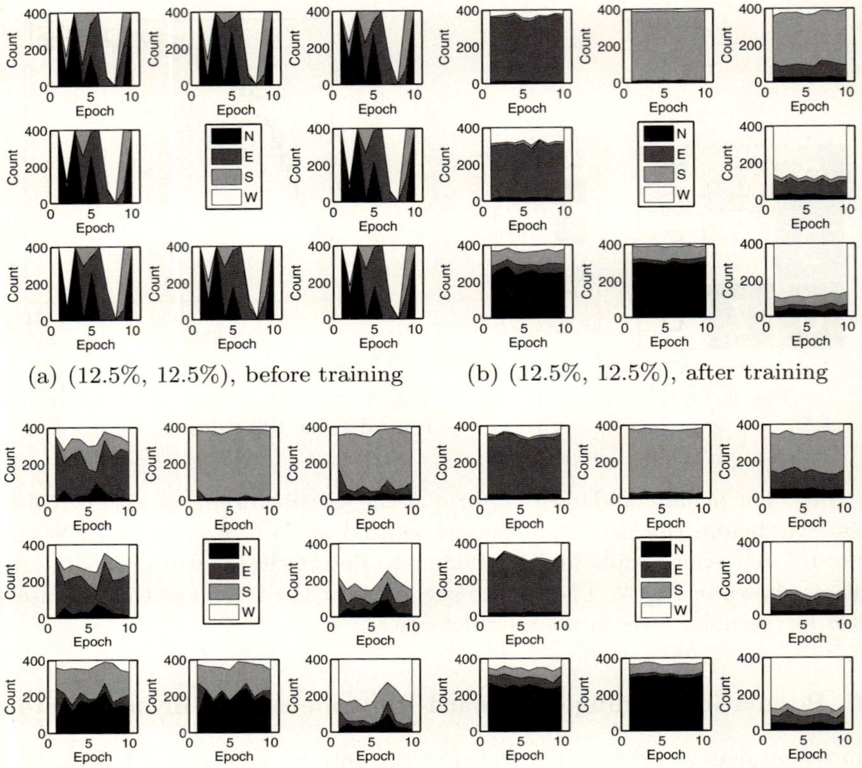

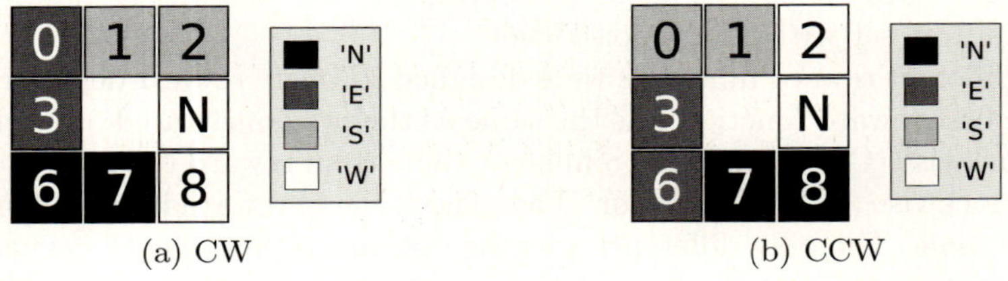

(a) (12.5%, 12.5%), before training (b) (12.5%, 12.5%), after training

(c) (12.5%, 50%), after training (d) (50%, 50%), after training

Fig. 5. Count of greedy actions induced by 400 test images for different combinations of training and test noise levels. The two numbers in the parenthesis indicate the noise levels in the training and testing phases.

(a) CW (b) CCW

Fig. 6. Optimal actions under the clockwise (CW) and counterclockwise (CCW) conditions

hierarchical clustering was employed. A distance matrix used for the hierarchical clustering was created from the Gram matrix using $D = g\mathbb{1}^{\top} + \mathbb{1}g^{\top} - 2G$, where g denotes the vectorized diagonal entries of G.

In the CW condition, as shown in Fig. 6(a), the following pairs had the same optimal action: ("1", "2"), ("0", "3"), ("6", "7"), ("8", "N") where "N" denotes

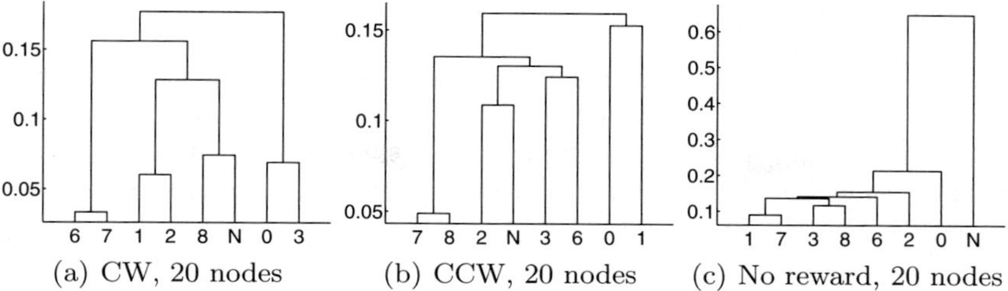

(a) CW, 20 nodes   (b) CCW, 20 nodes   (c) No reward, 20 nodes

Fig. 7. Hierarchically clustered median firing patterns of 20 hidden nodes

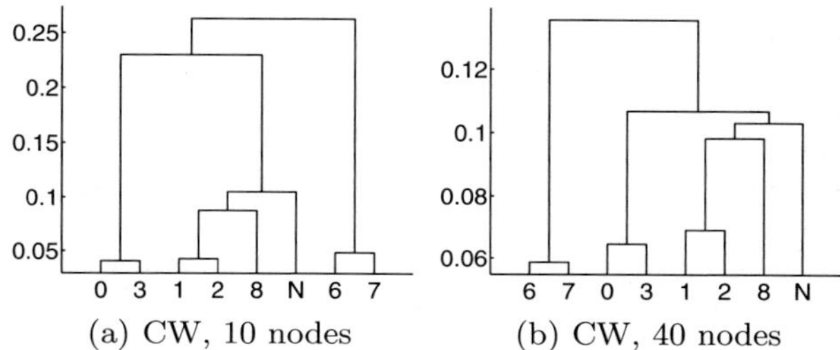

(a) CW, 10 nodes   (b) CW, 40 nodes

Fig. 8. Hierarchically clustered median firing patterns of 10 and 40 hidden nodes

the white noise. As shown in Fig. 7(a), each pair with the same optimal action formed a distinct cluster. In the CCW condition, the following pairs had the same optimal action: ("0", "1"), ("3", "6"), ("7", "8"), ("2", "N"). Further, all the pairs formed separate clusters. These aggregation patterns could not be explained by the common statistical regularity between digit images because the digit closest to "N" was "8" in the CW condition as opposed to "2" in the CCW condition.

As shown in Fig. 7(c), without any reward, perceptually similar "1" and "7" were grouped together; in the same manner, "3" and "8" were grouped together. This indicates that without a reward information, an agent extracts reward-independent statistical regularities of digits on the floor. This viewpoint is also supported by the large distance between "N," which did not possess any statistical regularities, and other digits.

4.3   Population Coding Invariant to the Size of the Hidden Layer

The size of the hidden layer was varied to examine its impact on the reward-driven representation of hidden states. Two new conditions with 10 and 40 hidden nodes were inspected, and the results were compared with the default condition with 20 hidden nodes shown in Fig. 7(a). Figs. 8(a) and (b) are not clearly different as compared to Fig. 7(a). Therefore, we can safely state that the reward-dependent representations formed by FE-based RL do not severely depend on the number of nodes in a hidden layer.

5   Conclusion

In this paper, we have shown that both reward-dependent and reward-invariant representations are robustly coded in the activation patterns of hidden nodes in the RBM. A new task called digit floor task has been introduced to test the properties of the FE-based RL. In future studies, we plan to integrate the FE-based RL and the hierarchical RBM [8].

References

1. Sutton, R.S., Barto, A.G.: Reinforcement Learning: An Introduction. MIT Press, Cambridge (1998)
2. Mahadevan, S.: Samuel Meets Amarel: Automating Value Function Approximation Using Global State Space Analysis. In: 20th National Conference on Artificial Intelligence, pp. 1000–1005. IEEE Press, New York (2005)
3. Keller, P.W., Mannor, S., Precup, D.: Automatic Basis Function Construction for Approximate Dynamic Programming and Reinforcement Learning. In: 23rd International Conference of Machine Learning, pp. 449–456. ACM Press, New York (2006)
4. Chrisman, L.: Reinforcement Learning with Perceptual Aliasing: The Perceptual Distinctions Approach. In: 10th National Conference on Artificial Intelligence, pp. 183–188. IEEE Press, New York (1992)
5. McCallum, A.: Overcoming Incomplete Perception with Util Distinction Memory. In: 10th International Conference on Machine Learning, pp. 190–196. Morgan Kaufmann, Amherst (1993)
6. Sallans, B., Hinton, G.E.: Using Free Energies to Represent Q-values in a Multiagent Reinforcement Learning Task. In: NIPS, pp. 1075–1081. MIT Press, Cambridge (2000)
7. Sallans, B., Hinton, G.E.: Reinforcement Learning with Factored States and Actions. Journal of Machine Learning Research 5, 1063–1088 (2004)
8. Hinton, G.E., Salakhutdinov, R.R.: Reducing the Dimensionality of Data with Neural Networks. Science 313(5786), 504–507 (2006)

Policy Gradients with Parameter-Based Exploration for Control

Frank Sehnke[1], Christian Osendorfer[1], Thomas Rückstieß[1],
Alex Graves[1], Jan Peters[3], and Jürgen Schmidhuber[1,2]

[1] Faculty of Computer Science, Technische Universität München, Germany
[2] IDSIA, Manno-Lugano, Switzerland
[3] Max-Planck Institute for Biological Cybernetics Tübingen, Germany

Abstract. We present a model-free reinforcement learning method for partially observable Markov decision problems. Our method estimates a likelihood gradient by sampling directly in parameter space, which leads to lower variance gradient estimates than those obtained by policy gradient methods such as REINFORCE. For several complex control tasks, including robust standing with a humanoid robot, we show that our method outperforms well-known algorithms from the fields of policy gradients, finite difference methods and population based heuristics. We also provide a detailed analysis of the differences between our method and the other algorithms.

1 Introduction

Policy gradient methods, so called because they search in policy space without using value estimation, are among the most effective optimisation strategies for complex, high dimensional reinforcement learning tasks [1,2,3,4]. However, a significant problem with policy gradient algorithms such as REINFORCE [5], is that the high variance in their gradient estimates leads to slow convergence. Various approaches have been proposed to reduce this variance [6,7,2,8].

In what follows we introduce an alternative method, called policy gradients with parameter-based exploration (PGPE), which replaces the search in policy space with a direct search in model parameter space. As with REINFORCE, the search is carried out by generating history samples, and using these to estimate the likelihood gradient with respect to the parameters. The advantage of PGPE is that a single parameter sample can be used to generate an entire action-state history, in contrast with policy gradient methods, where an action sample is drawn from the policy on every time step. This provides PGPE with lower variance history samples, and correspondingly lower variance gradient estimates. In addition, since PGPE estimates the parameter gradient directly, it can be used to train non-differentiable controllers.

The PGPE algorithm is derived in detail in Section 2. In Section 3, we test PGPE on three control experiments, and compare its performance with RE-INFORCE, evolution strategies (ES) [9], simultaneous perturbation stochastic adaptation (SPSA) [10], and natural actor critic (NAC) [4]. In Section 4 we

V. Kůrková et al. (Eds.): ICANN 2008, Part I, LNCS 5163, pp. 387–396, 2008.

analyse the relationship between PGPE and the other algorithms. In particular we carry out a range of experiments where we iteratively modify each of RE-INFORCE, SPSA and ES in such a way that they become more like PGPE, and evaluate the corresponding improvement in performance. Conclusions and directions for future work are presented in Section 5.

2   The Policy Gradients with Parameter-Based Exploration Method

In what follows we derive the PGPE algorithm from the general framework of episodic reinforcement learning in a Markovian environment. In doing so we highlight the differences between PGPE and policy gradient methods such as REINFORCE, and discuss why these differences lead to more accurate parameter gradient estimates.

Consider an agent interacting with an environment. Denote the state of environment at time t as s_t and the action at time t as a_t. Because we are interested in continuous state and action spaces (usually required for control tasks), we represent both a_t and s_t with real valued vectors. We assume that the environment is Markovian, i.e. that the current state-action pair defines a probability distribution over the possible next states $s_{t+1} \sim p(s_{t+1}|s_t, a_t)$. We further assume that the actions depend stochastically on the current state and some real valued vector θ of agent parameters: $a_t \sim p(a_t|s_t, \theta)$. Lastly, we assume that each state-action pair produces a scalar reward $r_t(a_t, s_t)$. We refer to a length T sequence of state-action pairs produced by an agent as a *history* $h = [s_{1:T}, a_{1:T}]$ (elsewhere in the literature such sequences are referred to as *trajectories* or *roll-outs*).

Given the above formulation we can associate a cumulative reward r with each history h by summing over the rewards at each time step: $r(h) = \sum_{t=1}^{T} r_t$. In this setting, the goal of reinforcement learning is to find the parameters θ that maximize the agent's expected reward:

$$J(\theta) = \int_H p(h|\theta)r(h)dh \tag{1}$$

An obvious way to maximise J is to use $\nabla_\theta J$ to carry out gradient ascent. Noting that the reward for a particular history is independent of θ, and using the standard identity $\nabla_x y(x) = y(x)\nabla_x \log x$, we can write

$$\nabla_\theta J(\theta) = \int_H p(h|\theta)\nabla_\theta \log p(h|\theta)r(h)dh \tag{2}$$

Since the environment is Markovian, and since the states are conditionally independent of the parameters given the agent's choice of actions, we can write $p(h|\theta) = p(s_1)\Pi_{t=1}^{T}p(s_{t+1}|s_t, a_t)p(a_t|s_t, \theta)$. Substituting this into Eq. (2) gives

$$\nabla_\theta J(\theta) = \int_H p(h|\theta) \sum_{t=1}^{T} \nabla_\theta p(a_t|s_t, \theta)r(h)dh \tag{3}$$

Clearly, integrating over the entire space of histories is unfeasible, and we therefore resort to sampling methods:

$$\nabla_\theta J(\theta) \approx \frac{1}{N} \sum_{i=1}^{N} \sum_{t=1}^{T} \nabla_\theta p(a_t^i | s_t^i, \theta) r(h^i) \tag{4}$$

where the histories h^i are chosen according to $p(h^i|\theta)$. The question then becomes one of how to model $p(a_t|s_t, \theta)$. In policy gradient methods such as REINFORCE, the parameters θ are used to determine a probabilistic *policy* $\pi_\theta(a_t|s_t) = p(a_t|s_t, \theta)$. A typical policy model would be a parametric function approximator whose outputs define the probabilities of taking different actions. In this case the histories can be sampled by choosing an action at each time step according to the policy distribution, and the final gradient can be calculated by differentiating the policy with respect to the parameters. However, the problem is that sampling from the policy on every time step leads to an excessively high variance in the sample over histories, and therefore in the estimated gradient.

PGPE addresses the variance problem by replacing the probabilistic policy with a probability distribution over the parameters themselves, that is

$$p(a_t|s_t, \rho) = \int_\Theta p(\theta|\rho) \delta_{F_\theta(s_t), a_t} d\theta, \tag{5}$$

where ρ are the hyperparameters determining the distribution over the parameters θ, $F_\theta(s_t)$ is the (deterministic) action chosen by the model with parameters θ in state s_t, and δ is the usual Dirac delta function. The advantage of this approach is that, because the actions are deterministic, an entire history can be generated using a single sample from the parameters, thereby reducing the variance in the gradient estimate. As an added benefit the gradient is estimated directly by sampling the parameters, which allows the use of non-differentiable controllers. The expected reward with hyperparameters ρ is:

$$J(\rho) = \int_\Theta \int_H p(h, \theta|\rho) r(h) dh d\theta \tag{6}$$

Differentiating this with respect to ρ and applying the log trick as before we get:

$$\nabla_\rho J(\rho) = \int_\Theta \int_H p(h, \theta|\rho) \nabla_\rho \log p(h, \theta|\rho) r(h) dh d\theta \tag{7}$$

Noting that h is conditionally independent of ρ given θ, we have $p(h, \theta|\rho) = p(h|\theta)p(\theta|\rho)$ and therefore $\nabla_\rho \log p(h, \theta|\rho) = \nabla_\rho p(\theta|\rho)$. Substituting this into Eq. (6) we get

$$\nabla_\rho J(\rho) = \int_\Theta \int_H p(h|\theta)p(\theta|\rho) \nabla_\rho \log p(\theta|\rho) r(h) dh d\theta \tag{8}$$

Again we approximate the above by sampling, this time by first choosing θ from $p(\theta|\rho)$, then running the agent to generate h from $p(h|\theta)$. In what follows we assume that ρ consists of a set of means μ_i and standard deviations σ_i that determine an independent normal distribution for each parameter θ_i in θ . Note that more complex forms for the dependency of θ on ρ could be used, at the expense of higher computational cost. Some rearrangement gives the following forms for the derivative of $\log p(\theta|\rho)$ with respect to μ_i and σ_i:

$$\nabla_{\mu_i} \log p(\theta|\rho) = \frac{(\theta_i - \mu_i)}{\sigma_i^2} \qquad \nabla_{\sigma_i} \log p(\theta) = \frac{(\theta_i - \mu_i)^2 - \sigma_i^2}{\sigma_i^3} \qquad (9)$$

Following Williams [5], we update each σ_i and μ_i in the direction of the gradient using a step size $\alpha_i = \alpha\sigma_i^2$, where α is a constant. Also following Williams we subtract a baseline b from the reward r for each history. This gives us the following hyperparameter update rules:

$$\Delta\mu_i = \alpha(r - b)(\theta_i - \mu_i) \qquad \Delta\sigma_i = \alpha(r - b)\frac{(\theta_i - \mu_i)^2 - \sigma_i^2}{\sigma_i} \qquad (10)$$

A possible objection to PGPE is that parameter space is generally higher dimensional than action space, and therefore has higher sampling complexity. However, recent results [11] indicate that this drawback was overestimated in the past. In this paper we present experiments where PGPE successfully trains a controller with more than 1000 parameters. Another issue is that PGPE, at least in its present form, is explicitly episodic, since the parameter sampling is carried out once per history. This contrasts with policy gradient methods, which can be applied to infinite horizon settings as long as frequent rewards can be computed.

3   Experiments

In this section we compare PGPE with SPSA, REINFORCE, NAC and ES, on three simulated control scenarios. These scenarios allow us to model problems of similar complexity to today's real-life RL problems [12,2].

In all experiments involving evolution strategies (ES) we used a local mutation operator. We did not examine correlated mutation and CMA-ES because both mutation operators add $n(n - 1)$ strategy parameters to the genome. Since we have more than 1000 parameters for the largest controller, this would lead to a prohibitive memory cost. In addition, the local mutation operator is more similar to the perturbations in PGPE, making it easier to compare the algorithms. All plots show the average results of 40 independent runs.

3.1   Pole Balancing

The first scenario is the standard pole balancing benchmark [11]. In this task the agent's goal is to maximize the length of time a movable cart can balance a pole upright in the center of a track. The agent's inputs are the angle and angular velocity of the pole and the position and velocity of the cart. The agent is represented by a linear controller with four inputs and one output. The simulation is

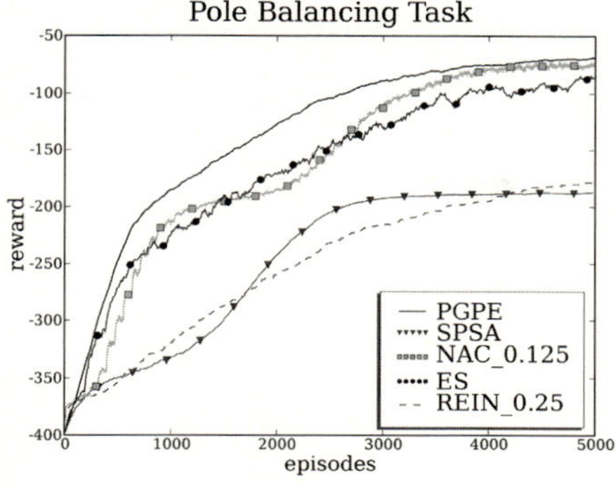

Fig. 1. PGPE compared to ES, SPSA, REINFORCE and NAC on the pole balancing benchmark

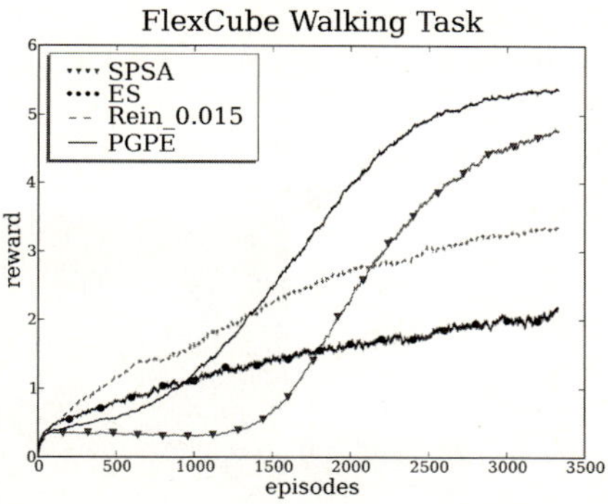

Fig. 2. PGPE compared to ES, SPSA and REINFORCE on the walking task

updated 50 times a second. The initial position of the cart and angle of the pole are chosen randomly. Figure 1 shows the performance of the various methods on the pole balancing task. All algorithms quickly learned to balance the pole, and all but SPSA eventually learned to do so in the center of the track. PGPE was both the fastest to learn and the most effective algorithm on this benchmark.

3.2 FlexCube Walking Task

The second scenario was a mass-particle system with 8 particles. The particles are modelled as point masses on the vertices of a cube, with every particle connected to every other by a spring (see Fig. 3). We refer to this scenario as the *FlexCube* framework. Though relatively simple, FlexCube can be used to perform sophisticated tasks with continuous state and action spaces. In this case the task is to make the cube "walk" — that is, to maximize the distance of its center of gravity from the starting point.

The agent can control the desired lengths of the 12 edge springs. Its inputs are the 12 current edge spring lengths, the 12 previous desired edge spring lengths (fed back from its own output at the last time step) and the 8 floor contact sensors in the vertices. The agent is represented by a Jordan network [13] with 32 inputs, 10 hidden units and 12 output units. Figure 2 shows the results on the walking task. All the algorithms learn to move the FlexCube. However, for reasons that are unclear to the authors, ES is only able to do so very slowly. PGPE substantially outperforms the other methods, both in learning speed and final reward. For a detailed view of the solutions in the walking task please refer to the video on http://www.pybrain.org/videos/icann08/.

3.3 Biped Robot Standing Task

The task in this scenario was to keep a simulated biped robot standing while perturbed by external forces. The simulation, based on the biped robot

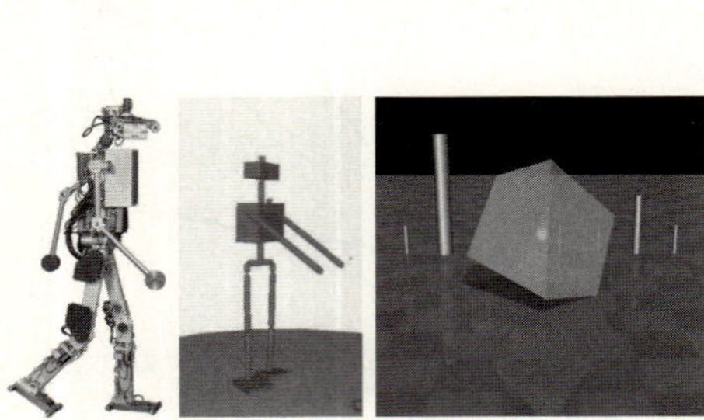

Fig. 3. The real Johnnie robot (left), its simulation (center) and the FlexCube (right)

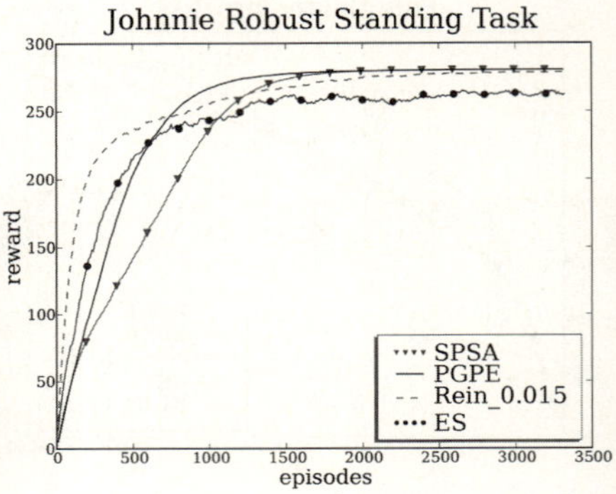

Fig. 4. PGPE compared to ES, SPSA and REINFORCE on the robust standing benchmark

Johnnie [14] was implemented using the Open Dynamics Engine. The lengths and masses of the body parts, the location of the connection points, and the range of allowed angles and torques in the joints were matched with those of the original robot. Due to the difficulty of accurately simulating the robot's feet, the friction between them and the ground was approximated with Coulomb friction. The framework has 11 degrees of freedom and a 41 dimensional observation vector (11 angles, 11 angular velocities, 11 forces, 2 pressure sensors in feet, 3 degrees of orientation and 3 degrees of acceleration in the head). The controller was a Jordan network [13] with 41 inputs, 20 hidden units and 11 output units.

The aim of the task is to maximize the height of the robot's head, up to the limit of standing completely upright. The robot is continually perturbed by random forces (see Figure 5) that would knock it over if it did not react.

As can be seen from the results in Fig. 4, the task was relatively easy, and all the methods were able to quickly achieve a high reward. REINFORCE learned especially quickly, and outperformed PGPE in the early stages of learning. However PGPE overtook it after about 500 training episodes. Figure 5 shows a typical scenario of the robust standing task with a reward outcome of 279. For more detailed views of the solution please refer to the video on http://www.pybrain.org/videos/icann08/.

3.4   Discussion

One general observation from our experiments was that the longer the episodes, the more PGPE outperformed policy gradient methods. This is not surprising, since the variance of the REINFORCE gradient estimates increases with the number of action samples. However it is an important benefit, given that most interesting real-world problems require much longer episodes than our experiments [1,2,12].

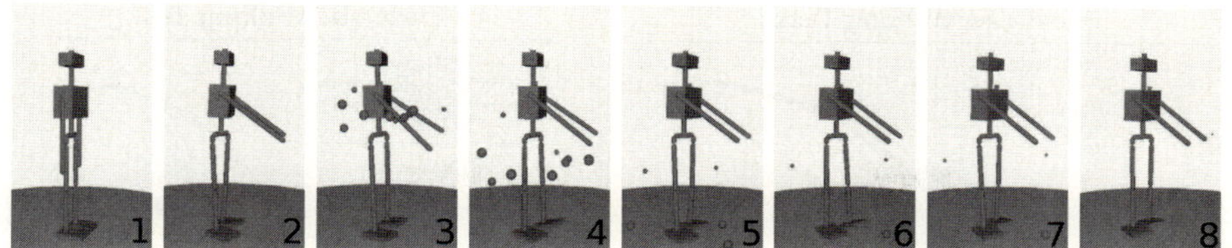

Fig. 5. From left to right, a typical solution which worked well in the robust standing task is shown: 1. Initial posture. 2. Stable posture. 3. Perturbation. 4. - 7. Backsteps right, left, right, left. 8. Stable posture regained.

4   Relationship to Other Algorithms

In this section we quantify the differences between PGPE and SPSA, ES and REINFORCE, and assess the impact of these differences on performance. Starting with each of other algorithms we incrementally alter them so that their behaviour (and performance) becomes closer to that of PGPE. In the case of SPSA we end up an algorithm identical to PGPE; for the other methods the transformed algorithm is similar, but still inferior, to PGPE.

4.1   From SPSA to PGPE

Two changes are required to transform SPSA into PGPE. First the uniform sampling of perturbations is replaced by Gaussian sampling, (with the finite differences gradient correspondingly replaced by the likelihood gradient). Second, the variances of the perturbations are turned into free parameters and trained with the rest of the model (initially the Gaussian sampling is carried out with fixed variance, just as the range of uniform sampling is fixed in SPSA). Figure 6 shows the performance of the three variants of SPSA on the walking task. Note that the final variant is identical to PGPE (solid line). For this task the main improvement comes from the switch to Gaussian sampling and the likelihood gradient (circles). Adding adaptive variances actually gives slightly slower learning at first, although the two converge later on. The original parameter update rule for SPSA is:

$$\theta_i(t+1) = \theta_i(t) - \alpha\frac{y_+ - y_-}{2\epsilon} \tag{11}$$

with $y_+ = r(\theta + \Delta\theta)$ and $y_- = r(\theta - \Delta\theta)$, where $r(\theta)$ is the evaluation function and $\Delta\theta$ is drawn from a Bernoulli distribution scaled by the time dependent step size $\epsilon(t)$, i.e. $\Delta\theta_i(t) = \epsilon(t) \cdot rand[-1, 1]$ In addition, a set of metaparameters is used to help SPSA converge. ϵ decays according to $\epsilon(t) = \frac{\epsilon(0)}{t^\gamma}$ with $0 < \gamma < 1$. Similarly, α decreases over time, with $\alpha = a/(t + A)^E$ for some fixed a, A and E [10]. The choice of initial parameters $\epsilon(0), \gamma, a, A$ and E is critical to the performance of SPSA. To switch from uniform to Gaussian sampling we simply modify the perturbation function to $\Delta\theta_i(t) = \mathcal{N}(0, \epsilon(t))$. We then follow the derivation of the likelihood gradient outlined in Section 2, to obtain the same

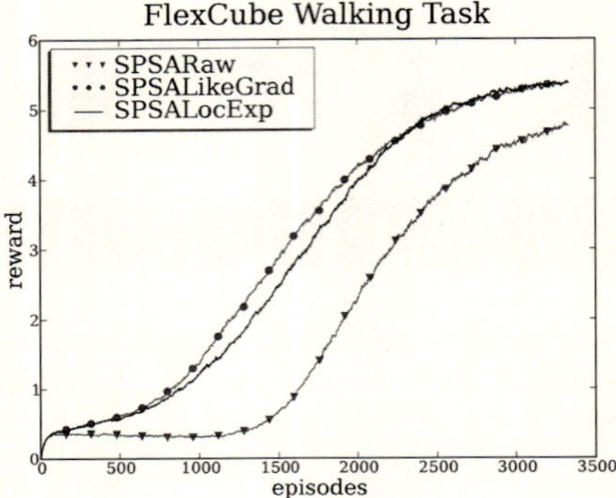

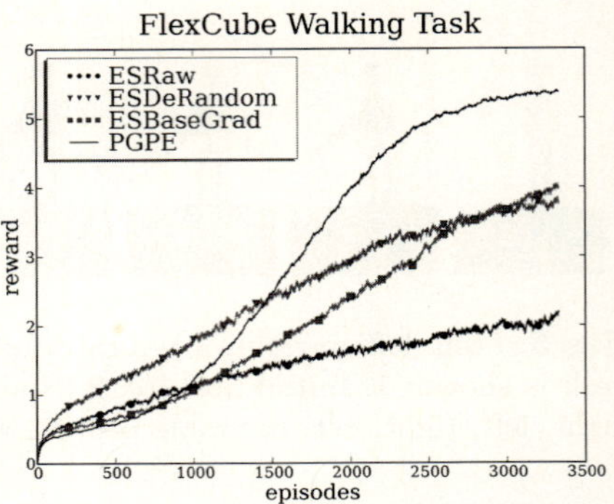

Fig. 6. Three variants of SPSA on the FlexCube walking task: the original algorithm (SPSARaw), the algorithm with normally distributed sampling and likelihood gradient (SPSALikeGrad), and with adaptive variance (SPSALocExp)

Fig. 7. Three variants of ES on the FlexCube walking task: the original algorithm (ESRaw), derandomized ES (ESDeRandom) and gradient following (ESBaseGrad). PGPE is shown for reference.

parameter update rule as used for PGPE (Eq. (10)). The correspondence with PGPE becomes exact when we calculate the gradient of the expected reward with respect to the sampling variance, giving us the standard deviation update rule stated in Eq. (10). As well as improved performance, the above modifications greatly reduce the number of free parameters in the algorithm. The SPSA metaparameters are now condensed to: a step size α_μ for updating the parameters, a step size α_σ for updating the standard deviations of the perturbations, and an initial value standard deviation σ_{init}. Furthermore, we found that the parameters $\alpha_\mu = 0.2$, $\alpha_\sigma = 0.1$ and $\sigma_{init} = 2.0$ worked very well for a wide variety of tasks.

4.2   From ES to PGPE

We now examine the effect of two modifications that largely bridge the gap between ES and PGPE. First we switch from standard ES to derandomized ES [15], which somewhat resembles the gradient based variance updates found in PGPE. We then change from using population based search to following a likelihood gradient. The results are plotted in Figure 7. As can be seen, both modifications bring significant improvements, although neither can match PGPE. While ES performs well initially, it is slow to converge to good optima. This is partly because, as well as having stochastic mutations, ES has stochastic updates to the standard deviations of the mutations, and the coupling of these two stochastic processes slows down convergence. Derandomized ES alleviates that problem by using instead a deterministic standard deviation update rule, based on the change in parameters between the parent and child. Tracking a population has advantages in the early phase of search, when broad, relatively undirected exploration is desirable. This is particularly true for the multimodal fitness spaces

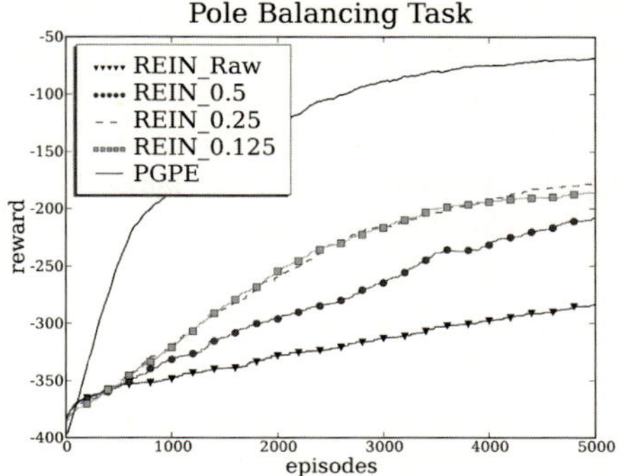

Fig. 8. REINFORCE on the pole balancing task, with various frequencies of action perturbation. PGPE is shown for reference.

typical of realistic control tasks. However in later phases convergence is usually more efficient with gradient based methods. Applying the likelihood gradient, the ES parameter update rule becomes:

$$\Delta\theta_i = \alpha \sum_{m=1}^{M} (r_m - b)(y_{m,i} - \theta_i), \tag{12}$$

where M is the number of samples and $y_{m,i}$ is parameter i of sample m.

4.3   From REINFORCE to PGPE

We previously asserted that the lower variance of PGPE's gradient estimates is partly due to the fact that PGPE requires only one parameter sample per history, whereas REINFORCE requires samples every time step. This suggests that reducing the frequency of REINFORCE perturbations should improve its gradient estimates, thereby bringing it closer to PGPE. Fig. 8 shows the performance of episodic REINFORCE with a perturbation probability of 1, 0.5, 0.25, and 0.125 per time step. In general, performance improved with decreasing perturbation probability. However the difference between 0.25 and 0.125 is negligible. This is because reducing the number of perturbations constrains the range of exploration at the same time as it reduces the variance of the gradient, leading to a saturation point beyond which performance does not increase. Note that the above trade off does not exist for PGPE, because a single perturbation of the parameters can lead to a large change in behaviour.

5   Conclusion and Future Work

We have introduced PGPE, a novel algorithm for episodic reinforcement learning based on a gradient based search through model parameter space. We derived the PGPE equations from the basic principle of reward maximization, and explained why they lead to lower variance gradient estimates than those obtained by policy gradient methods. We compared PGPE to a range of stochastic optimisation algorithms on three control tasks, and found that it gave superior performance in every case. Lastly we provided a detailed analysis of the relationship between PGPE and the other algorithms. One direction for future work would be to establish whether Williams' local convergence proofs for REINFORCE can be generalised to PGPE. Another would be to combine PGPE with recent improvements in policy gradient methods, such as natural gradients and base-line approximation [4].

Acknowledgement. This work is supported within the DFG excellence research cluster "Cognition for Technical Systems - CoTeSys", www.cotesys.org

References

1. Benbrahim, H., Franklin, J.: Biped dynamic walking using reinforcement learning. Robotics and Autonomous Systems Journal (1997)
2. Peters, J., Schaal, S.: Policy gradient methods for robotics. In: IROS-2006, Beijing, China, pp. 2219–2225 (2006)
3. Schraudolph, N., Yu, J., Aberdeen, D.: Fast online policy gradient learning with smd gain vector adaptation. In: Weiss, Y., Schölkopf, B., Platt, J. (eds.) Advances in Neural Information Processing Systems, vol. 18. MIT Press, Cambridge (2006)
4. Peters, J., Vijayakumar, S., Schaal, S.: Natural actor-critic. In: Gama, J., Camacho, R., Brazdil, P.B., Jorge, A.M., Torgo, L. (eds.) ECML 2005. LNCS (LNAI), vol. 3720, pp. 280–291. Springer, Heidelberg (2005)
5. Williams, R.: Simple statistical gradient-following algorithms for connectionist reinforcement learning. Machine Learning 8, 229–256 (1992)
6. Baxter, J., Bartlett, P.L.: Reinforcement learning in POMDPs via direct gradient ascent. In: Proc. 17th International Conf. on Machine Learning, pp. 41–48. Morgan Kaufmann, San Francisco (2000)
7. Aberdeen, D.: Policy-Gradient Algorithms for Partially Observable Markov Decision Processes. PhD thesis, Australian National University (2003)
8. Sutton, R., McAllester, D., Singh, S., Mansour, Y.: Policy gradient methods for reinforcement learning with function approximation. In: NIPS 1999, pp. 1057–1063 (2000)
9. Schwefel, H.: Evolution and optimum seeking. Wiley, New York (1995)
10. Spall, J.: An overview of the simultaneous perturbation method for efficient optimization. Johns Hopkins APL Technical Digest 19(4), 482–492 (1998)
11. Riedmiller, M., Peters, J., Schaal, S.: Evaluation of policy gradient methods and variants on the cart-pole benchmark. In: ADPRL 2007 (2007)
12. Müller, H., Lauer, M., Hafner, R., Lange, S., Merke, A., Riedmiller, M.: Making a robot learn to play soccer. In: Hertzberg, J., Beetz, M., Englert, R. (eds.) KI 2007. LNCS (LNAI), vol. 4667, pp. 220–234. Springer, Heidelberg (2007)
13. Jordan, M.: Attractor dynamics and parallelism in a connectionist sequential machine. In: Proc. of the Eighth Annual Conference of the Cognitive Science Society, vol. 8, pp. 531–546 (1986)
14. Ulbrich, H.: Institute of Applied Mechanics, TU München, Germany (2008), http://www.amm.mw.tum.de/
15. Hansen, N., Ostermeier, A.: Completely Derandomized Self-Adaptation in Evolution Strategies. Evolutionary Computation 9(2), 159–195 (2001)

A Continuous Internal-State Controller for Partially Observable Markov Decision Processes

Yuki Taniguchi, Takeshi Mori, and Shin Ishii

Graduate School of Informatics
Kyoto University
Gokasho, Uji, Kyoto, 611-0011, Japan
{yuki-t,tak-mori}@sys.i.kyoto-u.ac.jp,
ishii@i.kyoto-u.ac.jp

Abstract. In this study, in order to control partially observable Markov decision processes, we propose a novel framework called continuous state controller (CSC). The CSC incorporates an auxiliary "continuous" state variable, called an internal state, whose stochastic process is Markov. The parameters of the transition probability of the internal state are adjusted properly by a policy gradient-based reinforcement learning, by which the dynamics of the underlying unknown system can be extracted. Computer simulations show that good control of partially observable linear dynamical systems is achieved by our CSC.

1 Introduction

Many reinforcement learning (RL) techniques have been successfully applied to completely observable environments [6], which are often formulated as Markov decision processes (MDPs). However, many real-world problems such as autonomous control of robots cannot be formulated as MDPs, because noises or obstacles, which are introduced to the robot's sensors, prevent the agents (robots or robot controllers) from observing all of the state variables accurately. Such problems can be formulated as partially observable Markov decision processes (POMDPs) [3], and some RL methods employing belief states have been developed for solving POMDPs [8]. However, those techniques suffer from several difficulties even with effective approximations [2]; the model of the environment should be known, and even if we know the model, the dimensionality of the belief space is usually high. These difficulties make the POMDP-RL with belief states infeasible in real world problems.

So far, policy-gradient RL algorithms with finite state controllers (FSCs) have been proposed for solving POMDPs, called IState-GPOMDP and Exp-GPOMDP [1]. The FSC is a probabilistic policy which incorporates an internal state as an additional input, and the transition probability of the internal state is identified by the policy-gradient RL algorithms, together with the optimization of the policy. The essential dynamic characters of the target state space can be extracted by learning of the transition of the internal state, which is performed independently of the underlying dimensionality of the target state

V. Kůrková et al. (Eds.): ICANN 2008, Part I, LNCS 5163, pp. 397–406, 2008.

space. Because effective dimensionality for controlling a high-dimensional system is often much smaller than the dimensionality of the whole state space, as the dependence between the state variables increases, feature extraction by the policy-gradient RL algorithms with internal states can be more effective than directly reconstructing the true state space as done by the belief state-based methods. Actually, an existing study has shown the effectiveness of the IState-GPOMDP with the FSCs when applied to a partially observable multi-agent system through computer simulations [7].

Although the FSC shows good performance when applied to discrete-state dynamical systems, a direct application to more realistic problems whose state space is continuous should be intractable. In other words, essential features in such a continuous-state dynamical system would have continuous dynamics, which cannot be extracted and expressed by the FSCs.

To overcome this difficulty, in this study, we propose a novel framework by introducing a continuous internal-state transition model, called a continuous state controller (CSC), instead of the finite-state alternative (FSC). In this new framework, the parametric transition model of the internal state in the CSC can be learned by the IState-GPOMDP or Exp-GPOMDP together with the policy optimization. The parameters of the transition model are adjusted so as to maximize the average reward, and then the continuous dynamics of the essential features can be extracted within the framework of reward maximization. We apply this algorithm to control problems of linear dynamical systems under the assumption that some state variables cannot be observed. Computer simulations show that our continuous internal-state transition model could successfully extract the dynamical characters of the missing variables.

2 POMDPs with Internal-State Transition Models

We consider a finite POMDP, consisting of a set of true states $\mathcal{S}$, a set of observations $\mathcal{O}$, a set of actions $\mathcal{U}$, each of which is the output of the controller, and a set of scalar rewards $\mathcal{R}$. At time t, the true state $s_t \in \mathcal{S}$ cannot be perceived directly by the agent, but instead, the observation o_t, is drawn from the probability $P(o_t|s_t)$ conditioned on the state $s_t \in \mathcal{S}$ and then observed by the agent. The observation o_t is, for example, the true state variable lacking some dimensionalities or those disturbed by observation noise. It is usually difficult to realize an optimal control based only on such insufficient observations. According to our approach, an internal state $y_t \in \mathcal{Y}$ is then added to the POMDP definition above so as to become an additional input to the policy. When using a discrete internal state, such an augmented policy is called a finite state controller (FSC) and has been successfully used to solve POMDPs [7]. The stochastic process over the internal state is prescribed by the transition probability $P(y_{t+1}|y_t, o_t)$, where y_t and y_{t+1} are the internal states at time t and $t + 1$, respectively. In this case, the policy is the mapping from a pair of observation o_t and internal state y_t to the output u_t. Figure 1 represents the graphical model of a POMDP employing an internal state. The transition probability of the internal state $P(y_{t+1}|y_t, o_t)$

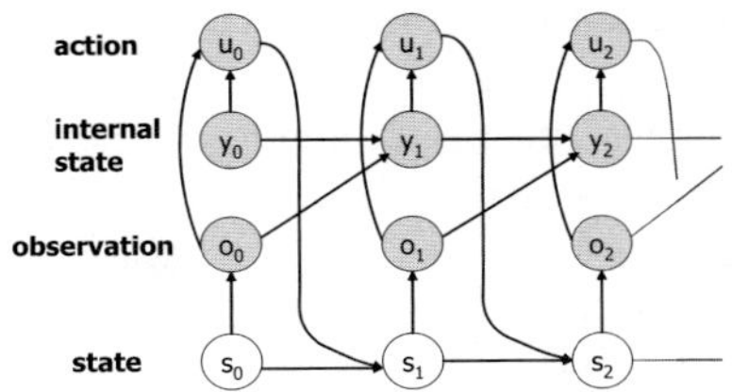

Fig. 1. Graphical model of a POMDP with an internal state. The current action u_t and the following internal state y_{t+1} depend on the current internal state y_t and the observation o_t.

embedded in the FSC is identified by the policy gradient algorithm together with the optimization of the policy $P(u_t|y_t, o_t)$. The important dynamic characters of the target state space are extracted by learning of the transition probability of the internal state.

3 Continuous State Controllers

Although the FSCs can achieve good performance in controlling partially observable discrete-state dynamic systems, the variables in the state space are often continuous in the real world, and then the essential dynamic characters can also have continuous dynamics. In this section, we propose a novel framework called continuous state controller (CSC), which has a continuous internal state and can express features with continuous dynamics. Now we explain how to apply the two policy-gradient RL algorithms with internal states which were formerly proposed by Aberdeen and Baxter [1], to our CSCs.

3.1 Continuous State Controllers for IState-GPOMDP

The IState-GPOMDP is a policy gradient-based RL method which does not seek to estimate the value function, but adjusts the policy parameters θ and the internal-state transition parameters ϕ directly to maximize the average reward:

$$\eta(\phi, \theta) := \lim_{T \to \infty} \frac{1}{T} E_{\phi,\theta} \left[\sum_{t=1}^{T} r_t \right], \tag{1}$$

where $E_{\phi,\theta}$ denotes the expectation with respect to the trajectory (s_0, y_0, o_0, u_0), $(s_1, y_1, o_1, u_1), \ldots$, prescribed by the parameters ϕ and θ, and r_t denotes the immediate reward. The internal state and the output of the controller are assumed to be drawn from the parameterized Gaussian distributions as

$$\begin{aligned}
p(y_{t+1}|y_t, o_t) &:= \mathcal{N}(y_{t+1}|f(y_t, o_t; \phi), \sigma_\phi^2) \\
p(u_t|y_t, o_t) &:= \mathcal{N}(u_t|g(y_t, o_t; \theta), \sigma_\theta^2),
\end{aligned} \tag{2}$$

where $f(y_t, o_t; \phi)$ and $g(y_t, o_t; \theta)$ are functions of (y_t, o_t) parameterized by ϕ and θ, respectively.

The following table shows the pseudo-code of the IState-GPOMDP applied to a continuous dynamical system.

```
0: repeat
1:    until the terminal condition, i.e., while t < T
2:        observe o_t from p(o_t|s_t).
3:        draw y_{t+1} from p(y_{t+1}|y_t, o_t, φ) and u_t from p(u_t|y_t, o_t, θ).
4:        update the estimation of the policy gradient Δ_t with respect to φ and θ.
5:        control the system by u_t.
6:        t + +.
7:    end
8:    αΔ_t is added to the parameters φ and θ, where α is the learning rate.
9: end
```

3.2   Continuous State Controllers for Exp-GPOMDP

In the previous subsection, to learn the CSCs by IState-GPOMDP, the internal state and the action are drawn from the distribution (2) at each time step. However, such a sampling process of the internal state can be omitted and be replaced by the propagation of the distribution as long as the distribution of the next internal state can be analytically propagated from that of the current internal state. Such a process can lead more efficient estimation of the policy gradient, and then the policy learning will be faster, because such a sampling noises would disappear due to the estimation. Exp-GPOMDP uses the propagation process for the policy gradient estimation.

If we assume the distribution of the internal state y_t at time t is $p(y_t) = \mathcal{N}(y_t; \mu, \sigma_t^2)$, the distribution of the next internal state y_{y+1} is obtained by the propagation and marginalization of the current internal state y_t as

$$p(y_{t+1}) = \int_{y_t} p(y_{t+1}|y_t, o_t; \phi)p(y_t)dy_t$$

$$= \int_{y_t} N(y_{t+1}; f(y_t, o_t; \phi), \sigma_\phi^2)N(y_t; \mu, \sigma_t^2)dy_t,$$

and similary, the distribution of action u_t is obtained by

$$p(u_t) = \int_{y_t} p(u_t|y_t, o_t; \theta)p(y_t)dy_t$$

$$= \int_{y_t} N(u_t; f(y_t, o_t; \theta), \sigma_\theta^2)N(y_t; \mu, \sigma_t^2)dy_t.$$

These calculations can be performed analytically as long as $f(y_t, o_t; \phi)$ and $f(y_t, o_t; \theta)$ are linear.

4   Computer Simulations

In this section, we apply the CSC to partially observable linear dynamical systems and evaluate its performance.

4.1   Linear Dynamical System

The first application is a linear quadratic regulator (LQR) task, depicted in Figure 2. The goal of this task is to make the ball stay around the origin. We define the dynamics of this system as

$$p(\mathbf{s}_{t+1}|\mathbf{s}_t, u_t) = \mathcal{N}(\mathbf{s}_{t+1}|\mathbf{A}\mathbf{s}_t + \mathbf{B}u_t, \boldsymbol{\Sigma}), \tag{3}$$

where $\mathbf{s}_t = (x_t, v_t)^T$ denotes the state vector composed of the position and velocity of the ball, u_t denotes the force applied to the ball at time t, and $\boldsymbol{\Sigma}$ is the variance of the state transition noise; $\boldsymbol{\Sigma} = \mathrm{diag}(1, 1) \times 10^{-3}$. $(^T)$ denotes the transpose. The matrices $\mathbf{A}$ and $\mathbf{B}$ denote the system parameters:

$$\mathbf{A} = \begin{bmatrix} 1 & \tau \\ 0 & 1 \end{bmatrix}, \mathbf{B} = \begin{bmatrix} 0 \\ \tau \end{bmatrix},$$

where the time constant is set at $\tau = 1/60s$. The agent observes the state and takes an action every time step. The rewards are given by $r(\mathbf{s}_t, u_t) = -(\mathbf{s}_t^T \mathbf{Q} \mathbf{s}_t + R u_t^2)$, where $\mathbf{Q} = \mathrm{diag}(0.025, 0.01)$ and $R = 0.01$, which formulate this problem as an underactuated system.

In our experimental setting, a single learning episode continues for 10 seconds, and each parameter is updated after every learning episode. If the absolute value of the position of the ball exceeds 10, the learning episode is terminated and the parameter is immediately updated. After that, the position and velocity are re-set to be around 0, and then a new learning episode starts.

4.2   Partially Observable Linear Dynamical System and Policy Parameterization

To evaluate the CSC's performance, we apply it to controlling a partially observable linear dynamical system, in which the velocity of the ball cannot be

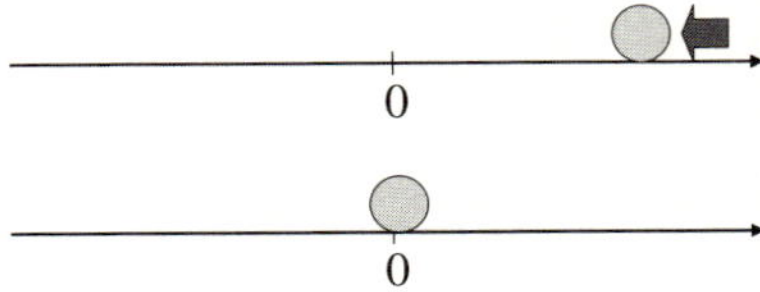

Fig. 2. An LQR task, a simple example of controlling a linear dynamical system. The goal of this task is to make the ball stay around the origin by controlling the force applied to the ball.

observed. It is impossible to control the system smoothly only by the currently observable position, so we introduce a continuous internal state as to compensate the missing velocity.

More concretely, in the completely observable LQR, it is easily controlled by learning the parameters (θ_1, θ_2) of the policy:

$$P(u_t|o_t) = \mathcal{N}(u_t|\theta_1 x_t + \theta_2 v_t, \sigma_\theta^2). \tag{4}$$

In the partially observable LQR, on the other hand, the velocity cannot be observed, so we replace it with the internal state of the policy:

$$P(u_t|o_t, y_t) = \mathcal{N}(u_t|\theta_1 x_t + \theta_2 y_t, \sigma_\theta^2), \tag{5}$$

and now the internal state transition is modeled as

$$P(y_{t+1}|o_t, y_t) = \mathcal{N}(y_{t+1}|\phi_1 x_t + \phi_2 y_t, \sigma_\phi^2), \tag{6}$$

where both σ_θ^2 and σ_ϕ^2 are set at 0.1^2. The parameters ϕ and θ are learned by the IState-GPOMDP or Exp-GPOMDP. The learning of σ_θ^2 and σ_ϕ^2 is would be possible, but they are fixed here. Incidentally, if they are too large, the converged reward drops to a lower value, and if they are too small, the policy gradient method dose not work.

4.3   Simulation Result

We applied the IState-GPOMDP with our CSC to the controlling problem of the partially observable linear dynamical system above. The initial values of ϕ_1 and ϕ_2 are $0 + \epsilon$ ($\epsilon \sim \mathcal{N}(0, 10^{-2})$), and those of θ_1 and θ_2 are $-20 + \epsilon$. Figure 3 shows the learning curve, which is averaged over 100 learning runs and smoothed over every 300 time step (5 seconds). The parameters were successfully learned so as to increase the averaged reward.

In Figure 4, panels (a) and (b) show the time-series of the position, the velocity and the internal state, and panels (c) and (d) show those of the acceleration and the internal state. Left panels, (a) and (c), are for before learning and right

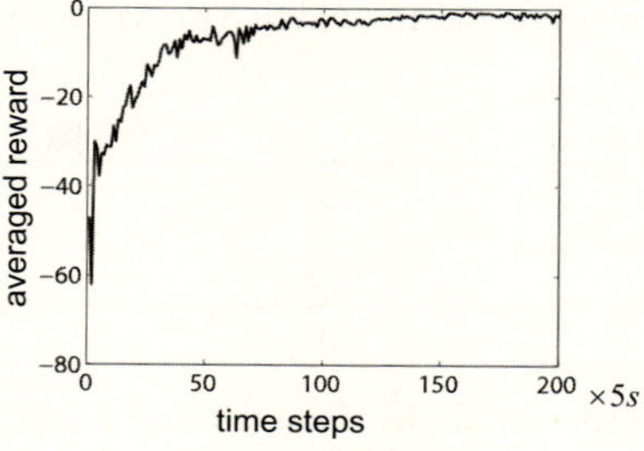

Fig. 3. The learning process

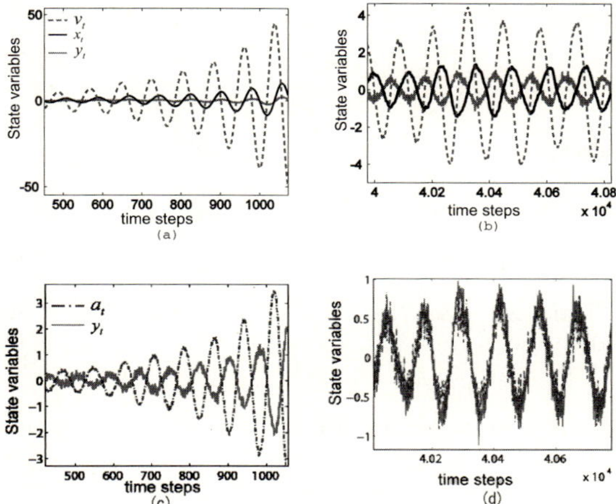

Fig. 4. (a),(b) The time-series before learning. (c),(d) The time-series after learning.

panels, (b) and (d), are for after learning. The internal state after learning came to reconstruct the acceleration rather than the velocity (Fig.4(d)). Because the average reward increased (Fig.3), the acceleration seems to be an important feature for controlling this system.

Since this system includes noises in the motion of the ball, some oscillatory behaviors remained even by the learned controller.

Next, we applied the Exp-GPOMDP with our CSC to the same problem. The parameters were initially set to the same values as in the previous experiment. Figure 5 shows the learning curves produced by the IState-GPOMDP and Exp-GPOMDP; panel (b) shows enlargement of the learning curves around when the parameters have converged. Although it seems that there is little difference in the learning curve between IState-GPOMDP and by Exp-GPOMDP (Fig.5(a)), the policy obtained by Exp-GPOMDP is superior to that by IState-GPOMDP (Fig.5(b)), possibly because the variance of policy gradient is smaller in Exp-GPOMDP than in IState-GPOMDP.

4.4   Alternative Internal State Transition Model

In the previous two experiments, the internal state after learning seems to imitate the acceleration rather than the velocity. This happens because the learning algorithm has made itself stable. In this problem, if the parameter ϕ_2 becomes more than 1, the internal state necessarily diverges. Unfortunately, the theoretically optimal value of ϕ_2 for imitating the velocity is close to 1. So the internal state transition learned to avoid such divergence in parameter learning, and instead, has become to reconstruct the acceleration in place of the velocity. In other words, if the parameter is learned to be close to 1, stochastic noises

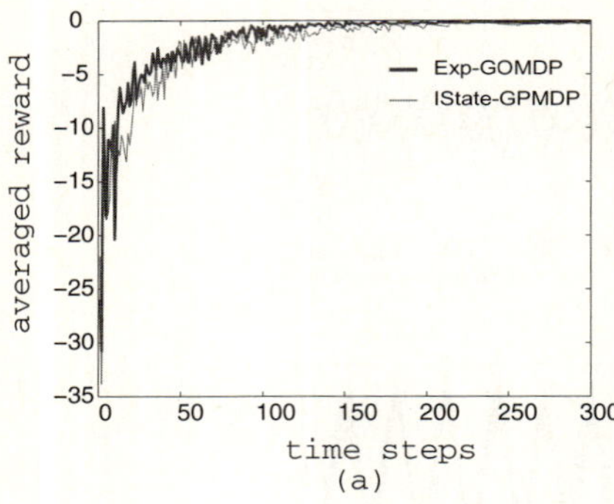
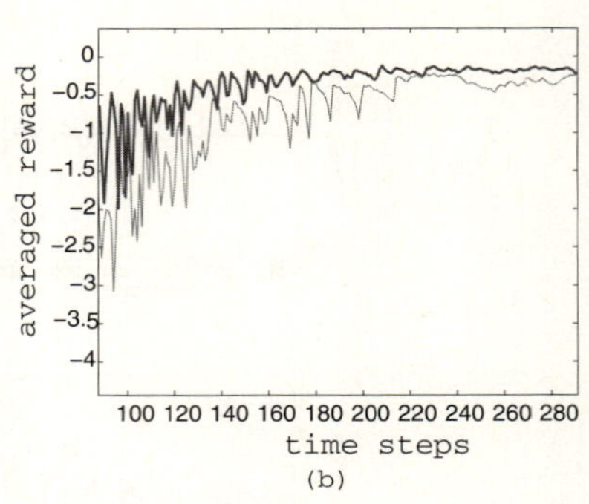

(a)   (b)

Fig. 5. The learning processes by IState-GPOMDP and Exp-GPOMDP

occasionally produce the divergent sequence of the internal state, leading to decrease in the average reward.

To solve the difficulty above, we introduce another internal transition model:

$$P(y_{t+1}|o_t, y_t) = \mathcal{N}(y_{t+1}|(1 - \exp(\phi_1))x_t + (1 - \exp(\phi_2))y_t, \sigma_\phi^2), \qquad (7)$$

where σ_ϕ^2 was set as 0.1^2. This model restricts the range of the coefficient of x_t to $[-\infty, 1)$, so as to avoid the divergence. We applied the IState-GPOMDP with our CSC whose internal state transition is modeled by (7), to the LQR problem. Figure 6 shows the time-series of the position, the velocity and the internal state after learning the parameters; as a consequence, (ϕ_1, ϕ_2) was converged to $(-2.7968, -4.0484)$ and (θ_1, θ_2) was converged to $(-10.4133, 0.1640)$. We can see that the internal state came to imitate the reversed velocity, suggesting that the internal state transition extracts the velocity information as an essential feature for controlling this task.

4.5   Cart-Pole Task

We applied our CSC to a cart-pole task, which is essentially a linear dynamical system but includes small nonlinearity. The dynamics of this system are defined by,

$$p(\mathbf{s}_{t+1}|\mathbf{s}_t, u_t) = \mathcal{N}(\mathbf{s}_{t+1}|\mathbf{A}\mathbf{s}_t + \mathbf{B}u_t, \mathbf{\Sigma}), \qquad (8)$$

where $\mathbf{s}_t = (x_t, \dot{x}_t, \alpha_t, \dot{\alpha}_t)^T$ denotes the state vector, whose components are the position and velocity of the cart, and the angle and angular velocity of the pole. u_t denotes the force applied to the cart at time t, and $\mathbf{\Sigma}$ is the variance of state transition noise; $\mathbf{\Sigma} = \mathrm{diag}(1, 1, 1, 1) \times 10^{-3}$. The matrices $\mathbf{A}$ and $\mathbf{B}$ denote the system parameters:

$$\mathbf{A} = \begin{bmatrix} 1 & \tau & 0 & 0 \\ 0 & 1 & 0 & 0 \\ 0 & 0 & 1 & \tau \\ 0 & 0 & \mu\tau & 1 \end{bmatrix}, \mathbf{B} = \begin{bmatrix} 0 \\ \tau \\ 0 \\ \mu\tau/g \end{bmatrix},$$

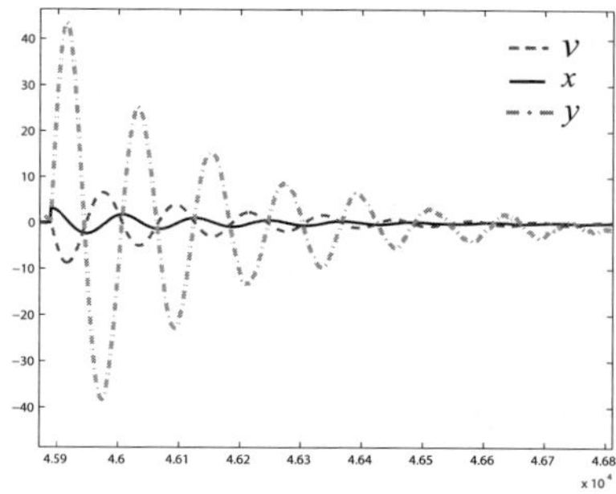

Fig. 6. The time-series of the position, velocity and internal state after learning

where $\tau = 1/60s$ and $\mu = 13.2s^{-2}$. The reward is given by $r(\mathbf{s}_t, u_t) = -(\mathbf{s}_t^T \mathbf{Q} \mathbf{s}_t + Ru_t^2)$, where $\mathbf{Q} = \mathrm{diag}(1.25, 1, 12, 0.25)$ and $R = 0.01$.

To make the environment partially observable, we assumed that the angular velocity cannot be observed and replaced it with the internal state of the policy:

$$P(u_t|o_t, y_t) = \mathcal{N}(u_t|\theta_1 x_t + \theta_2 \dot{x}_t + \theta_3 \alpha + \theta_4 y, \sigma_\theta^2), \tag{9}$$

and the internal state transition is modeled as

$$P(y_{t+1}|o_t, y_t) = \mathcal{N}(y_{t+1}|\phi_1 \alpha_t + \phi_2 y_t, \sigma_\phi^2), \tag{10}$$

where both σ_θ^2 and σ_ϕ^2 were set to 0.1^2. The parameters ϕ and θ were learned by IState-GPOMDP.

Like in the previous LQR task, a single learning episode continued for 10 seconds, and each parameter was updated after every learning episode. When the angle of the pole exceeded $\pi/18$, the episode ended, then the position, velocity, angle and angular velocity were reset to 0 to start the next episode.

Figure 7 shows the learning curve of the cart-pole task, which is over 10 learning runs and smoothed over every 10 episode, suggesting that the policy

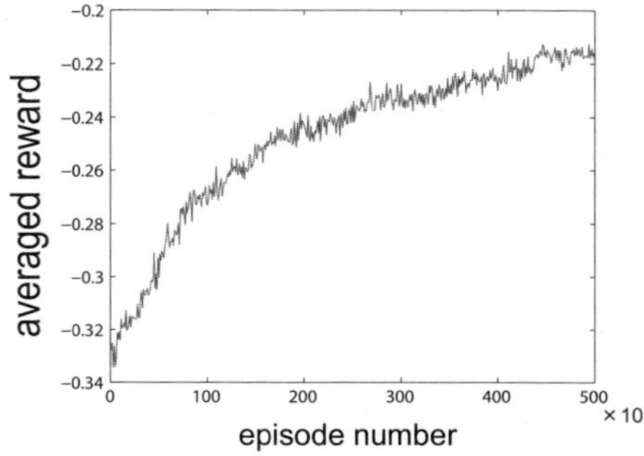

Fig. 7. The learning curve of the cart-pole task

and internal state transition were acquired to increase the reward. As the learning proceeded, the parameters converged and the reward approached its maximum value of 0, suggesting that our CSC achieved good controls for this partially observable cart-pole problem.

5   Concluding Remarks

In this article, we proposed the continuous state controller (CSC), which is a probabilistic policy incorporating continuous-state internal variables, and applied it to simple control tasks of partially observable linear dynamical systems. As a result, it could achieve good control by compensating missing state variables with the continuous internal state variables. The CSC we used in the experiments was, however, one-dimensional, so it is necessary to extend it to be applicable to more complicated or nonlinear dynamical systems. Such extension of our CSC will be shown in our future study. As well, thorough comparison with other methods is necessary in our future study.

References

1. Aberdeen, D., Baxter, J.: Scaling Internal State Policy-Gradient Methods for POMDPs. In: Proceedings of the 19th International Conference on Machine Learning, pp. 3–10 (2002)
2. Hauskrecht, M.: Value-function approximations for partially observable Markov decision processes. Journal of Artificial Intelligence Research 13, 33–99 (2000)
3. Kaelbling, L.P., Littman, M.L., Cassandra, A.R.: Planning and acting in partially observable stochastic domains. Artificial Intelligence 101, 99–134 (1998)
4. Littman, M.L.: Markov games as a framework for multi-agent reinforcement learning. In: Proceedings of the 11th International Conference on Machine Learning, pp. 157–163 (1994)
5. Stone, P., Veloso, M.: Multiagent Systems: A Survey from a Machine Learning Perspective. Autonomous Robotics 8(3) (2000)
6. Sutton, R., Barto, A.: An introduction to reinforcement learning. MIT Press, Cambridge (1998)
7. Taniguchi, Y., Mori, T., Ishii, S.: Reinforcement Learning for Cooperative Actions in a Partially Observable Multi-Agent System. In: de Sá, J.M., Alexandre, L.A., Duch, W., Mandic, D.P. (eds.) ICANN 2007. LNCS, vol. 4668, pp. 229–238. Springer, Heidelberg (2007)
8. Thrun, S.: Monte Carlo POMDPs: Advances in Neural Information Processing Systems, vol. 12, pp. 1064–1070 (2000)
9. Williams, R.J.: Simple statistical gradient-following algorithms for connectionist reinforcement learning. Machine Learning 8, 229–256 (1992)
10. Whitehead, S.D.: A complexity analysis of cooperative mechanisms in reinforcement leaning. In: Proc. of the 9th National Conf. on Artificial Intelligence, vol. 2, pp. 607–613 (1991)

Episodic Reinforcement Learning by Logistic Reward-Weighted Regression

Daan Wierstra[1], Tom Schaul[1], Jan Peters[2], and Juergen Schmidhuber[1,3]

[1] IDSIA, Galleria 2, 6928 Manno-Lugano, Switzerland
[2] MPI for Biological Cybernetics, Spemannstrasse 38, 72076 Tübingen, Germany
[3] Technical University Munich, D-85748 Garching, Germany

Abstract. It has been a long-standing goal in the adaptive control community to reduce the generically difficult, general reinforcement learning (RL) problem to simpler problems solvable by supervised learning. While this approach is today's standard for value function-based methods, fewer approaches are known that apply similar reductions to policy search methods. Recently, it has been shown that immediate RL problems can be solved by reward-weighted regression, and that the resulting algorithm is an expectation maximization (EM) algorithm with strong guarantees. In this paper, we extend this algorithm to the episodic case and show that it can be used in the context of LSTM recurrent neural networks (RNNs). The resulting RNN training algorithm is equivalent to a weighted self-modeling supervised learning technique. We focus on partially observable Markov decision problems (POMDPs) where it is essential that the policy is nonstationary in order to be optimal. We show that this new reward-weighted logistic regression used in conjunction with an RNN architecture can solve standard benchmark POMDPs with ease.

1   Introduction

In order to apply reinforcement learning (RL) to real-life scenarios it is often essential to deal with hidden and incomplete state information. While such problems have been discussed in the framework of partially observable Markov decision problems for a long time, this class of problems still lacks a satisfactory solution [1]. Most known methods to solve small POMDPs rely heavily on knowledge of the complete system, typically in the form of a belief-estimator or filter. Without such important information, the problem is considered intractable even for linear systems, and is not distinguishable from non-Markovian problems [2]. As a result, both POMDPs and non-Markovian problems largely defy traditional value function based approaches.

While policy search based approaches can be applied even with incomplete state information [3], they cannot yield an optimal solution unless the policy has an internal state [4]. As the internal state only needs to represent the features of the belief state and not all of its components, a function approximator with an internal state would be the ideal representation of a policy, and a recurrent

V. Kůrková et al. (Eds.): ICANN 2008, Part I, LNCS 5163, pp. 407–416, 2008.

neural network constitutes one of the few choices. It offers an internal state estimator as a natural component and is well-suited for unstructured domains.

However, the training of recurrent neural networks in the context of reinforcement learning is non-trivial as traditional methods often do not easily transfer to function approximators, and even if they do transfer, the resulting methods such as policy gradient algorithms do no longer employ the advantages of the strong results obtained for supervised learning. As a way out of this dilemma, we fall back onto a classical goal of reinforcement learning, i.e., we search for a way to reduce the reinforcement learning problem to a supervised learning problem where a multitude of methods exists for training recurrent neural networks. In order to do so, we re-evaluate the recent result in machine learning, that reinforcement learning can be reduced onto *reward-weighted regression* [5] which is a novel algorithm derived from Dayan & Hinton's [6] expectation maximization (EM) perspective on RL. We show that this approach generalizes from immediate rewards to episodic reinforcement learning to form Episodic Logistic Reward-Weighted Regression (ELRWR).

As a result, we obtain a novel, general learning method for memory-based policies such as recurrent neural networks in model-free partially observable environments, that is, a method that does not require prior knowledge of any of the dynamics of the problem setup. Using similar assumptions as in [5], we can show that episodic reinforcement learning can be solved as a utility-weighted nonlinear logistic regression problem in this context, which greatly accelerates the speed of learning. We obtain a reinforcement learning setup which is well-suited for training long short-term memory (LSTM) [7] recurrent neural networks, using the E-step of the algorithm to generate weightings for training the memory-capable LSTM network in the M-step. Intuitively, the network is trained to imitate or *self-model* its own actions, but with more successful episodes weighted more heavily than the unsuccessful ones, resulting in a convergence to an ever better policy. We evaluate ELRWR on a number of standard POMDP benchmarks, and show that this method provides a viable alternative to more traditional RL approaches.

2   Preliminaries

In this section, we will state our general problem, define our notation and briefly discuss long short-term memory (LSTM) recurrent neural networks.

2.1   Reinforcement Learning – Generalized Problem Statement

First, let us wintroduce the RL framework used in this paper and the corresponding notation. The environment produces a state g_t at every time step. Transitions from state to state are governed by a probability function $p(g_{t+1}|a_{1:t}, g_{1:t})$ unknown to the agent but dependent upon all previous actions $a_{1:t}$ executed by the agent and all previous states $g_{1:t}$ of the system. Let r_t be the reward assigned to the agent at time t, and let o_t be the corresponding observation produced by the environment. We assume that both quantities are governed by fixed distributions $p(o|g)$ and $p(r|g)$, solely dependent on state g.

In the more general reinforcement learning setting, we require that the agent has a memory of the generated experience consisting of finite episodes. Such episodes are generated by the agent's operations on the (possibly stochastic) environment, executing action a_t at every time step t, after observing o_t which depends solely on g_t. Observation o_t includes special 'observation' r_t (the reward). We define the *observed history*[1] h_t as the string or vector of observations and actions up to moment t since the beginning of the episode: $h_t = \langle o_1, a_1, o_2, a_2, \ldots, o_{t-1}, a_{t-1}, o_t \rangle$. The complete history H has finite length $T(H)$, and includes the unobserved states and is given by $H_T = \langle h_T, g_{1:T} \rangle$. At any time t, the statistic $R_t = (1 - \gamma) \sum_{k=t}^{T(H)} r_k \gamma^{t-k-1}$ denotes the *return* at time t where $0 < \gamma < 1$ denotes a discount factor.

The expectation of this return R_t at time $t = 1$ is also the measure of quality of our policy and, thus, the objective of reinforcement learning is to determine a policy which is optimal with respect to the expected future discounted rewards or expected return

$$J = E\left[R_1\right] = (1 - \gamma)\mathbf{E}\left[\sum_{t=1}^{T} \gamma^t r_t\right]. \tag{1}$$

An optimal or near-optimal policy in a non-Markovian or partially observable Markovian environment requires that the action a_t is taken depending on the *entire* preceding history. However, in most cases, we will not *need* to store the whole string of events but only sufficient statistics $S(h_t)$ of the events which we call the internal memory of the agent's past. Thus, a stochastic policy π can be defined as $\pi(a|h_t) = p(a|S(h_t); \theta)$, implemented as a recurrent neural network (RNN) with weights θ and stochastically interpretable output neurons implemented as a softmax layer. This produces a probability distribution over actions, from which actions a_t are drawn $a_t \sim \pi(a|h_t)$.

2.2 LSTM Recurrent Neural Networks

RNNs have attracted some attention in the past decade because of their simplicity and potential power. However, though powerful in theory, they turn out to be quite limited in practice due to their inability to capture long-term time dependencies – they suffer from the problem of *vanishing gradient* [8], the fact that the gradient signal vanishes as the error signal is propagated back through time. Because of this, events more than 10 time steps apart can typically not be related.

One method purposely designed to avoid this problem is long short-term memory (LSTM) [7], which constitutes a special RNN architecture capable of capturing long term time dependencies. The defining feature of this architecture is that it consists of a number of *memory cells*, which can be used to store activations

---

[1] Note that such histories are also called path or trajectory in the literature.

for an arbitrarily long time. Access to the memory cell is *gated* by units that learn to open or close depending on the context.

LSTM networks have been shown to outperform other RNNs on numerous time series requiring the use of deep memory [9]. Therefore, they seem well-suited for usage in POMDP algorithms for complex tasks requiring deep memory. In the context of reinforcement learning, RNNs are usually used to *predict value*, however, we use them to *control* an agent directly, to represent a controller's policy which receives observations and produces action probabilities at every time step.

Our LSTM networks are trained using backpropagation through time (BPTT) [10], where sequences of ⟨input, target, weighting⟩ samples are used to train the network to minimize a (weighted) error function.

3 Logistic Reward-Weighted Regression for Recurrent Neural Networks

Intuitively, it is clear that the general reinforcement learning problem is related to supervised learning problems as the policy should *match* previously taken motor commands such that episodes are more likely to be reproduced if they had a higher return. The network is trained to imitate or *self-model* its own actions, but with more successful episodes weighted more heavily than the unsuccessful ones, resulting in a convergence to an ever better policy. In this section, we will solidify this approach based on [5], and extend the previous results from a single-step, immediate reward scenario to the general episodic case.

We first discuss our basic assumptions and introduce reward-shaping. Subsequently, we show how a utility-weighted mean-squared error emerges from the general assumptions for an expectation maximization algorithm. Finally, we present the entire resulting algorithm.

3.1 Optimizing Utility-Transformed Returns

Let the return $R(H)$ be some measure of the total reward accrued during a history (e.g., $R(H)$ could be the average of the rewards for the average reward case or the future discounted sum for the discounted case), and let $p(H|\theta)$ be the probability of a history given policy-defining weights θ, then the quantity the algorithm should be optimizing is the expected return

$$J = \int_H p(H|\theta)R(H)dH. \tag{2}$$

This, in essence, indicates the expected return over all possible histories, weighted by their probabilities under policy π.

While a goal function such as found in Eq. (2) is sufficient in theory, algorithms which plainly optimize it have major disadvantages. They might be too aggressive when little experience is available, and converge prematurely to the best solution they have seen so far. On the opposite extreme, they might prove to be too passive and be biased by less fortunate experiences. Trading off such problems has been a long-standing challenge in reinforcement learning. However, in decision theory, such problems are surprisingly well-understood [11]. In that framework it is common to introduce a so-called utility transformation $u_\tau(R)$ which has to fulfill the requirement that it scales monotonically with R, is semi-positive and integrates to a constant. Once a utility transformation is inserted, we obtain an expected utility function given by

$$J_u(\theta) = \int p(H|\theta)u_\tau(R(H))\,dH. \tag{3}$$

The utility function $u_\tau(R)$ is an adjustment for the aggressiveness of the decision making algorithms, e.g., if it is concave, it's attitude is risk-averse while if it is convex, it will be more likely to consider a reward more than a coincidence. Obviously, it is of essential importance that this risk function is not manually tweaked but rather chosen such that its parameters τ can be controlled adaptively in accordance with the learning algorithm.

In this paper, we will consider one simple utility transformation function, the soft-transform $u_\tau(r) = \tau \exp(\tau r)$ also used in [5].

3.2   Expectation Maximization for Reinforcement Learning

Analogously as in [5,6], we can establish the lower bound

$$\log J_u(\theta) = \log \int q(H)\frac{p(H|\theta)u_\tau(R(H))}{q(H)}dH \tag{4}$$

$$\geq \int q(H)\log\frac{p(H|\theta)u_\tau(R(H))}{q(H)}dH \tag{5}$$

$$= \int q(H)\left[\log p(H|\theta) + \log u_\tau(R(H)) - \log q(H)\right]dH \tag{6}$$

$$= \mathcal{F}(q,\theta,\tau), \tag{7}$$

due to Jensen's inequality with the additional constraint $0 = \int q(H)dH - 1$. This points us to the following EM algorithm:

Proposition 1. *An Expectation Maximization algorithm for optimizing both the expected utility as well as the reward-shaping is given by*

$$E\text{-}Step: \quad q_{k+1}(H) = \frac{p(H|\theta)u_\tau(R(H))}{\int p(\tilde{H}|\theta)u_\tau\left(R(\tilde{H})\right)d\tilde{H}}, \tag{8}$$

$$M\text{-}Step\ Policy: \quad \theta_{k+1} = \arg\max_\theta \int q_{k+1}(H)\log p(H|\theta)dH, \tag{9}$$

$$M\text{-}Step\ Utility\ Adaptation: \quad \tau_{k+1} = \arg\max_\tau \int q_{k+1}(H)\log u_\tau(R(H))\,dH. \tag{10}$$

Proof. The E-Step is given by $q = \text{argmax}_q \mathcal{F}(q, \theta, \tau)$ while fulfilling the constraint $0 = \int q(H)dH - 1$. Thus, we have a Lagrangian $L(\lambda, q) = \mathcal{F}(q, \theta, \tau) - \lambda$. When differentiating $L(\lambda, q)$ with respect to q and setting the derivative to zero, we obtain $q^*(H) = p(H|\theta)u_\tau(R(H))\exp(\lambda - 1)$. We insert this back into the Lagrangian obtaining the dual function $L(\lambda, q^*) = \int q^*(H)dH - \lambda$. Thus, by setting $dL(\lambda, q^*)/d\lambda = 0$, we obtain $\lambda = 1 - \log\int p(H|\theta)u_\tau(R(H))\,dH$, and solving for q^* implies Eq (8). The M-steps compute $[\theta_{k+1}, \tau_{k+1}]^T = \text{argmax}_{\theta,\tau}\mathcal{F}(q_{k+1}, \theta, \tau)$. We can maximize $\mathcal{F}(q_{k+1}, \theta, \tau)$ for θ, τ independently, which yields Eqs. (9,10).

3.3 The Utility-Weighted Error Function for the Episodic Case

For every reinforcement learning problem, we need to establish the cost function $\mathcal{F}(q_{k+1}, \theta, \tau)$ and maximize it in order to derive an algorithm. For episodic reinforcement learning, we first need to recap the general settings. We can denote the probabilities $p(H|\theta)$ of histories H by

$$p(H|\theta) = p(\langle o_1, g_1 \rangle) \prod_{t=2}^{T(H)} p(\langle o_t, g_t \rangle | h_{t-1}, a_{t-1}, g_{1:t-1}) \pi(a_{t-1}|h_{t-1}) \tag{11}$$

which are dependent on an unknown initial state and observation distribution $p(\langle o_1, g_1 \rangle)$, and on unknown state transition function $p(g_{t+1}|a_{1:t}, g_{1:t})$. However, the policy $\pi(a_t|h_t)$ with parameters θ is known, where h_t denotes the history which is collapsed into the hidden state of the network.

It is clear that the expectation step has to follow by simply replacing the expectations by sample averages. Thus, we have

$$q_{k+1}(H_i) = \frac{u_\tau(R(H_i))}{\sum_{j=1}^{N} u_\tau(R(H_j))} \tag{12}$$

as E-step. We define $U_N = \sum_{j=1}^{N} u_\tau(R(H_j))$ as summed utility of all N histories.

The maximization or M-step of the algorithm requires optimizing θ such that $\mathcal{F}(q_{k+1}, \theta, \tau)$ is maximized. In order to optimize θ we realize that the probability of a particular history is simply the product of all actions and observations given subhistories (Eq. 11). Taking the log of this expression transforms this large product into a sum

$$\log p(H|\theta) = (\text{const}) + \sum_{t=1}^{T(H)} \log \pi(a_t|h_t) \tag{13}$$

where most parts are not affected by policy-defining parameters θ, i.e., are constant, since they are solely determined by the environment. Thus, when optimizing θ we can ignore this constant and purely focus on the outputs of the policy, optimizing the expression

$$\mathcal{F}(q_{k+1}, \theta, \tau) \propto \sum_{i=1}^{N} q_{k+1}(H_i)\log p(H_i|\theta) = \sum_{i=1}^{N} \frac{u_\tau(R(H_i))}{U_N} \sum_{t=1}^{T(i)} \log \pi(a_t^i|h_t^i),$$

$$\tag{14}$$

where a_t^i denotes an action from the complete history i at time t and h_t^i denotes the collapsed history i up to time-step t.

3.4   Logistic Reward-Weighted Regression for LSTMs

As we are interested in using recurrent neural networks as policies while avoiding the vanishing gradient problem, it is a logical choice that our policy $\pi(a_t|h_t)$ with parameters θ will be represented by a long short-term memory (LSTM) recurrent neural network. Here, we still condition on the the history h_t of our sequence up to time step t as it is collapsed into the hidden state of the network. We use a standard LSTM architecture where the discrete actions are drawn from a softmax output layer, that is, we have

$$\pi(a_t|h_t) = \frac{\exp(f(a_t, h_t))}{\sum_{a=1}^{A} \exp(f(a, h_t))} \tag{15}$$

for all A actions where the output of the neurons are $f(a_t, h_t)$. We can compute the cost function $\mathcal{F}(q_{k+1}, \theta, \tau)$ for this policy and obtain the utility-weighted conditional likelihood function

$$\mathcal{F}(q_{k+1}, \theta, \tau) = \sum_{i=1}^{N} \frac{u_\tau(R(H_i))}{U_N} \sum_{t=1}^{T(i)} \left(a_t^i f(a_t^i, h_t^i) - \log \sum_{a=1}^{A} \exp(f(a, h_t)) \right). \tag{16}$$

This optimization problem is equivalent to a weighted version of logistic regression [12]. As $f(a, h_t)$ is linear in the parameters of the output layer, these can be optimized directly. The hidden state related parameters of $f(a, h_t)$ can be optimized using backpropagation through time (BPTT) of the LSTM architecture. Both linear and nonlinear logistic regression problems cannot be solved in one single shot. Nevertheless, it is straightforward to show that the second-order expansion simply yields a linear regression problem which is equivalent to a Newton-Rapheson step on the utility-weighted conditional likelihood. As a result, we have an approximate regression problem

$$\mathcal{F}(q_{k+1}, \theta, \tau) \approx \sum_{i=1}^{N} \frac{u_\tau(R(H_i))}{U_N} \sum_{t=1}^{T(i)} \left(a_t^i - \pi(a_t^i|h_t^i) \right)^2, \tag{17}$$

which is exactly the utility-weighted squared error. Optimizing this expression by gradient descent allows us to use standard methods for determining the optimum. Nevertheless, there is a large difference in comparison to regular reward-weighted regression where the regression step can only be performed once – instead we can perform multiple BPTT training steps until convergence. In order to prevent overfitting we use the common technique of early stopping, assigning the sample histories to two separate batches for training and validation. While this supervised training scheme requires a relatively large demand in computation per sample history, it also reduces the number of episodes necessary for the policy to converge. Lastly, the update of τ optimizing Eq. 8 follows [5] as $\tau_{k+1} = \frac{\sum_{i=1}^{N} u_\tau(R(H_i))}{\sum_{i=1}^{N} u_\tau(R(H_i))R(H_i)}$ The complete algorithm pseudocode is displayed in Algorithm 1.

Algorithm 1. Episodic Logistic Reward-Weighted Regression

Initialize θ, training batch size N, $\tau = 1$, $k = 1$.
repeat
  for $i = 1 \ldots N$ **do**
    Sample episode $H_i = \langle o_1, a_1, o_2, a_2, \ldots, o_{T-1}, a_{T-1}, o_T \rangle$ using policy π.
    Evaluate return for $t = 1 : R(H_i)$.
    Compute utility of H_i as $u_\tau(R(H_i))$.
  end for
  Train weights θ of policy π until convergence with BPTT to minimize
$$\mathcal{F}(q_{k+1}, \theta, \tau) \approx \sum_{i=1}^{N} \frac{u_\tau(R(H_i))}{U_N} \sum_{t=1}^{T(i)} \left(a_t^i - \pi(a_t^i | h_t^i) \right)^2,$$
  using validation sample histories for early stopping.
  Recompute $\tau \leftarrow \dfrac{\sum_{i=1}^{N} u_\tau(R(H_i))}{\sum_{i=1}^{N} u_\tau(R(H_i))R(H_i)}$.
  $k \leftarrow k + 1$
until stopping criterion is met

---

4 Experiments

We experimented on 5 small POMDP benchmarks commonly used in the literature. The CheeseMaze, Tiger problem, Shuttle Docking benchmark and the 4x3Maze [13,14] are all classic POMDP problems which range from 2 to 11 states, with 2 to 7 observations. The last experiment was the T-maze [15], which was designed to test an RL algorithm's ability to correlate events far apart in history. It involves having to *learn* to remember the observation from the first time step until the episode ends. Its difficulty, depending on corridor lengths, can be adjusted. We investigated the T-Maze with corridor lengths 3, 5 and 7.

The policy was represented as an LSTM network, with input layer size dependent on the observation dimension, a hidden layer containing 2 LSTM memory cells, and a softmax output layer with size dependent on the number of actions applicable to the environment. The only tunable parameter, batch size N, was always set to 30, except for the CheeseMaze and the T-Maze, where it was set to 75. It was found that the algorithm is very robust to the particular setting of this parameter. One third of all batch sample episodes was used for validation in our early stopping scheme to prevent overfitting.

The specific settings for weighted supervised learning are of minor importance (assuming that the number of episodes determines performance), since we train every batch until (early stopping) convergence. Concretely, the LSTM network was initialized with weights uniformly distributed between -0.1 and 0.1. It was trained with BPTT using learning rate 0.002 and momentum 0.95, while weightings were used that are proportional to the self-adapting soft-transform $u_\tau(r) = \tau \exp(\tau r)$, but normalized such that the maximal weighting in every batch was always 1. All experiments consisted of 100 consecutive EM-steps, and were repeated 25 times to gain sufficient statistics.

Table 1. This table shows results averaged over 25 runs. Displayed are the average rewards and std. deviations obtained for the trained policy (Normal) after 100 EM steps, its greedy variant (Greedy) which always takes the learned policy's most likely action, the optimal policy manually calculated for each problem (Optimal), and a randomized policy (Random) as a reference. Shown results include statistics for T-Mazes with corridor lengths 3, 5 and 7.

Policy	Optimal	Random	Normal	Greedy
Tiger	6.7	-36	-6.5 ± 4.9	-5.7 ± 7.4
ShuttleDocking	1.69	-.31	$.709 \pm .059$	0.0 ± 0.0
4x3Maze	.27	.056	$.240 \pm .085$	$.246 \pm .092$
CheeseMaze	.257	.072	$.177 \pm .032$	$.212 \pm .057$
T-Maze3	1.0	.166	$.917 \pm .043$	1.0 ± 0.0
T-Maze5	0.666	.046	$.615 \pm .032$	$.662 \pm .021$
T-Maze7	0.5	.002	$.463 \pm .008$	$.484 \pm .090$

The results are shown in Table 1, which includes both the results for a random policy and the manually calculated optimal policy for each task as a reference. We can see that all problems converged quickly to a good solution, except for the Shuttle Docking benchmark where 13 out of 25 runs failed to converge to an acceptable solution. This might be due to the problem's inherently stochastic nature, which possibly induces the algorithm to converge prematurely. The T-Maze results are significantly less impressive than found in [15] and [4], where corridor lengths of 70 and 90 are reached. However, the result of solving T-Maze length 7 in less than 100 EM steps with batch size 75 constitutes a competitive result.

Good results were obtained without any fine tuning. This encourages us to expect that extensions of the approach will produce a rather general POMDP solver. Such extensions could include the properly re-weighted reuse of information from previous batches, resetting network weights for every EM step, and various improvements to the supervised learning scheme. Future research will include the investigation of the possibility of the use of value-functions and the time-specific reward-attributions to alleviate the credit assignment problem, by shifting responsibilities from entire sequences to single actions.

5   Conclusion

In this paper we introduced a novel, surprisingly simple EM-derived episodic reinforcement learning algorithm that learns from temporally delayed rewards. The method can learn to deal with partially observable environments by using long short-term memory, the parameters of which are updated using utility-weighted logistic regression as training method. The successful application of this algorithm to a number of POMDP benchmarks shows that reward-weighted regression is a promising approach for episodic reinforcement learning, even in non-Markovian settings.

Acknowledgments

This research was funded by SNF grants 200021-111968/1 and 200021-113364/1.

References

1. Kaelbling, L.P., Littman, M.L., Cassandra, A.R.: Planning and acting in partially observable stochastic domains. Artificial Intelligence 101 (1998)
2. Aoki, M.: Optimization of Stochastic Systems. Academic Press, New York (1967)
3. Baxter, J., Bartlett, P.: Infinite-horizon policy-gradient estimation. Journal of Artificial Intelligence Research 15, 319–350 (2001)
4. Wierstra, D., Foerster, A., Peters, J., Schmidhuber, J.: Solving deep memory pomdps with recurrent policy gradients. In: de Sá, J.M., Alexandre, L.A., Duch, W., Mandic, D.P. (eds.) ICANN 2007. LNCS, vol. 4668, pp. 697–706. Springer, Heidelberg (2007)
5. Peters, J., Schaal, S.: Reinforcement learning by reward-weighted regression for operational space control. In: Proceedings of the International Conference on Machine Learning (ICML) (2007)
6. Dayan, P., Hinton, G.E.: Using expectation-maximization for reinforcement learning. Neural Computation 9(2), 271–278 (1997)
7. Hochreiter, S., Schmidhuber, J.: Long short-term memory. Neural Computation 9(8), 1735–1780 (1997)
8. Hochreiter, S., Bengio, Y., Frasconi, P., Schmidhuber, J.: Gradient flow in recurrent nets: the difficulty of learning long-term dependencies. In: Kremer, S.C., Kolen, J.F. (eds.) A Field Guide to Dynamical Recurrent Neural Networks. IEEE Press, Los Alamitos (2001)
9. Schmidhuber, J.: RNN overview (2004),
   http://www.idsia.ch/~juergen/rnn.html
10. Werbos, P.: Back propagation through time: What it does and how to do it. Proceedings of the IEEE 78, 1550–1560 (1990)
11. Chernoff, H., Moses, L.E.: Elementary Decision Theory. Dover Publications (1987)
12. Kleinbaum, D.G., Klein, M., Pryor, E.R.: Logistic Regression, 2nd edn. Springer, Heidelberg (2002)
13. James, M.R., Singh, S., Littman, M.L.: Planning with predictive state representations. In: Proceedings 2004 International Conference on Machine Learning and Applications, pp. 304–311 (2004)
14. Bowling, M., McCracken, P., James, M., Neufeld, J., Wilkinson, D.: Learning predictive state representations using non-blind policies. In: ICML 2006: Proceedings of the 23rd international conference on Machine learning, pp. 129–136. ACM, New York (2006)
15. Bakker, B.: Reinforcement learning with long short-term memory. In: Advances in Neural Information Processing Syst., vol. 14 (2002)

Error-Entropy Minimization for Dynamical Systems Modeling

Jernej Zupanc

Faculty of Computer and Information Science,
University of Ljubljana, Slovenia

Abstract. Recent publications have presented many successful usages of elements from information theory in adaptive systems training. Error-entropy has been proven to outperform mean squared error as a cost function in many artificially generated data sets, but still few applications to real world data have been published. In this paper, we design a neural network trained with error-entropy minimization criterion and use it for dynamical systems modeling on artificial as well as real world data. Performance of this neural network is compared against the mean squared error driven approach in terms of computational complexity, parameter optimization, and error probability densities.

Keywords: information-theoretic learning, error-entropy minimization, neural network, industrial modeling, dynamical systems modeling, industrial forecasting.

1 Introduction

The ability of an adaptive system to provide good results is connected to the learning criterion it uses. One of the most frequently applied criteria for adaptive learning is mean squared error (MSE), which has been applied in many fields, where evaluation of a loss function is needed. This fact can be explained more with mathematical convenience than actual reflection of the true loss function in most situations. Even though MSE makes calculation relatively straightforward and simple, this criterion does not come without weaknesses. One of the major is that large errors are penalized more severely in comparison with small ones [1] which is due to the presumption that probability density function of error is Gaussian. In adaptive learning this property results in a heavy weighting of outliers which can in some cases yield a suboptimal performance of the trained model. This and other issues have led J. C. Principe et al. [2] to evolve an alternative approach to adaptive system training based on elements from information theory.

The idea behind Information-Theoretic Learning (ITL) is to use cost functions (see Fig.1) which directly manipulate information [2]. This allows us to extract more information from data than with usual approach using only second-order statistics [3]. The most evident measure of information is entropy. Even though it was first introduced by Shannon [4], many other definitions of entropy were

V. Kůrková et al. (Eds.): ICANN 2008, Part I, LNCS 5163, pp. 417–425, 2008.

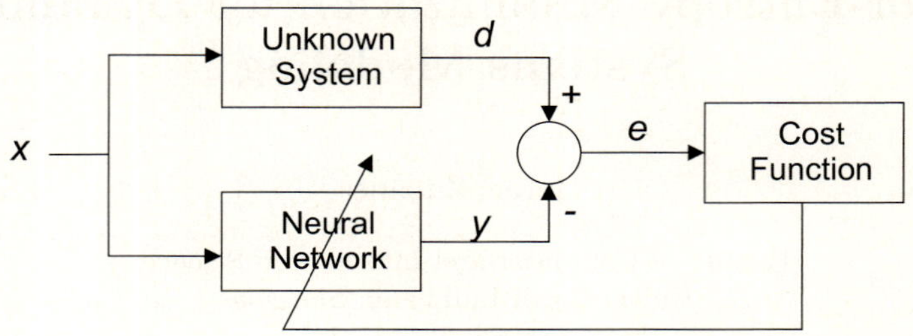

Fig. 1. Neural network prediction scheme in supervised learning

derived later, among which the Renyi's definition is worthy of a particular notice. When combined with the Parzen windowing a direct approximation of the data distribution entropy can be retrieved [5]. Erdogmus et al. [6] provided a derivation which links minimization of the Csiszar's divergence between the joint probability density functions (pdfs) of input-desired signals and input-output signals through minimizing Renyi's error-entropy. This way an information based learning criterion can be established to replace the MSE in adaptive system training. Erdogmus [3] also demonstrated the use of error-entropy minimization (EEM) in comparison with the MSE in solving artificial time series forecasting problems. Multi-layer perceptrons (MLPs) were trained with both criteria and results showed that the EEM trained network achieved a better fit to the pdfs of the desired samples.

The aim of our research was to demonstrate the usability of the ITL approach to adaptive learning on two different forecasting problems. The first is an artificially generated time series Logistic map and the second is a database containing measurements of polymer compound properties throughout an industrial mixing process. We have focused on ITL learning parameters optimization, various performance evaluators, and convergence problems.

The remainder of this paper is organized as follows. First an overview of the theoretical background is provided. Next we describe methods, data sets and parameters used for experiments. Section 3 describes experimental work together with results of the training. Finally a section for discussion and conclusion brings the paper to a close.

2   Theoretical Background

Information-Theoretic Learning uses Renyi's definition of entropy in combination with Parzen's nonparametric pdf estimation to compute entropy-based optimization criterion directly from samples. Renyi's entropy of order α over a discrete variable is defined as

$$H_{R_\alpha} = \frac{1}{1-\alpha} \log(\sum_{k=1}^{N} p_k^\alpha), \quad \alpha > 0, \alpha \neq 1 \ . \tag{1}$$

When $\alpha = 2$, this is called quadratic entropy, due to the quadratic form of the probability [2]. Because shape of pdf being estimated from samples can seldom be presented as a known distribution, Parzen's windowing [5] is used for nonparametric density estimation

$$\hat{f}(x) = \frac{1}{Nh} \sum_{i=1}^{N} K(\frac{x - X_i}{N}) \tag{2}$$

where N is the number of data points, K is the kernel function, and h is the kernel width which can also be called the smoothing parameter or bandwidth. It has been proved that the choice of kernel function itself does not assure a noticeable benefit in kernel efficiency [7], so other properties, like self-similarity and differentiability, were given more consideration. The Gaussian kernel

$$K_G(\mathbf{x}; \sigma) = \frac{1}{\sigma\sqrt{2\pi}} e^{-\frac{\mathbf{x}^\mathbf{T}\mathbf{x}}{2\sigma^2}} \tag{3}$$

suffices these criteria and is therefore a good choice for Parzen's density estimation. When combining (3) with (2) and (1) over a set of data $\mathbf{x} = \{\mathbf{x_i} \in \mathbf{R^m}\}$, Renyi's quadratic entropy with Parzen's density estimation using Gaussian kernel can be derived as

$$\hat{H}_{R_2} = -\log(\frac{1}{2\sqrt{2\pi}N^2\sigma^2} \sum_{i=1}^{N} \sum_{j=1}^{N} e^{-(\frac{x_i - x_j}{2\sigma^2})^2}) = -\log(\hat{V}(x)). \tag{4}$$

If the adaptive system is an MLP, the difference between d-desired output value and y-actual output value can further be defined as $e = d - y$ (see Fig.1). This error could be back-propagated to the weights through a derivative of MSE, but we chose to substitute adaptation criterion with error-entropy. This way, with steepest descent approach, the change of weights becomes

$$\Delta w = -\eta \frac{\partial \hat{H}}{\partial w} = +\eta \frac{\partial \hat{V}}{\partial w} . \tag{5}$$

In analogy with potential fields in physics, $\hat{V}$ was named information potential [2], and its derivative (6) information force. Because Renyi's quadratic entropy is a monotonic function of information potential [3], we can use maximization of the latter and further simplify our learning algorithm. When (4) is applied to the errors of the input-output mapping, information potential provides a learning criterion [2],[3],[6]. The derivation of information potential allows us to back-propagate information force

$$\frac{\partial \hat{V}(e)}{\partial w} = \frac{1}{2N^2\sigma^2} \sum_{i=1}^{N} \sum_{j=1}^{N} (e_j - e_i) \cdot K_G(e_i - e_j, 2\sigma^2) \cdot [\frac{\partial \hat{y}_j}{\partial w} - \frac{\partial \hat{y}_i}{\partial w}] \tag{6}$$

instead of MSE from the output mapping to the weights. The term $\frac{\partial \hat{y}}{\partial w}$ can be computed as in the standard back-propagation algorithm.

2.1 Methods

Two multi-layer perceptrons were investigated in detail. The MLP with EEM criterion was used next to a standard MLP which depended on MSE criterion. Both MLPs contained one layer of hidden neurons and a single output processing element, all with a sigmoid activation function. Back-propagation algorithm was used for adjustment of weights in both layers.

The MLPs used accept only a few parameters for training. The standard parameter for all neural networks using gradient descent is the learning rate η. In addition, the EEM driven MLP requires another parameter, the kernel width. This is due to the dependency on Parzen windowing which is required for error pdf estimation used in Renyi's entropy approximation. This parameter will be referred to as σ, which is also the default bandwidth parameter for Gaussian kernel in Parzen's windowing. In the proces of training no parameter adaptation approaches were used. Optimal parameters for both methods were experimentally acquired before the benchmark runs.

2.2 Sample Problems

As already stated in introduction, two sample problems were used for the purpose of providing comparisons. Logistic map is a time series based on recursive equation

$$x_t = r x_{t-1}(1 - x_{t-1}) , \quad x_0 = 0.01 . \tag{7}$$

When using an initial value $r = 4$, this time series is known to be chaotic. In our forecasting problem four successive data points in the time series were used to predict the fifth.

The second data set consists of real world data, collected in an industrial mixing process from polymer products industry. During the mixing of polymer compounds various variables are registered every second, the most important ones being: torque, temperature, rotation speed, and change of energy. Torque value is a good pointer of mixing progress, but is also highly nonlinear in time, with additional influence from other variables. Throughout the mixing process, it is desired that torque stays within boundaries (see Fig.2.2), which assures a high quality of the polymer compound after the mixing ends. Correction of torque during the mixing is possible only through modification of the rotation speed. Its effect on the change of torque is delayed, which is why the prediction of torque value is important. We have to use available data to predict torque in the foreseeable future. In case of expected error, the change of speed is performed to retain torque in tolerance.

Principal component analysis was used to extract most relevant data from the database acquired during months of mixing processes prior to our research. These data were then prepared for forecasting the value of torque at a specific time in the mixing process. Good results of our training could be used to intelligently manipulate the rotation speed and consequently raise the quality of polymer compounds and shorten average length of the mixing process.

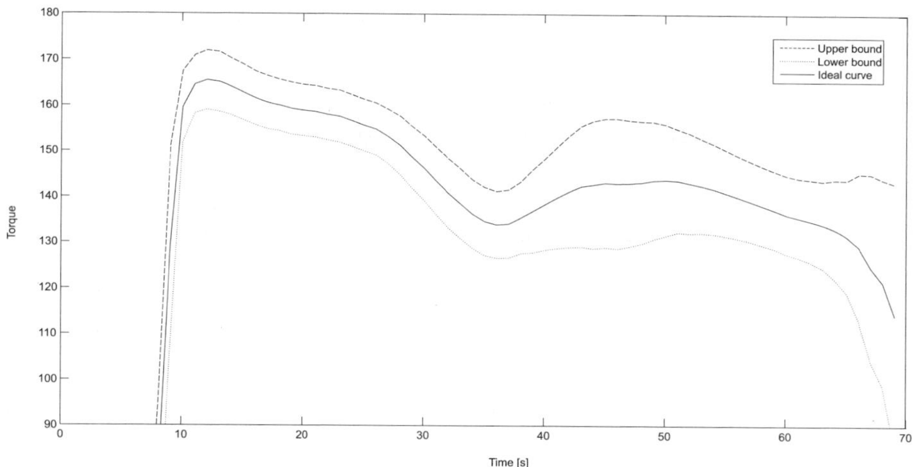

Fig. 2. Ideal value of torque with upper and lower bound

3   Experimental Work

We compared performances of two algorithms over two different data sets. The error-entropy driven multi-layer perceptron (EEM-MLP) was compared with multi-layer perceptron using mean squared error (MSE-MLP). Two data sets, the Logistic map time series (L-MAP) and torque forecasting data in polymer mixing process (MIX), were used. Parameters (σ and η) were not adapted during the training process, so their optimal values had to be found experimentally. For every data set each of the parameters had to be chosen separately to assure best convergence. After both algorithms had been tested to yield correct results and provide sufficient convergence, they were subjected to the benchmark data sets.

Data for each case (L-MAP, MIX) were divided into data subsets for training (200 samples) and testing (500 samples) the MLPs. To assure relevant results, 10 different data sets were randomly selected from the same databases, and 10 distinct initial random weight configurations were used for training. Both EEM-MLP and MSE-MLP used the same data sets to make the comparison of results relevant. Stopping criterion for the training process was defined as follows. Before each change in weights (batch mode) of the MLP, cumulative error on the training set was measured with the appropriate error criterion (MSE or IP). If the cumulative error on the training data set was smaller than all previous errors, the weights were buffered. The training process was stopped if for the last 2000 epochs the cumulative error of each epoch had not decreased below the previous minimal error. In this case, the last best buffered weights were used for testing. This stopping criterion was chosen because the error begun oscillating in some cases. If the error kept decreasing throughout the training, the process was stopped 5000 epochs after the beginning. This ad hoc criterion

Table 1. Learning rate values and MLP topologies used

	Learning rate		Topology	
	L-MAP	MIX	L-MAP	MIX
MSE-MLP	0.01	0.02	4-4-1	8-4-1
EEM-MLP	30	30	4-4-1	8-4-1

was used because our experiments showed that majority of most well trained models were obtained within this limit.

Both MLPs use learning rate parameter η for the steepest descent algorithm. All values for η were obtained experimentally and are given in Tab. 1 for both data sets. In addition to learning rate, the EEM-MLP accepts another parameter called kernel width σ. This parameter influences smoothness of the density approximation. While a large value of kernel size in Parzen's estimation could lead to over-smoothing of the probability density and a possible overlooking of the global optimum, a small kernel size could induce the training to get stuck in a local optimum. Through the experiments, $\sigma = 0.5$ was chosen as the optimal fixed size kernel width.

Topologies for both MLPs and experiments is given in Tab. 1 where $I - H - O$ stands for number of *input − hidden − output* elements. The topologies were also chosen through experiments. We chose topologies with smallest amount of neurons in the hidden layer that provided high convergence rate inside the 5000 epochs and good generalization.

The predictions of MLPs were sufficiently accurate in both experiments. The most noticeable difference is the time needed to complete one epoch. While complexity of the MSE-MLP is $O(n)$, the complexity of the EEM-MLP is $O(n^2)$ [2], which yields significant difference. Several approaches have been pointed in the literature [10], [11] where complexity of the EEM-MLP is reduced. Even

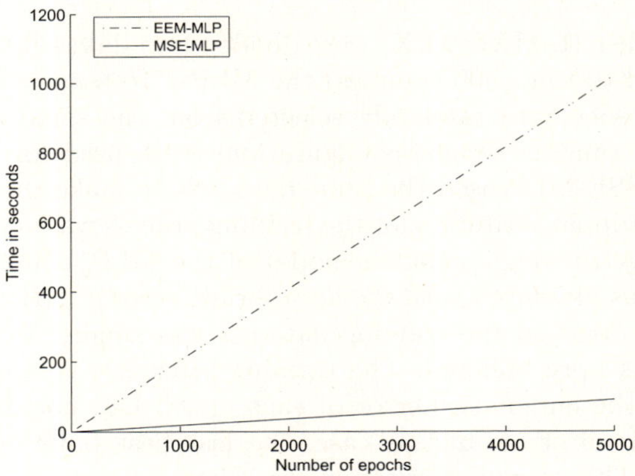

Fig. 3. Time needed to complete training epochs

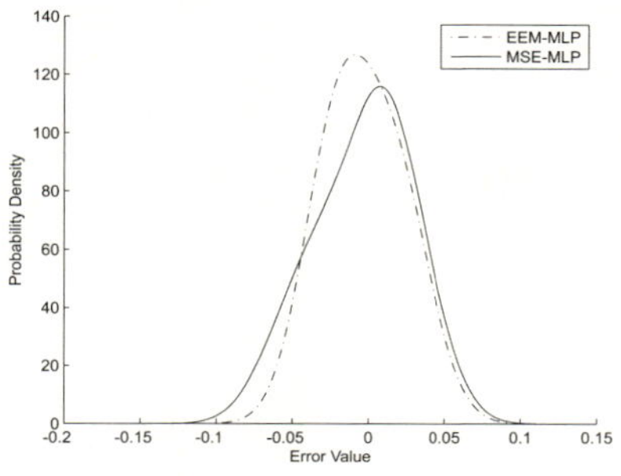 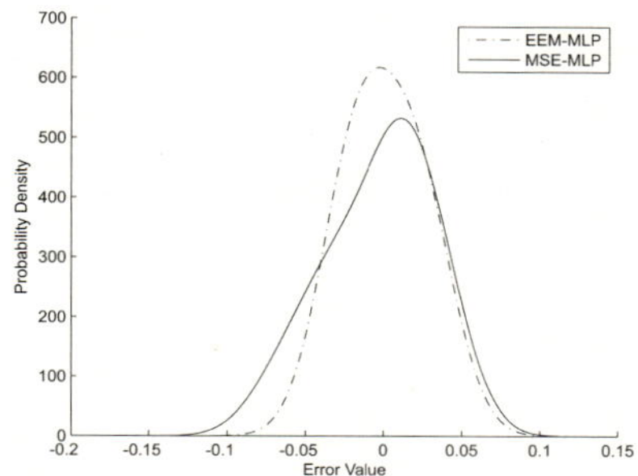

Fig. 4. Mean zero error pdfs for L-MAP on (a) train data set and (b) test data set

though the data sets were relatively small (200 input-output samples for each case), the difference in time needed to complete one training was notable (Fig. 3).

After all experiments were completed, we analyzed the models (weights of the neural networks) each algorithm provided. Algorithms were compared through mean absolute error (MAE) they achieved over train and test data sets, information potentials, mean squared errors and probability density functions of the errors. As expected, the EEM-MLP trained models were superior at the information potential criterion, and the MSE-MLP trained models provided better mean square error prediction. This is due to the nature of the algorithms where EEM-MLP maximizes the information potential (IP) and MSE-MLP minimizes the MSE to train the model. Because both (IP and MSE) criteria are evidently biased, error probability density functions were chosen to compare the MLP models on train and test data sets, together with a similar criterion, MAE. Error pdfs for trained MLPs over train and test L-MAP data sets are presented in Fig. 4(a) and (b). The bias of the EEM-MLP's output neuron was modified so the mean of error pdf is zero centered. [3]. The difference between the pdfs of both MLPs is notable in both figures. EEM-MLP provided a larger and more concentrated peak around zero error than MSE-MLP. This is most notable in Fig. 4(b) where the error pdf over the test data set is presented. When comparing Fig. 4(a) and (b), one can notice that the change of the EEM-MLP pdfs is not significant. Comparison of best MAE values the two MLPs provided are given in Tab. 2. It can be seen that the best EEM-MLP run outperformed the best MSE-MLP run in all cases.

A well known problem [6],[8],[9] with error-entropy minimization algorithms is selection of the kernel size for kernel density estimation. Because input-output

Table 2. Best MAE for both MLPs over MIX and L-MAP train and test data sets

	MIX		L-MAP	
	Train	Test	Train	Test
MSE-MLP	0.0703	0.0749	0.0227	0.0233
EEM-MLP	0.0672	0.0737	0.0156	0.0165

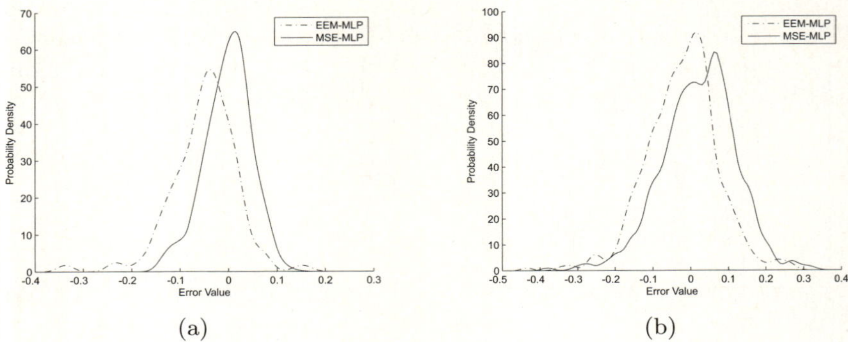

Fig. 5. Mean zero error pdfs for MIX on (a) train data set and (b) test data set

errors get smaller during training, kernel size should be adapted to allow a
finer estimation of the underlying density. Since our experiments were conducted
with a fixed kernel size, the EEM-MLP was more likely to oscillate, after an
optimum had been found, compared to MSE-MLP. On average the EEM-MLP
had convergence problems in about 20% while the MSE-MLP did not converge
in 5% of the runs. The training on the MIX data set also delivered interesting
results. Even though the EEM-MLP had similar convergence problems as with
L-MAP data set, the best run once again outperformed best MSE-MLP run
as seen in Tab. 2. The error pdfs reveal yet another feature. When comparing
the shapes of MSE-MLP error density in Fig. 5(a) and (b) one can notice that,
compared to EEM-MLP density, it has a higher and more concentrated peak
over the training data. When the two densities are compared in (b), it is evident
that here the MLP-MSE produced a lower and wider error density. This suggests
that the EEM-MLP provides a better generalization from the train to the test
data set, as the concentrated peak is presented on both figures.

4   Conclusion

In this paper we applied the error-entropy minimization criterion to training
and testing the multi-layer perceptron on an artificially generated and a real
world data set. It is shown that models trained with this criterion provide a good
equilibrium between the accuracy of prediction and generalization from the train
to the test data set. These properties are highly valued in industry applications
where off-line models are trained and later used for prediction. However the
original error-entropy algorithm has some drawbacks that need improvement. We
enlightened two areas where an optimization is required for a further expansion
of the error-entropy minimization criterion in real world problems. The most
noticeable is the computational inefficiency which almost renders training on
large data sets. The second area open for research is the pursue for an optimal
adaptive kernel size required for density estimation. Solution to this problem
would not only cease the oscillation and convergence problems but, together with

an adaptive learning rate, acquire the ability to find better optima in the steepest descent approach. With these optimizations, error-entropy minimization could be a very competitive choice for problems where fast and accurate computation has to be provided on large data sets.

References

1. Berger, J.O.: Statistical Decision Theory and Bayesian Analysis. Springer, Heidelberg (1985)
2. Principe, J.C., Xu, D., Fisher, J.W.: Information-Theoretic Learning. In: Haykin, S. (ed.) Unsupervised Adaptive Filtering, vol. I, pp. 265–319. Wiley, New York (2000)
3. Erdogmus, D., Principe, J.C.: An Error-Entropy Minimization Algorithm for Supervised Training of Nonlinear Adaptive Systems. IEEE Transactions on Signal Processing 50(7) (July 2002)
4. Shannon, C.E.: A Mathematical Theory of Communications. Bell Sys. Tech. J. 27, 379–423 (1948)
5. Parzen, E.: On Estimation of a Probability Density Function and Mode. Time Series Analysis Papers. Holden-Day, San Francisco (1967)
6. Erdogmus, D., Principe, J.C.: Generalized Information Potential Criterion for Adaptive System Training. IEEE Trans. Neural Networks 13, 1035–1044 (2002)
7. Silverman, B.W.: Density Estimation for Statistics and Data Analysis. Chapman and Hall, London (1993)
8. Morejon, R.A., Principe, J.C.: Advanced Search Algorithms for Information-Theoretic Learning with Kernel-Based Estimators. IEEE Transactions on Neural Networks 15 (July 2004)
9. Santos, J., Alexandre, L.A., Marques Sá, J.: The Error Entropy Minimization Algorithm for Neural Network Classification. In: Proceedings of the 5th International Conference on Recent Advances in Soft Computing, Nottingham, United Kingdom, pp. 92–97 (December 2004)
10. Erdogmus, D., Kenneth, E.H., Principe, J.C.: Online Entropy Manipulation: Stochastic Information Gradient. IEEE Signal Processing Letters 10(8) (August 2003)
11. Han, S., Rao, S., Principe, J.C.: Estimating the Information Potential with the Gauss Transform. In: Rosca, J.P., Erdogmus, D., Príncipe, J.C., Haykin, S. (eds.) ICA 2006. LNCS, vol. 3889, pp. 82–89. Springer, Heidelberg (2006)

Hybrid Evolution of Heterogeneous Neural Networks

Zdeněk Buk and Miroslav Šnorek

Department of Computer Science and Engineering, Faculty of Electrical Engineering,
Czech Technical University in Prague,
Karlovo namesti 13, 121 35, Prague, Czech Republic
`bukz1@fel.cvut.cz`
`http://cig.felk.cvut.cz`

Abstract. In this paper we are describing experiments and results of applications of the continual evolution algorithm to construction and optimization of recurrent neural networks with heterogeneous units. Our algorithm is a hybrid genetic algorithm with sequential individuals replacement, varibale population size and age-based probability control functions. Short introduction to main idea of the algorithm is given. We describe some new features implemented into the algorithm, the encoding of individuals, crossover, and mutation operators. The behavior of population during an evolutionary process is studied on atificial benchmark data sets. Results of the experiments confirm the theoretical properties of the algorithm.

1   Introduction

The Continual Evolution Algorithm (CEA) is a hybrid genetic algorithm mainly developed for recurrent neural networks training. In this algorithm the evolutionary and gradient-based optimizations are combined to create and optimize both the structure and weights setting of the network. Our recurrent neural network models are based on the general fully recurrent [2,3] (symetric fully connected oriented graph) topology. Examples of this networks are shown in figure 1. There are two tasks in construction of the network – first is the structure (topology and activation functions) of the network construction and second is the training process where the weights values of synapses in the network are set. Figure 1(a) shows the network where all neurons uses the same activation function and topology is represented by fully connected graph, figure 1(b) shows another example of the fully recurrent neural network, in this case each neuron uses different activation function. These two tasks – structure and weight values finding – are combined in our CEA Algorithm.

The following sections describe the main features of the CEA, evolution control mechanism used in the CEA, and mutation and crossover operators. The CEA algorithm is compared to the Differential Evolution (DE) algorithm [1] and pure gradien-based Realtime Recurrent Learning (RTRL) algorithm [2] in benchmark tasks which are described in section 3 (experiments).

V. Kůrková et al. (Eds.): ICANN 2008, Part I, LNCS 5163, pp. 426–434, 2008.

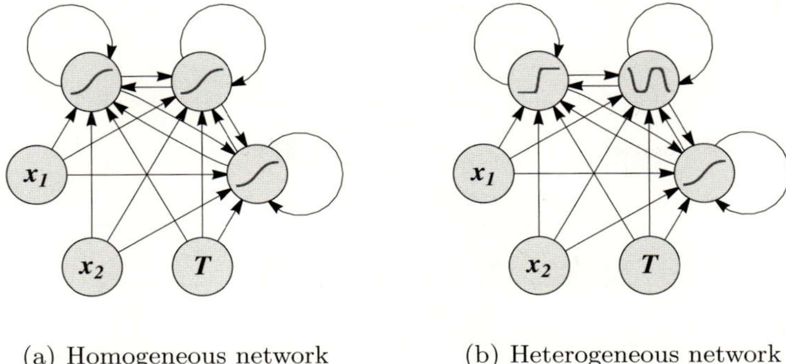

(a) Homogeneous network   (b) Heterogeneous network

Fig. 1. Examples of fully recurrent neural networks. Both networks have three neurons, two inputs (x_1, x_2) and threshold (T).

2   Continual Evolution Algorithm

Evolution process in CEA combines the gradient-based and genetic optimizations. The two-dimensional evolution in CEA is illustrated in figure 2. The CEA algorithm works with *variable population size* and it uses separated encoding of structure and parameters of models. Each individual in population in the CEA has *age parameter* and there is sequential replacement of individuals implemented – original individuals (parents) are kept in population and optimized together with new individuals (offsprings) – an example of population evolution and sequential replacement of individuals is shown in figure 3. The evolutionary process is controlled by a set of probability functions (depending on size of the population, quality of individuals, and their age).

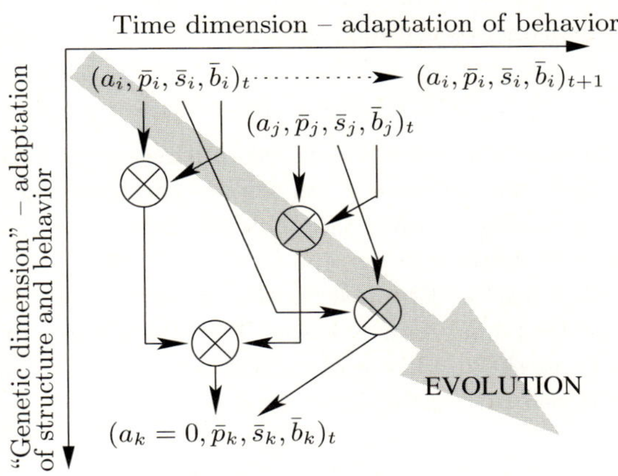

Fig. 2. Two-dimensional evolution in CEA

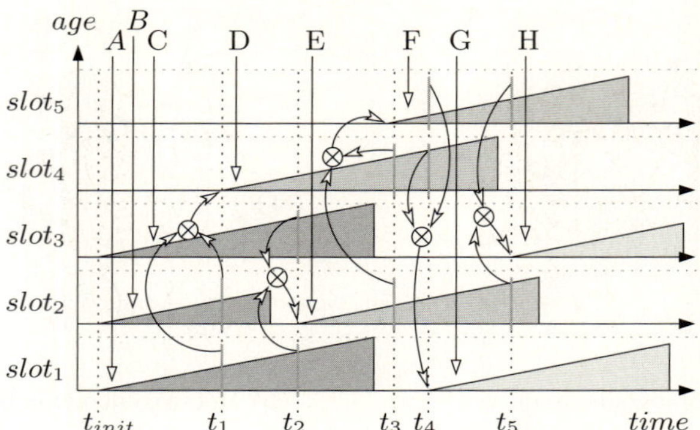

Fig. 3. Visualization of a CEA population example. This example shows the evolution of population with maximal size of five individuals, represented by five slots in the figure. The population is initialized with three individuals at time t_0 (two slots are empty in that moment). The vertical axis represents the age of each individual, maximal age is represented by horizontal dashed gray lines. Here can be seen that only some individuals reach this maximal age - it depends on some other factors such as probability, population size, and fitness value. At time t_1 the new individual (marked as D) is created as the offspring of the (parents) individuals A and B. The next reproduction processes come at t_2, t_3, t_4, and t_5. It is clear that the size of population (number of individuals) varies through time.

2.1 Evolution Control

The CEA is controlled by several auxiliary parameters that are computed for each individual in population. These parameters are used in the reproduction cycle. The first parameter is the *reproduction probability* which describes the probability, that the i-th individual of age a_i and quality F given by the fitness function, will be used for reproduction operation and that they will produce some new individual (offspring) to the next generation. The reproduction probability is defined as:

$$RP^*(\bar{x}_i) = RP^*(a_i, F(\bar{x}_i)), \tag{1}$$

where x_i is the i-th individual that we are computing the probability for, a_i is the age of this individual and function F represents the fitness function – so $F(\bar{x}_i)$ represents the fitness value (quality) of the i-th individual. Typical behavior of the reproduction probability can be seen in figure 4(b).

Parameter *death probability* represents the property that each individual has some maximal age that they can live for. The probability of survival of each individual depends on the quality and actual age of this individual. Here is an example how the death probability is defined:

$$DP^*(\bar{x}_i) = DP^*(a_i, F(\bar{x}_i)), \tag{2}$$

where x_i is the i-th individual that we are computing the probability for, a_i is the age of this individual and function F represents the fitness function – so $F(\bar{x}_i)$

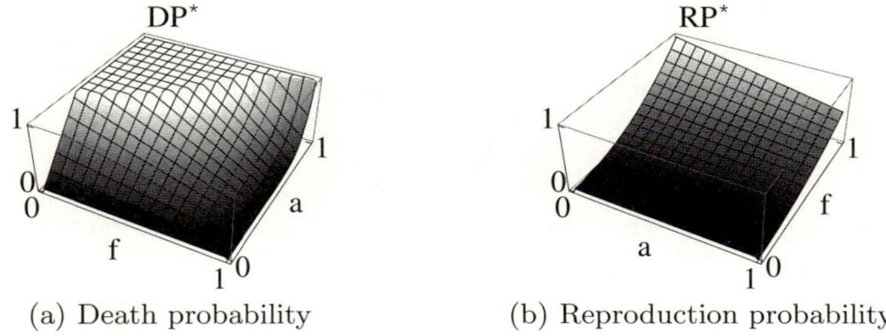

(a) Death probability   (b) Reproduction probability

Fig. 4. Example of the probability functions used in CEA algorithm

represents the fitness value (quality) of the i-th individual. Typical behavior of the death probability can be seen in figure 4(a).

All values signed by $*$ are so called *raw* values. So DP^* is the *raw death probability* and RP^* is the *raw reproduction probability*. The final values DP and RP, which the CEA works with, are computed from the raw values using the *balancing functions*. These functions represent the influence of the size of the population to this size itself – the bigger population will grow slowly (to some limit value, where no new individual will be born) and the smaller population will grow faster (for smaller populations the death probability is reduced and goes to zero – see examples below).

Final probabilities computation:

$$DP(\bar{x}_i) = BAL_{DP}(N, DP^*(\bar{x}_i)), \tag{3}$$

$$RP(\bar{x}_i) = BAL_{RP}(N, RP^*(\bar{x}_i)), \tag{4}$$

where N is the size of the actual population and the BAL_{DP} and BAL_{RP} are the general balancing functions. Examples of the balancing function is shown in figure 5.

2.2 Encoding of Individuals

This subsection describes the basic data structures used in CEA. In CEA the floating point vector is used for individuals encoding. This encoding vector is

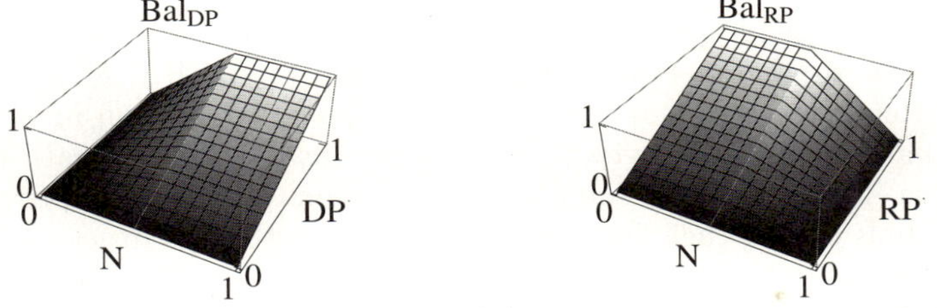

(a) Death probability balancing   (b) Reproduction probability balancing

Fig. 5. Example of the balancing functions used in CEA algorithm

additionally divided into logical parts, representing the structure and behavior of individual - topology and weights setting of neural network represented by the individual. The individual is represented by the following vector:

$$\bar{x}_i = (a_i, \bar{p}_i, \bar{s}_i, \bar{b}_i), \tag{5}$$

where a_i is the *age* of i-th individual, $\bar{p}_i$ is the initialization *parametric vector* (called instinct), $\bar{s}_i$ is the *structural* parameter and $\bar{b}_i$ is *behavioral vector* of i-th individual, which contains actual set of working parameters of the individual (at the beginning of evolution it is created as a copy of the $\bar{p}$ vector).

In our implementation and experiments we are using the floating point weight matrix for behavior encoding and adjacency matrix for topology of the network encoding. Part of the general $\bar{s}_i$ vector is also list of the activation functions used in the particular neural network. These functions are represented direcly as symbolic axpressions.

2.3   Mutation and Crossover Operators

In our experimental implementation we are using uniform crossover and random mutation of the adjacency and weight matrix. The crossover operator works with two individuals, but when new parametrical vector is being created the function combines four vectors (behavioral and parametrical) – $\bar{p}_1,\bar{b}_1$ and $\bar{p}_2,\bar{b}_2$. These parameters are uniformly combined.

3   Experiments

Because the CEA algorithm is under heavy research and developement, we are now solving mainly the benchmarking tasks with the recurrent neural networks. Most of the experiments we run with the simple learn to oscillate task [2].

On this task we study the behavior of the population evolution. We study the influence of the parameters of the CEA algorithm (elitism, population size, maximal age of the individuals, etc).

Figures 6 and 7 show selected results for simple sequence generators – the network topology and output values. Because the periodic signal used, the CEA algorithms found the periodic activation functions for the neurons – the sin-sigmoid function [11] is used in these solutions.

Figure 9 shows the example population evolution in CEA. It is the visuzalization of the fitness value of each individual in the population in time. It can be seen, that the better individual lives longer than the individuals with lower fitness value (lower quality). There can also be seen the quality improvements of the individuals during their life, only in some cases there is fitness value downgrade – these individuals have probably wrong topology, wich cannot be adapted to the problem.

Visualization of the fitness value of the best individual in the population for several runs of the algorithms is shown in figure 9.

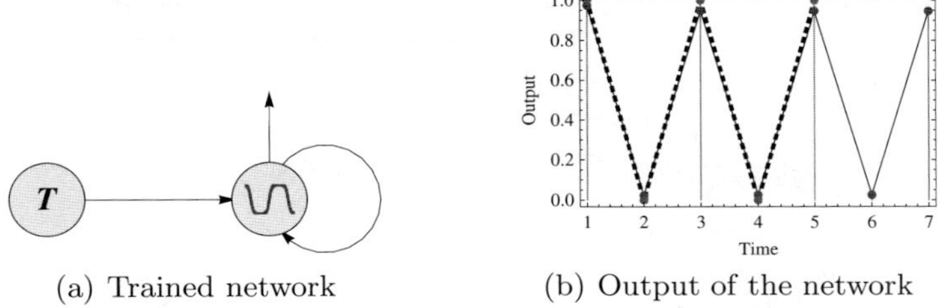

(a) Trained network   (b) Output of the network

Fig. 6. An example of the *learn to oscillate* experiment. The network was trained to generate simple 1010... periodic sequence. The graph shows the output value of the network in time and the training set (dashed line).

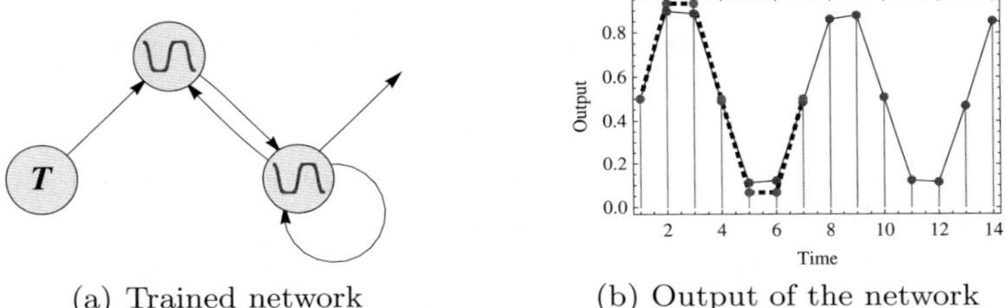

(a) Trained network   (b) Output of the network

Fig. 7. An example of the *learn to oscillate* experiment. The network was trained to generate sampled $sin(x)$ function. The graph shows the output value of the network in time and the training set (dashed line).

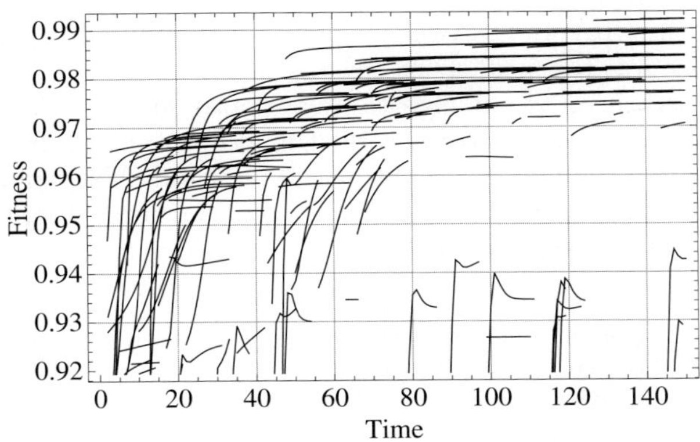

Fig. 8. Visualization of the population evolution in time. Fitness value of each individual is shown. The continual adaptation in time and variable time of life of each individual can be seen.

We have compared the CEA algorithm with Differential Evolution algorithm and pure gradient-based Realtime Recurrent Learning algorithm. All algorithms was tested on the same experiment (learn to oscillate) and some selected results

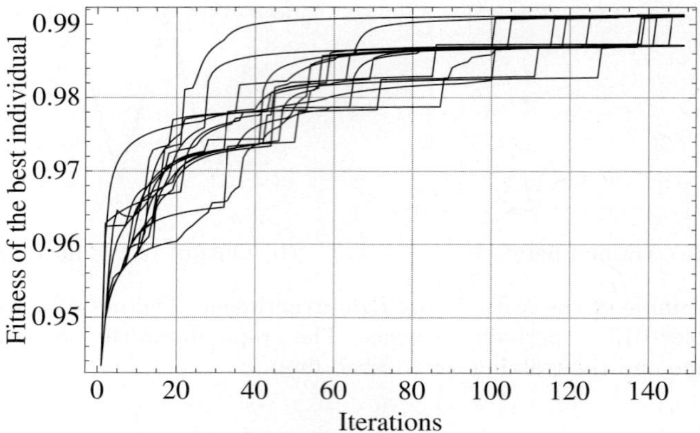

Fig. 9. Visualization of the fitness value of the best individual in population. Experiment was repeated 15 times.

are shown in table 1. These are approximate results got by several runs of the algorithm to get solutions with comparable quality. It can be seen that CEA algorithm is somewhere between the genetic (DE) and gradient (RTRL) methods.

Because of variable population size and combination of genetic and gradient-based methods in CEA there was highly reduced the number fitness evaluation needed for evolving good-quality individual (network). Differential evolution algorithm uses constant population size, so number of fitness evaluations is also constant for particular number of individuals and number of generation setting. Gradient-based algorithm (RTRL) works with only one network and it needed less steps than the CEA. In CEA the average size of population was 24 although the maximal size was the same as in DE and it was set to 100.

The RTRL algorithm is not able to create optimal topology of network. It always uses the fully connected network. DE also optimized the whole weight matrix representing the fully recurrent neural network. CEA reduced some connection - as an example see following numbers: for 6 neurons there are 42 weight values $(6 \times (6 + 1)$, one additional input for threshold value), best network from

Table 1. Selected algorithms comparison results measured on learn to oscillate benchmark experiment. Comparison of number of iterations needed in particular algorithms to train the network with comparable result (errors). DE-differential evolution, CEA-continual evolution algorithm, RTRL-real time recurrent learning. The table shows numbers for whole population and for the best individual for comparison. Shown values are approximate – they come from several runs of the algorithms.

	Total for whole algorithm run			For best solution (one individual)		
	DE	CEA	RTRL	DE	CEA	RTRL
No. of fitness evaluations	25 000	480	×	500	17	×
No. of gradient steps	×	9 000	500	×	340	500

one algorithm runs had 25 active connections, it means 17 connections was removed, which represents about 40% reduction.

Based on experiments we can say that we got better results with CEA for bigger state spaces - larger networks (larger maximal number of neurons). For smaller networks the DE algorithm served better results. RTRL algorithm is faster (mainly for very small networks) than DE or CEA, but RTRL is not able to optimize the topology of network.

4   Implementation

The algorithm and all experiments are implemented in Wolfram *Mathematica,* which allows us to use direct representations of the activation functions in symbolic form and many other features as a self modifying network structures, easy way to paralelization of the algorithm (it is not finished yet, but we are now testing some parts of the parallel CEA algorithm), etc.

5   Conclusion

The CEA algorithm allow us to construct recurrent neural networks with heterogeneous units, this algorithm needs less gradient steps for learning than the pure gradient learning algorithm (Realtime Recurrent Learning) and also less evolutionary steps (less fitness function evaluations) than the pure genetic algorithm (Differential Evolution).

The CEA algorithm is being developed and tested on experimental tasks now. The next step is to use this algorithm to create models of meteorological time series for temperature forecast. Another experiment being tested now is the 3D objects reconstruction process using CEA, where the algorithm is used to produce representation of 3D models instead of neural networks.

Acknowledgments

This work has been supported by the internal grant of Czech Technical University in Prague, IG: CTU0706813 - "Development of the Continual Evolution optimization algorithm and its application to the artificial neural networks adaptation" and the research program "Transdisciplinary Research in the Area of Biomedical Engineering II" (MSM6840770012) sponsored by the Ministry of Education, Youth and Sports of the Czech Republic.

Parts of the experiments used for the CEA tuning and research, mainly the visualizations used for population behavior study during evolution process, has been calculated using the ALTIX 3700 supercomputer at Supercomputing Centre of Czech Technical University in Prague.

References

1. Lampinen Jouni, A., Price Kenneth, V., Storn Rainer, M.: Differential Evolution A Practical Approach to Global Optimization (2005) ISBN 978-3-540-20950-8
2. Williams, R.J., Zipser, D.: A learning algorithm for continually running fully recurrent neural networks. Neural Computation 1(2), 270–280 (1989)
3. Williams, R.J.: Gradient-based learning algorithm for recurrent networks and their computational complexity. In: Chauvin, Y., Rumelhart, D.E. (eds.) Backpropagation: Theory, Architectures and Applications (1995)
4. Yao, X.: Evolving artificial neural networks (1999)
5. Buk, Z., Šnorek, M.: A brief introduction to the continual evolution algorithm. In: Proceedings of 6th International Conference APLIMAT 2007, Part III, Bratislava, pp. 71–76 (February 2007) ISBN: 978-80-969562-6-5
6. Buk, Z.: Survey of the computational intelligence methods for data mining. Technical Report DCSE-DTP-2007-01, Czech Technical University in Prague, FEE, CTU Prague, Czech Republic (2007)
7. Buk, Z., Šnorek, M.: Nature Inspired Methods for Models Creation and Optimization. In: Proceedings of 41th Spring International Conference MOSIS 2007, pp. 56–62. MARQ, Ostrava (2007)
8. Buk, Z., Šnorek, M.: Continual Evolution Algorithm for Building of ANN-based Models. In: Proceedings of the 6th EUROSIM Congress on Modelling and Simulation, Ljubljana, Slovenia (2007)
9. De Jong, K.A.: Evolutionary Computation: a unified approach. The MIT Press, Cambridge (2006)
10. Eiben, A.E., Smith, J.E.: Introduction to Evolutionary Computing. Springer, Heidelberg (2003)
11. Wong, K.-W., Leung, C.-S., Chang, S.-J.: Handwritten Digit Recognition Using Multi-Layer Feedforward Neural Networks with Periodic and Monotonic Activation Functions. In: 16th International Conference on Pattern Recognition (ICPR 2002). ICPR, vol. 3, p. 30106 (2002)

Ant Colony Optimization with Castes

Oleg Kovářík and Miroslav Skrbek

Department of Computer Science and Engineering, Faculty of Electrical Engineering,
Czech Technical University in Prague,
Karlovo namesti 13, 121 35, Prague, Czech Republic
kovaro1@fel.cvut.cz, skrbek@fel.cvut.cz
http://cig.felk.cvut.cz

Abstract. Ant Colony Optimization (ACO) is a nature inspired meta-heuristic for solving optimization problems. We present a new general approach for improving ACO adaptivity to problems, Ant Colony Optimization with Castes (ACO+C). By using groups of ants with different characteristics, known as castes in nature, we can achieve better results and faster convergence thanks to possibility to utilize different types of ant behaviour in parallel. This general principle is tested on one particular ACO algorithm: $\mathcal{MAX} - \mathcal{MIN}$ Ant System solving Symmetric and Asymmetric Travelling Salesman Problem. As experiments show, our method brings a significant improvement in the convergence speed as well as in the quality of solution for all tested instances.

1 Introduction

Stochastic algorithms derived from Ant Colony Optimization metaheuristic [1] proved to be successful in solving $\mathcal{NP}$-complete optimization problems. While using these algorithms, we often face a problem of slow convergence and of getting stuck in a local optima. To avoid these problems, several approaches were developed. We can utilize natural parallelism of ant algorithms and use multiple colonies to increase a probability of a faster solution discovery [2,3]. Also, approaches originally developed in the field of Evolutionary Computation, such as niching techniques [4], can be used to sustain a diversity in the (population based) ACO. Another option is to take into account the dependence of the optimal parameter settings on the concrete instance of the problem. Parameters can be adapted during the computation or derived from statistical evaluation of the solved problem instance (see [5] for survey of articles on these topics). We can also see this as a continuous optimization problem and solve it. There is a chance that the hierarchy of optimization problems might be able to reduce the dimension or the impact of parameters in the higher levels [6]. However, this process would be very time consuming indeed.

We tried another way to improve the ACO performance by using principles inspired by nature. In nature, ants of the same species are divided into a different number of castes, like queen(s), workers, soldiers etc. Each caste has unique capabilities depending on its size, body structure, ability to reproduce and encoded behavioural acts. This functional specialization called polyethism allows

V. Kůrková et al. (Eds.): ICANN 2008, Part I, LNCS 5163, pp. 435–442, 2008.

the whole colony to act as a sophisticated system with division of labour while keeping the individuals simple. Social insects are also often forming so called temporal castes because their behaviour together with physiology can change depending on their age. Various mechanisms of castes management across all social insect species are ranging from discrete castes to the continuous ones. These mechanisms include time-dependent, pheromone and environment driven changes in the structure of castes.

An ant colony has a distributed management which relies on a direct information exchange between the individual ants, and especially on indirect chemical communication mediated by pheromone trails. Diversification of castes provides a possibility to use different pheromones unique to each caste. There are documented species of ants which are using tens of different pheromones. Exhaustive description of real ants and ant colonies including more details on above mentioned attributes can be found in [7].

We imitated the natural principle of castes with different behaviour patterns. The castes feature helps the Ant Colony Optimization metaheuristic adapt to the particular problem instance. They also enable utilisation of different algorithm behaviours in different stages of computation. Our assumption was that this will lead to a better performance of the algorithm in average case as well as on special problem instances, where the standard Ant Colony Optimization algorithms would fail.

1.1 Ant Colony Optimization (ACO)

ACO is a metaheuristic approach to solve hard combinatorial optimization problems. It is inspired by the indirect communication in real ant colonies that is mediated by laying and following the pheromone trails. When searching for food, ants use a pheromone trail to mark their paths between the food and the nest. Other ants are more likely to follow the trails with greater amount of pheromone and if they find enough food, these trails are reinforced. Random deviations from their path also make it possible to gradually shorten the length of the path. Pheromone evaporation allows to forget less convenient paths.

ACO metaheuristic uses artificial ants and pheromone to imitate food searching behaviour of real ants. Algorithms built on this metaheuristic are stochastic as the ant movement is lead by probability function. They are iterative since they repeatedly simulate ant movements attended by pheromone laying and pheromone evaporating process:

```
procedure ACO_Meta_heuristic()
   while (termination_criterion_not_satisfied)
     ants_generation_and_activity()
     pheromone_evaporation()
     daemon_actions()     {optional}
   end while
end procedure
```

1.2 $\mathcal{MAX} - \mathcal{MIN}$ Ant System ($\mathcal{MMAS}$)

The first ACO algorithm was the Ant System (AS) designed to solve the Travelling Salesman Problem: for given set of cities and (symmetric or asymmetric) matrix of distances, we want to find a closed path with the requirement of visiting all cities with minimal total distance travelled. It is an $\mathcal{NP}$-complete discrete optimization problem often used as a benchmarking problem that is simple to figure out and visualize. It can be solved by using algorithms of different complexity including greedy and k-opt heuristics, very successful V-opt method (Lin-Kernighan-Johnson), nature inspired Genetic Algorithms, Simulated Annealing, Tabu Search and others. Last but not least, $\mathcal{MMAS}$ algorithm [8] proved to be successful in solving TSP.

$\mathcal{MMAS}$ uses ants with a memory containing partial tours between cities. The probability of moving from city i to allowed (unvisited) city j is given by

$$p_{ij}(t) = \begin{cases} \dfrac{[\tau_{ij}(t)]^{\alpha} \cdot [\eta_{ij}]^{\beta}}{\sum_{j \in allowed}[\tau_{ij}(t)]^{\alpha} \cdot [\eta_{ij}]^{\beta}} & \text{if } j \in allowed \\[2em] 0 & \text{otherwise} \end{cases} \tag{1}$$

where $\tau_{ij}(t)$ is the amount of pheromone between city i and j, $\eta_{ij} = 1/d_{ij}$ called visibility is computed from distance d_{ij} between cities, and α and β are parameters which allow the user to control the importance of the pheromone and visibility.

After ants constructed their paths, pheromone between every two cities is evaporated and then reinforced:

$$\tau_{ij}(t+1) = \rho \cdot \tau_{ij}(t) + \Delta\tau_{ij}(t) \tag{2}$$

where ρ is evaporation factor, and the amount of the pheromone to be added $\Delta\tau_{ij}(t)$ is a sum of amounts from all m ants:

$$\Delta\tau_{ij}(t) = \sum_{k=1}^{m} \Delta\tau_{ij}^{k}(t) \tag{3}$$

for

$$\Delta\tau_{ij}^{k}(t) = \begin{cases} Q/L^{k}(t) & \text{if } (i,j) \text{ is a member of the tour } T^{k}(t) \text{ of the } k\text{-th ant} \\[2em] 0 & \text{otherwise} \end{cases} \tag{4}$$

where Q is a constant value and $L^{k}(t)$ is the tour length of the k-th ant.

$\mathcal{MMAS}$ differs from AS in pheromone maintenance. It limits the possible amounts of pheromone to the interval $< \tau_{min}, \tau_{max} >$ to keep the probability of choosing each path grater than zero but not too big. This helps the algorithm to keep chances of escaping from local optima. Initial amounts are set to τ_{max} to boost up random search in the early iterations. There are several improvements to this algorithm of which we used probability q of choosing the nearest city instead of using eq. 5.

2   Proposed ACO with Castes (ACO+C)

2.1   General ACO+C

In the proposed modification of ACO metaheuristic, we utilised nature developed approach by simulating ant castes. The idea is to split the population of ants into several distinct groups. Ants from different groups have different sets of behavioural acts. For example, we can create "workers" which are moving along the strongest pheromone trails with only minor random changes. Or we can create "explorers" with high probability of random movements, problem heuristics exploitation or pheromone avoidance. As all ants are solving the problem together, explorers can help the algorithm escape from local optima in which the workers are stuck.

By assigning different pheromones to each caste and setting, we can achieve a parallel solving of the problem by several castes. To ensure that castes can exchange parts of the solution, a small probability of reacting to foreign pheromones can be set.

In order to create an instance of ACO+C, we need to specify:

- number of ants m
- number of castes $c \in \mathbb{N}, c \leq m$
- number of ants in each caste m_l for $l = 1..c$ and $m = \sum_{l=1}^{c} m_l$
- set of pheromones
- behavioural acts

where c and m_i can be constant or can change its values during the algorithm run.

2.2   Example: $\mathcal{MMAS}$+C

We designed a simple example of ACO+C improvement to test its potential. It extends one of the best performing ACO algorithms - $\mathcal{MMAS}$. To keep the results transparent, we used only one pheromone for all castes and we focused only on behavioural acts. They are represented by different values of α_l and β_l parameters for l-th caste (with m_l members). For ant from l-th caste, the probability of moving from city i to unvisited city j given by eq. 1 has changed to:

$$p_{ij}^l(t) = \begin{cases} \dfrac{[\tau_{ij}(t)]^{\alpha_l} \cdot [\eta_{ij}]^{\beta_l}}{\sum_{j \in allowed}[\tau_{ij}(t)]^{\alpha_l} \cdot [\eta_{ij}]^{\beta_l}} & \text{if } j \in allowed \\ \\ 0 & \text{otherwise} \end{cases} \tag{5}$$

This enables the algorithm to adapt to various types of solved TSP instances as well as to areas with different characteristics inside one TSP instance. We assumed that some instances would be better solvable with a stronger pheromone or heuristic preferences (α or β parameters) and some instances would be better solvable with smaller α or β values. After carrying out a couple of trial experiments, we decided to modify only one parameter to keep the algorithm simple. Parameter β has been chosen because of its great impact on algorithm performance. We also expected different optimal parameter settings for different stages of the computation. To support this expectation, we dynamically changed the number of ants in each caste depending on improvements of the solution.

3 Evaluation

The example of $\mathcal{MMAS}$ algorithm modification described in the previous section was tested on 8 instances of symmetric TSP and 3 instances of asymmetric TSP taken from TSPLIB [9]. Two sets of castes were designed. Set A (tbl. 1) with only one caste matches standard $\mathcal{MMAS}$ algorithm with recommended parameters. Set B (tbl. 2) consists of several castes with different probability of choosing shorter paths - parameter β.

Table 1. Configuration A (standard $\mathcal{MMAS}$)

caste	1
α	1.0
β	1.0

Table 2. Configuration B (10 castes with different value of the shortest distance heuristic parameter)

caste	0	1	2	3	4	5	6	7	8	9
α	1.0	1.0	1.0	1.0	1.0	1.0	1.0	1.0	1.0	1.0
β	9.0	8.0	7.0	6.0	5.0	4.0	3.0	2.0	1.0	-1.0

For each instance and for both configurations A and B (tbl. 1 and 2), the algorithm was executed 30 times. Average length of the shortest path $\overline{L}$, average iteration when the last improvement occurred $\overline{I}$ and their standard deviations were calculated. Stopping conditions for instance with n cities were met when no improvement occurred during last $S_{max} = 10n$ iterations. Number of ants was set to $m = n$ in all cases. Evaporation factor $\rho = 0.01$ and probability of choosing the nearest city instead of using eq. 5 was set to $q = 0.1$.

4 Results

Results of the tests are summarized in tbl. 3. In all cases, introduction of castes led to faster convergence and better average solutions. Average number of iterations needed for finding the best solution was reduced by 48% and average solutions were significantly closer to the optimum. To show differences between TSP instances, we counted average numbers of improvements of the solution made

Table 3. Comparison of configuration A (standard $\mathcal{MMAS}$) and B ($\mathcal{MMAS}$ with castes varying in parameter β). $\overline{I}$ is the number of iteration in which the best solution was found, $\overline{L}$ represents average length of the shortest path found, σ_L and σ_I are standard deviations of these values. The last two columns show deviations from the optimal path length. In experiments with configuration B, ants always found shorter path in smaller number of iterations at average.

Instance	Iterations				Lengths				> Opt.	
	A		B		A		B		A	B
	$\overline{I}$	σ_I	$\overline{I}$	σ_I	$\overline{L}$	σ_L	$\overline{L}$	σ_L	%	%
st70.tsp	**1723**	322	**1163**	372	684	7	679	4	**1.3**	**0.5**
eil76.tsp	**2081**	1055	**423**	474	544	5	540	3	**1.2**	**0.4**
rat99.tsp	**2220**	479	**943**	237	1225	15	1212	3	**1.2**	**0.1**
rd100.tsp	**2454**	539	**1060**	263	8023	95	7935	28	**1.4**	**0.3**
eil101.tsp	**3088**	890	**1577**	440	645	9	637	5	**2.5**	**1.3**
lin105.tsp	**2211**	490	**963**	277	14445	59	14388	32	**0.5**	**0.1**
pr107.tsp	**2981**	524	**1461**	414	44590	172	44446	101	**0.6**	**0.3**
bier127.tsp	**3143**	680	**1752**	742	119677	922	119049	389	**1.2**	**0.6**
ry48p.atsp	**1693**	561	**807**	464	14761	224	14546	51	**2.3**	**0.9**
ft70.atsp	**2796**	941	**1900**	825	39575	316	39289	176	**2.3**	**1.6**
ftv70.atsp	**1745**	675	**1373**	543	1992	38	1979	15	**2.2**	**1.5**

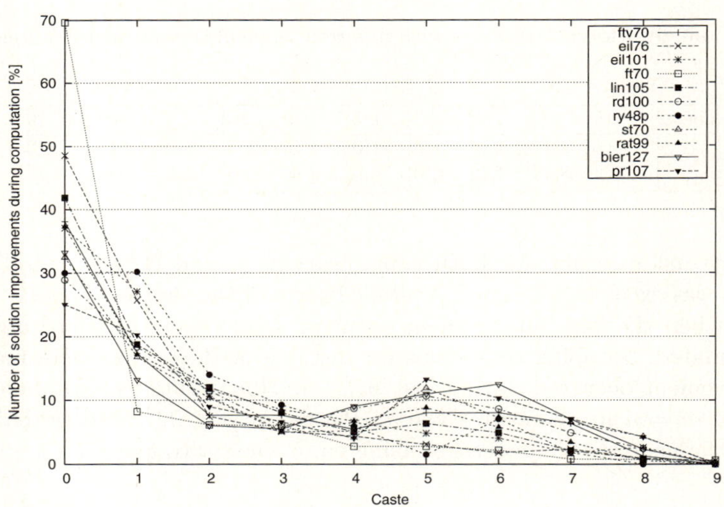

Fig. 1. Comparison of castes for configuration B (tbl. 2) and for all datasets. For each caste we can see average number of solution improvements. These improvements were made by ants from selected caste. We can see frequent utilization of castes with stronger greedy behaviour (on the left side). For some datasets there is also an increased hit rate in castes 5 and 6.

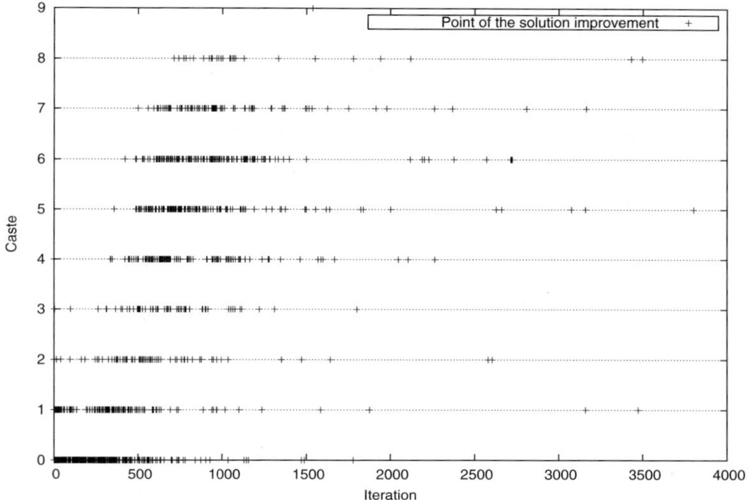

Fig. 2. Points of the solution improvement from 30 runs on the bier127 dataset. During the early stages of computation, castes with stronger preference for the shorter paths (castes with lower numbers) made the most of improvements. Their impact was gradually reduced while the solution advanced towards the optimum. Again configuration B (tbl. 2) was used.

by ants from distinct castes during the algorithm run. The resulting graph (see fig. 1) indicates the ability of the algorithm to adapt to problem instance specificity. Second graph (see fig. 2) shows points of improvements for the bier127.tsp instance. In the beginning of computation, castes with the highest impact of distance heuristic make the most of improvements. Later, when the algorithm converges to optimum, castes with more exploring behaviour are utilised.

5   Conclusion

We introduced Ant Colony Optimization with Castes (ACO+C) which is a general extension to algorithms based on ACO metaheuristic. The idea of dividing ants into several castes with different parameters was tested on the $\mathcal{MMAS}$ algorithm which follows the structure of ACO metaheuristic. We used 11 instances of Travelling Salesman Problem (8 symmetric, 3 asymmetric) from TSPLIB as benchmark data. Results for experimentally designed set of castes show better performance than standard $\mathcal{MMAS}$. The average number of iterations was reduced and average solution was shorter in all cases. These differences were significant according to Wilcoxon test.

We also showed that the different instances of TSP are more effectively solved by different castes and that the number of improvements made by ants from one caste can vary in different stages of computation.

These results are promising and we suggest at least two potential ways of further research. Firstly, more complex castes should be designed, for example castes for local search and castes with different types of pheromone. The second step could be optimization of castes developed by evolutionary computation techniques.

Acknowledgments

This work is supported by the internal grant of Czech Technical University in Prague, IG: CTU0807213 - "New modifications to ant algorithms" and the research program "Transdisciplinary Research in the Area of Biomedical Engineering II" (MSM6840770012) sponsored by the Ministry of Education, Youth and Sports of the Czech Republic.

References

1. Dorigo, M., Stützle, T.: The Ant Colony Optimization metaheuristic: Algorithms, Applications and Advances. In: Glover, F., Kochenberger, G. (eds.) Handbook of Metaheuristics. International Series in Operations Research and Management Science. Kluwer Academic Publishers, Dordrecht (2002)
2. Ellabib, I., Calamai, P., Basir, O.: Exchange strategies for multiple ant colony system. Inf. Sci. 177(5), 1248–1264 (2007)
3. Manfrin, M., Birattari, M., Stützle, T., Dorigo, M.: Parallel ant colony optimization for the traveling salesman problem. In: [10], pp. 224–234
4. Angus, D.: Niching for population-based ant colony optimization. In: E-SCIENCE 2006: Proceedings of the Second IEEE International Conference on e-Science and Grid Computing, Washington, DC, USA, p. 15. IEEE Computer Society, Los Alamitos (2006)
5. Pellegrini, P., Favaretto, D., Moretti, E.: On $\mathcal{MAX} - \mathcal{MIN}$ ant system's parameters. In: [10], pp. 203–214
6. Botee, H.M., Bonabeau, E.: Evolving ant colony optimization. Advanced Complex Systems 1, 149–159 (1998)
7. Hölldobler, B., Wilson, E.O.: The Ants. Belknap Press (March 1990)
8. Stützle, T., Hoos, H.: $\mathcal{MAX} - \mathcal{MIN}$ Ant System. Future Generation Computer Systems 16(8), 889–914 (2000)
9. Reinelt, G.: TSPLIB — a traveling salesman problem library. ORSA Journal on Computing 3(4), 376–384 (1991)
10. Dorigo, M., Gambardella, L.M., Birattari, M., Martinoli, A., Poli, R., Stützle, T. (eds.): ANTS 2006. LNCS, vol. 4150. Springer, Heidelberg (2006)

Neural Network Ensembles for Classification Problems Using Multiobjective Genetic Algorithms

David Lahoz and Pedro Mateo*

Dept. Métodos Estadísticos
Univ. Zaragoza - Spain
{davidla,mateo}@unizar.es

Abstract. In this work a Multiobjective Genetic Algorithm is developed in order to obtain an appropriate ensemble of neural networks. The algorithm does not use any back-propagation method. Furthermore, it considers directly the classification error instead of the mean square error. To obtain the multiobjective environment, the training pattern set is divided into subsets such that each one has its own error function and then, all the error functions are considered simultaneously. The proposed algorithm is found to be competitive with other current methods in the literature.

Keywords: Neural networks, Genetic algorithms, Multiple objective programming.

1   Introduction

Most of the recent works related to neural networks are devoted to the evolution of weights and/or structure of the networks by means of the use of ensembles. The idea of ensembles consists of evolving a set of different networks (in a dependent or independent way) in order to build a combined predictor with them (typically by weighted or unweighted voting). There are several ways to manage and subsequently combine ensembles. The papers [1,2,3,4,5,6,7,8,14] represent a small sample of these numerous works.

In this paper we develop a Multi-Objective Genetic Algorithm, MOGA, which permits us to train Neural Networks, NN, without using backpropagation methods. Furthermore, for the selected test problems only two hidden nodes are necessary in order to obtain the results presented in section 3, a smaller value than the usual in the literature. Therefore, similar or better values are obtained by means of simpler networks. The method proposed generalizes some ideas from [1]. In this work the training set was divided into two subsets, then, they were

____

* This research has been supported by the Spanish Ministry of Education and Science under grants MTM2006-15671 and research groups of *Gobierno de Aragón* E58 and E22.

V. Kůrková et al. (Eds.): ICANN 2008, Part I, LNCS 5163, pp. 443–451, 2008.

used to build a two-objective problem. After the training, the solutions of the Pareto frontier constitute the ensemble.

In our work, we take several stratified samples from the original training set and we construct a multivalued error function. Each component of this error function is associated with the classification error of each subset of training patterns. Then we develop a MOGA, which uses standard evolutionary operators: a non uniform mutation and a BLX-0.5 crossover operator and a specialized elitist selection operator.

As H. Abass established in [1] the use of a multiobjetive problem for training a set of networks enables the researchers to answer the following key open research questions in the literature. "1. On what basis should a network be included in, or excluded from the ensemble?. 2. How should the ensemble size be determined?. 3. How to ensure that the networks in an ensemble are different?" In this sense, our algorithm shares the same principles. Furthermore, the way we define the multiobjective environment in our work attempts to guarantee that subgroups of patterns having particular characteristics don't become diluted. If these patterns mainly belong to one or more of our subsets their effects become clear through their components of the classification error function.

Moreover, because we use a GA to train the network, one important advantage of our algorithm is that it has complete freedom to use any kind of error function. Other training methods, such as methods based on back- propagation, have less freedom to choose the error function to be minimized because this function has to fulfill some regularity conditions.

This paper is organized as follows: in section 2 the techniques of the MOGA and the NN used in this work are presented. The algorithms and their operators are shown in section 3, and the experiment in section 4. Finally, section 5 presents the conclusions.

2 Artificial NN's and Multi-Objective Genetic Algorithms

The NN model used is the MultiLayer Perceptron with two layers: the hidden and the output. The hidden node activation function is the logistic function, $f(x) = 1/(1 + exp(-x))$, and the output node activation function is the linear function. We will use the following notations: n_1, n_2 and n_3 correspond to the number of input, hidden and output nodes, respectively. Let X be the set of training patterns. Then, the pair $(\mathbf{X}^h, \mathbf{t}^h)$ represents the h-th pattern and its target value, where $\mathbf{X}^h = (x_1^h, \ldots, x_{n_1}^h)$. The targets take values in the set $\{(1, 0, \ldots, 0), (0, 1, 0, \ldots, 0), \ldots, (0, \ldots, 0, 1)\}$, that is, the component c-th is 1 and the rest are zeros if the pattern is in class c.

Each NN will be identified by means of its parameters: $w_{ij}^1, w_{ki}^2, \theta_i^1$ and $\theta_k^2, j = 1, \ldots, n_1, i = 1, \ldots, n_2, k = 1, \ldots, n_3$ where w's and θ's stand for the weights and bias, respectively. The output of the network for a pattern $(\mathbf{X}^h, \mathbf{t}^h)$ is $\tau^h = (\tau_1^h, \ldots, \tau_{n_3}^h)$ and the error of the network is defined by means of:

$$err = 1 - \frac{1}{p}\sum_{h=1}^{p}\mathbf{I}(\mathbf{t}^h,\tau^h) \tag{1}$$

where p is the number of patterns considered and $\mathbf{I} = 1$ if $argmax\ \tau^h = argmax\ \mathbf{t}^h$ and zero, otherwise ("winner-takes-all"). It is important to point out again that as we will use genetic algorithms for the network training, we are able to use the Classification Error Percentage (CEP) directly instead of using the usual Mean Square Error (MSE).

A Multi-objective Optimization Problem (MOP) can be defined in the following way:

$$\min \boldsymbol{f}(\boldsymbol{x}) = (f_1(\boldsymbol{x})\ldots f_J(\boldsymbol{x}))$$
$$\text{s. t:} \qquad \boldsymbol{x} = (x_1,\ldots,x_k) \in \mathcal{R}^k \tag{2}$$

Contrary to single objective optimization, in multi-objective optimization it is usually the impossibility of finding one optimal solution. Instead, algorithms for optimizing multiobjective problems try to find a family of points known as the Pareto optimal set. These points verify that there is no different feasible solution which strictly improves one component of the objective function vector without worsening at least one of the remaining ones.

GA's and their multi-objective version, MOGA's, are heuristic methods that imitate the behavior of natural evolution [10,11]. They allow us to work with a set of potential solutions to the problem, called population, instead of considering only one solution, as classical methods do. This fact constitutes an important advantage since they can converge, in a natural way, all the Pareto optimal set in only one execution of the algorithm.

3   The Algorithm

This section is divided into two subsections. In the first one we show the device used to create a multiobjective problem. In the second one the elements of the MOGA are shown.

3.1   Creating a Multi-objective Environment

To create a multi-objective environment we divide the set of training patterns. We build a separate and stratified partition of X, $X = \bigcup_{s=1}^{Q} X^s$ with $X^s \subseteq X$, $s = 1,\ldots,Q$ and $X^s \cap X^t = \emptyset, \forall s \neq t$. The h-th pattern of the subset X^s will be denoted by $(\mathbf{x}^{s,h},\mathbf{t}^{s,h})$, where $\mathbf{x}^{s,h}$ and $\mathbf{t}^{s,h}$ stand for the input and output variables, respectively. Moreover, we denote with $\tau^{s,h}$ the network output.

Then, given the parameters $w_{ij}^1, w_{ki}^2, \theta_i^1, \theta_k^2,\ i = 1,\ldots,n_2,\ j = 1,\ldots,n_1,\ k = 1,\ldots,n_3$ of a NN, Q error values are defined, one for each subset X^s, as:

$$err^s = CEP(X^s) = 1 - \sum_{h=1}^{|X^s|}\mathbf{I}(\mathbf{t}^{s,h},\tau^{s,h})/|X^s|,\quad s = 1,\ldots,Q$$

where $\mathbf{I} = 1$ if $argmax\ \tau^{s,h} = argmax\ \mathbf{t}^{s,h}$ and zero otherwise. Then, the multiobjective problem to be minimized is:

$$
\begin{aligned}
\min\quad &\mathbf{err} = (err^1, err^2, \ldots, err^Q)\\
\text{subject to}\quad &w_{ij}^1, w_{ki}^2, \theta_i^1, \theta_k^2 \in [-u, u],\\
&i = 1, \ldots, n_2,\ j = 1, \ldots, n_1,\ k = 1, \ldots, n_3.
\end{aligned}
\tag{3}
$$

3.2   Description of the Algorithm

As we have just said, GA's work with a set of potential solutions, called population. In our problem, each individual of the population corresponds to one neural network and it stores its parameters $w_{ij}^1, w_{ki}^2, \theta_i^1, \theta_k^2,\ i = 1, \ldots, n_2,\ j = 1, \ldots, n_1,\ k = 1, \ldots, n_3$. If we consider size_of_population neural networks in the population, it is coded as an array with $n_2 \times (n_1 + 1) + n_3 \times (n_2 + 1)$ rows and size_of_population columns.

Figure 1 shows the sketch of the proposed algorithm. The algorithm iterates on two levels. In the inner loop ($j = 1$ to $\#\ of\ iterations(i)$), the problem of equation (3) with $Q = i$ is solved. That is to say, the division of X into i subsets is carried out and the multi-objective problem (3) is established and solved. The solution of the problem is accomplished by means of a MOGA that incorporates the non-uniform mutation operator [10, p. 119] and the blend crossover operator BLX-0.5 [10, p. 109]. In a first stage, the mutation is applied and the mutated individuals are incorporated into the current population. Then the crossover operator is executed.

The following step consists of creating a fitness function to carry out the selection process. When a multi-objective function is considered, it is not possible to define a total order in the solution set, and, so as order to circumvent this fact, different approaches have been established in the literature. The approach we have taken [10, p. 40] consists of ranking all the individuals of the population $P(t)$ according to the Pareto ranking. It assigns a value equal to 1 to the efficient solutions (these solutions form the first Pareto layer, L_1), then these solutions are removed from the population. The efficient solutions of this new set are assigned rank 2 (second Pareto layer, L_2), and the process continues until there are no solutions left. The next step builds the new population by means of the selection operator (build $population\_aux(j + 1)$ from $population\_aux(j)$). We have proposed an elitist selection operator: it takes all the individuals of L_1, then all the individuals of L_2, and so on, until the new population is full. If all the elements of the last L_i considered cannot be incorporated, then two possibilities have to be considered depending on whether this last set is L_1 or not. If $L_i \neq L_1$, the available room is filled with some randomly selected elements of this set L_i. Otherwise, an iterative process that successively discards one of the two nearest solutions in L_1 is carried out. The process ends when the size of L_1 is equal to the desired size of the population. For this, the Euclidean distance between the vectors of parameters of all pairs of networks in the population are calculated. After this, the two nearest solutions are taken and the one having the biggest

```
for i = 1 to  max_#_of_subsets {
   build population_aux(1) from population(i − 1)
   for j = 1 to # of iterations(i)  {
      mutation of population_aux(j)
      recombination of population_aux(j) + new mutated elements
      evaluate and rank the obtained population
      build population_aux(j + 1) from population_aux(j)  }
   build population(i) from last population_aux
   # of iterations(i + 1) = # of iterations(i) × 1.25 }
build the final ensemble
```

Fig. 1. Sketch of the algorithm

CEP according to (1) is eliminated. If the size of L_1 is greater than the desired one, the process is repeated.

The previous steps are accomplished $\#of\ iterations(i)$ times. When the evolution of the i-objective problem is over, a set of solutions, $population(i)$, is stored. The sketch of this process is shown in Figure 2. It is a similar process to the one previously described but with some differences. First, for all pairs of similar efficient solutions (Euclidean distance smaller than 0.1), the ones having bigger CEP are successively eliminated. Then, depending on the number of remaining solutions, all of them are retained or only the $0.5 \times size\_of\_population$ networks with the smaller total CEP are kept. This process attempts to retain the most sparse efficient solutions obtained reaching the best total CEP.

Now the algorithm has to carry out a new iteration of the outer loop. The variable i is increased by 1, it implies that the training set is divided into one more subset, the multi-objective problem (3) has one more objective function. In order to start the inner loop again, a new initial population has to be established. For this purpose, we only need to complete $population(i − 1)$ with the appropriate number of random networks in order to get $size\_of\_population$ networks (remember that at the end of the inner loop, at the most $0.5 \times size\_of\_population$ solutions are stored). In the first iteration, when $i = 1$, all the population is formed with random networks.

To finish the description of the algorithm we have to show how the final ensemble is built. It is built following the process shown in Figure 2, but taking all the solutions kept at the end of the inner loop, $population(i)$, in the last five iterations, corresponding to the last five divisions. This value, 5, is a compromise value. We have looked for a value with a good behavior for all the test problems simultaneously. If we consider the test problems one by one, better final results could be obtained for each one.

4   The Experiment

In order to test the algorithm we have considered several common classification problems that are usual in the literature. We have also selected some representative papers, [12,13,14,15,16,17,18], and the results we have obtained are

Select the efficient solutions from the final auxiliary population.
Calculate the total CEP, according to (1), in the training set.
for each pair of networks having a Euclidean distance smaller
    than 0.1, the one having the biggest total CEP is eliminated.
if (number of remaining solutions $> 0.5 \times size\_of\_population$)
    The $0.5 \times size\_of\_population$ networks with smaller CEP are selected
else
    All remaining solutions are selected

Fig. 2. Obtention of the *population(i)* in iteration i

compared to theirs. Table 1 shows a brief description of the problems. It includes the name, the number of input values, classes, patterns, the architecture of the neural network we have used and the intervals where the variables, $w_{ij}^1, w_{ki}^2, \theta_i^1, \theta_k^2, \ i = 1, \ldots, n_2, \ j = 1, \ldots, n_1, \ k = 1, \ldots, n_3$, take values. More information about these problems can be found in the page *MLRepository.html* of the web site `http://mlearn.ics.uci.edu/`. It is important to point out that the number of hidden nodes of the neural network in all cases is two, a smaller value than the one usually used.

Table 1. Set of test problems considered

Name	No. inputs	No. classes	No. patterns	Structure n_1, n_2, n_3	Bounds of variables $(-u, u)$
Heart	13	2	270	13/2/2	[-1,1]
Wine	13	3	178	13/2/3	[-1,1]
Card	15	2	690	15/2/2	[-1,1]
Bcancer	9	2	699	9/2/2	[-1,1]
Ionosphere	33	2	351	33/2/2	[-1,1]
Pima	8	2	500	8/2/2	[-1,1]
Iris	4	3	150	4/2/3	[-1,1]

For all the test problems we have used a fixed set of parameters. These values have been set after several trials and, again, they constitute a compromise set of values to obtain a good global behavior for all problems. Specific values for specific problems could provide better results. The mutation probability considered is 0.2, the crossover probability is 0.6, for the first subset the number of iterations is 75, $\#of\ iterations = 75$, and they are increased in subsequent iterations by a factor of 1.25, the more subsets the more iterations, the population size is 50 ($size\_of\_population = 50$) for all problems and divisions and the maximum number of subsets considered is 10, $max\_\#\_of\_subsets = 10$.

In order to estimate the CEP we always use a ten fold crossvalidation which is repeated 5 times (except for the Pima data set in which a twelve fold crossvalidation is used instead). We choose one folder as the validation data set, another

Table 2. Classification Error Percentage(CEP) obtained in validation and training set. Majority Voting (MaVo), Simple Averaging (SiAv), Winner-Takes-All (WTA) and the error of the single NN with minimum CEP in *test set* are shown.

Name	VALIDATION SET				TRAINING SET		
	MaVo	SiAv	WTA	MinTest	MaVo	SiAv	WTA
Heart	16	16	16.4	17.9	16	16	16.4
Wine	2	2	2	4	0	0	0
Card	13.9	13.9	13.4	14.5	10.4	11.2	11.6
Bcancer	3.30	3.30	3.33	4.20	1.69	1.84	1.97
Ionosp.	12	13	13	14	5	6	7
Pima	22.6	22.6	22.8	23.45	20.0	21.1	21.7
Iris	3	3	3	3	1	1	1

one as the test data set, and the eight remaining folders form the training data set. The training set is used in the training process, the test set is only used to select a single NN for the minTest method (Table 2), and the validation set is used to validate the final ensembles, that is to say, to obtain the CEP's shown in Table 2.

The percentage of elements of each class that each folder receives is similar to the one in the set of original data. Table 2 shows the results obtained for every problem. The table shows the CEP for the validation and training set in the final ensemble calculated according to the majority voting, the simple averaging and the winner-takes-all methods. The table also shows the CEP for the network of the ensemble with minimum error in the test set. The capability of generalization of the algorithm can be seen in the moderate difference between the training error and the validation error. The method proposed is compared with other algorithms in Table 3 (all of them using a ten fold crossvalidation procedure for obtaining the results, except the Pima data set that uses a twelve crossvalidation procedure).

The first important issue to be noticed is that the use of ensembles provides better results than the use of only one neural network. Columns 2, 3 and 4 of Table 2 always show a better value than column 5. Secondly, we can see that the methods that use ensembles (MaVo, SiAv and WTA) show a similar behavior. If we compare these results with those in Table 3 we can point out the following facts. Our algorithm is not as good as the others for the Ionosphere data set; our best CEP is 12 and their values range from 6.6 to 9.5. This is due to the fact that we have fixed the same set of configuration parameters for all problems. As the Ionosphere data set has 33 input variables (many more input variables than all the other data sets) the fixed population size (50 neural networks) turns out to be insufficient for getting the properties of this data set. For the remaining data set we get better or equal CEP for Heart, Wine and Iris problems. For the Card data set we reach a CEP of 13.4 that improves the values provided in [12,14,17,18] but it worsens the results of [15,16]. For the Bcancer data set, we get better results than in [12,13] and we get worse results than in [16,17,18]. Finally, for the Pima data set we outperform [12,17,18] and we obtain worse

Table 3. CEP in previous works. The results shown are the best method among the algorithms tested in each one.

Name	[12]	[13]	[14]	[15]	[16]	[17]	[18]
Heart	19.7				20.4	16	16.2
Wine						2	2
Card	13.6		13.5	13	13	14.3	14.3
Bcancer	3.5	3.8			3.1	2.7	2.6
Ionosphere		6.6				7.5	9.5
Pima	24.4		22.1	22.3		24	23.7
Iris	5.7				6.9	4.8	2.4

CEP values than in [14,15]. Summarizing, as a whole, we obtain comparable or better values to those shown in Table 3. Furthermore, it should be taken into account that our algorithm is quite simple, in the sense of not needing complex mathematical procedures like gradient calculations, for instance. In the case of algorithms using NN approaches [14,15], our networks are less complex because we use only two hidden nodes whereas [14,15] use five hidden nodes.

5 Summary and Concluding Remarks

In this paper a new MOGA for training NN has been presented. Because of the characteristics of the algorithm we can directly minimize use the CEP instead of the more common MSE. The multi-objective environment is created by means of the division of the training data set in several subsets, this fact generalizes some ideas presented in [1].

Tables 2 and 3 show how our method is competitive with other methods for classification problems found in the literature (decision trees and NN) in spite of the fact that we have taken a fixed set of parameters for all the problems. If we set the parameters of the algorithm for some specific problem it would provide better results. Besides, the number of hidden neurons used in the examples is only two, a smaller value than the usual in the literature.

References

1. Abass, H.A.: Pareto neuro-evolution: constructing ensemble of neural networks using multi-objective optimization. In: IEEE Congress on Evolutionary Computation (CEC 2003), vol. 3, pp. 2074–2080. IEEE-Press, Canberra (2003)
2. Chandra, A., Yao, X.: DIVACE: Diverse and accurate ensemble learning algorithm. In: Yang, Z.R., Yin, H., Everson, R.M. (eds.) IDEAL 2004. LNCS, vol. 3177, pp. 619–625. Springer, Heidelberg (2004)
3. Chandra, A., Yao, X.: Evolutionary framework for the construction of diverse hybrid ensembles. In: 13thEuropean Symposium on Artificial Neural Networks (ESANN 2005), pp. 253–258. d-side pub., Bruges (2004)
4. Dietterich, T.G.: Ensemble methods in Machine Learning. In: 1st International Workshop on Multiple Classifier Systems, Cagliari, Italy, pp. 1–15 (2000)

5. García-Pedrajas, N., Hervás-Martinez, C., Ortiz-Boyer, D.: Cooperative coevolution of artificial neural network ensemble for pattern classification. IEEE Transactions on evolutionary computation 9(3), 271–302 (2005)
6. Gavin, J., Wyatt, H., Yao, X.: Diversity creation methods. A survey and categorisation. Information Fusion 6(1), 5–20 (2004)
7. Liu, Y., Yao, X.: Simultaneous training of negatively correlated neural networks in an ensemble. IEEE Transactions on systems, man, and cybernetics-part B: cybernetics 29(6), 716–725 (1999)
8. Islam, M., Yao, X., Murase, K.A.: A constructive algorithm for training cooperative neural network ensembles. IEEE Transactions on neural networks 14(4), 820–834 (2003)
9. Breiman, L.: Bagging predictors. Machine learning 24(2), 123–140 (1996)
10. Deb, K.: Multi-Objective Optimization using Evolutionary Algorithms. John Wiley and Sons, Chichester (2001)
11. Michalewicz, Z.: Genetic Algorithms + Data Structures = Evolution Programs, 3rd edn. Springer, Berlin (1996)
12. Webb, G.I.: Multiboosting: A technique for combining boosting and wagging. Machine learning 40(2), 159–196 (2000)
13. Friedman, J., Hastie, T., Tibshirani, T.: Additive logistic regresion: A statistical view of boosting. Annals Statistics 28(2), 337–407 (2000)
14. Liu, Y., Yao, X., Higuchi, T.: Evolutionary ensembles with negative correlation learning. IEEE Trans. Evol. Comput. 4, 380–387 (2000)
15. Liu, Y., Yao, X., Zhao, Q., Higuchi, T.: Evolving a cooperative population of neural networks by minimizing mutual information. In: IEEE congr. Evol. Comp. (CEC 2001), Seoul, Korea, pp. 384–389 (2001)
16. Dietterich, T.G.: An experimental comparison of three methods for constructing ensembles of decision trees: Bagging, boosting, ad randomization. Machine learning 40, 139–157 (2001)
17. Dzeroski, S., Zenko, B.: Is combining classifiers with stacking better than selecting the best one? Machine learning 54, 255–273 (2004)
18. Todorovski, L., Dzeroski, S.: Combining classifiers with meta decision trees. Machine learning, vol. 50, pp. 223–249. Springer, Heidelberg (2003)

Analysis of Vestibular-Ocular Reflex by Evolutionary Framework

D. Novák[1], A. Pilný[1], P. Kordík[1], Š. Holiga[1], P. Pošík[1],
R. Černý[2], and Richard Brzezny[2]

[1] Czech Technical University in Prague, Czech Republic
`xnovakd1@labe.felk.cvut.cz`
[2] 2nd Medical Faculty, Charles University, Czech Republic

Abstract. In this paper the problem of analysis of eye movements using sinusoidal head rotation test is presented. The goal of the method is to discard automatically the effect of the fast phase-saccades and consequently calculate the response of vestibular system in the form of phase shift and amplitude. The comparison of threshold detection and inductive models trained on saccades is carried out. After saccades detection we are left with discontinuous signal segments. This paper presents an approach to align them to form a smooth signal with the same frequencies that were originally present in the source signal. The approach is based on a direct estimation of the signal component parameters using the evolutionary strategy with covariance matrix adaptation. The performance of evolutionary approach is compared to least-square multimodal sinus fit. The experimental evaluation on real-world signals revealed that threshold saccades detection with combination of the evolutionary strategy is robust, scalable and reliable method.

1   Introduction

Vestibulo-ocular reflex (VOR) is responsible for maintaining retinal image stabilization in the eyes during relatively brief periods of head movement. By analyzing the VOR signal, physicians can recognize some pathologies of the vestibular organ which may result in e.g. failures of the balance of a patient. The principle of the frequency response measurement using servocontrolled rotating mechanism is relatively simple: the patient is situated in a chair which is then rotated in a defined way following a source signal-sine wave or a sum of sine (SOS) waves. This is called the head rotation test. Since the resulting eye signal is distorted by fast eye movements, so-called saccades, they must be removed from the signal.

The result of the stimuli is a prevailing pattern called nystagmus consisting of slow and fast phase - see Figure 2(a). First task of this work is the separation of slow and fast phases. Most previous algorithms used to detect fast phases were based on tresholding techniques (TH) [1]. We present a new approach based on Group of Adaptive Models Evolution (GAME) inductive modelling methodology [2].

The second task of this paper is a evolutionary based method with covariance matrix adaptation (CMA) for the direct estimation of the gain and phase lag of the individual sine components of the underlying SOS signal, i.e. for the measurement of several

V. Kůrková et al. (Eds.): ICANN 2008, Part I, LNCS 5163, pp. 452–461, 2008.

points of the frequency response at the same time. After the estimation, the VOR signal segments should match with the corresponding parts of the estimated SOS signal. This approach is compared to least-square method [1] fitted to sinusoid prototype.

The suggested methodology is first evaluated on artificially generated VOR-signal and next on the set of 7 healthy volunteers.

2   Methodology

A proposed methodology of VOR analysis is depicted in Figure 1. First, fast phases (or saccades as shown in Figure 2(a)) of eye movement signal S_{raw}, which is generated by vestibular organar (VO), are detected using GAME approach. After saccades detection, GAME topology selection is performed by inspecting FFT spectra of the reconstructed signal. GAMEs' results are compared to the thresholding method TH on artificially generated VOR signal using two evaluation criteria $E1$: performance rate in % and ration of false positive saccades detections to total number of saccades. Second, saccades are removed and eye signal S_{rem} is reconstructed by estimating vestibular organ's gains and phase shifts to a multimodal sinus technical signal $T(G_i)$. The evolutionary strategy with covariance matrix adaptation (CMA) is applied and its results are compared to linear square sinus fit (LSF) using again synthetic VOR signal. The performance $E2$ is evaluated by a ration of output gains of the processed signal $S(G_i)$ to input gains of the technical signal $T(G_i)$.

2.1   Data Acquisition

Eye movements were recorded using silver/silver chloride electrodes with a reference strip-chart recorder extended by digitized unit. For the eye movements capture the sampling frequency was set at 102.4 Hz that is consider to be enough for accurate capture of all eye movement characteristics. The rotating frequency of chair was constant throughout recording accelerating or decelerating and changing directions hence producing single and composed sinusoidal frequency signals with period $f_1 = 0.05$ Hz and $f_1 = 0.05$ Hz, $f_2 = 0.1$ Hz, resp. 7 patients have been measured, 5 with their head fixed, 2 without any head fixation.

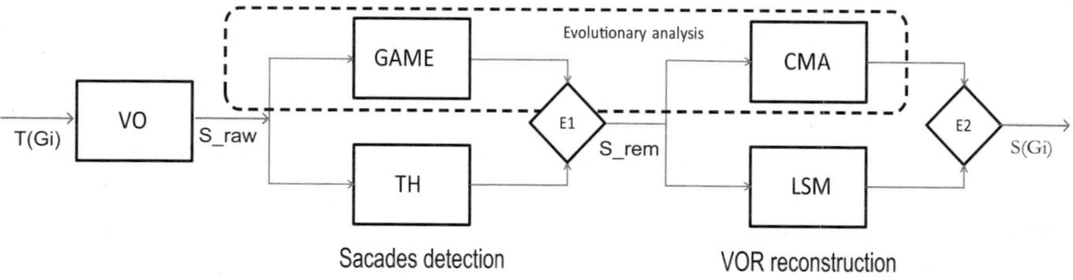

Fig. 1. Algorithm flow chart

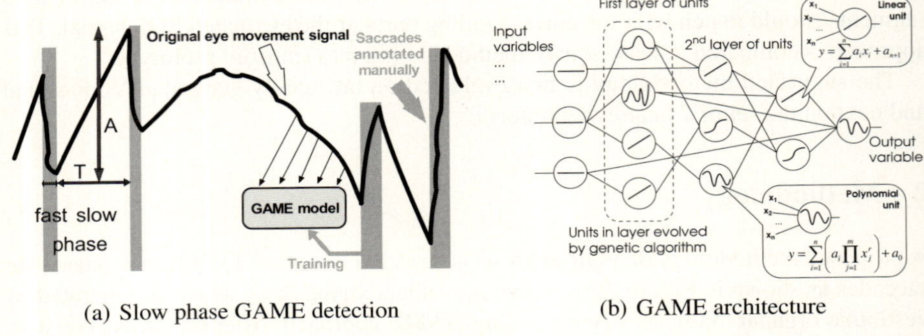

(a) Slow phase GAME detection   (b) GAME architecture

Fig. 2. (a): Slow phase velosity is defined as $SPV = A/T$. (b): GAME architecture.

2.2 Saccades GAME Detection

Group of Adaptive Models Evolution (GAME) [2] proceeds from the Group Model Data Handling (GMDH) theory. GMDH was designed to automatically generate model of the system in the form of polynomial equations. An example of inductive model created by GAME algorithm is depicted on the Figure 2(b). Similarly to Multi-Layered Perceptron (MLP) neural networks, GAME units (neurons) are connected in a feedforward network (model). The structure of the model is evolved by special niching genetic algorithm, layer by layer. Parameters of the model (coefficients of units' transfer functions) are optimized independently. Model can be composed from units of different transfer function type (e.g sigmoid, polynomial, sine, linear, exponential, rational, etc). Units with transfer function performing well on given data set survive the evolution process and form the model. Often, units of several different types survive in one model, making it hybrid. The eye movement signal is reconstructed in three steps: training data synthesis, identification of saccades and the signal reconstruction (removing the disturbances, filling out estimated values).

The first step is to prepare the training data. The key task is to annotate the saccades in the eye movement signal. For this purpose we implemented program which allows us to manually annotate the saccades in signal and save this information as a target signal for training - see Figure 2(a).

In the second step the data set is used to generate inductive models using the GAME method. Generated GAME model in fact acts as an automatic annotator of saccades. In the third step we reconstruct the signal using a script, which has two inputs (original eye movements signal and GAME model saccade estimation) and one output, the reconstructed signal. The method of reconstruction is based on information about signal character in the neighborhood of saccades. When saccade is estimated, the original signal is replaced with linear approximation of trends in the beginning and the end of saccade. The original eye movement signal with saccades (a), the output of the GAME model (b) and the reconstructed signal (c) is displayed on Figure 3.

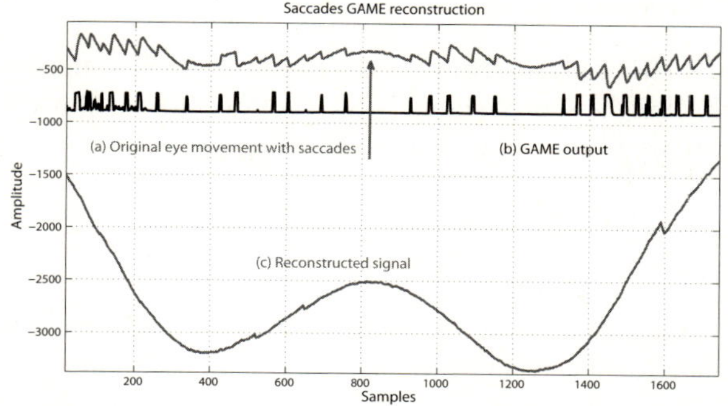

Fig. 3. The original signal (a), the output of GAME model (b) and reconstructed signal (c)

Thresholding approach. The GAME detection was evaluated against thresholding approach. Before the slow components were detected, the signal was preprocessed. First, the velocity was computed using derivative of horizontal eye movement channel. Afterwards the signal was filtered out implementing elliptic low pass filter. Taking advantage of signal statistical behaviour, the signal velocity was normalized in such a way that its mean was zero and standard deviation was one. In the next step the detection of slow phase was applied. It was based on horizontal velocity signal; the information of vertical signal was not regarded as useful. If we look at the typical example of eye movement development in time - see Figure 4(a), the prevailing pattern is nystagmus. In other words, the eye movement signal consists of series of triangles that are superposed on the isoline that is most similar to sinusoidal function. Artefacts presented in the signal such as noise are reduced by low pass filtering. Therefore after these assumptions the peaks of horizontal velocity signal detect the position of slow phases. Peaks were simply determined by setting up the threshold as it is shown in Figure 4(a).

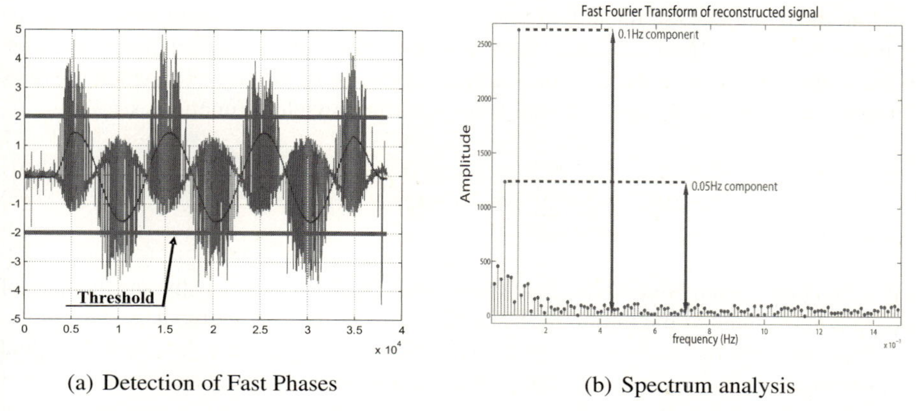

(a) Detection of Fast Phases   (b) Spectrum analysis

Fig. 4. (a): Thresholding method illustration. (b) FFT analysis after saccades reconstruction.

2.3 Signal CMA Reconstruction

The second part of VOR analysis focuses on signal reconstruction after fast phase removal. It is assumed that the source signal (which controls the rotation of the chair with the patient) is formed as a sum of sine waves (SOS):

$$y^S(t) = \sum_{i=1}^{n} a_i^S \sin(2\pi f_i t + \phi_i^S),$$

(1)

where $y(t)$ is the source signal and a_i, f_i and ϕ_i are the amplitude, the frequency and the phase shift of the individual sine components, respectively. The superscript S indicates the relation to the source signal. Furthermore, it is assumed that the output signal of the vestibular organ is of the same form as the input one, i.e. it contains only sine components with the *same frequencies* as the source signal but possibly with different amplitudes and phase shifts. It should be of the form

$$y(t) = \sum_{i=1}^{n} a_i \sin(2\pi f_i t + \phi_i).$$

(2)

If we knew the a_i and ϕ_i parameters of the output SOS signal components, we could calculate the amplification (a_i/a_i^S) and phase lag $(\phi_i - \phi_i^S)$ at individual frequencies and deduce the state of the vestibular organ.

Unfortunately, we do not have access to the output SOS signal described by Eq. 2. We have only the measured VOR signal, i.e. the segments of the output SOS signal that are left after filtering out the saccades from the eye-tracking signal. However, we can search for the unknown parameters a_i and ϕ_i of the output SOS signal by solving the optimization task described in the following text.

Minimizing Loss Function. Let m be the number of segments of the VOR signal at hand, $v_j(t)$, $j = 1 \ldots m$, be the actual j-th segment of the VOR signal and t_j^{ini} and t_j^{end} be the initial and the final time instants for the j-th signal segment. As stated above, we can find the parameters of the output SOS signal by searching the $2n$-dimensional space of points $\mathbf{x}$, $\mathbf{x} = (a_1, \phi_1, \ldots, a_n, \phi_n)$. Such a vector of parameters represents an estimate of the output SOS signal and we can compute the degree of fidelity with which the SOS corresponds to the VOR signal segments by constructing a loss function as follows:

$$L(\mathbf{x}) = \sum_{j=1}^{m} \sum_{i=t_j^{ini}}^{t_j^{end}} \left((v_j(i) - \bar{v}_j) - (y(i) - \bar{y}_j)\right)^2$$

(3)

where $\bar{v}_j$ is the mean value of the j-th VOR signal segment and $\bar{y}_j$ is the mean value of the current estimate of the output SOS signal related to the j-th segment

Subtracting the means $\bar{v}_j$ and $\bar{y}_j$ from the VOR signal segments $v_j(t)$ and SOS signal $y(t)$, respectively, we try to match the VOR signal segment to the corresponding part of the SOS signal. If they match, their difference is zero, otherwise it is a positive number quadratically increasing with the difference. This operation is carried out for all m VOR signal segments.

Optimization Method. The parameter vector $\mathbf{x}$ is projected to the loss function via the estimate of the SOS signal $y(i)$ (and via the mean values $y_j(i)$). The evolutionary strategy with covariance matrix adaptation was chosen to minimize the objective function $L(x)$. CMA is very recent and progressive stochastic optimization algorithm [3]. It maintains a D-dimensional normal distribution from which it samples new data points. The distribution is then in turn adapted based on the loss function values for these new points. The algorithm performs a kind of iterative principal component analysis of the selected perturbation vectors.

3   Results

First, the proposed approach will be evaluated on synthetic data to decide which of these two algorithms (evalutionary and threshold approach) is more suitable for estimating VOR clinical important paratemers as gain or phase shift.

3.1   Synthetic Data

The tests were carried out on six signals ($S1 - S6$) consisting of 1 to 6 sine components, i.e. the search was carried out in 2-, 4-, 6-, 8-,10- and 12-dimensional parameter spaces.

Generating VOR signal. First, for each sine component of the signal, the values of frequency and amplitude were generated. The ranges for individual parameters can be found in Table 1. This SOS signal then undergoes a disruption process which cuts it to individual segments with 'pauses' between them. This way the gaps created by filtering out the saccades are simulated. The segments are then placed to the same level-see Figure 5 where signal S_2 consisting of two sinus components is shown ($f_1 = 0.05, a_1 = 100, f_2 = 0.1, a_2 = 50$). Signal duration was set to $80s$, sampling frequency of synthetic signal to $102.4Hz$ and saccade duration to $0.1s$.

First, let us review the success rates of both algorithms when estimating the parameters of the SOS signal with the number of components ranging from 1 to 6. In Table 2, the ratio a_i/a_i^S of CMA and Least-Square Multimodal Sinus fit (LSF) is reported for all 6 testing sets. Figure 5 shows example of CMA reconstructed signals S_2.

Lest-square fit is sensitive to noise and outliers therefore the CMA approach significantly outperformed the least-square method. The CMA ratio a_i/a_i^S is equal to one in all cases. In each run, the algorithms were allowed to perform 10,000 evaluations of

Table 1. Settings for parameters of artificial VOR signals S1-S6

Frequency	Amplitude
f_i	a_i
S_1 $\langle 0.1 \rangle$	$\langle 50 \rangle$
S_2 $\langle 0.05, 0.1 \rangle$	$\langle 25, 50 \rangle$
S_3 $\langle 0.05, 0.1 \rangle$	$\langle 25, 50 \rangle$
S_4 $\langle 0.01, 0.04, 0.070.1 \rangle$	$\langle 25, 38, 50 \rangle$
S_5 $\langle 0.01, 0.033, 0.055, 0.076, 0.1 \rangle$	$\langle 25, 33, 38, 43, 50 \rangle$
S_6 $\langle 0.01, 0.028, 0.046, 0.064, 0.082, 0.1 \rangle$	$\langle 25, 30, 35, 40, 45, 50 \rangle$

Table 2. Comparision of artificial VOR signal gain estimation

		f1	f2	f3	f4	f5	f6
S1	CMA	1.00	-	-	-	-	-
	LSF	1.16	-	-	-	-	-
S2	CMA	1.00	1.00	-	-	-	-
	LSF	1.26	1.44	-	-	-	-
S3	CMA	1.00	1.00	1.00	-	-	-
	LSF	0.26	0.39	0.99	-	-	-
S4	CMA	1.00	1.00	1.00	1.00	-	-
	LSF	0.28	0.29	1.37	1.14	-	-
S5	CMA	1.00	1.00	1.00	1.00	1	-
	LSF	0.24	0.37	1.70	1.46	1.28	-
S6	CMA	1.00	1.00	1.00	1.00	1	1.00
	LSF	0.27	0.27	2.03	1.77	1.58	1.42

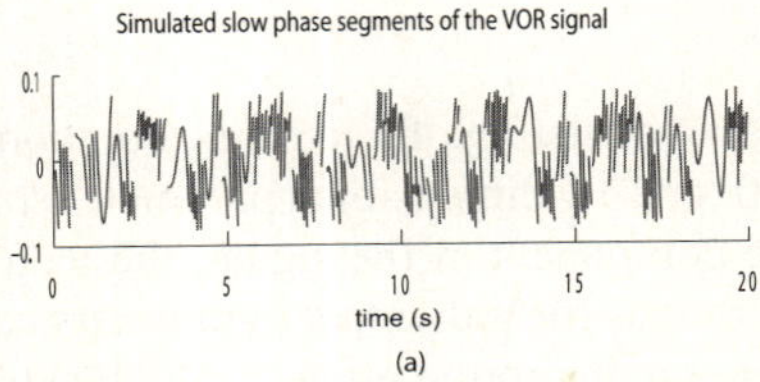

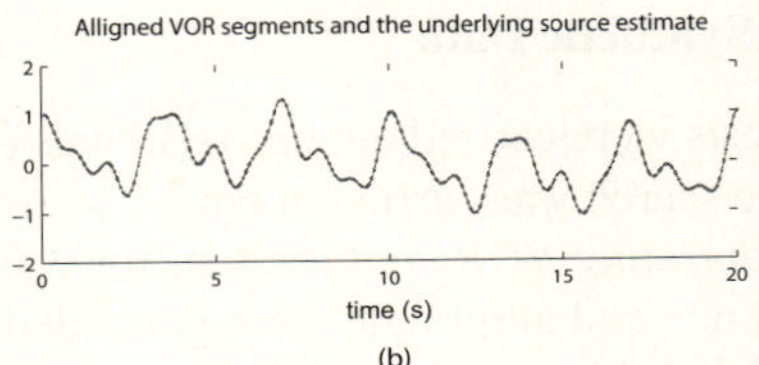

Fig. 5. (a) Simulated VOR signal with saccades removed. VOR signal segments aligned with the estimated SOS signal. (b) The parameters of the SOS signal are output of the algorithm.

the loss function and a particular run was considered to be successful if the algorithm found a parameter set with the loss function (3) value lower than 10^{-8}. Those learning parameters proved to be sufficient in order to precisely estimate the sinus components presented in the response of the vestibular organ.

3.2   Real-World Data

First of all, the fast phases were removed using GAME and thresholding approach (TH) for all 7 subjects $P1 - P7$. Before comparing the perfomance of both methods we must address the task of evaluating the most suitable GAME architecture.

GAME topology selection.   Three different models were used for saccades reconstruction. The model performance were evaluated on the base of spectrum analysis calculated by 4096 point FFT - one example is shown in Figure 4(b). Regarding neurons' transfer functions, Linear Config consists of linear units (linear and linearGJ neuron) only. NonLinConfig is composed of the followings transfer functions: PolySimpleNeuron,ExpNeuron, SigmNeuron, SinusNeuron a GaussianNeuron. The last NonLinearConfig ALL contains moreover polynomial units. 5 models were evaluated on the data of volunteer $P6$. The GAME topology comparison is depicted in Table 3 in

Table 3. Spectral results for three different GAMEs topology

c. model number	LinearConfig		NonLinearConfig		NonLinearConfig ALL	
	frequency[Hz]	amplitude	frequency[Hz]	amplitude	frequency[Hz]	amplitude
1	0,05	4,00E+005	0,05	1,60E+006	**0,05**	**1,20E+006**
	0,1	5,00E+005	0,1	1,50E+006	**0,1**	**1,16E+006**
	0,2	6,30E+004	0,18	8,50E-004	**0,22**	**2,60E+004**
	0,55	9,55E+004	0,23	4,00E-004	-	-
2	0,05	3,40E+005	0,05	1,30E+006	0,05	1,65E+006
	0,1	5,00E+005	0,1	1,32E+006	0,1	1,52E+006
	0,2	3,50E+004	0,15	8,00E+004	0,23	1,60E+004
	0,3	2,50E+004	0,23	3,50E+004	0,27	1,80E+004
	0,35	1,50E+004	0,3	2,50E+004	-	-
3	0,05	5,70E+005	0,05	9,90E+005	0,05	1,40E+006
	0,1	7,80E+005	0,1	1,04E+006	0,1	1,10E+006
	0,15	1,10E+005	0,28	5,00E+004	0,23	7,00E+004
	0,2	1,23E+005	-	-	0,28	4,00E+004
4	-	-	0,05	1,37E+006	0,05	1,08E+006
	0,1	3,40E+005	0,1	1,46E+006	0,1	1,13E+006
	0,23	1,00E+004	0,15	4,20E+004	0,23	1,13E+004
	0,35	2,00E+004	0,3	3,50E+004	0,28	2,80E+004
5	0,05	1,80E+005	0,05	1,23E+006	0,05	1,42E+006
	0,1	2,60E+005	0,1	1,18E+006	0,1	1,39E+006
	0,17	6,60E+004	0,15	5,70E+004	0,15	8,00E+004
	0,21	5,90E+004	0,23	2,80E+004	0,23	2,40E+004
	0,35	1,20E+004	0,35	3,40E+004	0,27	2,40E+004

terms of frequencies and their corresponding amplitudes found in FFT reconstructed spectrum. Only the fourth model with Liner Configuratin did not reveal 0.05 Hz component. GAME reconstruction also revealed other frequency components that were not presented in the SOS signal controlling the chair movements. However, their amplitudes are significantly less than amplitudes of 0.05 and 0.1 Hz components. In case of non-linear configurations, the difference is even in order of $10e^{-2}$. First model with NonLinearConfig ALLwas chosen for next analysis.

Next, the performance of saccades GAME and TH detection was evaluating using percentage rate of Sensitivity (Sn) and Specificity (Sp) for real signal $P6$. The saccades were annotated by four expert who determined the probability of saccades appearance. The average of those four annotation was taken as a final annotation. In case of GAME, $Sn = 87.8\%$, $Sp = 90.9\%$, in case of TH method, $Sn = xx$, $Sp = xx$. [!!Traditional thresholding techniques yielded better results in both performance criterions. GAME method requires more data in order to detect saccades more accurately.!!] Threshold method used also a priori information about the direction of chair rotation: when the chairs rotated to the right, only saccades exceeding positive threshold were processed (see Figure 4(a)) resulting in zero rate of false positives detections.

Furthermore, CMA reconstruction was carried out. One example of the resulting signal with one frequency component $f_1 = 0.05$ can be seen in Figure 6(a). We can see

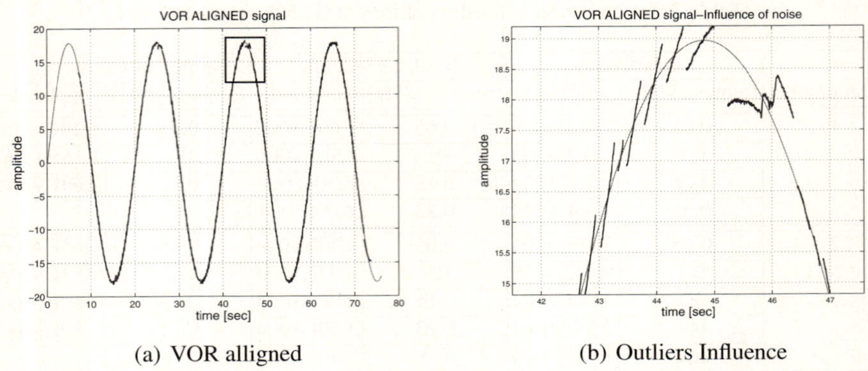

(a) VOR alligned   (b) Outliers Influence

Fig. 6. (a): Real-World VOR signal with slow components aligned. (b)Influence of noise in VOR aligned signal.

that all slow phases were precisely aligned in spite of some noisy segments that have been left after thresholding procedure. Even if not all fast phases were not detected producing some kind of noise, the evolutionary algorithms was able to cope with the outliers - see Figure 6(b).

Table 4 summarizes clinical parameters of real two frequency component signal ($f_1 = 0.05, f_2 = 0.1$): *(i)* gain G_i and phase shift ϕ_i for each component $f_i, i = 1, 2$, *(ii)* total slow phase velocity $SPV = A/T$ defined in Figure 2(a) when the chair rotates to the left SPV_{left} and right SPV_{Right} and total mean SPV.

CMA ratio estimation worked correctly with exception of phase shift reconstruction. The minus sign means that the CMA phase estimation was shifted by π compared to original phase value. Volunteers P_6 and P_7 did not have their head fixed during VOR measurement. SPV's parameters of $P7$ suggest that for the used frequency components 0.05 and 0.1Hz the fixation was not so relevant. However, head fixations proved to be important for P_6 only pointing out to big intra-personality variability in VOR data.

Table 4. Comparision of real VOR signal parameters estimation: GAME, CMA and LSF

	P_1	P_2	P_3	P_4	P_5	P_6	P_7
$G_1 CMA$	1.78	-3.25	-6.78	-4.70	-0.96	-6.52	-1.87
G_1^{LSF}	0.31	0.55	0.55	0.58	0.34	0.56	0.55
ϕ_1^{CMA}	1.42	-1.69	-1.31	-1.46	-1.66	-1.60	-1.97
ϕ_1^{LSF}	3.69	3.16	3.48	3.09	3.45	3.51	3.16
G_2^{CMA}	-0.63	1.45	2.97	2.65	0.41	3.20	0.90
G_2^{LSF}	0.25	0.48	0.55	0.54	0.30	0.55	0.48
ϕ_2^{CMA}	-1.88	1.14	1.43	1.31	1.14	1.32	0.92
ϕ_2^{LSF}	2.99	3.51	3.44	3.18	3.21	3.37	3.51
SPV_{left}	5.38	13.13	16.16	14.67	17.52	19.35	13.13
SPV_{rigth}	2.85	-9.41	-14.41	-10.14	-13.08	-22.39	-9.41
SPV_{mean}	4.03	11.14	15.23	12.25	15.15	20.97	11.14

There can be seen significant SPV's asymmetry of volunteer P1 when chair rotated to the left.

4 Conclusions

In this paper, a new method of VOR signal processing was introduced and experimentally evaluated both on artificially generated signals and on clinical real-world data. It relies on the right identification of the fast eye movements, saccades, that must be filtered out of the signal in advance. The simple and fast method based on GAME methodology for saccades removal was developed. The main advantage of the procedure lies in the simplicity of the basic algorithms and effectivity, even in the evaluation of irregular data such as patients suffering from vestibular organ disorders. However, a more examples for GAME learning is needed in order to make the GAME saccades detection suitable for use in the clinical situations.

Regarding phase and gain calculation, conventionally used methods interpolate the signal segments and carry out the Fourier transform to obtain the amplification and the phase shift on the original frequencies. On the contrary, the CMA proposed method directly estimates these parameters from the signal segments trying to align them with the underlying estimated mixture of sine waves.

Acknowledgment. The project was supported by the Ministry of Education, Youth and Sport of the Czech Republic with the grant No. MSM6840770012 entitled "Transdisciplinary Research in Biomedical Engineering II".

References

1. Gordon, J., Furman, J.M.R., Kamen, E.W.: System identification of the vestibulo-ocular reflex: application of the recursive least-squares algorithm. In: Proc. of the Fifteenth Annual Northeast Bioengineering Conference, pp. 199–200 (1989)
2. Kordik, P.: Fully Automated Knowledge Extraction using Group of Adaptive Models Evolution. PhD thesis, CTU Prague (2006)
3. Hansen, N., Ostermeier, A.: Completely derandomized self-adaptation in evolution strategies. Evolutionary Computation 9(2), 159–195 (2001)

Fetal Weight Prediction Models: Standard Techniques or Computational Intelligence Methods?

Tomáš Siegl, Pavel Kordík, Miroslav Šnorek, and Pavel Calda

Department of Computer Science and Engineering, FEE,
Czech Technical University, Prague, Czech Republic
sieglt@fel.cvut.cz, kordikp@fel.cvut.cz, snorek@fel.cvut.cz,
calda@gynstart.cz

Abstract. An accurate model of ultrasound estimation of fetal weight (EFW) can help in decision if the cesarean childbirth is necessary. We collected models from various sources and compared their accuracy. These models were mostly obtained by standard techniques such as linear and nonlinear regression. The aim of the comparison was to recommend a model best fitting to data measured for Czech population. Alternatively, we generated several linear and non-linear models by using our method GAME from the computational intelligence domain. GAME models can be serialized into simple equations that are understandable by domain experts. In this contribution, we show that automatically generated GAME models are at least as accurate (in terms of root mean squared error and standard deviations of predictions) as the best model computed by means of (time and expert skills demanding) standard techniques.

1   Introduction

During the ultrasound examination of fetus several parameters are measures from the screen. These parameters are subsequently used to the estimation of the fetal weight which is the key indicator if the cesarean childbirth is necessary for certain positions of the fetus (breech presentation).

The estimation of the fetal weight from measured parameters is based on most popular model (Hadlock et al, 1985 (8)) computed for US population. The aim of the study performed at First Medical School, Charles University, Department of Obstetrics and Gyneacology was to verify the validity of the model for Czech population. They compiled a data set of four hundred women with singleton pregnancies smaller or equal than 7 days before delivery in breech presentation. Input attributes are the measurements of bi-parietal diameter (BPD), abdominal circumference (AC), head circumference (HC), and femur length (FL) and output is the real weight of the child after delivery.

Our goal was to find the best fitting model of the fetal weight for this data set. We compiled several models for EFW published by renomed authors and compared their accuracy. Then we used our method Group of Adaptive Models

V. Kůrková et al. (Eds.): ICANN 2008, Part I, LNCS 5163, pp. 462–471, 2008.

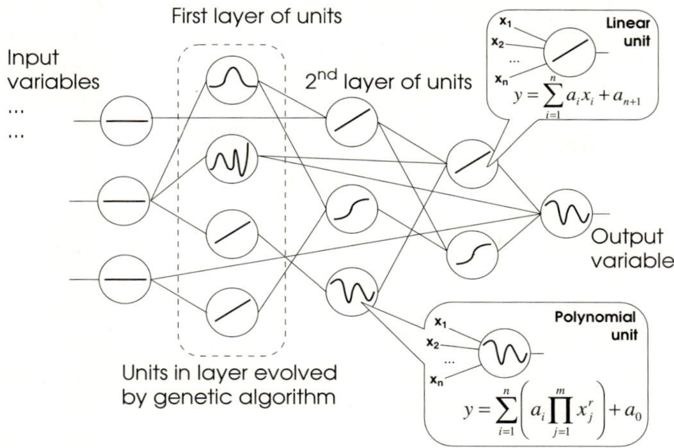

Fig. 1. The structure of the GAME model is evolved layer by layer using special genetic algorithm. While training an unit, selected optimization method adjusts coefficients $(a_1, ..., a_n)$ in the transfer function of the unit.

Evolution (GAME)[1] to compute the EFW model automatically. The GAME method is from the domain of Computational intelligence (CI) [1] and will be shortly described in the next section.

The problem of CI methods is that they are often treated as black box techniques and and these are not accepted readily by domain experts. The advantage of the GAME method is that models can be serialized into equations which are better accepted and can be used in spreadsheets. We find out, that models generated by the GAME method have high correlation coefficient as best of published models. The root mean squared error and standard deviation of the prediction indicate, that the GAME models are superior as shown bellow. In the last section, we discuss the possibility of overfitting that might occur, when model trains noise in the training data. GAME models prove to have good generalization abilities and no tendency to overfit the data was observed.

2   Group of Adaptive Model Evolution

Group of Adaptive Models Evolution (GAME) [6] proceeds from the Group Model Data Handling (GMDH) theory [4]. The GMDH was designed to automatically generate model of the system in the form of polynomial equations. The

---

[1] This research is partially supported by the grant Automated Knowledge Extraction (KJB201210701) of the Grant Agency of the Academy of Science of the Czech Republic and the research program "Transdisciplinary Research in the Area of Biomedical Engineering II" ($MSM6840770012$) sponsored by the Ministry of Education, Youth and Sports of the Czech Republic.

GAME algorithm respects basic ideas of the GMDH theory such as self organization and the principle of induction. It incorporates advanced optimization techniques to find best fitting model. The model can be composed from units of different transfer function type e.g sigmoid, polynomial, sine, linear, exponential, rational etc. and the best combination of units is selected by means of natural evolution. The model is constructed layer by layer and its complexity increases until the validation error of the model decreases. In each layer, niching genetic algorithm optimizes input connections, type and structure of the transfer functions of units. Best units from each niche are frozen and the construction of the model proceeds with next layers (see Figure 1).

The topology and parameters of the model can be also serialized into mathematical equations. The complexity of the model and its representation in form of equation depends on the character of the data set. However, if we need a simpler equation, we can restrict the competition to units with simple transfer functions (e.g. linear) or the output of the model can be replaced by any unit closer to the input layer. The accuracy of such simpler model will be lower so you need to find the balance between simplicity and accuracy of the model.

3   Data Acquisition

The study sample consisted of 400 women with singleton pregnancies less or equal than 7 days before delivery in breech presentation of the fetus. The measurements were peformed by 5 skilled sonographers, all using the same measurement method of fetal bioparameters. Acuson Antares (Siemens) and Logiq 9 (GE) premium class platforms were used in this study. Estimation of fetal weight was determined using measurements of bi-parietal diameter (BPD), abdominal circumference (AC), and femur length (FL). The measured data has been divided into two sets. First set has been used only for training and second one for evaluation of constructed models. During the learning phase, coefficient of units were optimized using part of the training set and remaining training vectors were used to select units with best generalization properties.

4   Experimental Setup

At first, we evaluated the performance of well known EFW models on our data set. The methodology used for the comparison reflects the need that domain experts understands our results. Therefore we use correlation, percentiles and other indicators, that are commonly used in medical journals. The primary objective of this chapter is to determine how accurately each of the methods predicted the actual observed birth weight and to compare them with each other. The accuracy of prediction was defined as the absolute difference between the predicted and observed weight. We also present estimated weight percentile table for each method which analyses weight categories where method fits.

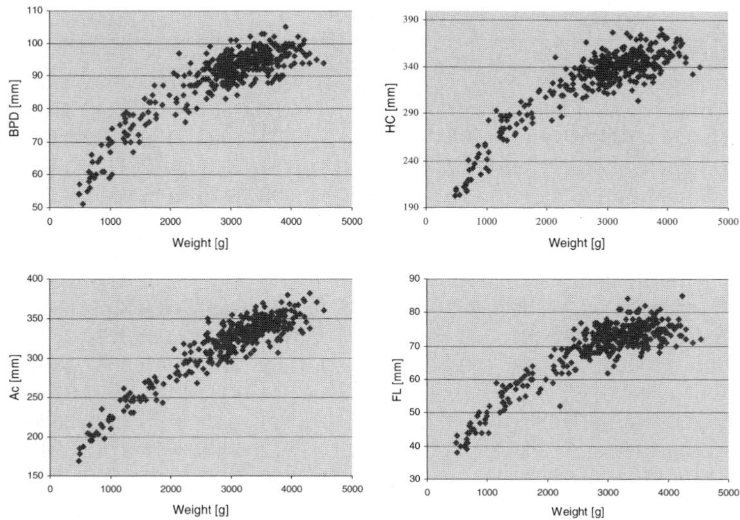

Fig. 2. Measured input data related to real birth weight

4.1   Comparison of Models Obtained by Standard Techniques

The table of percentiles (see Table 1) can indicate that a model is oriented just to certain weight category, ignoring the rest. We applied all models collected from the literature [10][11][12][13][14][15] to our data set and compared their percentiles. From the Table 1 we can observe that the equation (3) has all percentiles strongly biased against the reference data. Along with percentiles we compare other performance indicators of models (Model Correlation with Actual Birth Weight R^2, Mean absolute Error $\pm$, Standard deviation and the RMS Error). We sorted models according to their performance in the Table 2 and the best performing model on our data was Hadlock et al, 1985b (8). Some models (e.g. Hadlock (modification Nyberg et al, 2003a)) were well correlated with the target output, but their error was high. The reason is that they were biased towards our data. The Figure 3 shows why the worst model (as for the RMS error) is well correlated with the target variable. The regression line (predicted versus target output) should ideally have zero offset and slope one. The best performing model is getting close to this. The smaller distance of points from the regression line, the better. According to expectations, the most difficult range to predict corresponds with the cloud of data clustered around 3500g weight (see Figure 2). We list all models collected from literature so you can compare it with models obtained later by the GAME algorithm.

$$EFW = 0,23718 \times AC^2 + 0,0331 \times HC^2 \tag{1}$$

$$\log_{10} EFW = 1.335 - 0.0034 \times AC \times FL + 0.0316 \times BPD$$
$$+0.0457 \times AC + 0.1623 \times FL \tag{2}$$

Table 1. Basic statistic characteristics, Major Percentiles [g]

Method	5%	10%	50%	90%	95%
Reference data	937	1345	3080	3821	3966
Combs et al. 1993 (1)	1011	1557	3168	3644	3786
Hadlock (mod Nyhum 2003) (2)	904	1392	3156	3692	3807
Hadlock (mod Nyberg et al, 2003a) (3)	731	1031	2012	2278	2364
Hadlock (mod Nyberg et al, 2003b) (4)	908	1379	3111	3604	3773
Higginbottom et al, 1975 (5)	849	1230	2879	3589	3761
Jordaan, 1983 (6)	1289	1661	3069	3694	3879
Ott et al, 1986 (7)	973	1518	3218	3725	3842
Hadlock et al, 1985b (8)	894	1401	3168	3678	3836
Hsieh et al, 1987 (9)	1030	1429	3140	3779	3876
Shinozuka et al, 1987 (10)	967	1531	3190	3687	3802
Woo et al, 1985 (11)	843	1385	3120	3677	3785
Hadlock et al, 1985a (12)	894	1401	3168	3678	3836
Hadlock et al, 1984 (13)	912	1446	3178	3782	3945
Shepard et al, 1982 (14)	860	1385	3161	3800	3925

Table 2. Basic statistics characteristics, Model Correlation with Actual Birth Weight R^2, Mean absolute Error $\pm$ Standard deviation, RMS Error

Method	R^2	Mean Abs. Error [g] $\pm$ SD	RMS Error [g]
Hadlock et al, 1985b (8)	0.91	199 $\pm$ 171	261
Hadlock et al, 1985a (12)	0.91	202 $\pm$ 172	265
Shinozuka et al, 1987 (10)	0.91	211 $\pm$ 174	273
Combs et al. 1993 (1)	0.9	216 $\pm$ 173	276
Ott et al, 1986 (7)	0.91	215 $\pm$ 174	276
Hadlock (mod Nyberg et al, 2003b) (4)	0.9	212 $\pm$ 177	276
Hadlock (mod Nyhum 2003) (2)	0.91	202 $\pm$ 191	278
Woo et al, 1985 (11)	0.9	217 $\pm$ 181	282
Hadlock et al, 1984 (13)	0.89	232 $\pm$ 192	301
Shepard et al, 1982 (14)	0.88	233 $\pm$ 195	304
Hsieh et al, 1987 (9)	0.88	240 $\pm$ 190	306
Jordaan, 1983 (6)	0.87	260 $\pm$ 197	326
Higginbottom et al, 1975 (5)	0.86	313 $\pm$ 258	406
Hadlock (mod Nyberg et al, 2003a) (3)	0.91	1040 $\pm$ 447	1133

$$\log_{10} EFW = 1.5115 + 0.0436 \times AC + 0.1517 \times FL$$
$$-0.00321 \times FL \times AC + 0.0006923 \times HC \tag{3}$$

$$\log_{10} EFW = 1.304 + 0.05281 \times AC + 0.1938 \times FL - 0.004 \times FL \times AC \tag{4}$$

$$EFW = 0.0816 \times AC^3 \tag{5}$$

$$\log_{10} EFW = 2.3231 + 0.02904 \times AC + 0.0079 \times HC - 0.0058 \times BPD \tag{6}$$

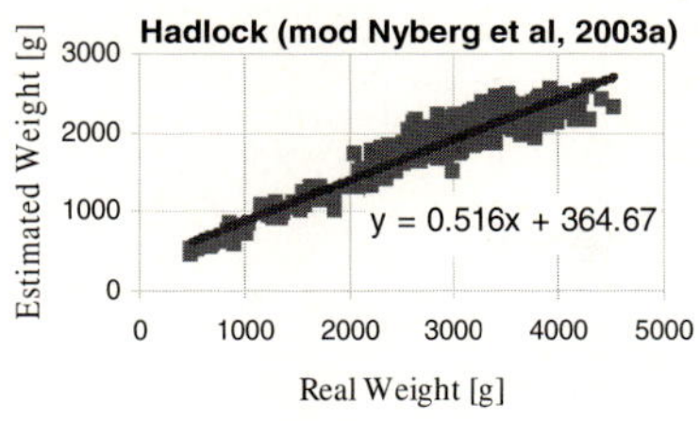

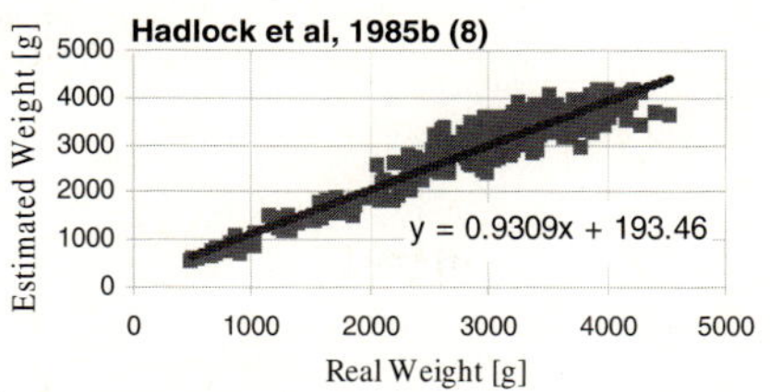

Fig. 3. Model (8) Correlation with Actual Birth Weight $R^2 = 0.91$

$$\log_{10} EFW = -2.0661 + 0.04355 \times HC + 0.05394 \times AC$$
$$-0.0008582 \times HC \times AC + 1.2594 \times \frac{FL}{AC} \times 1000 \qquad (7)$$

$$\log_{10} EFW = 1.326 - 0.00326 \times AC \times FL + 0.0107 \times HC$$
$$+0.0438 \times AC + 0.158 \times FL \qquad (8)$$

$$\log_{10} EFW = 2.7193 + 0.0094962 \times AC \times BPD - 0.1432 \times FL$$
$$-0.00076742 \times AC \times BPD^2 + 0.001745 \times FL \times BPD^2 \qquad (9)$$

$$EFW = 0.23966 \times AC \times AC \times FL + 1.623 \times BPD^3 \qquad (10)$$

$$\log_{10} EFW = 1.54 + 0.15 \times BPD + 0.00111 \times AC^2 - 0.0000764 \times BPD \times AC^2$$
$$+0.05 \times FL - 0.000992 \times FL * \times AC \qquad (11)$$

$$\log_{10} EFW = 1.335 - 0.0034 \times AC \times FL + 0.0316 \times BPD +$$
$$0.0457 \times AC + 0.1623 \times FL \qquad (12)$$

$$\log_{10} EFW = 1.182 + 0.0273 \times HC + 0.07057 \times AC - 0.00063 \times AC^2$$
$$-0.0002184 \times HC \times AC \qquad (13)$$

$$\log_{10} EFW = -1.7492 + 0.166 \times BPD + 0.046 \times AC - 0.002646 \times AC \times BPD \qquad (14)$$

5   GAME Models

In this section we generate several models ((15), (16), (17), (18)) by the GAME
algorithm and compare them with well known EFW models, which has been found
by linear and nonlinear regression methods by various authors in the past.

We loaded the data into the FAKE GAME open source application [2] and
generated models by using standard configuration (if not indicated differently)
of the GAME engine.

---

[2] http://sourceforge.net/projects/fakegame

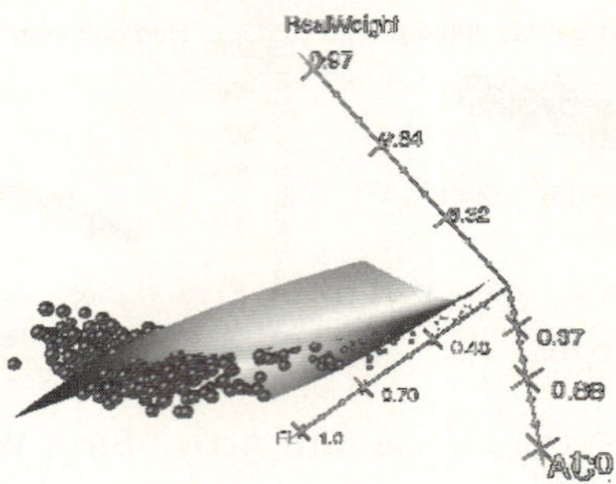

Fig. 4. An example of GAME model evolved on FL data. Regression hyperplane is smooth as expected.

Table 3. GAME models basic statistic characteristics, Major Percentiles [g]

Method	5%	10%	50%	90%	95%
(15)	950	1462	3149	3623	3779
(16)	906	1427	3236	3699	3800
(17)	937	1424	3145	3633	3741
(18)	886	1394	3173	3625	3720

Table 4. GAME models basic statistics characteristics, Model Correlation with Actual Birth Weight R^2, Mean absolute Error $\pm$ Standard deviation, RMS Error

Method	R^2	Mean Abs. Error [g] $\pm$ SD	RMS Error [g]
(17)	0.91	199 $\pm$ 168	261
(18)	0.91	203 $\pm$ 173	266
(16)	0.91	207 $\pm$ 169	267
(15)	0.91	209 $\pm$ 174	272

All generated models are simple and we also checked regression graphs of each model in GAME toolkit and see that every model has smooth progression (see 4) and approximate the output data set by hyperplane. Because the error is measured on testing data and the regression hyperplane is smooth, we can conclude that models are not overtrained [9] and have good generalization ability.

The first model was serialized into polynomial formula (just polynomial units were enabled and the penalization for complexity of the transfer function was applied to get simple formulae). The error of the model is therefore higher (Tab. 4) than that of models with more complex formulas obtained with the standard configuration of the GAME engine:

$$EFW = 0.0504 \times AC^2 - 16.427 \times AC + 38.867 \times FL + 284.074 \qquad (15)$$

$$EFW = -2061.54 + 1823.51 \times e^{0.006 \times BPD}$$
$$\times e^{\frac{1}{1.164 + 2073.08 \times e^{-0.017 \times AC - 0.035 \times FL}}}$$
$$\times e^{\frac{1}{-5.276 - 49.847 \times e^{3.125 \sin(0.455 - 0.006 \times BPD)}}} + \frac{1}{-0.271 - 221.677 \times e^{-0.024 \times HC}} \quad (16)$$

$$EFW = -7637.17 + 7870.09 \times e^{3.728 \times 10^{-6} \times (AC - 163)^2 + 0.0002 \times HC}$$
$$\times e^{\frac{1}{10.676 + 40011 \times e^{-0.096 \times BPD}}} + \frac{1}{7.557 + 6113.68 \times e^{-0.102 \times FL}} \quad (17)$$

Note that exponential and sigmoid units are very successful on this data set. Observed relationship of variables (Figure 4) is apparently nonlinear. To simplify generated equations, we transformed the output into logarithmic scale for the last model. Model produced by GAME does not contain exponential terms any more, but units with sine transfer function were more successful than polynomial units:

$$\log_{10} EFW = 2.18 + 0.0302 \times BPD + 0.0293 \times FL$$
$$-0.603 \sin(0.524 - 0.0526 \times AC)$$
$$-0.344 \sin(-0.029 \times AC - 0.117 \times FL + 0.946) \quad (18)$$

In case that experts prefer polynomial equation, Sine units can be easily disabled in the configuration of the GAME engine.

5.1   Ensemble of Three Models

Ensemble techniques are based on the idea that a collection of a finite number of models (eg. neural networks) is trained for the same task. Neural network ensemble [7] is a learning paradigm where a collection of a finite number of neural networks is trained for the same task. It originates from Hansen and Salamons work [8], which shows that the generalization ability of a neural network system can be significantly improved through ensembling a number of neural networks, i.e., training many neural networks and then combining their predictions. A single output can be created from a set of model outputs via simple averaging, in this case, the ensemble is called "simple ensemble".

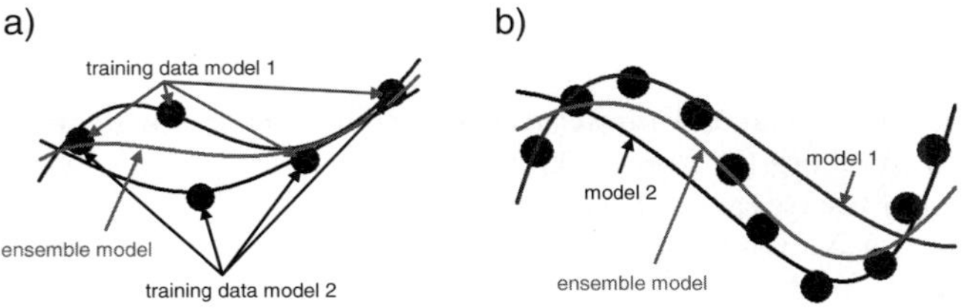

Fig. 5. Ensemble methods reduce variance a) and bias b) giving the resulting model better generalization abilities, thus improving accuracy on testing data

Table 5. GAME model ensemble basic statistic characteristics, Major Percentiles [g]

Ensemble	5%	10%	50%	90%	95%
GAME	952	1437	3182	3654	3781
Traditional methods	922	1440	3187	3693	3822

Table 6. GAME models ensemble basic statistics characteristics, Model Correlation with Actual Birth Weight R^2, Mean absolute Error $\pm$ Standard deviation, RMS Error

Ensemble	R^2	Mean Abs. Error [g] $\pm$ SD	RMS Error [g]
GAME	0.91	199 ± 160	260
Traditional methods	0.91	199 ± 161	261

5.2 Comparison of Ensemble Models

In this article we took three GAME models found earlier (15)(16)(17) and calculated same statistic characteristics and compared with previous results. We also created ensemble from three best tradition methods (8)(12)(10) and calculated statistic characteristics.   The ensemble of models has slightly better mean absolute error and the standard deviation than individual models.

6 Conclusion

All Fake Game models are at least good as best models found by statistical approach. We succeeded to find models with the same R^2, lower mean absolute error, lower RMS error and lower standard deviation than models found by traditional techniques. We also decreased mean absolute error, standard deviation and RMS error by using ensemble of three models which increases accuracy of estimation of fetal weight. The main advantage is that FAKE GAME models are not sensitive to noise located in input data. The GAME is very versatile, because generated models adapt to the character of the system modelled. The utilization of FAKE GAME methods brings additional knowledge about real world systems, reduces time needed for data preprocessing, the configuration of the data mining tools and the analysis of their results.

References

1. Wlodzislaw, D.: What is Computational Intelligence and what could it become? Springer, Heidelberg (in Press, 2007),
   `http://cogprints.org/5358/1/06-CIdef.pdf`
2. Kolmogorov, A.N.: On the Representations of Continuous Functions of Many Variables by Superposition of Continuous Functions of One Variable and Addition. Dokl. Akad. Nauk USSR 214, 953–956 (1957)
3. Hecht-Nielsen, R.: Kolmogorov's Mapping Neural Network Existence Theorem. In: Proceedings of the International Conference on Neural Networks, pp. 11–14. IEEE Press, New York (1987)

4. Ivakhnenko, A.G.: Polynomial theory of complex systems. IEEE Transactions on Systems, Man, and Cybernetics, SMC 1(1), 364–378 (1971)
5. Madala, Ivakhnenko, H.A.: Inductive Learning Algorithm for Complex System Modelling. CRC Press, Boca Raton (1994)
6. Kordik, P.: Fully automated knowledge extraction using Group of Adaptive Models Evolution. PhD thesis, Czech Technical University, Prague (March 2007), `http://neuron.felk.cvut.cz/game/doc.html`
7. Brown, R.: Diversity in Neural Network Ensembles. PhD thesis, The University of Birmingham, School of Computer Science, Birmingham B15 2TT, United Kingdom (January 2004)
8. Hansen, L., Salamon, P.: Neural network ensembles. IEEE Trans. Pattern Anal. Machine Intelligence 12(10), 993–1001 (1990)
9. Tetko, I.V., Livingstone, D.J., Luik, A.I.: Neural Network Studies. Comparison of Overfitting and Overtraining. J. Chem. Inf. Comput. Sci. 35, 826–833 (1995)
10. Colman, A., Maharaj, D., Hutton, J., Tuohy, J.: Reliability of ultrasound estimation of fetal weight in term singleton pregnancies. Journal of he New Zealand Medical Association 119(1241) (September 8, 2006)
11. Deter, R.L., Hadlock, F.P.: Use of ultrasound in the detection of macrosomia: A review. Journal of Clinical Ultrasound 13(8), 519–524 (1985)
12. Shepard, M.J., Richards, V.A., Berkowitz, R.L.: An evaluation of two equations for predicting fetal weight by ultrasound. American Journal of Obstetrics and Gynecology 142(1), 47–54 (1982)
13. Dudley, N.J.: A systematic review of the ultrasound estimation of fetal weight. Ultrasound Obstet Gynecol. 25, 80–89 (2005)
14. Hebbar, S.: Critical Evaluation of Various Methods of Estimating Fetal Weight by Ultrasound. J. Obstet Gynecol. Ind. 53(2), 131–133 (2003)
15. Nahum, G.G.: Estimation of Fetal Weight. eMedicine (May 13, 2007), `http://www.emedicine.com/med/topic3281.htm`

Evolutionary Canonical Particle Swarm Optimizer – A Proposal of Meta-optimization in Model Selection

Hong Zhang[1,2] and Masumi Ishikawa[1]

[1] Department of Brain Science & Engineering, Kyushu Institute of Technology
2-4 Hibikino, Wakamatsu, Kitakyushu 808-0196, Japan
{zhang,ishikawa}@brain.kyutech.ac.jp
[2] Department of Brain Science and Engineering,
Graduate School of Life Science & Systems Engineering,
Kyushu Institute of Technology, 2-4 Hibikino, Wakamatsu,
Kitakyushu 808-0196, Japan
Phone/Fax: +81-93-695-6112

Abstract. We proposed Evolutionary Particle Swarm Optimization (EPSO) which provides a new paradigm of meta-optimization for model selection in swarm intelligence. In this paper, we extend the technique of online evolutionary computation of EPSO to Canonical Particle Swarm Optimizer (CPSO), and propose Evolutionary Canonical Particle Swarm Optimizer (ECPSO) for optimizing CPSO. In order to effectually evaluate the performance of CPSO, a temporally cumulative fitness function of the best particle is adopted in ECPSO as the behavioral representative for entire swarm. Applications of the proposed method to a suite of 5-dimensional benchmark problems well demonstrate the effectiveness. Our experimental results clearly indicate that (1) the proper parameter sets in CPSO for solving various optimization problems are not unique; (2) the values of parameters in them are quite different from that of the original CPSO; (3) the search performance of the optimized CPSO is superior to that of the original CPSO, and to that of RGA/E except for the result to the Rastrigin's benchmark problem.

Keywords: canonical particle swarm optimization, evolutionary particle swarm optimization, temporally cumulative fitness, real-coded genetic algorithm, model selection.

1 Introduction

Particle Swarm Optimization (PSO) is a stochastic and population-based adaptive optimization algorithm proposed by Kennedy and Eberhart [5,9]. Due to its intuitive understandability, ease of implementation, and the ability in search, it has been widely applied to various disciplines in science, engineering, technology and application for solving large-scale, highly nonlinear, and multimodal optimization problems [4,7,15].

V. Kůrková et al. (Eds.): ICANN 2008, Part I, LNCS 5163, pp. 472–481, 2008.

During the last decade, many variants of PSO, e.g., Canonical Particle Swarm Optimizer (CPSO), Gregarious Particle Swarm Optimizer (G-PSO), and Adaptive Hierarchical Particle Swarm Optimizer (H-PSO) etc. were proposed [3,8,13,14,15,16]. Although the mechanisms of these variants of PSO are not complex with only few parameters to adjust, they, in general, can provide better results compared with other methods such as machine learning, neural network learning and genetic computation [10].

Needless to say, it is necessary to use a PSO model with higher performance balancing exploitation and exploration for efficiently solving a given optimization problem. However, how to appropriately determine the values of parameters in the PSO model is not yet completely solved, because the behavior of particle swarm is stochastic and dynamic in nature. Accordingly, model selection of PSO for improving the search performance is a hard but attractive and challenging problem.

Researchers paid much attention to the above problem, and proposed many algorithms and strategies for dealing with it [1,2,6]. In particular, to systematically estimate a proper parameter set in PSO model, Meissner et al. proposed an Optimized Particle Swarm Optimization (OPSO) method, which uses PSO for handling meta-optimization of PSO heuristics [11,12], and Zhang et al. recently proposed Evolutionary Particle Swarm Optimization (EPSO), in which the Real-coded Genetic Algorithm with Elitism strategy (RGA/E) is used for estimating appropriate values of parameters in PSO [17]. It should be noted that the different fitness functions are used in OPSO and EPSO for evaluating the performance of PSO in addition to difference in estimation methods.

Since a temporally cumulative fitness function of the best particle is adopted as the behavioral representative for entire swarm, theoretically EPSO gives a more reliable method for determining the appropriate values of parameters in PSO corresponding to a given optimization problem than that by OPSO with an instantaneous fitness function. It has also been demonstrated that EPSO has superior search performance than the original PSO [18,19].

Furthermore, the presumptive object of EPSO is limited not only to PSO, but also expendable to any variants of PSO and other stochastic optimization algorithms with adjustable parameters. In this paper, we extend the technique of online evolutionary computation of EPSO to CPSO, and propose Evolutionary Canonical Particle Swarm Optimizer (ECPSO) for optimizing CPSO without prior knowledge to efficiently solve various optimization problems.

The CPSO proposed by Clerc et al. [3] is a constricted version of PSO. It decreases the amplitude of the particle's oscillation by a constriction coefficient resulting in the quick convergence of the particles over time. Since it is based on a rigorous analysis of the dynamics of a simplified model of the original PSO, CPSO became highly influential in the field of PSO research. However, the major shortcomings of it is that some knowledge of the search space is required to efficiently guide swarm and adjust the balance between exploitation and exploration. This is also the reason why we pay much attention to it, and propose to optimize CPSO for handling this difficult.

To demonstrate the effectiveness of ECPSO, computer experiments on a suite of 5-dimensional benchmark problems are carried out. Based on the obtained results, we analyze the characteristics of the proposed method, and compare with the performance of the original CPSO, RGA/E, and EPSO in search.

The rest of the paper is organized as follows. Section 2 describes the original CPSO. Section 3 introduces an algorithm of ECPSO. Section 4 discusses the results of computer experiments applied to the given benchmark problems. Section 5 gives conclusions.

2 Overview of CPSO

Assuming N-dimensional search space, $\mathbf{S} \in \Re^N$, and a swarm of P particles, the position and velocity of the ith particle is an N-dimensional vector, $\boldsymbol{x}^i = (x_1^i, x_2^i, \cdots, x_N^i)^T$ and $\boldsymbol{v}^i = (v_1^i, v_2^i, \cdots, v_N^i)^T$, respectively. The constriction version of the original PSO [3] is defined by

$$\begin{cases} \boldsymbol{x}_{k+1}^i = \boldsymbol{x}_k^i + \boldsymbol{v}_{k+1}^i \\ \boldsymbol{v}_{k+1}^i = \chi\left(\boldsymbol{v}_k^i + c_1\boldsymbol{r}_1\otimes(\boldsymbol{x}_l^i-\boldsymbol{x}_k^i) + c_2\boldsymbol{r}_2\otimes(\boldsymbol{x}_g - \boldsymbol{x}_k^i)\right) \end{cases}$$

where χ is a constriction factor, c_1 is an individual confidence factor, and c_2 is a swarm confidence factor. $\boldsymbol{r}_1$ and $\boldsymbol{r}_2$ are two N-dimensional random vectors in which each element is uniformly distributed in $[0,1]$, the symbol, $\otimes$, is an element-by-element vector multiplication operator. $\boldsymbol{x}_l^i$ ($= arg \max_{k=1,2,\cdots} \{g(\boldsymbol{x}_k^i)\}$, where $g(\boldsymbol{x}_k^i)$ is the fitness value of the ith particle at time k) is the local best position, lbest, of the ith particle up to now, $\boldsymbol{x}_g (= arg \max_{i=1,2,\cdots} \{g(\boldsymbol{x}_l^i)\})$ is the global best position, gbest, among the whole particles, respectively.

The value of the constriction factor is typically obtained by $\chi = \dfrac{2\kappa}{|2-\phi-\sqrt{\phi^2-4\phi}|}$, for $\phi > 4$, where $\phi = c_1 + c_2$, and $\kappa = 1$. Different configurations of κ as well as a theoretical analysis of the derivation of the above formula, can be found in [3,4]. According to Clerc's constriction method, ϕ is commonly set to 4.1, then $c_1 = c_2 = 2.05$, and the constant multiplier χ is approximately 0.7298. Here we call them the values of parameters in the original CPSO.

3 Overview of ECPSO

ECPSO is a variant of EPSO that systematically estimates a proper parameter set of CPSO for efficiently solving a given optimization problem by online evolutionary computation. It can be extended easily to generate a CPSO model with higher performance instead of PSO used in EPSO [17].

Fig. 1 illustrates the flowchart of ECPSO. Specifically, the procedure of ECPSO is composed of two parts: The one is outer loop in which RGA/E is applied to simulating the survival of the fittest among individuals over generations for solving

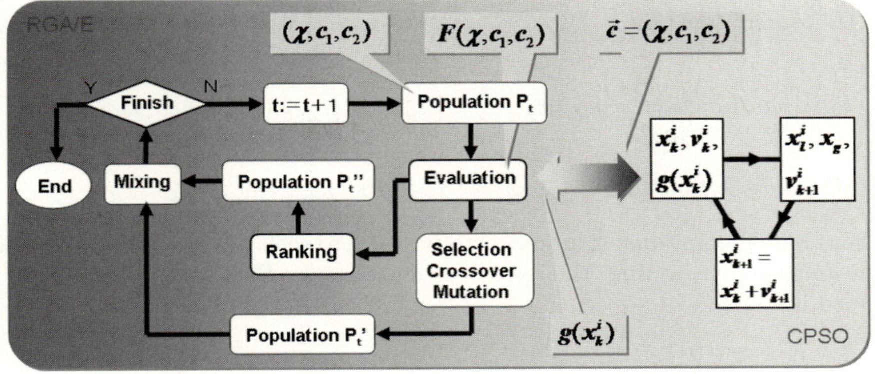

Fig. 1. Flowchart of ECPSO

real-valued optimization problems. While the CPSO finds a solution corresponding to a given optimization problem, it is used to optimize the values of parameters, i.e., χ, c_1, and c_2, in CPSO with online evolutionary computation. The other is inner loop in which CPSO runs. Namely the CPSO with the parameter values created by RGA/E is expected for achieving higher fitness than others.

As is well known, sometimes a swarm of particles searches well by chance, even if the parameter values they have are not suitable for efficiently solving the given optimization problem. In order to overcome this difficulty, here we adopt the following temporally cumulative fitness function for evaluating the performance of CPSO [18].

$$F(\chi, c_1, c_2) = \sum_{k=1}^{K} g(x_k^b)\big|_{\chi, c_1, c_2} \tag{1}$$

where x_k^b ($b = arg\,max_{i=1}^{P}\{g(x_k^i)\}$) is the position of the best particle at time k, and K is the maximum number of iterations in CPSO.

Since the fitness function, $F(\cdot)$, assesses the sum of instantaneous fitness function, $g(x_k^i)$, its variance is inversely proportional to the time span. Therefore, it leads to greatly inhibit the influence of noise in evaluation, and effectually to estimate CPSO models with superior search performance.

4   Computer Experiments

4.1   Benchmark Problems and Experimental Conditions

To facilitate comparison and analysis of the search performance of ECPSO, we use the following suite of benchmark problems for verification.

$Sphere$ function: $f_S(x) = \sum_{d=1}^{N} x_d^2$

$$Griewank\ \text{function:}\quad f_G(\boldsymbol{x}) = \frac{1}{4000}\sum_{d=1}^{N} x_d^2 - \prod_{d=1}^{N}\cos\left(\frac{x_d}{\sqrt{d}}\right) + 1$$

$$Rastrigin\ \text{function:}\quad f_{Ra}(\boldsymbol{x}) = \sum_{d=1}^{N}\left(x_d^2 - 10\cos(2\pi\,x_d)\right) + 10d$$

$$Rosenbrock\ \text{function:}\quad f_{Ro}(\boldsymbol{x}) = \sum_{d=1}^{N}\left[\left(100(x_{d+1} - x_d^2)\right)^2 + (x_d - 1)^2\right]$$

For finding an optimal solution corresponding to each benchmark problem, fitness function regarding the search environment, $\mathbf{S} \in (-5.12, 5.12)^N$, is defined by

$$g_*(\boldsymbol{x}) = \frac{1}{f_*(\boldsymbol{x}) + 1},$$

where the subscript $*$ stands for one of the followings: $S(Sphere$ function), $G(Griewank$ function), $Ra(Rastringin$ function), and $Ro(Rosenbrock$ function), respectively. Fig. 2 illustrates the fitness functions for 2-dimensional benchmark problems.

Table 1 gives the major parameters in ECPSO for estimating the appropriate values of parameters in CPSO. And initial conditions for CPSO search: the position of each particle is set randomly in the search space, and its velocity is set to zero.

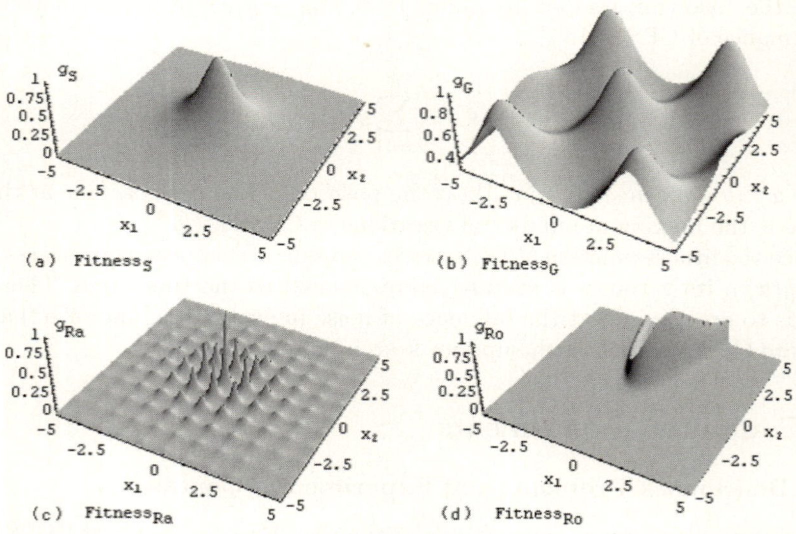

Fig. 2. The distribution of the fitness values according to each 2-dimensional benchmark problem. (a) Sphere function, (b) Griewank function, (c) Rastrigin function, (d) Rosenbrock function.

Table 1. The major parameters in ECPSO

Parameters	Value	Parameters	Value
The number of individuals, M	100	The number of superior individuals, s_n	1
The number of generation, G	20	Roulette wheel selection	–
The number of iterations, K	400	Probability of BLX-2.0 crossover, p_c	1.0
The number of particles, P	10	Probability of random mutation, p_m	1.0

4.2 Experimental Results

Computer experiments are carried out for the given 5-dimensional benchmark problems. Table 2 shows the resulting values of parameters in CPSO models and the frequency of models for each problem, respectively [1].

Table 2. The resulting values of parameters in CPSO models and the frequency of models for the given 5-dimensional benchmark problems (20 trials/per problem).

Problem	Optimized CPSO	Parameter			Frequency
		χ	c_1	c_2	
Sphere	a type	0.613±0.064	0	3.291±1.356	35.0%
	b type	0.552±0.114	1.783±1.870	3.562±1.646	65.0%
Griewank	a type	0.537±0.105	0	2.532±1.023	30.0%
	b type	0.601±0.132	1.773±1.343	2.080±1.274	70.0%
Rastrigin	a type	11.88±5.644	0	6.331±1.956	85.0%
	b type	8.809±1.031	0.326±0.332	5.010±1.791	15.0%
Rosenbrock	a type	0.876±0.000	0	0.446±0.405	25.0%
	b type	0.775±0.129	1.917±1.160	1.090±0.803	75.0%

Therefore, the following characteristics of ECPSO are observed.

- There are two types of the optimized CPSO models which all can find the optimal solution for each problem, although the frequencies of models contained within top 20 trails are different. Namely the one is that the value of parameter, c_1, is zero, and the other is that the value is not zero. In the meantime, the value of parameter, c_2, is never zero for any models under any problems. It clearly indicates that the swarm confidence factor plays an important role in CPSO search for efficiently solving the given problems.
- The estimated values of parameters, c_1 or c_2, for each problem are perfectly different, i.e., c_1 is not equal to c_2, and their values are also different from that in the original CPSO. This suggests that the appropriate values of parameters in CPSO are not only ones. Accordingly, there is necessity to individually estimate the appropriate values of parameters in CPSO by ECPSO.

---

[1] Computing environment: Intel(R) Xeon(TM); CPU 3.40GHz; Memory 2.00GB RAM; Computing tool: Mathematica 5.2; Computing time: about 9.8 minutes.

– The resulting value of the constriction facter, χ, drastically exceeds 1 only for Rastrigin's benchmark problem. This means that CPSO needs to have more randomization for enhancing the search performance to find an optimal solution or near-optimal solutions from the problem, because there are a lot of local minima in the search space (i.e., higher density case).

4.3   Evaluation of Performance

To identify the search performance of the resulting CPSO models, we measure the performance of each model by the average fitness value. Fig. 3 illustrates the frequency distribution of the **gbest**'s fitness values obtained by the optimized CPSO models with the average value of parameters in Table 2.

And Table 3 gives the search performance of the optimized CPSO models, i.e., the mean and standard deviation corresponding to each frequency distribution in Fig. 3. Comparison with the search performance of the optimized CPSO models in two types in Table 3 for each problem reveals that the average value of fitness of the model in b type is superior to that of the model in a type except for the Sphere's benchmark problem in which the optimal solution can be found even

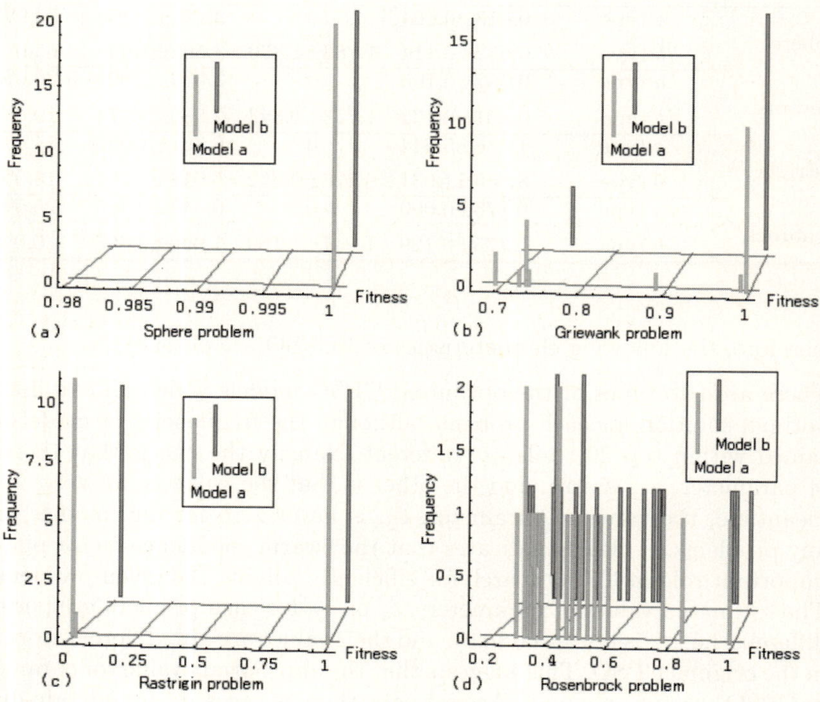

Fig. 3. The frequency distribution of the **gbest**'s fitness values for each 5-dimensional benchmark problem. (a) Sphere function, (b) Griewank function, (c) Rastrigin function, (d) Rosenbrock function.

Table 3. The search performance of the CPSO models optimized by ECPSO

Problem	Optimized CPSO	Fitness (mean±s.d.)
Sphere	a type	**1.0000**±0.0000
	b type	**1.0000**±0.0000
Griewank	a type	0.8855±0.1342
	b type	**0.9478**±0.1069
Rastrigin	a type	0.4202±0.4856
	b type	**0.5167**±0.4958
Rosenbrock	a type	0.5040±0.1853
	b type	**0.5352**±0.2624

without optimization. These results distinctly show that the basic construction of the estimated CPSO model in b type is relatively rational than the model in a type for efficiently solving the given problems.

4.4 Comparison for Performance

For illuminating the characteristics of ECPSO, here we compare with the search performance of other methods, i.e., the original CPSO, RGA/E, and EPSO. Table 4 shows the experimental results corresponding to each problem.

Table 4. The search performance of the original CPSO, ECPSO, RGA/E, and EPSO for each 5-dimensional benchmark problem (statistics over 20 trials)

Problem	Original CPSO	ECPSO	RGA/E	EPSO
Sphere	**1.0000**±0.0000	**1.0000**±0.0000	0.9980±0.0016	**1.0000**±0.0000
Griewank	0.9045±0.1302	0.9478±0.1069	0.7966±0.1175	**0.9829**±0.0129
Rastrigin	0.1210±0.0896	0.5167±0.4958	0.9616±0.0239	**1.0000**±0.0000
Rosenbrock	0.4210±0.2376	**0.5352**±0.2624	0.3723±0.1364	0.4694±0.2806

By comparison with the average values of fitness in Table 4, we observed that the search performance of ECPSO is superior to that of the original CPSO for each problem except for the result to the Sphere's benchmark problem which is a simple unimodal one. This result sufficiently indicates that the effectiveness of the proposed method, and model selection by ECPSO is of necessity for solving various optimization problems by CPSO.

And, we also observed that the search performance of ECPSO is also superior to that of RGA/E for the given problems except for the result to the Rastrigin's benchmark problem. This means that the optimized CPSO have higher search performance than RGA/E for the Sphere, the Griewank, and the Rosenbrock's benchmark problem, respectively. It also reflects that the performance of the optimized CPSO and the RGA/E depend on the characteristics of the given benchmark problems.

But the search performance of ECPSO is inferior to that of EPSO for the given problems except for the result to the Rosenbrock's benchmark problem. In theory, it is considered that the constriction factor in CPSO gives limitation for the velocity of the ith particle, $\boldsymbol{v}_{k+1}^{i}$, moving toward **gbest** and **lbest** at time $k + 1$ compared to PSO when the value of χ is smaller than 1.0. It seems to affect exploration in CPSO. Detailed examination on this, and experiments for the suite of high-dimensional benchmark problems will be done in the near future.

5   Conclusions

In this paper, we proposed Evolutionary Canonical Particle Swarm Optimization, ECPSO, for effectively solving various optimization problems. The crucial idea is to estimate the appropriate values of parameters in CPSO by online evolutionary computation, which was originally proposed in EPSO.

Applications of the proposed method, ECPSO, to a suite of 5-dimensional benchmark problems well demonstrate the effectiveness. Our experimental results clearly indicate that (1) the proper parameter sets in CPSO for solving various optimization problems are not unique; (2) the values of parameters in them are quite different from that of the original CPSO; (3) the search performance of the optimized CPSO is superior to that of the original CPSO, and to that of RGA/E except for the result to the Rastrigin's benchmark problem.

Most importantly, we confirmed that the proper parameter sets in CPSO for solving various optimization problems are not unique, and the values of parameters in them are quite different from the original CPSO. This emphasizes that the complexity in optimizing CPSO, and the importance of model selection of CPSO with ECPSO.

Applications of the proposed method to complex real-world problems are left for future study. It is expected that the fruitful results will be obtained in the disciplines of PSO research and meta-optimization.

Acknowledgment. This research was partially supported by the 21st century COE (Center of Excellence) program (#J19) granted to Kyushu Institute of Technology from the Ministry of Education, Culture, Sports, Science and Technology (MEXT), Japan.

References

1. Beielstein, T., Parsopoulos, K.E., Vrahatis, M.N.: Tuning PSO Parameters Through Sensitivity Analysis, Technical Report of the Collaborative Research Center 531 Computational Intelligence CI-124/02, University of Dortmund (2002)
2. Carlisle, A., Dozier, G.: An Off-The-Shelf PSO. In: The Workshop on Particle Swarm Optimization, Indianapolis, pp. 1–6 (2001)
3. Clerc, M., Kennedy, J.: The particle swarm-explosion, stability, and convergence in in a multidimensional complex space. IEEE Transactions on Evolutionary Computation 6(1), 58–73 (2000)

4. Clerc, M.: Particle Swarm Optimization. Iste Publishing Co., UK (2006)
5. Eberhart, R.C., Kennedy, J.: A new optimizer using particle swarm theory. In: The sixth International Symposium on Micro Machine and Human Science, Nagoya, Japan, pp. 39–43 (1995)
6. Eberhart, R.C., Shi, Y.: Comparing inertia weights and constriction factors in particleswarm optimization. In: The 2000 IEEE Congress on Evolutionary Computation, La Jolla, CA, vol. 1, pp. 84–88 (2000)
7. Eberhart, R.C., Shi, Y., Kennedy, J.: Swarm Intelligence. Morgan Kaufmann Publishers, CA (2001)
8. Janson, S., Middendorf, M.: A hierarchical particle swarm optimizer and its adaptive variant. IEEE Transactions on Systems, Man and Cybernetics, Part B 35(6), 1272–1282 (2005)
9. Kennedy, J., Eberhart, R.C.: Particle swarm optimization. In: The 1995 IEEE International Conference on Neural Networks, Piscataway, New Jersey, pp. 1942–1948 (1995)
10. Kennedy, J.: In Search of the Essential Particle Swarm. In: The 2006 IEEE Congress on Evolutionary Computations, Vancouver, BC, Canada, pp. 6158–6165 (2006)
11. Meissner, M., Schmuker, M., Schneider, G.: Optimized Particle Swarm Optimization (OPSO) and its application to artificial neural network training. BMC Bioinformatics 7(125) (2006)
12. Parsopoulos, K.E., Vrahatis, M.N.: Recent approaches to global optimization problems through Particle Swarm Optimization. Natural Computing 1, 235–306 (2002)
13. Pasupuleti, P., Battiti, R.: The Gregarious Recent Particle Swarm Optimizer (G-PSO). In: IEEE Congress on Evolutionary Computation, pp. 84–88 (2000)
14. Reyes-Sierra, M., Coello, C.A.C.: Multi-Objective Particle Swarm Optimizers: A Survey of the State-of-the-Art. International Journal of Computational Intelligence Research 2(3), 287–308 (2006)
15. Spina, R.: Optimisation of injection moulded parts by using ANN-PSO approach. Journal of Achievements in Materials and Manufacturing Engineering 15(1-2), 146–152 (2006)
16. Xie, X.-F., Zhang, W.-J., Yang, Z.-L.: A Dissipative Particle Swarm Optimization. In: The IEEE Congress on Evolutionary Computation (CEC 2002), Honolulu, USA, pp. 1456–1461 (2002)
17. Zhang, H., Ishikawa, M.: Evolutionary Particle Swarm Optimization (EPSO) – Estimation of Optimal PSO Parameters by GA. In: The IAENG International MultiConference of Engineers and Computer Scientists (IMECS 2007), Newswood Limited, Hong Kong, China, vol. 1, pp. 13–18 (2007)
18. Zhang, H., Ishikawa, M.: Designing Particle Swarm Optimization – Performance Comparison of Two Temporally Cumulative Fitness Functions in EPSO. In: 26th IASTED International Conference on Artificial Intelligence and Applications (AIA 2008), Innsbruck, Austria, pp. 301–306 (2008)
19. Zhang, H., Ishikawa, M.: Evolutionary Particle Swarm Optimization – Metaoptimization Method with GA for Estimating Optimal PSO Methods. In: Castillo, O., et al. (eds.) Trends in Intelligent Systems and Computer Engineering. Lecture Notes in Electrical Engineering, vol. 6, pp. 75–90. Springer, Heidelberg (2008)

Building Localized Basis Function Networks Using Context Dependent Clustering

Marcin Blachnik[1] and Włodzisław Duch[2]

[1] Silesian University of Technology, Electrotechnology Department,
Katowice, Krasińskiego 8, Poland
`marcin.blachnik@polsl.pl`
[2] Department of Informatics, Nicolaus Copernicus University,
Grudziądzka 5, Toruń, Poland
`Google: W. Duch`

Abstract. Networks based on basis set function expansions, such as the Radial Basis Function (RBF), or Separable Basis Function (SBF) networks, have non-linear parameters that are not trivial to optimize. Clustering techniques are frequently used to optimize positions of localized functions. Context-dependent fuzzy clustering techniques improve convergence of parameter optimization, leading to better networks and facilitating formulation of prototype-based logical rules that provide low-complexity models of data.

1 Introduction

Basis set expansions have been studied in approximation theory of functions since XIX century by such great mathematicians as Weierstrass, Chebyshev, Runge, Volterra, Lebesgue, Markov, Hermite, Fourier and many others. Multidimensional expansions into product of various orthogonal polynomials, splines, or Gaussian functions have been used in applied mathematics from its beginning. In neural networks radial basis function networks (RBF) [1] have gained great popularity, providing approximations of multidimensional functions by linear combinations of radial functions $\Phi(\mathbf{X}) = \sum_i W_i G(||\mathbf{X} - \mathbf{R}_i||)$. Although many radial functions exist (see the survey of neural transfer functions [2]) Gaussian functions are by far most popular. The n-dimensional Gaussian functions belong to a few types of radial functions that are separable, that is they may be presented as a product of n one-dimensional components. Such products may be identified either with "and" conditions in aggregation of fuzzy membership functions for different features, or with Naive Bayes estimation of multidimensional probability by a product of one-dimensional Gaussian distributions.

Separable basis function (SBF) networks, such as the Feature Space Mapping networks [3], use Gaussians or other one-dimensional functions to create approximations to multidimensional distributions $\Phi(\mathbf{X}) = \sum_I W_I \phi_{i1}(x_1)\phi_{i2}(x_2)\ldots\phi_{in}(x_n))$, where the multi-index $I = \{i1, i2, \ldots in\}$. In quantum mechanics and quantum chemistry [4] these one-dimensional functions are called orbitals and the total function to be approximated is called the wavefunction. SBF networks have many interesting properties, although they are not so popular as the RBF networks. In particular all similarity-based methods with additive distance functions are equivalent to SBF networks [5]. Such

V. Kurková et al. (Eds.): ICANN 2008, Part I, LNCS 5163, pp. 482–491, 2008.

representations may be justified by cognitive inspirations [6,3], with "intuitive" representation of knowledge based on similarity (measured by neural output) to prototypes, or a distance to peaks of probability density distributions that represent reference objects in feature spaces. Classification or approximation is then done using rules based on similarity to reference objects. In this paper only networks that use localized basis function (LBF) expansions (output is larger than some threshold value in a finite region of the input space) are considered, and in empirical tests only Gaussian Basis Functions (GBF) networks have been used. There is a tendency to call all basis set expansion networks RBF networks, even if the functions used are not radial. To avoid confusion the term LBF shall be used when localization is important, and more precise GBF when Gaussian functions are used.

In multilabel classification the goal is to discriminate between n-dimensional vectors $\mathbf{X} = [x_1, x_2, \ldots x_n]^T$ that carry c symbolic or numeric class labels $\mathbf{Y} = [y_1, y_2, \ldots y_c]$. In contrast to Bayesian formulation [1] class labels do not need to be interpreted as probabilities, and thus one vector may be a member of several classes. The classification system has some parameters that are optimized on a training dataset $\mathcal{D} = \{\mathbf{X}, \mathbf{Y}\}$ of N pairs of feature vectors $\mathbf{X}$ and their class labels $\mathbf{Y} = \mathbf{Y}(\mathbf{X})$, with the goal to optimize performance on a new data sampled from the same probability distributions.

In the LBF network expansion $g(\mathbf{X}; \mathbf{W}, \mathbf{R}) = \sum_i W_i G_i(\mathbf{X}, \mathbf{R}_i)$ linear coefficients $\mathbf{W}$ may be obtained using standard RBF techniques [1], but finding non-linear parameters $\mathbf{R}_i$, such as positions and dispersions of Gaussian functions, is more difficult. Most frequently RBF network training starts from clustering of the whole training set, placing the hidden units in the centers of obtained clusters. The hidden-output linear connections are then trained to minimize some cost function. This process can also be viewed as nonlinear mapping of the input data by localized functions to some image space, and then creating linear model in the image space. Kernel machines, in particular support vector machines (SVM) [7], perform this transformation only implicitly, but the idea is the same.

In the LBF networks used for classification problems clustering approach may lead to non-optimal position of hidden neurons, hence the number of hidden neurons required for sufficient accuracy may become unnecessarily large. Pruning techniques are used to remove less important neurons [1], and even to merge several neurons [8]. Pruning may be followed by re-optimization of parameters, adding more neurons, and repeating the whole cycle [9] to reduce the network and improve its generalization, but additional computational costs are involved.

SVM improves generalization by increasing the classification margin. Input vectors that are close to the SVM decision hyperplane are selected by the optimization procedure as Support Vectors (SV). Thus in contrast to other linear discrimination techniques that use all data vectors to define the hyperplane [10] in SVM only those vectors that lie near the decision border influence its final shape. In the Support Vector Neural Training algorithm [11] vectors that are far from the decision border (and therefore cause maximum or minimum activation of neurons) are gradually removed from the training set, leaving only support vectors close to the decision border. In this contribution context dependent clustering is used to restrict the subspace where the decision border is defined. Although SVM may be used for extraction of prototype-based rules (P-rules) [12] better results are obtained using context dependent clustering.

In the next section methods of building and optimizing LBF networks are discussed, in section 3 the basic context-dependent fuzzy C-means (CFCM) algorithm and the method of defining the context (section 4) are explained. Empirical results, presented in section 5, compare LBF networks based on the classical clustering and the context dependent clustering. The last section contains a discuss of the results and shows some potential applications of the proposed approach as a tool for creating prototype-based rule systems (P-systems).

2   Building LBF Network

Typical LBF network has only one hidden layer and an output layer. The LBF hidden layer nodes use localized transfer functions, for example radial distance-based activations $||\mathbf{X} - \mathbf{R}||$ between input vectors $\mathbf{X}$ and reference vectors $\mathbf{R}$. In the most common case neurons of the hidden layer have Gaussian transfer function. However, many other localized non-radial functions may also be used, and many radial functions are non-local [2], therefore LBF networks are not the same as the RBF networks.

The output layer usually implements a linear combination of signals generated by the hidden layer, leading to a decision function:

$$g(\mathbf{X}; \mathbf{W}, \mathbf{R}) = \sum_{i=1}^{K} W_i G_i(\mathbf{X}, \mathbf{R}_i) \tag{1}$$

where K is the number of hidden neurons and $\mathbf{W}$ are weights of the output layer. In case of a generalized Gaussian Basis Function expansion the particular parameterization of localized functions involves:

$$g(\mathbf{X}; \mathbf{W}, \mathbf{R}) = \sum_{i=1}^{K} W_i \exp\left(-||\mathbf{X} - \mathbf{R}_i||^2_{\Sigma_i}\right) \tag{2}$$

where $\mathbf{R}_i$ is the position of i-th prototype (location of the center of the hidden unit in the input space), and a local Mahalanobis distance function $|| \cdot ||_{\Sigma_i}$ with the covariance matrix $\mathbf{\Sigma}_i$ is calculated in some neighborhood around this center:

$$||\mathbf{X} - \mathbf{R}_i||^2_{\Sigma_i} = (\mathbf{X} - \mathbf{R}_i)^T \mathbf{\Sigma}^{-1} (\mathbf{X} - \mathbf{R}_i) \tag{3}$$

Covariance matrix could be replaced with a matrix of adaptive parameters learned from the training data, providing scaling and rotations for the shapes of local decision boarders. Although it is possible to rotate Gaussian function in n-dimensions using only n parameters [2] in practice almost always diagonal matrix $\mathbf{\Sigma}$ is used, providing distance scaling factors s_i. Training of such GBF networks requires adaptation of positions $\mathbf{R}_i$, scaling factors s_i and weights $\mathbf{W}$ of the output layers. As suggested in [13] there are three possible ways to build such networks:

1. One step process, when each training data vector is used to define one hidden unit, and training is resticted to the output layer;

2. Two step training, interchanging training of non-linear and linear parameters:
   - adjust location and scaling parameters of hidden units using some form of supervised clustering or decision trees, independently of the network outputs;
   - adjust output layer weights minimizing mean square error or another cost function:

$$E(\mathcal{D}; \mathbf{W}, \mathbf{R}) = \sum_{i=1}^{N} ||\mathbf{Y}_i - g(\mathbf{X}_i; \mathbf{W}, \mathbf{R})||^2 \tag{4}$$

   The output weights may then be determined using the pseudoinverse matrix:

$$\mathbf{W} = (\mathbf{G}^T \mathbf{G})^{-1} \mathbf{G}^T \mathbf{Y} \tag{5}$$

   where $G_{ij} = G(\mathbf{X}_i; \mathbf{R}_j)$ is the rectangular matrix $N \times K$ of similarities between the prototypes $\mathbf{R}$ and the training data $\mathbf{X}$.
3. Three step learning, where the first two steps described above are used for initialization, followed by simultaneous adaptation of all network parameters using gradient descend or other methods.

Unfortunately the error function $E(\mathcal{D}; \mathbf{W}, \mathbf{R})$ has multiple local minima making the gradient optimization quite difficult. Results may be significantly improved if a proper initialization of the network is found. Another problem facing the LBF networks is to determine the optimal number of hidden units. The simplest solution is to use all vectors to define a large number of hidden unites, but this leads to many problems with data overfitting, convergence difficulties due to ill-conditioned $\mathbf{G}$ matrix and waste of computational resources. Reduction of the number of hidden nodes to $K \ll N$ is a form of regularization that increases generalization of LBF networks. Other advantages of this approach include easier interpretation of the functions implemented by such networks. For example, in [3,14,15] relations between basis set expansion networks and fuzzy rule based systems (F-rules) have been analyzed, showing that for separable neural transfer functions (and thus a few types of radial basis functions) they are equivalent. Fuzzy rules are useful if they are comprehensible and generalize well, therefore a small number of neurons is highly desired. This is equally true for systems based on P-rules [16,5] that are equivalent to networks with different types of functions implemented by their hidden nodes.

Support Vector Machines provide good inspiration for construction of optimal shapes of decision borders that should be based on prototypes placed in a similar way as SVs around the border area. Classical (unsupervised) clustering algorithms analyze all data in the whole input space without information about their class distribution. Clustering each class independently may not be a good solution, leading to centers of clusters that may be far from optimal, as show in Fig. 1. Optimization methods from the LVQ family [17], or other gradient based methods used in the network training, cannot move such prototypes to areas where they could take an important role in the decision process.

Context-dependent clustering can be used with properly defined context to find much better positions of prototypes. Such approach has been previously applied with very good results for training of the prototype-based rule system with the nearest neighbor rule [18].

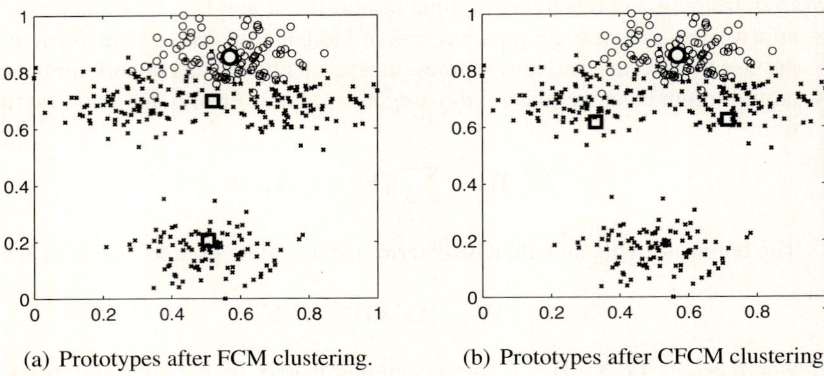

(a) Prototypes after FCM clustering.   (b) Prototypes after CFCM clustering.

Fig. 1. Comparison of prototype positions obtained from the FCM and CFCM clustering algorithms for the two class problem

3   Conditional Fuzzy C-Means

Context-dependent clustering methods may be realized in several ways, for example using the Conditional Fuzzy C-Means method (CFCM) [19], an extension of the standard Fuzzy C-Means (FCM) clustering algorithm [20]. CFCM clusters data under some external conditions defined for every training vector. These conditions are specified by an external variable z_k which is associated with each training vector $\mathbf{X}_k$, providing additional information that should be taken into account during clusterization. The strength of the clustering relations between the external variable z_k and the data vector $\mathbf{X}_k$ is defined by $f_k = \mu_A(z_k) \in [0, 1]$, where $\mu_A(z_k)$ may be interpreted as a membership function defined by particular context A. To find clusters that are near the decision border the external variable z is defined as an estimated distance between data vector $\mathbf{X}$ and the decision border.

FCM and CFCM are based on minimization of a cost function defined as:

$$J_m(\mathbf{U}, \mathbf{P}) = \sum_{i=1}^{K} \sum_{j=1}^{N} (u_{ji})^m \left\| \mathbf{X}_j - \mathbf{P}_i \right\|_A^2 \tag{6}$$

where K clusters are centered at $\mathbf{P}_i, i = 1..K$, $m > 1$ is a parameter, and the matrix $\mathbf{U} = (u_{ji})$ is a $K \times N$-dimensional membership matrix with elements $u_{ik} \in [0, 1]$ defining the degree of membership of the $\mathbf{X}_j$ vector in the i-th cluster. Conditional clustering is achieved by enforcing 3 conditions on the matrix $\mathbf{U}$:

1° Each vector $\mathbf{X}_j$ belongs to the i-th cluster to a degree that is between 0 to 1:

$$\bigvee_{1 \leq i \leq K} \bigvee_{1 \leq j \leq N} u_{ji} \in [0, 1] \tag{7}$$

2° Sum of the membership values of j-th vector $\mathbf{X}_j$ in all clusters is equal to f_j

$$\bigvee_{1 \leq j \leq N} \sum_{i=1}^{k} u_{ji} = f_j \tag{8}$$

3° Clusters must not be empty.

$$\bigvee_{1 \le i \le K} 0 < \sum_{j=1}^{N} u_{ji} < N \tag{9}$$

The cost function (6) is minimized under these conditions by placing centers at:

$$\bigvee_{1 \le i \le k} \mathbf{P}_i = \sum_{j=1}^{N} (u_{ji})^m x_j \left/ \sum_{j=1}^{N} (u_{ji})^m \right. \tag{10}$$

$$\bigvee_{\substack{1 \le i \le K \\ 1 \le j \le N}} u_{ji} = f_j \left/ \sum_{l=1}^{K} \left(\frac{\|\mathbf{X}_j - \mathbf{P}_i\|}{\|\mathbf{X}_j - \mathbf{P}_l\|} \right)^{2/(m-1)} \right. \tag{11}$$

4   Determining the Context

Conditional clustering has been used previously to generate P-rules [18]. Searching for optimal position of prototypes for P-rules is a difficult problem; moreover, a balance between the number of prototypes (simplicity of the rules) and the accuracy of the whole system is very important. Irrelevant prototypes, that is prototypes that do not have any influence on classification, should be removed. Such prototypes are usually far from decision borders. Therefore prototypes that lie close to vectors from one of the opposite classes should be preferred, as they should be close to the possible decision border.

To determine position of prototypes using conditional clustering algorithm an external coefficient z_k is defined, evaluating for a given vector $\mathbf{X}_k$ the ratio of a scatter between this vector and all vectors from the same class, divided by a scatter for all vectors from the remaining classes:

$$z_k = \sum_{j,\omega(\mathbf{X}_j)=\omega(\mathbf{X}_k)} \|\mathbf{X}_k - \mathbf{X}_j\|^2 \left/ \sum_{l,\omega(\mathbf{X}_l)\neq\omega(\mathbf{X}_k)} \|\mathbf{X}_k - \mathbf{X}_l\|^2 \right. \tag{12}$$

Here $\omega(\mathbf{X}_k)$ denotes class label of the vector $\mathbf{X}_k$. For quite inhomogeneous data summation may be restricted to local neighborhoods of the $\mathbf{X}_k$ vector. The z_k coefficients are then normalized to be in the range [0,1]:

$$z_k \Leftarrow \left(z_k - \min_k (z_k) \right) \left/ \left(\max_k (z_k) - \min_k (z_k) \right) \right. \tag{13}$$

Normalized z_k coefficients reach values close to 0 for vectors inside large homogenous clusters, and close to 1 if the vector z_k is near the vectors of the opposite classes and far from other vectors from the same class (for example, if it is an outlier). To avoid effects related to multiple density peaks in non-homogenous data distributions a cut-off point for distance of the order of standard deviation of the whole data could be introduced, although in our tests it was not necessary. These normalized coefficient determine the external variable which then is transformed into appropriate weights used as context or condition for CFCM clustering. Such weighting can be also understood as assigning appropriate linguistic term in the fuzzy sense. In the algorithm used below a Gaussian weighting was used:

$$f_k = \exp\left(\frac{z_k - \mu}{\sigma^2}\right) \tag{14}$$

with the best parameters in the range of μ =0.6–0.8 and σ =0.6–0.9, determined empirically for a wide range of datasets. The μ parameter controls where the prototypes will be placed; for small μ they are closer to the center of the cluster and for a larger μ closer to the decision borders. The range in which they are sought is determined by the σ parameter.

5   Experiments and Results

To verify the usefulness of proposed methodology for creating GBF networks several experiments have been performed, taking four rather different real-world benchmark datasets from the UCI repository [21]: Cleveland Heart Disease, Pima Indians Diabetes, Ionosphere and Appendicitis.

The primary goal of performed tests is to compare the effects of the context-based clustering with standard clustering algorithms on the initialization and performance of the GBF networks. The best results with the optimal number of neurons determined in the crossvalidation procedure are reported. Dispersions of the Gaussian neurons have been selected using a search procedure, and a search has also been made to find appropriate parameters for the context variable in the CFCM clustering case. In all tests the data has been normalized to keep each variable in the $[0, 1]$ range.

Generalization abilities of all methods applied to these datasets have been estimated using the 10-fold crossvalidation tests. Both accuracy of the model and its variance are collected in Table 5. Low variance shows good stability of the model, as some part of variance due to the partitioning of data by the crossvalidation procedure cannot be avoided. Stable models with low variance obtained in cross validation test should be preferred over more accurate but also more fluctuating ones. To select a good model a simple formula is used, minimizing estimation of error minus the standard deviation, as suggested in [22]; more sophisticated statistical tests give similar results.

In Table (5) results of GBF networks with FCM and CFCM initialization are compared with results of the SVM and P-rules. SVM calculations have been done using the *libsvm* implementation [23] of the algorithm with Gaussian kernels. C and σ parameters have been optimized using a grid search procedure to test performance of over 40 models for different values of these parameters. P-rules system (CFCM-LVQ type) is based on CFCM clustering followed by LVQ prototype optimization technique [18]. P-rules model has been designed to find the simplest possible data representation, therefore the number of selected prototypes is minimal. In addition it has build-in feature selection algorithm that reduces the complexity of each prototype, while for all other models feature selection hasn't been performed.

Table 1. Accuracy of the GBF network with k hidden neurons per class trained by FCM and CFCM algorithms, optimized SVM with Gaussian kernels and P-rules system

	GBF FCM		GBF CFCM		SVM		P-rules	
	acc±std	k	acc±std	k	acc±std	# SV	acc±std	# Proto
Appendicitis	10.55±3.50	3	10.55±3.75	3	11.46±3.57	30	16.23±5.80	2
Cleveland	16.12±2.50	5	15.13±2.39	5	16.79±1.98	132	17.2±2.15	2
Diabetes	22.14±1.56	55	21.88±1.54	45	21.75±1.52	429	23.04±1.17	2
Ionosphere	3.41±0.56	40	2.29±0.71	60	3.7±0.74	114	12.18±1.41	6

6 Discussion of Results and Further Perspectives

The context-dependent initialization of the localized basis set expansion networks proposed in this paper is a simple modification of the commonly used fuzzy C-means method. Experiments with GBF networks showed for all datasets and for all number of hidden neurons that the context-dependent clustering is more robust then the standard clustering algorithm used to initialize and train neural networks. The same approach is applicable to all basis set expansion networks that use localized functions with clearly defined centers. Surprisingly good results have been obtained in comparison to the SVM algorithm that is also based on Gaussian kernels, and is usually one of the best classifiers. The CFCM GBF networks proved to be not only more accurate, but they also provided models of low complexity. Much higher number of support vectors are selected by SVM, and for Gaussian kernels they are not always near the decision border. Although sparse SVM versions may be used for data understanding [24] results are not as good, and models are not so simple, as those obtained from the GBF networks trained with CFCM. The datasets analyzed in the previous section have been rather diverse, but all of them are relatively simple; this choice was intentional, as the main purpose was to check if simple data models with explanatory power may be generated. More complex data may not admit such simple models and thus there may be no advantage over the black box predictors.

Performance of all methods that calculate distances, includes clustering algorithms, are strongly affected by noise in high-dimensional problems. To avoid it feature selection or feature aggregation methods may be a necessary preliminary step. In contrast to SVM and most other kernel methods, basis expansion networks with full optimization of individual node parameters are capable of multi-resolution representation of probabilistic data distributions, providing complex shapes of decision border in selected areas of the input space where they are really needed, and simple borders in other areas. Small number of neurons has been sufficient for Appendicitis and Cleveland Heart datasets, but for the Ionosphere and Diabetes quite large number of neurons is needed. In case of Diabetes P-rules with two prototypes give only slightly worse result than 45 prototypes, thus the large number of additional prototypes lead to only a small correction of the decision borders. Tests are being conducted on more complex datasets to see how well CFCM training of GBF and other LBF networks will compare with other methods.

As suggested in Sec. 2 LBF networks may be used not only as black boxes, but also as tools for "intuitive" understanding of data structures, using similarity to justify the reasons for decisions made by the model. Separable Basis Function (SBF) networks may be interpreted as equivalent fuzzy system if the full covariance matrix mixing different features is not used. Such restriction does not occur in systems that extract prototype-based rules (P-rules), where the explanatory power of the model comes from the analysis of a set of characteristic examples taken as prototypes, and optimization of individual similarity measures [18,25,5,16]. The goal is to find minimal set of reference cases that allow for good generalization. P-rule systems may make decisions in several ways, but most often the single nearest prototype rule is used. In RBF, LBF or SBF networks the model is more complex, using instead of the $\max(\cdot)$ aggregation function a weighted sum of contributions from all hidden nodes. Such model is more flexible, but in effect more appropriate for approximation problems than for data understanding. Using appropriate normalization SBF networks may still be viewed as P-rule systems of the Takagi-Sugeno fuzzy model type.

In the CFCM approach to the LBF network initialization and training, all prototypes lie in the vicinity of decision borders. Compared to the standard training approach this leads to a reduction of unnecessary prototypes, which is highly desirable from the point of view of comprehensibility. However, model selection for LBF has to be done carefully, to avoid placing many prototypes close to the decision borders in cases when one prototype far from the border would be sufficient. This is not a big problem in P-systems with the nearest neighbor decision rule, but in the LBF networks the size of the support of localized functions has to be explicitly optimized. One drawback of the CFCM algorithm is the need to determine correct mean and sigma values in Eq. 14. Although default parameters work in most cases quite well, for some data a grid search for optimal values may be important. This may lead to an increase of computational complexity that could become a problem for a very large datasets, although for most problems learning is quite fast.

References

1. Bishop, C.M.: Pattern Recognition and Machine Learning. Springer, Heidelberg (2006)
2. Duch, W., Jankowski, N.: Survey of neural transfer functions. Neural Computing Surveys 2, 163–213 (1999)
3. Duch, W., Diercksen, G.H.F.: Feature space mapping as a universal adaptive system. Computer Physics Communications 87, 341–371 (1995)
4. Levine, I.: Quantum Chemistry, 5th edn. Prentice-Hall, Englewood Cliffs (1999)
5. Duch, W., Blachnik, M.: Fuzzy rule-based systems derived from similarity to prototypes. In: Pal, N.R., Kasabov, N., Mudi, R.K., Pal, S., Parui, S.K. (eds.) ICONIP 2004. LNCS, vol. 3316, pp. 912–917. Springer, Heidelberg (2004)
6. Pothos, M.E.: The rules versus similarity distinction. Behavioral and Brain Sciences 28, 1–49 (2005)
7. Schölkopf, B., Smola, A.: Learning with Kernels. Support Vector Machines, Regularization, Optimization, and Beyond. MIT Press, Cambridge (2001)
8. Jankowski, N.: Approximation and classification in medicine with incnet neural networks. In: Machine Learning and Applications, Workshop on Machine Learning in Medical Applications, Greece, pp. 53–58 (July 1999)
9. Adamczak, R., Duch, W., Jankowski, N.: New developments in the feature space mapping model. In: Third Conference on Neural Networks and Their Applications, Kule, Poland, pp. 65–70 (October 1997)

10. Webb, A.: Statistical Pattern Recognition. J. Wiley & Sons, Chichester (2002)
11. Duch, W.: Support vector neural training. In: Duch, W., Kacprzyk, J., Oja, E., Zadrożny, S. (eds.) ICANN 2005. LNCS, vol. 3697, pp. 67–72. Springer, Heidelberg (2005)
12. Blachnik, M., Duch, W.: Prototype rules from SVM. Springer Studies in Computational Intelligence, vol. 80, pp. 163–184. Springer, Heidelberg (2008)
13. Schwenker, F., Kestler, H., Palm, G.: Three learning phases for radial-basis-function networks. Neural Networks 14, 439–458 (2001)
14. Jang, J.S.R., Sun, C.: Functional equivalence between radial basis function neural networks and fuzzy inference systems. IEEE Transactions on Neural Networks 4, 156–158 (1993)
15. Kuncheva, L.: On the equivalence between fuzzy and statistical classifiers. International Journal of Uncertainty, Fuzziness and Knowledge-Based Systems 15, 245–253 (1996)
16. Duch, W., Grudziński, K.: Prototype based rules - new way to understand the data. In: IEEE International Joint Conference on Neural Networks, pp. 1858–1863. IEEE Press, Washington (2001)
17. Kohonen, T.: Self-organizing maps. Springer, Heidelberg (1995)
18. Blachnik, M., Duch, W., Wieczorek, T.: Selection of prototypes rules: context searching via clustering. In: Rutkowski, L., Tadeusiewicz, R., Zadeh, L.A., Żurada, J.M. (eds.) ICAISC 2006. LNCS (LNAI), vol. 4029, pp. 573–582. Springer, Heidelberg (2006)
19. Pedrycz, W.: Conditional fuzzy c-means. Pattern Recognition Letters 17, 625–632 (1996)
20. Hoppner, F., Klawonn, F., Kruse, R., Runkler, T.: Fuzzy Cluster Analysis. Wiley, Chichester (1999)
21. Merz, C., Murphy, P.: UCI repository of machine learning databases (1998-2004),
    `http://www.ics.uci.edu/~mlearn/MLRepository.html`
22. Jankowski, N., Grąbczewski, K.: Handwritten digit recognition – road to contest victory. In: IEEE Symposium on Computational Intelligence in Data Mining, pp. 491–498. IEEE Press, Los Alamitos (2007)
23. Lin, K., Lin, C.: A study on reduced support vector machines. IEEE Transactions on Neural Networks 14(6), 1449–1459 (2003)
24. Diederich, J. (ed.): Rule Extraction from Support Vector Machines. Springer Studies in Computational Intelligence, vol. 80. Springer, Heidelberg (2008)
25. Blachnik, M., Duch, W., Wieczorek, T.: Probabilistic distance measures for prototype-based rules. In: Proc. of the 12th Int. Conference on Neural Information Processing (ICONIP 2005), pp. 445–450. Taipei University Press, Taiwan (2005)

Adaptation of Connectionist Weighted Fuzzy Logic Programs with Kripke-Kleene Semantics

Alexandros Chortaras*, Giorgos Stamou, Andreas Stafylopatis, and Stefanos Kollias

School of Electrical and Computer Engineering
National Technical University of Athens
Zografou 157 80, Athens, Greece
{achort, gstam}@softlab.ntua.gr, {andreas,stefanos}@cs.ntua.gr

Abstract. Weighted fuzzy logic programs extend the expressiveness of fuzzy logic programs by allowing the association of a different significance weight with each atom that appears in the body of a fuzzy rule. The semantics and a connectionist representation of these programs have already been studied in the absence of negation; in this paper we first propose a Kripke-Kleene based semantics for the programs which allows for the use of negation as failure. Taking advantage of the increased modelling capabilities of the extended programs, we then describe their connectionist representation and study the problem of adapting the rule weights in order to fit a provided dataset. The adaptation algorithm we develop is based on the subgradient descent method and hence is appropriate to be employed as a learning algorithm for the training of the connectionist representation of the programs.

1 Introduction

The handling of uncertainty is an important requirement for logic programming languages because in several applications it is difficult to precisely determine the truth value of facts so as to use the expressivity of logic programming. Fuzzy logic has been suggested as a way to extend logic programming frameworks and provide them with the ability to reason with uncertain knowledge (e.g. [12], [9], [4]). In this context, *weighted fuzzy logic programs* (wflps) [1], [2] are a useful extension of such frameworks: by introducing additional atom weights, they allow each atom in a rule body to have a different importance in inferencing the head.

On the other hand, neural networks and connectionist models are architectures characterized by learning capabilities and fault tolerance. Therefore, the possibility to use them to represent logic programs (e.g. [7], [10]), perform symbolic reasoning and extract symbolic rules out of them is of considerable interest. When the rules include weights, a connectionist representation has a significant advantage: the connectionist model may be trained in order to learn the weights

---

* A. Chortaras is on a scholarship from the Alexander S. Onassis Public Benefit Foundation.

V. Kůrková et al. (Eds.): ICANN 2008, Part I, LNCS 5163, pp. 492–502, 2008.
© Springer-Verlag Berlin Heidelberg 2008

that make the rules fit best the data, so that the problem of extracting symbolic rules from the model is reduced to adapting the weights of the logic program.

A drawback of most of the suggested fuzzy logic programming languages is however that they do not allow the use of negation, which is essential if such languages are to be used with real applications. There are several alternative ways in which negation may be introduced; in this paper we extend wflps with negation as failure adopting the Kriple-Kleene semantics. This has the advantage that it allows a monotonous immediate consequence operator to be defined, a fact that allows us subsequently to effectively tackle the rule weight adaptation problem. Using an extension of the connectionist representation of definite wflps [2], we restate the weight adaptation problem [1] for the case of the Kriple-Kleene semantics and develop the details of a subgradient descent-based algorithm to be used for the training of the connectionist model.

In the rest of the paper, section 2 introduces the Kripke-Kleene semantics for wflps, section 3 describes their connectionist representation, section 4 defines the weight adaptation problem and algorithm and section 5 concludes the work.

2   Weighted Fuzzy Logic Programs with Negation

2.1   Syntax

A *fuzzy atom* A is the formula $p(u_1, \ldots, u_n)$ where p is a predicate and each u_i a variable or constant. The special atoms t, f, u and i represent absolute truth, falsehood, unknownness and inconsistency respectively. A *fuzzy literal* is a fuzzy atom A or its negation $\sim A$, which is interpreted as negation as failure.

Definition 1. *A* weighted fuzzy logic program $\mathcal{P}$ *is a finite set of* weighted fuzzy rules *that are clauses of the form*

$$w : B \leftarrow (w_1; L_1), \ldots, (w_n; L_n),$$

where $L_1, \ldots, L_n$ *are fuzzy literals and* B *is a fuzzy atom excluding* t, f, u *and* i, *such that any variable in* B *appears also in at least one* L_i. *The weight* $w \in [0, 1]$ *is the strength of the rule and all the other weights* w_i *are also in* $[0, 1]$.

For a rule R we write also $s(R)$ instead of w and $\boldsymbol{w}(R)$ instead of $(w_1, \ldots, w_n)$. An atom (rule) that contains only constants is *ground* and is denoted by $\underline{A}$ $(\underline{R})$. A ground rule obtained from a rule R by grounding its variables is an *instance* of R. A rule whose body includes only t, f, u and/or i is a *fuzzy fact*.

2.2   Kripke-Kleene Semantics

In the presence of negation, a useful interpretation of fuzzy logic programs can be obtained by mapping each ground atom to a truth interval $[v_1, v_2] \in \mathcal{B}$ with $\mathcal{B} \equiv [0, 1] \times [0, 1]$, instead of to a single truth value in $[0, 1]$. This corresponds to knowing only the interval within which the exact truth value of an atom lies. The interpretation of fuzzy logic programs in this way is based on bilattices [6]. A *bilattice* on $\mathcal{B}$ is a structure $(\mathcal{B}, \preceq_t, \preceq_k)$, where $\preceq_t$ is a *truth* and $\preceq_k$ a *knowledge* ordering. For two intervals $[a_i, b_i] \in \mathcal{B}$ the two orderings have as follows: $[a_1, b_1] \preceq_t [a_2, b_2]$ iff $a_1 \le a_2$ and $b_1 \le b_2$, and $[a_1, b_1] \preceq_k [a_2, b_2]$ iff $a_1 \le a_2$ and $b_1 \ge b_2$. The least and greatest elements of $\mathcal{B}$ w.r.t. $\preceq_t$ are $\perp_t = [0, 0]$ (false) and $\top_t = [1, 1]$ (true), and w.r.t $\preceq_k$ they are $\perp_k = [0, 1]$ (unknown) and $\top_k = [1, 0]$ (inconsistent) respectively. The join and meet operators with respect to $\preceq_t$ are $[a_1, b_1] \vee [a_2, b_2] = [\max\{a_1, a_2\}, \max\{b_1, b_2\}]$ and $[a_1, b_1] \wedge [a_2, b_2] = [\min\{a_1, a_2\}, \min\{b_1, b_2\}]$, and with respect to $\preceq_k$ they are $[a_1, b_1] \oplus [a_2, b_2] = [\max\{a_1, a_2\}, \min\{b_1, b_2\}]$ and $[a_1, b_1] \otimes [a_2, b_2] = [\min\{a_1, a_2\}, \max\{b_1, b_2\}]$ respectively. We also have that $\sim [a, b] = [1 - b, 1 - a]$ and any function f on $[0, 1]^n$ is extended on $\mathcal{B}^n$ by defining $f([a_1, b_1], \dots, [a_n, b_n]) = [f(a_1, \dots, a_n), f(b_1, \dots, b_n)]$.

Let $\mathcal{P}$ be a wflp, $P_\mathcal{P}$ the set of all the predicates that appear in $\mathcal{P}$ excluding t, f, u and i, $V_\mathcal{P}$ a superset of the set of constants that appear in $\mathcal{P}$, and $B_\mathcal{P} = B_\mathcal{P}(V_\mathcal{P})$ the set of all the ground atoms that can be constructed from the predicates in $P_\mathcal{P}$ and the constants in $V_\mathcal{P}$. Given $B_\mathcal{P}$, the *inference base* $R_{\underline{B}}$ of a ground atom $\underline{B}$ is the set of all the instances of the rules of $\mathcal{P}$ that consist of atoms from $B_\mathcal{P}$ and have $\underline{B}$ in their head. The *explicit base* $EB_\mathcal{P}$ of $\mathcal{P}$ is the set of all the atoms in $B_\mathcal{P}$ whose predicate appears only in the body of a rule in $\mathcal{P}$.

Definition 2. *An* interpretation I *with domain* $V_\mathcal{P}$ *of the wflp* $\mathcal{P}$ *is a mapping* $B_\mathcal{P} \to \mathcal{B}$. *An interpretation maps always* t *to* $\top_t$, f *to* $\perp_t$, u *to* $\perp_k$ *and* i *to* $\top_k$.

We denote the interval to which I maps $\underline{A}$ by $\underline{A}^I \equiv [\underline{A}^{bI}, \underline{A}^{tI}]$. The value of the body of rule $\underline{R} \equiv w : \underline{B} \leftarrow (w_1; \underline{L}_1), \dots, (w_n; \underline{L}_n)$ is given by a *weighted fuzzy conjunction operator* (wfco) [1], denoted by $\widetilde{\wedge}_{[\cdot]}$. In particular, the body value is $[\widetilde{\wedge}_{[w_1, \dots, w_n]}(C_1, \dots, C_n), \widetilde{\wedge}_{[w_1, \dots, w_n]}(D_1, \dots, D_n)]$, where $C_i = \underline{A}_i^{bI}$ and $D_i = \underline{A}_i^{tI}$ if $\underline{L} \equiv \underline{A}_i$, or $C_i = 1 - \underline{A}_i^{tI}$ and $D_i = 1 - \underline{A}_i^{bI}$ if $\underline{L} \equiv \sim\underline{A}_i$ for any atom $\underline{A}_i$. For the atom values in the body of $\underline{R}$ we write also $\boldsymbol{a}^I(\underline{R})$ instead of $(\underline{L}_1^I, \dots, \underline{L}_n^I)$. $\widetilde{\wedge}_{[\cdot]}$ is non-decreasing in its arguments a_i and an example is the operator

$$\widetilde{\wedge}_{[w_1, \dots, w_n]}^{std}(a_1, \dots, a_n) = \min_{i=1 \dots n} \max\{\bar{w} - w_i, \min\{\bar{w}, a_i\}\}, \text{ where } \bar{w} = \max_{i=1 \dots n} w_i.$$

Definition 3. *Given a wfco* $\widetilde{\wedge}_{[\cdot]}$, *a* t-*norm* T *and an* s-*norm* S, *an interpretation* I *with domain* $V_\mathcal{P}$ *is a* model *of the wflp* $\mathcal{P}$ *under* $(\widetilde{\wedge}_{[\cdot]}, T, S)$, *if*

$$S\left(\left\{T\left(\widetilde{\wedge}_{[\boldsymbol{w}(R)]}\left(\boldsymbol{a}^I(\underline{R})\right), s(R)\right)\right\}_{\underline{R} \in R_{\underline{B}}}\right) \preceq_t \underline{B}^I \tag{1}$$

for all $\underline{B} \in B_\mathcal{P}$ *such that* $R_{\underline{B}} \ne \oslash$ *and* $\perp_k \preceq_t \underline{B}^I$ *if* $R_{\underline{B}} = \oslash$.

The *immediate consequence* of I is the interpretation $F\Phi_\mathcal{P}(I)$, defined as $\underline{B}^{F\Phi_\mathcal{P}(I)} = S\left(\left\{T(\widetilde{\wedge}_{[\boldsymbol{w}(R)]}(\boldsymbol{a}^I(\underline{R})), s(R))\right\}_{\underline{R} \in R_{\underline{B}}}\right)$ if $R_{\underline{B}} \ne \oslash$ and $\underline{B}^{F\Phi_\mathcal{P}(I)} = \perp_k$ otherwise. If $\mathcal{I}_\mathcal{P}$ is the set of all the interpretations of $\mathcal{P}$, $F\Phi_\mathcal{P} : \mathcal{I}_\mathcal{P} \to \mathcal{I}_\mathcal{P}$ is the *immediate consequence operator*. It is monotonic in the lattice $(\mathcal{I}_\mathcal{P}, \preceq_k)$ and continuous if $\widetilde{\wedge}_{[\cdot]}$, T, S are continuous. Then, according to fixpoint theory it has a least fixpoint w.r.t. $\preceq_k$, which if $F\Phi_{\mathcal{P}}$ is continuous, is equal to $F\Phi_{\mathcal{P}}^{\uparrow\omega}$, where

$$F\Phi_{\mathcal{P}}^{\uparrow k} = \begin{cases} I_0 & \text{if } k = 0 \\ F\Phi_{\mathcal{P}}(F\Phi_{\mathcal{P}}^{\uparrow k-1}) & \text{if } k \text{ is a succesor ordinal} \\ \bigoplus_{l<k} F\Phi_{\mathcal{P}}^{\uparrow l} & \text{if } k \text{ is a limit ordinal } \omega, \end{cases} \tag{2}$$

and I_0 maps all atoms in $B_{\mathcal{P}}$ to $\perp_k$. It follows that $\mathcal{P}$ has a unique minimal model $FM_{\mathcal{P}} = F\Phi_{\mathcal{P}}^{\uparrow\omega}$ w.r.t. $\preceq_k$ under $(\widetilde{\wedge}_{[\cdot]}, T, S)$, which we define as the Kriple-Kleene model, or else the *intended meaning* of $\mathcal{P}$. In general, it may be impossible to reach $F\Phi_{\mathcal{P}}^{\uparrow\omega}$ in less than ω steps [3]. To overcome this difficulty we use a *fixpoint approximation*, which we quantify using a metric space on $\mathcal{I}_{\mathcal{P}}$, e.g. the function $\rho(I, J) = \max_{\underline{B} \in B_{\mathcal{P}}} d_{inf}(\underline{B}^I, \underline{B}^J)$ with $d_{inf}([a_1, b_1], [a_2, b_2]) = \max\{|a_1 - a_2|, |b_1 - b_2|\}$. Given such a metric it can be proved that if $F\Phi_{\mathcal{P}}^{\uparrow\omega}$ is unreachable in finite time, an approximation of it at any level of accuracy $\epsilon > 0$ may be computed in $n_0 < \omega$ iterations, i.e. $\rho(FM_{\mathcal{P}}, F\Phi_{\mathcal{P}}^{\uparrow n}) < \epsilon$ for all $n > n_0$.

3   Connectionist Representation

A connectionist model for the representation of a wflp $\mathcal{P}$ in the absence of negation was introduced in [2]. The proposed connectionist model has the benefit that its structure reflects the structure of $\mathcal{P}$ in a way that it has a well-defined symbolic interpretation. Here, we provide a rough descriptive account of the structure and the operation of the connectionist model, extended so that it works with interval instead of with point values.

Because $\mathcal{P}$ may in general not be propositional, the links of the connectionist model carry *complex named intervals*, i.e. multisets of pairs $(\boldsymbol{c}, [v_1, v_2])$, where $\boldsymbol{c}$ is a vector of constants (the arguments of a ground atom) and $[v_1, v_2]$ an interval in $\mathcal{B}$. The model consists of an input, a conjunction and a disjunction layer.

The conjunction layer consists of a set of *conjunctive neurons* N_c, one for each non-fact rule in $\mathcal{P}$. Each conjunctive neuron i has so many inputs as are the atoms of the body of the rule it represents, and each one of its inputs (which is connected to a disjunctive neuron j) is characterized by the weight w_{ij} of the respective atom. The neuron computes the value of the rule head by appropriately grounding the variables and combining the complex named intervals that appear in its inputs.

The disjunction layer consists of a set of *disjunctive neurons* N_d, one for each predicate in $P_{\mathcal{P}}$. Their role is to combine into one the possibly many intervals computed for the same ground atom by the conjunction layer neurons. Each input of a disjunctive neuron i is connected to the output of a conjunctive neuron j which computes a rule R whose head involves the predicate that the disjunctive neuron i represents. The weight w_{ij} of the input of i connected to j is the strength of R. Each disjunctive neuron has an additional input with weight 1, connected to a node of the input layer, which feeds into the connectionist model the information related with the facts of the program.

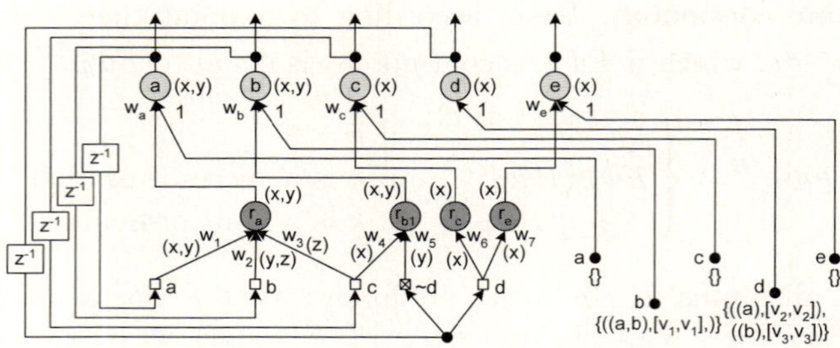

Fig. 1. The connectionist representation of the program of example 1

The connectionist model is recursive, and the output of disjunctive neuron i is connected through a unit delay node to an input of the (possibly many) conjunctive neurons j that represent the rules of the program that involve in their body the predicates that the disjunctive neuron i computes. If an atom in a rule body is negated, between the unit delay node and the conjunctive neuron a negation node that performs the interval negation operation is interpolated. The output of the model is the set of the complex named intervals that appear at the output of the disjunctive neurons, each characterized by the respective predicate name: the output at time point t is the set $S(t) = \{(p_1, out_1(t)), \ldots, (p_k, out_k(t))\}$, where p_i is the predicate that disjunctive neuron i represents and out_i its output.

By construction, the connectionist model of a wflp $\mathcal{P}$ is a connectionist implementation of $F\Phi_{\mathcal{P}}$, so that at time point t_i its output encodes interpretation $F\Phi_{\mathcal{P}}{}^{\uparrow i}$. Thus, the connectionist model computes $FM_{\mathcal{P}}$ at the same number of iterations that $F\Phi_{\mathcal{P}}$ needs to reach (an approximation of) its least fixpoint.

Example 1. The connectionist model that corresponds to the following program (the rules are given on the left and the facts on the right) is illustrated in Fig. 1.

$$w_a : a(x,y) \leftarrow (w_1; a(x,y)), (w_2; b(y,z)), (w_3, c(z)) \qquad v_1 : b(a,b) \leftarrow (1;t)$$
$$w_b : b(x,y) \leftarrow (w_4; c(x)), (w_5; \sim d(y)) \qquad v_2 : d(a) \leftarrow (1;t)$$
$$w_c : c(x) \leftarrow (w_6; d(x)) \qquad v_3 : d(b) \leftarrow (1;t)$$
$$w_e : e(y) \leftarrow (w_7; d(y))$$

4   Weight Adaptation

In presenting wflps we have assumed that the weights in the rules are known. If however we are given an appropriate dataset, this may be considered as an interpretation I of an unknown wflp which we want to learn so as to obtain a description, in the form of fuzzy rules, of the relations present in the dataset. Using some a priori knowledge we can build a logic theory to model the dataset in the form of a parametric wflp and leave the weights as parameters to be adapted based on the actual data, which should be in the form of a set of fuzzy facts $\mathcal{F}$. Due to the absence of non-fact rules in $\mathcal{F}$, the model of $\mathcal{F}$ is $F\Phi_{\mathcal{F}}^{\uparrow 1}$ and this is the interpretation I, on which the parametric wflp should be trained.

4.1   Weight Adaptation Problem

Following [1], a *parametric weighted fuzzy logic program* $\mathcal{P}(\boldsymbol{w})$ is a wflp in which (some of) the weights of its rules are unknown parameters. For some fixed ordering, we concatenate all the unknown weights in a parameter vector $\boldsymbol{w}$, whose size we denote by q. If w is (the name of) a weight parameter in $\boldsymbol{w}$, we denote its position in $\boldsymbol{w}$ by $q\langle w \rangle$. The weight adaptation problem has then as follows: Given a parametric wlfp $\mathcal{P}(\boldsymbol{w})$, an interpretation I for $\mathcal{P}(\boldsymbol{w})$ with domain $V_\mathcal{P}$, and a triple of operators $(\widetilde{\wedge}_{[\cdot]}, T, S)$, find an 'optimal' vector of weights $\hat{\boldsymbol{w}} \in [0,1]^q$ such that I is a model of $\mathcal{P}(\hat{\boldsymbol{w}})$ under $(\widetilde{\wedge}_{[\cdot]}, T, S)$. As we already mentioned, interpretation I can be considered to be $F\Phi_{\mathcal{F}}^{\uparrow 1}$ for an appropriate set of facts $\mathcal{F}$.

From definition 3, we know that an interpretation I with domain $V_\mathcal{P}$ is a model of $\mathcal{P}$ under $(\widetilde{\wedge}_{[\cdot]}, T, S)$, if inequality (1) holds for all ground atoms $\underline{B} \in B_\mathcal{P}$. Thus, if we consider a parametric program $\mathcal{P}(\boldsymbol{w})$ and write the corresponding set of inequalities letting the weights be unknown variables, we obtain a system $\Sigma(\boldsymbol{w})$ of $|B_\mathcal{P}|$ inequalities and q unknown weights: the vector $\boldsymbol{w}$. Then for some vector $\hat{\boldsymbol{w}} \in [0,1]^q$, I is a model of $\mathcal{P}(\hat{\boldsymbol{w}})$ if $\hat{\boldsymbol{w}}$ satisfies $\Sigma(\boldsymbol{w})$. In general, $\Sigma(\boldsymbol{w})$ has an infinite number of solutions. To choose a preferred one, we recourse to an error minimization criterion using some interval distance function d. We assume also that $\mathcal{P}(\boldsymbol{w})$ contains no facts and provide the following definition:

Definition 4. *The weight vector* $\boldsymbol{w}^*$ *is an* optimal fit *under* $(\widetilde{\wedge}_{[\cdot]}, T, S)$ *of the set of facts* $\mathcal{F}$ *with domain* $V_\mathcal{P}$ *to the parametric wflp* $\mathcal{P}(\boldsymbol{w})$, *if* $I = F\Phi_{\mathcal{F}}^{\uparrow 1}$ *is a model of* $\mathcal{P}(\boldsymbol{w}^*)$ *under* $(\widetilde{\wedge}_{[\cdot]}, T, S)$ *and it is also a minimizer of the function*

$$E(\boldsymbol{w}) = \frac{1}{2} \sum_{\underline{B} \in B_\mathcal{P}} \left(d\left(\underline{B}^I, \underline{B}^{M(\boldsymbol{w})} \right) \right)^2,$$

subject to $\boldsymbol{0} \leq \boldsymbol{w} \leq \boldsymbol{1}$, *where* $M(\boldsymbol{w}) = F\Phi_{\mathcal{P}'(\boldsymbol{w})}^{\uparrow \omega}$, $\mathcal{P}'(\boldsymbol{w}) = \mathcal{P}(\boldsymbol{w}) \cup \mathcal{P}_F$ *and* $\mathcal{P}_F$ *is the facts only program*

$$\mathcal{P}_F = \left\{ \underline{B}^{bI} : \underline{B} \leftarrow (1; i), \ \underline{B}^{tI} : \underline{B} \leftarrow (1; u) \mid \underline{B}^I \neq \perp_k \text{ and } \underline{B} \in EB_{\mathcal{P}(\boldsymbol{w})} \right\}.$$

Several interval distance functions d may be used in the above, e.g.

$$d_{euc}([a_1, b_1], [a_2, b_2]) = \sqrt{(a_1 - a_2)^2 + (b_1 - b_2)^2}, \tag{3}$$

$$d_{mid}([a_1, b_1], [a_2, b_2]) = \left| \frac{a_1 + b_1}{2} - \frac{a_2 + b_2}{2} \right|. \tag{4}$$

The weight adaptation problem may then be restated as finding an optimal fit of I to $\mathcal{P}(\boldsymbol{w})$, i.e. an optimizer $\boldsymbol{w}^*$ of the following problem $O(\boldsymbol{w})$:

$$\begin{aligned} &\text{minimize} \quad E(\boldsymbol{w}) \\ &\text{subject to} \quad \Sigma(\boldsymbol{w}) \text{ and } \boldsymbol{0} \leq \boldsymbol{w} \leq \boldsymbol{1}. \end{aligned} \tag{5}$$

As discussed in [1], because the satisfaction of the hard constraints $\Sigma(\boldsymbol{w})$ may in the case of noisy data lead to poor solutions, we may have to drop them to avoid obtaining poor solutions. We call the resulting simpler problem $O'(\boldsymbol{w})$, and in the sequel we will focus on it. We have to note however that in this case I will not in general be any more a model of $\mathcal{P}(\boldsymbol{w}^*)$ for an optimizer $\boldsymbol{w}^*$ or $O'(\boldsymbol{w})$.

In the case of wflps without negation it was proved in [1] that finding an optimizer of $O(\boldsymbol{w})$ is equivalent to independently finding optimizers for $|P_{\mathcal{P}}|$ smaller optimization problems, one for each $p \in P_{\mathcal{P}}$, whose goal function is the part of $E(\boldsymbol{w})$ that involves the ground atoms of predicate p. For interval valued wlfps this does not hold for any useful interval distance function and the entire optimization problem has to be directly considered.

The problem $O'(\boldsymbol{w})$ is in general non-convex and obtaining a global optimizer for it is computationally hard. In the following, we develop instead a gradient descent local optimization method to solve it, which can be used as a learning algorithm for the connectionist model described in section 3.

4.2   Adaptation Algorithm

An essential property of $E(\boldsymbol{w})$ is that in general it involves non-smooth functions. The local optimization of non-smooth functions relies on the use of subdifferentials and subgradients, which are generalizations of the gradient of a differentiable function. The *subdifferential* of an almost everywhere differentiable function $f : \mathbb{R}^n \to \mathbb{R}$ at $\boldsymbol{x}$ is defined as the convex hull $\partial f(\boldsymbol{x}) = \mathbf{conv}\{\lim_{\boldsymbol{y} \to \boldsymbol{x}} \nabla f(\boldsymbol{y}) \mid f$ is differentiable at $\boldsymbol{y}\}$, and any $\boldsymbol{\xi} \in \partial f(\boldsymbol{x})$ is a *subgradient* of f at $\boldsymbol{x}$.

Computing the subdifferential of an arbitrary function is in general not easy. However the following subgradient calculus rules [8] are helpful for some useful cases: 1) if f is differentiable at $\boldsymbol{x}$ then $\partial f(\boldsymbol{x}) = \{\nabla f(\boldsymbol{x})\}$; 2) $\partial f(\boldsymbol{x}) = -\partial(-f(\boldsymbol{x}))$; 3) if f_i are locally Lipschitz and subdifferentially regular (see [8]) then a)$\partial(f_1 + f_2)(\boldsymbol{x}) = \partial f_1(\boldsymbol{x}) + \partial f_2(\boldsymbol{x})$, b) if $f_1(\boldsymbol{x}), f_2(\boldsymbol{x}) \geq 0$ then $\partial(f_1 f_2)(\boldsymbol{x}) = f_2(\boldsymbol{x})\partial f_1(\boldsymbol{x}) + f_1(\boldsymbol{x})\partial f_2(\boldsymbol{x})$, c) $\partial \max_{i=1\ldots n} f_i(\boldsymbol{x}) = \mathbf{conv}\{\partial f_j(\boldsymbol{x}) \mid f_j(\boldsymbol{x}) = \max_{i=1\ldots n} f_i(\boldsymbol{x})\}$. The dual of the last equation holds for the min function.

If f is also convex, a minimizer $\boldsymbol{x}^*$ of it under the constraint that $\boldsymbol{x} \in D \subset \mathbb{R}^n$, can be obtained by applying the *projection subgradient descent* method [11], which given an initial point $\boldsymbol{x}^0 \in D$, is defined by the iteration

$$\boldsymbol{x}^{k+1} \leftarrow \Pi_D \left(\boldsymbol{x}^k - \alpha_k \frac{\boldsymbol{\xi}^k}{\|\boldsymbol{\xi}^k\|} \right) \quad \text{with} \quad \Pi_D(\boldsymbol{z}) = \arg\min_{\boldsymbol{x} \in D} \|\boldsymbol{z} - \boldsymbol{x}\|, \qquad (6)$$

where $\boldsymbol{\xi}^k$ is an arbitrary element of $\partial f(\boldsymbol{x}^k)$ and α_k a step factor. For convenience, in the following we denote an arbitrary element of $\partial f(\boldsymbol{x})$ by $[\partial f(\boldsymbol{x})]$. Because $-[\partial f(\boldsymbol{x})]$ is in fact not always a descent direction, a valid requirement for the step factor so that $\{f(\boldsymbol{x}^i)\}_{i=1}^{+\infty}$ converges to $f(\boldsymbol{x}^*)$ is that $\lim_{k \to +\infty} \alpha_k = 0$ and $\sum_{k=1}^{+\infty} \alpha_k = +\infty$.

Since $E(\boldsymbol{w})$ is non-smooth, the weight adaptation algorithm we propose consists in using the above subgradient descent iterative process as the learning rule of the proposed connectionist model in order to minimize $E(\boldsymbol{w})$. $E(\boldsymbol{w})$ is non-convex, but we follow here the standard practice in the field of optimization to apply methods developed for convex to non-convex functions [5]. The result is that the algorithm may get trapped into a local minimum or stationary point.

Because the semantics of $\mathcal{P}$ may be non-computable, we approximate also, for an appropriate n_0, the error $E(\boldsymbol{w})$ by the error function $\tilde{E}(\boldsymbol{w})$:

$$\tilde{E}(\boldsymbol{w}) = \frac{1}{2} \sum_{\underline{B} \in B_\mathcal{P}} \left(d \left(\underline{B}^I, \underline{B}^{F\Phi_{\mathcal{P}'(w)}^{\uparrow n_0}} \right) \right)^2 . \tag{7}$$

We can apply (6) to minimize (7) using as D the hypercube $[0,1]^q$ and as step factor (learning rate) an appropriate sequence, e.g. a sequence with $\alpha_0 < 1$ and $\alpha^k = \frac{\alpha_0}{\sqrt{j+1}}$, where j is the number of times that (6) did not lead to a decrease of the error function up to step k.

Going back to the connectionist model, we note that if $n_0 > 1$ it works recurrently. Given this remark, what remains to be computed is a subgradient of $\tilde{E}(\boldsymbol{w})$, which depends on the used distance function and the operators $(\widetilde{\wedge}_{[\cdot]}, T, S)$. In order to compute a subgradient, we can rewrite the subdifferential calculus properties by substituting $\partial f(\boldsymbol{x})$ with $[\partial f(\boldsymbol{x})]$. In the case of the max function (and similarly for the min), following the definition of the convex hull, we have $[\partial \max_{i=1...n} f_i(\boldsymbol{x})] = \sum_{i=1}^n \lambda_i [\partial f_i(\boldsymbol{x})]$, where by letting $R = \{j \mid f_j(\boldsymbol{x}) = \max_{i=1...n} f_i(\boldsymbol{x})\}$, the λ_is are such that $\sum_{i \in R} \lambda_i = 1$ (e.g. $\lambda_i = \frac{1}{|R|}$) and $\lambda_i = 0$ if $i \notin R$. We write also f^2 as $\max\{f, -f\}\max\{f, -f\}$ and $|f|$ as $\max\{f, -f\}$.

We now use these properties in order to compute a subgradient of $\tilde{E}(\boldsymbol{w})$. A general expression cannot be obtained; for this reason, here we will compute $[\partial \tilde{E}(\boldsymbol{w})]$ for the distance function d_{euc} and the triple of operators $(\widetilde{\wedge}_{[\cdot]}^{std}, \cdot, \max)$, for which it is easy to verify that the semantics are always exactly computable in finite time (i.e. for some $n_0 < \omega$). Because, unlike the case of differentiable functions, no chain rules of differentiation may be applied, we use directly the sudifferential calculus properties. This leads to a forward computation of the subgradient (instead of a backpropagation style computation).

For simplicity in the exposition we consider a propositional program, so that the complex named interval that carries each link of the connectionist model is a set with the sole element $(\langle\rangle, [v^b, v^t])$ where $\langle\rangle$ is the vector of dimension 0 and can be ignored. The output of a disjunctive neuron i that corresponds to atom $\underline{B}_i$ is $out_i(n_0) = [v_i^b, v_i^t] = \underline{B}_i^{F\Phi_{\mathcal{P}'(w)}^{\uparrow n_0}}$. Then for appropriate λ_1^{eb}, λ_2^{eb} and λ_1^{et}, λ_2^{et} that satisfy the necessary conditions, we obtain from (7) and (3) that

$$[\partial \tilde{E}(\boldsymbol{w})] = \sum_{i \in N_d} \left(|v_i^b - \underline{B}_i^{bI}|(\lambda_1^{eb} - \lambda_2^{eb})[\partial v_i^b] + |v_i^t - \underline{B}_i^{tI}|(\lambda_1^{et} - \lambda_2^{et})[\partial v_i^t] \right) . \tag{8}$$

Let J be the set of the conjunctive neurons connected to the inputs of disjunctive neuron i. The output of i is $[v_i^b, v_i^t]$ with $v_i^{\{b,t\}} = \max_{j \in J} w_{ij} v_j^{\{b,t\}}$ where $[v_j^b, v_j^t]$ is the output of conjunctive neuron $j \in J$. So for appropriate λ_j^ds we get

$$[\partial v_i^{\{b,t\}}] = \sum_{j \in J} \lambda_j^d \left(\boldsymbol{e}_{w_{ij}} v_j^{\{b,t\}} + w_{ij}[\partial v_j^{\{b,t\}}] \right) , \tag{9}$$

where $\boldsymbol{e}_{w_{ij}} = [\partial w_{ij}]$ is the unit vector in $[0,1]^q$ in the direction of $q\langle w_{ij}\rangle$.

Next, consider a conjunctive neuron i and let J be the set of the disjunctive neurons connected to its inputs. The output of i is $[v_i^b, v_i^t]$ with $v_i^{\{b,t\}} = \min_{j \in J} \max\{\bar{w}_i - w_{ij}, \min\{\bar{w}_i, u_j^{\{b,t\}}\}\}$ where $\bar{w}_i = \max_{j \in J} w_{ij}$ and either $[u_j^b, u_j^t] = [v_j^b, v_j^t]$ or $[u_j^b, u_j^t] = [1 - v_j^t, 1 - v_j^b]$, where $[v_j^b, v_j^t]$ is the output of neuron $j \in J$. As a result we have

$$[\partial v_i^{\{b,t\}}] = \sum_{j \in J} \lambda_j^{c3} \left(\lambda_1^{c2} \left([\partial \bar{w}_i] - e_{w_{ij}} \right) + \lambda_2^{c2} (\lambda_1^{c1} [\partial \bar{w}_i] + \lambda_2^{c1} [\partial u_j^{\{b,t\}}]) \right), \quad (10)$$

where $[\partial u_j^{\{b,t\}}] = [\partial v_j^{\{b,t\}}]$ or $[\partial u_j^{\{b,t\}}] = -[\partial v_j^{\{t,b\}}]$ (if a negation node is interpolated from j to i) and $[\partial \bar{w}_i] = \sum_{j \in J} \lambda_j^{c0} e_{w_{ij}}$ for appropriate λ_j^{ck} for $k = 1, 2, 3$.

Using equations (10), (9) and (8), we can compute $[\partial v_i^{\{b,t\}}]$ first for all the conjunctive neurons i, then for all the disjunctive neurons i, and finally $[\partial \tilde{E}(w)]$ in a forward manner, while simulating the connectionist model. Because of the forward computation, the fact that the connectionist model is recursive for $n_0 > 1$ does not pose any computational problems. At each iteration the same equations have to be used only by replacing $[\partial v_i^{\{b,t\}}]$ by $[\partial v_i^{\{b,t\}}(t+1)]$ and $[\partial u_j]$ by $[\partial u_j(t)]$ in equation (10). Then, $[\partial u_j(t)]$ is available because it has been computed during the previous iteration. For the first iteration $[\partial u_j(0)] = \mathbf{0}_q$ for all j, because for this iteration u_j is a constant that has been obtained from a fact.

Example 2. As a simple example consider the following program $\mathcal{P}(w)$

$$a(x) \leftarrow (w_1; \sim b(x)), (w_2; c(x))$$
$$b(x) \leftarrow (w_3; \sim a(x)), (w_4; d(x)),$$

in which we want to adapt the weight vector $w = (w_1, w_2, w_3, w_4)$.

We solved the problem by creating a synthetic interpretation I. Let p_i denote the value of atom $p^I(t_i)$ for some constant t_i. For $i = 1 \ldots 200$ we let $c_i, d_i = z_D \cdot [1, 1]$ where $z_D \sim U(0, 1)$. Using this data we created the program $\mathcal{P}_F$ and then computed the fixpoint of $\mathcal{P}(w) \cup \mathcal{P}_F$, by setting $w^t = (0.4, 1.0, 0.6, 0.9)$, in order to get the values of a_i, b_i. Finally, we disturbed the values of a_i, b_i by adding a noise element $\sigma_N z_N \cdot [1, 1]$ (and projecting the result to the interval $[0, 1]$) where $z_N \sim N(0, 1)$ and σ_N was a standard deviation parameter. The results of the adaptation using as initial weights the vector $w_0 = (0.25, 1.0, 0.5, 0.75)$ and the distance functions d_{euc} and d_{mid} are shown in table 1. We have taken the initial vector w_0 to be relatively close to w^t based on the assumption that underlies our whole approach, that we have a rough knowledge of the domain, which we want to *adapt* so that it fits better to the provided set of data.

As expected, the lower is the added noise, the closer is the computed weight vector $\hat{w}$ to w^t; taking w_0 to be close to w^t is crucial in this respect. When the data is too noisy the connectionist model locates a random local minimum. The performance was not significantly affected by the use of either d_{euc} or d_{mid}.

Table 1. Results for example 2

σ_N	d_{euc}		d_{mid}	
	$\hat{w}$	$E(\hat{w})$	$\hat{w}$	$E(\hat{w})$
0.00	$(0.372, 0.972, 0.600, 0.900)$	$4.97e^{-11}$	$(0.323, 0.972, 0.550, 0.900)$	$6.19e^{-3}$
0.10	$(0.360, 0.973, 0.587, 0.871)$	3.55	$(0.300, 0.973, 0.524, 0.871)$	1.77
0.25	$(0.285, 0.891, 0.556, 0.828)$	18.5	$(0.306, 0.892, 0.598, 0.827)$	9.21
0.50	$(0.386, 0.917, 0.276, 0.686)$	53.7	$(0.401, 0.917, 0.286, 0.686)$	26.8
1.00	$(0.045, 0.596, 0.426, 0.811)$	79.5	$(0.103, 0.624, 0.489, 0.811)$	39.7

5   Conclusions

We have introduced an extension of weighted fuzzy logic programs that allows
for the use of negation following a Kripke-Kleene based semantic interpretation.
Subsequently we have outlined a connectionist representation of the programs
and finally developed a subgradient descent algorithm in order to train it and
learn the weights of the rules using an appropriate set of data. Given the increased expressiveness of weighted fuzzy programs in modelling fuzzy datasets,
using connectionist models in order to learn logic programs is a significant
step in the field of integrating symbolism and connectionism and performing
connectionist-based reasoning. The suggested methodology has still to be evaluated with real data in order to determine its performance and effectiveness.
Several issues related with an efficient implementation also arise if we take into
account e.g. that according to the proposed algorithm each neuron has to carry
the entire subgradient vector of its output, which will in general be a sparse vector. The study of ways to escape from bad local minima as well as to take into
account the dropped constraints of the original problem (5), by including them
e.g. as penalty terms in the error function, is also an object of further work.

References

1. Chortaras, A., Stamou, G., Stafylopatis, A.: Adaptation of weighted fuzzy programs. In: Kollias, S.D., Stafylopatis, A., Duch, W., Oja, E. (eds.) ICANN 2006.
   LNCS, vol. 4132, pp. 45–54. Springer, Heidelberg (2006)
2. Chortaras, A., Stamou, G., Stafylopatis, A., Kollias, S.: A connectionist model for
   weighted fuzzy programs. In: IJCNN 2006: The 2006 International Joint Conference
   on Neural Networks, pp. 5362–5369 (2006)
3. Damásio, C.V., Medina, J., Ojeda-Aciego, M.: Termination of logic programs
   with imperfect information: applications and query procedure. Journal of Applied
   Logic 5(3), 435–458 (2007)
4. Damásio, C.V., Pereira, L.M.: Sorted monotonic logic programs and their embeddings. In: IPMU 2004: The 8th International Conference on Information Processing
   and Management of Uncertainty, pp. 807–814 (2004)
5. Eitzinger, C.: Nonsmooth training of fuzzy neural networks. Soft Computing 8,
   443–448 (2004)

6. Fitting, M.: Bilattices and the semantics of logic programming. Journal of Logic Programming 11, 91–116 (1991)
7. Hitzler, P., Holldöbler, S., Seda, A.K.: Logic programs and connectionist networks. Journal of Applied Logic 2(3), 245–272 (2004)
8. Kiwiel, K.C.: Methods of Descent for Nondifferentiable Optimization. Springer, Heidelberg (1985)
9. Lakshmanan, L.V.S., Shiri, N.: A parametric approach to deductive databases with uncertainty. IEEE Transactions on Knowledge and Data Engineering 13(4), 554–570 (2001)
10. Medina, J., Mérida-Casermeiro, E., Ojeda-Aciego, M.: A neural implementation of multiadjoint programming. Journal of Applied Logic 2(3), 310–324 (2004)
11. Shor, N.Z., Zhurbenko, N.G., Likhovid, A.P., Stetsyuk, P.I.: Algorithms of nondifferentiable optimization: Development and application. Cybernetics and Systems Analysis 39(4), 537–548 (2003)
12. Vojtáš, P.: Fuzzy logic programming. Fuzzy Sets and Systems 124, 361–370 (2001)

Neuro-fuzzy System for Road Signs Recognition

Bogusław Cyganek

AGH - University of Science and Technology,
Al. Mickiewicza 30, 30-059 Kraków, Poland
`cyganek@uci.agh.edu.pl`

Abstract. In this paper a hybrid neuro-fuzzy system for the real-time recognition of the road-signs is presented. For tracking an improvement to the continuously adaptive mean shift method is proposed. It consists in substitution of the probabilistic density for the especially formed membership function. Classification of binary pictograms of the detected signs is done with the kernel morphological neural network which is robust to noise, missing data, and small geometrical deformations of the patterns.

1   Introduction

Automatic recognition of road signs (RS) can increase traffic safety, hence there are many works devoted to RS detection and tracking [5-6]. A comparable example is a system proposed by Fang et al. [7]. It detects places with color and shape characteristic to the RSs with help of the neural networks. A fuzzy method is used for feature integration and detection. Detected objects are verified and tracked through consecutive frames based on the Kalman filter. However, the system is tuned to detect only shapes of a certain size and specific camera configuration. Also the applied neural networks are adjusted to detect only non distorted shapes.

Tracking can greatly increase systems abilities of detection of the RSs. The important properties of a tracking system are accuracy, speed and ability to track multiple objects at a time. A very common method is the already mentioned Kalman filter. However, this approach assumes a Gaussian distribution of data. The other approach which can deal with arbitrary PDFs was proposed by Fukunaga et al. [8]. It is called the mean shift method. Its version, called CamShift, was proposed by Bradski for face tracking [1]. Then, the method was analyzed and extended by Cheng et al. [2], as well as by Comaniciu et al. [3-4]. This version of the method is called a continuously adaptive mean shift. All of these methods utilize the underlying PDF. However, in many application a more practical approach is to use fuzzy measures instead of PDFs. Such an approach is proposed in this paper. Instead of PDFs membership functions conveying information characteristic to the tracked objects have to be constructed. This is discussed in section (3) of this paper.

Detected signs are then classified with help of the Morphological Neural Network (MNN). These comparatively less known networks exhibits robustness to

V. Kůrková et al. (Eds.): ICANN 2008, Part I, LNCS 5163, pp. 503–512, 2008.

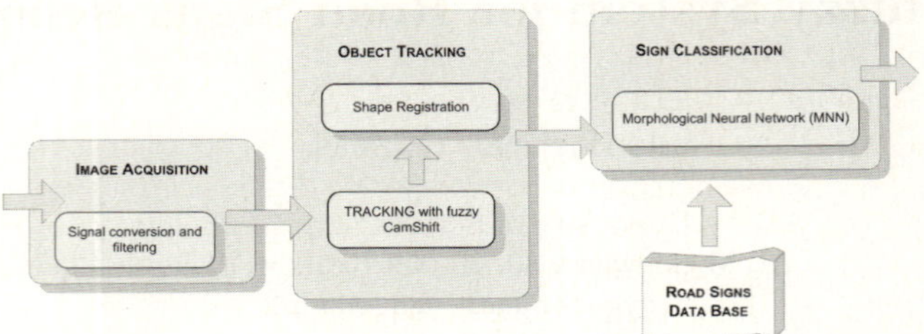

Fig. 1. Architecture of the neuro-fuzzy road sign recognition system. It consists of the image acquisition, object tracking, and the classification modules.

erosive/dilating noise and partially occluded patterns. They show also fast operation due to convergence in a single iteration. However, they require construction of the kernels specific to the prototype patterns, which in our case are pictograms of the detected signs. Robust performance of both, the detection and classification modules, made possible real-time operation of the presented system.

2 Architecture of the System

The system is an example of a hybrid architecture where many different modules has been connected together to perform a specific task. The big advantage of this approach is its flexibility, i.e. the system is opened, what means that the existing modules can be easily modified or even exchanged into new ones without affecting the other components of the system.

Fig. 1 presents architecture of the proposed system for recognition of RSs. Data processing starts with image acquisition, followed by the tracking and registration, and finally by recognition. The acquisition module provides color images. Object tracking is done with the novel fuzzy version of the continuously adaptive means shift method, described in the next section. Next, a detected shape is rectified by the affine warping module [5]. Finally, a sign is classified based on its binary pictogram. Classification is done with help of the morphological neural networks (MNN), trained for each group of signs. This is discussed in section (4) of this paper.

3 RS Tracking with the Fuzzy Continuously Adaptive Mean Shift Method

The mean shift algorithm is a non-parametric method of tracing mode of a PDF. It finds a location of the maximal probability associated with and existence of a searched object. For this purpose the algorithm climbs the gradient of a PDF. Thus a method needs to asses PDF of a searched object. In this paper, instead of PDF we propose to use properly designed fuzzy membership function in the mean shift method. Such a modification allows easier representation of the searched objects in terms of fuzzy logic. In this section we outline tracking with the fuzzy version of the continuously adaptive mean shift method.

The basic mean shift method assumes estimation of a density function from the sample points in an N dimensional space $\Re^N$. This is achieved at first step by a non parametric estimation of the PDF. Then the procedure climbs the estimated PDF gradient by recursive computation of the mean shift vector and translation of the centre of a mean shift kernel. For the mean shift procedure, the task of finding an object, based on its characteristic feature $\mathbf{v}$, is formulated as finding a discrete location $\mathbf{y}$ in the candidate image for which the associated density $s(\mathbf{y},\mathbf{v})$ is the most similar to the target density $d(\mathbf{v})$. However, the basic mean shift method relies on statically determined PDFs of the target and candidates. Further discussion on its properties and applications can be found for instance in the works by Comaniciu et al. [3-4].

A slightly different approach was proposed by Bradski in the CamShift method, originally designed for real-time face tracking [1]. It assumes repeated computations of densities in each frame, or its part. Next, the spatial moments are computed in the probability field to guide iterations aimed at finding the mode of the continuously updated PDF. This approach showed very useful in real-time tracking of objects and therefore a modified version of the CamShift is used in our system.

We outline the CamShift briefly. For a discrete function $f(x,y)$ its moments and central moments of order (a,b) are defined as follows [10]:

$$m_{ab} = \sum_{(x,y)\in U} x^a y^b f(x,y), \quad c_{ab} = \sum_{(x,y)\in R} (x-\bar{x})^a (y-\bar{y})^b f(x,y), \quad (1)$$

where the point with coordinates

$$\bar{x} = m_{10}/m_{00}, \quad \bar{y} = m_{01}/m_{00}, \tag{2}$$

is called a centroid (assuming that $m_{00} \neq 0$). Given a density function $f(x,y)$ the moment of inertia M of a rigid body U, around the line with a slope φ, passing through the centroid of U, is computed as follows:

$$M(\phi) = \sum_{(x,y)\in U} [(x-\bar{x})\sin\phi - (y-\bar{y})\cos\phi]^2 f(x,y). \tag{3}$$

Taking (1) into the above the following formula is obtained

$$M(\phi) = c_{20}\sin^2\phi - 2c_{11}\sin\phi\cos\phi + c_{02}\cos^2\phi. \tag{4}$$

The moment of inertia M is minimized with respect to φ when $\partial M(\varphi)/\partial\varphi = 0$, that is

$$\tan(2\phi) = 2c_{11}/(c_{20}-c_{02}), \quad \text{for } c_{20} \neq c_{02}, \tag{5}$$

which after simple expansion of $tan(2\varphi)$ leads to the following equation

$$\tan^2\phi + (c_{20}-c_{02})\tan\phi/c_{11} - 1 = 0, \quad \text{for } c_{11} \neq 0. \tag{6}$$

In general two values of φ are obtained from the above, one of which maximizes $M(\varphi)$. M_{min}/M_{max} provides the elongation factor of U. The line around which M is minimized is called the principal axis. It can be shown that this axis is collinear with the eigenvector corresponding to the largest eigenvector of the following tensor [10]

$$\mathbf{I} = \begin{bmatrix} c_{20} & c_{11} \\ c_{11} & c_{02} \end{bmatrix}. \tag{7}$$

An object U in an image, characterized by $\mathbf{I}$, can be approximated by an ellipse with the parameters α, β, and also with the slope φ [9]. These are as follows:

$$\alpha = \sqrt{\lambda_1}, \quad \beta = \sqrt{\lambda_2}, \text{ and } \phi = \frac{1}{2} ATAN2\left(2c_{11}, c_{20} - c_{02}\right), \tag{8}$$

where $ATAN2$ is an arcus-tangent function of a quotient of its two arguments, which is also defined when its second argument approaches zero ($\pm\pi/2$, based on a sign of its first argument), $\lambda_1 \geq \lambda_2$ are eigenvalues of $\mathbf{I}$ given as follows

$$\lambda_{1,2} = \frac{1}{2}\left[(c_{02} + c_{20}) \pm \sqrt{4c_{11}^2 + (c_{02} - c_{20})^2}\right]. \tag{9}$$

Further on the values α, β need to be scaled by a total 'mass', given by m_{00}, to reflect the relative size of an object in the image. Moreover, from practical reasons we wish to find and track a rectangle representing an ellipse [9]. Its parameters are computed from (8) and (9)

$$w = 2\sqrt{\lambda_1/m_{00}}, \quad l = 2\sqrt{\lambda_2/m_{00}}. \tag{10}$$

The found rectangles (w,l) have the same moment $\mathbf{I}$ as the object U. In practice, we allow a slight increase to the search window, and therefore the dimensions above are modified with the following formulas

$$w' = \rho w, \quad l' = \rho l, \tag{11}$$

where $1 < \rho < 2$ is a scaling parameter. In summary, the CamShift method for a single frame operates as follows [1][9]:

1. Set $t = 0$; $\rho = 1.8$; Set the stop thresholds τ_x, τ_y, τ_w, and τ_l, as well as the initial locations and dimensions of the regions of interest R_i. For each R_i do steps 2-4 until convergence;
2. $t = t + 1$; Compute the probability distributions in a region R_i;
3. Estimate the centroid $(\bar{x}, \bar{y})$ from (2) and dimensions (w', l') of R_i from (11);
4. Move R_i to a new centroid and new dimensions found in the previous step;
5. Check convergence conditions: if $(|\Delta\bar{x}| = |\bar{x}(t+1) - \bar{x}(t)| < \tau_x$, and $|\Delta\bar{y}| < \tau_y$, and $|\Delta w'| < \tau_w$, and $|\Delta l'| < \tau_l)$ then stop, else go to step 2.
6. Trim the exact positions of R_i, setting $\rho = 1$ in (11).

For each processed frame the final positions and sizes of the tracked regions R_i are stored in a special data structure. Then, each new frame is initialized with information from this data structure and the algorithm is repeated.

In CamShift the density function needs to be evaluated exclusively in the regions of interest R_i. Only at the first step of the algorithm the whole image has to be processed. This increases robustness of the method.

Nevertheless evaluation of the PDFs for multidimensional data is usually not easy. Therefore we propose to use fuzzy measures to be directly used in the CamShift method instead of probabilities. These show many desirable features, such as reduced memory and time requirements – frequently the membership functions are derived from the point samples in a form of the piecewise-linear functions. In our formulation, with each point P of an image a q-dimensional fuzzy measure $\mu_\xi = [\mu_1, \mu_2, \ldots, \mu_q]$ is assigned for a feature ξ. Then, in the CamShift algorithm (step 2) the probability in a region R_i is substituted for the fuzzy functions $g_1(\mu_\xi)$ or $g_2(\mu_\xi)$, as follows:

$$g_1(\mu_\xi) = \prod_{i=1}^{q} \mu_i. \tag{12}$$

$$g_2(\mu_\xi) = \min_{1 \le i \le q} (\mu_i), \tag{13}$$

In practice g_2 is easier for computations. A feature attributed to a pixel in an image can be color, shape, etc., for which its membership function can be inferred. It is not necessary to normalize μ_ξ since this is embedded in (10).

The best results of our system were obtained with fuzzy membership functions, given by (12) or (13), computed from the processed histograms of the H and S color channels, obtained from real traffic examples [6]. From the gathered data samples the histograms were built and then filtered with the Savitzky-Golay filter of 4th order which preserves statistical moments of that order. Such method has many desirable properties compared to a simple low-pass filtering, such as a moving average for instance. Then, the piecewise-linear membership functions were built from the histograms. To suppress noise all values which fell below a 10% threshold of the maximum value were arbitrarily set to 0.

4   Pictogram Classification with Morphological Neural Networks

Morphological neural networks were introduced by Ritter [12-13] over ten years ago. This relatively less known group of neural networks show many desirable properties such as high pattern capacity (compared for instance to the Hopfield network), resistance to the erosive/dilative type of noise, as well as convergence in only one step which allows fast response. These features inspired us to use of MNN in the presented RS recognition system.

Fig. 2a depicts a model of a morphological neuron, which is governed by the following equation

$$x_k[t+1] = F\left(p_k \bigvee_{i=1}^{N} r_{ik}(x_i[t] + w_{ik})\right), \tag{14}$$

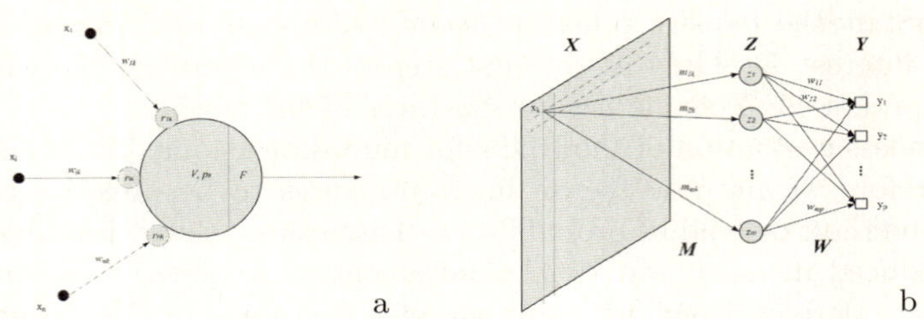

Fig. 2. Model of a morphological neuron (a) and network for image processing (b)

where r_{ik} denotes a pre-synaptic response which transfers excitatory (for $+1$) or inhibitory (for -1) influence of an *i-th* neuron, p_k conveys the post-synaptic response of a *k-th* neuron to the total input signal, $\vee$ denotes a *max* product, and finally $F(x)$ is a saturation function which takes on 1 for $x > 0$ and 0 for $x \leq 0$. In our realization all r_{ik} and p_k are positive. The max product $\vee$ for two matrices $\mathbf{A}_{pq}$ and $\mathbf{B}_{qr}$ is a matrix $\mathbf{C}_{pr}$, with elements c_{ij} defined as follows:

$$c_{ij} = \bigvee_{k=1}^{q} \left(a_{ik} + b_{kj} \right). \tag{15}$$

In analogous way the *min* function is defined. Fig. 2b presents structure of a MNN used for image classification. It is a version of a morphological associative memory in which a set of i input/output pairs is given as $(\mathbf{x}^1,\mathbf{y}^1)$, $\ldots$, $(\mathbf{x}^i,\mathbf{y}^i)$, where pattern $\mathbf{x}$ is a linear version of an image and $\mathbf{y}$ is binary version of a pattern's class. The first layer $\mathbf{X}$ consists of a binarized pictogram of a tracked sign. In our experiments its dimensions were warped to 256×256, which results in 8 kB. For binarization and sampling the method described in [5] is used. Vectors $\mathbf{y}$ decode (in the one-of-N code) the input classes, i.e. each different type of a road sign. Then, from pairs $(\mathbf{x}^i,\mathbf{y}^i)$ the matrices $\mathbf{X}=(\mathbf{x}^1,\ldots,\mathbf{x}^i)$ and $\mathbf{Y}=(\mathbf{y}^1,\ldots,\mathbf{y}^i)$ are created. In the simplest formulation with one layer, a matrix $\mathbf{W}$ or $\mathbf{M}$ of weights is determined from $\mathbf{X}$ and $\mathbf{Y}$ as follows [13]:

$$\mathbf{W_{XY}} = \bigwedge_{\alpha=1}^{i} \left(\mathbf{y}^\alpha \times (-\mathbf{x}^\alpha)' \right), \quad \mathbf{M_{XY}} = \bigvee_{\alpha=1}^{i} \left(\mathbf{y}^\alpha \times (-\mathbf{x}^\alpha)' \right), \tag{16}$$

where $\times$ denotes the morphological product of vectors which is analogous to the outer product in which multiplication is substituted for addition. Then the following holds

$$\mathbf{W_{XY}} \vee \mathbf{x}^\alpha = \mathbf{y}^\alpha, \quad \mathbf{M_{XY}} \wedge \mathbf{x}^\alpha = \mathbf{y}^\alpha, \tag{17}$$

even for erosively or dilatively distorted versions $\tilde{\mathbf{x}}^\alpha$ of $\mathbf{x}^\alpha$, for $\mathbf{W}$ and $\mathbf{M}$ respectively. The proofs of these statements can be found in [13][11][15]. In other words, a perfect recall is guaranteed if there are some pixels in a prototype image $\mathbf{x}$ which values are greater (lower) in this image from the maximum of all the remaining patterns $\mathbf{x}_i$ stored in that network. For binary images this means a unique pixel with value '1' in each of the prototypes $\mathbf{x}_i$. This in turn is connected with the concept of morphological independence and strong independence [13]. It also provides a methodology for the construction of the MNNs.

However, to make MNN robust to random, i.e. dilative *and* erosive noise, Ritter proposed a kernel MNN. The idea is to replace the associative memory $\mathbf{W}$ or $\mathbf{M}$ into a sequence of two memories $\mathbf{M}'$ and $\mathbf{W}'$ which are connected by a specially formed intermediate pattern $\mathbf{Z}$, called *a kernel* [12]. Structure of this network is depicted in Fig. 2b. It operates in accordance with the scheme: input-$\mathbf{M}'_{ZZ}$-$\mathbf{W}'_{ZY}$-output, i.e.:

$$\mathbf{W}'_{\mathbf{ZY}} \vee \underbrace{(\mathbf{M}'_{\mathbf{ZZ}} \wedge \tilde{\mathbf{x}}^{\alpha})}_{\mathbf{z}^{\alpha}} = \mathbf{y}^{\alpha}, \tag{18}$$

where $\tilde{\mathbf{x}}^{\alpha}$ is a randomly corrupted input pattern. $\mathbf{M}'_{ZZ}$ and $\mathbf{W}'_{ZY}$ in (18) are found in accordance with schemes (17), however for different matrices and after finding a kernel $\mathbf{Z}$. Construction of the kernel $\mathbf{Z}$ is outlined by the theorems given in [15][12]. In this work we follow this methodology. Fortunately our task is simplified since the input patterns, i.e. pictograms of the signs, are *binary* images only. Therefore a simple (i.e. not a strong) independence among patterns is sufficient for existence of a kernel. The independence among patterns means that each pattern has at least one component which is greater than a maximum computed over all other patterns. A very desirable at this point (but not absolutely necessary) is existence of a minimal representation which implies at most one non-zero entry in each kernel vector $\mathbf{z}^{\alpha}$. Our assumption here is that deformations of detected pictograms can be accommodated by the MNN. We know the 'true' pictograms of the signs, which are provided in a formal specification [14]. These constitute the reference binary patterns $\mathbf{x}^{\alpha}$ when creating the kernel $\mathbf{Z}$. A bit '1' is assigned to a pictogram symbols, '0' to the background. The reference patterns are then eroded to obtain the skeletons. These are verified to fulfill the independence conditions. This is done in checking one-against-all fashion. If the independence cannot be fulfilled then the checked patterns are tried to be corrected by removing pixels which cause failure of independence conditions. In our experiments only few patterns had to be modified this way. For MNN with binary patterns the kernels should comply with the following conditions [13]:

1. for all α: $\mathbf{z}^{\alpha} \leq \mathbf{x}^{\alpha}$ as well as
2. $\mathbf{z}^{\alpha} \wedge \mathbf{z}^{\beta} = \mathbf{0}$ and $\mathbf{z}^{\alpha} \not\leq \mathbf{x}^{\beta}$ (i.e. there is at least one component of $\mathbf{z}^{\alpha}$ which is greater from the corresponding component of $\mathbf{x}^{\beta}$) whenever $\alpha \neq \beta$.

Considering the above, the corrected patterns are used to create the kernel vectors $\mathbf{z}^{\alpha}$ in accordance with the method presented in [13].

Our method has some common points with the approach proposed by Raducanu et al. [11]. They proposed to use the morphological scale space, obtained with the zero shift structural element. The main assumption is that the characteristic points of objects are preserved in that domain. In effect the dilated or eroded scale space versions of the HNN exhibit robustness against two types of noise, i.e. dilative and erosive. However, the method of Raducanu requires construction of a set of MNNs.

5   Experimental Results

The experimental system is equipped with the Marlin-C33 cameras. These are connected by the IEEE 1394 link to the laptop computer with the Intel Duo core processor and 1GB of RAM. Software was written in C++ (Microsoft Visual 6.0). Fig. 3 depicts results of tracking and recognition of the "Children" sign which belongs to the group "A" of warning signs [14]. In this case the membership functions are constructed from the H and S channels of a subset of red color of the signs of this group. The membership function (13) is used in this case. The images in the first row of Fig. 3 show three consecutive frames of the video stream acquired from a moving car. The middle row visualizes the membership functions which are used in the fuzzy CamShift algorithm. These have direct influence on the detection results which are shown in the third row of Fig. 3. We noticed that due to reflections of the border of the windshield the spectra of color of the frame in Fig. 3c no longer fall into the specification obtained from the real color samples. In consequence the non zero membership field is strongly reduced which leads to the diminished tracking rectangle. In this case it does not contain the whole object and the pictogram cannot be recognized correctly. However, the previous two responses were correct which is sufficient in our system to verify a sign. We can assume even larger group of unanimous responses, however at a cost of time. The problem with proper segmentation is the most severe one in our system. Thus only daily conditions without shine and reflections usually produce acceptable results. More reliable color segmentation algorithms would improve the performance.

Fig. 4 presents results of tracking of the "40 km/h speed limit" (B-33) sign, acquired at daily conditions. For testing the (12) function was chosen at this time. Qualitative parameters of the system were measured in terms of Precision (P) vs. Recall (R) factors. Results of these measurements are presented in Table 1.

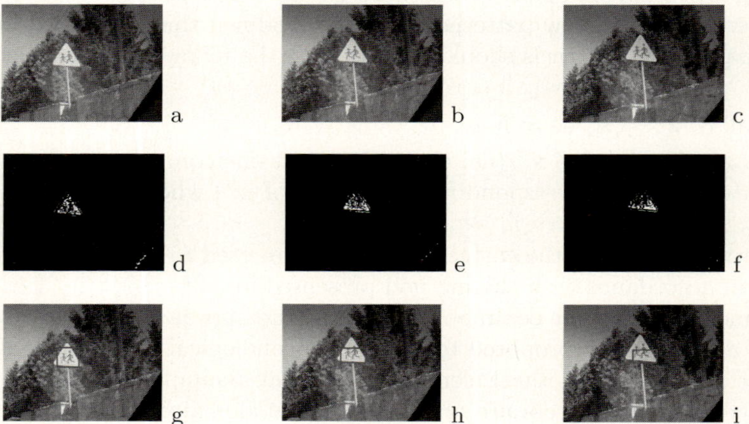

Fig. 3. Tracking and recognition of the "Children" sign (A-17) at daily conditions. Video frames (abc). Fuzzy membership functions of (13) (def). Tracked objects (ghi).

Table 1. Accuracy of recognition for different groups of signs: warning signs of group "A" and prohibition signs of group "B". Precision (P) versus recall (R).

Group	Warning signs (group "A")		Prohibition signs (group "B")	
	P	R	P	R
	0.89	0.90	0.80	0.88

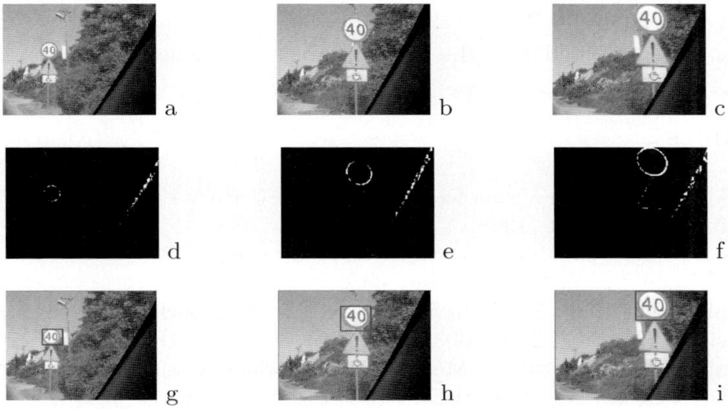

Fig. 4. Tracking of the "40 km/h speed limit" (B-33), daily conditions. Video frames (abc). Maps of the fuzzy membership functions (def). Detected and tracked signs (ghi).

The system is able to process 320×240 color frames in real time, i.e. at least 30 f/s. It is also resistant to noise, small lighting variations and some geometrical deformations of the signs due to the projective distortions of the acquisition system. The values of the recall parameter are affected mostly by performance of the detectors and in the lower degree by the MNN classifiers. However, the parameters depend on specific settings as well as type and quality of the input images. Only results for daily conditions are provided. Under night or worse weather conditions the recall parameter falls down due to large distortions.

6 Conclusions

The paper presents a hybrid system for recognition of the RSs. The front end constitutes a tracking module which performs the fuzzy adaptive mean shift. Two fuzzy membership functions (12) and (13) are built from the empirically gathered color samples. Both perform similarly, though (13) does not require multiplications. The most problems were caused by different characteristics of cameras used to gather reference color values and the other used during experiments. In practice this requires camera calibration or data correction. Binary pictograms of the detected signs are then classified with a kernel MNN. Such classifiers are robust of random noise, missing data, and small geometrical deformations. They have also very fast response time. Thanks to binary patterns the construction of the kernels for the MNN was simplified; Similarly, the memory requirements were moderated.

The method has been verified by experiments showing high accuracy and real-time processing abilities. The method can be easily extended to allow tracking of other objects based on their specific membership functions.

Acknowlegement

This work was supported from the Polish funds for the scientific research in 2008.

References

1. Bradski, G.: Computer Vision Face Tracking For Use in a Perceptual User Interface. Intel Technical Report (1998)
2. Cheng, Y.: Mean shift, mode seeking, and clustering. IEEE PAMI 15(6), 602–605 (1993)
3. Comaniciu, D., Meer, P.: Mean Shift: A Robust Approach Toward Feature Space Analysis. IEEE PAMI 24(5), 603–619 (2002)
4. Comaniciu, D., Ramesh, V., Meer, P.: The Variable Bandwidth Mean Shift and Data-Driven Scale Selection. IEEE ICCV 1, 438–445 (2001)
5. Cyganek, B.: Circular Road Signs Recognition with Soft Classifiers. Integrated Computer-Aided Engineering 14(4), 323–343 (2007)
6. Cyganek, B.: Soft System for Road Sign Detection. In: Advances in Soft Computing, vol. 41, pp. 316–326. Springer, Heidelberg (2007)
7. Fang, C.-Y., Chen, S.-W., Fuh, C.-S.: Road-Sign Detection and Tracking. IEEE Transactions on Vehicular Technology 52(5), 1329–1341 (2003)
8. Fukunaga, K., Hostetler, L.D.: The estimation of the gradient of a density function. IEEE Tr. Information Theory 21, 32–40 (1975)
9. Kim, K.I., Jung, K., Kim, J.H.: Texture-Based Approach for Text Detection in Images Using Support Vector Machines. IEEE PAMI 25(12), 1631–1639 (2003)
10. Klette, R., Rosenfeld, A.: Digital Geometry. Morgan-Kaufmann, San Francisco (2004)
11. Raducanu, B., Grana, M., Albizuri, F.X.: Morphological Scale Space and Associative Morphological Memories: Results on Robustness and Practical Applications. J. of Math. Imaging and Vision 19, 113–131 (2003)
12. Ritter, G.X., Sussner, P., Diaz, J.L.: Morphological Associative Memories. IEEE Transactions on Neural Networks 9(2), 281–293 (1998)
13. Ritter, G.X., Urcid, G., Iancu, L.: Reconstruction of Patterns from Noisy Inputs Using Morphological Associative Memories. J. of Math. Imaging 19, 95–111 (2003)
14. Road Signs and Signalization. Directive of the Polish Ministry of Infrastructure. Internal Affairs and Administration (Dz. U. Nr 170, poz. 1393) (2002)
15. Sussner, P.: Observations on Morphological Associative Memories and the Kernel Method. Neurocomputing 31, 167–183 (2000)

Neuro-inspired Speech Recognition with Recurrent Spiking Neurons

Arfan Ghani, T. Martin McGinnity, Liam P. Maguire, and Jim Harkin

Intelligent Systems Research Centre
University of Ulster, Magee Campus
Derry, BT487JL, N. Ireland, UK
`ghani-a1@ulster.ac.uk`

Abstract. This paper investigates the potential of recurrent spiking neurons for classification problems. It presents a hybrid approach based on the paradigm of *Reservoir Computing*. The practical applications based on recurrent spiking neurons are limited due to their non-trivial learning algorithms. In the paradigm of *Reservoir Computing*, instead of training the whole recurrent network only the output layer (known as readout neurons) are trained. These recurrent neural networks are termed as microcircuits which are viewed as basic computational units in cortical computation. These microcircuits are connected as columns which are linked with other neighboring columns in cortical areas. These columns read out information from each other and can serve both as reservoir and readout. The design space for this paradigm is split into three domains; front end, reservoir, and back end. This work contributes to the identification of suitable front and back end processing techniques along with stable and compact reservoir dynamics, which provides a reliable framework for classification related problems.

Keywords: Reservoir computing, liquid state machine, hybrid neuro inspired computing, speech recognition.

1 Introduction

The paradigm of reservoir computing is promising because it offers an alternative to the computational power of recurrent neural networks. It facilitates training in a recurrent neural network where linearly non separable low dimensional data is projected on a high dimensional space. The readout of the reservoir can be trained with partial information extracted from the reservoir which suffices to solve complex problems such as speech recognition. In this approach, the readout only observes the membrane potential of the spiking neurons at particular time steps which is far more efficient than fully quantifying the reservoir dynamics. It is due to this property that a relatively simple readout networks can be trained with meaningful internal dynamics of the reservoir. This paradigm appears to have great potential for engineering applications such as robotics, speech recognition, and wireless communication [8]. This paradigm can also be used for channel equalization of high speed data streams in wireless communication as suggested by [9].

V. Kůrková et al. (Eds.): ICANN 2008, Part I, LNCS 5163, pp. 513–522, 2008.

Since the inception of the theoretical foundations by Jaeger and Maass, various groups have focused on investigating different aspects of the paradigm for engineering applications, e.g., Skrownski et al investigated the paradigm of echo state networks for speech recognition applications where the HFCC (Human Factor Cepstral Coefficient) technique was investigated for front end processing and a HMM (Hidden Markov Model) classifier was used for back end processing. The overall performance was compared with the baseline HMM classifier. The main focus of the work was to investigate the noise robustness of the system based on echo state networks [4].Verstraeten et al analysed the classification accuracy of a reservoir with different benchmarks. In their study, both sigmoidal and LIF based reservoirs were tested for evaluating the memory capacity and overall classification accuracy was calculated based on different sizes of the reservoir. Moreover, different speech pre processing techniques were also elaborated and their robustness measured against the overall system performance [1]. In a recent study, Uysal et al investigated a noise robust technique by using phase synchrony coding [2]. However, in all of these studies, there are no specific guidelines regarding implementation and stability of reservoirs and all reported techniques significantly vary from each other.

Implementing stable reservoir is a challenging task and two important factors are the proper investigation of front and back end techniques. Regardless of the size of the reservoir or processing nodes, it is rather difficult to solve a problem without investigating a robust front end technique. This paper will investigate three main areas: a robust front end, a stable and compact reservoir, and an efficient back end engine for the task of speech recognition. The overall accuracy of the reservoir based classification technique will be compared with the baseline feedforward network.

2   Theoretical Background

In order to overcome the burden of training in recurrent networks, the paradigm of reservoir computing was introduced by Maass and Jaeger. This paradigm covers three main techniques in classification related problems: Echo State Machine [6]Backpropagation Decorrelation [11], and Liquid State Machine [5]. A common term is used for these techniques namely *reservoir computing* [1]. The fundamental motivation behind all these techniques is to overcome the computational burden of the recurrent neural network training. In the paradigm of reservoir computing, the partial response of a recurrent reservoir is observed from outside by a network trained by any suitable classification algorithm such as back propagation (see fig.1). It is much easier and computationally more efficient to train the output layer (feedforward network) or so called readout neurons, instead of the complete network of recurrent neurons. The mathematical theory of reservoir computing is based on the observation that if a complex recurrent neural circuit is excited by an input stream $u(t)$ and after some time s, such that when $t > s$, a liquid state $x\,(t)$ is captured, then it is very likely that this state will contain most of the information about recent inputs. According to the theory, it is not possible to

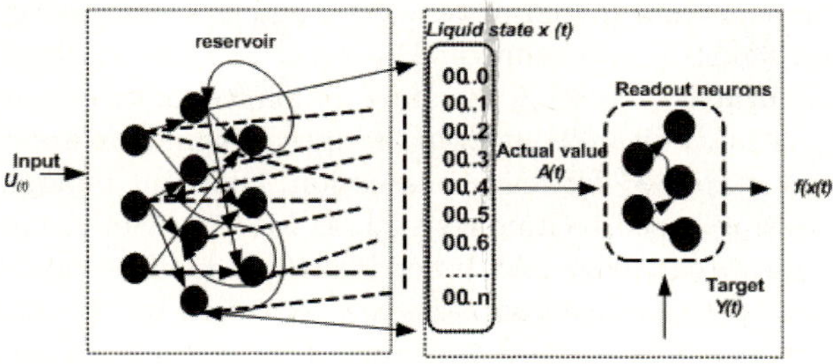

Fig. 1. An abstract overview of a neural reservoir

understand the neural code but it is not important because the liquid by itself serves as short term memory and the major task of learning depends on the state vector $x(t)$ which is exclusively used by the *readout* neurons. The reservoir transforms the input stream $u(t)$ to a high dimensional spatial state $x(t)$ [5].

The advantage of the neural reservoir is that it does not require a task specific connectivity and it does not require any specific code by which information is represented in the neural reservoir because only the readout neurons are trained [5]. Theoretical results imply that this is a universal state machine which has no limitation on the power of neural microcircuit as long as the reservoir and readouts fulfill the separation and approximation properties. A classification can be guaranteed by this paradigm if the dynamics of a reservoir exhibit a property of fading memory or echo state property. The concept of echo state property or fading memory is introduced because different states disappear over time. Theoretically, this paradigm appears to have no limits on the power of the neural microcircuit but there are no specific guidelines how to construct a stable or ordered recurrent neural reservoir, an appropriate front end and classification algorithm for backend readout neurons. In the following section, an experimental framework is proposed inspired by the paradigm of reservoir computing where the design space is split into three main areas: front end, back end and reservoir. Each one of these areas is analysed individually and then integrated for their performance evaluation. The experiment is based on the subset of isolated digit recognition (digits 0-9) dataset from the TI46 speech corpus [18].The dataset provides samples spoken by five different speakers with each digit being uttered four times by each speaker.This provides a total dataset of 200 speech samples (10 digits x 5 speakers x 4 utterances).

3   Front End (Pre-processing)

In the application of speech recognition, it is very important to detect the signal in the presence of a background noise in order to improve the accuracy of a system. A speech signal can be divided into three states: silence, unvoiced and voice. It is very important to remove the silence state in order to save the overall processing time and hence to improve the accuracy. In order to detect the silence part, an end-point detection technique is used where signal energy is calculated and a threshold value is determined. The total amount of data processing is minimised by accurately detecting the start and stop points of a sampled speech signal. Once the noise is filtered out from the speech signal, an appropriate speech coding technique is applied for feature selection. Different biologically plausible and signal processing based techniques such as MFCC (Mel Frequency Cepstral Coefficient), Lyon Passive Ear and Inner Hair Cell models have been reported in the literature and a detailed comparison is provided by [1]. These techniques provide good analysis of the signal but none of them offers an optimal solution. In this paper, a temporal based LPC (Linear Predictive Coding) technique is used which is one of the most useful methods for encoding a speech signal. In this method, current samples of the speech are predicted by the past p speech samples. Mathematically, this can be written as:

$$\tilde{x}(n) = a_1 x(n-1) + a_2 x(n-2) + ... + a_p x(n-p) \,. \tag{1}$$

Where $\tilde{x}(n)$ is the predicted signal value, $x(n-p)$ the previous observed value and a_p the predictor coefficient. The coefficients, $a_1...a_p$ remain constant while the objective is to estimate the next sample by linearly combining the most recent samples. Another important consideration is to minimise the mean square error between the actual sample and the estimated one. The error generated by this estimate can be calculated as:

$$e(n) = x(n) - \tilde{x}(n) \,. \tag{2}$$

Where $e(n)$ is the calculated error and $x(n)$ is the true signal value.

Speech is sampled at the rate of 12 KHz and the frame size is chosen as 30 ms and a frame rate of 20 ms. Autocorrelation coefficients were computed from the windowed frame where the Hamming window is used to minimise the signal discontinuities at the beginning and end of the frame. An efficient Levinson-Durbins algorithm is used to estimate the coefficients from a given speech signal. It is computationally expensive and not feasible to process all frames in the signal. Due to the various signal lengths the total numbers of frames are different. For this experiment, a total of four frames were selected for each spoken digit in linear distance from the start and end point of the signal, 7 coefficients per time frame over four frames and hence total 28 features per sample were processed. These feature vectors were found to be a good compromise between computational complexity and robustness. These frames were used for training and testing the baseline feed forward classifier. For the reservoir based approach, the same coefficients were used as inputs.

4   Reservoir Dynamics

In order to model a stable reservoir, it is very important that it should have two qualities: separation and approximation [5]. The approximation property refers to the capability of a readout function to classify the state vectors sampled from the reservoir. The separation property refers to the ability to separate

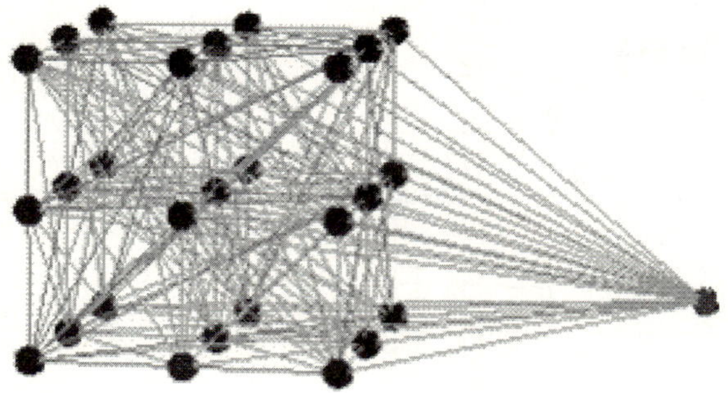

Fig. 2. This figure shows a 3D reservoir size of 27 neurons (3x3x3) fully connected with one input neuron

two different input sequences. This is important, because the readout network needs to have good classification accuracy. If the output responses of a reservoir for two different inputs are the same then the readout network will not be able to differentiate between the two patterns and thus will not be able to classify which pattern belongs to which class. In this study, reservoirs are generated in a stochastic manner where a 3D column is constructed which is a biologically plausible way to imitate microcolumnar structures in a neocortex (see fig.2) [5]. There have been other strategies proposed in the literature for their satisfactory performance based on empirical data and reported in [3],[4],[6],[1],[2]. The design space for constructing a stable reservoir is huge and depends on various important factors such as node type, probability of local and global connections and the size of the reservoir. Apart from the reservoir intrinsic dynamics, there are other factors which contribute to the overall performance of a reservoir such as input features which are used to perturb the reservoir. It is very important that a reliable front end is investigated so that the reservoir can effectively separate different input streams. Once a stable reservoir is constructed then different snapshots are recorded from the reservoir at different time steps and processed for approximating a readout function. If two reservoir states $xu(t)=(Ru)(t)$ and $xv(t)=(Rv)(t)$ for two different histories $x(.)$ and $v(.)$ are different then the reservoir dynamics are stable, otherwise they will be considered as chaotic. This property is desirable from a practical point of view because different input signals separated by the reservoir can more easily be classified by the readout neurons. A stable reservoir response to the input stimulus (digit 1) in terms of membrane voltages and spike times are shown in fig.3.

5   Backend Classification

A baseline feed forward MLP classifier for backend processing was investigated. For this experiment, the total dataset consisted of 200 samples, divided into two sets (training and testing), 20 samples (5 speakers x 4 utterances) for each

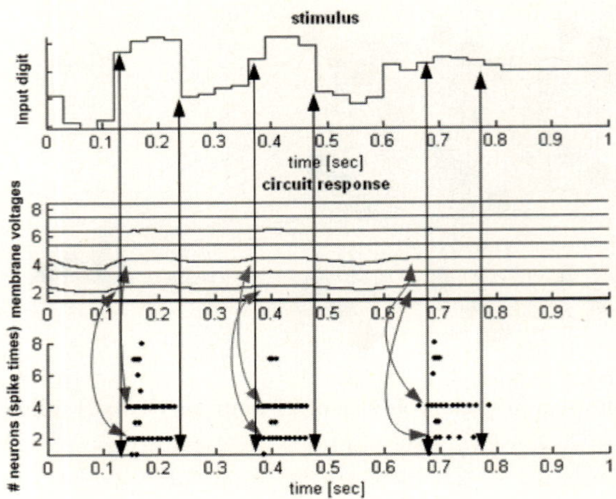

Fig. 3. This figure shows an input stimulus (digit 1) and response of a reservoir (8 neurons) in terms of membrane potentials and spike times

Table 1. Test accuracy in percentage %(total samples = 200, training samples =150, test samples = 50)

Hidden neurons	Digit 1	Digit 2	Digit 3	Digit 4	Digit 5	Digit 6	Digit 7	Digit 8	Digit 9	Digit 0
10	80	20	100	100	60	60	20	20	80	58
20	80	40	100	40	60	80	20	20	80	62
25	100	40	100	80	80	80	40	20	80	72
30	100	40	100	80	100	80	20	60	80	**76**
35	100	40	100	80	80	80	40	20	80	72

digit with 28 LPC features per sample were processed. In order to analyse the classification accuracy, different training sets and hidden layer neurons were investigated. In a series of experiments, the best results obtained in those trials are shown in table 1. The best performance achieved was limited to 76%.

6   Reservoir Based Classification

The inputs can be fed into the reservoir in two different ways: analog currents and spike trains. In this experiment, analog currents were used as inputs and features extracted through LPC technique were fed into reservoir as the post synaptic currents. In section 6.1, the input values were encoded in spike times and results were analysed. The reservoirs local and global connections were constructed in a random fashion and the reservoir by itself was not trained; rather, different reservoir states sampled every 0.25 seconds were recorded for backend classification by the readout neurons. A reservoir size of 8 neurons successfully classified the input data and the results are shown in table 2.

Table 2. Test performance with analog inputs and reservoir size of 8 neurons

Readout	Reservoir size	Readout network	Test accuracy %
Matlab RP	08	40-20-10	94.8
Matlab LM	**08**	**40-30-10**	**100**
Matlab BP	08	40-50-10	96

In these simulations the reservoir architecture and the number of neurons were remained fixed and internal states were analysed in response to different input stimuli. It is important to observe that reservoir internal states correspond to the input stimulus and could separate two different inputs. This is an important property which has to be verified for successful classification because readout neurons will have an exclusive access to the membrane voltages of the neurons in the reservoir, if the reservoir states were not significantly different from each other then readout will not be able to classify different inputs. Information processing in cortical neurons significantly depends on their local circuit connectivity. Many efforts have been made to investigate the neuronal wiring of cortical neurons and is still an active area of research [13],[14],[15],[16],[17]. The overall state of a reservoir very much depends on the connectivity of input neurons with the reservoir. For partially connected inputs, it is less likely that most of the neurons in the reservoir will have short term memory, however, for

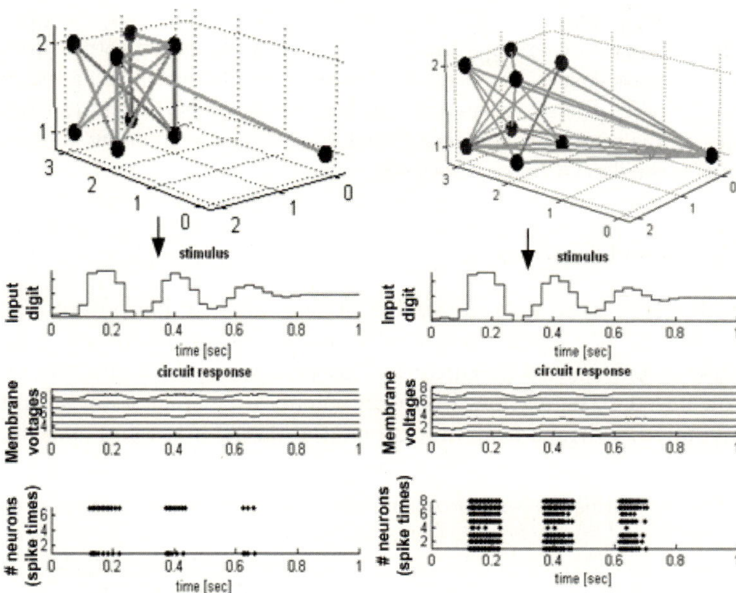

Fig. 4. This figure shows the effect of input connectivity on neural reservoir in terms of membrane potential and spike times

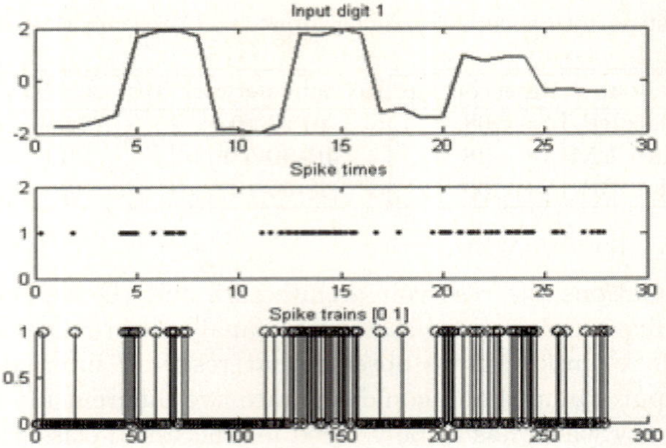

Fig. 5. This figure shows the input stimulus (digit 1) and corresponding spike times and spike trains

fully connected inputs, the probability of neurons having the short term memory increases many folds. In fig.4, a reservoir of size 8 neurons is shown where input stimulus (digit 0) is connected with the reservoir and states were recorded in terms of membrane voltages and spike times. An extremely low activity is observed when input connections were limited, however, the short term memory of a reservoir increases many folds with increased input connectivity. Input connectivity is an important design decision and affects the overall accuracy of the classifier because reservoir states are used as training vectors for readout neurons. If the reservoir states were not properly recorded then regardless of the size of the reservoir, the readouts will not be able to classify. A major advantage of this approach is the improved classification accuracy with a few neurons which overcomes the burden of bigger reservoirs.

6.1   Spike Based Coding

Squire and Kosslyn state that the timing of successive action potentials is irregular in the cortex [12]. To investigate this, another experiment was carried out where input analog values were converted into Poisson spike trains and processed for reservoir based classification. In order to generate a spike train, an interspike interval is randomly drawn from an exponential distribution and each successive

Table 3. Test performance with Poisson input spike trains and reservoir size of 8 neurons

Readout	Reservoir size	Readout network	Test accuracy %
Matlab RP	08	40-20-10	62
Matlab LM	**08**	**40-30-10**	**98**
Matlab BP	08	40-50-10	80.4

spiketime is calculated by the previous spiketime plus a randomly drawn inter-spike interval. In order to convert input features into Poisson spike trains, first negative analog values were converted into positive values and spike times were calculated (see fig.5). The spiketimes calculated through Poisson encoding were fed into the reservoir and states were recorded. The results are shown in table 3. The best accuracy achieved with Poisson encoding was limited to 98 %.

7   Conclusions

This work thoroughly investigated the theoretical framework of reservoir computing and extended by analysing the compact reservoir dynamics, front end pre processing and back end classification technique. The reservoir based recurrent neural architectures have proven to perform better on classification related tasks such as speech recognition, however, their performance can be increased if combined with feed forward networks. An alternative approach is proposed by utilising the idea of reservoir computing with efficient feature extraction technique and learning by rather simple feed forward network. This framework revealed a powerful alternative for recognition task and provided a significant improvement in terms of their performance and robustness. To the best of authors knowledge, none of the existing reservoir based techniques successfully classify the speech recognition problem with an extremely compact reservoir. This paper emphasised the modelling of a compact dynamic reservoir and empirically investigated the short term memory capacity and separation property for stable reservoirs. The results show that a compact and stable reservoir with efficient front end technique can solve significantly complex recognition task with simple readouts. It is obvious from these experiments that pre and post processing are important factors because reservoir computing can not guarantee to perform well if either the front or backend are not properly selected. Despite the promising results obtained through this investigation, a fundamental question remains open regarding the way signal pre-processing (feature extraction) is done in this study and other related work. In spiking neural networks, pre-processing through signal processing techniques may not be the best way to communicate with spiking neurons and this is the fundamental area which needs further investigation.

References

1. Verstraeten, D., Schrauwen, B., D'Haene, M., Stroobandt, D.: An experimental unification of reservoir computing methods. Neural Networks 20, 391–403 (2007)
2. Uysal, I., Sathyendra, H., Harris, J.G.: Spike based feature extraction for noise robust speech recognition using phase synchrony coding. In: International Symposium on Circuits and Systems, pp. 1529–1532 (2007)
3. Legenstein, R., Maass, W.: What makes a dynamically system computationally powerful? New Directions in Statistical Signal Processing: from Systems to Brain, pp. 127–154. MIT press, Cambridge (2007)
4. Skowronski, M.D., Harris, J.G.: Automatic speech recognition using a predictive echo state network classifier. Neural Networks 20(3), 414–423 (2007)

5. Maass, W., Natschläger, T., Markram, H.: Real-time computing without stable states: A new framework for neural computation based on perturbations. Neural Computation 14(11), 2531–2560 (2002)
6. Jaeger, H.: The echo state approach to analysing and training recurrent neural networks. Tech. Rep. Fraunhofer Institute for Autonomous Intelligent Systems: German National Research Center for Information Technology (GMD Report 148) (2001)
7. Ghani, A., McGinnity, T.M., Maguire, L.P., Harkin, J.G.: Analyzing the framework of 'Reservoir Computing' for hardware implementation. In: NIPS workshop on Echo State Networks, pp. 1–2 (2006)
8. Joshi, P., Maass, W.: Movement generation with circuits of spiking neurons. Neural Computation 17(8), 1715–1738 (2006)
9. Jaeger, H., Haas, H.: Harnessing nonlinearity: Predicting chaotic systems and saving energy in wireless communication. Science 304(5667), 78–80 (2004)
10. Joshi, P., Maass, W.: Movement generation and control with generic neural microcircuits. In: Ijspeert, A.J., Murata, M., Wakamiya, N. (eds.) BioADIT 2004. LNCS, vol. 3141, pp. 258–273. Springer, Heidelberg (2004)
11. Steil, J.J.: Backpropagation-Decorrelation: online recurrent learning with O(N) complexity. In: International Joint Conference on Neural Networks, vol. 1, pp. 843–848 (2004)
12. Squire, L.R., Kosslyn, S.M.: Findings and current opinion in cognitive neuroscience. The MIT Press, USA (1998)
13. Braitenberg, V., Schuz, A.: Anatomy of the Cortex: Statistics and Geometry. Springer, NY (1991)
14. Holmgren, C., Harkany, T., Svennenfors, B., Zilberter, Y.: Pyramidal cell communication within local networks in layer 2/3 of rat neocortex. J. Physiol. 551, 139–153 (2003)
15. Gupta, A., Wang, Y., Markram, H.: Organizing principles for a diversity of GABAergic interneuron and synapses in the neocortex. Science 287, 273–278 (2000)
16. Foldy, C., Dyhrfjeld-Johnsen, J., Soltesz, I.: Structure of cortical microcircuit theory. J. Physiol. 562, 47–54 (2005)
17. Yoshimura, Y., Dantzker, J.L.M., Callaway, E.M.: Excitatory cortical neurons form fine-scale functional networks. Nature 433, 868–873 (2005)
18. Doddington, G.R., Schalk, T.B.: Speech recognition: Turning theory to practice. IEEE Spectrum 18(9) (1981)

Predicting the Performance of Learning Algorithms Using Support Vector Machines as Meta-regressors

Silvio B. Guerra, Ricardo B.C. Prudêncio, and Teresa B. Ludermir

Center of Informatics, Federal University of Pernambuco
Pobox 7851 - CEP 50732-970 - Recife (PE) - Brazil
`silvio.guerra@gmail.com, rbcp@cin.ufpe.br,`
`tbl@cin.ufpe.br`

Abstract. In this work, we proposed the use of Support Vector Machines (SVM) to predict the performance of machine learning algorithms based on features of the learning problems. This work is related to the Meta-Regression approach, which has been successfully applied to predict learning performance, supporting algorithm selection. Experiments were performed in a case study in which SVMs with different kernel functions were used to predict the performance of Multi-Layer Perceptron (MLP) networks. The SVMs obtained better results in the evaluated task, when compared to different algorithms that have been applied as meta-regressors in previous work.

1   Introduction

Algorithm selection is an important aspect to the success of the Machine Learning applications [1]. This task is traditionally supported by empirically evaluating the candidate algorithms using the avaliable data, which can demand on expensive computational resources [2]. Algorithm selection can also be guided by expert rules, however such expert knowledge is not always easy to acquire, specially for new algorithms [3].

Considering the above context, different authors in literature have investigated the automatic acquisition of knowledge to predict the performance of learning algorithms, and to support algorithm selection, in a kind of Meta-Learning [1,2,3,4,5,6,7,8,9,10,11,12]. The knowledge in Meta-Learning is acquired from a set of meta-examples, in which each meta-example stores features of a learning problem solved in the past and information about the performance obtained by the candidate algorithms (the base-learners) on the problem. The knowledge in this case can be represented as a learning model (i.e., the meta-learner) that relates features of the problems and the performance of the learning algorithms.

A specific approach to Meta-Learning consists of using a regression algorithm to predict the value of a chosen performance measure (e.g., classification error) of the candidate algorithms based on the features of the problems. This approach is referred in the literature as Meta-Regression [10]. The meta-learner is a regression model that may be used to select the best candidate algorithm

V. Kůrková et al. (Eds.): ICANN 2008, Part I, LNCS 5163, pp. 523–532, 2008.

considering the highest predicted performance measure. Different regression algorithms have been applied in this context, such as decision trees, linear regression and instance-based algorithms [3,5,6].

In the current work, we investigated the use of Support Vector Machines (SVMs) [13] to Meta-Regression. In the literature, SVMs have been successfully applied to many different problems, achieving very competitive results when compared to other machine learning algorithms [14]. In the context of Meta-Regression, there is no previous work that evaluated the use of SVMs to predict the performance of learning algorithms.

Experiments were performed in a case study which consists of predicting the performance of Multi-Layer Perceptron (MLP) networks [15]. A set of 50 meta-examples was produced from the application of the MLP on 50 different learning problems. In the meta-level, we applied SVMs with different kernel functions (among polynomial and RBF kernels) to predict the value of the Normalized Mean of Squared Errors (NMSE) of the MLPs. The predictions of the NMSE were based on a set of 10 pre-defined features of the learning problems (e.g., number of training examples and correlation between attributes). The performance of the SVMs was evaluated in a leave-one-out experiment performed on the 50 meta-examples.

As a basis of comparison, we also performed experiments in the meta-level using three different benchmark regression algorithms which were already used in previous work to Meta-Regression: the M5 algorithm (decision trees), the linear regression, and the 1-nearest neighbor algorithm. The performed experiments revealed that the SVMs with simple polynomial kernels obtained better performance in the meta-regression task when compared to the benchmark algorithms. The experiments also revealed that good performance can also be achieved by using SVM with RBF kernels, depending, in this case, on an adequate choice of the kernel's parameters.

Section 2 brings a brief introduction about Meta-Learning, including the Meta-Regression approach. Section 3 presents details of the proposed work, as well as the case study. Section 4 brings the experiments and obtained results. Finally, section 5 presents some final conclusions and the future work.

2   Meta-Learning

According to [16], there are different interpretations of the term *Meta-Learning*. In our work, we focused on the definition of Meta-Learning as the automatic process of acquiring knowledge that relates the performance of learning algorithms to the features of the learning problems [1]. In this context, each *meta-example* is related to a learning problem and stores: (1) the features describing the problem, called *meta-features*; and (2) information about the performance of one or more algorithms when applied to the problem. The *meta-learner* is a learning system that receives as input a set of such meta-examples and then acquires knowledge used to predict the algorithms performance for new problems being solved.

The meta-features are, in general, statistics describing the training dataset of the problem, such as number of training examples, number of attributes, correlation between attributes, class entropy, among others [7,8]. In a strict formulation of Meta-Learning, each meta-example stores, as performance information, a class label which indicates the best algorithm for the problem, among a set of candidates [4]. In this case, the class label for each meta-example is defined by performing a cross-validation experiment using the available dataset. The meta-learner is simply a classifier which predicts the best algorithm based on the meta-features of the problem.

Although the strict Meta-Learning approach (as described above) has been applied by different authors (such as [2,4,9,17,18]), certain information loss may be introduced in the definition of the class labels associated to meta-examples. For instance, the performance of two algorithms may be very similar, and this information will be lost by merely recording the best algorithm as class label [5].

In order to overcome the above difficulty, the Meta-Regression approach [10] tries to directly predict the numerical value of accuracy (or alternatively the error rate) of each candidate algorithm. In this case, the meta-examples store as performance information the numerical values of accuracy obtained in previous problems. The meta-learner, in turn, is a regression model that may be used either to select the best candidate algorithm based on the highest predicted accuracy or to provide a ranking of algorithms based on the order of predicted accuracies.

In [3], the authors evaluated different algorithms as meta-regressors, including linear regression models, piecewise linear models, decision trees and instance-based regression. In [6], the authors used linear regression models to predict the accuracy of 8 classification algorithms, and the experiments revealed good results. In [5], the authors performed comparative experiments with both the strict Meta-Learning and Meta-Regression approaches, and observed that the latter one performed better when used to support algorithm selection.

3   Meta-Regression with SVMs

As seen in section 2, Meta-Regression is a flexible approach to predicting the performance of learning algorithms, supporting a more informative solution to algorithm selection.

In the current work, we investigate the use of Support Vector Machines (SVMs) in the context of Meta-Regression. SVM is a state-of-the-art algorithm in the Machine Learning field, successfully applied to different classification and regression problems [13,14]. Previous work in Meta-Regression has applied different regression methods to produce meta-learners (as seen in section 2), yielding relative success. However, to the best of our knowledge, there is no evaluation of the use of SVMs in this task.

Figure 1 brings the architecture of the Meta-Regression which summarizes our proposal. Each meta-example is composed by the meta-features of a learning task and the performance information derived from the empirical evaluation of the learning algorithm on the task. The set of generated meta-examples is given as input to the SVM algorithm, which will produce a regression model responsible

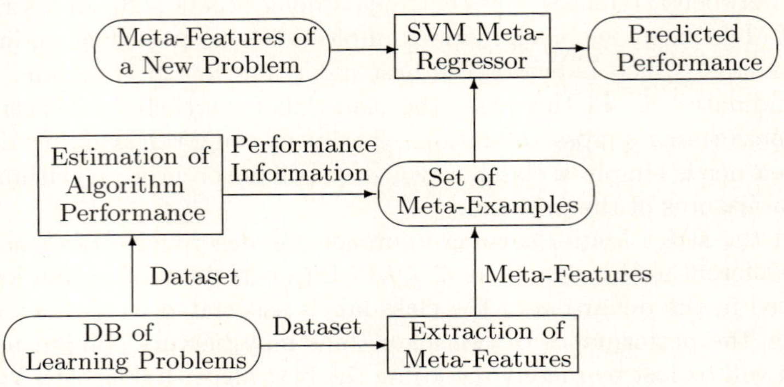

Fig. 1. Architecture of Meta-Regression

for predicting the algorithm's performance for new problems based on its meta-features.

A case study was performed in our work in a meta-learning task which corresponds to predict the performance of Multi-Layer Perceptron (MLP) networks [15] in regression problems. In this case study, we generated a set of 50 meta-examples from the application of the MLP in 50 regression problems (see section 3.1). Each meta-example is associated to a single problem and stores: (1) the value of 10 descriptive meta-features (see section 3.2); and (2) the test error obtained by the MLP when evaluated in the regression problem (see section 3.3). The set of meta-examples is given as input to a regression algorithm which will be able to predict the error of the MLP for new problems only based on the meta-attributes of the problems.

In the following sections, we provide details about the construction of the set of meta-examples, as well as the details on the SVMs used for Meta-Regression.

3.1   Datasets

In order to generate meta-examples, we collected 50 datasets corresponding to 50 different regression problems, available in the WEKA project[1]. On average, the collected datasets presented 4,392 examples and 13.92 attributes.

We observed in these datasets a large variability in both the number of training examples and the number of attributes. This variation may be convenient to Meta-Learning studies, since it is expected that the algorithms present significantly different patterns of performance, depending on the problem being solved.

The attributes in the original datasets were normalized to the [-1; +1] interval, aiming a better treatment by the MLP network. The order of the examples in each dataset were randomly permuted, in order to avoid eventual biases derived from the original dataset acquisition.

---

[1] These datasets are specifically the sets provided in the files *numeric* and *regression* available to download in http://www.cs.waikato.ac.nz/ml/weka/

3.2 Meta-features

In the developed work, a total number of 10 meta-features was used to describe the datasets of regression problems:

1. LogE - Log of the number of training examples;
2. LogEA - Log of the ration between number of training examples and number of attributes;
3. MinCT, MaxCT, MeanCT and StdCT - Minimum, maximum, mean and standard deviation of the absolute values of correlation between predictor attributes and the target attribute;
4. MinCAtr, MaxCAtr, MedCAtr and DesvCAtr - Minimum, maximum, mean and standard deviation of the absolute values of correlation between pairs of predictor attributes.

The meta-feature LogE is an indicator of the amount of data available for training, and LogEA, in turn, indicates the dimensionality of the dataset. The meta-features MinCT, MaxCT, MedCT and DesvCT indicate the amount of relevant information available to predict the target attribute. The meta-features MinCAtr, MaxCAtr, MedCAtr and DesvCAtr, in turn, indicate the amount of redundant information in the dataset. This set of meta-features was chosen by considering features adopted in previous work. As this set is probably non optimal, in future work, we will consider new features.

3.3 Performance Information

The final step to generate the meta-examples is to evaluate the performance of the MLP in the 50 collected regression tasks. From this evaluation, we can produce the performance information stored in the meta-examples which will correspond to the target attribute in the Meta-Regression task. In order to measure the performance of the MLP in each problem, the following methodology of evaluation was applied.

Each dataset was divided in the training, validation and test subsets, in the proportion of 50%, 25% and 25%, respectively. As usual, the training subset was used in the adjustment of the MLP's weights, the validation subset is used to estimate the generalization performance of the MLP during training, and the test subset is used to evaluated the performance of the trained MLP. In our work, we applied the standard Backpropagation (BP) algorithm [15] to train the MLP network[2]. The optimal number of hidden nodes[3] was empirically defined, by testing the values 1, 2, 4, 8, 16 and 32. For each number of hidden nodes, the MLP is trained 10 times with random initial weights. The stopping criteria of the training process followed the benchmarking rules provided in [20].

The optimal number of hidden nodes was finally chosen as the value in which the trained MLP obtained the lowest average NMSE (Normalized Mean of

---

[2] The BP algorithm was implemented by using the NNET Matlab toolbox [19]. Learning rates were defined by default.

[3] The MLP was defined with only one hidden layer.

Squared Errors) on the validation subset over the 10 training runs. The NMSE measure is defined as:

$$NMSE = \frac{\sum_{i=1}^{n}(y_i - \widetilde{y}_i)^2}{\sum_{i=1}^{n}(y_i - \overline{y})^2} \qquad (1)$$

In the equation, n is the number of examples in the validation subset, y_i and $\widetilde{y}_i$ are respectively the target and the predicted value for example i, and $\overline{y}$ is the average of the target attribute.

The performance information which is stored in the meta-example is the average NMSE obtained by the trained MLP (with optimal number of hidden nodes) on the test subset. According to [21], a property of NMSE which is interesting for Meta-Learning is that the values have no scale and are comparable across different datasets (which it would not be feasible if we had used a non-normalized error measure). Values of NMSE lower to 1 indicate that the MLP provided better predictions than the mean value. Values higher than 1 indicate that the MLP was not useful in the regression problem.

We highlight here that there are different algorithms to train MLPs, as well as other methodologies of training, that could have been used to improve performance. However, our aim in this work is not to achieve the best possible performance with MLPs but to *predict* the learning performance. Other learning algorithms to train MLPs (such as the Levenberg-Marquardt algorithm [22]) can be applied in the future as new case studies, possibly with more effective strategies to train the networks.

3.4   Meta-regressor

In our case study, the Meta-Regression task is to predict the NMSE measure of the MLP based on the features of the problems given as input. In our experiments, this task is dealt with SVM regressors.

An important aspect in the development of SVM is the chosen kernel function. In our work, we evaluated the use of two different types of kernel functions in the SVMs. First, we evaluated homogeneous polynomial kernels represented as:

$$K(x, x') = (x \cdot x')^p \qquad (2)$$

In the above equation, x and x' are instances in an attribute space. In our experiments, we used the values $p = 1$ (linear kernel) and $p = 2$ (quadratic kernel), which are the more common options for polynomial kernels. We also deployed in our work a Radial Basis Function (RBF) kernel in the form:

$$K(x, x') = e^{-\gamma \cdot \|x - x'\|^2} \qquad (3)$$

The RBF kernel is a non-linear function that, in comparison to the polynomial kernels, it is expected to handle more complex relationships between predictor and target attributes. In our experiments, we evaluated the values 0.01, 0.05 and 0.1 for the parameter γ.

Obviously, there are other possibilities of kernel functions (e.g., sigmoid kernels), as well as parameter settings that we intend to experiment in future work. Besides, automated procedures can be used in the future to define the values of these parameters [23].

In this work, we deployed the implementation of SVMs provided in the WEKA environment [24], which applies the sequential minimal optimization algorithm proposed by [13] for training the SVM regression model.

4   Experiments and Results

In this section, we evaluated the performance of the Meta-Regression process for the generated set of meta-examples. In our experiments, the SVM meta-regressor was evaluated by using a leave-one-out procedure. At each step, 49 meta-examples are used as the training set, and the remaining meta-example is used to test the trained meta-learner. This step is repeated 50 times, using at each time a different test meta-example. The meta-learning performance is then evaluated based on the predictions generated by the meta-learner in the test samples.

Two different criteria were used to evaluated the meta-learner: (1) the Normalized Mean of Squared Errors (NMSE) computed for the predictions of the meta-learner in the test meta-examples; and (2) the Correlation (COR) between the predictions generated by the meta-learner and the true values of the target attribute stored in the test meta-examples.

As a basis of comparison, the above methodology of experiments was also applied using three different methods as meta-regressors:

1. M5 algorithm: which was proposed by Quinlan [25] to induce regression trees;
2. 1-Nearest Neighbor (1-NN) algorithm: a special case of instance-based learning algorithm [26];
3. Linear Regression (LR) model.

All these methods were already used in previous work as meta-regressors (see, for instance, [3]). We highlight that these algorithms are representatives of different families of regression methods, providing different inductive biases for the learning problem being solved in the case study.

As it can be seen in Table 1, the SVM regressor with polynomial kernels (linear and quadratic) obtained better results on both evaluation measures when compared to the benchmark methods M5, 1-NN and LR. The best result among the polynomial kernels was obtained by the quadratic kernel, which also yielded a very competitive result (the second best one) when we include the SVM with RBF kernel in the comparison.

Considering the runs with RBF kernel, we observed that the performance of the SVM was very sensitive to the value of parameter γ. For $\gamma = 0.1$, the SVM obtained the best result over all evaluated algorithms and parameter settings. For $\gamma = 0.05$, the SVM also performed well compared to the other algorithms (yielding the third best result). However, for $\gamma = 0.01$, the SVM obtained a

Table 1. Results obtained by Meta-Regression in the leave-one-out experiment

	NMSE	COR
SVM (linear)	0.486	0.731
SVM (quadratic)	0.457	0.758
SVM (RBF, $\gamma = 0.1$)	0.374	0.794
SVM (RBF, $\gamma = 0.05$)	0.484	0.724
SVM (RBF, $\gamma = 0.01$)	0.595	0.701
M5	0.605	0.631
1-NN	0.539	0.710
LR	0.595	0.658

performance which was only better than the method M5. This result indicates that an adequate choice of the parameter of RBF kernels (i.e., the model selection) is an aspect that should be more carefully addressed in the case of SVM meta-regressors. Model selection is in fact a relevant topic of research in SVMs, which can be handled, for instance, by considering the characteristics of the data [13,21,27] and by deploying search techniques [23,28,29]. Such strategies will be considered in future work to model selection in our case study.

5   Conclusion

In this work, we presented the use of SVMs to Meta-Regression, which aims to predict the performance of learning algorithms based on the features of the learning problems being solved.

Experiments were performed in a case study which consisted in predicting the numerical performance of MLP networks when applied to regression tasks. A set of 50 meta-examples was generated in this case study in order to perform a comparative analysis of different meta-regressors. The results of a leave-one-out experiment revealed the viability of using SVMs in the investigated case study. The performance of the SVM meta-regressor was in general better than the performance obtained by the benchmark algorithms applied as a basis of comparison.

The current work has limitations that will be dealt with in future work. Although we applied, in the case study, polynomial and RBF kernels which are widely used in the literature, different other functions can be evaluated. Furthermore, we only tested few values for the kernel parameters, which could be optimized by using more sophisticated approaches. We also include as future work, the evaluation of SVM meta-regressors in new case studies related to other machine learning algorithms.

References

1. Giraud-Carrier, C., Vilalta, R., Brazdil, P.: Introduction to the special issue on meta-learning. Machine Learning 54(3), 187–193 (2004)
2. Kalousis, A., Hilario, M.: Representational issues in meta-learning. In: Proceedings of the 20th International Conferente on Machine Learning, pp. 313–320 (2003)

3. Gama, J., Brazdil, P.: Characterization of classification algorithms. In: Pinto-Ferreira, C., Mamede, N.J. (eds.) EPIA 1995. LNCS, vol. 990, pp. 189–200. Springer, Heidelberg (1995)

4. Aha, D.: Generalizing from case studies: A case study. In: Proceedings of the 9th International Workshop on Machine Learning, pp. 1–10. Morgan Kaufmann, San Francisco (1992)

5. Koepf, C., Taylor, C., Keller, J.: Meta-analysis: Data characterisation for classification and regression on a meta-level. In: Proceedings of the International Symposium on Data Mining and Statistics (2000)

6. Bensusan, H., Alexandros, K.: Estimating the predictive accuracy of a classifier. In: Proceedings of the 12th European Conference on Machine Learning, pp. 25–36 (2001)

7. Brazdil, P., Soares, C., da Costa, J.: Ranking learning algorithms: Using IBL and meta-learning on accuracy and time results. Machine Learning 50(3), 251–277 (2003)

8. Kalousis, A., Gama, J., Hilario, M.: On data and algorithms - understanding inductive performance. Machine Learning 54(3), 275–312 (2004)

9. Prudêncio, R.B.C., Ludermir, T.B.: Meta-learning approaches to selecting time series models. Neurocomputing 61, 121–137 (2004)

10. Koepf, C.: Meta-regression: performance prediction, pp. 89–106 (2006)

11. Prudêncio, R.B.C., Ludermir, T.B.: Active learning to support the generation of meta-examples. In: Proc. of the International Conference on Artificial Neural Networks, pp. 817–826 (2007)

12. Prudêncio, R.B.C., Ludermir, T.B.: Active selection of training examples for meta-learning. In: Proc. of the International Conference on Hybrid Intelligent Systems (2007)

13. Smola, A., Scholkopf, B.: A tutorial on support vector regression. Statistics and Computing 14(3), 199–222 (2004)

14. Wang, L.: Support Vector Machines: Theory and Applications. Springer, Heidelberg (2005)

15. Rumelhart, D.E., Hinton, G.E., Williams, R.J.: Learning representations by back-propagating errors. Nature 323, 533–536 (1986)

16. Vilalta, R., Giraud-Carrier, C., Brazdil, P.: Using meta-learning to support data-mining. International Journal of Computer Science Application I(31), 31–45 (2004)

17. Prudêncio, R.B.C., Ludermir, T.B., de Carvalho, F.A.T.: A modal symbolic classifier to select time series models. Pattern Recognition Letters 25(8), 911–921 (2004)

18. Leite, R., Brazdil, P.: Predicting relative performance of classifiers from samples. In: Proceedings of the 22nd International Conference on Machine Learning (2005)

19. Demuth, H., Beale, M.: Neural Network Toolbox: For use with MATLAB: User's Guide. The Mathworks (1993)

20. Prechelt, L.: A set of neural network benckmark problems and benchmarking rules. Technical Report 21/94, Fakultät für Information, Universitä Karlsruhe, Karlsruhe, Germany (Spetember 1994)

21. Soares, C., Brazdil, P., Kuba, P.: A meta-learning approach to select the kernel width in support vector regression. Machine Learning 54(3), 195–209 (2004)

22. Levenberg, K.: A method for the solution of certain non-linear problems in least squares. Quarterly Journal of Applied Mathmatics II(2), 164–168 (1944)

23. Chapelle, O., Vapnik, V., Bousquet, O., Mukherjee, S.: Choosing multiple parameters for support vector machines. Machine Learning 46(1), 131–159 (2002)

24. Witten, I.H., Frank, E.(eds.): WEKA: machine learning algorithms in Java. University of Waikato, New Zealand (2003)

25. Quinlan, J.: Learning with continuous classes. In: Proceedings of the Australian Joint Conference on Artificial Intelligence, pp. 343–348 (1992)
26. Aha, D., Kibler, D.: Instance-based learning algorithms. Machine Learning 6(3), 37–66 (1991)
27. Kuba, P., Brazdil, P., Soares, C., Woznica, A.: Exploiting sampling and meta-learning for parameter setting support vector machines. In: Garijo, F.J., Riquelme, J.-C., Toro, M. (eds.) IBERAMIA 2002. LNCS (LNAI), vol. 2527, pp. 217–225. Springer, Heidelberg (2002)
28. Cawley, G.: Model selection for support vector machines via adaptive step-size tabu search. In: Proceedings of the International Conference on Artificial Neural Networks and Genetic Algorithms, pp. 434–437 (2001)
29. Lessmann, S., Stahlbock, R., Crone, S.: Genetic algorithms for support vector machine model selection. In: Proceedings of the International Joint Conference on Neural Networks 2006, pp. 3063–3069 (2006)

Municipal Creditworthiness Modelling by Kohonen's Self-Organizing Feature Maps and Fuzzy Logic Neural Networks

Petr Hajek and Vladimir Olej

Institute of System Engineering and Informatics
Faculty of Economics and Administration
University of Pardubice
Studentska 84, 53210 Pardubice
Czech Republic
petr.hajek@upce.cz, vladimir.olej@upce.cz

Abstract. The paper presents the design of municipal creditworthiness parameters. Further, the design of model for municipal creditworthiness classification is presented. The model is composed of Kohonen's self-organizing feature maps and fuzzy logic neural networks, where the output of Kohonen's self-organizing feature maps represents the input of fuzzy logic neural networks. It is a feed-forward fuzzy logic neural network with three layers. Standard neurons are replaced by fuzzy neurons in the fuzzy logic neural network.

Keywords: Municipal creditworthiness, Kohonen's self-organizing feature maps, fuzzy logic neural networks, classification.

1 Introduction

Municipal creditworthiness [1] is the ability of a municipality to meet its short-term and long-term financial obligations. It is determined by factors (parameters) relevant to the assessed object. High municipal creditworthiness shows a low credit risk, while the low one shows a high credit risk. Municipal creditworthiness evaluation is currently being realized by methods (Scoring Systems, Rating and Rating-based Models) combining mathematical-statistical methods and expert opinion [1] and [2]. The output of the introduced methods, i.e. municipal creditworthiness, is represented either by a score (Scoring Systems) or by an assignment of the i-th object $o_i \in O$, $O = \{o_1, o_2, \ldots, o_i, \ldots, o_n\}$ to the j-th class $\omega_{i,j} \in \Omega$, $\Omega = \{\omega_{1,j}, \omega_{2,j}, \ldots, \omega_{i,j}, \ldots, \omega_{n,j}\}$.

Rating [2] is an independent expert evaluation based on complex analysis of all known risk parameters of municipal creditworthiness, however, it is considered to be rather subjective. Municipalities are classified into classes $\omega_{i,j} \in \Omega$ by Rating-based Models [2]. The classes $\omega_{i,j} \in \Omega$ are assigned to the municipalities by rating agencies. Only several municipalities of the Czech republic have assigned the class $\omega_{i,j} \in \Omega$.

V. Kůrková et al. (Eds.): ICANN 2008, Part I, LNCS 5163, pp. 533–542, 2008.

Therefore, the methods capable of processing and learning the expert knowledge, enabling their user to generalize and properly interpret, have proved to be most suitable. For example, fuzzy inference systems [3], and neural networks [4] and [5] are suitable for municipal creditworthiness evaluation. Neural networks are appropriate for municipal creditworthiness modelling due to their ability to learn, generalize and model non-linear relations. Nevertheless, the computational speed and robustness are retained. The use of natural language is typical for the municipal creditworthiness evaluation process. Moreover, the formulation of municipal creditworthiness' parameters in exact numbers does not correspond to reality. The previously mentioned problems can be realized by fuzzy logic [6]. Advantages of both methods can be gained by neuro-fuzzy systems [7].

The paper presents the design of municipal creditworthiness parameters. Only those parameters were selected which show low correlation dependences. Further, the paper presents the basic concepts of the Kohonen's Self-organizing Feature Maps (KSOFMs) and fuzzy logic neural networks (FLNNs). The contribution of the paper lies in the model design for municipal creditworthiness evaluation. The model realizes the advantages of both the unsupervised methods (combination of the KSOFM and K-means algorithm) and fuzzy logic (FLNNs).

2   Municipal Creditworthiness Parameters Design

In [8] common categories of parameters there are mentioned namely economic, debt, financial and administrative categories. The economic, debt and financial parameters are pivotal. Economic parameters affect long-term credit risk. The municipalities with more diversified economy and more favourable socio-economic conditions are better prepared for the economic recession. Debt parameters include the size and structure of the debt. Financial parameters inform about the budget implementation. Their values are extracted from the municipality budget. Based on the presented facts, the data matrix $\mathbf{P}$ can be designed (Table 1).

In Table 1 $o_i \in O$, $O = \{o_1, o_2, \ldots, o_i, \ldots, o_n\}$ are objects (municipalities), x_k is the k-th parameter, $x_{i,k}$ is the value of the parameter x_k for the i-th object $o_i \in O$, $\omega_{i,j}$ is the j-th class assigned to the i-th object $o_i \in O$, $\mathbf{p}_i = (x_{i,1}, x_{i,2}, \ldots, x_{i,k}, \ldots, x_{i,m})$ is the i-th pattern, $\mathbf{x} = (x_1, x_2, \ldots, x_k, \ldots, x_m)$ is the parameters vector. The design of parameters, based on previous correlation analysis and recommendations of notable experts, can be realized as presented in Table 2.

Table 1. Data matrix design

	x_1		x_k		x_m	
o_1	$x_{1,1}$	...	$x_{1,k}$	...	$x_{1,m}$	$\omega_{1,j}$
...	...	...	...	...	...	...
o_i	$x_{i,1}$	...	$x_{i,k}$	...	$x_{i,m}$	$\omega_{i,j}$
...	...	...	...	...	...	...
o_n	$x_{n,1}$	...	$x_{n,k}$	...	$x_{n,m}$	$\omega_{n,j}$

Table 2. Municipal creditworthiness parameters design

Parameters	
Economic	$x_1 = PO_r$, PO_r is population in the r-th year.
	$x_2 = PO_r/PO_{r-s}$, PO_{r-s} is population in the year r-s, and s is the selected time period.
	$x_3 = U$, U is the unemployment rate in a municipality.
	$x_4 = \sum_{i=1}^{k} (PZO_i/PZ)^2$, PZO_i is the employed population of the municipality in the i-th economic sector, i=1,2, ...,k, PZ is the total number of employed inhabitants, k is the number of the economic sector.
Debt	$x_5 = DS/OP$, $x_5 \in <0,1>$, DS is debt service, OP are periodical revenues.
	$x_6 = CD/PO$, CD is a total debt.
	$x_7 = KD/CD$, $x_7 \in <0,1>$, KD is short-term debt.
Financial	$x_8 = OP/BV$, $x_8 \in R^+$, BV are current expenditures.
	$x_9 = VP/CP$, $x_9 \in <0,1>$, VP are own revenues, CP are total revenues.
	$x_{10} = KV/CV$, $x_{10} \in <0,1>$, KV are capital expenditures, CV are total expenditures.
	$x_{11} = IP/CP$, $x_{11} \in <0,1>$, IP are capital revenues.
	$x_{12} = LM/PO$, [Czech Crowns], LM is the size of the municipal liquid assets.

3   Design of Model Based on Neuro-fuzzy Systems

Municipal creditworthiness evaluation is realized by a design of the model. Data pre-processing makes the suitable economic interpretation of results possible. Municipalities are assigned to clusters by unsupervised methods. The clusters are used as the inputs of the FLNN. Municipal creditworthiness modelling represents a classification problem.

3.1   Model Design and Data Pre-processing

The classes $\omega_{i,j} \in \Omega$, $\Omega = \{\omega_{1,j}, \omega_{2,j}, \ldots, \omega_{i,j}, \ldots, \omega_{n,j}\}$ of municipal creditworthiness are not known a priori. Therefore, it is suitable to realize the modelling of municipal creditworthiness by unsupervised methods. Based on the analysis of unsupervised methods, the combination of the KSOFM and K-means algorithm is a suitable method for municipal creditworthiness modelling [9]. This is also confirmed by the results of the KSOFM applications in the field of finance [10,11,12,13]. The results of the unsupervised methods are used as the inputs of the FLNN [7] in the model presented in Fig. 1. Supervised learning is realized by the FLNN, which makes the suitable economic interpretation of created clusters possible. The FLNN represents a method of generating rules R_u and fuzzy sets. This process can be also realized using other fuzzy classifiers [14] and [15]. Data pre-processing is carried out by means of data standardization. Thereby, the dependency on units is eliminated.

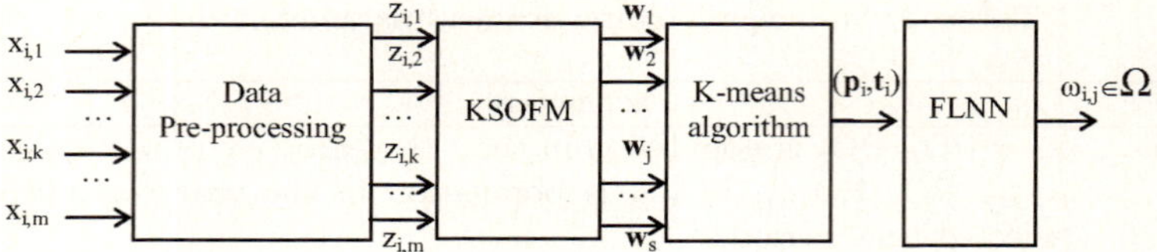

Fig. 1. The model design for the classification municipalities, where $\mathbf{p}_i=(x_{i,1},x_{i,2}, \dots,x_{i,k}, \dots,x_{i,m})$ is the i-th pattern, $z_{i,1},z_{i,2}, \dots,z_{i,k}, \dots,z_{i,m}$ are standardized values of parameters $\mathbf{x}=(x_1,x_2, \dots,x_k, \dots,x_m)$ for the i-th object $o_i \in O$, $O=\{o_1,o_2, \dots,o_i, \dots,o_n\}$, $(\mathbf{p}_i,\mathbf{t}_i)$ are patterns $\mathbf{p}_i$ assigned to classes $\omega_{i,j} \in \Omega$, $\Omega=\{\omega_{1,j},\omega_{2,j}, \dots,\omega_{i,j}, \dots,\omega_{n,j}\}$, $\omega_{i,j} \in \Omega$ is the output of the FLNN

3.2 Municipal Creditworthiness Modelling by Kohonen's Self-Organizing Feature Maps

The Kohonen's self-organizing feature maps [4] and [16] are based on competitive learning strategy. The input layer serves the distribution of the input patterns $\mathbf{p}_i$, i=1,2, $\dots$,n. The neurons in the competitive layer serve as the representatives (Codebook Vectors), and they are organized into topological structure (most often a two-dimensional grid), which designates the neighbouring network neurons. First, the Euclidean distances d_j are computed between the i-th pattern $\mathbf{p}_i$ and synapse weights $\mathbf{w}_{i,j}$ of all neurons in the competitive layer. The winning neuron j* (Best Matching Unit, BMU) is chosen, for which the distance d_j from the given pattern $\mathbf{p}_i$ is minimum. The output of this neuron is active, while the outputs of other neurons are inactive. The aim of the KSOFM learning is to approximate the probability density of the real input vectors $\mathbf{p}_i \in R^n$ by the finite number of representatives $\mathbf{w}_{j*} \in R^n$, where j=1,2, $\dots$,s. When the representatives $\mathbf{w}_{j*}$ are identified, the representative $\mathbf{w}_{j*}$ of the BMU is assigned to each vector $\mathbf{p}_i$. In the learning process of the KSOFM, it is necessary to define the concept of neighbourhood function, which determines the range of cooperation among the neurons, i.e. how many representatives $\mathbf{w}_{j*}$ in the neighbourhood of the BMU will be adapted, and to what degree. Gaussian neighbourhood function is in common use, which is defined as

$$h(j^*,j) = e^{\left(-\frac{d_E^2(j^*,j)}{\lambda^2(t)}\right)}, \tag{1}$$

where $h(j^*,j)$ is neighbourhood function, $d_E^2(j^*,j)$ is Euclidean distance of neurons j^* and j in the grid, $\lambda(t)$ is the size of the neighbourhood in time t. After the BMUs are found, the adaptation of synapse weights $\mathbf{w}_{i,j}$ follows. The principle of the sequential learning algorithm [16] is the fact, that the representatives $\mathbf{w}_{j*}$ of the BMU and its topological neighbours move towards the actual input vector $\mathbf{p}_i$ according to the relation

$$\mathbf{w}_{i,j}(t+1) = \mathbf{w}_{i,j}(t) + \eta(t) \times h(j^*,j) \times (\mathbf{p}_i(t) - \mathbf{w}_{i,j}(t)), \tag{2}$$

where $\eta(t) \in (0,1)$ is the learning rate. The batch-learning algorithm of the KSOFM [16] is a variant of the sequential algorithm. The difference consists in the fact that the whole training set O_{train} passes through the KSOFM only once, and only then the synapse weights $\mathbf{w}_{i,j}$ are adapted. The adaptation is realized by replacing the representative $\mathbf{w}_{j*}$ with the weighted average of the input vectors $\mathbf{p}_i$ [16].

3.3 Municipal Creditworthiness Modelling by Fuzzy Logic Neural Networks

The results of unsupervised learning are used as the inputs of the FLNN suitable for classification problem realization [7]. The FLNN parameters make the appropriate interpretation possible. It is a feed-forward FLNN with three layers. The input layer represents the input parameters $\mathbf{x}=(x_1,x_2, \ldots,x_k, \ldots,x_m)$, the hidden layer represents rules $R_1,R_2, \ldots,R_u, \ldots,R_N$ and the output layer denotes classes $\omega_{i,j} \in \Omega$. The rule R_u can be defined as follows [15]

$$R_u : \text{IF } x_1 \text{ is } A_1(x_1) \text{ AND } x_2 \text{ is } A_2(x_2) \text{ AND } \ldots \text{ AND } x_m \text{ is } A_m(x_m) \\ \text{THEN } \omega_{i,j} \in \Omega, \tag{3}$$

where $A_k(x_k)$, $k=1,2, \ldots,m$, are the values of linguistic variables appropriate for the corresponding fuzzy sets. Standard neurons are replaced by fuzzy neurons in the FLNN. Contrary to the standard neuron, the synapse weights among fuzzy neurons assign the value of membership function μ to the input value. The output value μ_y can be defined as

$$\mu_y = \mu_1(x_1) \otimes \mu_2(x_2) \otimes \ldots \otimes \mu_p(x_p), \tag{4}$$

where operator $\otimes$ represents one of the fuzzy operations (MIN,MAX) [17,18]. The membership function μ_y is defined as a continuous function, whose parameters are adapted in the learning process. The FLNN structure for a classification problem realization is presented as a feed-forward FLNN with three layers. Neurons in the hidden layer use t-norms as activation functions, neurons in the output layers use t-conorms [3]. The values of linguistic variable $A_k(x_k)$ appropriate for the corresponding fuzzy sets are represented by synapse weights $\mathbf{w}(x_k,R_u)$. Synapse weights $\mathbf{w}(R_u,\omega_{i,j})$ represent rules weights R_u. If $\mathbf{w}(R_u,\omega_{i,j}) \in [0,1]$, then every neuron in the hidden layer is connected with the only one neuron in the output layer. The aim of the designed classifier is to achieve a maximum classification accuracy S_{test} of the testing set O_{test}. At the same time, suitable interpretation of the designed model is demanded (i.e. low number N of rules R_u with the low number of variables in antecedent, the low number v_k of membership functions μ, etc.). The classifier is designed in two steps. First, the structure of the FLNN is determined. The structure is dependent on the number of clusters q created by the KSOFM. Then, parameters of the FLNN are identified in the process of learning. The rules R_u and fuzzy sets represent the parameters.

Let the FLNN has m inputs $\mathbf{x}=(x_1,x_2, \ldots,x_k, \ldots, x_m)$, $N<N_{max}$ neurons representing rules $R_1,R_2, \ldots,R_u, \ldots,R_N$ and q output neurons representing classes

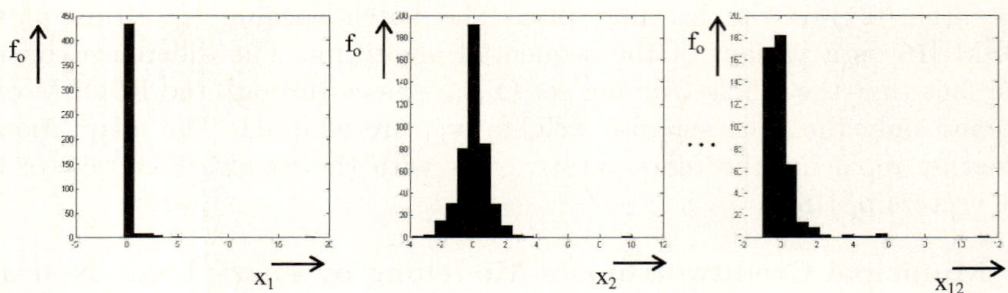

Fig. 2. Histograms of parameters $x_1, x_2, \ldots, x_m$, m=12, where f_o are the frequencies of municipalities $o_i \in O$

$\omega_{i,j} \in \Omega$. Next, let $O = \{(\mathbf{p}_1, \mathbf{t}_1), (\mathbf{p}_2, \mathbf{t}_2), \ldots, (\mathbf{p}_i, \mathbf{t}_i), \ldots, (\mathbf{p}_n, \mathbf{t}_n)\}$ is a set of n patterns, where $\mathbf{p}_i \in R^n$ is the vector of parameters values and $\mathbf{t}_i \in (0, 1)^q$ is the vector of patterns $\mathbf{p}_i$ assignments to classes $\omega_{i,j} \in \Omega$. Then, the learning algorithms of generating rules R_u and fuzzy sets are defined in [7].

4   Analysis of the Results

The sample of the input data matrix $\mathbf{P}$ of the KSOFM is shown in Fig. 2. The linear initialization, batch-training algorithm, number of representatives s=108, the starting size of the neighbourhood $\lambda(t)$=3, and the final size of the neighbourhood $\lambda(t)$=1 are the input parameters of the KSOFM. Using the KSOFM as such can detect the structure in the data. The K-means algorithm can be applied to the adapted KSOFM in order to find clusters $c_1, c_2, \ldots, c_q$ as it is presented in Fig. 3.

The K-means algorithm belongs to non-hierarchical algorithms of cluster analysis, where patterns $\mathbf{p}_1, \mathbf{p}_2, \ldots, \mathbf{p}_n$, are assigned to clusters $c_1, c_2, \ldots, c_r, \ldots, c_q$. The number of clusters q=6 is determined by indexes evaluating the quality of clustering [19]. Interpretation of clusters is realized by the values of parameters $x_1, x_2, \ldots, x_m$, m=12, for individual representatives $\mathbf{w}_{j*}$ (Fig. 4).

The learning rate η=0.1, number of epochs e=500 and maximum aggregation function represent the input parameters of the FLNN. Municipal creditworthiness

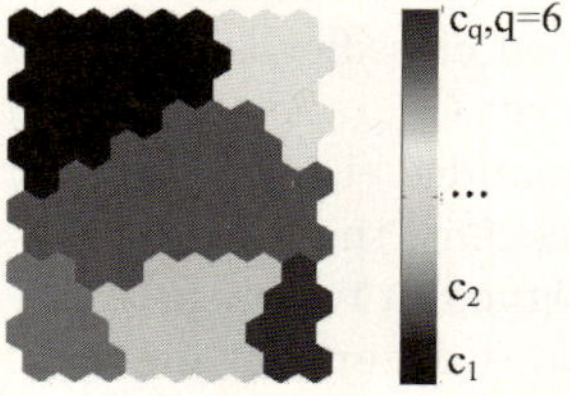

Fig. 3. Clustering of the KSOFM by K-means algorithm. Clustering process is realized in two levels. In the first level, n objects $o_i \in O$ are reduced to representatives $\mathbf{r}_1, \mathbf{r}_2, \ldots, \mathbf{r}_s$ by the KSOFM, the s representatives are clustered into q clusters.

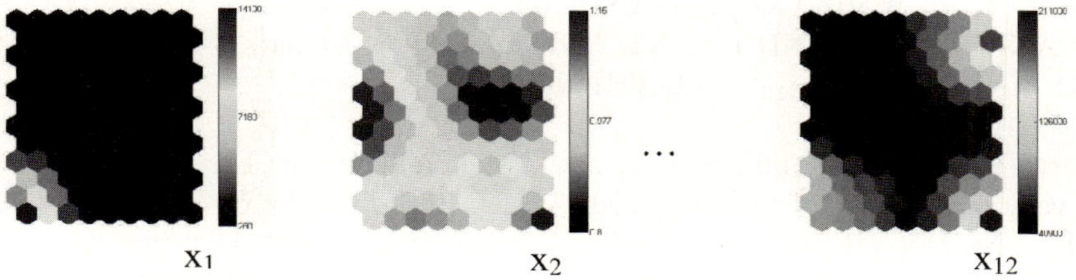

x_1 x_2 x_{12}

Fig. 4. Values of parameters $x_1, x_2, \ldots, x_m$, m=12, for individual representatives $\mathbf{w}_{j*}$

modelling is realized for different numbers v_k, shapes of membership functions, and for different numbers N of rules R_u. The set $O=\{o_1, o_2, \ldots, o_i, \ldots, o_n\}$ of objects (n=452 municipalities of Pardubice region, the Czech republic) is divided into training set O_{train} and testing set O_{test} The division of the set O is realized a number of times in order to get a representative sample of experiments. The quality of classification is measured by classification accuracy S_{test}, and its standard deviation σ_{test} for the testing set O_{test}.

The results of the classification for triangular membership functions are presented in Table 3. If the number v_k of membership functions increases, then the classification accuracy S_{test} increases too. Both an accurate classification and a low number N of rules R_u and v_k of membership functions should represent an optimum result. According to Table 3, the result can be acceptable, for which classification accuracy S_{test}=92.12%, standard deviation σ_{test}=3.56% and number N of rules R_u=165. A sample of the rules R_u for the number v_k=5 of membership functions is as follows

Table 3. The results of classification for triangular membership functions, S_{test} is the classification accuracy for the testing set O_{test}

v_k	S_{test} [%]	σ_{test} [%]	N
2	82.76	5.34	23
3	86.72	3.56	69
4	87.14	4.68	113
5	92.92	3.56	165
6	91.82	4.31	216
7	95.12	2.39	263
8	95.35	3.50	305
9	97.78	2.63	327

R_1: IF x_1 is VS AND x_2 is VS AND x_3 is VS AND x_4 is VS AND x_5 is VS AND x_6 is VS AND x_7 is VS AND x_8 is VS AND x_9 is VS FND x_{10} is VS AND x_{11} is VL AND x_{12} is VS THEN c_1

R_2: IF x_1 is VL AND x_2 is VS AND x_3 is VS AND x_4 is VS AND x_5 is VS AND x_6 is VS AND x_7 is M AND x_8 is VS AND x_9 is VS AND x_{10} is VS AND x_{11} is VL AND x_{12} is VS THEN c_1

...

R_{165}: IF x_1 is VS AND x_2 is VS AND x_3 is VS AND x_4 is VS AND x_5 is VS AND x_6 is VL AND x_7 is VS AND x_8 is M AND x_9 is S AND x_{10} is VS AND x_{11} is VS AND x_{12} is VL THEN c_6,

where VS is a very small, S is a small, M is a medium, L is a large and VL is a very large value of parameter x_k, k=12. Based on the created rules R_1,R_2, ...,R_N, N=165, clusters c_1,c_2, ...,c_q, q=6, can be labelled by classes $\omega_{i,j}\in\Omega$, where the class $\omega_{i,j}\in\Omega$, j=1 represents the best municipal creditworthiness, while class $\omega_{i,j}\in\Omega$, j=q, represents the worst one. This process has to be realized by an expert from the field of municipal creditworthiness evaluation. Then, the classes $\omega_{i,j}\in\Omega$, j=6, can be characterized as presented in Table 4.

Table 4. Descriptions of classes $\omega_{i,j}\in\Omega$

	Description
$\omega_{i,1}$	High ability of a municipality to meet its financial obligation. Very favourable economic conditions, low debt and excellent budget implementation.
$\omega_{i,2}$	Very good ability of a municipality to meet its financial obligation.
$\omega_{i,3}$	Good ability of a municipality to meet its financial obligation.
$\omega_{i,4}$	A municipality with stable economy, medium debt and good budget implementation.
$\omega_{i,5}$	Municipality meets its financial obligation only under favourable economic conditions.
$\omega_{i,6}$	A municipality meets its financial obligations with difficulty, the municipality is highly indebted.

The classification of the municipalities $o_i\in O$, i=452, into classes $\omega_{i,j}\in\Omega$, j=6, by the FLNN is shown in Fig. 5.

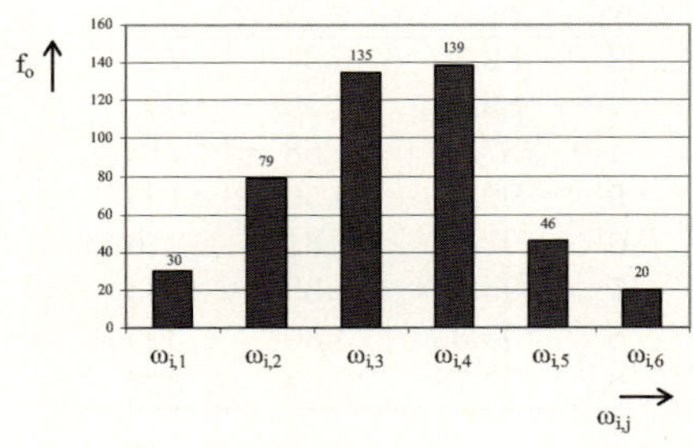

Fig. 5. Classification of the municipalities $o_i\in O$, i=452, into classes $\omega_{i,j}\in\Omega$, j=6, by the FLNN

5   Conclusion

The paper presents the design of municipal creditworthiness parameters. Further, the model is designed whereby municipal creditworthiness evaluation is realized.

First, the data are pre-processed. Next, municipal creditworthiness modelling is realized by the KSOFM. Number of the KSOFMs structures was designed with various input parameters in the process of modelling. The visualization of clusters by the KSOFM makes the proper economic interpretation of results possible. However, the knowledge presented to experts by means of rules is more suitable. Therefore, the outputs of the KSOFM are used as the inputs of the FLNN. The supervised classification by the FLNN makes the use of both fuzzy logic and neural networks possible. The goal of the learning process is achieved as the FLNN disposes of a high classification accuracy S_{test} with a low number v_k of membership functions and rules R_u, u=1,2, ...,N. The clusters can be labelled by an expert opinion based on the created rules. As result, the designed model is capable of processing and learning the expert knowledge and enabling its user to generalize at the same time. The designed model was carried out in NefClass (FLNN) and MATLAB (KSOFM) in MS Windows XP operation system.

Acknowledgements

The work is in part supported by the National Science Foundation of the Czech Republic, Grant No. 402/08/0849 and the Czech Ministry of Environment, Grant No. SP/4i2/60/07.

References

1. Olej, V., Hajek, P.: Modelling of Municipal Rating by Unsupervised Methods. WSEAS Transactions on Systems 6, 1679–1686 (2006)
2. Huang, Z.: Credit Rating Analysis with Support Vector Machines and Neural Networks: A Market Comparative Study. Decision Support Systems 37, 543–558 (2004)
3. Olej, V.: Modelling of Economic Processes based on Computational Intelligence. M and V, Hradec Kralove (2003)
4. Haykin, S.S.: Neural Networks: A Comprehensive Foundation. Prentice-Hall, Upper Saddle River (1999)
5. Kim, K.S., Han, I.: The Cluster-Indexing Method for Case-Based Reasoning Using Self-organizing Maps and Learning Vector Quantization for Bond Rating Cases. Expert Systems with Applications 21, 147–156 (2001)
6. Zadeh, L.A.: Fuzzy Sets. Information and Control 8, 338–353 (1965)
7. Nauck, D., Kruse, R.: A Neuro-Fuzzy Method to Learn Fuzzy Classification Rules from Data. Fuzzy Sets and Systems 89, 277–288 (1997)
8. Hajek, P.: Municipal Creditworthiness Modelling by Computational Intelligence Methods. Ph.D. Thesis, University of Pardubice (2006)
9. Hajek, P., Olej, V.: Municipal Creditworthiness Modelling by Clustering Methods. In: 10th International Conference on Engineering Applications of Neural Networks, Thessaloniki, Greece, pp. 168–177 (2007)
10. Deboeck, G., Kohonen, T.: Visual Explorations in Finance with Self-organizing Maps. Springer, London (1998)
11. Deboeck, G.: Financial Applications of Self-organizing Maps. Neural Network World, Neural and Massively-Parallel Computing and Information Systems 8, 213–241 (1998)

12. Back, B., Sere, K., Vanharanta, H.: Analyzing Financial Performance with Self-organizing Maps. In: IEEE International Joint Conference on Neural Networks, New York, USA, pp. 266–270 (1998)
13. Cottrell, M., de Bodt, E., Gregoire, P.: Financial Applications of the Self-organizing Map. In: 6th European Congress on Intelligent Techniques and Soft Computing, ELITE Foundation, Aachen, Germany, pp. 205–209 (1998)
14. Hoffmann, F., Baesens, B., Martens, J., Put, F., Vanthiemen, J.: Comparing a Genetic Fuzzy and a Neuro-Fuzzy Classifier for Credit Scoring. In: 5th International FLINS Conference on Computational Intelligent Systems for Applied Research, Gent, pp. 1067–1083 (2002)
15. Kuncheva, L.I.: Fuzzy Classifier Design. Springer, Berlin (2000)
16. Kohonen, T.: Self-organizing Maps. Springer, New York (2001)
17. Buckley, J.J., Hayashi, Y.: Hybrid Neural Networks can be Fuzzy Controllers and Fuzzy Expert Systems. Fuzzy Sets and Systems 60, 135–142 (1993)
18. Godjevac, J.: Comparison between Classical and Fuzzy Neurons. In: 2nd European Congress on Fuzzy Intelligent Technologies, Aachen, Germany, pp. 1326–1332 (1994)
19. Dunn, J.C.: Well Separated Clusters and Optimal Fuzzy Partitions. Journal of Cybernetics 4, 95–104 (1974)

Implementing Boolean Matrix Factorization

Roman Neruda[1], Václav Snášel[2], Jan Platoš[2], Pavel Krömer[2], Dušan Húsek[1],
and Alexander A. Frolov[3]

[1] Institute of Computer Science, Dept. of Neural Networks, Academy of Sciences of
Czech Republic, Pod Vodárenskou věží 2, 182 07 Prague, Czech Republic
`{roman,dusan}@cs.cas.cz`
[2] Department of Computer Science, VŠB – Technical University of Ostrava,
17. listopadu 15, 708 33 Ostrava, Czech Republic
`{vaclav.snasel,jan.platos,pavel.kromer.fei}@vsb.cz`
[3] Institute of Higher Nervous Activity and Neurophysiology, Russian Academy of
Sciences, Butlerova 5a, 117 485 Moscow, Russia
`aafrolov@mail.ru`

Abstract. Matrix factorization or factor analysis is an important task helpful in the analysis of high dimensional real world data. There are several well known methods and algorithms for factorization of real data but many application areas including information retrieval, pattern recognition and data mining require processing of binary rather than real data. Unfortunately, the methods used for real matrix factorization fail in the latter case. In this paper we introduce the background of the task, neural network, genetic algorithm and non-negative matrix facrotization based solvers and compare the results obtained from computer experiments.

1   Introduction

In order to perform object recognition (no matter which one) it is necessary to learn representations of the underlying characteristic components. Such components correspond to object-parts, or features. These data sets may comprise discrete attributes, such as those from market basket analysis, information retrieval, and bioinformatics, as well as continuous attributes such as those in scientific simulations, astrophysical measurements, and sensor networks.

Many applications in computer and system science involve analysis of large scale and often high dimensional binary datasets [8]. When dealing with such extensive information collections, it is usually very computationally expensive to perform some operations on the raw form of the data. Therefore, suitable methods approximating the data in lower dimensions or with lower rank are needed. In the following, we focus on the factorization of two-dimensional binary data (matrices, second order tensors).

The paper is structured as follows: first, a brief introduction to matrix factorization is given. In the following section, the basics of Evolutionary and Genetic Algorithms are presented. The rest of the paper brings description of Genetic Binary Matrix Factorization and summarizes performed computer experiments and conclusions drawn from our work.

V. Kůrková et al. (Eds.): ICANN 2008, Part I, LNCS 5163, pp. 543–552, 2008.

2   Matrix Factorization

Matrix factorization (or matrix decomposition) is an important task in data analysis and processing. A matrix factorization is the right-side matrix product in $A \approx F_1 \cdot F_2 \cdot \ldots \cdot F_k$ for the matrix A. The number of factor matrices depends usually on the requirements of given application area. Most often, $k = 2$ or $k = 3$. There are several matrix decomposition methods reducing data dimensions and simultaneously revealing structures hidden in the data. Such methods include Singular Value Decomposition (SVD) and Non-negative Matrix Factorization (NMF), which is our subject of interest in this research.

2.1   Non-negative Matrix Factorization

Non-negative matrix factorization (NMF) [1,9] is recently very popular unsupervised learning algorithm for efficient factorization of real matrices implementing the non-negativity constraint. NMF approximates real $m \times n$ matrix A as a product of two non-negative matrices W and H of the dimensions $m \times r$ and $r \times n$ respectively. Moreover, it applies that $r << m$ and $r << n$.

$$A \approx W \cdot H \tag{1}$$

There are several algorithms for NMF computation based on iterative minimization of given cost function [9]. The original NMF algorithm involved minimization of the Frobenius norm defined by formula (2).

$$\|A - WH\|_F^2 = \sum_{ij} |A_{ij} - (WH)_{ij}|^2 \tag{2}$$

Other investigated cost measures include square of the Euclidean distance between V and its approximation (4) or Kullback-Leibler divergence D (5). For every cost function, there are update rules (multiplicative or additive) applied iteratively in order to reduce the distance between original matrix V and its model [9].

$$\|A - WH\|^2 = \sum_{ij} (A_{ij} - (WH)_{ij}) \tag{3}$$

$$D(A \parallel WH) = \sum_{ij} \left(A_{ij} \log \frac{A_{ij}}{B_{ij}} - A_{ij} + B_{ij} \right) \tag{4}$$

Promising recent NMF algorithms are based on Gradient Descent Methods (GDM) or, extending the GDM, on Alternating Least Square computation [1]. NMF was reported to give good results in extracting features or concepts from processed data. Unfortunately, the common NMF algorithms excelling in NMF computation for real matrices are unsuitable for efficient factorization of binary matrices.

2.2 Boolean Matrix Factorization

Boolean matrix factorization (BMF) or Boolean factor analysis is the factorization of data sets in binary $(1,0)$ alphabet based on Boolean algebra. Boolean factor analysis is extremely important in computer applications since the natural data representation for computerized processing is binary. Binary factorization finds its application in data mining, information retrieval, pattern recognition, image processing or data compression [7].

The BMF can be defined in a similar manner as NMF [5,12,13]. Consider binary[1] matrix A of the dimension $m \times n$ as a Boolean product of two binary matrices W and H of the dimensions $m \times r$ and $r \times n$ respectively. Let r be a subject to $r << m$ and $r << n$. Then, BMF is searching for best W and H that approximate A:

$$A \approx W \otimes H \tag{5}$$

where $\otimes$ stands for Boolean matrix multiplication.

The urgency of BMF lies in the fact that computerized data are binary in its essence and BMF is intensively investigated. In [7], was introduced BMF algorithms based on formal concepts and blind search. Meeds [10] et al. presented BMF model for factorization of dyadic data, however, Meeds' decomposition features one non-binary (integer) factor.

2.3 Neural Network Boolean Factorization

Optimal solution of A decomposition according [7] by Blind search (brute force) and formal concepts is NP-hard problem and as such are not suitable for high dimensional data. On other side the classical linear methods could not take into account non-linearity of Boolean summation and therefore are inadequate for this task.

The *NBFA* is based on Hopfield-like neural network [2,3]. Used is the fully connected network of N neurons with binary activity (1 - active, 0 - nonactive). Each pattern of the learning set A^m is stored in the matrix of synaptic connections J' according to Hebbian rule:

$$J'_{ij} = \sum_{m=1}^{M} (A_i^m - q^m)(A_j^m - q^m), \ i,j = 1,\ldots,N, i{\neq}j, \ J'_{ii} = 0 \tag{6}$$

where M is the number of patterns in the learning set and bias $q^m = \sum_{i=1}^{N} A_i^m / N$ is the total relative activity of the m-th pattern. One special inhibitory neuron was added to N principal neurons of the Hopfield network. The neuron was activated during the presentation of every pattern of the learning set and was connected with all the principal neurons by bidirectional connections. To reveal factors the special two-run recall procedure was developed [3].

---

[1] Matrix A is binary iff $\forall ij : [a]_{ij} = 0 \vee [a]_{ij} = 1$.

3   Evolutionary Algorithms

Evolutionary algorithms (EA) are family of iterative stochastic search and optimization methods based on mimicking successful optimization strategies observed in nature [6,11]. EA operate with population (also known as pool) of artificial individuals (referred often as items or chromosomes) encoding possible problem solutions. Encoded individuals are evaluated using objective function which assigns a fitness value to each individual. Fitness value represents the quality (ranking) of each individual as solution of given problem. Competing individuals search the problem domain towards optimal solution [6].

3.1   Genetic Algorithms

Genetic Algorithms (GA) introduced by John Holland and extended by David Goldberg are wide applied and highly successful EA variant. Basic workflow of originally proposed standard generational GA is:

   I. Define objective function
  II. Encode initial population of possible solutions as fixed length binary strings and evaluate chromosomes in initial population using objective function
 III. Create new population (evolutionary search for better solutions):
    a. Select suitable chromosomes for reproduction (parents)
    b. Apply crossover operator on parents with respect to crossover probability to produce new chromosomes (offspring)
    c. Apply mutation operator on offspring chromosomes with respect to mutation probability. Add newly constituted chromosomes to new population
    d. Until the size of new population is smaller than size of current population go back to a.
    e. Replace current population by new population
 IV. Evaluate current population using objective function
  V. Check termination criteria; if not satisfied go back to III.

Many variants of standard generational GA have been proposed. The differences are mostly in particular selection, crossover, mutation and replacement strategy [6].

3.2   Genetic Binary Matrix Factorization

In this section, we to propose a Genetic Algorithm for Binary Matrix Factorization (Genetic BMF - GBMF). For that, we first analyze the factors that are to be found by the algorithm and define an algorithm suggesting initial values of the factors.

Binary factors. The factors W and H found by NMF algorithm by Lee and Seung can be straightforwardly interpreted. Columns of W are basis vectors of column space of A and columns of H are weights associated with the base vectors. In order to find out interpretation of the matrix factorization task for GA which might be different, consider a graph-like representation of a matrix:

$$A = \begin{pmatrix} 110 \\ 010 \\ 110 \end{pmatrix} \equiv \mathcal{G} \approx WH \qquad WH = \begin{pmatrix} 10 \\ 01 \\ 00 \end{pmatrix} \begin{pmatrix} 110 \\ 010 \end{pmatrix} \equiv \mathcal{G}' \qquad (7)$$

Then, the factorization can be seen as a task of finding tripartite graph $\mathcal{G}'$ that will exclusively preserve the arcs between the pairs of 'edge' vertices from $\mathcal{G}$ in the form of a two step long paths through a 'middle layer'. Intuitively, the number of vertices in middle layer corresponds to r in NMF. The graphs $\mathcal{G}$ and $\mathcal{G}'$ from 7 can be depicted as follows:

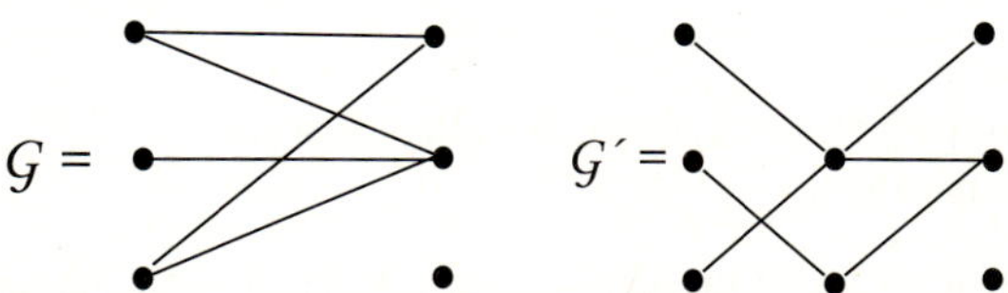

When adopting this notion of Boolean matrix factorization, the interpretation of factors W and H slightly differs from the interpretation of NMF factors. The rows of H are base vectors of row space of A and rows of W are associated weights.

Constructive algorithm for suggesting base vectors of boolean matrix row space (CAS). In order to provide the genetic algorithm with better than random initial population, a constructive algorithm for suggesting base vectors of matrix row space is defined:

Data: Binary data matrix A
Result: Initial suggestion of W and H
Initialization:
Compute the cardinality (number of 1 elements) of the rows of A;
Divide rows of A to classes A^C per cardinality;
Let W and H be 'empty';
Let $k = 0$;
repeat
    Let A_i^C be a random row from the row class with lowest cardinality;
    foreach *Base row $h_j \in H$* **do**
        if *h_j is not covered[a] by A_i^C* **then**
            Attach A_i^C to H as h_k;
            Set $W_{ik} = 1$;
            Increase k;
            Let $W_{ij} = 1$;
            Let $A_i^C = A_i^C - w_j$, where w_j is j-th column of matrix W;
        end
    end
until *A_i^C is zero vector* ;

Algorithm 1. CAS Algorithm

---

[a] i.e. $h_j - A_i^C = o$ where $o = (0, 0, \ldots, 0)$ is zero vector.

The rows of the matrix A can be then constructed using linear combinations of the rows of H.

Genetic algorithm for binary matrix factorization. We propose a genetic algorithm for BMF. It will exploit the initial factors constructed using the algorithm introduced in previous section. The objective function will be Hamming distance between reconstructed and original matrix. Crossover will aim to modify weights factor (matrix W) and mutation will primarily aim to alter basis vectors (matrix H). The algorithm can be summarized as follows:

Initialization:
Create initial population of N (WH) chromosomes;
Evaluate chromosomes;
repeat
    Select suitable chromosomes for reproduction (parents);
    Apply crossover on matrix W of selected parents;
    Mutate W with very small probability;
    Mutate H;
    Migrate offspring chromosomes to current population;
    Evaluate current population using objective function;
until *termination criteria are satisfied* ;

Algorithm 2. GenBMF

The evaluation of chromosomes in population is implemented as comparison of original matrix V to the product of W and H. The fitness function is based on Euclidean distance between V and WH:

$$f = \frac{1}{\sqrt{\sum_i \sum_j |(V[i,j] - WH[i,j])|}} \tag{8}$$

The termination criteria were based on specified threshold defining minimum acceptance of evolved solution and maximum number of generations processed. The maximum number of generations was set to 1000 and the minimum acceptance 0.3.

In this way, the algorithm explores different combinations of base vectors (via crossover) and simultaneously adjust the base vector suggestions. Evolutionary principles will be applied and the factor interpretation maintained.

4   Experimental Evaluation

This section provides summary on computer experiments conducted in order to verify described algorithms. The results are confronted with outputs of a reference NMF implementation.

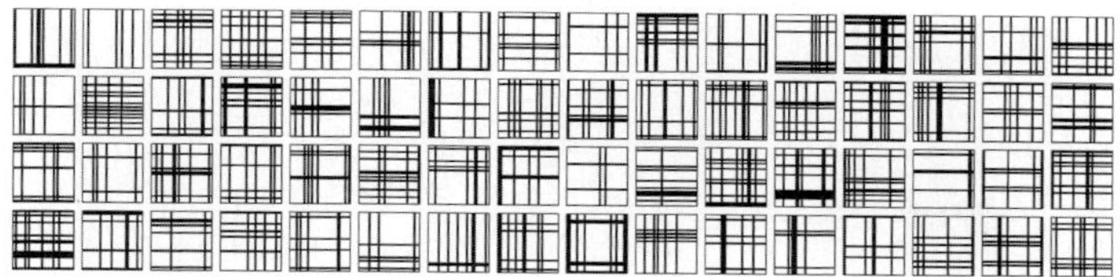

Fig. 1. Several images from generated collection

4.1 Evaluation of the NBFA Algorithm

For testing the NBFA algorithm was used a generic collection of 1600 32 ×
32 black-and-white images containing different combinations of horizontal and
vertical lines (bars). The probabilities of bars to occur in images were the same
and equal to 10/64, i.e. images contain 10 bars in average. An example of several
images from generated collection is depicted in Figure 1.

The decomposition of images into binary vectors by NBFA method was performed. Here factors contains only values $\{0, 1\}$ and model is Boolean. The factor
search was performed under assumption that the number of ones in factor is not
less than 5 and not greater than 200. Since the images are obtained by Boolean
summation of binary bars, it is not surprising, that NBFA is able to reconstruct
all bars as factors, providing an ideal solution, as we can see in Figure 2.

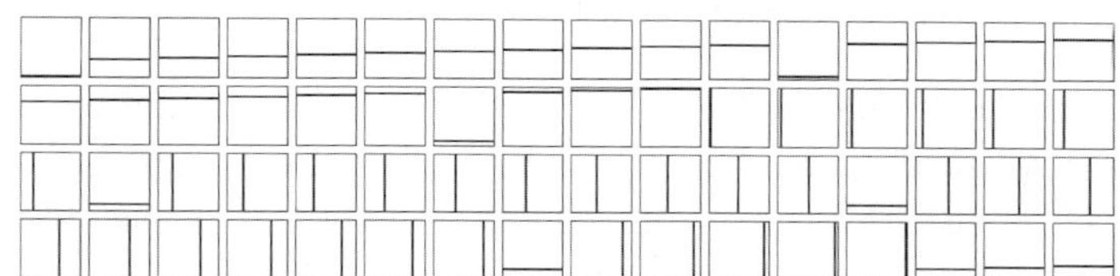

Fig. 2. First 64 factors retrieved by NBFA method

The method provides excellent factors. However, its drawback lies in its computational complexity.

4.2 Evaluation of the GBMF Algorithm

For testing the GBMF algorithm were used three different collections, each containing 1000 16 × 16 black-and-white bar images with the probalities 1/4, 1/8
and 1/16 respectively. Sample generated images are shown in Figure 3.

GBMF was implemented and run for some testing binary matrices. Black and
white images were chosen as representation of input and output binary matrices
for the ease of visual interpretation of the results.

The images in each collection were transformed into a single vector (16 × 16
image into a 1 × 256 row vector) and merged into a 1000 × 256 binary matrix.
The The matrix was then processed by GBMF with the following parameters: a

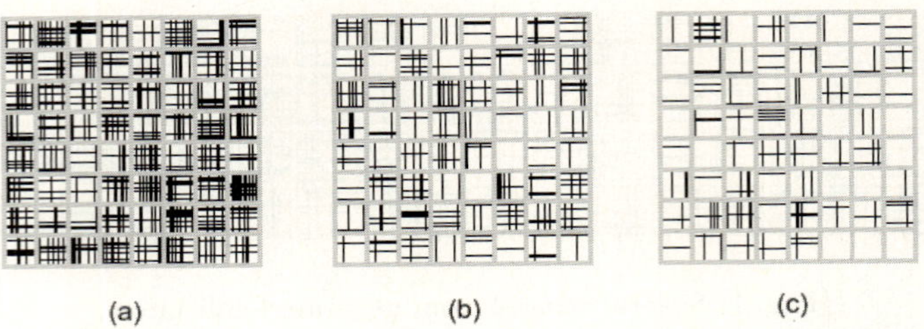

Fig. 3. Sample generated images in collections with bar probabilities 1/4 (a), 1/8 (b) and 1/16 (c)

population of 50 prospective factors, probability of crossover 0.9 and probability of mutation 0.2. The matrix dimension was reduced to 32, i.e. 32 factors were sought. GBMF was executed for 100 generations. The factors obtained for each testing collection are shown in Figure 4.

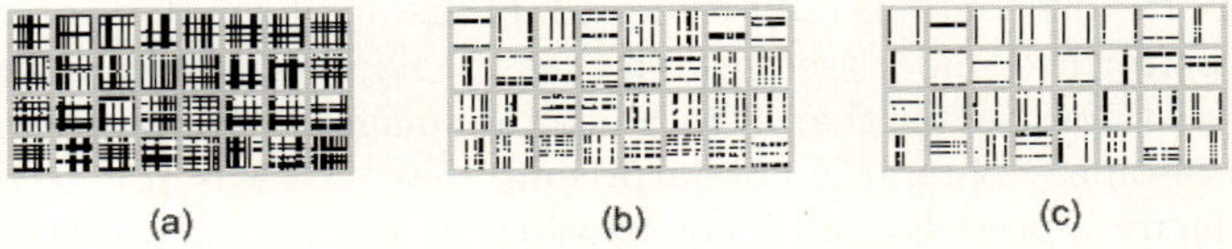

Fig. 4. Factors obtained by GBMF for collections with bar probabilities 1/4 (a), 1/8 (b) and 1/16 (c)

The factors feature some vertical and horizontal bars, usually more than one in each factor image. However, with reduction of the bar probability increases the precision of the algorithm and more appropriate factors were found.

4.3   Binary Factorization by NMF

We have implemented NMF and applied it on the binary matrix factorization task. The same collection as in 4.2 was processed by NMF. NMF was used to reduce find 32 factors of the 1000×256 input matrix. Obtained factors are shown in Figure 5.

The factors contain well discovered horizontal and vertical bars. However, some noise (gray points in the figures) is present since the output factor matrices are real and not binary.

5   Conclusions and Future Work

In this paper, we have investigated three methods for binary matrix factorisation. NBFA, based on neural networks, has showed an excellent ability to discover binary factors. Unfortunatelly, it is limited by its computational complexity.

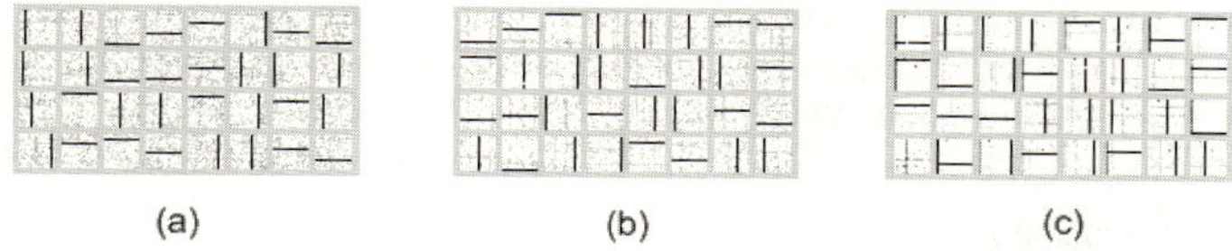

(a)   (b)   (c)

Fig. 5. Factors obtained by NMF for collections with bar probabilities 1/4 (a), 1/8 (b) and 1/16 (c)

GBMF, based on genetic algorithms, has been used for binary matrix factorization with good results. The obtained factors, however, usually contained more than one vertical or horizontal bar as the optimal factors would. With reduced probability of the bars (i.e. with reduced density of the processed matrix) were found better factors.

The last investigated method was NMF. It has proved its known excellence also in factorization of binary data but the obtained factors were real rather than binary.

The investigated methods are promising contributions in the field of factorisation of binary data. They have shown good ability to find factors of binary matrices but also space for significant improvements. Our future work aims to improve all three methods for binary matrix factorization - NBFA, GBMF and NMF.

References

1. Berry, M.W., Browne, M.L., Amy, N., Pauca, P.V., Plemmons, R.J.: Algorithms and applications for approximate nonnegative matrix factorization. Computational Statistics & Data Analysis 52(1), 155–173 (2007)
2. Frolov, A.A., Sirota, A.M., Húsek, D., Muraviev, I.P., Polyakov, P.J.: Binary factorization in hopfield-like neural networks: single-step approximation and computer simulations. Neural Network Word 14, 139–152 (2004)
3. Frolov, A.A., Húsek, D., Muravjev, P., Polyakov, P.: Boolean Factor Analysis by Attractor Neural Network. Neural Networks, IEEE Transactions 18(3), 698–707 (2007)
4. Frolov, A.A., Sirota, A.M., Húsek, D., Muraviev, I., Combe, P.: Binary Factorization in Hopfield-Like Neural Autoassociator: A Promising Tool for Data Compression. In: Pearson, D., Steele, N., Albrecht, R. (eds.) Artificial Neural Nets and Genetic Algorithms, pp. 58–62. Springer, Wien (2003)
5. Húsek, D., Moravec, P., Snášel, V., Frolov, A.A., Řezanková, H., Polyakov, P.: Comparison of Neural Network Boolean Factor Analysis Method with Some Other Dimension Reduction Methods on Bars Problem. In: Ghosh, A., De, R.K., Pal, S.K. (eds.) PReMI 2007. LNCS, vol. 4815, pp. 235–243. Springer, Heidelberg (2007)
6. Jones, G.: Genetic and evolutionary algorithms. In: von Rague, P. (ed.) Encyclopedia of Computational Chemistry. John Wiley and Sons, Chichester (1998)
7. Keprt, A., Snášel, V.: Binary Factor Analysis with Genetic Algorithms. In: Proceedings of 4th IEEE International Workshop on Soft Computing as Transdisciplinary Science and Technology, WSTST 2005, Muroran, Japan. LNCS/LNAI, pp. 1259–1268. Springer, Heidelberg (2005)

8. Koyutürk, M., Grama, A., Ramakrishnan, N.: Nonorthogonal decomposition of binary matrices for bounded-error data compression and analysis. ACM Trans. Math. Softw. 32(1), 33–69 (2006)
9. Lee, D.D., Seung, H.S.: Algorithms for non-negative matrix factorization. In: NIPS, pp. 556–562 (2000)
10. Meeds, E., Ghahramani, Z., Neal, R.M., Roweis, S.T.: Modeling dyadic data with binary latent factors. In: Schölkopf, B., Platt, J., Hoffman, T. (eds.) Advances in Neural Information Processing Systems, vol. 19, pp. 977–984. MIT Press, Cambridge (2007)
11. Mitchell, M.: An Introduction to Genetic Algorithms. MIT Press, Cambridge (1996)
12. Snášel, V., Húsek, D., Frolov, A.A., Řezanková, H., Moravec, P., Polyakov, P.: Bars Problem Solving - New Neural Network Method and Comparison. In: Gelbukh, A., Kuri Morales, Á.F. (eds.) MICAI 2007. LNCS (LNAI), vol. 4827, pp. 671–682. Springer, Heidelberg (2007)
13. Zhang, Z., Tao Li, T., Ding, C., Zhang, X.-S.: Binary Matrix Factorization with Applications. In: Proceedings of 2007 IEEE International Conference on Data Mining (ICDM 2007), pp. 391–400 (2007)

Application of Potts-Model Perceptron for Binary Patterns Identification

Vladimir Kryzhanovsky[1], Boris Kryzhanovsky[1], and Anatoly Fonarev[1,2]

[1] Center of Optical Neural Technologies of
Scientific Research Institute for System Analysis of
Russian Academy of Sciences
44/2 Vavilov Street, 119333 Moscow, Russian Federation
Vladimir.Krizhanovsky@gmail.com,iont.niisi@gmail.com
[2] CUNY City University of New York, Department of Engineering and Science
2800 Victory Blvd. SI, NY 10314

Abstract. We suggest an effective algorithm based on $q-$state Potts model providing an exponential growth of network storage capacity $M \sim N^{2S+1}$, where N is the dimension of the binary patterns and S is the free parameter of task. The algorithm allows us to identify a large number of highly distorted similar patterns. The negative influence of correlations of the patterns is suppressed by choosing a sufficiently large value of the parameter S. We show the efficiency of the algorithm by the example of a perceptron identifier, but it also can be used to increase the storage capacity of full connected systems of associative memory. Restrictions on S are discussed.

Keywords: identification, Potts model, $q-$state perceptron, storage capacity.

1 Introduction

The storage capacity of the Hopfield model is rather small. It allows one to storage $M \sim N/2\ln N$ randomized patterns only. These estimates are correct for randomized patterns, whose binary coordinates are independent random variables. If there are correlations between patterns, the recognizing ability of Hopfield model decreases drastically.

For a long time it was considered that the only way to overcome these difficulties was the sparse coding [1],[2]. This technique consists of random diluting of informative coordinates by a great number of spurious coordinates. Then with the aid of a special choice of the threshold and the level of the patterns activity, the storage capacity can be increased up to maximal value $M \sim (N/2\ln N)^2$. The last estimate is of theoretical interest only, since no distortions of the patterns are assumed. In the presence of distortion he sparse coding allows one to increase the storage capacity by 1.5-2 times only.

Another way to increase the binary storage capacity based on using of $q-$state models of neural nets was suggested in [3]-[4]. Q-state models were investigated in

V. Kůrková et al. (Eds.): ICANN 2008, Part I, LNCS 5163, pp. 553–561, 2008.

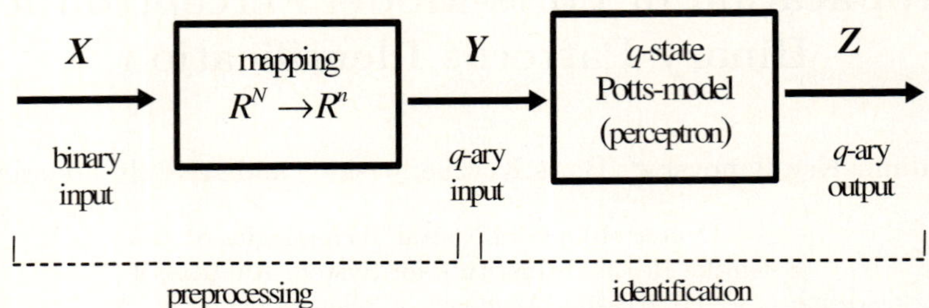

Fig. 1. Two-stage scheme of binary pattern identification

a lot of papers [5]-[15]. Among them the most well-known is the Potts spin-glass model [5]. The analysis of $q-$state models showed that they have an extremely large storage capacity $M \sim Nq^2/4\ln Nq$, which is q^2 greater than the same characteristic of the Hopfield model. In addition $q-$state models have an extremely large noise immunity and the ability for recognition in the presence of very large distortions. At the present $q-$state models of the associative memory are the best both with regard to the storage capacity and noise immunity. In the same time such high parameters of $q-$state models were practically not used up to now.

The situation changed when the algorithm of mapping of binary patterns into $q-$valued ones was proposed [3]. It was shown in [3] that such mapping allows one to use $q-$state neural networks for storing and processing of signals of any type and any dimension. Moreover, the mapping brings to nothing the main difficulty of all the associative memory systems, which is the negative influence of correlations between the patterns.

In the present paper we use the mapping algorithm [3] to create the identifier of binary patterns based on the Potts spin-glass model. The pattern identification includes two stages as shown schematically in Fig.1. At the first stage a preprocessing of the input binary signal is done. The preprocessing consists in mapping of a binary pattern $\mathbf{X}$ belonging to a N-dimensional configuration space into some internal $q-$nary pattern $\mathbf{Y}$ from the space of other dimension $(R^N \to R^n)$. In other words, we bring the black-white image $\mathbf{X}$ with the number of pixels N in one-to-one correspondence with the colored image $\mathbf{Y}$ with the less number of pixels $(n = N/r)$, but with the greater number of colors $(q = 2^r)$. The number $r > 1$ is called the mapping parameter. During the preprocessing two objects can be achieved at once. First, the mapping $\mathbf{X} \to \mathbf{Y}$ eliminates correlations between the input patterns. Second, for identification of patterns we can use a $q-$state neural network, whose storage capacity and noise immunity are much higher than in scalar models of the Hopfield type. For the purposes of illustration we use the most simple variant of the identification system based on the Potts-model perceptron.

The description of the $q-$state model properties is given in Sections 2. The mapping algorithm is described in Section 3. The description of binary patterns identification is given in Sections 4. Restrictions on the possibility of binary patterns identification discussed in Section 5.

2   Identification of the $q-$Nary Pattern

Let us have a set of n-dimensional $q-$nary patterns $\{\mathbf{Y}_\mu\}$:

$$\mathbf{Y}_\mu = (\mathbf{y}_1^{(\mu)}, \mathbf{y}_2^{(\mu)}, ..., \mathbf{y}_n^{(\mu)}) \tag{1}$$

where $\mathbf{y}_i^{(\mu)} \in \{\mathbf{e}_k\}^q$ and $\{\mathbf{e}_k\}^q$ is the set of basis vectors of the $q-$dimensional space R^q ($\mu = 1, ..., M; i = 1, ..., n; k = 1, ..., q$). It is supposed that each pattern $\mathbf{Y}_\mu$ is one to one associated with the identifier $\mathbf{Z}_\mu$:

$$\mathbf{Z}_\mu = (\mathbf{z}_1^{(\mu)}, \mathbf{z}_2^{(\mu)}, ..., \mathbf{z}_m^{(\mu)}) \tag{2}$$

where $\mathbf{z}_j^{(\mu)} \in \{\mathbf{e}_k\}^q$, $j = 1, ..., m$.

Usually, by the identifier $\mathbf{Z}_\mu$ one codes: i) a directive, produced as a result of the input $\mathbf{Y}_\mu$; ii) the number μ of the input vector, that allows one to reconstruct the pattern $\mathbf{Y}_\mu$ itself afterwards; iii) any other information related to $\mathbf{Y}_\mu$. We will examine the most simple case, when the number μ of the pattern $\mathbf{Y}_\mu$ is coding by the vector $\mathbf{Z}_\mu$: the sequence of numbers of the basis vectors of $q-$dimensional vector space along which the unit vectors $\mathbf{z}_1^{(\mu)}, \mathbf{z}_2^{(\mu)}, ..., \mathbf{z}_m^{(\mu)}$ are directed is just the number μ in the $q-$nary presentation. By $k_j = 1, ..., q$ we denote the number of the basis vector along which the vector $\mathbf{z}_j^{(\mu)}$ is directed. Then the number μ is defined by the expression:

$$\mu = 1 + \sum_{j=1}^{m} (k_j - 1)q^{j-1} \tag{3}$$

The problem is to create a network, which is able to reconstruct the identifier $\mathbf{Z}_\mu$ when presenting distorted pattern $\mathbf{Y}_\mu$. The Potts-model perceptron solving the problem of identification is shown in Fig.2. It consists of two layers of $q-$state neurons (n input and m output neurons). Each neuron of the input layer is connected with all the neurons of the output layer. The number of output neurons ($m = 1 + \log_q M$) is sufficient for coding the set of given patterns $\{\mathbf{Y}_\mu\}$. The interconnection matrix elements are given by the Hebb rule:

$$\mathbf{J}_{ji} = \sum_{\mu=1}^{M} \mathbf{z}_j^{(\mu)} \mathbf{y}_i^{(\mu)+} \tag{4}$$

where $\mathbf{y}_i^{(\mu)+}$ is the vector-row ($1 \leq i \leq n, 1 \leq j \leq m$).

Let an input image $\mathbf{Y} = (\mathbf{y}_1, \mathbf{y}_2, ..., \mathbf{y}_n)$ be a distorted copy of the l-th pattern $\mathbf{Y}_l$. The local field created by all the neurons from the input layer, which is acting on the j-th output neuron, have to be calculated as:

$$\mathbf{h}_j = \mathbf{h}_{j0} + \sum_{i=1}^{n} \mathbf{J}_{ji}\mathbf{y}_i, \quad \mathbf{h}_{j0} = -\frac{n}{q}\sum_{\mu=1}^{M} \mathbf{z}_j^{(\mu)} \tag{5}$$

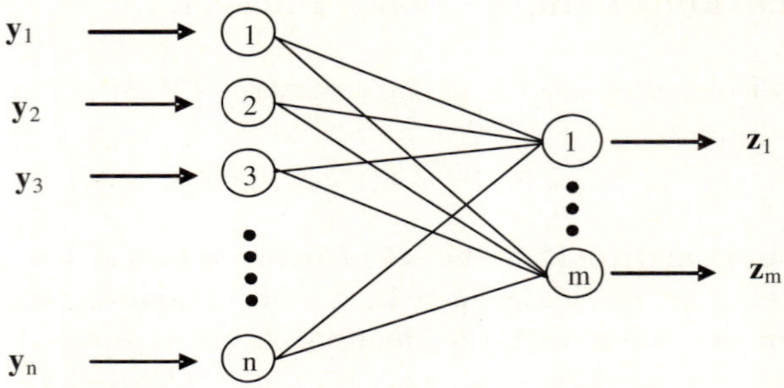

Fig. 2. The scheme of q–state Potts perceptron

Under the action of the local field $\mathbf{h}_j$ the j-th output neuron becomes aligned along the basis vector, whose direction is the most close to that of the local field. The calculation algorithm is as follows: i) the projections of the local field vector $\mathbf{h}_j$ onto all the basis vectors of the q–dimensional space are to be calculated; ii) the maximal projection is to be found (let it be the projection onto a basis vector $\mathbf{e}_{max}$); iii) the value $\mathbf{z}_i = \mathbf{e}_{max}$ is set to the output neuron. As it was shown in [5]-[7], under this dynamics all the components of the encoding output vector $\mathbf{Z}_l$ are retrieved reliable.

The reliability of the perceptron identifier, i.e. the probability P that the number of the input pattern is defined correctly, is given by the expression [14],[15]:

$$P = 1 - \frac{mq}{\gamma\sqrt{2\pi}} \exp\left(-\tfrac{1}{2}\gamma^2\right) \tag{6}$$

where

$$\gamma^2 = \frac{nq(q-1)}{2M}(1-b)^2 \tag{7}$$

Here γ is so called signal to noise ratio and b is the level of distortions in the input pattern (nb is the number of distorted pixels).

It follows from (6) that $P \to 1$ if $\gamma^2 > 2\ln mq$ and $n >> 1$. So, the maximal number of the patterns, which can be reliably identified by the perceptron, may be defined from the equality $\gamma^2 = 2\ln mq$ in the form:

$$M \sim \frac{nq(q-1)}{4\ln mq}(1-b)^2 \tag{8}$$

The expression (8) allows us to estimate the number of output neurons, which is necessary to enumerate the patterns written in the perceptron memory. Taking into account the equality $m = 1+\log_q M$, it is easy to obtain the desired estimate in the form:

$$m = 2 + \ln n / \ln q \tag{9}$$

We see that the number of output neurons is much less than the number of the input ones. If q is sufficiently large, as a rule, this value is of the order of or slightly larger than $3 \div 4$.

3   Mapping Algorithm

Here we describe the mapping algorithm of binary patterns into q−nary ones. Later the q−nary patterns are used for q−state Potts model construction.

Let $\mathbf{X} = (x_1, x_2, ..., x_N)$ be N-dimension binary vector, $x_i = \{0, 1\}$. We divide it mentally into n fragments containing r elements each ($n = N/r$). Each i-th fragment ($i = 1, ..., n$) can be considered as an integer $k_i - 1$ written down in the binary code ($1 \leq k_i \leq q$, $q = 2^r$). This fragment is associated with the vector $\mathbf{y}_i = \mathbf{e}_{k_i}$, where $\mathbf{e}_{k_i} \in \{\mathbf{e}_k\}^q$ is k_i-th basis vector in the space R^q. Thus, the vector $\mathbf{X}$ as a whole is put in one-to-one correspondence with a set of q-dimensional vectors, i.e. with an internal image $\mathbf{Y} = (\mathbf{y}_1, \mathbf{y}_2, ..., \mathbf{y}_n)$.

For example, the binary vector $\mathbf{X} = (01000000)$ can be split into two fragments of four elements: (0100) and (0000). The first fragment (it is "4" in the binary code, $k_1 = 5$) is associated with the vector $\mathbf{y}_1 = \mathbf{e}_5$ in the space of the dimension $q = 16$, and the second fragment (it is "0" in the binary code, $k_2 = 1$) is associated with the vector $\mathbf{y}_2 = \mathbf{e}_1$. The relevant mapping takes the form $\mathbf{X} = (0100000) \rightarrow \mathbf{Y} = (\mathbf{e}_5, \mathbf{e}_1)$.

It is important that the mapping is biunique and the binary vector $\mathbf{X}$ can be restored uniquely from its internal image $\mathbf{Y}$. It is even more that the mapping eliminates correlations between internal images. For example, let we have two 75% overlapping binary vectors $\mathbf{X}_1 = (10000001)$ and $\mathbf{X}_2 = (10011001)$. If now we use the mapping procedure with the parameter $r = 2$ ($n = 4, q = 4$) then we obtain two images $\mathbf{Y}_1 = (\mathbf{e}_3, \mathbf{e}_1, \mathbf{e}_1, \mathbf{e}_2)$ and $\mathbf{Y}_2 = (\mathbf{e}_3, \mathbf{e}_2, \mathbf{e}_3, \mathbf{e}_2)$, which are only overlapped by 50%. Using the same procedure with the mapping parameter $r = 4(n = 2, q = 16)$, we obtain two images $\mathbf{Y}_1 = (\mathbf{e}_9, \mathbf{e}_2)$ and $\mathbf{Y}_1 = (\mathbf{e}_{10}, \mathbf{e}_{10})$ that do not overlap completely. This means that our q−state perceptron can be effectively used for identification of correlated (similar) binary vectors.

4   Binary Pattern Identification

Here we describe the work of our model as a whole, i.e. the mapping of original binary patterns into internal images and identification of these images with the aid of q−state perceptron.

For a given mapping parameter r we apply the procedure for Sect.3 to the set of binary patterns $\{\mathbf{X}_\mu\}^M \in R^N$, $\mu \in \overline{1, M}$. As a result we obtain a set of n-dimensional q−nary internal images $\{\mathbf{Y}_\mu\}^M$, where $n = N/r$ and $q = 2^r$. These images can be considered as randomized ones. With the aid of these images we build the Potts-model perceptron as was described in Sect.2.

Now let us estimate the number of patterns, which can be recognized by our two-stage model (see Fig.1). Let the probability of distortions of coordinates of input binary pattern $\overline{\mathbf{X}}_\mu$ be $0 \leq p < 1/2$. Then mapping this vector into the q−nary representation, we obtain distorted internal image $\overline{\mathbf{Y}}_\mu$ with the level of distortion defined as $b = 1 - (1 - p)^r$.

The recognizing properties of Potts-model perceptron are given by the expressions (6)-(8) in which $n = N/r$, $q = 2^r$ and $1 - b = (1 - p)^r$ have to be

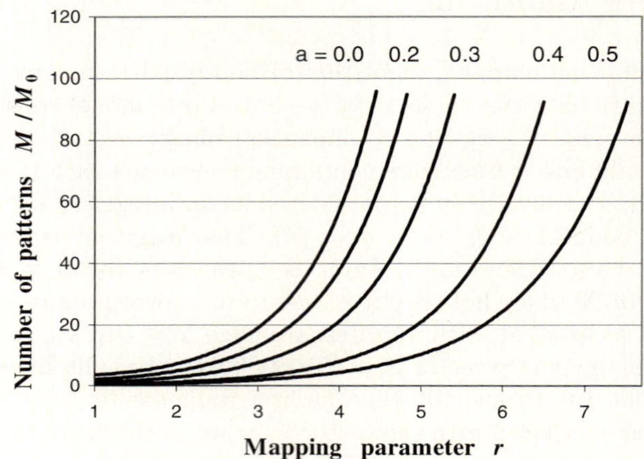

Fig. 3. The dependence of the storage capacity M on the mapping parameter r for the values of the bias $a = 0 \div 0.5$. The storage capacity of the Hopfield network M_0 is accept to be one.

substituted. In particular, the maximal number of images that can be identified by the perceptron constructed in such a way, is

$$M \sim \frac{N}{4r^2}[2(1-p)]^{2r} \tag{10}$$

The presence of the factor $(1-p)^r$ is due to the fact that even small distortions of the components of the binary vector $\mathbf{X}$ lead to very large distortions of the components of its $q-$nary image $\mathbf{Y}$. As a result, the number of patterns which can be recognized under such distortions also decreases. However, it follows from (10) that the value of M increases exponentially when the mapping parameter r increases. Indeed, let us write the mapping parameter in the normed form

$$r = S \log_2 N \tag{11}$$

For simplicity we set $p = 0$. Then the expression (10) takes the form

$$M \sim \frac{N^{2S+1}}{(2S \log_2 N)^2} \tag{12}$$

We see that the storage capacity of the described system increases exponentially when the parameter S increases. Already when $S \geq 0.5$, it becomes much greater than the storage capacity of any known one-level neural networks.

When $r = 1(q = 2)$ the expression (10) describes the functioning of perceptron based on the scalar Hopfield model. In this case the analysis of (10) shows that even if correlations are absent, the storage capacity does not exceed relatively small value $M_0 \sim N/2 \ln N$. Even if small correlations are introduced, the number of recognized patterns is still more less, and practically the network failed in realizing the functions of an associative memory.

When the mapping parameter r increases, the picture changes drastically. The network begins to operate as a $q-$nary model, i.e. its storage capacity increases significantly and the influence of the correlations goes down (see Fig.3 where the case of biased binary patterns is presented). In Fig.4 we show the decrease of the error of recognition when the parameter r increases, though the bias of the binary patters is large enough ($a = 0.4$) as well as the distortions of the images, which have to be recognized ($p = 0.3$). We see that when $r \leq 5$, the network failed in realizing the functions of an associative memory. In other words, it recognize not a single pattern ($P << 1$), since the number of standard images is too large ($M \geq N$). When r increases, the negative influence of correlations is suppressed completely, and for $r \sim 8 \div 10$ the network recognizes correlated patterns reliably, even for a rather large loading $M/N \geq 50$, which is unachievable for the networks of the Hopfield type.

5   Restrictions on the Storage Capacity

As it follows from (10)-(12) the storage capacity exponentially increases with r. However, r cannot increase unrestrictedly. Its limiting value is defined by the requirement of number of vector-neuron n being sufficiently large. Only in this case the expressions (6)-(9) obtained with central limit theorem are correct. Thus we need

$$n = \frac{N}{r} >> 1 \quad \Rightarrow \quad r << N \tag{13}$$

or

$$S << N/\log_2 N \tag{14}$$

This restriction is not very strong and allows one to get a very large storage capacity. For example, when $N \sim 10^3$ and $r = 10$, it can be assumed that the condition (13) is fulfilled. Then we have $M \sim 10^4 N/\ln N$. As we see this value exceeds the storage capacity of Hopfield model by four orders of magnitude.

In addition to the restriction (13), it is necessary to have at least one undistorted $q-$state neuron among the neurons of the internal image $\mathbf{Y}$. By other words, at least one of the pixels of internal image that has to be recognized, must be not distorted. This requirement has the form

$$n(1 - p)^r > 1 \quad \Rightarrow \quad (1 - p)^r > r/N \tag{15}$$

As it follows from Eq.(15), in the presence of a noise ($p \neq 0$), the value of mapping parameter r cannot exceed the critical value

$$r_c = \frac{\ln N}{|\ln(1 - p)|} \tag{16}$$

Correspondingly, the storage capacity cannot exceed the critical value

$$M_{\mathrm{cr}} \sim \frac{N^{2S_c - 1}}{(2S_c \log_2 N)^2} \tag{17}$$

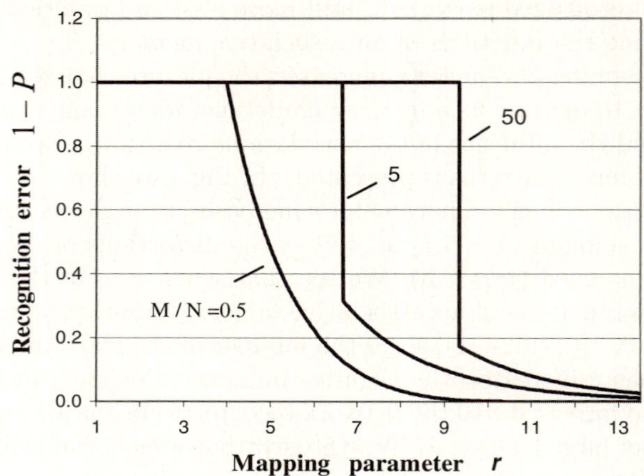

Fig. 4. The dependence of the error of recognition $(1-P)$ on the value of the parameter r: the loading parameter $M/N = 0.5, 5, 50$; $p = 0.3$; $a = 0.4$

where

$$S_c = |\ln(1 - p)|^{-1} \qquad (18)$$

When r increases inside region $1 \leq r \leq r_c$, the storage capacity M increases exponentially according to Eq.(12). However, when $r > r_c$, the network failed to recognize patterns. In this case, the basin of attraction becomes so small, that a distorted pattern falls out of its boundaries. On the other hand, we see that near the critical value r_c rather large storage capacity can be obtained.

6   Discussion of the Results and Conclusions

The algorithm described above can be used both for identification of binary and q−nary patterns (in the last case the preprocessing and the mapping are not necessary). Comparing the expression (8) with the results of the works [11]-[15], we see that for the perceptron algorithm the probability of the error recognition is n times less than in the case of the full connected neural networks. Actually this means that q−state perceptron is able to identify reliably the input vector and give out the correct directive, even when the full connected neural networks a priory give out incorrect output signals. Summarizing the presented results we can say that the proposed algorithm allows us to create the associative memory and identification systems with the storage capacity exponentially large with regard to the mapping parameter. These systems are able to work with sets of correlated binary patterns.

Applied for binary patterns recognition described model of neural network is able to store $M \sim N^{2S}$ patterns, where S can be chosen as $S >> 1$. For example, it is possible to store and recognize a large number of binary patterns: $M \sim N^6$ if $p = 0.1$; $M \sim N^3$ if $p = 0.2$; $M \sim N^{3/2}$ if $p = 0.3$.

Acknowledgements. The work in part supported by the Russian Foundation for Basic Research (#06-01-00109).

References

1. Perez-Vicente, C.J., Amit, D.J.: Optimized network for sparsely coded patterns. Journal of Physics A 22, 559–569 (1989)
2. Palm, G., Sommer, F.T.: Information capacity in recurrent McCulloch-Pitts networks with sparsely coded memory states. Network 3, 1–10 (1992)
3. Kryzhanovsky, B.V., Mikaelian, A.L.: An associative memory capable of recognizing strongly correlated patterns. Doklady Mathematics 67(3), 455–459 (2003)
4. Kryzhanovsky, B.V., Mikaelian, A.L., Fonarev, A.B.: Vector neural network identifing many strongly distorted and correlated patterns. In: Int. conf on Information Optics and Photonics Technology, Photonics Asia-2004, Beijing-2004. Proc. of SPIE, vol. 5642, pp. 124–133 (2004)
5. Kanter, I.: Potts-glass models of neural networks. Physical Review A 37(7), 2739–2742 (1988)
6. Cook, J.: The mean-field theory of a Q-state neural network model. Journal of Physics A 22, 2000–2012 (1989)
7. Vogt, H., Zippelius, A.: Invariant recognition in Potts glass neural networks. Journal of Physics A 25, 2209–2226 (1992)
8. Bolle, D., Dupont, P., Huyghebaert, J.: Thermodynamics properties of the $q-$state Potts-glass neural network. Phys. Rew. A 45, 4194–4197 (1992)
9. Wu, F.Y.: The Potts model. Review of Modern Physics 54, 235–268 (1982)
10. Nakamura, Y., Torii, K., Munaka, T.: Neural-network model composed of multidimensional spin neurons. Phys. Rev. E 51(2), 1538–1546 (1995)
11. Kryzhanovsky, B.V., Mikaelyan, A.L.: On the Recognition Ability of a Neural Network on Neurons with Parametric Transformation of Frequencies. Doklady Mathematics 65(2), 286–288 (2002)
12. Kryzhanovsky, B.V., Kryzhanovsky, V.M., Mikaelian, A.L., Fonarev, A.: Parametric dynamic neural network recognition power. Optical Memory&Neural Network 10(4), 211–218 (2001)
13. Kryzhanovsky, B.V., Litinskii, L.B., Fonarev, A.: Parametrical neural network based on the four-wave mixing process. Nuclear Instuments and Methods in Physics Research, A 502(2-3), 517–519 (2003)
14. Kryzhanovsky, B.V., Litinskii, L.B., Mikaelian, A.L.: Vector-neuron models of associative memory. In: Proc. of Int. Joint Conference on Neural Networks IJCNN 2004, Budapest-2004, pp. 909–1004 (2004)
15. Kryzhanovsky, B.V., Kryzhanovsky, V.M., Fonarev, A.B.: Decorrelating Parametrical Neural Network. In: Proc. of IJCNN Montreal-2005, pp. 1023–1026 (2005)

Using ARTMAP-Based Ensemble Systems Designed by Three Variants of Boosting

Araken de Medeiros Santos and Anne Magaly de Paula Canuto

Informatics and Applied Mathematics Department
Federal University of RN Natal, RN - Brazil, 59072-970
Tel.: +55-84-3215-3815
arakenmedeiros@gmail.com, anne@dimap.ufrn.br

Abstract. This paper analyzes the use of ARTMAP-based in structures of ensembles designed by three variants of boosting (Aggressive, Conservative and Inverse). In this investigation, it is aimed to analyze the influence of the RePART (Reward and Punishment ARTmap) neural network in ARTMAP-based ensembles, intending to define whether the use of this model is positive for ARTMAP-based ensembles. In addition, it aims to define which boosting strategy is the most suitable to be used in ARTMAP-based ensembles.

Keywords: Ensemble Systems, Boosting strategy, ARTMAP-Based neural networks, RePART.

1   Introduction

RePART (Reward/Punishment ART) is a more developed version of the simpler fuzzy ARTMAP which employs additional mechanisms to improve performance and operability, such as a reward/ punishment process, an instance counting parameter and variable vigilance parameter. The idea of using a reward/punishment process is to improve the performance of fuzzy ARTMAP networks. The idea of using a variable vigilance parameter is to decrease the complexity (category proliferation problem) of fuzzy ARTMAP models when used in applications with a large number of training patterns. The performance of RePART has been investigated in different pattern recognition tasks, as in [1, 2, 3, 4]. However, the performance of RePART only as individual classifiers has been investigated, not considering its performance in ensemble systems.

It is well known that substantial improvements can be obtained in difficult pattern recognition problems by combining or integrating the outputs of multiple classifiers. Classifier combinations (Multiple classifiers systems or ensembles) exploit the idea that different classifiers can offer complementary information about patterns to be classified, improving the effectiveness of the overall recognition process (see, for example, [5, 6]). In the literature, ensemble systems have been widely used for several pattern recognition tasks. In the last decade, for instance, a large number of papers [3, 6, 11] has proposed the combination of multiple classifiers as high performance classification systems, in several different areas.

In analyzing ARTMAP-based ensemble systems, this paper aims to analyze the influence of RePART in the accuracy of ensemble systems. These systems will be designed

V. Kůrková et al. (Eds.): ICANN 2008, Part I, LNCS 5163, pp. 562–571, 2008.

by three variants of boosting, which are Aggressive, Conservative and Inverse [11]. In [7], an initial investigation was performed. However, a smaller number of base classifiers and combination methods were used. In this paper, a wider analysis will be done and two system sizes will be used, which are: 12 and 48 base classifiers. In addition, ensemble systems were combined by five different combination methods.

2 ARTMAP-Based Models

The original binary ARTMAP network [8] incorporates two ART1 modules, ARTa (the input vector is received as input) and ARTb (the desired output vector is received as input) as well as a *map field* module. The latter module forms associations between categories from both ART Modules via a technique known as *outstar learning*.

An ARTMAP-based model employs a match-based learning process in which the learning is based on the similarity between the input pattern and the weights (templates) of the category neurons. Match-based learning is distinct from an error-based approach as employed in neural networks such as the standard Multi-Layer Perceptron in which the learning process is based on the error between the output of the network and the desired output (gradient descent methods). There are some significant advantages of match-based over error-based learning which may be characterized as easy knowledge extraction, no catastrophic forgetting and fast learning [5]. On the other hand, there are drawbacks in ARTMAP-based models. The main drawback is sensitivity to noise which can cause category proliferation during learning and misclassification during recalling.

- The Fuzzy ARTMAP Model: Fuzzy ARTMAP is a variant of the ARTMAP model which has the same abstract structure as ARTMAP. However, it makes use of some operations of fuzzy set theory [9]. Essentially, the fuzzy ARTMAP model works as follows. Input and desired output are clamped onto fuzzy ARTa and ARTb modules respectively. Their outputs are subsequently associated through the map field module [9].
- The ARTMAP-IC Model: The ARTMAP-IC neural model [10] basically uses the same learning process of ARTMAP model, with some improvements made in the testing phase. One of the changes is the use a distributed category method in order to define winner node in the recalling phase. In this method, instead of only one winner (winner-takes-all method), several neurons are described as winners (set of winners). These winners are used, along with the corresponding map field weights, to produce the final ARTMAP-IC result (map field winner neuron). Along with distributed code, ARTMAP-IC uses a category instance counting which is the frequency of activation of a category node during learning phase. During recall, instance counting is used with the output of the neuron (ARTa) in order to choose the winner category node.

2.1 RePART Model

RePART (reward/punishment ART) is a more developed version of the simpler Fuzzy ARTMAP which employs additional mechanisms to improve performance and operability of the Fuzzy ARTMAP learning and testing processes [3]. The main improvements are: a variable vigilance and a reward/punishment process.

Variable Vigilance: A vigilance parameter is used in ARTMAP-based networks as a parameter that defines the minimum similarity allowed between the clamped input pattern and the templates of the neurons during the learning phase. The choice of the vigilance parameter is very important to the performance of ARTMAP-based networks. If vigilance is set too low, the structure of the network is too small. In this case, a network tends to bias towards certain patterns and, as a consequence, over-generalizes. On the other hand, if the neural network structure is too large (high vigilance), there are too many variances of the pattern classes. This leads to the category proliferation problem and, in turn, it leads to over-training and poor generalization.

In order to smooth out the category proliferation problem, the use of variable vigilance was proposed. This vigilance starts high and decreases according to the complexity of the RePART architecture. The main idea behind the variable vigilance of the RePART model is to set individual vigilance during the training phase taking into account its average and frequency of activation as well as the number of category neurons associated with its corresponding class. The choice of individual vigilance is due to the fact that the category neurons have different behaviors, storing different numbers of patterns and, as a consequence, have different activation outputs. Also, in the some classes there is more intra-class variation than in other classes. This leads to a different number of category neurons being associated with each class, favoring some classes in the testing phase (for more details, see [2,5]).

The Reward and Punishment Process: In ARTMAP-based models, each ARTa (category) neuron stores weights which correspond to templates, which represent one or more patterns of a class. The activation of a category neuron detects the similarity between the input pattern and its template. The RePART model employs a distributed code in order to define the frequency of activation of a category node. In these respects, the RePART model is similar to the ARTMAP-IC model. RePART also uses the distributed category method defined in ARTMAP-IC. Thus, the use of a set of winners was preserved. In addition, a set of losers is added. During the recalling phase, each category neuron belongs either to the set of winners or the set of losers, according to its output. Thus, a reward/punishment mechanism was added to reward (set of winners) or punish (set of losers) neurons in the category layer. In this method, the more neurons of a class belong to the set of winners, the more likely the prediction is that the input pattern belongs to that class. The testing phase of RePART is summarized as:

1. When a testing pattern is clamped, the ARTa neurons produce their outputs;
2. All ARTa neurons are classified as winner or loser;
3. The winner node in the map field module is calculated, in the following way:
    (a) A reward is awarded to the category nodes of the winners group;
    (b) A punishment is applied to category nodes of the losers group;
4. The winning map field node is the top output one.

The magnitude of the reward depends on the position (ranking) of the category neuron in the set of winners, with the first winner (the highest output) having the biggest reward. An analogous process is performed with the punishment parameter. The intensity of punishment of a neuron depends on its position in the set of losers, in an inverse sense, in which the first loser has the lowest punishment.

As an ARTMAP-based model, after the map field winner node is chosen, an association between the map field winner node and the ARTb module winner node is created. Each map field neuron is linked only to its corresponding category ARTb module neuron in a 1-to-1 association.

3   Ensemble Systems

The need to have efficient computational systems that works with pattern recognition has motivated the interest in the study of ensemble systems [5, 6]. The main idea of using ensembles is that the combination of classifiers can lead to an improvement in the performance of a pattern recognition system. In the design of ensemble systems, two main issues are important, which are: the ensemble components and the combination methods that will be used. In relation to the first issue, the members of an ensemble are chosen and implemented. The correct choice of the set of base classifiers is fundamental to the overall performance of an ensemble. The ideal situation would be a set of base classifiers with uncorrelated errors and they would be combined in such a way as to minimize the effect of these failures. In other words, the base classifiers should be diverse among themselves. One way of obtaining diversity in ensemble systems is through the use of learning strategies, such as: Boosting, Bagging, among others.

Once a set of classifiers has been created, the next step is to choose an effective way of combining their outputs. The choice of the best combination method for an ensemble is very important and difficult to achieve. There are a great number of combination methods reported in the literature [1, 3, 5]. According to their functioning, there are two main strategies of combination methods, which are: fusion and selection [5].

- Fusion-based Methods: In this class, it is assumed that all classifiers are equally experienced in the whole feature space and the decisions of all classifiers are taken into account for any input pattern. There are a vast number of fusion-based methods reported in the literature, such as: Sum, Voting, Naive Bayes, k-NN, among others;
- Selection-based Methods: In this class, only one classifier is needed to correctly classify the input pattern. In order to do so, it is important to define a process to choose a member of the ensemble to make the decision. The choice is typically based on the certainty of the current decision. One of the main methods in classifier selection is Dynamic Classifier Selection (DCS) [5];

In this paper, five different fusion-based combination methods will be used, which are: decision trees (DT), k-NN (nearest neighbor), naive Bayesian (NB), Multilayer Perceptron (MLP) and sum (SUM).

3.1   The Boosting Strategy

As already mentioned, there are several ways to minimize the correlated error of the classifiers within an ensemble system. One way is through the use of distinct training set to the classifiers of an ensemble, using learning strategies. There are some learning strategies in the literature. One of the most used learning strategies is Boosting.

The Boosting process starts placing extra weight on the training patterns that represents the probability of a pattern to be misclassified (probability of misclassification) by a classifier [11]. The general idea of boosting is to develop a classifier team, D, sequentially training classifiers. At every step of boosting, the data set is sampled from the training data set Z. The weights are updates, putting more and more emphasis on patterns that have been misclassified. The more a training pattern is misclassified, the higher its probability of misclassification. Therefore, patterns with high probability of misclassification would then occur more often than those with low probability. Some patterns may not even occur in the training set at all although their probabilities of misclassification are not zero.

In this paper, three variants of Boosting are used, which were proposed in [11], and they can be described as follows:

- Aggressive: In this version, the weights of the correctly classified patterns are decreased whereas the weights of the incorrectly classified patterns are increased;
- Conservative: In this strategy, only the weights of the correctly classified patterns are decreased;
- Inverse: In this variant, unlike the conservative strategy, the weights of the incorrectly classified patterns are decreased;

4 Experimental Work

In order to analyze the influence of RePART in ensemble systems, an empirical comparison is performed. In this analysis, five combination methods are used. As it can be seen, four combination methods used in this paper are trainable methods (DT, k-NN, NB and MLP). In order to define the training set of these combination methods, a validation set will be created and used as the training set of the combination methods. The base classifiers and the ensemble systems were built by using the 10-fold cross-validation methodology. Thus, all accuracy results presented in paper refer to the mean over 10 different test sets. Also, in order to use a validation set, the datasets were divided into 11 folds of equal size (keeping the distribution of classes within each fold). From these sets, 10 of them are reserved to apply a 10-fold cross validation procedure and the remaining one is the validation set.

The base classifiers to be used in this investigation are: Fuzzy ARTMAP, ARTMAP-IC and RePART. Experiments were conducted using two different ensemble sizes, using 12 and 48 base classifiers. For each system size, the influence of using RePART is analyzed. To do this, the RePART models were incrementally added in the ensemble systems. In this sense, the percentage (%) of RePART models in the ensemble systems varied from 0 to 33, 66 and 100%. In ensemble systems with 33%, there are 67% of non-RePART models and 33% of RePART. For instance, for ensembles with 12 base classifiers, 0 means no RePART, 33 means 4 RePART and 8 non-RePART models, 66 means 8 RePART and 4 non-RePART models and 100 means 12 RePART models. As two different non-RePART models have been used, different ensemble systems are created by each configuration. For simplicity reasons, values shown here represent the average results of all possible configurations.

In this investigation, the hypothesis t-test has been used to analyze the statistical significance of the difference between the mean of the error rates, on independent test sets of the methods [11]. It is a test which involves testing two learned hypotheses on identical test sets. In order to perform the test, a set of samples (classifier results) from both algorithms should be used to calculate error mean and standard deviation. Based on the information provided, along with the number of samples, the significance of the difference between the two sets of samples, based on a degree of freedom (α), is defined. In this paper, the confidence level will be 95% $(\alpha = 0.05)$

Datasets. Two different datasets are used in this investigation, which are :

- Database A: It is a breast cancer dataset from UCI repository [12]. Instances were extracted from images of a fine needle aspirate (FNA) of a breast mass and they describe features of the cell nucleus. A total number of 9 attributes were used, in which are real-valued input features. Also, a total number of 480 instances were used, which are equally distributed into malignant and benign examples;
- Database B: This dataset was generated from Landsat Multi-Spectral Scanner image data by the Australian Centre for Remote Sensing. It thus contains 6435 patterns distributed into six classes [12].

4.1   Individual Classifiers

Before starting the investigation of the ensemble systems, it is important to analyze the accuracy of the individual classifiers. Table 1 shows the error rate and standard derivation of the individual classifiers employed in the ensembles with 12 and 48 base classifiers. The classifiers were created using the three variants of boosting, which are: aggressive, conservative and inverse. Values shown in Table 1 represent the average error rates of all executions of the base classifiers. According to the error rate provided by all classifiers, it can be seen that the RePART classifier has delivered the lowest error rate, followed by ARTMAP-IC and Fuzzy Artmap. Also, in general, RePART has delivered the lowest standard deviation of all methods, for both datasets. In relation to the error rate when varying the number of classifiers, Fuzzy ARTMAP and ARTMAP-IC had a similar behavior for both datasets, while RePART behaved differently in each dataset. Fuzzy ARTMAP and ARTMAP-IC increased the error rate when increasing the number of classifiers. On the other hand, RePART had a decrease in the error rate for dataset A, while it had a slight increase for dataset B.

4.2   Ensembles with Twelve Base Classifiers

First line of Table 2 illustrates the general error rate provided by the ensemble systems. Values shown in Table 2 are the average values of all three boosting strategies and all five combination methods. In showing Table 2, it is aimed to describe the general performance of the ensemble systems, when adding RePART as base classifiers.

Table 1. Error rate (ER) and ($\pm$) standard derivation (SD) of the individual classifiers

	FA	A-IC	RePART	FA	A-IC	RePART
	Database A			Database B		
12	9.37±6.59	7.41±5.80	4.46±3.71	21.50±2.74	20.95±1.83	19.99±2.01
48	9.80±6.61	7.64±6.00	4.37±3.61	23.25±6.25	21.74±5.65	20.01±2.07

Table 2. General error rates of the ensemble systems with 12 and 48 base classifiers

	Database A				Database B			
	0	33	66	100	0	33	66	100
12	5.86	5.46	5.05	4.51	17.01	16.55	16.21	16.79
48	6.21	5.56	4.98	4.56	17.21	16.79	16.69	16.70

In analyzing Table 2, it can be seen that the use of RePART has been positive to the accuracy of the ensemble systems, since the error rate of the ensemble systems decreased. This fact is more evident for dataset A, in which the more RePART is added to the ensemble systems, the less is the error rate of these systems. For dataset B, the RePART ensembles (100) have provided a slightly higher error rate, when compared with 33 and 66 ensembles. This is caused by the fact that when using two types of classifiers, there is a tendency to decrease diversity in the ensembles, increasing the accuracy of these systems. When comparing ensembles systems using only one type of classifier (0 and 100), the RePART ensembles (100) has provided a lower error rate than the non RePART ensembles (0). In order to verify whether the decrease in the error rate of the ensemble systems when using RePART is statistically significant, a hypothesis tests (t-test) was applied. In this analysis, ensembles without RePART (0) are compared with ensembles with RePART (33, 66 and 100). Apart from RePART ensembles (100) for dataset B, in all other 5 cases, the decrease in the error rate is statistically significant.

In analyzing the effect of adding RePART in the ensemble systems for each boosting strategy, Figure 1 shows the error rate of the ensemble systems, separated by the boosting strategy. Values shown in Figure 1 are the average values of all five combination methods. From Figure 1, it is possible to state that all three boosting strategies have a similar behavior, always decreasing the error rate when increase the percentage of RePART, for dataset A. For dataset B, there is a slight increase in the error rate when increasing from 66 to 100% of RePART. As already mentioned, this is caused by the decrease in the diversity of the ensembles, when using only RePART as base classifiers. When applying the statistical test, the conservative boosting has provided the largest number of statistically significant decreases, which was 4 (out of 6 - 33, 66 and 100 for datasets A and B), followed by Inverse and Aggressive (3).

The effects of using RePART in the ensemble systems, separated by the combination method, are shown in Figure 2. In analyzing Figure 2, it can be observed that, for dataset A, all combination methods had a similar behavior, decreasing the error rate when increasing the amount of RePART in the ensemble systems. Of the combination methods, the MLP combiner was the one which provided the highest decrease in the error rate for the RePART (100) and partially RePART (33 and 66) ensembles. When applying the statistical test, the MLP combiner has provided the largest number of statistically significant decreases, which was 5 (out of 6), followed by SUM (4), DT and NB (2) and k-NN (1).

4.3 Ensembles with Forty-Eight Base Classifiers

Second line of Table 2 illustrates the general error rate provided by the ensemble systems using 48 base classifiers. In analyzing Table 2, it can be seen that the use of

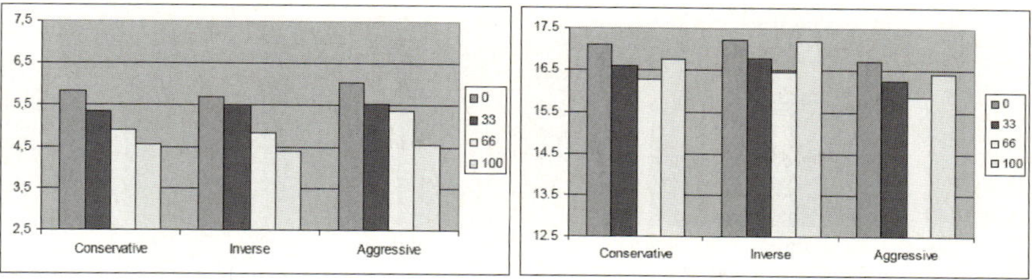

Fig. 1. The error rates of the ensemble systems with 12 base classifiers, separated by the boosting strategy and applied to datasets A (left) and B (right)

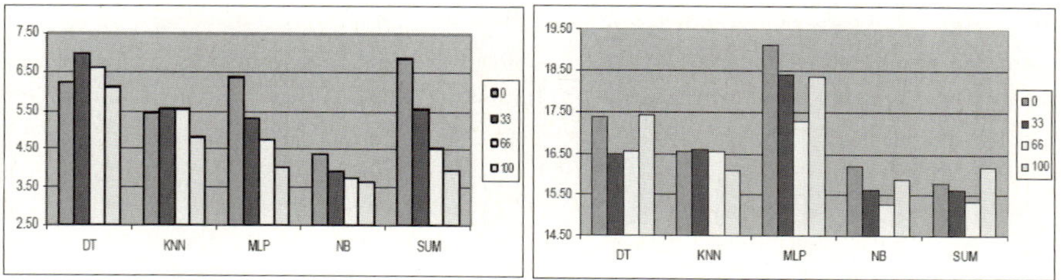

Fig. 2. The error rates of the ensemble systems with 12 base classifiers, separated by the combination method and applied to datasets A (left) and B (right)

RePART has been positive to the accuracy of the ensemble systems, since the error rate of the ensemble systems decreased when using ensembles, for both datasets. In applying the t-test, it could be observed that the decrease in the error rate is statistically significant, in all six cases (33, 66 and 100 for both datasets).

Figure 3 shows the error rate of the ensemble systems, separated by the boosting strategy. From Figure 3, it can be seen that all three boosting strategies have a similar behavior, always decreasing the error rate when increase the percentage of RePART, for both datasets (apart from inverse boosting for dataset B). When applying the statistical test, the aggressive boosting has provided the largest number of statistically significant decreases, which was 6, followed by Conservative (5) and Aggressive (2).

The effects of using RePART in the ensemble systems, separated by the combination method, are shown in Figure 4. Similar to the previous section, for dataset A, all combination methods had a similar behavior, decreasing the error rate when increasing the amount of RePART in the ensemble systems. On the other hand, for dataset B, some combination methods (SUM, NB and DT) have provided a increase in the error rate when using RePART ensembles (100), even when compared with non-RePART ensembles (0). The MLP combiner was the one which provided the highest decrease in the error rate for the RePART (100) and partially RePART (33 and 66) ensembles. When applying the statistical test, the MLP combiner has provided the largest number of statistically significant decreases, in all analyzed cases, followed by SUM, k-NN and NB (3) and DT (2).

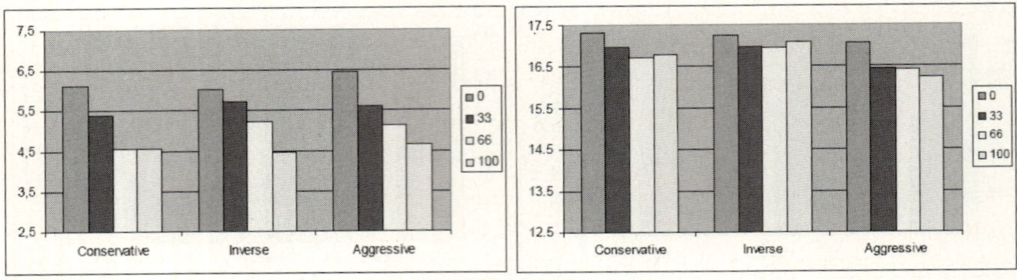

Fig. 3. The error rates of the ensemble systems with 48 base classifiers, separated by the boosting strategy and applied to datasets A (left) and B (right)

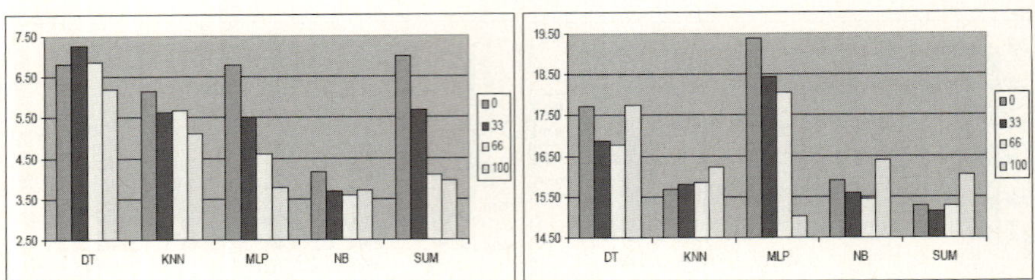

Fig. 4. The error rates of the ensemble system with 48 base classifiers, separated by the combination method and applied to datasets A (left) and B (right)

4.4   Analysis of the Results

After an analysis of the use of RePART in ensemble systems, it is possible to conclude that, in a general perspective, the use of RePART is positive to the ensemble systems, since it caused a decrease in the error rate of these systems, in most of the cases.

In relation to the ensemble size, the use of RePART had a bigger impact in ensembles with 48 base classifiers, having 36 statistically significant decreases (out of 54), while ensembles with 12 base classifiers had 28 statistically significant decreases. It is important to emphasize that although the use of more base classifiers was more beneficial for the use of RePART, for some combination methods, it caused a decrease in the performance of these systems for dataset B (increase in the error rate). This was compensated by the increase in the performance for the other combination methods and for the other dataset (A). In relation to the boosting strategy, the Conservative boosting had provided the best performance, providing 10 statistically significant decreases (out of 12), followed by Aggressive (9) and Inverse (5).

Of the combination methods, it could be noticed that the ensemble combined by MLP has provided the best performance, providing 11 statistically significant decreases (out of 12), followed by SUM (7), NB (5), DT (4) and k-NN (3).

5   Final Remarks

In this paper, an investigation of the influence of using RePART in ensemble systems was performed. In this analysis, three variants of boosting were applied to two different ensemble sizes (12 and 48). Also, five different combination methods were used and they were applied to two different datasets.

The results obtained in this paper have shown that the use of RePART was positive to the performance of ensembles, having statistically significant decreases in most of the cases (64 out of 108). In relation to the system size, the use of RePART in ensembles with 48 base classifiers has provided the largest number of statistically significant decreases. Of the boosting strategies, the largest number of statistically significant decreases was obtained by the Conservative boosting. Finally, of the combination methods, MLP is the best choice for combining the outputs of the classifiers. The MLP ensembles have provided the most accurate systems and the highest decrease in the error rate of the RePART ensembles, when compared with non-RePART ensembles.

References

1. Canuto, A., Howells, G., Fairhurst, M.: An investigation of the effects of variable vigilance within the repart neuro-fuzzy network. J. of Int. and Robotics Systems 29(4), 317–334 (2000)
2. Canuto, A., Fairhurst, M., Howells, G.: Improving artmap learning through variable vigilance. International Journal of Neural Systems 11(6), 509–522 (2001)
3. Canuto, A.: Combining neural networks and fuzzy logic for applications in character recognition. PhD thesis, University of Kent (2001)
4. Canuto, A., Santos, A.: A Comparative Investigation of the RePART Neural Network in Pattern Recognition Tasks. In: Proceedings of IEEE IJCNN 2004 (2004)
5. Kuncheva, L.I.: Combining Pattern Classifiers. Methods and Algorithms. Wiley, Chichester (2004)
6. Sharkey, A.J.C.: Multi-net System. In: Sharkey, A.J.C. (ed.) Combining Artificial Neural Nets: Ensemble and Modular Multi-net Systems, pp. 1–30. Springer, Heidelberg (1999)
7. Santos, A., Canuto, A.M.: Investigating the Influence of RePART in Ensemble Systems Designed by Boosting. In: IJCNN 2008 (accepted, 2008)
8. Carpenter, G., Grossberg, S., Reynolds, J.H.: Artmap: Supervised real-time learning and classification of nonstationary data by a self-organizing neural network. Neural Networks 4, 565–588 (1991)
9. Carpenter, G., Grossberg, S., Markunzo, M., Reynolds, J.H., Rosen, D.B.: Fuzzy artmap: A neural network architecture for incremental supervised learning of analog multidimensional maps. IEEE Transactions on Neural Networks 3, 698–713 (1992)
10. Carpenter, G., Markuzon, N.: Artmap-IC and medical diagnosis: instance counting and inconsistent cases. Neural Networks 11, 323–336 (1998)
11. Kuncheva, L., Whitaker, C.J.: Using diversity with three variants of boosting: Aggressive, conservative, and inverse. Multiple Classifier Systems (2002)
12. Blake, C.L., Merz, C.J.: UCI Repository of machine learning databases. University of California, Dep. of Inf. and Computer Science (1998),
    http://www.ics.uci.edu/~mlearn/MLRepository.html

Matrix Learning for Topographic Neural Maps

Banchar Arnonkijpanich[1], Barbara Hammer[2], Alexander Hasenfuss[2],
and Chidchanok Lursinsap[3]

[1] Khon Kaen University, Faculty of Science, Department of Mathematics, Thailand
[2] Clausthal University of Technology, Department of Computer Science, Germany
[3] Chulalongkorn University, Department of Mathematics, Thailand

Abstract. The self-organizing map (SOM) and neural gas (NG) constitute popular algorithms to represent data by means of prototypes arranged on a topographic map. Both methods rely on the Euclidean metric, hence clusters are isotropic. In this contribution, we extend prototype-based clustering algorithms such as NG and SOM towards a metric which is given by a full adaptive matrix such that ellipsoidal clusters are accounted for. We derive batch optimization learning rules for prototype and matrix adaptation based on a general cost function for NG and SOM and we show convergence of the algorithm. It can be seen that matrix learning implicitly performs minor local principal component analysis (PCA) and the local eigenvectors correspond to the main axes of the ellipsoidal clusters. We demonstrate the behavior in several examples.

1 Introduction

The self-organizing map (SOM) as proposed by Kohonen constitutes one of the most popular data inspection tools due to its intuitive, robust, and flexible behavior with numerous applications ranging from web and text mining up to telecommunications and robotics [12]. Data are represented by means of typical prototypes which are arranged on a fixed lattice structure, often a low-dimensional regular lattice such that visualization of data is easily possible. Neural gas (NG) as introduced by Martinetz transfers ideas of SOM towards a data optimum neighborhood structure such that a representation by prototypes and topographic mapping of data can be achieved without any constraints on the topology given by a prior lattice structure [15]. Both methods have in common that they extract typical prototypes from the data which are arranged according to the local data characteristics. Thereby, neighborhood integration accounts for a very robust behavior of the models. A variety of alternative prototype based clustering methods have been proposed including the popular k-means algorithm, mixture of Gaussians, or fuzzy clustering [3,9,23].

Standard NG and SOM are based on the Euclidean metric. Correspondingly, the clusters are isotropic with spherical class boundaries. Several methods extend SOM and NG towards more general distance measures, such as median or relational clustering which allows to process general dissimilarity data given by pairwise distances only [5,8]. These methods, however, use a fixed priorly

V. Kůrková et al. (Eds.): ICANN 2008, Part I, LNCS 5163, pp. 572–582, 2008.

chosen metric. For the Euclidean setting, a number of approaches which adapt the distance calculation to the data at hand have been proposed such that more general cluster shapes can be accounted for.

The methods [19,20] extend the setting towards a semi-supervised scenario and they adapt the metric such that the aspects which are relevant for the supervised labeling of data are emphasized in the visualization. These settings require additional knowledge for the adaptation of the metric parameters.

Methods which are solely based on the unsupervised training data for the neural map include the popular adaptive subspace SOM which extracts invariant features in terms of invariant local subspaces attached to the prototypes [13]. A similar idea is proposed in the approaches [14,17,22,1] where prototypes are enriched by vectors corresponding to the local main principal directions of the data. These methods combine principal component analysis (PCA) techniques such as Oja's rule and extensions thereof with a vector quantization scheme such as SOM, possibly using an enriched metric. Thereby, the approaches rely on heuristics, or the learning rules are derived separately for clustering and principal component analysis (PCA), respectively. Up to our knowledge, no approach which derives the learning rules from a uniform cost function exists.

Several alternatives propose to enrich the metric used for vector quantization towards a general adaptive form. One very elegant possibility uses a full matrix which can take individual dimension weighting and correlations into account. This scheme has been integrated into standard k-means algorithm and fuzzy extensions thereof as well as supervised prototype-based learning schemes such as learning vector quantization [18,11,21]. Interestingly, for k-means and fuzzy-k-means, matrix adaptation corresponds to an estimation of the local Mahalanobis distance of data. Hence the local principal directions are implicitly determined in these approaches and clusters are given by ellipsoids aligned along the local principal directions of data. Up to our knowledge, no extension of neural vector quantization schemes such as NG and SOM towards a general matrix exist.

In this approach, we extend NG and SOM towards a general adaptive matrix which characterizes the metric. We derive update rules of the parameters based on a uniform underlying cost function of NG and SOM in the variant as proposed by Heskes [10], relying on batch optimization schemes. The resulting update rules for the matrix correspond to a generalized Mahalanobis distance, hence the method can be linked to local PCA methods. We show convergence of the update rules, and we demonstrate the behavior in a variety of benchmark examples.

2   Topographic Neural Maps

Assume data points $x \in \mathbb{R}^m$ are distributed according to an underlying distribution P. The goal of prototype based clustering is to represent data by means of prototypes $w^i \in \mathbb{R}^m$, $i = 1, \ldots, n$, such that they represent the distribution P as accurate as possible. A data point $x \in \mathbb{R}^m$ is represented by the winning prototype $w^{I(x)}$ which is the prototype located closest to the data point, i.e. $I(x) = \operatorname{argmin}_i \{ d(x, w^i) \}$ measured according to the squared Euclidean distance

$d(\boldsymbol{x}, \boldsymbol{w}^i) = (\boldsymbol{x} - \boldsymbol{w}^i)^t(\boldsymbol{x} - \boldsymbol{w}^i)$. The mathematical objective of vector quantization is to minimize the quantization error, $E_{\mathrm{VQ}}(\boldsymbol{w}) = \frac{1}{2}\sum_{i=1}^{n}\int \delta_{i,I(\boldsymbol{x})}\cdot d(\boldsymbol{x}, \boldsymbol{w}^i)\, P(d\boldsymbol{x})$ where $\delta_{i,j}$ denotes the Kronecker function. Given a finite set of training examples $\boldsymbol{x}^1, \ldots, \boldsymbol{x}^p$, the popular k-means algorithm optimizes the corresponding discrete cost function directly by means of an EM scheme, subsequently optimizing data assignments and prototype locations. Since the cost function E_{VQ} is often multimodal, k-mean clustering gets easily stuck in local optima and multiple restarts are necessary to achieve a good solution.

NG and SOM offer alternatives which integrate neighborhood cooperation of the prototypes and, this way, achieve robustness of the results with respect to prototype initialization. In addition, neighborhood cooperation accounts for a topological arrangement of prototypes such that browsing and, in the case of SOM with low-dimensional lattice structure, visualization of data become easily possible. The cost function of NG is given by

$$E_{\mathrm{NG}}(\boldsymbol{w}) \sim \frac{1}{2}\sum_{i=1}^{n}\int h_\sigma(k_i(\boldsymbol{x})) \cdot d(\boldsymbol{x}, \boldsymbol{w}^i)\, P(d\boldsymbol{x}) \tag{1}$$

where

$$k_i(\boldsymbol{x}) = |\{\boldsymbol{w}^j \mid d(\boldsymbol{x}, \boldsymbol{w}^j) < d(\boldsymbol{x}, \boldsymbol{w}^i)\}| \tag{2}$$

is the rank of the prototypes sorted according to the distances and $h_\sigma(t) = \exp(-t/\sigma)$ is a Gaussian shaped curve with neighborhood range $\sigma > 0$. For vanishing neighborhood $\sigma \to 0$, the quantization error is recovered. NG is usually optimized by means of a stochastic gradient descent method. Alternatively, for a given finite data set as above, the corresponding discrete cost function $1/2 \cdot \sum_{i=1}^{n}\sum_{j=1}^{p} h_\sigma(k_i(\boldsymbol{x}^j)) \cdot d(\boldsymbol{x}^j, \boldsymbol{w}^i)$ can be optimized in a batch scheme in analogy to k-means using the update rules

$$k_{ij} = k_i(\boldsymbol{x}^j), \quad \boldsymbol{w}^i = \sum_j h_\sigma(k_{ij})\boldsymbol{x}^j \Big/ \sum_j h_\sigma(k_{ij}) \tag{3}$$

as pointed out in [5]. During training, the neighborhood range σ is annealed to 0 such that the quantization error is recovered in final steps. In intermediate steps, a neighborhood structure of the prototypes is determined according to the training data. This choice, on the one hand, accounts for a high robustness of the algorithm with respect to local minima of the quantization error, on the other hand, NG can be extended to extract an optimum topological structure from the data by means of Hebbian updates [16]. Due to its simple adaptation rule, the independence of a prior lattice, and the independence of initialization because of the integrated neighborhood cooperation, NG offers a simple and highly effective algorithm for data clustering.

SOM uses the adaptation strength $h_\sigma(nd(I(\boldsymbol{x}^j), i))$ instead of $h_\sigma(k_i(\boldsymbol{x}^j))$, where nd is a priorly chosen, often two-dimensional neighborhood structure of the neurons. A low-dimensional lattice offers the possibility to visualize data easily. However, if the primary goal is clustering, a fixed topology puts restrictions on the map and topology preservation can be violated. The original SOM does not possess a cost function in the continuous case and its mathematical investigation is difficult [4]. A slight change of the winner notation can be associated to a cost function as pointed out by Heskes [10]: SOM optimizes

$$E_{\text{SOM}}(\boldsymbol{w}) \sim \frac{1}{2} \sum_{i=1}^{n} \int \delta_{i,I^*(\boldsymbol{x})} \cdot \sum_{l=1}^{n} h_\sigma(nd(i,l)) \cdot d(\boldsymbol{x}, \boldsymbol{w}^l) \, P(d\boldsymbol{x}) \qquad (4)$$

if the winner is defined as $I^*(\boldsymbol{x}) = \text{argmin}_i \sum_{l=1}^{n} h_\sigma(nd(i,l))d(\boldsymbol{x}, \boldsymbol{w}^l)$. For vanishing neighborhood $\sigma \to 0$, the quantization error is recovered. Like NG, SOM is usually optimized by means of a stochastic gradient descent. For a given set of training data as beforehand, alternative batch optimization is possible. The neighborhood strength is thereby annealed $\sigma \to 0$ during training. This yields to a very fast adaptation towards a topology preserving map such that browsing and, in the case of a low dimensional lattice, data visualization become possible. Topological mismatches can occur due to two reasons, a mismatch of data and lattice topology or a too fast adaptation, as discussed in [7].

It has been proved in [5] that batch SOM and batch NG converge to a local optimum of the corresponding cost functions in a finite number of steps. Both methods offer clustering schemes which require only a few parameters: the number of clusters and the number of training epochs, such that robust methods are obtained which do not need a time consuming optimization of additional meta-parameters such as the learning rate.

3   Matrix Learning

Classical NG and SOM rely on the Euclidean metric which induces isotropic cluster shapes. General ellipsoidal shapes can be achieved by the generalized form

$$d_{\Lambda_i}(\boldsymbol{x}, \boldsymbol{w}^i) = (\boldsymbol{x} - \boldsymbol{w}^i)^t \Lambda_i (\boldsymbol{x} - \boldsymbol{w}^i) \qquad (5)$$

instead of the squared Euclidean metric where $\Lambda_i \in \mathbb{R}^{m \times m}$ is a symmetric positive definite matrix with $\det \Lambda_i = 1$. These constraints guarantee the metric property. A general matrix can account for correlations and non uniform scaling of the data dimensions. Since an optimum matrix Λ_i is not known before training, Λ_i is adapted according to the data. The following cost function results:

$$E_{\text{NG}}(\boldsymbol{w}, \Lambda) \sim \frac{1}{2} \sum_{i=1}^{n} \int h_\sigma(k_i(\boldsymbol{x})) \cdot d_{\Lambda_i}(\boldsymbol{x}, \boldsymbol{w}^i) \, P(d\boldsymbol{x}) \qquad (6)$$

for matrix NG whereby the assignments $k_i(\boldsymbol{x})$ are computed based on d_{Λ_i}, and Λ_i is restricted to symmetric positive definite forms with $\det \Lambda_i = 1$. The SOM cost function can be extended in a similar way.

We derive a batch optimization scheme for matrix NG based on this cost functions. We assume a finite number of data $\boldsymbol{x}^1, \ldots, \boldsymbol{x}^p$ and we consider the associated discrete cost function. We introduce hidden variables k_{ij} for $i = 1, \ldots, n$, $j = 1, \ldots, p$ which constitute a permutation of $\{0, \ldots, n-1\}$ for every fixed j.

Similar to batch NG, we extend the function to $E_{\mathrm{NG}}(\boldsymbol{w}, \Lambda, k_{ij}) = \sum_{ij} h_\sigma(k_{ij}) \cdot d_{\Lambda_i}(\boldsymbol{x}^j, \boldsymbol{w}^i)$ where the hidden variables k_{ij} take the place of $k_i(\boldsymbol{x}^j)$. Optimum assignments k_{ij} fulfill $k_{ij} = k_i(\boldsymbol{x}^j)$, hence $E_{\mathrm{NG}}(\boldsymbol{w}, \Lambda, k_{ij}) = E_{\mathrm{NG}}(\boldsymbol{w}, \Lambda)$ holds for optimum k_{ij}. Batch optimization, in turn, optimizes the hidden variables k_{ij} for fixed Λ and $\boldsymbol{w}$, and it determines optimum parameters Λ and $\boldsymbol{w}$ given fixed assignments k_{ij}. Now, we compute these optima.

Step 1: Optimization w.r.t k_{ij}: As already stated, given the constraints on k_{ij} and fixed $\boldsymbol{w}$ and Λ, optimum assignments fulfill the equation

$$k_{ij} = k_i(\boldsymbol{x}^j) \,. \tag{7}$$

Step 2: Optimization w.r.t. $\boldsymbol{w}^i$: Given fixed k_{ij}, the derivative of E_{NG} with respect to $\boldsymbol{w}^i$ into direction $\xi \in \mathbb{R}^m$ has the form $\sum_j h_\sigma(k_{ij})(\Lambda_i(\boldsymbol{w}^i - \boldsymbol{x}^j))^t \xi \overset{!}{=} 0$ set to 0 for all directions ξ. Therefore, $\sum_j h_\sigma(k_{ij})(\boldsymbol{w}^i - \boldsymbol{x}^j) = 0$ for all i, hence

$$\boldsymbol{w}^i = \frac{\sum_j h_\sigma(k_{ij})\boldsymbol{x}^j}{\sum_j h_\sigma(k_{ij})} \tag{8}$$

Step 3: Optimization w.r.t. Λ_i: We ignore the constraint of symmetry and positive definiteness of Λ_i for the moment. Taking into account the constraint $\det \Lambda_i = 1$, we obtain the Lagrange function $L = \frac{1}{2} \sum_{i=1}^n \sum_{j=1}^p h_\sigma(k_{ij}) \cdot d_{\Lambda_i}(\boldsymbol{x}^j, \boldsymbol{w}^i)$ $- \sum_{i=1}^n \lambda_i(\det \Lambda_i - 1)$ with Lagrange parameters $\lambda_i \in \mathbb{R}$. The derivative w.r.t Λ_i yields $\sum_j h_\sigma(k_{ij})(\boldsymbol{x}^j - \boldsymbol{w}^i)(\boldsymbol{x}^j - \boldsymbol{w}^i)^t - \lambda_i(\det \Lambda_i \cdot \Lambda_i^{-1}) \overset{!}{=} 0$, hence $\Lambda_i = \left(\sum_j h_\sigma(k_{ij})(\boldsymbol{x}^j - \boldsymbol{w}^i)(\boldsymbol{x}^j - \boldsymbol{w}^i)^t\right)^{-1} \lambda_i$ because $\det \Lambda_i = 1$. We set

$$S_i := \sum_j h_\sigma(k_{ij})(\boldsymbol{x}^j - \boldsymbol{w}^i)(\boldsymbol{x}^j - \boldsymbol{w}^i)^t \tag{9}$$

Using this definition, we have $S_i\Lambda_i = \lambda_i I$, I being the identity, and, hence, $\det(S_i\Lambda_i) = \det S_i = \lambda_i^n$, thus $\lambda_i = (\det S_i)^{1/n}$. Thus,

$$\Lambda_i = S_i^{-1}(\det S_i)^{1/n} \tag{10}$$

This is well defined if S_i is invertible, which holds if at least n linearly independent vectors $\boldsymbol{x}^j - \boldsymbol{w}^i$ exist. Further, the Hessian of the Lagrange function with respect to Λ_i yields the constant 0. Λ_i is symmetric because S_i is symmetric, and Λ_i is positive definite if S_i is. Using these equations, matrix NG is as follows:

> init $\boldsymbol{w}^i$ (e.g. using random data points)
> init Λ_i as identity I
> repeat until convergence
> > determine k_{ij} using (7)
> > determine $\boldsymbol{w}^i$ using (8)
> > determine Λ_i using (10)

The matrix S_i corresponds to the correlation of the data centered around prototype $\boldsymbol{w}^i$ and weighted according to its distance from the prototype. For vanishing neighborhood $\sigma \to 0$, the standard correlation matrix of the receptive field is obtained and the distance corresponds to the Mahalanobis distance. This causes a scaling of the principal axes of the data space by their inverse eigenvalues. Thus, ellipsoidal cluster shapes arise which main directions are centered along the local principal axes and the scaling is elongated along the main principal components. We conclude that matrix learning performs implicit local PCA and the main local principal components at prototype location $\boldsymbol{w}^i$ are given by the minor principal components of Λ_i (due to matrix inversion). Here, unlike a standard PCA, neighborhood cooperation is applied to both, prototype adaptation and matrix learning during batch training, such that a regularization of matrix learning is given which is beneficial e.g. for small clusters.

It can happen that the matrix S_i is not invertible. Therefore, we use the pseudoinverse of the matrix. Further, it might be advisable to include a regularization term by adding a small multiple of the unit matrix.

Matrix learning for SOM and k-means can be derived in a similar way by means of the respective cost functions. In the case of SOM, neighborhood cooperation is given by a prior lattice structure instead of the ranks.

4   Convergence

The cost functions of NG, SOM, and k-means can be uniformly stated as

$$E(\boldsymbol{w}, \Lambda) = \sum_{i=1}^{n} \sum_{j=1}^{p} f_1(k_{ij}(\boldsymbol{w}, \Lambda)) \cdot f_2^{ij}(\boldsymbol{w}, \Lambda) \tag{11}$$

where f_1 refers to the winner for SOM and k-means, and it is given as the neighborhood $h_\sigma(k_i(\boldsymbol{x}^j))$ for NG. f_2^{ij} denotes the squared distance $d_{\Lambda_i}(\boldsymbol{x}^j, \boldsymbol{w}^i)$ for NG and k-means, and it is given as $\sum_l h_\sigma(nd(i,l))d_{\Lambda_l}(\boldsymbol{x}^j, \boldsymbol{w}^l)$ for SOM.

Batch optimization substitutes the dependent values $k_{ij}(\boldsymbol{w}, \Lambda)$ by hidden parameters which are optimized under constraints within a finite set such that, for optimum values k_{ij} given fixed $\boldsymbol{w}$, Λ, the equality $k_{ij} = k_{ij}(\boldsymbol{w}, \Lambda)$ holds. Batch clustering in turn finds optimum values $\boldsymbol{w}$ and Λ, given fixed assignments k_{ij} and it determines new assignments k_{ij} given fixed parameters $\boldsymbol{w}$ and Λ. Consider the function

$$Q(\boldsymbol{w}, \Lambda, \boldsymbol{w}', \Lambda') = \sum_{i=1}^{n} \sum_{j=1}^{p} f_1(k_{ij}(\boldsymbol{w}, \Lambda)) \cdot f_2^{ij}(\boldsymbol{w}', \Lambda') \tag{12}$$

where $k_{ij}(\boldsymbol{w}, \Lambda)$ denotes the (unique) optimum assignment of k_{ij} given fixed $\boldsymbol{w}$ and Λ of the cost function. It holds $E(\boldsymbol{w}, \Lambda) = Q(\boldsymbol{w}, \Lambda, \boldsymbol{w}, \Lambda)$. Denote by $\boldsymbol{w}(t)$ and $\Lambda(t)$ the values derived in batch clustering in epoch t. Then we find $E(\boldsymbol{w}(t+1), \Lambda(t+1)) \leq Q(\boldsymbol{w}(t), \Lambda(t), \boldsymbol{w}(t+1), \Lambda(t+1))$ because $k_{ij}(\boldsymbol{w}(t+1), \Lambda(t+1))$ are optimum assignments for given $\boldsymbol{w}(t+1)$ and $\Lambda(t+1)$. Further, $Q(\boldsymbol{w}(t), \Lambda(t), \boldsymbol{w}(t+1), \Lambda(t+1)) \leq E(\boldsymbol{w}(t), \Lambda(t))$ because $\boldsymbol{w}(t+1)$ and $\Lambda(t+1)$ are chosen as optimum values for the assignments $k_{ij}(\boldsymbol{w}(t), \Lambda(t))$. Hence $E(\boldsymbol{w}(t+1),$

$\Lambda(t+1)) - E(\boldsymbol{w}(t), \Lambda(t)) \leq 0$ Thus, the cost function (12) is decreased in consecutive steps. Since the assignments k_{ij} stem from a finite set the algorithm converges in a finite number of steps towards a fixed point $\boldsymbol{w}^*$, Λ^*.

Assume that, for this fixed point, no data points $\boldsymbol{x}^j$ lie at the borders of receptive fields (which is fulfilled almost surely). Then, the cost function (11) is continuous at $\boldsymbol{w}^*$, Λ^*, and, since k_{ij} is discrete, the function $k_{ij}(\boldsymbol{w}, \Lambda)$ is constant in a vicinity of $\boldsymbol{w}^*$, Λ^*. Hence $E(\cdot)$ and $Q(\boldsymbol{w}^*, \Lambda^*, \cdot)$ are identical in a neighborhood of $\boldsymbol{w}^*$, Λ^*, i.e. local optima of Q correspond to local optima of E.

5   Experiments

We test the algorithm for an illustrative example and benchmarks from the UCI repository [2]. Larger experiments for realistic application are currently investigated, but beyond the scope of this paper. Initial neighborhood ranges between $n/2$ (for NG) and $n/12$ (for SOM) are used and the neighborhood is multiplicatively annealed to 0 during training. Training takes place for 100 epochs. We train k-means, neural gas, and SOM with two-dimensional rectangular neighborhood structure with and without matrix adaptation.

Checkerboard. In the checkerboard data set, data are arranged according to a checkerboard structure with almost circular and elongated clusters, respectively, as depicted in Fig. 1 (top and bottom, respectively). 25 prototypes are adapted using matrix NG, SOM, and k-means. In addition, we depict a characteristic isobar for every prototype. Obviously, NG and SOM yield perfect and robust results, whereas k-means suffers from local optima. In all cases, the major axes of the found ellipsoids corresponds to the local main principal component.

Spirals. The spirals data set is depicted in Fig. 2. It is trained using 34 prototypes with matrix learning and standard k-means, SOM, and NG, respectively. The results show that, for matrix learning the cluster shapes are adapted to the spiral, showing elongated ellipses at outer regions and circular shapes in the middle, while standard clustering without matrix adaptation cannot adjust the cluster forms to the given data as can be seen in Fig. 2 (bottom line). For both cases, the results of SOM and NG are virtually indistinguishable, whereas k-means suffers from local optima. We evaluate the results by their ability to classify the points correctly into two classes according to the two spirals. Thereby, we use posterior labeling of prototypes based on a majority vote. The classification accuracy is shown in Fig. 2, demonstrating the superior behavior of matrix learning and neighborhood integration.

UCI classification data. We consider 5 datasets from the UCI repository of machine learning [2]:

- Iris: The popular iris data set consists of 150 data points assigned to 3 classes. Data are numerical with 4 dimensions.

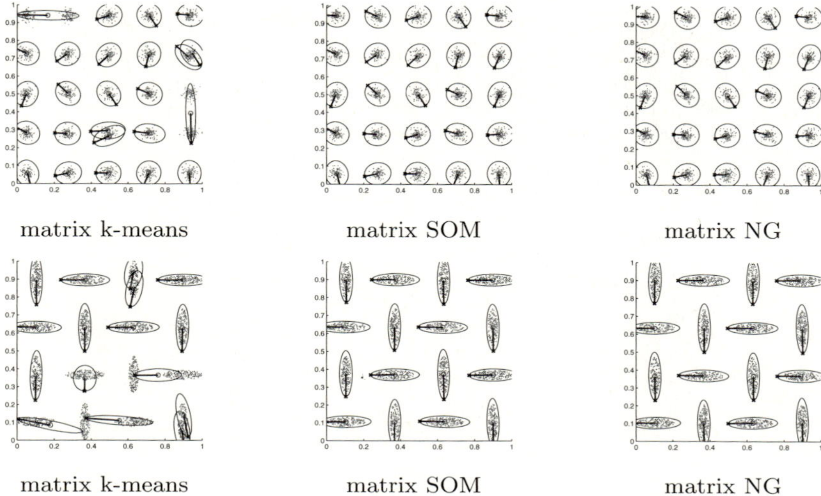

matrix k-means   matrix SOM   matrix NG

matrix k-means   matrix SOM   matrix NG

Fig. 1. Multimodal checkerboard data with circular shapes (top) and elongated clusters (bottom), respectively. The results of matrix k-means, SOM, and NG after 100 epochs are depicted.

- The Wisconsin diagnostic breast cancer data set contains 569 data of 2 classes represented by 32 dimensional vectors.
- The ionosphere data contains 351 examples in 2 classes, data are given as 34 dimensional vectors.

Class labels are available for the data sets, i.e. we can assign a label c_i to every data point $\boldsymbol{x}^i$ in the data. We evaluate the methods by the ability to group the data into the priorly given classes. For this purpose, we evaluate the classification accuracy obtained for posterior labeling of the clusters as before. For comparison, we consider the clustering which is obtained from the class-wise direct Mahalanobis distance given by the center and inverse correlation matrix of the priorly given classes.

The results (classification error and standard deviation over $50-100$ runs) of standard NG, SOM, and k-means and NG, SOM, and k-means with matrix learning are given in Tab. 1, whereby we use the same number of cluster centers as given classes. Standard clustering yields isotropic clusters whereas matrix learning adapts ellipsoidal shapes according to the data. Obviously, neighborhood integration improves the robustness and classification ability of the method. Further, matrix adaptation accounts for an improvement of 1-9% depending on the data set due to the larger flexibility of the metric. Interestingly, the result of a direct Mahalanobis distance is worse compared to matrix clustering in breast cancer and ionosphere data sets, although the first method uses the given class information and the same number of prototypes as for matrix clustering is available. This can be assigned to the fact that noise in the data set disrupts the

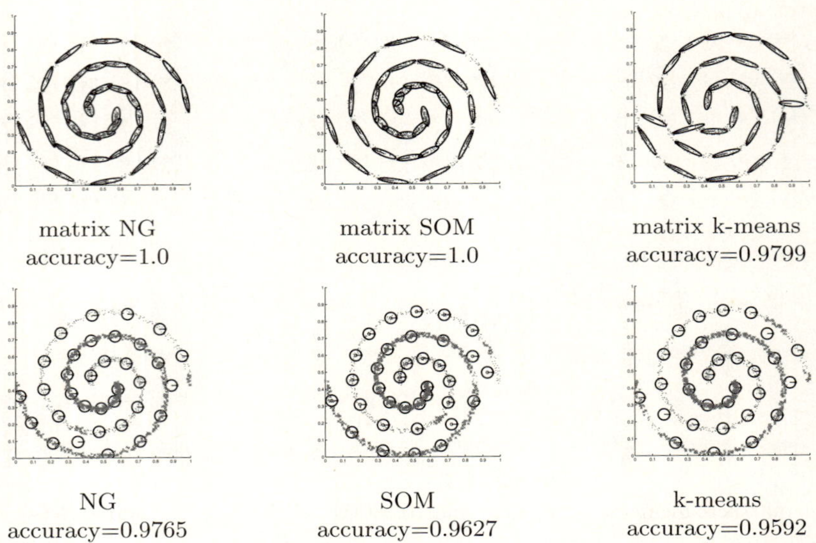

<table>
<tr><td align="center">matrix NG
accuracy=1.0</td><td align="center">matrix SOM
accuracy=1.0</td><td align="center">matrix k-means
accuracy=0.9799</td></tr>
<tr><td align="center">NG
accuracy=0.9765</td><td align="center">SOM
accuracy=0.9627</td><td align="center">k-means
accuracy=0.9592</td></tr>
</table>

Fig. 2. Results of matrix NG, SOM, and k-means (top) and standard NG, SOM, and k-means (bottom) after 100 epochs for the spirals data set. Obviously, k-means suffers from local optima. Further, the shape is better represented by matrix clustering as can be seen by the classification accuracy of the maps.

Table 1. Classification results of the clustering methods with and without matrix adaptation for various data sets from UCI. The mean classification accuracy and the standard deviation are reported.

	matrix k-Means	matrix SOM	matrix NG	direct Mahalanobis
iris	0.8606 ±0.1356	0.8909 ±0.1175	0.9147 ±0.0847	0.9800
breast cancer	0.8953 ±0.0226	0.9024 ±0.0245	0.9135 ±0.0192	0.8998
ionosphere	0.7320 ±0.0859	0.7226 ±0.0673	0.7197 ±0.0600	0.6410

	k-Means	SOM	NG	
iris	0.8499 ±0.0864	0.8385 ±0.0945	0.8867 ±0.0000	
breast cancer	0.8541 ±0.0000	0.8541 ±0.0000	0.8541 ±0.0000	
ionosphere	0.7074 ±0.0159	0.7114 ±0.0013	0.7097 ±0.0009	

relevant information. Unlike the direct Mahalanobis method, matrix clustering uses only those data for matrix computation which lies in the receptive field of the prototype, i.e. the geometric form is taken into account instead of the prior labeling, thus, better robustness with respect to outliers is obtained.

Naturally, the direct Mahalanobis method cannot be used in the case of unknown prior labeling or in the case of multimodal classes which should be represented by more than one prototype such as the two spirals data set. Matrix clustering can directly be used in these settings.

6 Conclusions

We presented an extension of neural based clustering such as NG and SOM towards a full matrix. This way, the metric can take scaling and correlations of dimensions into account. Based on batch optimization, an adaptation scheme for matrix learning has been proposed and convergence of the method has been proved. As demonstrated in experiments, matrix NG can improve the clustering performance compared to versions which rely on isotropic cluster shapes. The method has been demonstrated on several benchmark examples.

The found results support work to derive local PCA methods in previous approaches such as [17]. Unlike these previous methods, the presented approach relies on a single cost function and convergence is guaranteed. As demonstrated in the experiments, the convergence behavior of the method is very good in practical applications. Note that, once the connection of local PCA methods to matrix learning is formally established, the use of heuristics for the optimization steps in batch learning is justified, as long as they improve the respective cost terms. One possibility are online optimization methods which directly optimize the global cost function e.g. by means of a stochastic gradient descent. Alternatively, heuristics such as iterative PCA methods as presented in [17] can be included. These methods seem particularly important if the adaptation of full rank matrices is infeasible due to the data dimensionality. In this case, a restriction to a low rank adaptive approximation can be used. This can be formally justified by means of the same cost function by restricting to a low rank matrix.

References

1. Arnonkijpanich, B., Lursinsap, C.: Adaptive Second Order Self-Organizing Mapping for 2D Pattern Representation. In: IEEE-INNS IJCNN 2004, pp. 775–780 (2004)
2. Asuncion, A., Newman, D.J.: UCI Machine Learning Repository. School of Information and Computer Science. University of California, Irvine (2007), http://www.ics.uci.edu/~mlearn/MLRepository.html
3. Bottou, L., Bengio, Y.: Convergence properties of the k-means algorithm. In: Tesauro, G., Touretzky, D.S., Leen, T.K. (eds.) NIPS 1994, pp. 585–592. MIT, Cambridge (1995)
4. Cottrell, M., Fort, J.C., Pagès, G.: Theoretical aspects of the SOM algorithm. Neurocomputing 21, 119–138 (1999)
5. Cottrell, M., Hammer, B., Hasenfuss, A., Villmann, T.: Batch and median neural gas. Neural Networks 19, 762–771 (2006)
6. Duda, R.O., Hart, P.E., Storck, D.G.: Pattern Classification. Wiley, Chichester (2000)
7. Fort, J.-C., Letrémy, P., Cottrell, M.: Advantages and drawbacks of the Batch Kohonen algorithm. In: Verleysen, M. (ed.) ESANN 2002, pp. 223–230. D Facto (2002)
8. Hammer, B., Hasenfuss, A.: Relational Neural Gas. In: Hertzberg, J., Beetz, M., Englert, R. (eds.) KI 2007. LNCS (LNAI), vol. 4667, pp. 190–204. Springer, Heidelberg (2007)

9. Hartigan, J.A.: Clustering Algorithms. Wiley, Chichester (1975)
10. Heskes, T.: Self-organizing maps, vector quantization, and mixture modeling. IEEE Transactions on Neural Networks 12, 1299–1305 (2001)
11. Höppner, F., Klawonn, F., Kruse, R., Runkler, T.: Fuzzy Cluster Analysis. Wiley, Chichester (1999)
12. Kohonen, T.: Self-Organizing Maps. Springer, Heidelberg (1995)
13. Kohonen, T., Kaski, S., Lappalainen, H.: Self-organized formation of various invariant-feature filters in the Adaptive-Subspace SOM. Neural Computation 9(6), 1321–1344 (1997)
14. López-Rubio, E., Muñoz-Pérez, J., Gómez-Ruiz, J.A.: A principal components analysis self-organizing map. Neural Networks 17(2), 261–270 (2004)
15. Martinetz, T., Berkovich, S., Schulten, K.: 'Neural gas' network for vector quantization and its application to time series prediction. IEEE TNN 4(4), 558–569 (1993)
16. Martinetz, T., Schulten, K.J.: Topology representing networks. Neural Networks 7, 507–522 (1994)
17. Möller, R., Hoffmann, H.: An Extension of Neural Gas to Local PCA. Neurocomputing 62, 305–326 (2004)
18. Mochihashi, D., Kikui, G., Kita, K.: Learning an optimal distance metric in a linguistic vector space. Systems and computers in Japan 37(9), 12–21 (2006)
19. Peltonen, J., Klami, A., Kaski, S.: Improved learning of Riemannian metrics for exploratory analysis. Neural Networks 17, 1087–1100 (2004)
20. Schleif, F.-M., Villmann, T., Hammer, B.: Prototype based fuzzy classification in clinical proteomics. International Journal of Approximate Reasoning 47(1), 4–16 (2008)
21. Schneider, P., Biehl, M., Hammer, B.: Relevance matrices in LVQ. In: ESANN 2007, pp. 37–42 (2007)
22. Stones, L.Z., Zhang, Y., Jian, C.L.: Growing Hierarchical Principal Components Analysis Self-Organizing Map. In: Wang, J., Yi, Z., Żurada, J.M., Lu, B.-L., Yin, H. (eds.) ISNN 2006. LNCS, vol. 3971, pp. 701–706. Springer, Heidelberg (2006)
23. Zhong, S., Ghosh, J.: A unified framework for model-based clustering. Journal of Machine Learning Research 4, 1001–1037 (2003)

Clustering Quality and Topology Preservation in Fast Learning SOMs

Antonino Fiannaca[1,3], Giuseppe Di Fatta[2], Salvatore Gaglio[1,3], Riccardo Rizzo[1], and Alfonso Urso[1]

[1] ICAR-CNR, Consiglio Nazionale delle Ricerche, Palermo, Italy
[2] School of Systems Engineering, University of Reading, UK
[3] Dipartimento di Ingegneria Informatica, Universitá di Palermo, Italy

Abstract. The Self-Organizing Map (SOM) is a popular unsupervised neural network able to provide effective clustering and data visualization for multidimensional input spaces. Fast Learning SOM (FLSOM) adopts a learning algorithm that improves the performance of the standard SOM with respect to the convergence time in the training phase. In this paper we show that FLSOM also improves the quality of the map by providing better clustering quality and topology preservation of multidimensional input data. Several tests have been carried out on different multidimensional datasets, which demonstrate the superiority of the algorithm in comparison with the original SOM.

Keywords: SOM, FLSOM, Clustering.

1 Introduction

The Self-Organizing Map (SOM) algorithm is used to build a mapping of an input manifold from a high dimensional data space to a low-dimensional representation space. The Fast Learning SOM ($FLSOM$) [1] is an approach to speed up the learning process by introducing an optimization technique based on the Simulated Annealing (SA).

In this work we demostrate that the FLSOM algorithm is able to perform also a good clustering. We evaluate and compare the algorithm with a standard SOM over a number of artificial and real datasets. The experimental analysis shows that FLSOM provides better results in terms of both topology preservation and clustering criteria.

The paper has the following structure: the next section reports the basic SOM algorithm and describes the proposed one; the section 3 shows the advantage of using proposed algorithm for clustering process; the sections 4 and 5 report respectively the evaluation criteria and the experimental results. Finally some conclusions are reported.

2 SOM Adaptive Learning Rate Algorithm

2.1 Self-Organizing Maps

Self-Organizing Maps [2] are neural structures capable of building maps of the input data, which preserve neighborhood relationships. Although the SOM

V. Kůrková et al. (Eds.): ICANN 2008, Part I, LNCS 5163, pp. 583–592, 2008.

Table 1. SOM incremental learning algorithm

---

1. Initialize the N neurons with random weights.
2. Set step counter $t = 1$, epoch counter $p = 1$, maximum number of epochs $MaxP$.
3. While the stop condition $(p \geq MaxP)$ is not verified
   (a) Define $RndDataset$ as a list where input patterns are randomly ordered.
   (b) While $RndDataset$ is not empty
       i. Get a pattern $x(t)$ from $RndDataset$.
       ii. Find the best matching unit $bmu(x(t)) = \arg\left(\min_{i \in N} \|x(t) - w_i\|\right)$.
       iii. Update weights of bmu and its neighborhood with:
           $w_i(t+1) = w_i(t) + \alpha(t)h_{bmu,i}(t)\left[x - w_i(t)\right]$,
           where $h_{bmu,i}$ is the neighborhood gaussian function, kernel around the bmu.
       iv. Remove $x(t)$ from $RndDataset$.
       v. $t = t + 1$.
       vi. Update the value of learning rate $\alpha(t)$ and of neighbourhood radius $\sigma(t)$, used in the neighbourhood gaussian function h_{bmu}.
   (c) $p = p + 1$, $t = 1$.
4. End of learning after $MaxP$ epochs.

---

algorithm is well known, it seems appropriate to report some characteristics of the training method we have used. We have adopted the incremental learning algorithm [3] where the neural network is trained on-line in each epoch. The pseudocode of the basic learning algorithm is depicted in table 1. We refer to [2] for the details of the algorithm. Here we just highlight the update method of the two main parameters.

The learning rate factor $\alpha(t)$ and the neighborhood radius $\sigma(t)$ are decreasing functions of time and follow the same law, typically given by the formula:

$$\chi(t) = \chi_{MAX}\left(\frac{\chi_{MIN}}{\chi_{MAX}}\right)^{\left(\frac{t}{t_{max}}\right)}, \tag{1}$$

where $\chi = \alpha$ or $\chi = \sigma$.

In the standard SOM algorithm the values of α_{MAX} and α_{MIN} are constant. Whereas in the Fast Learning SOM these values change over time according to the evolution of training process. The adaptability of the learning process rate has shown to provide better performance in terms of the convergence time [1].

2.2   The Fast Learning SOM Algorithm (FLSOM)

The FLSOM algorithm is an adaptive version of the SOM algorithm, whose training function depends on a simulated annealing optimization process and the global evolution of the neural network.

Simulated Annealing (SA) is an optimization method, which is typically used for large scale problems where a global minimum may be hidden by many local

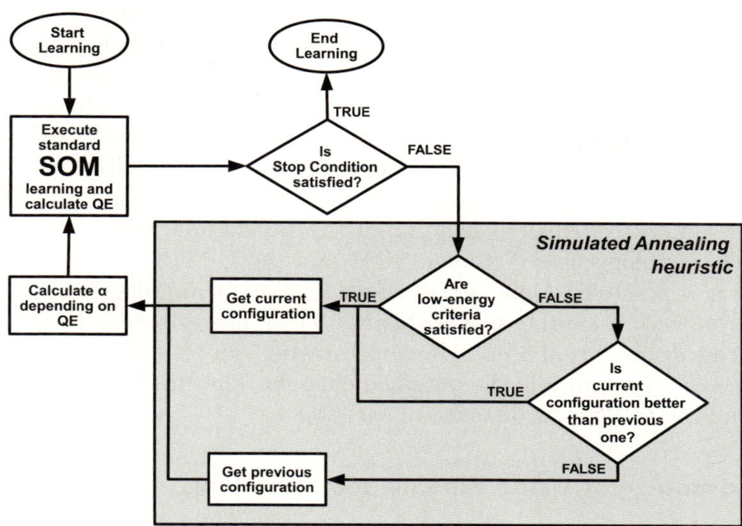

Fig. 1. FLSOM flow chart

minima. According to Metropolis algorithm [4], the SA algorithm replaces the configuration of a system by a random nearby configuration. The method applies a perturbation to the current solution with a probability that depends on a Boltzmann factor.

FLSOM uses this heuristic to improve the speed and the quality of the SOM learning process, preserving its unsupervised characteristic. An adaptive learning rate factor α is steered by the simulated annealing heuristic over the resolution of the map and depends on both t and the Quantization Error (QE).

The local QE of a SOM is defined as the euclidean distance between a data vector and its best matching unit according to: $QE = E\{\|x - m_c(x)\|\}$.

The detailed pseudocode of the adaptive algorithm is given in [1]. Figure 1 shows a general flow diagram of the algorithm. The gray area represents the adaptation of SA heuristic to the configuration of neural weights. The algorithm evaluates the configuration of Kohonen map generated by standard SOM, according to low-energy criteria introduced by [4]. The output of gray area is the configuration of neural network and the values of α_{MAX} and α_{MIN}.

3   SOMs for Clustering

Self Organizing maps are useful for 2-D visualization of high dimensional data; the data projection highlights the pattern clusters that can be visually detected, for example, using a U-Matrix representation [5]. For this reason, SOMs can be used for simultaneous clustering and visualization, or even for clustering via visualization.

Moreover, if we compare SOMs with both traditional vector quantization and projection methods (such as MDS [6]), we notice that SOMs present a relevant advantage: they provide a *topology approximation*. Neighbouring elements in the original space are projected to neighbouring grid points. As a consequence SOMs can be used to obtain topographic maps of the multidimensional data space[7,8].

In this work, we use SOMs for data clustering via visualization. In fact it is possible to obtain a cluster algorithm that neither needs a priori knowledge about the number of clusters as an input (as for example K-means),

Our methodology uses a trained SOM and the U-Matrix [9] representation: this way it is possible to look at the resulting map as an image where grayscale contours allow the identification of clusters. Then, using an automatic segmentation process derived from Seeded Region Growing, our framework can highlights, via boundaries recognition, the obtained clusters. The clustering result is evaluated using the techniques described in section 4.

3.1   Advantage of Using FLSOM for Clustering

Our experimental analysis shows that the SA heuristic offers good chances of finding a configuration with lower internal energy than the initial one. During the training phase, the SA heuristic assures the speed-up of the learning process through the increasing of the $\alpha(t)$ value. That implies neuron weights can reach a final configuration where similar patterns are collected faster than with the standard learning algorithm. At the end of the training process, each pattern fits in its related neuron better than in the standard SOM (as reported in [1]). As a consequence the algorithm provides a better compactness of similar patterns in specific areas, while preserving topologic relationships. Moreover, this property allows the generation of a lattice where distances among groups of homogeneous patterns are greater than the same distances calculated in a map trained with the standard learning algorithm.

Increasing of both inter-clusters dissimilarity and intra-clusters similarity encourages the adoption of FLSOM algorithm for clustering.

4   Evaluation Criteria

Two evaluation criteria are used to compare the SOM and FLSOM algorithms: topology preservation and clustering distribution. The former is computed at the end of the learning processes of each method. The latter is evaluated over clusters retrieved from the two trained maps by means of the same detection technique described in 3.

4.1   Topology Evaluation

The clustering, obtained by a self-organizing map algorithm, can be considered appropriate when two conditionos are satisfied. First, the dataset must be partitioned in the correct number of clusters. And, secondly, each element lies in the correct cluster in a consistent way, with respect to the distribution of elements in the original space. This property is called topology preservation. Several methods have been proposed to measure the topology violation of a trained SOM; the two most popular techniques are the Topographic Product (TP) [11] and the Topographic Function (TF) [12].

For our purpose, TF is not suitable for two reasons [13]. First, it may give misleading results because it does not differentiate between the adjacency of receptive fields in areas where the sample vectors are dense and in areas where they are sparse. Secondly, the comparison between the topographic functions of two different maps is very difficult.

In order to evaluate the topology preservation we use the Directional Product (DP) [14], which is an improved and computationally less expensive implementation of TP.

The maximum value of DP is 1.0 and higher values correspond to better topologies.

4.2   Clustering Evaluation

The results of the FLSOM clustering are compared with the results of the SOM standard algorithms, using a set of validation indexes. There are three approaches to evaluate cluster validity: internal criteria, external criteria and relative criteria [10]. Both external and relative criteria will not be taken in account, because they are based respectively on some user intuition over a specific structure (e.g. class labels), and on the evaluation among several results in terms of the same algorithm but with a different setting of parameters [15]. Internal criteria are easier to use because they can be obtained using the distribution of the output data.

More specifically, internal criteria are based on the minimization of intra-cluster similarity and the maximization of inter-cluster dissimilarity. These methods provide a measure of how close the elements are to their own cluster center ($compactness$) and how far the clusters are from each other ($separation$).

In order to evaluate clustering we use two indices: the $Overall\ Cluster\ Quality$ $index\ (Ocq(\xi))$ proposed in [15] and the $Scatter\text{-}Distance\ index\ (SD(\rho))$ developed in [16]. These two indices are built using the following tree quantities: $Compactness$, $Proximity$ and $Distance$.

The $Compactness$ of a cluster is defined as:

$$Compactness = \frac{1}{C} \sum_{i=1}^{C} \frac{var(c_i)}{var(X)},\qquad(2)$$

The $Proximity$ [15] is a function of the distance of the cluster centers and is defined as:

$$Proximity = \frac{1}{C(C-1)} \sum_{i=1}^{C} \sum_{j=1,j\neq i}^{C} \exp\left(-\left\|x_{c_i} - x_{c_j}\right\|\right),\qquad(3)$$

where C is the number of clusters generated on the data set X, $var(c_i)$ is the variance of the cluster c_i, $var(X)$ is the variance of the data set X and x_{c_i} and x_{c_j} are two cluster centers.

The *Distance* [16] is another function of the positions of the cluster centers:

$$Distance = \frac{D_{max}}{D_{min}} \sum_{k=1}^{C} \left(\sum_{z=1}^{C} \|x_{c_k} - x_{c_z}\| \right)^{-1}, \tag{4}$$

where $D_{max} = \max \left(\|x_{c_i} - x_{c_j}\| \right) \forall i, j \in \{1, 2, ..., C\}$ is the maximum distance between cluster centers. The $D_{min} = \min \left(\|x_{c_i} - x_{c_j}\| \right) \forall i, j \in \{1, 2, ..., C\}$ is the minimum distance between cluster centers.

Using these three quantities it is possible to build the $Ocq(\xi)$ as:

$$Ocq(\xi) = 1 - [\xi \times Compactness + (1 - \xi) \times Proximity], \tag{5}$$

where $\xi \in [0, 1]$ is the weight that balances cluster *compactness* and cluster *proximity*.

$SD(\rho)$ is defined as:

$$SD(\rho) = \rho \times Compactness + Distance, \tag{6}$$

where ρ is a weighting factor, introduced because the value of *compactness* could be in a different range than the value of *distance*.

Both indices use the *compactness* to evaluate the variance of clusters and dataset, whereas to evaluate the separation of clusters they use, respectively, the *Proximity* and the *Distance*. Values of the $Ocq(\xi)$ index are in the range $[0, 1]$ and greater values indicate a better result. Whereas low values of the $SD(\rho)$ index indicate a better quality of the clustering.

5 Experimental Results

In order to compare the algorithms, fourteen data set have been used to test the quality of clustering and the topologic preservation of the obtained maps.

5.1 Evaluation of the Proposed Algorithm

Experimental tests have been carried out over the datasets for both SOM implementations and the evaluation criteria have been computed. The results have been averaged over several runs in order to compare the performances. The statistical significance of the means has been evaluated using the t-Test.

The weights of all SOM networks are initialized with random values over the input space and all maps have a 80×80 square lattice. The training phase for each epoch is done with $\alpha_{MAX} = 0.75$, $\alpha_{MIN} = 0.15$, $\sigma_{MAX} = 7$, $\sigma_{MIN} = 2$, $\delta = 0.15$. In the FLSOM algorithm the values of the learning parameters are dynamically increased up to $\alpha_{MAX} = 1$ and $\alpha_{MIN} = 0.75$. These parameter values are those that give the best results for several datasets widely used in literature.

Table 2. Comparison between SOM and FLSOM

Clustering and Topology indices for three datasets					
	Two-C		**Blobs-3D**		**Iris**
	SOM	FLSOM	SOM	FLSOM	SOM	FLSOM
(Clustering parameters)						
Compatness	0.7679	**0.7671**	0.1501	**0.1416**	0.5168	**0.4820**
Distance	0.0561	**0.0559**	0.1042	**0.0917**	0.0887	**0.0884**
Proximity	0.0297	**0.0291**	0.0502	**0.0411**	0.0625	**0.0580**
(Clustering indices)						
SD ($\rho = 1$)	0.9269	**0.8338**	1.1656	**0.8527**	1.0979	**0.9307**
Ocq ($\xi = 0.5$)	0.5436	**0.5937**	0.3051	**0.4757**	0.4410	**0.5234**
(Topology index)						
Directional Product	1.0	1.0	0.9786	**0.9818**	0.9756	**0.9846**

5.2   Validation of the Proposed Learning Process

The validation of the FLSOM has been carried out using fourteen artificial datasets, thirteen artificial datasets and one real dataset. The thirteen artificial datasets are: the $Two - C$ two-dimensional dataset, providing two thousand input signals; the $Blobs - 3D$ constituted by 8 blobs in a 3D space, composed by three hundred input signal for each blob (see [1] for a picture of artificial datasets), then we built eleven datasets in a $\Re^8$ space assembling from 10 to 80 blobs of points on the vertices of the unitary hypercube. They provide one hundred input signals for each blobs (i.e. from 1000 to 8000 patterns). Finally, the real dataset is the well know "$Iris$, a dataset in four-dimensional space and provide one hundred fifty instances and there are three clusters, each one with 50 instances.

All the reported datasets are used for 50 runs of the learning algorithms. For each epoch of each SOM implementation, the means of Ocq, SD and DP have been calculated.

All artificial datasets are very simple collections of clusters, which are easily detected by both SOM implementation. Both algorithms have produced the correct number of clusters and each element is placed in the appropriate cluster. However, the analysis aims to show that the FLSOM algorithm does actually generate a clustering of a better quality. Our evaluation focuses on the intra and inter-cluster arrangement of elements.

5.3   Results

In table 2 the comparison between SOM and FLSOM in terms of both clustering and topology preservation is shown and better values, for each index and for each dataset, an highlighted with bold type. As shown in table 2 the value of ρ is 1.0, this means both members of equation 6 are weighted in the same manner; in equation 5 the assignement $\xi = 0.5$ is done in order to give equal weights to the two measure. Of course ranges could be different for each clustering parameter (i.e. *Compactness*, *Proximity* and *Distance*), because each value of the three clustering parameters depends on its own ranges that, in turn, depends on dataset and learning configuration. For this reason, for each dataset and for each clustering parameter, all obtained results have been normalized on their

Table 3. Gaps between SOM and FLSOM assessed for Blobs-3D dataset

	Clustering and Topology indices for all datasets				
Map Size	**20x20** SOM FLSOM	**30x30** SOM FLSOM	**50x50** SOM FLSOM	**80x80** SOM FLSOM	**100x100** SOM FLSOM
SD index	1.0739 **0.8582**	0.7015 **0.4607**	1.0916 **0.8148**	1.1656 **0.8527**	0.8446 **0.4255**
Ocq index	0.4329 **0.5104**	0.4925 **0.5863**	0.4970 **0.6031**	0.3051 **0.4757**	0.4982 **0.7134**
SD gaps	0.2157	0.2408	0.2768	0.3129	0.4190
Ocq gaps	0.0775	0.0938	0.1062	0.1706	0.2152

Table 4. Comparison between SOM and FLSOM for 8-dimensional datasets

Clustering and Topology indices for all 8-dimensional datasets								
Blobs	**rad**	**Learning Algorithm**	**DP**	**SD** $(\rho = 1)$	**Ocq** $(\xi = 0.5)$	**t-Test significance** Comp	Dist	Prox
10	0.1	FLSOM	**0.9853**	**0.6956**	**0.6705**	0.0094	0.0033	$1.1E^{-5}$
		SOM	0.9677	0.8223	0.45146			
10	0.2	FLSOM	**0.9889**	0.7129	**0.7111**	0.0026	0.0506	$1.0E^{-6}$
		SOM	0.9790	**0.5488**	0.5085			
10	0.3	FLSOM	**0.9885**	0.8947	**0.6354**	0.0060	0.2160	$1.0E^{-6}$
		SOM	0.9787	**0.6903**	0.4982			
15	0.1	FLSOM	**0.9828**	0.9493	**0.5464**	0.0018	0.2267	0.0037
		SOM	0.9768	**0.7211**	0.4968			
15	0.2	FLSOM	**0.9864**	**0.8676**	**0.4886**	0.8285	0.0008	$3.2E^{-5}$
		SOM	0.9831	1.3764	0.2920			
20	0.1	FLSOM	**0.9876**	**1.0267**	**0.4044**	0.0001	0.0029	0.0399
		SOM	0.9869	1.0624	0.3857			
20	0.2	FLSOM	**0.9821**	**1.0305**	**0.5318**	0.3436	0.7019	0.1026
		SOM	0.9816	1.7168	0.4542			
30	0.1	FLSOM	0.9824	**0.8196**	**0.4726**	0.0917	0.0165	0.0270
		SOM	**0.9832**	0.8541	0.3831			
65	0.1	FLSOM	**0.9772**	**0.7058**	**0.5791**	0.6880	0.0665	0.0354
		SOM	0.9757	0.9630	0.3374			
70	0.1	FLSOM	**0.9769**	**0.5874**	**0.6135**	0.4819	0.5805	0.7158
		SOM	0.9765	0.6319	0.6086			
80	0.1	FLSOM	0.9727	**0.5236**	**0.7233**	0.4947	0.3392	0.2972
		SOM	**0.9732**	0.6735	0.6500			

own range calculated considering executed runs. Finally, in the bottom of table, the index of topology preservation, *Directional Product*, is reported for each dataset.

Table 3 summaries the value of *SD* and *Ocq* indices for dataset *Blob-3D* versus number of neurons in Kohonen map. For each map size, better values are in bold type. As the table shows, not only the FLSOM works better than standard algorithm, but also gaps between indices, or likewise clustering quality, increase when maps increase.

Table 4 shows the comparison between SOM and FLSOM in terms of both Clustering and Topology for 8-dimensional datasets. Each row report results for a dataset. The first and the second columns show respectively the number of blobs (or resulting clusters) and the radius in the artificial datasets. The third column reports in the top the proposed algorithm and in the bottom the standard one. Next three columns show the values of respectively DP, SD and Ocq indices. Best results for them are highlighted with bold type. The proposed algorithm works better than the standard one the most of the time in terms of DP and SD, and it work better in terms of Ocq all the time. The last three columns report significance levels of t-Test, i.e. the probability of does not reject the null hypothesis, for respectively the *Compactness*, the *Distance* and the *Proximity* parameters. Worst results are underlined. T-Test significance confirms the most of the results are computed for different distributions. It should be no surprise that t-Test get worst result for datasets with several clusters (i.e. 70 - 80), in fact the distribution of the interaction amongs many blobs becomes less distinguished when a few features have to define several clusters.

6   Conclusions

In this paper FLSOM, a technique we proposed in the past for speeding up the execution time during SOM learning, has been tested for clustering. In order to compare the two neural networks a tool for automatic extraction of clusters based on SOM, U-Matrix visualization and segmentation techniques, has been implemented. Results of experimental tests, carried out on both artificial and real datasets, comparing the proposed method with standard SOM implementation, demonstrate the good performances obtained by FLSOM in terms of clustering and topology preservation.

References

1. Fiannaca, A., Di Fatta, G., Gaglio, S., Rizzo, R., Urso, A.: Improved SOM Learning using Simulated Annealing. In: de Sá, J.M., Alexandre, L.A., Duch, W., Mandic, D.P. (eds.) ICANN 2007. LNCS, vol. 4668, pp. 279–288. Springer, Heidelberg (2007)
2. Kohonen, T.: Self-Organizing Maps, 3rd edn. Springer, Berlin (2001)
3. Van Hulle, M.: Faithful Representations and Topographic Maps: From Distortion- to Information-Based Self-Organization. John Wiley, New York (2000)
4. Metropolis, N., Rosenbluth, A.W., Rosenbluth, M.N., Teller, A.H., Teller, E.: Equation of state calculations by fast computing machines. J. Chem. Phys. 21(6), 1087–1092 (1953)
5. Deboeck, G., Kohonen, T.: Visual Explorations in Finance using Self-Organizing-Maps. Springer, Heidelberg (1998)
6. Ripley, B.D.: Pattern Recognition and Neural Networks. Cambridge University Press, Cambridge (1996)
7. Kaski, S.: Data Exploration Using Self-Organizing Maps. PhD thesis, Helsinki University of Technology (1997)
8. Flexer, A.: On the Use of Self-Organizing Maps for Clustering and Visualization. In: Żytkow, J.M., Rauch, J. (eds.) PKDD 1999. LNCS (LNAI), vol. 1704, pp. 80–88. Springer, Heidelberg (1999)
9. Ultsch, A., Korus, D.: Integration of Neural Networks and Knowledge-Based Systems. In: Proceeding IEEE on International Conference on Neural Networks, pp. 425–426 (1995)

10. Halkidi, M., Vazirgiannis, M.: Clustering Validity Assessment: Finding the Optimal Partitioning of Data Set. In: Proc. of ICDM 2001, pp. 187–194 (2001)
11. Bauer, H.U., Pawelzik, K.R.: Quantifying the neighborhood preservation of self-organizing feature maps. IEEE Transaction on Neural Networks 3(4), 570–579 (1992)
12. Villman, T., Der, R., Hermann, M., Martinetz, T.: Topology preservation in self-organizing feature maps: Exact definition and measurement. IEEE Transaction on Neural Networks 8(2), 256–266 (1997)
13. Kiviluoto, K.: Topology Preservation in Self-Organizing Maps. In: Proc. of ICNN 1996, pp. 294–299 (1996)
14. Vidaurre, D., Muruzabal, J.: A Quick Assessment of Topology Preservation for SOM Structures. IEEE Transactions on Neural Networks 18(5), 1524–1528 (2007)
15. Yang, Y., Kamel, M.S.: An aggregated clustering approach using multi-ant colonies algorithms. Pattern Recognition 39(7), 1278–1289 (2006)
16. Halkidi, M., Batistakis, Y., Vazirgiannis, M.: Cluster validity methods: part II. SIGMOD Rec. 31(2), 40–45 (2002)

Enhancing Topology Preservation during Neural Field Development Via Wiring Length Minimization

Claudius Gläser, Frank Joublin, and Christian Goerick

Honda Research Institute Europe
Carl-Legien-Str. 30, D-63073 Offenbach/Main, Germany
`claudius.glaeser@honda-ri.de`

Abstract. We recently proposed a recurrent neural network model for the development of dynamic neural fields [1]. The learning regime incorporates homeostatic processes, such that the network is able to self-organize and maintain a stable operation mode even in face of experience-driven changes in synaptic strengths. However, the learned mappings do not necessarily have to be topology preserving. Here we extend our model by incorporating another mechanism which changes the positions of neurons in the output space. This algorithm operates with a purely local objective function of minimizing the wiring length and runs in parallel to the above mentioned learning process. We experimentally show that the incorporation of this additional mechanism leads to a significant decrease in topological defects and further enhances the quality of the learned mappings. Additionally, the proposed algorithm is not limited to our network model; rather it can be applied to any type of self-organizing maps.

1   Introduction

Self-organizing maps (SOMs) describe a group of methods in the domain of artificial neural networks. They provide learning algorithms for mapping high-dimensional input data onto a (discretized) low-dimensional output space, thereby pursueing two different goals. Firstly, SOMs perform vector quantization and therewith aim at minimizing the quantization error. Secondly, they strive for achieving topology preserving mappings by incorporating topological constraints into the vector quantization process.

For technical applications the topology preserving properties of SOMs are of particular interest. They have been used in the domains of data compression [2], medical diagnosis [3], or the monitoring of industrial processes [4]. Unfortunately, SOMs tend to privilege the minimization of the quantization error over topology preservation. For this reason, various mechanisms for improving topology preservation has been suggested [5,6,7].

Dynamic neural fields are special types of SOMs, which are particularly suited for modeling cortical development [8]. However, their use in technical applications is very limited, which is mainly due to stability issues. In order to circumvent this problem, we recently proposed a recurrent neural network model for the

V. Kůrková et al. (Eds.): ICANN 2008, Part I, LNCS 5163, pp. 593–602, 2008.

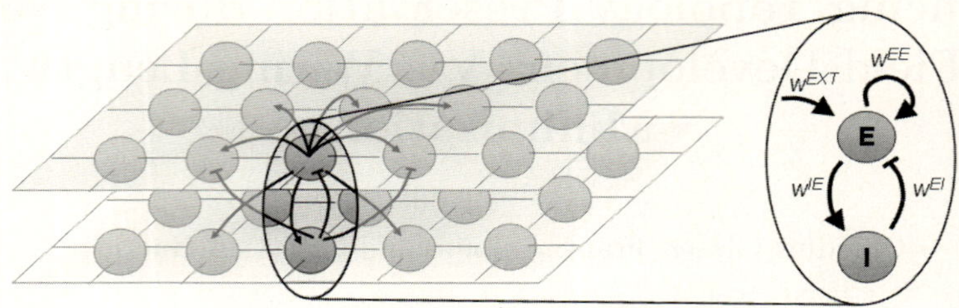

Fig. 1. The structure of the recurrent neural network model

homeostatic development of dynamic neural fields [1]. It is able to self-organize while maintaining a stable operation mode. However, the mappings learned by the model do not necessarily have to be topology preserving.

Here we will extend our model by incorporating another process exclusively dedicated to topology preservation. More precisely, we consider SOMs to be elastic nets with neighborhood relations between units being subject to change. We will show that a mechanism where units adapt their positions in the output space based on the concept of wiring length minimization is suitable for enhancing topology preservation. Furthermore, the method is not limited to our network model, rather it can be applied to any type of SOM.

The remainder is organized as follows. In section 2 we will shortly review our previously proposed network model. Next, we will propose a concept for enhancing topology preservation, compare it to existing approaches, and suggest a concrete implementation of it. After that, an experimental evaluation of the method will be given in section 4. The paper is finalized by a conclusion.

2   The Network Model

The recurrent neural network we presented in [1] is composed of excitatory units E and inhibitory units I, both being initially arranged on a 2-dimensional grid mimicking the neural tissue. The wiring of the network is show in Fig. 1. Afferent projections to excitatory units provide the input to the neural field. Furthermore, the units are laterally connected such that E-cells excite other E-cells as well as I-cells. In turn, E-cells receive inhibitory projections originating from I-cells.

The membrane potentials of excitatory and inhibitory units are denoted by the variables u and v, respectively. We will use i for specifying the unit located at position x_i of the cortical plane. The spatio-temporal evolution of the activity in the neural field following the presentation of a stimulus s can be described by the differential equations (1) and (2), where τ_E and τ_I are time constants, h_i^E and h^I are the resting potentials, and w_{ij}^* denotes the synaptic weight of a connection from unit j to unit i, where $* \in \{EE, EI, IE, EXT\}$ specifies the type of connection.

$$\tau_E \frac{du_i}{dt} = -u_i + \sum_j g(d_{ij}) \cdot w_{ij}^{EE} \cdot f(u_j) - \sum_j w_{ij}^{EI} \cdot f(v_j) + \sum_j w_{ij}^{EXT} \cdot s_j + h_i^E \quad (1)$$

$$\tau_I \frac{dv_i}{dt} = -v_i + \sum_j g(d_{ij}) \cdot w_{ij}^{IE} \cdot f(u_j) + h^I \quad (2)$$

Here, f is a monotonically increasing transfer function defining the relation between the membrane potential and the firing rate of a unit. We used a sigmoidal function with θ and γ denoting the threshold value and the gain factor, respectively. Additionally, we introduced a function g, which modulates the efficacy of excitatory lateral connections depending on the distance d_{ij} between the pre- and postsynaptic unit positions. The function g was chosen to follow the Gaussian distribution with a mean of 0.

$$f(z) = \frac{1}{1 + \exp\left(-\gamma(z - \theta)\right)} \quad (3)$$

Thus, we define that excitatory lateral connections between units within a local neighborhood are more efficient than those between far-distant units. It is, however, important to note that this is a fundamentally different approach from previously published network models. There the synaptic weight values of lateral connections are chosen as a function of the distance between the pre- and postsynaptic units, by which a topology on the neural tissue as well as within the feature space is defined. In contrast, we introduce a distance-dependent modulation of synaptic efficacy, but do not make any assumption on the synaptic weight values themselves. Following this argumentation, g could be understood as a physical constraint. The synapse location-dependent integration of excitatory synaptic input as it is carried out by passive dendrites is one possible interpretation [9].

A direct consequence of this is that large synaptic weight values could compensate for the distance-dependent modulation of connection efficacy. Thus, the mapping described by the neural field does not necessarily have to be topology preserving, that is nearby units having similar receptive fields. In the following we will propose an additional mechanism which improves the topology preservation of the learned mappings.

3   Topology Preservation

3.1   Existing Approaches

Several proposals have been made on how the topology preserving abilities of SOMs could be enhanced. Most of them rely on a fixed metric (fixed neighborhood relations) defined in the output space and try to dynamically adjust the width of the active neighborhood based on global heuristics or more sophisticated local measurements like input novelty [6], topology defects [5], or the degree of local folding [10]. Other approaches aim at achieving a reasonable balance between the two objective functions of minimizing the quantization error

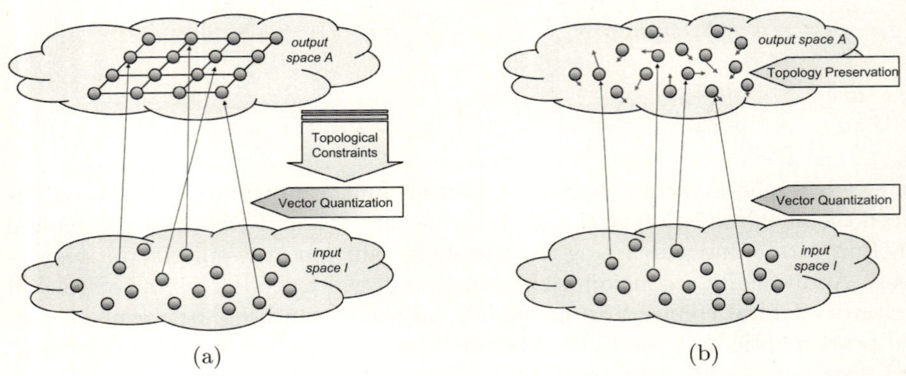

Fig. 2. An illustrative comparison between convential SOM learning algorithms (a) and the proposed system for enhancing topology preservation of SOMs (b)

and topological defects by differently weighting their influence to the learning algorithm [7]. Only a few approaches try to change the metric defined in the output space. They mostly concentrate on building tree-like neighborhood relations by hierarchically clustering codebook vectors [11].

3.2   Topology Preservation Via Wiring Length Minimization

According to Fig. 2(a) SOMs can be considered to be a group of methods for vector quantization, insofar as the codebook vectors of their units are adapted in order to minimize the quantization error. Thereby, topological constraints, which are defined by the units' neighborhood relations, are incorporated. By doing so, SOMs aim at finding a reasonable balance between the two objectives of minimizing the quantization error and minimizing topological defects of the mapping. Nevertheless, when mapping high-dimensional input data to a low-dimensional output space, the requirements described by the two objective functions most often cannot be simultaneously satisfied. In these cases, SOMs privilege vector quantization at the cost of an increase in topological defects.

We propose to release (or at least to relax) the topological constraints from the process of vector quantization by incorporating another process which is exclusively dedicated to topology preservation. This means that the objective function of minimizing topological defects is no longer implicitly defined by incorporating topological constraints into the vector quantization method; rather it is explicitly defined by a separate process running in parallel to vector quantization. Therefore, we propose to add another process to the usual SOM learning algorithms for vector quantization which changes the positions of the units in the output space based on the result of vector quantization, such that topology preservation is improved (see Fig. 2(b)). This means that, in the output space, a fixed metric is used (e.g. Euclidean distance), but the neighborhood relations between units are changed by adjusting their positions.

Therewith, SOMs can be seen as elastic networks of laterally connected units. Elastic means that individual units are placed in the output space, which is a continuous (physical) substrate with a fixed metric, and that individual units are able to dynamically adjust their positions on this substrate. More precisely, units are interconnected in a way, that the strength w_{ij} of the connection between two units i and j denotes the similarity between the codebook vectors w_i and w_j of the two units.

$$w_{ij} \quad \propto \quad \frac{1}{\| w_i, w_j \|^I} \tag{4}$$

There $\| \cdot \|^I$ is the distance metric of the input space I.

We propose that the process of adjusting the position of units in the output space is based on an individual unit's objective function of minimizing the weighted wiring length of its connections with other units of the SOM. Let x_i denote the position of a unit i in the output space and $\| \cdot \|^A$ be the distance metric of the output space. Then the objective function of a unit i for changing its position in the output space can be formulated according to (5), which is a purely local objective function. Alternatively, the problem can be formulated using a global objective function according to (6), which states to minimize the total weighted wiring length of the elastic network.

$$\sum_j w_{ij} \cdot \| x_i, x_j \|^A \quad \longrightarrow \quad \min \tag{5}$$

$$\sum_i \sum_j w_{ij} \cdot \| x_i, x_j \|^A \quad \longrightarrow \quad \min \tag{6}$$

3.3 Implementation of Wiring Length Minimization

If we apply the above mentioned framework for topology preservation via wiring length minimization to our network model described in section 2, we first have to chose appropriate values for the connection strengths of the elastic network. One possibility would be to chose them according to (4). However, in the case of developing neural fields with learned lateral connections as those featured by our network model, one can directly use the synaptic weight values of the within field connections. This is possible, because the synaptic weights of lateral connections are a measure for the similarity between the receptive fields of different units. If we do so, we obtain the following objective functions of individual units, where (7) and (8) hold for an excitatory unit i and an inhibitory unit i, respectively. Furthermore, d_{ij}^* with $* \in \{EE, EI, IE\}$ denotes the distance between units i and j according to the distance metric of the output space $\| \cdot \|^A$.

$$\sum_j w_{ij}^{EE} \cdot d_{ij}^{EE} + \sum_j w_{ji}^{EE} \cdot d_{ji}^{EE} + \sum_j w_{ji}^{IE} \cdot d_{ji}^{IE} \quad \longrightarrow \quad \min \tag{7}$$

$$\sum_j w_{ij}^{IE} \cdot d_{ij}^{IE} + \sum_j w_{ji}^{EI} \cdot d_{ji}^{EI} \quad \longrightarrow \quad \min \tag{8}$$

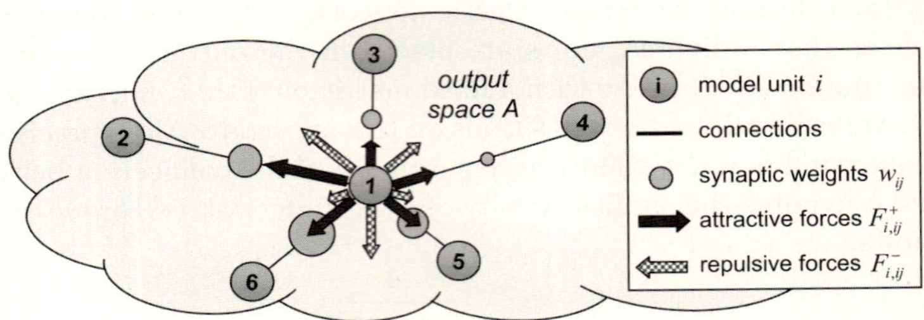

Fig. 3. The attractive and repulsive forces exerted on model units depend on the strengths of the connections as well as the distances between units

Now, multiple optimization techniques could be applied in order to minimize the wiring length. In example unit positions could be adapted via gradient descent or evolutionary strategies. Here, we follow another approach in which we interpret connections between units as springs with spring constants chosen according to the strengths w_{ij} of the connections. Following this argumentation, connections of the elastic net exert attractive forces on units in a way that the force $F_{i,ij}^{+}$ exerted on a unit i by a connection between individual units i and j increases with an increase in the connection strength w_{ij} or an increase in the distance d_{ij} between the two units and vice versa.

$$F_{i,ij}^{+} \quad \propto \quad w_{ij} \cdot d_{ij} \quad = \quad w_{ij} \cdot \| x_i, x_j \|^A \tag{9}$$

A trivial solution of (7) and (8) would be to set the distances between individual units to 0. The approach outlined above would converge to this solution as well. This would mean that all units coincide at the same position in the output space. For this reason, we additionally consider repulsive forces between units of the elastic net, such that the repulsive force $F_{i,ij}^{-}$ exerted on a unit i increases when the distance d_{ij} between the units i and j decreases and vice versa. The resulting physical interaction between units is illustrated in Fig. 3.

$$F_{i,ij}^{-} \quad \propto \quad \frac{1}{d_{ij}} \quad = \quad \frac{1}{\| x_i, x_j \|^A} \tag{10}$$

Additionally, it is important to note that in our model repulsive forces only exist between pairs of excitatory units as well as between pairs of inhibitory units, whereas there is no repulsive force between an excitatory and an inhibitory unit. This is reasonable when considering the output space to be the cortical tissue with excitatory and inhibitory units being located in different layers of it. Therewith the positions of excitatory and inhibitory units can be updated according to the following equations:

$$\Delta x_i^E \quad \propto \quad \sum_j F_{i,ij}^{+,EE} + \sum_j F_{i,ij}^{+,IE} + \sum_j F_{i,ij}^{+,EI} + \sum_j F_{i,ij}^{-,EE} \tag{11}$$

$$\Delta x_i^I \quad \propto \quad \sum_j F_{i,ij}^{+,IE} + \sum_j F_{i,ij}^{+,EI} + \sum_j F_{i,ij}^{-,II} \tag{12}$$

4 Experimental Results

In order to evaluate the presented method for improving the topology preservation of neural fields, we carried out an experiment in the domain of reference frame transformation. More precisely, we investigated the use of our recurrent neural network model for 1-dimensional eye-hand coordination. In order to robustly perform eye-hand coordination, an animal has to be able to transform between the different reference frames, an ability usually attributed to an intermodal body-calibration obtained in the early stages of development. The key aspect there is that simultaneously present stimuli become linked together and can later be used for the transformation from one modality into another. Here we want to use our network for modeling the calibration process during early self-exploration. Therefore, we have chosen three stimuli s^1, s^2, s^3 with $s^1, s^2 \in [-1, 1]$ and $s^3 = s^1 - s^2$, where s^1 and s^2 mimic the gaze and hand position in a body-centered reference frame, respectively, as well as s^3 representing the hand position in eye-centered coordinates.

Each of the three stimuli s^1, s^2, s^3 is represented by a population code composed of 21 neuron responses, resulting in a total of 63 inputs to the neural field. Target gaze- and hand-positions (s^1, s^2) were chosen randomly. The recurrent neural network is composed of 100 excitatory units and 100 inhibitory units, both initially being arranged on a 10x10 grid. The synaptic weight values of afferent projections to the field w_{ij}^{EXT} were initialized with small random values. Weights of lateral connections were initialized uniformly. In order to evaluate the effect of topology preservation via wiring length minimization (WLM), we compared the results obtained by two simulation runs; one run using WLM and one run not using WLM.

Firstly, we investigated if our interpretation of a neural field to be an elastic net with lateral connections exerting attractive forces on units is a suitable choice for minimizing wiring length. Therefore, the total weighted wiring length (see (6)) was calculated and normalized by the mean distance between units. Fig. 4(a) shows the temporal evolution of this measure for two developing neural fields, one using the process of WLM and one not using WLM. As can be seen the process of WLM eliminates the initial increase in total wiring length, which can be attributed to an initial rough adjustment of synaptic weight values. Later in training, the competitive nature of the learning regime incorporated by our network model induces a "die off" of many synapses, which results in a decrease in total weighted wiring length for both simulation runs. However, the neural field using WLM converges to a smaller value, which shows that our implementation is a suitable choice for WLM.

Secondly, the way how neurons change their positions in the output space will be illustrated. For the sake of simplicity we will concentrate on excitatory units. Therefore, Fig. 4(b) shows the starting and end positions of individual units. The neuron traces of four exemplarily chosen units are additionally shown. They illustrate that some units change their positions considerably. Furthermore, the traces indicate that position changes are large during the initial training phase, whereas units maintain their positions at the end of training.

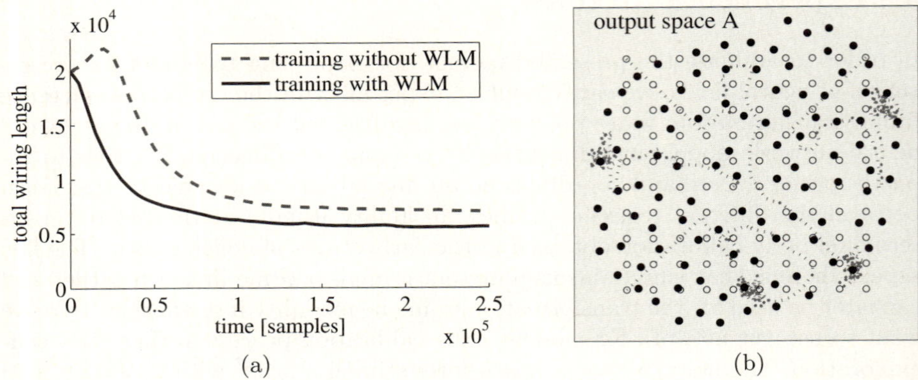

Fig. 4. (a) The total wiring length of the elastic net as a function of training time. (b) The positions of excitatory units before (open circles) and after training (filled circles) using WLM. Additionally, the traces of 4 exemplarily chosen units are shown (dotted lines).

The results presented so far focused on how WLM is achieved, however, they do not demonstrate the suitability of WLM for improving topology preservation. In order to do so we investigated the topology preserving properties of the mapping described by the neural field with a special emphasize on the effect of WLM. Therefore, we trained two neural fields: one using the learning regime incorporating WLM and one not using WLM. After training neighborhood relations were defined using Delaunay triangulation of the units' end positions. Finally, we plotted the position of the units' receptive fields (codebook vectors) in the input space (see Fig. 5 (a) and (b)), where the connections between units are plotted according to the neighborhood relations defined in output space. For a topology preserving mapping, this would result in a network, where adjacent

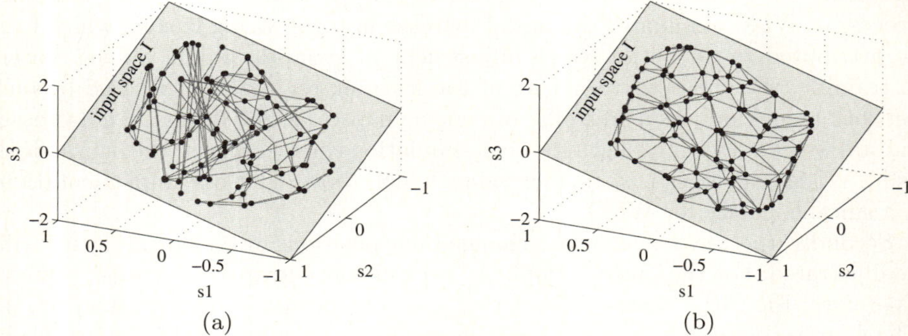

Fig. 5. The position of the units' receptive fields (codebook vectors) in input space as well as their neighborhood relations defined in output space for a neural field trained not using WLM (a) and one using WLM (b)

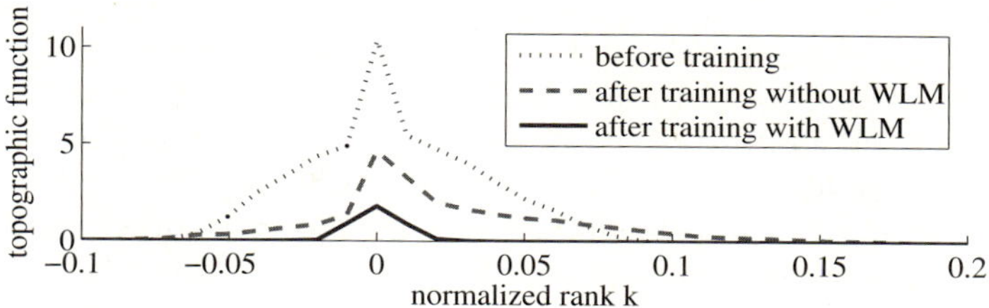

Fig. 6. The topographic function as a measure for the degree of topological defects is plotted for neural fields using WLM or not using WLM in the course of training

units in input space feature a connection according to the neighborhood relations defined in output space. As can be seen, this is the case for the neural field which was trained using WLM, whereas it is not the case when not using WLM.

In order to give a quantitative measure for the topology preserving properties of the learned mappings, we calculated the topographic function [12], which is a widely used method for characterizing the degree of topology preservation. Fig. 6 shows the corresponding plots, where the rank k determines the effective neighborhood range on which topology preservation was analyzed. More precisely, the topographic function for small $|k|$ describes the degree of local topological defects, whereas the topographic function for large $|k|$ describes the degree of global topological defects. Furthermore, a topographic function of 0 would be obtained when analyzing a mapping which is perfectly topology preserving. The plots of Fig. 6 show that the incorporation of WLM into the learning regime for the development of dynamic neural fields results in a significant decrease of topological defects on both a local scale and particularly on a global scale.

5   Conclusion

Our previous work [1] focused on the incorporation of recent advances in the understanding of how homeostatic processes regulate neuronal activity. There we could show that our network model is able to self-organize while maintaining a stable operation mode even in face of experience-driven changes. In this paper we extended our existing network model by another process, which enhances the topology preserving properties of the learned mappings. Therefore, we consider the neural field to be an elastic net with unit positions (and therewith neighborhood relations) being subject to change, where the objective function for changing the unit positions is the minimization of wiring length. With other words, units try to adapt their positions in the output space such that the distance to their neighbors in input space becomes minimized. For a concrete implementation we suggested to interpret connections between units as springs exerting forces on units. We experimentally showed that the incorporation of this additional mechanism for wiring length minimization significantly decreases the topological defects of the learned mapping. Furthermore, the self-regulatory properties of our network are not affected by this additional process.

The combination of these two algorithms further improves the quality of our network model and should ease its application in various domains. On the one hand, the self-regulatory nature of the model avoids stability problems, which one usually faces when using dynamic neural fields. On the other hand, the now incorporated enhancement of topology preservation broadens the domains to which the model can be potentially applied.

Additionally, the proposed mechanism for enhancing topology preservation is not limited to our network model. It is even not limited to dynamic neural fields. Rather it can be applied to various methods in the field of self-organizing maps, i.e. the widely used Kohonen-maps [13].

References

1. Gläser, C., Joublin, F., Goerick, C.: Homeostatic development of dynamic neural fields. In: Proc. International Conference on Development and Learning (2008)
2. Kangas, J.: Utilizing the similarity preserving properties of self-organizing maps in vector quantization of images. In: Proc. International Conference on Neural Networks, vol. 4, pp. 2081–2084 (1995)
3. Vörös, T., Keresztényi, Z., Fazekas, C., Laczkó, J.: Computer aided interactive remote diagnosis using self-organizing maps. In: Proc. 26th Annual International Conference of the IEEE Engineering in Medicine and Biology Society, vol. 5, pp. 3190–3193 (2004)
4. Alhoniemi, E., Hollmén, J., Simula, O., Vesanto, J.: Process monitoring and modeling using the self-organizing map. Integrated Computer-Aided Engineering 6(1), 3–14 (1999)
5. Herrmann, M.: Self-organizing feature maps with self-organizing neighborhood widths. In: Proc. International Conference on Neural Networks, pp. 2998–3003 (1995)
6. Phaf, R., Den Dulk, P., Tijsseling, A., Lebert, E.: Novelty-dependent learning and topological mapping. Connection Science 13(4), 293–321 (2001)
7. Kirk, J.S., Zurada, J.M.: Algorithms for improved topology preservation in self-organizingmaps. In: Proc. International Conference on Systems, Man, and Cybernetics, vol. 3, pp. 396–400 (1999)
8. Amari, S.: Dynamics of pattern formation in lateral-inhibition type neural fields. Biological Cybernetics 27, 77–87 (1977)
9. Williams, S., Stuart, G.: Role of dendritic synapse location in the control of action potential output. Trends in Neuroscience 26(3), 147–154 (2003)
10. Kiviluoto, K.: Topology preservation in self-organizing maps. In: Proc. International Conference on Neural Networks, vol. 1, pp. 294–299 (1996)
11. Kirk, J.S., Zurada, J.M.: A two-stage algorithm for improved topography preservation in self-organizing maps. In: Proc. International Conference on Systems, Man, and Cybernetics, vol. 4, pp. 2527–2532 (2000)
12. Villmann, T., Der, R., Herrmann, M., Martinetz, T.M.: Topology preservation in self-organizing feature maps: Exact definition and measurement. IEEE Transactions on Neural Networks 8(2), 256–266 (1997)
13. Kohonen, T.: Self-organized formation of topologically correct feature maps. Biological Cybernetics 43, 59–69 (1982)

Adaptive Translation: Finding Interlingual Mappings Using Self-Organizing Maps

Timo Honkela, Sami Virpioja, and Jaakko Väyrynen

Adaptive Informatics Research Centre
Helsinki University of Technology
P.O. Box 5400, FI-02015 TKK
Espoo, Finland

Abstract. This paper presents a method for creating interlingual word-to-word or phrase-to-phrase mappings between any two languages using the self-organizing map algorithm. The method can be used as a component in a statistical machine translation system. The conceptual space created by the self-organizing map serves as a kind of interlingual representation. The specific problems of machine translation are discussed in some detail. The proposed method serves in alleviating two problems. The main problem addressed here is the fact that different languages divide the conceptual space differently. The approach can also help in dealing with lexical ambiguity.

1   Introduction

In the following, the area of machine translation is discussed in broad terms. Specific problems that make high quality machine translation a difficult task are described.

1.1   Natural Language Understanding and Machine Translation

The language of a person is idiosyncratic and based on the subjective experiences of the individual. For instance, two persons may have a different conceptual or terminological density of the topic under consideration. A layperson, for instance, is likely to describe a phenomenon in general terms whereas an expert uses more specific terms. Moore and Carling [1] state that languages are in some respect like maps. If each of us sees the world from our particular perspective, then an individual's language is, in a sense, like a map of their world. Trying to understand another person is like trying to read the map of the other, a map of the world from another perspective [1]. If some persons speak the same language, many symbols in their vocabularies are the same. However, as discussed above, one cannot assume that the vocabularies of any two agents are exactly the same. These issues related to natural language understanding are even more challenging if communication in multiple languages is considered.

The term *machine translation* (MT) refers to computerized systems responsible for the production of translations with or without human assistance. It

V. Kůrková et al. (Eds.): ICANN 2008, Part I, LNCS 5163, pp. 603–612, 2008.

excludes computer-based translation tools which support translators by providing access to on-line dictionaries, remote terminology databanks, transmission and reception of texts, etc.

Development of high quality MT has proved to be a very challenging task. At best, the translation results can be reasonably good, for example, when the domain is small or a lot resources is used in developing the system. At worst, the results can be practically incomprehensible.

Another important challenging factor is that the development of a machine translation system requires a *lot of human effort*. This is true especially in the case of systems that are based on human-specified rules for morphology, syntax, lexical selection, semantic analysis and generation.

1.2   Translation Challenges

In the following, a list of seven *more detailed problems* are given. These problems need to be addressed in the development of MT systems and in comparisons of MT systems and approaches. A more philosophical consideration on the problems of translation is provided in [2].

1. Differences in conceptual mapping: Different languages divide the conceptual space differently, for example in the domain of physical space (cf. e.g. [3]) or colors (cf. e.g. [4]). The possibility of exact meaning preserving translation from one language to another can even be questioned on these grounds. In addition to overlapping areas of reference related to two words in different languages, there may be also words missing in some language for some particular phenomenon. One can, of course, introduce new words into a language in order to facilitate a one-to-one mapping between concepts in two languages would be an impossible task. Therefore, from the semantic and pragmatic point of view, it seems that both human and machine translation are always subject to some inaccuracy and there can be arguably several good ways to translate one expression to another language. This fact makes the evaluation of machine translation systems more difficult. For statistical MT this might also mean that it would be useful to have several different parallel translations available for a source language expression in order to facilitate modeling of the variety.

2. Lexical and syntactic ambiguity: In an MT system, one has to determine a suitable interpretation or to find the correct analysis of the syntactical structure for an expression in order to be able to find the corresponding expression in another language. This process is usually denoted as disambiguation. Traditionally lexical disambiguation refers to a process in which the "correct" meaning among several alternatives is determined. However, one can also consider the possibility that the underlying space of interpretations can be continuous (cf. e.g. [5]). In that case the disambiguation process would determine a more specific distribution or point in the conceptual space. However, it is nowadays much more commonplace in the MT systems to consider a word to have one or several distinct meanings.

3. Word, multi-word and sentence alignment: Meaning is often conveyed by phrases rather than by single words. In one language, an idea may be expressed

with one separate word, in another with a compound word, and in a third one with a multi-word phrase (collocation). It is also possible that there are one-to-many or many-to-many relationships in the sentence level. In an extreme case, one word in the source language might require several sentences in the target language in order to express approximately the same meaning in the cultural context of each language. It may also be that in one language some particular word classes are in use while in another language they are not. For instance, articles are used in many languages, in some others not. The alignment problem is a very actively researched topic (cf. e.g. [6]).

4. Differences in word order: Even for laypersons, a well known difference between languages is related to the word order. It is commonplace to categorize a language to have one basic word order, referring to the fact in which order subject, verb and object appear in a declarative sentence. In many languages, changes in word order occur due to topicalization or in questions. The basic word order is unmarked, i.e., it contains no extra information to the listener. In addition to the differences in the order of the basic constituents in a simple declarative sentence, there are many other points of difference, e.g., the order in which a noun and its attributes appear.

5. Inflectional word forms: Some languages have complex morphology, i.e., they have have inflectional word forms. The inflections may denote, e.g., the case of a noun or the tense of a verb. The existence of complex word forms relates closely to the alignment problem. In translation, each segment (morphemes) of one word form in the source language may correspond to one or several words in the target language. The order of the segments may very well be different from the order of the words (consider also the previous point).

6. Text segmentation: In some languages the words are not explicitly separated from each other in a text. This introduces another challenge: Before or associated with the alignment process, the system has to detect the word borders.

7. Speech-to-text and text-to-speech transformation: In the case of MT between spoken languages, the system has to perform transformation between the modalities unless the mapping is conducted directly without any textual intermediate representation. We are not aware of any such approach even if we consider it in principle possible.

Corpus-based methods have been introduced in order to deal, for instance, with the knowledge acquisition bottleneck. The availability of large corpora makes it possible to use various probabilistic and statistical methods to acquire automatically lexicon entries, parsing and disambiguation rules and other useful representations for a translation system. Next we describe the basic statistical translation model that is currently widely used.

1.3   Bayesian Models in Translation

Bayes' rule tells that

$$p(A|B) = p(B|A)p(A)/p(B), \tag{1}$$

where $p(A)$ is the prior probability or marginal probability of A, and $p(A|B)$ is the conditional probability of A, given B. Let us consider a situation in which we wish to translate Finnish sentences, f, into English sentences, e. Bayes' rule gives $p(e|f)p(f) = p(e, f) = p(f|e)p(e)$ and reduces to the basic equation of statistical machine translation: maximize $p(e|f) = p(f|e)p(e)$ over the appropriate e. This splits the translation problem into a translation model ($p(f|e)$) and a language model ($p(e)$). The decoding algorithm, given these models and a new sentence f, finds translation e.

In their classical paper, Peter Brown and his colleagues described a series of five statistical models of the translation process and gave algorithms for estimating the parameters of these models applying the basic equation [7]. During recent years, this statistical approach has had considerable successes, based on the availability of large parallel corpora and some further methodological developments (consider, e.g., [8,9]).

Translation or mapping between two language through use of the self-organizing map can be a viable solution for the problems 1 and 2 described above. The traditional Bayesian approach does not take these issues into account. Construction of maps of words based on the self-organizing map algorithm is next presented as an introductory theme. The self-organizing map is a natural means to build a conceptual space that can be used as a link between two or several languages.

1.4   Self-Organizing Map

The Self-Organizing Map (SOM) [10,11] defines an ordered mapping, a kind of projection from a set of given data items onto a regular, usually two-dimensional grid. A model m_i is associated with each grid node. These models are computed by the SOM algorithm. A data item will be mapped into the node whose model is most similar to the data item, i.e., has the smallest distance from the data item in some metric. The model is then usually a certain weighted local average of the given data items in the data space. But in addition to that, when the models are computed by the SOM algorithm, they are more similar at the nearby nodes than between nodes located farther away from each other on the grid. In this way the set of the models can be regarded to constitute a similarity graph, and structured 'skeleton' of the distribution of the given data items. [12]

1.5   Maps of Words

Charniak [13] presents the following scheme for grouping or clustering words into classes that reflect the commonality of some property.

- Define the properties that are taken into account and can be given a numerical value.
- Create a vector of length n with n numerical values for each item to be classified.
- Cluster the points that are near each other in the n-dimensional space.

Handling computerized form of written language rests on processing of discrete symbols. How can a symbolic input such as a word be given to a numeric algorithm? One useful numerical representation can be obtained by taking into account the sentential context in which the words occur. Before utilization of the context information, however, the numerical value of the code should not imply any order to the words. Therefore, it will be necessary to use uncorrelated vectors for encoding. The simplest method to introduce uncorrelated codes is to assign a unit vector for each word. When all different word forms in the input material are listed, a code vector can be defined to have as many components as there are word forms in the list. This method, however, is only practical in small experiments. With a vocabulary picked from a even reasonably large corpus the dimensionality of the vectors would become intolerably high. If the vocabulary is large, the word forms can be encoded by quasi-orthogonal random vectors of a much smaller dimensionality [14]. Such random vectors can still be considered to be sufficiently dissimilar mutually and not to convey any information about the meaning of the words. Mathematical analysis of the random encoding of the word vectors is presented in [15].

2   Experiments with Interlingual Mapping

To illustrate the idea of using the Self-Organizing Map in finding a mapping between vocabularies of two different languages, the results of two new experiments are reported in the following.

2.1   Situation Context

Like discussed in earlier sections, the maps of words are often constructed using the sentential contexts of words as input data. The result is that the more similar the contexts in which two words appear in the text, the closer the words tend to be on the map. Here this basic idea is extended to cover the notion of context in general: We consider the use of a collection of words in two languages, English and German, in a number of contexts. In this experiment, the contexts were real-life situations rather than some textual contexts.

Figure 1 presents the order of a number of words on a self-organizing map that serves simultaneously two purposes. First, it has organized different contexts to create a conceptual landscape (see, e.g., [5]). Second, the map includes a mapping between the English and German words used in the analysis.

The input for the map consists of words and their contexts. The German vocabulary includes 32 words (Advokat, Angler, Arzt, Autofahrer, ..., Zahnarzt) and the English vocabulary of 16 words (boss, dancer, dentist, director, etc.). For each word, there is a assessment by 10 to 27 subjects indicating the degree of suitability for the word to be used in a particular context. The number of contexts is 19. The resulting map is shown in Figure 1.

The map shows that those words in the two languages that have similar meaning are close to each on the map. In this particular experiment, the German subjects were usually using a larger vocabulary. Therefore, in many areas of the map, a particular conceptual area is covered by one English word (for instance,

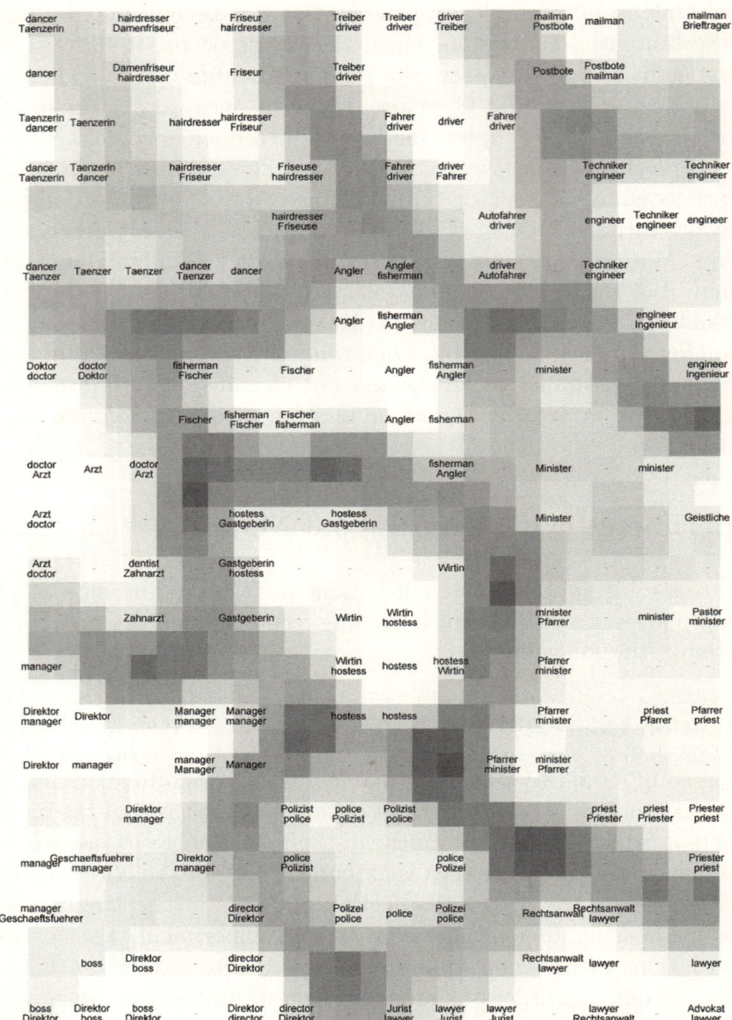

Fig. 1. A collection of German and English words positioned a conceptual landscape based on a Self-Organizing Map of contexts. The darker the shade of gray, the longer are the distances in the original input space. Thus, relatively light areas correspond to conceptual areas or clusters. The dots on the map denote empty prototypes, i.e., model vectors that are not the best match of any of the words under consideration.

"doctor" or "hairdresser") and by two or more German words (for instance, "Arzt" and "Doktor" or "Friseur", "Friseuse" and "Damenfriseur"). It is important to notice that the model covers both translation between languages and within languages. Namely, rather than dealing with German and English, the same model can be built for the language used in medical contexts by experts and laypersons.

2.2 Textual Context

In this experiment, a map of words was constructed using real-life sentential contexts of words as input data. The vocabulary was bilingual, as were the sentential contexts. The idea is to get words with similar contexts to appear close to each other on the map. With a bilingual vocabulary, we wish to find 1) a semantic ordering of words in the map, and 2) possible translation pairs of words from those close to each other on the map.

The words and their sentential contexts were obtained from the sentence-aligned English–German part of the parallel Europarl corpus (version 3) [16]. Each context spanned the words in the aligned regions in the corpus, covering 1,298,966 sentence pairs.

The input for the map consists of English and German words and their bilingual contexts. The most frequent 150 nouns were selected separately from both languages to get a vocabulary of 300 words. Note that the vocabulary does not contain sensible translations for all of its words. The contexts for the selected 300 words were calculated from the most frequent 3889 words in the two languages. In contrast to the previous experiment with feature contexts, each word was represented only with a single context vector.

The context variables were calculated as the number of co-occurrence counts of the words in the vocabulary with the context words, resulting with 300 word vectors with 3889 context variables. The variances of the context variables were normalized to one to make all the variables equally important. The L2 norm of the word vectors was normalized to one, which makes the Euclidian metric used by the SOM very closely related to the cosine similarity typically used with vector space models. The resulting map is shown in Figure 2 with the 300 words positioned to the closest map units.

The map shows the bilingual vocabulary evenly distributed to the map, with translations of words next to each other. The words are also ordered semantically, with related words close to each other (e.g., 'world', 'europe', 'country'). Inflections of the same word are typically also close to each other (e.g., 'Jahr', 'Jahre', 'Jahren'), with sometimes singular and plural forms a bit more separate. The map also shows several possible translations next to each other (e.g., 'aid', 'help', 'Hilfe'). Words without translation equivalents in the vocabulary are located to semantically near words (e.g., 'Lösung' near 'problem').

The results reported above are in an interesting contrast with another study in which words of two languages were presented in linguistic contexts [17]. Li and Farkas found out that the two languages were strictly separated on the map. The differences are explained by the selection of the word contexts: Li and Farkas used the distributions of preceding and following words in bilingual, intermixed sentences as semantic context, whereas we use "bag-of-words" representation of the full sentences in both languages.

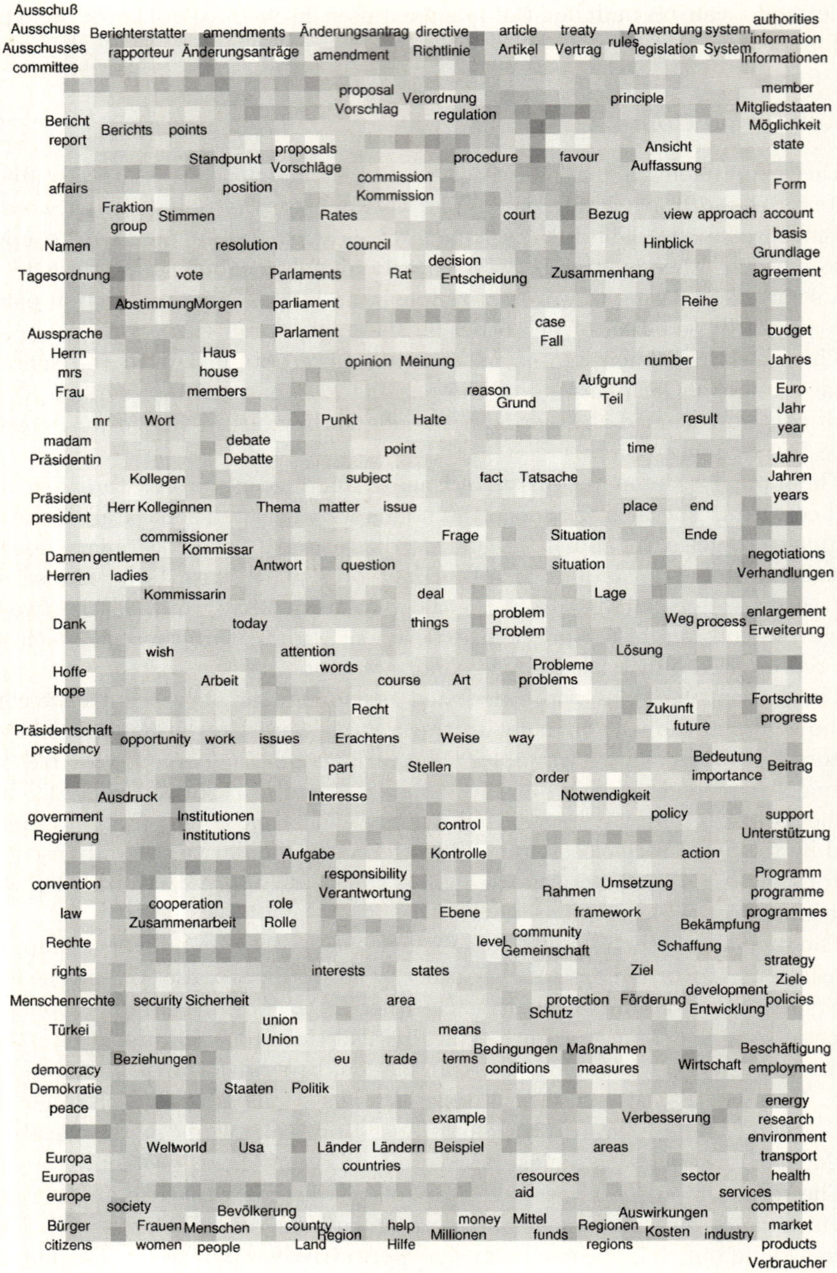

Fig. 2. German and English words on a Self-Organizing Map of contexts. The interpretation of the map is similar to that of Figure 1, except the map was taught with real-life bilingual sentential contexts, and for each word there is only one vector.

3   Conclusions and Discussion

The main difference between the approach outlined in the previous sections and the Bayesian method, in its commonly used form, is that the semantic or conceptual space is explicitly modeled in the SOM-based approach. Thus, the mapping between any two languages is based on an intermediate level of representation. This approach resembles, to some degree, the idea of using a knowledge-based interlingua in machine translation. The underlying philosophical assumptions about knowledge are, however, quite different. In a knowledge-based interlingua, the semantics of natural language expressions are typically represented as propositions and relations in symbolic hierarchical structures. The SOM can be used to span a continuous and multidimensional conceptual space in a data-driven manner. Moreover, the approach provides a natural means to deal with multimodal data [18] and, thus, deal with the symbol grounding problem [19].

References

1. Moore, T., Carling, C.: The Limitations of Language. Macmillan Press, Houndmills (1988)
2. Honkela, T.: Philosophical aspects of neural, probabilistic and fuzzy modeling of language use and translation. In: Proceedings of IJCNN 2007, International Joint Conference on Neural Networks (2007)
3. Bowerman, M.: The origins of children's spatial semantic categories: cognitive versus linguistic determinants. In: Gumperz, J., Levinson, S.C. (eds.) Rethinking linguistic relativity, pp. 145–176. Cambridge University Press, Cambridge (1996)
4. Berlin, B., Kay, P.: Basic Color Terms: Their Universality and Evolution. University of California Press (1991/1969)
5. Gärdenfors, P.: Conceptual Spaces. MIT Press, Cambridge (2000)
6. Ahrenberg, L., Andersson, M., Merkel, M.: A simple hybrid aligner for generating lexical correspondences in parallel texts. In: Proceedings of COLING-ACL 1998, pp. 29–35 (1992)
7. Brown, P.F., Pietra, S.A.D., Pietra, V.J.D., Mercer, R.L.: The mathematics of statistical machine translation: Parameter estimation. Computational Linguistics 19(2), 263–311 (1993)
8. Koehn, P., Och, F.J., Marcu, D.: Statistical phrase-based translation. In: NAACL 2003: Proceedings of the 2003 Conference of the North American Chapter of the Association for Computational Linguistics on Human Language Technology, Morristown, NJ, USA, pp. 48–54. Association for Computational Linguistics (2003)
9. Zhang, R., Yamamoto, H., Paul, M., Okuma, H., Yasuda, K., Lepage, Y., Denoual, E., Mochihashi, D., Finch, A., Sumita, E.: The NiCT-ATR Statistical Machine Translation System for IWSLT 2006. In: Proceedings of the International Workshop on Spoken Language Translation, Kyoto, Japan, pp. 83–90 (2006)
10. Kohonen, T.: Self-organizing formation of topologically correct feature maps. Biological Cybernetics 43(1), 59–69 (1982)
11. Kohonen, T.: Self-Organizing Maps. Springer Series in Information Sciences, vol. 30. Springer, Heidelberg (2001)
12. Kohonen, T., Honkela, T.: Kohonen network. Scholarpedia, p. 7421 (2007)
13. Charniak, E.: Statistical Language Learning. MIT Press, Cambridge (1993)

14. Ritter, H., Kohonen, T.: Self-organizing semantic maps. Biological Cybernetics 61(4), 241–254 (1989)
15. Kaski, S.: Dimensionality reduction by random mapping: Fast similarity computation for clustering. In: Proceedings of IJCNN 1998, International Joint Conference on Neural Networks, vol. 1, pp. 413–418. IEEE Service Center, Piscataway (1998)
16. Koehn, P.: Europarl: A parallel corpus for statistical machine translation. In: Proceedings of the 10th Machine Translation Summit, Phuket, Thailand, pp. 79–86 (2005)
17. Li, P., Farkas, I.: A self-organizing connectionist model of bilingual processing. In: Bilingual sentence processing, pp. 59–85. North-Holland, Amsterdam (2002)
18. Laaksonen, J., Viitaniemi, V.: Emergence of ontological relations from visual data with self-organizing maps. In: Proceedings of SCAI 2006, Scandinavian Conference on Artificial Intelligence, Espoo, Finland, pp. 31–38 (2006)
19. Harnad, S.: The symbol grounding problem. Physica D 42, 335–346 (1990)

Self-Organizing Neural Grove: Efficient Multiple Classifier System with Pruned Self-Generating Neural Trees

Hirotaka Inoue

Department of Electrical Engineering and Information Science,
Kure College of Technology,
2-2-11 Agaminami, Kure-shi, Hiroshima 737-8506, Japan
hiro@kure-nct.ac.jp

Abstract. Multiple classifier systems (MCS) have become popular during the last decade. Self-generating neural tree (SGNT) is one of the suitable base-classifiers for MCS because of the simple setting and fast learning. However, the computation cost of the MCS increases in proportion to the number of SGNT. In an earlier paper, we proposed a pruning method for the structure of the SGNT in the MCS to reduce the computation cost. In this paper, we propose a novel pruning method for more effective processing and we call this model as self-organizing neural grove (SONG). The pruning method is constructed from an on-line pruning method and an off-line pruning method. Experiments have been conducted to compare the SONG with an unpruned MCS based on SGNT, an MCS based on C4.5, and k-nearest neighbor method. The results show that the SONG can improve its classification accuracy as well as reducing the computation cost.

1   Introduction

Classifiers need to find hidden information in the given large data effectively and classify unknown data as accurately as possible [1]. Recently, to improve the classification accuracy, multiple classifier systems (MCS) such as neural network ensembles, bagging, and boosting have been used for practical data mining applications [2,3,4,5]. In general, the base classifiers of the MCS use traditional models such as neural networks (backpropagation network and radial basis function network) [6] and decision trees (CART and C4.5) [7].

Neural networks have great advantages of adaptability, flexibility, and universal nonlinear input-output mapping capability. However, to apply these neural networks, it is necessary to determine the network structure and some parameters by human experts, and it is quite difficult to choose the right network structure suitable for a particular application at hand. Moreover, they require a long training time to learn the input-output relation of the given data. These drawbacks prevent neural networks being the base classifier of the MCS for practical applications.

V. Kůrková et al. (Eds.): ICANN 2008, Part I, LNCS 5163, pp. 613–622, 2008.

Self-generating neural tree (SGNT) [8] have simple network design and high speed learning. SGNT are an extension of the self-organizing maps (SOM) of Kohonen [9] and utilize the competitive learning. The abilities of SGNT make it suitable for the base classifier of the MCS. In order to improve in the accuracy of SGNN, we proposed ensemble self-generating neural networks (ESGNN) for classification [10] as one of the MCS. Although the accuracy of ESGNN improves by using various SGNT, the computation cost, that is, the computation time and the memory capacity increases in proportion to the increase in number of SGNN in the MCS.

In an earlier paper [11], we proposed a pruning method for the structure of the SGNN in the MCS to reduce the computation cost. In this paper, we propose a novel MCS pruning method for more effective processing and we call this model as self-organizing neural grove (SONG). This pruning method is constructed from two stages. At the first stage, we introduce an on-line pruning method to reduce the computation cost by using class labels in learning. At the second stage, we optimize the structure of the SGNT in the MCS to improve the generalization capability by pruning the redundant leaves after learning. In the optimization stage, we introduce a threshold value as a pruning parameter to decide which subtree's leaves to prune and estimate with 10-fold cross-validation [12]. After the optimization, the SONG can improve its classification accuracy as well as reducing the computation cost. We use bagging [2] as a resampling technique for the SONG.

We investigate the improvement performance of the SONG by comparing it with an MCS based on C4.5 [13] using ten problems in UCI machine learning repository [14]. Moreover, we compare the SONG with k-nearest neighbor (k-NN) [15] to investigate the computational cost and the classification accuracy. The SONG demonstrates higher classification accuracy and faster processing speed than k-NN on average.

2 Constructing Self-Organizing Neural Grove

In this section, we describe how to prune redundant leaves in the SONG. First, we mention the on-line pruning method in learning of the SGNT. Second, we show the optimization method in constructing the SONG. Finally, we show a simple example of the pruning method for a two dimensional classification problem.

2.1 On-Line Pruning of Self-Generating Neural Tree

SGNT is based on SOM and implemented as a competitive learning. The SGNT can be constructed directly from the given training data without any intervening human effort. The SGNT algorithm is defined as a tree construction problem of how to construct a tree structure from the given data which consist of multiple attributes under the condition that the final leaves correspond to the given data.

Before we describe the SGNT algorithm, we denote some notations.

- input data vector: $e_i \in \mathbb{R}^m$.
- root, leaf, and node in the SGNT: n_j.

```
Input:
  A set of training examples E = {e_i}, i = 1, ... , N.
  A distance measure d(e_i,w_j).
Program Code:
  copy(n_1,e_1);
  for (i = 2, j = 2; i <= N; i++) {
    n_win = choose(e_i, n_1);
    if (leaf(n_win)) {
      copy(n_j, w_win);
      connect(n_j, n_win);
      j++;
    }
    copy(n_j, e_i);
    connect(n_j, n_win);
    j++;
    prune(n_win);
  }
Output:
  Constructed SGNT by E.
```

Fig. 1. SGNT algorithm

Table 1. Sub procedures of the SGNT algorithm

Sub procedure	Specification
$copy(n_j, \boldsymbol{e}_i/\boldsymbol{w}_{win})$	Create n_j, copy $\boldsymbol{e}_i/\boldsymbol{w}_{win}$ as $\boldsymbol{w}_j$ in n_j.
$choose(\boldsymbol{e}_i, n_1)$	Decide n_{win} for $\boldsymbol{e}_i$.
$leaf(n_{win})$	Check n_{win} whether n_{win} is a leaf.
$connect(n_j, n_{win})$	Connect n_j as a child leaf of n_{win}.
$prune(n_{win})$	Prune leaves if they have the same class.

- weight vector of n_j: $\boldsymbol{w}_j \in \mathbb{R}^m$.
- the number of the leaves in n_j: c_j.
- distance measure: $d(\boldsymbol{e}_i, \boldsymbol{w}_j)$.
- winner leaf for $\boldsymbol{e}_i$ in the SGNT: n_{win}.

The SGNT algorithm is a hierarchical clustering algorithm. The pseudo C code of the SGNT algorithm is given in Figure 1. In Figure 1, several sub procedures are used. Table 1 shows the sub procedures of the SGNT algorithm and their specifications.

In order to decide the winner leaf n_{win} in the sub procedure `choose(e_i,n_1)`, the competitive learning is used. If an n_j includes the n_{win} as its descendant in the SGNT, the weight w_{jk} $(k = 1, 2, \ldots, m)$ of the n_j is updated as follows:

$$w_{jk} \leftarrow w_{jk} + \frac{1}{c_j} \cdot (e_{ik} - w_{jk}), \quad 1 \le k \le m. \tag{1}$$

```
1 begin     initialize j = the height of the SGNT
2    do for each subtree's leaves in the height j
3       if the ratio of the most class ≥ the threshold value α,
4       then merge all leaves to parent node
5       if all subtrees are traversed in the height j,
6       then j ← j − 1
7    until j = 0
8 end.
```

Fig. 2. The merge phase

After all training data are inserted into the SGNT as the leaves, the leaves have each class label as the outputs and the weights of each node are the averages of the corresponding weights of all its leaves. The whole network of the SGNT reflects the given feature space by its topology. For more details concerning how to construct and perform the SGNT, see [8]. Note, to optimize the structure of the SGNT effectively, we remove the threshold value of the original SGNT algorithm in [8] to control the number of leaves based on the distance because of the trade-off between the memory capacity and the classification accuracy. In order to avoid the above problem, we introduce a new pruning method in the sub procedure `prune(n_win)`. We use the class label to prune leaves. For leaves that have the n_{win}'s parent node, if all leaves belong to the same class, then these leaves are pruned and the parent node is given the class.

2.2   Optimization of the SONG

The SGNT has the capability of high speed processing. However, the accuracy of the SGNT is inferior to the conventional approaches, such as nearest neighbor, because the SGNT has no guarantee to reach the nearest leaf for unknown data. Hence, we construct the SONG by taking the majority of plural SGNT's outputs to improve the accuracy.

Although the accuracy of the SONG is superior or comparable to the accuracy of conventional approaches, the computational cost increases in proportion to the increase in the number of SGNTs in the SONG. In particular, the huge memory requirement prevents the use of the SONG for large datasets even with latest computers.

In order to improve the classification accuracy, we propose an optimization method of the SONG for classification. This method has two parts, the merge phase and the evaluation phase. The merge phase is performed as a pruning algorithm to reduce dense leaves (Figure 2). This phase uses the class information and a threshold value α to decide which subtree's leaves to prune or not. For leaves that have the same parent node, if the proportion of the most common class is greater than or equal to the threshold value α, then these leaves are pruned and the parent node is given the most common class.

```
1 begin initialize α = 0.5
2    do for each α
3       evaluate the merge phase with 10-fold cross validation
4       if the best classification accuracy is obtained,
5       then record the α as the optimal threshold value
6       α ← α + 0.05
7    until α = 1
8 end.
```

Fig. 3. The evaluation phase

The optimum threshold values α of the given problems are different from each other. The evaluation phase is performed to choose the best threshold value by introducing 10-fold cross validation (Figure 3).

2.3 An Example of the Pruning Method for the SONG

We show an example of the pruning method for the SONG in Figure 4. This is a two-dimensional classification problem with two equal circular Gaussian distributions that have an overlap. The shaded plane is the decision region of class 0 and the other plane is the decision region of class 1 by the SGNT. The dotted line is the ideal decision boundary. The number of training samples is 200 (class0: 100,class1: 100) (Figure 4(a)).

The unpruned SGNT is given in Figure 4(b). In this case, 200 leaves and 120 nodes are automatically generated by the SGNT algorithm. In this unpruned SGNT, the height is 7 and the number of units is 320. In this, we define the unit to count the sum of the root, nodes, and leaves of the SGNT. The root is the node which is of height 0. The unit is used as a measure of the memory requirement in the next section. Figure 4(c) shows the pruned SGNT after the optimization stage in $\alpha = 1$. In this case, 159 leaves and 107 nodes are pruned away and 48 units remain. The decision boundary is the same as the unpruned SGNT. Figure 4(d) shows the pruned SGNT after the optimization stage in $\alpha = 0.6$. In this case, 182 leaves and 115 nodes are pruned away and only 21 units remain. Moreover, the decision boundary is improved more than the unpruned SGNT because this case can reduce the effect of the overlapping class by pruning the SGNT.

In the above example, we use all training data to construct the SGNT. The structure of the SGNT is changed by the order of the training data. Hence, we can construct the SONG from the same training data by changing the input order.

To show how well the SONG is optimized by the pruning algorithm, we show an example of the SONG in the same problem used above. Figure 5(a) and Figure 5(b) show the decision region of the SONG in $\alpha = 1$ and $\alpha = 0.6$, respectively. We set the number of SGNTs K as 25. The result of Figure 5(b) is a better estimation of the ideal decision region than the result of Figure 5(a). We investigate the pruning method for more complex problems in the next section.

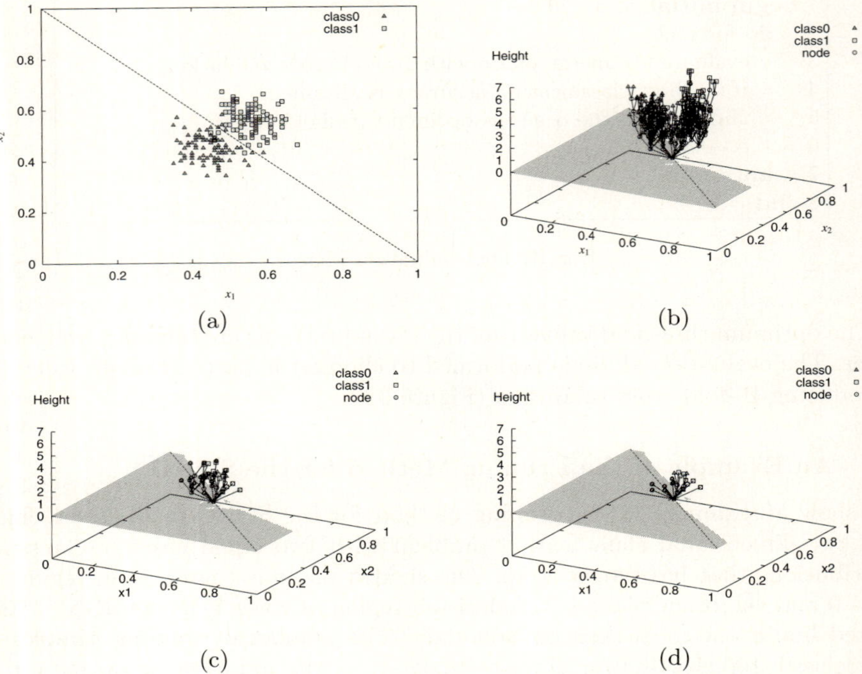

Fig. 4. An example of the SONG's pruning algorithm, (a) a two dimensional classification problem with two equal circular Gaussian distribution, (b) the structure of the unpruned SGNT, (c) the structure of the pruned SGNT ($\alpha = 1$), and (d) the structure of the pruned SGNT ($\alpha = 0.6$). The shaded plane is the decision region of class 0 by the SGNT and the dotted line shows the ideal decision boundary.

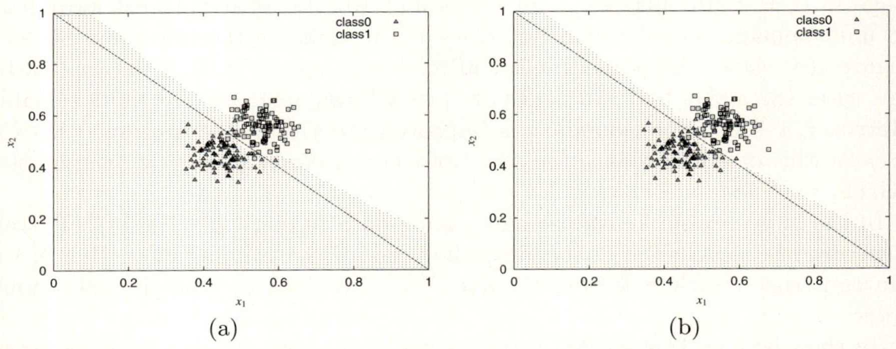

Fig. 5. An example of the SONG's decision boundary ($K = 25$), (a) $\alpha = 1$, and (b) $\alpha = 0.6$. The shaded plane is the decision region of class 0 by the SONG and the dotted line shows the ideal decision boundary.

3   Experimental Results

We investigate the computational cost (the memory capacity and the computation time) and the classification accuracy of the SONG with bagging for ten benchmark problems in UCI machine learning repository [14]. We evaluate how the SONG is pruned using 10-fold cross-validation for the ten benchmark problems. In this experiment, we use a modified Euclidean distance measure for the SONG and k-NN. Since the performance of the SONG is not sensitive in the threshold value α, we set the different threshold values α which are moved from 0.5 to 1; $\alpha = [0.5, 0.55, 0.6, \ldots, 1]$. We set the number of SGNT K in the SONG as 25 and execute 100 trials by changing the sampling order of each training set. All experiments in this section were performed on an UltraSPARC workstation with a 900MHz CPU, 1GB RAM, and Solaris 8.

Table 2 shows the average memory requirement and classification accuracy of 100 trials for the SONG. As the memory requirement, we count the number of units which is the sum of the root, nodes, and leaves of the SGNT. The average memory requirement is reduced from 65% to 96.6% and the classification accuracy is improved 0.1% to 2.9% by optimizing the SONG. This supports that the SONG can be effectively used for all datasets with regard to both the computation cost and the classification accuracy.

To evaluate the SONG's performance, we compare the SONG with an MCS based on C4.5. We set the number of classifiers K in the MCS as 25 and we construct both MCS by bagging. Table 3 shows the improved performance of the SONG and the MCS based on C4.5. The results of the SGNT and the SONG are the average of 100 trials. The SONG has a better performance than the MCS based on C4.5 for 6 of the 10 datasets. Although the MCS based on C4.5 degrades the classification accuracy for iris, the SONG can improve the classification accuracy for all problems. Therefore, the SONG is an efficient MCS on the basis of both the scalability for large scale datasets and the robust improving generalization capability for the noisy datasets comparable to the MCS with C4.5.

To show the advantages of the SONG, we compare it with k-NN on the same problems. In the SONG, we choose the best classification accuracy of 100 trials with bagging. In k-NN, we choose the best accuracy where k is 1,3,5,7,9,11,13,15, and 25 with 10-fold cross-validation. All methods are compiled by using gcc with the optimization level -O2 on the same workstation.

Table 4 shows the classification accuracy, the memory requirement, and the computation time achieved by the SONG and k-NN. Although there are compression methods available for k-NN [16], they take enormous computation time to construct an effective model. We use the exhaustive k-NN in this experiment. Since k-NN does not discard any training sample, the size of this classifier corresponds to the training set size. The results of k-NN correspond to the average measures obtained by 10-fold cross-validation that is the same experimental procedure of the SONG. Next, we show the results for each category.

First, with regard to the classification accuracy, the SONG is superior to k-NN for 8 of the 10 datasets and gives 1.6% improvement on average. Second, in

Table 2. The average memory requirement and classification accuracy of 100 trials for the bagged SGNT in the SONG. The standard deviation is given inside the bracket on classification accuracy ($\times 10^{-3}$).

Dataset	memory requirement			classification accuracy		
	pruned	unpruned	ratio	pruned	unpruned	ratio
balance-scale	107.68	861.18	12.5	0.866(6.36)	0.837(7.83)	+2.9
breast-cancer-w	30.88	897.37	3.4	0.97(2.41)	0.966(2.71)	+0.4
glass	104.33	297.75	35	0.714(13.01)	0.709(14.86)	+0.5
ionosphere	50.75	472.39	10.7	0.891(6.75)	0.862(7.33)	+2.9
iris	15.64	208.56	7.4	0.962(6.04)	0.955(5.45)	+0.7
letter	6197.5	27028.56	22.9	0.956(0.77)	0.955(0.72)	+0.1
liver-disorders	163.12	471.6	34.5	0.648(12.89)	0.636(13.36)	+1.2
new-thyroid	49.45	298.21	16.5	0.958(7.5)	0.957(7.49)	+0.1
pima-diabetes	204.4	1045.03	19.5	0.749(7.05)	0.728(7.83)	+2.1
wine	15	238.95	6.2	0.976(4.41)	0.972(5.57)	+0.4
Average	693.88	3181.96	16.9	0.869	0.858	+1.1

Table 3. The improved performance of the pruned MCS and the MCS based on C4.5 with bagging

Dataset	MCS based on SGNT			MCS based on C4.5		
	SGNT	MCS	ratio	C4.5	MCS	ratio
balance-scale	0.779	**0.866**	+8.7	0.795	0.827	+3.2
breast-cancer-w	0.956	**0.97**	+1.4	0.946	0.963	+1.7
glass	0.642	0.714	+7.2	0.664	**0.757**	+9.3
ionosphere	0.852	0.891	+3.9	0.897	**0.92**	+2.3
iris	0.943	**0.962**	+1.9	0.953	0.947	−0.6
letter	0.879	**0.956**	+7.7	0.880	0.938	+5.8
liver-disorders	0.59	0.648	+5.8	0.635	**0.736**	+10.1
new-thyroid	0.939	**0.958**	+1.9	0.93	0.94	+1
pima-diabetes	0.695	0.749	+5.4	0.749	**0.767**	+1.8
wine	0.955	**0.976**	+2.1	0.927	0.949	+2.2
Average	0.823	0.869	+4.6	0.837	**0.874**	+3

terms of the memory requirement, even though the SONG includes the root and the nodes which are generated by the SGNT generation algorithm, this is less than k-NN for all problems. Although the memory requirement of the SONG is totally used K times in Table 4, we release the memory of SGNT for each trial and reuse the memory for effective computation. Therefore, the memory requirement is suppressed by the size of the single SGNT. Finally, in view of the computation time, although the SONG consumes the cost of K times of the SGNT to construct the model and test for the unknown dataset, the average computation time is faster than k-NN. The SONG is slower than k-NN for small datasets such as glass, ionosphere, and iris. However, the SONG is faster than k-NN for large datasets such as balance-scale, letter, and pima-diabetes. In the

Table 4. The classification accuracy, the memory requirement, and the computation time of ten trials for the best pruned SONG and k-NN

Dataset	classification acc.		memory requirement		computation time (s)	
	SONG	k-NN	SONG	k-NN	SONG	k-NN
balance-scale	0.878	**0.888**	**109.93**	562.5	**0.82**	1.14
breast-cancer-w	**0.974**	0.969	**26.8**	629.1	**1.18**	1.25
glass	**0.758**	0.701	**91.33**	192.6	0.36	**0.08**
ionosphere	**0.912**	0.866	**51.38**	315.9	1.93	**0.2**
iris	**0.973**	0.96	**11.34**	135	0.13	**0.05**
letter	0.958	**0.96**	**6208.03**	18000	**208.52**	503.14
liver-disorders	**0.685**	0.653	**134.17**	310.5	**0.54**	0.56
new-thyroid	**0.972**	**0.972**	**45.74**	193.5	0.23	**0.05**
pima-diabetes	**0.764**	0.751	**183.57**	691.2	**1.72**	2.49
wine	**0.983**	0.977	**11.8**	160.2	0.31	**0.15**
Average	**0.885**	0.869	**687.41**	2119.1	**21.57**	50.91

case of letter, in particular, the computation time of the SONG is faster than k-NN by about 2.4 times. We need to repeat 10-fold cross validation many times to select the optimum parameters for α and k. This evaluation consumes much computation time for large datasets such as letter. Therefore, the SONG based on the fast and compact SGNT is useful and practical for large datasets. Moreover, the SONG has the ability of parallel computation because each classifier behaves independently. In conclusion, the SONG is a practical method for large-scale data mining compared with k-NN.

4   Conclusions

In this paper, we proposed a new pruning method for the MCS based on SGNT, which is called SONG, and evaluated the computation cost and the accuracy.We introduced an on-line and off-line pruning method and evaluated the SONG by 10-fold cross-validation. Experimental results showed that the memory requirement reduces remarkably, and the accuracy increases by using the pruned SGNT as the base classifier of the SONG. The SONG is a useful and practical MCS to classify large datasets. In future work, we will study an incremental learning and a parallel and distributed processing of the SONG for large scale data mining.

References

1. Han, J., Kamber, M.: Data Mining: Concepts and Techniques. Morgan Kaufmann Publishers, San Francisco (2000)
2. Breiman, L.: Bagging predictors. Machine Learning 24, 123–140 (1996)
3. Schapire, R.E.: The strength of weak learnability. Machine Learning 5(2), 197–227 (1990)
4. Quinlan, J.R.: Bagging, Boosting, and C4.5. In: Proceedings of the Thirteenth National Conference on Artificial Intelligence, Portland, OR, August 4–8, pp. 725–730. AAAI Press/ MIT Press (1996)

5. Rätsch, G., Onoda, T., Müller, K.R.: Soft margins for AdaBoost. Machine Learning 42(3), 287–320 (2001)
6. Bishop, C.M.: Neural Networks for Pattern Recognition. Oxford University Press, New York (1995)
7. Duda, R.O., Hart, P.E., Stork, D.G.: Pattern Classification, 2nd edn. John Wiley & Sons Inc., New York (2000)
8. Wen, W.X., Jennings, A., Liu, H.: Learning a neural tree. In: The International Joint Conference on Neural Networks, Beijing, China, November 3–6, vol. 2, pp. 751–756 (1992)
9. Kohonen, T.: Self-Organizing Maps. Springer, Berlin (1995)
10. Inoue, H., Narihisa, H.: Improving generalization ability of self-generating neural networks through ensemble averaging. In: Terano, T., Chen, A.L.P. (eds.) PAKDD 2000. LNCS (LNAI), vol. 1805, pp. 177–180. Springer, Heidelberg (2000)
11. Inoue, H., Narihisa, H.: Optimizing a multiple classifier system. In: Ishizuka, M., Sattar, A. (eds.) PRICAI 2002. LNCS (LNAI), vol. 2417, pp. 285–294. Springer, Heidelberg (2002)
12. Stone, M.: Cross-validation: A review. Math. Operationsforsch. Statist., Ser. Statistics 9(1), 127–139 (1978)
13. Quinlan, J.R.: C4.5: Programs for Machine Learning. Morgan Kaufmann, San Mateo (1993)
14. Blake, C., Merz, C.: UCI repository of machine learning databases (1998)
15. Patrick, E.A., Frederick, P., Fischer, I.: A generalized k-nearest neighbor rule. Information and Control 16(2), 128–152 (1970)
16. Zhang, B., Srihari, S.N.: Fast k-nearest neighbor classification using cluster-based trees. IEEE Transactions on Pattern and Machine Intelligence 26(4), 525–528 (2004)

Self-organized Complex Neural Networks through Nonlinear Temporally Asymmetric Hebbian Plasticity

Hideyuki Kato and Tohru Ikeguchi

Graduate school of Science and Engineering, Saitama University,
255 Shimo-Ohkubo Saitama 338–8570, Japan
kato@nls.ics.saitama-u.ac.jp

Abstract. Triggered by recent experimental results, the spike-timing-dependent plasticity (STDP) has been widely analyzed in the neuroscience. In this paper, we analyzed how spatial structure of neural networks will be organized through the STDP. In the experiments, we did use additive and multiplicative STDP rules as well as a nonlinear temporally asymmetric Hebbian rule. As a result, if the additive rule is applied, neural networks exhibit small-world properties. On the other hand, in the case of the multiplicative rule, the small-world networks are not constructed. In addition, we also found that the small-world properties of the neural networks become higher if the STDP rule is less dependent on current synaptic weights.

1   Introduction

In the brain, a huge number of neurons exist and are complexly connected each other to process mass information. Then, they must have constructed complex but specific structure. Using such complex structure, huge amounts of information are effectively processed.

Recent studies in the field of the neuroscience reveal that a new kind of synaptic plasticity is used in the brains: spike-timing-dependent synaptic plasticity (STDP), which has been experimentally observed in several regions of the brain from different kinds of species [1]. In Ref. [2], it is shown that long-term synaptic modification depends on firing timing between pre- and postsynaptic neurons, and that it arises in tens millisecond. More precisely, a synapse is strengthened if the postsynaptic spike follows the presynaptic spike. This synaptic modification is called the long-term potentiation. On the other hand, a synapse is weakened if the presynaptic spike follows the postsynaptic spike, which is called the long-term depression.

Since a seminal paper by Bi and Poo has been published [2], experimental studies have been devoted to analyzing the fundamental properties induced by the STDP. In the theoretical studies, many models of the STDP function have been proposed, and then their characteristics are analyzed numerically and theoretically. In 2000, Song et al. proposed an STDP function modeled approximately

V. Kůrková et al. (Eds.): ICANN 2008, Part I, LNCS 5163, pp. 623–631, 2008.

[3] from the data in Ref. [2]. It is shown that synapses are competitive and synaptic population forms a bimodal distribution through the STDP [3]. The model is called the additive STDP rule. On the other hand, Rossum et al. coincidentally proposed another STDP model. They experimentally and theoretically indicated that the synaptic distributions become unimodal in their model [4]. The model is called the multiplicative STDP rule.

Mathematical supports on how these synaptic distributions are consisted was given by the Fokker-Planck theorem [5]. However, it is not only interesting but also important to clarify what kinds of spatial network structure will emerge if the STDP is applied.

Until 1998, in the graph theory were mainly analyzed regular and random networks. Regular networks have specific structure or many clusters, then distances between any two nodes are large. On the other hand, random networks have small network radii but few clusters. However, it has been shown that real networks often exhibit small network radii although they still keep clusters. Watts and Strogatz recently proposed a new interesting concept, small-world networks [6]. They showed that the small-world networks have different characteristics from both regular and random networks: short characteristic path lengths but large clustering coefficients [6]. They also showed that small-world networks exist in the real world, for example, the co-acting relationship in movie films, the power grid networks, and the anatomical structure of *C. elegans*. In addition, neural networks often exhibit small-world properties. For example, such properties are discovered in macaque visual cortex [7], macaque cortex [8], and cat cortex [9]. Although these facts indicate that neural networks in the brains may use such small-world network properties for processing information, it has not yet been clarified how such properties are constructed or how the neural networks self-organize to evolve such functional complex networks.

Then, in this paper, to answer the above issues we analyzed how the self-organized neural network structure could emerge through the STDP from viewpoints of the complex network theory.

2   Neuron Model

To resolve the issues raised in this paper, we conduct numerical experiments with a mathematical neuron model. The Hodgkin-Huxley (HH) model is the most famous one as a mathematical neuron model [10]. The HH model is biologically meaningful, but computationally too complex for a large scale neural network. In contrast, the leaky integrate-and-fire (LIF) model is known as one of the simplest neuron models [11]. The LIF model is very simple and computationally effective. However, the LIF model is too simple to generate rich firing patterns. Then, in this paper, we used an Izhikevich's model as an element neuron in a neural network to realize different dynamics between excitatory and inhibitory neurons. The Izhikevich model we used is described by a two-dimensional ordinary differential equation. Then, it is computationally less effective than the LIF model but can generate rich firing patterns [12,13]. The model is defined as follows:

$$\dot{v} = 0.04v^2 + 5v + 140 - u + I, \tag{1}$$

$$\dot{u} = a(bv - u), \tag{2}$$

with the auxiliary after-spike resetting:

$$\text{if } v \geq +30\,[\text{mV}], \text{ then } \begin{cases} v \leftarrow c, \\ u \leftarrow u + d, \end{cases} \tag{3}$$

where v represents membrane potential of the neuron, u represents membrane recovery variable, and $\dot{} = d/dt$ where t represents time. The quantity I represents the sum of random external and synaptic current. Here, a, b, c, and d are parameters. Depending on the values of parameters in Eqs. (1)–(3) (a, b, c, and d), the neuron model can generate various kinds of firing patterns of neurons [12]. To investigate the fundamental mechanism of the STDP, we adopted a regular spiking type as the excitatory neurons and set parameters in Eqs (1)–(3) to $a = 0.02$, $b = 0.2$, $c = -65$, and $d = 8$ [12,13]. On the other hand, in the case of inhibitory neurons, we used a fast spiking neuron model, then, $a = 0.1$, $b = 0.2$, $c = -65$, and $d = 2$ [12,13].

3 Synaptic Modification

In the neural network, synaptic weights are modified by the STDP. The STDP rule was applied only to excitatory synapses, and the other inhibitory synapses were fixed [3,4,13]. Let us denote the firing times of pre- and postsynaptic neurons as t_{pre} and t_{post}, respectively. Then, we can describe the interspike interval between the pre- and postsynaptic neurons as $\Delta t = t_{\text{post}} - t_{\text{pre}}$. Using the variable, the STDP rule is defined as follows:

$$\Delta w = \begin{cases} \lambda f_+(w) \times K(\Delta t) & \text{if } \Delta t > 0, \\ -\lambda f_-(w) \times K(\Delta t) & \text{otherwise}, \end{cases} \tag{4}$$

where Δw represents an amount of synaptic modification [14]; w represents a synaptic weight normalized to $[0, 1]$. The parameter λ ($0 < \lambda \ll 1$) is a learning rate which scales the magnitude of an individual synaptic weight variance. In this paper, we set $\lambda = 0.1$ [13]. Then, the temporal filter $K(\Delta t)$ in Eq. (4) is defined as follows:

$$K(\Delta t) = e^{-\frac{|\Delta t|}{\tau}}. \tag{5}$$

In Eq. (5), the time constant τ of the exponential decay determines the temporal range of the learning window, and was set to 20 [ms] [3,4,13]. In addition, updating functions $f_+(w)$ and $f_-(w)$ in Eq. (4) are defined as follows:

$$f_+(w) = (1 - w)^\mu, \tag{6}$$

$$f_-(w) = \alpha w^\mu, \tag{7}$$

where $\alpha > 0$ denotes a degree of asymmetry between the scale of the potentiation and the depression, and was set to 1.2 [13]. Eqs. (6) and (7) indicate that synaptic weights can be updated in a nonlinear way with a parameter μ. If $\mu = 0$, Eqs. (6) and (7) show that $f_+(w)$ and $f_-(w)$ do not depend on the current synaptic weight. Then, Δw is also independent of the current weights. Hence, this rule is called an additive rule. On the other hand, if $\mu = 1$, $f_+(w)$ and $f_-(w)$ depend on the current synaptic weights linearly. In this case, this rule is called a multiplicative rule. If we set μ between 0 and 1, synaptic weights are modified nonlinearly depending on the current synaptic weights.

4   Methods

We used a neural network with $1,000$ neurons, and set the population rate of excitatory and inhibitory neurons to $8 : 2$. In the neural network, each neuron had 100 postsynaptic neurons which were randomly selected. There were no connections between any pairs of inhibitory neurons. In initial conditions, all excitatory synaptic weights were set to 6, and all inhibitory synaptic weights were set to -5. Excitatory synapses had $[1, 10]$ [ms] random axonal delays. On the other hand, inhibitory synapses had 1 [ms] axonal delays. We assumed that only excitatory synapses were modified by the STDP while the inhibitory synapses were maintained constantly [3,4,13]. The excitatory synaptic efficacy was set to $[0, 10]$. In addition, random Poissonian external inputs were applied to the neural network. The rate of inputs were 10 [Hz].

To characterize the structure of the neural network to which the STDP rule is applied for t [sec], we used two famous measures for the complex networks: a clustering coefficient C and a characteristic path length L (see below). It is important to analyze differences of the synaptic weights because the synaptic weights change through the STDP in the neural network. Then, we used C and L for weighted networks. However, the neural network is a weighted digraph, so that it is necessary to translate the neural network from the weighted digraph to the weighted graph.

A method for translating the neural network is described as follows: let us define w_{ij} as the synaptic weight from the ith to the jth neurons in the neural network. First, we compared w_{ij} with w_{ji}, and selected a larger weight. Namely, if $w_{ij} > w_{ji}$, we selected w_{ij}. Next, the selected weight w_{ij} was normalized by the maximum synaptic weight $w_{\max}$. Finally, we defined $\tilde{w}_{ij} = \tilde{w}_{ji} = w_{ij}/w_{\max}$ as a new weight between the ith and jth neurons. Applying such a process for all pairs of connected neurons, we translated the neural network from the weighted digraph to the weighted graph. We applied C and L to such a translated neural network.

The clustering coefficient C for the weighted networks is defined by

$$C = \frac{1}{N} \sum_i^N C_i \tag{8}$$

where N is the number of neurons in the neural network, and C_i is an individual clustering coefficient of the ith neuron [15,16,17,18]. The ith clustering coefficient C_i is defined as follows:

$$C_i = \frac{\sum_j^N \sum_{k \neq j}^N \tilde{w}_{ij} \tilde{w}_{ik} \tilde{w}_{jk}}{\sum_j^N \sum_{k \neq j}^N \tilde{w}_{ij} \tilde{w}_{ik}} \tag{9}$$

where $\tilde{w}_{ij}$ is a normalized synaptic weight between the ith and jth neurons and takes between 0 and 1 [19]. Because undirected networks were considered in Eq. (9), we set $\tilde{w}_{ij} = \tilde{w}_{ji}$.

The characteristic path length for the weighted networks is denoted by

$$L = \frac{1}{N(N-1)} \sum_i^N \sum_{j \neq i}^N d_{ij} \tag{10}$$

where d_{ij} is the shortest distance between the ith and jth neurons [17,18]. If the ith and jth neurons were directly connected, we defined an inverse of $\tilde{w}_{ij}$ as a distance between them, which means that $d_{ij} = 1/\tilde{w}_{ij}$. On the other hand, if the ith and jth neurons were not directly connected, we could not use the above definition. Then, in such a case, we used a different definition.

Now, let us explain how do we defined a distance between the ith and jth neurons which are not directly connected. In this case, there exist several paths from the ith neuron to the jth neuron. We assume here that there are M paths between the ith and jth neurons as shown in Fig. 1. In this case, we can denote the shortest distance d_{ij} as follows:

$$d_{ij} = \min_{k=1,2,\ldots,M} |P_{ij}(k)| \tag{11}$$

where $|P_{ij}(k)|$ is the kth path length between the ith and jth neurons. We also defined the distance between the mth and nth neurons which were directly connected and were included in the kth path $P_{ij}(k)$ as $l_{mn} = 1/\tilde{w}_{mn}$ (see Fig. 1). Using l_{mn}, the length of the kth path is defined

$$|P_{ij}(k)| = \sum_{(m,n)} l_{mn} \tag{12}$$

where (m,n) means that the summation of the right hand side is taken as $i, \ldots, m, n, \ldots, j$. Because the range of $\tilde{w}_{mn}$ is $[0,1]$, that of l_{mn} takes $[1, \infty]$. Then, it is possible to have $l_{mn} \to \infty$. In such a case, we defined the kth path length also goes to infinity.

The networks characterized by the measures are called the small-world networks. In the small-world networks, although L takes a small value which is as same as random networks, C takes a large value which is as same as regular (lattice) networks.

In addition, to investigate whether the neural network has the small-world property, it is necessary to compare with a regular network. In this paper, we defined a weight-averaged network as a regular network [20]. Namely, if $\bar{w}$ represents a weight in the regular network, $\bar{w}$ is written as

$$\bar{w} = \frac{1}{A} \sum_i^N \sum_{j>i}^N \tilde{w}_{ij} \tag{13}$$

where A is the number of connections whose weight $\tilde{w}_{ij}$ is larger than zero.

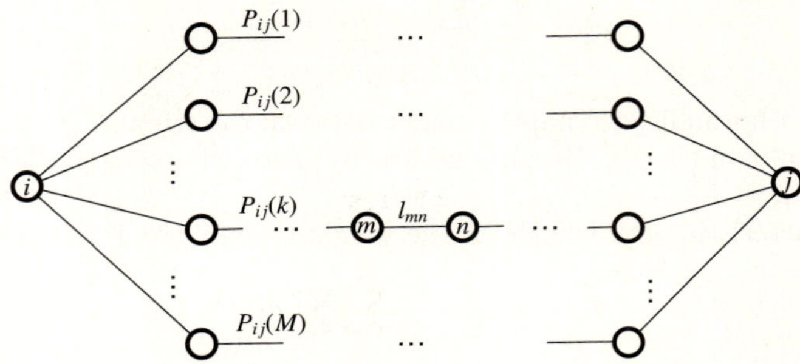

Fig. 1. This example shows that M paths exist between the ith and jth neurons. Circles and lines represent neurons and their connections, respectively. $P_{ij}(k)$ $(k = 1, 2, \ldots, M)$ represents the kth path between the ith and jth neurons. The variable l_{mn} is a distance between the mth and nth neurons which are directly connected and are included in the kth path.

In this paper, we described the clustering coefficient C and the characteristic path length L at t [sec] as $C_{NN}(t)$ and $L_{NN}(t)$. Furthermore, we described C and L of the regular network at t [sec] as $C_R(t)$ and $L_R(t)$, respectively. Thus, calculating $C(t) = C_{NN}(t)/C_R(t)$ and $L(t) = L_{NN}(t)/L_R(t)$, we investigated temporal evolution of the small-world properties of the neural network organized through the STDP from 0 to 600 [sec].

5   Results

Results are shown in Fig. 2. Each result was plotted at every 10 seconds. The step of the parameter μ was 0.1. All the results were averaged by 20 simulations.

From Fig. 2(a), in all cases of the parameter μ, the value of $C(t)$ decreases once from the initial condition. However, $C(t)$ recovers and increases, and then keeps about 1. In the case of $\mu = 0$, which corresponds to the additive STDP rule, the actual value of $C_{NN}(t)$ takes finally about 0.04, and the value of $C_R(t)$ takes as same as $C_{NN}(t)$, then, $C(t) \approx 1$. On the other hand, in the case of $\mu = 1$, which corresponds to the multiplicative STDP rule, $C_{NN}(t)$ takes finally about 0.08, and $C_R(t)$ also takes about 0.08. When μ is set to between 0 and 1, $C_{NN}(t)$ takes between 0.04 and 0.08. In addition, if μ becomes larger, the final value of $C_{NN}(t)$ gradually increases between 0.04 and 0.08. Then, the actual values of $C_{NN}(t)$ depend on the value of μ. However, comparing with $C_R(t)$, $C_{NN}(t)$ relatively takes a high value. Then, $C(t) \approx 1$ dose not depend on values of μ. According to these results, even if the parameter μ is set to any values, the neural network relatively remains many clusters like regular networks.

From Fig. 2(b), in the case of $\mu = 0$, $L(t)$ takes 1 at $t = 0$. The value of $L(t)$ rapidly decreases with the time from the initial condition, and then $L(t)$ takes about 0.004 at $t = 100$. After $t = 100$, $L(t)$ remains about 0.004. The actual value of $L_{NN}(t)$ takes about 3.08 at $t = 0$. From the initial condition, $L_{NN}(t)$ decreases

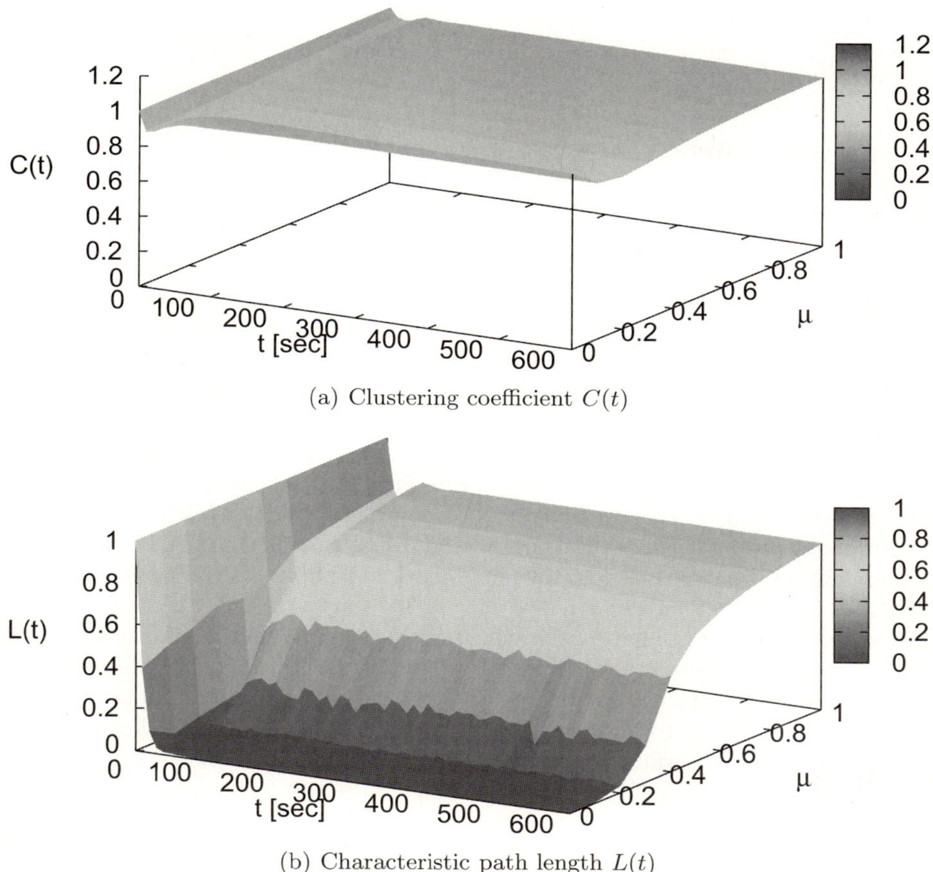

(a) Clustering coefficient $C(t)$

(b) Characteristic path length $L(t)$

Fig. 2. Temporal evolution of the small-world properties of the neural network organized through the STDP. Each result is plotted at every 10 seconds. The step of the parameter μ is 0.1. All the results were averaged by 20 simulations.

as time evolves, and $L_{NN}(t)$ finally takes about 2.18. In contrast, $L_R(t)$ takes same value with $L_{NN}(t)$ at $t = 0$. However, $L_R(t)$ increases with time evolution, and finally $L_R(t)$ takes about 450. Then, $L(t)$ becomes about 0.004. On the other hand, in the case of $\mu = 1$, $L(t)$ also takes 1 at $t = 0$. Then, $L(t)$ decreases as same as the case of $\mu = 0$ from the initial condition. However, in this case, $L(t)$ decreases to about 0.8. The actual value of $L_{NN}(t)$ takes about 3.08 in the initial condition and increases with the time. Finally, $L_{NN}(t)$ takes about 3.13. In addition, $L_R(t)$ takes the same value of $L_{NN}(t)$ in the initial condition. $L_R(t)$ increases with the time more than $L_{NN}(t)$, and finally takes about 3.9. Then, $L(t)$ takes about 0.8.

From both Figs. 2(a) and (b), if the parameter μ is set to a smaller value, it leads to $C(t) \approx 1$ and $L(t) \approx 0$. That is, the additive STDP rule leads the neural network to the highest small-world properties.

6 Conclusions

In this paper, we analyzed temporal evolution of the neural network structure organized through the STDP from viewpoints of complex networks. As a result, the clustering coefficient takes high value whenever any types of the STDP rule are applied to the neural network, that is, the neural network keeps many clusters with the learning. On the other hand, the characteristic path length takes different values depending on the types of the STDP rule. The characteristic path length becomes smaller as the STDP rule is less dependent on current synaptic weights. In other words, if the STDP rule is independent of synaptic weights, which corresponds to the additive STDP rule, the characteristic path length takes the smallest value. Then, when the dependence of the STDP rule on current synaptic weights is little, the neural network develops the highest small-world property. In addition, it was clarified that if the STDP rule is less dependent on synaptic weights, the small-world properties of the neural network becomes higher.

As future works, we analyze the self-organized neural network structure using different types of neurons and the STDP functions; for example, bursting neurons as the excitatory neurons. In addition, we analyze what kind of dynamics of the neural network leads to such complex structure, and the relationship between the structure and dynamics of the neural network. The research of TI is partially supported by Grant-in-Aid for Scientific Research (C) (No.20560352) from JSPS.

References

1. Abbott, L.F., Nelson, S.B.: Synaptic plasticity: taming the beast. Nature neuroscience supplement 3, 1178–1183 (2000)
2. Bi, G., Poo, M.: Synaptic Modifications in Cultured Hippocampal Neurons: Dependece on Spike Timing, Synaptic Strength and Postsynaptic Cell Type. The Journal of Neuroscience 18(24), 10464–10472 (1998)
3. Song, S., Miller, K.D., Abbott, L.F.: Competitive Hebbian learning through spike-timing-dependent synaptic plasticity. Nature Neuroscience 3(9), 919–926 (2000)
4. van Rossum, M.C.W., Bi, G.Q., Turrigiano, G.G.: Stable Hebbian Learning from Spike Timing-Dependent Plasticity. The Journal of Neuroscience 20(23), 8812–8821 (2000)
5. Rubin, J., Lee, D.D., Sompolinsky, H.: Equilibrium Properties of Temporally Asymmetric Hebbian Plasticity. Physical Review Letters 82(2), 364–367 (2001)
6. Watts, D.J., Strogatz, S.H.: Collective dynamics of 'small-world' networks. Nature 393, 440–442 (1998)
7. Hilgetag, C.C., Burns, G.A.P.C., O'Neill, M.A., Scannell, J.W., Young, M.P.: Anatomical connectivity defines the organization of clusters of cortical areas in the macaque monkey and the cat. Philosophical Transactions of The Royal Society B Biological Sciences 355(1393), 91–110 (2000)

8. Sporns, O., Tononi, G., Edelman, G.: Theoretical Neuroanatomy: Relating Anatomical and Functional Connectivity in Graphs and Cortical Connection Matrices. Cerebral Cortex 10(2), 127–141 (2000)
9. Shefi, O., Golding, I., Segev, R., Ben-Jacob, E., Ayali, A.: Morphological characterization of in vitro neuronal networks. Physical Review E 66, 021905 (2002)
10. Hodgkin, A.L., Huxley, A.F.: A quantitative description of membrane current and its application to conduction and excitation in nerve. Journal of Physiology 117, 500–544 (1952)
11. Troyer, T.W., Miller, K.D.: Physiological Gain Leads to High ISI Variability in a Simple Model of a Cortical Regular Spiking Cell. Neural Computation 9, 971–983 (1997)
12. Izhikevich, E.M.: Simple Model of Spiking Neurons. IEEE Transactions on Neural Networks 14(6), 1569–1572 (2003)
13. Izhikevich, E.M.: Polychronization: Computation with Spikes. Neural Computation 18, 245–282 (2006)
14. Gütig, R., Aharonov, R., Rotter, S., Sompolinsky, H.: Learning Input Correlations through Nonlinear Temporally Asymmetric Hebbian Plasticity. The Journal of Neuroscience 23(9), 3687–3714 (2003)
15. Barabási, A.L., Albert, R.: Statistical mechanics of complex networks. Reviews of Modern Physic 74(1), 47–97 (2002)
16. Newman, M.E.J.: The structure and function of complex networks. SIAM 45, 167–256 (2003)
17. Boccaletti, S., Latora, V., Moreno, Y., Chavez, M., Hwang, D.U.: Complex networks: Structure and dynamics. Physics Reports 424, 175–308 (2006)
18. Costa, L.D.F., Rodrigues, F.A., Travieso, G., Boas, R.R.V.: Characterization of complex networks: Asurvey of measurements. Advances in Physics 56(1), 167–242 (2007)
19. Saramäki, J., Kivelä, M., Onnela, J.P., KimoKaski, Kertész, J.: Generalizations of the clustering coefficient to weighted complex networks. Physical Review E 75(2), 027105 (2007)
20. Li, M., Fan, Y., Wang, D., Daqing Li, J.W., Di, Z.: Small-world effect induced by weight randomization on regular networks. Physics Letters A 364, 488–493 (2007)

Temporal Hebbian Self-Organizing Map for Sequences

Jan Koutník and Miroslav Šnorek

Computational Intelligence Group,
Department of Computer Science and Engineering,
Faculty of Electrical Engineering,
Czech Technical University in Prague
{koutnij,snorek}@fel.cvut.cz

Abstract. In this paper we present a new self-organizing neural network called Temporal Hebbian Self-organizing Map (THSOM) suitable for modelling of temporal sequences. The network is based on Kohonen's Self-organizing Map, which is extended with a layer of full recurrent connections among the neurons. The layer of recurrent connections is trained with Hebb's rule. The recurrent layer represents temporal order of the input vectors. The THSOM brings a straightforward way of embedding context information in recurrent SOM using neurons with Euclidean metric and scalar product. The recurrent layer can be easily converted into a stochastic automaton (Markov Chain) generating sequences used for previous THSOM training. Finally, two real world examples of THSOM usage are presented. THSOM was applied to extraction of road network from GPS data and to construction of spatio-temporal models of spike train sequences measured in human brain in vivo.

1 Introduction

In temporal sequences processing by neural networks recurrent connections utilize a context and form a short term memory similarly to sequential logical circuits [1]. It is of course possible to process temporal sequences using feed-forward neural networks only with satisfactory results but the context used through recurrent connections allows effective dealing with information stored in order of the input patterns or vectors.

In the field of self-organizing neural networks based on Kohonen's Self-organizing Map (SOM) [2] a couple of networks with recurrent connections were invented after the year 2000. The networks differ in fashion of incorporating the context as carefully described in [3].

Networks with single recurrent connection lead to a neuron itself only like Recurrent Self-organizing Map (RSOM) [4,5] and Temporal Kohonen Map (TKM) [6,7,8] represent a limited approach to the context utilization. Decoding of temporal and spatial part of the model created by the networks is not straightforward.

Another possibility is to use full context information stored in all network neurons in each neuron in further time step. The full context network such as

V. Kůrková et al. (Eds.): ICANN 2008, Part I, LNCS 5163, pp. 632–641, 2008.

Recursive SOM (RecSOM) [9] measures the context with Euclidean metric. A problem with dimensionality of the context proportional to a number of neurons in the network arises here.

The dimensionality problem can be solved by storing previous time step winner location like in SOM for Structured Data (SOMSD) [8]. The input of each neuron is enlarged by a number of neuron mesh indices. This causes the network to suffer from the dimensionality problem.

The most powerful variant of recurrent SOM is the Merging SOM (MSOM) [10], which contains two weight vectors. One stands for spatial input and second represents the context. The context in MSOM is calculated from weight vector of previous winner and its context. The MSOM combines ideas of TKM and SOMSD.

Our approach described in Section 2 is the most simple and straightforward recurrent extension of SOM. It respects the full context transferred to all neurons by full recurrent connections. The context is not processed by Euclidean metric like in RecSOM but with scalar product.

The paper is organized as follows. Section 2 contains a detailed description of THSOM. Section 3 shows an experiment with training THSOM with stochastic automaton output and extraction of the automaton back from trained THSOM and presents two experiments with real data processing using THSOM. Finally, conclusion is given in section 4.

2 Temporal Hebbian Self-Organizing Map (THSOM)

We introduce Temporal Hebbian Self-organizing Map (THSOM) [11] as the most simple SOM variant, which contains two maps – spatial map and temporal map. The temporal map is trained according to modified Hebbian rule. The spatial map uses standard Kohonen's training algorithm. THSOM utilizes whole context similarly to RecSOM but the context data are not measured by Euclidean metric and therefore THSOM suffers from dimensionality problem.

The THSOM architecture consists of one layer (grid) of neurons. The neurons are fully connected to input vector of dimension d, these connections form the *spatial map*. The neurons are connected to all neurons in the grid using recurrent temporal synapses (*temporal map*). The neurons contain two types of similarity measures, Euclidean metrics for measuring similarities in input space and scalar product for measuring similarities in activities of neurons in previous time step. THSOM architecture is depicted in Figure 1. Activation of THSOM neuron is calculated using the following equation:

$$v_i^{t+1} = (1 - \gamma)\left(1 - \sqrt{\frac{\sum_{j=1}^{D}(x_j^t - w_{ij}^t)^2}{D}}\right) + \gamma \sum_{k=1}^{n} y_k^t m_{ik}^t \qquad (1)$$

where v_i^{t+1} is activation (output) of i-th neuron in time $t+1$ (further time step), D is input vector dimension, x_j^t is j-th networks input in time t, w_{ij}^t is spatial weights for j-th input in time t, y_k^t is output of k-th neuron in time t, m_{ik}^t is

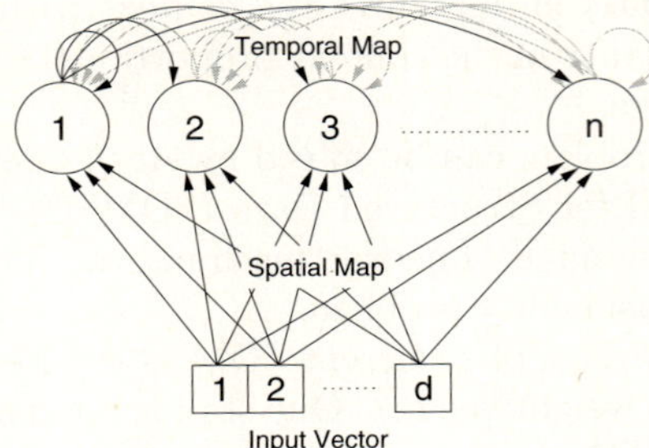

Fig. 1. THSOM Architecture. Neurons are depicted as circles. There are n neurons in one network layer. The input vector of dimension d is lead to the network via spatial map synapses. The spatial map makes full connection between the input vector and neurons. The temporal map makes a full connection among all neurons. Recurrent connections from one neuron to the other neurons are emphasized.

temporal weights for k-th neuron in time t (from neuron k to neuron i), n is number of neurons in network. Parameter γ controls balance between spatial and temporal maps. All neuron activations are normalized after the time step using equation:

$$y_i^{t+1} = \frac{v_i^{t+1}}{\max(v_i^{t+1})} \tag{2}$$

2.1 THSOM Training

THSOM is trained with sequences of input vectors. Training of a vector in the sequence involves three phases. First, the new BMU b for time $t+1$ is computed. Then, the spatial map is trained using SOM iterative training procedure with a respect to a neighborhood function. Finally, the temporal map is updated using the Hebb's rule.

Temporal weights of BMU b in time $t+1$ (after calculation of the new BMU) use modified discrete Hebbian learning rule according to the following formulas:

$$m_{bk}^{t+1} = \begin{cases} \min(\max(m_{bk}^t + \alpha(1 - m_{bk}^t + \beta), K_l), K_h) & \text{for } k = \text{argmax}_k(y_k^t) \\ \min(\max(m_{bk}^t - \alpha(m_{bk}^t + \beta), K_l), K_h) & \text{otherwise} \end{cases} \tag{3}$$

where y_k^t is an activation of neuron after previous time step, β and α control temporal synapses learning rate, α can start on 1.0 and is decreased in time for slow down oscillation around desired temporal weight value, K_l is a low boundary of temporal weights, usually ($K_l = 0$), K_h is a high boundary of temporal weights, usually ($K_h = 1$).

Input vectors should be in $\langle 0, 1 \rangle$ interval for balanced spatial and temporal map effect to BMU selection and therefore the proper function of THSOM neuron.

2.2   THSOM Function

THSOM function can be demonstrated on a simple example in which the network is fed with a sequence of two dimensional vectors generated as symbols by a Markov Chain (MC) using matrix $\mathbf{S}$. The MC has a transition matrix $\mathbf{A}$ and probabilities of states being initial states π.

$$\mathbf{S} = \begin{pmatrix} 0.0\ 0.0 \\ 0.0\ 1.0 \\ 1.0\ 1.0 \\ 1.0\ 0.0 \end{pmatrix} \qquad \mathbf{A} = \begin{pmatrix} 0.2\ 0.8\ 0.0\ 0.0 \\ 0.0\ 0.2\ 0.8\ 0.0 \\ 0.0\ 0.0\ 0.2\ 0.8 \\ 0.8\ 0.0\ 0.0\ 0.2 \end{pmatrix} \qquad \pi = \begin{pmatrix} 1.0 \\ 0.0 \\ 0.0 \\ 0.0 \end{pmatrix}$$

In each step a symbol represented by a row vector from $\mathbf{S}$ is added Gaussian noise with $\sigma = 0.02$.

After training, following spatial ($\mathbf{W}$) and temporal ($\mathbf{M}$) matrices were obtained:

$$\mathbf{W} = \begin{pmatrix} 0.99\ 0.0 \\ 0.0\ \ 1.0 \\ 1.0\ \ 0.99 \\ 0.01\ 0.0 \end{pmatrix} \qquad \mathbf{M} = \begin{pmatrix} 0.18\ 0.0\ \ 0.82\ 0.0 \\ 0.0\ \ 0.18\ 0.0\ \ 0.82 \\ 0.0\ \ 0.75\ 0.25\ 0.0 \\ 0.74\ 0.0\ \ 0.0\ \ 0.26 \end{pmatrix}$$

Figure 2 displays training of the model in the training phase. As we can see, the spatial map contains neurons put to same positions as positions of input vectors in $\mathbf{S}$ in the sequence. Spatial map $\mathbf{M}$ contains values close to 0.2 on the main diagonal. Values close to 0.8 are placed on column for a neuron – state from which a transition to actual neuron expressed by the matrix row is the most probable. The temporal weight matrix can be transformed into transition probability matrix of a Markov Chain using transposition and normalization as described in [12]. MC with randomly initialized transition matrix used for THSOM training and extraction of the MC back from temporal map example is given in Section 3.1.

3   Experimental Results

3.1   Extraction of Markov Chain from THSOM

Idea of extraction of Markov Chain from the temporal map was introduced in [13] and [11]. Equation 3 causes the temporal map to aggregate densities of transitions between symbols of an input sequence. Figure 3 contains a schema of mutual relationship of a Markov Chain and THSOM. MC with a transition matrix $\mathbf{A}$ generates a sequence of symbols, which is encoded to real vectors and presented to the THSOM in time steps. The THSOM builds spatial and temporal maps. Afterwards, the temporal map is converted back to matrix $\mathbf{A_2}$ similar to matrix $\mathbf{A}$ according to Manhattan metric.

In experimental results we used a Markov Chain to generate sequences that were presented to the network input, each vector from the sequence at one time step. The MC generates a symbol in each state. The symbols are generated

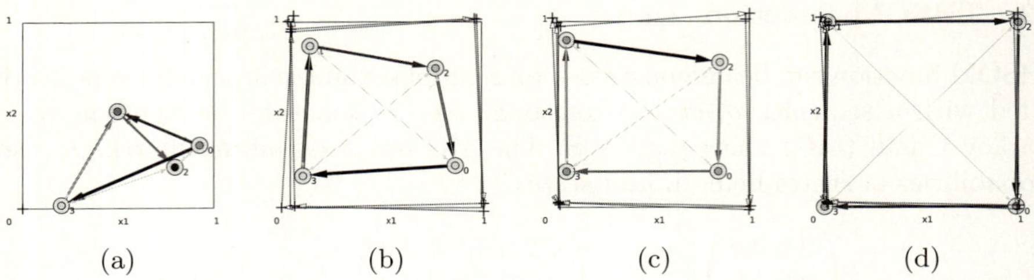

Fig. 2. Figures display building of spatio-temporal model. THSOM neurons are depicted as circles. The spatial map weights are represented by solid arrows. Arrow strength represents temporal weight value. Neuron inner circle color represents self-weight. Crosses represent input vectors. Consequences between input vectors are represented by dashed arrows. In each sub-figure, last 10 input vectors are included. Sub-figures display spatial and temporal maps after 0 – initialized (a), 10 (b), 100 (c) and 200 (d) steps.

equidistant in two dimensional Euclidean space, 2500 symbols in the sequence. Afterwards, we extracted the transition matrix from temporal weights and measure its distance to original transition matrix of the MC. We have tested reconstruction of transition matrices for Markov Chains with from 3 up to 16 states. Here is the original transition matrix of the MC with 9 states:

$$A = \begin{pmatrix} 0.0 & 0.0 & 0.1 & 0.0 & 0.0 & 0.1 & 0.0 & 0.3 & 0.5 \\ 0.1 & 0.1 & 0.1 & 0.0 & 0.0 & 0.0 & 0.2 & 0.2 & 0.3 \\ 0.2 & 0.1 & 0.0 & 0.0 & 0.0 & 0.2 & 0.2 & 0.1 & 0.2 \\ 0.1 & 0.1 & 0.0 & 0.0 & 0.2 & 0.1 & 0.0 & 0.0 & 0.5 \\ 0.0 & 0.2 & 0.2 & 0.1 & 0.1 & 0.1 & 0.1 & 0.0 & 0.2 \\ 0.1 & 0.2 & 0.0 & 0.1 & 0.1 & 0.0 & 0.2 & 0.2 & 0.1 \\ 0.1 & 0.0 & 0.1 & 0.1 & 0.0 & 0.1 & 0.2 & 0.1 & 0.3 \\ 0.0 & 0.2 & 0.0 & 0.0 & 0.2 & 0.2 & 0.0 & 0.2 & 0.2 \\ 0.1 & 0.2 & 0.0 & 0.1 & 0.0 & 0.1 & 0.1 & 0.1 & 0.3 \end{pmatrix}$$

Transition matrix extracted from the THSOM A_2 (Manhattan distance to A_1 divided by number of rows is 0.32):

$$A_2 = \begin{pmatrix} 0.0 & 0.0 & 0.054 & 0.0 & 0.0 & 0.05 & 0.0 & 0.325 & 0.571 \\ 0.039 & 0.128 & 0.064 & 0.0 & 0.0 & 0.0 & 0.238 & 0.185 & 0.346 \\ 0.431 & 0.236 & 0.0 & 0.0 & 0.0 & 0.251 & 0.066 & 0.0 & 0.017 \\ 0.079 & 0.084 & 0.0010 & 0.0 & 0.159 & 0.219 & 0.0 & 0.0 & 0.458 \\ 0.0 & 0.358 & 0.124 & 0.145 & 0.122 & 0.049 & 0.043 & 0.0 & 0.16 \\ 0.074 & 0.258 & 0.0010 & 0.09 & 0.116 & 0.0 & 0.241 & 0.098 & 0.063 \\ 0.097 & 0.0 & 0.137 & 0.088 & 0.0 & 0.05 & 0.318 & 0.136 & 0.175 \\ 0.0 & 0.168 & 0.0010 & 0.0 & 0.194 & 0.224 & 0.0 & 0.229 & 0.183 \\ 0.118 & 0.245 & 0.0 & 0.091 & 0.0 & 0.064 & 0.066 & 0.109 & 0.308 \end{pmatrix}$$

3.2   Clustering of GPS Traces

THSOM was applied to preprocessing and reconstruction of road mesh from GPS traces. The data were collected using in-car GPS receiver and consist of sequences of GPS measurement (2 dimensional vectors of latitude and longitude considered as Cartesian coordinates).

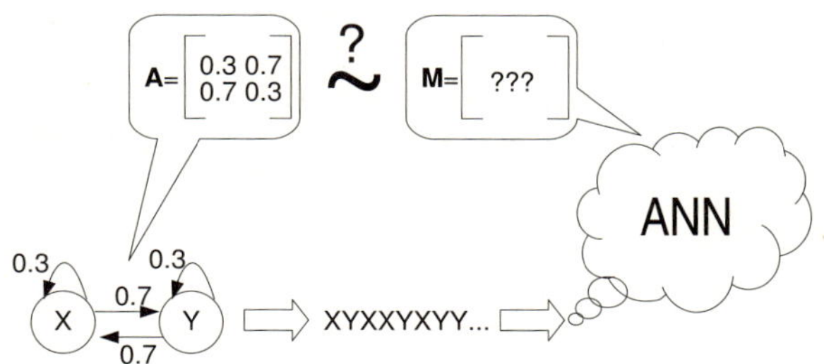

Fig. 3. THSOM is trained by symbol sequence generated by a Markov Chain. Afterwards, the MC will be extracted back from the neural network and compared to original MC.

Number of neurons was obtained experimentally. Less amount of neurons cause quantization in the input space and creates less and imprecise placed routable points. Higher number of neurons may put a cluster of neurons instead of a single point.

3.3   Spike Train Modelling

Parkinson disease surgery involves electric stimulation of deep brain cell called *subthalamic nucleus* (STN) [14]. Electrodes are put into human brain *in vivo* in order to reach STN cell under local anesthesia. During the surgery, signals detected on the electrodes are recorded in order to localize the STN. Sequences are composed from superposed patterns (pulses) in time called *spikes* The problem is to detect neurons in a background activity. It means to distinguish, which type of spike sequences occur on electrodes, because STN sends out characteristical sequences of spikes. Sequence of spikes is called *spike train.*

Spike trains are basically distinguished into three categories [15]: *irregular,* randomly occurring spikes, *burst* of spikes – period of relatively periodic spikes with a high frequency and *oscillation* at lower frequency.

Normally, features like *inter-spike interval* (ISI), *mean firing frequency* and *coefficient of variance* are extracted from spike trains and the whole spike sequence is classified according the features.

As we can see, classical analysis of spike trains involves statistics measured from recorded and reprocessed spike sequences globally. A spike train is considered as one sequence of either bursts or oscillations. Occurrence of bursts and oscillations influence first order statistics obtained from data. Statistics is used to model data even if the sequence is mixed from burst and oscillations but first

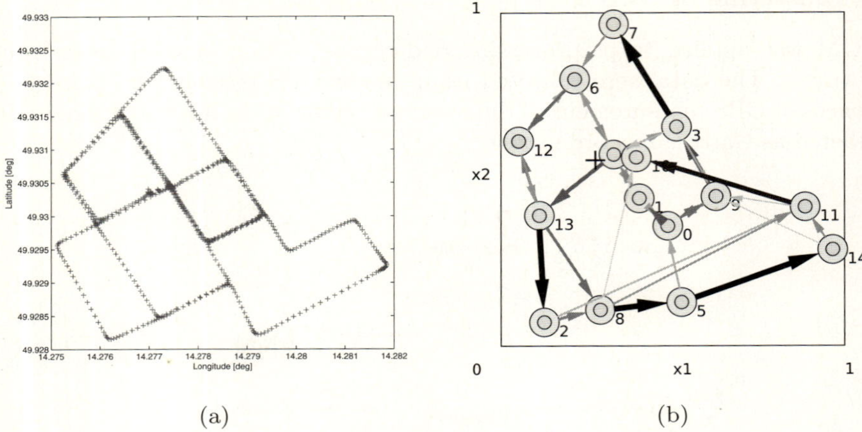

Fig. 4. GPS coordinates as measured by GPS receiver (a). Spatio-temporal map constructed by THSOM (b). Circles represent neurons – centroid. Arrows represent temporal weights between neurons. Weights lower than a threshold setup to 0.05 were removed from the plot and should not be considered as roads that can be used for routing. We can see that some roads are mapped in both directions (e.g., between neurons 12 and 13) and some road were recognized as single directional (e.g., between neurons 13 and 2).

order statistics and shape of distribution might not supply enough knowledge about the analyzed spike train.

We utilize THSOM model for construction of automaton model from spike trains. A Markov Chain is extracted from THSOM and from its visualization the spike train can be described as follows. If the transitions from state to state are non-zero numbers varying around $1/N$ where N is number of states, it points to an irregular spike train. Highly probable self-connections point to periodic discharge. The burst can be recognized as an oscillator represented by highly probable self-connection with the spatial weight of the neuron placed into the small ISI.

In the experiment THSOM models consisting from 3 to 20 neurons were created (different model for each spike train). Best results were achieved with networks having 5 neurons.

Table 1 contains features of signals obtained by conventional methods, whereas Table 2 contains features read from THSOM models visualization.

In experiments, THSOM models for each sequence of ISI have been created. THSOM networks with from 3 to 9 neurons were used. Best models were achieved those with 5 neurons. For each spike train 20 models were trained. Since THSOM is a stochastic method the resulting models are not the same after each of 20 training runs. The most frequent model was examined and additional features to those in Table 1 were judged from the Markov chains extracted from the THSOM temporal and spatial maps.

Table 1. Features extracted from ISI spike trains by conventional statistical methods. Features collected from 10 spike train data sets: number of spikes detected, mean ISI interval and its standard deviation, coefficient of variance, percentage of ISI shorter than 3 ms, mean frequency and its standard deviation, occurrence of a burst.

Dataset	6	9	11	13	15	17	18	21	26	27
# of spikes	73	145	105	193	217	207	107	387	281	201
$\overline{\text{ISI}}$ (ms)	123.4	68.4	92.9	51.4	45.6	45.2	84.4	25.3	34.8	49.5
σ ISI (ms)	200.9	56.2	121.4	66.0	71.0	89.1	199.1	18.9	32.6	42.7
CV	1.6	0.8	1.3	1.3	1.6	2.0	2.4	0.8	1.0	0.9
< 3 ms (%)	0	0	0	3.1	0	0	6.7	0.5	20.0	13.0
$\bar{f}$ (Hz)	8.1	14.6	10.8	19.5	22.0	22.1	11.9	39.5	28.7	20.2
σ f (Hz)	5.0	17.8	8.2	15.1	14.1	11.2	5.0	53.0	30.7	23.4
burst true	1	0	1	0	1	1	1	0	0	0

Each automaton in the figures has a number in all states. The number is the neuron position in the spatial one dimensional map and represents a result of trained spatial map.

Figure 5 contains raw spikes isolated by method described in this section (a) and Markov chains extracted from trained THSOM networks with 5 neurons (b). The Markov chain plot contains spatial position (spatial THSOM weight) in each state (neuron). The states are ordered as in one dimensional input space of inter-spike intervals from left to right. Transitions with a probability smaller than 0.1 are not included in the graphs. Transition probability is expressed by an opacity of the arrow representing the transition. There are 3 figures for selected data sets number 6, 9 and 11. All features extracted from the spike trains are summarized in Table 2.

We can see how Table 2 completed Table 1. Irregular spike discharges identified by Markov Chains lacking an oscillatory structure were found in spike train datasets labeled with false burst by conventional spike analysis method.

Figure 5(b) contains visualization of Markov Chain extracted from the TH-SOM trained with one of the spike trains.

Table 2. Features extracted from ISI spike trains by THSOM: mean burst ISI, mean burst frequency, mean ISI between two burst ($\overline{\text{BL}}$) and whether the spike train is irregular

Dataset	6	9	11	13	15	17	18	21	26	27
burst $\overline{\text{ISI}}$ (ms)	49	6	30	9	10	10	22	20	3	2
burst $\bar{f}$ (Hz)	20	167	33	111	100	100	46	50	333	500
$\overline{\text{BL}}$ (s)	1.46		0.45	0.1	0.19	0.37	1.73			
irregular	0	1	0	0	0	0	0	1	1	1

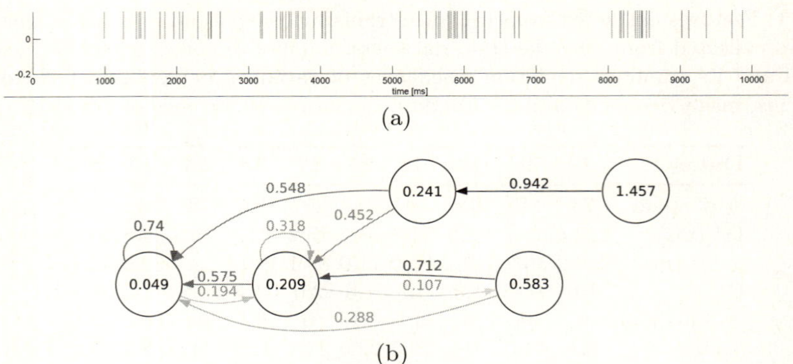

(a)

(b)

Fig. 5. Preprocessed spike train (a) and constructed Markov chain (b). We can see that there are two oscillators with mean frequencies of 20 and 4.8 Hz, which point to bursts. The bursts are separated with ISI with mean delay of 1.45 s. Delay between bursts of 0.68 s is less frequent.

4   Conclusion

We have introduced a recurrent extension of Self-organizing Map, which uses a complete context – activation of neurons in the previous network state for modelling of temporal sequences. The network has two layers of connections – spatial map and temporal map trained by Hebbian learning.

Separation of both maps allows easy extraction of Markov Chain from trained THSOM. We have shown experimentally that accumulated probabilities of transitions among input vectors generated by a Markov Chain can be converted back to transition Matrix of the Markov Chain.

We have described experiments with THSOM applied to reconstruction of road mesh from GPS records. The network was used to place neurons in spatial map into the crossroad centers. The temporal map represents the connections among the crossroads.

THSOM was applied to modelling of spike trains recorded in the human brain. Conventional statistical features were accompanied with description obtained from inspection of spatio-temporal model found in THSOM after training with sequences of time interval between spikes. Modelling in time domain has shown that conventional features like mean inter-spike interval tend to blur information stored in different parts of a single spike train.

Acknowledgement

This work has been supported by the research program "Transdisciplinary Research in the Area of Biomedical Engineering II" (MSM6840770012) sponsored by the Ministry of Education, Youth and Sports of the Czech Republic.

References

1. Schmidhuber, J., Bengio, Y.: Evaluating long-term dependency benchmark problems by random guessing (1997)
2. Kohonen, T.: Self-Organizing Maps, 3rd edn. Springer, Heidelberg (2001)
3. Hammer, B., Micheli, A., Neubauer, N., Sperduti, A., Strickert, M.: Self organizing maps for time series. In: Proceedings of WSOM 2005, pp. 115–122 (2005)
4. Varsta, M., Heikkonen, J., del R., Millan, J.: Context learning with the self organizing map. In: Proceedings of WSOM 1997, Workshop on Self-Organizing Maps, Espoo, Finland, June 4–6. Helsinki University of Technology, Neural Networks Research Centre, pp. 197–202 (1997)
5. Koskela, T., Varsta, M., Heikkonen, J., Kaski, K.: Temporal sequence processing using recurrent SOM. In: Proceedings of the 2nd International Conference on Knowledge-Based Intelligent Engineering Systems, Adelaide, Australia, vol. 1 (1998)
6. Varsta, M., Heikkonen, J., Lampinen, J.: Analytical comparison of the temporal kohonen map and the recurrent self organizing map. In: Verleysen, M. (ed.) European Symposium on Artificial Neural Networks (ESANN), pp. 273–280. D-facto Publications (2000)
7. Varsta, M.: Self-Organizing Map. In: Sequence Processing. PhD thesis, Helsinki University of Technology (2002)
8. Voegtlin, T.: Recursive self-organizing maps. Neural Netw. 15(8-9), 979–991 (2002)
9. Voegtlin, T.: Context quantization and contextual self-organizing maps (2000)
10. Strickert, M., Hammer, B.: Neural gas for sequences. In: Yamakawa, T. (ed.) Proceedings of the Workshop on Self-Organizing Networks (WSOM), Kyushu Institute of Technology, pp. 53–58 (2003)
11. Koutník, J.: Inductive modelling of temporal sequences by means of self-organization. In: Proceeding of Internation Workshop on Inductive Modelling (IWIM 2007), Ljubljana, CTU in Prague, pp. 269–277 (2007)
12. Koutník, J.: Self-organizing Maps for Modeling and Recognition of Temporal Sequences. PhD thesis, Czech Technical University in Prague (2008)
13. Koutník, J., Šnorek, M.: Neural network generating Hidden Markov Chain. In: Adaptive and Natural Computing Algorithms - Proceedings of the International Conference in Coimbra, pp. 518–521. Springer, Wien (2005)
14. Obeso, J.A., Guridi, J., Obeso, J.A., DeLong, M.: Surgery for parkinson's disease (editorial). Journal of Neurology, Neurosurgery and Psychiatry 62(1), 2–8 (1997)
15. Rodriguez-Oroz, M., Rodriguez, M., Guridi, J., Mewes, K., Chockkman, V., Vitek, J., DeLong, M.R., Obeso, J.A.: The subthalamic nucleus in Parkinson's disease: somatotopic organization and physiological characteristics. Brain 124(9), 1777–1790 (2001)

FLSOM with Different Rates for Classification in Imbalanced Datasets

Iván Machón-González and Hilario López-García

Universidad de Oviedo. Escuela Politécnica Superior de Ingeniería. Departamento de Ingeniería Eléctrica, Electrónica de Computadores y Sistemas. Edificio Departamental 2. Zona Oeste. Campus de Viesques s/n. 33204 Gijón/Xixón (Asturies). Spain

Abstract. There are several successful approaches dealing with imbalanced datasets. In this paper, the Fuzzy Labeled Self-Organizing Map (FLSOM) is extended to work with that type of data. The proposed approach is based on assigning two different values in the learning rate depending on the data vector membership of the class. The technique is tested with several datasets and compared with other approaches. The results seem to prove that FLSOM with different rates is a suitable tool and allows understanding and visualizing the data such as overlapped clusters.

1 Introduction

The problem consists of a dataset composed of two classes. One class has a great number of instances and it is known as the majority or negative class; whereas the other class is the minority or positive class since it collects a few number of samples. The membership to a certain class of each instance is known beforehand. When presenting a new testing data sample, the aim of the problem is to estimate which class it belongs to. This estimation uses the values of the attributes or variables of the sample. The main drawback is that a simple classifier for the most part tends to classify almost all the cases as negative.

There are several ways to solve the problem and many of them can be used combined to take advantage of each one. Two main approaches could be considered to deal with imbalanced datasets: sampling methods and approaches based on the algorithm. Within the sampling methods, two directions are possible: undersampling and oversampling.

Undersampling consists of reducing the number of samples of the negative class until both classes have approximately the same size. The most informative samples must be selected discarding only those ones which do not contribute any useful information. However, downsizing the negative class usually produces a loss of information and care must be taken in this key issue. Different techniques are possible in undersampling [1,2,3,4].

Oversampling approach [5,6] adds more samples to the positive class until both classes have approximately the same size. Repetition of the available instances could be considered but it would entail great risk of overfitting. Direct

V. Kůrková et al. (Eds.): ICANN 2008, Part I, LNCS 5163, pp. 642–651, 2008.

replacement of the old samples can be done. Also, new samples can be added to the positive class by generating interpolated data [6]. Obviously, a combination of undersampling and oversampling is possible [6,7].

The techniques based on the algorithm are designed to improve the classification taking advantage of its intrinsic performance. The imbalanced problem has been treated with neural networks of perceptrons [3], support vector machines [2,7,8], a variation of RBF networks extracting fuzzy rules [9] and assigning rates for the different classes [10] and prototypes [11].

This paper aims to improve the classification of instances in the typical two-classes problem of imbalanced datasets using a modification of the original FLSOM algorithm. The proposed approach is based on assigning two different values in the learning rate to the prototype vectors depending on the labeling data vector associated with the training data vector.

A large field of applications can be mentioned. Some examples of imbalanced applications are rarely seen diseases [12], text categorization [13], detection of fraudulent telephone calls [14] and oil spills detection [15]. The SOM algorithm is a useful technique for process monitoring and visualization and there are many applications in data mining [16] and in engineering [17,18].

The paper contains brief descriptions of FLSOM algorithm in section 2 and the metric to evaluate the classifier in section 3. The proposed modification of FLSOM is presented and tested in sections 4 and 5.

2   FLSOM Algorithm

FLSOM algorithm [19] is a version of SOM algorithm [20,17] where the prototype vectors are influenced by the labeling data vectors that define the clusters of the dataset. The classification task influences the values of the prototype vectors and both of them take place at the same time during the training.

Using SOM for classification tasks is an important feature to identify the clusters of the input data space, their relationships between each other and with the process variables. A classic method of classification after training consists of carrying out an assignment of each prototype vector to a certain cluster obtaining a new component plane. In this case, the clustering process is composed of two phases [21]. In the first phase, the SOM network is trained whereas in the second phase a partitive clustering algorithm, e.g. K-means, is applied to obtain several clustering structures of several numbers of clusters since the optimum number of clusters is unknown. The optimum clustering structure is chosen by means of a clustering validation index. This procedure is useful when the number of clusters and the classification (fuzzy or crisp distribution) of each data vector are not known beforehand [22,23]. However, these methods have the drawback of not influencing the training since they happen after it and don't modify the values of the prototype vectors.

This can be corrected using FLSOM. In this algorithm, the classification task influences the values of the prototype vectors and both of them take place at the same time during the training. In this way, FLSOM can be considered as a semisupervised algorithm. The training algorithm is based on an energy cost function of the SOM (E_{SOM}) proposed by Heskes in [24]. The cost function of equation (1) includes a term (E_{FL}) that represents the labeling or classification error of the prototype vectors y of the classification map with regard to the probabilistic vectors $\mathbf{x}$ supplied by the training dataset.

$$E_{\mathrm{FLSOM}} = (1 - \beta)\, E_{\mathrm{SOM}} + \beta\, E_{\mathrm{FL}}, \tag{1}$$

where $\beta \in [0, 1]$ reaches a compromise between E_{SOM} or the location of the prototype vectors $\mathbf{w}$ close to the data vectors $\mathbf{v}$ and E_{FL} or the approximation of the labeling prototype vectors y to the labeling data vectors $\mathbf{x}$.

Each labeling data vector $\mathbf{x}$, which is associated with a numerical data vector $\mathbf{v}$, is assigned in a fuzzy way according to a probabilistic membership to the clusters of the dataset. This stage is critical in the data preprocessing, but obviously the algorithm can be applied to crisp distributions.

A Gaussian kernel in the input data space is included within term E_{FL} so that the prototype vectors $\mathbf{w}$ close to the data vectors $\mathbf{v}$ determine the classification task. It is formulated in equation (2) with γ as kernel radius.

$$g_\gamma(\mathbf{v}, \mathbf{w_i}) = \exp\left(-\frac{\|\, \mathbf{v} - \mathbf{w_i}\, \|^2}{2\gamma^2}\right), \tag{2}$$

The learning rules are obtained by means of derivation of energy cost function (1) with respect to the numerical and labeling prototype vectors, $\mathbf{w_i}$ and $\mathbf{y_i}$, respectively. These learning rules are expressed in equations (3) and (4) considering using the squared Euclidean metric in the algorithm.

$$\Delta \mathbf{y_i} = \alpha(t)\, \beta\, g_\gamma(\mathbf{v}, \mathbf{w_i})\, (\mathbf{x} - \mathbf{y_i}), \tag{3}$$

$$\Delta \mathbf{w_i} = \alpha(t)\, (1 - \beta) \exp\left(-\frac{\|\, \mathbf{r}_i - \mathbf{r}_{i*}\, \|^2}{2\sigma(t)^2}\right)(\mathbf{v} - \mathbf{w_i}) \tag{4}$$

$$-\alpha(t)\, \beta\, \frac{1}{4\gamma^2}\, g_\gamma(\mathbf{v}, \mathbf{w_i})\, (\mathbf{v} - \mathbf{w_i})\, \|\, \mathbf{x} - \mathbf{y_i}\, \|^2$$

where $\alpha(t)$ is the learning rate, $\sigma(t)$ is the kernel radius and decreases monotonically and linearly, $\mathbf{r}_i$ and $\mathbf{r}_{i*}$ are the lattice coordinates of the updated neuron i and the winning neuron i^*, respectively.

3   Classifier Evaluation

Regarding classification of samples distributed in two classes, four different classes can be distinguished. TP and TN are the cases where the classification is correct dealing with positive and negative samples, respectively. FN is the case of a sample which is estimated as negative when is actually positive. Following the same reasoning, FP corresponds to the case of an actual negative

Table 1. Confusion matrix of the two-classes problem

	Estimated Positive	Estimated Negative
True Positive	True Positive (TP)	False Negative (FN)
True Negative	False Positive (FP)	True Negative (TN)

sample which is predicted as positive. Table 1 describes these four cases and is known as the confusion matrix.

The g-means metric [1] defined in (5) is usually used to measure the accuracy of the classification, where acc^+ and acc^- are commonly known as sensitivity and specificity, respectively. The sensitivity is related to the accuracy on the positive classes, whereas the specificity shows the accuracy on the estimation of the negative classes.

$$g = \sqrt{acc^+ \cdot acc^-} \tag{5}$$

$$acc^+ = \frac{TP}{TP + FN} \tag{6}$$

$$acc^- = \frac{TN}{TN + FP} \tag{7}$$

Calculating and distinguishing both values using the classification model is very important since poor models tend to have a good specificity but a bad sensitivity. The aim of g-means metric is to maximize the accuracy on each of the two classes by means of keeping them well balanced. However, strengthening one of those factors can be interesting in some problems. For example, detecting all the real cases of sickness is more important than diagnosing a healthy case as sick. In this case, the sensitivity must be as close to 1 as possible.

4   Modification of the Algorithm

The application of the original FLSOM algorithm would entail an unequal repartition of the number of the map units. A typical characteristic of the self-organizing algorithm is to employ a number of neurons proportional to the number of samples of the class in the search of the probability distribution of the input data. This should imply a poor positive class mapped with a small number of units and the sensitivity of the classifier would be low. For this reason, the original algorithm should be modified. In this paper, the proposed modification is based on assigning different rates to the prototype vectors in each iteration. Two values are possible depending on the labeling data vector $\mathbf{x}$ associated with the data vector $\mathbf{v}$ which is input in each iteration.

In this paper, learning rate $\alpha(t)$ is formulated according (8), where α_1 is a positive constant, t is the time step and t_{max} is the maximum number of time steps.

$$\alpha(t) = \frac{C\,\alpha_1}{99\frac{t}{t_{max}} + 1} \tag{8}$$

where $C = K^+$ if the labeling data vector $\mathbf{x}$ belongs to the positive class, otherwise $C = 1$. The values of K^+ must be adjusted obtaining a balance between sensitivity and specificity for each dataset. The modification consists of substituting the learning rate defined in (8) into the learning rules of FLSOM indicated in (3) and (4).

5   Experimental Testing

Nine datasets were used to check the algorithm performance in comparison with a Support Vector Machines (SVM) application [8], where SMOTE oversampling [6] and different error costs for different classes are considered. Furthermore, six datasets were considered to check the algorithm ability compared with a Paralell Perceptrons (PPs) implementation [3] that is based on an undersampling procedure. All the datasets were extracted from UCI repository of machine learning databases [25]. In both cases, the comparison was according to the g-means metric.

All the datasets have been randomly divided into two subsets, training and testing data. Training dataset is composed of 70% percent original samples and the rest of them (30%) belong to the testing dataset. The imbalance ratio of each dataset (ratio of negative to positive samples) has been preserved in both training and testing sets. The same proportion is taken in [8], whereas in [3] the size ratio between training and test is 90%-10%. Table 2 shows all these details. The number after the dataset name indicates the class considered as positive in those datasets that have more than 2 classes.

The original data vectors $\mathbf{v}$ were normalized according (9) to obtain a normalized set of data vectors $\mathbf{v_n}$; where $\mathbf{v}_j$, μ_j, σ_j are the original data, the mean

Table 2. Datasets used for experimental testing

Dataset	Variables	Negative samples (training)	Positive samples (training)	Imbalance ratio (training)	Negative samples (test)	Positive samples (test)	Imbalance ratio (test)
abalone19	7	2902	23	126	1243	9	138
car3	6	1162	49	24	497	20	25
glass7	9	130	21	6	55	8	7
hepatitis	18	66	14	5	27	5	5
hypothy3	21	1795	56	32	769	23	33
letter26	16	13487	514	26	5779	220	26
segment1	19	1386	231	6	594	99	6
sick	21	1702	149	11	729	63	12
soybean12	35	363	31	12	155	13	12
cancer	9	311	168	2	133	71	2
ionosphere	34	158	89	2	67	37	2
vowel0	12	630	63	10	270	27	10
vehicle1	18	444	149	3	190	63	3
diabetes	8	350	188	2	150	80	2

Table 3. Training parameters

Dataset	K^+	α_1	σ_{t_0}	$\sigma_{t_{max}}$	Map units	No. epochs
abalone19	30	0.1	4	1	30x30	100
car3	1	1	4	1	25x25	100
glass7	10	5	2	1	9x9	200
hepatitis	4	2.5	1	1	6x6	200
hypothy3	20	1	4	1	30x30	100
letter26	10	0.5	6	0.75	50x50	200
segment1	1	1	4	1	28x28	100
sick	6	1	4	1	30x30	100
soybean12	5	5	2	1	14x14	200
cancer	4	5	2	1	16x16	100
ionosphere	5	5	2	1	11x11	200
vowel0	3	5	3	1	18x18	100
vehicle1	2	5	3	1	17x17	200
diabetes	2	5	4	1	16x16	200

Table 4. Results. FLSOM versus SVMs.

	SVM			FLSOM		
	acc^+	acc^-	g-means	acc^+	acc^-	g-means
abalone	0.808	0.687	0.745	0.778	0.784	0.781
car	0.97	1	0.985	1	0.930	0.964
glass	0.808	1	0.899	0.875	0.982	0.927
hepatitis	0.708	0.833	0.768	0.800	0.852	0.826
hypothy	0.957	0.96	0.958	0.957	0.969	0.963
letter	0.997	0.966	0.982	0.982	0.956	0.969
segment	0.959	0.998	0.978	0.990	0.978	0.984
sick	0.865	0.874	0.870	0.857	0.886	0.872
soybean	1	1	1	1	0.987	0.994

value and the standard deviation of variable j, respectively. This data transformation leads to a normalized distribution with zero mean and unitary variance. In this way, all the variables are treated by the neural network with the same importance.

$$\mathbf{v}_{\mathbf{n}j} = \frac{\mathbf{v}_j - \mu_j}{\sigma_j} \tag{9}$$

Table 3 shows the parameters of the algorithm during for training. The map size or the number of prototype vectors of the FLSOM has been chosen as the half number of the training data vectors, except for the *letter* dataset which has been reduced due to the computing cost. Parameter β was chosen equal to 0.5 taking an intermediate weight between the approximation of the prototype vectors to the data vectors and the classification of the labeling prototype vectors. Parameter γ represents the kernel radius in the input space and it was considered

Table 5. Results. FLSOM versus PPs

	PP			FLSOM		
	acc^+	acc^-	g-means	acc^+	acc^-	g-means
cancer	0.972	0.961	0.966	0.972	0.977	0.975
ionosphere	0.718	0.943	0.823	0.703	0.970	0.826
glass	0.86	0.975	0.916	0.875	0.982	0.927
vowel0	0.881	0.988	0.933	1	0.970	0.985
vehicle1	0.611	0.859	0.725	0.778	0.658	0.715
diabetes	0.614	0.827	0.712	0.713	0.713	0.713

as a constant equal to 1 because it seems approximately the appropriate value for a normalized input data distribution with zero mean and unitary variance. This parameter is very sensitive and low values can produce maps that do not preserve the topographic property of the SOM. The radii $\sigma(t)$ decrease monotonically and linearly. The initial value σ_{t_0} is proportional to the map size since it must be high enough to allow the cooperative learning at the beginning of the training. The weights or values of the prototype vectors (w_i and y_i) were randomly initialized. The training was stopped at the number of epochs specified in Table 3 since the training error has an overall decreasing dynamic as in [3]. However, in [3] the number of epochs is 250 in all cases. The cost ratio is considered as the inverse of the imbalance ratio in [8]. In our work the value K^+ depends on the imbalance ratio but it deserves an adjustment to balance sensitivity and specificity.

The results are presented in tables 4 and 5 comparing the proposed algorithm with SVMs [8] and PPs [3], respectively. The algorithm is evaluated according g-means metric, but also specificity acc^- and sensitivity acc^+ are indicated. The FLSOM approach has a higher g-means metric than the SVM approach in six of a total number of nine datasets. Also it outperforms the rest of types of SVMs in four of the nine datasets, as it can be seen in [8]. Regarding the comparison

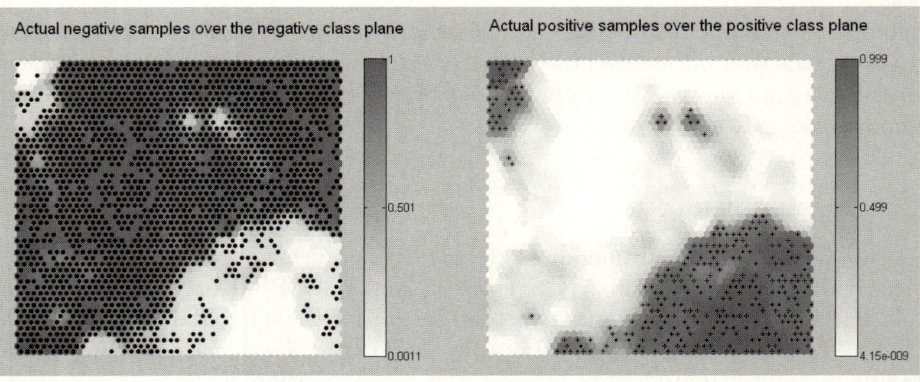

Fig. 1. Labeling planes for the *letter* training dataset

with the PPs approach [3], the proposed approach obtains higher values in five of the six datasets.

Several advantages are obtained by means of the use of this algorithm. The visualization of the clusters improves the data understanding. Fig. 1, e.g., shows the component planes of the classes, i.e., the labeling planes of the classifier trained with the letter dataset. Their values are between 0 and 1 and show the probabilistic membership of each map unit to each class. Therefore, the results are expressed in terms of probability. Also, the projection of the training dataset is shown in Fig. 1. Both positive and negative samples were separated and projected into different labeling planes for a better visualization. Actual positive samples are indicated with crosses whereas actual negative instances appear as dots. Regarding the sensitivity, the classification of the positive samples is very good. However, the specificity is not as good as the sensitivity because several instances are inside the positive region as it can be seen on the lower right side of the map. Those samples would be false positive. There are actual positive and negative instances overlapped in that region. The problem of classification also depends on the degree of data overlapping among the classes as the results suggest in [26]. In this way, using FLSOM can show the existence of overlapped clusters.

6   Conclusions

FLSOM is a version of SOM algorithm where the prototype vectors are influenced by the labeling data vectors that define the clusters of the dataset. This paper aims to improve the classification of instances in the typical two-classes problem of imbalanced datasets using a modification of the original FLSOM algorithm. The proposed approach is based on assigning two different values in the learning rate to the prototype vectors depending on the labeling data vector associated with the training data vector. The proposed algorithm was compared with two previous works in the topic literature. The results seem to prove that FLSOM algorithm with different rates is a interesting tool in order to deal with imbalanced datasets. Moreover, the classifier has the advantages of the SOM algorithm. The input space data can be visualized in the two-dimensional output space, watching the relationships between the variables and the classes and defining the possible clusters of misclassifying instances.

Acknowledgments

Our warmest thanks are expressed to the following for their financial support: Secretaría de Estado de Universidades e Investigación del Ministerio de Educación y Ciencia of Spain supporting the researching stay at Institut für Neuroinformatik of Universität Ulm with PR2007-0354 as agreement number.

References

1. Kubat, M., Matwin, S.: Addressing the Curse of Imbalanced Training Sets: One-sided Selection. In: Proc. 14th Int. Conf. on Machine Learning, pp. 179–186 (1997)
2. Vivaracho, C.: Improving SVM Training by Means of NTIL When the Data Sets Are Imbalanced. In: Esposito, F., Raś, Z.W., Malerba, D., Semeraro, G. (eds.) ISMIS 2006. LNCS (LNAI), vol. 4203, pp. 111–120. Springer, Heidelberg (2006)
3. Cantador, I., Dorronsoro, J.: Parallel Perceptrons, Activation Margins and Imbalanced Training Set Pruning. In: Marques, J.S., Pérez de la Blanca, N., Pina, P. (eds.) IbPRIA 2005. LNCS, vol. 3523, pp. 43–50. Springer, Heidelberg (2005)
4. Wilson, D.R., Martínez, T.R.: Reduction techniques for instance-based learning algorithms. Machine Learning 38, 257–286 (2000)
5. Guo, H., Viktor, H.L.: Learning from imbalanced data sets with boosting and data generation: the databoost-im approach. SIGKDD Explorations 6, 30–39 (2004)
6. Chawla, N.V., Bowyer, K.W., Hall, L.O., Kegelmeyer, W.P.: SMOTE: Synthetic Minority Over-Sampling Technique. Journal of Artificial Intelligence Research 16, 321–357 (2002)
7. Liu, Y., An, A., Huang, X.: Boosting Prediction Accuracy on Imbalanced Datasets with SVM Ensembles. In: Ng, W.-K., Kitsuregawa, M., Li, J., Chang, K. (eds.) PAKDD 2006. LNCS (LNAI), vol. 3918, pp. 107–118. Springer, Heidelberg (2006)
8. Akbani, R., Kwek, S., Japkowicz, N.: Applying support vector machines to imbalanced datasets. In: Boulicaut, J.-F., Esposito, F., Giannotti, F., Pedreschi, D. (eds.) ECML 2004. LNCS (LNAI), vol. 3201, pp. 39–50. Springer, Heidelberg (2004)
9. Soler, V., Roig, J., Prim, M.: Fuzzy Rule Extraction Using Recombined RecBF for Very-Imbalanced Datasets. In: Duch, W., Kacprzyk, J., Oja, E., Zadrożny, S. (eds.) ICANN 2005. LNCS, vol. 3697, pp. 685–690. Springer, Heidelberg (2005)
10. Veropoulos, K., Cristianini, N., Campbell, C.: Controlling the sensitivity of support vector machines. In: International Joint Conference on Artificial Intelligence, pp. 55–60 (1999)
11. Pazzani, M., Merz, C., Murphy, P., Ali, K., Hume, T., Brunk, C.: Reducing misclassification costs. In: Proc. 11th Int. Conf. on Machine Learning, pp. 217–225 (1994)
12. Woods, K., Doss, C., Bowyer, K.W., Solka, J., Priebe, C., Kegelmeyer, W.P.: Comparative evaluation of pattern recognition techniques for detection of microcalcifications in mammography. International Journal of Pattern Recognition and Artificial Intelligence 7, 1417–1436 (1993)
13. Zheng, Z., Wu, X., Srihari, R.: Feature Selection for Text Categorization on Imbalanced Data. SIGKDD Explorations 6(1), 80–89 (2004)
14. Fawcett, T., Provost, F.: Combining Data Mining and Machine Learning for Effective User Profile. In: Proceedings of the 2nd International Conference on Knowledge Discovery and Data Mining, pp. 8–13 (1996)
15. Kubat, M., Holte, R.C., Matwin, S.: Machine learning for the detection of oil spills in satellite radar images. Machine Learning 30, 195–215 (1998)
16. Lagus, K., Honkela, T., Kaski, S., Kohonen, T.: Self-organizing maps of document collections: a new approach to interactive exploration. In: Second International Conference on Knowledge Discovery and Data Mining, pp. 238–243 (1996)
17. Kohonen, T., Oja, E., Simula, O., Visa, A., Kangas, J.: Engineering applications of the self-organizing map. Proceedings of the IEEE 84, 1358–1384 (1996)
18. Simula, O., Kangas, J.: Process monitoring and visualization using self-organizing maps. Neural Networks for Chemical Engineers, 371–384 (1995)

19. Villmann, T., Seiffert, U., Schleif, F.-M., Brüss, C., Geweniger, T., Hammer, B.: Fuzzy Labeled Self-Organizing Map with Label-Ajusted Prototypes. In: Schwenker, F., Marinai, S. (eds.) ANNPR 2006. LNCS (LNAI), vol. 4087, pp. 46–56. Springer, Heidelberg (2006)
20. Kohonen, T.: Self-organizing maps, 3rd extended edn. 2001. Springer, Berlin (1995)
21. Vesanto, J., Alhoniemi, E.: Clustering of the Self-Organizing Map. IEEE Transactions Transactions on Neural Networks 11(3), 586–600 (2000)
22. López, H., Machón, I.: Self-organizing map and clustering for wastewater treatment monitoring. Engineering Applications of Artificial Intelligence 17(3), 215–225 (2004)
23. Machón, I., López, H.: End-point detection of the aerobic phase in a biological reactor using SOM and clustering algorithms. Engineering Applications of Artificial Intelligence 19(1), 19–28 (2006)
24. Heskes, T.: Energy functions for self-organizing maps. In: Oja, E., Kaski, S. (eds.) Kohonen Maps, pp. 303–316. Elsevier, Amsterdam (1999)
25. Newman, D.J., Hettich, S., Blake, C.L., Merz, C.J.: UCI repository of machine learning databases (1998), http://mlearn.ics.uci.edu/MLSummary.html
26. Prati, R., Batista, G., Monard, M.: Class Imbalances versus Class Overlapping: An Analysis of a Learning System Behavior. In: Monroy, R., Arroyo-Figueroa, G., Sucar, L.E., Sossa, H. (eds.) MICAI 2004. LNCS (LNAI), vol. 2972, pp. 312–321. Springer, Heidelberg (2004)

A Self-organizing Neural System for Background and Foreground Modeling

Lucia Maddalena[1] and Alfredo Petrosino[2]

[1] ICAR - National Research Council, Via P. Castellino 111, 80131 Naples, Italy
lucia.maddalena@na.icar.cnr.it
[2] DSA - University of Naples Parthenope, Centro Direzionale, Isola C/4, 80143 Naples, Italy
alfredo.petrosino@uniparthenope.it

Abstract. In this paper we propose a system that is able to detect moving objects in digital image sequences taken from stationary cameras and to distinguish wether they have eventually stopped in the scene. Our approach is based on self organization through artificial neural networks to construct a model of the scene background that can handle scenes containing moving backgrounds or gradual illumination variations, and models of stopped foreground layers that help in distinguishing between moving and stopped foreground regions, leading to an initial segmentation of scene objects. Experimental results are presented for color video sequences that represent typical situations critical for video surveillance systems.

Keywords: background modeling, foreground modeling, neural network, self organization, visual surveillance.

1 Introduction

Visual surveillance systems are demanded to automatically understand events happening at the monitored site, and this is achieved through several steps, including motion detection, object classification, tracking, activity understanding, and semantic description. For many of these tasks, it is necessary to build representations of the appearance of objects in the scene [3]. This paper focuses on two issues: how to construct a model of the scene background that supports sensitive detection of moving objects in the scene and how to build a model of the foreground objects that supports their tracking.

Aside from the intrinsic usefulness of being able to segment video streams into moving and background components, detecting moving objects provides a focus of attention for recognition, classification, and activity analysis, making these later steps more efficient, since only moving pixels need be considered [2]. The problem is known to be significant and difficult [10]. The most common and efficient method to tackle moving object detection for scenes from stationary cameras is background subtraction (see [8]), that is based on the comparison of the current sequence frame with a reference background model that includes information on the scene without moving objects. It is independent on the velocity

V. Kůrková et al. (Eds.): ICANN 2008, Part I, LNCS 5163, pp. 652–661, 2008.

of moving objects and it is not subject to the foreground aperture problem, but it is extremely sensitive to dynamic scene changes due to lighting and extraneous events.

Besides background modeling, also building representations for foreground objects is essential for tracking them and maintaining their identities throughout the image sequence [3]. Modeling the color distribution of homogeneous regions has been used successfully to track nonrigid bodies (e.g. [11]), and a variety of parametric and nonparametric statistical techniques have been adopted, ranging from single Gaussians (e.g. [11]), to mixture of Gaussians (e.g. [9]), to nonparametric techniques using histograms (e.g. [7]) or kernel density estimation [3].

Our aim is the construction of a system for motion detection based on the background and the foreground models automatically generated by a self-organizing method without prior knowledge of the pattern classes. The approach, firstly proposed for background modeling [5,6], consists in using biologically inspired problem-solving methods to solve motion detection tasks, typically based on visual attention mechanisms [1]. The aim is to obtain the objects that keep the users attention in accordance with a set of predefined features, including gray level, motion and shape features. Our approach defines a method for the generation of an active attention focus on a dynamic scene to be monitored for surveillance purposes. We adopt a self-organizing method for learning motion patterns represented by trajectories in the HSV space. By learning the trajectories and features of moving objects, the background and foreground models are built up. Such models, together with a layering mechanism, allow to construct a system able to detect motion and distinguish foreground objects into moving or stopped objects, even when they appear superimposed.

The paper is organized as follows. In Section 2 we sketch the architecture of the complete system for background and foreground modeling, while in Sections 3 and 4 we give a more detailed description of the approach adopted for background and stopped foreground modeling, respectively. In Section 5 we present some results obtained with the implementation of the proposed approach, while Section 6 includes concluding remarks.

2   The Complete System

The objective of our work is the construction of a system able to detect moving objects in digital image sequences taken from stationary cameras and to distinguish wether they have eventually stopped in the scene. Such goal is achieved through the estimate of a background model $\tilde{B}$, that represents only pixels not belonging to objects that are extraneous to the scene, and of a set of stopped foreground models $\tilde{S}^1, \ldots, \tilde{S}^L$, each of which represents only pixels belonging to one of the L objects that are extraneous to the scene and that appear as stopped in the scene.

The complete system for modeling background and stopped foreground is shown in Fig. 1. For each sequence frame I_t at time instant t, we first compare

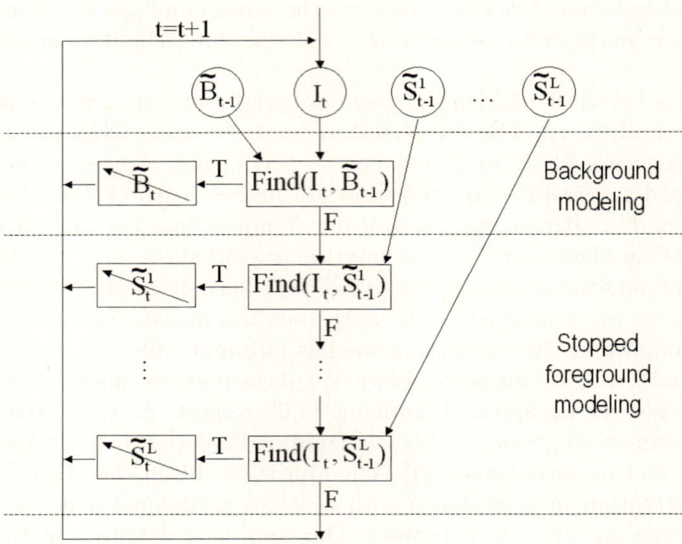

Fig. 1. Scheme of the complete system for background and stopped foreground modeling

it with the background model estimate $\tilde{B}_{t-1}$. If a match is found for the generic incoming pixel, then it means that it belongs to the scene background, and the background model is updated accordingly, giving a new estimate $\tilde{B}_t$ of the background model for the next iteration over time. If the match is not found, then it means that the pixel belongs to an object extraneous to the scene, and therefore we proceed in the search for a possible stopped foreground model to which the object belongs. To this end, we compare the incoming pixel with the first stopped foreground model estimate $\tilde{S}^1_{t-1}$. If a match is found, then the stopped foreground model is updated accordingly; otherwise we iterate the comparison with successive stopped foreground layers, and their eventual update. Pixels for which no match is found with any of the mentioned models constitute the moving foreground.

The described system architecture is independent of the type of model chosen for the background and for the stopped foreground objects, as well as of the criterion adopted for matching image pixels with such models and of the strategy for models update. In the following sections we describe our choices for such tasks, that lead to a self-organizing system based on artificial neural networks.

3　Modeling the Background by Self-organization

In our approach the background model is based on a self organizing neural network, inspired by Kohonen [4] and already adopted in [5,6]. The adopted artificial neural network is organized as a 2-D flat grid of neurons. Each neuron computes a function of the weighted linear combination of incoming inputs, with weights resembling the neural network learning, and can be therefore represented by a weight vector obtained collecting the weights related to incoming links. An incoming pattern is mapped to the neuron whose set of weight vectors is most similar to the pattern, and weight vectors in a neighborhood of such node are updated.

For each pixel we build a neuronal map consisting of $n \times n$ weight vectors, all represented in the HSV colour space, that allows to specify colours in a way that is close to human experience of colours. Each weight vector $c_i, i = 1, \ldots, n^2$, is therefore a 3D vector, initialized to the HSV components of the corresponding pixel of the first sequence frame I_0. The complete set of weight vectors for all pixels of an image I with N rows and M columns is represented as a neuronal map $\tilde{B}$ with $n \times N$ rows and $n \times M$ columns, where adjacent blocks of $n \times n$ weight vectors correspond to adjacent pixels in image I. An example of such neuronal map structure for a simple image I with $N = 2$ rows and $M = 3$ columns obtained choosing $n = 3$ is given in Fig. 2. The upper left pixel a of sequence frame I in Fig. 2-(a) has weight vectors $(a_1, \ldots, a_9)$ stored into the 3×3 elements of the upper left part of neuronal map $\tilde{B}$ in Fig. 2-(b), and analogous relations exist for each pixel of I and corresponding weight vectors storage.

By subtracting the current image from the background model, each pixel p_t of the t-th sequence frame I_t is compared to the current pixel weight vectors to determine if there exists a weight vector that matches it. The best matching weight vector is used as the pixel's encoding approximation, and therefore p_t is detected as foreground if no acceptable matching weight vector exists; otherwise it is classified as background. Matching for the incoming pixel p_t is performed

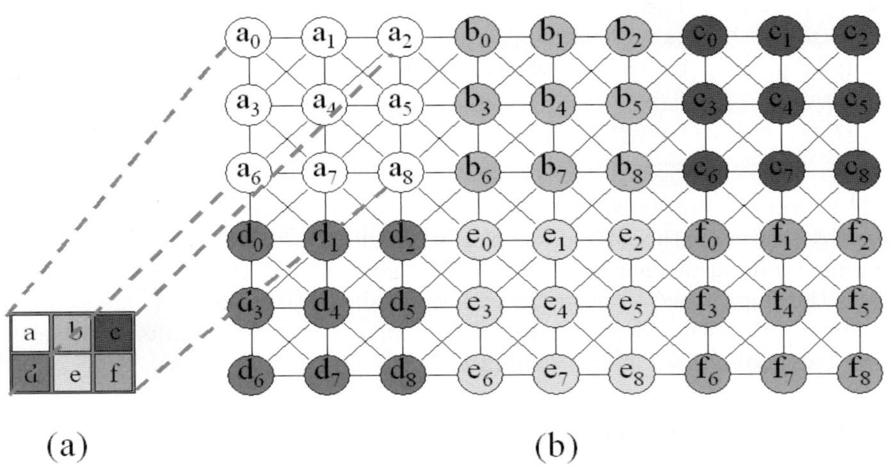

(a)   (b)

Fig. 2. A simple image (a) and the modeling neuronal map (b)

by looking for a weight vector c_m in the set $C = (c_1, \ldots, c_{n^2})$ of the current pixel weight vectors satisfying:

$$d(c_m, p_t) = \min_{i=1,\ldots,n^2} d(c_i, p_t) \leq \epsilon \tag{1}$$

where the metric $d(\cdot)$ and the threshold ϵ are suitably chosen.

The best matching weight vector $c_m = \tilde{B}_t(\overline{x}, \overline{y})$, together with all other weight vectors in a $n \times n$ neighborhood of the background model $\tilde{B}$, are updated according to weighted running average:

$$\tilde{B}_t(i,j) = (1 - \alpha_{i,j})\tilde{B}_{t-1}(i,j) + \alpha_{i,j}p_t, \quad \begin{array}{l} i = \overline{x} - \lfloor \frac{n}{2} \rfloor, \ldots, \overline{x} + \lfloor \frac{n}{2} \rfloor \\ j = \overline{y} - \lfloor \frac{n}{2} \rfloor, \ldots, \overline{y} + \lfloor \frac{n}{2} \rfloor \end{array} \tag{2}$$

Here $\alpha_{i,j} = \alpha \cdot w_{i,j}$, where α represents the learning factor, that depends on scene variability, while $w_{i,j}$ are Gaussian weights in the $n \times n$ neighborhood of c_m, that well correspond to the lateral inhibition activity of neurons. If the best match c_m is not found, the background model $\tilde{B}$ remains unchanged; such selectivity allows to adapt the background model to scene modifications without introducing the contribution of pixels not belonging to the background scene.

The background subtraction and update for an incoming pixel value p_t in sequence frame I_t allows to obtain the binary mask $M(p_t)$ and the estimated background model $\tilde{B}_t$ defined as in the following algorithm:

SOBS background subtraction and update algorithm

```
Initialize weight vectors C for pixel p_0 and store it into B̃_0
for t=1, LastFrame
   Find best match c_m in C to current sample p_t as in eqn. (1)
   if (c_m found) then
     M(p_t) = 0 //background
     update B̃_{t-1} in the neighborhood of c_m as in eqn. (2)
   else
     M(p_t) = 1 //foreground
```

The proposed approach has been shown to outperform several existing moving object detection methods, in particular in handling cases as moving backgrounds, gradual illumination changes, and cast shadows suppression [6]. There we present also a thorough discussion on the choice of all algorithm parameters and their impact on algorithm performance, and compare obtained results with several other background modeling techniques.

4   Modeling the Foreground by Self-organization

The binary mask M obtained by SOBS algorithm for each sequence frame I_t identifies foreground pixels, i.e. pixels that are not modeled by the background neuronal map $\tilde{B}$. However, due to the selective nature of SOBS background update process, objects that enter the scene and stop are always detected as moving foreground objects, even if they are not really *moving*. An example is given in Fig. 3, where both the walking man and the stopped bag are detected as moving foreground in the mask M computed by the SOBS algorithm (Fig. 3-(b)).

In order to distinguish between moving and stopped foreground objects, we introduce a layering mechanism, where each layer models a single stopped

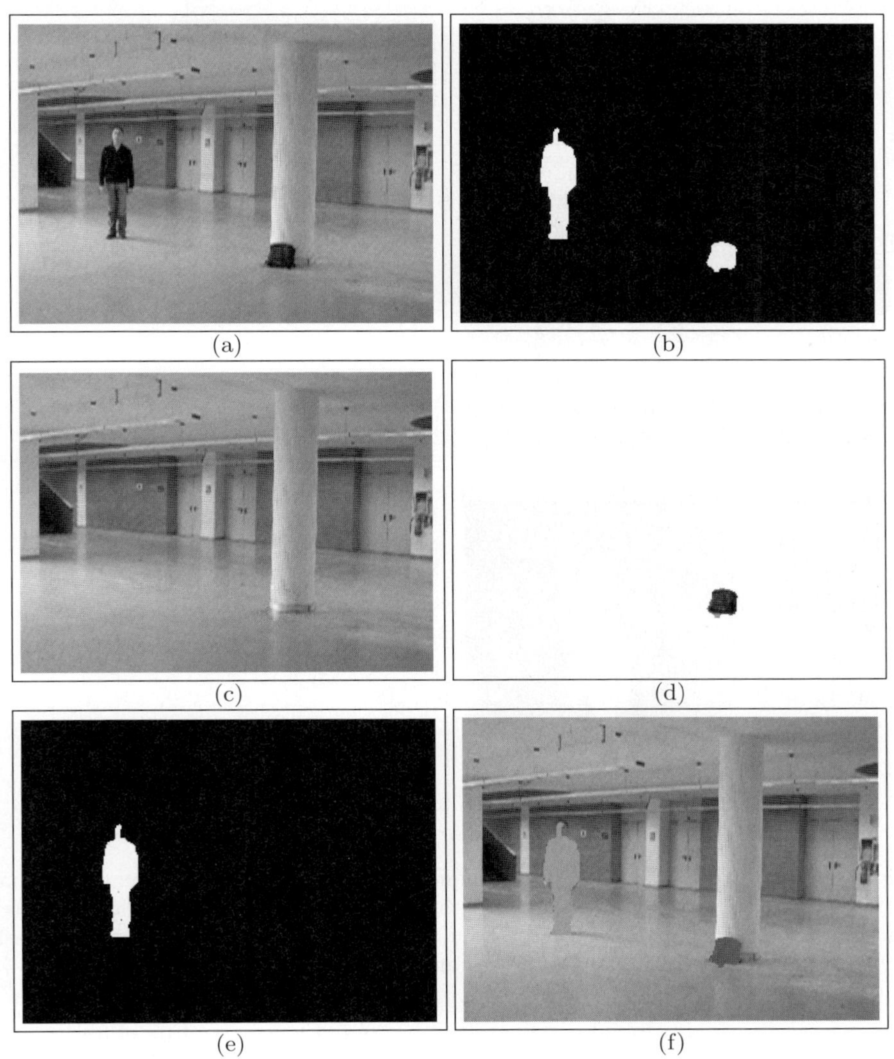

Fig. 3. Results of background and foreground modeling for Msa sequence: (a) original frame no. 208; (b) moving object detection mask M computed by SOBS algorithm; (c) background model $\tilde{B}$; (d) stopped foreground model $\tilde{S}^1$; (e) moving foreground mask F; (f) original frame with moving (green) and stopped (red) foreground objects

Fig. 4. Results of background and foreground modeling for Msa sequence: (a) original frame no. 350; (b) moving object detection mask M computed by SOBS algorithm; (c) background model $\tilde{B}$; (d) stopped foreground model $\tilde{S}^1$; (e) moving foreground mask F; (f) original frame with moving (green) and stopped (red) foreground objects

foreground object. If a pixel is detected as a moving foreground pixel for several consecutive frames, then it is considered as belonging to a stopped object, and it is inserted into a stopped foreground layer.

Stopped foreground pixels are modeled through a neuronal map analogous to the one described for background modeling and shown in Fig. 2. When a stopped foreground pixel is first detected, the corresponding weight vectors are initialized to the stopped foreground pixel itself. For subsequent sequence frames, weight vectors for stopped foreground pixels are updated in order to allow for adaptivity of the stopped foreground model. The updating follows selective running average of eqn. 2, where now $\tilde{B}$ indicates the stopped foreground model $\tilde{S}^i$ and pixel $(\overline{x}, \overline{y})$ of $\tilde{B}$ contains the best match c_m for the incoming stopped pixel p_t.

The foreground modeling algorithm for an incoming pixel value p_t in sequence frame I_t not belonging to the background model allows to obtain a binary mask F indicating moving foreground pixels and a set of estimated stopped foreground models $\tilde{S}_t = \left\{ \tilde{S}_t^i, i = 1, \ldots, L \right\}$ as follows:

Foreground modeling algorithm

```
Find best match c_m in S̃_{t-1} to current sample p_t as in eqn. (1)
if (c_m found in layer S̃^i_{t-1}) then
  F(p_t) = 0 //stopped
  update S̃^i_{t-1} in the neighborhood of c_m as in eqn. (2)
else
  F(p_t) = 1//moving
```

Modeling of stopped foreground layers allows to keep an up-to-date representation of all stopped pixels and to obtain an initial classification of scene objects, which can be used to support subsequent tracking and classification phases of the complete system. Moreover, it allows to distinguish moving and stopped foreground objects when it happens they are located in the same region of the image. Indeed, in this case the availability of an up-to-date model of all stopped objects allows to discern whether pixels that are not modeled as background belong to one of the stopped objects or to moving objects. An example is given in Fig. 4, where the stopped foreground model reported in Fig. 4-(d) allows to distinguish the stopped object (the bag on the floor) and the walking man, as shown in Fig. 4-(f).

5   Experimental Results

Experimental results for moving and stopped object detection using the proposed approach have been produced for several image sequences. Here we report results obtained for sequence Msa, which represents the typical critical video surveillance task of looking for suspect abandoned luggage. The Msa sequence is an indoor sequence consisting of 555 frames of 320×240 spatial resolution, acquired at a frequency of 30 frames/sec. The scene consists of a university hall, where a man comes in, leaves a bag on the floor, walks around, and then leaves. One representative frame together with obtained results is reported in Fig. 3. Here we report the original sequence frame no. 208 (Fig. 3-(a)) and the corresponding moving object detection mask M computed by the SOBS algorithm choosing $n = 3$ (Fig. 3-(b)). The detection mask shows that the walking man is perfectly detected. However the bag, which has been left by the man in previous frames, is still detected as an object extraneous to the background, due to the selective update of SOBS algorithm.

In Fig. 3 we also show the background model $\tilde{B}$ computed by the SOBS algorithm (Fig. 3-(c)), that appears as a quite accurate (enlarged) representation of the real background. We would remark that the background model $\tilde{B}$ is represented by a neuronal map whose size is $n^2 = 9$ times greater than that of the original image I; in the reported figures they appear to have the same size only for space constraints and for an easier comparison.

In Fig. 3-(d) we show the first stopped foreground layer $\tilde{S}^1$ obtained as explained in §4, that represents the bag left by the man in previous frames, while in Fig. 3-(e) we report the mask F containing only the moving foreground.

Final result is reported in Fig. 3-(f), where we show the original frame with moving foreground objects (in green) and stopped foreground objects (in red).

It can be observed that the neuronal maps modeling background (Fig. 3-(c)) and stopped foreground (Fig. 3-(d)), together with the moving foreground mask (Fig. 3-(e)), give a quite accurate and up-to-date representation of current sequence frame (Fig. 3-(a)), that can be used for an initial segmentation of the scene.

Another representative frame of our sequence together with obtained results is reported in Fig. 4. The sequence frame and the corresponding moving object detection mask computed by the SOBS algorithm are reported in Figs. 4-(a) and 4-(a). The detection mask, although quite accurate, does not allow to distinguish between the stopped object (the bag on the floor) and the walking man. However, the stopped foreground model, reported in Fig. 4-(d), allows to distinguish them, as shown by the moving foreground mask reported in Fig. 4-(e) and by the final result reported in Fig. 4-(f). Little confusion between stopped and moving pixels can be observed only under the man legs, due to the evident camouflage of the pants with the bag.

6   Conclusions

We have presented a self-organizing system for modeling background and foreground by learning motion patterns and so allowing foreground/background separation for scenes taken from stationary cameras, strongly required in video surveillance systems. Unlike existing systems that use individual flow vectors as inputs, our system learns in a self organizing manner motion trajectories. First, we describe a self-organizing neural network-based model of the scene background that can handle scenes containing moving backgrounds or gradual illumination variations, and achieves robust detection for different types of videos taken with stationary cameras. Second, we build a model of stopped foreground pixels, analogous to the one adopted for background, that helps in distinguishing between moving and stopped foreground regions, leading to an initial segmentation of scene objects. Experimental results on video sequences representing typical situations critical for video surveillance systems demonstrate the effectiveness of the proposed system.

References

1. Cantoni, V., Marinaro, M., Petrosino, A. (eds.): Visual Attention Mechanisms. Kluwer Academic/Plenum Publishers, New York (2002)
2. Collins, R.T., Lipton, A.J., Kanade, T., Fujiyoshi, H., Duggins, D., Tsin, Y., Tolliver, D., Enomoto, N., Hasegawa, O., Burt, P., Wixson, L.: A System for Video Surveillance and Monitoring, The Robotics Institute, Carnegie Mellon University, CMU-RI-TR-00-12 (2000)
3. Elgammal, A., Duraiswami, R., Harwood, D., Davis, L.S.: Background and Foreground Modeling Using Nonparametric Kernel Density Estimation for Visual Surveillance. Proceedings of the IEEE 90(7), 1151–1163 (2002)
4. Kohonen, T.: Self-Organization and Associative Memory, 2nd edn. Springer, Berlin (1988)
5. Maddalena, L., Petrosino, A.: A Self-Organizing Approach to Detection of Moving Patterns for Real-Time Applications. In: Mele, F., Ramella, G., Santillo, S., Ventriglia, F. (eds.) BVAI 2007. LNCS, vol. 4729, pp. 181–190. Springer, Heidelberg (2007)
6. Maddalena, L., Petrosino, A.: A Self-Organizing Approach to Background Subtraction for Visual Surveillance Applications. IEEE Trans. Image Process (July 2008) (to be published)
7. McKenna, S.J., Jabri, S., Duric, Z., Rosenfeld, A.: Tracking groups of people. Comput. Vision Image Understanding 80, 42–56 (2000)
8. Radke, R.J., Andra, S., Al-Kofahi, O., Roysam, B.: Image change detection algorithms: a systematic survey. IEEE Trans. Image Process. 14(3), 294–307 (2005)
9. Raja, Y., Mckenna, S.J., Gong, S.: Tracking color objects using adaptive mixture models. Image Vision Comput. 17, 225–231 (1999)
10. Toyama, K., Krumm, J., Brumitt, B., Meyers, B.: Wallflower: Principles and Practice of Background Maintenance. In: Proc. of the Seventh IEEE Conference on Computer Vision, vol. 1, pp. 255–261 (1999)
11. Wren, C., Azarbayejani, A., Darrell, T., Pentland, A.: Pfinder: Real-Time Tracking of the Human Body. IEEE Trans. on PAMI 19(7), 780–785 (1997)

Analyzing the Behavior of the SOM through Wavelet Decomposition of Time Series Generated during Its Execution

Víctor Mireles[1] and Antonio Neme[2]

[1] Universidad Nacional Autónoma de México
syats.vm@gmail.com
[2] Universidad Autónoma de la Ciudad de México
neme@nolineal.org.mx

Abstract. Cluster analysis applications of the SOM require it to be sensible to features, or groupings, of different sizes in the input data. On the other hand, the SOM's behavior while the organization process is taking place also exhibits regularities of different scales, such as periodic behaviors of different frequencies, or changes of different magnitudes in the weight vectors. A method based on the discrete wavelet transform is proposed for measuring the diversity of the scales of regularities, and this diversity is compared to the performance of the SOM. We argue that if this diversity of scales is high then the algorithm is more likely to detect differently sized features of data.

1   Introduction

The Self Organizing Map (SOM)[3], is an algorithm which forms a mapping of a finite set of points $V \subset \mathbb{R}^n$ into $G \subset \mathbb{N}^2$. This mapping gradually evolves through a process of self-organization from a completely random one, to one which preserves the distance relationships of the elements of V.

For this process of self organization to take place, each of the points of V is fed several times as inputs into the algorithm, which modifies the mapping according to a so called learning function depending on the input received. The algorithm goes on as follows(for further details see [5,4]): a two-dimensional grid G contains in each of its vertexes $a \in G$ a unit, or neuron, which stores an $n-$dimensional vector $v_a \in \mathbb{R}^n$, called its weight vector, which is initially random. Every time $x_i \in V$, an input vector or stimulus, is presented as input, a search is done to find which of the neurons' weight vectors is closer to the presented stimulus. This neuron $b \in G$ such that $d(v_b, x_i) \leq d(v_a, x_i) \ \forall a \in G$, with d a distance function in $\mathbb{R}^n$, is called the Best Matching Unit (BMU) and we say it has been activated by x_i. Once this $b \in G$ is found, the weight vectors of all neurons are adjusted according to the learning function

$$v_a(t + 1) = v_a(t) + \eta(t)h_{a,b}(t)(x_i - v_a(t)) \tag{1}$$

where $v_a(t)$ represents the weight vector of neuron a at time t, $\eta(t)$ is the learning rate parameter, and $h_{a,b}(t)$ is the neighborhood function which determines how

V. Kůrková et al. (Eds.): ICANN 2008, Part I, LNCS 5163, pp. 662–670, 2008.

much the neuron a is affected when neuron b is selected as BMU. Every time the whole set V has been presented, element by element, as input we say an epoch has elapsed. The algorithm repeats itself for several epochs until a stop criteria is met, usually that the weight vectors of all the neurons are modified less than a certain threshold.

Several applications to cluster analysis have been found for the SOM, be it for the classical algorithm [1] or for specially proposed modifications like that in [2]. Being sensible to clusters of different sizes (in terms of number of elements) is necessary for a good cluster analysis algorithm, more over, the sensitivity to features of a variety of scales is akin to all relatively interesting information processing algorithms. When we say an algorithm is sensible to a feature of a certain scale, we mean that its behavior is modified when a certain subset of all of its inputs has a certain property, and we define the scale of this feature as the number of elements of this subset. In the case of cluster analysis algorithms, a feature would be a cluster and its scale the number of input points belonging to this cluster. If this cluster is somehow detected by the algorithm it would yield a particular output that would not have occurred if the set of points forming the cluster did not comply with the relevant property, in this case, that the distance between them is, according to a certain threshold, smaller than the distance between points in the subset and any points not in this subset we are calling cluster.

However most algorithms, including the SOM, take inputs always of the same scale through out their execution. The inputs which the SOM takes and which determine the sort of modifications that are made to the mapping are always vectors $\in \mathbb{R}^n$, one of them at a time. Despite this, and owing to the memory properties of the map, which is altered accordingly to the input, the SOM algorithm is capable of distinguishing features of different scales in the input set of vectors. We can therefore say that the clusters to which the SOM is sensible are of a scale different to that of the individual points which are fed as inputs to it, moreover, the SOM is sensible to clusters of diverse scales that is, of diverse number of elements.

Clearly, a process must take place in the internal memory registers (weight vectors, status variables, etc) of the SOM for it to grasp differently scaled features in the input space. That is, if the SOM is to always take as input individual vectors in $\mathbb{R}^n$, it must somehow keep track of their properties, so that it can eventually verify if a certain property is valid for a subset (of more than one element) of the all the inputs. On the one hand, the neighborhood size is decreasing (or at least generally decreasing) in time. This means that the number of affected neurons, and therefore the number of memory registers (in the form of their weight vectors), which are affected when each individual input vector is presented varies with time. On the other hand, through out the self organization process it is not the same neurons which are affected by stimuli, that is, different BMU's are chosen for the same stimulus, nor is this unit's weight vector altered in the same magnitude, even if $\eta(t)$ were constant.

The intention of this work is to achieve better understanding of the behavior of the SOM algorithm, and which parts of it are responsible for its capacities as a cluster analysis tool. For this, a set of time series is generated during the execution of the SOM algorithm as a means of recording the map's modifications in time, namely, for each neuron we record the difference between its weight vector and the input vector which activated it as a BMU. These series are then analyzed to obtain a measure of the diversity of scales (magnitudes and frequencies) of the phenomena they record, this is done by analyzing the wavelet coefficients of the time series. We can then have, for each execution of the SOM, several measures of the diversity of scales of the modifications suffered by its memory registers. A relationship is investigated between the measures of a given execution and its performance in terms of the topographic error, a standard error measure for maps generated by the SOM which is argued to be related to the algorithms capacities in cluster analysis.

2   The Use of Wavelet Analysis

If we take a time series we would like to know the different scales of the information contained in it, in particular the periods of the variations this time series records. If the series' values oscillated between, say, -1 and 1, we could say that in the process which it records, there is an oscillation of period 2. However, in general, a time series is not produced by a single oscillating device. Yet, frequency domain analysis allows us to determine to what extent a time series seems to be generated by oscilating devices of several frequencies. Wavelet analysis[9,10] goes one step further, it gives us this frequency domain decomposition of a time series in a local manner, that is, it determines to what extent a certain portion of a time series resembles a function oscillating at a certain frequency independently of any such behavior in other, far away, parts of the series.

Mallat's algorithm [6] allows us to, given a time series $\{y_i\}$ $i \in \{1 \dots 2^M\}$, a mother wavelet Ψ and an scaling function ϕ, calculate a set $\{d_{j,i} \mid i \in \{1 \dots 2^{M-j}\}\}$ of detail coefficients organized in levels $j \in \{1 \dots M\}$ which give us information on how much the time series resembles the wavelet function $\Psi_{j,i}$ which has as support the i-th interval of size 2^j in the time series' time domain. This wavelet functions $\Psi_{j,i}$ are all translations and scallings of Ψ, the mother wavelet, which is a fast-decaying oscillating waveform with compact support. We have chosen possibly the simplest mother wavelet, that of Haar[7]. An often cited disadvantage of the Haar wavelet is its discontinuity, however due to the nature of the time series here analyzed, which are strongly discontinuous, and to the processes which generate them, we believe this is not so much of a drawback.

3   Methods

We take each of 4 data sets: Ionosphere (Dimension 34), Iris (Dimension 4), Codon Usage (Dimension 64) and Yeast gene expression(Dimension 7) and run on them the SOM algorithm 4000 times. The grid used is always of size 20×20 (that is $|G| = 400$), the neighbourhood is of bubble type and it is, just like the learning rate, exponentially decreasing. Executions of the algorithm on the same data set differ from each other in the initial weight vectors of the neurons, in the initial size of the decreasing neighborhood, and in the initial learning rate. The SOM stops when the BMU selection in an epoch is the same as for the previous one. Of the 4000 executions for each data set, 2000 are made taking the stimuli in the same order on every epoch, while in the other 2000, stimuli are taken in random order such that in every epoch all stimulli are taken once and only once. For each execution, let T denote the total number of epochs before the algorithm came to a stop, $|V|$ the length of the data set and n its dimension.

Each of these executions of the SOM algorithm gives us data which we analyze to determine the diversity of the scales of the phenomena there recorded. The data is extracted in the form of a set of 400 1-Dimensional series.

3.1   1D Series

This is a set of time-series rather than a single one. Given each of the 400 neurons, we are able to see when was it selected as a BMU. Every time it is selected as BMU, its weight vector is altered. The alteration depends on the one hand on the current learning rate, which varies with time. On the other hand, it depends on the difference between its weight vector and the input vector which activated it. This difference will tend to decrease as the map converges to its final steady configuration, but this tendency might be affected, and it usually is, if the neuron is selected as BMU by another input vector.

For each $a \in G$ we define $T_a \subset \mathbb{N}$ as the set of time instants in which a was selected as BMU. Let $E_a = |T_a|$, then we define the bijective function $S_a : \{1 \ldots E_a\} \to T_a$ such that $S_a(t) < S_a(s) \Leftrightarrow s < t$, this is in fact an ordering function. We can then define the 1D time series for a as:

$$D_a(t) = d(x(S_a(t)), v_a(S_a(t)))\ \ t \in \{1 \ldots E_a\} \tag{2}$$

where $d : \mathbb{R}^n \times \mathbb{R}^n \to \mathbb{R}$ is a distance function in $\mathbb{R}^n$ (in this case the euclidean distance), $x(s)$ is the stimulus for time s and $v_a(s)$ is the weight vector for neuron a in time s. In general E_a is different for every $a \in G$ and therefore the maximum level of wavelet decomposition each of these series accepts, varies.

3.2   Measuring the Diversity of Scales Present in the Time Series

Having generated these time series, it is first necessary to add zero's to them (zero-padding) at the end so that the length of each time series is in fact a power of two. We then proceed to calculate the wavelet transform using Haar wavelets. Once detail wavelet coefficients are calculated for the time series, we derive from them a set of two measures for each execution of the SOM algorithm.

Inverse of the range of the per-level mean of the coefficients for the 1D-series. For each $a \in G$, call N_a the largest level for which we can get a wavelet decomposition for D_a the 1D series corresponding to neuron a. Call $N_0 = max_{a \in G}\{N_a\}$ then we can define

$$M_k = mean_{a \in G_k, i \in \{1 \ldots 2^{E_a - k}\}}\{d_a(k, i)\}\ \ \ k \in \{1 \ldots N_0\} \tag{3}$$

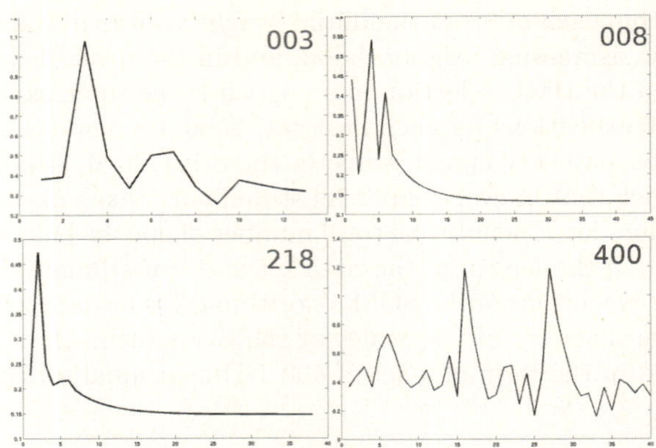

Fig. 1. 1D series for neurons 3, 8, 218 and 400 on an execution on the Iris set. Time on the x-axis, the corresponding values of the time series on the y-axis.

where $d_a(k, i)$ is the i'th k-level coefficient for series $D_a(t)$ and G_k is the set of neurons whose series accepts a wavelet transform of level k. We then define this measure as follows

$$IR = \frac{1}{max_{k \in \{1...N_0\}}\{M_k\} - min_{k \in \{1...N_0\}}\{M_k\}} \tag{4}$$

If there were a level whose mean coefficient is very different from the mean coefficient of another level this measure is very low. On the other hand, if the mean coefficient of all detail levels is quite the same, this measure is high. Therefore IR is a measure of the diversity of scales (periods) in the frequency domain.

Per neuron mean of the variance. For each $a \in G$ let var_a be the variance of the detail coefficients for the series $D_a(t)$. We then define this measure as:

$$NMV = \frac{1}{|G|} \sum_{a \in G} var_a \tag{5}$$

This is a measure of the diversity of scales in the space domain, specially for levels 1 and 2 of the wavelet decomposition, since these are better represented in the set of wavelet coefficients for a time series.

3.3   Measuring the Performance of the SOM

A common error measure for the maps generated through the SOM is the Topographic Error [11] (TE)

$$TE = \frac{1}{|V|} \sum_{x_i \in V} \delta(x_i, v_1, \ldots, v_{20 \times 20}) \tag{6}$$

Where, if we define the 2BMU for an input vector as the second neuron whose weight vector is more like it, we have

$$\delta(x_i, v_1, \ldots, v_{20\times20}) = \begin{cases} 1 & \text{if the BMU for } x_i \text{ is not adjacent to the 2BMU for } x_i \\ 0 & \text{otherwise} \end{cases}$$

If the SOM is to adequately recognize differently-sized clusters, we can expect each of these clusters to be mapped into a single neuron or a set of adjacent neurons, regardless of the radius of the cluster. Let us consider two points which are part of a same cluster, since they are close together in the input space, the BMU for one is likely to be, or at least be close to, the 2BMU for the other one. If they were mapped away from each other this wouldn't be the case, and therefore an increase into the topographic error would occur. If two vectors not belonging to the same cluster are mapped together, then the chances that the 2BMU for one of them is adjacent to its BMU are reduced, for one of its neighbours hasn't a weight vector which is similar to it. Therefore we can expect a low TE for a SOM that adequately recognized clusters, regardless of their size.

4   Results and Discussion

The results are summarized in the following graphs and further discussed below.

As can be seen in figures 2 and 3, especially in fig. 2, there is a generalized down-ward tendency in the error as the measures increase. It is however clear that the low error measure is not solely determined by the phenomena which determine the high measure. Rather than establishing that high IR or NMV

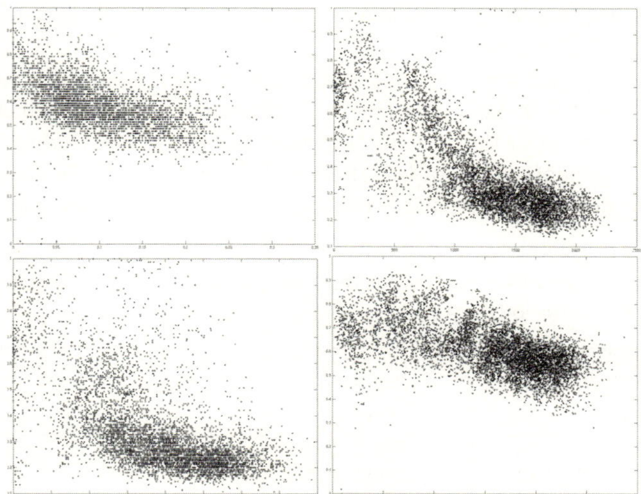

Fig. 2. Inverse of the range of the per-level mean of the coefficients for the 1D-series. Clockwise from top left: Codon, ion, yeast, iris. IR on the x-axis, topographic error TE on the y-axis.

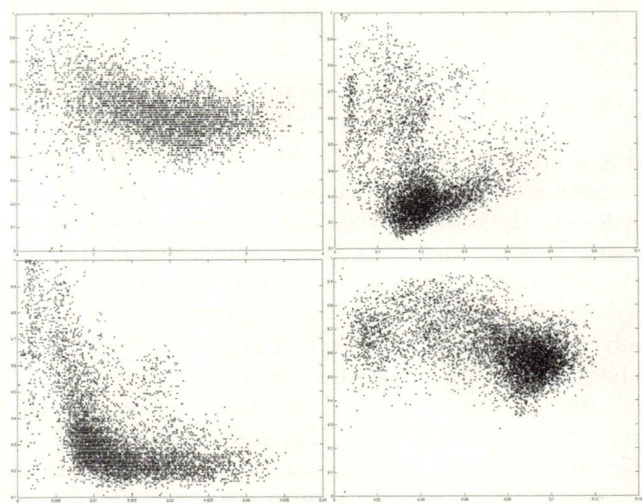

Fig. 3. Per neuron mean of the variance. Clockwise from top left: Codon, ion, yeast, iris. NMV on the x-axis, topographic error TE on the y-axis.

measures yield low error measures all the time, these results indicate that the likeness for a low error measure increases with the diversity of the scales of the phenomena present in the behavior of the SOM, we call this the main result of this work. This is supported by table 1 which compares the mean of each of the two measures for all the executions and the mean for the 400 (10%) best executions. In this table it can be seen how the mean of both measures is greater for the best 10% of the maps.

If the per-level averages of detail coefficients span a small interval (that is, a high IR), it means that no single level of wavelet decomposition has coefficients significantly larger than the rest, or that they all are, in relation to the absolute value of the coefficients, more or less similar. This means that the neuron's weight vector's difference with the stimuli exhibits in time a behavior which resembles

Table 1. Comparison of the means for our two measures between all the maps generated and the 10% with lowest TE

Data Set	Meassure	General Mean	Mean of the best 10% of maps
CODON	IR	40.2677	59.8557
CODON	NMV	0.0732	0.0791
ION	IR	1256.5	1712.1
ION	NMV	0.1998	.1767
IRIS	IR	7.5173	9.9331
IRIS	NMV	.0130	.0177
YEAST	IR	308.437	387.512
YEAST	NMV	0.0764	0.0841

Table 2. Correlation between the topographic error (TE) and the two parameters of our implementation of the algorithm: η_{inic}, initial learnig rate and U_{inic}, initial radius of the neighborhood

Data Set	Correlation with η_{inic}	Correlation with U_{inic}
CODON	.2710	-.2258
ION	.2127	-.3828
IRIS	.1760	-.2728
YEAST	.3901	-.3732

Table 3. Correlations between the two proposed measures and Topographic Error

Measure	Data Set	Correlation with TE
IR	CODON	-.4855
IR	ION	-.7896
IR	IRIS	-.6094
IR	YEAST	-.5865
NMV	CODON	-.4001
NMV	ION	-.2141
NMV	IRIS	-.5446
NMV	YEAST	-.4545

periodicity of different frequencies. The results showed if fig. 2 indicate that if there is a greatly dominant frequency, the topographic error is more likely to be high.

In the per neuron mean of the variance of coefficients (NMV) we see that in all data sets a low measure is more likely to give a high TE. This indicates that when, on average over all neurons, the distance between its weight vector and the input vector activating it resembles in the same degree, regardless of time, variations of different scales, the map is more likely to yield a high error measure. We can then say that the periodic nature of this time series must vary in time.

Considering TE a measure of the effectiveness of the organization process, it can be said that the above results indicate several features that this organization process is likely to have when being effective. This shreds some light on this yet to be understood process of self organization in the SOM.

We would like to see if large (in number of points) features in data require large and long in time processes within the SOM for them to be translated into features in the output map. If these sort of processes were taking place, a certain periodicity of small frequency would be present in the behavior of the SOM. The same could be said not only for large features in data, but for features of all sizes. If differently sized features were adequately presented in the final map a low TE would then result. Wavelet decomposition, due both to its multiresolution capabilities, and to the local manner of these, is the ideal tool for analyzing the behavior of the SOM algoirthm if we are trying to find out the scales of variations in its inner components. Our results suggest that a certain relation in fact exists between the diversity in scale of these variations and a good representation of differently-sized data in the final map.

The existence of this relation is more easily seen when we compare the correlations between topographic error and input parameters(Table 2) and the correlations between topographic error and our measures(Table 3).

References

1. Kraaijveld, M., Mao, J., Jain, A.: A non-linear projection method based on Kohonen's topology preserving maps. In: Proceedings of the 11th Int. Conf. on Pattern Recognition, pp. 41–45 (1992)
2. Flexer, A.: On the Use of Self-Organizing Maps for Clustering and Visualization. Intelligent Data Analysis 5, 373–384 (2001)
3. Kohonen, T.: Self-organized formation of topologically correct feature maps. Biological Cybernetics 43, 59–69 (1982)
4. Kohonen, T.: Self-Organizing Maps. Springer, Heidelberg (1995)
5. Haykin, S.: Neural networks a comprehensive foundation, 2nd edn. Prentice-Hall, Englewood Cliffs (1999)
6. Mallat, S.: A theory for multiresolution signal decomposition: the wavelet representation. IEEE Pattern Anal. and Machine Intell. 11(7), 674–693 (1989)
7. Haar, A.: Zur Theorie der orthogonalen Funktionensysteme. Math. Ann. 69, 331–371 (1910)
8. Walnut, D.: An Introduction to Wavelet Analysis. Birkhäuser, Basel (2001)
9. Daubechies, I.: Ten Lectures on Wavelets Society for Industrial and Applied Mathematics (1992)
10. Ogden, R.: Essential Wavelets for Statistical Applications and Data Analysis. Birkhauser, Basel (1997)
11. Kiviluoto, K.: Topology Preservation in Self-Organizing Maps. In: Proceedings of International Conference on Neural Networks, pp. 294–299 (1996)

Decreasing Neighborhood Revisited in Self-Organizing Maps

Antonio Neme[1], Elizabeth Chavez[2], Alejandra Cervera[3], and Victor Mireles[2]

[1] Universidad Autónoma de la Ciudad de México
[2] Universidad Nacional Autónoma de México
[3] Comisión Nacional Para el Conocimiento y Uso de la Biodiversidad,
México, D.F., Mex

Abstract. Decreasing neighborhood has been identified as a necessary condition for self-organization to hold in the self-organizing map (SOM). In the SOM, each best matching unit (BMU) decreases its influence area as a function of time and this area is always radial. Here, we present a model in which the BMU does not reduce its neighborhood, but the rest of the neurons exclude some BMUs from affecting them. In this model, what decreases as a function of time is the number of BMUs that affect each neuron, not the neighborhood of the BMUs. Each neuron identifies, from the set of BMUs that influenced it during each epoch, the farthest one and becomes refractory to it for the rest of the process. This BMU exclusion is not equivalent to the original decreasing neighborhood scheme. Even though the decreasing neighborhood condition is not totally maintained, self-organization remains, as shown by several experiments.

1   Introduction

The self-organizing map (SOM) is presented as a model of self-organization of neural connections, which is translated in the ability of the algorithm to produce organization from disorder [1]. One of the main properties of the SOM is the ability to preserve in the output map those topographical relations present in the input data [2], a very desirable property for data visualization and clustering. This property is achieved through a transformation of an incoming signal pattern of arbitrary dimension into a low-dimensional discrete map (usually one or two-dimensional) and by adaptively transforming data in a topologically ordered fashion [2,3]. Each input data is mapped to a single neuron in the lattice, to the one with the closest weight vector to the input vector, or best matching unit (BMU). The SOM preserves relationships during training through the learning equation, which establishes the effect each BMU has in any other neuron. Weight neurons are updated accordingly to:

$$w_n(t+1) = w_n(t) + \alpha_n(t)h_n(g,t)(x_i - w_n(t)) \tag{1}$$

where $\alpha(t)$ is the learning rate at time t and $h_n(g,t)$ is the neighborhood function from BMU neuron g to neuron n at time t. In general, the neighborhood function

V. Kůrková et al. (Eds.): ICANN 2008, Part I, LNCS 5163, pp. 671–679, 2008.
© Springer-Verlag Berlin Heidelberg 2008

decreases monotonically as a function of the distance from neuron g to neuron n. This decreasing property has been stated to be a necessary condition for convergence [4,5]. The SOM tries to preserve relationships in the input data by starting with a large neighborhood and reducing it during the course of training [2]. It has been reported as well that the learning factor α should be a decreasing function [5]. Both, neighborhood and learning parameter are reduced by an annealing scheme. several of these schemes have been proposed, although the form they take while decreasing is not critical [2,6].

The SOM has been studied as a model of cortical map formation [7]. However, the biological basis for the SOM is not completely plausible; for example, in the model, neurons rely on an homogeneous influence from activated neurons to all its neighbors, which contradicts physiological observations [8,9,10]. Active neurons do not equally affect all neurons within its influence area, neither are neurons equally affected by active neurons located at the same distance.

In this work, we propose a modification to the SOM algorithm such that BMUs does not decrease their neighborhood area, but neurons in the network become refractory to some BMUs. The refractory property states that the weight vector of a neuorn may not be modified by the influence of BMU even if it is located within its neighborhood. Those neurons that can not affect a given neuron are indeed the farthest BMUs during each epoch. Through the proposed scheme, we try to give the SOM a more biologically plausible behavior, and, as it is shown in the results, this modification leads to the formation of maps equivalent to those obtained by SOM.

2   The Modified Neighborhood Self-Organizing Map (MNSOM)

In the traditional SOM, each neuron decreases its area of influence as a function of time. In the discrete annealing scheme for neighborhood decreasing (bubble), each BMU reduces its area of influence by a certain value and all neurons located outside this area are not able to adapt its weight vectors. In the continuous neighborhood decrease, the affected neurons modify their weight vector as a function of a given kernel, mainly a Gaussian function [2].

In the model here introduced, hereafter modified neighborhood self-organizing map (MNSOM), the BMU does not decrease its area of influence (neighborhood), but the individual neurons may become refractory to some BMUs. Each BMU will try to affect all neurons within the specified neighborhood span, which does not decrease during the trainig process, but some neurons within this area of influence become refractory to the BMU.

Two sets of BMUs are defined in this model for each neuron i. The first one is the set of BMUs that affected i during the epoch t, $\gamma_i(t)$. The second set is composed by BMUs for whom i is refractory to, β_i. The main idea of the model is that i identifies a subset of $\gamma_i(t)$ that will enter β_i and thus, will not affect i for the remaining of the process. Once a BMU enters β_i, it can not be part of $\gamma_i(t)$ anymore.

How i selects BMUs from $\gamma_i(t)$ to enter β_i is critical to achieve good maps. At the beginning, β_i is empty. What we have found after several experiments is that the farthest BMUs on the first set are the best candidates to enter β_i. Once a BMU k enters β_i, it is not able to leave it. That is, i becomes refractory to k for the rest of the process.

The criteria here proposed is similar to the decreasing neighborhood scheme, as the farthest BMU will not affect anymore neuron i. However, these two schemes are not equivalent. Consider two neurons, i and j, located at the same distance from BMU k, and a BMU m located at a distance from both neurons i and j greater than that from k, but closer to j (see fig 1). In the SOM, once BMU k decreases its neighborhood, it will exclude, during the same epoch, both i and j. However, in the proposed model the behavior is different. Let $k \in \gamma_i(t)$ and $k \in \gamma_j(t)$. That is, k includes both i and j in its neighborhood. Now, suppose BMU m affected j but not i in epoch t, so $m \in \gamma_j(t)$ but $m \ni \gamma_i(t)$. As the criteria to leave $\gamma_i(t)$ is that of greater distance, i will select k from it and add to β_i, while j will select m and add it to β_j. Neuron i will be refractory to k while neuron j may still be affected by k but not by m.

At first glance, it is clear that non-radial areas of influence are formed, as k will affect j (as long as it is not the farthest BMU from j for an epoch) but will not affect i. Non-radial influence has been reported in several models [11,12,13]. However, this is not the only change in the map formation dynamics caused by the proposed scheme. Fig. 2 shows the average distance between neurons and the farthest BMU that affected them during an epoch and for several epochs. It is observed that, while in the SOM the average distance from neurons to the BMU decreases with time, in the MNSOM there is not such behavior.

Fig. 3 shows the lattice and the farthest BMU that affected neuron (0,0), upper left, at each epoch for both, SOM and MNSOM. The gray level of the BMU is dictated by the epoch in which it affected the referred neuron. Dark tones indicate an early influence (and thus, an early entering to β_i) while light

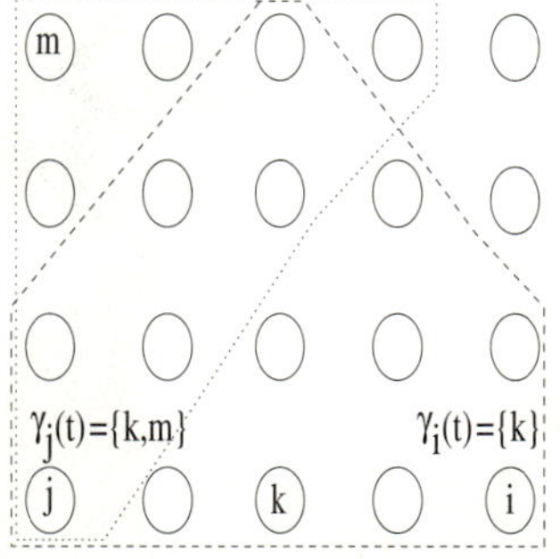

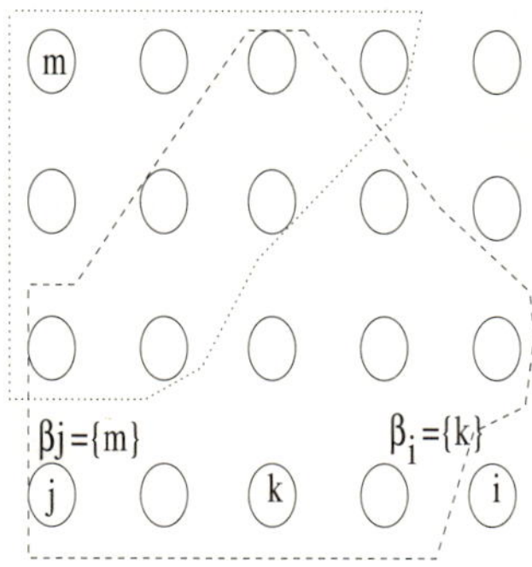

Fig. 1. Two BMUs's influence area. In the first image, neurons i and j became affected by one BMU (k) and two BMUs (k, m), respectively. According to the distance criteria, neuron i becomes refractory to k while neuron j becomes refractory to m (left image). As a result, non-radial influence areas emerge.

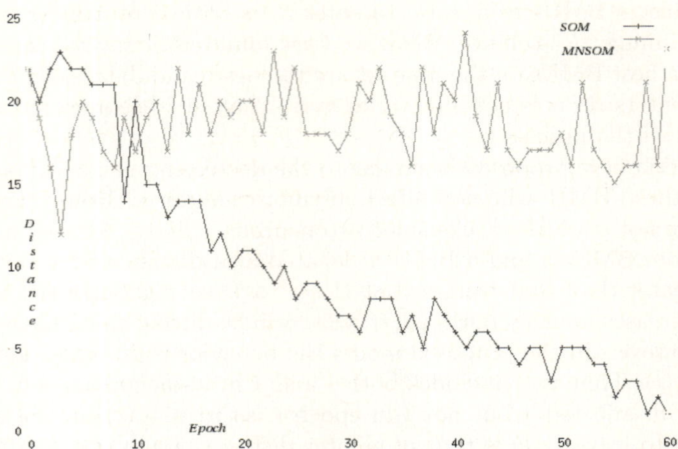

Fig. 2. Average distance between the farthest BMUs and neurons for SOM and MN-SOM. It is observed that the average distance decreases for SOM but remains for the MNSOM.

level represents a late influence on the neuron. It is observed that the number of BMUs that affected neuron (0,0) in the SOM is lower than the number of BMUs that affected the same neuron in the MNSOM, as neighborhood in the former case decreases exponentially, while in the later, BMUs are not impaired to affect it until they enter β_i.

In SOM, it is observed that the BMU tends to get closer to the affected neuron as a result of the decreasing neighborhood condition. The last BMU that affected

Fig. 3. Farthest BMU that affected neuron (0,0) (upper left) during the training process in a) SOM and b) MNSOM. The last BMU that affected the neuron is indicated by an arrow. The gray scale reflects the epoch: those BMUs with dark tones affected the neuron early while those BMUs in light levels affected it lately in the process.

neuron (0,0) is very close to it. In the MNSOM, the last BMU that affected it is not close. While in the original neighborhood annealing scheme the last BMU that affects a neuron is dictated only by $h_t(g, n)$, in the modified scheme the last BMU that will affect a neuron is that which has not already affected it, that is, the last BMU m in affecting a neuron i satisfies $m \ni \beta_i$.

Fig. 4 shows the average distance between the closest and farthest BMU for each epoch for all neurons. While the average distance tends to zero very fast in SOM, which means that only one BMU is affecting each neuron, the average distance in MNSOM presents a different behavior and tends to zero only at a lately stage during training. This effect is explained by the fact that in SOM the neighborhood is decreased symmetrically for all BMUs, while in MNSOM, the BMUs for whom neurons become refractory are not symmetrically distributed.

As there is an annealing scheme for neighborhood modification in the SOM, there is as well an annealing scheme for entering β_i, that is, for neurons to become refractory to BMUs. In the experiments here presented, the scheme is linear, that is, in every epoch each neuron becomes refractory to r BMUs. As the number of BMUs to which a neuron becomes refractory can not be greater than the number of neurons in the lattice, G, r should not be greater than G/number of epochs. If the number of epochs is larger than the number of neurons in the lattice, r should be set to one. From the experiments presented in the next section, we show that this r parameter is critical. The BMUs in $\gamma_i(t)$ are sorted from the closest one to the farthest one. After each epoch, the last r BMUs in $\gamma_i(t)$ are moved to β_i.

The traditional SOM lacks selectivity. In this model, neurons become refractory to specific BMUs, not to all BMUs within a given vicinity dictated by distance. A neuron i may be affected by BMUs at a distance $d(i, k)$ or greater,

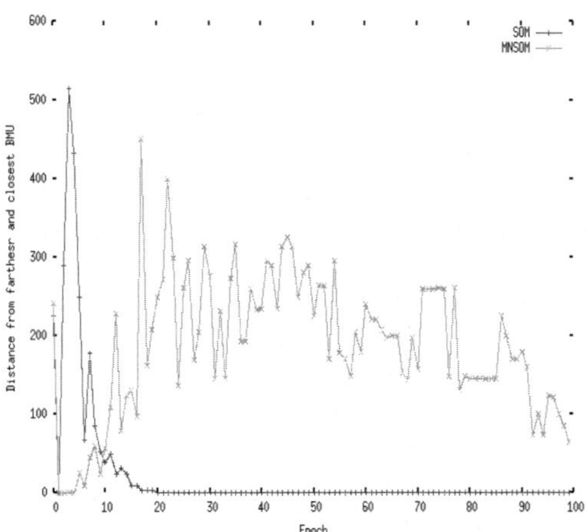

Fig. 4. Average distance between farthest and closest BMU for all neurons as a function of epoch. It is observed that in SOM, this distance decreases to zero, while in the MNSOM, this distance has a higher variance.

even if $k \in \beta_i$, which contrasts with the SOM, where a neuron i can not be affected by any BMU farther than the actual neighborhood span.

3   Results

In order to show how self-organization is possible in the presented model, several data sets were analyzed. The first set is a bidimensional spiral, for which the map folding is shown in fig 5. It is observed that the folding follows the spiral data set.

The second data set is an ordering of 4^k different inputs of dimension k, mapped over a $2^k \times 2^k$ grid. The idea is based on the Chaos Game Representation, from [15,16]. In this genome representation method, each sequence of k-length nucleotides is mapped to a certain cell of the grid and the relative frequency of each k-length nucleotide is represented as a gray scale over each cell. However, the cell ordering in the original representation suffers from discontinuity, as some adjacent cells represent very different sequences. For example, a cell represents

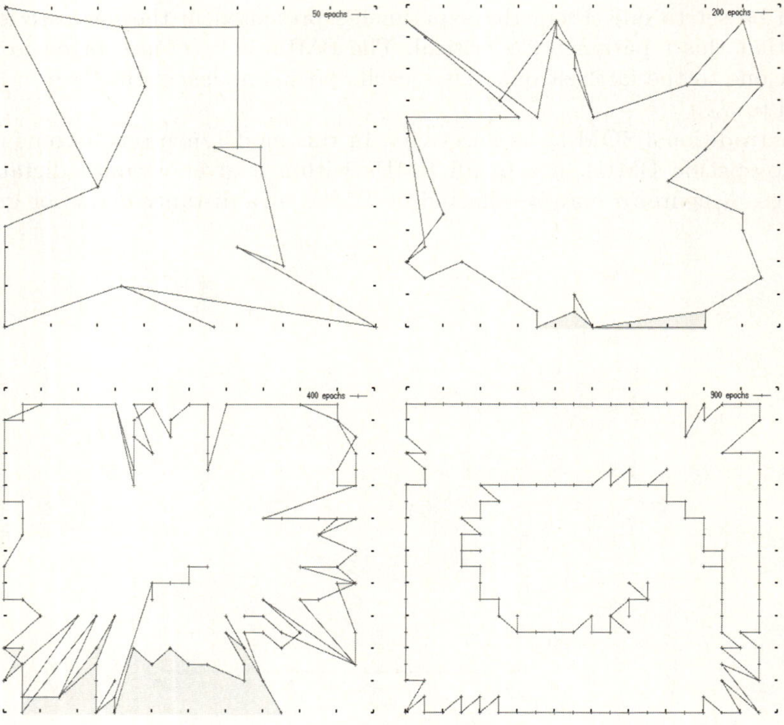

Fig. 5. Map folding for the the MNSOM for the spiral data set for 50, 200 400 and 900 epochs

TTA and one of its adjacent cells represents CCG. Discontinuities are measured as the number of different symbols over adjacent cells. In this example, the discontinuity between the two cells is 3, as $T \neq C$, $T \neq C$ and $A \neq G$. A perfect ordering would be that in which discontinuities over all adjacent cells are equal to 1, which, of course is not possible for several input vectors.

Applying the SOM algorithm to the input vectors leads to a new cell map, in which discontinuities are lower than in the original map. In the SOM, each neuron represents a cell to which the k-length sequence will be mapped. When the number of neurons (cells) equal the number of input vectors, there is also a lack of resolution, that is, several neurons do not map any input vector, while many of them are the BMU for several vectors. In the MNSOM model, the average discontinuity is as good as the SOM, but the number of unoccupied neurons is lower, as shown in table 1.

The iris data set (4-dimensional) was also analyzed with the MNSOM. Table 2 shows the ratio of topographic error (TE) [14] for the achieved maps through MNSOM and the topographic error obtained with SOM for these two data sets as a function of the influence span from BMUs. It is observed that, in general, the MNSOM maps are as good as those achieved through the SOM.

Although there is not a decreasing neighborhood in the MNSOM, it is important to identify how relevant is the neighborhood span in order to achieve

Table 1. Average distance between sequences in adjacent cells for several neighborhood spans (H). It is also shown the number of empty cells. Both measures were obtained from MNSOM. The average distance for the best map achieved by the traditional SOM is 3.32 with 143 empty cells. The lattice is 16x16 neurons and there are 256 input vectors, each four-dimensional.

H	Avg. distance	Empty cells
19	3.92	119
16	3.34	109
14	4.07	114
10	6.36	105

Table 2. Ratio of TE MNSOM maps and TE SOM for the iris data set for several epochs and number of BMUs entering β_i, r. The lattice is 20x20 neurons and the neighborhood for each BMU is 20. SOM map was trained 2000 epochs with an initial neighborhood of 20 and a final neighborhood of 0. $\alpha(0) = 0.05$ and $\alpha(final) = 0.00001$.

Epochs	TE ratio ($r = 2$)	TE ratio ($r = 4$)	TE ratio ($r = 6$)
200	1.04	0.78	1.01
400	0.83	0.76	1.04
800	0.96	0.81	0.66
1200	1.02	0.99	0.78
2000	1.01	0.99	1.03

Table 3. Ratio of TE MNSOM maps and TE SOM for the iris data set for 1000 epochs and $r = 2, 4, 6$. The lattice is 20x20 neurons. The neighborhood for each BMU (H) was varied to study the map quality.

H	TE ratio ($r = 2$)	TE ratio ($r = 4$)	TE ratio ($r = 6$)
15	1.01	0.99	0.95
10	0.97	1.0	1.0
5	0.99	1.0	1.0

good maps. Table 3 shows how the ratio of TE in MNSOM and that in SOM varies as a function of the neighborhood span. It is observed that the quality of the maps is similar to the quality of the SOM, even though the influence area of each BMU is kept very low (h = 5). The fact that the neighborhood area is not decreasing, but that neurons become refractory to distant BMUs explains this low error.

It is observed how r, the number of BMUs that enter β_i after each epoch, affects the quality of the maps.

4   Discussion and Conclusions

Although decreasing neighborhood has been identified as a necessary condition for self-organization to hold, we studied another possibility that leads to good maps. In the modification here presented, what decreases is not the BMU's neighborhood but the number of BMUs that affect each neuron. After each epoch, every neuron becomes refractory to some BMUs. Indeed, the BMUs to whom neurons become refractory are the farthest ones. However, this scheme is not equivalent to the original decreasing scheme.

In SOM, when a neuron becomes refractory to a BMU located at some distance, no BMU located at a distance greater than that may affect the neuron in sucesive epochs, as a consequence of the decreasing neighborhood. In the model here presented, this feature does not hold, as BMUs located at greater distances than those BMUs already unable to affect the neuron may still affect it.

The error measures of the MNSOM are equivalent to those of the SOM. However, we were not trying to improve the performance of the model, but to study the behavior when the neighborhood decreasing condition is not totally maintained. In this model, the dynamics of map formation are modified, but the resulting maps are similar. Although there is no mathematical proof, we have shown evidence that the maps obtained through this model are as good as those obtained with the SOM.

Finally, the biological plausibility of this model is also important, as in real neurons, there is a non-symmetrical influence between neurons. Selectivity in neural connections is present in this model, as a neuron becomes refractory to some BMUs, not to every one at a certain distance, as it occurs in the SOM.

References

1. Cottrell, M., Fort, J.C., Pagés, G.: Theoretical aspects of the SOM algorithm. Neurocomputing 21, 119–138 (1998)
2. Kohonen, T.: Self-Organizing maps, 3rd edn. Springer, Heidelberg (2000)
3. Ritter, H.: Self-Organizing Maps on non-euclidean Spaces Kohonen Maps. Oja, E., Kaski, S. (eds.), pp. 97–108 (1999)
4. Flanagan, J.: Sufficiente conditions for self-organization in the SOM with a decreasing neighborhood function of any width. In: Conf. of Art. Neural Networks. Conf. pub. No. 470 (1999)
5. Erwin, Obermayer, K., Schulten, K.: Self-organizing maps: Ordering, convergence properties and energy functions. Biol. Cyb. 67, 47–55 (1992)
6. Berglund, E., Sitte, J.: The parameterless Self-Organizing Map Algorithm. IEEE Trans. on Neural networks 17(2), 305–316 (2006)
7. Miikkulainen, R., Bednar, J., Choe, Y., Sirosh, J.: Computational maps in the visual cortex. Springer, New York (2005)
8. Rakic, S., Zecevic, N.: Emerging complexity of layer I in human cerebral cortex. Cerebral Cortex 13, 1072–1083 (2003)
9. Shefi, O., Golding, I., Segev, R., Ben-Jacob, E., Ayali, A.: Morphological characterization of in vitro neuronal networks. Phys. Rev. E 66, 21905(1–5) (2002)
10. Bailey, C., Kandel, E., SI, K.: The persistence of long-term memory. Neuron 44, 49–57 (2004)
11. Aoki, T., Aoyagi, T.: Self-organizing maps with asymmetric neighborhood function. Neural computation 19(9), 2515–2535 (2007)
12. Neme, A., Miramontes, P.: A parameter in the SOM learning rule that incorporates activation frequency. ICANN (1), 455–463 (2006)
13. Lee, J., Verleysen, M.: Self-organizing maps with recursive neighborhood adaption. Neural Networks 15, 993–1003 (2002)
14. Kiviluoto, K.: Topology preservation in Self-Organizing maps. In: Proc. ICNN 1996, IEEE Int. Conf. on Neural Networks (1996)
15. Almeida, J., Carrico, J., Maretzek, A., Noble, P., Fletcher, M.: Analysis of genomic sequences by chaos game representatio. Bioinformatics 17(5), 429–437 (2001)
16. Jeffrey, J.: Chaos game representation of sequences in Chaos and Fractals: a computer graphical journey. Pickover (1998)

A New GHSOM Model Applied to Network Security

E.J. Palomo, E. Domínguez, R.M. Luque, and J. Muñoz

Department of Computer Science
E.T.S.I.Informatica, University of Malaga
Campus Teatinos s/n, 29071 – Malaga, Spain
{ejpalomo,enriqued,rmluque,munozp}@lcc.uma.es

Abstract. The self-organizing map (SOM) have shown to be successful for the analysis of high-dimensional input data as in data mining applications such as network security. However, the static architecture and the lack of representation of hierarchical relations are its main drawbacks. The growing hierarchical SOM (GHSOM) address these limitations of the SOM. The GHSOM is an artificial neural network model with hierarchical architecture composed of independent growing SOMs. One limitation of these neural networks is that they just take into account numerical data, even though symbolic data can be present in many real life problems. In this paper a new GHSOM model with a new metric incorporing both numerical and symbolic data is proposed. This new GHSOM model is proposed for detecting network intrusions. An intrusion detection system (IDS) monitors the IP packets flowing over the network to capture intrusions or anomalies. One of the techniques used for anomaly detection is building statical models using metrics derived from observation of the user's actions. Randomly selected subsets that contains both attacks and normal records from the KDD Cup 1999 benchmark are used for training the proposed GHSOM. Experimental results are provided and compared to other hierarchical neural networks.

1 Introduction

Cluster analysis is the automatic assignment of a set of input data to a group or cluster. These input data are usually represented as features vectors of a high-dimensional space. The assignment involves a partition of the input patterns, where patterns with similar features are assigned to the same cluster or partition. The unsupervised learning methods try to identify the groups belonging to unlabeled input patterns. These methods are especially useful when we have no information about the input data.

The self-organizing map (SOM) constitutes an excellent tool for knowledge discovery in a data base, extraction of relevant information, detection of inherent structures in high-dimensional data and mapping these data into a two-dimensional representation space [1]. Since this mapping is adaptively done and preserves the topology of input data, they have been applied successfully in multiple areas. By preserving the topology of input data, the relations among input

V. Kůrková et al. (Eds.): ICANN 2008, Part I, LNCS 5163, pp. 680–689, 2008.

data are retained and the visualization of the clusters is made easy. Therefore, understanding of the structure of the data is provided. However, SOMs have some drawbacks. One of them arises because the network architecture has to been established in advance, and this task is usually difficult in a lot of problems of the real world due to the high dimensionality of the input data. Moreover, the architecture is not adapted during the learning process and, consequently, the map could not be an adequate representation of the input data. Another drawback is the hierarchical relations among input data. These hierarchical relations are difficult to detect and can lead to a worse understanding of the data.

A growing hierarchical SOM (GHSOM) intend to solve these difficulties. This neural network consists of several SOMs structured in layers, whose number of neurons, maps and layers are determined during the unsupervised learning process. Thus, the structure of the GHSOM is automatic adapted according to the structure of the data. However, the proposed GHSOM by [2], uses the Euclidean distance as metric to compare two input data. This can be useful for vectors with numerical features, but it is not appropriate for vectors where symbolic features are present. Since values from symbolic features do not have an order associated, it seems more suitable the use of a similarity measure instead of a distance measure. For that reason, a new GHSOM model based on a metric where symbolic data are considered is proposed in this paper. The new metric is based on the entropy and compares elements of symbolic features from their occurrence probabilities. The entropy concept is used to compute the representation quality of a unit and to check wheter the symbolic values are the same or not.

The proposed new GHSOM model is used to implement an Intrusion Detection System (IDS). An IDS usually has two approaches. The firs approach is known as misuse detection, which detect attacks storing the signatures of previously known attacks. This approach fails detecting unknown attacks or variants of known attacks. The second approach is known as the anomaly detection approach, where deviants from normal behaviour are considered attacks. According to this point of view, our IDS is an anomaly detection system, since detects new attack types and normal behaviour. Some anomaly detection systems using data mining techniques such as clustering, support vector machines (SVM) and neural network systems have been proposed [3,4,5]. Several IDS using a SOM have been done, however they have many difficulties detecting a wide variety of attacks with low false positive rates [6].

The proposed IDS detect anomalies or attacks by classifying IP connections into normal or anomalies connection records, and the type of attack if they are anomalies. In order to train our IDS based on GHSOM, the KDD Cup 1999 benchmark data has been used [7,8]. This data set has served as the first and only reliable benchmark data set that has been used for most of the research work on intrusion detection algorithms [9]. It contains a wide variety of simulated attacks in a military network environment. Since these connection records have both numerical and symbolic data, our GHSOM model results very appropriate to face this problem and yields good results in terms of detection rate and false positives rate.

The remainder of this paper is organized as follows. Section 2 provides a description of the GHSOM model. The new metric used in our GHSOM model is presented afterwards. In Section 3, we present an IDS built from our new GHSOM model, where symbolic data are found. Then, some experimental results are presented. Section 4 concludes this paper.

2   New GHSOM Model

The SOM constitutes an excellent tool for knowledge discovery and detection of inherent structures in high-dimensional data. However, it has some important drawbacks as mentioned before. One of them is the necessity of establish the network architecture in advance. Furthermore, the architecture is not adaptable during the learning process. The GHSOM intend to overcome these difficulties.

A GHSOM is a hierarchichal structure of several layers, whose number of neurons, maps and layers are determined during the unsupervised learning process. This neural network is initialized as a map with an initial number of neurons at the first layer. The training process is done in a top-down way, starting from the first-layer map. Each map is based on the growing grid [10], where a row or a column of neurons can be added during the training process to represent a set of data at a certain level of detail. After the map has reached this level of detail, each neuron is analyzed to check a measure of data heterogeneity. Neurons that represent heterogeneous data will be expanded in a new map at the next layer, where the input data will be represented in more detail. This process is repeated for each map, until each neuron does not require further expansion. Thus, the final architecture of the GHSOM mirrors the structure of the input data, therefore the number of neurons, maps or layers are depended on the input data. An example of the structure of a GHSOM is given in Fig. 1. This GHSOM has just two layers, where the first-layer map has 2x2 units (neurons). Layer 2 has two maps due to the expansion of two neurons from the first-layer. Note that the second map of this layer has grown more than the first one in order to improve the data representation. The layer 0 represents all the input data and is used to control the hierarchical growth process.

Growing of a GHSOM is controlled by two parameters that have to be setup in advance: τ_1, that is used to control the growth of a map; and τ_2, that is used to control the hierarchical growth, that is, the number of expanded maps. This control is based on the quantization error of a unit (qe), which is used to measure the representation quality of a unit. Usually, this quantization error has been used as the mean Euclidean distance between the weight vector of a unit i and the elements from input data that are mapped onto this unit i [2]. However, this metric is not suitable for input data where symbolic data are present. Since symbolic data set have not defined an order, it seems more appropriate to use a similarity measure rather than a distance measure.

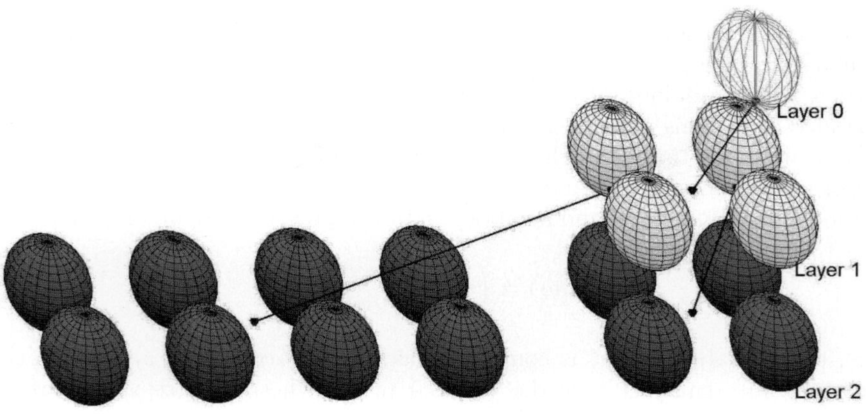

Fig. 1. Sample architecture of a GHSOM

In this paper, a new metric including both numerical and symbolic data is proposed. The Euclidean distance and the entropy is used as measure of error in the representation (quantization error) of a unit for numerical and symbolic data, respectively.

Let $x_j = \begin{pmatrix} x_j^n \\ x_j^s \end{pmatrix}$ be the jth input pattern, where x_j^n is the vector component of numerical features and x_j^s is the component of symbolic features. The error of a unit i (e_i) in the representation is defined as follows

$$e_i = \begin{pmatrix} e_i^n \\ e_i^s \end{pmatrix} = \sum_{x_j \in C_i} \begin{pmatrix} \|w_i^n - x_j^n\| \\ -p(x_j^s)\log(p(x_j^s)) \end{pmatrix} \tag{1}$$

where e_i^n and e_i^s are the error components of numerical and symbolic features, respectively, C_i is the set of patterns mapped onto the unit i, and $p(\cdot)$ is the probability of the element in C_i. The quantization error of the unit i is given by expression (2).

$$qe_i = |e_i| \tag{2}$$

The initial step for the training process is the calculation of the error at layer 0 as provided in (3). With w_0 we refer to the mean of the all input data I.

$$e_0 = \sum_{x_j \in I} \begin{pmatrix} \|w_0^n - x_j^n\| \\ -p(x_j^s)\log(p(x_j^s)) \end{pmatrix} \tag{3}$$

Note that the qe_0 measures the dissimilarity of all input data and it is used to control the growth of the neural network. Specifically, all units must represent their mapped subsets of data better than the unit 0, i.e., the quantization error of any unit (qe_i) must be smaller than a fraction (τ_2) of the quantization error of the unit 0 (qe_0) as specified in (4). The units not satisfying this condition are expanded in order to achieve a more detailed data representation.

$$qe_i < \tau_2 \cdot qe_0 \tag{4}$$

The newly created map from an expanded unit is trained according to the standard training procedure as continuously described. Each training iteration t starts with the random selection of an input pattern x mapped onto the upper expanded unit. This pattern is presented to the map and each unit determines its activation. In this case, the unit with the smallest euclidean distance and entropy to the input pattern is activated. This similarity measure between two vectors, where numerical and symbolic data are present, is defined as follows:

$$\delta(v_1, v_2) = \begin{pmatrix} \|v_1^n - v_2^n\| \\ -p(v_2^s)\log(p(v_2^s)) \end{pmatrix} \tag{5}$$

Here, the probability of v_2^s is computed taking into account v_1^s and v_2^s, so that the probability can just take the values 1 or 0.5 whether these values are the same or not, respectively. Therefore, the similarity between two symbolic values will be 0 if they are the same or 0.5 if they are different, that is, there is no similarity when the symbolic values are the same and the dissimilarity is always the same if they are different from each other regardless their values. This way, this similarity measure checks whether the symbolic values are the same or not. The activated unit is referred to as the winner, whose index is defined in (6).

$$r(t) = \arg\min_i |\delta(w_i, x(t))| \tag{6}$$

The winner is adapted according to the expression (7). Note that for numerical data the amount of adaptation is guided by a decreasing learning rate α and depends on a neighborhood function h_i. This neighborhood function affects to the number of units adapted and the adaptation proportion depending on the position of the unit. Here, we use a Gaussian neighborhood function with reduction of its neighborhood kernel in each iteration.

$$w_i(t+1) = \begin{pmatrix} w_i^n(t+1) \\ w_i^s(t+1) \end{pmatrix} = \begin{pmatrix} w_i^n(t) + \alpha(t)h_i(t)[x^n(t) - w_i^n(t)] \\ mode(C_i) \end{pmatrix} \tag{7}$$

For symbolic data, we neither use a neighborhood function nor a learning rate. This is because the values of symbolic features cannot be different from the existing values, that is, we cannot create new values for symbolic data. For that reason, to adapt $w_i^s(t+1)$, we choose the symbolic data value that constitutes the best representative of the input patterns mapped onto the unit i. This best representative is the mode of the set of the input patterns mapped onto the unit i (C_i). Note that to choose the winner (6), the input pattern in the iteration t is compared with the mode of the input patterns mapped onto each unit in the same iteration. This way, for symbolic data the winner will be the unit whose mode is the same that the input pattern $x^s(t)$.

After training a map m for λ iterations, the quantization error of each neuron is computed according to expression (2) and the mean of them (MQE_m) indicating the mean of the quantization errors of the map. In order to evaluate the stopping criterion for the growth of the map defined as $MQE_m < \tau_1 \cdot qe_u$, where qe_u is the qe of the corresponding unit u in the upper layer that was expanded.

If the MQE_m is bigger than a fraction τ_1 of qe_u, a row or a column of units will be inserted between two units, the unit with the highest quantization error e and its most dissimilar neighbor d. Then, this unit d is computed as follows:

$$d = \arg\max_i |\delta(w_i, w_e)| \qquad w_i \in \Lambda_e \tag{8}$$

where Λ_e is the set of neighbor units of e. Here, the probabilities of the all weight vectors are computed. When a row or a column of units is inserted, their weight vectors were initialized as the average of their corresponding neighbors as proposed in [11]. We use this method to initialize numerical features of weight vectors, but with symbolic features we use the mode rather than the mean, since we want that an inserted unit has the better existing symbolic features that represents its neighbouring environment.

When a map training and growing has finished, the stopping criterion of units expansion has to be evaluated. The qe measures the homogeneity or density of data of a unit, so the unit will be expanded in a new map if its qe is higher than a certain fraction (τ_2) of the qe_0, and the global stopping criterion of units expansion is defined as $qe_i < \tau_2 \cdot qe_0$. From the expanded unit, a new map of 2x2 units is created. To provide a global orientation of the indivual maps in the various layers of the hierarchy, a coherent initialization of the units of the new map was proposed in [11]. Thus, the weight vectors of units have to mirror the orientation of the neighbor units of its parent. The initialitazion proposed computed the mean of the parent and its neighbors in the respective directions. Again, we use this initialization method just for numerical features, but we propose other initialization for symbolic data based on the mode. The initialization of the symbolic features of a unit will be the mode of the parent and its neighbors in the respective directions, where the parent is taken into account twice. In this way, the parent has more weight than their neighbors, so that a symbolic feature of a unit will be initialized with the symbolic feature of the parent unless three neighbors has the same symbolic feature.

3 Experimental Results

In this paper, we have applied our new GHSOM model to solve network security problems. Concretely, we have built an Intrusion Detection System (IDS) to detect a wide variety of attacks in a network environment. To train our GHSOM, we are used the pre-processed KDD Cup 1999 benchmark data set created by MIT Lincoln Laboratory. The purpose of this benchmark was to build a network intrusion detector capable of distinguishing between intrusions or attacks, and normal connections. We have used the 10% KDD Cup 1999 training data set, which contains 494021 connection records, each with 41 features. This training data set contains 22 different attack types and normal type records. Among the training set features, three are symbolic, the related with protocol type, service and status of the connection flag; the rest are numerical.

For our experiments, we have selected two data subsets from the 494021 connection records: Set_100000 with 100000 connection records and Set_169000 with

169000. Both Set_100000 and Set_169000 contain the 22 attack types. We used a selection algorithm that, given the number of connection records and the number of different record types we want to select, tries to retrieve the data subset where the record types have the same distribution. But here it is not possible because the record types of training data have not the same distribution, so that the final distribution is adjusted according to the training data distribution. To train these two data subsets, we have chosen 0.1 as value for parameters τ_1 and τ_2, which have yielded two compact and very similar neural networks with just two layers. Set_100000 and Set_169000 generated just 16 neuronas during the training, although with different arrangement. The resulting GHSOM trained with Set_169000 is the same that the GHSOM showed in Fig. 1. Most of the related works, are only interested in knowing whether the input pattern is an attack or a normal connection record. Normal connection records that are identified as attacks, are called false positives. Otherwise, attacks identified as normals are called missed detections. However, we are also interested in knowing whether one attack type is identified as its proper type of attack instead another type of attack. That is, we want to know the global classification rate of our GHSOM taking into account the record type (group) in which an input pattern is classified. Taking into account this, the training results for Set_100000 and Set_169000 are given in Table 1. We can see that during the training with Set_100000, we achieve 99.98% detection rate and 3.03% false positive rate. The training with Set_169000 achieved 99.98% detection rate and a false positive rate of 5.74%, the same detection rate but with a high false positive rate. Regarding the global classification rate, we achieve 94.99% for Set_100000 and 95.09% for Set_169000.

Table 1. Training results of the proposed GHSOMs for different data subsets

Training Set	Detected(%)	False Positive(%)	Global Identified(%)
Set_100000	99.98	3.03	94.99
Set_169000	99.98	5.74	95.09

After the training, we test the trained GHSOMs by doing a simulation with the 10% KDD Cup 1999 training data set that contains 494021 connection records. The simulation is done without learning or growing. This way, we can see how the resulting GHSOMs classify a big set of input data. The simulating results are shown in Table 2. We achieve 99.99% and 99.91% detection rate and false positive rates of 5.41% and 5.11%, respectively. Regarding the global classifiaction rate, we achieve 97.09% for Set_100000 and 97.15% for Set_169000. Therefore, after the simulation we increased the detection rate until almost 100% and the global identified rate until 97% in both cases. The false positive rate is around the 5%, although we prefer higher false positive rates than lower detection rates.

Note that in this paper, we are using all the features of the connection records and the 22 training attack types both during the training and testing. In order to implement an IDS as presented in this paper, the hierarchical Kohonen net (K-Map) was proposed in [9]. This hierarchical K-Map has three static levels or

Table 2. Simulating results of the proposed GHSOMs with 494021 records for different data subsets

Training Set	Detected(%)	False Positive(%)	Global Identified(%)
Set_100000	99.99	5.41	97.09
Set_169000	99.91	5.11	97.15

layers, where each layer is a single K-Map or SOM. The best testing results of this IDS was 99.63% detection rate, but taking into account several limitations. They used a training set with 169000 connection records and 22 attack types as we used. However, it was tested with just three attack types while we used 22 attack types again. Their testing was also done for a pre-specified combination of 20 features, concretely, 4 for the first layer, 7 for the second and 9 for the third. Moreover, these three connected SOMs were established in advance using 48 neurons in each level, while we have used a GHSOM whose structure is determined during the training process and generated just 16 neurons. This way, we do not need further analyses from the problem domain to initialize neither the number of neurons or layers nor features or attacks subsets. The differencies between the features of these two IDS models are shown in Table 3. Furthermore, our good results are due to the distinction between symbolic and numerical features present in training data, that was considered in our new GHSOM model.

Table 3. Architectures and testing modes comparison results

	Layers	Neurons	Tested Attacks	Tested Features
Hierarchical K-Map	3	144	3	20
GHSOM	2	16	22	41

4   Conclusions

In this paper, we have proposed a new GHSOM model. This novelty model improves the GHSOM by taking into account the presence of symbolic data in input patterns. Since symbolic data have no an order associated, it seems more reasonable the use of a similarity measure rather than a distance measure. Self-organizing neural systems uses the Euclidean distance as distance measure. Here, we keep using this distance measure, but just for numerical data. Regarding to symbolic data, we use a new metric based on entropy. Since entropy constitues a measure of the amount of uncertainty associated with a data set, this measure can be used as the amount of "error" in the representation of symbolic data of a process unit. Moreover, entropy allows us to compare two symbolic values to see whether they are the same or not. This idea is used as similarity measure for symbolic data.

The new GHSOM model was used to implement an Intrusion Detection System (IDS). Training and simulating of our IDS was carried out using the KDD Cup 1999 benchmark data set, where numerical and symbolic data are found. Specifically, we used the 10% KDD Cup 1999 benchmark, which contains 494021 connection records. For the training, we selected two data subsets with 100000 and 169000 connection records, respectively. Then, we simulated them with the 494021 connection records from the 10% KDD Cup 1999 benchmark. During the simulation with both subsets, we were able to achieve a 99.99% and 99.91% (almost 100%) detection rate, with a false positive rate between 5.41% and 5.11%, respectively (see Table 2). The best results of the hierarchical Kohonen net (K-Map) proposed in [9] were 99.63% detection rate, but limiting the testing to three attack types, 20 pre-specified features and using 48 neurons in each of the three layers they used (see Table 3). Furthermore, the number of neurons and layers were established in advance, whereas our IDS implementation adapted its structure according to the input data and mirroring their hierarchical relations.

Acknowledgements

This work is partially supported by Spanish Ministry of Education and Science under contract TIC-07362, project name Self-Organizing Systems for Internet.

References

1. Kohonen, T.: Self-organized formation of topologically correct feature maps. Biological cybernetics 43(1), 59–69 (1982)
2. Rauber, A., Merkl, D., Dittenbach, M.: The growing hierarchical self-organizing map: Exploratory analysis of high-dimensional data. IEEE Transactions on Neural Networks 13(6), 1331–1341 (2002)
3. Lee, W., Stolfo, S., Chan, P., Eskin, E., Fan, W., Miller, M., Hershkop, S., Zhang, J.: Real time data mining-based intrusion detection. In: DARPA Information Survivability Conference and Exposition II, vol. 1, pp. 89–100 (2001)
4. Maxion, R., Tan, K.: Anomaly detection in embedded systems. IEEE Transactions on Computers 51(2), 108–120 (2002)
5. Tan, K., Maxion, R.: Determining the operational limits of an anomaly-based intrusion detector. IEEE Journal on Selected Areas in Communications 21(1), 96–110 (2003)
6. Ying, H., Feng, T.J., Cao, J.K., Ding, X.Q., Zhou, Y.H.: Research on some problems in the kohonen som algorithm. In: International Conference on Machine Learning and Cybernetics, vol. 3, pp. 1279–1282 (2002)
7. Lee, W., Stolfo, S., Mok, K.: A data mining framework for building intrusion detection models. In: IEEE Symposium on Security and Privacy, pp. 120–132 (1999)
8. Stolfo, S., Fan, W., Lee, W., Prodromidis, A., Chan, P.: Cost-based modeling for fraud and intrusion detection: results from the jam project. In: DARPA Information Survivability Conference and Exposition, 2000, DISCEX 2000. Proceedings, vol. 2, pp. 130–144 (2000)

9. Sarasamma, S., Zhu, Q., Huff, J.: Hierarchical kohonenen net for anomaly detection in network security. IEEE Transactions on Systems Man and Cybernetics Part B-Cybernetics 35(2), 302–312 (2005)
10. Fritzke, B.: Growing grid - a self-organizing network with constant neighborhood range and adaptation strength. Neural Processing Letters 2(5), 9–13 (1995)
11. Dittenbach, M., Rauber, A., Merkl, D.: Recent advances with the growing hierarchical self-organizing map. In: 3rd Workshop on Self-Organising Maps (WSOM), pp. 140–145 (2001)

Reduction of Visual Information in Neural Network Learning Visualization

Matúš Užák, Rudolf Jakša, and Peter Sinčák

Center for Intelligent Technologies
Department of Cybernetics and Artificial Intelligence
Technical University of Košice, Slovakia
`uzak@neuron.tuke.sk`, `jaksa@neuron.tuke.sk`, `peter.sincak@tuke.sk`

Abstract. Visualization of neural network learning faces the problem of overwhelming amount of visual information. This paper describes the application of clustering methods for reduction of visual information in the response function visualization. If only clusters of neurons are visualized instead of direct visualization of responses of all neurons in the network, the amount of visually presented information can be significantly reduced. This is useful for user fatigue reduction and also for minimization of the visualization equipment requirements. We show that application of Kohonen network or Growing Neural Gas with Utility Factor algorithm allows to visualize the learning of moderate-sized neural networks. Comparison of both algorithms in this task is provided, also with performance analysis and example results of response function visualization.

1   Introduction

Visualization of artificial neural networks learning allows researchers to observe processes running inside the neural network during the learning. This is useful for learning algorithms analysis, for estimation of the progress of learning, or to allow interaction with the learning algorithm [1].

Several methods are available for the visualization of the learning of neural network. Craven and Shavlik [2] provide the survey of visualization methods including Hinton diagrams, bond diagrams, hyperplane diagrams, response-function plots, and trajectory diagrams.

Response function visualization method shows responses of individual neurons to given range of inputs of network. The neural network is fed with all possible input combinations from the given range (for inst. $x_1 \in \langle 0, 1 \rangle$, $x_2 \in \langle 0, 1 \rangle$), while activations of all hidden neurons and all outputs are collected and displayed. Two inputs to a neural network with 100 samples per each input produce 10000 combinations. Responses to these 10000 input patterns on every neuron are displayed as 100×100 pixel pictures where the color of every pixel represents actual response to given single input combination (See Fig. 1). This method is best suited for two-input networks when two-dimensional plots are drawn [3][1].

---

[1] Lang and Witbrock were among the first to use response function plots (on the famous two-spirals task, which is two-dimensional).

V. Kůrková et al. (Eds.): ICANN 2008, Part I, LNCS 5163, pp. 690–699, 2008.

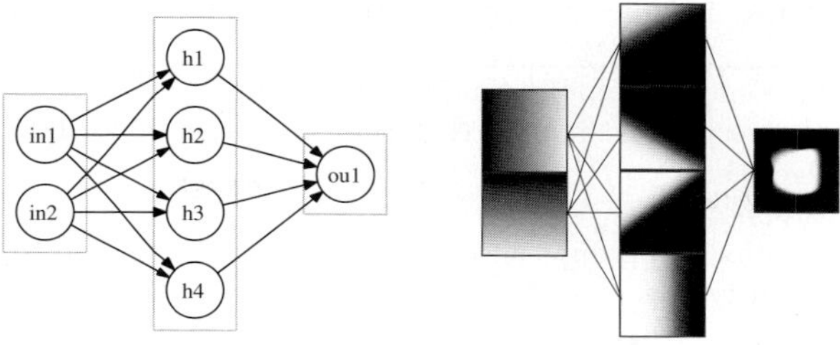

Fig. 1. Responses of neurons in Multilayer Perceptron to the inputs from $\langle 0, 1\rangle \times \langle 0, 1\rangle$

Visualization of the responses of all neurons in a neural network with 100 neurons, and the 100×100 pixel resolution requires $100 \times 100 \times 100 = 1\,million$ pixels, which is the full resolution of 1280×1024 pixel display. The 1600×1200 pixel display contains approximately $2\,million$ pixels which in our example provides the capacity to display 200 neurons. The size of the network can limit the ability to visualize responses of neurons directly due to exceeding the display possibilities.

Further problem with visualization of responses of larger number of neurons is the human ability to follow the changes in large number of pictures simultaneously. In the interactive evolutionary computation image processing scenarios (IEC-IP) usually 12 images are presented to the observer in parallel [4]. This is approximately 10 times less then in our 100 neuron neural network visualization example. If we accept the IEC-IP task to be similarly demanding to the human user as the response function visualization, the size of neural network can easily limit human ability to follow the visualization.

To deal with the described limits on visualization capacity clustering methods can be employed. If clusters of 10 neurons each (neurons which have similar responses) are constructed and only the representatives of these clusters are visualized the visualization requirements will be reduced 10 times.

2   Clustering of Neuron Responses

Information flood by the visualization of responses of neurons of larger neural networks can be reduced using clustering of these responses and by further processing of the representatives of the clusters only. Clustering might be performed in several different manners. The method from Table 1 will be used and Fig. 2 illustrates the whole clustering process. A scale of clustering algorithms is available to perform the task of clustering. We will use the Kohonen network [5], the more advanced Growing Neural Gas algorithm [6], and the Growing Neural Gas with the Utility Factor algorithm [7].

The Kohonen network is an adaptive clustering method introduced in 1982 [5,8]. The space in which the clusters are formed is geometrically aligned and has to be predefined. Order of clusters or their similarity can be extracted from this network. The survey of clustering algorithms for application in the Interactive

Table 1. Procedure of clustering of neuron responses

1. Compute the responses of all neurons in the network,
2. propagate the responses in full resolution (for inst. 100×100 pixel) into the clustering algorithm,
3. find the clusters,
4. divide all neurons into groups according to their membership to clusters,
5. get a representative response for every cluster (from the clustering algorithm),
6. visualize only the representative responses.

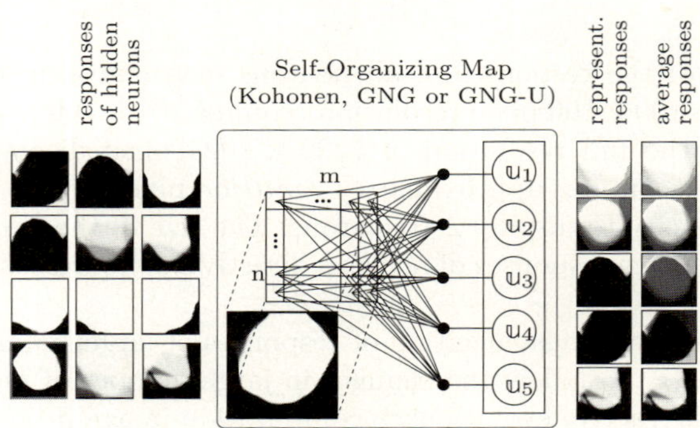

Fig. 2. Response function plots of the hidden neurons are used as the input data for the clustering algorithm. Outputs of the clustering algorithm (Kohonen,GNG,GNG-U) are centers of the clusters, which can be visualized as representative responses directly. We can also construct the representative response as the average of responses of all members of the cluster.

Evolutionary Computation (IEC) [9] finds Kohonen network to produce good distribution of clusters (for IEC purposes[2]) with acceptable computational load.

Growing Neural Gas algorithm (GNG) by Fritzke (1995) [6] provides some advancements in the adaptation qualities over basic Kohonen network. The dynamics of the adaptation is more suitable for tracking non-stationary data distribution. Also the geometrical alignment of the output space is self-constructed and dynamically adapted by the algorithm. The Growing Neural Gas with Utility Factor algorithm (GNG-U) by Fritzke (1997) [7] is a further advancement of the GNG algorithm with improved adaptivity to the changing data distribution. The GNG-U algorithm is designed to effectively remove clusters which are no longer useful and to construct new ones. The experimental comparison of GNG and GNG-U can be found in [11]. The dynamics of neural network learning can show sudden changes therefore the GNG-U algorithm should be suitable for clustering of responses of neurons during the learning. The Kohonen network has only two control parameters however, while GNG (GNG-U) has six (seven) user adjustable parameters.

---

[2] The process of visualization and manual intervention in IEC applications [10] is similar to the usage of visualization of neural network learning.

3   Description of the Clustering Algorithms Used

In this section we will briefly describe Kohonen network [5,8], GNG [6] and GNG-U [7] algorithms which will be used for clustering of responses of neurons.

The Network $\mathbb{N} = \{u_1, u_2, \ldots, u_N\}$ consists of N units u_i and has M inputs x_j. Each unit u_i has an associated reference vector $\overline{w}_i \in \mathbb{R}^M$, indicating its position in the M-dimensional input space. Input patterns $\overline{x}^p \in \mathbb{R}^M$ are drawn from a finite training data set $\mathbb{P} = \{\overline{x}^1, \overline{x}^2, \ldots, \overline{x}^P\}$. For a given input signal $\overline{x}^p$ the winner $u_* = \arg\min_{u_i \in \mathbb{N}} \|\overline{x}^p - \overline{w}_i\|$ among the units in $\mathbb{N}$ is the unit with the reference vector closest to $\overline{x}^p$.

In the Kohonen network the reference vectors $\overline{w}_i$ are updated by rule (1) for every input pattern $\overline{x}^p$.

$$\Delta \overline{w}_i = \gamma \Lambda_{*i}(\overline{x}^p - \overline{w}_i), \qquad \Lambda_{ij} = e^{-\frac{\|\overline{u}_i - \overline{u}_j\|}{r(t)}} \tag{1}$$

Parameter γ is the learning rate. The neighborhood function Λ_{ij} defines neighborhood relation of units i and j. The neighborhood radius $r(t)$ converges to zero during the training time $t \in \mathbb{T}$.

In the GNG network the reference vector of the winner $\overline{w}_*$ and also reference vectors $\overline{w}_o$ of its N_o direct topological neighbors[3] are updated by rule (2). For every input pattern $\overline{x}^p$, $\overline{w}_*$ and $\overline{w}_o$ are shifted by γ_* and γ_o fractions of their distances from $\overline{x}^p$ towards the $\overline{x}^p$.

$$\Delta \overline{w}_* = \gamma_*(\overline{x}^p - \overline{w}_*), \qquad \Delta \overline{w}_o = \gamma_o(\overline{x}^p - \overline{w}_o) \tag{2}$$

A new unit k is inserted into GNG network $\mathbb{N}$ after every π_k presentations of training data-set $\mathbb{P}$. It is inserted between the unit k_1 with maximal accumulated error e_i and its neighbor with the maximal error k_2. The original connection is replaced by two new connections to the new unit k. The errors of units k_1 and k_2 are reduced by the parameter α and the error of the new unit k is set to the new error of k_1. Generally, the error e_i is specific for every unit i and it is adapted for every winner as $\Delta e_* = \|\overline{x}^p - \overline{w}_*\|$. Errors of all units are subjected to exponential decay $\Delta e_i = -\beta e_i$ after each adaptation step.

The maximal number of clusters in GNG is limited. Also if any connection becomes older than τ_c, the edge is removed from the network. If the unit will lose all its connections it is removed too. The age of connections is incremented for all edges of the winner u_* by one. The age of connection between winner u_* and 2nd winner[4] u_{*2} (which is created if necessary) is set to zero.

The GNG-U algorithm adds a rule, that unit i with the minimal utility value[5] u_i is removed from the network when $e_{max}/u_i > \eta$, where η is a constant and e_{max} is the maximal error among all units in the network. Utility value u_i is specific for every unit and it is adapted for every winner as $\Delta u_* = \|\overline{x}^p - \overline{w}_{*2}\| - \|\overline{x}^p - \overline{w}_*\|$. Utility values of all units are subjected to exponential decay $\Delta u_i = -\beta u_i$ after each adaptation step.

---

[3] Neighboring nodes in the graph that represents the structure of GNG-U clusters.

[4] The second nearest unit to the current input $\overline{x}^p$.

[5] Symbol u_i designates the i-th network unit, vector $\overline{u}_i$ designates the position of i-th unit in the network structure, whereas scalar u_i designates its utility value.

4   Evaluation Criteria for Clustering

We will evaluate the performance of studied clustering algorithms subjectively visually ans also quantitatively. In the tasks having visual character and being targeted toward a human user like ours the subjective evaluation is used commonly. Human preferences, intuition, psychological aspects or kansei [10] take a role in the subjective evaluation.

The quantitative evaluation is based on calculation of cluster errors in the network. Let $\mathbb{P}_i = \{\overline{x}_i^1, \overline{x}_i^2, \ldots, \overline{x}_i^{P_i}\}$ be a set of all P_i input patterns for which the unit i is the winner. Let the *cluster error* J_i of unit i be defined as the sum of differences between its reference vector and all inputs $\overline{x}_i$ from $\mathbb{P}_i$. Let the *network error* $J_\mathbb{N}$ be defined as the sum of cluster errors of all units in $\mathbb{N}$.

$$J_i = \sum_{p \in \mathbb{P}_i} ||\overline{x}_i^p - \overline{w}_i||, \qquad J_\mathbb{N} = \sum_{i \in \mathbb{N}} J_i \tag{3}$$

Cluster error provides overview of the representation capabilities of units in $\mathbb{N}$. Large error for a given unit i indicates that its reference vector is not representative enough for inputs in $\mathbb{P}_i$. Network error provides valuable information about the progress of training process. It is useful to compare different clustering algorithms, their parameters or network topologies suitable for given task.

5   Comparison of Clustering Performance

The objective of the experiments presented is to illustrate implementation of the clustering methods for response function visualization, to show the example results and to evaluate performance of Kohonen network and GNG algorithms in this visualization task. The trained neural network (the learning of which will be visualized) is a Multilayer Perceptron Network (MLP) with two hidden layers. The Backpropagation learning algorithm is used to train the MLP. The learning task is a two-class classification of geometrical structures. The first variant is the circle-in-the-square task, where examples from inside the circle are the first class and the outside of circle is the second class (See Fig. 3 top). The second variant is the two-entangled-spirals task (See Fig. 3 bottom), where the body of the first spiral is the first class and the second spiral forms the second class.

Kohonen network (and also GNG) clustering will be applied two times: on the first layer of MLP, and on the second layer of MLP. The Backpropagation algorithm on MLP network is stopped in the half of the training process and the actual responses of neurons are re-sampled as training data for clustering to allow equal start for both clustering algorithms. Parameters of all networks are in Table 2. The parameters of Kohonen network were chosen experimentally. The GNG parameters were chosen according to [6], except for π_k and τ_c, which were tuned experimentally.

The examples of the responses of neurons on the 1st and 2nd hidden layers of MLP on the Circle and the Spirals task and the progress of clustering are

Table 2. Parameters of algorithms

Kohonen network parameters

$r(t)$	$\gamma(t)$	inputs
$1 \rightarrow 0$	$1 \rightarrow 0$	100×100

task	layer	patterns	outputs
circle	1	20	10×1
	2	10	4×1
spirals	1	40	15×1
	2	20	10×1

GNG parameters

τ_c	α	β	π_k	γ_*	γ_0
50	0.5	0.995	2	0.2	0.006

task	layer	pats.	max. clusters
circle	1	20	10
	2	10	4
spirals	1	40	15
	2	20	10

Backpropagation parameters

γ	patterns	circle top.	spirals top.
0.25	10000	2-20-10-1	3-40-20-1

GNG-U parameters

τ_c	α	β	π_k	γ_*	γ_0	η
120	0.5	0.0005	10	0.05	0.0006	3

Table 3. Cluster errors of Kohonen network and GNG clustering

	1st MLP layer				2nd MLP layer			
	Circle		Spirals		Circle		Spirals	
J_i	Koh	GNG	Koh	GNG	Koh	GNG	Koh	GNG
J_1	18.72	19.09	98.78	79.98	0.00	23.31	0.00	69.30
J_2	0.00	22.82	125.38	35.68	33.41	48.74	0.00	0.69
J_3	0.00	64.84	72.08	43.07	19.75	2.98	29.25	26.97
J_4	18.31	43.46	0.00	37.06	81.74	21.04	18.53	0.07
J_5	25.76	17.17	40.72	61.11			0.00	0.00
J_6	0.00	4.19	0.00	34.45			0.00	29.47
J_7	31.88	0.00	57.64	55.10			26.96	16.24
J_8	16.29	9.14	59.52	28.52			22.64	17.73
J_9	8.75	1.69	48.43	29.88			17.12	4.22
J_{10}	22.82	14.09	8.09	69.37			0.00	2.24
J_{11}			0.00	3.66				
J_{12}			85.55	0.00				
J_{13}			0.00	58.23				
J_{14}			0.00	15.88				
J_{15}			0.00	4.36				
$J_\mathbb{N}$	142.55	196.51	596.21	556.36	134.91	96.07	114.52	166.94

in Fig. 3. The quantitative comparison of cluster errors on the 1st and 2nd hidden layers of MLPs for both studied tasks is in Table 3. Interesting are the cluster errors equal to zero for both studied algorithms. In this case for the particular node u_i, $\mathbb{P}_i$ was a single member set (or empty set). The results for both algorithms are similar but the GNG algorithm needs more training cycles to achieve these results (similar to those of Kohonen network).

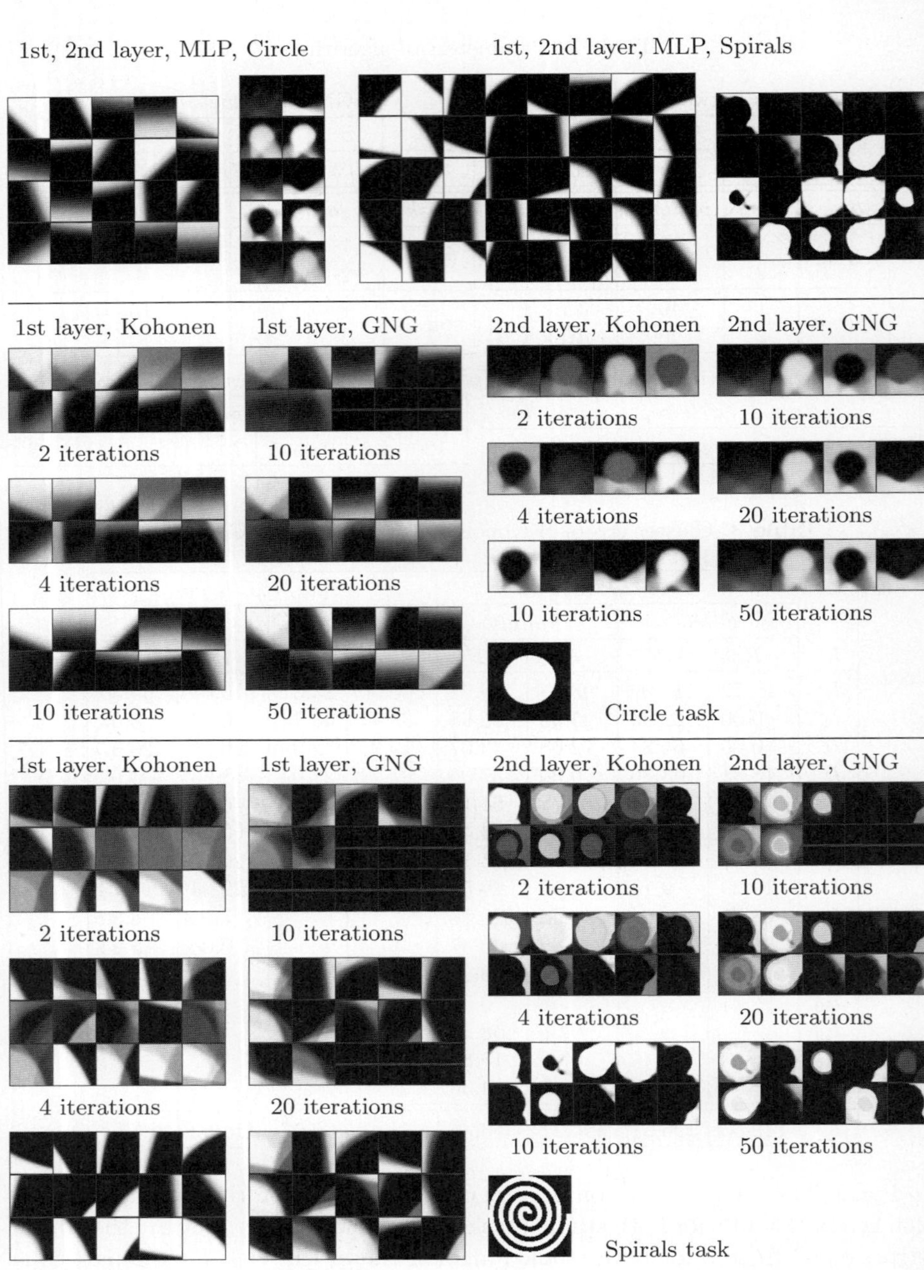

Fig. 3. Progress of adaptation of clusters of responses of neurons in both layers of MLP. Representative responses are displayed.

Table 4. Simultaneous MLP and Clustering adaptation

Backpropagation learning of MLP	*Kohonen or GNG clustering of responses*
1. Initialize weights of the MLP network,	1. Initialize weights of the Kohonen (GNG, GNG-U) network,
2. update weights of MLP network according to the Backpropagation algorithm for every training pattern,	2. every 10 training cycles of MLP compute the actual responses of neurons,
3. repeat the step 2.	3. update weights (10 learning cycles) of Kohonen (GNG, GNG-U) network using the last obtained responses from the MLP network,
	4. repeat from the step 2.

6   Simultaneous MLP and Clustering Adaptation

In this experiment both algorithms – MLP training and also clustering did run continuously during the whole experiment. MLP was initialized with the same set of weights in experiments with Kohonen network and GNG, GNG-U algorithms. Clustering was performed independently in parallel with the Backpropagation algorithm always considering the actual responses of neurons from the MLP network (See Table 4). The course of training for both algorithms is 10 cycles MLP then 10 cycles clustering then 10 cycles MLP, etc.

The parameters of Kohonen network must decrease in the training process. The Kohonen network in this experiment is therefore reinitialized[6] (weights of Kohonen network set to random values) after every 10-th training iteration. The GNG algorithm is capable to shift cluster centers continuously, exploiting the ability to remove the unused nodes. Because all GNG parameters are constant during the training process, the state of GNG is not reinitialized between subsequent runs. Moreover the performance improvement using the Utility Factor presented in GNG-U algorithm will be tested. The parameters of GNG-U are chosen according to [7], except for π_k and τ_c, which were tuned experimentally (see Table 2 for actual values).

The plots of the network error progress during the training are in Fig. 4 and Fig. 5. The big spikes on both figures result from the described reinitialization of Kohonen network. The course of error grows first. This is the result of initial MLP training which makes the task of clustering continuously harder. Later, the courses stabilize as the MLP training as well as clustering algorithms stabilize themselves. Interesting was the fact that in average only one edge removal was performed by GNG during clustering of neurons responses in MLP in both tasks. In the experiments on the simple task (Fig. 4), Kohonen network performs better. In the more difficult task (Fig. 5 left) the GNG algorithm performs better. The performance advantage of GNG-U versus GNG varies between tasks.

---

[6] Another approach suggested by Kohonen is to re-adjust the topology of the network at different stages during the training process, based on the minimal spanning tree between the reference vectors $\overline{w}_i$ [12].

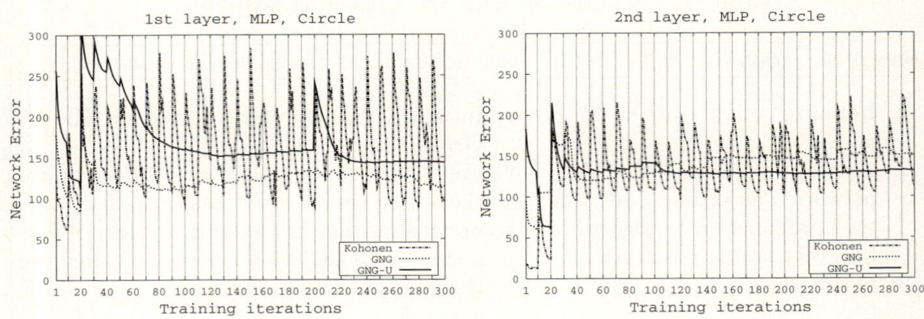

Fig. 4. Display of network error in the training process on the responses of both hidden layers of MLP on the Circle task. Weights of MLP are simultaneously updated by the Backpropagation algorithm.

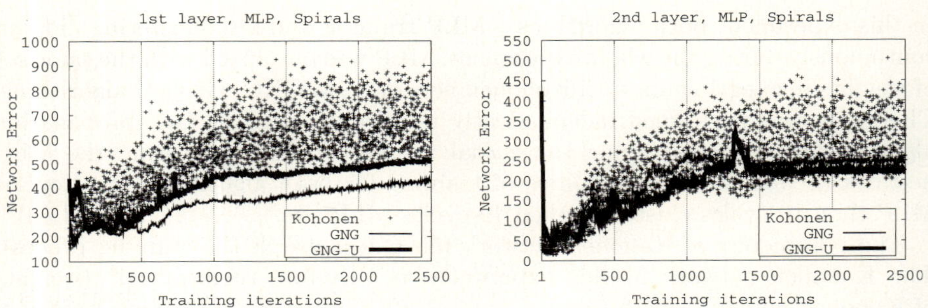

Fig. 5. Display of network error in the training process on the responses of both hidden layers of MLP on the Spirals task. Weights of MLP are simultaneously updated by the Backpropagation algorithm.

7   Conclusion

Visual information reduction in the visualization of neural network learning using Kohonen or GNG-U clustering is possible. The benefit of information reduction is not only minimization of the visualization equipment requirements but mainly an attempt to reduce the user fatigue present in systems based on human–machine interaction as studied in the field of IEC. Based on quantitative evaluation of our experiments both studied algorithms are suitable for clustering of response function plots. GNG-U is able to adapt continuously to a rapidly changing distribution of MLPs responses while Kohonen network is able to accomplish clustering cycle in shorter time. We prefer the visual output of Kohonen network. Usage of Kohonen network is also more comfortable as the user is asked to tune up only two adjustable parameters compared to the seven parameters present in the GNG-U algorithm.

Our future work is focused on development of methods for interactive learning of neural networks [1], where the visual information will be based on clusters representatives provided by the Kohonen network. We also suggest a possibility of exploitation of the information about similarly responding neurons for a neural network structure reduction method. The purpose is to improve the generalization performance of the neural network.

References

1. Užák, M., Jakša, R.: Framework for the interactive learning of artificial neural networks. In: Kollias, S.D., Stafylopatis, A., Duch, W., Oja, E. (eds.) ICANN 2006. LNCS, vol. 4131, pp. 103–112. Springer, Heidelberg (2006)
2. Craven, M.W., Shavlik, J.W.: Visualizing learning and computation in artificial neural networks. International Journal on Artificial Intelligence Tools (1), 399–425 (1991)
3. Lang, K.J., Witbrock, M.J.: Learning to tell two spirals apart. In: Proceedings of the 1988 Connectionist Models Summer School, pp. 52–59 (1988)
4. Jakša, R., Takagi, H.: Tuning of image parameters by interactive evolutionary computation. In: Proc. of 2003 IEEE International Conference on Systems, Man & Cybernetics (SMC 2003), pp. 492–497 (2003)
5. Kohonen, T.: Self-organized formation of topologically correct feature maps. Neurocomputing: foundations of research, pp. 509–521 (1982)
6. Fritzke, B.: A growing neural gas network learns topologies. In: Tesauro, G., Touretzky, D.S., Leen, T.K. (eds.) Advances in Neural Information Processing Systems, vol. 7, pp. 625–632. MIT Press, Cambridge (1995)
7. Fritzke, B.: A self-organizing network that can follow non-stationary distributions. In: Gerstner, W., Hasler, M., Germond, A., Nicoud, J.-D. (eds.) ICANN 1997. LNCS, vol. 1327, pp. 613–618. Springer, Heidelberg (1997)
8. Kohonen, T.: Automatic formation of topological maps of patterns in a self organizing system. In: Oja, E., Simula, O. (eds.) 2nd Scand. Conference on Image Analysis, pp. 214–220 (1981)
9. Pratihar, D., Hayashida, N., Takagi, H.: Comparison of mapping methods to visualize the EC landscape. In: 5th International Conference on Knowledge-Based Intelligent Information Engineering Systems & Allied Technologies (KES 2001), pp. 223–227 (2001)
10. Takagi, H.: Interactive evolutionary computation: Fusion of the capabilities of EC optimization and human evaluation. Proceedings of the IEEE 89(9), 1275–1296 (2001)
11. Holmström, J.: Growing Neural Gas - experiments with GNG, GNG with Utility and supervised GNG. Master's thesis, Uppsala University (August 2002)
12. Kangas, J., Kohonen, T., Laaksonen, J.: Variants of self-organizing maps. IEEE Transactions on Neural Networks 1, 93–99 (1990)

Heuristiscs-Based High-Level Strategy for Multi-agent Systems*

Péter Gasztonyi[1] and István Harmati[2]

[1] Department of Control Engineering and Information Technology
Budapest University of Technology and Economics,
Magyar Tudósok krt. 2/B313, Budapest, Hungary H-2117
peti@fw.hu
[2] Department of Control Engineering and Information Technology
Budapest University of Technology and Economics,
Magyar Tudósok krt. 2/B422, Budapest, Hungary H-2117
harmati@iit.bme.hu

Abstract. In this paper, a high-level strategy concept is presented for robot soccer, based on low level heuristic inference methods, rather than explicit rule-based strategy. During tactical positioning, no strict role set is assigned for the agents, instead a fitting point of the role-space is selected dynamically. The algorithm for this approach applies fuzzy logic. We compute fields-of-quality, regarding some relevant aspects of the scenario, and integrate them into one decision-field, according to given strategic parameters (used as weights). The most relevant locations are derived from the decision-field through subtractive clustering, and the agents are allocated to these locations, as their desired positions, according to their significance and their cost of reaching the given target. If an agent is in a position to manipulate the ball, an appropriate action is being selected for it. The simulation and experiments prove that the proposed approach can be efficient in dynamically changing environment or against opponents of different strategies.

1 Introduction

The control and motion-planning of multiagent systems typically call for modular managing, thus the robot soccer strategies follow the same approach. Most solutions define about 3 levels for the modular control system, and these levels are responsible for different scope of phisical movement or strategical positioning. A typical realization is described in [4] and [7]. These systems operate with deterministic rule based strategies, while others try to optimize the role assignment dinamically [2]. However, these algorithms are outworn by newer conceptions, involving holistic approach [5] or interpreting the roles less strictly [3]. These

* The research was partly supported by the Hungarian Science Research Fund under grant OTKA K 71762 and by the János Bolyai Research Scholarship of the Hungarian Academy of Sciences.

V. Kůrková et al. (Eds.): ICANN 2008, Part I, LNCS 5163, pp. 700–709, 2008.

approaches let the agents to take their tasks according to global states, or to define their positions in a more transitional way ('area of responsibility').

We intend to eliminate explicit role assignment, but a hierarchical control system is needed to control the agents' movements. Similarly to the generic concepts, we follow the control logics in Fig. 1.

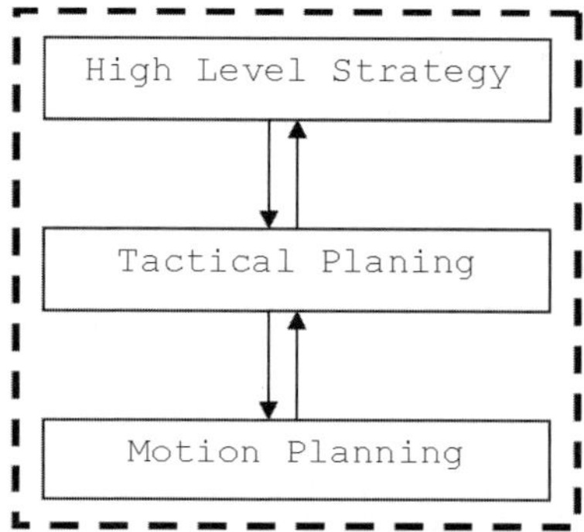

Fig. 1. The modular control system for robot soccer

The specific objective of the layers are:

- *High Level Strategy* (HLS): coordinating the strategy; setting up actual strategic goals; distributing area and responsibility to players
- *Tactical Planning* (TP): dynamic path planning for agents (handling joined movements); carrying out basic skills and special actions
- *Motion Planning* (MP): low level path planning for the given robot type, with no obstacles supposed in space; generating control signs.

The objective of high level strategies is to provide optimal dislocation for the team in the dynamic environment, beside the obstructive behaviour of other team's agents on the same field. Optimal dislocation in our context means that robots go to suitable positions to get the ball and make a goal. Optimal dislocation of the agents has a contribution to the dominance of the team, and assists to achieve own objectives - and to have the opponent fail theirs. However, since the optimal solution of the problem is not available, this paper aims to integrate some known partial heuristic solutions related to the subject. Taking simplifications, e.g. using fuzzy logic, which fits well to the problem, can be effective in the solution.

In Sect. 2, the problem is shown in details, and the used configuration is specified along with the outline of the strategy logic system. After this overview, the main concept of the solution is presented in Sect. 3, including the action-selection function of the agents. Finally, based on the above introduced knowledge, some aspect of the team coordinaton is discussed in Sect. 4, and finally, we draw the conclusions.

2 Preliminaries

In our framework, the mobile agents are realized by differential driven robots. These robots are described with their transition equations:

$$\dot{x} = \frac{r}{2}(u_\mathrm{r} + u_\mathrm{l})\cos\phi$$
$$\dot{y} = \frac{r}{2}(u_\mathrm{r} + u_\mathrm{l})\sin\phi \tag{1}$$
$$\dot{\phi} = \frac{r}{L}(u_\mathrm{r} - u_\mathrm{l})$$

where x and y are the coordinates, ϕ is the orientation of the agent, while u is the angular speed of a wheel, r is the wheel diameter and L is the distance between the wheels.

As we do not have extra ball manipulator devices on-board, the simplified exterior of the robots is a symmetric cylindered case. However, this type of cover is easy to apply and also to simulate; it affects the logic rather badly, due to expecting a quite precise maneuvering from the agents - as we will see this problem a bit more detailed in Sect. 3. This setup results in a soccer team whose members try to bump into the ball such a way that the ball is preferred to be driven into the opposite team's cage - which goal cage is a hole in the boarding of the squared field. As of the similarities with the physics of the billiards game, this model is referred to as 'billiard ball model'. According to the simulation, the 3D cylindrical robots are projected into a 2D circle, and the collisions are simulated in the 2D plane, too.

One team can consist arbitrary number of robots, but the algorithm is implemented for a typical of 3..5 agents. These robots have a maximum angular velocity for their wheels, thus limiting both the linear and turning speeds. They also have a weight parameter, which is counted during collision calculations. As part of the billiard ball model, these calculations are detailed in [1]. According to the current model, the ball friction is constant: the ball loses 0.8% of its current speed in every simulation cycle.

The modular arhitecture of the team-coordination logic is based on the following considerations:

- The drafting of the main goal includes the following: a) score goals, b) more than the opponent. This suggest a parallel realization of two independent algorithms for the different aims (scoring and defending), which are fundamentally competing, but can also have a validity attribute according to the current game situation, so in the given scenario, one can decide whether to attack or defend. Furthermore, these layers (and so the calculations) might be connected on the recognition that the one team's offence chances are the other's threats.
- The "scoring" also requires a special action, namely, manipulating the ball in a desired direction (let it be called a "kick"). A kick is a detached action from the class of different types of positional-purpose moving, because its execution is strongly affected not only by the intention, but also the current

scenario. Thus, it has to be calculated separately and at a higher priority level (before other calculations).
- When not actually trying to score ("kicking to the goal"), but simply positioning, an appropriate logic is needed for setting up an order among the agents of the team and decide the targets themselves.
- An interchangeable interface is recommended for the agents' control and communication, as to support not fully homogeneous teams, or two homogeneous teams of different type of robots, too. In these cases, the motion control of certain robot types will be different, of course, but remains compatible to be built into the team.
- Besides these internal physical constraints, one might face other limitations, e.g. the presence of other objects - which might also be a subject to handle.

3   Fundamentals of the HLS Layer

The basic concept of the HLS in subject is to avoid explicit fix rule-based strategy and force improvisative play in order to be able to adapt to different opponents or to those who are dinamically changing their strategies. The main idea is to take the ball trajectory design as the primary object (instead of positioning the players directly, according to a certain formation or any heuristics). This can be done by positioning the players, of course, as to form the vertices of the ball trajectory in question - but as from our aspect, the available reasonable ball trajectories, in theory, will not depend on any predefined formation.

One could say that the ball trajectory can not be planned due to the chaotic nature of the game, which comes from the indeterministic nature of the opponent's moves. In our case, however, the ball's dynamics are much faster than players' moves, thus these trajectories can be planned for a short while, indeed. Since the players have limited speeds and they have to bump against the ball to kick it, the required (potential) vertices are spread across the field in limited ranges, but this aspect also makes possible to combine some different quality measurement into one decision-field. The algorithm counts some low-level quality-fields on the mesh, according to properties which are drafted by general strategical demands. To determine the suitable ball trajectories, we need to combine the low-level fields to a decision-field. Different possible trajectories come from different reasons, and the integration to a common field is performed through the following *strategic parameters*, which weight the individual quality-fields for a linear combination:

- **Shooting Rate:** The preference of shooting on goal.
- **Pressure:** The preference for agents to act on the opponent's side. The intention is to keep far the opposite robots with the ball from the goal cage.
- **Possession:** The preference of safe kick sequences (where the possibility of the opponent will touch the ball is minimal).
- **RiskTake:** A real value between -1..+1, that affects the default general parameters of actions, e.g. the range of safe reach for an agent.

- **ActiveSubtractRadii:** The radius parameter of subtractive clustering, when deriving *positioning* target points from the decision field. Thus, it affects the admissible proximity for agents.
- **PassingSubtractRadii:** The similar radius parameter when deriving a target for a *kick*.
- **ActiveSubtractPoints:** The number of target points to be returned when running the subtractive function. It is desired to calculate more points than the number of team members, as advanced distribution functions can benefit from this.

The quality-fields are realized by fuzzy membership functions defined on the field mesh. Some of these meshes, which depend only on the field itself, are constant (like the *ShootingArea* field), while other fields, which depend on the certain scenario, are dynamically calculated (like *ShotClear* or *PassClear* fields, which tell the visibility of the mesh points as the source or target, to or from a given point). Examples are pictured in Fig. 2 and 3.

The 'good places' (of where to touch the ball, where to kick, and thus, where to move by the others) can be determined through the obtained decision-field; and a secondary forthcoming target can also be chosen to refine the actual action

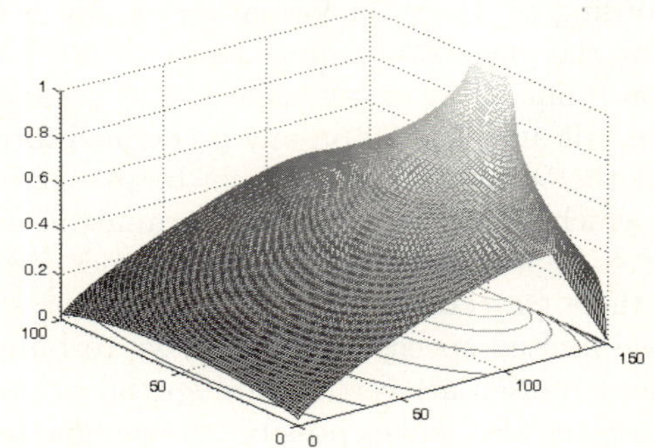

Fig. 2. *ShootingArea* field tells the constant visibility of the goal cage from the mesh field points, *without* considering any agents

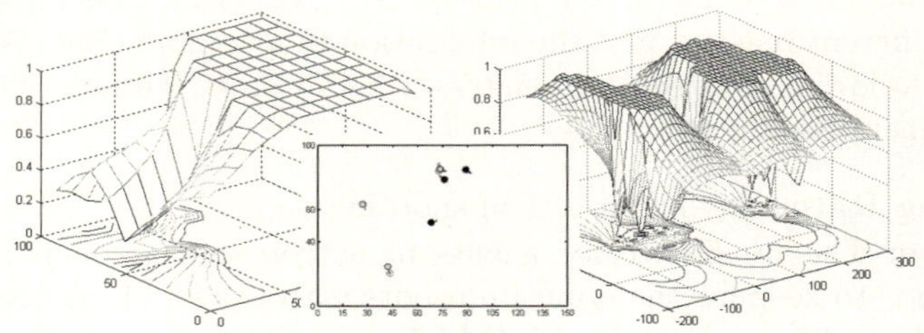

Fig. 3. *ShotClear* field tells the actual visibility of the mesh points from the goal cage, considering *only* the opponent agents. Simple field (L), the simulation scenario(C) and the extended mirrored field (R) are shown.

and prepare an 'indirect pass', that is, to aim the kick *not* in the *actual* direction of a teammate, but to a point where it shall be at the time *when* the ball reaches it. This procedure can apply only to the players who have a reasonable chance to perform the kick in a limited time, considering both the technical part of the kick and the opponents' dislocation. Once the kick action is fixed, the other agents can move according to this decision.

Other main fields can be constructed of 'where-to-move' for the players, which is derived from the possible-ball-trajectory-vertices-field, through a simple time-cost function of players to move there. This can involve more possible vertices, and more distribution algorithms can be used to allocate targets to players. A sample quality field for passing, and a subtractive allocation procedure is shown in Fig. 4.

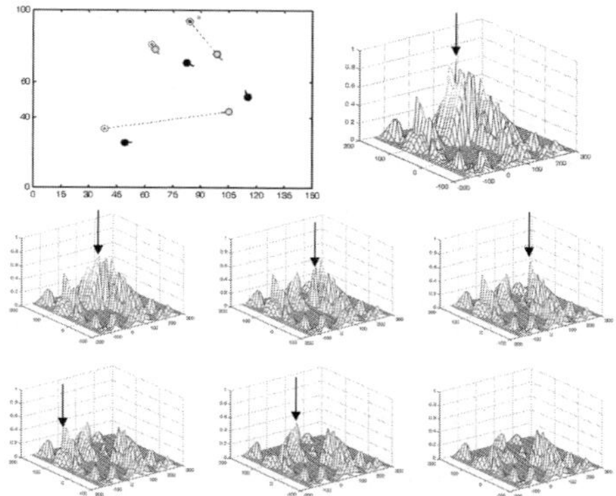

Fig. 4. Passing quality-field and subtractive allocation procedure

The formula of one subtraction step for one discretized grid point $P=(x,y)$ of the field is as follows:

$$Q(P) = \max(0, Q(P)$$
$$-Q_{\max} \cdot \exp(-0.5 \cdot \quad (\frac{(x_{\max} - x, y_{\max} - y)}{\sigma})^2)) \tag{2}$$

where $(x_{\max} - x, y_{\max} - y)$ is the distance of P from the location of the maximum value of $Q_{\max}$ in the field, while sigma is a constant parameter.

Although it has been mentioned in Sect. 2, that the one team's offence chances are the other's threats, thus they could be treated as the dual pairs of each other, despite in the realization, this method was shown to be wasting too much of resources, therefore simplifications had to be applied in the defence logic. The connection between the offense logic and the simplified defense logic remains the recognition that the ball, if untouched, moves following a straight line. Thus, it can only change its direction if a player kicks it or it is bumped into the wall. The effect of collisions with the wall is calculated through an appropriate reflection transformation (depending on physics modeling). Altogether, if a self player is clearly able to kick the ball before the opponent (or a "mirror instance") reaches it, then that action will be chosen by the algorithm. The only real threat to be prepared to is if an opponent agent is likely to manipulate ball before a self teammate could do the same. Assuming that it is going to kick the ball as soon as possible (which is likely the most efficient action), the defence can be prepared to close the possible ways from the kicking point to the goal cage or to other potential kickers of the opponent. This likelihood (the possibility that the opponent reaches the ball first) is linked to a parameter which partly describes the seriousness of the current threat - the other part is that *where* in the field is that situation. Fig. 5 shows an example of localizing these threats of possible opponent kicks.

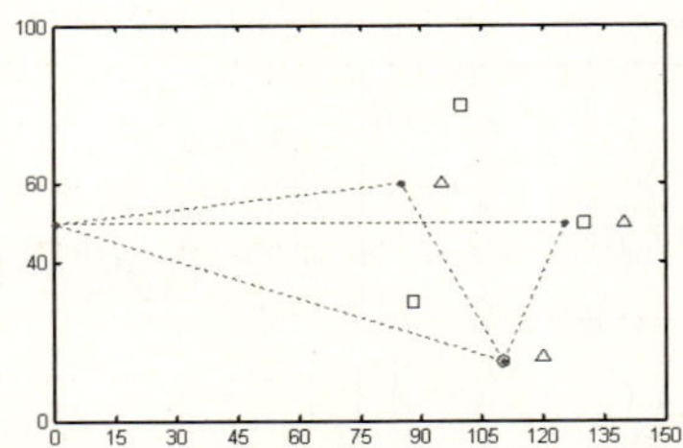

Fig. 5. Localizing threats of opponent kicks. Squares are teammates, triangles are opponents, and the ball is estimated to be kicked by the lower opponent at *(110,15)*. Dashed lines are some possible paths of ball to own goal. The the right vertex is well covered by a teammate.

4   Team Coordination

The purpose of positioning is to create a condition in which the likelihood of accomplishing the currently ordered task is the possible maximum. Positioning in itself is not necessary to be treated as a skill - indeed, we position to be in position to perform a real skill in the future.

It is easier, however, to assign certain roles to players, which means that we discretize the responsibility and many attributes for the players according to multiple components. By tying up a player to a single role, we simply lose the overlapped information between the real situation's demands and the possible actions, determined by that certain role. This is because the actual situation can transit to another where completely different role assignment may occur. In this case, sharp changes in roles would have to be carried out to some players, which is clearly not the most economic way of coordinating the players and the team. Instead, we want to take every opportunitiy and the relevant threats into consideration by their appropriate importance in every moment. This effort can be fulfilled by the earlier described use of properly weighted multi-purpose positioning, which results in smoother transitions between the defined roles. Moreover, if the defined selectable purposes of positioning do not correspond directly to a role set, rather are components of conventional 'roles', then the agents are available to pick the most fitting continuous point of the purpose-space for the actual situation.

This aspect implies even the omission of a fixed Goalkeeper role, e.g. as if the conditions allow to involve all players in an offence, the most efficient dislocation will be chosen, considering the appointed strategic and tactical parameters.

Once the position is solved this way, if one agent has the chance to manipulate the ball, a direct skill action (kick) is called. If more agents has the opportunity to kick, more of them may be investigated for a target and for a secondary purpose to refine the first aim. The kicker selection method uses fuzzy algorithm, which fits well to the problem and can be easily expanded. Fuzzy rules for the kicker selection may look like:

- if *distanceToBall* is **near** and *relativeSpeed* is **approaching** and *potential-ShootAngle* is **front** then *playerIsShooter* is **high**,
- if *distanceToBall* is **medium** or *relativeSpeed* is **invariant** then *playerIs-Shooter* is **medium**,

where

$$\mu_{\text{distanceToBall}} = \frac{distanceToBall_{\min}}{distanceToBall} \tag{3}$$

and

$$\mu_{\text{potentialShootAngle}} = \cos(\frac{potentialShootAngle}{2}) \tag{4}$$

and *relativeSpeed* is the ball velocity vector projected to the direction towards the player, while *potentialShootAngle* is the estimated angle between the ball's moving direction and the direction from the expected kick-point to the related target.

When the potential kickers are selected with their available targets along with the secondary options, these choices will logically form a set of precedence graph of actions. If these nodes are ranked, the best action-chain can be found, through an appropriate weight-function, which is affected by the actual values of the strategical parameters - like risk-taking, shooting-rate, etc.

The selected action-chain determines the kicker and its chosen target, along with the secondary intention concerning a possible next kick, which would refine the execution of the forthcoming one. As a consequence of the billiard ball model, proper calculations are needed for the optimal motion planning for a correct kick: a) the main option is whether to kick the ball in one touch or try to stop it first (which would result in 2 kicks indeed) b) the second task is to plan the vertices for a safe bump, i.e. to be in the right position in the right time, avoiding any collisions with the ball before the planned one. Detailed considerations and calculations on this problem can be found in [1].

As the kick action is a well defined maneuver, it is executed at the TP level. Additionally, for the time-cost estimations on HLS level, the kick action function should have an alternative interface for these quick-response queries.

In case of no kicker can be selected clearly, e.g. due to the opponent's proximity to the ball, 'some agent of the self team should take more part in the defensive tasks', that is, to gain more affinity to parry the detected threats. This fuzzy phrase can be easily implemented with the utilization of some properties describing the situation in the field. The affinities in question are described with two parameters: *maxAff*, which determines the maximum affinity in the team, and *maxExp*, an exponential which affects the distribution ratio of the individual affinity values among the agents. With these parameters, the individual defensive affinities are calculated by:

$$defAff(1) = maxAff \tag{5}$$

for the agent that can be allocated to the most serious threat by the lowest cost, in the given situation, and

$$defAff(j) = maxAff \cdot \exp(maxExp \cdot \frac{j-1}{M-1}) \tag{6}$$

for the other agents, where M is the number of members in a team.
The properties, as the inputs of the fuzzy inference system, are:

- *pressure*: the current value of the formerly mentioned strategic parameter,
- *kicklead*: an estimated lead in simulation time cycles for the self players to reach the ball before the opponent,
- *sumcritic*: a 'danger-meter' value, calculated from the properties of the detected threats,
- *average_q_active*: a 'chance-meter' value, calculated from the quality of possible targets, those are derived from the decision maps.

Finally, the fuzzy rules to determine the *maxAff* and *maxExp* parameters:

- if *pressure* is **high** and *kicklead* is **high** then *most_defensive* is **verylow**
- if *pressure* is **low** then *most_defensive* is **mid**
- if *kicklead* is **none** then *most_defensive* is **high**
- if *kicklead* is **high** then *most_defensive* is **none**
- if *sumcritic* is **high** then *most_defensive* is **high**
- if *pressure* is **low** then *affinity_distribution* is **split**
- if *kicklead* is **none** then *affinity_distribution* is **equalized**
- if *kicklead* is **high** then *affinity_distribution* is **split**
- if *sumcritic* is **high** then *affinity_distribution* is **balanced**
- if *sumcritic* is **low** then *affinity_distribution* is **split**
- if *average_q_active* is **poor** then *most_defensive* is **high**
- if *average_q_active* is **poor** then *affinity_distribution* is **balanced**
- if *average_q_active* is **promising** and *sumcritic* is **high** then *affinity_distribution* is **split**.

Then the output *most_defensive* is binded to *maxAff*, and *affinity_distribution* to *maxExp*.

5 Conclusion

A new high level strategy has been proposed for robotic soccer game. It has been shown that our concept can be very scalable and results a quite smooth positioning, while even being compatible with regular role-based strategies, if no interleaving is allowed and roles are fully discretized. Simulations showed that the combination of fuzzy rule based strategy and induvidual quality fields leads to a potentially strong high level coordination. One real challenge is to find an optimal tuning algorithm for the strategic parameters, which would assure the proper tracking speed of a varying tactic used by an advanced opponent. A possible solution can be some kind of reinforcment learning technique.

References

1. Gasztonyi, P.: Magasszintü csapatstratégiák fejlesztése robotfocihoz, M.Sc. Thesis, Budapest University of Technology and Economics, Hungary (2006)
2. Sng, H.L., Sen Gupta, G., Messom, C.H.: Strategy for Collaboration in Robot Soccer. In: IEEE DELTA 2002, New Zealand, pp. 347–351 (2002) ISBN: 0-7695-1453-7
3. Tews, A., Wyeth, G.: Multi-Robot Coordination in the Robot Soccer Environment, Uneversity of Queensland, Australia
4. Bowling, M.: Multiagent Learning in the Presence of Agents with Limitations, Ph.D. Thesis, Carnegie Mellon University, Pittsburgh, PA, USA, CMU-CS-03-118
5. Jennings, A., Shah, C., Jani, H, Upadhya, M., Kapadia, M.: Team Yarra: holistic soccer, RMIT University, Melbourne, Australia
6. Stone, P., Veloso, M.: Task Decomposition and Dynamic Role Assignment for RealTime Strategic Teamwork, Computer Science Department, Carnegie Mellon University
7. Aparício, M.: Design and Implementation of a Population of Cooperative Autonomous Robots, M.Sc. Thesis, Universidade Técnica de Lisboa, Portugal (2000)
8. Lantos, B.: Fuzzy Systems and Genetic Algorithms, Müegyetemi Kiadó (2002) ISBN: 963 420 706 5
9. Wang, T., Liu, J., Jin, X.: Minority Game Strategies for Dynamic Multi-Agent Role Assignment, Hong Kong Baptist University, Kowloon Tong, Hong Kong
10. Russel, S.J., Norvig, P.: Artificial Intelligence: A Modern Approach. Prentice-Hall, Englewood Cliffs (2002)
11. Tatlidede, U., Kaplan, K., Köse, H., Akin, H.L.: Reinforcement Learning for Multi-Agent Coordination in Robot Soccer Domain, Bogazici University, Istanbul, Turkey (2005)

Echo State Networks for Online Prediction of Movement Data – Comparing Investigations

Sven Hellbach[1], Sören Strauss[1], Julian P. Eggert[2],
Edgar Körner[2], and Horst-Michael Gross[1]

[1] Ilmenau University of Technology, Neuroinformatics and Cognitive Robotics Labs,
POB 10 05 65, 98684 Ilmenau, Germany
`sven.hellbach@tu-ilmenau.de`
[2] Honda Research Institute Europe GmbH, Carl-Legien-Strasse 30,
63073 Offenbach/Main, Germany
`julian.eggert@honda-ri.de`

Abstract. This paper's intention is to adapt Echo State Networks to problems being faced in the field of Human-Robot Interactions. The idea is to predict movement data of persons moving in the local surroundings by understanding it as time series. The prediction is done using a black box model, which means that no further information is used than the past of the trajectory itself. This means the suggested approaches are able to adapt to different situations. For experiments, real movement data as well as synthetical trajectories (sine and Lorenz-attractor) are used. Echo State Networks are compared to other state-of-the-art time series analysis algorithms, such as Local Modeling, Cluster Weighted Modeling, Echo State Networks, and Autoregressive Models. Since mobile robots highly depend on real-time application.

Keywords: Echo State Networks, Time Series Analysis, Prediction, Movement Data, Robot, Motion Trajectories.

1 Introduction

For autonomous robots, like SCITOS [1], it is important to predict their own movement as well as the motion of people and other robots in their local environment, for example to avoid collisions or to evolve a proactive behavior in Human-Robot-Interaction (HRI). Hence, further actions can be planned more efficiently.

Most publications in this field focus on optimal navigation strategies [2,3]. This paper, however, suggests to spend more effort into prediction of the motion of the dynamic objects instead. Often, only linear approximations or linear combinations are used to solve this problem.

The approach presented here is the interpretation of movement trajectories as time series and their prediction into the future. For performing this prediction an assortment of time series analysis algorithms was implemented and tested.

Echo State Networks have proven to be able to predict chaotic time series with a comparatively high accuracy [4]. Because of the fact that unexpected behavior can

V. Kůrková et al. (Eds.): ICANN 2008, Part I, LNCS 5163, pp. 710–719, 2008.

occur, movement data can also be understood as chaotic time series. Echo State Networks should be able to predict such data well. Their usage for the prediction of movement data has – to our best knowledge – not been investigated yet.

For an application of Echo State Networks, the movement trajectory of a person in the vicinity of the mobile robot, typically gained from a person tracker like [5] needs to be presented as a time series. For this reason, the given trajectory of the motion is now interpreted as $\mathcal{T}$ with values $\mathbf{s}_i$ for time steps $i = 0, \ldots, n-1$:

$$\mathcal{T} = (\mathbf{s}_0, \mathbf{s}_1, \ldots, \mathbf{s}_{n-1}) \tag{1}$$

Each $\mathbf{s}_i$ can be assumed as the tracked object's position, e. g. in a three dimensional Cartesian state space $\mathbf{s}_i = (x_i, y_i, z_i)^T$. Basically, the prediction for each future point on the trajectory is done iteratively for up to 500 time steps (about 8.3 seconds of motion using a sampling rate of 60 Hz).

The prediction in general takes place with the so-called black box model which means that no further background information is used than the past trajectory itself. The aspired prediction shall follow the trajectory's characteristics, which can be found in their past. Furthermore, no explicit model is given, to be able to freely adapt to new types of trajectories, i. e. new situations.

The following section at first briefly explains the principles of Echo State Networks and introduces different versions. In section 3, experiments and comparisons of the results on the movement data are presented, while the last section concludes this paper.

2   Echo State Networks

It is commonly known, that Neural networks are well suited for function approximation tasks. For the specific task of predicting time series, Echo State Networks (ESNs) are often used recently[4].

2.1   Principle

ESNs have some specific features which differ from "standard" neural networks: The hidden layer consists of neurons which are randomly connected (see Fig. 1). If the connectivity is low, this layer provides independent output trajectories. For this reason, the hidden layer is also called *reservoir*. Furthermore, there are neurons which are connected to circles in the reservoir, so that past states "echo" in the reservoir. That is the reason why only the current time series value $\mathbf{s}_n$ is needed as input.

The weights of the reservoir determine the matrix $\mathbf{W}^r$. In [4], it is mentioned that the spectral radius *spec* of this matrix[1] is an important parameter and must not have values above 1.0 to guarantee stable networks. The randomly initialized reservoir matrix $\mathbf{W}^r$ can easily be adapted to a matrix with a desired spectral radius. However, [6] argued that networks with a spectral radius close or slightly

---

[1] The spectral radius of a matrix equals to the largest Eigenvalue.

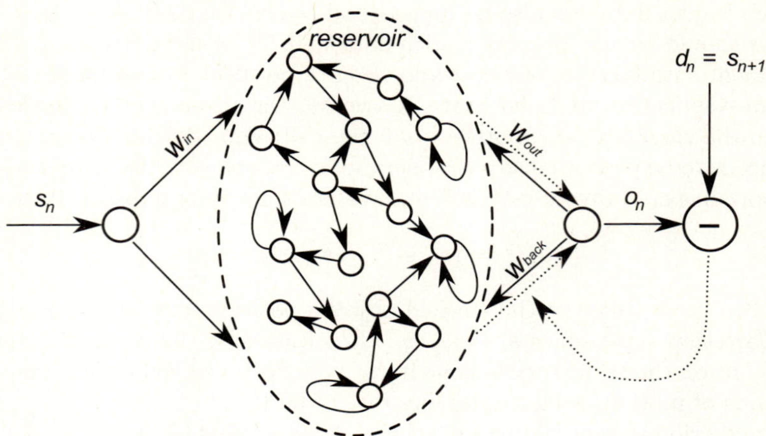

Fig. 1. The design of Echo State Networks has some characteristic features. In addition to the randomly connected reservoir r_n, the training algorithm is pretty simple for neural networks: Only the output weights $\mathbf{w}_{out}$ are adapted.

above 1.0 may lead to better results. Both possibilities are evaluated for their suitability for motion prediction.

Furthermore, the sparseness of the reservoir matrix plays an important role. A sparse reservoir matrix means that most of the weights in the reservoir are set to zero. This can be interpreted as the reservoir being decomposed into subsets, which are responsible for basic signals being overlaid by the output layer. As suggested in [4] and [6] both sources, about 80% of the weights are set to zero.

Another characteristic of ESNs is that only the output weights $\mathbf{w}_{o}ut$ are adapted and learned. All other weights (input, reservoir, feedback) are chosen randomly and stay static.

2.2   Training and Application

For training, the network is initialized randomly, and the training time series is used as network input step by step. The internal states $\mathbf{r}_n$ are calculated by using the following recursive equation:

$$\mathbf{r}_n = f(\mathbf{W}^r \cdot \mathbf{r}_{n-1} + \mathbf{w}_{in} \cdot \mathbf{s}_n + \mathbf{w}_{back} \cdot \mathbf{o}_{n-1}) \tag{2}$$

$\mathbf{r}_n$ describes the internal state at time step n. $\mathbf{W}^r$ stand for the reservoir matrix, while $\mathbf{w}_{in}$ and $\mathbf{w}_{back}$ are the weights at the respective edges (See Fig. 1), while f is the transfer function of the reservoir neurons which can be the Fermi-function or the hyperbolic tangent.

From a predefined starting point, the internal states $\mathbf{r}_n$ can be combined to a matrix $\mathbf{R}$. The starting point should be around the time step 100 or later to overcome possible bad initial values in the network. The adaption step for the output weights $\mathbf{w}_{out}$ is a linear regression using this matrix and the vector of the related output values $\mathbf{o}$:

$$\mathbf{w}_{out} = (\mathbf{R}^T\mathbf{R})^{-1}\mathbf{R}^T\mathbf{o}. \tag{3}$$

After weight adaption, the network can be applied for prediction. Thereto, the network is fed again with the whole trajectory data as input, this time step by step. If the prediction is taking place (i. e. reaching the last known point in time) the output is fed back to the input. So, the last network output is used as the next input to be able to generate more than one prediction step. In our experiments, up to 800 prediction steps are generated.

2.3   Enhancements

In [6] some additional Echo State Network features are introduced, like an online adapting rule and a plasticity rule to adapt the Fermi transfer function parameters in the reservoir (*intrinsic plasticity*). Furthermore, additional weights such as a direct input-output ($\mathbf{w}_{dir}$) weight and a loop at the output neuron ($\mathbf{w}_{rec}$) are suggested. Apart from the online rule, all other of those enhancements were evaluated and tested.

Intrinsic plasticity is performed online. It helps to adjust the reservoir transfer functions for better adapting to the current prediction task. It takes place before starting the learning of the output weights and shouldn't last longer than 200 time steps, otherwise predictions could get instable. Unfortunately, intrinsic plasticity has the effect that the eigenvalues and thus the spectral radius of the reservoir matrix increases.

Since in Echo State Network a huge number of parameters can be adjusted, a more automated process would be reasonable, especially, for those network weights, which are not changed during the regular training process, i. e. the $\mathbf{w}_{in}$, $\mathbf{w}_{back}$, and W^r. This paper suggests to use multiple instances of the network, as a kind of simple stochastical search in the parameter space. All instances are trained using the same input data, after initializing the fixed weights diferrently (in a random manner). During the training process, the output of each network is compared with the corresponding values of the training trajectory. The network showing the best prediction results for the yet unknown training data is then selected for further application.

3   Motion Prediction

The algorithms presented in this paper are intended to be used for motion prediction to enable a mobile robot evolving a proactive behavior in HRI. To be comparable and reproduceable, however, movement data taken from the University of Glasgow is used [7]. This benchmark data is available as 3D coordinate representation for each limb of a human performing a certain action, e. g. walking (see Fig. 2). Using this data is even more challenging, because several basic motions are combined (i. e. intrinsic movement, e. g. of the foot combined with the walking direction). The data set consists of 25 trajectories containing 1,500 up to 2,500 sampled points.

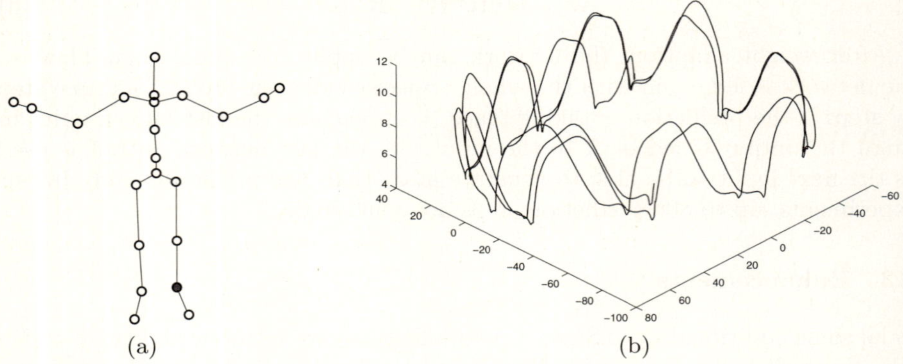

Fig. 2. Example of movement data from the University of Glasgow. Shown are the body points from which data is available (a) and an exemplary trajectory of the movement of the left ankle while walking in circles (b).

3.1   Test Conditions

Trajectories. Besides the movement data coming from the University of Glasgow, periodical and "standard" chaotic time series are used. All time series are three-dimensional.

The movement data has a resolution of 60 time steps per second, so that an average prediction of about 500 steps means a prediction of 8.3 seconds into the future. Present movement prediction techniques last considerably shorter.

The periodical trajectories consists of up to three superimposed sine waves, with each dimension being independent from the others.

As chaotic time series the Lorenz-Attractor is used. It is a simple system of differential equations where the single dimensions are not independent. This time series is a typically chaotic one, so small changes in a state leads to dramatic differences after a short time period.

Quality Measures. For comparing the prediction results, some kind of quality measures are necessary. The used quality measures are based on the normalized mean square error $NMSE$. Hence, the standard mean square error is normalized using the variance σ^2 of the time series.

$$NMSE = \frac{1}{N \cdot \sigma^2} \sum_{i=1}^{N} (\mathbf{s}_i^{pred} - \mathbf{s}_i^{orig})^2 = \frac{MSE}{\sigma^2} \tag{4}$$

Since the trajectories are three-dimensional and dimensions with greater difference are supposed to be more important, the highest variance of all dimensions is used as normalization.

Two different kinds of the defined measure are used. The first one, the short term error STE, is responsible for evaluating a short period of the prediction. It uses the first $N = 75$ predicted output values (which means 1.25 sec) with a weighting of $\frac{1}{f}$ of the f-th prediction step. Since some of the algorithms show the tendency to drift away, the performance is furthermore evaluated using the long term error LTE, which uses all prediction steps with a weighting of $\frac{1}{\sqrt{f}}$.

3.2   Reference Algorithms

Time Series Analysis Algorithms. Echo State Networks are compared to other state-of-the-art time series analysis algorithms, to be able to assess not only their absolute performance. They have been reimplemented in MatLab, following the methods described in the respective papers. Again, the black box model is used to have a similar starting point for all approaches.

1. **Autoregressive Models:** Assume a linear relation in the observed time series which means that any time series value can be determined by using a linear combination of p previous values. Different approaches can be used to determine these linear coefficients. The Wiener Filter, the Durbin-Levinson algorithm and the Yule-Walker equations were used to calculate the coefficients. For further details see [8] and [9].
2. **Local Modeling:** This algorithm tries to find similar states of the observed trajectory. Therefore, the usually low dimensional time series is transformed in a higher dimensional space, the so-called embedding space. Details from this algorithm can be found in [10] and [11].
3. **Cluster Weighted Modeling:** This approach is similar to the Local Modeling but from a probabilistic point of view. Hence, the embedding space is clustered with Gaussians. For more details see [10] and [11], too.

Trivial Comparison Algorithms. The first algorithm, defining the baseline, is a simple repetition of the last observed time series value and is called repetetive algorithm in the following. Also a linear algorithm is used as reference. This algorithm simply does a linear approximation based on the last two points of the time series. The result of the better one is used as reference. Both algorithms have to be outperformed clearly to get useful predictions.

3.3   Test Results

The following tests are to demonstrate the advantages and disadvantages of the Echo State variants and the other time series analysis algorithms presented within this paper. For the application of the algorithms, a lot of parameters had to be specified. The paramter values presented in the following are chosen after extensive tests, which cannot be not discussed here, because of space limitations.

Comparison of the Echo State Networks versions. Since, the literature provides slightly different variants of Echo State Networks, two different ones are evaluated here. On the one hand, networks with a structure from [4], called in the following Jaeger networks and on the other hand, networks with a structure from [6] (Steil networks). For both networks, the spectral radius is set differently. While Jaeger [4] uses $spec = 0.8$, with Steil networks it is set to $spec = 1.0$.

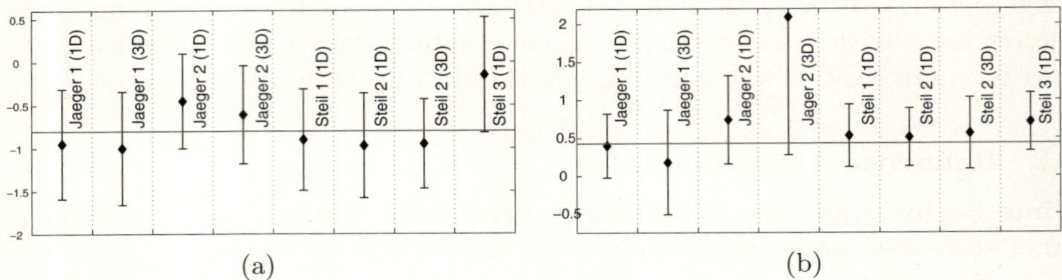

(a)   (b)

Fig. 3. The graphs show the *STE* (a) and *LTE* (b) plotted for each of the Echo State Network version tested on 1D and 3D movement data. The ordinate uses a logarithmic scale. Hence, lower values mean a better prediction. The error bars represent the standard deviation from the mean. The different tests are labeled with "Jaeger" and "Steil" using the networks presented in [4] and [6] respectively. For Jaeger networks, two versions are tested. On the one hand, parameters, like number of neurons, spectral radius, and sparseness of the reservoir, where set to fixed values. On the other hand, those parameters are obtained randomly. For Steil networks, the number of neurons is increased (25, 100, 250). Additionally, version 3 of Steil network uses input s_{n-1} and s_{n-2} as input (not only s_{n-1} as for all other tests).

Both networks are evaluated on real motion data (see Fig. 3). As already mentioned, the motion data is available as a trajectory in 3D Cartesian space. These 3D points are used directly as input for the network (labeled "3D" in Fig. 3), or they are split into three 1D time series, predicted independently with three networks (labeled "1D" in Fig. 3).

It is recommended in [6] to initialize all Steil network weights to 0.05. Since only weights to the output layer are adapted during training process, all other weights stay at 0.05. Actually, this value could not be confirmed with the test on movement data. It could be shown for both network versions, that the the feedback weights $\mathbf{w}_{back}$ must be scaled very low (about 10^{-20}) to guarantee stable networks. Furthermore, the input weights $\mathbf{w}_{in}$ are set to values of about 10^{-5}. For all other weights the influence of the chosen values is not that significantly.

Steil networks have additional weights to the output layer ($\mathbf{w}_{dir}, \mathbf{w}_{rec}$). These weights can be included in the learning process as it is suggested in [6]. Unfortunately, this leads to instable networks, so that these weights were not learned for predicting the movement data. These weights should be scaled low about 10^{-20} as well, because they have a similar function like the feedback weights $\mathbf{w}_{back}$.

Jaeger [4] suggests to use 50 up to 2000 neurons for the reservoir. In the prediction of movement data, the size of the reservoir is set to lie between 25 and 250 neurons. However, a higher number of neurons doesn't lead to significant better results.

Steil [6] adviises to apply Intrinsic Plasticity (adaption of transfer function parameters) for the first 200 time steps to improve classification results of the network. Those benchmark results were gained by applying the online learning rule. Since only offline learning rule is used here, the results could not be

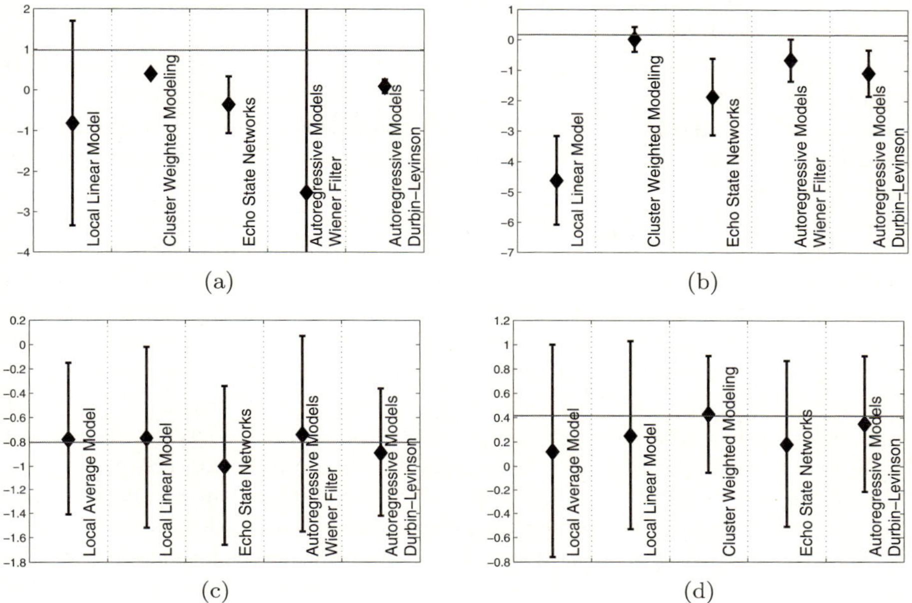

Fig. 4. The graphs shows the STE (b), (c) and LTE (a), (d) plotted for each of the algorithms tested with sine (a), Lorenz-attractor data (b), and movement data (c), (d). The ordinate uses a logarithmic scale. Hence, lower values mean a better prediction. The error bars represent the standard deviation from the mean. For the STE all results lie relatively close together while the reference algorithm can only be beaten clearly by the Echo State Networks. Longer predictions show more differences in the results of the algorithms. Also the mean errors are higher than STE, as expected in longer predictions. The reference is beaten more clearly in general. Local Average Models and Echo State Networks show the best results.

confirmed. In both types of networks, Intrinsic Plasticity seems to have only minor effects when the offline learning rule is applied.

Additionally, Steil networks were extended in a TDNN-like fashion, using more than only the last point of the trajectory as input (labeled "Steil 3" in Fig. 3). It can be observed that this leads to better predictions in the very first steps (about 5) but may destabilize the prediction in the following steps. In general, it leads to worse results for the chosen quality measures as they include 75 prediction steps.

Comparison to the reference algorithms. In the prediction of sine trajectories, the Autoregressive Models show the best results in the mean for STE and LTE (see Fig. 4(a)). Note that these algorithms can build up to high values, so that the standard deviation in this case is very high.

With worse mean errors the standard deviation is also lower. The Local Models and Echo State Networks lead also to quite good prediction results, while the used reference (data repetition of linear approximation) is beaten clearly by all prediction algorithms.

For predicting the chaotic Lorenz-Attractor (see Fig. 4(b)) the Local Linear Model leads to the best results. Echo State Networks perform also well – especially with higher number of neurons. Here, the reference algorithms are outperformed clearly, as well. The standard deviation of the prediction quality is relatively high.

For the prediction of real movement data, the Echo State Networks lead to the best results for the STE as it is shown in Fig. 4(c), while for long term prediction Local Models have slightly better results (Fig. 4(d)). The AR models perform barely better than the reference. Here the Durbin-Levinson algorithm achieves the best prediction quality. Cluster Weighted Models show the worst performance and their mean errors stay even behind the simple reference algorithms.

It seems that the usage of the additional weights in the Steil networks $(\mathbf{w}_{dir}, \mathbf{w}_{rec})$ destabilizes the prediction especially over a long term, because of the fact that these networks have considerably worse prediction results.

In general, the difference between each of the algorithms and to the reference is much smaller than for the predictions of the sine or Lorenz-Attractor trajectories. Nevertheless, the best algorithms still beat the simple references clearly (as expected) and are able to predict movement trajectories, consisting of several hundred hypothetical steps, very well.

It can be concluded that the prediction of movement data seems to be a harder problem than predicting standard chaotic trajectories, such as those generated by the Lorenz-Attractor. This is caused by unique unexpected and unpredictable behavior, which can be observed in the movement data. Therefore, the choice of the number of neurons in the ESN reservoir, for example, has only a minor effect. In tests, the difference in the prediction results of movement data between 25 and 250 neurons were insignificant. It can be presumed that the structure of the movement data does not allow a higher accuracy in the prediction unlike other chaotic time series [4].

4   Conclusion

The intention of this paper was to connect the well-known fields of time series prediction and movement data handling in a new way. It was possible to show, that some of the algorithm are preforming well, while predicting movement data. Generally, it can be said that movement data behaves different than periodical sine and chaotic Lorenz-Attractor time series.

The tested algorithms show good results on movement data. However, some improved versions of the algorithms, which show good results for sine and Lorenz-attractor time series, show no benefit for movement data.

Echo State Networks and Local Models turned out to be suitable algorithms for movement prediction.

Autoregressive Models and again ESNs are able to predict fast enough for an online application without any further adaption. From the current point of view Echo State Networks are the "winning" approach which are able to solve the problem. Hence, further analysis should put the focus on this approach and on the additional improvements, which have not yet been tested.

The other algorithms can be upgraded as well. Local Models could be a good alternative to ESNs, if they could be accelerated without deterioration of quality. Besides, the enhanced versions of the AR Models such as ARMA or ARIMA models could be tested. Furthermore, the usage of an irregular embedding is imaginable.

As a next step, an adequate proactive navigation and interaction strategy exploiting the prediction results needs to be investigated. One drawback for predicting motion data is the fact that human beings may perform unexpected motion. Since the discussed algorithms rely on the past characteristics, it is possible to use them for detection of such unexpected behavior (as some kind of "saliency"-cue) for improving robot human interaction.

References

1. Schröter, C., Böhme, H.J., Gross, H.M.: Memory-efficient gridmaps in rao-blackwellized particle filters for slam using sonar range sensors. In: Proc. European Conference on Mobile Robots (2007)
2. Pett, S., Fraichard, T.: Safe navigation of a car-like robot within a dynamic environment. In: Proc. of European Conf. on Mobile Robots, pp. 116–121 (2005)
3. Owen, E., Montano, L.: Motion planning in dynamic environments using the velocity space. In: Proc. of RJS/IEEE IROS, pp. 997–1002 (2005)
4. Jaeger, H., Haas, H.: Harnessing nonlinearity: predicting chaotic systems and saving energy in wireless telecommunication. Science, 78–80 (April 2004)
5. Scheidig, A., Müller, S., Martin, C., Gross, H.M.: Generating person's movement trajectories on a mobile robot. In: Proc. of International Symposium on Robots and Human Interactive Communications, pp. 747–752 (2006)
6. Steil, J.J.: Online reservoir adaptation by intrinsic plasticity for backpropagation-decorrelation and echo state learning. Neural Networks 20, 353–364 (2007)
7. http://paco.psy.gla.ac.uk/data_ptd.php
8. Wiener, N.: Extrapolation, Interpolation, and Smoothing of Stationary Time Series. Wiley, Chichester (1949)
9. Shumway, R.H., Stoffer, D.S.: Time Series Analysis and Its Applications. Springer Texts in Statistics (2000)
10. Abarbanel, H., Parlitz, U.: Nonlinear Analysis of Time Series Data. In: Handbook of Time Series Analysis, pp. 1–37. WILEY-VCH, Chichester (2006)
11. Engster, D., Parlitz, U.: Local and Cluster Weighted Modeling for Time Serie Prediction. In: Handbook of Time Series Analysis, pp. 38–65. WILEY-VCH, Chichester (2006)
12. Gross, H.M., Richarz, J., Müller, S., Scheidig, A., Martin, C.: Probabilistic multi-modal people tracker and monocular pointing pose estimator for visual instruction of mobile robot assistants. In: Proc. World Congress on Comp. Intelligence (2006)

Comparison of RBF Network Learning and Reinforcement Learning on the Maze Exploration Problem

Stanislav Slušný, Roman Neruda, and Petra Vidnerová

Institute of Computer Science
Academy of Sciences of the Czech Republic
Pod vodárenskou věží 2, Prague 8, Czech Republic
{slusny,roman,petra}@cs.cas.cz

Abstract. An emergence of intelligent behavior within a simple robotic agent is studied in this paper. Two control mechanisms for an agent are considered — a radial basis function neural network trained by evolutionary algorithm, and a traditional reinforcement learning algorithm over a finite agent state space. A comparison of these two approaches is presented on the maze exploration problem.

1   Introduction

Reactive and behavior based systems deal with agents of low levels of cognitive complexity in complex, noisy, and uncertain environments. The focus is on the intelligent behaviours that arise as a result of an agent's interaction with its environment. The ultimate goal of the process is to develop an embodied and autonomous agent with a high degree of adaptive possibilities [9]. Two main approaches to tackle this problem are currently the traditional reinforcement learning (RL) [13] and evolutionary robotics (ER) [8].

Both these approaches fall into the same category of learning algorithms that are often used for tasks where it is not possible to employ more specific supervised learning techniques. Designing an agent control mechanism is a typical example of such a problem where an instant reward of agent actions is not available. We are usually able to judge positive or negative behaviour patterns of an agent (such as finding a particular spot in a maze or hitting a wall) and evaluate it on the coarser time scale. This information is used by different learning algorithms of reinforcement type to strengthen successful partial behavior patterns, and in the course of adaptation process, to develop an agent solving a given task.

The ER approach attacks the problem through a self-organization process based on artificial evolution [5]. Control mechanisms of an agent are typically based on a neural network which provides direct mapping from agent sensors to effectors. Most of the current applications use traditional multi-layer perceptron networks [4]. In our approach we utilize local unit network architecture called radial basis function (RBF) network which has competitive performance, more learning options, and (due to its local nature) better interpretation possibilities [11,12].

V. Kůrková et al. (Eds.): ICANN 2008, Part I, LNCS 5163, pp. 720–729, 2008.

2   Reinforcement Learning

Let us consider an embodied agent that is interacting with the environment by its sensors and effectors. The essential assumption of RL is that the agent has to be able to sense rewards coming from the environment. Rewards evaluate taken actions, agent's task is to maximize them. There has been several algorithms suggested so far. We have used the Q-learning algorithm, which was first breakthrough of RL [14].

The next important assumption is that agent is working in discrete time steps. Symbol S will denote finite discrete set of states and symbol A set of actions. In each time step t, agent determines its actual state and chooses one action. Therefore, agent's life can be written as a sequence $o_0 a_0 r_0 o_1 a_1 r_1 \ldots$ where o_t denotes observation through the sensors, $a_t \in A$ action and finally symbol $r_t \in R$ represents *reward*, that was received at time t.

The most serious assumption of RL algorithms is the *Markov property*, which states, that agent does not need history of previous observations to make decision. The decision of the agent is based on the last observation o_t only. When this property holds, we can use theory coming from the field of *Markov decision processes* (MDP).

The direct implication of Markov property is the equality of states and observations. The strategy π, which determines what action is chosen in particular state, can be defined as function $\pi : S \rightarrow A$, where $\pi(s_t) = a_t$.

Now, the task of the agent is to find optimal strategy π^*. Optimal strategy is the one, that maximalizes expected reward. In MDP, single optimal deterministic strategy always exists, no matter in what state has the agent started.

The quantity $V^\pi(s_t)$ is called discounted cumulative reward. It is telling us, what reward can be expected, if the agent starts in state s_t and follows policy π,

$$V^\pi(s_t) = r_t + \gamma r_{t+1} + \gamma^2 r_{t+2} + \ldots = \sum_{i=0} \gamma^i r_{t+i}. \tag{1}$$

Here $0 \leq \gamma < 1$ is a constant that determines the relative value of delayed versus immediate rewards.

Optimal strategy π^* can now be defined as

$$\pi^* = argmax_\pi V^\pi(s), \forall s \in S \tag{2}$$

To simplify the notation, let us write $V^*(s)$ instead of symbol V^{π^*}, value function corresponding to optimal strategy π^*.

$$V^*(s) = max_\pi V^\pi(s) \tag{3}$$

The Q-learning algorithm was the first algorithm to compute optimal strategy π^* [14]. The key idea of the algorithm is to define the so-called *Q-values*. $Q^\pi(s, a)$ is the expected reward, if the agent takes action a in state s and then follows policy π.

$$Q^\pi(s, a) = r(s, a) + \gamma V^\pi(s'), \tag{4}$$

where s' is the state, in which agent occurs taking action a in state s ($s' = \delta(s, a)$).

1. Let S be the finite set of states and A finite set of actions.
   $\forall s \in S, a \in A : Q(s,a) = 0$
2. Process sensors and obtain state s
3. Repeat:
   - Choose and carry out action a
   - Receive reward r
   - Obtain new state s'
   - $Q(s,a) \leftarrow r + \gamma \max_{a'} Q(s',a')$
   - $s \leftarrow s'$

Algorithm 1. Q-learning

Q-learning algorithm (Algorithm 1) guarantees convergence to optimal values of $Q^*(s,a)$, if Q-values are represented without any function approximations (in table), rewards are bounded and every state-action pair is visited infinitely often. To fulfill the last condition, every action has to be chosen with non-zero probability. Probability $P(a|s)$ of choosing action a in state s is defined as [6]

$$P(a_i|s) = k^{Q(s,a_i)} / \sum_j k^{Q(s,a_j)}, \tag{5}$$

where constant $k > 0$ determines exploitation-exploration rate. Big values of k will make agent to choose actions with above average values. On the other hand, small values will make agent to choose actions randomly. Usually, learning process is started with small k, that is slightly increasing during the course of learning.

Optimal values V^*s can be obtained from $Q^*(s,a)$ by the equality

$$V^*(s) = \max_{a'} Q(s,a'). \tag{6}$$

3   Evolutionary Robotics

Evolutionary robotics combines two AI approaches: neural networks and evolutionary algorithms. The control system of the robot is realized by a neural network, in our case an RBF network. It is difficult to train such a network by traditional supervised learning algorithms since they require instant feedback in each step, which is not the case for evolution of behavior. Here we typically can evaluate each run of a robot as a good or bad one, but it is impossible to assess each one move as good or bad. Thus, the evolutionary algorithm represent one of the few possibilities how to train the network.

The *RBF network* [10,7,1], used in this work, is a feed-forward neural network with one hidden layer of *RBF units* and linear output layer. The network function is given by Eq. (8).

The evolutionary algorithms (EA) [5,3] represent a stochastic search technique used to find approximate solutions to optimization and search problems. They use techniques inspired by evolutionary biology such as mutation, selection, and crossover. The EA typically works with a population of *individuals* representing abstract representations of feasible solutions. Each individual is assigned a *fitness* that is a measure of how good solution it represents. The better the solution is, the higher the fitness value it gets. The population evolves toward better solutions. The evolution starts from a population of completely random individuals and iterates in generations. In each generation, the fitness of each individual is evaluated. Individuals are stochastically selected from the current population (based on their fitness), and modified by means of operators *mutation* and *crossover* to form a new population. The new population is then used in the next iteration of the algorithm.

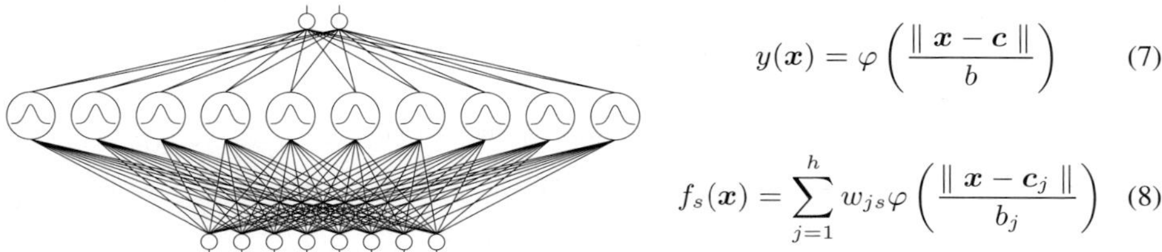

$$y(\boldsymbol{x}) = \varphi\left(\frac{\|\,\boldsymbol{x} - \boldsymbol{c}\,\|}{b}\right) \qquad (7)$$

$$f_s(\boldsymbol{x}) = \sum_{j=1}^{h} w_{js}\varphi\left(\frac{\|\,\boldsymbol{x} - \boldsymbol{c}_j\,\|}{b_j}\right) \qquad (8)$$

Fig. 1. RBF Network. Eq. (7) evaluates the output of a hidden RBF unit, f_s (8) is the output of the s-th output unit. φ is an activation function, typically Gaussian function $\varphi(s) = e^{-s^2}$.

In our approach, the evolutionary algorithm is responsible for modification of weights of an RBF network. The weights are encoded using a floating-point encoding, so an individual is a vector of floating-point values of network weights. The number of hidden units is determined in advance and does not undergo the evolutionary process. Typical evolutionary operators for the floating point encoded genotype have been used, namely the uniform crossover and the mutation which performs a slight additive change in the parameter value. A standard tournament selection is used together with a small elitist rate parameter [11]. Detailed discussions about the fitness function are presented in the Sec. 4.

4   Experimental Framework

In order to compare performance and properties of described algorithms, we conducted simulated experiment. Miniature robot of type e-puck[2] was trained to explore the environment and avoid walls. E-puck is a mobile robot with a diameter of 70 mm and a weight of 50 g. The robot is supported by two lateral wheels that can rotate in both directions and two rigid pivots in the front and in the back. The sensory system employs eight "active infrared light" sensors distributed around the body, six on one side and two on other side. In "passive mode", they measure the amount of infrared light in the environment, which is roughly proportional to the amount of visible light. In "active mode" these sensors emit a ray of infrared light and measure the amount of reflected light. The closer they are to a surface, the higher is the amount of infrared light measured. The e-puck sensors can detect a white paper at a maximum distance of approximately 8 cm. Sensors return values from interval $[0, 4095]$. Effectors accept values from

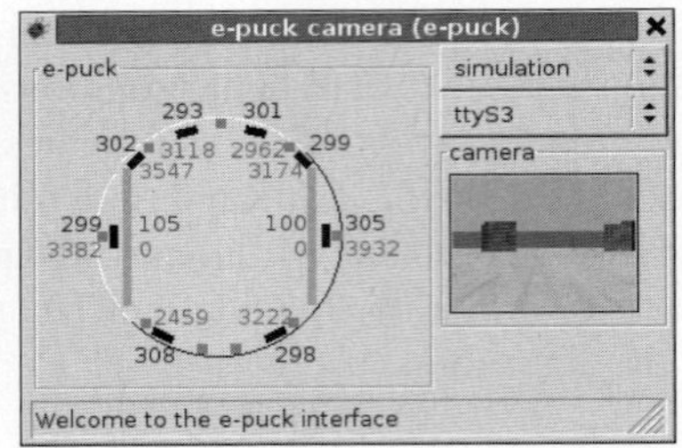

Fig. 2. Miniature e-puck robot

Table 1. Sensor values and their meaning

Sensor value	Meaning
0-50	NOWHERE
51-300	FEEL
301-500	VERYFAR
501-1000	FAR
1001-2000	NEAR
2001-3000	VERYNEAR
3001-4095	CRASHED

interval $[-1000, 1000]$. The higher the absolute value, the faster is the motor moving in either direction.

Without any further preprocessing of sensor's and effector's values, the state space would be too big. Therefore, instead of raw sensor values, learning algorithms worked with "perceptions". Instead of 4095 raw sensor values, we used only 5 perceptions (see Tab. 1). Effector's values were processed in similar way: instead of 2000 values, learning algorithm was allowed to choose from values $[-500, -100, 200, 300, 500]$. To reduce the state space even more, we grouped pairs of sensors together and back sensors were not used at all.

Agent was trained in the simulated environment of size 100 x 60 cm and tested in more complex environment of size 110 x 100 cm. We used Webots [15] simulation software. Simulation process consisted of predefined number of steps. In each simulation step agent processed sensor values and set speed to the left and right motor. One simulation step took 32 ms.

In first experiment, we have used Q-learning algorithm as described in Section 2. Each state was represented by a triple of perceptions. For example, the state [NEAR, NOWHERE, NOWHERE] means, that the robot sees a wall on its left side only. Action was represneted by a pair [left speed, right speed]. Learning process was divided into episodes. Each episode took at most 800 simulation steps. At the end of each episode, agent was moved to one from 5 randomly chosen positions. Episode could be finished earlier, if agent hit the wall. The learning process was stopped after 10000 episodes. Parameter γ was set to 0.3.

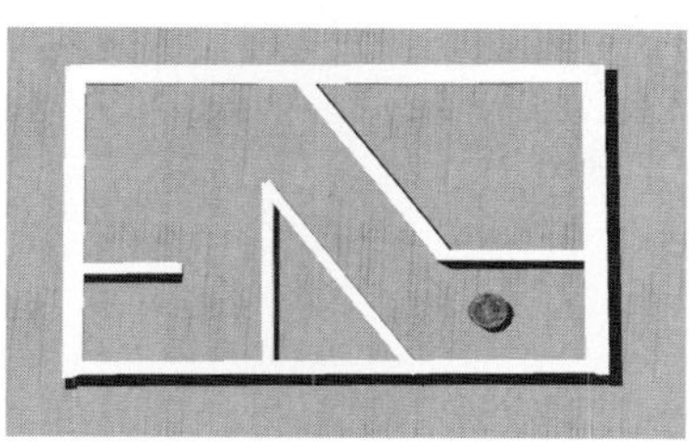 

Fig. 3. Simulated environments for agent training and testing: a) Agent was trained in the simulated environment of size 100 x 60 cm. b) Simulated testing environment of size 110 x 100 cm.

In the second experiment the evolutionary RBF networks were applied to the same maze-exploration task (see Fig. 3).

The network has 3 input units, 5 hidden Gaussian units, and 2 output units. The three inputs correspond to the coupled sensor values (two left sensors, two front sensors, two right sensors), which are preprocessed in the way described in Tab. 1. The two outputs correspond to the left and right wheel speeds and before applying to robot wheels they are rounded to one of 5 valid values. Fitness evaluation consists of two trials, which differ by agent's starting location (the two starting positions are in the opposite ends of the maze). Agent is left to live in the environment for 800 simulation steps, each simulation step corresponds to 32ms.

In each step, a three-component score is calculated to motivate agent to learn to move and to avoid obstacles:

$$T_{k,j} = V_{k,j}(1 - \sqrt{\Delta V_{k,j}})(1 - i_{k,j}). \tag{9}$$

First component $V_{k,j}$ is computed by summing absolute values of motor speed (scaled to $\langle -1, 1\rangle$) in the k-th simulation step and j-th trial, generating value between 0 and 1. The second component $(1 - \sqrt{\Delta V_{k,j}})$ encourages the two wheels to rotate in the same direction. The last component $(1 - i_{k,j})$ encourage obstacle avoidance. The value $i_{k,j}$ of the most active sensor (scaled to $\langle 0, 1\rangle$) in k-th simulation step and j-th trial provides a conservative measure of how close the robot is to an object. The closer it is to an object, the higher is the measured value in range from 0.0 to 1.0. Thus, $T_{k,j}$ is in range from 0.0 to 1.0, too.

In the j-th trial, score S_j is computed by summing normalized trial gains $T_{k,j}$ in each simulation step:

$$S_j = \sum_{k=1}^{800} \frac{T_{k,j}}{800}. \tag{10}$$

To stimulate maze exploration, agent is rewarded, when it passes through one of pre-defined zones. There are three zones located in the maze. They can not be sensed by an agent. The reward $\Delta_j \in \{0, 1, 2, 3\}$ is given by the number of zones visited in the j-th trial. The fitness value is then computed as follows:

$$Fitness = \sum_{j=1}^{2}(S_j + \Delta_j).\tag{11}$$

5   Experimental Results

5.1   Rules Inducted by Reinforcement Learning Algorithm

Table 2 contains states with biggest and smallest Q-values and their best action. The states with biggest Q-values contain mostly perception NOWHERE. On the other side, states with smallest Q-values contain perception CRASHED.

Learned behaviour corresponds to obstacle avoidance behaviour. The most interested are the states, which contain perception "NEAR". Expected rules "when obstacle left, then turn right" can be found. States without perception "NEAR" were evaluated as safe — even if bad action was chosen in this state, it could be fixed by choosing good

Table 2. 5 states with biggest and smallest Q-values and their best actions

State			Action	Q-value
left	front	right		
NOWHERE	NOWHERE	VERYFAR	[500, 300]	5775.71729
NOWHERE	NOWHERE	NOWHERE	[300, 300]	5768.35059
VERYFAR	NOWHERE	NOWHERE	[300, 500]	5759.31055
NOWHERE	NOWHERE	FEEL	[300, 300]	5753.71240
NOWHERE	VERYFAR	NOWHERE	[500, 100]	5718.16797
CRASHED	CRASHED	CRASHED	[300, 500]	-40055.38281
CRASHED	NOWHERE	CRASHED	[300, 300]	-40107.77734
NOWHERE	CRASHED	VERYNEAR	[300, 500]	-40128.28906
FAR	VERYNEAR	CRASHED	[300, 500]	-40210.53125
NOWHERE	CRASHED	NEAR	[200, 500]	-40376.87891

Table 3. Rules represented by RBF units (listed values are original RBF network parameters after discretization)

Sensor			Width	Motor	
left	front	right		left	right
VERYNEAR	NEAR	VERYFAR	1.56	500	-100
FEEL	NOWHERE	NOWHERE	1.93	-500	500
NEAR	NEAR	NOWHERE	0.75	500	-500
FEEL	NOWHERE	NEAR	0.29	500	-500
VERYFAR	NOWHERE	NOWHERE	0.16	500	500

action in next state. Therefore, these actions do not tell us a lot about agent's behaviour. On the other side, action with perception VERYNEAR leaded to the crash, usually. Agent was not able to avoid the collision.

5.2   Rules Extracted from RBF Networks

The experiment with evolutionary RBF network was repeated 10 times, each run lasted 200 generations. In all cases the successful behaviour was found, i.e. the evolved robot was able to explore the whole maze without crashing to the walls.

Table 3 shows parameters of an evolved network with five RBF units. We can understand them as rules providing mapping from input sensor space to motor control.

Table 4. Rules and their frequencies. Rules are actual rules used by RBF network in train arena.

	Sensor			Motor	
	left	front	right	left	right
250	FEEL	NOWHERE	NOWHERE	500	500
172	NOWHERE	NOWHERE	NOWHERE	100	500
105	VERYFAR	NOWHERE	NOWHERE	500	500
98	FEEL	NOWHERE	FEEL	500	500
68	NOWHERE	NOWHERE	FEEL	100	500
67	FAR	NOWHERE	NOWHERE	500	300
60	FEEL	FEEL	NOWHERE	500	300
51	NEAR	NOWHERE	NOWHERE	500	100
20	VERYFAR	FEEL	NOWHERE	500	300
18	NOWHERE	FEEL	NOWHERE	300	300
15	NEAR	FEEL	NOWHERE	500	-100
15	FAR	FEEL	NOWHERE	500	-100
10	FAR	FAR	NOWHERE	500	-500
9	NEAR	FAR	NOWHERE	500	-500
8	NOWHERE	FEEL	FEEL	300	300
6	FEEL	VERYFAR	NOWHERE	500	-100
5	VERYFAR	FAR	NOWHERE	500	-500
5	NEAR	VERYFAR	NOWHERE	500	-500
3	NOWHERE	VERYFAR	FEEL	500	100
3	NEAR	NEAR	NOWHERE	500	-500
2	FEEL	FAR	NOWHERE	500	-100
2	FEEL	FAR	FEEL	500	-100
1	VERYNEAR	NEAR	NOWHERE	500	-500
1	VERYNEAR	FAR	NOWHERE	500	-500
1	VERYFAR	VERYFAR	NOWHERE	500	-500
1	VERYFAR	NEAR	FEEL	500	-500
1	NOWHERE	FAR	FEEL	500	100
1	FEEL	NEAR	FEEL	500	-500
1	FAR	VERYFAR	NOWHERE	500	-500
1	FAR	NEAR	NOWHERE	500	-500

Table 5. Most important rules from Tab. 4 and their semantics

Sensor			Motor		
left	front	right	left	right	
FEEL	NOWHERE	NOWHERE	500	500	straight movement
NOWHERE	NOWHERE	NOWHERE	100	500	turning left
VERYFAR	NOWHERE	NOWHERE	500	500	straight movement
FEEL	NOWHERE	FEEL	500	500	straight movement
NOWHERE	NOWHERE	FEEL	100	500	turning left
FAR	NOWHERE	NOWHERE	500	300	turning right
FEEL	FEEL	NOWHERE	500	300	turning right
NEAR	NOWHERE	NOWHERE	500	100	turning right

However, these 'rules' act in accord, since the whole network computes linear sum of the corresponding five gaussians.

The following two tables Tab. 4 and Tab. 5 show rules from actual run of the robot in the train arena. The nine most frequently used rules are shown in the latter table. It can be seen that this agent represents a typical evolved left-hand wall follower. Straight movement is a result of situations when there is a wall far left, or both far left and right. If the robot sees nothing, it rotates leftwise (rule 2), The front collision is avoided by turning right, as well as a near proximity to the left wall (rules 6–8).

The evolved robot was then tested in the bigger testing maze. It behaved in a consistent manner, using same rules, demonstrating generalization of the behaviour trained in the former maze.

6   Conclusion

We have presented experiments with RL and ER algorithms training a robot to explore a maze. It is known from the literature, and from our previous works, that this problem is manageable by both RL and ER learning with different neural representations of the control mechanism. Usually, in such a successful learning episode, an agent with general behavioral pattern emerges that is able to explore previously unseen maze in an efficient way.

In this work we have focused on comparison of rules derived by traditional RL approach and by the evolved neural networks. We have chosen the RBF network architecture with local processing units. These networks are easily to interpret in terms of rules than traditional perceptron networks. A simple analysis shows that both RL and ER resulted in a rules that are reasonable, and easy to interpret as higher-level behaviour traits. The RL approach shows rational obstacle avoidance, while the neuro-evolution approach comes with more compact individuals that can be clearly classified as left-hand wall followers (or right-hand wall followers, respectively).

As we have seen, different learning approaches can lead to different behaviours. Agents trained by evolutionary algorithms usually show simple behaviours. Often, changing basic environment constraints (dimensions of environment, for example) can make learned strategy fail([8]). In our experiment, learned strategy is simple (it can be described by several rules) but effective. Agent learned by Q-learning algorithm showed more complex behaviour. It can cope with situations, in which agent trained by ER would fail. However, effective wall following strategy was not discovered.

In our further work, we would like to take advantages of both approaches. The basic learning mechanism will be evolutionary algorithm. Behavioral diversity could be maintained by managing population of agents that use different learning approaches (RBF networks and Reinforcement learning). As we have shown, different learning algorithms can lead to different behavioral patterns. In both algorithms used, experience can be expressed as a set of rules. Taking this into account, genetic operators could be designed to allow simple rules exchange mechanisms.

Acknowledgements

R. Neruda and P. Vidnerova have been supported by the Grant Agency of the Czech Republic project no. 201/08/1744; S. Slusny has been supported by the Grant Agency of the Charles University project no. GAUK 7637/2007.

References

1. Broomhead, D.S., Lowe, D.: Multivariable functional interpolation and adaptive networks. Complex Systems 2, 321–355 (1988)
2. E-puck, online documentation, http://www.e-puck.org
3. Fogel, D.B.: Evolutionary Computation: The Fossil Record. MIT/ IEEE Press (1998)
4. Haykin, S.: Neural Networks: a comprehensive foundation, 2nd edn. Prentice-Hall, Englewood Cliffs (1999)
5. Holland, J.: Adaptation In Natural and Artificial Systems. MIT Press, Cambridge (1992)
6. Mitchell, T.: Machine Learning. McGraw-Hill, New York (1997)
7. Moody, J., Darken, C.: Fast learning in networks of locally-tuned processing units. Neural Computation 1, 289–303 (1989)
8. Nolfi, S., Floreano, D.: Evolutionary Robotics — The Biology, Intelligence and Techology of Self-Organizing Machines. MIT Press, Cambridge (2000)
9. Pfeifer, R., Scheier, C.: Understanding Intelligence. MIT Press, Cambridge (2000)
10. Poggio, T., Girosi, F.: A theory of networks for approximation and learning. Technical report, Massachusetts Institute of Technology, Cambridge, MA, USA, A. I. Memo No. 1140, C.B.I.P. Paper No. 31 (1989)
11. Slušný, S., Neruda, R.: Evolving homing behaviour for team of robots. In: Computational Intelligence, Robotics and Autonomous Systems. Massey University, Palmerston North (2007)
12. Slušný, S., Neruda, R., Vidnerová, P.: Evolution of simple behavior patterns for autonomous robotic agent. In: System Science and Simulation in Engineering, pp. 411–417. WSEAS Press (2007)
13. Richard Sutton, S., Andrew Barto, G.: Reinforcement Learning: An Introduction. MIT Press, Cambridge (1998)
14. Watkins, C.J.C.H.: Learning from delayed rewards. Ph.D. thesis (1989)
15. Webots simulator. On-line documentation, http://www.cyberbotics.com/

Modular Neural Networks for Model-Free Behavioral Learning

Johane Takeuchi, Osamu Shouno, and Hiroshi Tsujino

Honda Research Institute Japan Co., Ltd.
8-1 Honcho, Wako-shi, Saitama 351-0188, Japan
`{johane.takeuchi,shouno,tsujino}@jp.honda-ri.com`

Abstract. Future robots/agents will need to perform situation specific behaviors for each user. To cope with diverse and unexpected situations, model-free behavioral learning is required. We have constructed a modular neural network model based on reinforcement learning and demonstrated that the model can learn multiple kinds of state transitions with the same architectures and parameter values, and without pre-designed models of environments. We recently developed a modular neural network model equipped with a modified on-line modular learning algorithm, which is more suitable for neural networks and more efficient in learning. This paper describes the performances of constructed models using the probabilistically fluctuated Markov decision process including partially observable conditions. In the test transitions, the observed states probabilistically fluctuated. The new learning model is able to function in those complex transitions without specific adjustments for each transition.

Keywords: Reinforcement learning, On-line learning, Partially observable environment.

1 Introduction

Animals flexibly learn appropriate response to various situations in order to survive in the real world even without problem formulations by others. Such capabilities will be required for future symbiotic robots/agents that will be expected to support humans in houses, offices, and so on. They will autonomously perform situation associated behaviors/dialogs and carry out tasks for users in the real world. Appropriate behavioral learning systems will be required to achieve the situated behavior/dialog controls because real situations are diverse and volatile. Current machine learning, however, covers only isolated single problems that are appropriately modeled by the designers of the learning systems. We are striving to extend machine learning to real-environmental problems where it is difficult to formulate environmental models. We describe the practical importances of our studies with an example form the engineering viewpoints of creating future robots/agents.

The behavior/dialog control system of future robots/agents will be comprised of a model-based system, which enables the products to work incipiently. The

V. Kůrková et al. (Eds.): ICANN 2008, Part I, LNCS 5163, pp. 730–739, 2008.

model-based systems will likely fail to perform situation specific behaviors because they are not inherently grounded to every user's situations. Every users' situations are often both outside of our expectations and changeable. These will make robots/agents commercially uncompetitive, and moreover, not truly symbiotic. As a commercial mass product, however, it is impossible to adjust every system sold whenever robots encounter unknown situations. A possible option is somehow updating the model within the system. Another idea for overcoming these shortcomings is introducing an on-line learning system that bonds situations with behaviors. We believe that such on-line learning is applicable even for updating model-based systems, for example, to find something wrong in the current model. What kinds of requirements do we need to create such a learning system? The learning system should function using only its experiences. In other words, the system must be autonomous and situated. Reinforcement learning (RL) [1] is a learning algorithm that makes a map between observed states and actions, which is very close to our purposes. The standard implementations of RL are, however, neither autonomous nor situated. In general, RL requires task-specific reward functions that are carefully tailored by designers. One has to adjust the system to make them function, so RL is not fully autonomous. In the standard implementations of RL, value functions are represented by means of tables. These tables are usually also designed by designers. We should inquire about the appropriate representation method that situated its inputs. By satisfying these conditions, the learning system will cope with diverse real situations without adjustments by hand so that the learning system can autonomously complement behavior/dialog controllers.

As a first step, we have suggested construction of an RL system capable of functioning in several kinds of state transitions without hand-wired adjustments for each transition. This is crucial in order to achieve learning systems coping with real situations where several kinds of state transitions potentially emerge. In our previous model [2], we introduced internal rewards to reduce hand-wired factors of reward functions. Several researches have attempted to introduce internal rewarding mechanisms for artificial learning algorithms (e.g., refer to [3]). Our model adopted simple prediction error-based reward generators. In addition, the system consisted of neural networks equipped with a modular reinforcement learning algorithm. The modular system that decomposes the observed state space in a self-organizing manner stabilizes the internal rewards calculated from prediction errors. We found that the combined system with both modular networks and the internal reward generators led to performance that converged to the optimal sequences of actions in multiple tested transitions without any specific adjustments for each transition. The tested transitions included not only a Markov decision process (MDP) but also an MDP under partially observable conditions without prior knowledge of the transition.

In this paper, we mainly examine two recently introduced factors. One is small negative rewards to every transition in addition to the prediction based internal rewards. Another is the the algorithms of on-line modular decompositions. In the previous studies, we applied the modular reinforcement learning algorithm

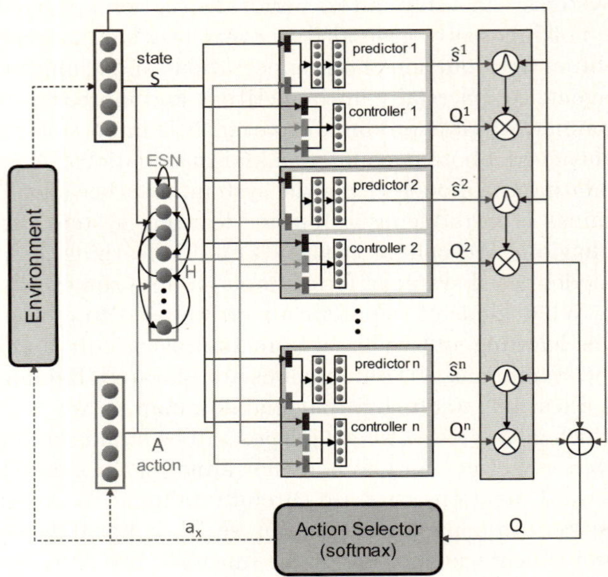

Fig. 1. A schematic diagram of the modular neural network model. The predictor network in each module predicts the current state $\widehat{s_t}^n$ from the previous observed state s_{t-1} and the previous action a_{t-1}. Each predictor network contains one hidden layer (six nodes). The outputs of the controller layer in each module (Q^n) are Q-function vector values for primitive actions. The inputs of each controller layer are the observed state s_t (S), the previous action a_{t-1} (A), and the history information h_{t-1} (H). The history information is represented with an echo state network (ESN).

suggested by Doya, *et al.* [4], which was effective in on-line learning. We refer to the algorithm in this paper as MOSAIC (MOdule Selection And Identification Control [5]). We extended the algorithm to be more suitable for neural networks and to be more efficient at learning. We demonstrated the learning efficiencies on these two factors using a probabilistically fluctuated state transition that can be expected to be encountered in real situations and more difficult to learn than those in our previous studies.

2   Learning Model

Tabular representations are inadequate to approximate value functions in several kinds of transitions without predefinitions of observed states. To combine function approximation with a system to generate novelty rewards, we used neural networks for constructing the models. Figure 1 shows a schematic diagram of the constructed model. The observed state, the previous action, and the recurrent layer were input sources of the controller layer in each module. The history information was represented by means of an echo state reservoir that was a randomly connected recurrent neural network [6]. Each state was represented by a fifteen-dimensional normalized vector consisting of random values. The previous actions were represented by ten-dimensional unit vectors. We used a softmax function for the action selector. For the controller in each module, we adopted a gradient-descent SARSA(λ) algorithm [1]. It has been pointed out that the combination of off-policy learning and function approximation potentially causes divergence [1]. On the other hand, on-policy learning with a linear approximation such as gradient-descent SARSA(λ) hardly met this difficulty. In addition, we utilized a non-linear approximation with two-layered neural networks. Each node of output layers implemented a non-linear function for its outputs; $y_i = \tanh(a\mathbf{w}_i \cdot \mathbf{s})$, which cannot approximate the value function that $|v| > 1$. This is, however, not the severe problem in this study. By these limitations of output values, we could suppress the divergence of approximations. Connecting weights $\mathbf{w}$ were updated for module n as follows.

$$\mathbf{w}^n_{t+1} = \mathbf{w}^n_t + \alpha\delta^n_t\mathbf{e}^n_t,$$

where

$$\delta^n_t = r^{ex}_t + r^{in}_t + b + \gamma Q(s_t, a_t) - Q(s_{t-1}, a_{t-1}), \qquad (1)$$
$$\mathbf{e}^n_{t+1} = \gamma\lambda\mathbf{e}^n_t + \rho^n_t\nabla_{\mathbf{w}^n_t}Q^n(s_t, a_t).$$

r^{ex}_t is an external reward at time t, r^{ex}_t is an internal one, γ is the discount factor ($\gamma \in [0, 1]$), α is an update state constant and λ is an eligibility trace constant. In this paper, we introduced the term b that was a small negative reward. The term $Q(s_t, a_t)$ was calculated from all modules as,

$$Q = \sum_k \rho^k_t Q^k \qquad (2)$$

where ρ^k_t is the responsibility signal [4] calculated by

$$\rho^k_t = C\exp(-(\Delta^k_t)^2/\sigma^2) \qquad (3)$$
$$\Delta^k_t = \|s_t - \widehat{s_t}^k\|,$$

where $\widehat{s_t}$ is the predicted state for time t and s_t is the observed state at time t. The intrinsic reward was calculated as follows;

$$r^{in}_t = \begin{cases} d_t & (d_t > 0) \\ 0 & (d_t \le 0), \end{cases} \qquad (4)$$
$$d_t = \Delta_t - Ave_n[\Delta], (\Delta_t = \min_k \Delta^k_t),$$

where Ave_n is the calculation of the moving average over n times. Therefore, when the error signal exceeds the current average by more than zero, internal rewards are given non-zero values. The learning algorithm of the prediction network in each module is a conventional back-propagation algorithm,

$$\mathbf{w}^n_{t+1} = \mathbf{w}^n_t + \epsilon\rho^n_t\nabla\left[\frac{1}{2}(s_t - \widehat{s_t}^n)^2\right].$$

The prediction networks do not contain any recurrent connections.

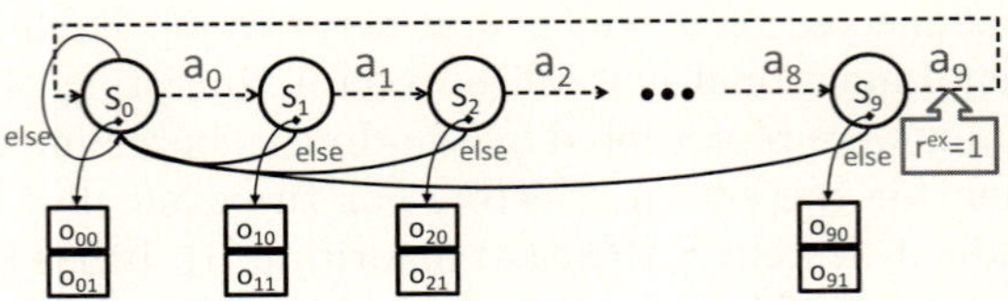

Fig. 2. Test continuing MDP environment. An external reward $r^{ex} = 1$ is earned after sequential events. Dashed lines indicate probabilistic transitions.

In our previous study, we applied the MOSAIC RL [4] to neural networks engaging in on-line learning of both predictions and controllers, which could cause interferences between modules or concentrations to fewer modules. Another modular RL algorithm for neural networks was suggested by Nishida *et al.* [7] using the modular network self-organization map (mnSOM) that is an extension of the conventional Kohonen's SOM [8]. In the mnSOM each SOM node extended to a functional module such as the module in our model, in which the winner node was the module that has the lowest prediction error at a time. By combining the mnSOM with parameter-less SOM proposed in [9], we made the mnSOM algorithm on-line learning. We have two options to calculate Q in the mnSOM for RL. One is 'winner takes all' so $Q = Q^k$ where the module k has the lowest prediction error (we will refer to this as SOM). Another is using (2) which is called SOM+MOSAIC in later sections.

3   Results

This section demonstrates the efficiencies of our proposed model in two kinds of transitions. All parameters of RL were fixed through all simulations, for example, $\gamma = 0.9$, $\lambda = 0.1$, and $\alpha = 0.3$ (for plain SARSA, $\alpha = 0.1$, please see later descriptions). The duration of the moving average in (4) was 100. The parameter of responsibility signals in (3) was set at $(1/\sigma)^2 = 20$.

3.1   Probabilistically Fluctuated Markov Environment

Figure 2 shows the MDP environment for the test. The number of action sets was ten, that is $\{a_0, a_1, \cdots , a_9\}$, and the number of states was also ten. At initial state s_0, if the agent selected actions sequentially in the order of the subscriptions, external rewards were earned, otherwise, the agent received no positive external reward. We engaged in continuing tasks so the transition repeated from the initial state even after earning an external reward. We applied these kinds of transitions for tests in our previous studies not only deterministic ones but also partially probabilistic ones. In this paper, we tested the constructed model using a full probabilistic transition with probabilistic fluctuations. The probability transition from s_0 to s_1 with action selection a_0 is 0.3. Other transitions of the correct sequence to earn external rewards were set as $p = 0.9$. In addition, the observed states were probabilistically changed. For example, in state s_0, the agent could

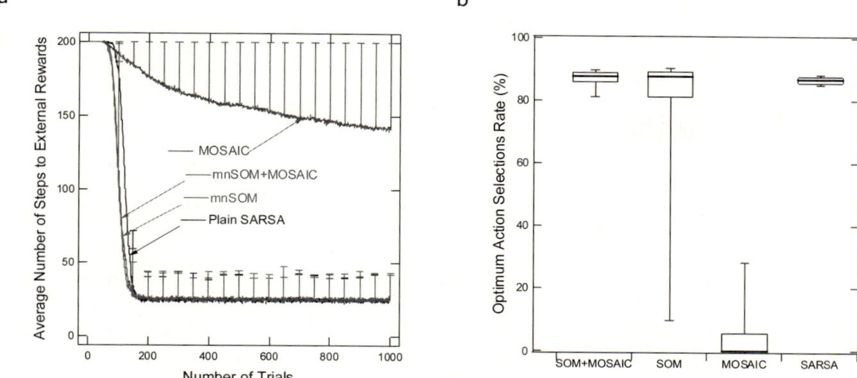

Fig. 3. a) The learning curves in the MDP environment. The ordinates indicate the number of steps to earn external rewards before time runs out (200 steps). The abscissas indicate the number of trials the agent had attempted in each set. Each curve is the average of 500 trial sets. For networked models, all input sources were connected to the controllers in each module, the number of modules was 25, and internal rewards were activated. b) Boxplot of the optimum action selections rate of 500 trial-sets. The median rates are represented by the middle horizontal lines. Interquartile ranges (25th–75th percentile) are denoted by height of box. Whiskers denote 10th–90th percentiles.

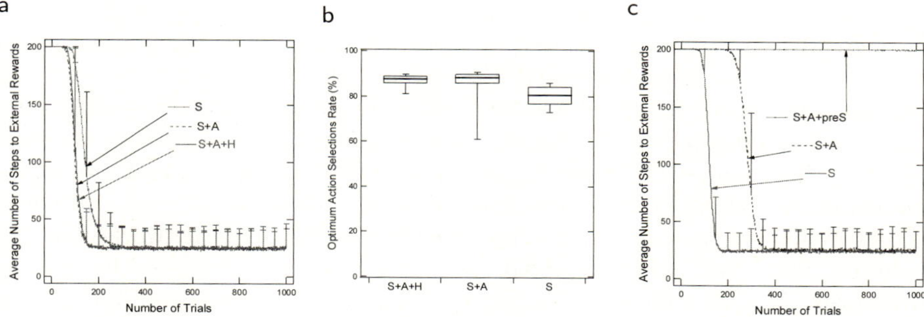

Fig. 4. a) The learning curves of the networked model (SOM+MOSAIC) in the MDP environment. 'S' denotes that the input source of the controllers in each module is the sensory signal. In a similar way, 'S+A' is combinations of sensory signal and previous action information and 'S+A+H' means that all sources including history information represented by the recurrent network are connected. b) The optimum action selections rate for (a). c) The learning curves of plain SARSAs in the MDP environment. 'preS' means the previous sensor inputs.

observe a state signal o_{00} or o_{01} with a probability of 0.5. Each state s_n was represented by o_{n0} and o_{n1}, so the number of observed states was twenty in this environment. All state signals o_{nx} were fifteen dimensional vectors that were randomly generated, therefore, o_{n0} and o_{n1} for the same s_n were not correlated

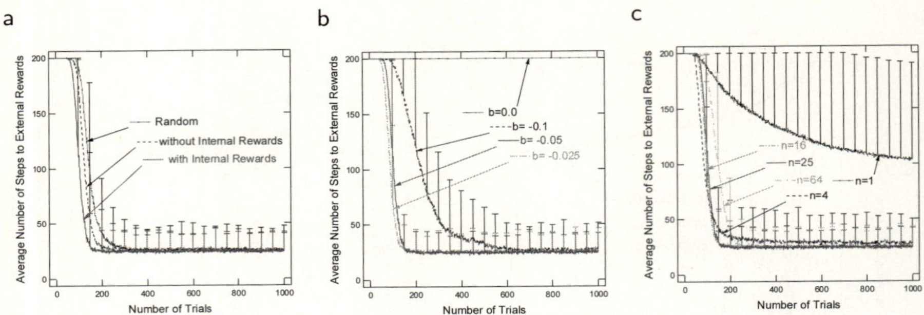

Fig. 5. The learning curves of the networked model (SOM+MOSAIC) in the MDP environment. a) Effects of internal rewards. b) Effects of negative rewards. c) Effects of changing the number of modules.

except coincidentally. We intended to make these test transitions close to real situations. The test transitions used were a simple model of actually possible situations in which different observed states could indicate the same state.

Figure 3a shows learning curves for the average performance of 500 trial sets (each set consisting of 1000 trials) with changing module algorithms including a plain tabular SARSA. Each trial set was simulated with different state vector values. Each trial started with the last states of the adjacent trials, and stopped either when the agent was rewarded, or when time ran out. As we described in the previous section, every transition in all cases cost negative rewards $b = -0.05$. The case of MOSAIC was impaired in this environment. For not fluctuated and shorter transitions, MOSAIC could learn as described in our previous study [2]. The cases including SOM, SOM+MOSAIC and plain SARSA had almost the same average learning speed. For the plain SARSA, we required that the table directory represented all observing states o_{nx}. On the other hand, networked models should make their representations from indirect observation of vector signals in self-organizing manner from *tabula rasa*. Still the networked models including SOM and SOM+MOSAIC achieved good performances. In order to investigate detailed differences between them, we investigated optimum selections of actions in each trial (Fig. 3b). 'Optimum selections of action' means that the agent chose only correct actions to earn external rewards in each trial. We did not confirm the significant difference between the the cases of SOM and SOM+MOSAIC by means of statistical non-parametric tests (we used Kruskal-Wallis test and post hoc Nemenyi test throughout this paper) with significant level 0.05. The others were different with significant level 0.01.

The above results for the networked models contained several parameters to be examined. The first one is the inputs of each controller. The results are shown in Fig. 4a. Adding input sources of controllers had little effect on the performances. The cases of S+A+H and S+A were rather better in this figure. The optimum selection rates depicted in Fig. 4b shows that the interquartile of the case S+A+H was the narrowest in this figure. The statistical test with significant level 0.01 showed that the cases and S+A+H and S were different

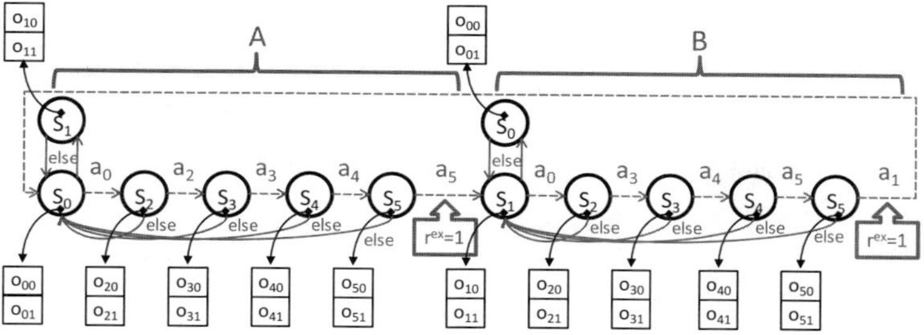

Fig. 6. Test continuing partially observable environment. The optimum action selection when the agent starts from the state s_0 is $a_0 \rightarrow a_2 \rightarrow a_3 \rightarrow a_4 \rightarrow a_5$ (A), thereby receiving r^{ex} and then $a_0 \rightarrow a_3 \rightarrow a_4 \rightarrow a_5 \rightarrow a_1$ (B) whereby the agent receives another r^{ex}.

from each other. We did not confirm the differences between the case of S+A+H and S+A. As a matter of course, SARSA with table representation could be impaired adding input sources, which are indicated by Fig. 4c. These properties of the networked model are important in order to treat several kinds of transitions without changing representations. The next parameter to be examined is internal rewards (Fig. 5a). In our previous study, internal rewards were required to achieve learning these learning tasks. Although internal rewards improved the performance, the case without internal rewards still functioned. The newly introduced factor, negative rewards, was more effective in the results. Figure 5b shows that the case of $b = 0.0$ resulted in no learning progress. We lastly examined the effects of the number of modules, which is shown in Fig. 5c. In these parameter sets and environment, $n = 25$ and $n = 16$ were best. Still $n = 4$ and $n = 64$ could function. As a whole, the system was not particularly sensitive to the extra input sources and the number of modules.

3.2 Partially Observable Environment

We also investigated differences in performance using the partially observable environments (Fig. 6). In this case, if the agent succeeded in earning an external reward, the unobservable state of the environment changed to another state. Thus, because of the unobservable state, the state transitions were changed and the agent had to select another sequence of actions to earn another external reward. The number of *true states* was six and the number of observed states was twelve. Although the number of states was less than the MDP case, the same observed states could be different for action selections in this case. Each trial ended when the agent received two external rewards so that the shortest number of transitions was ten, this being the same as the MDP case. The agent had to pass two probabilistic *alternating* transitions of $p = 0.3$ to earn two external rewards. If it failed to transit this portion, the state transited to s_1 as shown in the portion A in Fig. 6 (s_0 for portion B); thus, before attempting again to transit to s_2, the agent was required to output an arbitrary action to transit to

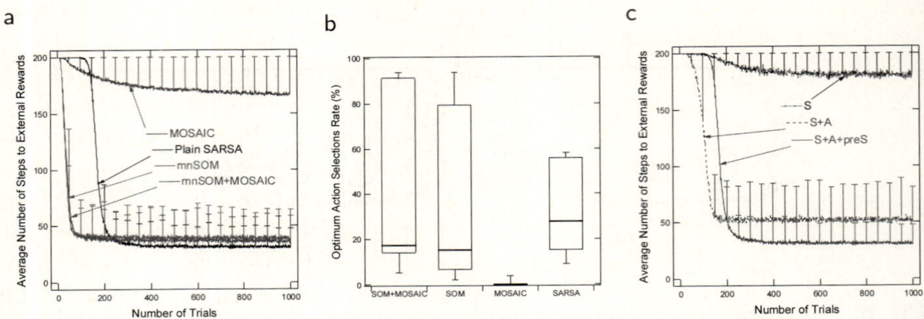

Fig. 7. a) The learning curves in the HOMDP environment. For SARSAs, the table representations were combinations of observing state, previous actions, and previous observed state. b) The optimum action selections rate for (a). c) The learning curves of the networked model (SOM+MOSAIC) in the HOMDP environment.

s_0 (s_1 for the case shown in the portion B). The memory information to solve the partially observability was previous actions at transitions from s_3 to s_5. When the agent was in state s_2, the agent could not distinguish the hidden states from the previous action because the previous actions were the same in both hidden states. History information was inevitably necessary to distinguish the situation. We refer to this environment as HOMDP (High Order MDP).

Figure 7a shows the learning curves of networked models and the plain SARSA. The fastest results were obtained for the cases of SOM and SOM+MOSAIC. The learning speed of the plain SARSAs was slower than SOM and SOM+MOSAIC, but the the plain SARSA converged to shorter steps. These results of the plain SARSA are corollary. The table representation of this case was large (12 states × 6 actions × 12 previous states) so learning required a longer time. The representations of previous states were, however, more direct in comparison with the representation of the recurrent neural network. Hence, learning could be more stable than networked model cases. The optimum selections rate of actions (Fig. 7b) in the case of SOM+MOSAIC distributed to higher rate values than the case of others. We confirmed that they were different by meas of the statistical test with significance level 0.01 except between the cases of plain SARSA and SOM+MOSAIC. The learning curves with different input sources shown in Fig. 7c reflected the characteristics of this test environment. The modular networked model could function in both MDP and HOMDP without adjusting the architectures and parameters.

4   Discussion

The results indicated that small negative rewards were more efficient than novelty based internal rewards. It has been said that negative rewards could shorten the steps to earn rewards. This may depend on the initial settings. In those cases, the initial values of connection weights were nearly zero, then, the initial outputs of controllers were also nearly zero. If we set the initial outputs to large negative values, learning will fail. We, however, confirmed that the initial negative values up to -0.4 hardly degraded the performances in the plain SARSA with $b = -0.05$. We speculate that using both temporal internal rewards that are always non-negative values and negative rewards at every step can be efficient in real situations, which was partially demonstrated in the results. We will carry out detailed studies on these aspects.

SOM+MOSAIC and SOM achieved better performance than MOSAIC in the tested environments. The performance of SOM+MOSAIC was relatively better than SOM at least in the HOMDP environment. During on-line learning, the system encountered only states of the first step such as s_0 and s_1. Afterwards, other states could be observed step by step. For MOSAIC, the differences of observed states at initial and final likely causes interferences and concentrations of functioned modules in the tested environments. On the other hand, SOM can properly distribute modules to each observed state in most cases, but only winner modules take all to calculate Q, so moving the winner node for each state can cause impairments. By calculating Q using responsibility signals, the system becomes more robust.

The constructed neural network model achieved comparable performances to plain SARSAs that were adjusted according to each test environment. We demonstrated that the network representation using SOM and MOSAIC could cope with complex MDP environment including partially observable conditions without modeling environments in advance.

References

1. Sutton, R.S., Barto, A.G.: Reinforcement Learning: An Introduction. MIT Press, Cambridge (1998)
2. Takeuchi, J., Shouno, O., Tsujino, H.: Modular neural networks for reinforcement learning with temporal intrinsic rewards. In: Proceedings of 2007 International Joint Conference on Neural Networks (IJCNN). CD-ROM (2007)
3. Schmidhuber, J.: Self-motivated development through rewards for predictor errors/ improvements. In: Developmental Robotics 2005 AAAI Spring Symposium (2005)
4. Doya, K., Samejima, K., Katagiri, K.-I., Kawato, M.: Multiple model-based reinforcement learning. Neural Computation 14, 1347–1369 (2002)
5. Haruno, M., Wolpert, D.M., Kawato, M.: Mosaic model for sensorimotor learning and control. Neural Computation 13, 2201–2220 (2001)
6. Jaeger, H.: The 'echo state' approach to analysing and training recurrent neural networks. Gmd report 148, German National Research Center for Information Technology (2001)
7. Nishida, S., Ishii, K., Furukawa, T.: An online adaptation control system using mnsom. In: King, I., Wang, J., Chan, L.-W., Wang, D. (eds.) ICONIP 2006. LNCS, vol. 4232, pp. 935–942. Springer, Heidelberg (2006)
8. Kohonen, T.: Self-Organizing Maps. Springer, Heidelberg (1995)
9. Berglund, E., Sitte, J.: The parameterless self-organizing map algorithm. IEEE transactions on neural networks 17(2), 305–316 (2006)

From Exploration to Planning

Cornelius Weber and Jochen Triesch

Frankfurt Institute for Advanced Studies, Johann Wolfgang Goethe University,
Ruth-Moufang-Straße 1, Frankfurt am Main, Germany
{c.weber,triesch}@fias.uni-frankfurt.de
http://fias.uni-frankfurt.de/neuro

Abstract. Learning and behaviour of mobile robots faces limitations. In reinforcement learning, for example, an agent learns a strategy to get to only one specific target point within a state space. However, we can grasp a visually localized object at any point in space or navigate to any position in a room. We present a neural network model in which an agent learns a model of the state space that allows him to get to an arbitrarily chosen goal via a short route. By randomly exploring the state space, the agent learns associations between two adjoining states and the action that links them. Given arbitrary starting and goal positions, route-finding is done in two steps. First, an activation gradient spreads around the goal position along the associative connections. Second, the agent uses state-action associations to determine the actions leading to ascend the gradient toward the goal. All mechanisms are biologically justifiable.

1   Introduction

Neuro- and computer scientists are trying to formulate and implement models of human intelligence since decades. A model for learning goal directed actions, reinforcement learning [1], is among the most influential. While these algorithms were primarily developed in the field of machine learning, they have behavioral and algorithmic counterparts in neurobiology, such as the reward prediction error that may be coded by dopamine neurons in the basal ganglia [2,3].

Reinforcement learning requires an extensive randomized exploration phase. During this phase, the agent visits every state and makes any possible transition from one state to another usually several times. Despite this experience, the agent does not learn a general model of the environment that would allow it to plan a route to any goal. Instead, for any position it learns an action that leads to the one trained goal only, or, in the case of multiple goal problems [4], to a fixed set of trained goals.

The topic of this paper are learning forward- and inverse models when the agent explores the environment (state space); no goal is present, and no goal-related value function, as it is often used in reinforcement learning, is assigned to the states in this phase. Further, a planning algorithm is presented that uses these trained models to allow the agent then to get from any position in the environment to any other.

V. Kůrková et al. (Eds.): ICANN 2008, Part I, LNCS 5163, pp. 740–749, 2008.
© Springer-Verlag Berlin Heidelberg 2008

Forward- and Inverse Models. The brain seems to use forward- and inverse models in abundance [5,6]. Such models can be involved in the control of simple movements, but they are also implicated in higher level functions such as action understanding by mirror neurons [7,8] or a representation of the "self" [9].

Computational models usually train the forward- and inverse models in a phase in which random motor actions are taken, which is termed "motor babbling". This concept from infant language development [10] is established in developmental robotics [11,12] for learning body control ("embodiment").

In our quest to represent the agent's actions in the entire environment by internal models, we will not build one sophisticated model of a complex action. Instead, we will build many tiny models, each representing only two neighboring states and the action linking them. All together, they will cover all connected state pairs.

During exploration, the agent learns three kinds of internal models. *(i)* Associative models of the environment represent connectedness of states (Fig. 1). *(ii)* Inverse models calculate the action necessary to move from one state to a desired neighboring state (Fig. 2). *(iii)* Forward models predict a future state from a current state and chosen action (Fig. 3).

All weights are trained by an error rule in which the difference between the weight's prediction and the actual outcome determines the scale and the sign of a learning step. We have used such a learning rule for a predictive model interpretation of V1 complex cells in the visual cortex [13].

Planning. The main idea of planning is to use inverse- and forward models repetitively, so that a trajectory between two states can be formed even if they are distant. The other idea that actually comes first, is to form a kind of energy landscape that is maximized toward the goal. We will use associative weights to spread a "goal hill" as neural activation vector around the goal. Its slope will then guide the trajectory that forms using the inverse- and forward models' weights, starting at the agent position and arriving at the goal.

The spread of the "goal hill" along the connections that associate connected states reminds of the pheromone trails that some ants lay when travelling from a food source back home [14]. This process is much more efficient than to build up a value function as reinforcement learning does, even when using the learnt world model (as opposed to acting in the real world), as in Dyna-Q [1]. This work recapitulates with neural networks part of the work of [15] who explains multiple learning paradigms observed in animal behaviour with formal logic.

2   Methods

2.1   Exploration

During exploration, the agent moves around randomly in the environment and learns the relations between states and actions[1]. At a given time step let the agent

---

[1] In models of purposive actions, the agent moves so as to acquire specifically useful information [16,17,18,19]. For us, however, simple, purely random exploration suffices.

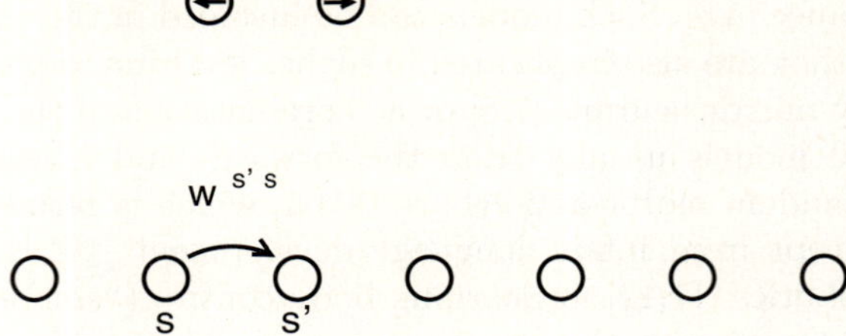

Fig. 1. The model consists of a state space (shown below) and action units (top). For a one-dimensional state space as shown here, only two actions (move left or right) exist. The weights $W^{s's}$ associate neighboring states s and s' irrespective of action taken. For planning, they will be used to find a route between the goal and the agent.

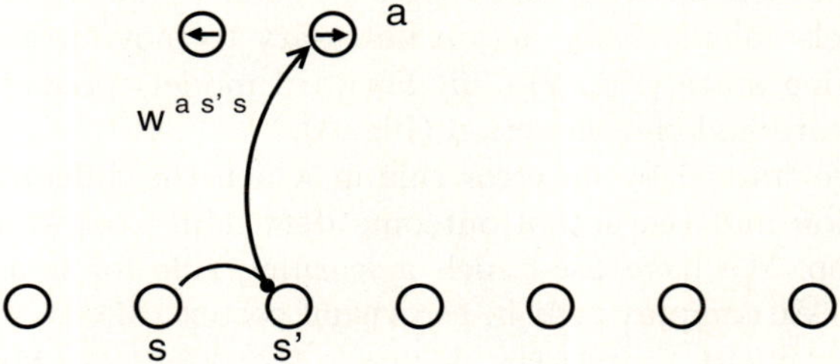

Fig. 2. The weights $W^{as's}$ "postdict" the action a that has been taken in state s when arriving at state s'. For planning, they will be used to identify the action necessary to get to a desired state.

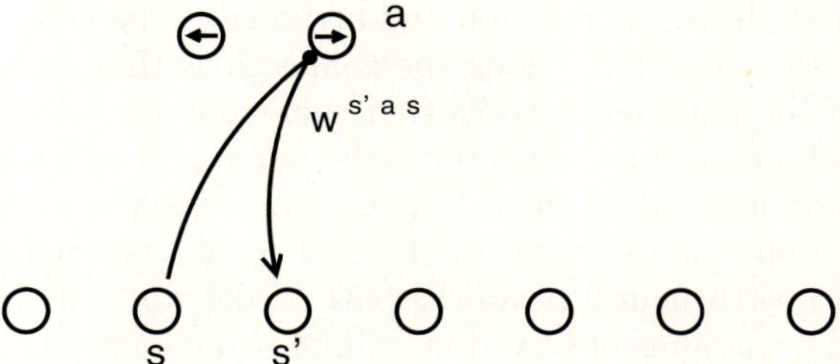

Fig. 3. The weights $W^{s'as}$ predict the next state s' given the current state s and the chosen action a. During planning they will allow mental simulation, without the agent having to move physically.

be in state s and perform action a that leads it to state s'. The environment forms a grid world, and the agent is at one node of the grid. Hence s is a vector with an entry for any node, and all entries are zero except the one corresponding to the position of the agent, which is 1. a is a vector with one component for each action, and the component denoting the chosen action is 1, the others are zero. We assume there is no noise. After sufficient exploration of the environment, we assume that every possible triplet (s, a, s') has been sampled a few times. That is, from every state, every possible action has been taken and every directly reachable (neighboring) state has been reached from it.

To build a model of how actions are related to states, the agent learns three sets of weights. First, $W^{s's}$ tells how likely it is to get to state s' from state s, irrespective of (averaging over) the chosen actions (Fig. 1). Second, $W^{as's}$ tells, which action a has been used to get from s to s' (Fig. 2). Third, $W^{s'as}$ tells, which state s' the agent will arrive at after performing action a in state s (Fig. 3).

The inverse model $W^{as's}$ and the forward model $W^{s'as}$ combine dual input (state-state and state-action, respectively) to generate their output. To activate such a weight, both input conditions must be met, hence it acts as a logical AND gate. Since each state and every action are described by one specific unit being active, the product of the two inputs forms this AND gate. Units that take a sum over such products are called $\Sigma\Pi$ (Sigma-Pi) units.

If s has N elements and a has K elements, then $W^{s's}$ has N^2 elements and $W^{as's}$ and $W^{s'as}$ have KN^2 elements each. Note that most elements of these matrices will be zero, because from a given state only a small number of other states can be reached directly.

At any point during exploration the agent moves from state s using action a to state s'. This triplet (s', s, a) has all the values necessary to learn the three weight types. Each weight type learns to predict its respective outcome value by adjusting the weight incrementally to reduce the prediction error.

For the associative weights $W^{s's}$ the prediction of s' from s is

$$\tilde{s}'_i = \sum_j w^{s's}_{ij} s_j. \tag{1}$$

They are trained to reduce the prediction error $\|s' - \tilde{s}'\|$ according to

$$\Delta w^{s's}_{ij} = \epsilon\,(s'_i - \tilde{s}'_i)\,s_j \tag{2}$$

with learning rate ϵ. Since these weights do not consider the action being taken, they compute an average over the possible reachable states rather than a precise prediction of the actual next state.

The $W^{as's}$ make a "postdiction" $\tilde{a}$ of the action a that was taken in s before the agent arrived at s':

$$\tilde{a}_k = \sum_{ij} w^{as's}_{kij} s'_i s_j. \tag{3}$$

Note that $w^{as's}_{kij}$ is distinguished from $w^{as's}_{kji}$, that is, the direction of movement matters. They are trained to reduce the error $\|a - \tilde{a}\|$ as follows:

$$\Delta w^{as's}_{kij} = \epsilon\,(a_k - \tilde{a}_k)\,s'_i s_j. \tag{4}$$

Here the $\tilde{a}_k$ are continuous values while a_k is 1 for the action that is taken during the random exploration and zero otherwise.

The $W^{s'as}$ predict the actual next state s' by taking into account both s and a:

$$\hat{s}'_i = \sum_{kj} w^{s'as}_{ikj} \, a_k \, s_j. \tag{5}$$

They are trained to reduce the prediction error $\|s' - \hat{s}'\|$ as follows:

$$\Delta w^{s'as}_{ikj} = \epsilon \, (s'_i - \hat{s}'_i) \, a_k \, s_j. \tag{6}$$

2.2   Planning

After learning the weights in the exploration phase, a trajectory between any two states can be planned. The planning process consists of two phases. In the first phase, the unit in state space at the goal position is set active. Activation spreads far around the goal using the associative weights $W^{s's}$. It is important that this emerging activation hill decreases monotonously away from the goal and that the agent position happens to be at its slope. Hence, along a line from the agent to the goal there will be an activation gradient that the agent follows to reach the goal. This gradient ascent is the second phase of the planning process.

Spreading the Goal Hill (Phase 1). Let the goal be at unit g in the state space, so we initialize our "hill" h of activation at the goal with $h_i = 1$, for unit i which is the goal position g, and $h_i = 0$, for all other units.

We spread the activations along those pathways captured in the associative weights $W^{s's}$ by repeating the following three equations for all state space units

$$h'_i = \sum_{j} w^{s's}_{ij} \, h_j \tag{7}$$

$$h''_i = h_i + h'_i \tag{8}$$

$$h_i = scale(h''_i) \tag{9}$$

until the goal hill reaches the position of the agent. Eq. 7 carries the goal unit's activation one step along all connections $\{w^{s's}_{ij}\}$. Eq. 8 adds up the arriving activations so that they do not become too small away from the goal. In Eq. 9, $scale(.)$ is a non-local function that scales all activities by a constant so that the largest value is 1. It prevents the activations from diverging to large values.

Note that the activation hill spreads primarily along the $W^{s's}$ connections that denote movements *away* from the goal, instead of toward the goal. This is no problem if actions from state unit j to unit i have been taken as often as actions into the other direction during exploration. The $W^{s's}$ ought to be used in the opposite direction by the goal hill as compared to the movements.

Hill Ascent (Phase 2). We start by placing the agent to unit m by defining $f_i = 1$, at position $i = m$, and $f_i = 0$, elsewhere.

Next we activate the action units from combined input of h and f:

$$a_k = \sum_{ij} w_{kij}^{a\,s'\,s}\, h_i\, f_j. \tag{10}$$

Since f_j is non-zero only at unit m, only those h_i are effective where i is reachable from m. (This is because all $w_{kij}^{a\,s'\,s}$ are zero if state i cannot be reached from state j by any action.) Each of the effective h_i corresponds to one action, and since the h_i nearest to the goal is the largest, the corresponding action unit will be most activated. (This is assuming that the weights are approximately equal.) A winner-take-all function then sets the maximally activated action unit to 1, others to zero.

Finally, we either perform the action of the selected unit in the world, or we mentally simulate it by predicting the next state as follows:

$$f_i' = \sum_{kj} w_{ikj}^{s'\,a\,s}\, a_k\, f_j \tag{11}$$

A winner-take-all function over the f_i' then determines the next position of the agent. We write this new position as f_i and repeat from Eq. 10 until the agent reaches the goal.

3   Results

First, we tested the model successfully on a plain grid world (not shown). The agent explored the grid and was then able to navigate to several given goal positions. Next, we put some obstacles into the grid world and created the maze shown in Fig. 4 a). The agent explored the space by randomly activating one of its four motor units to move north, west, south or east. Whenever the agent aimed into a wall it was placed to the current position, instead.

All weights were initialized with zero values. Self-connections of $W^{s's}$ were not allowed to grow. 25000 individual movements in the maze, and hence learning steps, were performed with a learning rate of $\epsilon = 0.01$ for all weights. Longer exploration would lead to more homogeneous weights, but was not necessary.

As a result of learning, Fig. 4 b) shows the weights $W^{s's}$ of a few units, those that are boxed in Fig. 4 a). Units on a wall have only zero weights. The other units have non-zero weights from all accessible neighboring fields. The weights are noisy, because the transitions between states s and s' have occurred with different frequencies for different state pairs during exploration.

The planning process is shown in Fig. 5. The two phases, spreading of the "goal hill" and hill ascent are done for a fixed number iterations each, here 48, to cover at least the distance between the agent and the goal. In the first phase, the activation hill h spreads from the goal position throughout the entire maze, avoiding the walls, as shown in Fig. 5 a) and c) for two slightly different goal positions. In the second phase, shown in Fig. 5 b) and d), the agent "travels" without actual interaction with the real world, but only by simulating its movement using Eqs. 10 and 11. Avoiding the wall positions, it navigates toward the goal along a short path. Planning is fully deterministic.

a)   b)

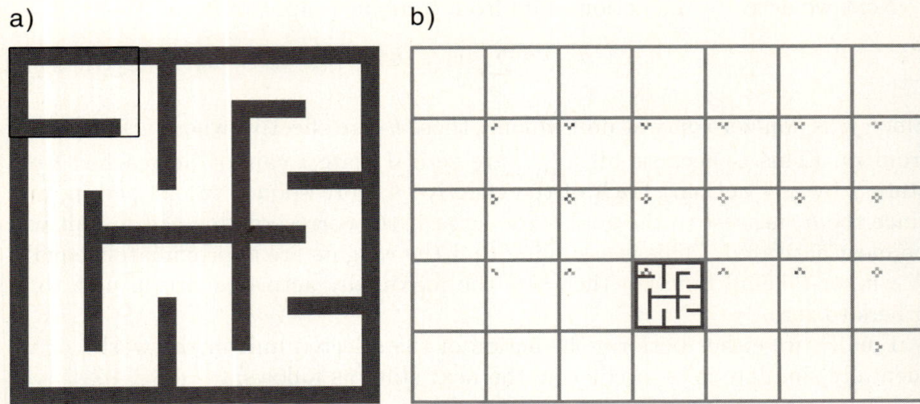

Fig. 4. a) The grid-world, with impenetrable fields shown in red color. b) A part of the associative weight matrix $W^{s's}$. Each little square shows the input weights (receptive field) of one unit; on one unit's input the maze is overlayed. Blue denotes strong weights. Units have no self-connections as seen in the white square "between" a unit's weights to the neighbors. Only units in the upper left part, corresponding to the box in a), are shown.

4   Discussion

Reinforcement Learning. Our small maze example suggests that our planning algorithm greatly enhances reinforcement learning. Are there any cases of goal-finding that can be learnt by reinforcement learning but not by planning?

Planning requires access to the internal representation of the entire state space, including the goal position. Before the agent actually reaches the goal, it might not be possible for it to imagine the goal and to seed the "goal hill" for initializing phase 1 of the planning process. For example, we can plan to navigate a maze if we know the layout, but we have to resort to other strategies if we have only local information.

Experiments which are typical for reinforcement learning often combine incoherent stimuli like a sound, a light and food. In these experiments, the forming of the associative weights $W^{s's}$ may become difficult. Reinforcement learning does not require direct links between neighboring states, but only links from all state space units to a critic and actor units.

In experiments with rats who have to find a hidden platform in a water maze, the rats learn new positions of the platform faster after several blocks of learning. A reinforcement leaning model explained these experiments using a "direction computer" that was not neurally specified [20]. We suggest an alternative explanation, that the rats learn the goals faster, because they have acquired internal models of the water maze.

Robustness. In Fig. 5 d) we can see that the agent misses the goal narrowly. The reason, we conjecture, is that the "goal hill" has become very flat around the

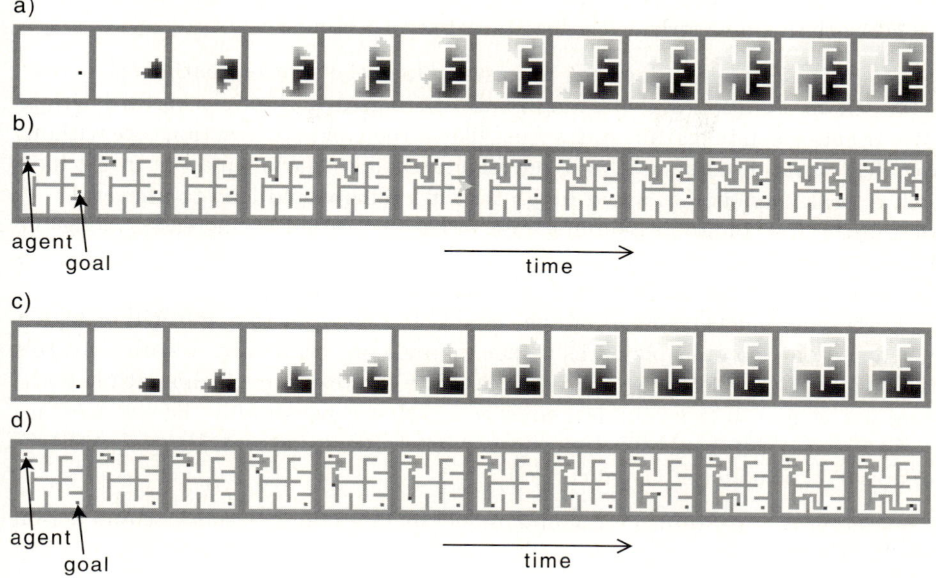

Fig. 5. Network activations in the planning phase. a) and c) show the "goal hill" h spreading over the entire maze during 48 time steps. Only every 4th time step is displayed. To make small activations better visible, the square root of h_i is taken four times. b) and d) show the agent position (dark blue point). The starting- and goal positions are marked by a red point. The agent's trajectory is marked light blue.

goal, while there is noise in the $W^{as's}$ weights that are responsible to choose the action together with information from the gradient of the hill (Eq. 10). Generally for a hill of limited height, the wider it is, the narrower becomes its slope, making it harder to identify[2].

The weights $W^{as's}$ and $W^{s'as}$ are not expected to be uneven in a deterministic scenario such as our grid world. However, in our simulation the weights are noisy because we have stopped learning well before convergence (neither have we annealed the learning rates).

A possible remedy to deal with a flat slope is to spread the "goal hill" several times, and let the agent ascend it only when it perceives changes in the hill's activations. Another remedy may be to use the timing of arriving spikes (cf. [21]). The idea is to let spikes (or bursts) travel away from the hill center along the $W^{s's}$ and to let the agent move toward the direction from which it first receives a spike.

A noisy environment as well as a probabilistic description of the model are to be tackled in the future. Furthermore, since the implementation is neurally and

---

[2] Note that in Eq. 10 the *relative* activations h around the agent are important for determining the action. A slope with a fixed relative gradient could range indefinitely, however, the values would become infinitesimally small.

not table-based, it may not require discrete states. Hence, we may test in future whether the model works with distributed representations.

Short Path. The agent does not necessarily take the shortest path to the goal.

In the exploration and learning phase, the agent may prefer certain actions over others when in certain states (just like in the maze some actions are without an effect when directed at a wall, but in a realistic scenario, an agent might have just a weak tendency to avoid rough terrain). This would lead to uneven $W^{s's}$, and in the first phase of planning, the "goal hill" would be increased along the typically preferred paths. This will lead to — wanted or unwanted — preferences for well experienced paths during goal seeking as well.

Furthermore, when spreading the "goal hill", the input to a unit will be higher when arriving via multiple paths rather than via one path only. Accordingly, this would lead to a preference to travel along "safe" states from which multiple paths lead to the goal. If this is undesired, schemes that consider only the strongest input to spread the "goal hill" are conceivable (cf. Riesenhuber & Poggio's "MAX" operator in a model of visual cortex).

Embedding Architecture. Our simple model includes only state space and action units. Allowing for a variable goal, it is distinguished from a purely reactive sense-action mechanism. However, the question of who sets the goal is not in the scope of the model. This might come from sensory input, for example if an attractive object, such as a fruit, is seen in the visual field. It might as well be determined by a more abstract thought process on a cognitive level.

A challenging question also in reinforcement learning is, how does the state space emerge from the raw sensory stimuli. Models of sensory preprocessing that should fill this gap are mostly trained in an unsupervised fashion, so sensory representations are not optimized for action or planning. Some models learn to represent sensory input following the requirements of reinforcement learning (e.g. [22,23]). Consequentially, one might consider whether sensory processing could be designed to aid planning processes.

Acknowledgments

We acknowledge financial support by the European Union through projects FP6-2005-015803 and MEXT-CT-2006-042484 and by the Hertie Foundation. We thank Felipe Gerhard for a useful discussion, and Cristina Savin, Prashant Joshi, Constantin Rothkopf and the two anonymous reviewers for valuable feedback on the document.

References

1. Sutton, R., Barto, A.: Reinforcement Learning: An Introduction. MIT Press, Cambridge (1998)
2. Ungless, M., Magill, P., Bolam, J.: Uniform inhibition of dopamine neurons in the ventral tegmental area by aversive stimuli. Science 303, 2040–2042 (2004)

3. Tobler, P., Fiorillo, C., Schultz, W.: Adaptive coding of reward value by dopamine neurons. Science 307(5715), 1642–1645 (2005)
4. Foster, D., Dayan, P.: Structure in the space of value functions. Machine Learning 49, 325–346 (2002)
5. Davidson, P., Wolpert, D.: Widespread access to predictive models in the motor system: A short review. Journal of Neural Engineering 2, 8313–8319 (2005)
6. Iacoboni, M., Wilson, S.: Beyond a single area: motor control and language within a neural architecture encompassing broca's area. Cortex 42(4), 503–506 (2006)
7. Miall, R.: Connecting mirror neurons and forward models. Neuroreport 14(16), 2135–2137 (2003)
8. Oztop, E., Wolpert, D., Kawato, M.: Mirror neurons: Key for mental simulation? In: Twelfth annual computational neuroscience meeting CNS, p. 81 (2003)
9. Churchland, P.: Self-representation in nervous systems. Science 296, 308–310 (2002)
10. Plaut, D.C., Kello, C.T.: The emergence of phonology from the interplay of speech comprehension and production: A distributed connectionist approach. In: The emergence of language. B. MacWhinney (1998)
11. Metta, G., Panerai, F., Manzotti, R., Sandini, G.: Babybot: an artificial developing robotic agent. In: SAB (2000)
12. Dearden, A., Demiris, Y.: Learning forward models for robots. In: IJCAI, pp. 1440–1445 (2005)
13. Weber, C.: Self-organization of orientation maps, lateral connections, and dynamic receptive fields in the primary visual cortex. In: Dorffner, G., Bischof, H., Hornik, K. (eds.) ICANN 2001. LNCS, vol. 2130, pp. 1147–1152. Springer, Heidelberg (2001)
14. Dorigo, M., Birattari, M., Stützle, T.: Ant colony optimization. Computational Intelligence Magazine, IEEE 1(4), 28–39 (2006)
15. Witkowski, M.: An action-selection calculus. Adaptive Behavior 15(1), 73–97 (2007)
16. Schmidhuber, J.: Developmental robotics, optimal artificial curiosity, creativity, music, and the fine arts. Connection Science 18(2), 173–187 (1991)
17. Herrmann, J., Pawelzik, K., Geisel, T.: Learning predictive representations. Neurocomputing 32-33, 785–791 (2000)
18. Oudeyer, P., Kaplan, F., Hafner, V., Whyte, A.: The playground experiment: Task-independent development of a curious robot. In: AAAI Spring Symposium Workshop on Developmental Robotics (2005)
19. Der, R., Martius, G.: From motor babbling to purposive actions: Emerging self-exploration in a dynamical systems approach to early robot development. In: SAB, pp. 406–421. Springer, Berlin (2006)
20. Foster, D., Morris, R., Dayan, P.: A model of hippocampally dependent navigation, using the temporal difference learning rule. Hippocampus 10, 1–16 (2000)
21. Van Rullen, R., Thorpe, S.: Rate coding versus temporal order coding: What the retinal ganglion cells tell the visual cortex. Neur. Comp. 13, 1255–1283 (2001)
22. Roelfsema, P., van Ooyen, A.: Attention-gated reinforcement learning of internal representations for classification. Neur. Comp. 17, 2176–2214 (2005)
23. McCallum, A.: Reinforcement Learning with Selective Perception and Hidden State. PhD thesis, U. of Rochester (1995)

Sentence-Level Evaluation Using Co-occurences of N-Grams

Theologos Athanaselis, Stelios Bakamidis, Konstantinos Mamouras,
and Ioannis Dologlou

Institute for Language and Speech Processing
Artemidos 6 and Epidavrou, Maroussi, Athens, 15125
Greece
tathana@ilsp.gr
www.ilsp.gr

Abstract. This work presents an evaluation method of Greek sentences
with respect to word order errors. The evaluation method is based on
words' reordering and choosing the version that maximizes the number
of trigram hits according to a language model. The new parameter of
the proposed technique concerns the incorporation of unigram probability. This probability corresponds to the frequency of each unigram to be
posed in the first and in the last position of the training set sentences.
The comparative advantage of this method is that it works with a large
set of words, and avoids the laborious and costly process of collecting
word order errors for creating error patterns.

Keywords: Language model, word order errors, permutations filtering,
Hellenic National Corpus.

1   Introduction

Automatic grammar checking is traditionally done by manually written rules,
constructed by computer linguists. Methods for detecting grammatical errors
without manually constructed rules have been presented before. Atwell [1] uses
the probabilities in a statistical part-of the speech tagger, detecting errors as
low probability part of speech sequences. Chodorow and Leacock [2] suggested
an unsupervised method for detecting grammatical errors by inferring negative
evidence from edited textual corpora. Bigert and Knutsson [3] presented how a
new text is compared to known correct text and deviations from the norm are
flagged as suspected errors. Sjöbergh [4] introduced a method of grammar errors
recognition by adding errors to a lot of (mostly error free) unannotated text and
by using a machine learning algorithm.

Natural language generation forms an important component of many natural
language applications e.g. man-machine interface, automatic translation, text
generation, etc. Bag generation is one of the natural language generation methods. Given a sentence, we cut it into words, place these words in a bag and try
to recover the sentence from the bag [5]. Bag generation has been considered

V. Kůrková et al. (Eds.): ICANN 2008, Part I, LNCS 5163, pp. 750–758, 2008.

within the statistical and rule-based paradigms of computational linguistics and each has handled this problem differently [6]. Bag generation has received particular attention in lexicalist approaches to machine translation, as exemplified by Shake-and-Bake generation [7]. The Shake-and-Bake generation algorithm of [8] combines target language signs using the technique known as generate and test. In effect, an arbitrary permutation of signs is input to a shift-reduce parser which tests them for grammatical well-formedness. If they are well-formed, the system halts indicating success. If not, another permutation is tried and the process is repeated. The complexity of this algorithm is $O(N!)$ because all the permutations $N!$ ($N!$ for any input of size N) may have to be explored to find the correct answer, and indeed must be explored in order to verify that there is no answer.

The paper is organized as follows. The architecture of the entire system and a description of each component follow in section 2. Section 3 shows how permutations are filtered by the proposed method. Section 4 specifies the method that is used for searching valid trigrams in a sentence. The results of using HNC (Hellenic National Corpus) [9] experimental scheme are presented in section 5. Finally, the conclusions are made in section 6.

2 Description of the System

2.1 The Issue

It is straightforward that the best way to reconstruct a sentence with word order errors is to reorder the words. However, the question is how it can be achieved without knowing the POS tag of each word. Many techniques have been developed in the past to cope with this problem using a grammar parser and rules. However, the success rates reported in the literature are in fact low. A way to reorder the words is to use all the possible permutations. The crucial drawback of this approach is that, given a sentence with length N words, the number of all permutations is $N!$. This number is very large and seems to be restrictive for further processing.

2.2 The Components

The novelty of the proposed method consists in the use of a technique for filtering the $N!$ permutations. The process of repairing sentences with word-order errors incorporates the followings tools.

- a simple and efficient confusion matrix technique
- a language model's trigrams and bigrams

Consequently, the correctness of each sentence depends on the number of valid trigrams. Therefore, this method evaluates the correctness of each sentence after filtering, and provides as a result, a sentence with the same words but in the correct order.

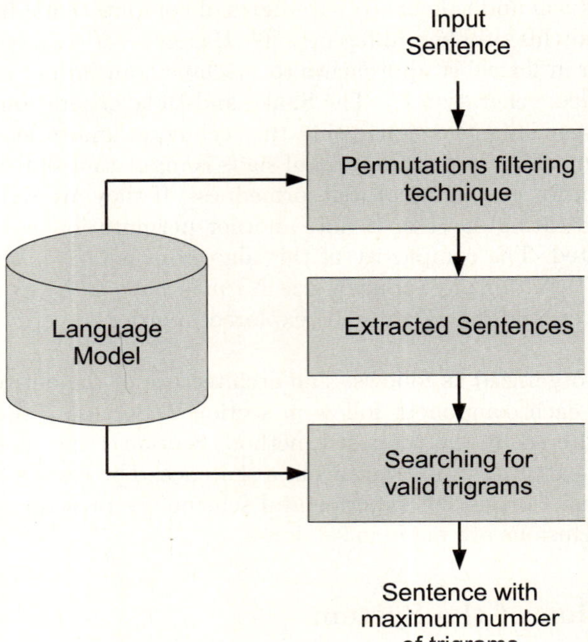

Fig. 1. System's Components

3   Reducing the Search Space

The question here is how feasible it is to deal with all the permutations for
sentences with a large number of words. Therefore, a filtering process of all
possible permutations is necessary. The filtering involves the construction of a
confusion matrix $N \times N$ in order to extract possible permuted sentences. Given a
sentence $a = \left(w[1], w[2], \ldots, w[N-1], w[N] \right)$ with N words, a confusion matrix
$A \in \mathbb{R}^{N \times N}$ can be constructed, as shown in the Table 1.

The size of the matrix depends on the length of the sentence. The objective of
this confusion matrix is to extract the valid bigrams according to the language

Table 1. Construction of the confusion matrix $N \times N$ for a given sentence

$word$	$w[1]$	$w[2]$	$w[3]$	...	$w[N-1]$	$w[N]$
$w[1]$	$P[1,1]$	$P[1,2]$	$P[1,3]$	...	$P[1,N-1]$	$P[1,N]$
$w[2]$	$P[2,1]$	$P[2,2]$	$P[2,3]$	...	$P[2,N-1]$	$P[2,N]$
$w[3]$	$P[3,1]$	$P[3,2]$	$P[3,3]$	...	$P[3,N-1]$	$P[3,N]$
$\vdots$	$\vdots$	$\vdots$	$\vdots$	$\ddots$	$\vdots$	$\vdots$
$w[N]$	$P[N,1]$	$P[N,2]$	$P[N,3]$	...	$P[N,N-1]$	$P[N,N]$

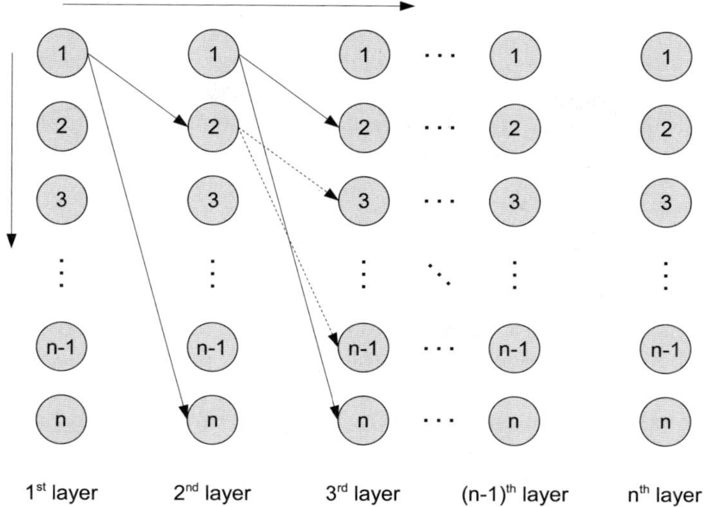

Fig. 2. Illustration of the lattice with n layers and n states

model. The element $P[i, j]$ indicates the validity of each pair of words $(w[i], w[j])$ according to the list of the language model's bigrams. Note that, the number of valid bigrams is lower than the size of the confusion matrix, which is $N(N-1)$, since not all possible pairs of words are valid according to the language model. In order to generate permuted sentences using the valid bigrams, all the possible word sequences must be found. This is the search problem and its solution is the domain of this filtering process.

As with all search problems there are many approaches. In this paper a left to right approach is used. To understand how it works the permutation filtering process, imagine a network of N layers with N states. N is the number of words the the sentence contains. Each layer corresponds to a position in the sentence. Each state is a possible word. All the states on layer 1 are then connected to all possible states on the second layer and so on according to the language model. The connection between two states (i, j) of neighboring layers $(k-1, k)$ exists when the bigram $(w[i], w[j])$ is valid. This network effectively visualizes the algorithm to obtain the permutations. Starting from any state in layer 1 and moving forward through all the available connections to the Nth layer of the network, all the possible permutations can be obtained. No state should be "visited" twice in this movement.

In order to further reduce the number of permutations the use of unigram probability is necessary. This probability indicates the frequency of each unigram being positioned in the first or in the last layer. The unigram probability is calculated by estimating the probabilities

$$P[W_1 \text{ is the first word of the sentence}]$$

and

$$P[W_n \text{ is the last word of the sentence}]$$

from the training set sentences. It is clear that some of the sentence's words cannot be placed at the first or the last layer since they are not used as beginning or end of well-formed sentences. Therefore, across the network, the states on the first and the last layer that correspond to words with low probability are ignored.

4   Evaluating Sentences

The prime function of this approach is to decompose any input sentence into a set of trigrams. To do so, a block of words is selected. In order to extract the trigrams of the input sentence, the size of each block is typically set to 3 words, and blocks normally overlap over two words. Therefore, an input sentence of length N, includes $N - 2$ trigrams. The second step of this method involves the search for valid trigrams for each sentence. In the third step of this method the number of valid trigrams per each permuted sentence is calculated. The idea is that the more shared sub-strings (in our case trigrams) the permuted sentence has with the HNC corpus, the more correct the sentence is. Although some of the sentence's trigrams may be typically correct, it is possible they are not included into the list of LM's trigrams due to the data sparsity problem. The plethora of LM's trigrams relies on the quality of corpus.

The criterion for ranking all the permuted sentences is the number of valid trigrams. The system provides as an output, a sentence with the maximum number of valid trigrams. In case where two or more sentences have the same number of valid trigrams a new distance metric should be defined. This distance metric is based on the logarithmic probability of the trigrams. The total log probability is computed by adding the log probability of each trigram, whereas the probability of non valid trigrams is assigned to a very big negative number (-100000). Therefore the sentence with the maximum probability is the system's response.

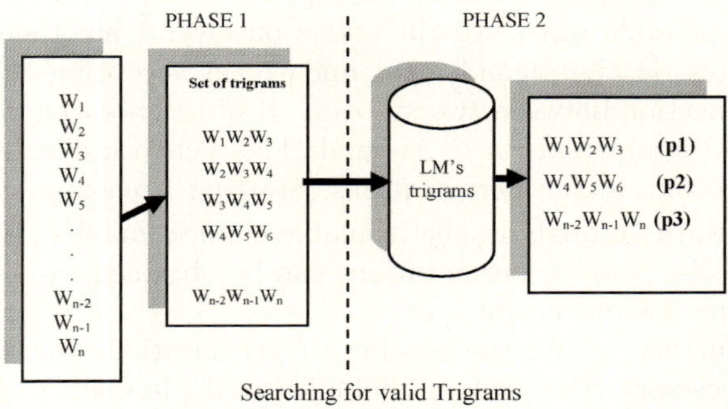

Fig. 3. The architecture of the subsystem for searching valid trigrams according to the LM. The (p_i) corresponds to the log probability of each valid trigram.

5   Experimentation

5.1   Language Model Characteristics

One major problem with standard N-gram models is that they must be trained from some corpus, and because any particular training corpus is finite, some perfectly acceptable N-grams are bound to be missing from it [10]. That is, the N-gram matrix for any given training corpus is sparse; it is bound to have a very large number of cases of putative "zero probability N-grams" that should have some non zero probability. Some part of this problem is endemic to N-grams; since they can not use long distance context, they always tend to underestimate the probability of strings that happen not to have occurred nearby in their training corpus. There are some techniques that can be used in order to assign a non zero probability to these zero probability N-grams. In this work, the language model has been trained using free Internet data from newspapers and consists of N-grams ($N = 2, 3$) with Good-Turing discounting [11] and Katz back off [12] for smoothing. Training data contains about 9 million sentences and 120 million words. The table below shows the number of unigrams, bigrams, and trigrams for the Greek version of language model.

5.2   Experimental Scheme

The experimentation involves a test set of 753120 sentences. Test sentences have been selected randomly from HNC (Hellenic national Corpus). They have variable length with minimum 7 words and maximum 12 words. The 95% of the test words belong to the vocabulary of training data. For experimental purposes our test set consists of sentences with no word order errors. For every one of the test sentences the system suggests the ten best sentences according to the ordering defined in the previous section. The goal of this experimentation is to show that the place of the input sentence among the 10-best sentences is improved using the unigram probability.

Note that the test sentences are not included in the training set of the statistical language model that is used as tool for the proposed method. Table 3 shows the number of sentences that have been used for experimentation and their distribution over the variable 'words per sentence'.

5.3   Experimental Results

Figure 4 depicts the capability of the system to provide the input sentence among the 10-best list. Also it is shown that using the unigram probability

Table 2. The number of different elements of LM

LM Elements	Number
Unigrams	425,562
Bigrams	7,652,952
Trigrams	7,029,801

Table 3. The number of sentences with different length

words per sentence	# test sentences	percentage (%)
7	164,410	21.83
8	146,036	19.39
9	130,471	17.32
10	116,789	15.51
11	103,214	13.70
12	92,200	12.24
total	753,120	

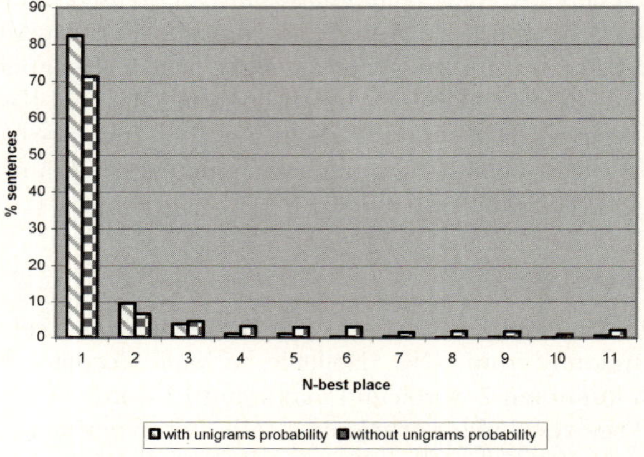

Fig. 4. The percentage of test sentences in different places into the N-best list (N=10)

the percentage of sentences that are in the first place increases about 10%. The x-axis corresponds to the place of the correct sentence into this list. The last position (11) indicates that the correct sentence is out of this list.

Table 4 depicts the differences in the number of permutations for sentences with length from 7 to 12 words. The point is that the number of permutations that are extracted with the filtering process is significantly lower than the corresponding value without filtering. It is obvious that the performance of filtering

Table 4. The mean value of permutations for HNC sentences

words	with filtering	no filtering
7	324	5,040
8	1,332	40,320
9	8,227	362,880
10	45,092	3,628,800
11	352,678	39,916,800
12	1,502,661	479,001,600

process depends mainly on the number of valid bigrams. This implies that the language model's reliability affects the outcome of the system and especially of the filtering process.

6   Conclusions

The proposed method is effective in repairing erroneous sentences. Therefore the method can be adopted by a grammar checker as a word order repairing tool. The findings from the experimentation show that 82.41% of the total sentences have been suggested as the sentences with the high score using unigrams' probability.

On the other hand, when no unigrams' probability is incorporated to the system the 72% of the total sentences have been suggested as the sentences with the high score. The system could not provide an appropriate response for 1.15% of the test sentences and 2.13% respectively. The incorrect output of the system can be explained considering that some words are not included into the training data vocabulary, hence some of the sentences' trigrams are considered as invalid.

References

1. Atwell, E.S.: How to detect grammatical errors in a text without parsing it. In: Proceedings of the 3rd Conference of the European Chapter of the Association for Computational Linguistics, EACL, pp. 38–45 (1987)
2. Chodorow, M., Leacock, C.: An unsupervised method for detecting grammatical errors. In: Proceedings of the 2000 Conference of the North American Chapter of the Association for Computational Linguistics, NAACL, pp. 140–147 (2000)
3. Bigert, J., Knutsson, O.: Robust error detection: A hybrid approach combining unsupervised error detection and linguistic knowledge. In: Proceedings of the 2nd Workshop on Robust Methods in Analysis of Natural language Data, ROMAND, pp. 10–19 (2002)
4. Sjöbergh, J.: Chunking: An unsupervised method to find errors in text. In: Proceedings of the 15th Nordic Conference of Computational Linguistics, NODALIDA (2005)
5. Brown, P.F., Cocke, J., Della Pietra, S., Della Pietra, V.J., Jelinek, F., Lafferty, J.D., Mercer, R.L., Roossin, P.S.: A statistical approach to machine translation. Computational Linguistics 16(2), 79–85 (1990)
6. Chen, H.H., Lee, Y.S.: A corrective training algorithm for adaptive learning in bag generation. In: Proceedings of International Conference on New Methods In Language Processing, Manchester, pp. 248–254 (September 1994)
7. Beaven, J.L.: Shake-and-bake machine translation. In: [13]
8. Whitelock, P.: Shake-and-bake translation. In: [13], pp. 784–790
9. ILSP: Hellenic National Corpus, http://hnc.ilsp.gr/en/
10. Jelinek, F.: Continuous speech recognition by statistical methods. IEEE Proceedings 64(4), 532–556 (1976)
11. Good, I.J.: The population frequencies of species and the estimation of population parameters. Biometrika 40(3-4), 237–264 (1953)

12. Katz, S.M.: Estimation of probabilities from sparse data for the language model component of a speech recogniser. IEEE Transactions on Acoustics, Speech and Signal Processing 35(3), 400–401 (1987)
13. COLING: International Conference on New Methods In Language Processing. In: Proceedings of International Conference on New Methods In Language Processing, Nantes, France, COLING (1992)

Identifying Single Source Data for Mixing Matrix Estimation in Instantaneous Blind Source Separation[*]

Pau Bofill

Dept. d'Arquitectura de Computadors, Universitat Politècnica de Catalunya (UPC)
Campus Nord, Mòdul D6, 08036 Barcelona, Spain
pau@ac.upc.edu

Abstract. This paper presents a simple yet effective way of improving the estimate of the mixing matrix, in instantaneous blind source separation, by using only reliable data.

The paper describes how the idea of detecting single source data is implemented by selecting only the data which remain for two consecutive frames in the same spatial signature. Such data, which are most likely to belong to a single source, are then used to accurately identify the spatial directions of the sources and, hence, the mixing matrix.

The paper also presents a refined histogram procedure which improves on the potential function method to estimate the mixing matrix, in the two dimensional case (two sensors).

The approach was experimentally evaluated and submitted to the first Stereo Audio Source Separation Evaluation Campaign (SASSEC), with good results in matrix estimation both for development and test data.

Keywords: blind source separation; single source data; matrix estimation; potential function method.

1   Introduction

An *instantaneous* or *linear* mixture is modeled by

$$x = As \; , \tag{1}$$

where s is an N-dimensional column vector corresponding to the sources, x is an M-dimensional column vector corresponding to the observations and A is an $M \times N$ mixing matrix. The problem of *blind source separation* [1], consists in finding the solution to the above equation when A and s are both *unknown* (see [2,3] for a general survey on source separation). When the components of A are filters rather than coefficients, the mixture is *convolutive*, and the problem is much harder.

---

[*] Supported by the Ministry of Science and Technology of Spain (contract TIN2007-60625).

V. Kůrková et al. (Eds.): ICANN 2008, Part I, LNCS 5163, pp. 759–767, 2008.

When $N > M$ (more sources than sensors), the separation problem is *underdetermined*. Solving for the mixing matrix, then, is not enough, because the system is overcomplete. When sources are sparse [4], which is usually the case for music and speech sources in the frequency domain, they are likely to be disjoint. Then, the observation vectors belong mostly to one source or another, and the separation is possible.

Usually the separation is performed in two steps. In the first step the mixing matrix is estimated, and in the second step the observation vectors are decomposed into their source components. Estimation of the mixing matrix is usually performed either with histograms [5] or by clustering [6]. Source decomposition is usually performed either by time-frequency masking [7] or by solving the system with constraints [8,9].

No matter if the system is underdetermined or not, the goal of this paper is to improve the mixing matrix estimation step, by considering only the data that belong mostly to a single source [10]. In our approach, such data are identified because their spatial signature remains aproximately constant along two consecutive time frames. This paper focuses on the instantaneous case for stereo ($M = 2$) speech and music signals, although the basic idea should in principle be applicable to the general case as well.

The approach was tested on all the instantaneous mixtures of the Stereo Audio Source Separation database (SASSEC) [11].

Section 2 describes how the spatial signature of the sources is in fact embedded in the scatter-plot of the observation data. Section 3 describes our approach to identifying single source data. Section 4 shows a procedure for inferring the directions of the sources (i.e, the mixing matrix) from the scatter-plot of the data, which is an improvement of the potential function approach in [5]. Section 5 describes our evaluation experiments with the SASSEC database. Finally, Section 6 is devoted to conclusions.

2   The Magnitude Scatter-Plot of the Left vs. Right Channels: The Spatial Signature of the Sources

In the case of time-domain signals, Eqn. (1) holds for each data sample x_i and source vector s_i at time instant i, with a common mixing matrix A. Nevertheless, for audio signals the separation is much easier in the short-time discrete frequency transform domain (STDFT), where sources are much sparser. In this case, observation vectors x^{tk} are the complex-valued STDFT coefficients at each time frame t and frequency bin k. Their corresponding source vectors will be denoted by s^{tk}. By linearity, the mixing matrix is still A (in the convolutive case, though, the coefficients of the mixing matrix are complex as well).

Expanding A along its column vectors a_j, Eqn. (1) can be rewritten as

$$x^{tk} = \sum_{j=1}^{N} a_j s_j^{tk} \ . \tag{2}$$

When only a single source (source j^*) is active, the observation vector is $\boldsymbol{x}^{tk} \simeq \boldsymbol{a}_{j^*} s_{j^*}^{tk}$. Then, $\boldsymbol{x}^{tk}$ is *proportional in magnitude* to $\boldsymbol{a}_{j^*}$

$$|\boldsymbol{x}^{tk}| \simeq \boldsymbol{a}_{j^*} |s_{j^*}^{tk}| \ , \tag{3}$$

with $|z|$ the modulus or magnitude of complex value z. The $\boldsymbol{a}_j$'s are the *basis vectors* of the mixture, and they define the *spatial signature* of the sources in observation space.

In the binaural case (i.e., $M = 2$), let v_r and v_l, the right and left channels, be the cartesian coordinates of vector $\boldsymbol{v}$, and let $\mathcal{L}_v$ and θ_v, length and angle, be its polar coordinates. If we draw the *scatter-plot* of the magnitudes (i.e., if we plot the magnitude of the left channel vs the magnitude of the right channel), all observation vectors belonging only to source j^* will be aligned along a stright line defined by the angle $\theta_{j^*} = atan(a_{lj^*}/a_{rj^*})$.

The actual problem, though, is the other way around. The basis vectors are unkown and we have to inferr them from the magnitude scatter-plot of the observation data. Since

$$|\boldsymbol{x}^{tk}| = \left| \sum_{j=1}^{N} \boldsymbol{a}_j s_j^{tk} \right| \ , \tag{4}$$

most observation vectors will be mixtures of two or more sources, and their position in the magnitude scatter-plot will depend both on the coefficients of the mixture and on the relative phases of their components. Yet, if we could provide data points belonging mostly to separated sources, then, each subset of data would cluster along their corresponding source direction. By measuring these directions $\theta_{\boldsymbol{a}_j}$ on the scatter-plot, and assuming unit length (i.e., $\mathcal{L}_{a_j} = 1$), we would recover the columns of the mixing matrix. The unit length assumption is no loss of generality, because of the *scaling ambiguity*. Since both $\boldsymbol{A}$ and $\boldsymbol{s}$ in (1) are unknown, the observation $\boldsymbol{x}$ is invariant to arbitrary but inverse scale factors applied to $\boldsymbol{A}$ and $\boldsymbol{s}$, respectively.

In practice, we do not have single source data. All observation vectors are somehow a mixture of all the sources. But when the sources are *sparse*[1] enough, the likelihood that they are *disjoint* increases, and many data belong 'mostly' to a single source.

The next Section suggests an approach for reliably identifying single source data.

3 Identifying Single Source Data: Vectors That Keep Their Direction along Consecutive Frames

The rationale of this paper is to estimate the mixing matrix by retrieving the basis vectors from single source data scatter-plots. Our approach to finding such

---

[1] Sparsity is usually modelled by a laplacian distribution, but in the context of this paper, it is enough to consider that a signal is sparse when smaller values are much more frequent than larger ones.

data is the following. Let us assume two observation vectors $\boldsymbol{x}^{tk}$ and $\boldsymbol{x}^{(t+1)k}$, belonging to two consecutive time frames, t and $t+1$, of the same frequency bin k. Let's assume, first, that both observation vectors belong to a single source j^*. Then, no matter how much $s_{j^*}^{tk}$ and $s_{j^*}^{(t+1)k}$ differ from each other, both observation vectors will lie on the spatial direction of $\boldsymbol{a}_{j^*}$.

On the other side, if the observation vectors are actually a mixture of two or more sources (i.e., sources j^1 and j^2), then

$$|\boldsymbol{x}^{tk}| = \left|\boldsymbol{a}_{j^1} s_{j^1}^{tk} + \boldsymbol{a}_{j^2} s_{j^2}^{tk}\right| \ , \tag{5}$$

and $\theta_{|\boldsymbol{x}^{tk}|}$ and $\theta_{|\boldsymbol{x}^{(t+1)k}|}$ will not be the same, unless both s_{j^1} and s_{j^2} remain constant along both frames. Since the last condition is very unlikely, *we decide that observation vectors which keep their direction along two consecutive frames belong to a single source.*

In practice, then, before estimating the mixing matrix we select only *single source* data satisfying

$$\left|\theta_{|\boldsymbol{x}^{tk}|} - \theta_{|\boldsymbol{x}^{(t+1)k}|}\right| < \Theta \ , \tag{6}$$

with Θ a threshold to be set experimentally.

Besides, and in order to avoid the distorsion of poorly informed data, we also filter out the data with magnitudes shorter than a fraction $R < 1$ of the strongest datum. Thus, we further select *significant* data satisfying

$$|\boldsymbol{x}^{tk}| > R \ \max |\boldsymbol{x}^{tk}| \ . \tag{7}$$

The rationale for doing so is that absolute disturbances have much more effect on the angle of shorter data.

Figures 1(a) and 1(b) show the magnitude scatter-plot of a mixture of 4 male speech signals from the SASSEC database, before and after data selection, respectively. As can be seen, the spatial signature of the sources is much more clear in the second case. Next Section describes a particular procedure for estimating the mixing matrix from the (preferably filtered) magnitude scatter-plot.

4 Estimating the Mixing Matrix from the Magnitude Scatter-Plot: Finding the Relevant Peaks of a Weighted Polar Histogram

The following procedure is an improvement of the potential funtion based procedure in [5], designed for stereo instantaneous mixtures which, of course, works much better on data filtered as in the previous Section.

We first built a weighted polar histogram of the magnitudes of the observation data on a grid with $p = 1 \cdots P$ bins uniformly distributed between $\Theta_{p_1} = \theta_{\min} + \Delta/2$ and $\Theta_{p_P} = \theta_{\max} - \Delta/2$, with $\Delta = \frac{\theta_{\max} - \theta_{\min}}{P}$, $\theta_{\min} = \min(\theta_{|\boldsymbol{x}^{tk}|})$ and $\theta_{\max} = \max(\theta_{|\boldsymbol{x}^{tk}|})$. The height $H(p)$ of each bin is then computed as

$$H(p) = \sum_{tk} \mathcal{L}_{|\boldsymbol{x}^{tk}|}, \text{for all} \ \ tk \ \ \text{such that} \ \ \theta_p - \Delta/2 \le \theta_{|\boldsymbol{x}^{tk}|} < \theta_p + \Delta/2 \ . \tag{8}$$

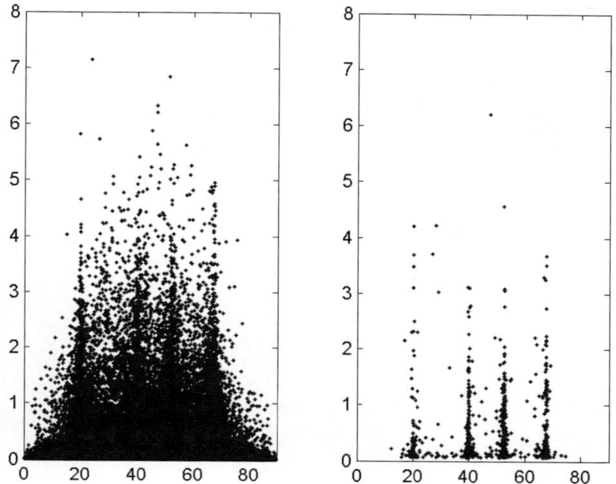

Fig. 1. The magnitude scatter-plot of a mixture of 4 male speech signals, from the SASSEC database. (a) Unfiltered data. (b) Only signal source and significant data. The plot shows $\mathcal{L}_{|x^{tk}|}$ vs $\theta_{|x^{tk}|}$.

That is, each observation vector is assigned by angle to each corresponding bin, with a weight corresponding to its length, thus putting more emphasis on the most significant data.

Then, the histogram is smoothed using a cosine kernel function, as follows. Let the kernel be

$$\phi(\theta) = 1/2 + 1/2 \cos\left(\frac{2\pi\theta}{\theta_{\max} - \theta_{\min}}\right) \quad \text{for} \ \ |\theta| < (\theta_{\max} - \theta_{\min})/2 \ , \quad (9)$$
$$= 0 \ \ \text{otherwise.}$$

The smoothed potential field $F(p)$ at each grid point p is then computed as the sum of it's neighbour histogram bins, weighted by a stretched version of the kernel. That is,

$$F(p) = \sum_q \phi(\lambda(\theta_p - \theta_q))H(q) \ . \quad (10)$$

The scaling parameter λ controls the width of the kernel. With $\lambda = 1$, the kernel spreads over the whole input angle range, and with $\lambda = 10$ it spreads over %10 of the range (Fig. 2). In that way, the smoothness of the potential function F can be controlled, in order to avoid too many or too few peaks. The parameter λ is again set experimentally.

Finally, a local maximum search procedure is applied to $F(p)$ to find the peaks, which should correspond to the spatial directions of the sources.

The resolution of the above procedure is Δ, which depends on the range of the data. Using this adaptive resolution scheme makes it possible to discriminate

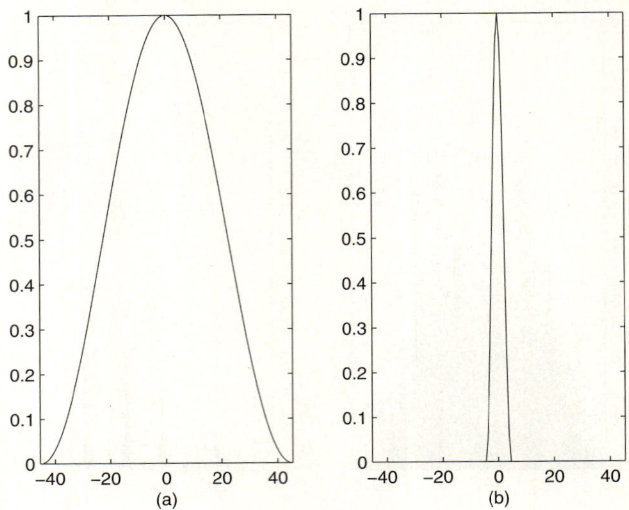

Fig. 2. The kernel function $\phi(\theta)$ over an input angle range of 90 degrees. (a) For $\lambda = 1$. (b) For $\lambda = 10$.

between the sources, even when they have been mixed very close to each other. In fact, the separability of the sources depends on their intrinsic disjointness, rather than on the proximity of the source directions.

5   Experimental Evaluation on the SASSEC Data

The procedure of the previous Section with the data selection of Section 3 was tested on all the instantaneous mixtures of the SASSEC Campaign. Development data consisted on the stereo underdetermined mixture of the following signal sets: four female voices, four male voices, three music instruments without drums, and three music instruments, one of them being drums. Test data was a different realization of the same four mixture types.

Different combinations were tried for identifying single source data, but the one that worked best was the one described in Section 3. We also tried selecting data which kept their direction along three or more consecutive frames, but that yielded too few data, and some data that did not correspond to true sources. This issue is related with the length T of the time window in the STDFT analysis. A shorter window yields observation vectors that are more persistent along frames, but they have wider spectral peaks, resulting in a higher overlap between sources. We also considered persistency over contiguous frequency bins, taking advantadge of the width of the spectral peaks, but results were weaker. Exploring the actual structure of sound (harmonicity, etc.) is another approach that might prove useful.

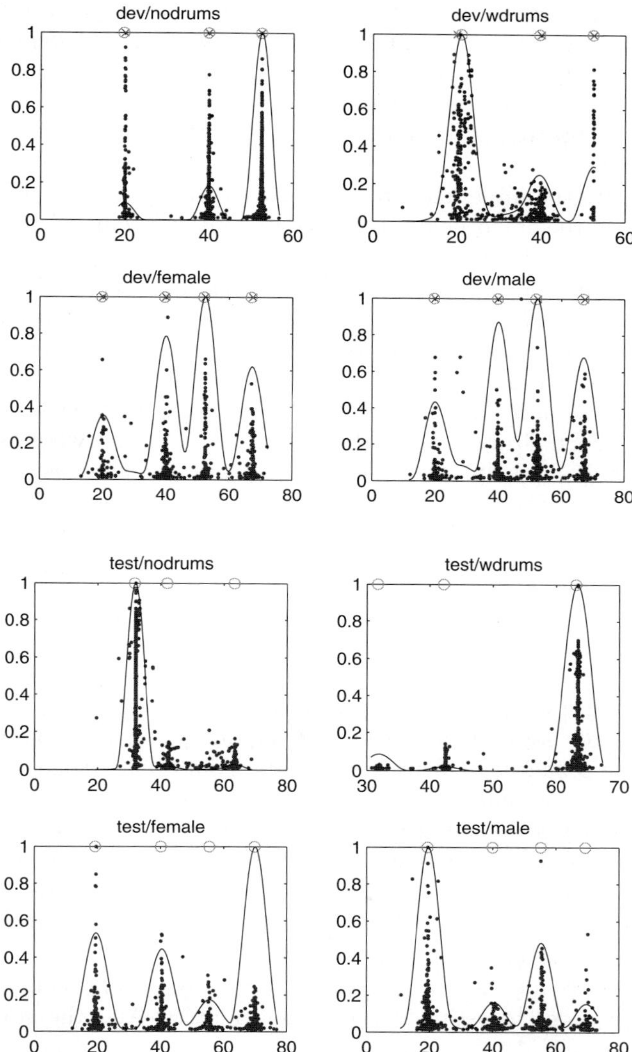

Fig. 3. Magnitude scatter-plots (dots), and smoothed potential functions F (curves) vs angle θ for all the development and test data (see text)

In practice, test data proved to be harder than development data, and the parameters settings that worked well for the later didn't allways work for the former. Yet, making use of the fact that the number of sources was known, we could find a unique set of parameters that retrieved exactly the desired number of sources for all mixtures, both development and test. All signals were 160,000 samples in length, with a sampling frequency of 16KHz. The STDFT analysis was performed with a hanning window of 512 to 2048 samples in length (depending on the signal). Consecutive frames were overlapped in 60% of their length. Parameters were set to the following values: $\Theta = 0.05$ degrees, $R = 1\%$, $P = 120$ and $\lambda = 4$.

With this parameter setting the estimate of the spatial directions was correct for all mixtures, up to grid resolution (for test data results were confirmed by optical inspection).

Figure 3 shows the filtered scatter-plots of the magnitudes of the observation data (dots), and the smoothed potential functions F's (curves) for all development (the four on the left) and test mixtures (the four on the right). Considering the peaks in detail, in some cases the field function is very well informed and develops very tall maxima, whereas in other situations the peak is identified with difficulties. Yet, the final parameter setting described above, scopes a wide range of mixture combinations.

Once the mixing matrix was identified, the sources were reconstructed using Second Order Cone Programming (SOCP) [12,8] over the whole set of observation data. Overall results of the separation scored average, as compared with other algorithms on the same data sets of the SASSEC campaign. Although the matrix estimation step was performed without error, other algorithms performed better for the source reconstruction task [11].

6   Conclusion

In this paper we have presented a particular realization of Xie et al.'s idea of using single source data for estimating the mixing matrix in a source separation task [10]. Our approach is to identify observation data that lie on the same spatial signature for two consecutive time frames. Then, we presented an improvement of Bofill & Zibulevsky's potential function approach [5], so as to identify the dominant directions of the scatter-plot of the observation vectors, corresponding to columns of the mixing matrix. Sources were then reconstructed using Second Order Cone Programming (SOCP), and results were submitted to the Stereo Audio Source Evaluation Campaing (SASSEC).

The single source idea is a simple but effective way of improving the reliability of the data, in order to estimate the mixing matrix. And the stability of the spatial signature for identifying single source data is a criterium that can be of use for source separation tasks in the general case (any number of sources and sensors and instantaneous or convolutive mixtures). In fact, filtering in reliable data should improve any mixing matrix estimation technique.

The main improvement in the potential function based method is to rescale the kernel function to the range of input angles, so that the resolution of the procedure is adaptive to the spreadness of the sources. This is effective, since the separability of the sources is intrinsic to their disjointness, rather than their closeness to each other.

Experimental results on SASSEC scored average, for the source reconstruction task, but results were always correct in the estimate of the mixing matrices.

In conclusion, the use of single source data significantly enhances the detection of source directions.

References

1. Jutten, C., Herault, J.: Blind Separation of Sources, an Adaptive Algorithm Based on Neuromimetic Architecture. Signal Processing 24(1), 1–10 (1991)
2. Makino, S., Lee, T.W., Sawada, H. (eds.): Blind Speech Separation. Springer, Heidelberg (2007)
3. O'Grady, P.D., Pearlmutter, B.A., Rickard, S.T.: Survey of Sparse and Non-Sparse Methods in Source Separation. International Journal of Imaging Systems and Technology (IJIST) 15, 18–33 (2005)
4. Lewicki, M., Sejnowski, T.: Learning Overcomplete Representations. Advances in neural information processing systems 10, 556–562 (in press, 1998)
5. Bofill, P., Zibulevsky, M.: Underdetermined Blind Source Separation using Sparse Representations. Signal Processing 81, 2353–2362 (2001)
6. Zibulevsky, M., Kisilev, P., Zeevi, Y., Pearlmutter, B.: Blind source separation via multinode sparse representation. In: Advances in Neural Information Processing Systems, vol. 14, pp. 1049–1056 (2002)
7. Blin, A., Araki, S., Makino, S.: Underdetermined blind separation of convolutive mixtures of speech using time-frequency mask and mixing matrix estimation. IEICE Trans. Fundamentals E88-A(7), 1693–1700 (2005)
8. Bofill, P.: Underdetermined Blind Separation of Delayed Sound Sources in the Frequency Domain. In: Fanni, A., Uncini, A. (eds.) Neurocomputing, Special issue: Evolving Solution with Neural Networks, vol. 55(3-4), pp. 627–641 (October 2003)
9. Winter, S., Hiroshi, S., Makino, S.: On Real and Complex Valued l_1-norm minimization for overcomplete blind source separation. In: IEEE Workshop on Apps. of Sig. Proc. to Audio and Acoustics, pp. 86–89 (October 2005)
10. Xie, Xiao, Fu: A novel approach for undedetermined blind speech sources separation. In: Proc. DCDIS Impulsive Dynamical Systems, pp. 1846–1853 (2005)
11. Vincent, E., Sawada, H., Bofill, P., Makino, S., Rosca, J.: First Stereo Audio Source Separation Evaluation Campaign: Data, Algorithms and Results. In: Davies, M.E., James, C.J., Abdallah, S.A., Plumbley, M.D. (eds.) ICA 2007. LNCS, vol. 4666, pp. 552–559. Springer, Heidelberg (2007)
12. Lobo, M.S., Vandenberghe, L., Boyd, S., Lebret, H.: Applications of Second-order Cone Programming. In: Linear Algebra and its Applications, vol. 284, pp. 193–228 (1998)

ECG Signal Classification Using GAME Neural Network and Its Comparison to Other Classifiers

Miroslav Čepek[1], Miroslav Šnorek, and Václav Chudáček

Dept. of Computer Science and Engineering, Karlovo nám. 13, 121 35 Praha 2, Czech Republic
`cepekm1@fel.cvut.cz`

Abstract. Long term Holter monitoring is widely applied to patients with heart diseases. Many of those diseases are not constantly present in the ECG signal but occurs from time to time. To detect these infrequent problems the Holter long time ECG recording is recorded and analysed. There are many methods for automatic detection of irregularities in the ECG signal. In this paper we will comapare the Support Vector Machine (SVM), J48 decision tree (J48), RBF artificial neural network (RBF), Simple logistic function and our novel GAME neural network for detection of the Premature Ventricular Contractions. We will compare and discuss classification performance of mentioned methods. There are also very many features which describes the ECG signal therefore we will try to identify features important for correct classification and examine how the accuracy is affected with only selected features in training set.

1 Introduction

Long term Holter [1] monitoring is widely applied to patients with heart problems such as arrhythmias. Identification of heart beats with unusual timing or unusual electrocardiogram (ECG) morphology can be very helpful for early diagnosis of hearts with pathological electrophysiology.

Many different methods have been proposed to solve the crucial problem of long-term Holter recordings evaluation which actually is transformed into the problem of discrimination between normal "N" and premature ventricular "V" contrations. Solving this problem is of major importance for the identification and diagnostic of heart problems. An advanced system with automatic classification and analysis capabilities is needed in order to process the large amounts of data the Holter measurement records.

Much research effort has been put till today to examine and classify data, which are usually based on beat-shape description parameters [2,3] , shape descriptive parameters transformed with the KarhunenLoeve coefficients [4], and hermite polynomials [5]. Some other works use frequency-based features [5] or time-frequency features [6] and features obtained from heartbeat interval measurements [7] to identify cardiac arrhythmia. Different methods which deal with this problem range from linear discriminants [5,3], neural networks [5] and self-organised maps [8], to other methods as in [8].

In any case, for the comparison of different approaches the setup of the experiments, where the type of the training and the selection of testing sets are defined, is of high importance. There are two main setups to be considered: training based on a local learning set [8,2] and on a global learning set [8,9].

V. Kůrková et al. (Eds.): ICANN 2008, Part I, LNCS 5163, pp. 768–777, 2008.

When the local training is used, the classifier is fitted at least partially to the data given by the measurements on the patient on which the testing will be performed. Therefore this approach requires a cardiologist to annotate part of the signal in advance before the automatic classification can be started.

On the other hand the global classification is independent of a patient and therefore it is expected to be much more robust in classifying new recordings because it does not require an annotation of a cardiologist in advance. But the drawback of this method is that it usually yields worse results in comparison with the locally trained classifiers since the morphology of N and V beats differ not only from patient to patient, but also according to the position of the ECG leads, even on the same person.

In addition to this, published works differ also in the results presentation. Some papers use accuracy measures [9] this is common for the classical AI field. Others use as measures the sensitivity and specificity [8] that are more suitable for the classification in the medical field. While the accuracy gives an overall insight of the correctness of the methods, the sensitivity and the specificity evaluate the methods from the clinical point of view; this is more suitable for decision making in the situations dealing with uneven numbers of beats for each classification group.

In this paper we attempt to provide a thorough investigation of different pattern analysis techniques using a global classifier on two different datasets. The results are then compared by the means of sensitivity and specificity.

In our work we will use two different databases with Holter ECG recordings. One of them will be MIT-BIH and the second AHA (American Heart Association) databases [10,11]. These will be described in details in Section 6. Both data bases have undergone the same preprocessing steps, where the same features have been computed. In Section 7, the results are compared using the well established measures of sensitivity and specificity and we will discuss our results.

2   GAME Neural Network

In many applications it is important to find optimal model of unknown system (for example in classification, prediction, approximation, etc.). Such model can be found using two different approaches – deductive and inductive. The GAME artificial neural network (ANN) is based on inductive approach. This means that parameters and also structure of the ANN are parts of a learning process (the parameters are selected and the NN is constructed from some minimal blocks during the learning process).

The GAME ANN extends the concept of GMDH network [12,13]. The GMDH allows only one type of minimal block (neurons with one transfer function). On the other hand in GAME ANN there are neurons with many different transfer functions (linear, sigmoid, polynomial, etc...). The GAME has a feed-forward structure [14] as illustrated in figure 1.

The GAME NN is built during the training phase from scratch. Each new layer is constructed in the following way: at first a large number of new neurons is generated. Neurons differ in the transfer function, the number of inputs and in the number of neurons in previous layer that the new neuron is connected to.

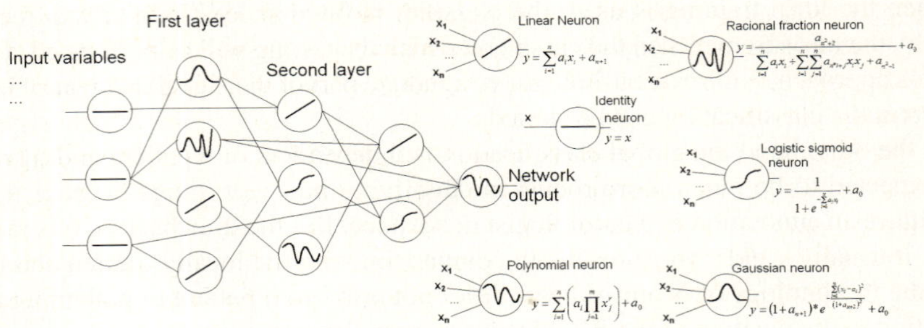

Fig. 1. Example of GAME artificial neural network with different types of neurons

The next step is to find the optimal setup of internal parameters and the best connections to neurons in previous layer. To do this, GAME uses an advanced genetic algorithm. The population consists of neurons in each new layer and each neuron is coded to a genome.

In the beginning of the genetic algorithm, the neurons are generated randomly (derivation of the first generation). The neurons are evaluated using a validation set and their accuracy corresponds to neuron's fitness. Then neurons are selected to cross-over according to their fitness. The cross-over primary involves only the part of genome which encodes the connections to the previous layer. If both neurons employed in cross-over have the same transfer function, their structural parameters are also "crossed-over". When the cross-over is finished all neurons optimise their transfer function using a properly assigned method. After that the genetic algorithm is repeated for a number of generations until the accuracy of the population exceeds a specified threshold.

All neurons in new layer are evaluated using separated testing set and the worst neurons are deleted from the layer. Then the layer is "frozen" and the algorithm continues with creation of a new layer. This is repeated until a neuron with a satisfactory accuracy is found and this neuron is the output of the network.

2.1 GAME Method Setup

The number of inputs is equal to number of features extracted from the data – 14. For the main experiment neuron with following transfer function are allowed: linear, polynomial, exponential, sigmoid, sinus, Gaussian.

Neurons may be connected to any neuron in previous layers or any input.

For the second experiment only neurons with following transfer functions are allowed: linear, polynomial and sigmoid.

3 Other Classification Methods

3.1 Decision Tree J48

This technique is based on natural and intuitive idea to classify objects through a sequence of question in which the next question depends on answer to current question.

This very useful for categorical and ordinal data, but can be extended also to continuous variables. Such sequence of questions can be displayed as a direct decision tree. Root node and inner nodes are the questions and leaf nodes are classes to which evaluated object belongs. The links are answers to the these questions.

The most popular approaches is to select question are based on information entropy or information gain. For more details see [15].

3.2   Simple Logistics

The linear logistic regression perform a least-squares of a set of parameters $\beta_0 \ldots \beta_k$ to a numeric target variable to form a model

$$f(x) = \beta_0 + \sum_{i=1}^{k} x_i \beta_i$$

When we obtain the value $f(x)$ it says nothing about class the x belongs to. One method is to use posterior class probabilities $P(G = jjX = x)$ for the J classes via linear functions in x while at the same time ensuring they sum to one and remain in $< 0; 1 >$. The probabilities are computed as

$$P(G = j|X = x) = \frac{e^{F_j(x)}}{\sum_{k=1}^{J} e^{F_k(x)}}$$

where $\sum_{k=1}^{J} e^{F_k(x)} = 0$.

The $F_j(X) = \beta_j^T * x$ and it is usually by maximum likehood estimates for the parameters β_j. And the assigned class is is the one with the highest posterior probability [15].

3.3   Support Vector Machines

Support Vector Machines (SVMs) are learning systems that are trained using an algorithm based on optimisation theory [16,17]. The SVM solution finds a hyperplane in the feature space that keeps the empirical error small while maximising the margin between the hyperplane and the instances closest to it. New pattern x is classified to either one of the two categories (in case of dichotomizing problems $y_i \in \{-1, 1\}$):

$$f(x) = sign(\sum_{i=1}^{n} y_i a_i K(x, x_i) + b)$$

where b is a threshold parameter. The coefficients are found by solving a maximisation quadratic programming problem which is "controlled" by a penalty factor C and are assigned to each of the training patterns x_i. The points for which $a_i > 0$, are called Support Vectors and are the points lying closest to the hyperplane.

The kernel function K implicitly performs the mapping from the input space to the feature space. Among others, the most popular kernels are the polynomial, the radial basis function networks and the two-layer perceptrons. In our experimental procedure we employ the radial basis function machines (the width s, which is common to all kernels, is specified also apriori by the user [18])

$$K(x, x_i) = exp(-\frac{1}{2s^2}\|x - x_i\|^2), i = 1 \ldots l$$

3.4   Radial Basis Function ANN

Radial Basis Function (RBF) ANN is trained with a fast supervised learning algorithm which is suitable both for regression and classification. It consists of one input, one output and one hidden layers, which contain RBF neurons. Each RBF neuron is described by a transfer function

$$y = exp(-\frac{X_i - C_i}{s^2})$$

that represents a d-dimensional Gaussian "bump" (where d is the dimension of the input vectors), with the centre at a point C_i and a width s. Output neurons calculate the weighted sum of RBF neurons output. Training of the network is divided into two phases. In the first phase, C_i and s are set for each RBF neuron. The second phase adjusts weights of output neurons. The details can be found in [19].

4   Specificity and Sensitivity

Specificity and Sensitivity are measures of binary classifier quality. Primary use of this measures is in medicine where it says how successful is the method in diagnosing healthy and ill people. But nowadays they are also used in other fields.

In Specificity and Sensitivity analysis are also important terms true/false positive and true/false negative. The **true positives** are contraction correctly classified as abnormal ("V" contractions), **true negatives** are contraction correctly classified as normal ("N"contractions). **False positives** are contractions incorrectly classified as abnormal and **false negatives** are contraction incorrectly classified as normal.

$$Specificity = \frac{Number of True Negatives}{Number of True Negatives + Number of False Positives}$$

$$Sensitivity = \frac{Number of True Positives}{Number of True Positives + Number of False Negatives}$$

5   Feature Ranking

The GAME can be also used for feature ranking. As the genetic algorithm searches for the optimal setup of the units it also picks inputs which contributes the most to classifiaction accuracy. If we take finished model, and count units connected to each input, we get the proportional significance of each input.

We compare results of the GAME to some selected feature ranking methods implemented in the Weka software. In this software the feature selection process is divided into two parts. The first part selects subset of features and the second part evaluates quality of selected subset. The subset of feature selection methods used were:

- **BestFirst** – performs greedy hill climbing with backtracking.
- **Ranker** – is not actual search method, but is a ranking scheme for individual attributes. It sorts attributes by their individual significance.

Following subset feature evaluators were used:

- **CfsSubsetEval** – calculates the predictive ability of each feature individually and the degree of redundancy among them [15].
- **InfoGainAttributeEval** – evaluates features by measuring their information gain with respect to the class [15].

6   Heart Beat Databases and Their Preprocessing

6.1   Premature Ventricular Contraction

Premature ventricular contractions occur when there is an ectopic focus in the ventricle that sends out its own electrical impulse, pre-empting the next sinus beat. These abnormal and early (premature) impulses from the ventricles interrupt the normal rhythm. This premature impulse spreads across the ventricles, causing then to contract, and sometimes spreads backwards up the AV node to the atria as well.

The PVC is different from a normal sinus beat because the areas of heart muscle are stimulated in an abnormal order; the ventricles are stimulated first. Further, because the route of electrical stimulation is not normal, the impulse does not pass over the ventricles in the regular, organised way. The result may be a heart beat that is less effective in pumping blood. In addition, since the PVC comes early, the left ventricle is not properly filled, so the cardiac output from these premature beats is lower than a normal sinus beat. If one feels the pulse at the wrist in someone with PVC's, the premature beats might generate a noticeably weaker pulse, or sometimes no pulse at all. The description is taken from [20].

6.2   Heart Beat Databases

For training and testing the previously described methods, the **MIT-BIH arrhythmia database** [10] and the **AHA (American Heart Association) database** [11] are used. From both databases, thirty minute long segments annotated by experts are used. Since we focus on the discrimination between ventricular and normal beats, only beats labelled as 'V' or 'N' are selected for the classification purposes.

6.3   Data Preprocessing

During preprocessing the raw signal was splited into individual heart contractions and each contraction is described by set of about 40 parameters. The figure 2 shows some features computed.

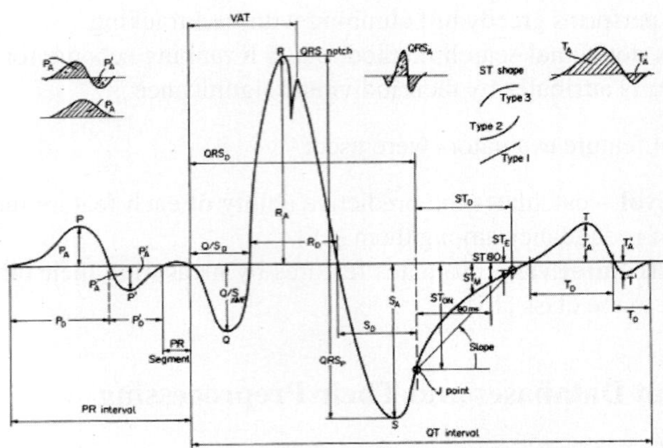

Fig. 2. Selected features computed from ECG signal

7   Results

In this section we will present our results. First we will present results for the all features. In the next subsection we will present features selection results and in the last part of this section we will show the results of classification methods for reduced data.

All results presented in this section were obtained as average value of given parameter. Averages are calculated from corresponding parameters for testing sets. Testing set for

- AHA and MIT as training set – was recording from both databases.
- MIT as training set – only recordings from AHA database.
- AHA as training set – only recordings from MIT database.

7.1   Full Features Classification Result

The Table 1 shows final results for full data. Results shows that all methods have very good average specificity – more than 90 %. This means that all methods are good in recognising normal beats as normal. This is not quite surprising because majority of beats in moth databases are normal heart contractions.

On the other hand the sensitivity is much poorer. This means that all methods have problem with correct recognising ventricular contractions. Many normal beat are also recognised ventricular beats. Almost all methods have sensitivity less than 60% which is very poor result. The exception is the GAME neural network which achieves very good sensitivity – more than 87%. This is – with respect to other methods – an excellent result. The average accuracy is also quite good – model have usually average accuracy more than 85%. This good result is spoiled by sensitivity as described above.

7.2   Feature Selection Results

In this particular dataset there are over 40 different features. We were also curious if all these features are needed and relevant for the correct classification. Even some features may be misleading and may harm the model accuracy.

We performed feature ranking using 10 GAME neural network models and feature selection methods implemented in Weka software. Results obtained from methods were naturally slightly different but the first six features were always the same.

- Prematurity – shows how much given beat sooner or later than expected.
- intRR – distance of R-R peaks of two subsequent beats.
- intQRS – length of the QRS complex.
- areaQRS_abs – area under the QRS complex. Area above the zero is added to area below the zero.
- areaQRS_real – area under the QRS complex. Area below the zero is subtracted from area above the zero.
- STarea80 – area under the curve between S and T peaks.

The premature ventricular contraction is heart contraction which comes before its time and more it is distorted (see 6.1 or [20]). All features selected have some relation to length of the contraction or its the most important parameters. Therefore we believe that selected features are relevant.

7.3   Reduced Features Classification Result

Results of classifiers for the data with selected features are presented in Table 2.

Table 1. Results for testing data and all attributes. All values are in %. Values shows percentage of correctly recognised instances therefore higher number is better. Spec is abbreviation from Specificity, Sens from Sensitivity and Acc from accuracy.

	AHA as Training set			MIT as Training set			Both as Training set		
Method name	Spec [%]	Sens [%]	Acc [%]	Spec [%]	Sens [%]	Acc [%]	Spec [%]	Sens [%]	Acc [%]
J48 Tree	90	30	90	90	26	59	100	48	92
RBF NN	96	22	74	93	8	17	98	44	88
Simple Log	99	39	93	92	51	81	99	51	92
SVM	99	40	93	94	48	82	99	52	94
GAME NN	95	97	96	90	87	83	94	96	93

Table 2. Results for testing data with selected features. All values are in %. Values shows percentage of correctly recognised instances therefore higher number is better. Spec is abbreviation from Specificity, Sens from Sensitivity and Acc from accuracy.

	AHA as Training set			MIT as Training set			Both as Training set		
Method name	Spec [%]	Sens [%]	Acc [%]	Spec [%]	Sens [%]	Acc [%]	Spec [%]	Sens [%]	Acc [%]
J48 Tree	99	31	93	99	46	81	100	48	93
RBF NN	99	39	93	99	46	82	100	47	92
Simple Log	99	43	94	99	49	79	100	48	91
SVM	99	44	94	99	53	86	100	49	91
GAME NN	93	93	93	86	86	74	92	92	90

The results show great improvement of the results for the J48 Decision Tree, RBF Neural network, Simple Logistics and SVM methods (all methods implemented in Weka). Results of method were improved is up to 10% in specificity and up to 40% sensitivity. This means that ability of recognising premature ventricular beats was improved greatly. The reason for improvement is that attributes with irrelevant or noisy information are removed. Methods then need not to work with the noisy inputs and achieves better results.

On the other hand, the accuracy of the GAME network was slightly lowered. The reason for this is that the GAME network is able create better combinations of attributes and is able to get additional information even for those noisy inputs.

8   Conclusion

In this paper we presented classification of the Holter long term ECG recordings using several well known methods (Support Vector Machine, Decision Tree, Logistic Regression function, RBF artificial neural network) and novel GAME neural network. The best method is shown to be the GAME neural network. It achieves very good and well balanced results for specificity and sensitivity measures. This means that the GAME network is good in recognising both normal and ventricular contractions.

Other methods tends to recognise only normal contractions and very often ventricular contraction are also classified as normal. In the extreme case almost all ventricular contractions are classified as normal and omits the ventricular contractions (for example see RBF NN row in Table 1).

In this work we also examined features which are important for classification. We identified following features: Prematurity, intRR, intQRS, areaQRS_abs, areaQRS_real, STarea80. Description of these features can be found in section 7.2. These features corresponds to what we expect and we believe that they are correct.

The results of the methods for the selected features are better. This is caused by removal of irrelevant and noisy inputs. The accuracy is higher about 5% (see Table 2).

Acknowledgement

This research is partially supported by the grant Automated Knowledge Extraction (KJB201210701) of the Grant Agency of the Academy of Science of the Czech Republic and the research program "Transdisciplinary Research in the Area of Biomedical Engineering II" (MSM6840770012) sponsored by the Ministry of Education, Youth and Sports of the Czech Republic.

References

1. Holter, N.J.: New method for heart studies. Science 134, 1214–1220 (1961)
2. de Chazal, P., O'Dweyr, O., Reilly, R.B.: Automatic classification of heartbeats using ECG morp. heartbeat intervals features. IEEE Trans. on Biom. Eng. 51(7), 1196–1206 (2004)

3. Cueasta-Frau, D., Perez-Cortes, J.C., Andreu-Garcia, G.: Clustering of electrocardiograph signals in computer-aided holter analysis. Computer methods and programs in Biomedicine 72, 179–196 (2003)
4. Moody, G., Mark, R.: Qrs morphology representation and noise estimation using the karhunen-loeve transform. Computer Cardiology 16, 269–272 (1989)
5. Lagerholm, M., Peterson, C., Edenbrandt, L., Sornmo, L.: Clustering ECG complexes using hermite functions and self-organizing maps. IEEE Trans. on Biom. Eng. 47, 838–848 (2000)
6. Christov, I., Herreto, G.G., Kraseva, V., Jekova, I., Gotchev, A., Egiazarian, K.: Comparative study of morphological and time-frequency ECG descriptos for heartbeat classification. Medical Engineering & Physics 28, 876–887 (2006)
7. Tsipouras, M.G., Voglis, C., Lagaris, I.E., Fotiadis, D.I.: Cardiac arrythmia classification using support vector machines. In: The 3rd European Medical and Biological Engineering Conference (2005)
8. Bortolan, G., Jekova, I., Christov, I.: Comparison of four methods for premature ventricular contraction and normal beat clustering. Computers in Cardiology 32, 921–924 (2005)
9. Hu, Y.H., Palreddy, S., Tompkins, W.J.: A patient-adaptable ECG beat classifier using a mixture of experts approach. IEEE Trans. on Biom. Eng. 44, 891–900 (1997)
10. Goldberger, A.L., Amaral, L., Glass, L., Hausdorf, J.M., Ivanov, P.C., Mark, R.G., Mietus, J.E., Moody, G., Peng, C.K., Stanley, H.E.: Physiobank, physiotoolkit and physionet: Compnents of a new research resource for complex physiologic signals. Circulation 101(23), 215–220
11. The American Heart Association. AHA-Database. AHA Database Series 1, AHA database series, 1 edn. (1997)
12. Muller, J.A., Lemke, F.: Self-Organising Data Mining, Berlin (2000)
13. Madala, H., Ivakhnenko, A.: Inductive Learning Algorithm for Complex System Modelling. CRC Press, Boca Raton (1994)
14. Pavel KordÃk: Fully Automated Knowledge Extraction using Group of AdaptiveModels Evolution. PhD thesis, Czech Technical University, Prague (March 2007)
15. Witten, I., Frank, E.: Data Mining Practical Machine Learning Tools and Techniques. Elsevier, Amsterdam (2005)
16. Burges, C.J.C.: A tutorial on support vector machines for pattern recognition. Data Mining and Knowledge Discovery 2(2), 121–167 (1998)
17. Muller, K.R., Mika, S., Ratsch, G., Tsuda, K., Scholkopf, B.: An introduction to kernel-based learning algorithms. IEEE Trans. on Neural Networks 2(2), 181–201 (2001)
18. Haykin, S.: Neural Networks: A Comprehensive Foundation. Prentice-Hall, Englewood Cliffs (1999)
19. Buhmann, M.D., Ablowitz, M.J.: Radial Basis Functions: Theory and Implementations. Cambridge University, Cambridge (2003)
20. Morganroth, J.: Premature ventricular complexes. diagnosis and indications for therapy. Journal of American Medical Association (1984)

Predictive Modeling with Echo State Networks[*]

Michal Čerňanský[1] and Peter Tiňo[2]

[1] Faculty of Informatics and Information Technologies, STU Bratislava, Slovakia
[2] School of Computer Science, University of Birmingham, United Kingdom
cernansky@fiit.stuba.sk, P.Tino@cs.bham.ac.uk

Abstract. A lot of attention is now being focused on connectionist models known under the name "reservoir computing". The most prominent example of these approaches is a recurrent neural network architecture called an echo state network (ESN). ESNs were successfully applied in several time series modeling tasks and according to the authors they performed exceptionally well. Multiple enhancements to standard ESN were proposed in the literature. In this paper we follow the opposite direction by suggesting several simplifications to the original ESN architecture. ESN reservoir features contractive dynamics resulting from its' initialization with small weights. Sometimes it serves just as a simple memory of inputs and provides only negligible "extra-value" over much simple methods. We experimentally support this claim and we show that many tasks modeled by ESNs can be handled with much simple approaches.

1   Introduction

Echo state network (ESN) [1] is a novel recurrent neural network (RNN) architecture based on a rich reservoir of potentially interesting behavior. The reservoir of ESN is the recurrent layer formed of a large number of sparsely interconnected units with non-trainable weights. ESN training procedure is a simple adjustment of output weights to fit training data. ESNs were successfully applied in several sequence modeling tasks and performed exceptionally well [2,3]. Also some attempts were made to process symbolic time series using ESNs with interesting results [4].

On the other side part of the community is skeptic about ESNs being used for practical applications [5]. There are many open questions, as noted by the author of ESNs [6]. It is still unclear how to prepare the reservoir with respect to the task, what topologies should be used and how to measure the reservoir quality for example.

The key feature of ESNs is so called "echo-state" property: under certain conditions ESN state is a function of finite history of inputs presented to the network - the state is the "echo" of the input history. It simply means that ESNs are based on contractive dynamics where recurrent units reflect the history of inputs presented to the network. The most recent input has the most important influence to the current network state and this influence gradually fades out.

To study properties responsible for excellent performance of ESN architecture we propose several simplified models. The first we call "feedforward echo state network"

---

[*] This work was supported by the grants VG-1/0848/08 and VG-1/0822/08.

V. Kůrková et al. (Eds.): ICANN 2008, Part I, LNCS 5163, pp. 778–787, 2008.

(FF-ESN). In FF-ESN, a cascaded reservoir topology with triangular recurrent weight matrix is employed and so each FF-ESN can be unfolded into an equivalent feedforward network. By using this architecture we try to reveal that recurrent nature of the ESN reservoir is not the key architectural feature, as it usually is in recurrent multilayer perceptron trained by common algorithm such as backpropagation through time. The next simplified model is ESN-like architecture with reservoir composed of units organized in one chain, hence only elements on recurrent weight matrix subdiagonal have nonzero values. This model benefits from the nonlinear combination of network inputs and this combination is achieved by a very straightforward manner. Finally we have tested performance of linear autoregressive model represented by architecture with linear hidden units organized in chain and thus forming tapped delay line.

2   Models

2.1   Echo State Networks

Echo state networks represent a new powerful approach in recurrent neural network research [1,3]. They belong to the class of methods known under the name "reservoir computing". Reservoir of ESN is formed of large sparsely interconnected and randomly initialized recurrent layer composed of huge number of standard sigmoid units. As a readout mechanism ESN uses standard output layer and to extract interesting features from dynamical reservoir only output connections are modified during learning process. A significant advantage of this approach over standard RNNs is that simple linear regression algorithms can be used for adjusting output weights. When $\mathbf{u}(t)$ is an input vector at time step t, activations of hidden units $\mathbf{x}(t)$ are updated according to

$$\mathbf{x}(t) = f\left(\mathbf{W}^{\mathrm{in}} \cdot [\mathbf{u}(t), 1] + \mathbf{W} \cdot \mathbf{x}(t-1) + \mathbf{W}^{\mathrm{back}} \cdot \mathbf{y}(t-1)\right), \tag{1}$$

where f is the hidden unit's activation function, $\mathbf{W}$, $\mathbf{W}^{\mathrm{in}}$ and $\mathbf{W}^{\mathrm{back}}$ are hidden-hidden, input-hidden, and output-hidden connections' matrices, respectively. Activations of utput units are calculated as

$$\mathbf{y}(t) = f\left(\mathbf{W}^{\mathrm{out}} \cdot [\mathbf{u}(t), \mathbf{x}(t), 1]\right), \tag{2}$$

where $\mathbf{W}^{\mathrm{out}}$ is output connections' matrix.

Echo state property means that for each internal unit x_i there exists an echo function e_i such that the current state can be written as $x_i(t) = e_i(u(t), u(t-1), \ldots)$, the network state is an "echo" of the input history [1]. The recent input presented to the network has more influence to the network state than an older input, the input influence gradually fades out. So the same input signal history $u(t), u(t-1)$, will drive the network to the same state $x_i(t)$ in time t regardless the network initial state. Echo states are crucial for successful operation of ESNs, their existence is usually ensured by rescaling recurrent weight matrix $\mathbf{W}$ to specified spectral radius λ. This can be achieved by simply multiplying all elements of a randomly generated recurrent weight matrix with λ/λ_{max}, where λ_{max} is the spectral radius of the original matrix.

2.2 Feed-Forward ESN Model

Taking into account the contractive dynamics of ESN exhibited through existence of "echo states", the network output can be seen as a complex nonlinear function of the input history with finite length since the influence of inputs fades out exponentially in time and inputs presented in earlier time steps can be ignored. By removing recurrent connections we propose modified model with the reservoir of units connected in a feed-forward manner. Units in a reservoir can be indexed and their activities depend only on activities of units with smaller indices. No cycles are present in the graph with nodes representing units and edges representing connections. FF-ESN is shown in the Fig. 1a. Technically these connections are still recurrent ones because units are fed by activities from previous time steps. But this network can be easily transformed into regular feed-forward network by the process identical to the RNN unfolding in time when using backpropagation through time learning algorithm (see Fig. 1b).

Exactly the same training process as is commonly used in training regular ESNs can be used. The only difference is how a recurrent weight matrix is generated. Initial values for biases, recurrent and backward weights falls to the same ranges as described for regular ESNs. Recurrent weight matrix is rescaled to the required spectral radius λ and the matrix is then made lower triangular by keeping only elements below diagonal. To force the ESN to keep longer history of inputs in activities every unit i was connected to the previous one $i-1$ through the weight $w_{i,i-1}$ of chosen constant value, in our experiments we used the value of spectral radius λ.

2.3 Tapped Delay Line Models

Further simplification of ESN architecture resulted in the model with hidden unites organized into a tapped delay line. Hence the only existing recurrent weights are $w_{i,i-1}$ connecting every hidden unit to its predecessor, excluding the first hidden unit. The first tapped delay line model variant labeled TDL-I (tapped delay line with inputs) also allows input and threshold connections for hidden units. TDL-I uses hidden units with nonlinear activation function and was constructed in order to prove our supposition that very simple nonlinear combination of inputs can be responsible for stunning performance of ESNs on some tasks. The second variant is even greater simplification. Its a

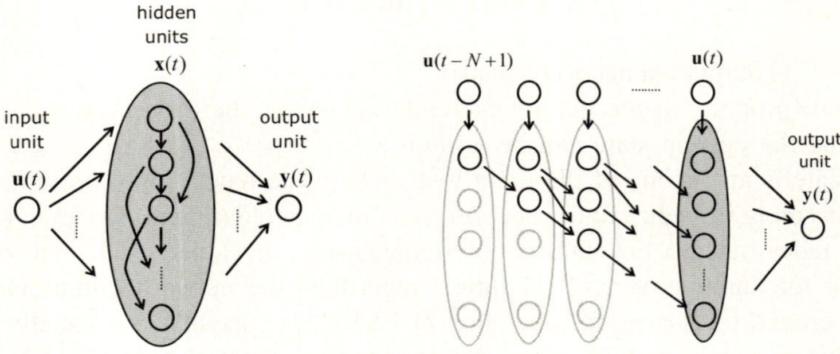

Fig. 1. (a) Modified "feed-forward" ESN architecture. (b) Feed-forwad ESN unfolded in time into regular feed-forward network.

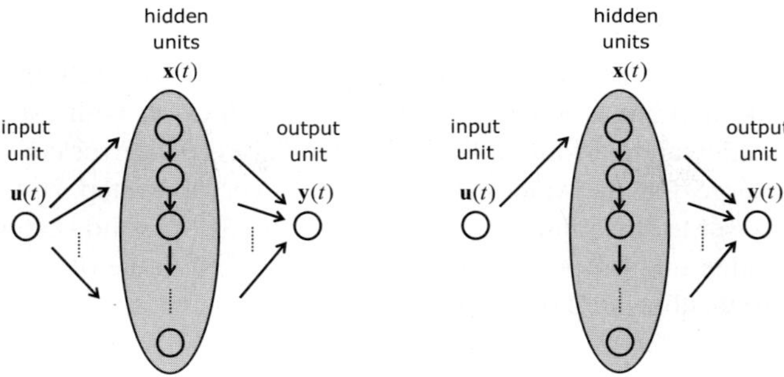

Fig. 2. (a) Architecture with reservoir orginized as TDL with nonlinear units connected to input. (b) Architecture with reservoir formed of linear units organized to TDL.

TDL with linear hidden units and only the first hidden unit is connected to the input. All connections are set to 1.0. It supposes that the process being modeled can be handled with auto-regressive model. Both models are represented in Fig. 2.

3   Method

We trained ENS, FF-ESN, TDL-I and TDL models on several time series modeling tasks. We provide mean square error (MSE) results for several datasets taken from published works. For each model and each dataset we have calculated means and standard deviations over 10 simulation runs with the best meta-parameters we were able to find (only number of reservoir units was chosen to match the published experiments).

We used both standard training as proposed in [1] and training using recursive least squares [3] for each of four architectures and each dataset. Since we used output units with linear activation function for all architectures target values were directly used in the least squares fit. All time series are single-dimensional hence all architectures had only one input and one output unit.

The standard training consists of first running the network on the train set and collecting activities of input and hidden units. Initial "washout" activities are thrown away and remaining values forming matrix $\mathbf{X}$ are used for least squares fitting. Output weights are found by solving $\mathbf{X}\mathbf{w} = \hat{\mathbf{y}}$ for $\mathbf{w}$ using singular value decomposition. $\mathbf{w}$ stands for the vecor of output unit weights. It is formed by concatenating output unit threshold, input-output weights and hidden-output weights. Matrix $\mathbf{X}$ has rows corresponding to time steps. The first value in each row is the constant 1.0 corresponding to the output unit threshold followed first by input values and then hidden activities collected from the corresponding time step. Vector $\hat{\mathbf{y}}$ is the vector of target values from the training set.

Recursive least squares training starts after initial "washout" steps. Output weights are updated every time step according to the following equations:

$$\mathbf{k}(t) = \frac{\mathbf{P}(t-1)\mathbf{v}(t)}{\mathbf{v}^T(t)\mathbf{P}(t-1)\mathbf{v}(t) + \gamma}, \tag{3}$$

$$\mathbf{P}(t) = \gamma^{-1}\left(\mathbf{P}(t-1) - \mathbf{k}(t)\mathbf{v}^T(t)\mathbf{P}(t-1)\right), \tag{4}$$

$$\mathbf{w}(t) = \mathbf{w}(t-1) + \mathbf{k}(t)[\hat{y}(t) - y(t)], \tag{5}$$

where $\mathbf{k}$ stands for the innovation vector calculated in every time step. $\hat{y}$ and y correspond to the desired and calculated output unit activities. $\mathbf{w}$ is the vector of output weights (threshold, weights from input and hidden units). $\mathbf{P}$ is error covariance matrix initialized with large diagonal values and updated in every time step. Forgetting parameter γ is usually set to the value smaller or equal to 1.0. $\mathbf{v}$ is the vector of activities of input and hidden units form actual time step. The first value of $\mathbf{v}$ is constant 1.0 corresponding to the threshold of the output unit.

4   Results

4.1   Mackey-Glass Chaotic Time Series

Mackey-Glass (MG) system is a chaotic time series where ESNs showed excellent performance exceeding other approaches by several orders [1,3]. MG system is defined by the differential equations $\partial x/\partial t = \big(0.2(x-\tau)\big)/\big(1 + x(t-\tau)^{10} - 0.1x(t)\big)$ and we used $\tau = 17$ as in [3]. We have generated 10000 values using MG system. Than values were shifted by adding -1 and tanh function was used to squash values into appropriate interval as in [3]. The first 5000 values were used for the training set and remaining 5000 values were used for testing. Initial 1000 from the training set were used only for forward propagation and corresponding hidden units activities were discarded from the adaptation process.

Models with 1000 hidden units were trained for the next value prediction task. Input weights and thresholds for ESN, FF-ESN and TDL-I were initialized from $(-0.2, 0.2)$ interval. Recurrent weights for ESN and FF-ESN were created with 1.0% probability and then recurrent weight matrix was rescaled to spectral radius $\lambda = 0.70$ for ESN and 0.65 for FF-ESN. "Backbone" recurrent connections $w_{i,i-1}$ were set to 0.65 for both FF-ESN and TDL-I model. RLS forgetting parameter γ was set to 1.0. Resulting MSE means together with standard deviations in parenthesis are shown in the Tab. 1. The best performance was achieved using ESN, with FF-ESN and TDL-I models inferior performance was obtained, but within the same order of magnitude. This is in slight contrast from our previous findings, where FF-ESNs performed comparably well on MG generation task [7]. FF-ESN architecture is more sensitive to the network initialization parameters and more thorough parameter searching process could result in some performance improvement. Quite surprising is a very good prediction performance of TDL-I model. Although this model is extremely simple and has far less connections than ESN or FF-ESN, its resulting performance was comparable. Simple linear TDL

Table 1. The next value prediction results for Mackey-Glass time series

	One Step Least Square Fit	Recursive Least Squares
ESN	9.82×10^{-16} (9.02×10^{-17})	3.30×10^{-9} (1.95×10^{-10})
FF-ESN	1.84×10^{-15} (2.13×10^{-16})	1.77×10^{-9} (1.62×10^{-10})
TDL-I	1.88×10^{-15} (2.49×10^{-16})	3.74×10^{-9} (1.19×10^{-10})
TDL	1.18×10^{-8}	1.32×10^{-8}

model is not able to achieve comparable performance to other models that can exploit nonlinearity through hidden units' activation functions when combining inputs. RLS results for all models are severely inferior to results obtained by standard training, since numerical accuracy is a key factor to achieve results of 10^{-15} order. The required precision is lost when doing recursive updates. On the other hand it's hard to imagine real-life application where results of similar precision would be achievable.

4.2   Nonlinear Communication Channel

We used the same nonlinear channel model as in [3]. First the a sequence $d(n)$ of symbols transmitted through the channel was generated by randomly choosing values from $\{-3, -1, 1, 3\}$. Then $d(n)$ values were used to form $q(n)$ sequence by $q(n) = 0.08d(n+2) - 0.12d(n+1) + d(n) + 0.18d(n-1) - 0.1d(n-2) + 0.09d(n-3) - -0.05d(n-4) + 0.04d(n-5) + 0.03d(n-6) + 0.01d(n-7)$ what represents linear filter equation. Nonlinear transformation is then applied to $q(n)$ sequence to produce row corrupted signal $u(n)$ by $u(n) = q(n) + 0.0036q(n)^2 - 0.11q(n^3) + v(n)$ where $v(n)$ represents the zero mean Gaussian noise. In our experiments no noise was added. The task was to predict value $d(n-2)$ when $u(n)$ was presented to the network. We also shifted $u(n)$ signal by adding value of 30 like in [3].

Models were trained on 4000 values and then the performance was evaluated on the next 6000 values. Initial 100 values from the training set were not used in adaptation process. All models had 47 hidden units (as in [3]), input weights for ESN, FF-ESN and TDL-I models were generated from $(-0.025, 0.025)$ interval. Recurrent weight were generated with 20% probability and the recurrent weight matrix was then rescaled to spectral radius $\lambda = 0.5$. The same value was used for FF-ESN and TDL-I backbone connections. RLS forgetting parameter γ was set to 1.0. Results are given in the Tab. 2. Results for ESN, FF-ESN and TDL-I models are very similar for both standard training and RLS training, although slight degradation of performance can be observed for RLS. Simple architecture of TDL-I model is sufficient to obtain good results. Although performance of linear TDL model is much inferior, it is still comparable to other models.

4.3   NARMA System

The 10th order nonlinear autoregressive moving average (NARMA) system from [8] defined as $d(n+1) = 0.3d(n) + 0.05d(n)\left[\sum_{i=0}^{9} d(n-i)\right] + 1.5u(n-9)u(n) + 0.1$ was used. Values for input sequence $u(n)$ were generated uniformly from interval $[0, 0.5]$

Table 2. The next value prediction results for nonlinear communication channel

	One Step Least Square Fit	Recursive Least Squares
ESN	0.022 (0.0016)	0.028 (0.010)
FF-ESN	0.022 (0.0017)	0.025 (0.0039)
TDL-I	0.018 (0.0024)	0.026 (0.0027)
TDL	0.052	0.053

Table 3. Prediction results for NARMA system

	One Step Least Square Fit	Recursive Least Squares
ESN	0.00091 (0.00017)	0.00084 (0.000084)
FF-ESN	0.00090 (0.00011)	0.00096 (0.00027)
TDL-I	0.00132 (0.00021)	0.00141 (0.00023)
TDL	0.00189	0.00189

and the task was to predict $d(n)$. Training set was composed of 2200 values and first 200 values were used as initial washout steps. Next 2000 values were used to train models. Testing was performed on the next 2000 values. No noise was added to the sequence.

We trained models with hidden layer formed of 100 units. Input weight were initialized from $(-0.1, 0.1)$ interval. Recurrent weights were created with 5% probability and recurrent weight matrix was rescaled to have spectral radius $\lambda = 0.95$ (for ESN and FF-ESN model). Backbone connections (for FF-ESN and TDL-I models) were set to the same value. RLS forgetting factor was set to $\gamma = 1.0$. Results are shown in Tab. 3. Results reveal only slight differences between standard and RLS training. Both ESN and FF-ESN models achieved similar performance. Significantly worse but yet comparable performance was achieved with TDL-I and TDL models.

4.4 Predictive Modeling Problem

The next task is called predictive modeling taken form [9]. Time series of 10000 values was generated using $sin(n + sin(n))$ for $n = 1, 2, \ldots$, the first 1000 values initial transient activities are thrown away, training is performed on the next 4000 values and predictive performance is evaluated on the next 5000 values.

Models with 600 units were tested, small $\lambda = 0.35$ value was used to rescale recurrent weight matrix for ESN and FF-ESN models. The same value was also used for backbone connections of FF-ESN and TDL-I models. Recurrent weights for ESN and FF-ESN were created with 10% probability. Input weights for ESN, FF-ESN and TDL-I were created from $(-0.25, 0.25)$ interval. Hidden units had no threshold connections. Results are shown in Tab. 4. We were able to train models using nonlinear activation function to the same level of performance. Linear TDL showed significantly poorer performance. ESN model in [9] was trained with suboptimal parameters, scaling recurrent weights to relatively small spectral radius resulted into significant prediction improvement. We were not able to train models to the comparable level of performance with RLS, much better results were obtained using standard one-step linear regression.

Table 4. The next value prediction results for predictive modeling problem

	One Step Least Square Fit	Recursive Least Squares
ESN	0.00046 (5.20×10^{-5})	0.018 (0.0019)
FF-ESN	0.00055 (6.22×10^{-5})	0.0074 (0.00066)
TDL-I	0.00058 (6.52×10^{-5})	0.015 (0.00081)
TDL	0.056	0.056

4.5 Multiple Superimposed Ocillator

Multiple superimposed oscillators time series was used for the next value prediction problem in [9,10]. Dataset values are created as $sin(0.2n) + sin(0.311n)$ for $n = 1, 2, \ldots$. Sequence of 1000 values was generated, the first 100 values were used for initial transient steps. Training was performed on the following 600 values and predictive performance was evaluated on the next 300 values.

We provide results for models with 400 hidden units. Input weight were initialized from $(-0.2, 0.2)$ interval. Recurrent weights were created with 10% probability and recurrent weight matrix was rescaled to have spectral radius $\lambda = 0.5$ for ESN for FF-ESN models. Also backbone connections for FF-ESN and TDL-I models were set to 0.5. RLS forgetting factor was set to $\gamma = 1.0$. Results are shown in Tab. 5. Results reveal that MSO problem can be modeled by simple linear autoregressive model and hence linear TDL outperforms other models exploiting nonlinearity by several orders of magnitude. All models were trained to significantly better performance than in [9] since they are equipped with linear output units. Because of smaller λ values hidden units operate in linear part of their activation functions and models achieve better results than if higher λ values were used.

4.6 IPIX Radar

IPIX radar data are taken from [9] where the dataset of 2000 values was used as noisy nonlinear real-life prediction task. The first 200 values were used as initial transient inputs, the training was performed using the next 800 values and predictive performance was evaluated on the remaining 1000 values.

Models of 80 units were tested. Input weight were initialized from $(-0.2, 0.2)$ interval. Recurrent weights were created with 10% probability and recurrent weight matrix was rescaled to have spectral radius $\lambda = 0.9$ (for ESN and FF-ESN model). Backbone connections (for FF-ESN and TDL-I models) were set to the same value. RLS forgetting factor was set to $\gamma = 1.0$. Results are shown in Tab. 6. We have achieved slightly

Table 5. The next value prediction results for multiple superimposed oscillator problem

	One Step Least Square Fit	Recursive Least Squares
ESN	1.30×10^{-26} (1.11×10^{-26})	5.64×10^{-12} (2.27×10^{-12})
FF-ESN	6.30×10^{-27} (4.79×10^{-27})	1.34×10^{-12} (3.14×10^{-13})
TDL-I	7.45×10^{-27} (3.59×10^{-26})	5.35×10^{-12} (1.37×10^{-12})
TDL	1.04×10^{-28}	1.90×10^{-24}

Table 6. The next value prediction results for IPIX Radar data

	One Step Least Square Fit	Recursive Least Squares
ESN	0.00079 (0.000043)	0.00071 (0.000039)
FF-ESN	0.00074 (0.000052)	0.00070 (0.000019)
TDL-I	0.00075 (0.000051)	0.00073 (0.000016)
TDL	0.00084	0.00079

better accuracy as in [9] for ESN, FF-ESN and TDL-I models. Also there was no difference between standard training and RLS training performance. Linear TDL model achieved slightly worse results, but still comparable to results of other models.

5   Conclusion

Several ESN model simplifications were suggested in this work and multiple simulations of different tasks taken from literature were performed. Summary of results is shown in Tab. 7. FF-ESN architecture with lower-triangular recurrent weight matrix is only a minor modification to ESN and on most datasets achieved comparable results to standard ESN. Surprisingly this is also the case of TDL-I model with hidden units organized into tapped delay line. This model is based on more simple and more straightforward way of combining inputs presented to the network and performed very well in comparing with standard ESN on multiple tasks. Experiments with TDL model with linear hidden units forming tapped delay line revealed that some tasks used in the community can be simply handled as linear autoregressive models. One should be aware that for some tasks ESN reservoir can serve just as simple memory of inputs. We encourage researchers to compare results obtained using ESN models with standard techniques such as autoregressive moving average model. Using appropriate parameters for ESN reservoir preparation can result into important performance improvement. For most of the tasks from [9] we were able to find better parameters and results differ significantly. Nevertheless searching for good parameters is difficult and time demanding process and more appropriate parameters may still remain undiscovered.

Many improvements were suggested for ESNs in the literature: decoupled echo state networks [9], refined version of the training algorithm [3], working with enhanced states [8], using leaky-integrator units [1] etc. In this work we have tried to simplify ESN architecture to better understand where the performance gains originate. It seems that for some of tasks taken from other papers, no complex reservoir architecture is required to achieve comparable results. Interestingly enough, iterative RLS training was not as

Table 7. Summary of MSE results

One Step Least Squares Fit	ESN	FF-ESN	TDL-I	TDL
Mackey Glass Chaotic Time Series	9.82×10^{-16}	1.84×10^{-15}	1.88×10^{-15}	1.18×10^{-8}
Nonlinear Communication Channel	0.022	0.022	0.018	0.052
NARMA System	0.00091	0.00090	0.00132	0.00189
Predictive Modeling Problem	0.00047	0.00055	0.00058	0.056
Multiple Superimposed Oscillators	1.30×10^{-26}	6.30×10^{-27}	7.47×10^{-27}	1.04×10^{-28}
IPIX Radar	0.00079	0.00074	0.00075	0.00084
Recursive Least Squares	**ESN**	**FF-ESN**	**TDL-I**	**TDL**
Mackey Glass Chaotic Time Series	3.30×10^{-9}	1.77×10^{-9}	3.74×10^{-9}	1.32×10^{-8}
Nonlinear Communication Channel	0.028	0.025	0.026	0.053
NARMA System	0.00084	0.00096	0.00141	0.00189
Predictive Modeling Problem	0.018	0.0074	0.015	0.056
Multiple Superimposed Oscillators	5.64×10^{-12}	1.34×10^{-12}	5.35×10^{-12}	1.90×10^{-34}
IPIX Radar	0.00071	0.00070	0.00073	0.00079

accurate as standard one-step training algorithm. Extremely high precision observed when using ESNs on some tasks is achieved by using one step linear regression for training.

References

1. Jaeger, H.: The "echo state" approach to analysing and training recurrent neural networks. Technical Report GMD 148, German National Research Center for Information Technology (2001)
2. Jaeger, H.: Adaptive nonlinear system identification with echo state networks. In: Becker, S., Thrun, S., Obermayer, K. (eds.) Advances in Neural Information Processing Systems, vol. 15, pp. 593–600. MIT Press, Cambridge (2003)
3. Jaeger, H., Haas, H.: Harnessing nonlinearity: predicting chaotic systems and saving energy in wireless communication. Science 304(5667), 78–80 (2004)
4. Frank, S.L.: Learn more by training less: Systematicity in sentence processing by recurrent networks. Connection Science (in press, 2006)
5. Prokhorov, D.: Echo state networks: Appeal and challenges. In: Proceedings of International Joint Conference on Neural Networks, IJCNN 2005, Montreal, Canada, pp. 1463–1466 (2005)
6. Jaeger, H.: Reservoir riddles: Suggestions for echo state network research. In: Proceedings of International Joint Conference on Neural Networks IJCNN 2005, Montreal, Canada, pp. 1460–1462 (2005)
7. Čerňanský, M., Makula, M.: Feed-forward echo state networks. In: Proceedings of International Joint Conference on Neural Networks IJCNN 2005, Montreal, Canada, pp. 1479–1482 (2005)
8. Jaeger, H.: Adaptive nonlinear system identification with echo state networks. In: Proceedings of Neural Information Processing Systems NIPS 2002, Vancouver, Canada (2002)
9. Xue, Y., Yang, L., Haykin, S.: Decoupled echo state network with lateral inhibition. IEEE Transactions on Neural Network (January 2007) (in press)
10. Wierstra, D., Gomez, F.J., Schmidhuber, J.: Modeling systems with internal state using evolino. In: GECCO 2005: Proceedings of the 2005 conference on Genetic and evolutionary computation, pp. 1795–1802. ACM, New York (2005)

Sparse Coding Neural Gas for the Separation of Noisy Overcomplete Sources

Kai Labusch, Erhardt Barth, and Thomas Martinetz

University of Lübeck - Institute for Neuro- and Bioinformatics
Ratzeburger Alle 160 23538 Lübeck - Germany

Abstract. We consider the problem of separating noisy overcomplete sources from linear mixtures, i.e., we observe N mixtures of $M > N$ sparse sources. We show that the "Sparse Coding Neural Gas" (SCNG) algorithm [1] can be employed in order to estimate the mixing matrix. Based on the learned mixing matrix the sources are obtained by orthogonal matching pursuit. Using artificially generated data, we evaluate the influence of (i) the coherence of the mixing matrix, (ii) the noise level, and (iii) the sparseness of the sources with respect to the performance that can be achieved on the representation level. Our results show that if the coherence of the mixing matrix and the noise level are sufficiently small and the underlying sources are sufficiently sparse, the sources can be estimated from the observed mixtures.

1   Introduction

Suppose we are given a number of observations $X = (\mathbf{x}_1, \ldots, \mathbf{x}_L)$, $\mathbf{x}_j \in \mathbb{R}^N$ that are a linear mixture of a number of sparse sources $S = (\mathbf{s}_1, \ldots, \mathbf{s}_M)^T = (\mathbf{a}_1, \ldots, \mathbf{a}_L)$, $\mathbf{s}_i \in \mathbb{R}^L$ and $\mathbf{a}_j \in \mathbb{R}^M$:

$$\mathbf{x}_j = C\mathbf{a}_j + \epsilon_j \quad \|\epsilon_j\| \leq \delta \,. \tag{1}$$

Here $C = (\mathbf{c}_1, \ldots, \mathbf{c}_M), \mathbf{c}_j \in \mathbb{R}^N$ denotes the mixing matrix. We require $\|\mathbf{c}_j\| = 1$ without loss of generality. The vector $\mathbf{a}_j = (s_{1,j}, \ldots, s_{M,j})^T$ contains the contribution of the sources $\mathbf{s}_i$ to the mixture $\mathbf{x}_j$. Additionally a certain amount of additive noise ϵ_j is present. Is it possible to estimate the sources $\mathbf{s}_i$ only from the mixtures $\mathbf{x}_j$ without knowing the mixing matrix C? In the past, a number of methods have been proposed that can be used to estimate the $\mathbf{s}_i$ and C knowing only the mixtures $\mathbf{x}_j$ assuming $\|\epsilon_j\| = 0$ and $M = N$ [2]. Some methods assume that the sources are statistically independent [3]. More recently the problem of estimating an overcomplete set of sources has been studied which arises if M is larger than the number of observed mixtures N [4,5,6,7]. Fewer approaches have been proposed for source separation under the presence of noise [8]. An overview of the broad field of blind source separation and independent component analysis (ICA) can be found in [9].

In [1] we have proposed the "Sparse Coding Neural Gas" (SCNG) algorithm in order to learn sparse overcomplete data representations under the presence

V. Kůrková et al. (Eds.): ICANN 2008, Part I, LNCS 5163, pp. 788–797, 2008.

of additive noise. Here we show how a slightly modified version of the same algorithm can be employed to tackle the problem of source separation in a noisy overcomplete setting. We do not make assumptions regarding the type of noise but our method requires that the underlying sources $\mathbf{s}_i$ are sufficiently sparse, in particular, it requires that the $\mathbf{a}_j$ have to be sparse and that the noise level δ as well as the number of sources M is known.

1.1   Source Separation and Orthogonal Matching Pursuit

Recently some properties of the orthogonal matching pursuit algorithm (OMP) [10] with respect to the obtained performance on the representation level have been shown [11]. These results provide the theoretical foundation that allows us to apply OMP to the problem of source separation. We here briefly discuss the most important aspects with respect to our work.

Our method does not require that the sources $\mathbf{s}_i$ are independent but it requires that only few sources contribute to each mixture $\mathbf{x}_j$, i.e, that the $\mathbf{a}_j$ are sparse. However, an important observation is that if the underlying components $\mathbf{s}_i$ are sparse and independent, for a given mixture $\mathbf{x}_j$ the vector $\mathbf{a}_j$ will be sparse too.

In order to apply the OMP algorithm to problem (1) let us assume that we know the mixing matrix C. Let us further assume that we know the noise level δ. Let $\mathbf{a}_j$ be the vector containing a small number k of non-zero entries such that

$$\mathbf{x}_j = C\mathbf{a}_j + \epsilon_j \quad \|\epsilon_j\| \leq \delta \tag{2}$$

holds for a given observation $\mathbf{x}_j$. OMP can be used to estimate $\mathbf{a}_j$ by solving

$$\|\mathbf{x}_j - C\mathbf{a}_j^{\mathrm{OMP}}\| \leq \delta . \tag{3}$$

It can be shown that

$$\|\mathbf{a}_j^{\mathrm{OMP}} - \mathbf{a}_j\| \leq \Lambda_{\mathrm{OMP}}\, \delta \tag{4}$$

holds if the smallest entry in $\mathbf{a}_j$ is sufficiently large and the number of non-zero entries in $\mathbf{a}_j$ is sufficiently small. Let

$$H(C) = \max_{1 \leq i,j \leq M, i \neq j} |\mathbf{c}_i^T \mathbf{c}_j| \tag{5}$$

be the mutual coherence H of the mixing matrix C. The smaller $H(C)$, N/M and k are, the smaller Λ_{OMP} becomes and the smaller $\min(\mathbf{a}_j)$ is allowed to be [11]. Since (4) only holds if the smallest entry in $\mathbf{a}_j$ is sufficiently large, OMP has the property of local stability with respect to (4) [11]. Furthermore it can be shown that under the same conditions $\mathbf{a}_j^{\mathrm{OMP}}$ contains only non-zeros that also appear in $\mathbf{a}_j$ [11]. An even globally stable approximation of $\mathbf{a}_j$ can be obtained by methods such as basis pursuit [11,12].

1.2 Optimised Orthogonal Matching Pursuit

The SCNG algorithm is based on an approximation technique that is closely related to OMP and called "Optimised Orthogonal Matching Pursuit" (OOMP) [13].

Given an observed mixture $\mathbf{x}_j$ the algorithm iteratively constructs $\mathbf{x}_j$ out of the columns of the mixing matrix C. The indices of the columns of C that have been used already are collected in the set U. Initially $U = \emptyset$. The number of elements in U, i.e, $|U|$, equals the number of iterations that have been performed so far by the OOMP algorithm. The columns of C that are indexed by U are denoted by C^U. We obtain the approximation of the noise term ϵ_j, i.e., the residual ϵ_j^U, by removing the projection of $\mathbf{x}_j$ to the subspace spanned by C^U from $\mathbf{x}_j$. The residual is initialized with $\epsilon_j^U = \mathbf{x}_j$. The algorithm runs until $\|\epsilon_j^U\| \leq \delta$. A temporary mixing matrix R is used whose columns are orthogonal to the subspace spanned by C^U. Initially $R = (\mathbf{r}_1, \ldots, \mathbf{r}_l, \ldots, \mathbf{r}_M) = C$. Note that R can be obtained by removing the projection of the columns of C to the subspace spanned C^U from C and setting the norm of the residuals $\mathbf{r}_l$ to one. In each iteration the algorithm determines the column $\mathbf{r}_l$ of R with $l \notin U$ that has maximum overlap with respect to the current residual ϵ_j^U

$$l_{\text{win}} = \arg \max_{l, l \notin U} (\mathbf{r}_l^T \epsilon_j^U)^2 . \tag{6}$$

Then, in the construction step, the orthogonal projection with respect to $\mathbf{r}_{l_{\text{win}}}$ is removed from the columns of R and ϵ_j^U

$$\mathbf{r}_l = \mathbf{r}_l - (\mathbf{r}_{l_{\text{win}}}^T \mathbf{r}_l)\mathbf{r}_{l_{\text{win}}} \tag{7}$$

$$\epsilon_j^U = \epsilon_j^U - (\mathbf{r}_{l_{\text{win}}}^T \epsilon_j^U)\mathbf{r}_{l_{\text{win}}} . \tag{8}$$

After the projection has been removed l_{win} is added to U, i.e., $U = U \cup l_{\text{win}}$. The columns $\mathbf{r}_l$ with $l \notin U$ might be selected in the subsequent iterations of the algorithm. The norm of these columns is set to unit length. If the stopping criterion $\|\epsilon_j^U\| \leq \delta$ has been reached, the final entries of $\mathbf{a}_j^{\text{OOMP}}$ can be obtained by recursively collecting the contribution of of each column of C during the construction process taking into account the normalization of the columns of R in each iteration.

Due to the normalization of the columns of R after the orthogonal projection has been performed, the selection criterion (6) ensures that the norm of the residual ϵ_j^U obtained by (8) is minimal. Hence, the OOMP algorithm can provide an approximation of $\mathbf{a}_j$ containing even less non-zeros than the approximation provided by OMP.

2 Learning the Mixing Matrix

We now consider the problem of estimating the mixing matrix $C = (\mathbf{c}_1, \ldots, \mathbf{c}_M)$ from the mixtures $\mathbf{x}_j$ provided that we know the noise level δ and the number of underlying sources M. As a consequence of the sparseness of the underlying sources $\mathbf{s}_i$, we are looking for a mixing matrix C that minimizes the number of non-zero entries of $\mathbf{a}_j^{\mathrm{OOMP}}$, i.e., the number of iteration steps of the OOMP algorithm, given a noise level of δ

$$\min_C \frac{1}{L} \sum_{j=1}^{L} \|\mathbf{a}_j^{\mathrm{OOMP}}\|_0 \quad \text{subject to} \quad \forall j \; : \; \|\mathbf{x}_j - C\mathbf{a}_j^{\mathrm{OOMP}}\| \leq \delta \; . \tag{9}$$

Here $\|\mathbf{a}_j^{\mathrm{OOMP}}\|_0$ denotes the number of non-zero entries in $\mathbf{a}_j^{\mathrm{OOMP}}$. In order to minimize (9), we perform an update of R and C prior to the construction step (7) and (8) in each iteration of the OOMP algorithm. In order to reduce the total number of iterations, this update step minimizes the norm of the residual that is going to be obtained in the current iteration. The norm of the residual becomes small if

$$(\mathbf{r}_{l_{\mathrm{win}}}^T \epsilon_j^U)^2 \tag{10}$$

is large. Hence, we have to consider the optimization problem

$$\max_{\mathbf{r}_{l_{\mathrm{win}}}} \sum_{j=1}^{L} (\mathbf{r}_{l_{\mathrm{win}}}^T \epsilon_j^U)^2 \quad \text{subject to} \quad \|\mathbf{r}_{l_{\mathrm{win}}}\| = 1. \tag{11}$$

An optimization of (11) can be achieved by using Oja's rule [14], which is

$$\mathbf{r}_{l_{\mathrm{win}}} = \mathbf{r}_{l_{\mathrm{win}}} + \alpha \, y(\epsilon_j^U - y \, \mathbf{r}_{l_{\mathrm{win}}}) \tag{12}$$

with $y = \mathbf{r}_{l_{\mathrm{win}}}^T \epsilon_j^U$ and learning rate α. Instead of updating only the winning column of R, i.e, $\mathbf{r}_{l_{\mathrm{win}}}$, we employ the soft competitive learning approach of the "Neural Gas" (NG) algorithm [15] in order to update each column of R that might be selected in the next iteration. We determine the sequence

$$-\left(\mathbf{r}_{l_0}^T \epsilon_j^U\right)^2 \leq \ldots \leq -\left(\mathbf{r}_{l_k}^T \epsilon_j^U\right)^2 \leq \ldots \leq -\left(\mathbf{r}_{l_{M-|U|}}^T \epsilon_j^U\right)^2 \quad , \; l_k \notin U \; . \tag{13}$$

Combining Oja's rule with the soft-competitive update of the NG algorithm, we obtain

$$\Delta \mathbf{r}_{l_k} = \Delta \mathbf{c}_{l_k} = \alpha_t e^{-k/\lambda_t} y \left(\epsilon_j^U - y \, \mathbf{r}_{l_k}\right) \tag{14}$$

Here α_t and λ_t are the learning rate resp. neighbourhood size at time t:

$$\lambda_t = \lambda_0 \left(\lambda_{\mathrm{final}}/\lambda_0\right)^{t/t_{\mathrm{max}}} \tag{15}$$

$$\alpha_t = \alpha_0 \left(\alpha_{\mathrm{final}}/\alpha_0\right)^{t/t_{\mathrm{max}}} \; . \tag{16}$$

Equation (14) corresponds to the update of the NG algorithm with the distance measure

$$D(a, b) = \|a - b\| \tag{17}$$

in the input space replaced by

$$D(a, b) = \frac{\langle a, b \rangle}{\|a\|\|b\|} \tag{18}$$

for determining the rank in the sequence (13). For $t \to t_{\max}$ one obtains equation (12) as update rule. Note that (14) accumulates the updates of all iterations in the learned mixing matrix C. Due to the orthogonal projection (7) and (8) performed in each iteration, these updates are pairwise orthogonal. Furthermore, note that the columns of the original matrix emerge in random order in the learned mixing matrix and due to Oja's rule the learned columns might be multiplied by -1. The entire SCNG method is shown in Algorithm 1.

Algorithm 1. The sparse coding neural gas algorithm for source separation

---

initialize $C = (\mathbf{c}_1, \ldots, \mathbf{c}_M)$ using uniform random values
for $t = 0$ to $t_{\max}$ **do**
  select random sample $\mathbf{x}$ out of X
  set $\mathbf{c}_1, \ldots, \mathbf{c}_M$ to unit length
  calculate current size of neighbourhood: $\lambda_t = \lambda_0 \left(\lambda_{\text{final}}/\lambda_0\right)^{t/t_{\max}}$
  calculate current learning rate: $\alpha_t = \alpha_0 \left(\alpha_{\text{final}}/\alpha_0\right)^{t/t_{\max}}$
  set $U = \emptyset$, $\epsilon^U = \mathbf{x}$ and $R = (\mathbf{r}_1, \ldots, \mathbf{r}_M) = C = (\mathbf{c}_1, \ldots, \mathbf{c}_M)$
  while $\|\epsilon^U\| > \delta$ **do**
    determine $l_0, \ldots, l_k, \ldots, l_{M-|U|}$ with $l_k \notin U$:

$$-(\mathbf{r}_{l_0}^T \epsilon^U)^2 \leq \ldots \leq -(\mathbf{r}_{l_k}^T \epsilon^U)^2 \leq \ldots \leq -(\mathbf{r}_{l_{M-|U|}}^T \epsilon^U)^2$$

    for $k = 1$ to $M - |U|$ **do**
      with $y = \mathbf{r}_{l_k}^T \epsilon^U$ update $\mathbf{c}_{l_k} = \mathbf{c}_{l_k} + \Delta_{l_k}$ and $\mathbf{r}_{l_k} = \mathbf{r}_{l_k} + \Delta_{l_k}$ with

$$\Delta_{l_k} = \alpha_t e^{-k/\lambda_t} y \left(\epsilon^U - y\, \mathbf{r}_{l_k}\right)$$

      set $\mathbf{r}_{l_k}$ to unit length
    end for
    determine $l_{\text{win}} = \arg \max_{l \notin U} (\mathbf{r}_l^T \epsilon^U)^2$
    remove projection to $\mathbf{r}_{l_{\text{win}}}$ from ϵ^U and R:

$$\epsilon^U = \epsilon^U - (\mathbf{r}_{l_{\text{win}}}^T \epsilon^U)\mathbf{r}_{l_{\text{win}}}$$

$$\mathbf{r}_l = \mathbf{r}_l - (\mathbf{r}_{l_{\text{win}}}^T \mathbf{r}_l)\mathbf{r}_{l_{\text{win}}} \quad , l = 1, \ldots, M$$

    set $U = U \cup l_{\text{win}}$
  end while
end for

---

3   Experiments

In order to evaluate the performance of the SCNG algorithm with respect to the reconstruction of the underlying sources, we performed a number of experiments on artificial data. We generated sparse underlying sources $S = (\mathbf{s}_1, \ldots, \mathbf{s}_M)^T = (\mathbf{a}_1, \ldots, \mathbf{a}_L), \mathbf{s}_i \in \mathrm{I\!R}^L, \mathbf{a}_j \in \mathrm{I\!R}^M$. This was done by setting up to k entries of the $\mathbf{a}_j$ to uniformly distributed random values in $[-1, 1]$. For each $\mathbf{a}_j$ the number of non-zero entries was obtained from a uniform distribution in $[0, k]$. The noise

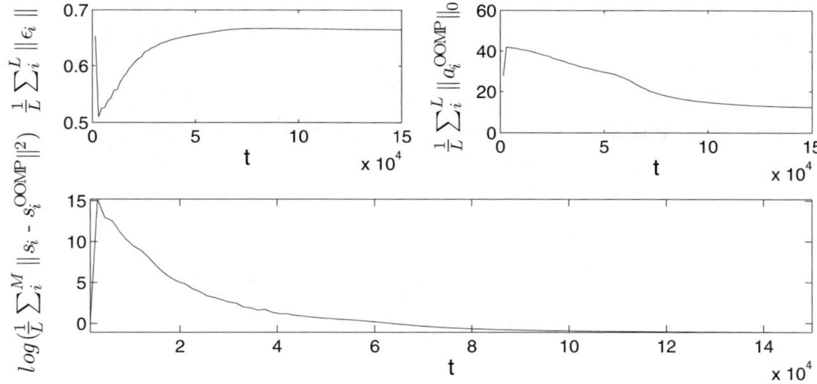

Fig. 1. The figure shows the convergence of the SCNG algorithm. Top left: The mean norm of the residual ϵ_j^U of the last iteration of the OOMP algorithm. Middle: Mean number of iterations performed by the OOMP algorithm until $\epsilon_j^U \leq \delta$. Right: Logarithmic plot of the mean squared distance between the estimated sources s_i^{OOMP} and the true sources s_i. We used $M = 100$, $N = 50$, $NR = 0.01$, $H(C) = 0.4$, $k = 15$ and $\delta = 0.7$.

was generated by adding $E = (\mathbf{n}_1, \ldots, \mathbf{n}_M)^T = (\mathbf{e}_1, \ldots, \mathbf{e}_L), \mathbf{n}_i \in \mathrm{I\!R}^L, \mathbf{e}_j \in \mathrm{I\!R}^M$ containing small uniformly distributed random values in $[-1, 1]$ such that

$$\mathbf{x}_i = C(\mathbf{a}_j + \mathbf{e}_j) = C\mathbf{a}_j + \epsilon_j \ . \tag{19}$$

The noise parameter δ was obtained as

$$\delta = \frac{1}{L} \sum_{j=1}^{L} \|\epsilon_j\| \ . \tag{20}$$

We scaled the S and thereby the $\mathbf{a}_j$ so that $var(CS) = 1$. The amplitude of the values in E was chosen such that

$$\frac{var(CE)}{var(CS)} = NR \ . \tag{21}$$

In order to obtain a random mixture matrix $C \in \mathrm{I\!R}^{N \times M}$ with coherence z, we repeatedly chose a matrix from a uniform distribution in $[-1, 1]$ until $H(C) = z$. Then, the norm of the columns of the mixture matrix was set to unit length.

In our experiments, we study the error on the representatition level. This means that for each observation $\mathbf{x}_j$, we evaluate the difference between the original contributions of the underlying sources, i.e., $\mathbf{a}_j$ to $\mathbf{x}_j$, and the contributions $\mathbf{a}_j^{OOMP}$ that were estimated by the OOMP algorithm on the basis of the mixing matrix C that was learned by the SCNG algorithm

$$\frac{1}{L} \sum_{j=1}^{L} \|\mathbf{a}_j - \mathbf{a}_j^{OOMP}\|_2^2 = \frac{1}{L} \sum_{i=1}^{M} \|\mathbf{s}_i - \mathbf{s}_i^{OOMP}\|_2^2. \tag{22}$$

Here $S^{\text{OOMP}} = (\mathbf{s}_1^{\text{OOMP}}, \ldots, \mathbf{s}_M^{\text{OOMP}})^T = (\mathbf{a}_1^{\text{OOMP}}, \ldots, \mathbf{a}_L^{\text{OOMP}})$ are the underlying sources obtained from the OOMP algorithm. In order to evaluate (22) we have to assign the entries in $\mathbf{a}_j^{\text{OOMP}}$ to the entries in $\mathbf{a}_j$ which is equivalent to assigning the original sources $\mathbf{s}_i$ to the estimated sources $\mathbf{s}_i^{\text{OOMP}}$. This problem arises due to the random order in which the columns of the original mixing matrix appear in the learned mixing matrix. For the assignment we perform the following procedure:

1. Set $I_{\text{orig}} : \{1, \ldots, M\}$ and $I_{\text{learned}} : \{1, \ldots, M\}$.
2. Find and assign $\mathbf{s}_i$ and $\mathbf{s}_j^{\text{OOMP}}$ with $i \in I_{\text{orig}}, j \in I_{\text{learned}}$ such that

$$\frac{|\mathbf{s}_j^{\text{OOMP}} \mathbf{s}_i^T|}{\|\mathbf{s}_i\| \|\mathbf{s}_j^{\text{OOMP}}\|} \text{ is maximal.}$$

3. Remove i from I_{orig} and j from I_{learned}.
4. If $\mathbf{s}_j^{\text{OOMP}} \mathbf{s}_i^T < 0$ set $\mathbf{s}_j^{\text{OOMP}} = -\mathbf{s}_j^{\text{OOMP}}$.
5. Proceed with (2) until $I_{\text{orig}} = I_{\text{learned}} = \emptyset$.

For all experiments we used $L = 10000$ and $\alpha_0 = 0.1$, $\alpha_{\text{final}} = 0.0001$ for the learing rate as well as $\lambda_0 = M/2$ and $\lambda_{\text{final}} = 0.01$ for the neighbourhood size. We repeated all experiments 10 times and report the mean result over the 10 runs. The number of learning iterations of the SCNG algorithm was set to $t_{\text{max}} = 15 * 10000$.

In our first experiment, we evaluated the convergence of the SCNG algorithm over time in case of $N = 50$ observations of $M = 100$ underlying sources with up to $k = 15$ non-zero entries. The result is shown in Figure 1. The norm of the residual of the final iteration ϵ_j^U of the OOMP algorithm converges to δ. The number of iterations of the OOMP algorithm converges to the mean number of non-zero entries. At the same time also the error on the representation level is minimized. The 3 underlying sources that were estimated best as well as 3 of the mixtures from which they were obtained are shown in Figure 2.

In the next experiment, we set $N = 20$, $M = 40$ and $NR = 0.1$ and varied the coherence $H(C)$ of the mixing matrix as well as the sparseness k of the underlying components. The result is shown in Figure 3. The sparser the sources are and the smaller the coherence of the mixing matrix is, the better the obtained performance is. Then, we fixed $H(C) = 0.6$ and $k = 5$ and varied the overcompleteness by setting $M = 20, \ldots, 80$. From Figure 3 it can be seen that only slightly varying performance is obtained though the overcompleteness strongly increases. Furthermore we varied N from 10 to 50 and set $M = 2N$ as well as $k = \lceil N/10 \rceil$. Figure 3 shows that almost the same performance is obtained, i.e., the obtained performance does not depend on the number of sources and observations if the fraction N/M as well as the sparseness of the sources is constant.

The result of the next experiment is shown in Figure 4. We set $N = 20$, $M = 40$ and $H(C) = 0.6$ and varied the noise level as well as the sparseness of the sources. As expected, the more noise is present and the less sparse the sources are, the lower the obtained performance is. In a last experiment, we took the

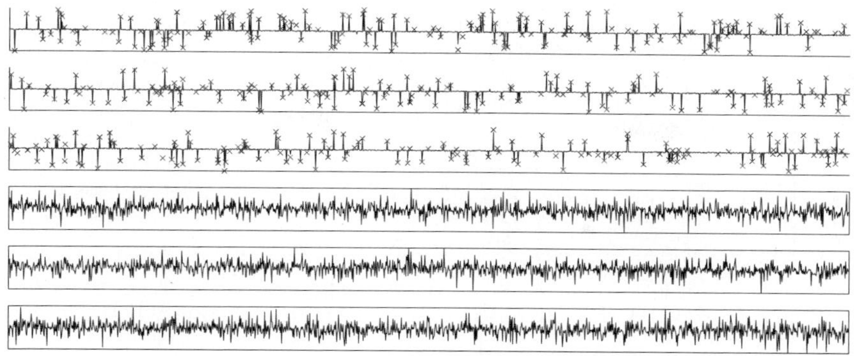

Fig. 2. The upper part of the figure shows the 3 out of $M = 100$ underlying sources that were reconstructed best from the $N = 50$ mixtures observed. 3 examples of the observations are shown in the lower part. In the upper part the solid line depicts $\mathbf{s}_i + \mathbf{n}_i$ whereas the crosses depict $\mathbf{s}_i^{\mathrm{OOMP}}$ that were obtained by applying the OOMP to the mixtures using the mixing matrix that was learned by the SCNG algorithm. The coherence of the mixture matrix was set to $H(C) = 0.4$. The data was generated using $k = 15$, $NR = 0.01$. We used $\delta = 0.7$.

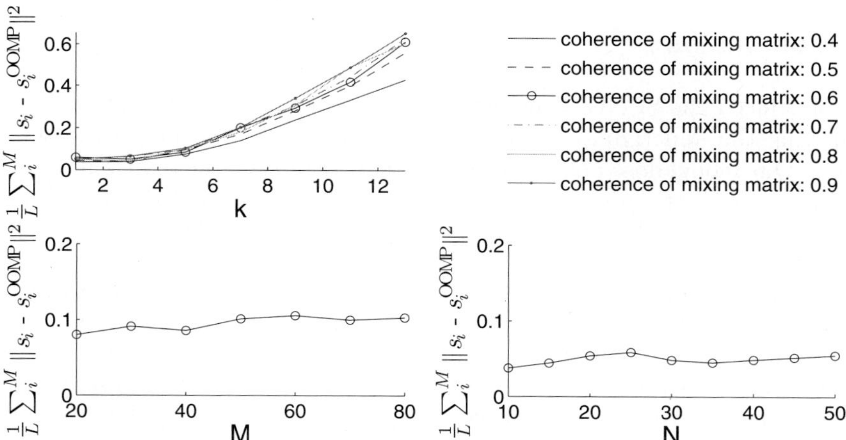

Fig. 3. Top left: The influence of decreasing sparseness and increasing coherence of the mixing matrix with respect to the reconstruction error is shown. We used $N = 20$, $M = 40$ and $NR = 0.1$. Bottom left: The obtained reconstruction error using $M = 20, \ldots, 80$ and $k = 5$. Bottom right: The obtained reconstruction error for $N = 10, \ldots, 50$ with $M = 2N$ and $k = \lceil N/10 \rceil$.

same parameters, set $k = 5$ and studied the obtained reconstruction performance depending on the coherence of the mixing matrix and the noise level. The result is also shown in Figure 4. It can be seen that in our experimental setting the

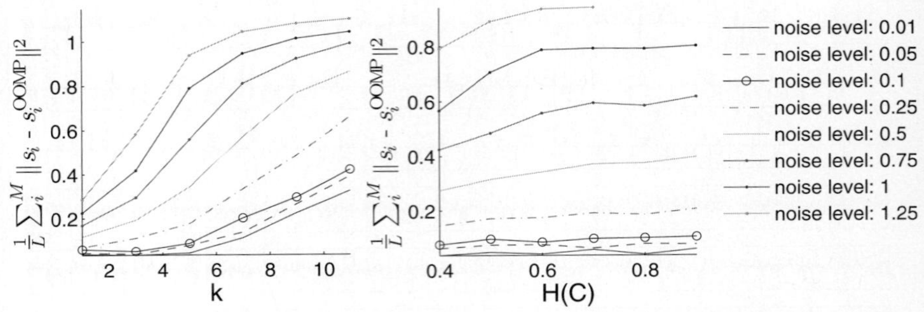

Fig. 4. Left: We used $N = 20, M = 40$. The coherence of the mixing matrix was set to 0.6. The reconstruction performance depending on the noise level as well as on the sparseness is shown. Right: The sparseness parameter k was set to 5. The obtained reconstruction error depending on the noise level and the coherence of the mixing matrix is shown.

noise level has an strong impact on the performance. The influence of the noise cannot be compensated by the coherence of the mixing matrix.

4   Conclusion

We introduced the SCNG algorithm in order to tackle the difficult problem of estimating the underlying sources of a linear mixture in a noisy overcomplete setting. Our model does not make assumptions regarding the distribution of the sources or the distribution of the noise. However, the method requires that the sources are sparse and that the noise level as well as the number of the underlying sources are known or can be estimated.

Based on the mixing matrix that was learned by the SCNG algorithm, we evaluated the performance on the representation level by employing the OOMP algorithm in order to obtain the sources from the observations. We analyzed the performance with respect to the reconstruction of the original sources that can be achieved. We studied the influence of the coherence of the mixing matrix, the noise level and the sparseness of the underlying sources. If the sources are sufficiently sparse and the coherence of the mixing matrix and the noise level are sufficiently small, the SCNG algorithm is able to learn the mixing matrix and the sources can be reconstructed. We also evaluated the influence of the overcompleteness with respect to the obtained performance. Our results show that sufficiently sparse sources can be reconstructed even in highly overcomplete settings. In order to improve the performance on the representation level computationally more demanding methods such as basis pursuit might be used.

References

1. Labusch, K., Barth, E., Martinetz, T.: Learning Data Representations with Sparse Coding Neural Gas. In: Verleysen, M. (ed.) Proceedings of the 16th European Symposium on Artificial Neural Networks, pp. 233–238. D-Side Publishers (2008)
2. Bell, A.J., Sejnowski, T.J.: An information-maximization approach to blind separation and blind deconvolution. Neural Computation 7(6), 1129–1159 (1995)
3. Hyvärinen, A.: Fast and robust fixed-point algorithms for independent component analysis. IEEE Transactions on Neural Networks 10(3), 626–634 (1999)
4. Hyvarinen, A., Cristescu, R., Oja, E.: A fast algorithm for estimating overcomplete ica bases for image windows. In: Proceedings of the International Joint Conference on Neural Networks, IJCNN 1999, vol. 2, pp. 894–899 (1999)
5. Lee, T.W., Lewicki, M., Girolami, M., Sejnowski, T.: Blind source separation of more sources than mixtures using overcomplete representations. IEEE Signal Processing Letters 6(4), 87–90 (1999)
6. Lewicki, M.S., Sejnowski, T.J.: Learning Overcomplete Representations. Neural Computation 12(2), 337–365 (2000)
7. Theis, F., Lang, E., Puntonet, C.: A geometric algorithm for overcomplete linear ICA. Neurocomputing 56, 381–398 (2004)
8. Hyvärinen, A.: Gaussian moments for noisy independent component analysis. IEEE Signal Processing Letters 6(6), 145–147 (1999)
9. Hyvarinen, A., Karhunen, J., Oja, E.: Independent Component Analysis. Wiley-Interscience, Chichester (2001)
10. Pati, Y., Rezaiifar, R., Krishnaprasad, P.: Orthogonal matching pursuit: Recursive function approximation with applications to wavelet decomposition. In: Proceedings of the 27 th Annual Asilomar Conference on Signals, Systems (November 1993)
11. Donoho, D.L., Elad, M., Temlyakov, V.N.: Stable recovery of sparse overcomplete representations in the presence of noise. IEEE Transactions on Information Theory 52(1), 6–18 (2006)
12. Chen, S.S., Donoho, D.L., Saunders, M.A.: Atomic decomposition by basis pursuit. SIAM Journal on Scientific Computing 20(1), 33–61 (1998)
13. Rebollo-Neira, L., Lowe, D.: Optimized orthogonal matching pursuit approach. IEEE Signal Processing Letters 9(4), 137–140 (2002)
14. Oja, E.: A simplified neuron model as a principal component analyzer. J. Math. Biol. 15, 267–273 (1982)
15. Martinetz, T., Berkovich, S., Schulten, K.: "Neural-gas" Network for Vector Quantization and its Application to Time-Series Prediction. IEEE-Transactions on Neural Networks 4(4), 558–569 (1993)

Mutual Information Based Input Variable Selection Algorithm and Wavelet Neural Network for Time Series Prediction

Rashidi Khazaee Parviz, Mozayani Nasser, and M.R. Jahed Motlagh

Computer Eng. Department, Iran University of Science and Technology
p_rashidi@comp.iust.ac.ir, Mozayani@iust.ac.ir, Jahedmr@iust.ac.ir

Abstract. In this paper we have presented an Integrated Wavelet Neural Network (WNN) model and Mutual Information (MI)-based input selection algorithm for time series prediction. Based on MI the proper input variables, which describe the time series' dynamics properly, will be selected. The WNN Prediction model uses selected variables and predicts the future. This model utilized for time series prediction benchmark in NN3 competition and sunspot data. Comprehensive results show that integrated Mutual information based input variable selection algorithm and wavelet network based prediction model, which uses selected variable from lagged value, outperforms other models in prediction of time series.

1 Introduction

Predicting the future in chaotic, high dimensional and nonlinear dynamic systems is a complicated task and needs a lot of efforts [1,2]. To do an efficient prediction, it is necessary to develop advanced algorithms and prediction models. One way, to have a good prediction model in high dimensional and complicated problems, is providing proper data. To do this, relevant variables from input data set must be selected. Relevant input variables selection is one of the most important problems in modeling and prediction tasks. The objectives of input selection algorithms are finding a subset of inputs from original input dataset, which describe system dynamics properly [2,3]. In the prediction tasks proper inputs means selecting the most valuable variables from high dimensional input variables which have maximum dependency with prediction variable and have minimum redundancy. Ding and Peng [4] presented a feature selection algorithm based on minimum redundancy. Peng et al. [5] proposed max-dependency, max-relevance, and min-redundancy criteria and used first order feature selection algorithm for variable Selection. Yousefi et al.[1] presented greedy input selection algorithm for feature selection and used LLNF (Local Linear Neuro Fuzzy) Model for time series prediction. Vahabei et al. [6] used greedy input selection algorithm and LLNF for load forecasting. The LLNF model as general function approximator can be used as a general framework to predict the main patterns of the time series due its great performance in prediction of nonlinear and chaotic time series[1].

V. Kůrková et al. (Eds.): ICANN 2008, Part I, LNCS 5163, pp. 798–807, 2008.

Most of recently published papers used Mutual Information (MI) as similarity measure between variables [1], [4]-[6]. J. Hao focused on MI-based input selection algorithms in her thesis [7] and presented efficient MI estimation algorithms and time series prediction models. Battiti [8] used MI in data analysis and presented feature selection algorithm based on this measure. Compared to the other similarity measures like correlation coefficient, partial correlation coefficient and Linear Regression, MI is very efficient method to estimating the dependency and redundancy between variables in nonlinear problems [1],[6]-[9].

In this paper we proposed an MI based input selection algorithm and wavelet network Model for time series prediction. Wavelet networks are very efficient tools in nonparametric estimation [10] and they have been used for time series prediction successfully [1,6,7]. Chen et al. [11] used Local Linear Wavelet network for time series prediction. A new class of wavelet network structure, which has been presented by Billings et al. [12], will be used as WNN structure in this paper. The proposed model will be used for prediction of the sunspot Dataset [14] and 3rd time series of NN3 forecasting competition time series [13]. Simulation results shows that the proposed model outperforms other models with fewer nodes in hidden layer.

The paper is organized as follow: Mutual Information will be described in section 2. Section 3 deals with wavelet network structure. Proposed model will be presented in section 4. Prediction results are presented in section 5. Finally some conclusions are given.

2   Mutual Information

The MI can be used to evaluate the dependencies between random variables. The MI between two variables, say X and Y, is the amount of information obtained from X in the presence of Y and vice versa. In the time series prediction problem, if Y is the output and S is a subset of the input variables X, the MI between S and Y is one criterion for measuring the dependence between selected inputs and outputs. Thus, we have to select subset S from input set X, which gives maximum MI, to predict the output Y. In information theory, MI can be calculated based on entropy [15]. In this paper discrete variables are used, For discrete variables Px,y represent the joint probability density function between x and y, Px and Py Represent the marginal density functions of x and y respectively.

The entropy of X is defined by Shannon [15] as:

$$H(X) = -\sum_{-\infty}^{+\infty} P_x(x) \log(P_x(x)) \tag{1}$$

Where log is the natural logarithm and then, the information is measured in natural units. The remaining uncertainty of X is measured by the conditional entropy as:

$$H(X|Y) = -\sum_{-\infty}^{+\infty} P_y(y) * \sum_{-\infty}^{+\infty} P_x(x|Y=y) \log P_x(x|Y=y)) \tag{2}$$

The joint entropy is defined as:

$$H(X,Y) = -\sum_{-\infty}^{+\infty} P_{x,y}(X,Y) \log P_{x,y}(X,Y) \tag{3}$$

The MI between variables X and Y is defined [16] as:

$$MI(X,Y) = H(X) - H(X|Y) = H(X) + H(Y) - H(X,Y) \tag{4}$$

From eq. (1) to (4), MI is computed as:

$$MI(X,Y) = -\sum_{-\infty}^{+\infty}\sum_{-\infty}^{+\infty} P_{x,y}(x,y) \log(\frac{P_{x,y}(x,y)}{P_x(x)P_y(y)}) \tag{5}$$

For computing the MI, only estimations of the probability density functions $P_{x,y}$, P_x and P_y are required. J. Hao [7] presented four different MI estimation methods namely: Histogram based estimator, kernel based estimator, k-NN based estimator and the high dimensional volume based estimator. Here we have used Histogram based estimator method for the state of simplicity.

3   Wavelet Network

In terms of wavelet transformation theory, wavelets are defined in the following form:

$$\Psi_i(x) = \frac{1}{\sqrt{d_i}}\Psi(\frac{x-t_i}{d_i}); t_i, d_i \in \Re, i \in Z \tag{6}$$

As shown in (6), wavelets are a family of functions generated from one single function $\Psi(X)$, Mother Wavelet, by the operation of dilation and translation operators. The x represents inputs to the WNN model, and t_i, d_i are translation and scaling parameters, respectively.

The neural network structure, which uses wavelet function as its activation function, is named as Wavenet or wavelet neural network (WNN). Recently developed WNN structure [17] has a linear term instead of its usual structure (Fig. 1.a). Adding linear term to network helps it to make proper function estimation [17].

The output of a new WNN Structure(Fig. 1.a) is given by:

$$f(x) = \sum_{i=1}^{m} w_i\Psi_i(x) + Cx + b \tag{7}$$

Where $\Psi(x)$ is the wavelet activation function of ith unit of the hidden layer and w_i is the weight of connection link, which connecting the ith unit of the hidden layer to the output layer unit, C and b are linear term parameters. Note that for the n-dimensional input space, the multivariate wavelet basis function can be calculated by the tensor product of n single wavelet basis functions as follow:

$$\Psi(x) = \Pi_{j=1}^n \phi(x_j) = \Pi_{j=1}^n \phi(\frac{x_j - t_j}{d_j}) \tag{8}$$

The two parameters d_j and t_j can either be predetermined based on the wavelet transformation theory or be determined by a training algorithm. Localization property of WNN is an interesting feature, which helps the model to predict time series efficiently [11]. The localization of hidden unit is determined by adjusting those parameters. Localized activation of the hidden layer unit is an intrinsic feature of the basis function networks, so that the connection weights associated with the units can be viewed as locally accurate piecewise constant models and their validity for a given input is indicated by the activation functions.

4   Proposed Integrated Prediction Model

In this section proposed input variable selection algorithm and integrated prediction model are presented.

4.1   MI-Based Input Variable Selection Algorithm

Here we present input selection algorithm, which uses MI as similarity measure. This algorithm was presented by Battiti in 1994 for data analysis [8]. First order feature selection algorithm was presented by Peng et al. in 2005 for feature selection [5] and greedy input selection algorithm was presented by Yousefi et al. in 2007 [1]. The main goal of this algorithm is to select a subset of features S from X (inputs), which has minimum redundancy and has maximum relevance with the target class c (output). This algorithm computes $MI(x_i, x_j)$ and $MI(x_i, c)$, where x_i and x_j are individual inputs and c is the target class (output). The algorithm is as follow:

1. First compute $MI(xi, c)$ between input variables and target class.
2. Set S to be empty and L to contain all input variables.
3. Select input variable with largest $MI(xi, c)$ and add it to the selected subset S , then remove it from L.
4. For each remaining inputs:
   (a) Select input variable, which maximizes :

$$|MI(x_i, c) - \frac{1}{|S|} \sum_{x_j \in S} MI(x_i, x_j)| \tag{9}$$

   (b) Add selected input in 4.a to the S and remove it from L.
5. If $|S|$ is less than the desired number of variables then go to step 4.
6. Use variables in S as the selected inputs and exit.

4.2   Integrated Prediction Model

Block diagram in Fig. 1.b shown the integrated, Mutual Information based input selection algorithm and WNN, model for time series prediction.

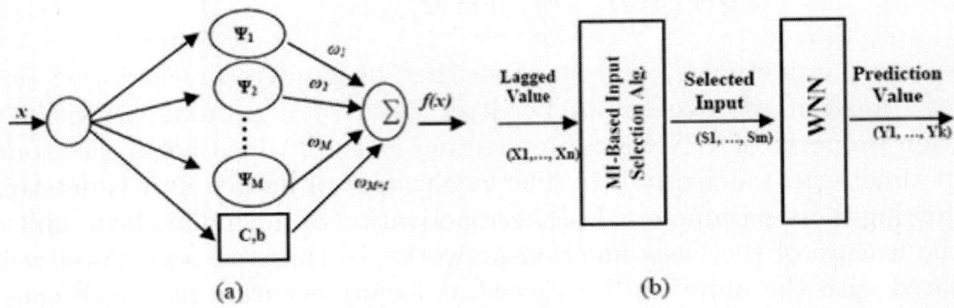

Fig. 1. a) New structure of Wavelet Neural Network (WNN). b) Integrated prediction Model.

As shown in Fig. 1.b at first, the proper Input will be selected by using the Input selection algorithm described in section 4.1. Then, the selected inputs are feed to WNN.

5   Results

For evaluating our method we used two datasets used in [1]: the 3rd times series of NN3 forecasting competition and sunspot data.

5.1   Prediction of the 3rd Time Series of NN3 Forecasting Competition

The first experimentation is adopted from [1] which was used for performance analysis of LLNF prediction model. It was used for prediction of the 3rd time series of reduced data set from NN3 forecasting competitions' time series. In this subsection, prediction performance of proposed model on this dataset is presented. This time series could be downloaded from [13]. The time series includes 125 data. Test set includes the last 13 record of dataset and other records are used for training.

At first MI based input selection algorithm, which presented in section 4.1, selects proper inputs for WNN model (fig. 1.b). This algorithm selects 5 most relevant input variables from 20 lagged values of time series. Table. 1 summarizes the selected inputs based on the proposed method in comparison with other methods such as correlation analysis and gamma test [18].

As shown in Table 1, the five out of twenty input variables are selected to create a five dimensional input vector. After creating the training and test datasets, the orthogonal least square (OLS) algorithm [19] is applied to initialize the parameters of the WNN model. Proper initialization methods help to estimate model parameters correctly and increase its efficiency progressively [20].

Table 1. Selected Input for 3rd time series of reduced data set from NN3 forecasting competitions' time series

Algorithms	5 recent	Gamma Test	Correlation analysis	Proposed Alg.
Selected input	X(t-1)	X(t-1)	X(t-1)	X(t-1)
	X(t-2)	X(t-2)	X(t-11)	X(t-2)
	X(t-3)	X(t-5)	X(t-13)	X(t-11)
	X(t-4)	X(t-13)	X(t-12)	X(t-10)
	X(t-5)	X(t-19)	X(t-18)	X(t-12)

In this paper, the Normalized Mean Square Error (NMSE) is used as the error index with following definition:

$$NMSE = \left(\frac{\sum_{i=1}^{n}(y-\widehat{y})^2}{\sum_{i=1}^{n}(y-\overline{y})^2}\right) \tag{10}$$

y, $\widehat{y}$ and y are observed data, predicted data and average of observed data respectively. Prediction Results of the proposed integrated WNN Model Summarized in Table2. Table 2 shows prediction result of LLNF model in [1] respectively.

Table 2. Prediction Result of 3rd time series from reduced data set from NN3 forecasting competitions' time series

Algorithms	WNN Model			LLNF Model[1]		
	Node Count	Training Error	Test Error	Node Count	Training Error	Test Error
5 recent value	5	0.042	0.045	7	0.3053	0.0916
Correlation analysis	2	0.163	0.078	6	0.0105	0.2067
Gamma Test	3	0.096	0.056	6	0.1995	0.2129
Proposed Alg.	7	0.0037	0.047	4	0.0916	0.1151

Comparison of prediction results shows that the proposed integrated prediction model gives better results than the LLNF prediction Model.

Fig. 2 shows the results of one-step-ahead prediction model in training phase and Fig. 3 shows the results of one-step-ahead prediction model with different input selection algorithms in the test phase. It is shown in Table 2 and Fig. 3 that the proposed input selection algorithm gives the minimum training error, minimum test error and helps the prediction model to forecast efficiently.

5.2   Solar Activity Forecasting

To show the proposed model's efficiency, we applied it to another benchmark dataset, the sunspot numbers. The sunspot number is a good measure of solar activity and is computed according to the Wolf formulation:

$$R = k(10g + s) \tag{11}$$

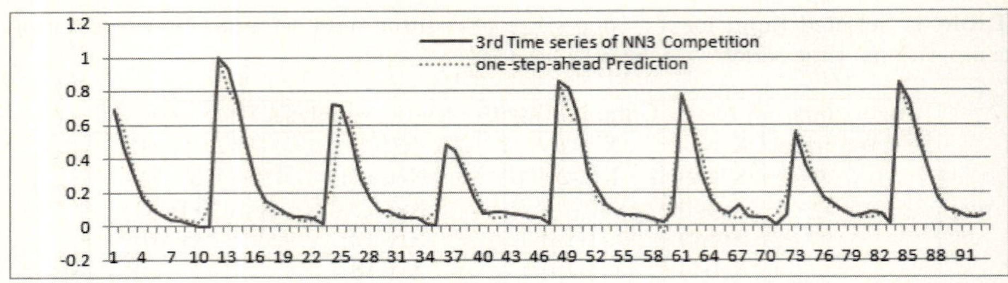

Fig. 2. Training Result of the 3rd time series of reduced data set from NN3 forecasting competitions' time series with WNN Model

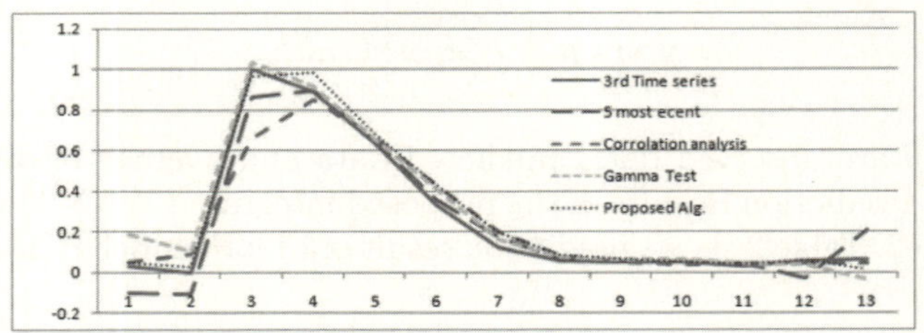

Fig. 3. Test Result of the 3rd time series of reduced data set from NN3 forecasting competitions' time series with WNN Model

Where g is the number of sunspot groups, s is the total number of spots in all groups and k is a variable scaling factor which is related to the conditions of observation. The monthly and yearly average number of sunspots is accessible through several web sites from the sunspot Index, for example Data Center in Belgium [14].

Yousefi et al. [1] used yearly sunspot numbers time series in their paper. We used the same data in order to have a good comparison with their model and also to analyze our proposed model. This dataset has 303 yearly data, from 1700 to 2002. The first 254 years is used for training and the last 49 yearly data is used for testing. The proposed input selection algorithm selects 5 best inputs out of 15 lagged values. Table 3 shows the results of the four different input variables selection methods.

Table 3. Selected Inputs for solar activity time series

Algorithms	5 recent	Gamma Test	Correlation analysis	Proposed Alg.
Selected input	X(t-1)	X(t-1)	X(t-1)	X(t-1)
	X(t-2)	X(t-2)	X(t-9)	X(t-2)
	X(t-3)	X(t-3)	X(t-10)	X(t-5)
	X(t-4)	X(t-5)	X(t-11)	X(t-10)
	X(t-5)	X(t-10)	X(t-12)	X(t-11)

Table 4. Prediction Result of solar activity time series via proposed model

Algorithms	WNN Model			LLNF Model[1]		
	Node Count	Training Error	Test Error	Node Count	Training Error	Test Error
5 recent value	9	0.0819	0.1023	6	0.1095	0.1196
Correlation analysis	2	0.1769	0.1519	8	0.1420	0.2200
Gamma Test	13	0.0559	0.0976	9	0.0981	0.1083
Proposed Alg.	3	0.1073	0.1180	3	0.1136	0.1159

Table 4 shows prediction results of four different input selection method and WNN Model. This Table summarizes the prediction result of LLNF model presented in [1] respectivly. Prediction result shows that integrated WNN prediction model with input selection algorithm gives better result than LLNF prediction model.

Fig. 4 and 5 show training and test result of the WNN model with different input selection methods. It has been show in these figures that the WNN model learns the training data correctly and gives correct prediction in the test phase.

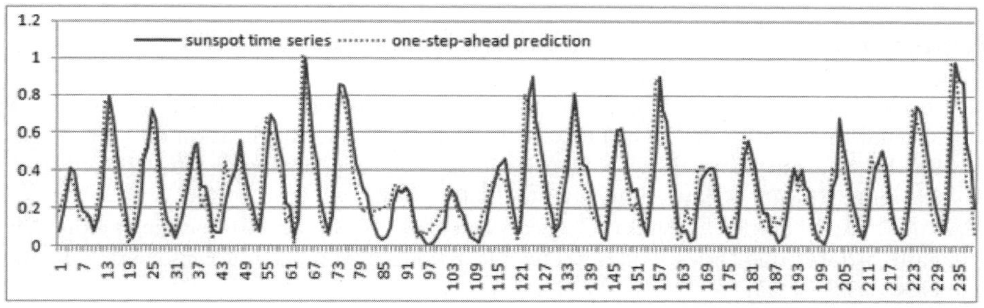

Fig. 4. Training result of Sunspot yearly number

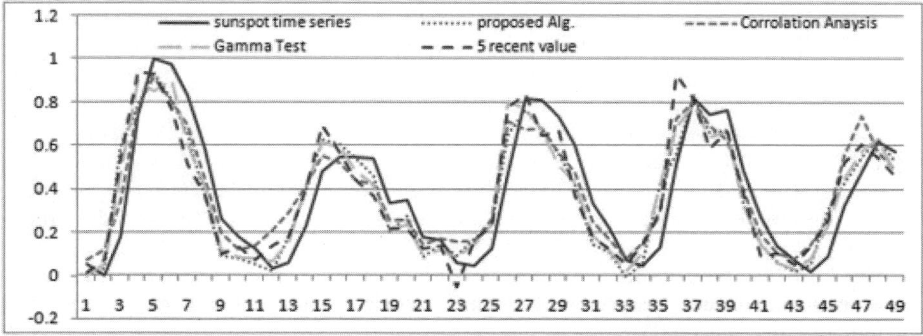

Fig. 5. Test result of Sunspot yearly number

In comparison with prediction results of the first benchmark-3rd time series of NN3 forecasting competition- the second prediction result shows a relative improvement, although it doesn't provide a significant meaning. So for each time series, it is necessary to find best input selection algorithms and prediction models.

6   Conclusion

In this paper, to select the proper inputs for the prediction model, an MI-based input selection algorithm was presented. Then the integrated prediction model based on this input selection algorithm and WNN was proposed. Training and testing results of the two benchmark show that this model can be an efficient prediction model and outperform LLNF Prediction Model. On the other hand, the proposed input selection model, which uses MI as similarity measure, can select relatively relevant input variables.

The prediction results of the two benchmarks show that WNN based prediction model outperforms the LLNF model. It is worth noting that these two models have relatively close results on the second benchmark which means that finding a general predicator needs lots of efforts in the future.

References

1. Rezaei Yousefi, M.M., Mirmomeni, M., Lucas, C.: Input Variables Selection Using Mutual Information for Neuro Fuzzy Modeling with the Application to Time Series Forecasting. In: Proceedings of International Joint Conference on Neural Networks, Orlando, Florida, USA, August 12-17 (2007)
2. Yoon, H., Yang, K., Shahabi, C.: Feature subset selection and feature ranking for multivariate time series. IEEE Trans. On Knowledge and Data Engineering 17(9), 1186–1198 (2005)
3. Guyon, I., Elisseeff, A.: An introduction to variable and feature selection. Journal of Machine Learning Research 3, 1157–1182 (2003)
4. Ding, C., Peng, H.: Minimum redundancy feature selection from microarray gene expression data. Journal of Bioinformatics and Computational Biology 3(2), 85–205 (2005)
5. Peng, H., Long, F., Ding, C.: Feature selection based on mutual information: criteria of max-dependency, max-relevance, and min-redundancy. IEEE Trans. on Pattern Analysis and Machine Intelligence 27(8), 1226–1238 (2005)
6. Vahabei, A.H., Rezaei Yousefi, M.M., Araabi, B.N., Barghinia, S., Ansarimehr, P., Lucas, C.: Mutual Information based Input selection in Neuro-Fuzzy modeling for short-term load forecasting of Iran National Power System. In: 2007 International IEEE Conference on control and Automation, Guangzhou, China, May 30-June 1 (2007)
7. Hao, J.: Input selection using mutual information - applications to time series prediction, Helsinki University of Technology, MS thesis, Dep. of Computer Science and Engineering (2005)
8. Battiti, R.: Using mutual information for selecting features in supervised neural net learning. IEEE Trans. On Neural Networks 5, 537–550 (1994)

9. Al-Ani, A., Deriche, M.: An optimal feature selection technique using the concept of mutual information. In: Int. Symposium on Signal Processing and its Applications (ISSPA), Kuala Lumpur, Malaysia, pp. 477–480 (2001)
10. Zhang, Q.: Using Wavelet Network in Nonparametric Estimation. IEEE, Trans. Neural Networks 8(2) (March 1997)
11. Chen, Y., Yang, B., Dong, J.: Time-series prediction using a local linear wavelet neural network. Neurocomputing 69, 449–465 (2006)
12. Billings, S.A., Wei, H.-L.: A New Class of Wavelet Networks for Nonlinear System Identification. IEEE Transactions On Neural Networks 16(4) (July 2005)
13. http://www.neural-forecasting-competition.com/datasets.htm
14. http://sidc.oma.be/sunspot-data/
15. Shannon, C.E.: A Mathematical theory of communication. The Bell System Technical 27, 379–423, 623–656 (1948)
16. Cover, T., Thomas, J.: Elements of information theory. Wiley, New York (1991)
17. Banakar, A., Azeem, M.F.: Generalized Wavelet Network Model and its Application in Time Series prediction. In: IEEE International Joint conference on Neural Networks (2006)
18. Reyhani, N., Hao, J., Ji, Y., Lendasse, A.: Mutual information and gamma test for input selection. In: European Symposium on Artificial Neural Networks, Bruges, Belgium (2005)
19. Chen, S., Billings, S.A., Luo, W.: Orthogonal least squares methods and their application to nonlinear system identification. Int. J. Control 50, 1873–1896 (1989)
20. Oussar, Y., Dreyfus, G.: Initialization by Selection for Wavelet Network Training. Nerocomputing 34, 131–143 (2000)

Stable Output Feedback in Reservoir Computing Using Ridge Regression*

Francis Wyffels**, Benjamin Schrauwen, and Dirk Stroobandt

Electronics and Information Systems Department, Ghent University, Belgium
`Francis.wyffels@UGent.be`

Abstract. An important property of Reservoir Computing, and signal processing techniques in general, is generalization and noise robustness. In trajectory generation tasks, we don't want that a small deviation leads to an instability. For forecasting and system identification we want to avoid over-fitting. In prior work on Reservoir Computing, the addition of noise to the dynamic reservoir trajectory is generally used. In this work, we show that high-performing reservoirs can be trained using only the commonly used ridge regression. We experimentally validate these claims on two very different tasks: long-term, robust trajectory generation and system identification of a heating tank with variable dead-time.

1 Introduction

Reservoir Computing is a recently developed technique for very quickly training recurrent networks which can be used for trajectory generation, system identification and many other temporal tasks. To accomplish this, Reservoir Computing uses an untrained dynamic system (the reservoir), where the desired function is implemented by a linear, memory-less mapping from the full instantaneous state of the dynamical system to the desired output. Only this linear mapping is learned. These techniques were originally introduced as Liquid State Machines [1] or Echo State Networks [2], but are now commonly referred to as Reservoir Computing [3]. In this work we consider only analog sigmoidal neurons, but the techniques can be broadly applied.

Two important aspects of learning methods in general, is robustness and generalization. When Reservoir Computing is used for trajectory generation, one wants it to autonomously generate stable trajectories for a very long time (optimally forever), even when disturbed by noise. It is possible that a small deviation of the trajectory gets amplified and thus leads to instability. Also if we want to do forecasting or system identification over-fitting becomes important and the reservoir needs to have good generalization capabilities.

In [2] two techniques for improving the performance of Reservoir Computing are introduced: adding noise to the output feedback and state noise injection. The

---

* This research is partially funded by FWO Flanders project G.0317.05 and the Photonics@be Interuniversity Attraction Poles program (IAP 6/10), initiated by the Belgian State, Prime Minister's Services, Sience Policy Office.
** Corresponding author.

V. Kůrková et al. (Eds.): ICANN 2008, Part I, LNCS 5163, pp. 808–817, 2008.

two methods have a regulatory influence, and are able to improve the stability of the trained network although it is not clearly understand why this works [4].

A big advantage of Reservoir Computing is the use of linear regression methods for training the readout. These techniques are well understood and regularization has been thoroughly studied. One way to overcome over-fitting, called ridge regression, also known as Tikhonov regression [5,6], is based on the idea of adding an additional cost term to the least squares optimization so that the norm of the parameters is also kept small. This is related to weight decay, generally used during on-line learning. In [7] a first attempt to use ridge regression in Reservoir Computing has been made for a 2D trajectory generation task. Although ridge regression was not optimally applied: only one fixed regularization parameter was selected.

In this work we will compare these different regularization schemes on two very different tasks: the long-term robust generation of a 2D trajectory, and the identification of a complex system with variable dead-time.

2   Reservoir Computing and Regularization

Although there are some variations on the description of Reservoir Computing, we will use the following equations:

$$\mathbf{x}[n+1] = f\left(W_{\mathrm{res}}^{\mathrm{res}}\mathbf{x}[n] + W_{\mathrm{inp}}^{\mathrm{res}}\mathbf{u}[n] + W_{\mathrm{out}}^{\mathrm{res}}\mathbf{y}[n] + W_{\mathrm{bias}}^{\mathrm{res}}\right)$$
$$\widehat{\mathbf{y}}[n+1] = W_{\mathrm{res}}^{\mathrm{out}}\mathbf{x}[n+1] + W_{\mathrm{inp}}^{\mathrm{out}}\mathbf{u}[n] + W_{\mathrm{bias}}^{\mathrm{out}}, \tag{1}$$

with $\mathbf{x}[n]$ denoting the reservoir's state, $\mathbf{u}[n]$ the input and with $\mathbf{y}[n]$ the desired output at time step n. The actual reservoir output is denoted $\widehat{\mathbf{y}}[n]$. All the weight matrices to the reservoir (denoted $W_{\star}^{\mathrm{res}}$) are fixed and randomly created, while all connections to the output ($W_{\star}^{\mathrm{out}}$) are trained. The non-linearity f we use, is a hyperbolic tangent (sigmoidal), but other common used units are linear, spiking and threshold logic neurons. The reservoir weights $W_{\mathrm{res}}^{\mathrm{res}}$ are scaled such that the dynamic system is operating at the edge-of-stability by normalizing the spectral radius of this matrix [2].

As mentioned in the introduction, training is done by minimizing the error between the desired output and the reservoir output. Simple least squares linear regression methods are generally used.

The generation of time-series (e.g. prediction tasks, trajectory generation,...) is performed by training the reservoir to predict the next time step of the time-series given only the current time-step. The training is done using teacher forcing as presented in equation (1). When we want to use the trained readout for generating time-series, we feed-back the reservoir output into the reservoir.

Note that the initial state is determined by first warming up the reservoir by teacher forcing it with a part of the time-series [8]. In practise this is often done by starting the free-run from the last state of the training sequence.

2.1   Dynamic Noise Injection

The majority of the current work on Reservoir Computing uses the injection of noise in the reservoir during training to improve robustness and generalization [2,4]. This effectively "simulates" what would happen if a small error is made by the linear mapping, and so forces the network to be able to recover from its own mistakes (e.g. learning a broad trajectory).

Two quite similar techniques have been used [2,4]: (1) adding noise with equal variance to all reservoir nodes, or (2) adding noise to the output feedback. This can be written as:

$$\mathbf{x}[n+1] = f\left(W_{\text{res}}^{\text{res}}\mathbf{x}[n] + W_{\text{inp}}^{\text{res}}\mathbf{u}[n] + W_{\text{out}}^{\text{res}}\left(\mathbf{y}[n] + \nu_1[n]\right) + W_{\text{bias}}^{\text{res}} + \nu_2[n]\right),(2)$$

where ν_1 is the output feedback noise and ν_2 is the state noise. In both cases, the noise variance has to be tuned. The only difference between these techniques is the difference of noise variance on the different reservoir states.

In [2] it was shown that when noise is inserted into the network, this leads to an increase of stability for the Mackey-Glass generation task. It was also noticed that the more state noise that is used, the more stable the reservoir is, but at the same time the precision of the prediction decreases.

2.2   Regularising the Readout

The most commonly used scheme for regularization is ridge regression [5]. An extra term, dependent on the readout weights ω is added to the linear regression cost function:

$$J_{\text{ridge}}(\omega) = \frac{1}{2}\left(A\omega - B\right)^T\left(A\omega - B\right) + \frac{1}{2}\lambda\left\|\omega\right\|_2^2, \tag{3}$$

This leads to the following matrix solution:

$$\omega_{\text{ridge}} = \left(A^T A + \lambda I\right)^{-1} A^T B, \tag{4}$$

Matrix A consists of the concatenation of all inputs to the readout, also known as the reservoir states. Matrix B contains the desired outputs.

Ridge regression uses an extra parameter λ, the regularization parameter, which determines the allowed readout weights norm. Note however that this parameter has no absolute meaning. It depends on the magnitude of the correlation matrix $A^T A$, and thus this parameter needs to be optimized for each specific reservoir and cannot blindly be reused or fixed to some arbitrary value.

Because the regularization parameter has no absolute meaning, often the effective number of parameters γ is calculated which gives insight how the generalization performance relates to the expected training set error for nonlinear systems [9]. With σ_i the ith eigenvalue of the $p \times p$ matrix $A^T A$ the number of effective parameters γ can be calculated by

$$\gamma = \sum_i^p \frac{\sigma_i}{\sigma_i + \lambda}, \tag{5}$$

3   Experimental Setup

We will compare different regularization schemes by means of two different tasks. The first task subjects the generation of the figure-eight which is not trivial to solve with recurrent neural networks [7,10]. The second task is the identification of a nonlinear dynamical system with a variable dead time. The reservoir has to be trained to predict the temperature of the outlet of a heating tank in connection to a variable cold water inlet. But, as an additional difficulty, the outlet temperature that has to predicted, is measured after the water flows in long a small tube which introduces a variable dead time. Both experiments are explained in more detail in the following two subsections.

3.1   Trajectory Generation: Figure-Eight

For this task, a reservoir is trained to generate the two output coordinates of the figure-eight which consists of largely 200 sampling points. One of the difficulties of learning a figure-eight lies in the cross-over point which corresponds with a many-to-one-mapping. Fortunately, as been seen in different tasks such as robot localization [11,12], a reservoir is good at memorizing its position.

For this task a reservoir consisting of 100 sigmoidal neurons were randomly connected to each other with a connectivity of 100%. The spectral radius of the reservoir was set to 0.99 and each neuron had an auxiliary input with a bias sampled from the uniform interval $[0; 0.1]$ which makes the reservoir slightly more nonlinear. We used Butterworth filter neurons [13] with following cut-off frequencies: f_{low} uniformly spread from 0 to 0.049 $\times\pi\mathrm{rad/sample}$ and $f_{\mathrm{high}} = f_{\mathrm{low}} + 0.02 \times \pi\mathrm{rad/sample}$. Additional, we will consider all experiments using leaky integrator neurons [2] with a leak-rate set to 0.05. There was feedback from the output to all the neurons with a randomly choosen scaling factor of -0.005 or 0.005. Each time, the network was trained using 3 figure-eight examples and after a warming up period of 400 samples the reservoir was left running freely 99,000 samples what corresponds with nearly 450 figure-eight periods. During freerun we will do both adding Gaussian distributed state noise (zero mean and variance 0.0001) and adding no state noise. The addition of state noise in freerun is used to test wether the generated figure-eight is dynamically stable (in the sense of being a periodic attractor) [7].

The values of the parameters introduced by the regularization schemes (e.g. state noise in equation 2, regularization parameter in equation 4) are dependent of the experiment and will be discussed in section 4.

3.2   The Identification of a Dynamical System with Variable Dead-Time

This task concerns the prediction of the temperature of the water outlet of a heating tank in connection to a variable cold water inlet. The outlet temperature is measured after flowing through a long tube. Because the system is closed (the output flow is equal to the variable input flow) the system has a variable dead-time which is an additional difficulty in predicting the outlet tempearture at the right moment. The heating tank contains a heating element with a constant power, the outlet temperature is controlled by varying the input flow of cold water. A full description of the plant can be found in [14]. The reservoir needs to predict the (variable) delayed output temperature given an input flow of cold water. Due to the slow dynamics of the heating tank system, a time-step of 4 s will be used.

A reservoir of 400 randomly connected neurons was created, and a (arbitrary choosen) connection rate of 90% was used. Just like the previous task, the spectral radius was set to 0.99 and each neuron had an auxiliary input with a constant bias sampled from the uniform distribution $[0; 0.9]$ which gives the reservoir more nonlinear properties. Additional there was feedback from from the output to all the neurons with a scaling factor of 0.01 this because the mean output was too big compared to the input. The feedback was needed to increase the (fading) memory of the reservoir which gives the reservoir extra knowledge about its current state. Due small input values, the input was scaled up by factor 10. For this task, Butterworth filter neurons were used which cut-off frequencies were set in following manner: f_{low} uniformly spread from 0 to 0.02 $\times \pi\text{rad/sample}$ and $f_{\text{high}} = f_{\text{low}} + 0.01 \times \pi\text{rad/sample}$. The network was trained using 8,000 samples of random input-output examples extracted by simulation. Next, the reservoir was left predicting 3,000 samples in connection to its input, 1,000 samples were discarded for warming up.

4   Experimental Results

4.1   Comparing Ridge Rigression and Dynamic Noise Injection

We will first compare the performance of both ridge regression and dynamical noise injection during training mode. Therefore, the two previously discussed tasks were considered using both schemes. The results are presented in Figure 1. For the generation of the figure-eight we considered two cases. In the first case the figure-eight generation was done without adding noise during freerun mode which is denoted by a solid black line. For the second case of the figure-eight generation we added Gaussion distributed noise with zero mean and variance 0.0001 during freerun which tests the dynamical stability of the generated trajectory [7]. This is denoted by a black dotted line. The results of the system identification task are indicated with dots. Each experiment was repeated 25 times, each time using a randomly connected reservoir. Both regularization methods were used separatedly from each other, each time calculating the NMSE for a broad range of the parameter (e.g. regularization parameter or state noise variance).

At the top left of Figure 1 the averaged NMSE is given for the figure-eight generation task using ridge regression. Because the regularization parameter has no absolute meaning we averaged over the effective number of parameters as presented in equation 5. At the top right, the same experiment is shown but now with state noise injection during training. When no noise was injected during

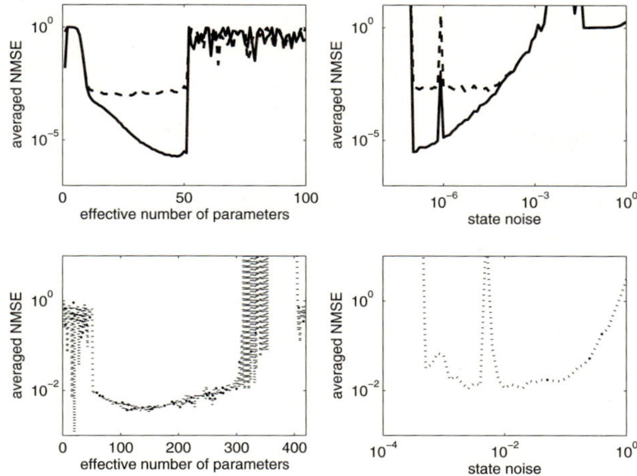

Fig. 1. Ridge regression versus state noise injection during training: on the left the averaged NMSE for three tasks is shown (at the top, solid line: generation figure-eight without noise injection during freerun mode, dotted line: generation figure-eight with noise injection in freerun mode, at the bottom, dots: system identification task). On the right the same tasks are shown but now with state noise injection during training. We see better overall performance using the ridge regression method. One can also observe that more regularization (less number of effective parameters) is needed when dynamical noise injection during freerun mode is performed.

freerun mode a factor three improvement was obtained by using ridge regression instead of state noise injection during freerun mode. When noise was injected during freerun, the increase in performance by using ridge regression was less spectacular but still more than 30%. Moreover the regularization parameter can be optimized much faster than the state noise parameter because only one simulation of the reservoir is needed. Only the readout weights have to be trained with different settings for the regularization parameter. Another interesting observation is that the optimal effective number of parameters shifts towards the left when noise injection during freerun is used. Thus less effective parameters are used which means more regularization is done (a bigger value of the regularization parameter has to be chosen), thus more efforts are made to stay stable. The figures are given for reservoirs using band-pass neurons but we came to the same conclusions by using leaky integrator neurons.

In this paper we also investigated what happens if both methods, ridge regression and noise injection during training mode, are combined. Therefore we used noise injection during training, ranging the variance of the noise in $10^{[0:-0.1:-15]}$[1], and for each value of the noise variance we ranged the regularization parameter over the same interval calculating each time the NMSE over 30 figure-eight periods. Then, for each value for the variance of the noise, an optimal regularization

---

[1] Which means that the variance ranged from 1 to 10^{-15}, each time by decreasing the exponent with -0.1.

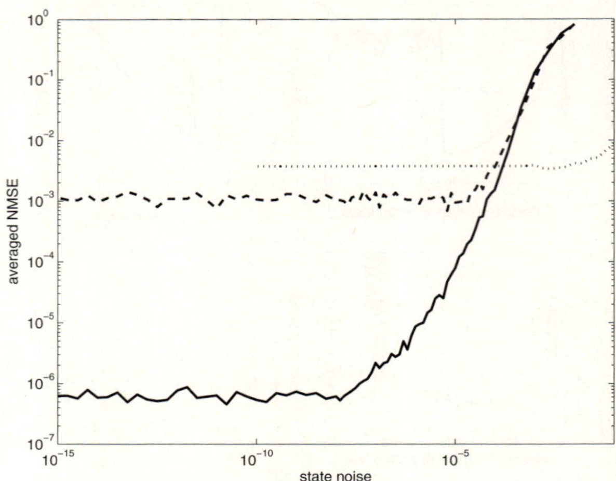

Fig. 2. Averaged NMSE is shown for two different tasks in function of the state noise injection during training mode, each time using an optimal regularization parameter. For the figure-eight generation task we can see that adding state noise has no beneficial influence on the performance. For the system identification task we see that a slightly improved performance can be obtained when state noise injection during training is combined with ridge regression.

parameter was chosen, based on the NMSE calculations. Finally simulation was redone during 30 full rides along the figure-eight, using the optimal regularization parameter and a new NMSE value was calculated with the generated output of the reservoir. All these steps were repeated using 20 randomly connected reservoirs. For the two different tasks, results are given in Figure 2. For the figure-eight generation task, one can see that when the state noise injection during training becomes small, the NMSE remains constant. For the system identification task we see that state noise injection during training and ridge regression can be combined in order to get the best performance. But the performance boost is rather weak while the optimization of the state noise in combination with ridge regression is a computationally intensive task.

4.2   The Capabilities of Reservoir Computing Using Ridge Regression

We have seen that when ridge regression is used, no state noise injection is needed in the training phase for the figure-eight generation task. Better performance is obtained with ridge regression. An additional benefit is that for optimizing the regularization parameter, you don't need to simulate the reservoir with different state noise configurations. You only need to calculate the readout weights according equation (4) using different settings of the regularization parameter.

   In order to show the capabilities of ridge regression we put Reservoir Computing through a severe test: the figure-eight generation task with noise addition

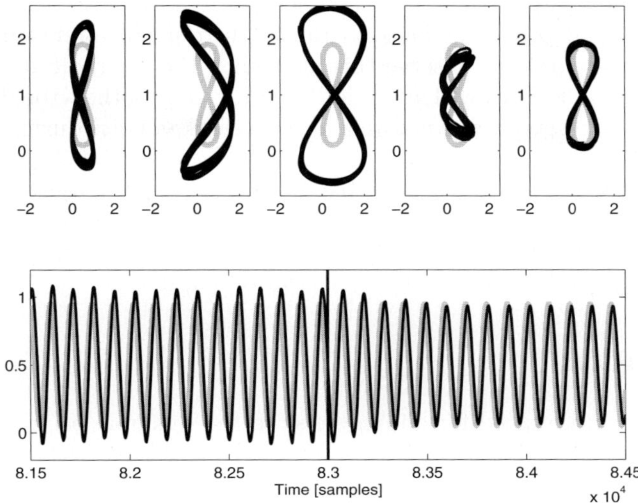

Fig. 3. At the top, five examples of the figure-eight generated with Reservoir Computing using ridge regression are shown. In order to test the dynamical stability Gaussian noise was added to the reservoir states. The shown figure-eights (solid black lines) are approximately 10 full rides along the figure-eight after running more than 400 figure-eight periods. In the background, the teacher eight is shown in gray. At the bottom, one of the two outputs generated by the reservoir is given. Here at sample 83,000 the variance of the Gaussian noise in freerun mode was set from 0.001 to zero. We see some convergence back to the desired signal.

during freerun to the reservoir states sampled from a Gaussian distribution with zero mean and variance 0.001. This task was repeated 25 times using another randomly created reservoir. Each time the optimal regularization parameter λ was searched in the range $10^{[-4:-0.25:-14]}$. Only the last 2,500 samples of the freerun phase were considered presenting approximately the 10 full rides along the figure-eight when already 440 figure-eight periods were autonomously generated by the reservoir. We have seen that 17 out of 25 runs gave proper figure-eights, some with slightly bigger amplitude. Five examples of this are shown in Figure 3. In none of the 25 runs the amplitude of the output exploded, but sometimes the two outputs weren't properly coupled anymore. Although the examples in Figure 3 were generated using a reservoir of band-pass neurons, similar results were seen using leaky integrator neurons.

We have also observed that the reservoir is able to correct itself when the noise variance is reduced which is also illustrated in Figure 3. One of the two outputs is shown when the state noise is switched off. For that, a new experiment was done, in which the variance of the Gaussian noise added during freerun was set to zero (and zero mean) after freely running 83,000 samples with Gaussian noise injection with variance 0.001 (and zero mean). On Figure 3 samples 82,000 to 84,000 are shown. From sample 82,000 to 83,000 we see some deviation from its desired behavior, actually we see the generated output "dancing" around the desired output. From sample $83,000$, when the noise is set to zero, one can see that the signal converges back to the learned signal after some transient behavior. We were able to observe this also when the reservoir states were fed with a constant value. But, we have to add that for some of the wrongly generated figure-eights, the reservoir was not able to recover when noise during freerun was set to zero.

By using ridge regression we were able to generate stable trajectories. Because all the readout weights are kept small, the reservoir becomes robust against noise. Moreover, fast optimization is possible because the simulation of the reservoir during training is invariant to the regularization parameter.

5 Conclusions

In previous work, the need for adding noise to the dynamic trajectory of reservoirs was assumed to be needed in order to get stable trajectory generation. In this work we clearly showed that high performing reservoirs can be trained using only the very simple and commonly used ridge regression regularization scheme on the readout. Both short-term high precision and/or long-term stable trajectories can be generated in this way.

This was demonstrated on two very different tasks: long term generation of a stable 2D trajectory, and the identification of a complex system with variable dead-time. On both tasks the superiority of ridge regression was demonstrated. Ridge regression not only performs better, but also leads to less inter-trial variance; is much better understood and forms the basis of e.a. kernel methods. Moreover, the regularization parameter can be optimized much faster. These results show that regularizing the linear readout not only keeps the weight small, but also regularizes the trained trajectory in state-space. Thanks to the well understood theory behind the ridge regression scheme, these findings can in future work lead to a better theoretical understanding of the effect of readout regularization on the regularization of the state-space attractor.

References

1. Maass, W., Natschläger, T., Markram, H.: Real-time computing without stable states: A new framework for neural computation based on perturbations. Neural Computation 14(11), 2531–2560 (2002)
2. Jaeger, H.: The "echo state" approach to analysing and training recurrent neural networks. Technical Report GMD Report 148, German National Research Center for Information Technology (2001)
3. Verstraeten, D., Schrauwen, B., D'Haene, M., Stroobandt, D.: An experimental unification of reservoir computing methods. Neural Networks 20, 391–403 (2007)
4. Jaeger, H.: Tutorial on training recurrent neural networks, covering BPTT, RTRL, EKF and the "echo state network" approach. Technical Report GMD Report 159, German National Research Center for Information Technology (2002)
5. Hoerl, A., Kennard, R.: Ridge regression: Biased estimation for nonorthogonal problems. Technometrics 12, 55–67 (1970)

6. Tikhonov, A., Arsenin, V.: Solution of Ill-posed Problems. Winston & Sons (1977)
7. Jaeger, H., Lukosevicius, M., Popovici, D., Siewert, U.: Optimization and applications of echo state networks with leaky-integrator neurons. Neural Networks 20, 335–352 (2007)
8. Feldkamp, L., Prokhorov, D., Eagen, C., Yuan, F.: Enhanced multi-stream Kalman filter training for recurrent neural networks. In: Nonlinear modeling: Advanced black-box techniques edn. Kluwer, Dordrecht (1998)
9. Moody, J., Hanson, S., Lippmann, R.: The effective number of parameters: An analysis of generalization and regularization in nonlinear systems. Advances in Neural Information Processing Systems 4, 847–854 (1992)
10. Zegers, P., Sundareshan, M.K.: Trajectory generation and modulation using dynamic neural networks. IEEE Transactions on Neural Networks 14, 520–533 (2003)
11. Antonelo, E., Schrauwen, B., Dutoit, X., Stroobandt, D., Nuttin, M.: Event detection and localization in mobile robot navigation using reservoir computing. In: de Sá, J.M., Alexandre, L.A., Duch, W., Mandic, D.P. (eds.) ICANN 2007. LNCS, vol. 4669, pp. 660–669. Springer, Heidelberg (2007)
12. Antonelo, E.A., Schrauwen, B., Stroobandt, D.: Experiments with reservoir computing on the road sign problem. In: Brazilian Congress on Neural Networks (CBRN) (2007)
13. Wyffels, F., Schrauwen, B., Verstraeten, D., Stroobandt, D.: Band-pass reservoir computing. In: Proceedings of the International Joint Conference on Neural Networks (accepted, 2008)
14. Cristea, S., de Prada, C., De Keyser, R.: Predictive control of a process with variable dead-time. In: CD-Proceedings of the 16th IFAC World Congress (2005)

Spatio-temporal Summarizing Method of Periodic Image Sequences with Kohonen Maps

Mohamed Berkane, Patrick Clarysse,
and Isabelle E. Magnin

CREATIS-LRMN, CNRS UMR 5220, Inserm U630,University of Lyon,
Lyon, France
{Mohamed.Berkane,Patrick.Clarysse,Isabelle.Magnin}@creatis.insa-lyon.fr
http://www.creatis.insa-lyon.fr

Abstract. This work relates to the study of periodic events in medical imaging. Currently, biological phenomena exhibiting a periodic behaviour such as the heart motion are observed through the continuous recording of signals / images. Indeed, due to various reasons, cycle duration may slightly vary in duration and magnitude. It is important for understanding to be able to extract the meaningful information from the mass of acquired data. This paper presents a new neural-based method for the extraction of summarized cycle from long and massive recordings. Its concept is simple and it can be implemented on a hardware architecture to make the process very fast. The proposed method is demonstrated on sequences of noise-free and noisy synthetic images eventually in the presence of artefacts.

Keywords: Neural network, periodic motion, image sequence, summarized sequence.

1   Introduction

Computer vision systems can be used to analyze motion in natural scenes based on dedicated simple to very sophisticated algorithms. In this paper, we consider the particular and important class of periodic and quasi-periodic events. Such phenomena can often be observed in biomedical imaging. We will give three examples. The first one relates to the analysis of the periodic heart motion from dynamic imaging in Magnetic Resonance Imaging (MRI), ultrasounds or X-rays Computed Tomography (CT). For instance, cardiac ultrasonic imaging results in temporal image sequences that depict the evolution of the hearts anatomy over cardiac cycles thanks to a frame rate of, at least, 25 images per second [1]. For diagnosis purpose, the clinician intends to infer an average behaviour on one cardiac cycle from this continuous image sequence. The Electrocardiographic (ECG) signal is another example where specific cardiac events have to be extracted from a continuous signal recording over several tenths of cardiac cycles. A third example issues from the imaging of brain functions in MRI. In conventional experiments, human subjects are asked to periodically repeat a mental or

V. Kůrková et al. (Eds.): ICANN 2008, Part I, LNCS 5163, pp. 818–826, 2008.

active action while imaging their brain in order to detect the activated regions. Such a study usually results in continuous recording of sequences of action and rest periods which have to be jointly analyzed to determine the most reactive regions.

These examples illustrate two different event natures; cardiac motion is autonomous and reflex controlled while the other is provoked and triggered by an commandment. Note that the observed phenomena are not strictly periodic that is cycles can slightly differ in duration. However, all the three cases result in a considerable amount of signal/image sequences from which meaningful information must be retrieved. In particular, we are interested on the one hand to characterize the average status/motion of the heart over the cardiac cycle and determine eventual anomalies and, on the other hand, to detect prominent activated regions in the brain. The idea behind the acquisition of multiple event cycles is that both acquisition system imperfections and temporal variability of the observed phenomenon can be somehow taken into account when extracting the useful information.

We propose a connectionist approach to pre-process the image sequence in order to extract the dominant behaviour. Neural networks represent an important set of methods for analysis and data processing characterized by the combination of simple elements. Neural based methods differ mainly by their topologies and the algorithms used for training. Their operating mode relies on the distribution of the weights on the network connections, the data propagation strategy during the training phase and the exploitation phase. The proposed approach relies on Kohonen maps (or self organizing maps) [2, 3]. Kohonen maps are unsupervised networks with competitive learning which not only learn how to model the space of the inputs from prototypes, but also build a map with one or two dimensions allowing to structure this space. A Kohonen map based operator is designed to summarize dominant moving patterns in quasi-periodic image sequences. It is applied to one of the above introduced cardiac example to demonstrate the interest of the proposed approach.

2   Proposed Method

We consider as input to the problem a sequence of signals or images which sample several periods of an observed pseudo-periodic phenomena. In the context of this paper, image sequences representing the motion of the heart during several cardiac cycles will be studied in the experiments.

The architecture of the proposed neural network consists in three layers: an input layer, an intermediate layer which represents a one-dimensional Kohonen map and an output layer (**fig 1**). Each neuron of the input layer is connected to all neurons of the Kohonen map. In the same way, each neuron of the Kohonen map is connected to all neurons of the output layer. Cycles of the image sequence are iteratively presented to the input layer.

Let $e_1..e_i..e_n$ the input neurons, where n represents the number of pixels per image; let $c_1 \ldots c_t \ldots c_T$ be the neurons of the self-organizing (SO) map with T the number of images per sequence cycle (at least an average) and let $s_1 \ldots s_i \ldots s_n$ the output neurons. The connections between the neurons of the SO map and the input neurons are weighted by: w_i^t with $i=1..n$ and $t=1..T$. The set of w_i^t forms a matrix noted W. Similarly, the connections between the neurons of the SO map and the output neurons are represented through the weight matrix P with element p_i^t with $i=1..n$ and $t=1..T$.

The network is initialized by assigning the weight of each neuron of the Kohonen map with the pixel gray level value of the images of the first cycle. The training consists in iteratively presenting images and searching for the winning neuron which presents the closest values to the input values (fig 1, up). For the updating of the weights of the winner, we applied the Winner Take All (WTA) strategy. In other words, there is no interaction between the winner and its neighbours. Once the winner is selected, it will be penalized for the next search steps during the processing of the current cycle. The purpose of the weight modification of this neuron is to converge towards values, most adequately representing the corresponding image, according to measurements issued from the acquisition of the several cycles. At the end of the process, the neurons of the self organizing map carry an image sequence, called *summarized sequence* (fig 2), which will be further exploited in a second phase to deduce the dominant motion in the sequence.

3   Implementation

3.1   Initialization Phase

During the initial phase, the intensity values of the pixels of the images of the first cycle, noted $I_1^t \ldots I_i^t \ldots I_N^t$ with $t=1..T$, are assigned to the connection weights w_i^t between the neurons of the SO map and the input neurons:

$$w_1^1 \leftarrow I_1^1 \ldots w_1^t \leftarrow I_1^t \ldots w_1^n \leftarrow I_1^n$$
$$w_i^1 \leftarrow I_i^1 \ldots w_i^t \leftarrow I_i^t \ldots w_i^n \leftarrow I_i^n$$
$$w_n^1 \leftarrow I_n^n \ldots w_n^t \leftarrow I_n^t \ldots w_n^n \leftarrow I_n^n \tag{1}$$

3.2   Learning Phase

The whole training set is composed of the images of all the cycles following cycle 1. The first values presented to the input neurons are the intensity values of the first image of the second cycle, then the second image, and so on and so forth until the image T of the last cycle. The global shift between the weights carried by neurons connections of the SO map and the neuron input values is computed according to equation (2):

$$\Delta_t = \frac{T}{\sum_{j=1}^{T} \sum_{i=1}^{n} \left(w_i^t - e_i \right)^2} \sum_{i=1}^{n} \left(w_i^t - e_i \right)^2 + s(t) \tag{2}$$

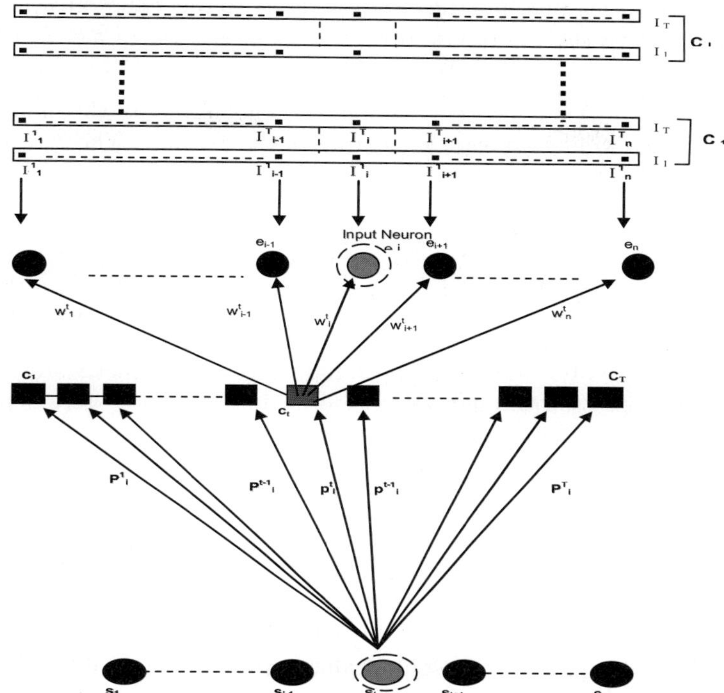

Fig. 1. Topology of the proposed network: black squares represent the neurons of the Kohonen map and the plain circles represent neurons belonging either to the input or output layers

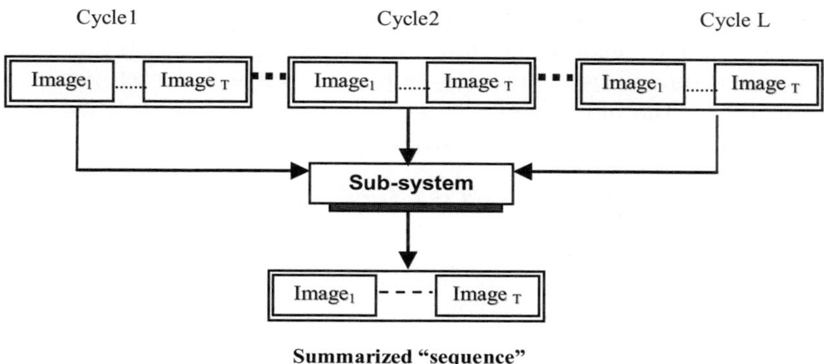

Fig. 2. Inputs and outputs of the first phase of the proposed system

With $s(t)$ a penalty function attributed to each winner neuron in order to minimize the likelihood it is again selected as the winner during the processing of the same cycle. The winner neuron is the one which presents the minimal difference, or equivalently, which presents the closest weights to the values presented at the input neurons. Lets set:

$$\Delta_{t^*} = Min\left(\Delta_t\right) \tag{3}$$

Δ_{t^*} is the minimal value corresponding to the neuron t^* of the map, which is considered as the winner among the neurons of the SO map. The weights of this neuron are modified according to:

$$w_i{}^{t^*} = w_i{}^{t^*} + f_i\left(\Delta_{t^*}\right) \ for \ i = 1\ldots n \tag{4}$$

$$f_i\left(\Delta_{t^*}\right) = \left(w_i{}^{t^*} - e_i\right)/\lambda \ for \ i = 1\ldots n \tag{5}$$

The result of this phase, is a weight matrix, W, which represent a kind of consensus sequence of T images of the L input cycles. In other words, the periodical motion expressed through L cycles of T images will be represented by a sequence W, namely the *summarized sequence*.

4   Exploitation

This phase consists in exploiting matrix W to estimate the dominant motion at each pixel. Matrix P represents the weights of connections between the neurons of the SO map and the neurons of the output layer. This matrix have as elements p_i^t with *t=1..T* and *i =1..n* (T and n being respectively: the number of images per sequence and the number of pixels per image). The value p_i^t represents the intensity value of pixel i in image t. This implies that the matrix P represents the intensity values of all pixels in all images of the sequence resulting from the training phase. To detect the motion at each pixel, a differential method is used. The intensity change between image *t-1* and the image *t*, is deduced by comparing, locally, the values of the connection weights between the neurons of the output layer and those of the chart. So, each neighbour answering positively to the comparison will be balanced by a positive value. For each pixel, we will obtain for result, the most probable displacement of the pixel. This process is repeated for all the neurons of the output layer.

$$if \ \left|p_i^{t+1} - p_i^t\right| > \epsilon \ for \ i = 1\ldots T$$
$$diff_i \leftarrow diff_i + 1 \tag{6}$$

5   Results

Our approach has been experimented on synthetic sequences which present the advantage to provide a perfectly known reference.

5.1   Synthetic Image Sequences

Synthetic image sequences of a beating heart have been created to test our approach. They are based on an analytical model of a beating heart observed

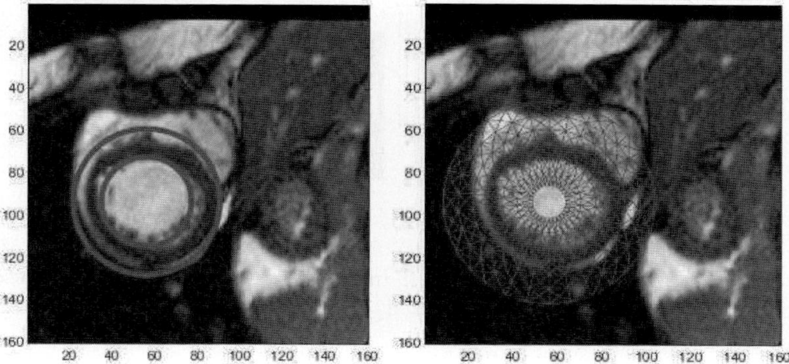

Fig. 3. Principle of the generation of the synthetic cardiac sequence. (a) Concentric circles delimit the deformation area in a MR image of the left ventricular myocardium (b) Superimposition over the image of a Finite Element mesh to which the analytical motion is applied.

through an image slice [4]. Two concentric circles, delimiting the left ventricular myocardium, are positioned on a real 2D Magnetic Resonance Image (MRI) (fig. 3). The ring is regularly partitioned into triangular elements to which the transformation is applied. The motion law integrates several components: global rigid and local rotations, myocardial thickening, elliptization. It is designed to avoid discontinuities at the edges of the grid and to obtain a complete smooth image sequence over the entire cardiac cycle. Two periodic sequences have been generated with 7 cardiac cycles. A first one was obtained by adding at each cycle a variable Gaussian noise (seq1 is illustrated in fig. 4). A second sequence was obtained by adding both Gaussian noise and various artefacts (seq2, fig. 5). Each sequence consists in 7 periods (or cycles), each composed of 22 images. Our neural method is then applied to the sequences to extract the *summarized sequences*.

5.2   Experiments

The quality of the summarized sequence is evaluated by measuring the signal-to-Ratio (SNR) according the equation (7) relatively to the known reference sequence. Table 1 give the SNR values and the corresponding graphical representation in fig. 6 for the 7 cycles and the summarized sequence for both image sequences. Test 8 gives the result when applying a simple mean operator which consists in computing average intensity value at each pixel throughout the sequence. Test 9 gives the result for the summarised sequences from the noisy sequences with $\lambda=2$ and $s(t)= 1$ and from the noisy and artefacted sequences with $\lambda=3.1$ and $s(t)= 1$.

$$SNR = 10\log\left[\frac{\sum_{xy}\left(I_0\left(x,y\right)\right)^2}{\sum_{xy}\left(I_r\left(x,y\right) - I_0\left(x,y\right)\right)^2}\right] \tag{7}$$

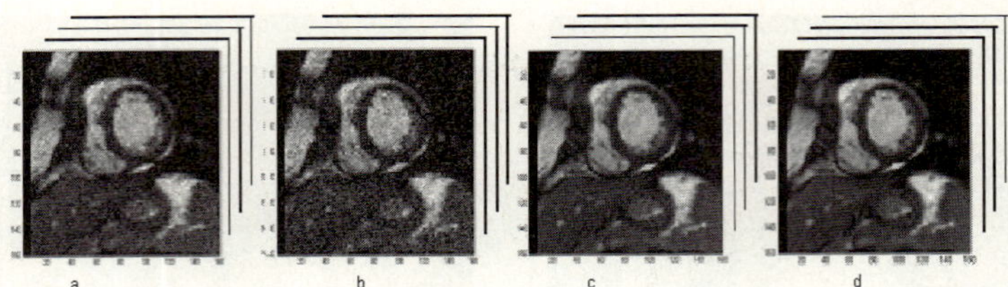

Fig. 4. First images of some cycles of the corrupted sequence (a) 1st Image of seq1-6 with Gaussian noise $\sigma=30$ and $\mu=0$ (b) 1st Image of seq1-7 with Gaussian noise $\sigma=35$ and $\mu=0$ (c) 1st image of the Summarized sequence (d) 1st image of the reference sequence

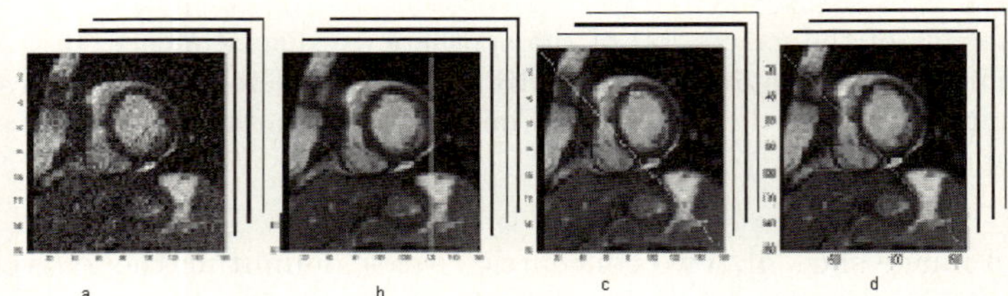

Fig. 5. First images of selected cycles of the perturbed sequence with addition of artefacts (a) 1st Image of seq2-1 with Gaussian noise $\sigma=35$ et $\mu=0$, with a diagonal black line (b) 1st Image of seq2-3 with Gaussian noise $\sigma=5$ et $\mu=0$, with a vertical white line (c) 1st Image of seq2-6 with Gaussian noise $\sigma=5$ et $\mu=0$, with a diagonal white line (d) 1st Image of the summarized sequence

A selected image pixel located at (160, 75) is tracked in various cycles of the average sequence, in the summarized sequence and in the reference sequence. The associated intensity variation is shown in fig. 7.

Table 1. Signal to noise ratio of tested sequences

Label	Sequences	SNR noisy sequences	Sequences	SNR noisy with artefact sequences
1	Sequence 1-1	24.9	Sequence 2-1	15.3
2	Sequence 1-2	19.2	Sequence 2-2	9.3
3	Sequence 1-3	15.9	Sequence 2-3	9.2
4	Sequence 1-4	13.6	Sequence 2-4	14.8
5	Sequence 1-5	11.9	Sequence 2-5	15.1
6	Sequence 1-6	10.6	Sequence 2-6	15.0
7	Sequence 1-7	9.5	Sequence 2-7	18.5
8	**Average sequence**	20,8	**Average sequence**	20,5
9	**Summarized sequence1**	22.4	**Summarized sequence2**	21

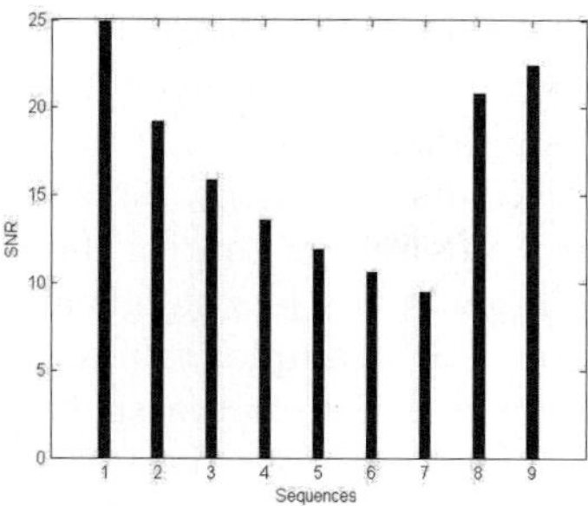 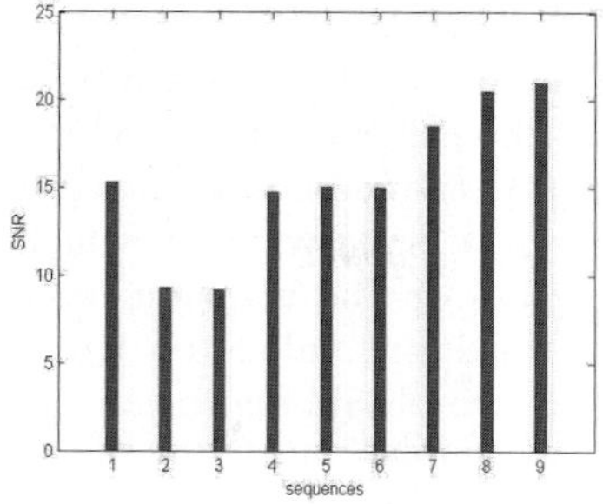

Fig. 6. Bar graph of the SNR. The sequence order follows the label number in Table 1. (Left) Noise corrupted sequence. (Right) Noise corrupted sequence with artefacts.

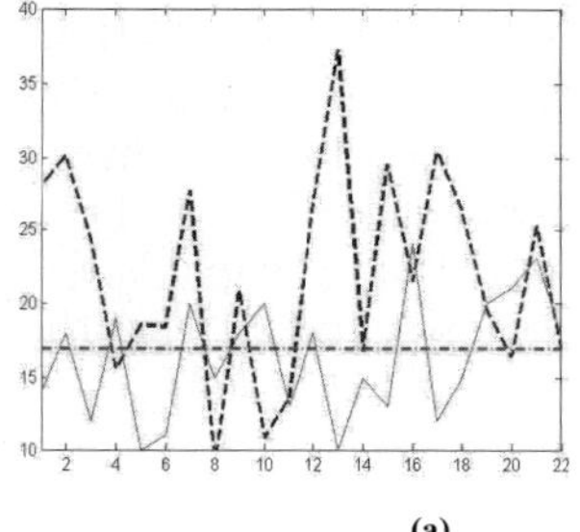 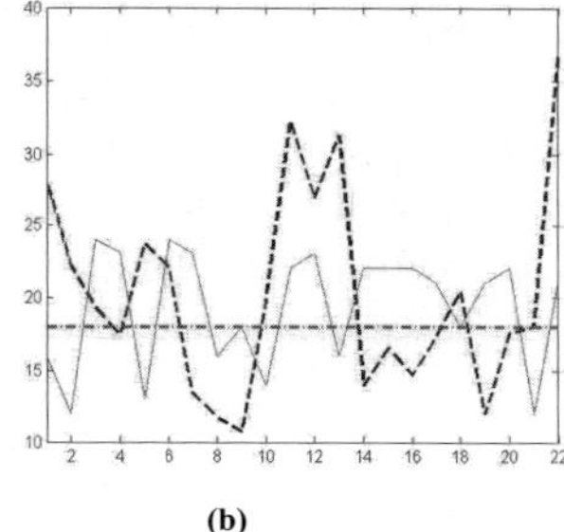

(a)   (b)

Fig. 7. (a): Intensity tracking of a selected pixel located at (160, 75) for the 22 images for the sequence of reference (dash-dot horizontal line), for the summarized sequence in solid green line and in dashed blue line for the same pixel in average sequence. (b): Another selected random pixel.

6   Discussion and Conclusion

Two synthetic multi-cycle periodic sequences have been considered to illustrate the proposed method. They represent various variations in the cycle repetition through artefacts and various added noise levels that could perturb the image acquisition process of the observed biological phenomena. As shown in Fig. 6 and table 1, the summarized sequences obtained an improved SNR compared to individual cycles and therefore result in an improvement of image quality with the summarized sequence. Also, the grey level invariance assumption during motion is better respected in the summarized sequence (Fig. 7). The possibility of having a competition within neurons during the training phase allows, notably, a shift in the order of the sequence images in case of the presence of slight variations in the cycle duration from one sequence to another.

Regarding the learning, we can use the neighborhood of order Nb with Nb a positive or zero. The use of $Nb = 0$ means that there is no effect on the neuron neighbours, therefore, the weight of the neighbours does not change. This is the strategy with the WTA algorithm (Winner Take All). In the case $Nb \geq 1$ one must define the effect of entries on the neuron neighbours. this is what is done with WTM algorithm (Winner Take Most). Further the work will be devoted to the exploitation of the proposed method based one a WTM algorithm in order to validate the generalization capacity of this approach.

Considering the method's parameters, we can tailor our approach to deal with various cases. λ is a parameter which makes it possible to control the influence of input values on the winner neuron. The larger λ the less significant the influence of the input values on winner neuron. For example, taking $\lambda = (l+1)$ with $l=1..L$, we obtain the mean operator which is therefore a special case of our approach. Function $s(t)$ in equation (2), which penalises the winner for the same cycle, is another parameter to control the network evolution. Note also the possibility of implementing the neural network in hardware, making possible the development of a high speed process. The periodic variations of biological phenomena as observed for instance with the heart, may imply important difficulties for the cardiovascular disease diagnostic or the realization of a surgical gesture. The proposed approach aims at providing a reference consensus cycle from the observation of many. It is mainly based on the learning capabilities of the neural networks and SO maps in particular. In the context of the other example on brain activation, introduced in section 1, the principle is the same as the method would provide one summarized activation sequence from the observation of several.

References

1. Ledesma-Carbayo, M., Kybic, J., Desco, M., Santos, A., Sühling, M., Hunziker, P., Unser, M.: Spatio-temporal nonrigid registration for ultrasound cardiac motion estimation. IEEE Transactions on Medical Imaging 24, 1113–1126 (2005)
2. Kohonen, T., Kaski, S., Lagus, K., Salojarvi, J., Honkela, J., Paatero, V., Saarela, A.: Self organization of a massive document collection. IEEE Transactions on Neural Networks 11(3), 574–585 (2000)
3. Kohonen, T.: Self-Organizing Maps. Springer, Berlin (2001)
4. Clarysse, P., Basset, C., Khouas, L., Croisille, P., Friboulet, D., Odet, C., Magnin, I.E.: 2D spatial and temporal displacement field fitting from cardiac MR tagging. Medical Image analysis 3, 253–268 (2000)
5. Dekker, A.H.: Predicting Periodic Phenomena with Self-Organising Maps. In: MODSIM International Congress on Modelling and Simulation, Australia, New Zealand, December 2007, pp. 2293–2298 (2007)

Image Classification by Histogram Features Created with Learning Vector Quantization

Marcin Blachnik[1,2] and Jorma Laaksonen[2]

[1] Silesian University of Technology, Electrotechnology Department,
Katowice Krasiskiego st. 8, Poland
[2] Helsinki University of Technology, Adaptive Informatics Research Centre,
P.O. Box 5400, FI-02015 TKK, Finland
marcin.blachnik@polsl.pl, jorma.laaksonen@tkk.fi

Abstract. Histogram-based data analysis is one of the most popular solutions for many problems related to image processing such as object recognition and classification. In general a histogram preserves more information from the first-order statistics of the original data than simple averaging of the raw data values. In the simplest case, a histogram model can be specified for a specific image feature type independently of any real image content. In the opposite extreme, the histograms can be made dependent not only the actual image contents, but also on the known semantic classes of the images. In this paper, we propose to use the Learning Vector Quantization (LVQ) algorithm in fine-tuning the codebook vectors for more efficient histogram creation. The performed experiments show that the accuracy of the Interest Point Local Descriptors (IPLD) feature for image classification can be improved by the proposed technique.

1 Introduction

In image analysis many different problems can be identified, but most of them are related to one of two categories: classification where the goal is prediction of presence or absence of a certain class (object) in the test image; and detection where the goal is prediction and selection of the bounding box of a certain object. In this paper the first problem will be considered based on histogram features created from low-level visual descriptor vectors extracted from interest points detected in the images.

Histogram-based image descriptors have proofed to be one of the most useful methods for image representation in classification tasks. They allow high compression of an image represented as single vector with relatively small set of attributes. Histogram representation can be used for many different types of features extracted from images. Some of the most popular types are color histograms [1], and SIFT [2] or IPLD [3] feature histograms. Histogram and interest point based features allowed winning the PASCAL NoE[1] Visual Object Classes Challenge 2006[2].

---

[1] http://www.pascal-network.org
[2] http://www.pascal-network.org/challenges/VOC/voc2006/

V. Kůrková et al. (Eds.): ICANN 2008, Part I, LNCS 5163, pp. 827–836, 2008.

There exist many ways of creating various types of histograms from simple 1-D to spatial. However, the basic approach is common and based on clustering features extracted from images and selecting special *reference vectors* also called *prototypes* or *codebook vectors*. These prototypes will then act as the cluster centers of the histogram bins in the high-dimensional feature space. The image-independent, fixed prototypes will then be further utilized for creating a content-describing histogram for each image to be classified. The number of low-level vector samples assigned to each prototypes in the nearest neighbor mapping is used as a new vector describing single image. Each histogram can then be normalized by the number of samples originating from the image, leading to an estimate of a discrete probability distribution.

In the simplest case, a histogram model can be specified for a specific image feature type independently of any real image content. A more evolved technique is to calculate the histogram prototypes from an existing image collection. In that case, one might not have any information available concerning the categories of objects in the images. If such category information is available, the creation of histograms can be based on clustering raw image features extracted from each object category independently. In such a case, however, no knowledge is gained from the class distribution, which may lead to low quality of final classification task of images set.

One of the methods that can utilize statistical inter-class knowledge is the Learning Vector Quantization (LVQ) algorithm [4]. The LVQ algorithm optimizes the positions of the prototypes (codebook vectors) according to the distances between samples from different classes, leading to improved discriminative abilities. We assume that this better discrimination will benefit also the process of histogram creation. The goal of this paper is to present improvements in creating histograms that can be achieved by applying a supervised codebook optimization algorithm – such as the LVQ algorithm – on features extracted from interest points of the images.

The rest of the paper is organized as follows: Section 2 first discusses the general question of discriminative prototype selection. Our approach for it is then described in detail, briefly introducing the Fuzzy C-Means clustering and the LVQ algorithm. The next Section 3 describes our data and experiments, evaluating SVM classifier on IPLD feature histograms, and presents a comparison of results obtained with and without LVQ tuning of the histogram codebooks. The last Section 4 concludes the paper and discusses the obtained results, pointing out possible future research directions.

2   Selecting Prototype Positions

Statistical pattern recognition algorithms are typically based on finding similarities and dissimilarities between training samples and using this knowledge for further reasoning. In image analysis this is equivalent to comparing representations of images with different depicted objects in order to evaluate similarities between them. This process, however, can be very complicated, especially when one considers the huge number of pixels of a single image to be analyzed.

A commonly-used solution for the image representation problem is to extract low-level visual features from the image – or some special interest points in it – and to either average or create a histogram of those features. For preparing the histograms, some computationally simple methods – such as k-means clustering – are used to first create the codebook of prototypes and then to extract the actual histogram feature vectors. Each single image is then analyzed according to the low-dimensional histogram vectors with more complex classification methods like the Support Vector Machine (SVM).

In classification tasks, histograms are usually created in one of two principal ways. First, they can be obtained by clustering the whole dataset; second, they can be obtained clustering each semantic object class independently. Similar concept is well known also in similarity-based methods of learning classification problems, where these two approaches are also called *post-supervised* and *pre-supervised*, respectively [5].

Simple clustering algorithms are often used for selecting prototypes (codebook vectors) which are further used to create histograms of the images. However, clustering does not produce any knowledge of the inter-class distribution of the classes. This may lead to poor quality of obtained histograms, and sub-optimal classification results. The same problem appears in classical classification problems, as reported eg. by Blachnik et al. [6] and Kuncheva [7]. Moreover, while creating histograms the goal is to find the least possible number of prototypes because too high a number can decrease the quality and generalization ability of the histograms. This appears when just a few samples from each image are used per one prototype.

The aim in selecting good prototypes is to look for prototypes that assure maximum discrimination abilities. In traditional nearest-neighbor classification the same goal has been sought for by using the Learning Vector Quantization (LVQ) algorithm [4]. The LVQ optimization tends to place prototypes of one class (say positive) inside the centers of homogeneous clusters of that class, far from representatives of the second (negative) class, and vice versa. After the LVQ training, the obtained prototypes represent the nearest-neighbor decision borders between the classes, providing good discrimination abilities.

In the LVQ training the main problem is the proper initialization of the codebook vectors. However, this can be solved sufficiently by using prototypes obtained after clustering each class separately. In the presented approach for LVQ-based histograms, Fuzzy C-Means algorithm was used to find initial position of the codebook vectors, and these prototypes were then optimized by the LVQ training.

2.1 Fuzzy C-Means Clustering

One examples of modern clustering methods is the Fuzzy C-Means algorithm (FCM) [8]. It is an extension of the classical k-means clustering algorithm. Instead of binary $\{0, 1\}$ membership of a vector $\mathbf{x}$ to the cluster represented by reference vector $\mathbf{p}_i$, such membership is in FCM a continuous function in bounds $[0, 1]$.

FCM belongs to batch clustering methods, where prototypes are iteratively updated after processing of the whole dataset. Batch clustering methods require a special partition matrix $\mathbf{U}$ of size $N \times k$, where N is the number of vectors and k is the number of clusters. This matrix of elements u_{ji} represent membership value of assigning vector $\mathbf{x}_j$ to cluster i defined by a prototype $\mathbf{p}_i$. FCM is based on the minimization of a cost function defined as:

$$J_m(\mathbf{U}, \mathbf{P}) = \sum_{i=1}^{k} \sum_{j=1}^{N} (u_{ji})^m \left\| \mathbf{x}_j - \mathbf{p}_i \right\|^2 \tag{1}$$

where $m > 1$ is a fuzziness parameter, typically $m = 2$, and $\mathbf{P}$ is a matrix whose columns are the prototype vectors $\mathbf{p}_i$.

For obtaining good clustering results, the partition matrix $\mathbf{U}$ has to fulfill three conditions:

1° each vector $\mathbf{x}_j$ belongs to the i-th cluster to a degree between $[0, 1]$:

$$\bigvee_{1 \le i \le k} \bigvee_{1 \le j \le N} u_{ji} \in [0, 1] \tag{2}$$

2° the sum of membership values of j-th vector $\mathbf{x}_j$ in all clusters is equal to 1

$$\bigvee_{1 \le j \le N} \sum_{i=1}^{k} u_{ji} = 1 \tag{3}$$

3° no clusters are empty.

$$\bigvee_{1 \le i \le k} 0 < \sum_{j=1}^{N} u_{ji} < N \tag{4}$$

Cost function (1) is minimized under these conditions by [8]:

$$\bigvee_{1 \le i \le k} \mathbf{p}_i = \sum_{j=1}^{N} (u_{ji})^m \mathbf{x}_j \left/ \sum_{j=1}^{N} (u_{ji})^m \right. \tag{5}$$

$$\bigvee_{\substack{1 \le i \le k \\ 1 \le j \le N}} u_{ji} = \left(\sum_{l=1}^{k} \left(\frac{\left\| \mathbf{x}_j - \mathbf{p}_i \right\|}{\left\| \mathbf{x}_j - \mathbf{p}_l \right\|} \right)^{2/(m-1)} \right)^{-1} \tag{6}$$

The above-described clustering method applied to classification problems does not optimize position of the codebook vectors according to inter-class distributions. This drawback can be improved by prototype optimization algorithms used in classification problems.

2.2 LVQ Training

The Learning Vector Quantization (LVQ) is a family of neurally-motivated algorithms which during the training phase optimize the positions of codebook vectors to the maximize classification accuracy. This type of training can be seen as a special form of the k-NN or rather nearest prototype classifier (NPC) [9]. The family of LVQ networks consist of many algorithms, where the most basic one is called LVQ1 and based on iterative optimization of the position of the prototype vectors.

Generally there exist two approaches to the training phase: batch and online learning. In the first approach the positions of the codebook vectors are updated after presentation of the whole training set, while in the online mode the codebook vectors $\mathbf{p}_i$ are updated after presentation of each training instance using the formula:

$$\mathbf{p}_i(t+1) = \mathbf{p}_i(t) + \alpha(t)\left(\mathbf{x}(t) - \mathbf{p}_i(t)\right), \text{ if class}(\mathbf{p}_i) = \text{class}(\mathbf{x}(t))$$
$$\mathbf{p}_i(t+1) = \mathbf{p}_i(t) - \alpha(t)\left(\mathbf{x}(t) - \mathbf{p}_i(t)\right), \text{ if class}(\mathbf{p}_i) \neq \text{class}(\mathbf{x}(t)) \tag{7}$$

$$i = \operatorname*{argmin}_{a=1,\dots,l} \|\mathbf{x}(t) - \mathbf{p}_a\| \tag{8}$$

where t is the current iteration step, i denotes the index of the prototype which is the closest one to the current training vector $\mathbf{x}(t)$ according to formula (8), and $\alpha(t)$ is the learning rate, a monotonically decreasing function which takes values between $[0, 1]$.

Self-performed experiments and other researchers experience proofs that the second approach – online learning – is more stable and less sensitive to outliers. It also imposes smaller memory requirements.

3   Experiments and Results

Datasets used in this experiments were taken from the Visual Object Classes Challenge 2007 (VOC2007) dataset [10]. It includes 2493 training images and 2500 images used for validation. In each image, may appear one or more of $n = 20$ object classes, such as *person, train, cat, dog, cow, bicycle*, etc.

Instead of average precision (AP), the measure suggested by the authors of the challenge, we decided to use receiver operator characteristic (ROC) curves and especially the area under curve (AUC) measure to determine the accuracy of classification. The switch between these accuracy measures was motivated by the problem that the AP measure depends on the number of intervals used to determine the value, and changing the number of intervals may affect the order of otherwise similar results.

The Support Vector Machine (SVM) was used as the final classifier with SVM-light implementation of T. Joachims [11]. The SVM used in the experiments had Gaussian kernels, and the SVM parameters such as C and γ (the kernel width) were optimized by using the greedy search procedure.

3.1   IPLD Features

The Interest Point Local Descriptors (IPLD) [3] feature extraction technique first detects a set of interest points in an image and then calculates features describing the local neighborhoods of those points. The interest points have in our experiments been the union of points with strongest responses to the Harris–Laplace corner detector and the difference of Gaussians (DoG) keypoint filtering. The typical number of interest points found in each image has been around one thousand.

The features extracted around the interest points are based on the SIFT [2] descriptors. They are comprised of edge orientation histograms with eight directions. There are a total of 16 histograms extracted around the interest point in a 4×4 grid. The total dimensionality of the feature vectors is thus 128.

3.2 Training Strategy

The process we used to create a classifier can be divided into several steps:

1. IPLD interest point detection and feature extraction from images
2. randomly sampling 300 interest points from each image
3. grouping feature vectors from images of the *studied class*, and those from the remaining classes as the *background class*
4. dataset normalization, so each variable has 0 mean and variance equal to 1.
5. clustering the studied class and the background class independently into 150 clusters per class with the Fuzzy C-Means algorithm
6. prototype optimization using labeled data and the LVQ algorithm
7. creating the histogram feature vector for each image
8. training and optimizing SVM classifier based on the obtained histogram features

First of all each image of the VOC2007 database was processed and the IPLD interest points and features were calculated from it. The mean number of interest points per image was around 1000, and from that number 300 interest points were randomly selected for further analysis.

In the classical approach used for selecting codebook vectors, each class is after sampling clustered independently. This requires keeping $300 \times n_i$ vectors in memory, where n_i is the number of images in the i-th class. On the average each class of the VOC2007 dataset had around 140 images, except the *person* class which had 1025 images.

The number of samples is especially important for the LVQ training which requires labeled data and combination of samples from different classes into the training set. This may create a problem with the dataset size of $(\sum n_i) \cdot 300$ samples. The class-wise dataset combination can be done either as *one against all* strategy or *one against one*. In the first option, all vectors of all the other classes except the studied class, labeled as $C_1 = 1$, are grouped together and labeled as $C_2 = -1$. In the one against one strategy, $n - 1$ datasets are required to create a codebook for single class because prototypes are generated to discriminate pairs of classes. From these two approaches, the one against all strategy was selected here because it has linear complexity to the number of classes. Moreover, to avoid imbalanced representation of vectors in the labeled LVQ training set, the

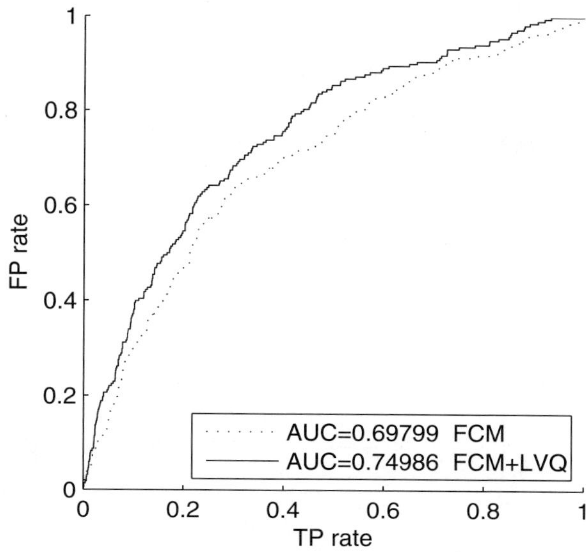

Fig. 1. ROC curve for the *dog* class obtained by using FCM clustering alone and using FCM+LVQ optimization

data was resampled to obtain an equal number of samples in the studied and background classes. In the same time the memory complexity was also reduced.

The training vectors in both the studied and background classes were clustered independently by using the Fuzzy C-Means algorithm to obtain 150 prototypes per class. The prototypes were then further optimized by using the LVQ1 training on the training data. Histograms were then calculated for all the images in both the training and validation data sets. The histograms were normalized be the number of samples in each image to produce proper discrete probability estimates. The prepared histogram features from the training set images were used for training the SVM classifier and those from the validation set images were used for evaluating its accuracy.

3.3   Classification Results

In order to evaluate the performance of the proposed approach, three classes, *dog*, *cat* and *person*, were selected from the dataset. For each of these three classes we ran a separate experiment and built a classifier by following the steps described in the previous section.

Accuracy measures such as the ROC curve and its AUC value were evaluated by using the validation image set of the challenge. The obtained ROC curves are presented in Figures 1, 2 and 3 for each class, *dog*, *cat* and *person*, respectively. One can clearly see that the ROC curves of the LVQ-optimized histogram features are always better than those obtained without LVQ optimization.

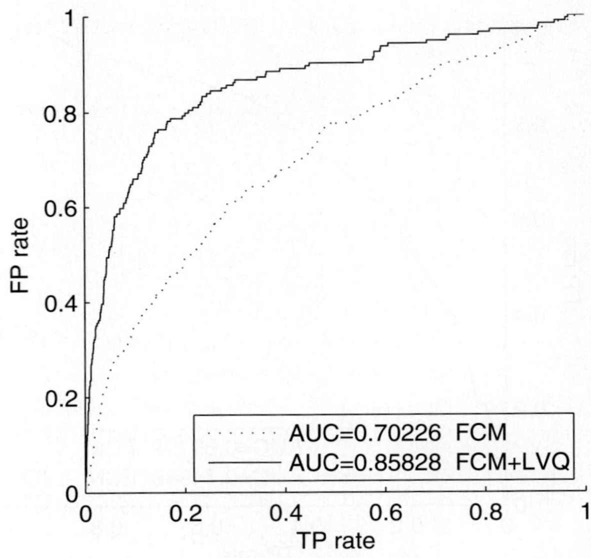

Fig. 2. ROC curve for the *cat* class obtained by using FCM clustering alone and using FCM+LVQ optimization

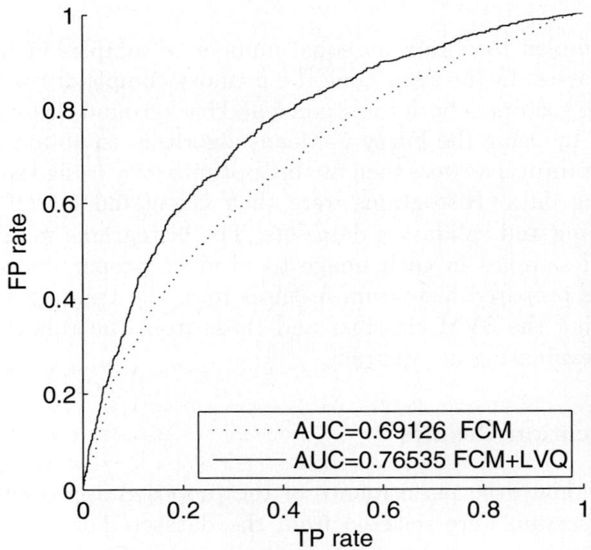

Fig. 3. ROC curve for the *person* class obtained by using FCM clustering alone and using FCM+LVQ optimization

The AUC values and the SVM accuracies obtained during the training with five-fold cross-validation are presented in Table 1. The "FCM" results have been been obtained with the Fuzzy C-Means clustering alone, whereas the

Table 1. AUC value on validation data and after 5xCV training of the SVM classifier

	dog		*cat*		*person*	
	AUC	5xCV	AUC	5xCV	AUC	5xCV
FCM	0.6980	0.6837	0.7023	0.7150	0.6913	0.7080
FCM+LVQ	0.7499	0.8457	0.8583	0.9179	0.7654	0.8010

results in the "FCM+LVQ" row have included LVQ1 training after the class-wise clustering.

The presented results show clear improvement from the classical approach based on only independent class clustering. The positions of the prototypes optimized by the LVQ1 algorithm allow obtaining more discriminative histograms. Therefore, the difference in accuracy between just clustering and clustering followed by LVQ is significant. This phenomenon is visible both for the final validation results in the "AUC" columns of Table 1 and for the cross-validation results during the training.

4   Conclusions and Future Directions

The proposed approach for selecting prototypes was in this paper applied only for IPLD histograms of images while it can be used as well for other types of visual features and other totally different data domains. However, the applicability of the technique will need further investigation.

An important issue that should be solved is the optimization of the number of prototypes selected for each class. Our current approach was based on a manually selected value. However, selecting an appropriate number of codebook vectors may further improve the quality of the histograms and enhance their discrimination abilities. The codebook size selection problem might perhaps be solved by using the Race algorithm proposed by Blachnik in [6] or by Dynamic LVQ solution developed by Bermejo in [12]. Instead of the initial optimization of the number of prototypes, feature selection algorithms could also be used during the SVM training for pruning any useless codebook vectors.

One more possible improvement that might lead to achieving good prototype sets is based on the utilization of the Relevance Vector Machine (RVM) [13]. The RVM belongs to the group of sparse Bayesian methods and tends to automatic selection of a small number of prototypes that lie far from the decision borders. This property might be desired for codebooks used for creating histograms.

All in all, we have shown in this paper that supervised tuning of histogram prototypes is a promising approach for creating efficient features for classification. The LVQ algorithm is a simple and well-known method for this purpose, but not the only one. Future research will include further investigations concerning both the unsupervised selection of the initial prototypes and their supervised tuning.

References

1. ISO/IEC: Information technology - Multimedia content description interface - Part 3: Visual, 15938-3:2002(E) (2002)
2. Lowe, D.G.: Distinctive image features from scale-invariant keypoints. International Journal of Computer Vision 60(2), 91–110 (2004)
3. Dorko, G., Schmid, C.: Selection of scale-invariant parts for object class recognition. In: Proceedings of Ninth IEEE International Conference on Computer Vision, vol. 1, p. 634. IEEE Computer Society, Los Alamitos (2003)
4. Kohonen, T.: Self-Organizing Maps. Springer, Heidelberg (2001)
5. Kuncheva, L., Bezdek, J.: Presupervised and postsupervised prototype classifier design. IEEE Transactions on Neural Networks 10(5), 1142–1152 (1999)
6. Blachnik, M., Duch, W., Wieczorek, T.: Selection of prototypes rules: context searching via clustering. In: Rutkowski, L., Tadeusiewicz, R., Zadeh, L.A., Żurada, J.M. (eds.) ICAISC 2006. LNCS (LNAI), vol. 4029, pp. 573–582. Springer, Heidelberg (2006)
7. Kuncheva, L., Bezdek, J.: Nearest prototype classification: Clustering, genetic algorithms or random search? IEEE Transactions on Systems, Man, and Cybernetics C28(1), 160–164 (1998)
8. Hoppner, F., Klawonn, F., Kruse, R., Runkler, T.: Fuzzy Cluster Analysis. Wiley, Chichester (1999)
9. Kuncheva, L., Bezdek, J.: An integrated framework for generalized nearest prototype classifier design. International Journal of Uncertainty 6(5), 437–457 (1998)
10. Everingham, M., Van Gool, L., Williams, C.K.I., Winn, J., Zisserman, A.: The PASCAL Visual Object Classes Challenge 2007 (VOC2007) Results (2007), http://www.pascal-network.org/challenges/OC/voc2007/ workshop/index.html
11. Joachims, T.: Learning to Classify Text Using Support Vector Machines. Kluwer Academic Publishers, Dordrecht (2002)
12. Bermejo, S.: Learning with nearest neighbor classifiers. PhD thesis, Technical University of Catalonia (2000)
13. Tipping, M.: The relevance vector machine. In: Advances in Neural Information Processing Systems. Morgan Kaufmann, San Francisco (2000)

A Statistical Model for Histogram Refinement

Nizar Bouguila and Walid ElGuebaly

CIISE, Concordia University, Montreal, Canada
bouguila@ciise.concordia.ca, w_elgue@encs.concordia.ca

Abstract. Color histograms have been widely used successfully in many computer vision and image processing applications. However, they do not include any spatial information. In this paper, we propose a statistical model to integrate both color and spatial information. Our model is based on finite multiple-Beta-Bernoulli mixtures. The results show that our model achieves good performance in a specific image classification problem (city vs. landscape).

1 Introduction

Multimedia data are undergoing an important expansion in terms of volume and variety. Color histograms are one of the most important and used techniques for images representation. This can be explained by their trivial computation and the fact that histograms are invariant to translation and rotation about the viewing axis, and provide a stable object recognition in the presence of occlusions and over views change [1]. However, histograms do not include any spatial information which is an important issue in human visual perception [2]. Indeed, images with completely different appearance and spatial colors organization may have similar histograms. Many approaches have been proposed to consider the spatial information [3,4]. A successful approach to cure this problem and introduce the spatial information was proposed by Pass and Zabih [5]. Their technique is called histogram refinement and imposes additional constraints on histograms. These constraints are introduced by subdividing the set of pixels having a given color into classes based on local features. A possible subdivision can be based on pixels spatial coherence where a pixel is considered as coherent if it is a part of a large group of connected pixels of the same color, and incoherent if not [5]. In this paper, we propose a statistical framework to model the histogram refinement technique. Our framework is based on finite multiple-Beta-Bernoulli mixture models.

In the next section, we present our statistical model. In Section II, we propose an algorithm to estimate the parameters of our statistical model. In Section III, we describe the experimental results in a specific image classification problem (city vs. landscape). Finally, Section IV concludes the paper.

2 The Model

Let I a $K \times L$ image considered to be a set of pixels $\{X_{lk}, l = 1, \ldots, L; k = 1, \ldots, K\}$, where X_{lk} is the pixel in position (l, k) of image I. The colors in I

V. Kůrková et al. (Eds.): ICANN 2008, Part I, LNCS 5163, pp. 837–846, 2008.

are quantized into C colors $c_1, \ldots, c_C$. The color histogram $H = (h_{c_1}, \ldots, h_{c_C})$ is a discrete vector of counts in which each h_c represents the number of pixels of color c. According to its histogram, a given image can be represented by the following probability distribution

$$p(I|\boldsymbol{\pi}) = \prod_{c=c_1}^{c_C} \prod_{l=1}^{L} \prod_{k=1}^{K} \pi_c^{\delta(X_{lk}=c)} = \prod_{c=c_1}^{c_C} \pi_c^{h_c} \tag{1}$$

which is a multinomial distribution with parameters $\boldsymbol{\pi} = \{\pi_{c_1}, \ldots, \pi_{c_C}\}$ and $\delta(X_{lk} = c)$ is an indicator function. However, recent developments have shown that this choice is inappropriate in many cases and applications (See [6,7], for instance, for interesting discussions about the drawbacks of the multinomial assumption). Better models are given by the multinomial Dirichlet and the multinomial generalized Dirichlet obtained by taking Dirichlet [8] and generalized Dirichlet [9] distributions, respectively, as priors. The multinomial Dirichlet with parameters $\boldsymbol{\eta} = (\eta_{c_1}, \ldots, \eta_{c_C})$ is given by [6,10]

$$p(I|\boldsymbol{\eta}) = \frac{\Gamma(\sum_{c=c_1}^{c_C} \eta_c)}{\Gamma(\sum_{c=c_1}^{c_C}(\eta_c + h_c))} \prod_{c=c_1}^{c_C} \frac{\Gamma(\eta_c + h_c)}{\Gamma(\eta_c)} \tag{2}$$

The multinomial generalized Dirichlet distribution with parameters $\boldsymbol{\beta} = (\phi_1, \beta_1, \ldots, \phi_{c_C}, \beta_{c_C})$ is given by [7,11]

$$p(I|\boldsymbol{\beta}) = \prod_{c=1}^{c_C-1} \frac{\Gamma(\phi_c + \beta_c)}{\Gamma(\phi_c)\Gamma(\beta_c)} \prod_{c=1}^{c_C-1} \frac{\Gamma(\phi_c')\Gamma(\beta_c')}{\Gamma(\phi_c' + \beta_c')} \tag{3}$$

where $\phi_c' = \phi_c + h_c$ and $\beta_c' = \beta_c + h_{c+1} + \ldots + h_{c_C}$ for $c = 1, \ldots, c_C - 1$.

By considering the histogram refinement technique, a given image can be represented by a color coherence vector (CCV) [5,12] $\langle (f_{c_1}, h_{c_1} - f_{c_1}), \ldots, (f_{c_C}, h_{c_C} - f_{c_C}) \rangle$ which is a vector of pairs, one for each color. In each pair $(f_c, h_c - f_c)$, f_c represents the number of pixels having color c and coherent (i.e the pixel is a part of a large group of connected pixels of the same color). Thus, an image can be viewed as a collection of samples from a multiple-Bernoulli distribution. Indeed, we can assume that we sample from a multiple-Bernoulli distribution once for each pixel in the image, where each binary trial corresponds to the event that a pixel with a given color is coherent or not. Modeling the image in this manner gives us the following

$$p(I|\boldsymbol{\pi}) = \prod_{k=1}^{K} \prod_{l=1}^{L} \prod_{c=c_1}^{c_C} \left[\left(\pi_c\right)^{r_{kl}} \left(1 - \pi_c\right)^{1-r_{kl}} \right]^{\delta(X_{lk}=c)} = \prod_{c=c_1}^{c_C} \pi_c^{f_c} (1 - \pi_c)^{h_c - f_c} \tag{4}$$

where $r_{kl} = 1$ if pixel X_{lk} is coherent and 0 if not. However, as the multinomial this distribution gives poor estimates and suffers from zero counts as it is based only on the frequencies. A solution is to consider a smoothing prior to $\boldsymbol{\pi}$. We choose a multiple-Beta distribution which is actually a conjugate prior [13]

$$p(\boldsymbol{\pi}|\boldsymbol{\alpha}) = \prod_{c=c_1}^{c_C} \frac{\Gamma(\alpha_c + \beta_c)}{\Gamma(\alpha_c)\Gamma(\beta_c)} \pi_c^{\alpha_c - 1}(1 - \pi_c)^{\beta_c - 1} \tag{5}$$

where $\boldsymbol{\alpha} = (\alpha_{c_1}, \beta_{c_1}, \ldots, \alpha_{c_C}, \beta_{c_C})$ is the hyperparameters vector. Then, an image I can be represented by a multiple-Bernoulli-Beta distribution as follows

$$p(I|\boldsymbol{\alpha}) = \int_{\boldsymbol{\pi}} p(I|\boldsymbol{\pi})p(\boldsymbol{\pi}|\boldsymbol{\alpha})d\boldsymbol{\pi} = \prod_{c=c_1}^{c_C} \frac{\Gamma(\alpha_c + \beta_c)\Gamma(\alpha_c + f_c)\Gamma(\beta_c + h_c - f_c)}{\Gamma(\alpha_c)\Gamma(\beta_c)\Gamma(\alpha_c + \beta_c + h_c)} \tag{6}$$

3   Model Learning

Given a set of N images $\mathcal{I} = \{I_1, \ldots, I_N\}$, where each image is represented by a coherence vector $\langle (f_{ic_1}, h_{ic_1} - f_{ic_1}), \ldots, (f_{ic_C}, h_{ic_C} - f_{ic_C}) \rangle$. Each image can be modeled by a finite multiple-Beta-Bernoulli mixture model

$$p(I|\Theta) = \sum_{j=1}^{M} p_j p(I|\boldsymbol{\alpha}_j) \tag{7}$$

where p_j $(0 < p_j \leq 1$ and $\sum_{j=1}^{M} p_j = 1)$ are the mixing proportions and $\Theta = (p_1, \ldots, p_M, \boldsymbol{\alpha}_1, \ldots, \boldsymbol{\alpha}_M)$. The loglikelihood corresponding to our mixture model given by Eq. 7 is

$$\log p(\mathcal{I}|\Theta) = \log \prod_{i=1}^{N} p(I_i|\Theta) = \sum_{i=1}^{N} \log \sum_{j=1}^{M} p_j p(I_i|\boldsymbol{\alpha}_j) \tag{8}$$

A known method to estimate the parameters is the ML approach given by $\hat{\Theta}_{ML} = \arg\max_{\Theta}\{\log p(\mathcal{I}|\Theta)\}$. Note that the maximization is done with respect to the constraints over the mixing parameters. The usual choice for obtaining ML estimation of mixture parameters is the EM algorithm which is a general approach in the presence of incomplete data [14]. In EM, the "complete" data are considered to be $Y_i = \{I_i, \boldsymbol{Z}_i\}$, where $\boldsymbol{Z}_i = (Z_{i1}, \ldots, Z_{iM})$ with $Z_{ij} = 1$ if I_i belongs to class j and 0 otherwise, constituting the "missing" data. The relevant assumption is that the density of an image I_i given $\boldsymbol{Z}_i$ is given by $\prod_{j=1}^{M} p(I_i|\boldsymbol{\pi}_j)^{Z_{ij}}$. The resulting *complete-data log-likelihood* is

$$L(\mathcal{I}, \Theta, \mathcal{Z}) = \sum_{i=1}^{N} \sum_{j=1}^{M} Z_{ij} \log \left(p_j p(I_i|\boldsymbol{\alpha}_j) \right) \tag{9}$$

The EM algorithm produces a sequence of estimates $\{\Theta^t, t = 0, 1, 2 \ldots\}$ by applying two steps in alternation

1. **E-step:** Compute $\hat{Z}_{ij}^{(t)} = \dfrac{p(I_i|\boldsymbol{\alpha}_j^{(t)})p_j^{(t)}}{\sum_{j=1}^{M} p(I_i|\boldsymbol{\alpha}_j^{(t)})p_j^{(t)}}$ given the parameter estimates from the initialization.

2. **M-step:** Update the parameter estimates according to

$$\Theta^{(t+1)} = \arg\max_\Theta \left\{ \sum_{i=1}^N \sum_{j=1}^M \hat{Z}_{ij}^{(t)} \log(p(I_i|\boldsymbol{\alpha}_j^{(t)})p_j^{(t)}) \right\}$$

The quantity $\hat{Z}_{ij}$ represents the conditional expectation of Z_{ij} given the image I_i and parameter vector Θ. The value Z_{ij}^* of $\hat{Z}_{ij}$ at a maximum of Eq. 9 is the conditional probability that observation i belongs to class j (the *posterior* probability). The EM algorithm is widely used in the case of finite mixture models estimation. However, it highly depends on initialization and it suffers from a local maxima problem because of the multimodal nature of the likelihood when we deal with mixture models [14]. Different extensions were proposed to overcome this problem [14] and one of the most successful extension of the EM was the deterministic annealing method [15] which has been used to avoid the initialization dependence and poor local maxima. Deterministic annealing is obtained by modifying the E-step in which we compute the following parameterized variant of the original posterior probability and is given by [15]

$$\hat{w}_{ij}^{(t)} = \frac{\left(p(I_i|\boldsymbol{\alpha}_j^{(t)})p_j^{(t)}\right)^\tau}{\sum_{j=1}^M \left(p(I_i|\boldsymbol{\alpha}_j^{(t)})p_j^{(t)}\right)^\tau} \tag{10}$$

where $\tau = \frac{1}{T}$ and T corresponds to the *computational temperature*. The algorithm starts at high temperature which is lowered during the iterations. In the M-step we update the parameter estimates according to

$$\Theta^{(t+1)} = \arg\max_\Theta \left\{ L(\mathcal{I},\Theta,\mathcal{Z},T) = \sum_{i=1}^N \sum_{j=1}^M \hat{w}_{ij}^{(t)} \log(p(I_i|\boldsymbol{\alpha}_j^{(t)})p_j^{(t)}) \right\} \tag{11}$$

Using the ML approach, we obtain the following estimates for the mixing parameters:

$$p_j = \frac{1}{N} \sum_{i=1}^N \hat{w}_{ij} \tag{12}$$

For the estimation of the $\boldsymbol{\alpha}_j$, we have used a Newton-Raphson approach, since a closed-form solution does not exist. Indeed,

$$\frac{\partial L(\mathcal{I},\Theta,\mathcal{Z},T)}{\partial \alpha_{jc}} = \sum_{i=1}^N \hat{w}_{ij}[(\Psi(\alpha_{jc}+\beta_{jc})-\Psi(\alpha_{jc}))+(\Psi(\alpha_{jc}+f_{ic})-\Psi(\alpha_{jc}+\beta_{jc}+h_{ic}))] \tag{13}$$

$$\frac{\partial L(\mathcal{I},\Theta,\mathcal{Z},T)}{\partial \beta_{jc}} = \sum_{i=1}^N \hat{w}_{ij}[(\Psi(\alpha_{jc}+\beta_{jc})-\Psi(\beta_{jc}))+(\Psi(\beta_{jc}+h_{ic}-f_{ic})-\Psi(\alpha_{jc}+\beta_{jc}+h_{ic}))] \tag{14}$$

where $\Psi(.)$ is the digamma function. By computing the second and mixed derivatives of $L(\mathcal{I},\Theta,\mathcal{Z},T)$ we obtain

$$\frac{\partial^2 L(\mathcal{I},\Theta,\mathcal{Z},T)}{\partial \alpha_{jc_1}\partial \alpha_{jc_2}} = \begin{cases} \sum_{i=1}^N \hat{w}_{ij}(\Psi'(\alpha_{jc}+f_{ic})-\Psi'(\alpha_{jc}+\beta_{jc}+h_{ic})) \\ +(\Psi'(\alpha_{jc}+\beta_{jc})-\Psi'(\alpha_{jc}))\sum_{i=1}^N \hat{w}_{ij} & \text{if } c_1=c_2=c \\ 0 & \text{otherwise} \end{cases} \tag{15}$$

$$\frac{\partial^2 L(\mathcal{I},\Theta,\mathcal{Z},T)}{\partial\beta_{jc_1}\partial\beta_{jc_2}} = \begin{cases} \sum_{i=1}^{N}\hat{w}_{ij}(\Psi'(\beta_{jc}+h_{ic}-f_{ic})-\Psi'(\alpha_{jc}+\beta_{jc}+h_{ic})) \\ +(\Psi'(\alpha_{jc}+\beta_{jc})-\Psi'(\beta_{jc}))\sum_{i=1}^{N}\hat{w}_{ij}\ \text{if}\ c_1=c_2=c \\ \qquad\qquad 0 \qquad\quad \text{otherwise} \end{cases} \tag{16}$$

$$\frac{\partial^2 L(\mathcal{I},\Theta,\mathcal{Z},T)}{\partial\alpha_{jc_1}\partial\beta_{jc_2}} = \begin{cases} (\Psi'(\alpha_{jc}+\beta_{jc}))\sum_{i=1}^{N}\hat{w}_{ij}-\sum_{i=1}^{N}\hat{w}_{ij}(\Psi(\alpha_{jc}+\beta_{jc}+h_{ic})) \\ \qquad\qquad\quad \text{if}\ c_1=c_2=c \\ \qquad\qquad 0 \qquad\quad \text{otherwise} \end{cases}$$
$$\tag{17}$$

where Ψ' is the trigamma function. Then, the hessian matrix has a block-diagonal structure

$$H(\boldsymbol{\alpha}_j)=\text{block-diag}\{H_1(\alpha_{jc_1},\beta_{jc_1}),\ldots,H_{c_C}(\alpha_{jc_C},\beta_{jc_C})\} \tag{18}$$

where

$$H_c(\alpha_{jc},\beta_{jc})=\begin{pmatrix} \frac{\partial^2 L(\mathcal{I},\Theta,\mathcal{Z},T)}{\partial^2\alpha_{jc}} & \frac{\partial^2 L(\mathcal{I},\Theta,\mathcal{Z},T)}{\partial\alpha_{jc}\partial\beta_{jc}} \\ \frac{\partial^2 L(\mathcal{I},\Theta,\mathcal{Z},T)}{\partial\beta_{jc}\partial\alpha_{jc}} & \frac{\partial^2 L(\mathcal{I},\Theta,\mathcal{Z},T)}{\partial^2\beta_{jc}} \end{pmatrix} \tag{19}$$

and we have [16, Theorem 8.8.16]

$$H(\boldsymbol{\alpha}_j)^{-1}=\text{block-diag}\{H_1(\alpha_{jc_1},\beta_{jc_1})^{-1},\ldots,H_{c_C}(\alpha_{jc_C},\beta_{jc_C})^{-1}\} \tag{20}$$

We remark that $H_c(\alpha_{jc},\beta_{jc})$ can be written as

$$H_c(\alpha_{jc},\beta_{jc})=D+\gamma\boldsymbol{a}\boldsymbol{a}^{tr} \tag{21}$$

where $D=\text{diag}\Big[-\Psi'(\alpha_{jc})\sum_{i=1}^{N}\hat{w}_{ij}+\sum_{i=1}^{N}\hat{w}_{ij}\Psi'(\alpha_{jc}+f_{ic}),\ -\Psi'(\beta_{jc})\sum_{i=1}^{N}\hat{w}_{ij}+$

$\sum_{i=1}^{N}\hat{w}_{ij}\Psi'(\beta_{jc}+h_{ic}-f_{ic})\Big]$, $\gamma=\Psi'(\alpha_{jc}+\beta_{jc})\sum_{i=1}^{N}\hat{w}_{ij}-\sum_{i=1}^{N}\hat{w}_{ij}\Psi'(\alpha_{jc}+$

$\beta_{jc}+h_{ic})$, $\boldsymbol{a}^{tr}=\mathbf{1}$ and $\gamma\neq(\sum_{k=1}^{2}\frac{a_k^2}{D_{kk}})^{-1}$. Then, the inverse of the matrix $H_c(\alpha_{jc},\beta_{jc})$ is given by [16, Theorem 8.3.3]:

$$H_c(\alpha_{jc},\beta_{jc})^{-1}=D^*+\delta^* a^* a^{*tr} \tag{22}$$

where

$$D^*=D^{-1}=\text{diag}[1/D_1,1/D_2]$$
$$=\text{diag}\Bigg[\frac{1}{-\Psi'(\alpha_{jc})\sum_{i=1}^{N}\hat{w}_{ij}+\sum_{i=1}^{N}\hat{w}_{ij}\Psi'(\alpha_{jc}+f_{ic})}, \tag{23}$$
$$\frac{1}{-\Psi'(\beta_{jc})\sum_{i=1}^{N}\hat{w}_{ij}+\sum_{i=1}^{N}\hat{w}_{ij}\Psi'(\beta_{jc}+h_{ic}-f_{ic})}\Bigg]$$

$$a^{*tr}=(a_1/D_1,a_2/D_2)=(1/D_1,1/D_2) \tag{24}$$

$$\delta^* = -\gamma\left(1 + \gamma(1/D_1 + 1/D_2)\right)^{-1} \tag{25}$$

Given a set of initial estimates, Newton-Raphson method can now be used. The iterative scheme of the Newton-Raphson method is given by the following equation:

$$\boldsymbol{\alpha}_j^{(t+1)} = \boldsymbol{\alpha}_j^{(t)} - H(\boldsymbol{\alpha}_j^{(t)})^{-1}\frac{\partial L(\mathcal{I},\Theta,\mathcal{Z},T)}{\partial \boldsymbol{\alpha}_j^{(t)}} \tag{26}$$

For the selection of the number of clusters, we have used the minimum description length (MDL) [1] criterion [19], given by

$$MDL(M) = \log p(\mathcal{I}|\hat{\Theta}) - \frac{N_p \log N}{2} \tag{27}$$

where $N_p = M(1 + 2c_C)$ is the number of parameters in the model.

Algorithm
For each candidate value of M:

1. Set $\tau \leftarrow \tau_{min}$ ($\tau_{min} \ll 1$), choose an initial estimate $\Theta^{(0)}$ and set $t \leftarrow 0$
2. Iterate the two following steps until convergence:
    (a) E-Step: Compute $\hat{w}_{ij}^{(t)}$ using Eq. 10
    (b) M-Step: Update the $p_j^{(t)}$ using Eq. 12 and the $\boldsymbol{\alpha}_j^{(t)}$ using Eq. 26
3. Increase τ ($\tau \leftarrow \tau \times \Delta$)
4. If $\tau \leq 1$, set $t \leftarrow t + 1$, go to step 2.
5. Calculate the associated MDL using Eq. 27.
6. Select the optimal model M^* such that: $M^* = \arg\max_M MDL(M)$

Experimentally, we have concluded that τ_{min} set to 0.04 is enough which is the same confirmation reached in [20]. For the temperature, we used $\Delta = 5$ which is a choice that gives good results and corresponds to 3 phases as in [20].

4   Experimental Results: Image Classification (City vs. Landscape)

The main goal of this section is to show the importance of the introduction of spatial information through the comparisons of mixture of multinomial Dirichlet and multinomial generalized Dirichlet [2], used to model the color information, and our model based on mixtures of multiple-Beta-Bernoulli used to model both the color and spatial information. Images are represented in the RGB [3] color

---

[1] Of course other mixture selection criteria may be used (See [17,18] for a discussion).

[2] See [21,6] and [7,11] for details about multinomial Dirichlet and multinomial generalized Dirichlet mixtures estimation.

[3] We have also tested the representation of the image colors in the HSV and LAB spaces and we did not remark much changes in the results.

Fig. 1. Sample images from each group. Row 1: Landscape images, Row 2: City images.

space. The color space is discretized in 64 distinct colors in the image. A pixel is considered coherent if the size of its connected components exceeds the fixed value of 300. Following [5], we have scaled all images to contain a total number of pixels equal to 38,978, so a region is considered as coherent if its area is about 1% of the image.

With the amount of digital information growing rapidly, the need for efficient automatic images categorization techniques has increased. Images categorization facilitates navigation and content-based images retrieval, and at the same time provides tools for continual maintenance as images categories grow in size. In this section, we propose a categorization approach based on our statistical model. We mainly focus on a specific interesting organization problem: city images vs. landscape images for which coherence vectors were shown to give good results [22]. Our experimental study is conducted on an image database consisting of 30,000 images (15,000 city images and 15,000 landscape images) collected from various sources. Figure 1 shows examples of images from both classes. Note that compared to city images, landcape images have relatively constant colors.

Our main problem in this section can be defined as follows: Given an input image, assign it to either the city or landscape class. The assignment is based on a set of training images which are already labeled. All training images are passed through the color coherence vectors computation stage, and then through the mixture's parameters estimation stage, in which the color coherence vectors are modeled as multiple-Beta-Bernoulli mixtures. After this stage, city and

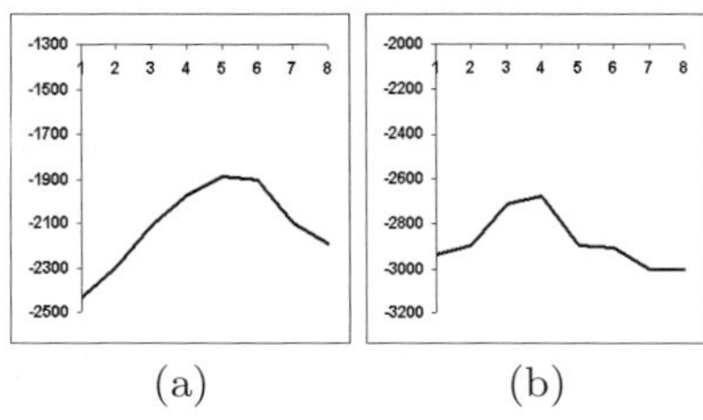

(a)   (b)

Fig. 2. Number of clusters found for the two training classes using the MDL criterion. (a) Landscape, (b) City.

Table 1. Classification accuracies for training and testing sets

	Training set	Testing set
Multiple-Bernoulli mixture	98.54	92.46
Multinomial generalized Dirichlet mixture	94.93	89.79
Multinomial Dirichlet mixture	92.89	85.32

Table 2. Confusion matrices for multiple-Beta-Bernoulli mixtures. (a) Training set, (b) Test set.

	City	Landscape
City	4918	82
Landscape	64	4936

(a)

	City	Landscape
City	9680	320
Landscape	434	9566

(b)

landscape classes are represented by M_{city}- and $M_{landscape}$-components multiple-Beta-Bernoulli mixtures $p(city|\Theta_{city})$ and $p(landscape|\Theta_{landcape})$, respectively. Finally, the classification stage uses the Bayesian decision rule, to determine to which class each image will be assigned, given as following: image I is assigned to class "City" if $p(I|\Theta_{city}) > p(I|\Theta_{landscape})$ and to class "landscape", otherwise. In our experiments we have used 10,000 images for training (5,000 city images and 5,000 landscape images). Figure 2 shows the number of clusters found for each training class when we apply our algorithm with the MDL criterion. Table 1 shows the classification results for the training and test sets, respectively, when applying multinomial Dirichlet, multinomial generalized Dirichlet and multiple-Beta-Bernoulli mixtures. The best accuracies of 98.54% and 92.46% (which corresponds to 146 and 754 misclassified images for the training and test sets, respectively) were obtained with the multiple-Beta-Bernoulli mixture. It is clear also that the best results (over 98% and 92% for the training and test sets, respectively) were obtained using multiple-Beta-Bernoulli mixtures, which show that the spatial color distribution can be effective to discriminate

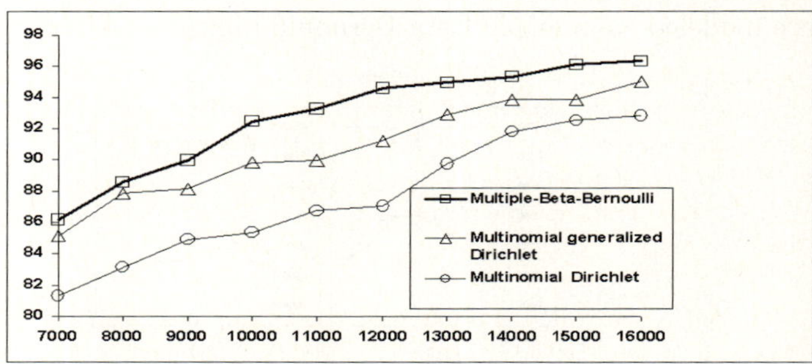

Fig. 3. Accuracy, using Multiple-Beta-Bernoulli mixture, as a function of the number of images in the training set

between landscape and city images. Tables 2.a and 2.b show the confusion matrices for both the training and test sets using multiple-Bernoulli mixtures. Figure 3 shows the classification accuracies for the test set as a function of the number of images in the training set. According to this figure, increasing the number of images in the training set improves the classification accuracy which is actually an expected result showing that our model has good capacities to learn when additional images are introduced.

5   Conclusion

A statistical model to add spatial constraints to the image histogram has been proposed. Each image pixel with a given color is considered to be coherent or not and then modeled as a multiple-Beta-Bernoulli mixture. All the model parameters are estimated using the ML approach through the DAEM algorithm mixed with a Newton-Raphson step. To assess the capabilities of this model, experiments have been carried out in a specific image classification problem.

Acknowledgment

The completion of this research was made possible thanks to the Natural Sciences and Engineering Research Council of Canada (NSERC), a NATEQ Nouveaux Chercheurs Grant, and a start-up grant from Concordia University.

References

1. Swain, M., Ballard, D.: Color Indexing. International Journal of Computer Vision 7(1), 11–32 (1991)
2. Treisman, A., Paterson, R.: A Feature Integration Theory of Attention. Cognitive Psychology 12(1), 97–136 (1980)
3. Bouguila, N.: Spatial Color Image Databases Summarization. In: Proc. of the IEEE International Conference on Acoustics, Speech, and Signal Processing (ICASSP 2007), Honolulu, HI, USA, vol. 1, pp. I–953–I–956 (2007)
4. Bouguila, N., ElGuebaly, W.: A Generative Model for Spatial Color Image Databases Categorization. In: Proc. of the IEEE International Conference on Acoustics, Speech, and Signal Processing (ICASSP 2008), Las Vegas, Nevada, USA, pp. 821–824 (2008)
5. Pass, G., Zabih, R.: Histogram Refinement for Content-Based Image Retrieval. In: Proc. of the 3rd IEEE Workshop on Applications of Computer Vision (WACV), pp. 96–102 (1996)
6. Bouguila, N., Ziou, D.: Unsupervised Learning of a Finite Discrete Mixture: Applications to Texture Modeling and Image Databases Summarization. Journal of Visual Communication and Image Representation 18(4), 295–309 (2007)
7. Bouguila, N.: Clustering of Count Data Using Generalized Dirichlet Multinomial Distributions. IEEE Transactions on Knowledge and Data Engineering 20(4), 462–474 (2008)

8. Bouguila, N., Ziou, D., Vaillancourt, J.: Unsupervised Learning of a Finite Mixture Model Based on the Dirichlet Distribution and its Application. IEEE Transactions on Image Processing 13(11), 1533–1543 (2004)
9. Bouguila, N., Ziou, D.: A Hybrid SEM Algorithm for High-Dimensional Unsupervised Learning Using a Finite Generalized Dirichlet Mixture. IEEE Transactions on Image Processing 15(9), 2657–2668 (2006)
10. Bouguila, N., Ziou, D.: Improving Content Based Image Retrieval Systems Using Finite Multinomial Dirichlet Mixture. In: Proc. of the IEEE Workshop on Machine Learning for Signal Processing (MLSP 2004), Sao Luis, Brazil, pp. 23–32 (2004)
11. Bouguila, N., ElGuebaly, W.: On Discrete Data Clustering. In: Washio, T., Suzuki, E., Ting, K.M., Inokuchi, A. (eds.) PAKDD 2008. LNCS (LNAI), vol. 5012, pp. 503–510. Springer, Heidelberg (2008)
12. Pass, G., Zabih, R., Miller, J.: Comparing Images Using Color Coherence Vectors. In: Proc. of the Fourth ACM International Conference on Multimedia, pp. 65–73 (1997)
13. Metzler, D., Lavrenko, V., Croft, W.B.: Formal Multiple-Bernoulli Models for Language Modeling. In: Proc. of the 27th Annual International ACM Conference on Research and Development in Information Retrieval (SIGIR), pp. 540–541. ACM Press, New York (2004)
14. McLachlan, G.J., Krishnan, T.: The EM Algorithm and Extensions. Wiley-Interscience, New York (1997)
15. Ueda, N., Nakano, R.: Deterministic Annealing EM Algorithm. Neural Networks 11, 271–282 (1998)
16. Graybill, F.A.: Matrices with Applications in Statistics. Wadsworth, California (1983)
17. Bouguila, N., Ziou, D.: Unsupervised Selection of a Finite Dirichlet Mixture Model: An MML-Based Approach. IEEE Transactions on Knowledge and Data Engineering 18(8), 993–1009 (2006)
18. Bouguila, N., Ziou, D.: High-Dimensional Unsupervised Selection and Estimation of a Finite Generalized Dirichlet Mixture Model Based on Minimum Message Length. IEEE Transactions on Pattern Analysis and Machine Intelligence 29(10), 1716–1731 (2007)
19. Rissanen, J.J.: Modeling by Shortest Data Description. Automatica 14, 445–471 (1978)
20. Elkan, C.: Clustering Documents with an Exponential-Family Approximation of the Dirichlet Compound Multinomial Distribution. In: Proc. of the 23rd International Conference on Machine Learning (ICML 2006), pp. 289–296. ACM Press, New York (2006)
21. Bouguila, N., Ziou, D., Vaillancourt, J.: Novel Mixtures Based on the Dirichlet Distribution: Application to Data and Image Classification. In: Perner, P., Rosenfeld, A. (eds.) MLDM 2003. LNCS, vol. 2734, pp. 172–181. Springer, Heidelberg (2003)
22. Vailaya, A., Jain, A.K., Zhang, H.J.: On Image Classification: City vs. Landscape. In: Proc. of the IEEE Workshop on Content-Based Access of Image and Video Libraries, pp. 3–8 (1998)

Efficient Video Shot Summarization Using an Enhanced Spectral Clustering Approach

Vasileios Chasanis, Aristidis Likas, and Nikolaos Galatsanos

Department of Computer Science, University of Ioannina,
45110 Ioannina, Greece
{vchasani,arly,galatsanos}@cs.uoi.gr

Abstract. Video summarization is a powerful tool to handle the huge amount of data generated every day. At shot level, the key-frame extraction problem provides sufficient indexing and browsing of large video databases. In this paper we propose an approach that estimates the number of key-frames using elements of the spectral graph theory. Next, the frames of the video sequence are clustered into groups using an improved version of the spectral clustering algorithm. Experimental results show that our algorithm efficiently summarizes the content of a video shot producing unique and representative key-frames outperforming other methods.

Keywords: Video summarization, Key-frame extraction, Spectral clustering, Global k-means.

1   Introduction

Due to the rapid increase in the amount of video data generated nowadays, appropriate video indexing and browsing tools are required to efficiently store videos in large databases. Such indexed databases assist the efficient retrieval of any video in modern video retrieval systems. The most popular indexing and summarization method is based on key-frame extraction. More specifically, a video shot, which is the smallest physical segment of a video (defined as an unbroken sequence of frames recorded from one camera) can be sufficiently summarized using its most representative frames, which are the key-frames. A good summarization of a shot contributes to an efficient indexing and browsing of large video databases. The key-frame extraction problem seeks to find the most representative frames of a shot. However any such algorithm should fulfil some requirements. Firstly, the key-frames should represent the whole video content without missing important information and secondly, these key-frames should not be similar, in terms of video content information, thus containing redundant information.

Several approaches have been proposed for the key-frame extraction problem. We assume that a video is already segmented into shots using any shot detection algorithm. The simplest methods choose the first, last and median frames of a shot or a combination of the previous ones to describe the content of a shot [7].

V. Kůrková et al. (Eds.): ICANN 2008, Part I, LNCS 5163, pp. 847–856, 2008.

In [9] the optical flow is computed and the local minima of a motion metric are selected as key-frames. In [2] it is proposed to form a trajectory from the feature vectors for all frames within a shot. The magnitude of the second derivative of this feature trajectory with respect to time is used as a curvature measure in this case. As key-frames the local minima and maxima of this magnitude are selected. In [12] multiple frames are detected using unsupervised clustering based on the visual variations in shots. A main drawback of this algorithm is the determination of the appropriate number of key-frames to represent each shot which depends on the threshold parameter that controls the density of the clusters. A variant of this algorithm is presented in [6] where the final number of key-frames depends on a threshold parameter which defines whether two frames are similar. In [3] the extraction of the key-frames is based on detecting curvature points within the curve of cumulative frame differences and suggest quality measures to evaluate the summaries.

In our approach the frames of a video sequence are clustered into groups using an improved version of the typical spectral clustering method [5] that uses the global k-means algorithm [4] in the clustering stage after the eigenvector computation. Moreover, the number of key-frames is estimated using results from spectral graph theory. The rest of the paper is organized as follows: In section 2 we present our key-frame extraction algorithm. In section 3 a method for the estimation of the number of clusters is proposed and in section 4 we present quality measures for the evaluation of the summary of a shot. In section 5 we present numerical experiments and compare our method with three other approaches. Finally in section 6 we provide conclusions and suggestions for further study.

2   Key-Frames Extraction

To deal with the problem of key frame extraction two major issues must be addressed. Firstly the number of key-frames must me estimated and second an efficient algorithm must produce unique key-frames that will summarize the content of the shot. In our work for each frame a 16-bin HSV normalized histogram is used, with 8 bins for hue and 4 bins for each of saturation and value.

2.1   The Typical Spectral Clustering Algorithm

To perform key-frame extraction the video frames of a shot are clustered into groups using an improved spectral clustering algorithm. Then, the medoid of each group, defined as the frame of a group whose average similarity to all other frames of this group is maximal, is characterized as a key-frame. The main steps of the typical spectral clustering algorithm [5] are described next. Suppose there is a set of objects $S = s_1, s_2, \ldots, s_N$ to be partitioned into K groups.

1. Compute similarity matrix $A \in \mathbb{R}^{N \times N}$ for the pairs of objects of the data set S.
2. Define D to be the diagonal matrix whose (i, i) element is the sum of the A's i-th row and construct the Laplacian matrix $L = I - D^{-1/2}AD^{-1/2}$.

3. Compute the K principal eigenvectors $x_1, x_2, \ldots, x_K$ of matrix L to build an $N \times K$ matrix $X = [x_1\, x_2 \ldots\, x_K]$.
4. Renormalize each row of X to have unit length and form matrix Y so that:

$$y_{ij} = x_{ij}/(\sum_j x_{ij}^2)^{1/2} \ . \tag{1}$$

5. Cluster the rows of Y into K groups using k-means.
6. Finally, assign object s_i to cluster j if and only if row i of the matrix Y has been assigned to cluster j.

In what concerns our key-frame extraction problem, suppose we are given a data set $H = H_1, \ldots, H_N$ where H_n is the feature vector (normalized color histogram) of the n-th frame. The distance function we consider is the Euclidean distance between the histograms of the frames. As a result each element of the similarity matrix A is computed as follows:

$$a(i, j) = 1 - \sqrt{\sum_{h \in bins} (H_i(h) - H_j(h))^2} \ . \tag{2}$$

2.2   Global k-Means

Furthermore, in the fifth step of the spectral clustering algorithm instead of using the typical k-means approach, we have used the very efficient global k-means algorithm [4]. Global k-means in an incremental deterministic clustering algorithm that overcomes the important initialization problem of the typical k-means approach. This initialization problem has been found to be severe in the case of frame clustering, significantly affecting the quality of the key-frames. Using the global k-means, the obtained key frames usually provide a sensible representation of shot content. Next we briefly review the global k-means algorithm. Suppose we are given a data set $X = x_1, \ldots, x_N$, $x_n \in R^d$ to be partitioned into K disjoint clusters $C_1, C_2, \ldots, C_K$.

This algorithm is incremental in nature. It is based on the idea that the optimal partition into K groups can be obtained through local search (using k-means) starting from an initial state with i) the k-1 centers placed at the optimal positions for the (k-1)-clustering problem and ii) the remaining k-th center placed at an appropriate position within the dataset. Based on this idea, the K-clustering problem is incrementally solved as follows. Starting with k=1, find the optimal solution which is the centroid of the data set X. To solve the problem with two clusters, the k-means algorithm is executed N times (where N is the size of the data set) from the following initial positions of the cluster centers: the first cluster center is always placed at the optimal position for the problem with k=1, whereas the second center at execution n is initially placed at the position of data x_n. The best solution obtained after the N executions of k-means is considered as the solution for k=2. In general if we want to solve the problem with k clusters, N runs of the k-means algorithm are performed, where each run n starts with the k-1 centers initially placed at the positions corresponding to the solution obtained for the (k-1)-clustering problem, while the k-th center is initially placed at the position of data x_n. A great benefit of this algorithm is that it provides the solutions for all k-clustering problems with $k \leq K$.

3   Estimation of Number of Key-Frames Using Spectral Graph Theory

As already mentioned in the introduction, the number of key-frames cannot be predetermined due to the different content of each shot. In our approach we attempt to estimate the number of the key-frames using results from the spectral graph theory.

Assume we wish to partition dataset S into disjoint subsets $(S_1, \ldots, S_K)$, and let $X = [X_1, \ldots, X_K] \in \mathbb{R}^{N \times K}$ denote the partition matrix, where X_j is the binary indicator vector for set S_j such that:

$$
\begin{array}{lll}
X(i,j) = 1 & : & if\ i \in S_j \\
X(i,j) = 0 & : & otherwise
\end{array}
\tag{3}
$$

The optimal solution [10] is defined as:

$$
\begin{array}{c}
\max_{X}\ trace(X^T L X) \\
s.t.\ X^T X = I_K \text{ and } X(i,j) \in \{0,1\}
\end{array}
\tag{4}
$$

where L is the Laplacian matrix defined in section 2.1. The spectral clustering algorithm (for K clusters) provides solution to the following optimization problem:

$$
\begin{array}{c}
\max_{Y}\ trace(Y^T L Y) \\
s.t.\ Y^T Y = I_K
\end{array}
\tag{5}
$$

Relaxing Y into the continuous domain turns the discrete problem into a continuous optimization problem. The optimal solution is attained at $Y = U_K$, where the columns u_i of $U_k, i = 1, \ldots, K$, are the eigenvectors corresponding to the ordered top K largest eigenvalues λ_i of L. Since it holds that [11]:

$$
\lambda_1 + \lambda_2 + \ldots + \lambda_K = \max_{Y^T Y = I_K}\ trace(Y^T L Y) \,,
\tag{6}
$$

the optimization criterion that also quantifies the quality of the solution for K clusters and its corresponding difference for successive values of K are respectively given by:

$$
\begin{array}{l}
sol(K) = \lambda_1 + \lambda_2 + \ldots + \lambda_K \\
sol(K+1) - sol(K) = \lambda_{K+1}
\end{array}
\tag{7}
$$

When the improvement in this optimization criterion (i.e. the value of the λ_{K+1} eigenvalue) is below a threshold, improvement by the addition of cluster $K+1$

is considered negligible, thus the estimate of the number of clusters is assumed to be K. The threshold value that is used in all our experiments was fixed to $Th=0.005$ with very good results.

4 Summary Evaluation

A difficult issue of the key-frame extraction problem is related to the evaluation of the extracted key-frames, since it is rather subjective which frames are the best representatives of the content of a shot. There are several quality measures that can be used to evaluate the efficiency of the algorithms. In [3], two quality measures are used. The first is the Fidelity measure proposed in [1] and the second is the Shot Reconstruction Degree measure proposed in [8].

4.1 Average Fidelity

The Fidelity measure compares each key-frame with other frames in the shot. Given the frame sequence $F = \{F_1, F_2, \ldots, F_N\}$ and the set of key-frames $KF = \{KF_1, KF_2, \ldots, KF_{N_{kf}}\}$ the distance between the set of key-frames KF and a frame F_n is defined as:

$$d(F_n, KF) = \min_j \ Diff(F_n, KF_j), \ j = 1, 2, \ldots, N_{kf} \ , \tag{8}$$

where N_{kf} is the number of key-frames and $Diff(F_i, F_j)$ a distance measure between two frames F_i and F_j. The Fidelity measure is computed as:

$$\mathrm{Fidelity}(F, KF) = \mathrm{MaxDiff} - d_{all}(F, KF) \ , \tag{9}$$

where MaxDiff is a constant representing the largest possible value that the frame difference measure can assume and $d_{all}(F, KF)$ is as follows:

$$d_{all}(F, KF) = \max_n \ d(F_n, KF), \ n = 1, 2, \ldots, N \ . \tag{10}$$

However as mentioned in [8], Fidelity cannot capture well the dynamics of a shot since it focuses on global details. For that reason we compute the Average Fidelity which is computed using the average of the minimal distances between the key frame set and the video shot and is given from the following equation:

$$\mathrm{Average \ Fidelity}(F, KF) = \mathrm{MaxDiff} - \frac{1}{N} \sum_{n=1}^{N} d(F_n, KF) \ . \tag{11}$$

4.2 Shot Reconstruction Degree

Given the set of key-frames, the whole frame sequence of a shot can be reconstructed using an interpolation algorithm. The better the reconstructed video sequence approximates the original sequence, the better the set of key-frames summarizes the video content. More specifically, given the frame sequence F, the set of key-frames KF and a frame interpolation algorithm IA(), we can reconstruct any frame from a pair of key-frames in KF [8]:

$$\tilde{F}_n = \text{IA}(KFn_j, KFn_{j+1}, n, n_j, n_{j+1}), \ n_j \leq n < n_{j+1} \ . \tag{12}$$

The Shot Reconstruction Degree (SRD) measure is defined as follows:

$$SRD(F, KF) = \sum_{n=0}^{N-1} (Sim(F_n, \tilde{F}_n)) \ , \tag{13}$$

where Sim() is given from the following equation:

$$Sim(F_n, \tilde{F}_n) = \log(MaxDiff/Diff(F_n, \tilde{F}_n)) \ , \tag{14}$$

where $Diff(F_i, F_j)$ is a distance measure between two frames F_i and F_j and MaxDiff the largest possible value that the frame difference measure can assume.

5   Experiments

In this section we present the application and evaluation of our key-frame extraction algorithm compared to three other algorithms.

5.1   Data

In our experiments we used seven frame sequences. The first frame sequence describes an action of a comedy movie that takes place in an office. The other six sequences are taken from sports. Three of them describe three attempts in the NBA Slam Dunk Contest and the other three a goal attempt in a football match taken from three individual cameras. In *Table 1* we present the characteristics of the video data set.

5.2   Comparison

We compare the proposed approach to three other methods. The first one is the simple k-means algorithm. For each shot we perform 20 iterations of the k-means algorithm keeping the iteration with the minimum clustering error. The number of clusters in k-means algorithm is assumed to be the same with one selected

Table 1. Characteristics of video data set

	FRAME SEQUENCE						
	F_1	F_2	F_3	F_4	F_5	F_6	F_7
No. FRAMES	633	144	145	146	225	300	172
GENRE	Comedy	Basketball	Basketball	Basketball	Football	Football	Football

using the proposed estimation algorithm of section 3. The second technique is presented in [6], as a variant of the method presented in [12]. Initially, the middle frame of the video sequence is selected as the first key-frame and added to the empty set of key-frames KF. Next, each frame in the video sequence is compared with the current set of key-frames. If it differs from every key-frame in the current set, then it is added into the set as a new key-frame. This algorithm uses a threshold to discriminate whether two frames are similar or not. In our experiments this threshold parameter is set to such a value that the number of key-frames extracted is the same as in our algorithm. Finally, the third technique is the spectral clustering algorithm employing the simple k-means algorithm.

5.3   Evaluation

To evaluate the results of the extracted key-frames we use the metrics mentioned in section 4. More specifically in *Tables 2-3* we present the performance results for the Average Fidelity and SDR measures respectively. To compute the SDR we use a simple linear interpolation algorithm on the frame's features. It is clear that our approach provides the best summarization of each shot compared to the other methods and the best reconstruction of the original video sequence from the extracted key-frames.

Table 2. Comparative results of the tested key-frame extraction algorithms using Average Fidelity

AV.FIDELITY	FRAME SEQUENCE						
ALGORITHM	F_1	F_2	F_3	F_4	F_5	F_6	F_7
Our method	0.973	0.9437	0.9507	0.9559	0.9687	0.9546	0.978
K-means	0.9549	0.9278	0.9344	0.948	0.9467	0.931	0.9654
Method in [6]	0.9616	0.8913	0.9268	0.9405	0.955	0.9424	0.9672
Typical Spectral Algorithm	0.9619	0.9235	0.9253	0.9462	0.9625	0.9318	0.9675

Table 3. Comparative results of the tested key-frame extraction algorithms using SDR

SDR	FRAME SEQUENCE						
ALGORITHM	F_1	F_2	F_3	F_4	F_5	F_6	F_7
Our method	1859.66	425.87	511.58	527.32	855.96	860.01	711.18
K-means	1533.34	369.87	430.78	356.46	808.24	753.75	648.71
Method in [6]	1693.1	292.43	374.23	340.89	758.23	813.1	642.97
Typical Spectral Algorithm	1620.6	362.64	431.32	393.02	780.33	791.2	663.15

5.4   Representation

As already mentioned in section 3.2 a great benefit of the global k-means algorithm is that it provides the solutions for all intermediate k-clustering problems with $k \leq K$. In *Fig. 1* we give an example of the extracted key-frames of a video shot with object and camera motion. Moving from the top to the bottom of

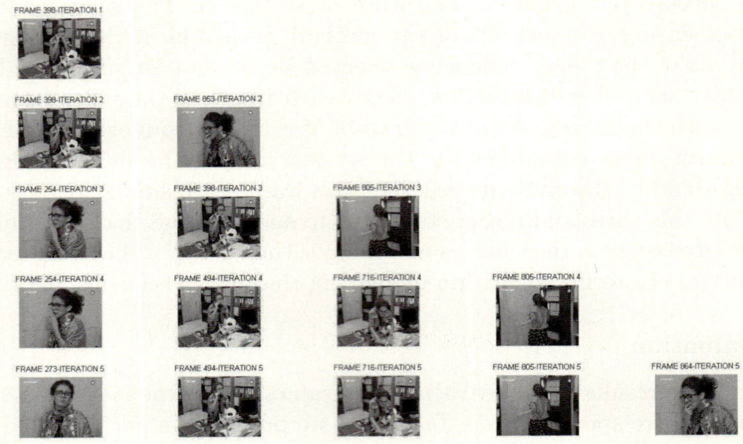

Fig. 1. Key-frame extraction of a shot using the proposed method($N_{kf} = 5$)

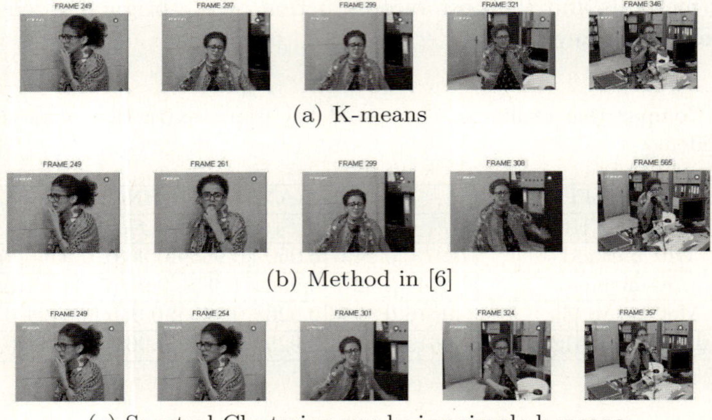

(a) K-means

(b) Method in [6]

(c) Spectral Clustering employing simple k-means

Fig. 2. Results for the key-frame extraction algorithms used for comparison

this figure we show all solutions until the selected number of key-frames N_{kf}=5 is reached. The shot that we used contains 633 frames. It shows a woman in an office set-up. This shot can be semantically divided into 5 sub-shots. a) The woman stands against a door eavesdropping and then rushes to her office to pick up the phone that is ringing; b) she talks on the phone, c) lays the receiver of the phone down with a visible effort not to make any noise, d) she rushes back to the door, and e)she continues eavesdropping. In *Fig. 2* we provide the key-frames extracted performing the simple k-means algorithm, the algorithm in [6] and the typical spectral clustering algorithm. All algorithms fail to provide a solution

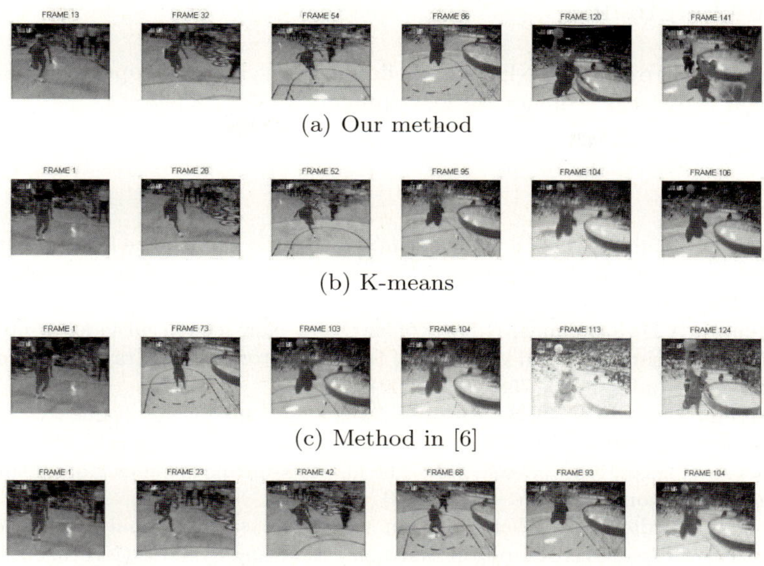

(a) Our method

(b) K-means

(c) Method in [6]

(d) Spectral Clustering employing simple k-means

Fig. 3. Key-frame extraction algorithms in comparison in basketball sequence

adequately describing the visual content of the shot, whereas our approach provides a sensible solution. More specifically, they do not produce any frames for sub-shots (c), (d) and (e) and instead produce multiple frames for sub-shot (a). In contrast the proposed approach produces key frames for all sub-shots. In *Fig. 3* we provide the key-frames for these four algorithms for a video shot containing 145 frames and describing a slam dunk attempt. It becomes clear that our algorithm summarizes the attempt from the beginning to the end, whereas the other three fail to describe the end of the action.

6   Conclusions

In this paper a new method for key-frame extraction is proposed. Key-frames are extracted using a spectral clustering method employing the global k-means algorithm in the clustering procedure. Furthermore, the number of key-frames is estimated using results from the spectral graph theory, by examining the eigenvalues of the similarity matrix corresponding to pairs of shot frames. Appropriate quality measures indicate that our method outperforms traditional techniques and provides efficient summarization and reconstruction of a video sequence from the extracted key-frames. In future work, we will try to improve the performance of our method by examining other features for the frames, such as motion and edge histograms.

Acknowledgments

This research project (PENED) is co-financed by E.U.-European Social Fund (75%) and the Greek Ministry of Development-GSRT (25%).

References

1. Chang, H.S., Sull, S., Lee, S.U.: Efficient Video Indexing Scheme for Content-Based Retrieval. IEEE on Transactions Circuits and Systems Video Technology 9(8), 1269–1279 (1999)
2. Doulamis, A.D., Doulamis, N.D., Kollias, S.D.: Non-sequential video content representation using temporal variation of feature vectors. IEEE Transactions on Consumer Electronics 46(3), 758–768 (2000)
3. Gianluigi, C., Raimondo, S.: An innovative algorithm for key frame extraction in video summarization. Journal of Real-Time Image Proceesing 1(1), 69–88 (2006)
4. Likas, A., Vlassis, N., Verbeek, J.J.: The global k-means clustering algorithm. Pattern Recognition 36(2), 451–461 (2003)
5. Ng, A.Y., Jordan, M.I., Weiss, Y.: On spectral clustering: Analysis and an algorithm. In: Proceedings Neural Information Processing Systems (NIPS 2001) (2001)
6. Rasheed, Z., Shah, M.: Detection and representation of scenes in videos. IEEE Transactions on Multimedia 7(6), 1097–1105 (2005)
7. Rui, Y., Huang, T.S., Mehrotra, S.: Exploring video structure beyond the shots. In: Proceedings of IEEE International Conference on Multimedia Computing and Systems (ICMCS), Texas, USA, pp. 237–240 (1998)
8. Tieyan, L., Zhang, X., Feng, J., Lo, K.T.: Shot reconstruction degree: a novel criterion for key frame selection. Pattern Recognition Letters 25, 1451–1457 (2004)
9. Wolf, W.: Key frame selection by motion analysis. In: Proceedings of IEEE International Conference on Acoustics, Speech, and Signal Processing, pp. 1228–1231 (1996)
10. Xing, E.P., Jordan, M.I.: On semidefinite relaxation for normalized k-cut and connections to spectral clustering. Technical Report CSD-03-1265, Computer Science Division, University of California, Berkeley (2003)
11. Zha, H., Ding, C., Gu, M., He, X., Simon, H.: Spectral relaxation for k-means clustering. In: Neural Information Processing Systems (NIPS 2001) (2001)
12. Zhuang, Y., Rui, Y., Huang, T.S., Mehrotra, S.: Adaptive key frame extraction using unsupervised clustering. In: Proceedings of IEEE International Conference on Image Processing, pp. 866–870 (1998)

Surface Reconstruction Techniques Using Neural Networks to Recover Noisy 3D Scenes

David Elizondo, Shang-Ming Zhou, and Charalambos Chrysostomou

Centre for Computational Intelligence,
School of Computing, De Montfort University,
The Gateway, Leicester LE1 9BH, UK
Phone: +44(0) 116-207-8471; Fax: +44(0)116-207-8159
{elizondo,smzhou}@dmu.ac.uk,
charalambos_chrysostomou@hotmail.com

Abstract. This paper presents a novel neural network approach to recovering of 3D surfaces from single gray scale images. The proposed neural network uses photometric stereo to estimate local surfaces orientation for surfaces at each point of the surface that was observed from same viewpoint but with different illumination direction in surfaces that follow the Lambertian reflection model. The parameters for the neural network are a 3x3 brightness patch with pixel values of the image and the light source direction. The light source direction of the surface is calculated using two different approaches. The first approach uses a mathematical method and the second one a neural network method. The images used to test the neural network were both synthetic and real images. Only synthetic images were used to compare the approaches mainly because the surface was known and the error could be calculated. The results show that the proposed neural network is able to recover the surface with a highly accurately estimate.

Keywords: Neural networks, 3D surfaces, reconstruction, noisy scenes, light sources.

1 Introduction

Shape from shading algorithms try to recover the 3D shape from a single gray-scale image. The recovered surface can be expressed by depth Z, the surface normal or surface gradient (p, q). In mathematical methods there are two main categories of algorithms for computing shape from a shaded image, global methods and local methods. In the global methods the shape is recovered by minimizing a cost function concerning some constraints such as smoothness. In such approaches the variational calculus is used to iteratively compute the shape, which is globally consistent. In the local methods, the shape is recovered by using local constraints in relation to the surface being recovered, or concerning the reflectance map being used. The common assumptions about surface shape and reflectance are that the surface is locally spherical, and the reflectance map is linear in surface shape. The local methods, on the other hand, are more complex, however they provide more accurate recovered surfaces.

V. Kůrková et al. (Eds.): ICANN 2008, Part I, LNCS 5163, pp. 857–866, 2008.

Some researchers have proposed to apply the neural network schemes to surface reconstruction and image approximation. Wei and Hirzinger used a feed forward back-propagation neural network to reconstruct surfaces from shading [1], in which the p and q are finite point of values. Mostafa, Yamany and Farag proposed to integrate dense depth map calculated from shading algorithm combined with a sparse range data to reconstruct 3D surfaces [2]. In [3], a neural network model was developed to learn localized brightness patterns for reconstructing surface from shading. Cheng used neural network based photometric stereo to estimate local surfaces orientation for surfaces that are observed from same viewpoint nevertheless with different illuminate direction [4]. Zhou, Li and Xu proposed a Hopfield neural network for image approximation [5], which has a great potential of being applied to surface reconstruction. However, most of the current methods for source reconstruction focus on the objects in scenes that are free of noise. Through some experiments it was found that for the object in the scenes with added noises the estimating results about light source directions are impacted much so that the inaccurate surface reconstruction results are obtained. In the interest of reconstructing 3D Lambertian surface from a single image of a noisy scene, we propose a novel neural network based approach in this paper.

The proposed neural network scheme uses photometric stereo to estimate local surfaces orientation for surfaces. This is done at each point of the surface which is observed from the same viewpoint but with a different illumination direction. The surfaces follow the Lambertian reflection model. The parameters for the neural network are a 3x3 brightness patch with pixel values of the image and the light source direction of scenes with noises. The light source direction of the surface is calculated using two different approaches, a mathematical one and a neural network based one. The motivation behind choosing a neural network scheme lies in that complex relationships in data can be discovered; neural network can solve both linear and non linear problems, which is ideal for image reconstruction problems where the solution is roughly unknown. Furthermore, nearly all (synthetic or real) shapes from shading images suffer from noise. Therefore, a neural network approach is a suitable solution the problem of recovering of 3D images. Only synthetic images were used to compare the approaches mainly because the surface was known and the error could be calculated. The results show that the proposed neural network method is efficient in recovering a highly accurate estimate of the desired surface from single images of noisy scenes.

The orgnisation of this paper is as follows. Section 2 proposed a neural network approach to estimating light source direction from images in noisy scenes. Then based on the light source estimation result, 3D surfaces are to be recovered by a proposed neural network model in Section 3. Section 4 presents the experimental results on noisy synthetic images and real images. Section 5 concludes this paper.

2   Neural Network Approach to Estimating Light Source Direction in Noisy Scenes

In this paper, the estimation of illumination direction and albedo from a noisy scene is conducted under the following three assumptions:

- All illumination originates from one light source in the scene;

- All objects have Lambertian reflectance surfaces;
- The surface normal of the objects in the scene complies with normal distribution.

The first two assumptions are made in many existing approaches as well. Although these assumptions are seemingly simple and limit the applications, the current research indicates that estimating illumination direction and properties based on these assumptions is a difficult task. In this section a neural network based approach is presented to estimate the light source direction from image in noisy scenes and reduce the error produced by the noise, in which some scene features about noisy image are used as inputs to the neural network.

2.1 Scene Features of Noisy Image

Given a noisy image, the following features are extracted by some direct mathematical formula [6], these features will work as the inputs of the neural network to recover the original slant and tilt of light source.

The first feature of a noisy image is the average of of the image brightness:

$$E_1 = \frac{\sum_y \sum_x R_{x,y}}{T} \tag{1}$$

where $R_{x,y}$ is the image intensity value at pixel (x, y), T denotes the total number of pixels in the image.

The second feature of a noisy image is the average of the squared image brightness:

$$E_2 = \frac{\sum_y \sum_x (R_{x,y}{}^2)}{T} \tag{2}$$

The third and fourth features of a noisy image are the averages of the horizontal and vertical components of the direction of the noisy image spatial gradient separately:

$$E_x = \frac{\sum_y \sum_x (R_{x,y} - R_{x,y-1})}{T} \tag{3}$$

$$E_y = \frac{\sum_y \sum_x (R_{x,y} - R_{x-1,y})}{T} \tag{4}$$

The fifth and sixth features are the contaminated slant (CS) angle and contaminated tilt (CT) angle extracted from a noisy image.

Given the Lambertian reflectance model whose surface reflect light in all directions, the slant angle θ_i is the angle between the camera axis and the light source direction, while the tilt angle ϕ is the angle between the x-axis and the projection light in same direction. The CT is computed by:

$$Ct = tan^{-1} \left(\frac{E_x}{E_y} \right) \tag{5}$$

while the CS is given by:

$$Cs = \begin{cases} 1 & if\ \alpha \geq 1 \\ cos^{-1} \left(\frac{4E_1}{g} \right) & if\ \alpha < 1 \end{cases} \tag{6}$$

where

$$g = \sqrt{6\pi^2 E_2 - 48 E_1^2} \tag{7}$$

and

$$\alpha = \frac{4E_1}{g} \tag{8}$$

2.2 Structure of the Proposed Neural Network

Given a noisy scene, where the information available to us is only the features extracted in the above subsection, our task is to recover the original slant angle and tilt angle of the light source, and calculate the light source direction. It is known in the signal processing community that it is rather challenging to recover the original signal from data highly contaminated by noise without prior knowledge. So we propose a neural neural network model with six input neurons and two output neurons to fufil this task. The six input neurons are merely "fan-out" units accepting the six features extracted above from noisy image, no processing takes place in these units. The sigmoid activation function is used in this feedforward network with one hidden layer of units.

Then the most important issue is how to effectively train the network from a noisy scene for recovery of the original slant and tilt of light source. To this end, we need to create the training data. Given N initial images from scenes that are less contaminated by noise, and their corresponding slant s_i and tilt angles t_i of light source are calculated in a similar way as did by the (6) and (5) separately, let $y^{(i)} = (s_i, t_i)^T$. Then Gaussian noises are added to these N images, and six above scene features are extracted from the noise added images, let $x^{(i)} = (E_1^i, E_2^2, E_x^i, E_y^i, Cs_i, Ct_i)^T$. Then the data set $\left\{ (x^{(i)}, y^{(i)}) \right\}_{i=1}^{N}$ is used to train the neural network based on the backpropagation algorithm. Such networks can learn arbitrary associations by using differentiable activation functions. A theoretical foundation of backpropagation can be found in [7] and [8].

The rationale of the neural network's merit in reducing noise lies in that the output of every neuron in the hidden layer and output layer is obtained by having to perform the weighted averaging of inputs from previous layer, which acts as a lowpass filter. So the noise from the input layer will be cancelled much through the connections between input layer and hidden layer, and reduced further through the connections between hidden layer and output layer.

2.3 Estimation of Light Source Direction

Given a noisy image, the recovered slant s and tilt t can be obtained by applying the trained neural network to the features. Then the light source direction in (S_x, S_y, S_z) can be estimated as $S_x = cos(t) \cdot s$, $S_y = sin(t) \cdot s$, and $S_z = cos(s)$.

3 Neural Network Approach to Recover 3D Surfaces from Noise Images

In this paper a further neural network is proposed to reconstruct 3D Lambertian surfaces from images and compared with already available mathematical methods. The novel neural network is a multilayer feed-forward back-propagation neural network.

3.1 Neural Network Structure

In this method, the height z for a particular point of a 3D surface can be derived from a brightness patch, (bp), and the given light source s as described in equation 9.

$$z_{(x,y)} = f_I(f_L(bp|s)) \tag{9}$$

where f_L is the function that adjust the image, in this case the brightness pattern, to the light source. f_I is the function that tries to compute the surface heights taking into consideration the adjusted height brightness pattern to the light source for a particular pixel taking into consideration also the neighborhood pixels. The idea behind taking also neighborhood pixel values to reconstruct the surface height for every surface point is to reduce noise existing in the image and to help identify "optical illusions" that may also exist in the image. The novel neural network is expected to outperform mathematical methods since many of these methods are highly influenced by noise and can only find useful applications to synthetic images were many surface and image variables are known.

The proposed feedforward neural network model with 3 layers uses photometric stereo to estimate local surfaces orientation for surfaces at each point of the surface that was observed from the same viewpoint but with different illumination direction. The final outputs of the neural network are:

$$y = f^3(LW^{3,2}f^2(LW^{2,1}f^1(IW^{1,1}p + b^1) + b^2) + b^3)$$
$$= tansig(LW^{3,2}tansig(LW^{2,1}tansig(IW^{1,1}p + b^1) + b^2) + b^3) \tag{10}$$

where $f^i(i = 1, 2, 3)$ correspond to the *tansig* activation functions, IW to the input weights, p to the neural network input vector and b^1 to the first layer bias, $LW^{2,1}$ to the hidden weights of the second layer, α^1 to the neural network output of the input layer and b^2 is the second layer bias with dimensions 20×1, $LW^{3,2}$ to the hidden weights of the third layer, α^2 to the neural network outputs of the second layer and b^3 to the third layer bias with dimensions 1×1. Twenty neurons are used for the first layer and the output of the first layer is a 20×1 matrix representing the relationships between the intensity values and the and the light source direction at each surface patch. Twenty neurons are also used for the second layer, and the output of the layer is also a 20×1 matrix representing the modified intensities to the given light source. Twenty neurons were also used for the third layer, and the output of the layer is a single value representing the height of the surface for a particular point.

The proposed neural network has as input an intensity patch and the light source direction. With the light source provided, the neural network should have the ability to recover the desired surface. Furthermore, the neural network model is also expected to be able to adjust the surface patch orientation according to the light source direction.

As shown in Figure 1 the above input for the neural network patch is exactly the same. The main difference is that the light source direction is given. For the above brightness patch the light source direction is $(s_x = 1, s_y = 0, s_z = 1)$. That means that the light source direction is on the right side of the surface in relation to the viewpoint direction. As expected the plain must be brighter on the right, thus the surface orientation may appear wrongly. The proposed neural network is expected to identify these

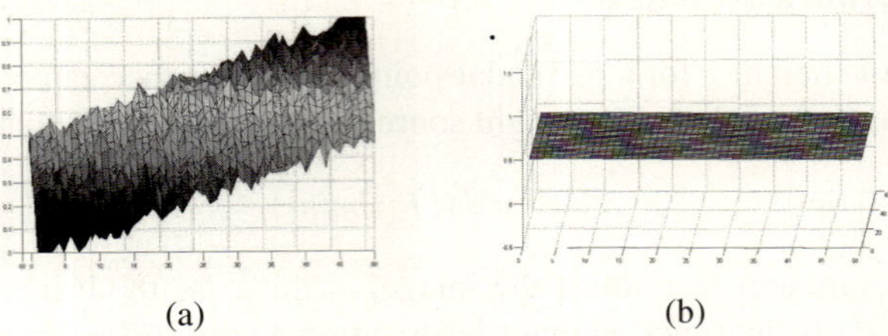

(a)   (b)

Fig. 1. Brightness Patch for Light Source

"optical illusions" and to recover the real surface. A similar neural network approach was developed by Cheng [4] with the difference that the light source is given as opposed to being calculated as in the approach described in this paper. Another difference in the structure of the neural networks is that in the model described by Cheng the network tries to recover all the available points in the brightness patch given. In this paper the network tries to recover only particular surface points needed by the neural network. Also the neural network described in this paper moves the brightness patch by one point after a surface point is recovered. The method described by Cheng tries to estimate how to split the image into a set of brightness patches and to derive a set of surface patches that are then combined to create the final surface. The proposed method is expected to perform better as it can handle noise and "optical illusions" better with no loss of data due to the overlapping of points in the surface, as opposed to the strategy proposed by Cheng were the average value is taken instead.

Another neural network based method for 3D image reconstruction, proposed by Wei and Hirzinger [1], tries to estimate the surface using p and q values for each surface point separately only for a finite number of points. This approach is considered to have low noise reduction as it connects the desire point only with one neighbor point.

A neural network based approach proposed by Arie and Nandy [3] uses brightness patches to reconstruct the surface. This method uses the patches to train a neural network to learn surface patterns according to light source direction. This approach is consider to have highly accurate results. Therefore, this idea was adopted and used with a major modification. The proposed neural network tries to move the brightness patch only one point at a time to the x or the y direction of the surface contrary with Arie and Nandy method that moves towards the next brightness patch.

3.2   Training Data

For this neural network the three images with known surface were used for training. The training data represents a surface of a sphere from the same viewpoint but with different light source directions like $S_x = 0$, $S_y = 0$ and $S_z = 1$; $S_x = 0.5$, $S_y = 0.5$ and $S_z = 1$; $S_x = -0.5$ $S_y = -0.5$ and $S_z = 1$. In order to enable the neural network to learn to be adaptive to noise Gaussian noise was added to the images with mean equal to zero and variance equal to 0.001. Then the images were reshaped from a $N \times M$ matrix to a $12 \times NM$ with the light source direction added to the data. The input vector of the neural network consisted of 3x3 matrix of the surrounded information. Information provided by this matrix included the point to be recovered, the height and the light source direction. The data was normalized on the interval [-1,1] prior to training the neural network.

3.3 Training Algorithm

The goal of the neural network training process is to produce a neural network that not only works well and produces small errors with the training data but also responds well to previously unseen testing data. When the network performs as well on the training data as on new data then the network generalizes well. In this paper, the update of the weights of the neural network was based on the Levenberg-Marquards optimization algorithm. This approach minimizes the combination of squared errors and weights and determines a correct combination for the neural network to generalize well. This process is called Bayesian regularization, details of the Bayesian regularization training procedure can be found in [9] [10].

4 Experimental Results

In this section, the proposed neural network model is assessed by recovering 3D surfaces of some images in noisy scenes. However, as mentioned in the previews sections, the surface and the light source direction have to satisfy some assumption, for example in the scene only one light source must exist or the surface must follow the Lambertian reflectance model. These restrictions make it very difficult to find good training or testing sets for the neural network. To overcome these restrictions in our experiments we used a combination of synthetic and real images where the light source direction, albedo and the real 3D surface is known.

Synthetic images. The synthetic images described in [11,12] were created by using the original depth maps from range data obtained from a range laser finder (see [11]). To create realistic synthetic images, the virtual objects have to be shaded consistently under the real illumination condition of the scene. To this end, the forward discrete approximation of depth is used to calculate the surface gradient, and shaded images are generated using the Lambertian reflectance model with different light conditions and source directions of the same surface, which can referred to as a merit of synthetic images in comparison with real images. It is known that real images usually do not perfectly meet the above assumptions of reflectance model due to errors that can not be corrected or measured.

In this paper, four synthetic surfaces were generated accordingly, they are: Sphere, synthetic Vase, Penny and Mozart. For example, Figure 2 show the synthetic sphere image taken from [11] and the noisy version obtained from one light direction. These images correspond to the same surface, from the same view point direction, but with different light source directions. Their corresponding noisy images were obtained from the same view point with different light source directions separately.

Real images. Furthermore three real images and their noisy real versions were also used for training and testing the neural network model, they are: Lenna, Pepper and Real Vase. For example, the Figure 3 shows the Pepper image and its corresponding noisy version with one light direction.

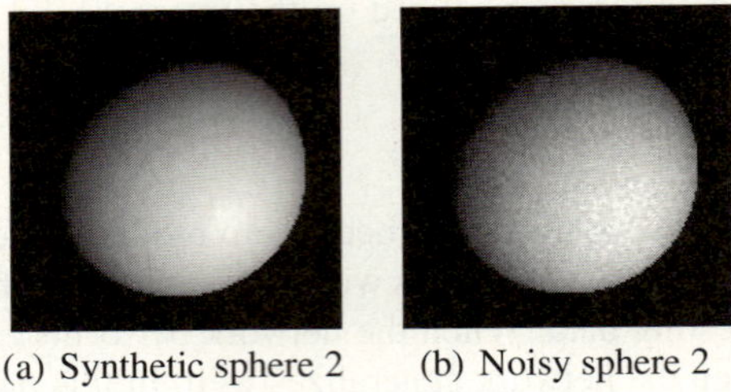

(a) Synthetic sphere 2   (b) Noisy sphere 2

Fig. 2. Synthetic Sphere 2 Image

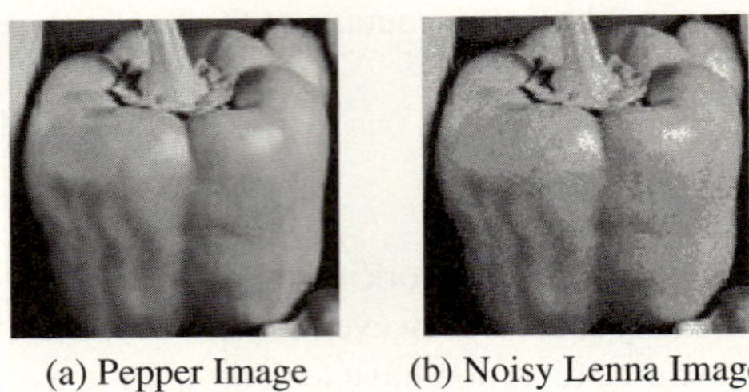

(a) Pepper Image   (b) Noisy Lenna Image

Fig. 3. Real Pepper Images

4.1   The Performance of the Neural Network

The Levenberg-Marquardt training algorithm with Bayesian regularization, described above, was implemented to build the proposed neural network model. The data that was used for training was three images of a spherical surface from the same viewpoint direction but different light source directions. Furthermore, Gaussian noise was added to the training data. Figures 4 and 5 show the training of the proposed neural network. The learning parameters and the number of hidden neurons were determined after several training tests were ran. Table 1 shows the training results for the different choices of the numbers of hidden units.

The image used for testing is the vase surface with light source direction (1,0,1). This image was chosen to test the performance of the neural network in different light source direction from the ones given as training input so that the neural network generalizes well. The decision that has to be made is for the number of the neurons that for each

Table 1. Error results obtained with different number of hidden units

Neurons	Training Error	Testing Error	Validation Error	Testing Image
15	179.59	64.72	65.76	0.10
20	175.88	63.80	66.07	0.09
25	105.90	57.40	58.04	0.11
30	93.38	48.17	50.50	0.13

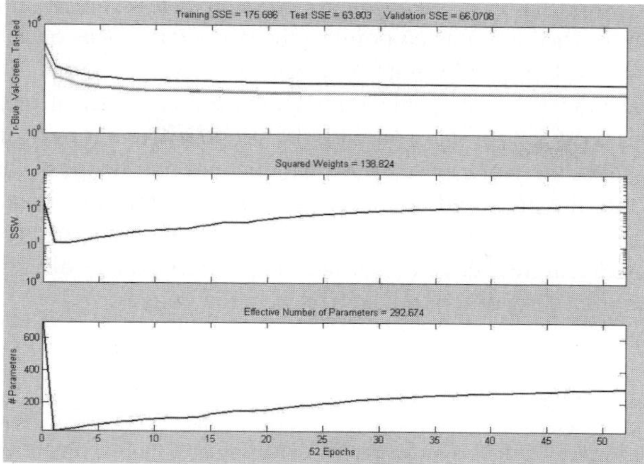

Fig. 4. Neural Network Training

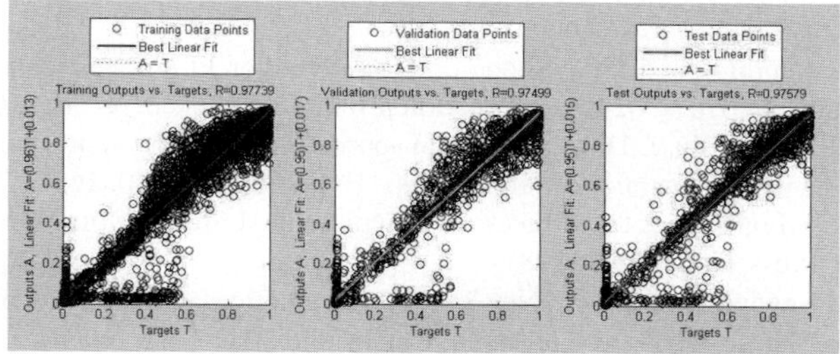

Fig. 5. Neural Network Regression

layer were the network generalize for all surfaces. As seen in Table 1 the best results were obtained using 20 hidden neurons. The three real images, Lenna, Pepper and Real Vase as shown above were used for testing. As the real depth of the surfaces is unknown no evaluation can be done for estimating the performance of the algorithms. The light source direction for the above three surfaces was specified with the images, please refer to [11] for more details regarding the real images. The light source directions for the real images are: $S_x = 1.5, S_y = 0.866, S_z = 1$ for Lenna; $S_x = 0.766, S_y = 0.642, S_z = 1$ for Pepper; and $S_x = 0.5, S_y = 0.5, S_z = 1$ for Real Vase. Hence the proposed neural network approach is efficient in recovering the 3D surface from sing images in the noisy scenes.

5   Conclusions

This paper developed a novel neural network that attempts to recover 3D surfaces from a single gray scale image in noisy scenes and compare it with existing mathematical approaches. The proposed neural network uses photometric stereo to estimate local surfaces orientation for surfaces at each point of the surface that was observed from same viewpoint but with different illumination direction in surfaces that follow the Lambertian reflection model. In this paper three existing mathematical method were also implemented and used for comparison with the proposed neural network. The images used to test the neural network were both synthetic and real images. The experimental results showed that the proposed neural network is efficient in estimating the desired surface with high accuracy. The proposed neural network outperformed the other approach with more accurate recovered surfaces if the light source direction was not the same as the viewpoint direction or noise existed in the images.

References

1. Wei, G.Q., Hirzinger, G.: Learning shape from shading by a multilayer network. IEEE Transactions On Neural Networks 7, 985–994 (1996)
2. Mostafa, M.G.H., Yamany, S.M., Farag, A.A.: Integrating shape from shading and range data using neural networks. In: CVPR, pp. 2015–2020 (1999)
3. Ben-Arie, J., Nandy, D.: A neural network approach for reconstructing surface shape from shading. ICIP (2), 972–976 (1998)
4. Cheng, W.C.: Neural-network-based photometric stereo for 3d surface reconstruction. Neural Networks, 2006. In: IJCNN 2006. International Joint Conference, pp. 404–410 (2006)
5. Zhou, S.M., Li, H.X., Xu, L.D.: A variational approach to intensity approximation for remote sensing images using dynamic neural networks. Expert Systems 20, 163–170 (2003)
6. Grimson, W.: From Images to Surfaces: A Computational Study of the Human Early Visual System. MIT Press, Cambridge (1981)
7. Werbos, P.J.: Beyond Regression: NewTools for Prediction and Analysis in the Behavioral Sciences. PhD Thesis, Harvard University, Cambridge, MA, USA (1974)
8. Rumelhart, D., Hinton, G., Williams, R.: Learning internal representations by error propagation. In: Anderson, J., Rosenfeld, E. (eds.) Neurocomputing
9. Dan Foresee, F., Hagan, M.: Gauss-newton approximation to bayesian learning. In: International Conference on Neural Networks,1997, vol. 3, pp. 1930–1935 (1997)
10. MacKay, D.: Bayesian interpolation. Neural Computation 4(3), 415–447 (1992)
11. Zhang, R., Tsai, P.S., Cryer, J.E., Shah, M.: Shape from shading: a survey. IEEE Trans. Pattern Anal. Mach. Intell. 21(8), 690–706 (1999)
12. Zhang, R., Tsai, P.S., Cryer, J.E., Shah, M.: A survey of shape from shading methods. Technical Report CS-TR-97-15, Computer Science Dept., Univ. of Central Florida (1997)

A Spatio-temporal Extension of the SUSAN-Filter

Benedikt Kaiser[1] and Gunther Heidemann[2]

[1] University of Karlsruhe, Institute for Process Control and Robotics
Building 40.28, Kaiserstr. 12, D-76128 Karlsruhe, Germany
[2] University of Stuttgart, Intelligent Systems Group
Universitätsstr. 38, D-70569 Stuttgart, Germany

Abstract. This paper proposes a detector for spatio-temporal interest points. Interest point detection is a common technique in computer vision to extract salient regions and represent them by a single point for further processing. But while many algorithms exist for static images, there is hardly any method to obtain interest points from image sequences for the representation of salient motion. Here we introduce SUSANinTime, an extension of the well known SUSAN algorithm from 2D to 2D+1D, where the third dimension is time. While SUSAN-2D extracts edge- and corner points, SUSANinTime detects basic events such as turning points of object trajectories. To find out the type and saliency of the detected events, we analyze the second order statistics of the spatio-temporal volume surrounding the interest points in real world image sequences.

1   Introduction

The detection of actions and events in image sequences is a key ability in Computer Vision. Examples are the recognition of complex hand movements for Human-Machine Interaction, or the particularly difficult task to recognize interactions of a human with objects, such as taking and filling a cup or dialing a number. The development of generic, non-specialized solutions for the decomposition and classification of such action sequences is still an unsolved problem.

For the decomposition of static scenes into meaningful components, *interest points* (IPs) are a standard technique to filter out and represent areas which appear relevant at the signal level. Applications are image retrieval [1], active vision [2], object recognition [3], or image compression [4] to name but a few. It is therefore promising to transfer the concept to the spatio-temporal domain.

The so far most common approach to detect and classify actions and events is computation of the *optical flow*. However, computations are costly, and the evaluation of the optical flow for actions requires an idea what is being searched for. That is, a model (in whatever form) of the action in question must exist. As an alternative to this branch of Computer Vision, we investigate the generalisation of IPs, since IP-methods are usually context-free. Though certainly not the entire complexity of actions in natural image sequences can be covered by spatio-temporal IPs, the approach aims at detecting at least a relevant and

V. Kůrková et al. (Eds.): ICANN 2008, Part I, LNCS 5163, pp. 867–876, 2008.

usable subset of *basic events*. Such events might be, e.g., acceleration, rotation, turning points in object trajectories, or approach of two objects (a closing gap).

IPs in static images are points which are "salient" or "distinctive" as compared to their neighbourhood. To be more precise, an IP is not necessarily an isolated salient pixel, rather, the IP at a pixel location stands for a salient patch of the image. Most algorithms for IP detection are aimed at the detection of corners or edges in grey value images [5,6,7,8,9,10]. Methods of this kind which detect edges or corners are particularly promising for the spatio-temporal case, since 3D-corners may indicate the turning points of 2D-corners in a temporal sequence. Thus they would indicate saliency both in the geometrical sense and in the sense of an event.

Laptev and Lindeberg [11] have shown how the concept of spatial edge- and corner-based IPs can be extended to the spatio-temporal domain for the HARRIS detector [6]. They extend the detection of IPs from the eigenvalues of the 2D auto-correlation matrix of the signal to the 3D matrix in a straight forward approach, in addition, they propose a scale space approach to deal with different spatial and temporal scales. But since the 2D- and 3D-HARRIS detector depends on the often problematic computation of grey value derivatives, we chose to extend another IP-detector to the spatio-temporal domain: The SUSAN detector proposed by Smith and Brady [8], which detects edges and corners merely from the grey values without the use of derivatives.

In the following, first the spatial SUSAN detector will be shortly described (section 2), then the "naive" extension of SUSAN to the temporal axis (SUSAN-3D) is introduced in section 3. Since tests of SUSAN-3D reveal several drawbacks of the straight forward extension from 2D to 3D (which presumably hold also for the 3D-version of HARRIS), a modified version (SUSANinTime) is introduced in section 4. SUSANinTime is tested on real world image sequences, and section 5 applies second-order statistics to identify typical events that can be detected. The final section 6 gives a conclusion and outlook.

2   The Original SUSAN Approach

Smith and Brady have proposed an approach to detect edges and corners, i.e., one- and two-dimensional image features [8]. While most algorithms of this kind rely on the (first) derivatives of the image matrix, the SUSAN-detector relies on the local binarisation of grey values. To compute the edge- or corner strength of a pixel (called the "nucleus"), a circular mask A around the pixel is considered. By choice of a brightness difference threshold ϑ, an area within the mask is selected which consists of pixels similar in brightness to the nucleus. This area is called USAN ("Univalue Segment Assimilating Nucleus"). To be more precise, let $I(r)$ denote the grey value at pixel r, n the area (i.e. # pixels) of the USAN, and r_0 the nucleus. Then

$$n(r_0) = \sum_{r \in A} c(r, r_0), \qquad \text{with} \quad c(r, r_0) = \begin{cases} 1 \text{ for } |I(r) - I(r_0)| \leq \vartheta \\ 0 \text{ for } |I(r) - I(r_0)| > \vartheta . \end{cases} \tag{1}$$

The response of the SUSAN-detector at pixel r_0 is given by

$$R(r_0) = \begin{cases} g - n(r_0) & \text{for } n(r_0) < g \\ 0 & \text{else}, \end{cases} \tag{2}$$

where g is called the *geometric threshold*. For edge detection, a suitable value is $g = \frac{3}{4} n_{max}$, for corner detection $g = \frac{1}{2} n_{max}$. It can be shown that these values are optimal in certain aspects for a particular signal-to-noise ratio.

Smith and Brady [8] obtain a saliency map which indicates edge and corner strength as the inverted USAN area for each nucleus pixel. IPs are then found as the local minima of the USAN area, thus the name SUSAN (= Smallest Univalue Segment Assimilating Nucleus).

Fig. 1 illustrates how the USAN-area and thus the response of the detector changes according to the circular mask being positioned on a corner, an edge or a homogeneous region. To find the local direction of an edge and to localize

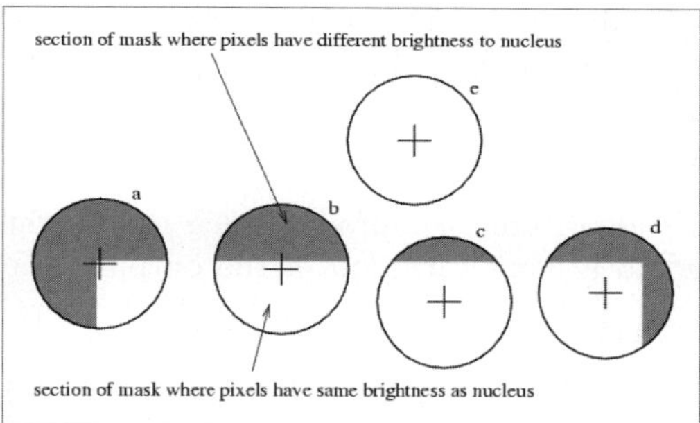

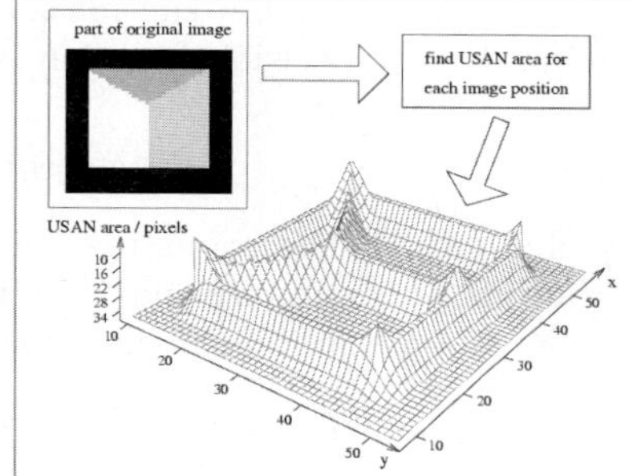

Fig. 1. Left: Five circular masks: The white areas are the USAN (with respect to the nucleus marked by a cross). From [8]. Right: A three-dimensional plot of the response of the SUSAN-detector to a test image (from [8]).

corners precisely, geometrical features of the USAN have to be exploited, see [8] for details.

The SUSAN-approach is well suited for a fast detection of one- and two-dimensional basic image features with the benefit that both localization precision and the implicitly built-in noise reduction are robust to changes of the size of the circular mask.

3   The SUSAN-3D-Detector

In this section we introduce the extension of the normal SUSAN-2D-detector to the spatio-temporal ("3D") domain [12].

The generalization of an isotropic circular mask in two dimensions is a sphere in three dimensions. But since the spatial coordinates are independent of time, a (rotationally symmetric) ellipsoid around the time axis is betters suited for event detection since it allows suitable scaling (note the same physical event may, e.g., be captured using different frame rates). However, also other 3D-shapes with a circular cross section come into question. We have investigated two algorithms using different 3D-masks M_E and M_Z:

$$M_E(x,y,t) = \begin{cases} 1 \text{ if } \frac{x^2+y^2}{R_{xy}^2} + \frac{t^2}{R_t^2} < 1, \\ 0 \text{ otherwise} \end{cases} \tag{3}$$

$$M_Z(x,y,t) = \begin{cases} 1 \text{ if } \frac{x^2+y^2}{R_{xy}^2} \wedge -R_t \leq t \leq R_t, \\ 0 \text{ otherwise} \end{cases} \tag{4}$$

where R_{xy} denotes the radius in the x-y-plain, and R_t the extension of the mask on the temporal $t-$axis. In the same way as the SUSAN-2D-detector is applied to each spatial point, now the mask must be applied to each spatio-temporal point of an image sequence, and the grey value of the nucleus is compared to the other pixels within the mask to obtain the USAN-volume. In place of the original binary decision function, we use the improved version as proposed by Smith and Brady:

$$c(r, r_0) = e^{-(\frac{I(r)-I(r_0)}{t})^6}. \tag{5}$$

By this means, the robustness of the algorithm is improved since now slight variations of luminance may not lead to a large variation of the output. The USAN-volume can be calculated as

$$V(r_0) = \sum_r c(r, r_0) M(r_{0x} + r_x, r_{0y} + r_y, r_{0t} + t), \tag{6}$$

with $r_0 = (r_{0x}, r_{0y}, r_{0t})^T$ and $r = (r_x, r_y, r_t)^T$.

Fig. 2 illustrates the way SUSAN-3D processes different motion events (in only one spatial dimension). The x-axis points from left to right, the t-axis upwards. The USAN is the white area within the mask. While Fig. 2(a) shows an edge element at rest, Fig. 2(b) depicts motion at constant velocity. The result of $V = 0.5$ (of the area) is the same in both cases. Figs. 2(c) and 2(d) depict a stop-event and acceleration, respectively. For these cases, values V clearly below 0.5 are to be expected. The smallest value V is to be expected in the case depicted in Fig. 2(e), which shows a turning point. Fig. 2(f) shows either a sudden appearance of the object or motion at a velocity to high to be resolved. Again, the volume is $V = 0.5$.

Evaluation of the "naive" SUSAN-3D-detector in earlier work [13] has shown that processing the 3D-USAN values in the manner of the conventional SUSAN-2D-detector "salient" events can be detected, such as acceleration and turn around (values the smaller the stronger curvature). However, rest, constant motion, and sudden appearance (all three $V = 0.5$) can not be discriminated. While this is still satisfactory for rest and constant motion ($V = 0.5$ being larger than for acceleration and turn-around), sudden appearance should get the smallest

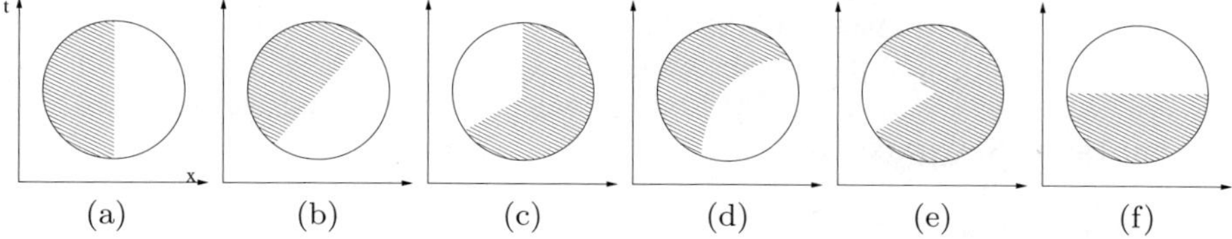

Fig. 2. Examples for different kinds of motion. The USANs are shown in the $x - t$-plain: (a) Rest, (b) constant velocity, (c) stop-event, (d) acceleration, (e) turn-around, (f) sudden appearance [13].

value (i.e. highest saliency output). Further, a major drawback of the simple SUSAN-3D-detector is the implicit coupling of geometrical and dynamical features. Even worse, static features are over-weighted compared to dynamic ones.

This can be shown using artificial image sequences, using as a mask a cylinder of $2R_{xy} = 7$ pixels and $2R_t = 7$ frames, yielding a total volume of 224. The brightness threshold is set to 27 as in the 2D-version. The test sequence shows squares moving in different ways. Fig. 3, left, shows, from top to bottom, the SUSAN-3D response maps for a square accelerating from left to right; a square moving at high constant velocity; a moving one with low constant velocity; and a square at rest. Obviously, the response of SUSAN-3D is mainly governed by geometrical, not dynamical features. Therefore, an alternative approach is proposed: SUSANinTime.

4   Improving the Naive Approach

To eliminate the drawbacks of SUSAN-3D outlined above, one way would be to employ additional features for correcting the detector response. But then the simple and elegant idea of using the USAN-volume as a feature for IP-detection would be more or less abandoned. For this reason, we will outline an alternative approach, which applies the SUSAN principle only on the temporal dimension.

The first step is computation of the USAN-area within a cylindrical volume around the nucleus. The single x-y-slices of the cylindrical mask are evaluated to find the USAN-areas for every frame, these values are saved in a 1D-array (`areas[]`). Then the SUSAN principle is applied to the 1D-array `areas[]` in the following way: The USAN-area at the current time is considered to be a (second) nucleus value (`nucleus2`). Note the second nucleus value is an *area*, not a grey value. Then the array `areas[]` is binarised with respect to the `nucleus2` value, and the final detector response is the sum of the now binarised array.

```
SUSANinTime(x, y, t)
    nucleus <- getpixel(x,y,t)
    FOR t' FROM -R TO R DO
        areas[t'] <- 0
        FOR y' FROM -r TO r DO
```

```
        FOR x' FROM -r TO r DO
            IF mask[x'][y'][t'] = 1 THEN
                pixel <- getpixel(x + x', y + y', t + t');
                areas[t'] <- areas[t'] + c_1(nucleus, pixel)
    nucleus2 <- areas[0]
    value <- 0
    FOR t' FROM -R TO R DO
        value <- value + c_2(nucleus2, areas[t'])
    RETURN value
```

Here, `mask[x][y][t]` takes a value of 1 if `x,y,t` is inside the volume covered by the detector, else 0. c_1 and c_2 denote the thresholding functions for the spatial binarisation of the x-y-slices and the binarisation of the `areas[]` array, respectively.

The idea of the SUSANinTime algorithm is to give a high response to such space-time volumes which exhibit a high activity, where "activity" is defined as a high temporal variation of the USAN-area. The SUSANinTime-detector computes for which span of time the USAN-areas are still within an environment of the nucleus value. This time span takes a maximum for zero velocity and decreases with increasing velocity. Thus, it becomes also clear that the return value of SUSANinTime does not allow measurement of acceleration.

Though the SUSANinTime-algorithm can not classify space-time events in categories velocity / acceleration, it has nevertheless highly useful properties. Fig. 3, right, shows the response of the SUSANinTime detector, where the input sequence is the same as for Fig. 3, left. The detector response is approximately proportional to the velocity, for stationary regions, the detector is "blind" (here, small intensity values denote a high response). In contrast to Fig. 3, left, the corners of the squares yield no stronger response than the edges, though being geometrically more salient. So the response is determined by the dynamics, not stationary

Fig. 3. Left: Response of SUSAN-3D at a single moment for different image sequences displaying a square moving from left to right, see text. Right: Response of SUSANinTime to the same sequences [13].

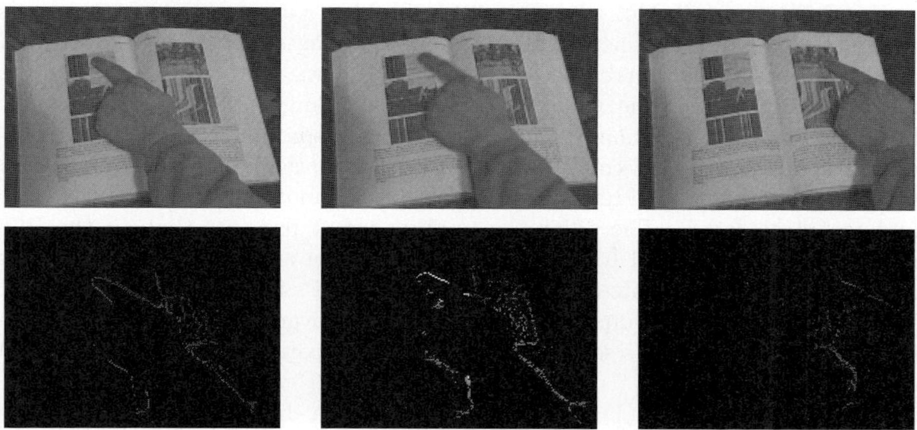

Fig. 4. Response of the SUSANinTime detector for a real image sequence. The pointing event can be clearly detected in the second frame, see text [13].

features — a major advantage, since now geometrical features detected by separate modules can be included in a final saliency map in a well-defined way.

First experiments on real world image sequences have shown that the algorithm yields robust results. Fig. 4 shows the output of the SUSANinTime detector for three frames of a sequence. Different intensities of the map reflect the different velocities. At first, the hand accelerates in a movement to the right. The pointing event is a turning point in the trajectory, i.e., stopping and accelerating again. The turning point is clearly visible in the second frame. In the last frame, the hand stops. Note that the appearance or disappearance of (otherwise) static structures in the background is also detected as motion.

5   Statistical Event Analysis by PCA

Spatio-temporal IPs can be evaluated by applying the detector to image sequences and visualizing the results, looking for coincidence of IPs and semantically meaningful physical events. However, observations of this kind will remain qualitative, not only because the number of IPs that can be investigated in detail remains limited, but also because it depends on human judgement. We are therefore interested in a more principled way to find the "prototypical" events to which the SUSANinTime-detector is sensitive. We therefore compute the principal components (PCs) of the space-time volumes surrounding a large number of IPs found in a real world sequence.

Principal component analysis (PCA) leads to a data representation by the orthogonal eigenvectors of the covariance matrix. Those corresponding to the largest eigenvalues represent the directions of highest variance. If the SUSANinTime-detector is indeed most sensitive to elementary visual events, such events should be among the eigenvectors of high eigenvalues.

For the computation of the eigenvectors, a large number of data samples is necessary to obtain sufficient statistics. The training sequence is 100 seconds, comprising parts of different television shows, news, and movies (at 25fps). The evaluated space-time volumes are ellipsoidal, multiplied (not convolved!) by a 3D Gaussian of comparable size to attenuate border effects. The procedure is similar to the PC-computation in [14], where the PCs of natural (static) scenes are computed. We refer to this paper for the motivation of our procedure and technical details. The difference to the current paper is that (*i*) the PC-computation is extended to the temporal dimension and (*ii*) that the samples are not grabbed at random but are centered at the IPs found by SUSANinTime. The eigenvectors were computed directly from the covariance matrix of the sample volumes. Fig. 5 shows some of the eigenvectors corresponding to the largest eigenvalues.

The first eigenvector (1) is the mean of the data distribution and therefore mainly reflects the 3D-Gaussian. It can be interpreted as a constant image with neither spatial nor temporal structure. Eigenvector (2) represents the first and most simple event: A "temporal edge", i.e. the transition from black to white without spatial structure ("switching on a lamp"). Note that the orientation of the eigenvectors is arbitrary, so the black–to–white event is equivalent to a white–to–black event. Eigenvectors (3) and (5) are the spatial equivalent to (2), a stationary horizontal and vertical edge. Further, eigenvectors (4) and (8) are the two ways to increase the complexity of event (2) by one step: (4) is a "temporal bar", i.e. the bright spot disappears for a short while and re-appears (lamp is switched off–on or on–off). Eigenvector (8) is the first which combines spatial and temporal structure: An edge "flips".

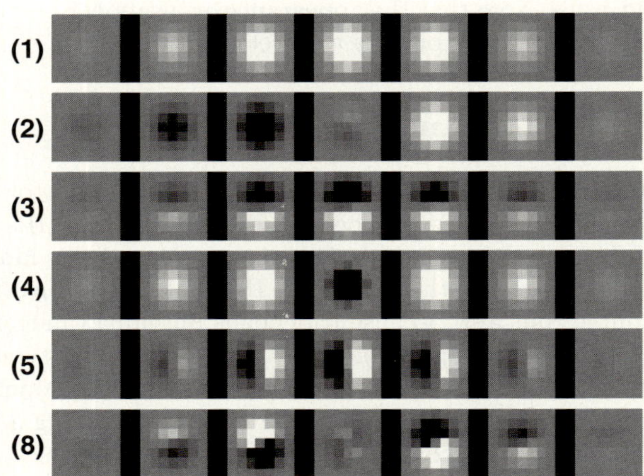

Fig. 5. Some of the eigenvectors with largest eigenvalues of IP-centered 7x7x7 space–time volumes extracted from a 100 second image sequence. The number is the rank of the eigenvalue. See text.

Calculation of the first 20 PCs shows that a significant number differs from the PCs of static scenes known from [14]: 13 PCs represent events, only 7 are static patterns. Further, PCs representing events complement each other, e.g., PC (8) is supplemented by another PC showing a flipping edge perpendicular to (8). Likewise, while PCs (2) and (4) are the temporal patterns "off–on" and "on–off–on", respectively, further PCs exist representing "on–off–on–off" etc. From the amount of dynamic PCs it can be concluded that SUSANinTime does indeed filter out spatio-temporal events. However, due to limited computational resources, we can only compute 7x7x7 x-y-t volumes. For this resolution, PC computation exceeding 20 makes little sense since the patterns become more complex and cannot be sufficiently resolved — note that spatial frequency of the PCs increases with decreasing eigenvalues, since the spectral power of natural images falls with spatial frequency f according to a power law $1/f^p$ [15]. Complex events can therefore not yet be found. We will discuss the consequences of the current investigations in the next section and give an outlook how further experiments can better analyze event detection.

6   Conclusion

We have proposed a spatio-temporal extension of the SUSAN algorithm for event detection. While the naive extension of SUSAN to a third dimension (time) is unsatisfactory since static object features (edges and corners) lead to higher responses in the computed sequence of saliency maps than the dynamical features, the SUSANinTime approach yields promising results. In real world sequences of hand movement, semantically meaningful events such as the turning point of a pointing gesture or turning points in object manipulation can be detected.

Since tests of this kind inevitably remain qualitative, we are searching for a more principled evaluation. We have therefore computed the PCs of IP-centered space-time volumes from real world sequences. The idea is to get the events which are typically caught by the detector encoded by the PCs with the highest eigenvalues. This hope has been fulfilled insofar as the 20 PCs of highest eigenvalues do indeed encode basic events. Moreover, comparison to the PCs of static images from [14] shows that only few of the static patterns are found among the high eigenvalues, i.e. the detector has a high selectivity for dynamic structures. However, due to limited computational power, only little spatial and temporal resolution is possible, thus limiting also the number of PCs that can be evaluated with reasonable accuracy.

The first step in future work is therefore increasing spatial and temporal resolution. But even more important is improving the interpretation of the PCs. We will therefore compare the PCs of IP-centered space-time volumes with those of randomly picked ones, following the idea presented in [3]. By this means we can compare the PC-based representation of natural motion in general to the specific events captured by the detector. We expect that the PCs of IP-centered space-time volumes are more or less identical to those of random volumes, but with a different order of eigenvalues which puts in front the eigenvectors to which the detector is sensitive. Within the spatio-temporal subspace spanned by these eigenvectors lie the events the detector can filter out of a sequence. With this novel technique, a variety of other 2D+1D extensions of known 2D IP-detectors can be evaluated.

References

1. Tian, Q., Sebe, N., Lew, M.S., Loupias, E., Huang, T.S.: Image Retrieval Using Wavelet-Based Salient Points. J. of Electronic Imaging 10(4), 835–849 (2001)
2. Backer, G., Mertsching, B., Bollmann, M.: Data and Model-Driven Gaze Control for an Active-Vision System. IEEE Trans. on Pattern Analysis and Machine Intelligence 23(12), 1415–1429 (2001)
3. Heidemann, G.: Focus-of-Attention from Local Color Symmetries. IEEE Trans. on Pattern Analysis and Machine Intelligence 26(7), 817–830 (2004)
4. Privitera, C.M., Stark, L.W.: Algorithms for Defining Visual Regions-of-Interest: Comparison with Eye Fixations. IEEE Trans. on Pattern Analysis and Machine Intelligence 22(9), 970–982 (2000)
5. Moravec, H.P.: Towards Automatic Visual Obstacle Avoidance. In: Proc. 5th Int'l Joint Conf. on Artificial Intelligence, Cambridge, Massachusetts, USA, pp. 584–587 (1977)
6. Harris, C., Stephens, M.: A Combined Corner and Edge Detector. In: Proc. 4th Alvey Vision Conf., pp. 147–151 (1988)
7. Schmid, C., Mohr, R.: Local Grayvalue Invariants for Image Retrieval. IEEE Trans. on Pattern Analysis and Machine Intelligence 19(5), 530–535 (1997)
8. Smith, S., Brady, J.: SUSAN – A New Approach to Low Level Image Processing. Int'l J. of Computer Vision 23(1), 45–78 (1997)
9. Zheng, Z., Wang, H., Teoh, W.: Analysis of Gray Level Corner Detection. Pattern Recognition Letters 20, 149–162 (1999)
10. Zitová, B., Kautsky, J., Peters, G., Flusser, J.: Robust detection of significant points in multiframe images. Pattern Recognition Letters 20(2), 199–206 (1999)
11. Laptev, I., Lindeberg, T.: Space-time Interest Points. In: Proc. ICCV 2003, pp. 432–439 (2003)
12. Heidemann, G., Kaiser, B., Bax, I., Bekel, H., Ritter, H.: Spatiotemporal Events and Action Sequences. Technical report, Bielefeld Univ., Neuroinformatics Group (2005)
13. Kaiser, B., Heidemann, G.: Context-Free Detection of Events. In: Ersbøll, B.K., Pedersen, K.S. (eds.) SCIA 2007. LNCS, vol. 4522, pp. 223–232. Springer, Heidelberg (2007)
14. Hancock, P.J.B., Baddeley, R.J., Smith, L.S.: The Principal Components of Natural Images. Network 3, 61–70 (1992)
15. Tolhurst, D.J., Tadmor, Y., Chao, T.: Amplitude spectra of natural images. Ophthalmic & Physiological Optics 12(2), 229–232 (1992)

A Neighborhood-Based Competitive Network for Video Segmentation and Object Detection

R.M. Luque Baena[1], E. Dominguez[1], D. López-Rodríguez[2], and E.J. Palomo[1]

[1] Department of Computer Science, University of Málaga, Málaga, Spain
{rmluque,enriqued,ejpalomo}@lcc.uma.es
[2] Department of Applied Mathematics, University of Málaga, Málaga, Spain
dlopez@ctima.uma.es

Abstract. This work proposes an unsupervised competitive neural network based on adaptive neighborhoods for video segmentation and object detection. The designed neural network is proposed to form a background model based on subtraction approach. The synaptic weights and the adaptive neighborhood of the neurons serve as a model of the background and are updated to reflect the statistics of the background. The segmentation performance of the proposed neural network is examined and compared to mixture of Gaussian models. The proposed algorithm is parallelized on a pixel level and designed to enable efficient hardware implementation to achieve real-time processing at great frame rates.

1   Introduction

The goal of the video object segmentation is to separate pixels corresponding to foreground from those corresponding to background. This task is complex by the increasing resolution of video sequences, continuing advances in the video capture and transmission technology. As a result, research into more efficient algorithms for real-time object segmentation continues unabated.

The process of modeling the background by comparison with the frames of the sequence is often referred to as background subtraction. These methods are widely exploited for moving object detection in videos. Detection of moving is the next step of information extraction in many computer vision applications, such as video surveillance and people tracking. In these applications, robust tracking of objects is required for a reliable and effective moving object detection. While the fast execution and flexibility in different scenarios (indoor, outdoor) or different light conditions should be considered basic requirements to be met, precision is another important goal. In fact, a precise moving object detection makes tracking more reliable and faster.

Adaptive models are typically used by averaging the images over time [1,2,3] creating a background approximation. While these method are effective in situations where objects move continuously, they are not robust to scenes with many moving objects particularly if they move slowly. Toyama et al. [4] propose the Wallflower algorithm in which background maintenance and background subtraction are carried out at pixel, region, and frame levels. Haritaoglu et al. [5] build a statistical model by representing each pixel with three values: its minimum and maximum intensity, and the maximum intensity difference between consecutive frames, which are updated periodically.

V. Kůrková et al. (Eds.): ICANN 2008, Part I, LNCS 5163, pp. 877–886, 2008.

McKenna et al. [6] use an adaptive background model with color and gradient information to reduce the influences of shadows and unreliable color cues.

Wren et al. [7] used a multiclass statistical model based on Gaussian distributions. But the background model is a single Gaussian per pixel. A modified version modeling each pixel as a mixture of Gaussians is proposed by Stauffer and Grimson [8]. This approach is robust to scenes with many moving objects and lighting changes, but it is only able to achieve real time processing of small video formats (120x160 pixels).

Neuronal networks have been widely used for images segmentation. There exists numerous works in the fields of typewriting recognition [9] and medical images segmentation [10,11], but many of these neural techniques are not suitable for real time applications by their long time of computation in the training process. The development of a parallelized object segmentation approach, which would allow for object detection in real time for complex video sequences, is the focus of this paper. Neural networks posse intrinsic parallelism which can be exploited in a suitable hardware implementation to achieve fast segmentation of foreground objects.

In this work an unsupervised competitive neural network is proposed for objects segmentation in real time. The proposed approach employs is based on competitive neural network to achieve background subtraction. A new unsupervised competitive neural network based on neighborhood is designed to serve both as an adaptive model of the background in a video sequence and a classifier of pixels as background or foreground. The segmentation performance of the proposed neural network is qualitatively examined and compared to mixture of Gaussian models.

2   A Neural Model for Color Video Segmentation

Let us consider a competitive network for each pixel in the video. Each of these networks models the RGB space associated to the corresponding pixel. Thus, every pattern is formed by the three values (R,G,B) for the pixel. Then, the sequence of input patterns learned by each network consists on the different (R,G,B) values of the same pixel in every video frame.

The target of the network is to classify the input pattern (for the specified pixel) at each frame as foreground or background. Each neuron is used to model only one class. It can be noted that our model allows the use of many neurons to represent multimodal classes.

Our model performs a clustering based on a crisp-fuzzy hybrid neighborhood. For each neuron, there is a value r_j, representing the crisp neighborhood of the corresponding synaptic weight w_j: $\mathcal{N}_j = \{x : \|x - w_j\| \leq r_j\}$. The fuzzy neighborhood of neuron j is given by a membership function μ_j defined over the entire input space, and taking values in the interval $(0, 1)$.

Usually, the membership function present in this model is of the form: $\mu_j(x) = e^{-k_j(\|x-w_j\|-r_j)}$

The value of the parameter k_j is related to the slope of the function. The higher its value, the higher the slope is. For a great value of k_j, the fuzzy neighborhood of w_j will be more concentrated around $\mathcal{N}_j$. The effect (on the corresponding h_j) of increasing the value of k_j is shown in Fig. 1.

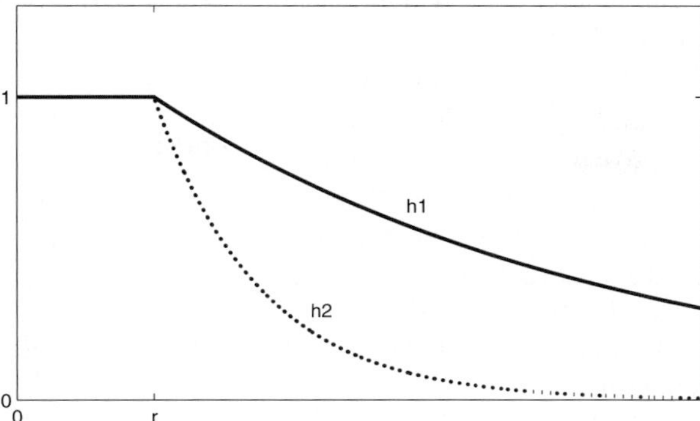

Fig. 1. Comparison between the synaptic potentials h_1 and h_2 of two neurons, such that $k_2 > k_1$, supposing $r_1 = r_2 = r$

With the help of this parameter, we can model a mechanism of consciousness, necessary to avoid dead neurons. When a neuron is activated and updated many times, its crisp neighborhood usually englobes a very high percentage of the patterns associated to the neuron. In this case, patterns outside this neighborhood are not likely to belong to the corresponding category. Thus, the membership function of the fuzzy neighborhood should be sharp, with a high slope.

If a neuron, j, is rarely activated, then $\mathcal{N}_j$ does not represent accurately the patterns associated to the neuron. Then, the membership of the fuzzy neighborhood should express the actual fuzziness present in the data and therefore assign higher values to patterns outside $\mathcal{N}_j$.

Thus, a good way to define k_j is proportional to the number of times that neuron j has been activated, n_j.

The use of this hybrid neighborhood allows us to better assign an input pattern to a class or category:

- If input pattern x belongs to $\mathcal{N}_j$, the network assumes that the best matching category for x is the associated to w_j.
- If input pattern x does not belong to any crisp neighborhood, its most likely category is represented by the neuron achieving the maximum value of the membership function for this particular input.

The fuzzy neighborhood also allows to include other mechanisms in the learning phase, such as neuron consciousness.

The items above provide us with a reasonable way to select the winning neuron for each input pattern, which is the one to be updated at the current iteration, after the input pattern is received by the network. We define the synaptic potential received by neuron j when pattern x is presented to the network as

$$h_j(x) = \begin{cases} 1, & \text{if } x \in \mathcal{N}_j \\ \mu_j(x), & \text{otherwise} \end{cases} \tag{1}$$

The winning neuron (with index denoted as $q(x)$) is the one receiving the maximum synaptic potential:

$$h_{q(x)} = max_j \, h_j(x)$$

To break ties (more common when two crisp neighborhoods, corresponding to different neurons, are overlapped), we consider the neuron which has been activated more times as the winning neuron. That is, if, for every j, n_j counts the number of times neuron j has been winner, and pattern x belongs to $\mathcal{N}_{j_1} \cap \mathcal{N}_{j_2}$ (overlapped neighborhoods), then $q(x)$ is defined as k for which $n_k = max\{n_{j_1}, n_{j_2}\}$. In case $n_{j_1} = n_{j_2}$, a random neuron in $\{j_1, j_2\}$ is selected as winner.

The learning rule used by each network to model the input space is the standard competitive learning rule:

$$w_{q(x)}(t+1) = w_{q(x)}(t) + \lambda \cdot (x - w_{q(x)}(t))$$

which can be viewed as the stochastic gradient descent technique to minimize the squared error function (also named distortion function):

$$F(W) = \sum_i \|x_i - w_{q(x_i)}\|^2 \tag{2}$$

where W represents a matrix whose rows are the w_j and λ is the so-called learning rate parameter, usually decreasing to 0.

However, an undesirable effect takes place under some conditions: the overlapping of two or more neighborhoods. Our model can be extended to solve this problem. We can consider a new formulation in order to avoid such overlaps.

Let us denote $Q(x) = \{j \neq q(x) : x \in \mathcal{N}_j\}$ the set of the indices j such that $x \in \mathcal{N}_j \cap \mathcal{N}_{q(x)}$, i.e., neuron j and $q(x)$ overlap at x.

In this case, neurons j with $j \in Q(x)$ may be repelled from the data, getting away from the winning neuron (which gets closer to the data) and reducing the overlap. Therefore, one should maximize the quantity $\|x - w_j\|^2$ with $j \in Q(x)$.

On the other hand, an overlap indicates that the crisp neighborhood of the winning neuron is not big enough to englobe all its associated patterns. Thus, a way to solve this problem is to increase the value of r_j, that is, to maximize (to some extent) the value of the radius r_j^2.

Let us find a learning rule with these properties, starting from an objective function, similar to Eq. (2), to be minimized.

Note that there is overlap if, and only if, $Q(x)$ is not empty. Define $\delta_{Q(x)} = 1$ if $Q(x) \neq \varnothing$, otherwise $\delta_{Q(x)} = 0$.

Then, the objective function to be optimized can be expressed in the following terms:

$$F(W, r) = \sum_i \left[\|x_i - w_{q(x_i)}\|^2 - \sum_{j \in Q(x_i)} \|x_i - w_{q'(x_i)}\|^2 - \delta_{Q(x_i)} r_{q(x_i)}^2 \right] \tag{3}$$

The minus sign expresses the fact that the second and third terms must be maximized. It can be observed that the repulsion of neurons $j \in Q(x_i)$, and the increase of the radius associated to the winner, only take place when an overlap occurs.

Note that, in this new formulation, the value r_j of the radius of $\mathcal{N}_j$ is treated as an adaptive parameter, and is automatically recalculated after an overlap is detected.

The learning rule associated to the new model can be obtained by applying the stochastic gradient descent technique to optimize the function given in Eq.(3). This technique has been used in other works [12] to develop new competitive learning rules from a family of objective functions.

To apply this technique, let us consider a single term of the objective function $F(W, r)$ (denote $x = x_i$, for simplicity):

$$T(W, r) = \frac{1}{2} \left(\|x - w_{q(x)}\|^2 - \sum_{j \in Q(x)} \|x - w_{q'(x)}\|^2 - \delta_{Q(x)} r_{q(x)}^2 \right)$$

Then, the learning rule is derived by differentiating the above expression with respect to w_j and r_j, and making $w_j(t+1) = w_j(t) - \lambda \frac{\partial T}{\partial w_j}$ and $r_j(t+1) = r_j(t) - \lambda \frac{\partial T}{\partial r_j}$.

$$\frac{\partial T}{\partial w_j} = \begin{cases} w_j - x, & \text{if } j = q(x) \\ x - w_j, & \text{if } j \in Q(x) \\ 0, & \text{otherwise} \end{cases}$$

$$\frac{\partial T}{\partial r_j} = \begin{cases} -r_j, & \text{if } (j = q(x)) \wedge (\delta_{Q(x)} = 1) \\ 0, & \text{otherwise} \end{cases}$$

Thus, we obtain the learning rule:

$$w_{q(x)}(t+1) = w_{q(x)}(t) + \lambda(x - w_{q(x)}(t)) \tag{4}$$

$$w_j(t+1) = w_j(t) - \lambda(x - w_j(t)) \quad \text{if } j \in Q(x) \tag{5}$$

$$r_{q(x)}(t+1) = (1+\lambda)r_{q(x)}(t) \quad \text{if } \delta_{Q(x)} = 1 \tag{6}$$

This constructive way of deducing this learning rule (by using the stochastic gradient descent method) ensures the minimization of the objective function of our model, given in Eq. (3).

3   Results

In this section the performance of the proposed neural approach is evaluated by testing our system on several image sequences. This video sequences are associated to different types of scenes obtained by means of a web camera or downloaded from internet. We have used very heterogeneous scenes, with different sizes and compressed by different techniques. These sequences also presented different features, from diverse kind of lighting to distinct objects in motion (people, vehicles) in order to conduct a more comprehensive study of the proposed method. Some of the scenes have been used and referenced in the literature.[1]

The most remarkable aspects evaluated in this paper, which decide whether our proposal may be used in a real-world environment, are reliability and robustness. Therefore, our system has been compared with other standard algorithms designed to perform

---

[1] Some of them are available at http://cvrr.ucsd.edu/aton/shadow

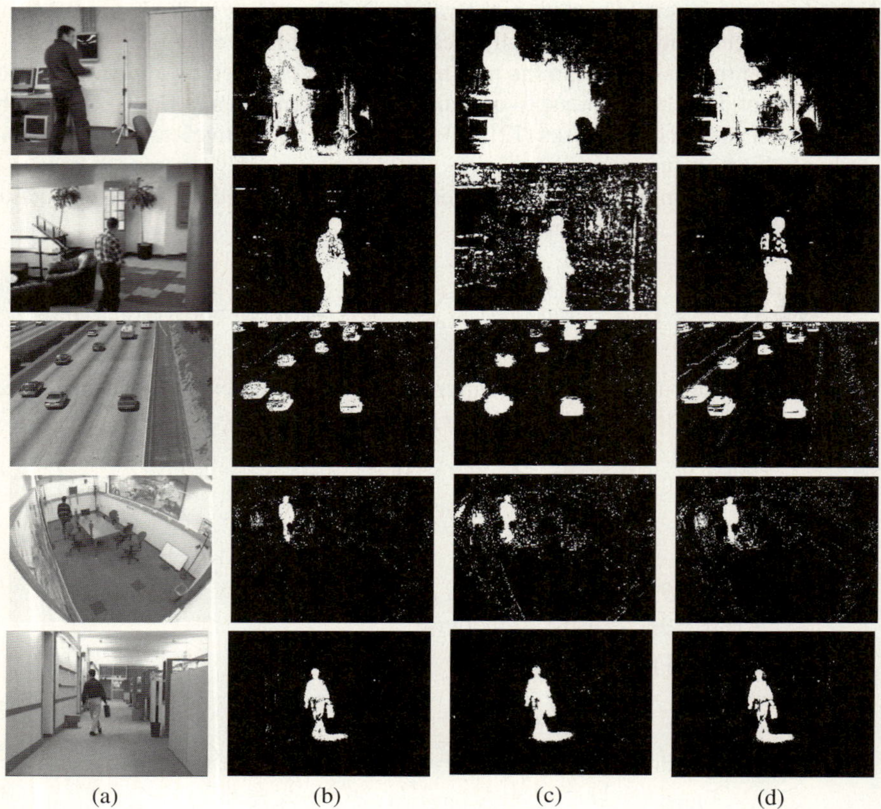

(a)   (b)   (c)   (d)

Fig. 2. Results of applying three segmentation methods to several frames. (a) column shows the capture frames for each scene in raw form; (b) proposed method; (c) mixture of Gaussians method; (d) background subtraction.

object segmentation in video sequences. Concretely, we compare with one of the most cited techniques in the specialized literature, the mixture of Gaussians model (MoG) [13]. This statistical model is based on a mixture of Gaussian distributions, able to solve correctly sudden illumination changes, multimodal backgrounds and deal with both indoor and outdoor scenes. In this work, we have set the number of normal distributions to $K = 3$. The mean of each of these distributions is initially computed as the gray-level value of the corresponding pixel in the first frame of the sequence. The other parameter, the standard deviation, is initialized to 25. Additionally, the results of another simpler technique (Background Subtraction, BS) have also been analyzed. This algorithm [3] consists on subtracting the processed frame from a background model previously computed. A fixed threshold value has been established for the whole set of analyzed scenes.

In our neural approach, fine-tuning of the model initial parameters (number of neurons, learning rate, initial radius r_j) is made before the segmentation process. These

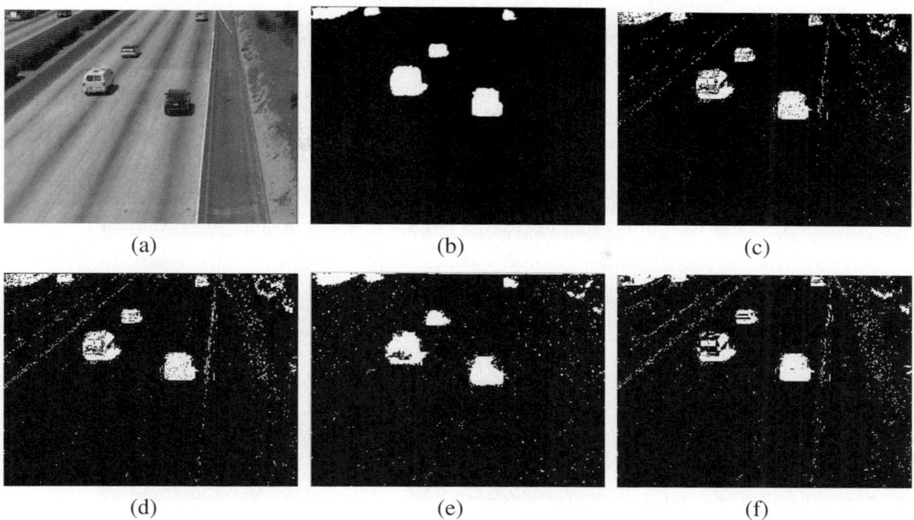

(a)   (b)   (c)

(d)   (e)   (f)

Fig. 3. Comparison among the studied techniques: (a) a traffic frame obtained in raw form; (b) ground truth; (c) neural network approach with repulsion; (d) neural network approach without repulsion; (e) mixture of gaussians method; (f) background subtraction method

Table 1. Comparative analysis of the success rate among the studied methods for the sequence observed in Fig. 3

Method	% Matching	% FP	% FN
NCM+	98.255	0.750	23.822
NCM	97.713	1.427	21.347
MoG	97.336	1.842	20.894
BS	97.125	1.772	27.355

parameters remain the same for all the scenes analyzed. Three neurons have been used to model the scene including both background and foreground, although more neurons can be added for a better representation of multimodal backgrounds. The B most activated neurons are used to model the background, whereas the rest of neurons correspond to foreground objects. This value B is computed as the amount of neurons whose number of activations $n_{a_1}, \ldots, n_{a_B}$ verify $\frac{n_{a_1} + \ldots + n_{a_B}}{N} > T$ for a prefixed threshold T, where N is the total number of activations of all neurons, as proposed in [13]. In this work, we have used $T = 0.7$. The learning rate is decreased after each frame until stabilizing at a fixed value, while the radius of the neural neighborhood is initialized to 25. Both the simple model with neighborhoods (NCM) and the extended model (NCM+), in which a repulsion mechanism is used to avoid the overlapping of two neighborhoods (see Eqs. (4)-(6)), have been tested.

Figure 2 shows comparison results among different techniques for several video sequences. It can be noted a better silhouette detection of the objects using our neural approach, removing also great amount of noise regarding to the others algorithms.

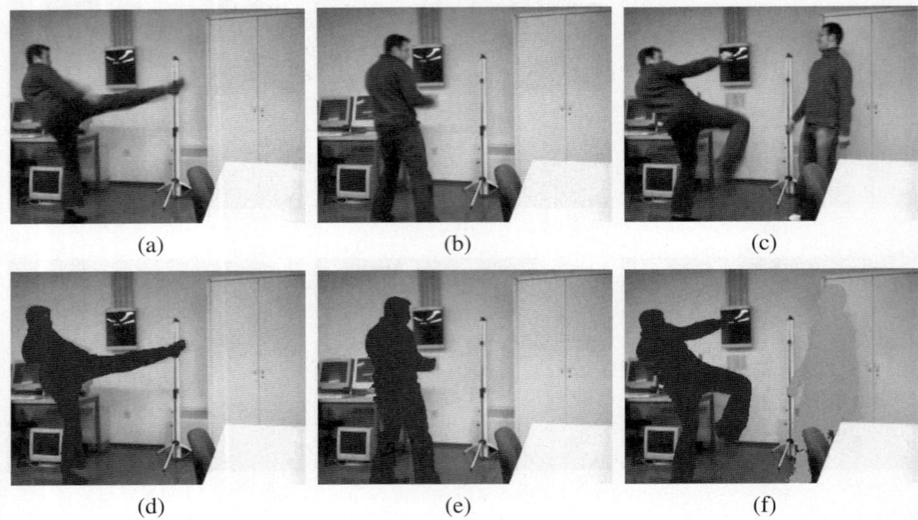

Fig. 4. Final results of the proposed method. In (a), (b) and (c) we can observe three frames of a scene in raw form. (d), (e) and (f) show the segmentation results using our neural approach (NCM+), after applying shadow detection and morphological operations.

A quantitative comparison among the different algorithms is shown in Table 1. Three measures have been defined to evaluate the performance of each method. A false positive rate (FP) shows the number of wrong background pixels and a false negative rate (FN) is used to inform the number of misleading foreground pixels. The success rate, indicating the accuracy of the corresponding algorithm, is presented in the column labelled '% Matching'. It can be observed that our method gets better results than the rest of analyzed methods, although the false negative rate is not as good as in the MoG model. Nevertheless, our model defines more clearly the shape of the object. Therefore, making a post-processing phase using morphological operators, we can achieve a smaller error rate.

Final results of our method, showing a quite good segmentation, can be observed in Fig. 3. If these results are to be used in a subsequent tracking phase, others post-processing mechanisms are needed to achieve reliable solutions:

– Objects in motion in the scene usually can cast shadow on the background which could affect to the overall segmentation as occurs in Fig. 3. In this case, shadow pixels should be detected. Some techniques have been proposed in the literature [14,15,16,17] to this end. In our case, the method proposed in [18] has been implemented, based on the proportionality property between the shadow and the background in the RGB color space.
– As mentioned before, the results have also been improved by applying morphological operators after the segmentation process.

Figure 4 shows the final result of our approach, after applying a shadow detection method and enhancing the obtained frame by using morphological operations. As we can observe, this segmentation is effective enough to be used in a subsequent tracking phase.

Another advantage of the system is its efficiency (in terms of computational cost) compared with other techniques. It is possible to develop our neural network approach in a parallel hardware implementation, in which the segmentation time can be improved using a typical field programmable gate array (FPGA). Therefore, it will require a lower number of iterations to achieve the overall segmentation. Thus our method is able to perform in real-time.

4   Conclusions and Future Work

In this work, we have developed a new competitive neural model focusing on object detection and segmentation in video scenes. This new model is based on an unsupervised learning performed to model the RGB space of each pixel.

The main contribution of this model is the use of adaptive neighborhoods that incorporate some mechanisms of repulsion and consciousness. The neighborhoods used in the model are combination of crisp and fuzzy ones. The crisp neighborhood tries to model the vicinity of the synaptic weight vector in the RGB space, whereas the fuzzy one helps to avoid the apparition of dead neurons, i.e., neurons that never activate.

The segmentation accuracy of the proposed neural network is compared to mixture of Gaussian models. In all the performed comparisons, our model achieved better results in terms of success rate and false positive rate, whereas the false negative rate is, at least, comparable to the obtained by the other studied methods.

The proposed algorithm is parallelized on a pixel level and designed to enable efficient hardware implementation to achieve real-time processing at great frame rates.

Other applications of this model will be studied, including the incorporation of this neural network to a remote sensing system performing people and vehicles tracking on closed scenes.

Acknowledgements

This work is partially supported by Junta de Andalucía (Spain) under contract TIC-01615, project name Intelligent Remote Sensing Systems.

References

1. Cucchiara, R., Grana, C., Piccardi, M., Prati, A.: Detecting moving objects, ghosts, and shadows in video streams. IEEE Transactions on Pattern Analysis and Machine Intelligence 25(10), 1337–1342 (2003)
2. Koller, D., Weber, J., Huang, T., Malik, J., Ogasawara, G., Rao, B., Russell, S.: Towards robust automatic traffic scene analysis in real-time. In: Proceedings of the International Conference on Pattern Recognition (1994)

3. Lo, B., Velastin, S.: Automatic congestion detection system for underground platforms. In: Proceedings of 2001 International Symposium on Intelligent Multimedia, Video and Speech Processing, 2001, pp. 158–161 (2001)
4. Toyama, K., Krumm, J., Brumitt, B., Meyers, B.: Wallflower: Principles and practice of background maintenance. In: Proceedings of the 1999, 7th IEEE International Conference on Computer Vision (ICCV 1999), pp. 255–261 (1999)
5. Haritaoglu, I., Harwood, D., Davis, L.: W4: real-time surveillance of people and their activities. IEEE Transactions on Pattern Analysis and Machine Intelligence 22(8), 809–830 (2000)
6. McKenna, S., Jabri, S., Duric, Z., Rosenfeld, A., Wechsler, H.: Tracking groups of people. Computer Vision and Image Understanding: CVIU 80(1), 42–56 (2000)
7. Wren, C., Azarbayejani, A., Darrell, T., Pentl, A.: Pfinder: Real-time tracking of the human body. IEEE Transactions on Pattern Analysis and Machine Intelligence 19(7), 780–785 (1997)
8. Stauffer, C., Grimson, W.: Adaptive background mixture models for real-time tracking. In: IEEE Computer Society Conference on Computer Vision and Pattern Recognition (1999)
9. Zabavin, N., Kuznetsova, M., Luk'yanitsa, A., Torshin, A., Fedchenko, V.: Recognition of handwritten characters by means of artificial neural networks. Journal of Computer and Systems Sciences International 38(5), 831–834 (1999)
10. Hall, L., Bensaid, A., Clarke, L., Velthuizen, R., Silbiger, M., Bezdek, J.: A comparison of neural network and fuzzy clustering techniques in segmenting magnetic resonance images of the brain. IEEE Transactions on Neural Networks 3(5), 672–682 (1992)
11. Pham, D., Xu, C., Prince, J.: Current methods in medical image segmentation. Annual Review of Biomedical Engineering 2, 315 ± (2000)
12. Mérida-Casermeiro, E., López-Rodríguez, D., Galán-Marín, G., Ortiz-de Lazcano-Lobato, J.M.: Improved production of competitive learning rules with an additional term for vector quantization. In: Beliczynski, B., Dzielinski, A., Iwanowski, M., Ribeiro, B. (eds.) ICANNGA 2007. LNCS, vol. 4431, pp. 461–469. Springer, Heidelberg (2007)
13. Stauffer, C., Grimson, W.: Learning patterns of activity using real-time tracking. IEEE Transactions on Pattern Analysis and Machine Intelligence 22(8), 747–757 (2000)
14. Cucchiara, R., Grana, C., Piccardi, M., Prati, A., Sirotti, S.: Improving shadow suppression in moving object detection with hsv color information. In: IEEE Intelligent Transportation Systems Conference Proceedings, pp. 334–339 (2001)
15. Mikic, I., Cosman, P., Kogut, G., Trivedi, M.: Moving shadow and object detection in traffic scenes. In: International Conference on Pattern Recognition, ICPR. IEEE Computer Society, Los Alamitos (2000)
16. Martel-Brisson, N., Zaccarin, A.: Learning and removing cast shadows through a multidistribution approach. IEEE Transactions on Pattern Analysis and Machine Intelligence 29(7), 1133–1146 (2007)
17. Prati, A., Cucchiara, R., Mikic, I., Trivedi, M.: Analysis and detection of shadows in video streams: a comparative evaluation. In: Proceedings of the IEEE Computer Society Conference on Computer Vision and Pattern Recognition, CVPR, pp. 571–576 (2001)
18. Horprasert, T., Harwood, D., Davis, L.S.: A statistical approach for real-time robust background subtraction and shadow detection. In: Proceedings of IEEE International Conference on Computer Vision (1999)

A Hierarchic Method for Footprint Segmentation Based on SOM

Marco Mora Cofre[1], Ruben Valenzuela[1], and Girma Berhe[2]

[1] Les Fous du Pixel
Image Processing Research Group
Department of Computer Science, Catholic University of Maule
Casilla 617, Talca, Chile
`mora@spock.ucm.cl, rvalenzuela@lfdp-iprg.net`
`http://ganimides.ucm.cl/mmora/`
[2] Department of Computer Science, University of Luxembourg, Campus Kirchberg
6, rue Richard Coudenhove-Kalergi, L-1359 Luxembourg
`girma.berhe@uni.lu`

Abstract. In this study we propose a new approach for solving the problem of segmenting the footprint in color images. Previous studies have presented direct and supervised methods for segmenting the footprint pattern. This new approach proposes, in comparison to the previous methods, a hierarchic segmentation method, the use of different color models to represent the image pixels, and the non-supervised classification based on SOM. The characteristics of the method allow a robust footprint segmentation with a high level of autonomy.

1   Introduction

The footprint segmentation problem is relevant for the construction of systems to recognize people [1], and for the computer-assisted detection of foot diseases such as the diabetic foot [2] , the flat foot and the cave foot [3].

When a person stands, the portion of the foot in contact with the ground is called footprint, and the zone not touching the surface is called foot vault [4]. To visualize the footprint, an instrument called podoscope is commonly used. This instrument can visualize the footprint based on the optic principle [2] or based on electronic pressure-based sensors [5].

The literature provides some segmentation methods for the footprint. In [6] a supervised method for segmenting the footprint based on conventional operations for image treatment is proposed. In [7], neural networks with supervised training are proposed, and in [8] the problem is solved by using more sophisticated techniques for image processing such as the anisotropic filtering and active contours. The previous methods have two characteristics in common: all try by a series of treatments to directly segment the footprint from the rest of the image, and all present low level of autonomy.

This study is part of a research with the goal of designing and implementing a computer system for detection of foot diseases, at a low cost, and robust enough to

V. Kůrková et al. (Eds.): ICANN 2008, Part I, LNCS 5163, pp. 887–896, 2008.

be used in the massive diagnosis of diseases, with low human intervention. Considering the previous points, this article presents a new footprint segmentation method. Instead of a direct segmentation, a hierarchical approximation for the footprint segmentation is proposed. This approximation consists in successive segmentations of the image in order to progressively separate its components, to the point of being able to differentiate the footprint from the rest of the foot image. The method has different stages. Initially the image is simplified by a color reduction. Secondly, the background pixels are separated from the footprint pixels. The two previous stages are based in a non-supervised neuronal network SOM [9]. A third step is the separation of the footprint pixels from the pixels of the full foot image. For that, the distance between the classes in different color models are analyzed [10], and a non-supervised classification by SOM is applied again. Finally, the quality of the segmentation is improved by applying a median filter and by eliminating small bodies based on segmentation by size [11].

This study has the following structure: the second section presents a non-automated method based in umbrals for the footprint detection. In the third section all the stages of the proposed method for the footprint segmentation are explained in detail. The fourth section shows the analysis of the results for this new method. The study ends with the conclusions.

2   A Non-automated Method for Footprint Segmentation

As it is said in [7], simple techniques such as the umbral detection do not allow a good segmentation of the footprint in a grey-scale image because of a problem called metamerism [12]. Metamerism is a phenomenon for grey-scale images, where pixels with different values in the color image have associated the same value in the gray-scale image.

Figure 1 shows different footprint images. Figure 1(a) is a drawing showing different zones of a footprint, zone 1 is the footprint, and zone 2 corresponds to the vault of the foot. Figure 1(b) shows a grey-scale image of a footprint. The metamerism effect is shown in figure 1(c): the red circles show the footprint and vault pixels with a similar grey level, but with different colors in the image in figure 1(d). Other problem with this type of images in gray scale is the existence of limit zones between the footprint and the vault with low contrast, which make difficult to segment the footprint by using the information provided by the color image.

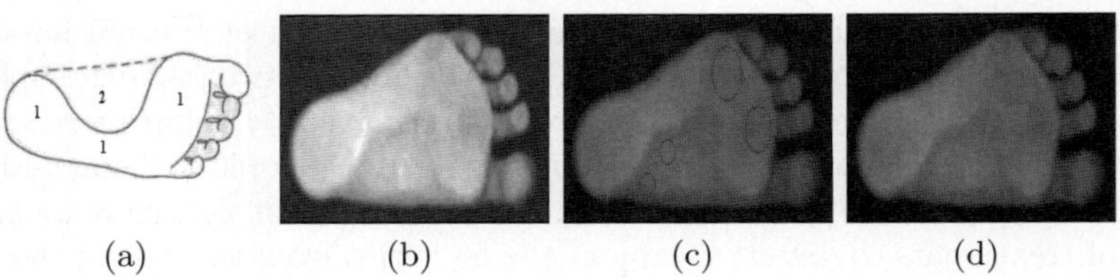

(a)   (b)   (c)   (d)

Fig. 1. Zones and images of the foot plant

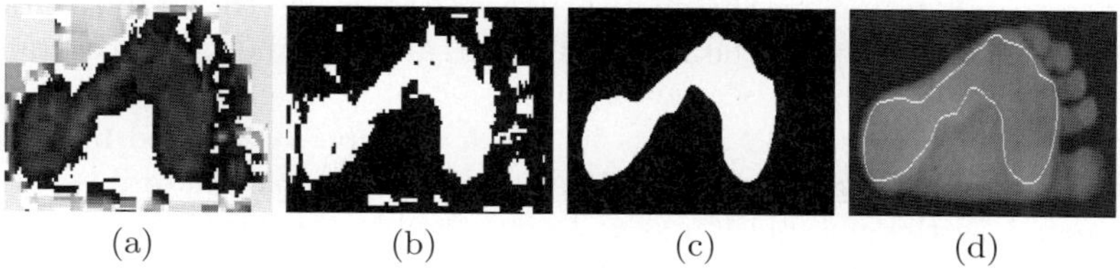

(a)   (b)   (c)   (d)

Fig. 2. Stages in the channel H based method

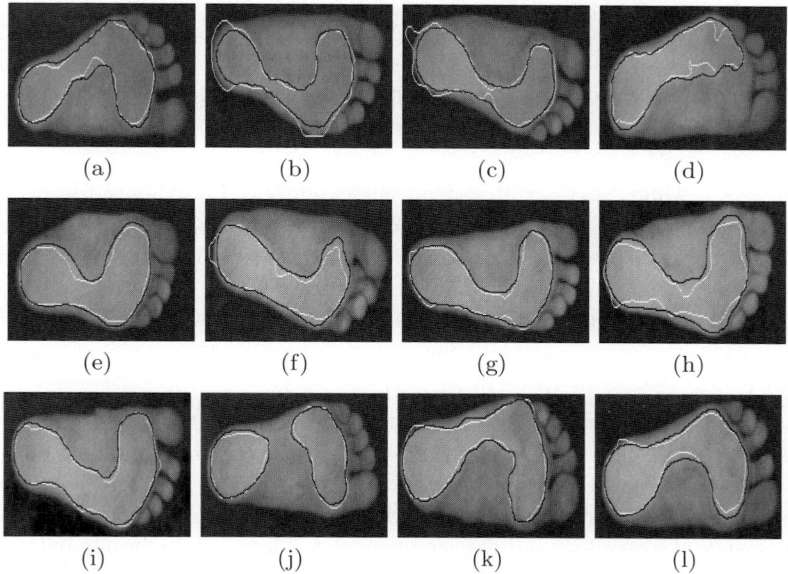

(a)   (b)   (c)   (d)
(e)   (f)   (g)   (h)
(i)   (j)   (k)   (l)

Fig. 3. Results of the channel H based method

A color model has the specification of a tridimensional space with coordinates and a subspace of this system where each color is represented by a unique point. In this way, a digital color image is made of 3 matrices, each one representing a channel of the color model [11]. A simple method although not automated for the segmentation of the footprint consists in finding a channel of a color model properly representing the footprint, manually determining for each image the range of values for the footprint, segmenting the footprint considering the defined umbrals, and finally improving the quality of the obtained footprint by classic techniques such as the segmentation by size and the softening by the application of median filter [11]. In this study, the most traditional color models have been used (RGB, CIE XYZ, CMY), chromatic models (HSV, HSL, YIQ, YUV, YcbCr), and the models known as uniforms (CIE Lab, CIE Luv, CIE Lch) [10].

It has been determined that the channel H from the HSV model allows this type of segmentation. Figure 2 shows in detail the stages in the method based on the treatment of channel H. Figure 2(a) shows the channel H, figure 2(b) shows the segmentation by thresholds, figure 2(c) shows the segmentation by size and improvements with the median, and figure 2(d) shows the contour over the original image.

Figure 3 shows a set of 12 images in which the non-automated method was tried. Each figure shows a footprint image, the white color describes the contour detected by the channel H method, and the black color shows the contour segmented by a specialist. Even though the method is simple, in some cases the detected contour is far away from the real one. Additionally, the thresholds segmenting the footprint are particular for each image and it is very unlikely that an automated method can define them with precision.

3   Hierarchic Method for Footprint Segmentation Based on SOM

This section presents the details for our hierarchic and automated approach to segment the footprint. The stages in our proposal are the following: (1) Simplification of the image by color reduction based on SOM, (2) segmentation of the foot plant image based on SOM, (3) segmentation of the footprint based on SOM, and (4) improvement of the footprint segmentation.

3.1   Color Reduction Applied on the Original Image

Our segmentation approach is based on non-supervised classification. Then, it is convenient to have distant patterns for classes. In the case of having a high number of patterns, they can cover the interspaces, reducing then the distance among classes. To avoid the above, we propose to simplify the image by reducing the amount of colors.

Methods to reduce colors have been proposed based on variants of SOM [13,14]. We adopted the classic scheme of SOM for classifying patterns, a row vector of neurons, which update only the winner neuron and has as many output neurons as the number of color classes desired to obtain.

Figure 4 shows a sequence of images obtained from the color reduction by SOM. Figures 4(a), 4(b), 4(c) and 4(d) show the respective reduction to 5, 10, 20 and 25 colors. The figure shows that, with the progressive increase of colors, the footprint

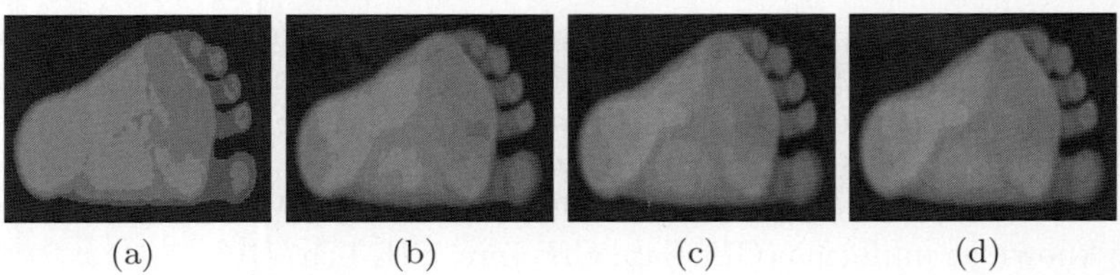

(a)   (b)   (c)   (d)

Fig. 4. Color reduction

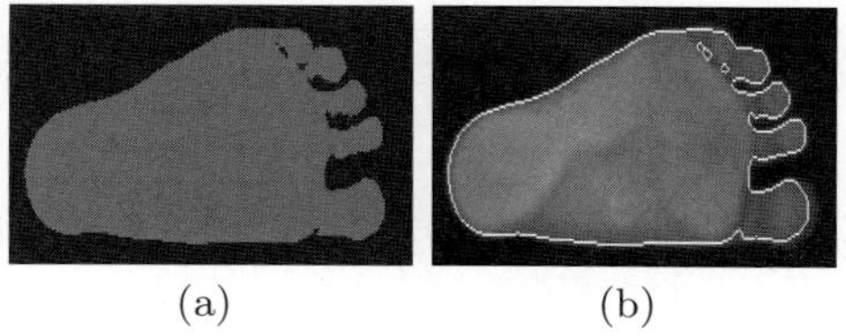

Fig. 5. Foot plant image segmentation

starts to distinguish with more clarity. Our research has shown that 20 colors provide a good footprint representation. It is necessary to also consider that as the number of neurons increases, the computing time increases as well.

3.2 Foot Plant Image Segmentation

In our hierarchic approach, a first segmentation level implies the separation of the foot image from the background of the image. The idea is to treat the background of the image as a class, and the foot image (vault and footprint together) as the second class. The goal is to detect the pixels from the image belonging to the footprint in order to focus the second segmentation stage on them.

In this case we adopted the same scheme implemented for the color reduction, this time considering two output neurons. Figure 5 shows the result for the reduction of the foot image to two colors. Specifically, figure 5(a) shows the resulting image with two colors, and figure 5(b) shows the contour obtained over the image in initial colors. An adequate separation between the image background and the foot image has been obtained.

3.3 Segmentation of the Footprint Based on Som

The second level of our hierarchic approach to segmentation is the pixel extraction for the footprint from the total pixels from the foot image. Two new classes are considered, one for the footprint class and the other for the vault class.

The image is then transformed in different color models, in order to search for two channels from a model representing the pixels of the footprint class and the ones for the vault class the farthest away possible, being able to segment by non-supervised segmentation. All the color models in section 2 have been used, but figure 6 only shows a selection of them. The three first figures: 6(a), 6(b) and 6(c) show the dispersion diagram for the channels of model chromatic model HSV, the following three figures 6(d), 6(e) and 6(f) show the diagrams for the Lab uniform model, and figures 6(g), 6(h), 6(i) show the diagrams for the channels of the Luv uniform model. From the figures it can be noted that the uniform models are the ones better separating the footprint class and the vault class, specifically, the channels a-b from the CIE Lab model and the channels u-v from the CIE Luv model.

To determine the pair of channels better representing the characteristics of separation, in other words, the plane with the footprint class and the vault

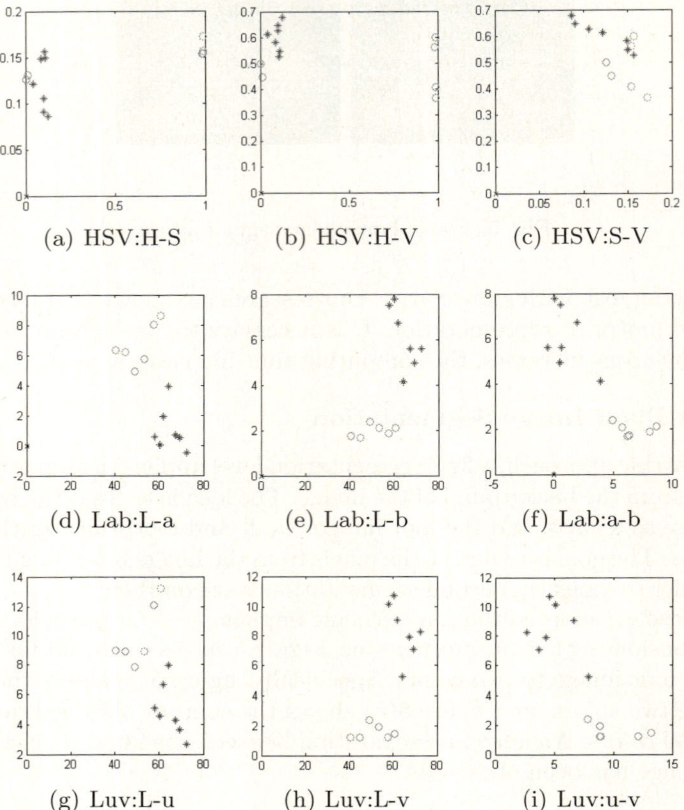

Fig. 6. Foot plant image and its color models

class more separated, we used two measures. The first measure is the Euclidian distance between the centers of the classes (DE). In this case, a normalized distance is used to compare with respect to the biggest distance found between two patterns in each plane (DEN), in order to be able to compare the distance between centers among different pairs of channels. We also used the Λ of Wilks (LW) [15], which is the coefficient between the intra-group dispersion and the total dispersion (sum of the dispersion intra and inter group). Values close to zero of Λ mean that the groups are distant, and values close to 1 mean that the groups are close. Table 1 for the 12 images shows the DEN for the plane a-b, the DEN for plane u-v, the Λ for plane a-b and plane u-v. The table shows that the Euclidian distance between the centers is bigger in the plane a-b, but the Λ in the majority of the cases is a little smaller in the plane u-v. From the table can be concluded that plane u-v has slightly better characteristics to separate the footprint and vault patterns than plane a-b. Despite of the previous and based on our approach, both planes provide a good pattern classification.

Table 1. Distance evaluation between classes

No	DEN a-b (Lab)	DEN u-v (Luv)	Λ a-b (Lab)	Λ u-v(Luv)
1	0.71	0.70	0.5952	0.5996
2	0.70	0.70	0.4476	0.4475
3	0.87	0.84	0.3258	0.3237
4	0.76	0.76	0.2541	0.2576
5	0.83	0.80	0.3365	0.3453
6	0.85	0.83	0.4250	0.4220
7	0.67	0.66	0.3632	0.3714
8	0.71	0.67	0.2531	0.2507
9	0.71	0.66	0.6060	0.6035
10	0.67	0.66	0.3486	0.3489
11	0.67	0.62	0.0798	0.0793
12	0.68	0.66	0.5050	0.5030

To separate the footprint pixels from the vault pixels, we have implemented a non-supervised SOM neuronal network but instead of calculating the distance between the entry pattern and the weight vectors for the neurons by using the traditional Euclidian distance, we used the cosine distance. This distance allowed to better differentiate the footprint pattern and the vault pattern. Figure 7(a) shows the non-supervised classification of the footprint pixels.

3.4   Improvement of the Footprint Segmentation

From figure 7(a) it can be observed that the footprint obtained from the segmentation is no soft and the image presents small bodies disconnected from the footprint. To improve the results obtained a segmentation by size and a median filter are used. Figure 7(b) shows the segmentation by size, figure 7(c) shows the softening by median, and figure 7(d) shows the contour obtained on top of the image in original color.

Figure 8 shows the footprint obtained for the 12 images used for the non-automated method in section 2. The white curves show the contour resulting from the automated segmentation based on SOM, and the black curves show the contour traced by a specialist. It can be observed that the contours obtained by our automated approximation are very close to the contours traced by the specialist.

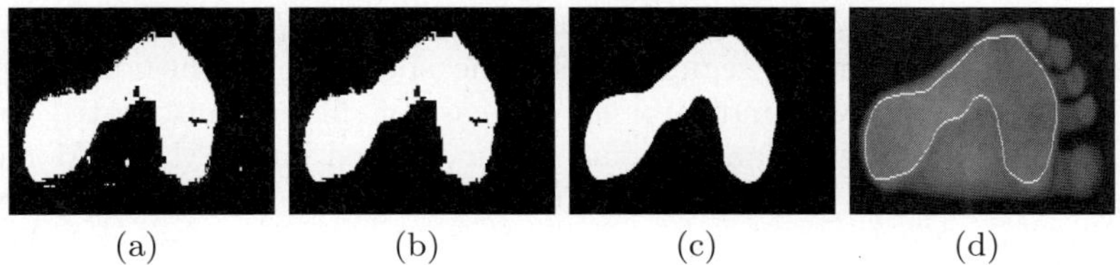

(a)   (b)   (c)   (d)

Fig. 7. Improvement of the footprint segmentation

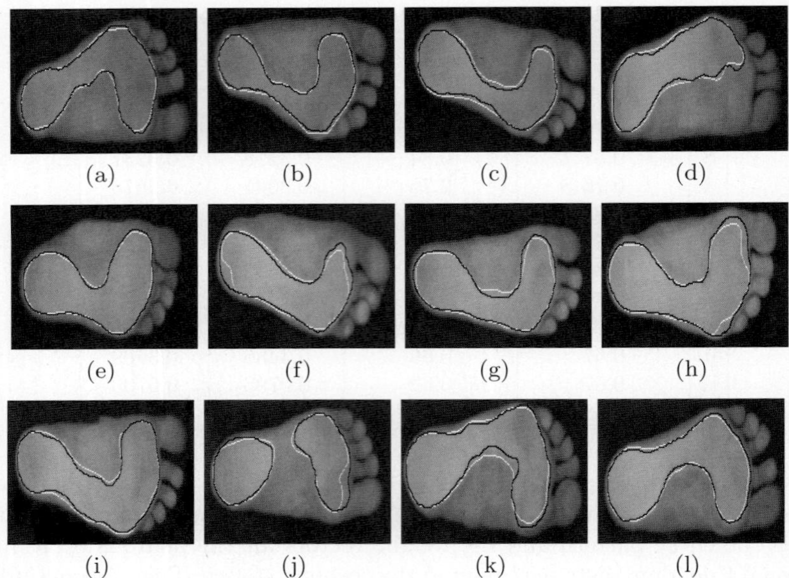

Fig. 8. Results of the SOM based method

4   Analysis of the Results Obtained by the Method

To numerically evaluate the results provided by our method based on SOM we used the Pratt figure of merit (FOM) [16]. The FOM indicator is a measure of similarity between two curves with values between 0 and 1. If the curves are similar the FOM is close to 1, and if the curves are very different its value is close to 0. The formula for FOM is the following:

$$FOM = \frac{1}{max(I_I, I_E)} \sum_{i=1}^{I_A} \frac{1}{(1 + a * d^2(i))}$$

where I_I and I_E represent the number of pixels for the ideal contour (traced by the specialist) and the contour obtained, respectively. a is a factor of scale (usually $1/9$). $d^2(i)$ is the distance between a pixel i from the contour obtained to its closest in the ideal contour. It is important to notice that this indicator can be used only if the ideal contour is available.

Considering the contour segmented by the specialist as the ideal contour, we can calculate the FOM indicator for the case of the non-automated method (FOM method channel H) and for the method based on SOM (FOM method SOM). Table 2 shows the indicators FOM for both methods for each one of the 12 images. By checking each row in the table, it can be proved that our method based on SOM has its FOM indicators closer to 1 than the method based in channel H. This way we proved numerically that our automated method produces better results than the non-automated method.

Table 2. evaluation of the methods based on FOM

N Image	FOM chanel H based method	FOM SOM based method
1	0.7480	0.8676
2	0.6545	0.8282
3	0.6936	0.7177
4	0.7420	0.7504
5	0.6232	0.8548
6	0.6369	0.6535
7	0.7392	0.7576
8	0.4506	0.7332
9	0.7741	0.7681
10	0.7800	0.7445
11	0.7460	0.7120
12	0.6553	0.8009

5   Conclusions

This article has presented a non-supervised method based on SOM for the footprint segmentation in color images. Our proposal presents several innovations compared to methods presented in previous studies. In first place, an image simplification by color reduction was implemented. Secondly, there was a separation of the background pixels from the pixels of the foot image. Finally, the footprint pixels and the vault pixels were separated by non-supervised classification. The first, the second and the fourth stages of the method are totally based on a SOM non-supervised neuronal network.

Because of the hierarchic approach and the non-supervised classification, the proposed method presents a great autonomy and provides a viable alternative to be implemented for the footprint segmentation stage in a system for foot disease detection.

References

1. Nakajima, K., Mizukami, Y., Tanaka, K., Tamura, T.: Footprint-based personal recognition. IEEE Transactions on Biomedical Engineering 47(11), 1534–1537 (2000)
2. Morsy, A., Hosny, A.: A new system for the assessment of diabetic foot planter pressure. In: Proceedings of 26th Annual International Conference of the IEEE EMBS, pp. 1376–1379 (2004)
3. Mora, M., Jarur, M.C., Sbarbaro, D.: Automatic diagnosis of the footprint pathologies based on neural networks. In: Beliczynski, B., Dzielinski, A., Iwanowski, M., Ribeiro, B. (eds.) ICANNGA 2007. LNCS, vol. 4432, pp. 107–114. Springer, Heidelberg (2007)
4. Valenti, V.: Orthotic Treatment of Walk Alterations (in spanish), 1st edn. Panamerican Medicine (1979)
5. Patil, K., Bhat, V., Bhatia, M., Narayanamurthy, V., Parivalan, R.: New online methods for analysis of foot preesures in diabetic neuropathy. Frontiers Me. Biol. Engg. 9, 49–62 (1999)

6. Chu, W., Lee, S., Chu, W., Wang, T., Lee, M.: The use of arch index to characterized arch height: a digital image processing approach. IEEE Transaction on Biomedical Engineering 42(11), 1088–1093 (1995)
7. Mora, M., Sbarbaro, D.: A robust footprint detection using color images and neural networks. In: Sanfeliu, A., Cortés, M.L. (eds.) CIARP 2005. LNCS, vol. 3773, pp. 311–318. Springer, Heidelberg (2005)
8. Mora, M., Jarur, M., Pavesi, L., Achu, E., Drut, H.: A pattern recognition approach to diagnose foot plant pathologies: From segmentation to classification. In: Bellazzi, R., Abu-Hanna, A., Hunter, J. (eds.) AIME 2007. LNCS (LNAI), vol. 4594, pp. 378–387. Springer, Heidelberg (2007)
9. Kohonen, T.: Self-Organizing Maps. Springer, Heidelberg (2001)
10. Cheng, H., Jiang, X., Sun, Y., Wang, J.: Color image segmentation: advances and prospects. Pattern Recognition 34(9), 2259–2281 (2001)
11. Gonzalez, R., Woods, R.: Digital image processing, 2nd edn. Addison-Wesley Longman Publishing Co., Inc., Boston (2001)
12. Gevers, T., Groen, F.: Segmentation of color images. In: Proceedings of 7th Scandinavian Conference on Image Analysis 1991 (SCI 1991) (1991)
13. Atsalakis, A., Paparmakos, N.: Color reduction and estimation of the number of dominant colors by using a self-growing and self-organized neural gas. Engineering Applications of Artificial Intelligence 19(7), 769–786 (2006)
14. Papamarkos, N., Atsalakis, A., Strouthopoulos, C.: Adaptive color reduction. IEEE Transaction on System, Man and Cybernetics, Part B 32(1), 44–56 (2002)
15. Mardia, K., Kent, J., Bibby, J.: Multivariate Analysis, 1st edn. Academic Press, London (1979)
16. Pratt, W.: Digital image processing, 2nd edn. John Wiley and Sons, Chichester (1991)

Co-occurrence Matrixes for the Quality Assessment of Coded Images

Judith Redi[1], Paolo Gastaldo[1], Rodolfo Zunino[1], and Ingrid Heynderickx[2]

[1] Dept. of Biophysical and Electronic Engineering (DIBE), Genoa University
Via Opera Pia 11a, 16145 Genoa, Italy
{judith.redi,paolo.gastaldo,rodolfo.zunino}@unige.it
[2] Philips Research Laboratories Prof. Holstlaan 4 - 5656 AA Eindhoven NL
and Delft Technical University, Mekelweg 4 2628 CD Delft NL
ingrid.heynderickx@philips.com

Abstract. Intrinsic nonlinearity complicates the modeling of perceived quality of digital images, especially when using feature-based objective methods. The research described in this paper indicates that models from Computational Intelligence can predict quality and cope with multi-dimensional data characterized by complex perceptual relationships. A reduced-reference scheme exploits Support Vector Machines (SVMs) to assess the degradation in perceived image quality induced by three different distortion types: JPEG compression, white noise, and Gaussian blur. First, an objective description of the images is obtained by exploiting the co-occurrence matrix and its features; then, the SVM supports the nonlinear mapping between the objective description and the quality evaluation. Experimental results confirm the validity of the approach.

Keywords: objective quality assessment, neural networks.

1   Introduction

Efficient evaluation of image quality is a crucial issue to improve present-day technologies used for signal storage and communication [1]. Thus, there is a great interest in developing reliable methods for assessing the quality of digital images. Sub-jective testing methods [2] require human assessors to judge the overall quality of a set of images. Those approaches are time-consuming, and very expensive. Moreover, they give an empirical, consistent representation of the perceptual phenomenon, but do not lead to a deterministic model of the evaluation process. Conversely, objective assessment methods [1,3] aim at the automated estimation of perceived quality by using numerical features extracted from images. These approaches bypass human assessors but, to be effective, they must provide results consistent with data obtained from subjective experiments.

Objective methods are usually classified with respect to the amount of information available from the original, unprocessed image. No-reference approaches [1] assess perceived quality in the absence of the original image. Full reference methods [1,3] assume to have both the original and the processed image available, and relate perceived quality to the difference between these two. Typically,

V. Kůrková et al. (Eds.): ICANN 2008, Part I, LNCS 5163, pp. 897–906, 2008.

these approaches are extended by including analytical models of the human visual system (HVS) in order to improve their accuracy and reliability [4,5]. Finally, reduced-reference schemes [1,6] only use a limited set of features from the original image; they can e.g. access the quality of noise-affected images.

The research presented in this paper follows a reduced-reference objective approach, since only some characteristics of the original image (i.e. a feature vector) are needed; connectionist techniques tackle the crucial issue of reliably mapping the objective description of an image into the associated quality evaluation. The specific objective of the research is to predict the change in quality as a consequence of any type of distortion on the original image. Co-occurrence matrixes provide an effective paradigm to extract an objective description from the images. Previous research [7,8,9] showed that features derived from the co-occurrence matrix [10] are able to provide an accurate prediction of quality. A Support Vector Machine (SVM) [11] supports the underlying objective model, which is learnt and adjusted empirically on actual data drawn from subjective testing. The SVM operates on the vectors constructed from objective features, which are extracted from the processed image and the corresponding original signal, respectively. A quality score attributed to the processed image is the eventual output. The basic advantage of a connectionist approach is the ability to deal effectively with multidimensional data as e.g. originating from complex perceptual relationships, without an explicit mathematical model of the underlying phenomenon. Indeed, SVMs represent one of nowadays most effective methodologies for tackling classification and regression problems in complex, nonlinear data distributions [11]. The present study shows that the SVM-based reduced-reference model is effective in assessing the degradation in perceived image quality as a consequence of image distortions. Three different sources of distortion are considered: JPEG compression, white noise, and Gaussian blur. The second release of the LIVE database [12] provided the experimental domain, within which the proposed framework is evaluated. The obtained results confirm the general validity of the SVM-based approach, also considering the substantial difference in the three image distortion types.

2 Support Vector Machines for Regression Problems

A SVM is used in this research to map feature-based image descriptions into scalar values that represent the perceived image quality. As such, the overall objective-quality assessment model can be regarded as a regression problem, in which learning evolves according to an empirical sample.

Each training pattern is encoded by a d-dimensional vector, $\mathbf{x}$, and is labelled by a target value, y_i. The vector, $\mathbf{x}$, contains a feature-based description of an image, while the value y_i represents the (normalized) quality score associate with that image. Thus the training set includes n_p patterns, and can be written as $Z = \{(x_i, y_i); i = 1, .., n_p; x_i \in \Re^d, y_i \in [-1, +1]\}$. The empirical problem formulation implies that the actual function, $\gamma(\mathbf{x})$, which maps a feature-based vector into its corresponding quality score, y, is unknown, hence the learning process should infer the function $\hat{\gamma}(\mathbf{x}) : \Re^d \mapsto \Re$ that best approximates $\gamma(\mathbf{x})$. This usually is quite a complicated task, especially in the case of quality assessment, which is a perceptual phenomenon, not fully understood yet.

SVMs regression models approximate the target function for an input vector, x, as:

$$\hat{\gamma}_{svm}(\mathbf{x}) = \sum_{i=1}^{n_{sv}} (\alpha_i - \alpha_i^*) y_i K(\tilde{\mathbf{x}}_i, \mathbf{x}) + b \tag{1}$$

where α_i, α_i^* are positive parameters and b is a bias. The patterns $\{(\tilde{\mathbf{x}}_i, y_i), i = 1, ..., n_{sv}\}$ used are a subset of the training set and are called support vectors. Expression (1) shows that $\hat{\gamma}_{svm}(\mathbf{x})$ is a series expansion having kernel function $K(\cdot, \cdot)$ as a basis and involving part or all of the training patterns. As a result of this well-known kernel trick, expression (1) also shows that one can handle inner products of patterns in the transform space, yet disregarding the specific mapping of each single pattern. In the following, the use of the Radial Basis Function (RBF) kernel will be assumed throughout the paper:

$$K(\mathbf{x}_1, \mathbf{x}_2) = \exp(-||\mathbf{x}_1 - \mathbf{x}_2||^2 / \sigma^2) \tag{2}$$

The coefficients α_i, α_i^* and b in expression (1) must be adjusted in compliance with the input sample distribution so as to minimize some cost function measuring the deviation resulting from the approximation. To this end, Vapnik [11] suggested the use of ϵ-insensitive loss functions, which penalize the error whenever the absolute approximation error remains smaller than ϵ. A complete treatment of Vapniks approach is beyond the scope of this paper; here, it is sufficient to realize that the SVM optimization task can be reformulated as a minimization problem in a Hilbert space, such that SVM learning is equivalent to solve:

$$\min_{\alpha, \alpha^*} \tfrac{1}{2} \sum_{i,j=1}^{n_p} \beta_i \beta_j K(\mathbf{x}_i, \mathbf{x}_j) - \sum_{i=1}^{n_p} \beta_i y_i + \sum_{i=1}^{n_p} \psi_i \epsilon$$
$$0 \leq \alpha_i, \alpha_i^* \leq C \tag{3}$$
$$\sum_{i=1}^{n_p} \beta_i = 0$$

where $\beta_i = (\alpha_i - \alpha_i^*)$, $\psi_i = (\alpha_i + \alpha_i^*)$ and C is a fixed regularization term that rules the trade-off between accuracy and complexity.

SVMs offer both theoretical and practical advantages during training. First of all, the training problem (3) is a constrained quadratic programming problem (CQP) on a convex cost function, which allows one to reach the global minimum by using polynomial-complexity Quadratic Programming (QP) algorithms [11]. Secondly, the optimization problem involves inner products between pairs of patterns, hence the curse of dimensionality does not enter the complexity of the training process explicitly. Finally, kernel-based representations allow SVMs to handle arbitrary distributions that may not be linearly separable in the data space.

When considering application effectiveness, one aims to predict the approximation error on overall run-time data not used during training. From this viewpoint, a crucial reason of the SVM success is the fact that the complexity of the learning machine can be controlled through the Structural Risk Minimization principle [11]. This principle leads to upper bounds of the generalization ability of a trained SVM, albeit in a statistical way. In particular, the eventual generalization performance of a SVM depends on the specific setting of the kernel hyper-parameters. Both theoretical [11,13] and empirical [14] approaches have been proposed in the literature to define the generalization limits. Theoretical bounds to an SVMs run-time error use worst-case derivations, and therefore prove impractical in real-world domains. Conversely, comparisons of empirical methods [14] show that they can yield effective and tight predictions of the SVMs performance. In the research presented in this paper, the hyper-parameter tuning problem involves the pair $\{C, \sigma\}$, and an empirical approach involving k-fold cross validation [14] provided the final setting of the hyper-parameters values.

3 Images Objective Description for Quality Prediction

The research presented in this paper adopts a reduced-reference paradigm to address the quality assessment of distorted images. The overall problem can be set formally as follows. Let $I^{(n)}$ be an original, reference image, and $\bar{I}^{(n,r)}$ an image resulting from applying some impairment to $I^{(n)}$ with a distortion strength, r. Denote $q^{(n)}$ and $q^{(n,r)}$ as the perceptual quality scores obtained from a subjective evaluation of $I^{(n)}$ and $I^{(n,r)}$, respectively. Likewise, $\mathbf{x}^{(n)}$ and $\mathbf{x}^{(n,r)}$ denote the feature-based description of $I^{(n)}$ and $\bar{I}^{(n,r)}$, respectively. The fact that in our approach only $\mathbf{x}^{(n)}$ is needed for the quality estimation, and not the full information of the original image $I^{(n)}$, clarifies the classification of reduced reference paradigm. The loss/gain in quality that results from somehow distorting the image $I^{(n)}$ can be estimated in two ways: either directly, by computing the distance, $d_S(q^{(n)}, q^{(n,r)})$, between the subjective scores associated with the two images, or indirectly, by observing the objective vectors, $\mathbf{x}^{(n)}$ and $\mathbf{x}^{(n,r)}$, and predicting the difference in quality from the comparison between the two vectors. In the framework adopted in this paper (fig. 1), a connectionist model supports the latter predictive task. The objective descriptors $\mathbf{x}^{(n)}$ and $\mathbf{x}^{(n,r)}$ feed a SVM, which is trained to estimate the quantity $d_S(q^{(n)}, q^{(n,r)})$, and therefore reproduces the difference in quality between the original and the distorted image. Hence, the prediction of image quality is decoupled into two tasks: 1) the design of the objective metric (i.e. feature vector), which, when chosen properly, results in an effective descriptive basis for the images to assess their quality, and 2) the design of the mapping function, which may be highly nonlinear and even mimic unknown perceptual mechanisms.

3.1 Feature-Based Description of the Image

The present research exploits features derived from the co-occurrence matrix to construct an objective description of the image. A co-occurrence matrix, C_a, measures local correlation among gray tones within an image sub-region, a, including $H_a \times W_a$ pixels. In particular, the matrix element $C_{i,j}^{(a)}(\lambda, \Theta)$ describes a joint occurrence of lu-minance levels, and counts the pairs of pixels in a that:

1) have gray levels i and j, and 2) are separated by λ radial units at an angle Θ to the horizontal axis. Formally, C_a is defined as:

$$C_{i,j}^{(a)}(\lambda,\Theta) = \left| \left\{ (m,n), \begin{array}{c} 0 \leq m \leq W_a - 1 \\ 0 \leq n \leq H_a - 1 \end{array} \; s.t. \; \begin{array}{c} a_{m,n} = i \\ a_{m+\Delta h,n+\Delta v} = j \end{array} \right\} \right| \; i,j = 0,...,N_g \tag{4}$$

where Δh and Δv are the horizontal and vertical displacements in the Θ direction, respectively:

$$\Delta h = \lambda \lceil \cos \Theta \rceil, \Delta v = \lambda \lceil \sin \Theta \rceil, 0 < \Theta \leq \frac{\pi}{2} \tag{5}$$

$$\Delta h = \lambda \lfloor \cos \Theta \rfloor, \Delta v = \lambda \lceil \sin \Theta \rceil, \frac{\pi}{2} < \Theta \leq \pi \tag{6}$$

Although C_a is a $N_g \times N_g$ matrix, where N_g is the number of luminance quantization levels, a co-occurrence matrix can be characterized by using a set of scalar descriptors, which were formalized in [10]. All of those features are implicitly indexed by the image region, a, from which the matrix is calculated. The research presented here adopts a subset, $\Phi = \{f_u; u = 1, \cdots, N_f\}$, of $N_f = 10$ features that have already been tested to be successful [8]. The feature set Φ supports the objective vector, $\mathbf{x}_u^{(n,r)}$, that characterizes an image, $\bar{I}^{(n,r)}$ and is used in two steps. First, the image is split into non-overlapping square regions, each holding $N_a \times N_a$ pixels. For each block, the co-occurrence matrix and the related features f_u are computed. The block size, N_a, plays a crucial role in the reliability of the image description, and the image size should be taken into account to ensure proper sampling by a sufficient number of measures. Moreover, the features derived from the co-occurrence matrix $C^{(a)}(\lambda,\Theta)$ may suffer from border effects. The percentage of pixels that do not enter the computation of $C_{i,j}^{(a)}(\lambda,\Theta)$ decreases as N_a increases, hence one should avoid to use small block sizes; a typical setting is $N_a > 8$. The first step of the vector-computation procedure yields as many values for a given feature as the number of blocks.

The second step compresses the information thus obtained, and aggregates block-level information into one objective vector that characterizes the whole image. Such a global-level statistical reduction is consistent with the actual measuring procedure, since human assessors usually generate one overall quality score per image. Hence, local-level, block-based information must somehow be reduced into one objective vector per image, which has to be further mapped onto a single quality score. The approach adopted in this paper uses percentiles as statistical descriptors for estimating the distribution of a feature f_u over the image. Similar approaches have already been proved to be effective for quality assessment of compressed [8] and uncompressed images [7]. Thus, for a parameter setting λ^*, Θ^*, the image $\bar{I}^{(n,r)}$ is represented by a set of N_f objective vectors, $\mathbf{x}_{u,(\lambda^*,\Theta^*)}^{(n,r)}$, $u = 1,..,N_f$, which contain detail-related information. In the remainder of this paper, the index pairs (n,r) and (λ^*,Θ^*) will be omitted wherever possible, in order to simplify the notation. The procedure to construct the objective vectors can be summarized as follows:

Algorithm 1. Objective vectors construction

1: **procedure** BUILDVECTORS(a picture $\bar{I}^{(n,r)}$, a descriptive feature f_u and a parameter setting $(\lambda*, \Theta^*)$)

2:     (Block-level feature extraction)
   – Split $\bar{I}^{(n,r)}$ into N_b non-overlapping blocks (size N_a) to obtain the set: $B = \{b_m; m = 1, ..., N_b\}$
   – For each block $b_m \in B$ compute the associate co-occurrence matrix $C^{(m)}(\lambda^*, \Theta^*)$
   – For each matrix $C^{(m)}$ compute the values $x_{u,m}$ of feature $f_{u'}$ and obtain the set of values

$$X_u = \{x_{u,m}; m = 1, ..., N_b\} \tag{7}$$

3:     (Feature integration into one image descriptor)
   – Compute a percentile-based description of the sample set X_u; let p_a be the α-th percentile:

$$\phi_{\alpha,u} = p_\alpha(X_u) \tag{8}$$

   – Assemble the objective descriptor vector, $\mathbf{x}_u^{(n,r)}$, for the feature f_u on the image $\bar{I}^{(n,r)}$ as:

$$\mathbf{x}_u^{(n,r)} = \{\phi_{\alpha,u}^{(n,r)}; \alpha = 0, 20, 40, 60, 80, 100\} \tag{9}$$

4: **end procedure**

3.2   Reduced-Reference Quality Assessment by Exploiting SVMs

The reduced-reference approach to image quality assessment exploits 1) the consistency of the co-occurrence matrix as a descriptive basis for images [7,8,9], and 2) the effectiveness of SVMs in tackling regression problems for nonlinear relationships. According to the scheme presented in Fig. 1, objective vectors derived from co-occurrence matrixes feed the SVM. The relationship between the subjectively evaluated quality scores and the objectively predicted estimates of the quality is established in the SVM-training process, which adjusts the model parameters so as to minimize the error between the subjective measured quality gain/losses, $d_S(q^{(n)}, q^{(n,r)})$, and the objectively predicted quality change, $\hat{d}_S(q^{(n)}, q^{(n,r)})$. The former quantities must be measured on panels of human assessors and, incidentally, constitute the expensive as-pect of the overall methodology.

More specifically, the first step in the neural-based quality estimator (see Figure 1) consists of the feature extractor modules, which process both the reference image $I^{(n)}$ and the processed signal $\bar{I}^{(n,r)}$ to yield the 6-dimensional vectors $\mathbf{x}_u^{(n)}$ and $\mathbf{x}_u^{(n,r)}$, computed according to 9. From the output of the feature extractor modules the objective stimulus to the neural component is obtained simply by combining the two vectors:

$$\mathbf{z}_u^{(n,r)} = \lfloor \mathbf{x}_u^{(n)}, \mathbf{x}_u^{(n,r)} \rfloor \tag{10}$$

The resulting 12-dimensional vector $\mathbf{z}_u^{(n,r)}$ feeds an SVM-based estimator. It should be noted that each vector $\mathbf{z}_u^{(n,r)}$ per feature f_u feeds its own SVM-based estimator. Therefore, the complete system for objective quality prediction

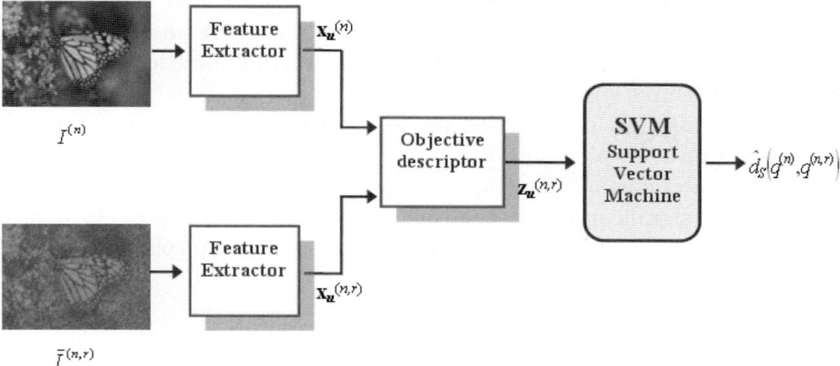

Fig. 1. The proposed neural-based quality estimator

just replicates as many independent SVMs (also trained independently) as the number of employed features. These estimators are further integrated within an ensemble structure [15]. Research on ensemble methods for quality prediction [9,15] showed that plain averaging is the most effective method for integrating the predictions of all estimators. This approach can be formalized as:

$$\hat{d}_S(q^{(n)}, q^{(n,r)}) = \frac{1}{K} \sum_k \hat{d}_S(q^{(n)}, q^{(n,r)})_{f=f_u} \tag{11}$$

4 Experimental Results

The second release [12] of the public database provided by the Laboratory for Im-age and Video Engineering (LIVE) of the University of Texas at Austin has been used as testbed for this study. The database includes test images obtained by applying different kinds of distortion to 29 original images (with a spatial resolution of either 480x720 pixels or 768x512 pixels). In the present experiment, three different types of distortion have been considered: JPEG compression, white noise in the RGB compo-nents, and Gaussian blur. In total, the dataset included 159 JPEG images, 145 images containing noise, and 145 blurred images. The LIVE database includes for each image a quality score obtained from a panel of human assessors that took part in a subjective quality experiment. People were asked to assess the perceived quality of the original and distorted image on a 6-points continuous scale, which was subsequently linearly remapped onto the range $[1, 100]$. Consequently, the remapped data of the original and distorted image were subtracted and averaged in order to obtain a Difference Mean Opinion Score (DMOS) per processed image [12].

4.1 Experimental Set-Up

For the experiment reported here, the features derived from the co-occurrence matrix were computed on the luminance signal only. The radius amplitude, λ, was set at a fixed value of $\lambda = 1$, while the angle, Θ, was set at $\Theta = 0$. This choice aimed at minimizing computational complexity and was subsequently supported experimentally, as it was found that larger values of λ did not alter the statistical distribution of features significantly.

The unsupervised feature-selection procedure presented in [9] was applied to the set of features ϕ, to pick out those descriptors that were the most informative over the three types of distortion. Many features were eventually discarded being either not significantly informative or mutually correlated. As a result, two objective measures, i.e. diagonal energy and IMC, were selected. They are defined as follows:

$$\texttt{diagonal energy} = \frac{\sum_{i,j:i=j}[C^{(a)}]^2}{\sum_{i,j}[C^{(a)}]^2} \tag{12}$$

$$\texttt{IMC} = \frac{\sum_{i,j} C^{(a)} \log C^{(a)} - \sum_{i,j} C^{(a)} \log[C^{(a)}]^2}{C_i^{(a)}}; C_i^{(a)} = \sum_j C_{i,j}^{(a)} \tag{13}$$

To complete the system set-up it was necessary to tune the kernel hyper-parameters to control the generalization ability (see section 2). To this purpose, the k-fold cross-validation technique was applied, which proved [14] to be the best method in terms of accuracy in predicting generalization performance (and therefore in terms of effectiveness in driving model selection). Based on the results of the cross-validation, the following values were selected for the experiment: for the JPEG compressed images: $C = 100$, and $\sigma = 0.5$, for the noisy images: $C = 100$, and $\sigma = 1$ and for the blurred images: $C = 10$, and $\sigma = 0.5$.

4.2   Results

To evaluate the overall performance of the quality estimator a k-fold test method was applied, being an effective strategy to obtain reliable results in cases where few data are available. The three datasets (i.e. each based on a different distortion) contained several distorted versions of the same 29 original images. Each dataset was divided into 5 different folds, each containing all the distorted versions of a few original images, in such a way that none of the 29 originals belonged to more than one fold. In each experimental run, the training set consisted of 23 originals (24 in case of run number 5) that were fed to the SVMs, and the remaining six originals (5 for run number 5) were employed to test the models generalization ability in quality estimation. Three different indicators were used to evaluate the systems accuracy, defined as the discrepancy between the objectively predicted quality change and the subjectively evaluated quality change : $\hat{d}_S(q^{(n)}, q^{(n,r)})$ and the subjectively evaluated quality change $d_S(q^{(n)}, q^{(n,r)})$:

- the mean prediction error, μ_{err}, between d_S and $\hat{d}_S$;
- the mean value of the absolute prediction error, $\mu_{|err|}$;
- the root mean square (RMS) error between d_S and $\hat{d}_S$.

Table 1. Test results for the K-Fold experiments on the White Noise dataset

White Noise	Run#1	Run#2	Run#3	Run#4	Run#5	Avg
μ_{err}	-0.052	-0.014	0.021	0.036	0.034	0.005
$\mu_{|err|}$	0.123	0.082	0.060	0.081	0.085	0.087
RMS	0.153	0.100	0.078	0.128	0.102	0.113

Table 2. Test results for the K-Fold experiments on the JPEG Compression dataset

Jpeg Compression	Run#1	Run#2	Run#3	Run#4	Run#5	Avg
μ_{err}	-0.007	0.019	0.045	0.042	-0.086	0.003
$\mu_{|err|}$	0.138	0.141	0.071	0.136	0.135	0.125
RMS	0.197	0.180	0.107	0.159	0.188	0.167

Table 3. Test results for the K-Fold experiments on the Gaussian Blur dataset

Gaussian Blur	Run#1	Run#2	Run#3	Run#4	Run#5	Avg
μ_{err}	-0.003	-0.081	0.040	0.160	-0.165	-0.010
$\mu_{|err|}$	0.151	0.175	0.143	0.186	0.209	0.173
RMS	0.180	0.219	0.160	0.230	0.264	0.211

Tables 1, 2 and 3 report for each distortion type the values of the previously mentioned indicators per experimental run. The overall generalization error can be esti-mated by averaging the indicators over the five runs. The values in the tables refer to a quality range normalized to $[-1, 1]$.

Tables 1, 2 and 3 show how the proposed framework is able to attain a satisfactory performance in quality prediction, despite the fact it is employed with three very different kinds of distortion. Assessing the quality gain/loss of noisy images, the systems mean absolute prediction error, averaged over the five folds, is 0.09, which results in an estimation accuracy of 95.5% Table 2 shows that the mean absolute prediction error $\mu_{|err|}$ for JPEG images is 0.13, which is comparable to the accuracy obtained in past studies [8,9]. Finally, table 3 illustrates that the accuracy for the blurred images is somewhat lower with an average $\mu_{|err|}$ of 0.17. Nonetheless, also for blurred images the model is able to guarantee an averaged error lower than 9%, which in most applications can be considered more than acceptable

References

1. Wang, Z., Sheikh, H.R., Bovik, A.C.: Objective video quality assessment. In: Furth, B., Marques, O. (eds.) The Handbook of Video Databases: Design and Applications. CRC Press, Boca Raton (2003)
2. International Telecommunication Union: Methodology for the subjective assessment of the quality of television pictures (1995)

3. Wang, Z., Sabir, M.F., Bovik, A.C.: A statistical evaluation of recent full reference image quality assessment algorithm. IEEE Trans. Image Processing 15, 3441–3452 (2006)
4. Yeh, E.M., Kokaram, A.C., Kingsbury, N.G.: Perceptual distortion measure for edgelike artifacts in image sequences. In: Proc. The 3rd International Conference on Human Vision and Electronic Imaging, SPIE, pp. 160–172 (1998)
5. Karunasekera, S.A., Kingsbury, N.G.: A distortion measure for blocking artifacts in images based on human visual sensitivity. IEEE Trans. Image Processing 4, 713–724 (1995)
6. Wang, Z., Simoncelli, E.P.: Reduced-reference image quality assessment using a wavelet-domain natural image statistic model. In: Proc. SPIE Human Vision and Electronic Imaging X, vol. 5666, pp. 149–159 (2005)
7. Gastaldo, P., Zunino, R., Heynderickx, I., Vicario, E.: Objective quality assessment of displayed images by using neural networks. Signal Processing: Image Communication 20, 643–661 (2005)
8. Gastaldo, P., Zunino, R.: Neural networks for the no-reference assessment of perceived quality. Journal of Electronic Imaging 14, 033004/1–11 (2005)
9. Gastaldo, P., Parodi, G., Redi, J., Zunino, R.: No-reference quality assessment of JPEG images by using cbp neural networks. In: de Sá, J.M., Alexandre, L.A., Duch, W., Mandic, D.P. (eds.) ICANN 2007. LNCS, vol. 4669, pp. 564–572. Springer, Heidelberg (2007)
10. Haralick, R., Shanmugam, K., Dinstein, I.: Textural features for image classification. IEEE Trans. On Systems, Man and Cybernetics SMC-3, 610–621 (1973)
11. Vapnik, V.: Statistical Learning Theory. Wiley, New York (1998)
12. Sheikh, H.R., Wang, Z., Cormack, L., Bovik, A.C.: Live image quality assessment database, http://live.ece.utexas.edu/research/quality
13. Bartlett, P., Boucheron, S., Lugosi, G.: Model selection and error estimation. Machine Learning 48, 85–113 (2002)
14. Anguita, D., Ridella, S., Rivieccio, F., Zunino, R.: Hyperparameter tuning criteria for support vector classifiers. Neurocomputing 55, 109–134 (2003)
15. Kittler, J., Hatef, M., Duin, R.P.W., Matas, J.: On combining classifiers. IEEE Trans. Pattern Analysis and Machine Intelligence 20, 226–239 (1998)

Semantic Adaptation of Neural Network Classifiers in Image Segmentation

Nikolaos Simou, Thanos Athanasiadis, Stefanos Kollias,
Giorgos Stamou, and Andreas Stafylopatis

Department of Electrical and Computer Engineering,
National Technical University of Athens,
Zographou 15780, Greece
{nsimou,thanos}@image.ntua.gr, stefanos@cs.ntua.gr

Abstract. Semantic analysis of multimedia content is an on going research area that has gained a lot of attention over the last few years. Additionally, machine learning techniques are widely used for multimedia analysis with great success. This work presents a combined approach to semantic adaptation of neural network classifiers in multimedia framework. It is based on a fuzzy reasoning engine which is able to evaluate the outputs and the confidence levels of the neural network classifier, using a knowledge base. Improved image segmentation results are obtained, which are used for adaptation of the network classifier, further increasing its ability to provide accurate classification of the specific content.

1   Introduction

The usage of semantic analysis in multimedia applications is currently a field of extensive research [9] that also forms recent R&D activities of European IST projects, such as Acemedia, Muscle, K-Space, X-Media, Mesh. Moreover, machine learning techniques are also used in the field to handle specific aspects related to learning classification or adaptation. In this paper, we show that both technologies can be interweaved to provide improved performance segmentation of static or moving images.

In the following, we describe the overall architecture used for semantic adaptation of a neural network classifier in image or video segmentation. The architecture of the proposal is illustrated in Figure 1.

An image, or a video frame is initially processed by a segmentation algorithm [1] which partitions it in a number of regions, that may have a symbolic interpretation. Standard MPEG-7 low level visual features are extracted from these regions forming the input of the adaptable neural network classifier, which assigns a semantic label and a confidence value to each segment. The obtained classification results are then processed by the application of a semantic-based segmentation algorithm, which aims to refine the initial labels and the derived segmentation masks. Finally, neighboring regions that share common semantic labels and meet certain criteria are merged to form a more meaningful segmentation of the image.

V. Kůrková et al. (Eds.): ICANN 2008, Part I, LNCS 5163, pp. 907–916, 2008.

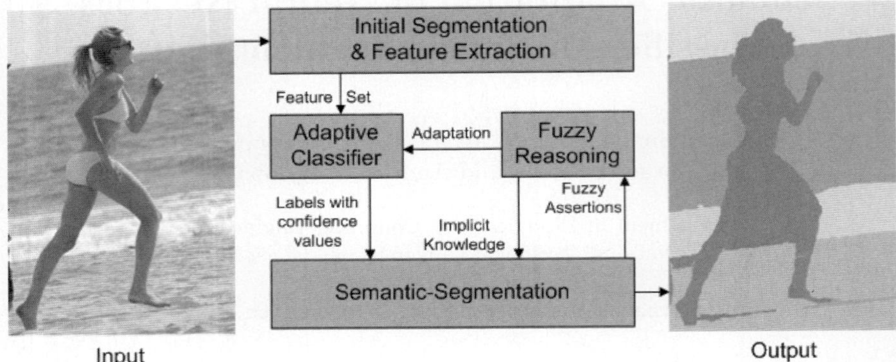

Fig. 1. The semantic adaptation architecture

In particular, the region-associated semantic labels and degrees of confidence are refined by the fuzzy reasoning engine FiRE[1]. FiRE is based on the expressive description logic (DL) f-$\mathcal{SHIN}$ [11]. The segments of the image are represented as DL-individuals, participating in the domain concepts to a given degree, and together with their spatial relations, they comprise the fuzzy assertion component (ABox) of the knowledge base. The terminology (TBox) is defined by using the domain concepts, declaring more general and complex concepts regarding both the segments and the image.

Using such a representation, implicit knowledge about segments can be extracted. This inferred knowledge either assigns them to higher concepts or corrects labels that have been mistakenly assigned by the classifier. These results constitute a source of information that can be used:

- to feed a semantic segmentation algorithm merging the updated segments and producing an improved segmentation mask
- as input to the adaptable neural network classifier

This classifier uses the semantically corrected results from reasoning, for adaptation purposes, so as to improve:

- its knowledge of the specific domain
- its performance over the next videos frames or images of similar content.

This reasoning adaptation cycle can be repeated more than once, depending on the image, or video frame, complexity.

The rest of the paper is organized as follows. In the next section the semantically adaptive neural network classifier is presented. Section 3 introduces the fuzzy knowledge base that was used for the refinement of segments and their confidence values. Section 4 presents the algorithm which performs the semantic image segmentation task. Finally, the last section presents some preliminary

---

[1] FiRE can be found at http://www.image.ece.ntua.gr/FiRE together with installation instructions and examples.

results of the proposed architecture. Furthermore, conclusions and suggested further work are provided in Section 6.

2   The Semantically Adaptable Classifier

The neural network classifier accepts an input vector $\overline{x}_i$ containing the features extracted from each region which determined by the initial segmentation phase, and categorizes it to one of, say, p available region classes ω_i.

The output vector $\overline{y}(\overline{x}_i)$ is

$$\overline{y}(\overline{x}_i) = \left[p^i_{\omega_1} p^i_{\omega_2} \cdots p^i_{\omega_p} \right]^T \tag{1}$$

where $p^i_{\omega_j}$ denotes the probability that the ith region belongs to the jth class.

The neural network is initially trained to perform the classification task using a specific training set, say $S_b = \left\{ \left(\overline{x'}_1, \overline{d'}_1 \right), \cdots, \left(\overline{x'}_{m_b}, \overline{d'}_{m_b} \right) \right\}$, where vectors $\overline{x'}_i$ and $\overline{d'}_i$ with $i = 1, 2, \cdots, m_b$ denote the i input training vector and the corresponding desired output vector consisting of p elements. In the present case the input features are the low-level descriptors for every image segment. These are the MPEG-7: Scalable Color, Homogeneous Texture, Edge Histogram and Region Shape. The computed feature vector is employed by the neural network for the generation of the initial hypotheses regarding the segments semantic labels.

Then, the network classifier is applied to a new video frame or image. Whenever the network performance is estimated as non very accurate, or erroneous, a slightly different network weight set should be estimated. This can be established through a network adaptation procedure.

Let $\overline{w}_b$ include all weights of the network before adaptation, and $\overline{w}_a$ the new weight vector, which is obtained after adaptation is performed. To perform the adaptation, a training set S_c is formed including features of say m_c regions the semantic label of which has been refined or modified by the fuzzy reasoning engine; $S_c = \left\{ \left(\overline{x}_1, \overline{d}_1 \right), \cdots, \left(\overline{x}_{m_c}, \overline{d}_{m_c} \right) \right\}$ where $\overline{x}_i$ and $\overline{d}_i$ with $i = 1, 2, \cdots, m_c$ correspond to the i input and the desired output data to be used for adaptation. The adaptation algorithm that is activated, whenever such a need is detected, computes the new network weights $\overline{w}_a$, minimizing the following error criteria with respect to weights,

$$E_a = E_{c,a} + \eta E_{f,a}$$

$$E_{c,a} = \frac{1}{2} \sum_{i=1}^{m_c} \left\| \overline{z}_a(\overline{x}_i) - \overline{d}_i \right\|_2$$

$$E_{f,a} = \frac{1}{2} \sum_{i=1}^{m_b} \left\| \overline{z}_a(\overline{x'}_i) - \overline{d'}_i \right\|_2 \tag{2}$$

where $E_{c,a}$ is the error performed over training set S_c ("current" knowledge), $E_{f,a}$ the corresponding error over training set S_b ("former" knowledge); $\overline{z}_a(\overline{x}_i)$

and $\overline{z}_a(\overline{x}'_i)$ are the outputs of the adapted network, corresponding to the input vectors $\overline{x}_i$ and $\overline{x}'_i$ respectively, of the network consisting of weights $\overline{w}_a$. Similarly $\overline{z}_b(\overline{x}_i)$ would represent the output of the network, consisting of weights $\overline{w}_b$, when accepting vector $\overline{x}_i$ at its input. Parameter η is a weighting factor accounting for the significance of the current training set compared to the former one and $\|\cdot\|_2$ denotes the L_2-norm.

The goal of the training procedure is to minimize $E_{f,a}$ and estimate the new network weights $\overline{w}_a$. The adopted algorithm has been proposed by the authors in [5][6] leads and provides an analytical and tractable solution for estimating $\overline{w}_a$, through linearization of the non-linear activation function of the neurons.

Equation (2) indicates that the new network weights are estimated taking into account both the current and the previous network knowledge. To stress, however, the importance of current training data in (2), the first term is replaced by the constraint that the actual network outputs are equal to the desired ones. Assuming that weight adaptation refers to small increments that is

$$z_a(\overline{x}_i) = d_i, i = 1, \cdots, m_c, \forall \overline{x} \in S_c \tag{3}$$

Moreover, minimization of the second term of (2), which expresses the effect of the new network weights over the data set S_b, can be considered as minimization of the absolute difference of the error over the data in S_b with respect to the previous and the current network weights. This means that the weight increments are minimally modified, with respect to the following error criterion

$$E_S = \|E_{f,a} - E_{f,b}\|_2 \tag{4}$$

with $E_{f,b}$ defined similarly to $E_{f,a}$, with $\overline{z}_a$ replaced by $\overline{z}_b$ in (2).

It can be shown [8] that (4) takes the form of

$$E_S = \frac{1}{2}(\Delta\overline{w})^T \cdot K^T \cdot K \cdot \Delta\overline{w} \tag{5}$$

where the elements of matrix K are expressed in terms of the previous network weights w_b and the training data in S_b. The error function defined by (5) is convex since it is of squared form. The gradient projection method has been used to estimate the weight increments. [5] [6]

Detection of the regions in which the output of the neural network classifier is not appropriate and, consequently activation of the adaptation is required, is achieved through a comparison of the semantic label and confidence value produced by the fuzzy reasoning engine with the one estimated by the original neural network classifier. Whenever the difference in these values is significant, adaptation is activated. Following to this, the adapted network can be applied to similar images or consequent video frames contributing to the improvement of the obtained segmentation results.

3   The Fuzzy Reasoning Engine

This section presents the operation of the fuzzy reasoning engine together with the fuzzy knowledge base that have been used for the adaptation of the neural network classifier.

Description Logics (DLs) [3] are a family of logic-based knowledge representation formalisms designed to represent and reason about the knowledge of an application domain in a structured and well-understood way. Recently, DLs have been extended to accommodate imperfect information [12,10].

As pointed out in the fuzzy DL literature, fuzzy extensions of DLs involve only the *assertion* of individuals to concepts and the semantics of the new language. Hence, FiRE which is the reasoner used, supports fuzzy $\mathcal{SHIN}$ using an alphabet of distinct concept names ($\mathbf{C}$), role names ($\mathbf{R}$) and individual names ($\mathbf{I}$). The $\mathcal{SHIN}$ constructors regarding concepts are disjunction ($C_1 \sqcup C_2$), conjunction ($C_1 \sqcap C_2$), negation ($\neg C$), full existential quantification ($\exists R.C$) and value restrictions ($\forall R.C$).Furthermore the $\mathcal{SHIN}$ language permits the hierarchy of roles as well as the use of transitive and inverse roles.

The semantic analysis module evaluates the spatial relations for each region, providing information for their location relatively to their neighboring regions. Additionally, the adaptive classifier estimates a degree of participation for each region in some trained labels.

Hence the alphabet of our fuzzy knowledge base consists of the relations representing the roles in our terminology and forming the following set:

$Roles = \{above - of, below - of, left - of, right - of\}$.

as well as the concepts that the adaptive classifier may estimate, that are :

$Concepts = \{Sky, Building, Person, Rock, Tree, Vegetation, Sea, Grass,$
$Ground, Sand, Trunk, Dried - plant, Pavement, Boat, Wave\}$

The set of individuals consist of the segments, and the images.

Using these sets, we have defined a terminology that refines some concepts with the aid of the regions spatial relations.

For the specific architecture axioms which correct mistaken estimations of analysis are defined for further adaptation purposes. For example Sea is

Table 1. Knowledge Base (*TBox*)

$\mathcal{T} = \{$SEA $\equiv$ Sea $\sqcap$ (($\exists$right $-$ of.(Sea $\sqcup$ Wave)) $\sqcup$ ($\exists$left $-$ of.(Sea $\sqcup$ Wave)) $\sqcup$($\exists$above $-$ of.(Sea $\sqcup$ Wave)) $\sqcup$ ($\exists$below $-$ of.(Sea $\sqcup$ Wave $\sqcup$ Sky))), SAND $\equiv$ Sand $\sqcap$ (($\exists$right $-$ of.(Sand $\sqcup$ Wave)) $\sqcup$ ($\exists$left $-$ of.(Sand $\sqcup$ Wave)) $\sqcup$($\exists$above $-$ of.(Sand $\sqcup$ Wave)) $\sqcup$ ($\exists$below $-$ of.(Sand $\sqcup$ Wave $\sqcup$ Sea))), WAVE $\equiv$ Wave $\sqcap$ ($\exists$right $-$ of.(Sea $\sqcup$ Wave)) $\sqcup$ ($\exists$left $-$ of.(Sea $\sqcup$ Wave)) $\sqcup$($\exists$above $-$ of.(Sea $\sqcup$ Wave)) $\sqcup$ ($\exists$below $-$ of.(Sea $\sqcup$ Wave))),
$\mathcal{R} = \{$above $-$ of, below $-$ of, left $-$ of, right $-$ of, below $-$ of$^-$ = above $-$ of, left $-$ of$^-$ = right $-$ of$\}$

re-defined as SEA and is specified by the concept Sea assigned by the classifier and by a neighboring criterion concept which requires neighbors to be either one of Wave, Sea or Sky.

The main reasoning services provided by crisp reasoners are *entailment* and *subsumption*. These services are also available in FiRE together with greatest lower bound queries which take the advantage of the fuzzy element. Since a fuzzy $ABox$ might contain many positive assertions for the same individual, without forming a contradiction, it is of interest to compute what is the best lower and upper truth-value bounds of a fuzzy assertion. The term of *greatest lower bound* (GLB) of a fuzzy assertion w.r.t. a knowledge base has been defined in [12].

In this case, a variation of greatest lower bound reasoning service is used for the semantic refinement of the labels provided by the neural network classifier. Since the classifier is trained, we assume a correct estimation of the region label but with a mistaken confidence value. Hence, we first compute the GLB of the region of interest to the concept of interest (i.e. SEA). We then evaluate the GLB of the region of interest to the neighbor criterion concept of the concept of interest (if SEA is the concept of interest then neighbor criterion concept is $((\exists \mathsf{right} - \mathsf{of}.(\mathsf{Sand} \sqcup \mathsf{Wave})) \sqcup (\exists \mathsf{left} - \mathsf{of}.(\mathsf{Sand} \sqcup \mathsf{Wave})) \sqcup ...)$. If this bound is greater that the value that was originally assigned to that concept then the region value is refined, differently it remains as assigned. For example, if a region has been assigned by the classifier as Sea to degree 0.8, and it is also "below-of" a region assigned as Sky to a degree 0.9, then due to the SEA axiom defined in the terminology (Table 1), the Sea value will be refined to 0.9.(Note that if the Sky value was 0.7 then the Sea value would have remained as assigned) This value will form the desired input for the adaptation of the neural network classifier.

4   Semantically Adaptive Image Segmentation

In this section we examine how a variation of a traditional segmentation technique, the Recursive Shortest Spanning Tree, also known as RSST [7], can be used to integrate and apply the results provided by the adaptive reasoning mechanism. RSST is a bottom-up segmentation algorithm that begins from the pixel level and iteratively merges similar neighboring regions until certain termination criteria are satisfied. It uses an internal graph representation of image regions, like the Attributed Relation Graph (ARG) [4]. In the beginning, all edges of the graph are sorted according to a criterion, e.g. color dissimilarity of the two connected regions using Euclidean distance of the color components. The edge with the least weight is found and the two regions connected by that edge are merged. After each step, the merged region's attributes (e.g. region's mean color) is re-calculated. RSST will also re-calculate weights of related edges and resort them, so that in every step the edge with the least weight will be selected. This process goes on recursively, until termination criteria are met. Such criteria may vary, but they usually are either the number of regions, or a threshold on the distance.

We modify this algorithm to operate on the fuzzy sets in a similar way as if they worked on low-level features (such as color, texture, etc.). This variation follows in principle the algorithmic definition of the traditional RSST, though a few adjustments were considered necessary and were added. S-RSST aims to improve the usual oversegmentation results by incorporating region labeling in the segmentation process [2]. The modification of the traditional algorithm to S-RSST lies on the definition of the two criteria: (a) The dissimilarity criterion between two adjacent regions a and b (vertices v_a and v_b in the graph), based on which the graph's edges are sorted and (b) the termination criterion.

For the calculation of the similarity between two regions, two approaches have been examined. The first one is based on the definition of a metric between two fuzzy sets, those that correspond to the candidate concepts of the two regions. This dissimilarity value is computed according to the following formula and is assigned as the weight of the respective graph's edge e_{ab}:

$$w(e_{ab}) = 1 - \sup_{c_k \in C} (t - norm(\mu_a(c_k), \mu_b(c_k))) \tag{6}$$

where a and b are two neighboring regions and $\mu_a(c_k)$ is the degree of membership of the concept $c_k \in C$ in the fuzzy set L_a.

Let us now examine one iteration of the S-RSST algorithm. Firstly, the edge e_{ab} with the least weight is selected, then regions a and b are merged. Vertex v_b is removed completely from the ARG, whereas v_a is updated appropriately. This update procedure consists of the following two actions:

1. Re-evaluation of the degrees of membership of the labels fuzzy set in a weighted average (w.r.t. the regions' size) fashion.
2. Re-adjustment of the ARG edges by removing edge e_{ab} and re-evaluating the weight of the affected edges.

This procedure continues until the edge e^* with the least weight in the ARG is bigger than a threshold: $w(e^*) > T_w$. This threshold is calculated in the beginning of the algorithm, based on the histogram of all weights of the set of all edges.

5   Results

In this section, certain results of the semantically adaptive architecture evaluated on real images are presented. As described in Section 1, an image is initially processed by the low-level segmentation algorithm that produces the segmented mask together with the input features for the adaptive neural network classifier. The classifier produces region-associated labels and degrees of confidence. These values pass through the semantic segmentation module and form the input for the fuzzy reasoning engine. Fuzzy reasoning provides refinement of some regions values according to which classifier adaptation is performed.

Figure 2 presents for some images, the initial output of the classifier and the semantically adaptive segmentation results.

It can be seen that based on the implicit knowledge provided by the fuzzy reasoner, semantic adaptation of the neural network classifier is achieved and used in improving the performance of the image segmentation module. The neural

Fig. 2. (a) Original Image (b)Segmentation based on the original neural network classifier (c)Semantic segmentation using the adapted NN classifier

network accepts an input vector of 5 elements composed of the MPEG7 Scalable Color, Homogeneous Texture, Edge Histogram and Region Shape features and provides 15 outputs, corresponding to the fifteen concepts which form the Concepts alphabet of the fuzzy reasoning engine mentioned in Section 3. Based on pruning a two hidden layer architecture was formed composed of ten and six neurons respectively. Segmentation of a data set about 200 image results in a training set of 4000 regions (i.e feature vectors) which were used for training, while 50 more images were used for testing.

As indicatively shown in Figure 2 the results are very promising. The fuzzy reasoning engine propagates the confidence values of region labels, which have "correct" spatial relations according to the fuzzy knowledge base, to the neighboring regions. These semantically corrected values are used for adaptation of the classifier in order to improve its knowledge of the specific domain and also its performance.

6   Conclusions

In this paper we have presented an architecture used for semantic adaptation of a neural network classifier in image or video segmentation. The proposed architecture combines techniques used for semantic multimedia analysis together with an adaptive classifier. A semantic segmentation algorithm and a fuzzy reasoning engine provide semantically corrected results that are used by the classifier for adaptation.

An evaluation of our architecture was made using images, presenting very promising results and a strong potential. The improved performance of adapted classifier on segments estimation could be also successfully used for segment indexing. Future work includes evaluation of the architecture using video frames and various domains.

Acknowledgment

This research was supported by the European Commission under contract FP6-027026 K-SPACE.

References

1. Adamek, T., O'Connor, N., Murphy, N.: Region-based segmentation of images using syntactic visual features. In: Proc. Workshop on Image Analysis for Multimedia Interactive Services, WIAMIS 2005, Montreux, Switzerland, April 13-15 (2005)
2. Athanasiadis, T., Mylonas, P., Avrithis, Y., Kollias, S.: Semantic image segmentation and object labeling. IEEE Trans. on Circuits and Systems for Video Technology 17(3), 298–312
3. Baader, F., McGuinness, D., Nardi, D., Patel-Schneider, P.F.: The Description Logic Handbook: Theory, implementation and applications. Cambridge University Press, Cambridge (2002)
4. Berretti, S., Del Bimbo, A., Vicario, E.: Efficient matching and indexing of graph models in content-based retrieval. IEEE Trans. on Circuits and Systems for Video Technology 11(12), 1089–1105 (2001)
5. Doulamis, N., Doulamis, A., Kollias, S.: On-line retrainable neural networks: Improving performance of neural networks in image analysis problems. IEEE Transactions on Neural Networks 11, 1–20 (2000)
6. Ioannou, S., Kessous, L., Caridakis, G., Karpouzis, K., Aharonson, V., Kollias, S.: Adaptive on-line neural network retraining for real life multimodal emotion recognition. In: Kollias, S.D., Stafylopatis, A., Duch, W., Oja, E. (eds.) ICANN 2006. LNCS, vol. 4131, pp. 81–92. Springer, Heidelberg (2006)

7. Morris, O.J., Lee, M.J., Constantinides, A.G.: Graph theory for image analysis: An approach based on the shortest spanning tree. Inst. Elect. Eng. 133, 146–152 (1986)
8. Park, D., EL-Sharkawi, M.A., Marks II., R.J.: An adaptively trained neural network. IEEE Transactions on Neural Networks 2, 334–345 (1991)
9. Stamou, G., Kollias, S.: Multimedia Content and the Semantic Web: Methods, Standards and Tools. John Wiley & Sons Ltd, Chichester (2005)
10. Stoilos, G., Stamou, G., Pan, J.Z., Tzouvaras, V., Horrocks, I.: Reasoning with very expressive fuzzy description logics (2007)
11. Stoilos, G., Stamou, G., Tzouvaras, V., Pan, J.Z., Horrocks, I.: The fuzzy description logic f-shin. In: A International Workshop on Uncertainty Reasoning For the Semantic Web, 2005 (2005)
12. Straccia, U.: Reasoning within fuzzy description logics. Journal of Artificial Intelligence Research 14, 137–166 (2001)

Partially Monotone Networks Applied to Breast Cancer Detection on Mammograms

Marina Velikova[1], Hennie Daniels[2,3], and Maurice Samulski[1]

[1] Department of Radiology, Radboud University Nijmegen Medical Centre
Nijmegen, The Netherlands
`{m.velikova,m.samulski}@rad.umcn.nl`
[2] Rotterdam School of Management, Erasmus University, Rotterdam
[3] Center for Economic Research, Tilburg University, Tilburg, The Netherlands
`daniels@uvt.nl`

Abstract. In many prediction problems it is known that the response variable depends monotonically on most of the explanatory variables but not on all. Often such partially monotone problems cannot be accurately solved by unconstrained methods such as standard neural networks. In this paper we propose so-called MIN-MAX networks that are partially monotone by construction. We prove that this type of networks have the uniform approximation property, which is a generalization of the result by Sill on totally monotone networks. In a case study on breast cancer detection on mammograms we show that enforcing partial monotonicity constraints in MIN-MAX networks leads to models that not only comply with the domain knowledge but also outperform in terms of accuracy standard neural networks especially if the data set is relative small.

1 Introduction

In many prediction problems it is known that the response variable depends monotonically on most of the explanatory variables but *not* on all. Common sense suggests that the house price has a monotone increasing dependence on the number of rooms and the total house area. However, such monotone dependency on the number of floors does not necessarily hold; for example, some expensive houses (such as villas) may have only one floor, whereas cheaper houses may have three floors. As shown in [1], the age of abalone shellfish is monotone in its length and height but not necessarily monotone in the shell weight. Finally, the likelihood for cancer of a mammographic finding is in monotone relationships with most of its mammographic characteristics (e.g., contrast, size) but not in all (e.g., distance to chest).

In this paper, we study the incorporation of such partial monotonicity constraints in neural networks. We discuss previous studies that deal with monotonicity in neural networks and we present some new contributions to the field:

- We prove the universal approximation capabilities of three-layer neural networks with combinations of minimum and maximum functions over linear

V. Kůrková et al. (Eds.): ICANN 2008, Part I, LNCS 5163, pp. 917–926, 2008.

units for non-monotone and partially monotone functions. This is a generalization of the results obtained by Sill on monotone functions [2].

- In the case of partially monotone problems, we demonstrate that models based on partially monotone networks are more accurate and more interpretive than models based on standard neural networks.

The paper is organized as follows. In the next section, we introduce notations and definitions related to monotone problems and models, distinguishing between total and partial monotonicity. In Section 3, we discuss a class of three-layer neural networks with a combination of minimum and maximum operators over linear functions, so-called MIN-MAX networks. In addition to monotone MIN-MAX networks proposed by Sill in [2] here we describe the unconstrained and partially monotone networks and we prove their universal approximation capabilities. In a case study on breast cancer detection on mammograms, we demonstrate that partially monotone MIN-MAX networks outperform standard neural networks for partially monotone problems. Section 4 concludes the paper.

2   Monotone Prediction Problems and Models

Let $\mathcal{X} = \prod_{i=1}^{k} \mathcal{X}_i$ be an *input space* represented by k attributes (features). A particular point $\mathbf{x} \in \mathcal{X}$ is defined by the vector $\mathbf{x} = (x_1, x_2, \ldots, x_k)$, where $x_i \in \mathcal{X}_i$, $i = 1, 2, \ldots, k$. Furthermore, a totally ordered set of labels $\mathcal{L}$ is defined. In the discrete case, we have $\mathcal{L} = \{1, 2, \ldots, \ell_{max}\}$ where ℓ_{max} is the maximal label. Note that ordinal labels can be easily quantified by assigning numbers from 1 for the lowest category to ℓ_{max} for the highest category. In the continuous case, we have $\mathcal{L} \subset \Re$ or $\mathcal{L} \subset \Re^+$. Unless the distinction is made explicitly, the term *label* is used to refer generally to the dependent variable irrespective of its type (continuous or discrete).

Next a function f is defined as a mapping $f : \mathcal{X} \to \mathcal{L}$ that assigns a label $\ell \in \mathcal{L}$ to every input vector $\mathbf{x} \in \mathcal{X}$. In prediction problems, the objective is to find an approximation $\hat{f}$ of f as close as possible, e.g., in L_1 or L_2 norm. In particular, in regression we try to estimate the average dependence of ℓ given $\mathbf{x}$, $\mathcal{E}[\ell|\mathbf{x}]$, whereas in classification, we look for a discrete mapping function represented by a classification rule $r(\ell_{\mathbf{x}})$ assigning a class ℓ to each point $\mathbf{x}$ in the input space.

In reality, the information we have about f is mostly provided by a data set $D = (\mathbf{x}^n, \ell_{\mathbf{x}^n})_{n=1}^{N}$, where N is the number of points, $\mathbf{x} \in \mathcal{X}$ and $\ell_{\mathbf{x}} \in \mathcal{L}$. In other words, $X = \{\mathbf{x}^n\}_{n=1}^{N}$ is a set of k independent variables represented by an $N \times k$ matrix, and $L = \{\ell_{\mathbf{x}^n}\}_{n=1}^{N}$ is a vector with the values of the dependent variable. In this context, D corresponds to a mapping $f_D : X \to L$ and we assume that f_D is a close proximity of f. Ideally, f_D is equal to f over X, which is seldomly the case in practice due to the noise present in the data.

Hence, our ultimate goal in prediction problems is restricted to obtaining a close approximation $\hat{f}_{M_D}$ of f by building a prediction model M_D from the given data D. The main assumption we make here is that f exhibits monotonicity properties with respect to the input variables and therefore, $\hat{f}_{M_D}$ should also obey these properties in a strict fashion.

One can distinguish between two types of problems, and their respective models, concerning the monotonicity properties. The distinction is based on the set of input variables, which are in monotone relationships with the response:

1. *Totally monotone prediction problems (models)*: f $(\hat{f}_{M_D})$ depends monotonically on *all* variables in the input space.
2. *Partially monotone prediction problems (models)*: f $(\hat{f}_{M_D})$ depends monotonically on *some* variables in the input space but *not* all.

2.1   Total Monotonicity

In monotone prediction problems, we assume that D is generated by a process with the following properties

$$\ell_{\mathbf{x}} = f(\mathbf{x}) + \epsilon \tag{1}$$

where f is a monotone function, and ϵ is a random variable with zero mean and constant variance σ_ϵ^2. Monotonicity of f on $\mathbf{x}$ is defined on all independent variables by

$$\mathbf{x}_1 \geq \mathbf{x}_2 \Rightarrow f(\mathbf{x}_1) \geq f(\mathbf{x}_2) \tag{2}$$

where $\mathbf{x}_1 \geq \mathbf{x}_2$ is a partial ordering on $\mathcal{X}$ defined by $x_i^1 \geq x_i^2$, for $i = 1, 2, \ldots, k$. The pair $(\mathbf{x}_1, \mathbf{x}_2)$ is called comparable, and if the relationship defined in (2) holds, it is also a monotone pair. Note that even though f is monotone, the data generated by (1) is not necessarily monotone due to the random effect of ϵ.

2.2   Partial Monotonicity

In partially monotone problems, we have $\mathcal{X} = \mathcal{X}^m \times \mathcal{X}^{nm}$ with $\mathcal{X}^m = \prod_{i=1}^{m} \mathcal{X}_i$ and $\mathcal{X}^{nm} = \prod_{i=m+1}^{k} \mathcal{X}_i$ for $1 \leq m < k$. Furthermore, we have a data set $D = (\mathbf{x}^m, \mathbf{x}^{nm}, \ell_{\mathbf{x}})^N$, where $\mathbf{x}^m \in \mathcal{X}^m$, $\mathbf{x}^{nm} \in \mathcal{X}^{nm}$, and N is the number of observations. A data point $\mathbf{x} \in D$ is represented by $\mathbf{x} = (\mathbf{x}^m, \mathbf{x}^{nm})$; the label of $\mathbf{x}$ is $\ell_{\mathbf{x}}$. We assume that D is generated by the following process

$$\ell_{\mathbf{x}} = f(\mathbf{x}^m, \mathbf{x}^{nm}) + \epsilon, \tag{3}$$

where f is a monotone function in $\mathbf{x}^m$ and ϵ is a random error. In regression problems, ϵ has zero mean, whereas in classification problems ϵ is a small probability that the assigned class is incorrect.

The partial monotonicity constraint of f on $\mathbf{x}^m$ is defined by

$$\mathbf{x}^{nm} = \mathbf{x}'^{nm} \text{ and } \mathbf{x}^m \geq \mathbf{x}'^m \Rightarrow f(\mathbf{x}) \geq f(\mathbf{x}'). \tag{4}$$

Henceforth, we call $\mathcal{X}^m$ the *set of monotone variables* and $\mathcal{X}^{nm}$ the *set of non-monotone variables*. By *non-monotone* we mean that monotone dependence is not known a priori. Although, we do not constrain the size of the two sets, our main assumption for the problems considered in this paper is that we have only a small number of non-monotone variables (e.g., < 5), and a large number of monotone variables. Thus, monotonicity plays an important role in the data generating process, and needs to be preserved.

Though the distinction between the two types of problems (models) is also made in the literature (e.g., by [3]), we want to emphasize that the terms "totally" and "partially" refer to the set of inputs for which monotone relationships hold with respect to the target–not to the monotonicity property as such.

3   Partially Monotone Networks

3.1   Description

The majority of previous studies deal with the construction of totally monotone neural networks ([4]). The only relevant work on partially monotone networks we have found is by Armstrong ([5]). The networks described here are based on the three-layer (two hidden-layer) architecture introduced by Sill in [2]. The input layer is connected to the first hidden layer consisting of a set of linear units (hyperplanes), which are combined into several groups. The number of units in each group is not necessarily the same. Corresponding to each group there is a second hidden-layer unit, which computes the maximum over all first-layer units within the group. The final output unit is defined as the minimum over all groups. Due to the network architecture, it is easy to constrain the weights on the monotone variable(s) to be non-negative, and thus to obtain a partially monotone model. In the remainder of the paper we refer to three-layer neural networks with a combination of minimum and maximum operators over linear functions as MIN-MAX networks.

We will prove that MIN-MAX networks have universal approximation capabilities. A general abstract proof for the universal function approximation capabilities of unconstrained MIN-MAX networks has been communicated to us through e-mail by William Armstrong. Here we will give a more constructive proof. We also follow the grid representation scheme used by Sill in the proof for the universal approximation capabilities of three-layer monotone networks ([2]). But first a formal description of the network architecture is given.

Let R denote the number of nodes in the second hidden layer, which is equal to the number of groups in the first hidden layer. The outputs of the groups are denoted by $g_1, g_2, \ldots, g_R$. Let h_r denote the number of hyperplanes within group $r, r = 1, 2, \ldots, R$. The parameters (weights) of the hyperplanes in r are k-dimensional vectors denoted by $\mathbf{w}_{(r,1)}, \mathbf{w}_{(r,2)}, \ldots, \mathbf{w}_{(r,h_r)}$; the matrix of all weights and biases is denoted by W. Then, the output at group r is defined by:

$$g_r(x) = \min_j \left(\mathbf{w}_{(r,j)} \cdot \mathbf{x} + \theta_{(r,j)} \right), \quad 1 \le j \le h_r, \tag{5}$$

where θ is a bias term. The final output $O_\mathbf{x}$ of the network for an input $\mathbf{x}$ is:

$$O_\mathbf{x} = \max_r g_r(x),$$

or in classification problems

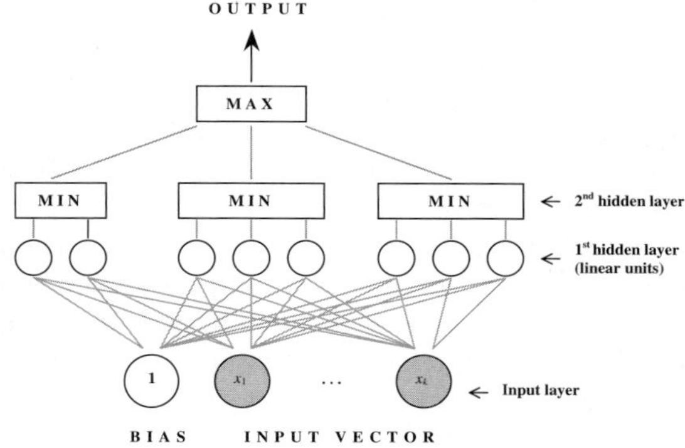

Fig. 1. Architecture of three-layer neural network with MIN and MAX operators

$$O_{\mathbf{x}} = \max_r \sigma(g_r(x)),$$

where σ is the sigmoid function. This architecture is depicted in Fig. 1.

To guarantee that the network output is monotone in a particular input, all weights for that input to the first hidden layer are constrained to be non-negative (non-positive), if increasing (decreasing) monotonicity is desired for that input. Here, the parameters in (5) are enforced to be non-negative by taking an appropriate transformation such as $w = z^2$, where z is a free parameter.

In the following two theorems we show that unconstrained and partially monotone MIN-MAX networks have universal approximation capabilities.

Theorem 1. *Let $\mathcal{X}$ denote a closed bounded domain of k inputs (dimensions) and f be a continuous bounded function mapping $\mathcal{X}$ to $\Re^+$. Then, for any $\epsilon > 0$ there exists a function O_x that can be implemented by a MIN-MAX network such that $|f(\boldsymbol{x}) - O_x| < \epsilon$, for any $\boldsymbol{x} \in \mathcal{X}$.*

Proof. Let $\epsilon > 0$. Define an equispaced grid of points on $\mathcal{X}$ such that the spacing between grid points along each dimension is δ, $\delta > 0$, and it is taken such that for any grid point $\mathbf{s}$ and for any $\mathbf{x}$ with $\|\mathbf{x} - \mathbf{s}\|_\infty < \delta$, we have $|f(\mathbf{x}) - f(\mathbf{s})| < \epsilon$. Here $\|\cdot\|_\infty$ denotes the max-norm distance between two points:

$$\|\mathbf{x} - \mathbf{x}'\|_\infty = \max_i |x_i - x_i'|, \quad i = 1, 2, \ldots, k.$$

Corresponding to each grid point $\mathbf{s}$, we assign a group consisting of $2k + 1$ hyperplanes: one hyperplane is the constant output $h_{\mathbf{s}} = f(\mathbf{s})$. In addition, along each dimension d we place two hyperplanes:

$$h_{\mathbf{s}}^+ (x_d) = \frac{2}{\delta} f(\mathbf{s}) (x_d - s_d + \delta) \quad \text{and} \quad h_{\mathbf{s}}^- (x_d) = \frac{2}{\delta} f(\mathbf{s}) (s_d - x_d + \delta).$$

Furthermore, we denote

$$\underline{f} = \min_{\mathbf{s}} f(\mathbf{s}) \quad \text{and} \quad \overline{f} = \max_{\mathbf{s}} f(\mathbf{s}), \tag{6}$$

where $\mathbf{s}$ are the grid points close to $\mathbf{x}$, i.e., $\|\mathbf{x} - \mathbf{s}\|_\infty < \delta$. The number of these points is at most 2^k. We will show that

$$\underline{f} \leq O_\mathbf{x} \leq \overline{f}. \tag{7}$$

First we prove the left-hand side of the inequality in (7). It is clear that for any point $\mathbf{x}$, $\mathbf{x} \in \mathcal{X}$, there always exists at least one grid point $\mathbf{q}$ such that $\|\mathbf{x} - \mathbf{q}\|_\infty \leq \delta/2$. Then, by the definition of the hyperplanes the group's output (minimum over the hyperplanes) for $\mathbf{q}$ at $\mathbf{x}$ is $f(\mathbf{q})$. Hence, the final network's output $O_\mathbf{x}$ (maximum over all the groups' outputs) at $\mathbf{x}$ is $\geq f(\mathbf{q}) \geq \underline{f}$.

Next we proceed with the proof of the right-hand side of the inequality in (7). We consider two sets of grid points and we show that their group's outputs are $\leq \overline{f}$. First for any grid point $\mathbf{s}$ with $\|\mathbf{x} - \mathbf{s}\|_\infty < \delta$, the group's output for $\mathbf{s}$ at $\mathbf{x}$ lies in the range $(0, f(\mathbf{s})]$, and therefore it is $\leq \overline{f}$. Now we take a grid point $\mathbf{t}$ with $\|\mathbf{x} - \mathbf{t}\|_\infty \geq \delta$. In other words, there is at least one dimension d for which $|x_d - t_d| \geq \delta$. If $x_d - t_d \geq \delta$, then $h_\mathbf{t}^-(x_d) \leq 0$; if $t_d - x_d \geq \delta$, then $h_\mathbf{t}^+(x_d) \leq 0$. Since at each group we compute the minimum over all the hyperplanes, the group associated with $\mathbf{t}$ would produce an output at $\mathbf{x}$ that is $\leq 0 < \overline{f}$. Since the group's outputs at $\mathbf{x}$ for all grid points are $\leq \overline{f}$, it follows that $O_\mathbf{x} \leq \overline{f}$.

Finally, by the construction of the grid, for all points $\mathbf{s}$, $\|\mathbf{x} - \mathbf{s}\|_\infty < \delta$, we have $|f(\mathbf{x}) - f(\mathbf{s})| < \epsilon$. Thus, $|f(\mathbf{x}) - \underline{f}| < \epsilon$ and $|f(\mathbf{x}) - \overline{f}| < \epsilon$. Hence, we have $|f(\mathbf{x}) - O_\mathbf{x}| < \epsilon$, which completes the proof. $\square$

Remark. The theorem can be generalized to mappings $g : \mathcal{X} \to \mathfrak{R}$. First note that any function g can be represented as $g = f - C$, where f is a positive function and C is a constant. Next, given Theorem 1 for any $\epsilon > 0$ we can always find a MIN-MAX network's approximation $\hat{f}$ of f such that $|f - \hat{f}| < \epsilon$. Finally, we apply $\hat{g} = \hat{f} - C$ to obtain the approximation $\hat{g}$ of g.

Theorem 2. *Let $\mathcal{X}$ denote a closed bounded domain of k inputs (dimensions) and f be a continuous bounded function mapping $\mathcal{X}$ to $\mathfrak{R}^+$ that is monotonically increasing in x_k. Then, for any $\epsilon > 0$ there exists a partially monotone function O_x that can be implemented by a MIN-MAX network with weights on the kth input constrained to be non-negative, such that $|f(x) - O_x| < \epsilon$, for any $x \in \mathcal{X}$.*

Proof. The proof follows the same line of reasoning as the proof of Theorem 1. The difference is that now to each grid point $\mathbf{s}$, along the monotone dimension k, we place only one hyperplane: $h_\mathbf{s}^+(x_k) = \dfrac{2}{\delta} f(\mathbf{s})(x_k - s_k + \delta)$. Thus, at any grid point we have $2k$ hyperplanes in total. We show again that

$$\underline{f} \leq O_\mathbf{x} \leq \overline{f},$$

for $\underline{f}$ and $\overline{f}$ defined in (6). The left-hand side of the inequality is proved analogously to the unconstrained case. To prove the right-hand side, we show that for all grid points the group's output is $\leq \overline{f}$.

First, for any grid point $\mathbf{s}$ with $\|\mathbf{x} - \mathbf{s}\|_\infty < \delta$, the group's output (minimum over the hyperplanes) at $\mathbf{x}$ lies in the range $(0, f(\mathbf{s})]$, so it is $\leq \overline{f}$. Next, we take any grid point $\mathbf{q}$ with $q_k - x_k \geq \delta$ or $|x_d - q_d| \geq \delta$, for at least one non-monotone dimension d, $d = 1, 2, \ldots, k-1$. Then, by the definition of the group hyperplanes, the group's output for $\mathbf{q}$ at $\mathbf{x}$ is $\leq 0 < \overline{f}$. Finally, we consider any grid point $\mathbf{t}$ with $x_k - t_k \geq \delta$ and $|x_d - t_d| < \delta$, for $d = 1, 2, \ldots, k-1$. Then, by the definition of the hyperplanes, the group's output for $\mathbf{t}$ at $\mathbf{x}$ lies in the range $(0, f(\mathbf{t})]$. Note that there always exists at least one grid point $\mathbf{s}$ with $\|\mathbf{x} - \mathbf{s}\|_\infty < \delta$ such that $t_k < s_k$ and $t_d = s_d$, for $d = 1, 2, \ldots, k - 1$. Then by monotonicity of f on the kth dimension, we have $f(\mathbf{t}) \leq f(\mathbf{s}) \leq \overline{f}$.

Since the group's outputs at $\mathbf{x}$ for all grid points are $\leq \overline{f}$, it follows that $O_{\mathbf{x}} \leq \overline{f}$. Following the same line of reasoning as in the proof for the unconstrained networks, we can show that $|f(\mathbf{x}) - O_{\mathbf{x}}| < \epsilon$. $\qquad\square$

Remark. The proof is easily generalized for more than one monotone inputs. It is also analogous if f is monotonically decreasing in one or more inputs. Then, along the monotone dimension(s) we take h^- instead of h^+.

3.2 Case Study on Breast Cancer Detection on Mammograms

A screening mammographic exam usually consists of four images, corresponding to each breast scanned in two views - mediolateral-oblique (MLO) taken under 45^0 angle and craniocaudal (CC) taken top-down (Figure 2). In reading mammograms, radiologists judge for the presence of a *lesion* (a physical cancerous object present on mammograms) by comparing both views and breasts. The general rule is that a lesion is to be observed in both views.

Except the multi-view dependencies, in the domain of breast cancer exists also another type of knowledge, namely monotone relationships between some of the mammographic characteristics of the lesion and its likelihood for cancer. For example, the more the contrast (brightness) or the spiculation (star-like) shape of the lesion the higher the cancer likelihood. Linear texture is another characteristic that has a decreasing monotone effect on the likelihood–cancerous lesions tend to have non-linear texture structures. At the same time, the lesion location regarding the skin and the chest does not necessarily exhibit such monotonicity dependency as ordering cannot be applied.

Computer-aided detection (CAD) systems, however, often do not incorporate such domain knowledge. Hence, the correlations in the lesion characteristics are

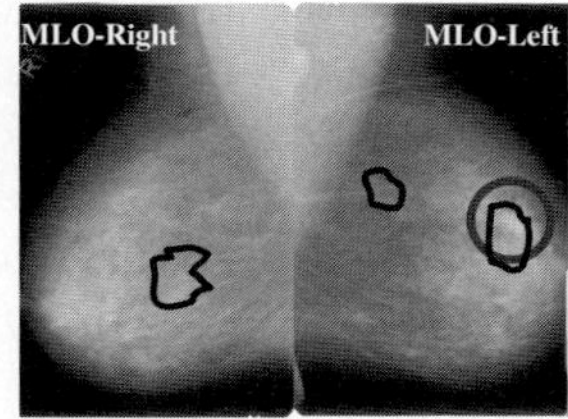

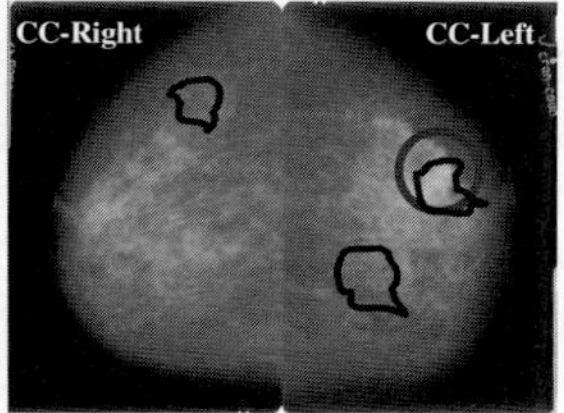

Fig. 2. MLO and CC projections of a left and right breast. The circles denote a cancerous object (so-called lesion) and the free-shaped areas are regions detected as suspicious by a computer-aided detection system.

ignored and the breast cancer detection can be obscured due to the lack of consistency in lesion marking. As a result CAD leads to a high number of false positive detected regions (see Figure 2). This limits the usability and the trust in the performance of such systems.

Here we apply partially monotone MIN-MAX networks to account for such monotone relationships between the mammographic characteristics of the regions detected by the UMCN CAD system ([6]). We use actual screening data of 28 patients including 12 cancerous cases. All cases contained both MLO and CC views. All cancerous cases contain one lesion visible in at least one view, which was verified by pathology reports to be malignant. For each mammographic view (image) we select 3 regions detected by the CAD system and every region from MLO view is linked with every region in CC view, thus obtaining 504 links in total. We define a link, L_{ij} for $i, j = 1, 2, 3$, as *cancerous* if at least one of the linked regions is cancerous; otherwise the link is labeled as *normal*. Thus we obtain a binary class variable *link*. Next each region is described by 8 continuous features automatically extracted by the CAD system, which tend to be relatively invariant across the views. These features include two measures for suspiciousness computed by CAD and lesion characteristics such as spiculation, size, contrast, linear texture and location coordinates - distance to skin and distance to chest. In summary, our data contain 504 observations (links) with 16 features (2 regions × 8 characteristics) and 1 class variable. Given the domain knowledge we consider the location measures (4 in total per observation) as non-monotone variables whereas the remaining 12 variables are monotone.

On the data thus constructed we apply partially monotone MIN-MAX networks (PartMon) and standard neural networks with weight decay (NNet) as a benchmark method for comparison. We use different combinations of parameters for MIN-MAX networks: groups × planes - 2 × 2 (4n), 3 × 3 (9n), and for standard neural networks: hidden nodes - 4, 9; weight decay - 0.000001 (1e-6), 0.0001 (1e-4). The evaluation of each network performance is done using two-fold cross validation: the dataset is split into two subsets with equal number of observations and proportion of cancerous cases. Each half is used as a training set and as a test set. Although we have a discrete class variable the result from both types of networks is a continuous likelihood measure that a link is cancerous, M_{canc}. Therefore the comparison analysis is done using Receiver Operating Characteristic (ROC) curve, a standard performance measure in the radiologists' practice ([7]). Figure 3 presents the classification outcome per link with the respective Area Under the Curve (AUC) measures for all network architectures.

While the link results are very promising from a radiologists' point of view it is more important to look at the case level performance. The measure for a case being cancerous is computed by taking the maximum link likelihood out of all 18 links for the case, i.e., $M_{canc}(Case) = \max_{i,j,br} M_{canc}(L_{ij}^{br})$, for $i, j = 1, 2, 3$ and $br = $ left, right. Except the standard AUC measure, we compute another performance indicator for both methods (PartMon and NNet): percentage (number) of correctly detected lesions in the 12 cancerous cases, i.e., we count how many

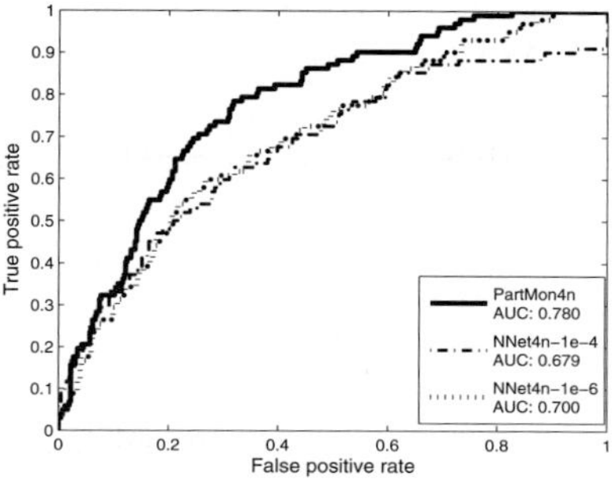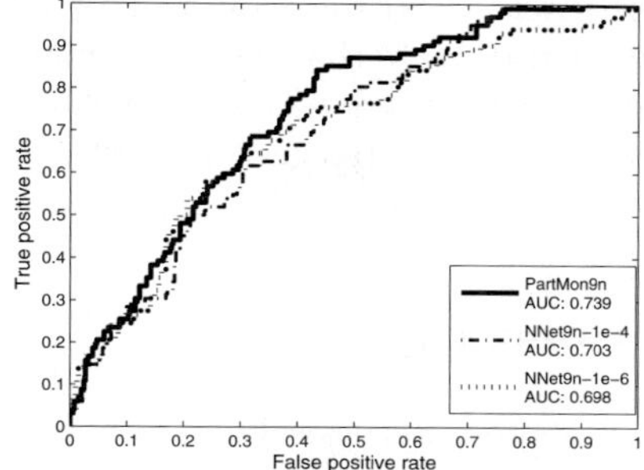

Fig. 3. ROC analysis on the performance of partially monotone MIN-MAX networks (PartMon) and standard neural networks with weight decay (NNet) on breast cancer data based on different architectures

Table 1. Performance of PartMon and NNet on the breast cancer data at a case level

Method	AUC	Perc. (#) correctly detected lesions in 12 cancerous cases
PartMon4n	**0.755**	**100%** (12)
PartMon9n	0.693	92% (11)
NNet4n-1e-4	0.729	83% (10)
NNet4n-1e-6	0.724	75% (9)
NNet9n-1e-4	0.646	67% (8)
NNet9n-1e-6	0.719	67% (8)

times the selected link with maximum likelihood corresponds to the lesion in the case. The results are presented in Table 1.

The results at link and case level demonstrate that the incorporation of partial monotonicity constraints leads to models that not only comply with the expert knowledge but also improve the detection rate of breast cancer. Considering different network architectures, we see that the overall performance of both PartMon and NNet tends to decrease with the increase of the number of hidden nodes. This implies that the flexibility of larger networks leads to overfitting of the training data. The best performance at a link and case level is achieved for partially monotone MIN-MAX networks with 2 groups and 2 hyperplanes. These relatively small networks are capable of detecting more cancers at low false positive (FP) rates (< 0.6). This leads to improved case performance as well as to 100% correctly detected lesions, i.e., for all 12 cancerous cases PartMon4n yields the maximum likelihood for a link corresponding to the lesion in the case. For PartMon9n the improvement at a link level is observed for higher FP rates (> 0.4), which could explain the decreased case performance observed in lower AUC and 1 missed lesion. Overall, we can say that partially monotone models improve not only the distinction between cancerous and normal cases but also the lesion marking on the mammogram.

We have conducted analogous experiments with housing data to predict the house price based on a number of house characteristics. The results obtained again demonstrate that partially monotone MIN-MAX networks outperform significantly the standard neural networks in terms of classification accuracy (averaged mean squared errors: PartMon= 0.03 and NNet= 0.07, $p = 0.0\%$) and stability across architectures and data samples.

4   Conclusion

In this paper, we described three-layer neural networks with a combination of min and max functions over the hidden layers (MIN-MAX networks) without and with partial monotonicity constraints, enforced on some of the weight parameters. We proved their universal function approximation capabilities, as for totally monotone networks it has already been proved by Sill in an earlier study. In a real case study on mammographic breast cancer detection, we demonstrated the superior performance of partially monotone MIN-MAX networks over standard neural networks.

Acknowledgments. This work is partially funded by the Netherlands Organisation for Scientific Research under BRICKS/FOCUS grant number 642.066.605.

References

1. Velikova, M.: Monotone models for prediction in data mining. PhD thesis, Tilburg University, Tilburg, The Netherlands (2006)
2. Sill, J.: Monotonic networks. In: Advances in Neural Information Processing Systems (NIPS), vol. 10, pp. 661–667. MIT Press, Cambridge (1998)
3. Tuy, H.: Monotonic optimization: problems and solution approaches. SIAM Journal on Optimization 11(2), 464–494 (2000)
4. Daniels, H., Velikova, M.: Derivation of monotone decision models from noisy data. IEEE Trans. on Systems, Man and Cybernetics, Part C 36(5), 705–710 (2006)
5. Armstrong, W.W., Thomas, M.M.: Adaptive logic networks. In: Handbook of Neural Computation, vol. 10, pp. 1–14. Oxford University Press, Oxford (1997)
6. van Engeland, S., Karssemeijer, N.: Combining two mammographic projections in a computer aided mass detection method. Medical Physics 34(3), 898–905 (2007)
7. Hanley, J.A., McNeil, B.J.: The meaning and use of the area under a Receiver Operating Characteristic (ROC) curve. Radiology 143, 29–36 (1982)

A Neuro-fuzzy Approach to User Attention Recognition

Stylianos Asteriadis, Kostas Karpouzis, and Stefanos Kollias

National Technical University of Athens,
School of Electrical and Computer Engineering,
Image, Video and Multimedia Systems Laboratory,
GR-157 80 Zographou, Greece
stiast@image.ece.ntua.gr, {kkarpou,stefanos}@cs.ntua.gr

Abstract. User attention recognition in front of a monitor or a specific task is a crucial issue in many applications, ranging from e-learning to driving. Visual input is very important when extracting information regarding a user's attention when recorded with a camera. However, intrusive equipment (special helmets, glasses equipped with cameras recording the eye movements, etc.) impose constraints on users spontaneity, especially when the target group consists of under aged users. In this paper, we propose a system for inferring user attention (state) in front of a computer monitor, only with the usage of a simple camera. The system can be used for real time applications and does not need calibration in terms of camera parameters. It can function under normal lighting conditions and needs no adaptation for each user.

Keywords: Head Pose, Eye Gaze, Facial Feature Detection, User Attention Estimation.

1   Introduction

For the estimation of a user's attention (state) in front of a computer monitor, two key issues are the movements (rotational and translational) of his head, as well as the directionality of his eye gaze. Not a lot of work has been published offering a combination of the two criteria, in order to evaluate the attentiveness or non-attentiveness of a person in front of a camera, especially without the need of specially designed hardware. In the current work, a top to down approach is presented, from head detection to fusion of biometrics using a specially trained neuro-fuzzy system, for evaluating the attentive state of the user.

There are techniques around the issue of head pose estimation which use more than one camera, or extra equipment ([1],[2],[3],[4],[5]), techniques based on facial feature detection ([6],[7]), suffering from the problem of robustness, or techniques that estimate the head pose ([8],[9]) using face bounding boxes, requiring the detected region to be aligned with a training set. In eye gaze estimation systems, the eye detection accuracy plays a significant role depending on the application (higher for security, medical and control systems). Some of the existing eye gaze

V. Kůrková et al. (Eds.): ICANN 2008, Part I, LNCS 5163, pp. 927–936, 2008.

techniques ([10],[11]), estimate the iris contours (circles or ellipses) on the image plane and, using edge operators, detect the iris outer boundaries. While not a lot of work has been done towards combining eye gaze and head pose information, in a non-intrusive environment, the proposed method uses a combination of the two inputs, together with other biometrics to infer the user's attention in front of a computer monitor, without the need of any special equipment apart from a simple web camera facing the user.

The structure of the paper is the following: Details on features used for User Attention Recognition are discussed in Section 2. Firstly, we discuss our facial feature detection method. Then, head pose and eye gaze estimation are analytically described. In Section 3, the training and testing of a Neuro-Fuzzy system is presented for characterizing the state of a user at frequent time intervals. Conclusions and future work are discussed in Section 4.

2   Feature Extraction for User Attention Recognition

For estimating a user's attention to the monitor of a computer, estimating his head pose, his eye gaze and his distance from the monitor are essential elements. To this aim, a neuro-fuzzy inference system was built, using as input the above metrics, as will be discussed hereafter. The system first detects the face of the user, when he is facing the camera frontally. Facial features are subsequently detected and tracked. Based on the facial feature movements, decisions regarding the user's head position and eye gaze are taken. The user's distance from the monitor is also estimated.

2.1   Facial Feature Detection

For face detection, the Boosted Cascade method described in [12] is employed, as it gives robust and real time results. The method output provides with the face region in which, usually, there exists background which can be asymmetric with regards to the right and left part of the region. An accurate estimate of the face region is important, since, for facial feature detection, blocks of predefined size will be used and, for this reason, accurate face detection is necessary. A postprocess method used in [13] was employed here for refining the face position in the frame.

For eye center localization, an approach based on [13] was used. For the detection of the eye corners (left, right, upper and lower) a technique similar to that described in [14] is used: Having found the eye center, a small area around it is used for the rest of the points to be detected.

In the current work, the point between the nostrils has also been used, as will be seen later. To this aim, nostrils have been detected at the frontal position of the user. Starting from the middle point between the eyes, an area of length equal to the inter-ocular distance is considered and the two darkest points are found to be the nostrils.

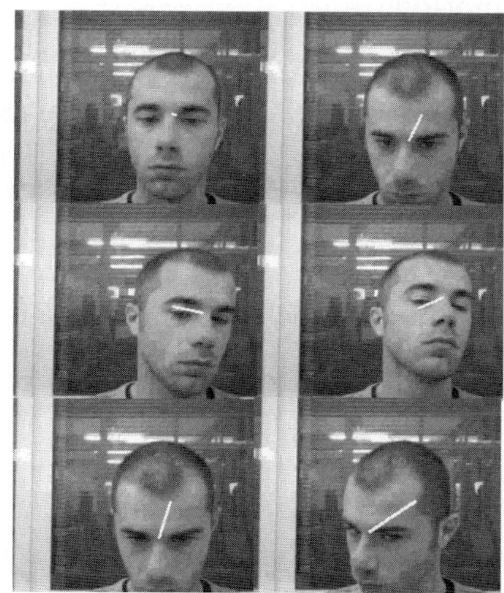

Fig. 1. Various head poses captured in front of a computer monitor using a simple web-camera. The white line (head pose vector) corresponds to the displacement of the point in the middle of the eyes with regards to the frontal pose.

2.2   Feature Tracking and Head Pose Estimation

Eye center tracking in a video sequence is done using a three-level Lucas-Kanade algorithm and, if only rotational movements of the face are considered, the movement of the middle point between the tracked eye centers can give a good insight of the face pose. This is achieved by subtracting the 2D image coordinates of its position at the frontal view of the face from those corresponding to a head rotation. The result can further be normalized by the inter-ocular distance in order the results to be scale independent. The resulting vector, from now on will be referred to as head pose vector (see Fig. 1).

To handle real application problems, where the user moves parallel and vertical to the camera plane, a series of rules has been extracted. If the user moves parallel to the image plane, the fraction between the inter-ocular distance and the vertical distance between the eyes and the nostrils remains almost constant. In this case, no rotation of the face is considered and, thus, the frontal pose is determined. Also, rapid rotations, apart from occluding some of the features (and, consequently, tracking is lost) make it difficult for the visible features to be tracked. In such cases, when the user comes back to his frontal position, the vector corresponding to pose estimation reduces in length and stays fixed for as long as the user is looking at the monitor. In these cases, the algorithm can re-initialize by re-detecting the face and the facial features.

2.3   Eye Gaze Estimation

For gaze detection, the areas defined by the four points around each eye are used. Prototype eye areas depicting right, left, upper and lower gaze directionality (see Fig. 2) are used to calculate mean greyscale images corresponding to each gaze direction. The areas defined by the four detected points around the eyes, are then correlated to these images.

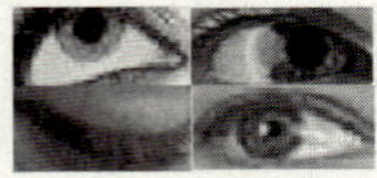

Fig. 2.   Prototype eye patches of different eye directionality

The normalized differences of the correlation values of the eye area with the left and right, as well as upper and lower mean gaze images are calculated:

$$H_r = \frac{(R_{r,l} - R_{r,r})}{max(R_{r,l}, R_{r,r}, R_{r,u}, R_{r,d})} \tag{1}$$

$$V_r = \frac{(R_{r,u} - R_{r,d})}{max(R_{r,l}, R_{r,r}, R_{r,u}, R_{r,d})} \tag{2}$$

$$H_l = \frac{(R_{l,l} - R_{l,r})}{max(R_{r,l}, R_{r,r}, R_{r,u}, R_{r,d})} \tag{3}$$

$$V_l = \frac{(R_{l,u} - R_{l,d})}{max(R_{r,l}, R_{r,r}, R_{r,u}, R_{r,d})} \tag{4}$$

Where $R_{i,j}$ is the correlation of the i (i=left,right) eye with the j (j=left, right, upper, lower) mean grayscale image. The normalized value of the horizontal and vertical gaze directionalities (conventionally, angles) are then the weighted mean:

$$H = ((2 - l) \cdot H_r + l \cdot H_l)/2 \tag{5}$$

$$V = ((2 - l) \cdot V_r + l \cdot V_l)/2 \tag{6}$$

where l is the fraction of the mean intensity in the left and right areas. This fraction is used to weight the gaze directionality values so that eye areas of greater luminance are favored in cases of shadowed faces. This pair of values (H,V) constitutes the gaze vector, as it will be called from now on in this paper. Typical cases of a person whose head is rotated frontally in relation to the monitor but his eyes are moving are shown in Fig. 3. The black line corresponds to the gaze vector whose magnitude declares the degree of the gaze away from the center of the field of view of the user.

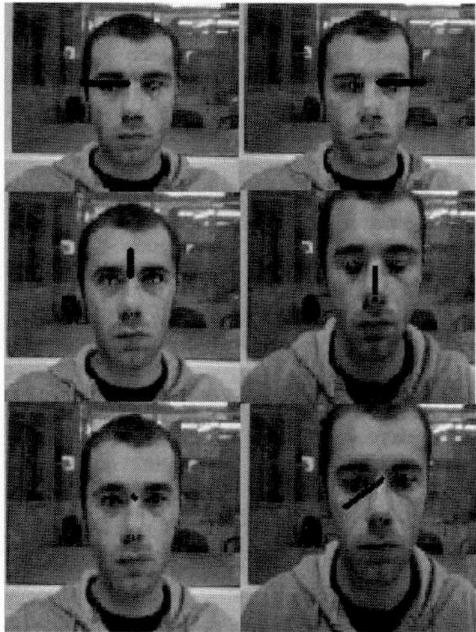

Fig. 3. Various gaze instances captured in front of a computer monitor using a simple web-cam. The black line (gaze vector) shows the magnitude of the gaze vector, as well as its directionality.

3   Neuro-fuzzy System Design and Evaluation

The target group of this research involves 20 learners at a certain level of education - they are 9 years old +/- 18 months. The selection of these children is based on the fact that they have not experienced problem of comprehension and use of language but they face barrier to learning caused by learning problems. One of the challenges that we had to face was the age of the participants of the experiment. As they are children, they are more active in the duration of the video recording. In addition, it was a unique opportunity to test the facial feature detection method that we use, as well as the eye-gaze and head-pose estimation algorithms to this difficult target group. The experimental tests took place in a familiar learning environment for the children: their school. In this framework, we collected videos of children from a Greek and a Danish school. We installed a web camera in the computer of the school and we asked each of the children to read an electronic document which was displayed in a 17' monitor.

As mentioned above, every time specific geometric criteria are fulfilled, the algorithm re-initializes. The first frame has to be a face looking at the camera frontally and, at this point, the pose vector has length equal to zero. As the person changes his head pose, this vector increases in length (see Fig. 4). Also, as a person's eyes look at different directions, the gaze vector changes in magnitude (see Fig. 5) [1].

---

[1] For privacy reasons, the videos used were from adults in our laboratory.

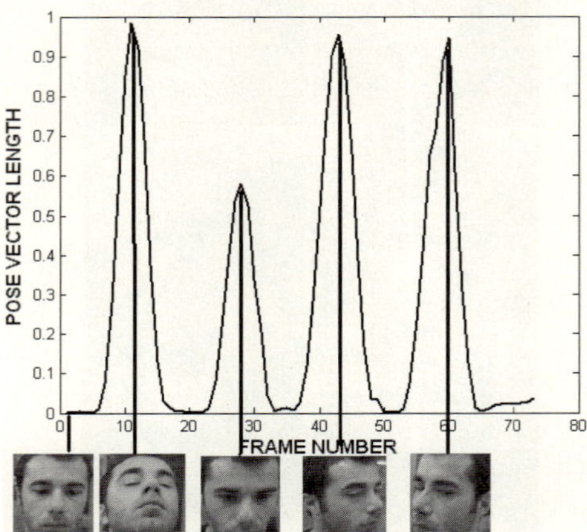

Fig. 4. Pose changes during a video of a person in front of a monitor

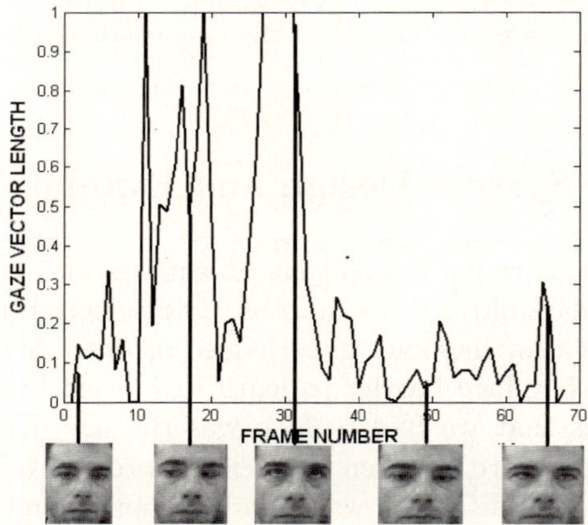

Fig. 5. Gaze changes during a video of a person not moving his head in front of a monitor

In order to estimate a user's attention, based on children's video files, a Sugeno-type fuzzy inference system was built. The metrics used in our case were the head pose vector, the gaze vector and the inter-ocular distance fractions between consecutive frames. We used 10 video segments, between 800 and 1200 frames each, and examined the means of the above metrics within video samples of 50 frames. Thus, for our experiments we had a total of 200 samples. The lengths of the videos were chosen so that, in each shot, all states were equally allocated. For training and testing, a leave-one-out approach was followed.

The pose and gaze vector magnitudes (see Fig. 4, 5) are values between zero and one, while the inter-ocular distance fraction was calculated as a fraction of

Table 1. Neuro-Fuzzy system decision accuracy

State	Average Error	%success
attentive	0.04	100
non-attentive	0.24	72
TOTAL	0.117	87.7

the inter-ocular distance measured every time the algorithm was re-initialized. Thus, the values of the inter-ocular distance between consecutive frames follows the distribution shown in Fig. 6. It can be noticed that the mean is shifted on the left. This is due to the fact that, as a face rotates right and left, the projected inter-ocular distance is reduced. In fact, these values were used, as will be seen later, together with pose and gaze vectors, for inferring a person's attention. The output's values, based on the annotation of the database, were either 1, declaring those time segments when the child was attentive, and 0 for those periods when the child was not paying attention to the monitor.

Prior to training, our data were clustered using the sub-cluster algorithm [15]. This algorithm, instead of using a grid partition of the data, clusters them and, thus, leads to fuzzy systems deprived of the curse of dimensionality. For clustering, many radius values for the cluster centers were tried and the one that gave the best trade-off between complexity and accuracy was 0.1 for all inputs. The number of clusters created by the algorithm determines the optimum number of the fuzzy rules. After defining the fuzzy inference system architecture, its parameters (membership function centers and widths), were acquired by applying a least squares and back-propagation gradient descent method [16].

After training, the output surface using the pose vector length and gaze vector length are shown in Fig. 7, and the respective surface using the pose vector length

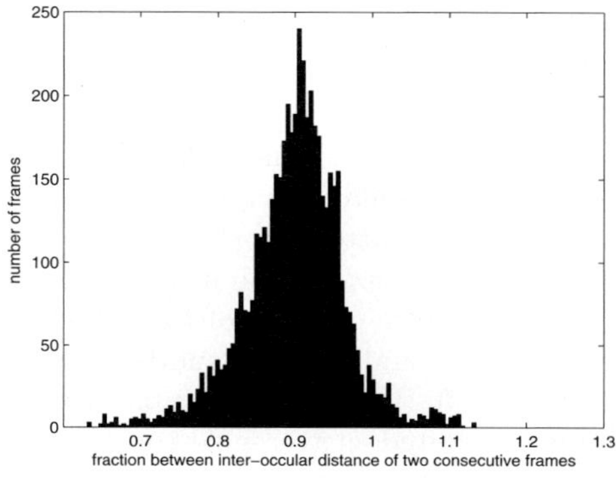

Fig. 6. Distribution of the fraction of the inter-ocular distance between consecutive frames

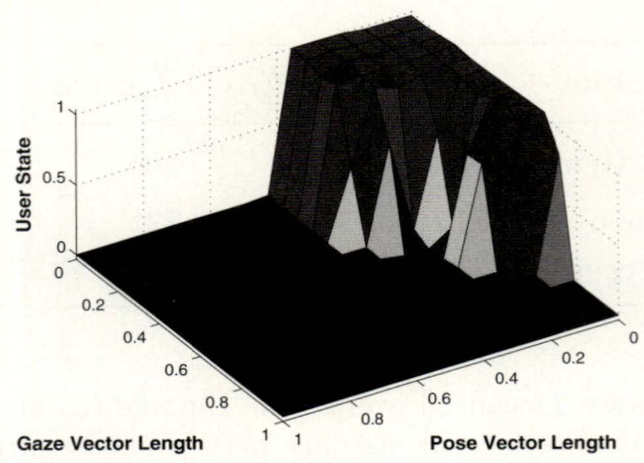

Fig. 7. Output surface using the pose vector length and gaze vector length

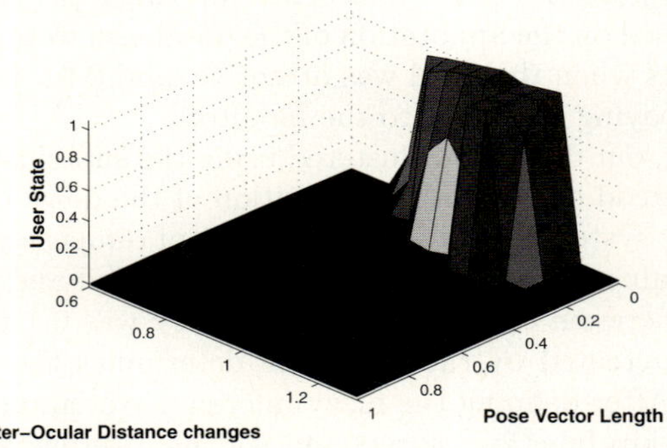

Fig. 8. Output surface using the pose vector length and inter-ocular distance fraction between successive frames

and inter-ocular distance fraction as inputs is shown in Fig. 8. It can be seen from the above figures that, for small values of the pose vector length and the gaze vector length, the output takes values close to one (attentive), while, as gaze or pose vector lengths increase in magnitude, the output's value goes to zero. Similarly, it can be seen that , for small values of the pose vector length, and for small changes of the inter-ocular distance (the fraction takes values close to 1), the user can be considered as attentive (the output is close to one). Large values of the pose vector length mean that the user is non-attentive (the output is close to zero), while sudden, large changes of the inter-ocular distance mean that the user is non-attentive. Usually, these sudden changes occur when the user rotates rapidly right and left. The following table summarizes the results of our approach. In the column denoted as average error, the average absolute error between the output and the annotation is reported, while in the second column, a crisp decision is considered, meaning that a state is inferred to be attentive if the output value is above 0.5 and non-attentive if it is below 0.5.

4 Conclusions and Future Work

In this work, the steps of a method for user attentiveness evaluation in front of a computer monitor was presented. The advantages of the method are that it does not require any special setup in terms of hardware, it runs real-time, it is user independent and can work efficiently under uncontrolled lighting conditions. A neuro-fuzzy system was developed for the evaluation of the state of the user. The results of using a neuro-fuzzy system showed that, discriminating between attentiveness and non-attentiveness does not have to be a crisp decision but takes into account the magnitude of the biometrics it uses in order to characterize a user with a certain degree of certainty. Future work shall include more biometric measurements (mouth, hands movements, frowning of the eyebrows) in order to develop a system that can discriminate among more states (tiredness, frustration, distraction).

Acknowledgments

This work was partially funded by European Commission IST Project 'AGENT-DYSL' (under contract FP6 IST-2005-2.5.11 e-Inclusion-034549) and by IST Project 'FEELIX',(under contract FP6 IST-045169).

References

1. Voit, M., Nickel, K., Stiefelhagen, R.: Multi-view head pose estimation using neural networks. In: Second Canadian Conference on Computer and Robot Vision (CRV), Victoria, BC, Canada, pp. 347–352. IEEE Computer Society, Los Alamitos (2005)
2. Mao, Y., Suen, C.Y., Sun, C., Feng, C.: Pose estimation based on two images from different views. In: Eighth IEEE Workshop on Applications of Computer Vision (WACV), Washington, DC, USA, p. 9. IEEE Computer Society, Los Alamitos (2007)
3. Beymer, D., Flickner, M.: Eye gaze tracking using an active stereo head. In: IEEE Computer Society Conference on Computer Vision and Pattern Recognition (CVPR), Madison, WI, USA, vol. 2, pp. 451–458. IEEE Computer Society, Los Alamitos (2003)
4. Meyer, A., Böhme, M., Martinetz, T., Barth, E.: A single-camera remote eye tracker. LNCS (LNAI), pp. 208–211. Springer, Heidelberg (2006)
5. Hennessey, C., Noureddin, B., Lawrence, P.D.: A single camera eye-gaze tracking system with free head motion. In: Proceedings of the Eye Tracking Research & Application Symposium (ETRA), San Diego, California, USA, pp. 87–94. ACM, New York (2006)
6. Gee, A., Cipolla, R.: Non-intrusive gaze tracking for human-computer interaction. In: Int. Conference on Mechatronics and Machine Vision in Pract., Toowoomba, Australia, pp. 112–117 (1994)
7. Gourier, N., Hall, D., Crowley, J.: Estimating face orientation from robust detection of salient facial features. In: International Workshop on Visual Observation of Deictic Gestures (ICPR), Cambridge, UK (2004)

8. Seo, K., Cohen, I., You, S., Neumann, U.: Face pose estimation system by combining hybrid ica-svm learning and re-registration. In: 5th Asian Conference on Computer Vision, Jeju, Korea (2004)

9. Stiefelhagen, R.: Estimating Head Pose with Neural Networks - Results on the Pointing 2004 ICPR Workshop Evaluation Data. In: Pointing 2004 Workshop (ICPR), Cambridge, UK (August 2004)

10. Daugman, J.: High confidence visual recognition of persons by a test of statistical independence. IEEE Trans. Pattern Anal. Mach. Intell. 15(11), 1148–1161 (1993)

11. Deng, J.Y., Lai, F.: Region-based template deformation and masking for eye-feature extraction and description. Pattern Recognition 30(3), 403–419 (1997)

12. Viola, P.A., Jones, M.J.: Rapid object detection using a boosted cascade of simple features. In: Conference on Computer Vision and Pattern Recognition (CVPR), Kauai, HI, vol. 1, pp. 511–518 (December 2001)

13. Asteriadis, S., Nikolaidis, N., Pitas, I., Pardàs, M.: Detection of facial characteristics based on edge information. In: Second International Conference on Computer Vision Theory and Applications(VISAPP), Barcelona, Spain

14. Zhou, Z.H., Geng, X.: Projection functions for eye detection. Pattern Recognition 37(5), 1049–1056 (2004)

15. Chiu, S.L.: Fuzzy Model Identification Based on Cluster Estimation. Journal of Intelligent and Fuzzy Systems 2(3) (1994)

16. Jang, J.S.R.: ANFIS: Adaptive-Network-Based Fuzzy Inference System. IEEE Transactions on Systems, Man, and Cybernetics 23, 665–684 (1993)

TriangleVision: A Toy Visual System

Thomas Bangert

15 Bence House, Rainsborough Avenue, SE8 5RU, London, UK
Tel.: +44 (20) 8691 6039
`txbangert@hotmail.com`

Abstract. This paper presents a simple but fully functioning and complete artificial visual system. The triangle is the simplest object of perception and therefore the simplest visual system is one which *sees* only triangles. The system presented is *complete* is the sense that it will *see* any triangle presented to it as visual input in bitmap form, even triangles with *illusory contours* that can only be detected by inference.

Keywords: computer vision, image analysis, Kanizsa figures.

1   Introduction

The human visual system is extraordinarily complex[1]. It is so complex that little progress has been made in understanding its function. Furthermore, very little progress has been made in artificially (computationally) replicating functionality of visual systems found in humans and other organisms, even organisms with much simpler visual systems.

The most advanced artificial visual systems have recently been put to the test in the DARPA Grand Challenge[2]. The Grand Challenge is a competition sponsored by the United States Defense Advanced Research Projects Agency in which the entrants are autonomous vehicles that must navigate a route across difficult desert terrain. This is a task that is primarily a visual one, and one which many animals are able to perform with ease, using visual systems that are varied in design but generally much less sophisticated than the human visual system. In 2004 when the Grand Challenge was first held none of the vehicles completed the race and when it was held in the following year only 5 out of the 23 finalists were able to complete the race. This is despite the fact that all the vehicles were required to do was to follow an unpaved desert road from the start position to the destination; something a human driver is able to do with ease. The 5 vehicles that completed the race did not complete the race because of their ability to *see* the road, but because entrants were given a detailed map of the route in advance and allowed the use of the Global Positioning System to determine in real time their precise location on the map. The vehicles therefore did not need sophisticated visual systems to navigate the route. The only visual information the vehicles needed was detection of obstacles (such as rocks or potholes). To detect obstacles the entrants employed an array of technically sophisticated and expensive sensors such as laser rangefinders and radar. These sensors performed dedicated tasks,

V. Kůrková et al. (Eds.): ICANN 2008, Part I, LNCS 5163, pp. 937–950, 2008.

primarily *obstacle detection*. None of the entrants employed what could be called a more general visual system, which would require sophisticated processing but allow the use of simple and inexpensive sensors.

Visual systems vary in complexity. Some organisms such as humans have a very sophisticated visual system whereas other organisms have much simpler visual systems[1]. Even the simplest of the visual systems are not well understood. One way to classify the complexity of a visual system is by the complexity of the geometric objects the visual system is capable of representing. Circles may be said to be more complex than polygons and therefore a visual system that is capable of processing line arcs as well as straight lines may be said to be more complex. Rather than attempting to computationally reproduce the function of the most complex natural visual systems, the approach taken here is to construct a visual system of minimal complexity.

It is assumed that the basis of any visual system is geometry, and the simplest of geometric objects is the triangle. Therefore the simplest of all possible visual systems is one which *sees* only triangles. This paper presents such a visual system. The visual system presented is limited to representing triangles but it is capable of detecting all triangles from a given visual input, even meta-triangles (triangles whose edges are themselves composed of triangles) and *illusory* triangles, which can only be detected by inference.

1.1   Completeness

We may say that a visual system is complete when it is able to detect from a given visual input all that can effectively be represented by that visual system. That is, that there is no other visual system with an identical imaging model that is able to extract additional objects from the same visual input.

1.2   PostScript

PostScript is a system whose purpose is to represent text and graphical images for reproduction on printed pages. To do this it employs a set of geometric primitives and rules in respect to how they are used. This set of rules is called the *imaging model* of the system.[3] The creators of the PostScript imaging model claim it to be "very similar to the model we instinctively adopt when drawing by hand"[4]. It is suggested that when we *instinctively* draw by hand that the visual system of our brain is employed to carry out the processing necessary for the task. Therefore when seeking to develop an artificial visual system it may well be useful to take inspiration from existing models that in some way reflect the visual system of the human brain.

When we draw by hand we must draw on a piece of paper or canvas. It is always the case that we draw *on* something, and this something may be seen as the framework (or space) of that activity. The canvas does not form a part of the drawing itself but establishes the space on which the drawing is set out. In this same sense, PostScript posits an abstract drawing space, which is referred to as the *current page*. It is a two dimensional space (or canvas) upon which

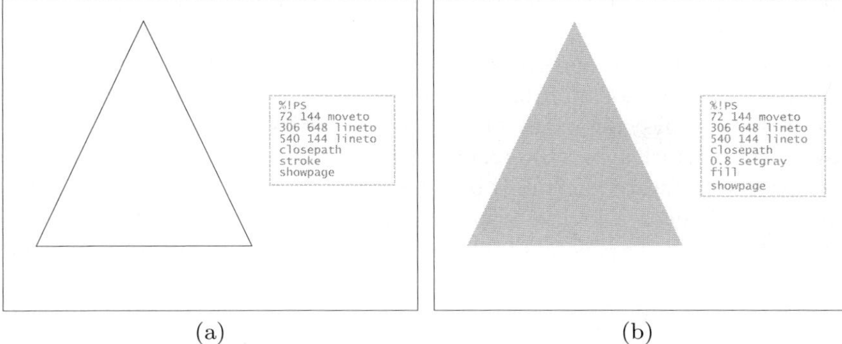

Fig. 1. A triangle, and its PostScript code (a). The line width is set automatically by PostScript to the default. It may also be set to any arbitrary width with the `setlinewidth` command. A PostScript triangle with grey texture and no line (b).

PostScript *painting operators* draw. This space is a Cartesian plane where any arbitrary point can be uniquely addressed by the Cartesian coordinate system. Postscript output is confined to the first quadrant of this plane and therefore the bottom left corner of the page is defined as the origin for a PostScript page.

The basic PostScript geometric operators are coordinate points and lines. A line is drawn by setting two points (by giving the X-Y coordinate) and giving the line draw command. More complex geometric objects may be drawn by repeating this process. Figure 1(a) gives an example of how a triangle is drawn using PostScript.

PostScript represents objects such as triangles in a rigorous formal way. Lines are drawn between set points and then the area enclosed by the path set out by the lines is used to establish a shape. PostScript can represent very complex shapes, but we will restrict ourselves to the simplest of its geometric shapes, the triangle. Under the PostScript imaging model the exterior of a shape can have its lines stroked (that is drawn) and the shape can be filled with a texture. Figure 1(a) gives an example of a conventional triangle with stroked lines. Figure 1 gives an example of a triangle with a simple gray texture and no stroked lines. Lines can also be more complex by being dashed. Shapes can also be defined that have no stroked lines and no texture as shown in figures 5(a) and 5(b).

2   Prolog

The principal function of an artificial visual system is to induce detailed and precise geometry from a visual input that is incomplete, inconsistent and poor in quality. Prolog (PROgramming in LOGic) is the only programming language that is non-deterministic and allows the representation of entities which do not have to be instantiated at the outset[5]. It allows for the establishment of relations between as yet unestablished entities and without having to provide further detail will work through all the possibilities and determine the values that allow an entity to be instantiated once sufficient grounds for it are found[6]. These are the kinds of processes inherent to visual systems and therefore Prolog is a natural choice for implementing the processing framework of an artificial visual system.

Most Prolog implementations allow what might be called a logical space or framework into which new statements of fact may be asserted or from which previously asserted statements may be retracted. It is therefore possible in Prolog to directly implement a logical visual space set out purely by the geometrical entities of that visual space and the logical relationships between them. Once seeded with the facts that represent raw sensor data such a system would operate on the raw facts by way of recursive transformation. This system would transform the initial perception-objects into entities of increasing abstractness until the visual space reaches a steady state that is populated entirely by the geometrical entities of the visual space. Recursion is inherent to Prolog and it is a process that operates from a set of rules by which entities are replaced by new entities and where these new entities in turn become subject to the same process and may in turn be replaced.

3   Retinal Processing

The retina is the only part of the human visual system that is to some degree understood[1]. While it may well be possible to design visual systems that are superior to the visual systems found in natural organisms the fact that visual systems have so far proven to be in practice intractable suggests that artificial visual systems should take inspiration from natural visual systems.

The first layer of the retina (see figure 6) is a simple sensor array which responds to light. Assuming rate coding, each sensor produces a value representing light intensity. This function is well understood and has been replicated with considerable success for a variety of functions, notably in digital cameras. A digital camera simply stores the sensor light intensity values at a certain point in time in a two dimensional array, which can then be read back and viewed from the camera's memory store.

While photography as a whole may be said to have been inspired by the study of the human visual system, specifically the retina, the function replicated by the camera is merely the first layer of the retina, this being a simple sensor array. However, the retina (unlike the digital camera) does not send the output of the sensor array directly to the brain. The retina incorporates an additional layer that carries out pre-processing function(s) and the neural circuitry for this function is arranged immediately behind the sensor array (actually in the mammalian retina the physiology is inverted). This pre-processing circuitry is not complex, has been studied in considerable detail and its function is to some degree understood.

In the mammalian retina what the pre-processing circuitry does is to arrange the individual light sensors into groups. Each group operates as receptor unit, has a set function and produces a unitary output. These groups are generally

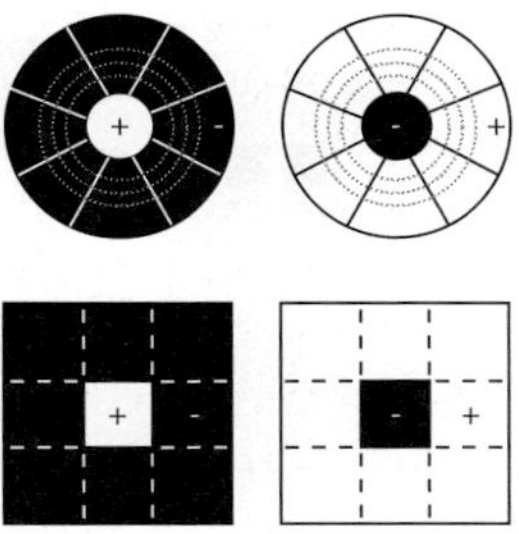

Fig. 2. Receptive Fields. The receptive fields of natural visual systems are circular and vary in size. This has been simplified to receptive fields with square zones of a fixed size of 3 by 3 pixels.

referred to as receptive fields. Receptive fields do not produce a value for light intensity but rather a value for a change in the intensity between two areas of the field. Receptive fields are roughly circular in organization and have two distinct areas; a circular centre and a a concentric surround (see figure 2). A receptive field measures the difference in light intensity between these two areas and gives as output the value of the contrast. There are two distinct types of receptive fields: on-centre and off-centre. The former gives a contrast value when the light intensity is greater in the centre than the surround (the center is *on* and the surround is *off*)and the latter gives a contrast value when the light intensity in the surround is greater than in the centre (the center is *off* and the surround is *on*). It is these contrast values that are sent from the retina to the brain. It is important to note that these contrast values are the only visual information that the brain receives.

While the function of receptive fields within the visual system is not fully understood, it is generally accepted that receptive fields are a first step in the process of edge detection. How edges are detected or elicited from images has been a matter of study for approximately half a century and has elicited many rather complex models[7]. This complexity has in part been driven by the empirical observation that receptive fields have a concentric surround. This concentricity is paradoxical to the otherwise straight forward function of edge contrast detection.

For our simplified visual system we have adopted a simple computational model of receptive fields, a model that is directly useful for edge contrast detection. It is proposed that the empirical findings in respect of receptive fields are correct but that there are alternatives to the concentric surround hypothesis that are consistent with the empirical findings. It is proposed that the concentric surround is not a whole unit but is divided into segments. It is assumed that the surround is composed of 8 units, each being equal in size to the center. Each surround unit calculates a contrast value in comparison to the centre. This is then integrated to a single output. A receptive field therefore calculates 8 separate contrast values internally and takes the maximum of these values as its output. A receptive field of this type when presented with the conventional experimental stimuli would give an output identical to the conventional model of a receptive field.

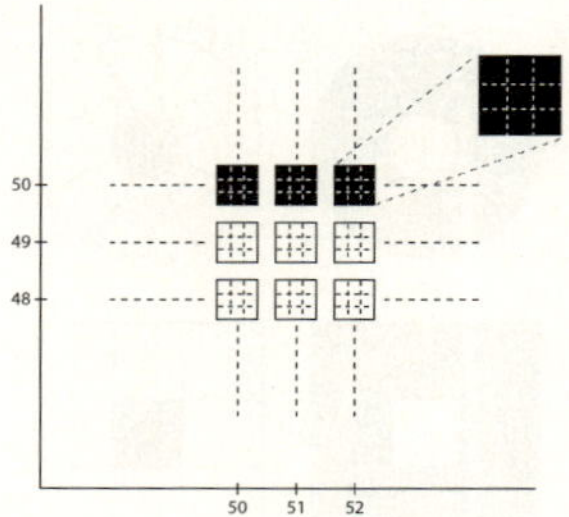

Fig. 3. A bitmap image divided into receptive field zones of 3 by 3 pixels. The pixels shown illustrate a horizontal black/white contrast zone.

4   The Visual System

The system is designed to take any PostScript produced raster (bitmap) image and produce from it geometric representation that may be used to accurately recreate the original PostScript code. The original PostScript code may be arbitrarily complex, but limited to the only geometric form allowed by our system, the triangle.

4.1   Receptive Fields and Points

Receptive fields of the human visual system have a number of different sizes. For the sake of simplicity the visual system presented here employs receptive fields of a single fixed size. It is impractical to perform edge detection pixel by pixel and therefore the receptive field size was set to 3 pixels. The total size of a receptive field is therefore 9 pixels (square). To achieve the best resolution only the center of the receptive fields are mapped to the bitmap input rather than both the center and surround. This requires surrounds to overlap onto the centers of other receptive fields. This effectively sets the receptive fields to 3 by 3 pixels. If the sensor array is (for example) 1350 by 900 then the number of receptive fields would be 450 by 300. The center of each receptive field receives a fixed input of 9 pixels and the surround is a set of 8 zones also of 9 pixels each. The receptive fields are in a fixed location but to allow for some overlap and flexibility each receptive field may shift one pixel in any direction to maximize available contrast. Receptive fields come in two types: center-on and center-off. This means that for every group of 3 by 3 pixels there are two receptive fields. All receptive fields are autonomous of each other.

   The function of a receptive field under this system is to assert a contrast value fact when its contrast threshold (a global value) is breached. Facts are asserted as Prolog statements. For example, in figure 3 the center-off receptive fields at addresses (50,50), (51,50), (52,50) will activate and find a contrast of 255. As a result they will assert the following Prolog statements:

```
point((50,50),255).
point((51,50),255).
point((52,50),255).
```

The center-on receptive fields at (50,49),(51,49), (52,49) will also activate and assert the corresponding points, but with inverse contrast.

```
point((50,49),-255).
point((51,49),-255).
point((52,49),-255).
```

4.2 Edges

Edges are defined as 3 or more adjacent points, and this allows direction to be extracted from points that are individually directionless. Points are considered adjacent in the directions allowed by the receptive field. They may be adjacent horizontally, vertically or diagonally (two directions of diagonality). For the points asserted from figure 3 the following edge would be asserted.

```
edge((50,50), (52,50), 255,
  [[(50,50),(52,50),[(51,50)]],[(50,49),(52,49),[(51,49)]]]).
```

The edge is from the (50,50) to (52,50) and has a contrast of 255. All the points associated with the edge are retracted when the edge is asserted. The retracted points are stored as a list with the asserted edge. The edge may at some point be retracted, and if so the edge points stored in the list would be re-asserted.

The fragment of the code that asserts edges is given below, in simplified form. Ignoring some of the minor issues such as handedness, an edge is asserted if the edge detector finds a set of edge points adjacent to a set of edge points parallel to it (linear equations must match) but with inverse contrast . When an edge has been found, the edge points are retracted and the edge is asserted. The simplified code fragment presented below is given as representative of the general approach used by the system.

```
edge :-
  edge_detector(PointA,PointB,Contrast,EdgePointsAB),
  edge_detector(PointC,PointD,-Contrast,EdgePointsCD),
  linear_eqn(PointA,PointB,Slope,ConstantAB),
  linear_eqn(PointC,PointD,Slope,ConstantCD),
  ConstantAB>ConstantCD, nextTo(ConstantAB,ConstantCD),
  ContrastEdge is Contrast * sign(Slope),
  retractor(EdgePointsAB), retractor(EdgePointCD),
  assertz(edge(PointA,PointB,ContrastEdge,
          [EdgePointsAB,EdgePointsCD])).
```

4.3 Lines

A line is defined as two edges that are parallel and *reasonably* close to each other. For the sake of simplicity we ignore that lines have a thickness. A simplified code fragment is given below.

```
line :-
  edge(PointA,PointB,Contrast,EdgeAB),
  edge(PointC,PointD,-Contrast,EdgeCD),
  linear_eqn(PointA,PointB,Slope,ConstantAB),\pagebreak
  linear_eqn(PointC,PointD,Slope,ConstantCD),
  reasonably_close(ConstantAB,ConstantCD),
  voidcheck(PointA,PointB,PointC,PointD,[EdgeAB,EdgeCD]),
  retractor(EdgeAB), retractor(EdgeCD),
  assertz(line(PointA,PointB,Contrast,[EdgeA,EdgeB])).
```

4.4   Verteces

A vertex is defined as the point where two non-parallel lines meet. Vertexes are pseudo-objects that are computed from lines or edges at the point where they meet. They may be used to assert objects but they will not prevent objects from being asserted if they fall within their bounds. As they do not contradict assertion they are retracted if they fall within the boundaries of an object that is asserted.

4.5   Triangles

A triangle is a set of 3 matching lines or vertexes. Again for the simplified sample code, we ignore that lines have thickness and that triangles may have texture.

```
triangle :-
  line(PointA,PointB,Contrast,LineAB),
  line(PointB,PointC,Contrast,LineBC),
  line(PointC,PointA,Contrast,LineAC),
  trianglegeometry(PointA,PointB,PointC),
  voidcheck(PointA,PointB,PointC,[LineAB,LineBC,LineAC]),
  retractor(LineAB), retractor(LineBC), retractor(LineAC),
  assertz(triangle(PointA,PointB,PointC,
        Contrast,[LineAB,LineBC,LineAC])).
```

4.6   Processing Flow

As shown by the simplified Prolog code the visual system sets out the space into which objects may be asserted and the geometric nature of that space. This object-space is then seeded by the receptive fields to which the visual input is presented. This introduces low-level facts into the visual system. The processing flow of the visual system is to recursively replace these fact statements with ever more high level replacements. Points are replaced with edges, edges with lines and finally lines with the end product objects, in this case triangles. The system may be said to complete when it reaches a steady state, where no further objects are replaced or where there is circular object replacement.

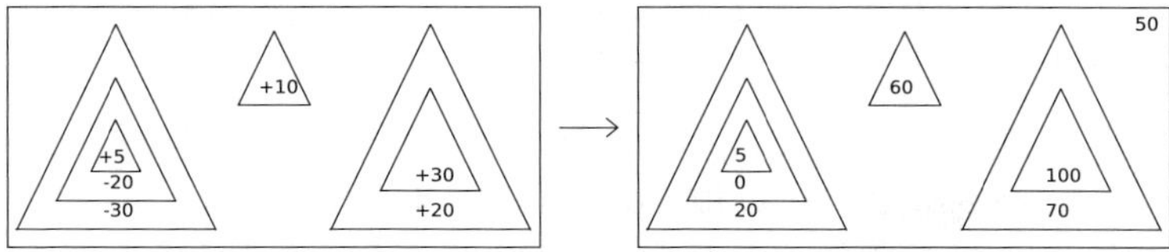

Fig. 4. Contrast colour normalization. The left panel shows objects with the raw contrast values as derived from edges and lines. The right panel shows the same objects with normalized colour values. The right panel has a colour value for the canvas as well as the object, and this is determined by negating the negative raw contrast.

5   Colour

The system defines colour as a simple luminance value. As with natural visual systems, this may be extended by the addition of two further colour contrast values, but at its simplest colour may be said to be a simple luminance value. As visual input is restricted to the output of receptive fields, colour must be computed purely from relative contrast values. These contrast values feed through edges, lines and finally to the final objects of the visual system. Contrast values may be negative as well as positive. Once the visual system completes (that is, reaches a steady state from a given initial sensory input) then normalized colour values can be calculated.

The simplest contrast colour model is derived simply by normalizing negative contrast values. This is done by calculating the baseline colour of white from the lowest negative contrast value. The colour of the canvas is then simply set to the absolute value of the lowest contrast (as final colour values cannot be negative, the lowest contrast value plus the canvas value must equal zero). Figure 4 gives an example how colours are calculated. It is assumed (with respect to the system) that colour is represented by values from 0 to 255. The colour white (full luminance) is defined to have the value of 0 and 255 is defined as black (no luminance).

Colours are recursively computed from the contrasts of objects in relation to their background. The default background is the canvas (current page). The colour of the canvas is set to the difference between positive and negative contrast values. For example, if the canvas has a single black triangle, then the canvas would have a colour of 0 (white) and the triangle would have a colour of 255 (black). For a more illustrative example, consider a canvas with two separate triangles, one with a contrast value of -100 and another with a contrast value of +155. The lowest contrast value is -100 and therefore the normalization value would be +100. The final colours for the two triangles would as a result be 0 (white) and 255 (black), and the canvas would be 100 (light gray). Figure 4 give a slightly more complex example which involves recursion. The nested triangles give a combined minimum contrast value of -50, and therefore the normalization value would be +50.

6   Kanizsa Figures and Illusory Contours: An Extended Example

One of the strengths of the visual system is that it provides the theoretical basis for perception of illusory contours and implicit objects (Kanizsa figures)[8][9]. When presented with figure 5(a) human observers generally see three notched black triangles. However, when the black triangles are rotated so as to obscure their tips a human observer will generally now see a single large white triangle in place of the small triangular notches. Such figures are called subjective figures or illusory figures because they are visible only on the basis of how they obscure other figures. Moreover, such figures have an interesting aside in that despite there being no empirical difference in luminance a human observer will report that the subjective figure is brighter than its surroundings. The area where the brightness changes (there being no change in the original image itself) is called an illusory contour (or edge).

Figure 5(a) shows three black on white triangles with white on black triangular notches on one side. Under the visual system presented here an object will not complete (be *recognized*) if any of the objects that fall within its borders are unaccounted for (that is, contradict it). The white on black notches lead to the assertion of two white on black edges and a white on black vertex, and these will prevent the larger black on white edges from completing a triangle. There is also a break in the line, which also prevents the triangle from completing. It will fail because a bridging line cannot be asserted to complete one of the lines for the triangle (lines cannot simply be asserted as needed). Therefore the black on white triangle will fail. However, the two white on black edges and the vertex are sufficient to allow the small white on black triangle to be asserted. It has no line and has a contrast colour of white on black (the contrast colour of the two edges). Assertion of the small white triangle now removes the objects that initially blocked completion of the larger black on white triangle. A bridging line can now be asserted because the space is a void (covered by the small white on black triangle) where anything can be asserted as needed. This allows line completion, and as there are no other objects that remain to contradict it the black on white triangle completes and is asserted.

Figure 5(b) differs from figure 5(a) only in that the black triangles have been rotated so that the tip of the triangle is obscured rather than the base. In this case the small white on black triangles will complete as before. However, completion of the 3 small white on black triangles does in this instance not allow the larger black on white triangles to complete. This is because the vertexes lie outside of the void left by the asserted triangles. The entire image therefore fails to complete. Failure leads to backtracking and eventually to the retraction of the 3 small white on black triangles. Examining alternatives, the three vertexes formed by the white on black edges line up correctly to form a large triangle which completes without complications. The void left by the assertion of a large triangle now allows double bridging lines to be asserted to complete the black on white triangles.

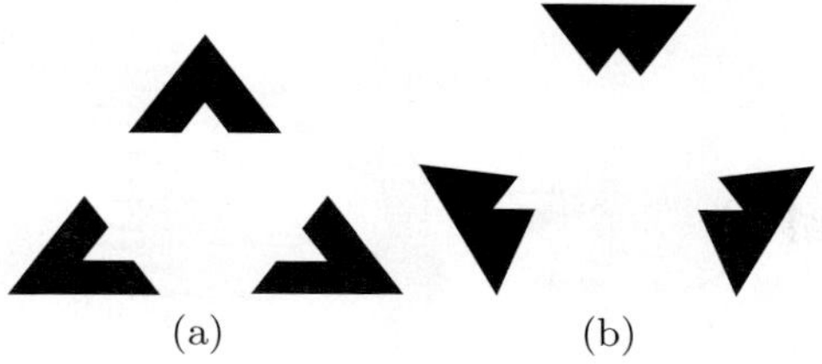

Fig. 5. Three black on white triangles with triangle notches (a). The same notched triangles as in figure (a), but rotated so that the triangular notch obscures the tip of the triangle. Most human observers will now perceive an additional white triangle, whose 3 vertexes cover the tips of the 3 black triangles (b).

It is important to note that colours are relative contrast values rather than absolute. These values are calculated from the contrast values that are passed down from edges and then adjusted globally when the image completes. The large triangle in Figure 5(b) is asserted on the basis of 3 vertexes (which have no contrast value) and without any associated edges or lines. This means there is no contrast value to assign to the triangle. The solution to this is to simply set the contrast value to the default(-1). Without the induced triangle the colour normalization would result in a white canvas with 3 black triangles. However, the induced white triangle has a contrast colour of -1, which makes it the lowest colour contrast value on the canvas, and is therefore assigned as reference white. As a result the background canvas is set to a colour of 1. In this way the colour of the inferred Kanizsa triangle may be seen to be *whiter than white*[8].

7   Related Work

One of the few researchers who has made an attempt to develop the framework of a visual system is Marr[10]. Marr attempts to give an overview of the entire human visual system. While Marr's work is ambitious in scope it fails in providing a detailed and functional model of visual systems. Marr's work has not led to the development of effective artificial visual systems.

8   Discussion

Conventional edge detection algorithms take a bitmap image as an input and produce a bitmap image of the same dimensions as output. While the visual input for the human visual system is equivalent to a bitmap image, the human visual system does not do pixel based global transforms. All retinal sensors are directly organized into receptive fields in the retinal itself via a small set of specialized neurons, which lie directly behind the visual sensory neurons. As can be seen from Figure 6 the neural structure of the retina is not overly complex,

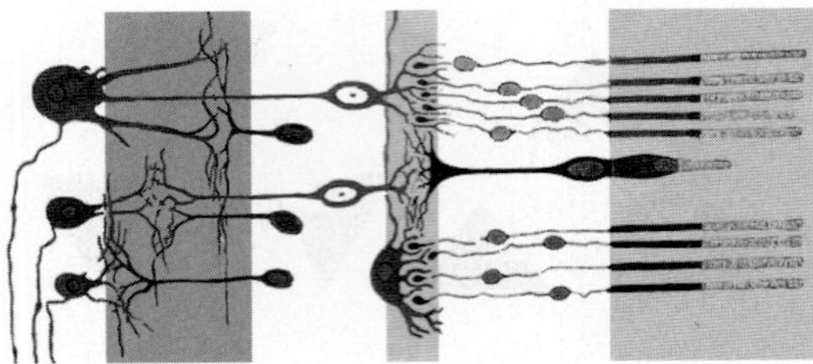

Fig. 6. Structure of the human retina

with sensory neurons organized by a small number of specialist intermediary neurons, whose output is sent to the brain[11]. While it is not universally agreed what the function of the specialist neurons involved is, it is generally accepted that they organize the individual sensor neurons into receptive fields, and that the output of receptive fields is then used to carry out some form of edge detection. This is made more precise in this paper by setting out the dedicated function of calculating segment contrast values and integrating them into individual receptive fields.

One of the corollaries of the segmented contrast hypothesis is an explanation for why two types of receptive field are found in natural visual systems. As the two types of receptive field are inversely related, they will operate as a pair – that is, an edge will always attract both types of receptive field, one for each side (or hand) of the edge. It might appear from this that only a single type of receptive field is necessary to carry out edge detection. Evolution rarely indulges in redundancy which serves no useful function, and therefore it follows that the second inverse receptive field is likely to have some necessary function for the visual system. Indeed this function becomes apparent when we seek to compute edges from points and lines from edges. Edges are derived from sets of adjacent points. Edges have direction (or angle), which is information that individual receptive fields do not generate. It can, however, be derived from the direction in which a set of two or more points is adjacent. An edge is a contrast difference boundary, and this boundary can be of two types, positive contrast or negative contrast. This handedness (or chirality) is also information that is not generated by individual receptive fields. It is also not derivable from a set of adjacent points. Indeed it cannot be resolved from the information provided by a single type of receptive field. A second type of receptive field, one which provides inverse contrast values, allows resolution of whether an edge contrast is positive or negative. This therefore sets out the reason why two types of receptive field are necessary and as a result commonly found in natural visual system.

Using the precise imaging model set out by PostScript has allowed the rigorous definition of the representational space of the visual system. One of the primary features of the visual system is to formally describe how the abstract representational space of the visual system is linked to the space of the visual input. The space set out by the receptive fields is equivalent to a bitmap and is a two dimensional integer space. The PostScript imaging model allows this two dimensional integer space of the visual input to be transformed into a space for abstract objects by simply converting integers to real numbers. This space of real numbers is then represented by the use of Prolog, which effectively transforms what was intially a two dimensional bitmap into a set of logic statements. The coordinate system of the initial visual input is maintained but objects are represented by abstract logic statements. These logic statements exist within a logical space, in which geometric relationships recursively transform known facts into hypothetical solutions. This is a logical search space, and the visual processing that may be said to be done in this space has very little relation to the pixels and bitmap that compose the original input. Visual processing activity consists solely of the inference mechanisms of mathematical logic and the rules of geometry. Nevertheless, the final objects are set out using the coordinates that match the coordinates of the initial bitmap input and the abstract objects may be accurately rasterized by translation into PostScript.

9   Future Work

The visual system is not yet fully implemented in Prolog. However, the logic for the system is complete and simply needs to be implemented.

PostScript has a very comprehensive imaging model that is designed to be able to represent a significant subset of what the human visual system is able to represent. PostScript is of course limited to two dimensional static images whereas the human visual system functions in four dimensions; 3 dimensional space as well as time. Nevertheless, a visual system that would incorporate all of the PostScript imaging model (Bézier curves in particular) would be a significant step in producing a more general visual system that would approach the functionality of the human visual system.

The system is at present designed to process bitmap images that are of good quality, without noise or distortion. At present minor noise such as random pixels are simply ignored because they are not detected by the receptive fields. Minor distortion is also corrected by the geometry margin of error set into the geometry rules. It is unclear at present how significant distortion or noise will be dealt with. It is expected that this is to be a significant area of research.

10   Conclusion

TriangleVision is a toy visual system, but one which within its limited parameters is a complete visual system. Its visual field is restricted to triangles, but in this limited visual environment it may be said to genuinely *see* what is presented to it. As the visual environment is further restricted to triangles produced by the PostScript imaging model, we are able to formally define what it means to *see* as the ability to reconstruct PostScript code from a bitmap image. This ability can be accurately measured by simply comparing the original PostScript code that produced the raster output with the PostScript code produced by the system from the bitmap input.

The ability of the visual system presented here to provide a theoretical model for illusory contour and Kanizsa figure perception is a strong indication that this artificial visual system operates in a way that is directly related to the way the human visual system functions. The contrast model of colour employed is able to fully account for the phenomenon of illusory contours.

References

1. Hubel, D.H.: Eye, Brain, and Vision. Scientific American Library, New York (1988)
2. Thrun, S., Montemerlo, M., Dahlkamp, H., Stavens, D., Aron, A., Diebel, J., Fong, P., Gale, J., Halpenny, M., Hoffmann, G., Lau, K., Oakley, C., Palatucci, M., Pratt, V., Stang, P., Strohband, S., Dupont, C., Jendrossek, L.E., Koelen, C., Markey, C., Rummel, C., van Niekerk, J., Jensen, E., Alessandrini, P., Bradski, G., Davies, B., Ettinger, S., Kaehler, A., Nefian, A., Mahoney, P.: Winning the DARPA grand challenge. Journal of Field Robotics (2006) (accepted for publication)
3. Adobe: PostScript Language Reference Manual, 2nd edn. Addison-Wesley Professional, Boston, MA, USA
4. Adobe: PostScript Language Tutorial and Cookbook. Addison-Wesley Longman Publishing, Boston, MA, USA (1986)
5. Clocksin, W., Mellish, C.: Programming in Prolog. Springer, Berlin (1984)
6. Giannesini, F., Kanoui, H., Pasero, R., van Caneghem, M.: Prolog. Addison-Esley, Wokingham (1986)
7. Grossberg, S., Mingolla, E.: Neural dynamics of form perception: Boundary completion, illusory figures, and neon color spreading. Psychological Review 92(2), 173–211 (1985)
8. Kanizsa, G.: Subjective contours. Scientific American 234(4), 48–52 (1976)
9. Kanizsa, G.: Organization in Vision: Essays on Gestalt Perception. Praeger Publishers, New York (1970)
10. Marr, D.: Vision. W.H. Freeman and Company, New York (1982)
11. Cajal, S.R.y.: Histologie du Système Nerveux de l'Homme et des Vertébrés. Maloine, Paris (1911)

Face Recognition with VG-RAM Weightless Neural Networks

Alberto F. De Souza[1], Claudine Badue[1], Felipe Pedroni[1], Elias Oliveira[2],
Stiven Schwanz Dias[1], Hallysson Oliveira[1], and Soterio Ferreira de Souza

[1] Departamento de Informática
[2] Departamento de Ciência da Informação
Universidade Federal do Espírito Santo
Av. Fernando Ferrari, 514, 29075-910 - Vitória-ES, Brazil
alberto@lcad.inf.ufes.br

Abstract. Virtual Generalizing Random Access Memory Weightless Neural Networks (VG-RAM WNN) are effective machine learning tools that offer simple implementation and fast training and test. We examined the performance of VG-RAM WNN on face recognition using a well known face database—the AR Face Database. We evaluated two VG-RAM WNN architectures configured with different numbers of neurons and synapses per neuron. Our experimental results show that, even when training with a single picture per person, VG-RAM WNN are robust to various facial expressions, occlusions and illumination conditions, showing better performance than many well known face recognition techniques.

1 Introduction

Computerized human face recognition has many practical applications, such as access control, security monitoring, and surveillance systems, and has been one of the most challenging and active research areas in computer vision for many decades [1]. Even though current machine recognition systems have reached a certain level of maturity, the recognition of faces with different facial expressions, occlusions, and changes in illumination and/or pose is still a hard problem.

A general statement of the problem of machine recognition of faces can be formulated as follows: given an image of a scene, identify or verify one or more persons in the scene using a database of faces. In identification problems, given a face as input, the system reports back the identity of an individual based on a database of known individuals; whereas in verification problems, the system confirms or rejects the claimed identity of the input face. In both cases, the solution typically involves segmentation of faces from scenes (face detection), feature extraction from the face regions, recognition, or verification. In this work, we examined the recognition part of the identification problem only.

Many methods have been used to tackle the problem of face recognition. One can roughly divide these into (i) *holistic* methods, (ii) *feature-based* methods, and (iii) *hybrid* methods [1]. Holistic methods use the whole face region as the raw input to a recognition system. In feature-based methods, local features, such

V. Kůrková et al. (Eds.): ICANN 2008, Part I, LNCS 5163, pp. 951–960, 2008.

as the eyes, nose, and mouth, are first extracted and their locations and local statistics (geometric and/or appearance) are fed into a classifier. Hybrid methods use both local features and the whole face region to recognize a face.

Among holistic approaches, eigenfaces [2] and fisher-faces [3,4] have proved to be effective in experiments with large databases. Feature-based approaches [5,6,7,8] have also been quite successful and, compared to holistic approaches, are less sensitive to facial expressions, variations in illumination and occlusion. Some of the hybrid approaches include the modular eigenface method [9], the Flexible Appearance Model method [10], and a method that combines component-based recognition with 3D morphable models [11]. Experiments with hybrid methods showed slight improvements over holistic and feature-based methods.

In this work, we evaluated the performance of virtual generalizing random access memory weightless neural networks (VG-RAM WNN [12]) on face recognition using the AR Face Database [13]. There are many face databases freely available for research purposes (see [1] for a comprehensive list); we chose the AR Face Database because we were interested in face recognition under different facial expressions, types of occlusion and illumination conditions, and this database has face images of the same people encompassing all these variations. We evaluated two VG-RAM WNN architectures, one holistic and the other feature-based, each implemented with different numbers of neurons and synapses per neuron. We compared the best VG-RAM WNN performance with that of: (i) a holistic method based on principal component analysis (PCA) [2]; (ii) feature-based methods based on non-negative matrix factorization (NMF) [5], local non-negative matrix factorization (LNMF) [6], and line edge maps (LEM) [7]; and (iii) hybrid methods based on weighted eigenspace representation (WER) [9] and attributed relational graph (ARG) matching [8]. We selected these for comparison because they are representative of some of the best methods for face recognition present in the literature. Our results showed that, even training with a single face image per person, VG-RAM WNN outperformed all mentioned techniques under all face conditions tested.

This paper is organized as follows. Section 2 introduces VG-RAM WNN and Section 3 describes how we have used them for face recognition. Section 4 presents our experimental methodology and experimental results. Our conclusions and directions for future work follow in Section 5.

2   VG-RAM WNN

RAM-based neural networks, also known as n-tuple classifiers or weightless neural networks, do not store knowledge in their connections but in Random Access Memories (RAM) inside the network's nodes, or neurons. These neurons operate with binary input values and use RAM as lookup tables: the synapses of each neuron collect a vector of bits from the network's inputs that is used as the RAM address, and the value stored at this address is the neuron's output. Training can be made in one shot and basically consists of storing the desired output in the address associated with the input vector of the neuron [14].

lookup table	X_1	X_2	X_3	Y
entry #1	1	1	0	class 1
entry #2	0	0	1	class 2
entry #3	0	1	0	class 3
	↑	↑	↑	↓
input	1	0	1	**class 2**

Fig. 1. VG-RAM WNN neuron lookup table

In spite of their remarkable simplicity, RAM-based neural networks are very effective as pattern recognition tools, offering fast training and test, in addition to easy implementation [12]. However, if the network input is too large, the memory size becomes prohibitive, since it must be equal to 2^n, where n is the input size. Virtual Generalizing RAM (VG-RAM) weightless neural networks (WNN) are RAM-based neural networks that only require memory capacity to store the data related to the training set [15]. In the neurons of these networks, the memory stores the input-output pairs shown during training, instead of only the output. In the test phase, the memory of VG-RAM WNN neurons is searched associatively by comparing the input presented to the network with all inputs in the input-output pairs learned. The output of each VG-RAM WNN neuron is taken from the pair whose input is nearest to the input presented—the distance function employed by VG-RAM WNN neurons is the *hamming distance*. If there is more than one pair at the same minimum distance from the input presented, the neuron's output is chosen randomly among these pairs.

Figure 1 shows the lookup table of a VG-RAM WNN neuron with three synapses (X_1, X_2 and X_3). This lookup table contains three entries (input-output pairs), which were stored during the training phase (entry #1, entry #2 and entry #3). During the test phase, when an input vector (input) is presented to the network, the VG-RAM WNN test algorithm calculates the distance between this input vector and each input of the input-output pairs stored in the lookup table. In the example of Figure 1, the *hamming distance* from the input to entry #1 is two, because both X_2 and X_3 bits do not match the input vector. The distance to entry #2 is one, because X_1 is the only non-matching bit. The distance to entry #3 is three, as the reader may easily verify. Hence, for this input vector, the algorithm evaluates the neuron's output, Y, as class 2, since it is the output value stored in entry #2.

3   Face Recognition with VG-RAM WNN

As stated in the Introduction, in this work we examined the recognition part of the identification problem only. Face segmentation is performed semi-automatically and, thanks to the properties of the VG-RAM WNN architectures employed, explicit feature extraction (e.g., line edge extraction; eye, nose, or mouth segmentation; etc.) is not required, even though in one of the two VG-RAM WNN architectures studied some neurons specializes in specific regions of the faces and, because of that, we say it is feature-based. The other VG-RAM WNN architecture studied is holistic.

3.1   Holistic Architecture

The holistic architecture has a single bidimensional array of $m \times n$ VG-RAM WNN neurons, N, where each neuron, $n_{i,j}$, has a set of synapses $W = \{w_1, \ldots, w_{|W|}\}$, which are randomly connected to the network's bidimensional input, Φ, of $u \times v$ inputs, $\phi_{i,j}$ (Figure 2). This random interconnection pattern is created when the network is built and does not change afterwards.

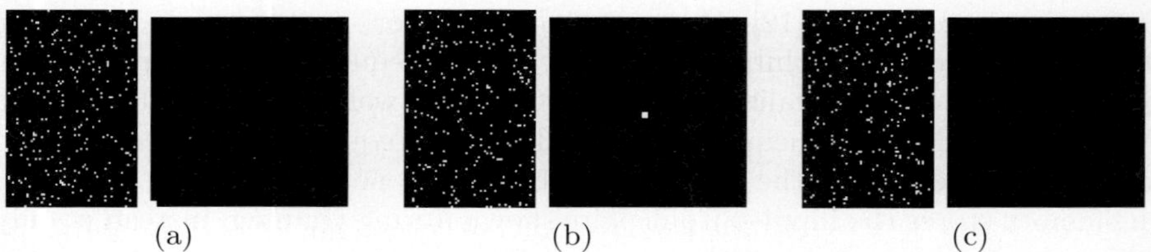

(a)   (b)   (c)

Fig. 2. The synaptic interconnection pattern of the holistic architecture. (a) Left, input Φ: in white, the elements $\phi_{i,j}$ of the input Φ that are connected to neuron $n_{0,0}$ of N via $w_1, \ldots, w_{|W|}$; right, neuron array N: in white, the neuron $n_{0,0}$ of N. (b) Left: in white, the elements $\phi_{i,j}$ of Φ connected to $n_{\frac{m}{2}, \frac{n}{2}}$; right: in white, the neuron $n_{\frac{m}{2}, \frac{n}{2}}$ of N. (c) Left: in white, the elements of Φ connected to $n_{m,n}$; right: in white, the neuron $n_{m,n}$.

VG-RAM WNN synapses can only get a single bit from the input. Thus, in order to allow our VG-RAM WNN to deal with images, we use *minchinton cells* [16]. In the proposed VG-RAM WNN architectures, each neuron's synapse, w_t, forms a minchinton cell with the next, w_{t+1} ($w_{|W|}$ forms a minchinton cell with w_1). The type of the minchinton cell we have used returns 1 if the synapse w_t of the cell is connected to an input element, $\phi_{i,j}$, whose value, $x_{i,j}$, is larger than the value of the element $\phi_{k,l}$, $x_{k,l}$, to which the synapse w_{t+1} is connected (i.e., $x_{i,j} > x_{k,l}$); otherwise, it returns zero.

The input face images, I, of $\xi \times \eta$ pixels (Figure 3(a)), are rotated, scaled and cropped (Figure 3(b)), so that the face in the image fits within the VG-RAM WNN input, Φ. The rotation, scaling and cropping are performed semi-automatically, i.e., the position of the eyes are marked manually and, based on this marking, the face in the image is computationally adjusted to fit into Φ. Before being copied to Φ, the transformed image is filtered by a Gaussian filter to smooth out artifacts produced by the transformations (Figure 3(c)).

During training, the face image I is transformed and filtered, and its pixels are copied to the VG-RAM WNN's input Φ and all $n_{i,j}$ neurons' outputs, $y_{i,j}$, are set to the value of the label associated with the face (an integer). All neurons are then trained to output this label with this input image. During testing, each face image I is also transformed, filtered, and copied to the VG-RAM WNN's

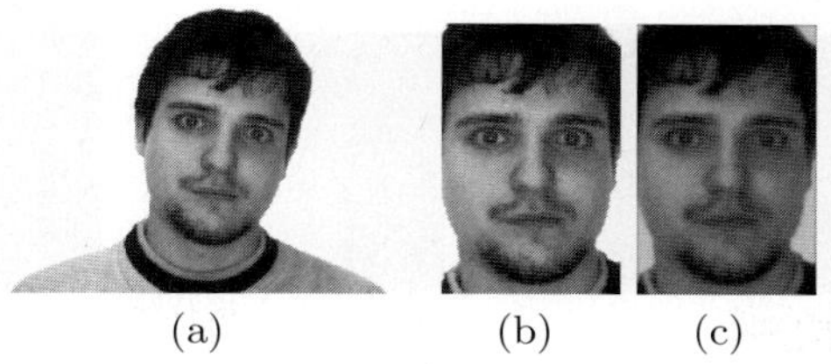

(a)   (b)   (c)

Fig. 3. Face image and its preprocessing. (a) Original image; (b) rotated, scaled and cropped image; and (c) filtered image.

input Φ. After that, all neurons' outputs $y_{i,j}$ are computed and the number of neurons outputting each label is counted. The network's output is the label with the largest count.

3.2   Feature-Based Architecture

As the holistic architecture, the feature-based architecture has a single bidimensional array of $m \times n$ VG-RAM WNN neurons, N, where each neuron, $n_{i,j}$, has a set of synapses, $W = \{w_1, \ldots, w_{|W|}\}$, which are connected to the network's bidimensional input, Φ, of $u \times v$ inputs. The synaptic interconnection pattern of each neuron $n_{i,j}$, $\Omega_{i,j}(W)$, is, however, different (Figure 4). In the feature-based architecture, $\Omega_{i,j}(W)$ follows a bidimensional Normal distribution centered at ϕ_{μ_k,μ_l}, where $\mu_k = \frac{i.u}{m}$ and $\mu_l = \frac{j.v}{n}$, i.e., the coordinates k and l of the elements of Φ to which $n_{i,j}$ connects via W follow the probability density functions:

$$\varphi_{\mu_k,\sigma^2}(k) = \frac{1}{\sigma\sqrt{2\pi}} e^{-\frac{(k-\mu_k)^2}{2\sigma^2}} \tag{1}$$

$$\varphi_{\mu_l,\sigma^2}(l) = \frac{1}{\sigma\sqrt{2\pi}} e^{-\frac{(l-\mu_l)^2}{2\sigma^2}} \tag{2}$$

where σ is a parameter of the architecture. This interconnection pattern mimics that observed in many classes of biological neurons, and is also created when the network is built and does not change afterwards.

A comparison between Figure 2 and Figure 4 illustrates the difference between the interconnection patterns of the holistic and feature-based architectures. In the feature-based architecture (Figure 4), each neuron $n_{i,j}$ monitors a region of the input Φ and, therefore, specializes in the face features that are mapped to that region. On the other hand, each neuron $n_{i,j}$ of the holistic architecture monitors the whole face (Figure 2).

As in the holistic architecture, in the feature-based architecture each neuron's synapse, w_t, forms a minchinton cell with the next, w_{t+1}, and, before training or testing, the input face images, I, are rotated, scaled, cropped, filtered and only then copied to the VG-RAM WNN input Φ. Training and testing are performed the same way as in the holistic architecture.

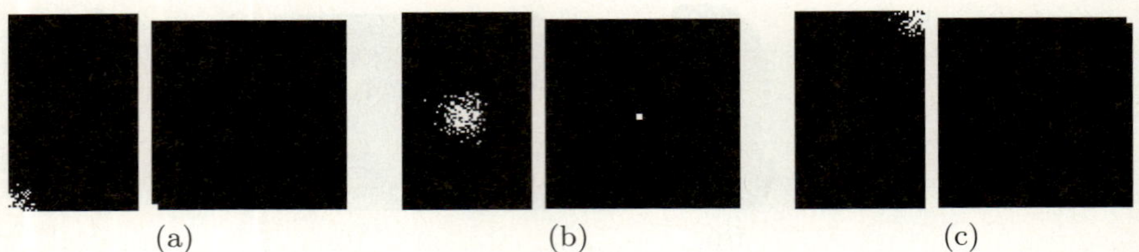

(a)   (b)   (c)

Fig. 4. The synaptic interconnection pattern of the feature-based architecture. (a) Left, input Φ: in white, the elements $\phi_{i,j}$ of the input Φ that are connected to neuron $n_{0,0}$ of N via $w_1, \ldots, w_{|W|}$; right, neuron array N: in white, the neuron $n_{0,0}$ of N. (b) Left: in white, the elements $\phi_{i,j}$ of Φ connected to $n_{\frac{m}{2},\frac{n}{2}}$; right: in white, the neuron $n_{\frac{m}{2},\frac{n}{2}}$ of N. (c) Left: in white, the elements of Φ connected to $n_{m,n}$; right: in white, the neuron $n_{m,n}$.

4 Experimental Evaluation

We used the AR Face Database [13] to evaluate the performance of VG-RAM WNN on face recognition. This face database contains over 4,000 color images corresponding to 135 people's faces (76 men and 59 women). Images feature frontal view faces with different facial expressions, illumination conditions, and occlusions (sun glasses and scarf). The 768×576 pixels pictures were taken under strictly controlled conditions, but no restrictions on wear (clothes, glasses, etc.), make-up, hair style, etc. were imposed to participants.

In order to facilitate the comparison with other results in the literature, we used only the following subset of image types of the AR Face Database [13]: neutral expression, smile, anger, scream, left light on, right light on, all side lights on, wearing sun glasses, and wearing scarf. These can be divided into four groups (see Figure 5): (i) normal (neutral expression); (ii) under expression variation (smile, anger, scream); (iii) under illumination changes (left light on, right light on, all side lights on); and (iv) with occlusion (wearing sun glasses, wearing scarf).

We randomly selected 50 people from the database to tune the parameters of the VG-RAM WNN architectures (25 men and 25 women). We used one normal face image of each person to train (50 images), and the smile, anger, wearing sun

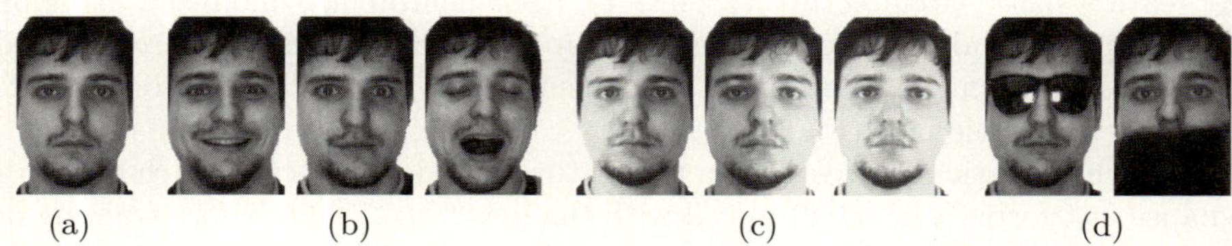

(a)   (b)   (c)   (d)

Fig. 5. The AR face database: (a) normal (neutral expression); (b) under expression variation (smile, anger, scream); (c) under illumination changes (left light on, right light on, all side lights on); and (d) with occlusion (wearing sun glasses, wearing scarf)

glasses, and wearing scarf to evaluate the architectures (200 images) while varying their parameters. In the following subsections, we describe the experiments we performed to tune the parameters of the architectures.

4.1   Holistic Architecture Parameter Tunning

The holistic architecture has three parameters: (i) the number of neurons, $m \times n$; (ii) the number of synapses per neuron, $|W|$; and (iii) the size of the network input, $u \times v$. We tested networks with: $m \times n$ equal to 2×2, 4×4, 16×16, 32×32 and 64×64; number of synapses per neuron equal to 32, 64, 128 and 256; and $u \times v$ equal to 128×200 (we did not vary $u \times v$ to reduce the parameter search space). Figure 6(a) presents the results of the experiments we carried out to tune the parameters of the holistic architecture. As Figure 6(a) shows, the performance, i.e., the percentage of correctly recognized faces (recognition rate) of the holistic architecture grows with the number of neurons and synapses per neuron; however, as these numbers increase, the gains in performance decrease forming a plateau towards the maximum performance. The simplest configuration in the plateau has around 16×16 neurons and 64 synapses.

4.2   Feature-Based Architecture Parameter Tunning

The feature-based architecture has four parameters: (i) the number of neurons; (ii) the number of synapses per neuron; (iii) the size of the network input; and (iv) σ (see Section 3.2). We tested networks with: $m \times n$ equal to 2×2, 4×4, 16×16, 32×32 and 64×64; number of synapses per neuron equal to 32, 64, 128 and 256; $u \times v$ equal to 128×200; and σ equal to 10 (we did not vary $u \times v$ and σ to reduce the parameter search space).

Figure 6(b) presents the results of the experiments we conducted to tune the parameters of the feature-based architecture. As Figure 6(b) shows, the performance of the feature-based architecture also grows with the number of

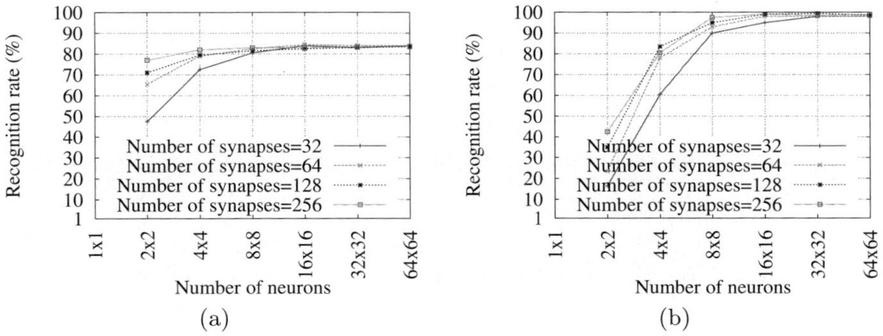

Fig. 6. Performance tunning: (a) holistic architecture and (b) feature-based architecture

neurons and synapses per neuron, and again reaches a plateau at about 32×32 neurons and 128 synapses. However, it is important to note that, in this case, the plateau is very close to 100% accuracy (99.5%).

4.3   Comparison with Other Techniques

We compared the performance of the holistic and feature-based VG-RAM WNN architectures with that of other techniques. For that, we took the best VG-RAM WNN architectures configurations (holistic: 16×16 neurons and 64 synapses per neuron; feature-based: 32×32 neurons and 128 synapses per neuron), trained them with the normal face image of all people in the database (135 images), and tested them with the remaining face images of Figure 5 of all people in the database (135 images of each face image category). Table 1 summarizes this comparison, showing one technique on each line, grouped by type, and the corresponding performance for each face image category on each column.

As Table 1 shows, the VG-RAM WNN holistic architecture (VWH) performed better than all holistic and feature-based techniques examined in all face image categories. It also performed better than the hybrid techniques, except for the categories with occlusion and single side illumination. That was expected, since occlusions and single side illumination compromise eventual similarities between the input patterns learned by the VWH neurons and those collected by its synapses from a partially occluded or illuminated face. Nevertheless, it is

Table 1. The recognition rate on the AR Face Database. PCA: principal component analysis [2] (results obtained from [8]); VWH: VG-RAM WNN holistic architecture; NMF: non-negative matrix factorization [5] (results from [8]); LNMF: local non-negative matrix factorization [6] (results from [8]); LEM: line edge maps [7] (results from [7] with only 112 people of the AR Face Database); VWF: VG-RAM WNN feature-based architecture; WER: weighted eigenspace representation [9] (results from [9] with only 50 people of the AR Face Database); and ARG: attributed relational graph matching [8] (results from [8]).

Type	Technique	Category							
		Smile	Anger	Scream	Glasses	Scarf	Left light	Right light	All side lights
HOL[a]	PCA	94.1%	79.3%	44.4%	32.9%	2.2%	7.4%	7.4%	2.2%
	VWH	**98.5%**	**97.8%**	**91.1%**	**66.7%**	**25.2%**	**97.8%**	**95.6%**	**95.6%**
FBA[b]	NMF	68.1%	50.4%	18.5%	3.7%	0.7%	N/A[d]	N/A	N/A
	LNMF	94.8%	76.3%	44.4%	18.5%	9.6%	N/A	N/A	N/A
	LEM	78.6%	92.9%	31.3%	N/A	N/A	92.9%	91.1%	74.1%
	VWF	**99.3%**	**99.3%**	**93.3%**	**85.2%**	**98.5%**	**99.3%**	**98.5%**	**99.3%**
HYB[c]	WER	84.0%	94.0%	32.0%	80.0%	82.0%	N/A	N/A	N/A
	ARG	97.8%	96.3%	66.7%	80.7%	85.2%	98.5%	96.3%	91.1%

[a]HOL: holistic techniques. [b]FBA: feature-based techniques. [c]HYB: hybrid techniques. [d]N/A: not available.

important to note the overall performance achieved by VWH, which is better than that of several other relevant techniques from literature.

The VG-RAM WNN feature-based architecture (VWF) performed better than all other techniques examined in all categories, in many cases for a significant margin. This performance is the result of two factors. First, each VWF (or VWH) synapse collects the result of a comparison between two pixels, executed by its corresponding minchinton cell. Our best VWF has 128 synapses per neuron and 32×32 neurons. Therefore, during test, 131072 ($128\times32\times32$) such comparisons are executed on an input face image and the results are checked against equivalent results learned from training images. This amount of pixel comparisons allows not only high discrimination capability but also generalization. Second, thanks to the characteristics of the VWF architecture, i.e., its synaptic interconnection pattern, each VWF neuron monitors a specific region of the face only, which reduces the overall impact of occlusions and varying illumination conditions on recognition performance.

5   Conclusions and Future Work

In this work, we presented an experimental evaluation of the performance of Virtual Generalizing Random Access Memory Weightless Neural Networks (VG-RAM WNN [12]) on face recognition. We presented two VG-RAM WNN face recognition architectures, one holistic and the other feature-based, and examined its performance with a well known face database, the AR Face Database. This database is challenging for face recognition systems because it has images with different facial expressions, occlusions, and varying illumination conditions. The best performing architecture (feature-based) showed robustness in all image conditions and better performance than many other techniques from literature, even when trained with a single sample per person.

In future works, we will examine the performance of VG-RAM WNN with other databases and use it to tackle other problems associated with face recognition systems, such as face detection, face alignment, face recognition in video, etc.

Acknowledgements

We would like to thank *Receita Federal do Brasil, Conselho Nacional de Desenvolvimento Científico e Tecnológico* — CNPq-Brasil (grants 308207/2004-1, 471898/2004-0, 620165/2006-5, 309831/2007-5), *Financiadora de Estudos e Projetos* — FINEP-Brasil (grants CT-INFRA-PRO-UFES/2005, CT-INFRA-PRO-UFES/2006), and *Fundação Espírito Santense de Tecnologia* — FAPES-Brasil (grant 37711393/2007) for their support to this research work.

References

1. Zhao, W.Y., Chellappa, R., Phillips, P.J., Rosenfeld, A.: Face recognition: A literature survey. ACM Computing Surveys 35(4), 399–458 (2003)
2. Turk, M., Pentland, A.: Eigenfaces for recognition. Journal of Cognitive Neuroscience 3, 71–86 (1991)

3. Belhumeur, P.N., Hespanha, J.P., Kriegman, D.J.: Eigenfaces vs. fisherfaces: Recognition using class specific linear projection. IEEE Trans. Pattern Anal. Mach. Intell. 19(7), 711–720 (1997)
4. Etemad, K., Chellappa, R.: Discriminant analysis for recognition of human face images. Journal of the Optical Society of America A 14(8), 1724–1733 (1997)
5. Lee, D.D., Seung, S.H.: Learning the parts of objects by non-negative matrix factorization. Nature 401, 788–791 (1999)
6. Li, S.Z., Hou, X.W., Zhang, H., Cheng, Q.: Learning spatially localized, parts-based representation. In: Proceedings of Computer Vision and Pattern Recognition, pp. 207–212 (December 2001)
7. Gao, Y., Leung, M.K.: Face recognition using line edge map. IEEE Transactions Pattern Analysis and Machine Learning 24(6), 764–779 (2002)
8. Park, B.G.: Face recognition using face-arg matching. IEEE Trans. Pattern Anal. Mach. Intell. 27(12), 1982–1988 (2005)
9. Martínez, A.M.: Recognizing imprecisely localized, partially occluded, and expression variant faces from a single sample per class. IEEE Transactions on Pattern Analysis and Machine Intelligence 24(6), 748–763 (2002)
10. Lanitis, A., Taylor, C.J., Cootes, T.F.: Automatic face identification system using flexible appearance models. Image Vision Computing 13(5), 393–401 (1995)
11. Huang, J., Heisele, B., Blanz, V.: Component based face recognition with 3d morphable models. In: Proceedings of the International Conference on Audio and Video-Based Person Authentication, pp. 27–34 (2003)
12. Aleksander, I.: From WISARD to MAGNUS: a Family of Weightless Virtual Neural Machines. In: RAM-Based Neural Networks, pp. 18–30. World Scientific, Singapore (1998)
13. Martinez, A., Benavente, R.: The AR face database. Technical Report #24, Computer Vision Center (CVC), Universitat Autònoma de Barcelona (1998)
14. Aleksander, I.: Self-adaptive universal logic circuits. IEE Electronic Letters 2(8), 231–232 (1966)
15. Ludermir, T.B., Carvalho, A.C.P.L.F., Braga, A.P., Souto, M.D.: Weightless neural models: a review of current and past works. Neural Computing Surveys 2, 41–61 (1999)
16. Mitchell, R.J., Bishop, J.M., Box, S.K., Hawker, J.F.: Comparison of Some Methods for Processing Grey Level Data in Weightless Networks. In: RAM-Based Neural Networks, pp. 61–70. World Scientific, Singapore (1998)

Invariant Object Recognition with Slow Feature Analysis

M. Franzius, N. Wilbert, and L. Wiskott

Institute for Theoretical Biology, Humboldt-Universität zu Berlin, Germany

Abstract. Primates are very good at recognizing objects independently of viewing angle or retinal position and outperform existing computer vision systems by far. But invariant object recognition is only one prerequisite for successful interaction with the environment. An animal also needs to assess an object's position and relative rotational angle. We propose here a model that is able to extract object identity, position, and rotation angles, where each code is independent of all others. We demonstrate the model behavior on complex three-dimensional objects under translation and in-depth rotation on homogeneous backgrounds. A similar model has previously been shown to extract hippocampal spatial codes from quasi-natural videos. The rigorous mathematical analysis of this earlier application carries over to the scenario of invariant object recognition.

1   Introduction

Sensory signals convey information about the world surrounding an animal. However, a visual signal can change dramatically even when only a single object is slightly moved or rotated. The visual signal from the retina, for example, varies strongly when distance, position, viewing angle, or lighting conditions change. A high-level representation of object identity in the brain should, however, remain constant or *invariant* under these different conditions. How could the brain extract this abstract information from the highly variable stimuli it perceives? How can this task be learned from the statistics of the visual stimuli in an unsupervised way?

The primate visual system is organized hierarchically with increasing receptive field sizes, increasing stimulus specificity and invariance towards to top of the (ventral) hierarchy, where many neurons are coding for objects with an extreme amount of invariance to position, angle, scale etc.

Many models of invariant object recognition have been proposed in the last decades [1]. However, many approaches fail or have not been shown to work for natural stimuli and complex transformations like in-depth rotations. But invariant object recognition is only one task the (primate) brain has to achieve in order to successfully interact with the environment. We do not only need to extract the identity of an object ("what is seen?") independently of its position and view direction, we also want to extract the position of an object ("where is it?") independently of its identity or viewing angle. The relative rotational

V. Kůrková et al. (Eds.): ICANN 2008, Part I, LNCS 5163, pp. 961–970, 2008.

angle of a viewed object can be just as crucial ("does the tiger look at me?"). In principle, we might want a representation of any aspect (i.e., size, viewing angle, lighting direction etc.) independently of all the others and optimally, all these tasks should be solved with a single computational principle. In the following we will refer to the configuration of position and angles relative to the viewer as the *configuration* of an object (sometimes also called a pose). We will call a 2D image of an object in a specific configuration a *view*. In general, the process of extracting the configuration of an object from a view is very hard to solve, especially in the presence of a cluttered background and many different possible objects. In this work we use high-dimensional views of complex objects but restrict the problem to the cases of only one object present at any time point and a static homogeneous background. A good model should also *generalize* to previously unseen configurations, i.e., it should learn the relevant statistics of transformations rather than just memorizing specific views. It should also generalize to new objects, e.g., it should successfully estimate the position and orientation[1] of an object that was never shown before.

We apply here a hierarchical model based on the learning principle of *temporal slowness*. The structure of the resulting representation solely depends on the statistics of presentation of the training views. This information is encoded in an analytically predictable way and very simple to decode by linear regression. Except for minor changes, the model used here is identical to that used earlier for the modeling of place cells and head direction cells in the hippocampal formation from quasi-natural videos [2].

2   Methods

2.1   Stimulus Generation

The model was trained and tested with image sequences containing views of different objects. OpenGL was used to render the views of the objects as textured 3D-models in front of a white homogenous background. To prevent the model from relying on simple color cues (especially for object classification) we only used grayscale views for the results presented here. Two different object classes were used that are described below. For each object class the model was trained with five objects. In the testing phase we added five new objects, which the model had never seen during training.

In the first experiment the objects were clusters of textured spheres as shown in Figure 1 (a), which provide a difficult but generic task. The same six textured spheres (textures from VisTex database [3]) were used in different spatial arrangements for all the objects. For each object the spheres were randomly fixed on a hexagonal lattice of size $2 \times 2 \times 3$. As the examples in Figure 1 (a) illustrate, identifying the rotation angles and identities for such objects is quite difficult even for human observers. Using spheres has the advantage that the outline does

---

[1] Generally, there is no canonical "0°"-view of an object, thus a random phase offset of absolute phase for a new object is to be expected.

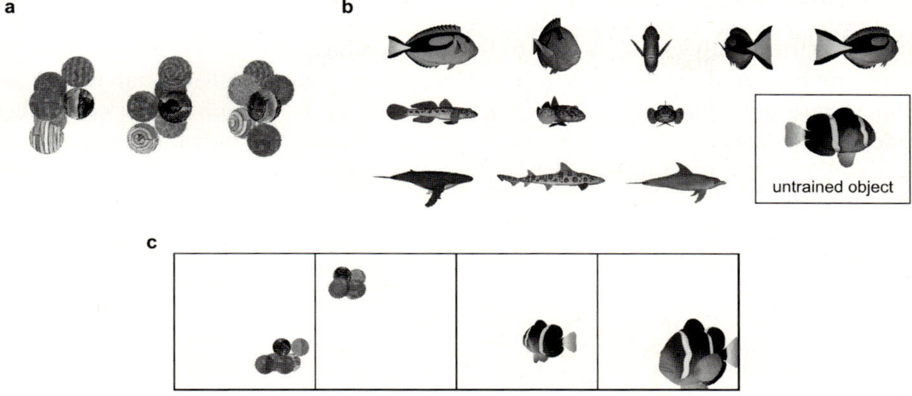

Fig. 1. Objects used for stimulus generation. Part (a) shows the sphere objects (each cluster of 6 spheres is one object). The first two views show the same object under different in-depth rotation angles, while the third view shows a different object. In part (b) the five fish objects used for training are shown with examples for the effect of in-depth rotation. The fish model on the bottom right is one of the five untrained fish used for testing. Part (c) shows some examples for the training and testing images.

not give a simple clue for the in-plane rotation angle. In the second experiment, models of different fish (see Figure 1 (b)) were used to provide more natural stimuli from a single object class (all models taken from [4]).

For sphere objects, the x-coordinate, the y-coordinate (vertical image coordinate), the in-depth rotation angle ϕ_y and the in-plane rotation angle ϕ_z were chosen as configuration variables. x and y range from 0 to 1, ϕ_y and ϕ_z from 0° to 360°. Another configuration variable was the object identity ranging from one to ten, with objects one to five being the training objects. So the transformations the model had to learn consisted of translations in the plane, rotations along the y and z axes (with the in-depth rotation coming first) and changes of object identity. For the fish objects, the configuration variables were x, y, z, ϕ_y and object identity. So compared to the sphere objects we added translations in depth along the z-axis and removed the in-plane rotations. A pure z-translation changes both the object size and the position in the frame, due to the perspective projection.

The configurations for the training sequences were generated as a random walk procedure like in [2]. To generate a configuration in the sequence we add a random term to the current spatial and angular velocities of the currently shown object. By adjusting the magnitude of the velocity updates one can effectively choose the timescales of the changes which are relevant for SFA. The position and angles are then updated according to these velocities. In each step the object identity was changed with low probability ($p = 0.001$). A blank frame was inserted in between if a switch took place. This procedure adds some realism, because in natural scenes a change in object identity without any changes in other configuration variables generally does not occur.

2.2 Slow Feature Analysis

Slow Feature Analysis solves the following learning task: Given a multidimensional input signal we want to find instantaneous scalar input-output functions that generate output signals that vary as slowly as possible but still carry significant information. To ensure the latter we require the output signals to be uncorrelated and have unit variance. In mathematical terms, this can be stated as follows:

Definition 1 (Optimization problem). *Given a function space $\mathcal{F}$ and an I-dimensional input signal $\mathbf{x}(t)$ find a set of J real-valued input-output functions $g_j(\mathbf{x}) \in \mathcal{F}$ such that the output signals $y_j(t) := g_j(\mathbf{x}(t))$*

$$minimize\ \Delta(y_j) := \langle \dot{y}_j^2 \rangle_t \tag{1}$$

under the constraints

$$\langle y_j \rangle_t = 0 \quad \textit{(zero mean)}, \tag{2}$$

$$\langle y_j^2 \rangle_t = 1 \quad \textit{(unit variance)}, \tag{3}$$

$$\forall i < j : \langle y_i y_j \rangle_t = 0 \quad \textit{(decorrelation and order)}, \tag{4}$$

with $\langle \cdot \rangle_t$ and $\dot{y}$ indicating temporal averaging and the derivative of y, respectively.

Equation (1) introduces the Δ-value, which is a measure of the temporal slowness of the signal $y(t)$. It is given by the mean square of the signal's temporal derivative, so that small Δ-values indicate slowly varying signals. The constraints (2) and (3) avoid the trivial constant solution and constraint (4) ensures that different functions g_j code for different aspects of the input.

It is important to note that although the objective is slowness, the functions g_j are instantaneous functions of the input, so that slowness cannot be enforced by low-pass filtering. Slow output signals can only be obtained if the input signal contains slowly varying features that can be extracted instantaneously by the functions g_j. Note also that for the same reason, once trained, the system works fast, not slowly.

In the computationally relevant case where $\mathcal{F}$ is finite-dimensional the solution to the optimization problem can be found by means of Slow Feature Analysis [5,6]. This algorithm, which is based on an eigenvector approach, is guaranteed to find the global optimum. Biologically more plausible learning rules for the optimization problem, both for graded response and spiking units exist [7,8]. If the function space is infinite-dimensional, the problem requires variational calculus and will in general be difficult to solve. In [2] it has been shown that the optimization problem for the high-dimensional visual input can be reformulated for the low-dimensional configuration input. This provides very useful predictions for the SFA-output of our model (even though it is only based on the finite-dimensional SFA algorithm). We use these predictions for the postprocessing of our model output (see 2.4). For the detailed predictions and their derivations see [9] or [2].

2.3   Network Architecture

The computational model consists of a converging hierarchy of layers of SFA nodes (see Figure 2). Each SFA node finds the slowest output features from its input according to the SFA algorithm and performs the following sequence of operations: linear SFA for dimensionality reduction, quadratic expansion with subsequent additive Gaussian white noise (with a variance of 0.05), another linear SFA step for slow-feature extraction, and clipping of extreme values at ± 64 (Figure 2). Effectively, a node implements a subset of full quadratic SFA. The clipping removes extreme values that can occur on test data very different from training data.

In the following, the part of the input image that influences a node's output will be denoted as its receptive field. On the lowest layer, the receptive field of each node consists of an image patch of 10 by 10 grayscale pixels that jointly cover the input image of 128 by 128 pixels. The nodes form a regular (i.e., non-foveated) 24 by 24 grid with partially overlapping receptive fields. The second layer contains 11 by 11 nodes, each receiving input from 4 by 4 layer 1 nodes with neighboring receptive fields, resembling a retinotopical layout. The third layer contains 4 by 4 nodes, each receiving input from 5 by 5 layer 2 nodes with neighboring receptive fields, again in a retinotopical layout. All 4 by 4 layer 3 outputs converge onto a single node in layer 4, whose output we call SFA-output. The output of each node consists of the 32 slowest outputs (called units), except for the top layer where 512 dimensions are used. Thus, the hierarchical organization of the model captures two important aspects of cortical visual processing: increasing receptive field sizes and accumulating computational power at higher layers.

The layers are trained sequentially from bottom to top (with slightly faster varying views for the top layers). For computational efficiency, we train only one node with stimuli from all node locations in its layer and replicate this node throughout the layer. This mechanism effectively implements a weight-sharing constraint. However, the system performance does not critically depend on this mechanism. To the contrary, individually learned nodes improve the overall performance.

For our simulations, we use 100,000 time points for the training of each layer. Since training time of the entire model on a single PC is on the order of multiple days, the implementation is parallelized and training times thus reduced to hours. The simulated views are generated from its configuration (position, angles, and object identity). The network is implemented in Python using the MDP toolbox [10], and the code is available upon request.

2.4   Feature Extraction with Linear Regression

The function space available to the model is large but limited (a subset of polynomials of degree $2^4 = 16$), which will generally lead to deviations from the theoretical predictions. This causes linear mixing of features in the SFA-output which is practically impossible to avoid in complex applications. Therefore we extracted the individual configuration features in a separate post-processing step.

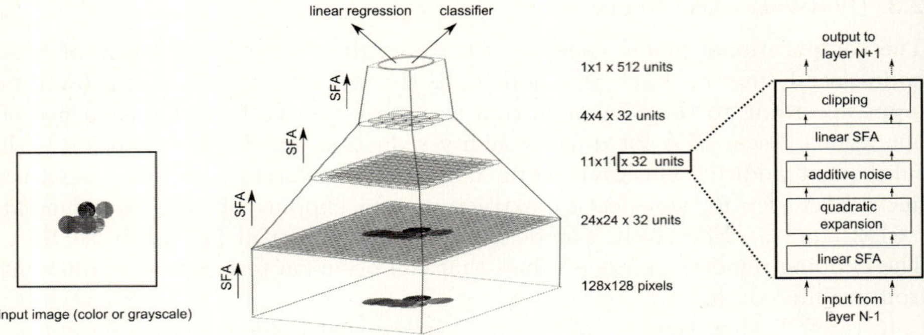

Fig. 2. Model architecture and stimuli. An input image is fed into the hierarchical network. The circles in each layer denote the overlapping receptive fields for the SFA-nodes and converge towards the top layer. The same set of steps is applied on each layer, which is visualized on the right hand side.

Our main objective here was to show that the relevant features were indeed extracted from the raw image data. The easiest way to do this is by calculating a multivariate linear regression of the SFA-output against the known configuration values. While the regression procedure is obviously supervised, it nevertheless shows that the relevant signals are easily accessible. Extracting this information from the raw image data linearly is not possible (especially for the angles and object identity). One should note, that the dimension of the model output is smaller than the raw image data by two orders of magnitude. We also checked that the nonlinear expansions without SFA reduces the regression performance to almost chance level.

Since the predicted SFA solutions are cosine functions of the position values (see [9]), one has to map the reference values correspondingly before calculating the regression. The results from the regression are then mapped with the inverse function. For those SFA solutions that code for rotational angles, the theory predicts both sine and cosine functions of the angle value. Therefore we did regressions for both mappings and then calculated the angles via the arctangent. If multiple objects are trained and separated with a blank (see 2.1), the solutions will in general be object dependent. This rules out global regressions for all objects. The easiest way around this is to perform individual regressions for all objects. While this procedure sacrifices the object invariance it does not affect the invariance under all other transformations.

The object identity of N different objects is optimally encoded (under the SFA objective) by up to $N - 1$ uncorrelated step functions, which are invariant under all other transformations. In the SFA-output this should lead to separated clusters for the trained objects. For untrained objects, those SFA-outputs coding for object identity should take new values, thereby separating all objects. To explore the potential classification ability of the model we applied two very simple classifiers on the SFA-output: a k-nearest-neighbour and a Gaussian classifier.

3   Results

The stimulus set used for testing consisted of 100,000 views (about 10,000 per object), which were generated in the same way as the training data. Half of this data was used to calculate the regressions or to train the classifiers, the other half for testing[2].

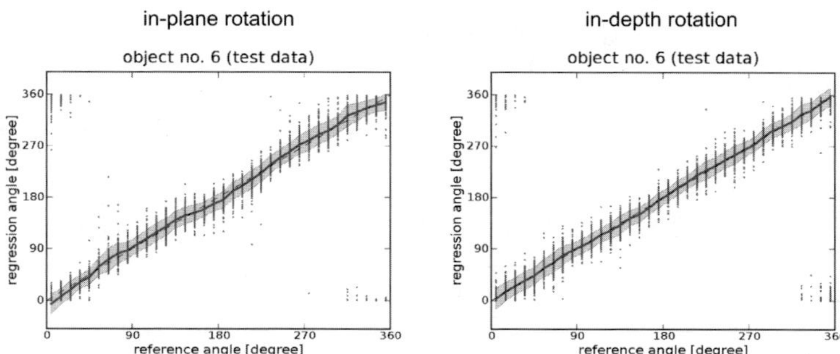

Fig. 3. Regression results for position and angle. For the first untrained sphere object the feature values that were calculated with linear regression are plotted against the correct reference values. The green dots are data points, the red line is the mean and the gray area shows $\pm$ one standard deviation around the mean value. The regression of the x coordinate was based on all five training objects.

Position and Rotational Angles. To extract the x and y object coordinates from the model SFA-output we used multivariate linear regression, as described in Section 2.4. As one can see in Figure 3 and Table 1, this works well for the sphere objects, with a standard deviation of 5% for trained objects and 7% for untrained objects. For the trained fish we achieve the same performance as for the trained sphere objects. For untrained fish the additional size variations from the z-transformations take their toll and pull the performance down to 14% for the y-coordinate.

To extract the in-plane and in-depth rotation angles for the spheres, an individual regression for each object was calculated. This still requires invariance of the representations under the other transformations (including the other rotation type). As the results in Table 2 show, both rotational angles were extracted with about $15°$ standard deviation.

For the fish, the standard deviations are about twice as large as for the spheres, mostly due to systematic errors. The fish models have a very similar front and back view and therefore the model has difficulties to differentiate between those two views. When taking the mean absolute error instead (see Figure 2), the performance gap to the spheres is smaller because of the reduced influence of outliers. As for the angles, individual regressions for all objects were used to calculate the z coordinate for the fish objects (z-transformations were not used for the sphere objects).

---

[2] The detailed results are available at www.nikowilbert.de/supplements.

Table 1. Standard deviations for the coordinate regressions. The values are given in percent relative to the coordinate range. The chance level is 28.9%. For example, the x-coordinate of trained sphere objects on test data was reconstructed with an RMSE of 5%.

	Spheres		Fish	
	trained	new	trained	new
x	5%	7%	5%	12%
y	5%	7%	5%	14%

Table 2. Mean standard deviations for the angles and the z coordinate over all ten objects. For the fish we give in ϕ_y^* the mean absolute error for ϕ_y, since a large part of the error is systematic due to the 180° pseudo-symmetry. The chance level is 104°.

Spheres ϕ_z	ϕ_y	Fish ϕ_y	ϕ_y	z
14.8°	15.8°	36.4°	23.4°	0.11%

Classification. To quantify the classification ability of the model, two classifiers were used on the SFA-output. The classifiers were trained with about 5000 data points per object (i.e., half of the data, as for the regressions). A random fraction of the remaining data points (about 150 per object) was then used to test the classifier performance. The k-nearest-neighbour classifier generally performed with about 96% hit rate (see Table 3). As expected, the Gaussian classifier performed not as well, with a hit rate between 88% and 96%.

Table 3. Classifier hit rates in percent. The columns labeled with "KNN" refer to the k-nearest-neighbour classifier ($k = 8$), while those with "G" refer to the Gaussian classifier. Chance level is 10% for all objects and 20% on the sets of training or test objects.

	trained objects		test objects		all objects	
	KNN	G	KNN	G	KNN	G
spheres	96.4	88.3	96.6	95.7	95.0	89.2
fish	97.7	94.6	99.2	88.1	96.7	88.8

4   Discussion

In the previous section we have shown how the hierarchical model learns independent representations of object position, angle, and identity from quasi-natural stimuli. The simultaneous extraction of all these different features is one novelty of our model. This capability is made possible by the invariance of one extracted feature with respect to the others (only limited with respect to object identity).

Related Work. The slowness principle has been applied in many models of the visual system in the past [11,12,13,5,6,2]. VisNet is the most prominent example of hierarchical feed-forward neural network models of the primate ventral visual system [1]. Weights in this model are adapted according to the trace rule [11,14,15], which is closely related to Slow Feature Analysis [8]. Like our model, VisNet can learn a position-invariant (or view-invariant) but object-specific code. In contrast to our model, only a single invariant representation is extracted (i.e., object identity) and the other parameters (e.g., object position, rotation angles, lighting direction) are discarded. The ability of our model to code for more than object identity in a structured way is therefore the greatest functional difference.

Conclusion and Outlook. The model proposed here learns invariant object representations based on the statistics of the presented stimuli during the training phase. Specifically, representations of transformations that occur slowly or seldom are formed, while information on other transformations that occur most often or quickly are discarded. An advantage of such a system is that no invariances have to be "hard-wired". Instead we use only generic SFA units on all network layers. The unsupervised learning of the system can be described by a comprehensive mathematical analysis that predicts specific output signal components. However, if many transformations occur simultaneously and on similar timescales, solutions tend to mix. In this case a final step of linear regression or a simple classifier reconstructs the relevant transformations.

We show that the system generalizes well to previously unseen configurations (i.e., shows no overfitting for test positions and viewing angles of objects) and to previously unseen objects. However, the system behavior for completely different configurations, like for two simultaneously presented objects, cannot be predicted with our current theory. The performance of the network for cluttered background also remains to be tested.

References

1. Rolls, E.T., Deco, G.: The Computational Neuroscience of Vision. Oxford University Press, New York (2002)
2. Franzius, M., Sprekeler, H., Wiskott, L.: Slowness and sparseness lead to place, head-diretion and spatial-view cells. Public Library of Science (PLoS) Computational Biology 3(8), 166 (2007)
3. Picard, R., Graczyk, C., Mann, S., Wachman, J., Picard, L., Campbell, L.: Vision texture (2002),
   `http://vismod.media.mit.edu/vismod/imagery/VisionTexture/vistex.html`
4. Toucan Corporation: Toucan virtual museum (2005),
   `http://toucan.web.infoseek.co.jp/3DCG/3ds/FishModelsE.html`
5. Wiskott, L., Sejnowski, T.: Slow feature analysis: Unsupervised learning of invariances. Neural Computation 14(4), 715–770 (2002)
6. Berkes, P., Wiskott, L.: Slow feature analysis yields a rich repertoire of complex cell properties. Journal of Vision 5(6), 579–602 (2005),
   `http://journalofvision.org/5/6/9/`

7. Hashimoto, W.: Quadratic forms in natural images. Network: Computation in Neural Systems 14(4), 765–788 (2003)
8. Sprekeler, H., Michaelis, C., Wiskott, L.: Slowness: An objective for spike-timing-plasticity? PLoS Computational Biology 3(6), 112 (2007)
9. Wiskott, L.: Slow feature analysis: A theoretical analysis of optimal free responses. Neural Computation 15(9), 2147–2177 (2003)
10. Berkes, P., Zito, T.: Modular toolkit for data processing (version 2.0) (2005), `http://mdp-toolkit.sourceforge.net`
11. Földiák, P.: Learning invariance from transformation sequences. Neural Computation 3, 194–200 (1991)
12. Stone, J.V., Bray, A.: A learning rule for extracting spatio-temporal invariances. Network: Computation in Neural Systems 6, 429–436 (1995)
13. Kayser, C., Einhäuser, W., Dümmer, O., König, P., Körding, K.: Extracting slow subspaces from natural videos leads to complex cells. In: Dorffner, G., Bischof, H., Hornik, K. (eds.) ICANN 2001. LNCS, vol. 2130, pp. 1075–1080. Springer, Heidelberg (2001)
14. Rolls, E.T.: Neurophysiological mechanisms underlying face processing within and beyond the temporal cortical visual areas. Philosophical Transactions of the Royal Society 335, 11–21 (1992)
15. Wallis, G., Rolls, E.T.: Invariant face and object recognition in the visual system. Progress in Neurobiology 51(2), 167–194 (1997)

Analysis-by-Synthesis by Learning to Invert Generative Black Boxes

Vinod Nair, Josh Susskind, and Geoffrey E. Hinton

Department of Computer Science
University of Toronto, Toronto, Ontario, Canada

Abstract. For learning meaningful representations of data, a rich source of prior knowledge may come in the form of a generative black box, e.g. a graphics program that generates realistic facial images. We consider the problem of learning the *inverse* of a given generative model from data. The problem is non-trivial because it is difficult to create labelled training cases by hand, and the generative mapping is a black box in the sense that there is no analytic expression for its gradient. We describe a way of training a feedforward neural network that starts with just one labelled training example and uses the generative black box to "breed" more training data. As learning proceeds, the training set evolves and the labels that the network assigns to unlabelled training data converge to their correct values. We demonstrate our approach by learning to invert a generative model of eyes and an active appearance model of faces.

1 Introduction

"Analysis-by-synthesis" is the idea of explaining an observed data vector (e.g. an image) in terms of a compact set of hidden causes that generated it. A *generative model* specifies how the underlying causes produce the data vector. A *recognition model* is the inverse mapping – it infers the causes from a given data vector. In coding terms, the recognition and generative models are the encoder and decoder, respectively, and the hidden causes represent a *code vector*. The composite of the two models implements the identity function: recognition of a data vector followed by generation from the resulting code vector should reconstruct the original data vector.

In this paper we consider the following problem: given a training set of data vectors and a generative model for that data, learn the corresponding recognition model. For example, suppose that we have a face dataset and a graphics program that can generate realistic images of faces. This program may have a set of real-valued inputs (e.g. pose, lighting, facial muscle activations) that can be smoothly varied to create any face. The task is to learn a recognition model that infers from a face image the graphics inputs that will accurately reconstruct that image. The inputs to the generative model can be seen as a compact, high-level, generative representation of the data vector.

Note that this is a *new type of problem* that existing learning algorithms are not designed to solve. Here we assume that an arbitrary generative model of the data is given as part of the problem, and the goal is to learn the inverse of

V. Kůrková et al. (Eds.): ICANN 2008, Part I, LNCS 5163, pp. 971–981, 2008.

that particular model. In contrast, algorithms such as PCA, factor analysis, and ICA simply assume specific parametric forms for the generative model and fit the parameters to the data. A nonlinear autoencoder learns separate generative and recognition models simultaneously by also assuming the generative model to be of a specific form with an analytic gradient. As explained later, our problem is more difficult than the ones solved by these standard methods. Our main contribution here is a learning algorithm that solves the above problem.

There are two main motivations for this work. First, it provides a general way of incorporating domain knowledge, via a generative model, into the learning of a data representation. Generative models are a natural way of expressing complex prior knowledge. For example, in speech, knowledge about how the vocal system produces sounds can be expressed as an articulatory synthesizer. In modelling face images, knowledge about facial muscles and skin can be expressed as a physics-based graphics model. The knowledge contained in such models gives their inputs (i.e. the code vector) high-level semantics, so inverting these models creates useful high-level representations of the data. Such representations can capture the true degrees of freedom in the data much better than those learned by generic algorithms like PCA or ICA.

Second, our work allows the automatic transfer of research advances in generative modelling to the learning of compact data representations. Progress in graphics and speech synthesis can be used directly to learn better representations of images and speech data. For example, the graphics community is rapidly advancing the realism and detail of physically-based graphics models for faces [1], [2]. Successfully inverting them will result in new facial image representations that can significantly improve applications such as face recognition, facial image coding and expression analysis.

To make our learning algorithm as general as possible, we do not assume any knowledge of the internal details of the generative model. It is a "black box" function that can be evaluated as many times as we like, but no additional information about it – in particular its gradient – is given. So the learning of the recognition model is de-coupled from the internal details of the generative model. Thus, any of the wide variety of generative models proposed in the literature for different types of data can be used without changing the learning algorithm. The alternative to not using the black box assumption is to design a separate learning algorithm for each inversion problem, which is undesirable.

Overview of our approach: Learning to invert a nonlinear[1], deterministic generative black box is difficult. Assuming real-valued code and data vectors, we want to learn a regression mapping from data space to code space. There are three problems: first, only the regression function's inputs (data vectors) are given for learning – the corresponding target outputs are unknown. If they were known, then the problem reduces to standard supervised learning. For a complex generative model, inferring a code vector that reconstructs a given data vector is nontrivial.

---

[1] The linear case is easy: elements of the generative matrix can be discovered trivially by evaluating it on the standard basis vectors.

Second, the black box's gradient is unknown, so learning cannot be done by propagating the gradient of the data reconstruction error through the black box. Estimating it numerically by finite differencing is too inefficient. If the black box gradient were known, then learning becomes similar to that of an autoencoder, except the generative model is fixed.

Third, codes corresponding to the real data occupy only a small volume in code space. For example, consider a face model that simulates muscles with springs. Humans cannot independently control each facial muscle, so the muscle activations are dependent on each other. Therefore only a small subspace of possible spring states correspond to valid facial configurations. This makes it impractical to naively take *random* samples from code space, generate data vectors from those samples using the black box, and learn the recognition model from the resulting input-output pairs. Such an approach will waste almost all the capacity of the recognition model on "junk" training cases far away from the real data that we are interested in modelling.

Our approach solves all three problems. The basic idea is to use the black box itself to generate synthetic data vectors, but from codes sampled in a more intelligent way than random sampling. To restrict the learning to the small portion of code space that is actually relevant, a single code vector corresponding to a real training data vector is assumed to be given. Starting from this single labelled example, the algorithm computes a set of nearby code vectors from which the synthetic data vectors are generated. The recognition model is trained by standard supervised learning on the resulting input-output pairs. As learning proceeds, the sampling procedure produces code vectors from an increasingly broader distribution. The details are in section 2. As the algorithm "breeds" its own labelled training cases, we refer to it as *breeder learning*.

We use breeder learning to invert two different generative black boxes, one for images of eyes (section 3.1) and the other for faces (section 3.2). In the former case, we got the graphics program from its authors [3] and simply used it as a subroutine in our algorithm without looking at its internal implementation details. This shows the generality of the algorithm, as well as the usefulness of de-coupling the design details of the generative model from the learning.

2   Breeder Learning

Figure 1 summarizes the learning algorithm. The algorithm requires a single code vector, chosen by hand, that produces a data vector close to the ones in the training set. We call this code vector the *prototype*. The inputs to the algorithm are the prototype, the generative black box, and the training set X of data vectors. The final output is a recognition model trained to accurately infer from a data vector its corresponding code vector. We choose a feedforward neural network with a single hidden layer of sigmoid units to implement the recognition model. An important advantage of neural networks is that, unlike batch learning methods such as support vector machines, they can be trained online on a dynamically generated, ever-changing dataset, as is the case here.

Algorithm for training a recognition network R_w parameterized by weight vector w:

Given: Training set X of n data vectors $\{x_1, x_2, ..., x_n\}$, generative black box G, prototype code vector p.

Initialization: Set output biases of R_w using p, and the remaining weights to samples from a zero-mean Gaussian with a small standard deviation.

Weight update computed using the ith (unlabelled) training case x_i:
Let y_i be the code vector inferred from x_i using the current recognition network R_w.
1. $y_i = R_w(x_i)$.
2. Perturb y_i randomly to create y_i'. (Note: The exact perturbation method is specified later.)
3. $x_i' = G(y_i')$.
4. Supervised learning on (x_i', y_i'):
   (a) $y_i'' = R_w(x_i')$.
   (b) $E = \|y_i' - y_i''\|^2$.
   (c) $w \leftarrow w - \eta \frac{\partial E}{\partial w}$.

Fig. 1. Summary of the breeder learning algorithm

The biases of the recognition network's output units are initialized in such a way that the network outputs the prototype in the absence of any input. The remaining weights of the neural network are initialized to small random values, which prevents the input from having much effect on the output initially. So, early on in training, the network (mostly) ignores its input and outputs codes in the vicinity of the prototype.

Once initialized, the weights are updated iteratively. Each iteration has four major steps: 1) The "real" data vectors in X are propagated through the current recognition network to infer their corresponding code vectors. 2) These code vectors are then perturbed with noise. (The exact perturbation method depends on whether the output units are linear or sigmoid. It will be specified later when the applications are discussed.) 3) The generative model is applied to the perturbed codes to produce their corresponding synthetic data vectors. The noisy codes and the data vectors generated from them form a correct set of training pairs for the network. 4) The weights are updated by the negative gradient of the squared error loss (in code space) calculated for the synthetic pairs.

Because of how the network is initialized, it first learns to invert the generative model in a small neighbourhood around the prototype. The early noisy codes will be minor variants of the prototype, so the synthetic training pairs will not be very diverse. At this point the network can correctly infer the codes for only a small subset of X, i.e., those that are near the prototype's data vector. Randomly perturbing the outputs allows the network to discover codes slightly farther away from the prototype. Training on them expands the region in code space that the network can correctly handle.

In subsequent iterations, the network will correctly infer the codes for a few more real data vectors. Perturbing their codes generates new ones that are even farther away from the prototype. As learning progresses, the synthetic training pairs become increasingly diverse, as the codes come from a larger region in code space. The network eventually learns to handle the entire region of code space corresponding to the real data vectors.

Some notes about the algorithm:

• The underlying assumption is that Euclidean distance in code space is a more semantically meaningful way of assessing similarity than any generic distance metric in data space. Therefore small random perturbations in code space should produce semantically similar data vectors that may nevertheless have a large Euclidean distance between them.

• The algorithm is not doing a naive random search in code space. Instead, it uses the current recognition network itself to produce new codes to learn on. So the network's ability to correctly generalize to previously unseen data vectors allows the algorithm to discover codes that correspond to real data vectors much more efficiently than a random search.

• The training set of code-data pairs is generated on the fly, and interestingly, it depends on the trained model itself. It starts off as a mostly homogeneous set and becomes more diverse as learning progresses.

• There is no attempt to filter the synthetic code-data pairs by removing those data vectors that are highly dissimilar from the real data vectors. Generic similarity metrics in data space can be highly misleading, and such filtering is likely to make the learning worse. Without filtering the network will occasionally learn on "junk" training cases. But such cases are unlikely to be exactly re-generated many times, so they are quickly forgotten.

• The use of a single prototype code vector amounts to assuming that there are no well-separated modes in code space corresponding to real data. If this assumption is false, it is possible to formulate a mixture version of the algorithm that can handle far-apart modes. It would require creating one prototype per mode.

Random-code learning: A simpler alternative to breeder learning is to 1) sample an isotropic Gaussian centred on the prototype code vector and with a fixed variance, 2) generate synthetic data vectors from these samples, and 3) train the recognition network on the resulting pairs. In our experiments (section 3) this alternative consistently performs worse than breeder learning. If the Gaussian's variance is too large, many of the sampled codes will correspond to junk data vectors. If it is too small, it will almost never see valid codes that happen to be far away from the prototype. So the particular way in which breeder learning creates new codes is crucial for its success and cannot be replaced by a naive random search.

3   Results

The rest of the paper describes two applications of breeder learning: inverting a generative model for eye images [3], and inverting an active appearance model for faces [4]. In both cases we learn a generative representation for real images by taking a generative model from the literature and simply "plugging it in" as the black box into breeder learning. These applications show 1) the generality of our algorithm, and 2) its usefulness for exploiting an existing generative model to learn a compact, high-level representation of the data.

3.1   Inverting a 2D Model of Eye Images

The generative black box is a 2D model of eye images proposed by Moriyama et al. [3]. They use knowledge about the eye's anatomy to define a high-level generative representation of eye images. Since breeder learning does not need to know the model's internal details, we explain them only briefly here. See [3] for a full description.

The inputs to the black box are parameters that specify high-level properties of the eye, such as gaze direction and how open the eyelid is. The model uses the inputs to first compute a set of polygonal curves that represent the 2D shape of the sclera, iris, upper eyelid, lower eyelid, and the corners of the eye. Once the shape is computed, we use a simple texture model to generate a 32×64 grayscale image from it (see examples in the even columns of figure 3). In total there are eight inputs to the black box, all normalized to be in the range $[0, 1]$. (These inputs affect only the shape; the texture model is fixed.) Given this black box and a training set of real eye images, we use breeder learning to learn the corresponding recognition model.

Dataset and training details: We use 1272 eye images collected from faces of people acting out different expressions. We normalize all images to be 32×64, and apply histogram equalization to remove lighting variations. See the odd-numbered columns of figure 3 for example images. Since the eye images come from faces with many different expressions and ethnicities, they contain a wide variety of shapes and represent a difficult shape modelling task. We select the prototype code to be the vector with all components set to 0.5, which is the midpoint of each code dimension's range of possible values. From the set of 1272 images, 872 are used for training, 200 for validation and 200 for testing.

The recognition network has 2048 input units ($32 \times 64 = 2048$ pixels), 100 sigmoid units in the hidden layer, and 8 sigmoid units in the output layer. A code vector is randomly perturbed during learning by adding zero-mean Gaussian noise with a standard deviation of 0.25 to the total input of each code unit. Training is stopped when the root mean squared error (RMSE) of the validation images is minimized. The recognition network trained by breeder learning achieves its best performance on the validation set after about 1900 epochs.

Figure 2 shows the RMSE achieved by breeder learning on the validation set as training proceeds. Random-code learning is unable to improve the RMSE beyond a certain value and starts overfitting because it only sees training cases from a limited region around the prototype. We tried various values for the variance of random-code learning, and the results shown are for the one that gave the best performance on the validation set.

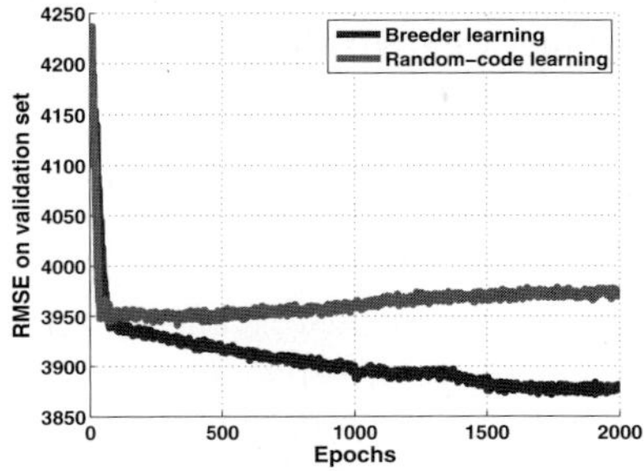

Algorithm	Test RMSE
Breeder learning	3902.56
Random-code learning	3977.37

Fig. 2. Validation set RMSE (left graph) during training, and test set RMSE (above table) after training, for breeder and random-code learning algorithms on the eye dataset

Figure 3 shows examples of test images reconstructed by the recognition network trained with breeder learning. The inferred boundaries of the sclera and iris regions are superimposed on the real image. Notice that the network is able to correctly infer the codes for eyes with significantly different shapes. This is despite using a very limited texture model in the black box.

3.2 Inverting an Active Appearance Model of Faces

We now consider inverting an active appearance model (AAM) of face images [4]. The AAM is a popular nonlinear generative model that incorporates high-level knowledge about facial shape and texture to learn a low-dimensional representation of faces. Unlike the eye model, here the black box itself is learned from data, but this difference is irrelevant from the point of view of breeder learning.

Our implementation of the AAM follows Cootes et al. [4]. Again, we only give a brief overview of it here. It consists of separate PCA models for facial shape and texture, whose outputs are combined via a nonlinear warp to generate the face image. As in [4], we apply PCA again to the shape and texture representations of the training images to produce a higher-level "appearance" model of faces. The warp makes the AAM's output a nonlinear function of its input.

We first train the AAM using a set of face images, and then use it as a *fixed* generative black box for breeder learning. The face images are of size 30×30, and

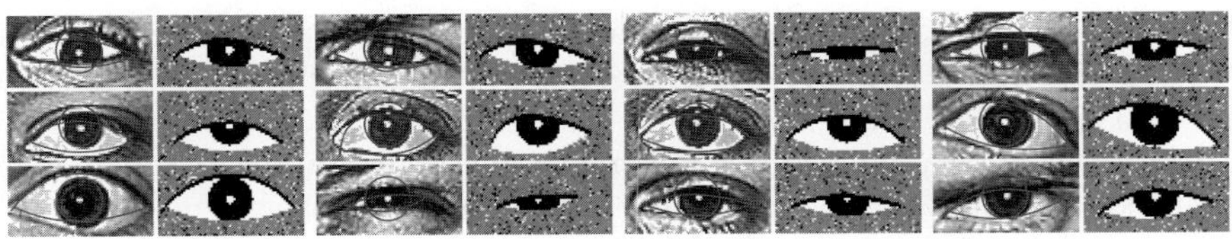

Fig. 3. Test image reconstructions computed by the recognition network trained with breeder learning

the AAM's appearance representation (i.e., the code vector) is chosen to be 60-dimensional. So for the purposes of breeder learning, we treat the AAM as a black box that takes 60 real-valued inputs and produces a 30 × 30 face image as output. (Note that the AAM learning procedure itself computes the codes for its *training* images as part of learning, so they are known, but we do not use them when learning the recognition network. The correct codes for the *test* images are truly unknown.)

Dataset and training details: We use 400 frontal faces (histogram-equalized) containing different expressions and identities. The dataset is split into 300 training images, 50 validation images, and 50 test images. None of the identities in the test set appear in the training and validation sets, so at test time, the recognition network has to generalize correctly to unseen identities, which is a difficult task. Note that only the 300 training and 50 validation images are used in the learning of the AAM itself.

The recognition network has 900 input units, a hidden layer of 100 sigmoid units, and 60 linear output units. We select the origin of the code space, corresponding to the face with the mean shape and mean texture, as the prototype code. Since the network's output units are linear, the code vectors are perturbed during learning by adding zero-mean Gaussian noise (with 0.1 standard deviation) directly to the outputs. The recognition network trained by breeder learning achieves its best performance on the validation set after slightly fewer than 3400 epochs.

Figure 4 shows the RMSE results. Interestingly, the best reconstruction error achieved by breeder learning on the validation set is below that of the AAM itself (dashed line in the graph). This means that the net is able to find codes that are better in the squared pixel error sense than the ones found by the AAM learning. Example reconstructions of test faces are shown in figure 5. The recognition task here is quite challenging as the network has to generalize to new identities. In most cases, the network reconstructs the face with approximately the correct expression and identity. In contrast, the reconstructions computed by the network learned with random-code learning are visually much worse and most of them resemble the face corresponding to the prototype code.

3.3   Iterative Refinement of Reconstructions with a Generative Network

So far recognition has been treated as a purely *feedforward, bottom-up* computation. A key property of analysis-by-synthesis is the use of top-down knowledge in the generative model to improve inference via a feedback loop that minimizes an error measure in data space itself. In our case implementing such a feedback loop requires knowing the gradient of the generative black box. But once a fully-trained recognition network is available, an alternative approach becomes possible.

The idea is to approximate the function implemented by the black box with a *generative neural network* [5]. This network emulates the black box: it takes a code vector as input and computes the corresponding data vector as output. Once such a generative network is trained, it can be used to compute the gradient

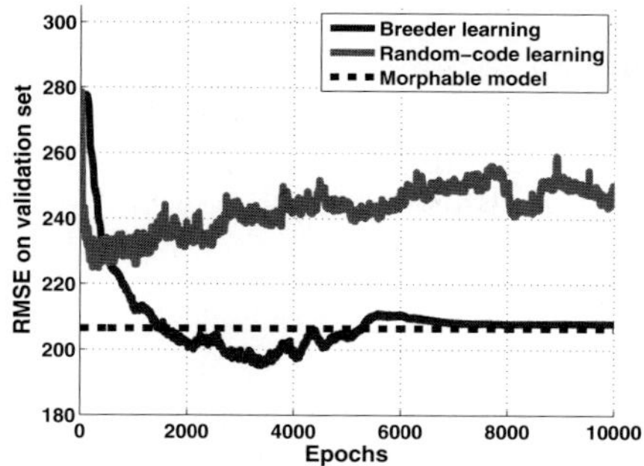

Algorithm	Test RMSE
Breeder learning	200.17
Random-code learning	229.31

Fig. 4. Validation set RMSE (left graph) during training, and test set RMSE (above table) after training, for breeder and random-code learning algorithms on the face dataset

of the data reconstruction error with respect to the code. As a result, inference now becomes a gradient-based iterative optimization problem that minimizes reconstruction error in data space.

Training the generative network is possible only because a fully-trained recognition model is already available. It provides the generative network approximately correct target codes for the training images. Given these code-image pairs, training reduces to a standard supervised learning task. The recognition network restricts the learning to the small part of data space that contains the real data, thus making it practical.

Figure 6 summarizes the closed-loop version of the recognition procedure. The initial code is computed by a bottom-up pass through the recognition network as before. But unlike in open-loop recognition, this initial estimate is subsequently refined by gradient descent on the squared error between the data vector and its reconstruction. The iterations continue until the squared error stops improving.

We learned a generative network to emulate the AAM and then used it to refine the reconstructions of faces. The average improvement in squared pixel

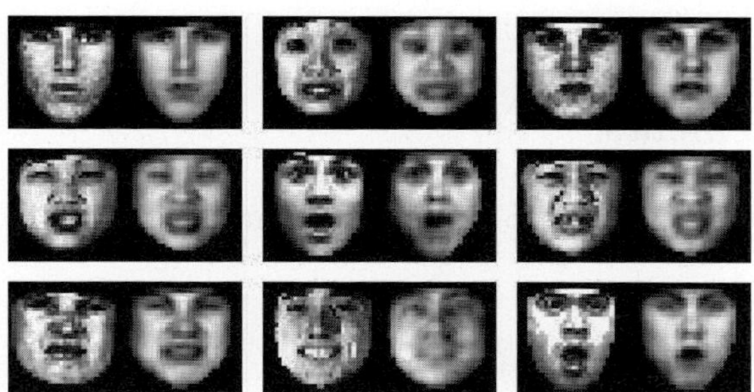

Fig. 5. Test image reconstructions computed by the recognition network trained with breeder learning

---

Algorithm for closed-loop recognition of data vector x:

Given: Generative black box G, recognition network R_w, generative network $G_{w'}$.

Initialization: $y = R_w(x)$.

For each refinement iteration:

1. $x' = G(y)$.
2. $E = \|x - x'\|^2$.
3. Compute $\frac{\partial E}{\partial y}$ by backpropagation through $G_{w'}$.
4. $y \leftarrow y - \eta \frac{\partial E}{\partial y}$.

Fig. 6. The closed-loop recognition algorithm using a generative neural network

error for the validation and test sets are 6.28% and 5.41%, respectively. It should be emphasized that the closed-loop recognition algorithm is used only as a way of fine-tuning the initial open-loop code estimate, which is already a very good solution. This side-steps the issue of whether a generic distance metric such as Euclidean distance can be used to correctly measure similarity in data space. Here we use Euclidean distance to measure *local* similarity only, i.e. to decide how an already good solution can be made a little bit better. So minimizing Euclidean distance can be sensible for fine-tuning, even if it is prone to get stuck in shallow local minima when starting from a random solution.

4   Conclusions

Breeder learning is a new tool for engineers building recognition models. By taking advantage of the rich domain knowledge in a generative model, it can learn much higher-level representations of data than those learned by standard methods such as PCA or ICA. Inverting complex physically-based generative models is one of the most promising future applications of breeder learning. Successfully inverting them will result in improved data representations for applications such as classification and compression.

Acknowledgements

This work was supported by NSERC, CIFAR, and FQRNT.

References

1. Lee, Y., Terzopoulos, D., Waters, K.: Realistic modeling for facial animation. In: SIGGRAPH (1995)
2. Sifakis, E., Neverov, I., Fedkiw, R.: Automatic determination of facial muscle activations from sparse motion capture marker data. In: SIGGRAPH (2005)

3. Moriyama, T., Kanade, T., Xiao, J., Cohn, J.F.: Meticulously detailed eye region model and its application to analysis of facial images. IEEE Transactions on Pattern Analysis and Machine Intelligence 28(5), 738–752 (2006)
4. Cootes, T.F., Edwards, G.J., Taylor, C.J.: Active appearance models. IEEE Transactions on Pattern Analysis and Machine Intelligence 23(6), 681–685 (2001)
5. Jordan, M., Rumelhart, D.: Forward models: Supervised learning with a distal teacher. Cognitive Science 16, 307–354 (1992)

A Bio-inspired Connectionist Architecture for Visual Classification of Moving Objects

Pedro L. Sánchez Orellana and Claudio Castellanos Sánchez[*]

Laboratory of Information Technologies of Centre for Research and Advanced Studies
LTI Cinvestav - Tamaulipas, Ciudad Victoria, Tamaulipas, México
{psanchez,castellanos}@tamps.cinvestav.mx

Abstract. In this paper we propose a bio-inspired connectionist architecture for visual classification of moving objects in articulated and non-articulated ones. It is based on a previous model called CONEPVIM, which uses the behaviour of simples cells in the visual primary cortex (V1) to estimate motion. Besides, on general new ideas on the area of neurophysiology for detecting the articulated moving objects and understanding the processing of the visual information from the earlier vision. We tested our architecture with real images without illumination constraints and any previous knowledge.

Keywords: Bio-inspired connectionist architecture, visual classification, articulated/non-articulated moving objects.

1 Introduction

Detection of moving humans has become an important task for several applications such as: surveillance [1] or some health applications to detect abnormalities when the person walks [2]. It is also a main part of more elaborated systems like gait recognition systems. However, this is not a trivial task because it implies the isolation of the person from the environment, which sometimes it is not static making this task even more difficult. In order to solve this problem several methodologies have been developed, most of them uses the background subtraction to recover what is moving in the images, some others more complex methods are reported in [3,4], but they become harder to implement at working with environments without restrictions.

So the main disadvantages of working with the classic techniques mainly has to do with the inconvenience of working with complex environments. To overcome them a different type of methodologies came up, such as the so called bio-inspired. They consist on algorithms based or inspired on functions of some areas in the brain, so they try to take advantage of the capability of the human brain to recognise objects even in complex scenes, and isolate them from other moving things.

The paper is organised in four stages. In the first one, the related work to bio-inspiration is described to show the context where we want to establish

* This research was partially funded by project number 51623 from "Fondo Mixto Conacyt-Gobierno del Estado de Tamaulipas".

V. Kůrková et al. (Eds.): ICANN 2008, Part I, LNCS 5163, pp. 982–990, 2008.

our architecture. For the second one, some biological foundations are shown to understand how does the methodology extracts the information to solve the problem of human being motion recognition. In the third, our architecture is presented and the parts that composes it are described. And finally get with the experiments, results and the conclusions and perspectives.

2   Related Work

Even though the bio-inspiration for pedestrian detection is reasonably a new approach to solve the problem, we can still find some good works on the topic, for example, Laxmin et al. [5] present a neural network model based on the capabilities of the human beings to recognise motion of other human based just on partial information. Lange et al. [6] presented experiments that indicates that for recognising a human shape in motion it is important to consider the spatial and temporal structure of the shape, it means, it is important to keep up with a specific motion pattern. Also there has been some works that tries to explain the way the information is processed in the brain by describing some interactions between cells; for example, Casile et al. [7] showed that the detection of the movement might be explained by some mechanisms in cortical areas of some macaques. To prove this, a statistical analysis (ANOVA) was applied to the training images to extract the most representative information and finally a neurophysiological model was created to test these features.

However, according to recent works, the interpretation of the reality in our brain is formed along the whole visual pathway where the information is distributed into processes, adding in each of them necessary information to create the conception of reality. This new discoveries are the bases for a new type of bio-inspiration, where the objective is not only to take the parts of the information to get the final conclusion, but also to understand what happens in the middle stages. For example, Romo [8] considers the information that the brain receives (by the senses) as just one part of the final result, so the final result contains also information added by the brain. This translated to the case of the conception of the shape of a human body at a specific moment in the image sequence depends not only on what we extract from the image, but also on the information that the brain uses to complement it.

This new perspective has some advantages with respect to the traditional way of bio-inspiration because it gives a different perspective to understand certain areas in the brain. In this context Castellanos Sánchez [9] used the known information of the primary visual cortex (V1), and the dorsal pathway to create a model (called CONEPVIM) to estimate the movement of objects from sequences of images. This model is composed by two stages, the first one models the simple cells in V1 to extract the spatial-temporal characteristics of the images. The responses of the first stage feeds the second one, that considers the neurons in MT and MST to estimate velocity, direction and whether the motion in the scene was caused by the motion of the observer or by motion of objects.

To solve the problem of distinguishing between the motion of articulated and non-articulated objects we will keep on with the methodology of Castellanos Sánchez by using the first stage of his model and also, the optic of Romo, about the way the brain constructs the reality, to answer a couple of up-to-now open questions. What is the best representation for modelling the pedestrian detection? and what does really happens in the intermediate stages in the brain? In other words, how does the brain completes the information in each stage? For this we will start describing some biological foundation for our proposed architecture bio-inspired in the behaviour of the simple cells in V1, more specifically parts the visual cortex.

3   Biological Bases

Recent research on computational neuroscience has provided an improved understanding of human brain functionality and bio-inspired models have been proposed to mimic the computational abilities of the brain for motion perception and understanding. For example, there are some inspired in the functioning of (V1) with a strong neural cooperative-competitive interactions that converge to a local, distributed and oriented auto-organisation [10,11,12]. Some others are inspired by the middle temporal area (MT) with the cooperative-competitive interactions between V1 and MT and an influence range [13,14]. And the others are inspired by the middle superior area (MST) for the coherent motion and ego-motion [15,16].

In this work our analysis will be focused on the responses of the neurons in V1, modelled by means of a causal spatio-temporal Gabor-like filters. The filter has the capability to extract the motion information in a very local and distributed form. This characteristic of the filter was used to understand the way an articulated object (human) moves. For us, the movements described by an articulated object produces different reactions in these neurons compared to the responses to a non articulated ones. So to explain this idea we used some works realised by Murray [17], where the composition of the human walking was described as several pieces that form the stepping of the humans. For instance, one whole step is composed by three major movements, one forward left step, forward right step and finally another forward left step, but the interesting thing comes in the middle of this movements, while one foot is moving the other one remains static.

This of course, in terms of the responses of the neurons produces a different patterns of neurons activations which are unique for the articulated objects. Mainly because the non articulated ones never change their structure while moving. Once that the foundations were mentioned we will describe the methodology to solve the problem.

4   Architecture

The proposed architecture for visual detection of moving objects is divided basically in three stages. The first stage is the acquisition of real images; the second one is the casual spatio-temporal treatment of the images, by means of spatial-temporal

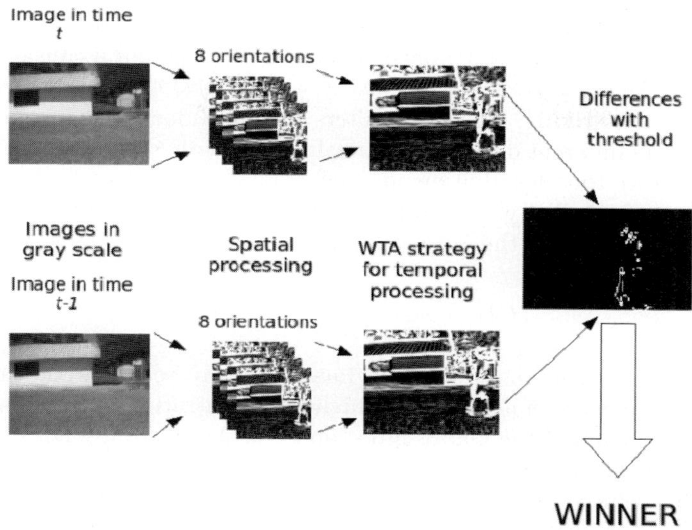

Fig. 1. Architecture for recognition of biological motion, based of the first module of CONEPVIM and the ideas described by Romo

Gabor-like filters, and the third is the analysis of the responses of the simple cells by means of a method of the type WTA, to measure the responses of the filter.

In other words, we combine the responses of the simple cells (estimated with the first module of the CONEPVIM model) and the ideas proposed by Romo, to get to the understanding of how the processing in each stage of the pathways contributes to the conceptualisation of the reality, rather than just describing properties of the perceived information.

In the next subsections we show a more detailed description of the three stages of the methodology (see figure 1), starting with the image capturing.

4.1   Image Acquisition

In order to prove our methodology we constructed a database of sequences of images. For this database our goal was to keep them as natural as possible, so we used a simple webcam which has an automatic adjustment of the contrast. This characteristic is not useful when the base for the Gabor filters is the measurement of the intensities of the pixels in the images. Besides, the position of the camera was static.

The next step in our model is application of the first module of CONEPVIM model.

4.2   First Module

The first stage of the model depicted by spatial filtering and temporal processing of figure 1, which performs a causal spatio-temporal filtering. It models the magnocellular cells seen as motion sensors that depend on the gradient of image intensity and on its temporal derivatives [18,19,20]. This filtering is performed in two steps (see equation 1) : a Gabor-like spatial filtering and a causal temporal processing [21,22].

For the spatial filtering, Gabor-like filters are implemented as image convolution kernels in Θ different directions. I usually work with $\Theta = 8$ for simplicity. In order to integrate the contributions of each orientations we applied a principle of neural networks called winner takes all (WTA) and calculate the maximum response for each pixel in the 8 orientations.

$$H(t) = max_\theta \int G_\theta * I(x, y)dt \tag{1}$$

where $I(\cdot, \cdot)$ are the intensities of the image in gray scale, so the maximum response of the neurons will indicate a higher concentration of the information, G are the coefficients of the Gabor filter, and $*$ is the convolution.

Then the causal temporal processing involves the temporal computation of $H(t)$. In other words, it simply measures the absolute difference between $H(t)$ and $H(t-1)$, that belongs to images in consecutive parts of the sequence. This difference is rated by using a threshold ϵ because not all the differences are useful, they might be product of a distortion in the acquisition stage. This process is repeated for all the images of the sequence as follows:

$$D(t) = \begin{cases} \mid H(t) - H(t-1) \mid, & \mid H(t) - H(t-1) \mid >= \epsilon \\ 0 & , \quad \mid H(t) - H(t-1) \mid < \epsilon \end{cases} \tag{2}$$

This principle is valid under the strong hypothesis of a very high sampling frequency to ensure a local motion detection and an immediate constant local speed. For more details on this filtering see [23].

Once we get the responses grouped by time, it means the conjunction of the Θ orientations grouped by the temporal filter, we use a methodology for counting on the responses of the neurons in time t. It will be described in the next subsection.

4.3 Responses Analysis

As it was explained in the last section, in the temporal processing of $D(t)$ we use a threshold to determine which of the neurons are responding to the changes in the images. Although, not all the changes in the scene can be counted on because it is necessary to consider the error that are obtained in the capturing stage (such as automatic contrast or auto focus characteristics).

We are searching for different patterns of activations in the neurons, depending of the type of movement that is presented. Our theory is that due to the way an articulated object moves (might be a human or an animal) it is possible to discriminate it from other type of moving objects that are not articulated (vehicles). For this we counted on the responses and rated them according to the amount of activation that they show.

For counting the responses we estimated the number of activation of the cells splitting them into 40 parts of responses, this value was determined empirically. The number of responses was multiplied by the information that it represented, this was to rate the information. For us, not all the responses can be treated the same it depends on the amount of activations rather than the magnitude of them. Besides it was divided into the proportion of the maximum object that for us is possible to recognise, lets say that a human occupies at most the 33.3% of the space in the image, for a bigger value we would not be able to recognise the body because of missing information. The function of this maximum size of the object is to show a proportion of activations in the neurons based on the maximum expected value and it is described as follow:

$$Winner(t) = \frac{\sum_i^n R_i \cdot P_i}{Q} \tag{3}$$

where R_i is the set of the active neurons in the i-esim percentile, P_i is the percentile, and $Q = \frac{length(I)}{3}$ is the maximum supposed size of an object in the image I.

This architecture was tested with a database of real images. The experiments are described in the following section.

5   Experiments and Results

For the experiments we used a several sequences of real images taken with a webcam, this to test the methodology even with the lowest conditions for making the calculations (due to the function of self-adjustment of contrast in the webcam). The images also where taken at a very low rate, 15 fps, which causes not to see the changes in the responses of the filter very accurately.

The experiments are separated in two, in the first one we tested the algorithm with sequences were just a person was walking in front of the camera. For each one of the subjects a total of 6 sequences were taken (each one with a different orientation with respect to the camera). For the second experiment we recorded a total of 4 vehicles, and for each one of them 4 sequences were taken also varying the angle with respect to the camera.

The results of the experiments are shown in the figure 2, where it can be seen that the amount of responses of the neurons is completely different for both the sequences of the human (articulated object) and the vehicle (non articulated object). For the case of the human, the activations of the neurons along the time (x axis) are not in a uniform growing up style, which is caused by moving one articulation while the other one still static. These values contrast diametrically with the graphics of the vehicle where the winner responses are more smoother due to the way a car moves, the moving structure of the car never change along the time.

The behaviour of the winner neurons is reflected on the spikes that appear in the graphics, the show the behaviour (in terms of responses) of the active neurons along the sequence.

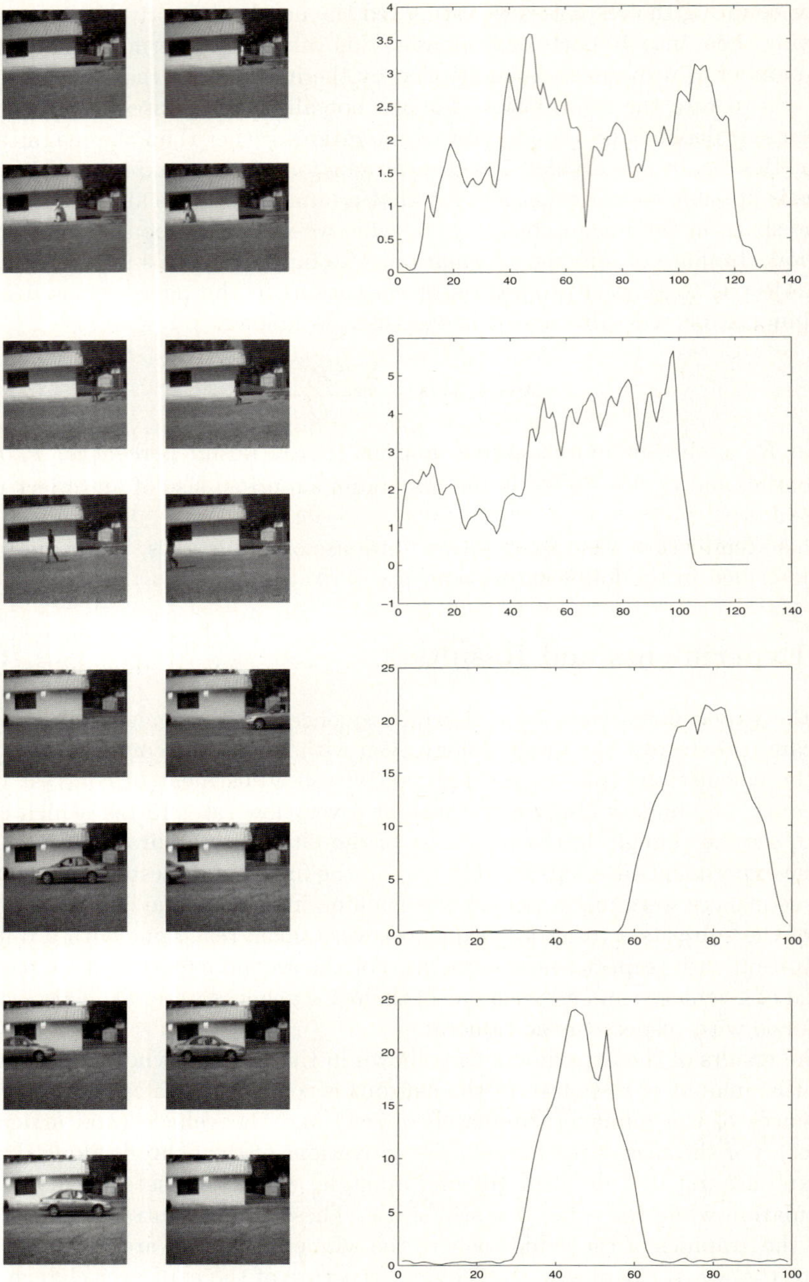

Fig. 2. Real sequences of images used in this work: Carlos, Miguel, Carro1 and Carro2 in each two rows, four images of each sequence, respectively. In the second column their winners in the time, respectively.

6   Conclusions and Future Work

This work permitted us to explore the capabilities of the earlier stages in the visual pathway, the results are very interesting considering the possibility to distinguish between articulated objects and not articulated ones. However, it is important to mention that the architecture might show the same behaviour for other types of articulated objects (cats or dogs) so the next step for us is to test it with different types of sequences.

In order to test in extremely conditions the methodology the sequences were taken with a simple webcam in a outdoor scenery without any kind of constraints, if fact the scene was not very static due to wind and other climatological conditions, although the responses of the filter also reflects this difficulties we managed to reduces the effects by means of the threshold.

Our objective is to developer a bio-inspired system that allow us to recognise persons by their gait. For this we take as inspiration the methodologies proposed by Giese et al. [24], Jhuang et al. [25] and Castellanos [9] where some clues about how the processing of the images is done in the brain. It is important to say that these methodologies were done for recognising peoples in sequences. Although, they also have to isolate the human shape from the environment. Traditionally this methodologies extracts the moving sequences by modelling the simple cells in V1 with Gabor filters and feeding upper processing stages with the responses of the filter. So the detection of the human moving object is done in the upper stages.

Compared to these methodologies our proposal considers that the conceptualisation of the objects in the brain is a process in which every stage on the visual pathway contributes to create the idea of the object that is being seen.

References

1. Xu, H., Jiwei, L., Lei, L., Zhiliang, W.: Gait recognition considering directions of walking. IEEE Cybernetics and Intelligent Systems, 1–5 (2006)
2. Kirtley, C., Smith, R.: Application of multimedia to the study of human movement. multimedia tools and applications. Multimedia Tools and Applications 14, 259–268 (2001)
3. Yoo, J.H., Nixon, M.S., Harris, C.J.: Model-driven statistical analysis of human gait motion. In: ICIP 2002 International Conference on Image Processing, vol. 1, pp. 285–288 (2002)
4. Veres, G., Gordon, L., Carter, J., Nixon, M.: What image information is important in silhouette-based gait recognition. In: IEEE Computer Vision and Pattern Recognition conference (2004)
5. Laxmi, V., Carter, J.N., Damper, R.I.: Biologically-inspired human motion detection. In: 10th European Symposium on Artificial Neural Networks, pp. 95–100 (2002)
6. Lange, J., Lappe, M.: The role of spatial and temporal information in biological motion perception. Advances in Cognitive Psychology, University of Muenster, Germany 3, 419–428 (2007)

7. Casile, A., Giese, M.: Critical features for biological motion. Journal of Vision 3(9), 85 (2003)
8. Romo, V.R.: Neural correlate of subjective sensory experience gradually builds up across cortical areas. PNAS 103, 14266–14271 (2006)
9. Castellanos Sánchez, C.: Neuromimetic indicators for visual perception of motion. In: 2nd International Symposium on Brain, Vision and Artificial Inteligence, vol. 103, pp. 134–143 (2007)
10. Fellez, W.A., Taylor, J.G.: Establishing retinotopy by lateral-inhibition type homogeneous neural fields. Neurocomputing 48, 313–322 (2002)
11. Latham, P.E., Nirenberg, S.: Computing and stability in cortical networks. Neural Computation, 1385–1412 (2004)
12. Moga, S.: Apprendre par imitation: une nouvelle voie d'apprentissage pour les robots autonomes. PhD thesis, Université de Cergy-Pontoise, Cergy-Pontoise, France (September 2000)
13. Simoncelli, E.P., Heeger, D.J.: A model of neural responses in visual area MT. Vision Research 38(5), 743–761 (1998)
14. Mingolla, E.: Neural models of motion integration and segmentation. Neural Networks 16, 939–945 (2003)
15. Pack, C., Grossberg, S., Mingolla, E.: A neural model of smooth pursuit control and motion perception by cortical area MST. Technical Report CAS/CNR-TR-99-023, Boston University, Department of Cognitive and Neural Systems and Center for Adaptive Systems, 677 Beacon St, Boston, MA 02215 (September 2000)
16. Zemel, R.S., Sejnowski, T.J.: A model for encoding multiple object motions and self-motion in area MST of primate visual cortex. The Journal of Neurosciences 18(1), 531–547 (1998)
17. Murray, M.: Gait as a total pattern of movement. American Journal in Physics and Medicine 46, 290–332 (1967)
18. Pollen, D., Ronner, S.: Phase relationships between adjacent simple cells in the visual cortex. Science 212, 1409–1411 (1981)
19. Newsome, W.T., Gizzi, M.S., Movshon, J.A.: Spatial and temporal properties of neurons in macaque MT. Investigative Ophtalmology and Visual Science Supplement 24, 106 (1983)
20. Hammond, P., Reck, J.: Influence of velocity on directional tuning of complex cells in cat striate cortex for texture motion. Neuroscience Letters 19, 309–314 (1981)
21. Torres Huitzil, C., Girau, B., Castellanos Sánchez, C.: On-chip visual perception of motion: A bio-inspired connectionist model on FPGA. Neural Networks 18(5-6), 557–565 (2005)
22. Castellanos Sánchez, C., Girau, B., Alexandre, F.: A connectionist approach for visual perception of motion. In: Smith, L., Hussain, A., Aleksander, I. (eds.) Brain Inspired Cognitive Systems (BICS 2004), BIS3-1:1–7 (September 2004)
23. Castellanos Sánchez, C.: Neuromimetic connectionist model for embedded visual perception of motion. PhD thesis, Université Henri Poincaré (Nancy I), Nancy, France, Bibliothèque des Sciences et Techniques (October 2005)
24. Giese, M., Poggio, T.: Neural mechanisms for the recognition of biological movements and action. Nature Reviews Neuroscience 4, 179–192 (2003)
25. Jhuang, H., Serre, T., Wolf, L., Poggio, T.: A biologically inspired system for action recognition. In: Proceedings of the Eleventh IEEE International Conference on Computer Vision (in press, 2007)

A Visual Object Recognition System Invariant to Scale and Rotation

Yasuomi D. Sato, Jenia Jitsev, and Christoph von der Malsburg

Frankfurt Institute for Advanced Studies (FIAS), Johann Wolfgang Goethe University, Ruth-Moufang-Str. 1, 60438, Frankfurt am Main, Germany

Abstract. We here address the problem of scale and orientation invariant object recognition, making use of a correspondence-based mechanism, in which the identity of an object represented by sensory signals is determined by matching it to a representation stored in memory. The sensory representation is in general affected by various transformations, notably scale and rotation, thus giving rise to the fundamental problem of invariant object recognition. We focus here on a neurally plausible mechanism that deals simultaneously with identification of the object and detection of the transformation, both types of information being important for visual processing. Our mechanism is based on macrocolumnar units. These evaluate identity- and transformation-specific feature similarities, performing competitive computation on the alternatives of their own subtask, and cooperate to make a coherent global decision for the identity, scale and rotation of the object.

1 Introduction

We here address the problem of visual object recognition invariant to change in scale and orientation. This problem is far from being resolved for varying stimuli of natural complexity [1]. Following Rosenblatt [2], invariance is often assumed to be achieved feature-by-feature, by connecting all cells of an input domain that correspond to transformed versions of a given feature type to one master neuron to represent presence or absence of that feature type. Recognition is then based on the set of feature types that occur in the image. This approach is called *feature-based recognition*. When changes in scale and orientation are among the allowed transformation, and when neurobiologically realistic feature types such as Gabors [3,4,5] are used, this approach runs into the difficulty that the main distinction between different features —scale and orientation— is completely lost, so that recognition breaks down. There are feature types that are immune to this difficulty [6], but these features are hardly able to express the texture of objects, such is necessary, for instance, for face recognition.

An alternative approach, which we follow here, is to recognize the transformation between input image and stored representation and to use it to map input and stored features into each other. This way, the full information contained in the features is maintained. In addition, as has been pointed out [7], knowing the transformation parameters of the image is of functional value itself, information

V. Kůrková et al. (Eds.): ICANN 2008, Part I, LNCS 5163, pp. 991–1000, 2008.

that is lost in the feature-specific approach [8]. The model we present here continues own earlier work [9,10] in showing that not only the correct transformation but also one of several stored object models can be recognized. It is a simple proof of principle, assuming that a pair of corresponding center points in image and stored model has already been found (thus ignoring the translation invariance problem for the time being) and limiting itself just to the set of features centered on these points. The advantage is the simplicity of the model, and we will address in later publications full translation invariance and representation of complete objects instead of just one local feature set.

Subsystem Coordination.   In our approach, not only object identification but also the image transformations are of interest. Although these different dimensions are in principle independent of each other, as it is not possible to predict one given the values of others, they conjointly determine image structure and must be disentangled during the recognition process. Thus, it is necessary to know transformation parameters to recognize an object, and conversely, it is helpful to know the identity of an object before attempting to estimate transformation parameters. It is therefore necessary to find mechanisms to coordinate subtasks such that they cooperatively achieve the global recognition task.

Central elements of our system are *control columns*, which implement the recognition subtasks and have the ability to control the information flow between the domains [11]. We have three separate control columns for the representation of image scale, image orientation and object identity, each consisting of a number of minicolumns for the representation of different transformation parameters or object identities. Each of these columns attempts to arrive at a decision already before getting any information from the other subsystems, by gathering evidence in the form of *identity-* or *transformation-specific* feature similarities. In the course of the global decision process, the three control columns get clearer and clearer signals conveying the local decisions from the others, quickly converging to a global consensus. As our experiments show, the feature similarities gathered by the three subsystems initially in the absence of any information from the others contain enough information to set the process in motion.

2   System Setup, Domain Organization

Our system is based on two domains. A *Gallery Domain* G contains representations of N different object images G_g, $g \in \{0, 1, \ldots, N-1\}$. These serve as stored representations of the different objects we want to recall from memory. An *Input Domain* I receives afferent input containing sensory signals generated by test images, taken at different scale and/or orientation. The images in each domain are represented by sets of Gabor wavelet responses *(Jets)* [12]. From each image, one jet $\boldsymbol{J}^{\mathrm{L}}$, L = I or G, is extracted from a central point (as shown in Fig.1). Jet components are defined as convolutions of the image with a family of Gabor functions. Orientation and spatial frequency are sampled discretely. Wavelet orientation can take the values $\theta = \pi k/n$, $k \in \{0, 1, \ldots, (n-1)\}$, n being the total number of possible orientations, while wavelet scale is sampled

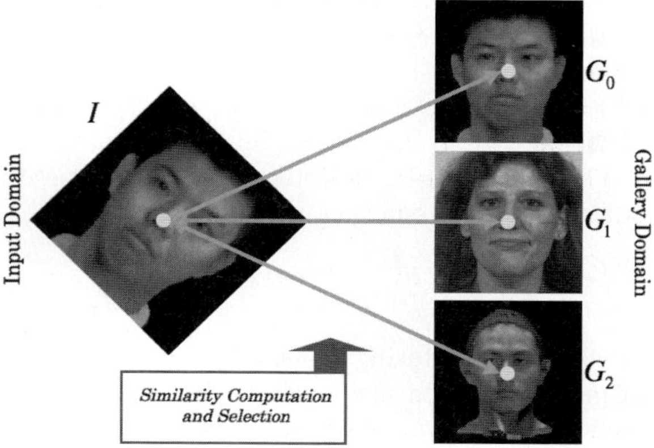

Fig. 1. Different objects are stored as reference views in memory *(Gallery domain)*. One of these is selected when a transformed version of it is presented in the *input domain*. For simplicity, we here represent images by a single set of Gabor features *(Jet)* all centered on a single point. Transformation-specific similarities of the features are evaluated to obtain the correct match. Competitive evaluation across all transformation-specific similarities provides the identity of the most suitable candidate in the gallery.

as $a = a_0^{l_{\mathrm{L}}}$, $0 < a_0 < 1$, with $\mathrm{L} \in \{\mathrm{I}, \mathrm{G}\}$ and $l_{\mathrm{I}} \in \{0, 1, \ldots, m_d + m_f + m_u - 1\}$, $l_{\mathrm{G}} \in \{m_d, \ldots, m_d + m_f - 1\}$. If wavelets of $m_f > 1$ different scales are stored for objects in G, and as many wavelet scales are to be compared to data in I in spite of different image scale, I must cover a larger range of scales, extended by m_u and m_d steps above and below the central range.

This log-polar sampling of the frequency domain of the feature space allows us to deal straight-forwardly with scale and rotation transformations. Selecting one of the images G_g as reference, the scale transformation can be interpreted as a radial shift of the spatial frequency components in the jet in I relative to the reference jet in G_g feature, while the rotation corresponds to a circular permutation of the orientation components around the center point. Some additional notations are used here for the size and rotation transformations applied to the object image in I. Size changes of the image I relative to the image G_g are designated with the scale factor of $[1/a_0]^s$ where $s \in \{-m_d, \ldots, -1, 0, +1, \ldots, +m_u\}$. Similarly, the rotation change of the I image is given by $\pi \times [r/n]$, $r \in \{0, \ldots, n-1\}$.

As in [9,10,13], we model the representation of features by a pair of cortical feature macrocolumns, one in I, one in G. The macrocolumn in the I domain consists of a set of $N_{\mathrm{I}} = (m_d + m_f + m_u) \times n$ minicolumns, with activities $\{P_0^{\mathrm{I}}, \ldots, P_{N_{\mathrm{I}}-1}^{\mathrm{I}}\}$, each minicolumn representing a single jet component. Similarly, the feature jet for the image G_g in the G domain is represented by a macrocolumn containing $N_{\mathrm{G}} = m_f \times n$ minicolumns. The equation system for the activations in the domain L, $\mathrm{L} = \mathrm{I}$ or G, is thus given by

$$\frac{dP_\alpha^{\mathrm{L}}}{dt} = f_\alpha(\boldsymbol{P}^{\mathrm{L}}) + \kappa_{\mathrm{L}} J_\alpha^{\mathrm{L}}, \qquad (\alpha = 0, \ldots, N_{\mathrm{L}} - 1), \qquad (1)$$

where κ_{L} controls the strength of the Gabor jet input in the domain L. The function $f_\alpha(\cdot)$ has the form

$$f_\alpha(\boldsymbol{P}^{\mathrm{L}}) = a_1 P_\alpha^{\mathrm{L}} \left(P_\alpha^{\mathrm{L}} - \nu(t) \max_{\beta=0,\ldots,N_{\mathrm{L}}-1} \{P_\beta^{\mathrm{L}}\} - (P_\alpha^{\mathrm{L}})^2 \right), \tag{2}$$

where $a_1 = 25.0$. The function $\nu(t)$ describes a cycle in time during which the feedback is periodically modulated:

$$\nu(t) = \begin{cases} (\nu_{max} - \nu_{min})\frac{t-T_k}{T_1} + \nu_{min}, & (T_k \leqslant t < T_k + T_1) \\ \nu_{max}, & (T_k + T_1 \leqslant t < T_k + T_1 + T_{relax}). \end{cases} \tag{3}$$

Here, T_1 [ms] is the time period during which ν increases while T_k, $k = 1, 2, 3, \ldots$, are the periodic times at which a new cycle begins. T_{relax} is a relaxation time after ν has reached its maximum. As parameter values we chose $\nu_{min} = 0.4$, $\nu_{max} = 1$, $T_1 = 36.0$ [ms] and $T_{relax} = T_1/6$. While $\nu(t)$ is growing, one minicolumn after the other is turned off, depending on the size of the coefficients in Eq.(1), until only one minicolumn, the one with the strongest input, remains active at the end of the cycle.

3   Transformation-Specific Feature Similarities

To decide whether an object g in domain G matches the image I under unknown orientation or scale, we compute a jet similarity that is specific for one

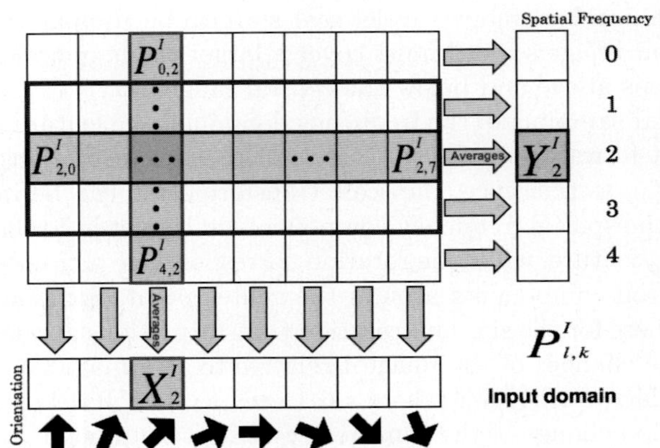

Fig. 2. Marginal feature vectors. A single macrocolumn in the input domain I, which consists of minicolumns representing $(m_d + m_f + m_u) \times n$ jet components, is mapped onto the log-polar coordinate of the domain. The columns (orientations) and rows (spatial frequencies) are then grouped together. As an example, $m_d = m_u = 1$, $m_f = 3$ and $n = 8$ are set up. The minicolumn activities $(P^{\mathrm{I}}_{l,k})$ are averaged groupwise over respective scales and orientations. Then, the averaged activity responses $(X^{\mathrm{I}}_k$ and $Y^{\mathrm{I}}_l)$ are aligned as vectors of grouped orientation and scale responses. The right and bottom panels respectively represent the computed vectors for scale and orientation.

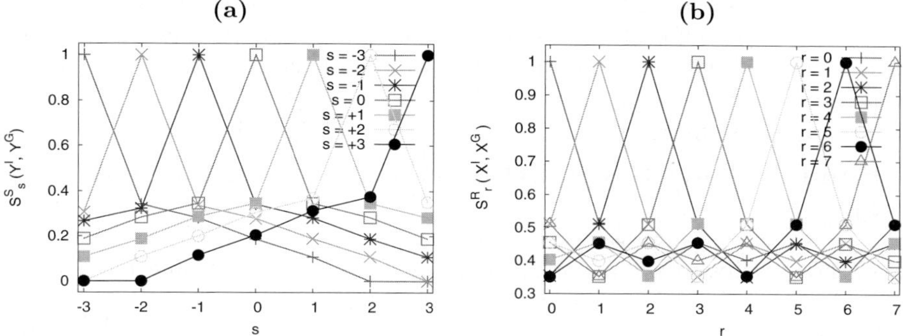

Fig. 3. Transformation-specific feature similarities for (a) scale and (b) rotation, plotted as function of scale (s) or rotation (r). To measure the similarities, one of the stored objects was taken from the gallery and transformed to present it in I. Different curves correspond to different relative transformation. Note that each transformation-specific similarity peaks for the correct relative transformation parameter. System parameters are $\kappa_I = \kappa_G = 0.5$ and $\kappa = 1.5$, $m_f = 4$, $m_u = m_d = 3$ and $n = 8$.

transformation parameter (s or r) but is averaged ("marginalized") over the other. In preparation, we compute marginalized feature vectors (see Fig. 2)

$$X_k^I = \frac{1}{|l_I|} \sum_{l \in l_I} P_{l,k}^I, \tag{4}$$

$$Y_l^I = \frac{1}{n} \sum_{k=0}^{n-1} P_{l,k}^I, \tag{5}$$

where l is the wavelet scale and k the wavelet orientation parameter for the I or G domain. These quantities are combined into vectors $\boldsymbol{X}^I = (X_0^I, \ldots, X_{n-1}^I)^T$ and $\boldsymbol{Y}^I = (Y_0^I, \ldots, Y_{m_d+m_f+m_u-1}^I)^T$, called *marginalized feature vectors*. Analogously, for domain G, we compute vectors $\boldsymbol{X}_g^G = (X_0^{G_g}, \ldots, X_{n-1}^{G_g})^T$ and $\boldsymbol{Y}_g^G = (Y_{m_d}^{G_g}, \ldots, Y_{m_d+m_f-1}^{G_g})^T$ with components $X_k^{G_g} = [1/m_f] \sum_{l \in l_G} P_{l,k}^{G_g}$ and $Y_l^{G_g} = [1/n] \sum_{k=0}^{n-1} P_{l,k}^{G_g}$.

With these definitions, we are ready to compute similarities that are specific for one stored object and one relative transformation parameter, s or r, and are marginalized for the other, r or s:

$$S_s^S(\boldsymbol{Y}^I, \boldsymbol{Y}_g^G) = \frac{\sum_{l \in l_G} Y_{l+s}^I Y_l^{G_g}}{\sqrt{\sum_{l \in (l_G+s)} \left(Y_l^I\right)^2 \sum_{l \in l_G} \left(Y_l^{G_g}\right)^2}}, \tag{6}$$

$$S_r^R(\boldsymbol{X}^I, \boldsymbol{X}_g^G) = \begin{cases} \dfrac{\sum_{k=0}^{n-1} X_{k-r+n}^I X_k^{G_g}}{\sqrt{\sum_{k=0}^{n-1} (X_k^I)^2 \sum_{k=0}^{n-1} (X_k^{G_g})^2}}, & \text{if } k < r \\[2ex] \dfrac{\sum_{k=0}^{n-1} X_{k-r}^I X_k^{G_g}}{\sqrt{\sum_{k=0}^{n-1} (X_k^I)^2 \sum_{k=0}^{n-1} (X_k^{G_g})^2}}, & \text{else} \end{cases}. \tag{7}$$

It might be expected that these feature similarities are without structure, but our experiments show that for a given stored object, relative scale of the image can be estimated without knowing orientation, and relative orientation can be estimated without knowing scale, see Fig. 3.

4   Simultaneous Recognition of Transformation and Object

We are now ready to set up our system of control columns, one for Scale S, one for rotation R, and one to represent object identity in the gallery G. The numbers of minicolumns in these three subsystems are $N_S = (m_d + 1 + m_u)$, $N_R = n$ and $N_G = N$, respectively. Designating activities of minicolumns by P_γ^C, with $(C, \gamma) \in \{(S, s), (R, r), (G, g)\}$, columnar dynamics is described by equations analogous to Eq. (1):

$$\frac{dP_\gamma^C}{dt} = f_\gamma(\boldsymbol{P}^C) + \kappa F_\gamma^C, \tag{8}$$

where κ controls the strength of the influence of the input F_γ^C to the respective control column. These quantities F_γ^C represent the mutual influence of the three control columns on each other and may be interpreted as weighted similarities that are specific for identity, scale and orientation:

$$F_g^G = \sum_s P_s^S S_s^S(\boldsymbol{Y}^I, \boldsymbol{Y}_g^G) + \sum_r P_r^R S_r^R(\boldsymbol{X}^I, \boldsymbol{X}_g^G), \tag{9}$$

$$F_s^S = \sum_g P_g^G S_s^S(\boldsymbol{Y}^I, \boldsymbol{Y}_g^G), \tag{10}$$

$$F_r^R = \sum_g P_g^G S_r^R(\boldsymbol{X}^I, \boldsymbol{X}_g^G). \tag{11}$$

These formulae may be seen as dynamical analogs to Bayesian expressions, if similarities are set in relation to conditional probabilities and minicolumnar activities are taken as analogous to probably estimates for values g, s and r. Fig. 4 gives a simplified schematic for the interaction between two control columns for scale and identity with only three scales and two stored objects. Fig. 5 shows results from a simulation of a full system with three stored objects (facial images, the input being a transformed version of one of these images). At the end of a single ν cycle, the system has decided on a relative scale, on a relative orientation and on one of the three stored objects. The small matrices in that figure represent the mapping of jet components of different scale (left matrix array) and of different orientation (right matrix array), time running from top to bottom.

5   Discussion and Conclusion

We are aware that the work presented here is very incomplete and in need of extension. Object images need to be represented more completely than by

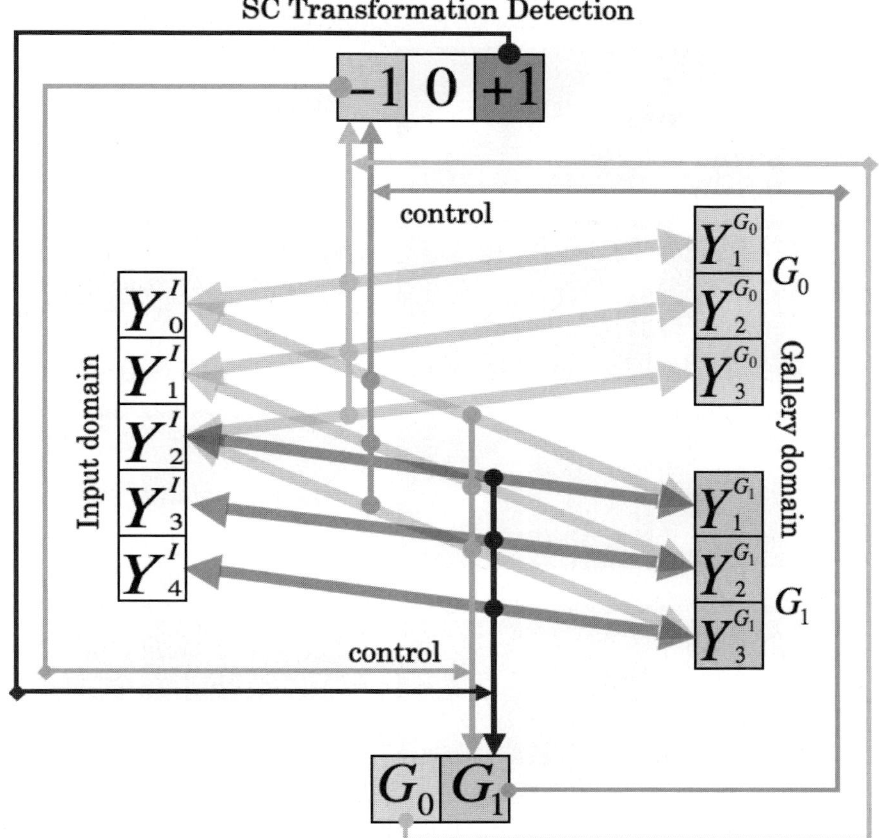

Fig. 4. Coordination between scale and identity control columns (schematic). Only three scales and two object identities and no orientation column are shown. Vertical arrows symbolize feature similarities, long arrows around the periphery stand for the modulation of similarities by minicolumnar activities, see Eqs. (9)–(11), and implement the coordination between the control columns.

one Gabor jet. Translation invariance is to be included instead of starting the recognition process by already giving two correspondence points, and galleries of serious size are to be considered. On the basis of some experience we believe, however, that no fundamental obstacles will be encountered on this path.

The first mechanism coming to mind when facing the problem of object recognition is probably template matching: take a literal, two-dimensional model of a candidate object and move it over the picture until it fits precisely. Historically, this idea soon went out of fashion as it seemed unwieldy: there are just too many dimensions to be gotten right — candidate object, position, size and

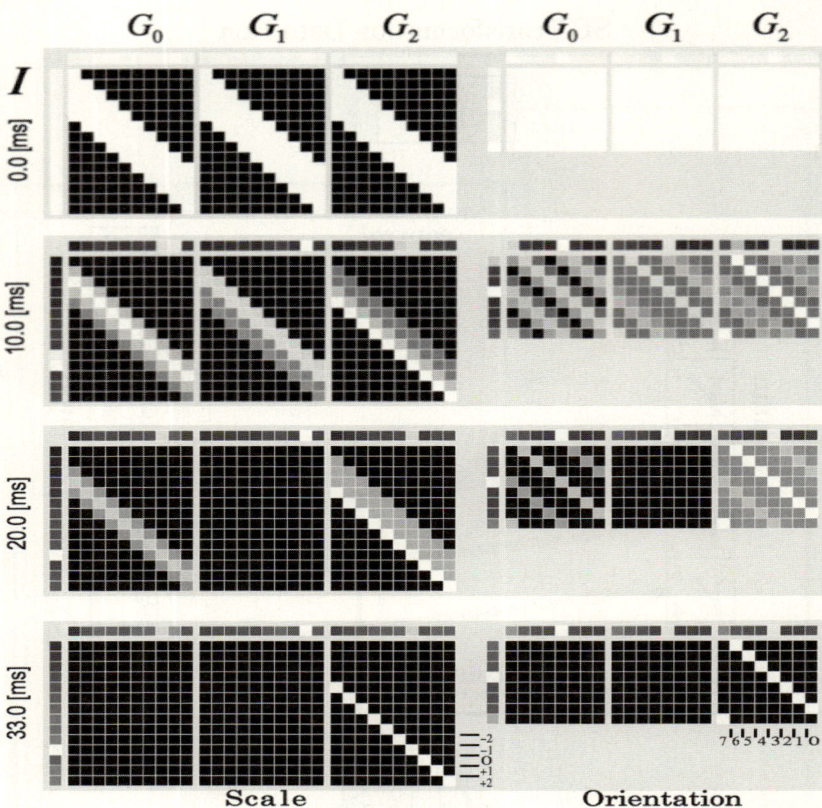

Fig. 5. Time course of the evolution of scale and orientation maps between the domains I and G over a ν cycle (time running from top to bottom) for a system with jet size of $m_f = 10$, $m_u = m_d = 2$ and $n = 8$. A transformed version of object G_2 with $[1/a_0]^2$ and $157.5°$ is presented in the I domain. The small matrices in the array at left represent the maps between wavelet components of marginalized feature vectors in I and G for different relative scale, each column corresponding to one stored object, white pixels corresponding to active links (only some connections are active initially, due to the limited range set by the parameters m_d and m_u). The right array of matrices is analogous for relative orientation. Over the ν cycle, the matrices contract to one diagonal for one object, corresponding to one active unit in each control column at the end of the cycle, corresponding to identity G_2, relative scale $s = +2$ and relative orientation $r = 7$. System parameters are $\kappa_I = \kappa_G = 1.0$ and $\kappa = 2.1$.

orientation, not even to speak of deformation, illumination and other image variations — before a match can be scored. Now, the correspondence-based approach we are pursuing can be seen as a version of template matching. However, the experiments performed here and in other work of our group raise the hope that not all dimensions must be gotten right precisely before a match is found, but that several of these dimensions can be estimated on the basis of marginal information, and that these estimates can subsequently be refined efficiently by coordination between subsystems. Our dynamic formulation may be seen in analogy to a Bayesian approach, in which, given an image, a total probability for object identity is to be optimized by combining partial observations with the help of conditional probabilities. The whole brain is probably to be seen as a large network of subsystems, each of which is faced with uncertainty due to incomplete information but able to reduce this uncertainty by mutual coordination, so that a reliable interpretation of the global situation arises. We hope that our work will prove of value on the way to this kind of formulation of the brain problem. Possibly, concrete steps in that direction will have to be taken when ramping up an object recognition system to realistic numbers of stored models. It will then be necessary again to divide and conquer, by estimating different dimensions of the universe of object shapes independently and letting the corresponding subsystems coordinate their activity into a global estimate of the exact shape.

Acknowledgments

This work is funded by the Hertie Foundation and by the EU project Daisy, FP6-2005-015803.

References

1. Serre, T., Wolf, L., Bileschi, S., Riesenhuber, M., Poggio, T.: Robust Object Recognition with Cortex-Like Mechanisms. IEEE Transactions on Pattern Analysis and Machine Intelligence 29(3), 411–426 (2007)
2. Rosenblatt, F.: Principles of Neurodynamics: Perceptrons and the Theory of Brain Mechanisms. Spartan Books, Washington (1961)
3. Gabor, D.: Theory of Communication. J. IEE 93, 429–459 (1946)
4. Daugmann, J.G.: Uncertainty relation for resolution in space, spatial frequency, and orientation optimized by two-dimensional visual cortical filters. J. Opt. Soc. Am. A 2, 1160–1169 (1985)
5. Jones, J., Palmer, L.: An evaluation of the two-dimensional Gabor filter model of simple receptive fields in cat striate cortex. J. Neurophysiol. 58, 1233–1258 (1987)
6. Lowe, D.G.: Distinctive Image Features from Scale-Invariant Keypoints. International Journal of Computer Vision 60(2), 91–110 (2004)
7. Arathorn, D.W.: Map-Seeking circuits in Visual Cognition — A Computational Mechanism for Biological and Machine Vision. Standford Univ. Press, Stanford (2002)
8. Serre, T., Kouh, M., Cadieu, C., Knoblich, U., Kreiman, G., Poggio, T.: A Theory of Object Recognition: Computations and Circuits in Feedforwad Path of the Ventral Stream in Primate Visual Cortex. AI Memo 2005-036/CBCL Memo 259. Massachusetts Inst. of Technology, Cambridge (2005)
9. Sato, Y.D., Wolff, C., Wolfrum, P., von der Malsburg, C.: Dynamic Link Matching between Feature Columns for Different Scale and Orientation. In: Ishikawa, M., Doya, K., Miyamoto, H., Yamakawa, T. (eds.) ICONIP 2007, Part I. LNCS, vol. 4984, pp. 385–394. Springer, Heidelberg (2008)

10. Jitsev, J., Sato, Y.D., von der Malsburg, C.: A Neural System for Scale and Orientation Invariant Correspondence Finding. In: Proc. COSYNE 2008, Salt Lake City, Utah, USA (2008)
11. von der Malsburg, C.: Dynamic Link Architecture. In: The handbook of brain theory and neural networks, pp. 329–331. MIT Press, Cambridge (1998)
12. Wiskott, L., Fellous, J.-M., Krüger, N., von der Malsburg, C.: Face recognition by elastic bunch graph matching. IEEE Transactions on Pattern Analysis and Machine Intelligence 19(7), 775–779 (1997)
13. Lücke, J., von der Malsburg, C.: Rapid Correspondence Finding in Networks of Cortical Columns. In: Kollias, S.D., Stafylopatis, A., Duch, W., Oja, E. (eds.) ICANN 2006. LNCS, vol. 4131, pp. 668–677. Springer, Heidelberg (2006)
14. Freedman, D.J., Riesenhuber, M., Poggio, T., Miller, E.K.: Categorical representation of visual stimuli in the primate prefrontal cortex. Science 291, 312–316 (2001)
15. Freedman, D.J., Riesenhuber, M., Poggio, T., Miller, E.K.: Visual categorization and the primate prefrontal cortex: Neurophysiology and behavior. J. Neurophys. 88, 930–942 (2002)

Recognizing Facial Expressions: A Comparison of Computational Approaches

Aruna Shenoy[1], Tim M. Gale[1,2], Neil Davey[1],
Bruce Christiansen[1], and Ray Frank[1]

[1] School of Computer Science, University of Hertfordshire,
United Kingdom, AL10 9AB
`{a.1.shenoy,t.gale,r.j.frank,n.davey}@herts.ac.uk`
[2] Department of Psychiatry, Queen Elizabeth II Hospital, Welwyn Garden City
Herts, AL7 4HQ UK

Abstract. Recognizing facial expressions are a key part of human social interaction,and processing of facial expression information is largely automatic, but it is a non-trivial task for a computational system. The purpose of this work is to develop computational models capable of differentiating between a range of human facial expressions. Raw face images are examples of high dimensional data, so here we use some dimensionality reduction techniques: Linear Discriminant Analysis, Principal Component Analysis and Curvilinear Component Analysis. We also preprocess the images with a bank of Gabor filters, so that important features in the face images are identified. Subsequently the faces are classified using a Support Vector Machine. We show that it is possible to differentiate faces with a neutral expression from those with a smiling expression with high accuracy. Moreover we can achieve this with data that has been massively reduced in size: in the best case the original images are reduced to just 11 dimensions.

Keywords: Facial Expressions, Image Analysis, Classification, Dimensionality Reduction.

1 Introduction

According to Ekman and Friesen [1] there are six easily discernible facial expressions: anger, happiness(smile), fear, surprise, disgust and sadness, apart from neutral. Moreover these are readily and consistently recognized across different cultures [2]. In the work reported here we show how a computational model can identify facial expressions from simple facial images. In particular we show how smiling faces and neutral faces can be differentiated. Data presentation plays an important role in any type of recognition. High dimensional data is normally reduced to a manageable low dimensional data set. We perform dimensionality reduction and classification using Linear Discriminant Analysis and also dimensionality reduction using Principal Component Analysis (PCA) and Curvilinear Component Analysis (CCA). PCA is a linear projection technique and it may be more appropriate to use a non linear Curvilinear Component Analysis (CCA) [3]. The Intrinsic Dimension (ID) [4], which is the true dimension of the data, is often much less than the original dimension of the data. To use this efficiently, the actual dimension of the data must be estimated. We use the Correlation Dimension to estimate

V. Kůrková et al. (Eds.): ICANN 2008, Part I, LNCS 5163, pp. 1001–1010, 2008.

the Intrinsic Dimension. We compare the classification results of these methods with raw face images and of Gabor Pre-processed images [5],[6]. The features of the face (or any object for that matter) may be aligned at any angle. Using a suitable Gabor filter at the required orientation, certain features can be given high importance and other features less importance. Usually, a bank of such filters is used with different parameters and later the resultant image is a L2 max (at every pixel the maximum of feature vector obtained from the filter bank) superposition of the outputs from the filter bank.

2   Background

We begin with a simple experiment to classify two expressions: neutral and smiling. We use Linear Discriminant Analysis (LDA) for dimensionality reduction and classification. We also use a variety of other dimensionality reduction techniques, a Support Vector Machine (SVM) [7] based classification technique and these are described below.

2.1   Linear Discriminant Analysis (LDA)

For a two class problem, LDA is commonly known as Fisher Linear discriminant analysis after Fisher [8] who used it in his taxonomy based experiments. Belhuemer was the first to use the LDA on faces and used it for dimensionality reduction [9] and it can be used as a classifier. LDA attempts to find the linear projection of the data that produces maximum between class separation and minimum within class scatter. In the simple example shown in Figure 1, a projection on to the vertical axis separates the two classes whilst minimizing the within class scatter. Conversely, a projection onto horizontal axis does not separate the classes. Formally the algorithm can be described as follows. The between class scatter covariance matrix is given by:

$$\mathbf{S_B} = (\mathbf{m_2} - \mathbf{m_1})(\mathbf{m_2} - \mathbf{m_1})^{\mathbf{T}} \tag{1}$$

The within class covariance matrix is given by:

$$\mathbf{S_W} = \sum_{i=1}^{C_j} \sum_{n \epsilon C_k} (\mathbf{X^n} - \mathbf{m_i})(\mathbf{X^n} - \mathbf{m_i})^{\mathbf{T}} \tag{2}$$

where m_1 and m_2 are the means of the datasets of the class 1 and 2 respectively. C is the number of classes and C_k is the k_{th} class. The eigenvector solution of $\mathbf{S_w^{-1}S_B}$ gives the projection vector which in the context of face image classification is known as the Fisher face.

2.2   Gabor Filters

A Gabor filter can be applied to images to extract features aligned at particular orientations. Gabor filters possess the optimal localization properties in both spatial and frequency domains, and they have been successfully used in many applications [10]. A Gabor filter is a function obtained by modulating a sinusoidal with a Gaussian function. The useful parameters of a Gabor filter are orientation and frequency. The Gabor

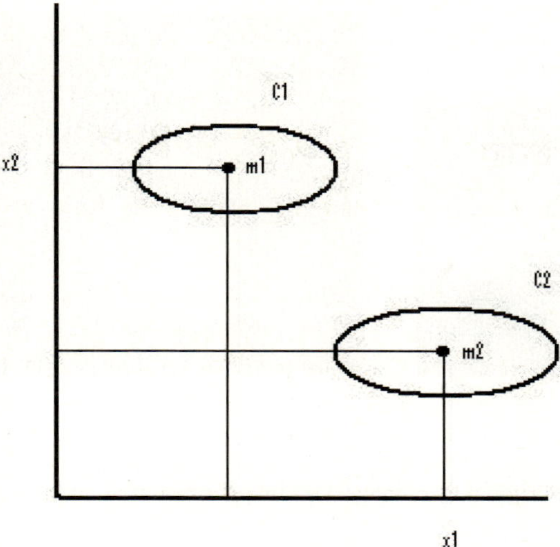

Fig. 1. The figure shows the classes which are overlapping along the direction of x1. However, they can be projected on to direction x2 where there will be no overlap at all.

filter is thought to mimic the simple cells in the visual cortex. The various 2D receptive field profiles encountered in populations of simple cells in the visual cortex are well described by an optimal family of 2D filters [11]. In our case a Gabor filter bank is implemented on face images with 8 different orientations and 5 different frequencies. Recent studies on modeling of visual cortical cells [12] suggest a tuned band pass filter bank structure. Formally, the Gabor filter is a Gaussian (with variances S_x and S_y along x and y-axes respectively) modulated by a complex sinusoid (with centre frequencies U and V along x and y-axes respectively) and is described by the following equation 3

$$g(x,y) = \frac{\exp\left[-\frac{1}{2}\left[\left(\frac{x}{S_x}\right)^2 + \left(\frac{y}{S_y}\right)^2\right] + 2\delta j(Ux + Vy)\right]}{2\delta S_x S_y} \tag{3}$$

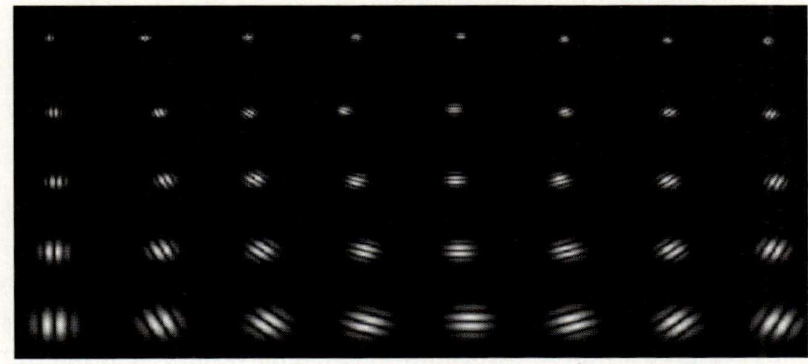

Fig. 2. Gabor filters: Real part of the Gabor kernels at five scales and eight orientations

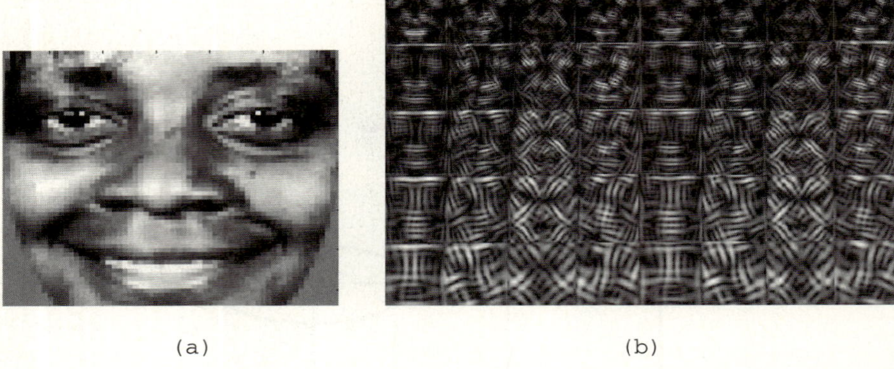

(a)   (b)

Fig. 3. *(a)* Original face image,*(b)* Forty Convolution outputs of Gabor

The variance terms and dictates the spread of the band pass filter centered at the frequencies U and V in the frequency domain. This filter is complex in nature.

A Gabor filter can be described by the following parameters: The S_x and S_y of the Gaussian explain the shape of the base (circle or ellipse), frequency (f) of the sinusoid, orientation (θ) of the applied sinusoid Figure 2 shows examples of various Gabor filters. Figure 3b) shows the effect of applying a variety of Gabor filters shown in Figure 2 to the sample image shown in Figure 3a). Note how the features at particular orientations are exaggerated.

An augmented Gabor feature vector is created of a size far greater than the original data for the image. Every pixel is then represented by a vector of size 40 and demands dimensionality reduction before further processing. So a 64 x 64 image is transformed to size 64 x 64 x 5 x 8. Thus, the feature vector consists of all useful information extracted from different frequencies, orientations and from all locations, and hence is very useful for expression recognition.

Once the feature vector is obtained, it can be handled in various ways. We simply take the L2 max norm for each pixel in the feature vector. So that the final value of a pixel is the maximum value found by any of the filters for that pixel. The L2 max norm Superposition principle is used on the outputs of the filter bank and the Figure 4b) shows the output for the original image of Figure 4 a).

2.3   Curvilinear Component Analysis

Curvilinear Component Analysis (CCA) is a non-linear projection method that preserves distance relationships in both input and output spaces. CCA is a useful method for redundant and non linear data structure representation and can be used in dimensionality reduction. CCA is useful with highly non-linear data, where PCA or any other linear method fails to give suitable information [3]. The D-dimensional input X should be mapped onto the output p-dimensional space Y. Their d-dimensional output vectors y_i should reflect the topology of the inputs x_i. In order to do that, Euclidean distances

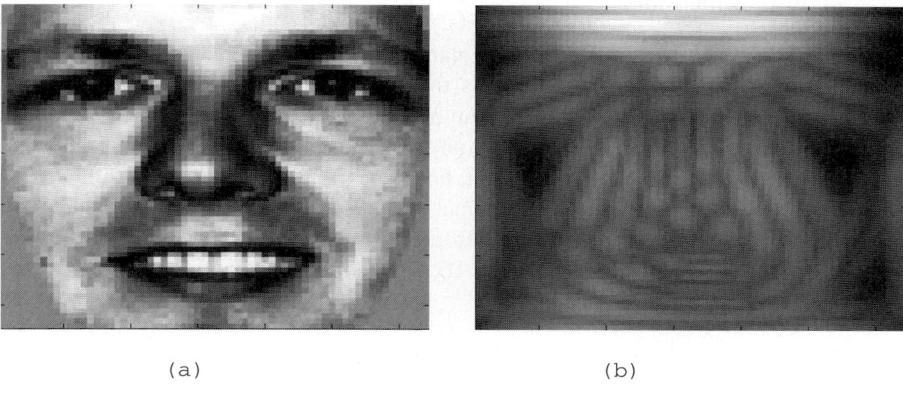

(a)   (b)

Fig. 4. a) Original Image used for the Filter bank b)Supposition Output(L2 max norm)

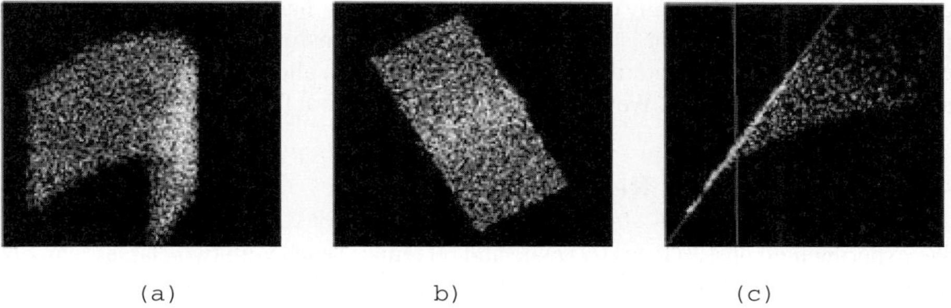

(a)   b)   (c)

Fig. 5. (a) 3D horse shoe dataset (b) 2D CCA projection (c) plot

between the x_i 's are considered. Corresponding distances in the output space y_i's is calculated such that the distance relationship between the data points is maintained.

CCA puts more emphasis on maintaining the short distances than the longer ones. Formally, this reasoning leads to the following error function:

$$E = \frac{1}{2} \sum_{i=1}^{N} \sum_{j=1}^{N} \left[d_{i,j}^{X} - d_{i,j}^{Y} \right]^2 F_\lambda \left[d_{i,j}^{Y} \right] \ \forall j \neq i \tag{4}$$

where $d_{i,j}$ and $d_{i,j}^{Y}$ are the Euclidean distances between the points i and j in the input space X and the projected output space Y respectively and N is the number of data points. $F_\lambda \left[d_{i,j}^{Y} \right]$ is the neighbourhood function, a monotonically decreasing function of distance. In order to check that the relationship is maintained a plot of the distances in the input space and the output space $(d_y - d_x)$plot is produced. For a well maintained topology,d_y should be proportional to the value of d_x at least for small values of d_y's. Figure 5 shows CCA projections for the 3D data horse shoe data. The $(d_y - d_x)$ plot shown is good in the sense that the smaller distances are very well matched [5].

2.4   Intrinsic Dimension

One problem with CCA is deciding how many dimensions the projected space should occupy, and one way of obtaining this is to use the intrinsic dimension of the data manifold. The Intrinsic Dimension (ID) can be defined as the minimum number of free variables required to define data without any significant information loss. Due to the possibility of correlations among the data, both linear and nonlinear, a D-dimensional dataset may actually lie on a d-dimensional manifold ($D \geq d$). The ID of such data is then said to be d. There are various methods of calculating the ID; here we use the correlation Dimension [8] to calculate the ID of face image dataset.

3   Classification Using Support Vector Machines

A number of classifiers can be used in the final stage for classification. We have concentrated on the Support Vector Machine. Support Vector Machines (SVM) are a set of related supervised learning methods used for classification and regression. SVM's are used extensively for many classification tasks such as: handwritten digit recognition [14] or Object Recognition [15]. A SVM implicitly transforms the data into a higher dimensional data space (determined by the kernel) which allows the classification to be accomplished more easily. We have used the LIBSVM tool [7] for SVM classification.

4   Experiments and Results

We experimented on 120 faces (60 male and 60 female) each with two classes, namely: neutral and smiling (60 faces for each expression). The images are from The FERET dataset [16] and some examples are shown in Figure 6. The training set was 80 faces (with 40 female, 40 male and equal numbers of them with neutral and smiling). Two test sets were created. In both test sets the number of each type of face is balanced. For example, there were 5 smiling male faces and 5 smiling female faces. The first set has images which are easily discernible smiling faces. The second test set has smiling and neutral faces, but the smiling faces are not easily discernible. With all faces aligned based on their eye location, a 128 x 128 image was cropped from the original (150 x 130). The resolution of these faces is then reduced to 64 x 64.

A LDA projection was made onto the Fisher face shown in Figure 7. The two test sets were then classified by using the nearest neighbor in the test set in the projection space. The results are as in Table 1. Figure 7 shows the Fisher face obtained by performing the LDA on the training data set. For PCA reduction we use the first few principal components which account for 95% of the total variance of the data, and project the data onto these principal components. This resulted in using 66 components of the raw dataset

Table 1. Classification accuracy of raw faces using LDA

Accuracy %	Test Set 1	Test Set 2
LDA	95	75

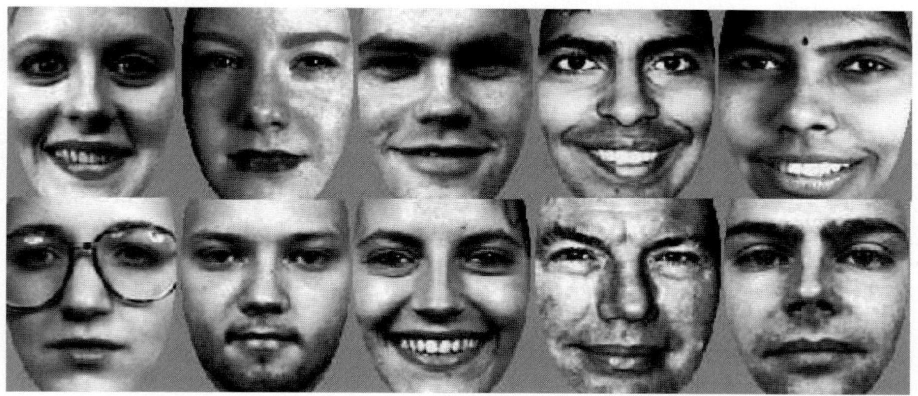

Fig. 6. Example FERET images used in our experiments which are cropped to the size of 128 x 128 to extract the facial region and reduced to 64 x 64 for all experiments

and 35 components in the Gabor pre-processed dataset. As CCA is a highly non-linear dimensionality reduction technique, we use the intrinsic dimensionality technique and reduce the components to its Intrinsic Dimension. The Intrinsic Dimension of the raw faces was approximated as 14 and that of Gabor pre- processed images was 11. The classification results are shown in Table 2. Figure 8 shows the Eigenfaces obtained by the PCA technique.

After dimensionality reduction a standard SVM (with Gaussian kernel) was used to classify the images. The parameters of the SVM were optimized using 5-fold validation.

The results of the classification are as in Table 2. The PCA, being a linear dimensionality reduction technique, did not do quite as well as CCA. With CCA there was good generalization, but the key point to be noted here is the number of components used for the classification. The CCA makes use of just 14 components with raw faces and just 11 components with the Gabor pre-processed images to get good classification results.

This suggests that the Gabor filters are highlighting salient information which can be encoded in a small number of dimensions using CCA. Some examples of misclassifications are shown in Figure 9. The reason for these misclassifications is probably due to

Table 2. SVM Classification accuracy of raw faces and Gabor pre-processed images with PCA and CCA dimensionality reduction techniques

SVM Accuracy %	Test Set 1	Test Set 2
Raw Faces(64x64)	95	80
Raw with PCA66	90	75
Raw with CCA14	90	80
Gabor pre-processed Faces (64x64)	95	80
Gabor with PCA35	70	60
Gabor with CCA11	95	80

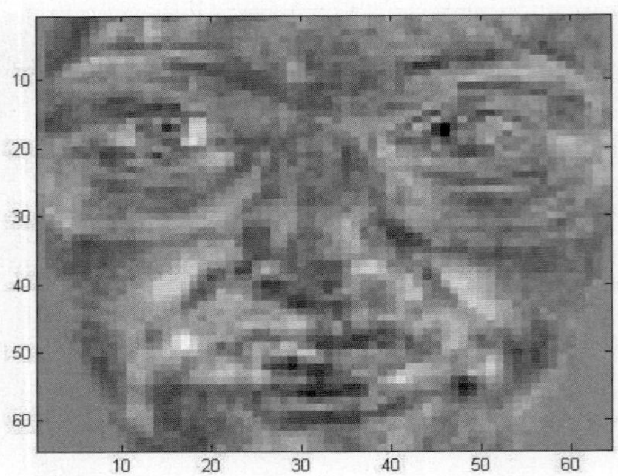

Fig. 7. Fisher face obtained for a dataset with two classes namely, Neutral and Smiling

Fig. 8. The first 5 eigenfaces of the complete data set

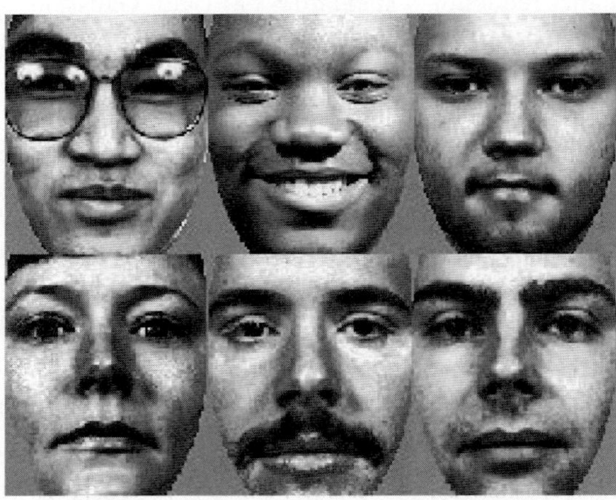

Fig. 9. Examples of the misclassified set of faces. Top row shows smiling faces wrongly classified as neutral. Bottom row shows neutral faces wrongly classified as smiling.

the relatively small size of training set. For example, the mustachioed face in the middle of the bottom row is misclassified as smiling. The only mustachioed face in the training set is of the same man smiling.

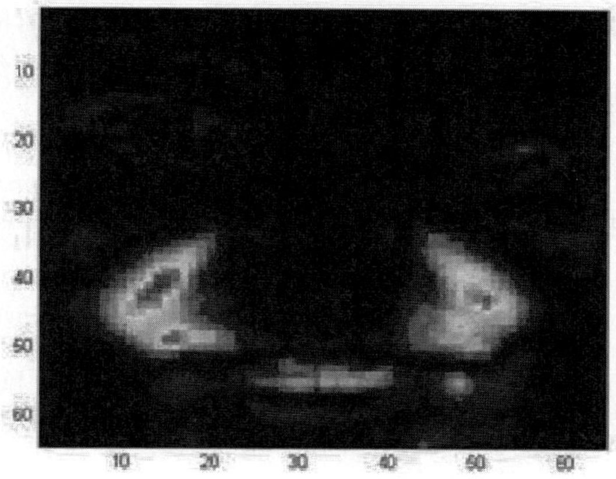

Fig. 10. Encoding face

5   Conclusion

Identifying facial expressions is a challenging and interesting task. Our experiment shows that identification from raw images can be performed very well. However, with a larger data set, it may be computationally intractable to use the raw images. It is therefore important to reduce the dimensionality of the data. Performing classification using LDA was a trivial task and the result was very impressive. It is interesting to see the effect size for each pixel in the image.

In other words which pixels discriminate most between smile and neutral faces can be seen and the result of this analysis is shown in Figure 10. The Creasing of the cheeks is diagnostic of smiling faces; teeth may also be an important indicator, though to a lesser extent. A linear method such as PCA does not appear to be sufficiently tunable to identify features that are relevant for facial expression characterization. Though the result of classification with LDA is impressive, for large datasets with face images, PCA needs to be done prior to the LDA. However, on performing Gabor preprocessing on the images and following it with the CCA, there was good generalization in spite of the massive reduction in dimensionality. The most remarkable finding in this study is that the facial expression can be identified with just 11 components found by CCA. Future work will include extend the experiment to a larger data set and for other expressions.

References

1. Ekman, P., Friesen, W.V.: Constants across cultures in the face of the emotion. Journal of Personality and Social Psychology 17 (1971)
2. Batty, B., Taylor, M.J.: Early processing of the six basic facial emotional expressions. Cognitive Brain Research 17 (2003)
3. Demartines, P., Herault, D.J.: Curvilinear component analysis: A self-organizing neural network for nonlinear mapping of data sets. IEEE Transactions on Neural Networks 8(1), 148–154 (1997)

4. Grassberger, P., Proccacia, I.: Measuring the strangeness of strange attractors. Physica D 9 (1983)
5. Jain, A.K., Farrokhnia, F.: Unsupervised texture segmentation using Gabor filters. Pattern Recognition 24(12) (1991)
6. Movellan, J.R.: Tutorial on Gabor Filters (2002)
7. Chang, C.-C., Lin, C.-J.: LIBSVM: a library for support vector machines (2001)
8. Fisher, R.A.: The use of mutliple measures in anatomical problems. Ann. Eugenics. 7, 179–188 (1936)
9. Belhumeur, Kriegman: Eigenfaces vs. Fisherfaces: recognition using class specific linear projection. IEEE Transactions on Pattern Analysis and Machine Intelligence 19(7), 711–720 (1997)
10. Zheng, D., Zhao, Y., Wang, J.: Features Extraction using A Gabor Filter Family. In: Proceedings of the sixth Lasted International conference, Signal and Image processing, Hawaii (2004)
11. Daugman, J.G.: Uncertainty relation for resolution in space, spatial frequency and orientation optimized by two dimensional visual cortical filters. Journal of Optical Society of America A 2(7) (1985)
12. Kulikowski: Theory of spatial position and spatial frequency relations in the receptive fields of simple cells in the visual cortex. Biological Cybernetics 43(3),187–198 (1982)
13. Smith, L.I.: Tutorial on Principal Component Analysis (2002)
14. Cortes, C., Vapnik, V.: Support Vector Networks. Machine Learning 20, 273–297 (1995)
15. Blanz, V., et al.: Comparison of view-based object recognition algorithms using realistic 3D models. In: Proc. Int. Conf. on Artificial Neural Networks, pp. 251–256 (1996)
16. Philips, P.J., et al.: The FERET evaluation methodology for face recognition algorithms. Image and Vision Computing 16(5), 295–306 (1998)

A Probabilistic Prediction Method for Object Contour Tracking

Daniel Weiler[1], Volker Willert[2], and Julian Eggert[2]

[1] Darmstadt University of Technology, Darmstadt D-64283, Germany
[2] Honda Research Institute Europe GmbH, Offenbach D-63073, Germany

Abstract. In this paper we present an approach for probabilistic contour prediction in an object tracking system. We combine level-set methods for image segmentation with optical flow estimations based on probability distribution functions (pdf's) calculated at each image position. Unlike most recent level-set methods that consider exclusively the sign of the level-set function to determine an object and its background, we introduce a novel interpretation of the value of the level-set function that reflects the confidence in the contour. To this end, in a sequence of consecutive images, the contour of an object is transformed according to the optical flow estimation and used as the initial object hypothesis in the following image. The values of the initial level-set function are set according to the optical flow pdf's and thus provide an opportunity to incorporate the uncertainties of the optical flow estimation in the object contour prediction.

1   Introduction

In this paper we propose an object contour tracking approach based on level-set methods for image segmentation and correlation-based patch-matching methods for optical flow estimation. Using level-set methods for object detection enables us to overcome the problems imposed by nonrigid object deformations and object appearance changes. In tracking applications with dynamic template adaption these changes lead to template drift and in applications without template adaption to a decreased robustness. Utilising probabilistic optical flow for the prediction of the object contour constitutes a non-parametric prediction model that is capable of representing nonrigid object deformation as well as complex and rapid object movements and thus providing a segmentation method with a reliable initial contour that leads to a robust and quick convergence of the level-set method even in the presence of a comparably low camera frame rate. Furthermore, we introduce a novel interpretation of the value of the level-set function. Unlike most recent level-set methods that consider exclusively the sign of the level-set function to determine an object and its surroundings, we use the value of the level-set function to reflect the confidence in the predicted initial contour. This yields a robust and quick convergence of the level-set method in those sections of the contour with a high initial confidence and a flexible and

V. Kůrková et al. (Eds.): ICANN 2008, Part I, LNCS 5163, pp. 1011–1020, 2008.

mostly unconstrained (and thus also quick) convergence in those sections with a low initial confidence.

The segmentation occurs by means of level-set methods [1,2,3,4,5], which separate all image pixels into two disjoint regions [1] by favouring homogeneous image properties for pixels within the same region and dissimilar image properties for pixels belonging to different regions. The level-set formalism describes the region properties using an energy functional that implicitly contains the region description. Minimising the energy functional leads to the segmentation of the image. The formulation of the energy functional dates back to e.g. Mumford and Shah [2] and to Zhu and Yuille [3]. Later on, the functionals were reformulated and minimised using the level-set framework e.g. by [4]. Among all segmentation algorithms from computer vision, level-set methods provide perhaps the closest link with the biologically motivated, connectionist models as represented e.g. by [6]. Similar to neural models, level-set methods work on a grid of nodes located in image/retinotopic space, interpreting the grid as having local connectivity, and using local rules for the propagation of activity in the grid. Time is included explicitly into the model by a formulation of the dynamics of the nodes activity. Furthermore, the external influence from other sources (larger network effects, feedback from other areas, inclusion of prior knowledge) can be readily integrated on a node-per-node basis, which makes level-sets appealing for the integration into biologically motivated system frameworks.

Optical flow estimation, i.e. the evaluation of the pixel-motion in a sequence of consecutive images, yielded two prominent solution classes: namely gradient-based differential [7,8] and correlation-based patch-matching [9,10] algorithms. While the former is based on the *gradient constraint equation* that utilises spatiotemporal derivatives of the image intensity and thus requires nearly linear image intensity resulting in velocities smaller than one pixel per frame and a high frame rate of the camera, the latter uses similarity or distance measures between a small patch of an image and its shifted counterpart that leads to discrete velocities and comparatively high computational costs.

In [11] a comprehensive survey of object tracking algorithms is given. Depending on the vision task, object tracking algorithms are based on several object representations , object detection strategies and prediction methods for the object location . Nonrigid object deformation (e.g. walking person), complex and rapid object movements (e.g. playing children), entire object appearance changes (e.g. front side vs. back side) and object occlusions form some of the numerous challenges in the field of object tracking.

In this paper we propose an approach that combines level-set segmentation algorithms and optical flow estimation methods to form a tracking system. With that combination we are able to overcome some of the principle problems the approaches exhibit, when utilised separately (e.g.: initial level-set function, local optima of the energy functional, aperture problem). The paper is organised as follows. In Sect. 2.1 and 2.2 we describe the level-set method applied for image segmentation and the probabilistic optical flow estimation used for the prediction of the initial object contour, respectively. Section 3 introduces the proposed probabilistic prediction method for object contour tracking. In Sect. 3.1 we suggest a novel interpretation of the value of the initial level-set function. An approach for level-set based object contour tracking based on a *parametric* prediction model is introduced in Sect. 3.2, and extended by a *non-parametric* prediction model in Sect. 3.3. The results of the proposed algorithms are presented in Sect. 4. A short discussion finalises the paper.

2   Level-Set Segmentation and Optical Flow Estimation

2.1   Standard Level-Set Based Region Segmentation

Level-set methods are front propagation methods. Starting with an initial contour, a figure-background segregation task is solved by iteratively moving the contour according to the solution of a partial differential equation (PDE). The PDE is often originated from the minimisation of an energy functional [2,3].

Compared to "active contours" (snakes) [12], that also constitute front propagation methods and explicitly represent a contour by supporting points, level-set methods represent contours implicitly by a level-set function that is defined over the complete image plane. The contour is defined as an iso-level in the level-set function, i.e. the contour is the set of all locations, where the level-set function has a specific value. This value is commonly chosen to be zero, thus the inside and outside regions can easily be determined by the Heaviside function $H(x)$ [1].

The proposed object contour tracking framework is based on a standard two-region level-set method for image segmentation. In a level-set framework, a level-set function $\phi \in \Omega \mapsto \mathbb{R}$ is used to divide the image plane Ω into two disjoint regions, Ω_1 (background) and Ω_2 (object), where $\phi(x) > 0$ if $x \in \Omega_1$ and $\phi(x) < 0$ if $x \in \Omega_2$. A functional of the level-set function ϕ can be formulated that incorporates the following constraints:

- Segmentation constraint: the data within each region Ω_i should be as similar as possible to the corresponding region descriptor ρ_i.
- Smoothness constraint: the length of the contour separating the regions Ω_i should be as short as possible.

This leads to the expression

$$E(\phi) = \nu \int_{\Omega} |\nabla H(\phi)| dx - \sum_{i=1}^{2} \int_{\Omega} \chi_i(\phi) \log p_i \, dx \qquad (1)$$

with the Heaviside function $H(\phi)$ and $\chi_1 = H(\phi)$ and $\chi_2 = 1 - H(\phi)$. That is, the χ_i's act as region masks, since $\chi_i = 1$ for $x \in \Omega_i$ and 0 otherwise. The first term acts as a smoothness term, that favours few large regions as well as smooth region boundaries, whereas the second term contains assignment probabilities $p_1(x)$ and $p_2(x)$ that a pixel at position x belongs to the outer and inner regions Ω_1 and Ω_2, respectively, favouring a unique region assignment.

---

[1] $H(x) = 1$ for $x > 0$ and $H(x) = 0$ for $x \leq 0$.

Minimisation of this functional with respect to the level-set function ϕ using gradient descent leads to

$$\frac{\partial \phi}{\partial t} = \delta(\phi) \left[\nu \operatorname{div} \left(\frac{\nabla \phi}{|\nabla \phi|} \right) + \log \frac{p_1}{p_2} \right] . \tag{2}$$

A region descriptor $\rho_i(\mathbf{f})$ that depends on the image feature vector $\mathbf{f}$ serves to describe the characteristic properties of the outer vs. the inner regions. The assignment probabilities $p_i(x)$ for each image position are calculated based on an image feature vector via $p_i(x) := \rho_i(\mathbf{f}(x))$. The parameters of the region descriptor $\rho_i(\mathbf{f})$ are gained in a separate step using the measured feature vectors $\mathbf{f}(x)$ at all positions $x \in \Omega_i$ of a region i.

2.2   Probabilistic Optical Flow Estimation

The characteristic motion pattern of an object in an image sequence $\mathbf{I}^{1:t}$ at time t is given by the optical flow $\mathbf{V}^t$ within the region that constitutes the object. The optical flow $\mathbf{V}^t = \{\mathbf{v}_x^t\}$ is the set of velocity vectors $\mathbf{v}_x^t$ of all pixels at every location x in the image $\mathbf{I}^t$ at time t, meaning that the movement of each pixel is represented with one velocity hypothesis. This representation neglects the fact that in most cases the pixel movement cannot be unambiguously estimated due to different kinds of motion-specific correspondence problems (e.g. the aperture problem) and noisy data the measurement is based on.

As has been suggested and discussed by several authors [10], velocity probability density functions (pdf's) are well suited to handle several kinds of motion ambiguities. Following these ideas we model the uncertainty of the optical flow $\mathbf{V}^t$ as follows:

$$P(\mathbf{V}^t|Y^t) = \prod_x P(\mathbf{v}_x^t|Y^t) \qquad \text{with} \quad Y^t = \{\mathbf{I}^t, \mathbf{I}^{t+1}\} , \tag{3}$$

where the probability for the optical flow $P(\mathbf{V}^t|Y^t)$ is composed of locally independent velocity pdf's $P(\mathbf{v}_x^t|Y^t)$ for every image location x. $P(\mathbf{v}_x^t|Y^t)$ can be calculated using several standard methods, for details refer e.g. to [10]. These pdf's fully describe the motion estimations available for each position x, taking along (un)certainties and serving as a basis for the probabilistic prediction method for object contour tracking as proposed in Sect. 3.3.

3   Probabilistic Prediction Method for Contour Tracking

3.1   Interpretation of the Value of the Initial Level-Set Function

In general, level-set methods evaluate exclusively the sign of the level-set function to determine an object and its surroundings. The exact value of the level-set function is not considered by most approaches. Signed-distance functions are a common means of regulating the value of the level-set function, as they enforce the absolute value of the gradient of the level-set function to be one.   For the approach we propose in this paper (explained in detail in the next section), it is required to extend the common understanding of the values of the level-set function. Considering the front propagation and gradient descent nature of the applied level-set method for image segmentation, the height of the level-set function influences the time (number of iterations) until the occurrence of a zero crossing (change of region assignment). In particular for numerical stability a maximum time step value is required. Thus, sections of the contour exhibiting large values of the level-set function in their neighbourhood generally move slower than those with smaller values. Following that idea, a steep gradient of the *initial* level-set function for a segmentation algorithm yields a slow deformation of the contour, whereas a flat gradient leads to a mostly unconstrained and quick deformation. Altogether this results in the possibility to control the velocity of the propagated front, embedded entirely and without any algorithmic changes in a standard level-set framework for image segmentation.

3.2   Level-Set Based Segmentation in Image Sequences

Building an iterative level-set based object tracker, a trivial approach would be the usage of the final level-set function of the preceding image ϕ^{t-1} as the initial level-set function $\hat{\phi}^t$ of the current image. To accelerate the convergence of the minimisation process one might also use a smoothed version of the level-set function:

$$\hat{\phi}^t = K_\sigma * \phi^{t-1} \tag{4}$$

The performance of this approach depends on the velocity and deformations of the tracked object. While the approach will succeed in tracking the object in the presence of small movements and deformations, it is likely to fail under huge deformations or large object movements.

To circumvent the above mentioned problem, tracking algorithms include a prediction stage that estimates the object position in the next frame. Introducing a first order prediction method in our level-set based framework would consider the last two segmentation results χ_2^{t-1} and χ_2^{t-2}, measure the transformation between them and predict the current initialisation of the image segmentation algorithm on the basis of the measured transformation. A parametric approach, based on a similarity[2] transformation $\mathbf{A}$, requires the estimation $\boldsymbol{F}$ of four parameters, namely the translation vector $\boldsymbol{t} = (t_x, t_y)^T$, the rotation ω and scale s, comprised in a state vector $\boldsymbol{s} = (t_x, t_y, \omega, s)^T$. In a level-set framework the object translation might be estimated by the translation of the centre of gravity of the inside masks χ_2^{t-1} and χ_2^{t-2}, the rotation by the evaluation of the principal component[3] of the two masks and the scale by the square root of the mask area ratio.

$$\hat{\phi}^t = \mathbf{A}(\phi^{t-1}, \boldsymbol{s}^{t-1}) \qquad \text{with} \quad \boldsymbol{s}^{t-1} = \boldsymbol{F}(\chi_2^{t-1}, \chi_2^{t-2}) \tag{5}$$

---

[2] Similarity transformations constitute a subgroup of affine transformations where the transformation matrix A is a scalar times an orthogonal matrix.

[3] Here the principal component is the eigenvector to the largest eigenvalue of the covariance matrix of the positions of the points within the masks χ_2^{t-1} and χ_2^{t-2}.

In contrast to the previous approach with "zero order" prediction, even objects with high velocities can be tracked, as long as they move to some extent in accordance with the assumed similarity transformation model. Object movements that violate the prediction model, in particular high dynamic movements, again lead to failure.

To cope with high dynamic movements, higher order prediction models might be exploited, but they still underlie the limitation to movements that approximately follow the assumed model. Another approach includes the measurement of the real motion of all pixels (optical flow), belonging to the object, thus providing a means to accurately estimate the object position in the next frame, even in the presence of high dynamic movements. In this way, the prediction is not based on previous frames $Y^{t-2} = \{\mathbf{I}^{t-2}, \mathbf{I}^{t-1}\}$ only, but also on the current frame $Y^{t-1} = \{\mathbf{I}^{t-1}, \mathbf{I}^t\}$. Extending the above approach by the measurement of optical flow leads to the estimation of the state vector $\boldsymbol{s} = (t_x, t_y, \omega, s)^T$ from the flow field $\mathbf{V}^{t-1}$, that might be achieved by a regression analysis $\boldsymbol{R}$.

$$\hat{\phi}^t = \mathbf{A}(\phi^{t-1}, \boldsymbol{s}^t) \qquad \text{with} \quad \boldsymbol{s}^t = \boldsymbol{R}(\mathbf{V}^{t-1}, \chi_2^{t-1}) \tag{6}$$

Although the actual pixel velocities within the object are measured and used for an accurate prediction of the object position, a similarity transformation model is used for the prediction of the contour of the object. Strong deformations of the object will still lead to an imprecise initialisation of the image segmentation algorithm that might decrease the speed of convergence and the robustness of the segmentation. In the next section a purely non-parametric approach is introduced that comprises both a non-parametric estimation of the object position and a non-parametric estimation of the object deformation.

3.3   Probabilistic Prediction Method

In the following we propose an extension of the object tracking algorithm, developed in the previous section, that incorporates the optical flow measurement not only in the estimation of the object position, but also in determining the accurate deformation of the object. The optical flow $\mathbf{V}^t$ already contains all information required. Utilising an image processing warp[4] algorithm $\mathbf{W}_v$ that moves each pixel within an image according to a given vector field, enables us to purely non-parametrically predict an initial level-set function $\hat{\phi}^t$ for the segmentation of the current image $\mathbf{I}^t$.

$$\hat{\phi}^t = \mathbf{W}_v(\phi^{t-1}, \mathbf{V}^{t-1}) \tag{7}$$

If the optical flow estimation provides an additional confidence measure $\mathbf{C}$ a modulation of the prediction will lead to large values of the initial level-set function at locations with high confidence and to small values at locations with low confidence. Thus the flexibility of the moving contour, as introduced in Sect. 3.1 is adapted by the confidence of the optical flow estimation.

$$\hat{\phi}^t = \mathbf{W}_v(\phi^{t-1}, \mathbf{V}^{t-1}, \mathbf{C}) \tag{8}$$

---

[4] Backward warping yielded best results. For the backward warping, also the velocity vectors need to be measured back in time.

In a last step, to introduce an even more robust and faster convergence of the proposed algorithm, the entire velocity pdf $P(\mathbf{V}^t|Y^t)$ is exploited in the prediction stage to determine not only an accurate initial region $\hat{\chi}_2^t$, but also provide an optimal slope (see Sect. 3.1) of the initial level-set function $\hat{\phi}^t$. Utilising a weighted warping algorithm $\mathbf{W}_p$ that moves each pixel within an image not only in one direction, but in all possible directions and overlays all moved pixels weighted by the probability $P(\mathbf{V}^t|Y^t)$ for the given pixel and direction, enables us to determine both the optimal initial region and the optimal slope of the initial level-set function $\hat{\phi}^t$.

$$\hat{\phi}^t = \mathbf{W}_p(\phi^{t-1}, P(\mathbf{V}^{t-1}|Y^{t-1})) \quad \text{with} \quad \hat{\phi}^t(x) = \sum_{v_{x'}^t} P(v_{x'}^{t-1}|Y^{t-1}) \cdot \phi^{t-1}(x - v_{x'}^{t-1}) \tag{9}$$

Altogether the proposed approach keeps the motion ambiguities of the optical flow estimation and yields a flat gradient of the initial level-set function at those sections of the contour where the information from the optical flow is ambiguous and offers only low confidence, leading to a mostly unconstrained and quick convergence. To the contrary, in regions of the contour where the optical flow has a high confidence, the predicted initial level-set function exhibits a steep gradient, enforcing only little change to the contour. The proposed approach enables a smooth transition between the prediction algorithm and the level-set image segmentation method. Thus, the deformation of the contour is locally controlled depending on which algorithm is superior. In sections of the contour with little structure and thus only small confidence in the optical flow measurement, the segmentation method will drive the contour evolution, whereas in sections, where the optical flow estimation is very accurate, the impact of the segmentation method on the contour deformation is suppressed and dominated by the prediction algorithm.

4   Main Results

In order to show the performance of the proposed approach three exemplary test image sequences were chosen. First, a sequence was artificially created with known ground truth by moving an object in front of a background. The movement was strictly based on similarity transformations, i.e. the transformations of the object exhibit exclusively translation, rotation and scale. Second, two real world examples were chosen. One outdoor scene with a driving car and an indoor scene with a high dynamically moving object.

Figure 1 (top row) shows two initial level-set functions, indicating the same initial figure-background condition of a circle in the middle of the image for the image segmentation algorithm and thus leading to the same segmentation result (bottom row, left). The only difference of the initial level-set functions are the steepness of their gradients at the contour, yielding different numbers

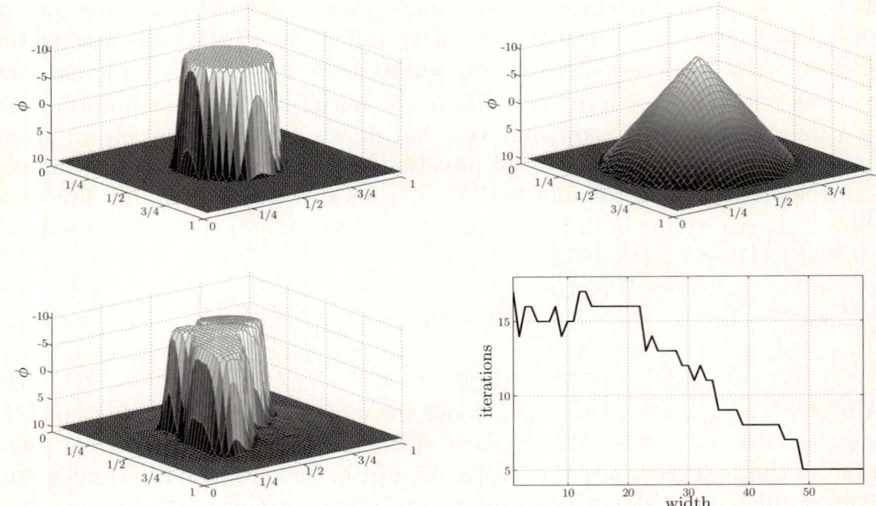

Fig. 1. Top row: Two different initial level-set functions ϕ for the same figure-background regions with steep (left) and flat (right) gradient. Bottom row: Final level-set function, after segmentation (left) and number of iterations until convergence of the segmentation algorithm, plotted for different widths of the contour of the initial level-set function (right).

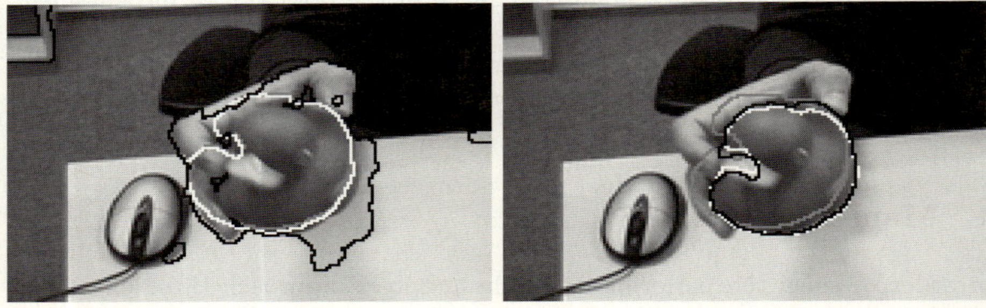

Fig. 2. Detailed view of the real-world image sequence with high dynamic motion, overlaid with segmentation results: previous segmentation result χ_2^{t-1} (grey), current segmentation initial condition $\hat{\phi}^t$ (white) and current segmentation result χ_2^t (black). Shown are identical frames left and right, but different prediction approaches: first order prediction based on the last two segmentation results χ_2^{t-2} and χ_2^{t-1} (5), yielding a segmentation initial that leads to unrobust tracking, as the segmentation algorithm is stuck in a local minimum (left) and probabilistic optical flow based measurement (9), being able to track the high dynamically moving object (right).

of iterations until convergence of the segmentation algorithm. Figure 1 (bottom row, right) shows the number of iterations until convergence of the segmentation algorithm, depending on the width (steepness of the gradient) of the contour of

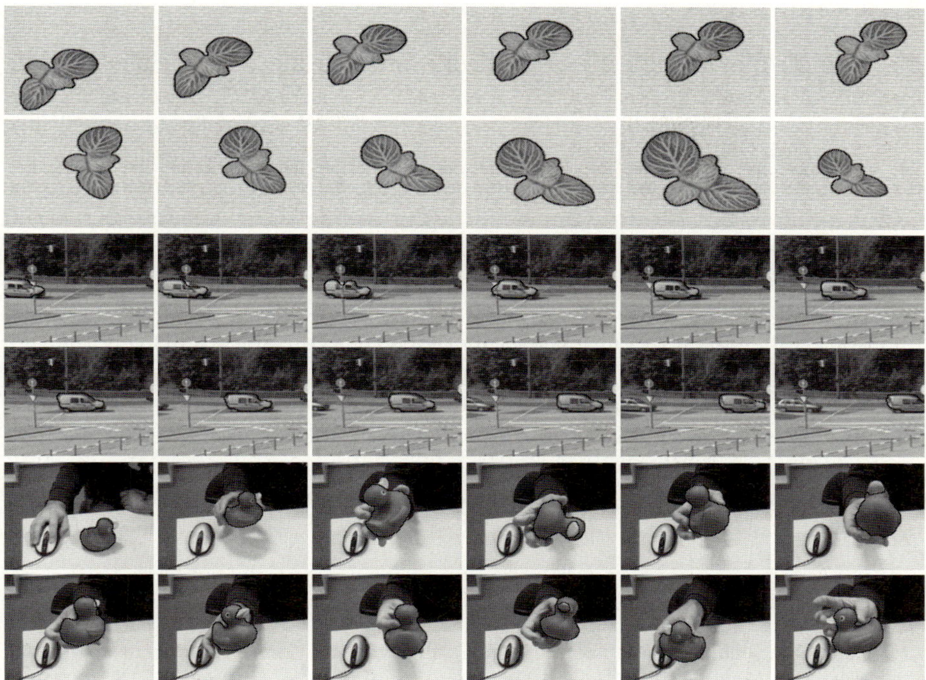

Fig. 3. Overview of used test sequences. Artificial test sequence (top), outdoor car sequence (middle), indoor sequence with high dynamic motion (bottom). The images are overlaid with the segmentation result χ_2^t (black). Shown are the first (upper left), last (lower right) and ten intermidiate frames of the sequence.

the initial level-set function. In Fig. 2 one frame of the real-world test image sequence with high dynamic motion is shown in detail. The images are overlaid with segmentation results: previous segmentation result χ_2^{t-1} (grey), current segmentation initial condition $\hat{\phi}^t$ (white) and current segmentation result χ_2^t (black). Figure 2 shows two times the same frame, but processed with different prediction approaches. Whereas Fig. 2 (left) shows the results of a method with first order prediction based on the last two segmentation results χ_2^{t-2} and χ_2^{t-1} (5) yielding a segmentation initial condition that leads to unreliable tracking (the segmentation algorithm gets stuck in an unfavourable local minimum as the initial condition is already too far away from the desired final contour), Fig. 2 (right) shows an approach based on the probabilistic optical flow measurement (9), that is able to track the high dynamically moving object. Figure 3 shows an overview of the used test sequences: an artificial test sequence (top), an outdoor car sequence (middle) and an indoor sequence with high dynamic motion (bottom). The sequences are overlaid with the segmentation result χ_2^t (black) using the prediction method introduced in Sect. 3.3 (9).

5   Conclusions

We presented an approach for object contour tracking, based on a level-set method for image segmentation and a correlation-based patch-matching method for probabilistic optical flow estimation. Utilising the probabilistic optical flow for the prediction of the object contour constitutes a non-parametric prediction model that is capable of representing nonrigid object deformation as well as complex and rapid object movements, thus providing the segmentation method with a reliable initial contour that leads to a robust and quick convergence of the level-set method.

Furthermore we introduced a novel interpretation of the value of the level-set function. Unlike most recent level-set methods that consider exclusively the sign of the level-set function to determine an object and its surroundings, we use the value of the level-set function to reflect the confidence in the predicted initial contour. This yields a robust and quick convergence of the level-set method in those sections of the contour with a high initial confidence and a flexible, mostly unconstrained and quick convergence in those sections with a low initial confidence.

References

1. Osher, S., Sethian, J.A.: Fronts propagating with curvature-dependent speed: Algorithms based on Hamilton-Jacobi formulations. J. Cmpt. Phys. 79, 12–49 (1988)
2. Mumford, D., Shah, J.: Optimal approximation by piecewise smooth functions and associated variational problems. Commun. Pure Appl. Math. 42, 577–685 (1989)
3. Zhu, S.C., Yuille, A.L.: Region competition: Unifying snakes, region growing, and bayes/MDL for multiband image segmentation. PAMI 18(9), 884–900 (1996)
4. Chan, T., Vese, L.: Active contours without edges. IEEE Trans. Image Process. 10(2), 266–277 (2001)
5. Kim, J., Fisher, J.W., Yezzi, A.J., Çetin, M., Willsky, A.S.: Nonparametric methods for image segmentation using information theory and curve evolution. In: International Conference on Image Processing, Rochester, New York, vol. 3, pp. 797–800 (September 2002)
6. Grossberg, Stephen, Hong, Simon: A neural model of surface perception: Lightness, anchoring, and filling-in. Spatial Vision 19(2-4), 263–321 (2006)
7. Horn, B.K.P., Schunck, B.G.: Determining optical flow. Artif. Intell. 17(1-3), 185–203 (1981)
8. Beauchemin, S.S., Barron, J.L.: The computation of optical flow. ACM Comp. Surv. 27(3), 433–467 (1995)
9. Simoncelli, E.P., Adelson, E.H., Heeger, D.J.: Probability distributions of optical flow. In: Proc Conf on Computer Vision and Pattern Recognition, Mauii, Hawaii. IEEE Computer Vision and Pattern Recognition (CVPR), pp. 310–315 (1991)
10. Weiss, Y., Fleet, D.: Velocity likelihoods in biological and machine vision. In: Probabilistic Models of the Brain: Perception and Neural Function, pp. 77–96 (2002)
11. Yilmaz, A., Javed, O., Shah, M.: Object tracking: A survey. ACM Comput. Surv. 38(4), 13 (2006)
12. Kass, M., Witkin, A., Terzopoulos, D.: Snakes: Active contour models. International Journal for Computer Vision 1(4), 321–331 (1988)

Author Index

Printing: Mercedes-Druck, Berlin
Binding: Stein+Lehmann, Berlin

Lecture Notes in Computer Science

Sublibrary 1: Theoretical Computer Science and General Issues

For information about Vols. 1– 4960
please contact your bookseller or Springer

Vol. 5234: V. Adve, M.J. Garzarán, P. Petersen (Eds.), Languages and Compilers for Parallel Computing. XV, 354 pages. 2008.

Vol. 5228: A. Barbero (Ed.), Coding Theory and Applications. VIII, 197 pages. 2008.

Vol. 5226: D.-S. Huang, D.C. Wunsch II, D.S. Levine, K.-H. Jo (Eds.), Advanced Intelligent Computing Theories and Applications. XXVII, 1273 pages. 2008.

Vol. 5217: M. Dorigo, M. Birattari, C. Blum, M. Clerc, T. Stützle, A.F.T. Winfield (Eds.), Ant Colony Optimization and Swarm Intelligence. XV, 416 pages. 2008.

Vol. 5216: G.S. Hornby, L. Sekanina, P.C. Haddow (Eds.), Evolvable Systems: From Biology to Hardware. XV, 444 pages. 2008.

Vol. 5215: F. Cassez, C. Jard (Eds.), Formal Modeling and Analysis of Timed Systems. X, 295 pages. 2008.

Vol. 5214: H. Ehrig, R. Heckel, G. Rozenberg, G. Taentzer (Eds.), Graph Transformations. XIII, 523 pages. 2008.

Vol. 5204: C.S. Calude, J.F. Costa, R. Freund, M. Oswald, G. Rozenberg (Eds.), Unconventional Computing. X, 259 pages. 2008.

Vol. 5203: S.W. Golomb, M.G. Parker, A. Pott, A. Winterhof (Eds.), Sequences and Their Applications - SETA 2008. XII, 421 pages. 2008.

Vol. 5201: F. van Breugel, M. Chechik (Eds.), CONCUR 2008 - Concurrency Theory. XIII, 524 pages. 2008.

Vol. 5191: H. Umeo, S. Morishita, K. Nishinari, T. Komatsuzaki, S. Bandini (Eds.), Cellular Automata. XVI, 577 pages. 2008.

Vol. 5172: S. Dolev, T. Haist, M. Oltean (Eds.), Optical SuperComputing. IX, 129 pages. 2008.

Vol. 5171: A. Goel, K. Jansen, J.D.P. Rolim, R. Rubinfeld (Eds.), Approximation, Randomization and Combinatorial Optimization. XII, 604 pages. 2008.

Vol. 5170: O.A. Mohamed, C. Muñoz, S. Tahar (Eds.), Theorem Proving in Higher Order Logics. X, 321 pages. 2008.

Vol. 5169: J. Fong, R. Kwan, F.L. Wang (Eds.), Hybrid Learning and Education. XII, 474 pages. 2008.

Vol. 5168: E. Luque, T. Margalef, D. Benítez (Eds.), Euro-Par 2008 – Parallel Processing. XXVIII, 964 pages. 2008.

Vol. 5165: B. Yang, D.-Z. Du, C.A. Wang (Eds.), Combinatorial Optimization and Applications. XII, 480 pages. 2008.

Vol. 5164: V. Kůrková, R. Neruda, J. Koutník (Eds.), Artificial Neural Networks - ICANN 2008, Part II. XXIX, 986 pages. 2008.

Vol. 5163: V. Kůrková, R. Neruda, J. Koutník, (Eds.), Artificial Neural Networks - ICANN 2008, Part I. XXIX, 1026 pages. 2008.

Vol. 5162: E. Ochmański, J. Tyszkiewicz (Eds.), Mathematical Foundations of Computer Science 2008. XIV, 626 pages. 2008.

Vol. 5160: J.S. Fitzgerald, A.E. Haxthausen, H. Yenigun (Eds.), Theoretical Aspects of Computing - ICTAC 2008. XI, 455 pages. 2008.

Vol. 5156: K. Havelund, R. Majumdar, J. Palsberg (Eds.), Model Checking Software. X, 343 pages. 2008.

Vol. 5148: O. Ibarra, B. Ravikumar (Eds.), Implementation and Applications of Automata. XII, 289 pages. 2008.

Vol. 5147: K. Horimoto, G. Regensburger, M. Rosenkranz, H. Yoshida (Eds.), Algebraic Biology. XII, 245 pages. 2008.

Vol. 5133: P. Audebaud, C. Paulin-Mohring (Eds.), Mathematics of Program Construction. X, 423 pages. 2008.

Vol. 5132: P.J. Bentley, D. Lee, S. Jung (Eds.), Artificial Immune Systems. XIV, 436 pages. 2008.

Vol. 5130: J. von zur Gathen, J.L. Imaña, Ç.K. Koç (Eds.), Arithmetic of Finite Fields. X, 205 pages. 2008.

Vol. 5126: L. Aceto, I. Damgård, L.A. Goldberg, M.M. Halldórsson, A. Ingólfsdóttir, I. Walukiewicz (Eds.), Automata, Languages and Programming, Part II. XXII, 730 pages. 2008.

Vol. 5125: L. Aceto, I. Damgård, L.A. Goldberg, M.M. Halldórsson, A. Ingólfsdóttir, I. Walukiewicz (Eds.), Automata, Languages and Programming, Part I. XXIII, 896 pages. 2008.

Vol. 5124: J. Gudmundsson (Ed.), Algorithm Theory – SWAT 2008. XIII, 438 pages. 2008.

Vol. 5123: A. Gupta, S. Malik (Eds.), Computer Aided Verification. XVII, 558 pages. 2008.

Vol. 5117: A. Voronkov (Ed.), Rewriting Techniques and Applications. XIII, 457 pages. 2008.

Vol. 5114: M. Bereković, N. Dimopoulos, S. Wong (Eds.), Embedded Computer Systems: Architectures, Modeling, and Simulation. XVI, 300 pages. 2008.

Vol. 5104: F. Bello, E. Edwards (Eds.), Biomedical Simulation. XI, 228 pages. 2008.

Vol. 5103: M. Bubak, G.D. van Albada, J. Dongarra, P.M.A. Sloot (Eds.), Computational Science – ICCS 2008, Part III. XXVIII, 758 pages. 2008.

Vol. 5102: M. Bubak, G.D. van Albada, J. Dongarra, P.M.A. Sloot (Eds.), Computational Science – ICCS 2008, Part II. XXVIII, 752 pages. 2008.

Vol. 5101: M. Bubak, G.D. van Albada, J. Dongarra, P.M.A. Sloot (Eds.), Computational Science – ICCS 2008, Part I. XLVI, 1058 pages. 2008.

Vol. 5092: X. Hu, J. Wang (Eds.), Computing and Combinatorics. XIV, 680 pages. 2008.

Vol. 5090: R.T. Mittermeir, M.M. Sysło (Eds.), Informatics Education - Supporting Computational Thinking. XV, 357 pages. 2008.

Vol. 5084: J.F. Peters, A. Skowron (Eds.), Transactions on Rough Sets VIII. X, 521 pages. 2008.

Vol. 5083: O. Chitil, Z. Horváth, V. Zsók (Eds.), Implementation and Application of Functional Languages. XI, 272 pages. 2008.

Vol. 5073: O. Gervasi, B. Murgante, A. Laganà, D. Taniar, Y. Mun, M.L. Gavrilova (Eds.), Computational Science and Its Applications – ICCSA 2008, Part II. XXIX, 1280 pages. 2008.

Vol. 5072: O. Gervasi, B. Murgante, A. Laganà, D. Taniar, Y. Mun, M.L. Gavrilova (Eds.), Computational Science and Its Applications – ICCSA 2008, Part I. XXIX, 1266 pages. 2008.

Vol. 5065: P. Degano, R. De Nicola, J. Meseguer (Eds.), Concurrency, Graphs and Models. XV, 810 pages. 2008.

Vol. 5062: K.M. van Hee, R. Valk (Eds.), Applications and Theory of Petri Nets. XIII, 429 pages. 2008.

Vol. 5059: F.P. Preparata, X. Wu, J. Yin (Eds.), Frontiers in Algorithmics. XI, 350 pages. 2008.

Vol. 5058: A.A. Shvartsman, P. Felber (Eds.), Structural Information and Communication Complexity. X, 307 pages. 2008.

Vol. 5050: J.M. Zurada, G.G. Yen, J. Wang (Eds.), Computational Intelligence: Research Frontiers. XVI, 389 pages. 2008.

Vol. 5045: P. Hertling, C.M. Hoffmann, W. Luther, N. Revol (Eds.), Reliable Implementation of Real Number Algorithms: Theory and Practice. XI, 239 pages. 2008.

Vol. 5038: C.C. McGeoch (Ed.), Experimental Algorithms. X, 363 pages. 2008.

Vol. 5036: S. Wu, L.T. Yang, T.L. Xu (Eds.), Advances in Grid and Pervasive Computing. XV, 518 pages. 2008.

Vol. 5035: A. Lodi, A. Panconesi, G. Rinaldi (Eds.), Integer Programming and Combinatorial Optimization. XI, 477 pages. 2008.

Vol. 5029: P. Ferragina, G.M. Landau (Eds.), Combinatorial Pattern Matching. XIII, 317 pages. 2008.

Vol. 5028: A. Beckmann, C. Dimitracopoulos, B. Löwe (Eds.), Logic and Theory of Algorithms. XIX, 596 pages. 2008.

Vol. 5022: A.G. Bourgeois, S.Q. Zheng (Eds.), Algorithms and Architectures for Parallel Processing. XIII, 336 pages. 2008.

Vol. 5018: M. Grohe, R. Niedermeier (Eds.), Parameterized and Exact Computation. X, 227 pages. 2008.

Vol. 5015: L. Perron, M.A. Trick (Eds.), Integration of AI and OR Techniques in Constraint Programming for Combinatorial Optimization Problems. XII, 394 pages. 2008.

Vol. 5011: A.J. van der Poorten, A. Stein (Eds.), Algorithmic Number Theory. IX, 455 pages. 2008.

Vol. 5010: E.A. Hirsch, A.A. Razborov, A. Semenov, A. Slissenko (Eds.), Computer Science – Theory and Applications. XIII, 411 pages. 2008.

Vol. 5008: A. Gasteratos, M. Vincze, J.K. Tsotsos (Eds.), Computer Vision Systems. XV, 560 pages. 2008.

Vol. 5004: R. Eigenmann, B.R. de Supinski (Eds.), OpenMP in a New Era of Parallelism. X, 191 pages. 2008.

Vol. 5000: O. Grumberg, H. Veith (Eds.), 25 Years of Model Checking. VII, 231 pages. 2008.

Vol. 4996: H. Kleine Büning, X. Zhao (Eds.), Theory and Applications of Satisfiability Testing – SAT 2008. X, 305 pages. 2008.

Vol. 4988: R. Berghammer, B. Möller, G. Struth (Eds.), Relations and Kleene Algebra in Computer Science. X, 397 pages. 2008.

Vol. 4985: M. Ishikawa, K. Doya, H. Miyamoto, T. Yamakawa (Eds.), Neural Information Processing, Part II. XXX, 1091 pages. 2008.

Vol. 4984: M. Ishikawa, K. Doya, H. Miyamoto, T. Yamakawa (Eds.), Neural Information Processing, Part I. XXX, 1147 pages. 2008.

Vol. 4981: M. Egerstedt, B. Mishra (Eds.), Hybrid Systems: Computation and Control. XV, 680 pages. 2008.

Vol. 4978: M. Agrawal, D.-Z. Du, Z. Duan, A. Li (Eds.), Theory and Applications of Models of Computation. XII, 598 pages. 2008.

Vol. 4975: F. Chen, B. Jüttler (Eds.), Advances in Geometric Modeling and Processing. XV, 606 pages. 2008.

Vol. 4974: M. Giacobini, A. Brabazon, S. Cagnoni, G.A. Di Caro, R. Drechsler, A. Ekárt, A.I. Esparcia-Alcázar, M. Farooq, A. Fink, J. McCormack, M. O'Neill, J. Romero, F. Rothlauf, G. Squillero, A.Ş. Uyar, S. Yang (Eds.), Applications of Evolutionary Computing. XXV, 701 pages. 2008.

Vol. 4973: E. Marchiori, J.H. Moore (Eds.), Evolutionary Computation, Machine Learning and Data Mining in Bioinformatics. X, 213 pages. 2008.

Vol. 4972: J. van Hemert, C. Cotta (Eds.), Evolutionary Computation in Combinatorial Optimization. XII, 289 pages. 2008.

Vol. 4971: M. O'Neill, L. Vanneschi, S. Gustafson, A.I. Esparcia Alcázar, I. De Falco, A. Della Cioppa, E. Tarantino (Eds.), Genetic Programming. XI, 375 pages. 2008.

Vol. 4967: R. Wyrzykowski, J. Dongarra, K. Karczewski, J. Wasniewski (Eds.), Parallel Processing and Applied Mathematics. XXIII, 1414 pages. 2008.

Vol. 4963: C.R. Ramakrishnan, J. Rehof (Eds.), Tools and Algorithms for the Construction and Analysis of Systems. XVI, 518 pages. 2008.

Vol. 4962: R. Amadio (Ed.), Foundations of Software Science and Computational Structures. XV, 505 pages. 2008.

Vol. 4961: J.L. Fiadeiro, P. Inverardi (Eds.), Fundamental Approaches to Software Engineering. XIII, 430 pages. 2008.